MW01257020

THERMAL INSULATION BUILDING GUIDE

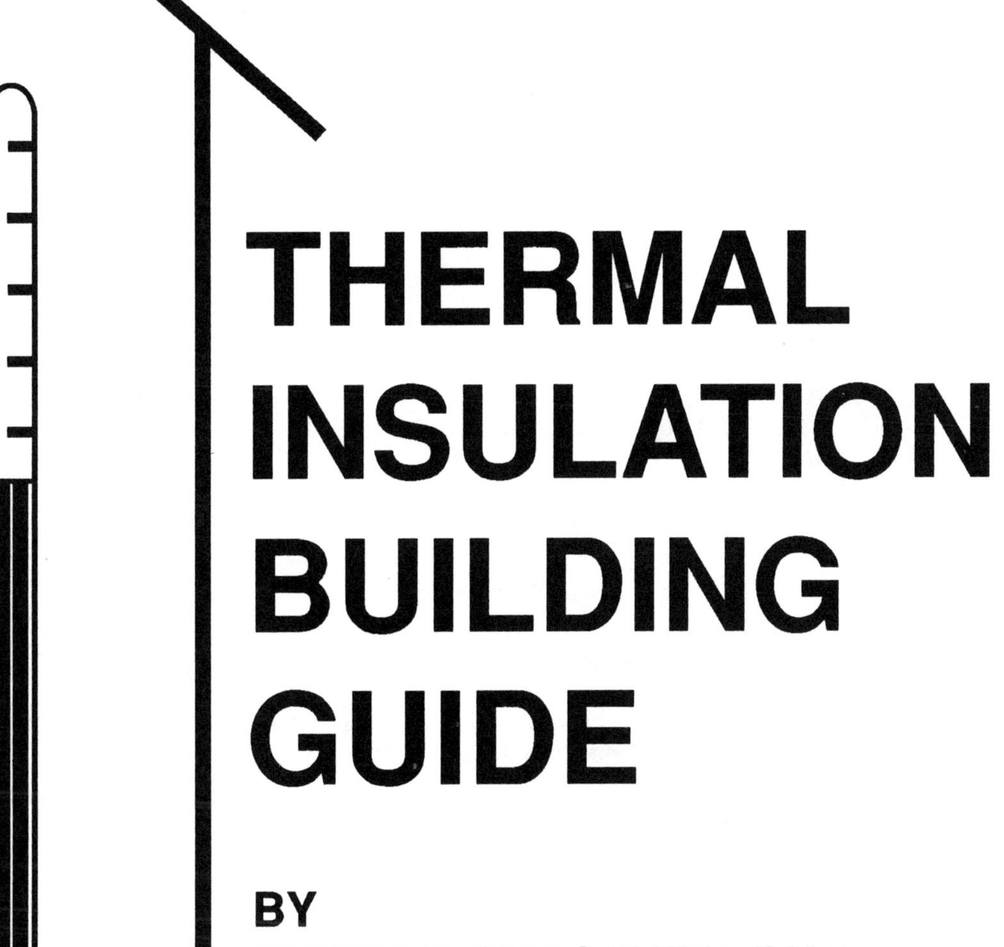

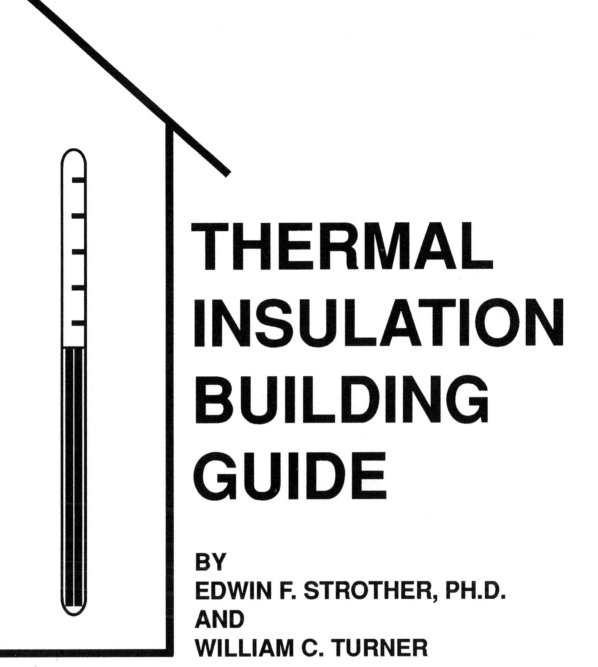

THERMAL INSULATION BUILDING GUIDE

BY
EDWIN F. STROTHER, PH.D.
AND
WILLIAM C. TURNER

Robert E. Krieger Publishing Company
Malabar, Florida
1990

Original Edition 1990

Printed and Published by
ROBERT E. KRIEGER PUBLISHING COMPANY, INC.
KRIEGER DRIVE
MALABAR, FL 32950

Library of Congress Cataloging-in-Publication Data

Strother, Edwin.
Thermal insulation building Guide.
Bibliography: p.
Includes index.
1. Insulation (Heat) 2. Building--Energy conservation.
I. Turner, William C., 1942- . II. Title.
TH1715.T885 1984 693.8'32 82-16232
ISBN 0-88275-985-X

10 9 8 7 6 5 4 3 2

CONTENTS

Publisher's Note

In 1980 William C. Turner and John F. Malloy brought out the first in a series of books intended to give complete information on the subject of thermal insulation. The series was designed to develop the difficult concepts first. Thus, 1980 saw the publication of HANDBOOK OF THERMAL INSULATION DESIGN ECONOMICS, and 1981 saw the release of THERMAL INSULATION HANDBOOK. In the interim came John Malloy's death; however, William Turner pressed on to finish the series with THERMAL INSULATION BUILDING GUIDE. Regrettably the project was still incomplete at the time of his death. After several false starts, we were able to procure the services of Dr. Edwin F. Strother who has brought the original concept to completion. We are pleased to present this volume as the capstone of the original series.

LIST OF TABLES

(Note: E.U. = English Units; M.U. = Metric Units)

1 Fundamentals of Heat Transfer

It is well known that human beings are sensitive to temperature and temperature variations. In fact, a large portion of all human activities and concerns are associated directly with those factors which affect thermal comfort. These factors include shelter, clothing, heating, air conditioning, and thermal insulation. While thermal insulation is the subject of this book, it will, nevertheless, often be necessary to deal with the thermal and insulating characteristics of the other factors. This being so, it is important that one has a conceptual understanding of heat, energy, and temperature. In many instances, these terms are used incorrectly. *Energy* is defined as the capacity of acting; power, the rate of doing work.

Energy can exist in various forms, such as (1) *Gravitational Potential Energy* — energy stored between two or more masses under each other's mutual gravitational attractions; (2) *Mechanical Potential Energy* — energy stored in a mechanical system (like a spring) which is under tension, compression, torsion, or shear forces; (3) *Chemical Potential Energy* — energy which is chemically stored within matter and which may be released during an exothermic chemical reaction; and (4) *Kinetic Energy* — energy associated with any mass. The sum of all kinetic and potential energies within a solid, liquid, gas, or plasma, can be collectively referred to as *Internal Energy*.

Work can be thought of as a transient form of mechanical energy and *Heat* as a transient form of internal energy. More precisely, heat is energy in transition or transfer from one body to another by virtue of a temperature difference existing between the bodies. Heat energy transferred from one body to another body at a lower temperature by physical contact is heat transfer by *conduction*. Heat energy transmission through space from a hotter body to a colder body is by *radiation*. *Convection* occurs when the internal thermal energy of matter is transferred from one location to another by the physical movement of that matter. This form of heat transfer is closely associated with mass transfer (usually liquids or gases). Problems involving convection are, therefore, more difficult to handle mathematically than those involving either pure conduction or radiation. But since the movement of air over a heated surface affects the amount of heat transferred by conduction and radiation, the mathematical formulas treat convection as if it were a separate form of heat transfer.

The convection of gases or liquids can be classified as (1) *Natural (or Free) Convection*, (2) *Forced Convection*, or (3) a combination of both. *Natural (or Free) Convection* occurs when differential heating portion of the fluid rises due to its lower density while the cooler portions, having greater density, tend to sink. *Forced Convection* is relative motion of a liquid or a gas caused by some external agent. (In a gas this external agent may be a fan or blower; in a liquid the external force may be supplied by a pump.)

Energy Measurement Units. One of the major difficulties in an applied science like engineering thermodynamics is that of transforming units of measure between the English and the Metric Systems. In the English system a common unit of energy is the British thermal unit (Btu) which is defined as the quantity of heat required to raise the temperature of one pound of water one degree Fahrenheit at or near 32°F. The Btu is related to other energy units as follows:

$$\text{One Btu} = 778 \text{ ft.-lb.} = 252 \text{ calories} = 1055 \text{ joules} = 2.928 \times 10^{-4} \text{ kilowatt-hrs.}$$

In the Metric system, heat energy is measured in joules or kilowatt-hours (kW-hr.) where

$$\text{One joule} = 2.778 \times 10^{-4} \text{ watt-hr.} = 9.48 \times 10^{-4} \text{ Btu} = 0.7376 \text{ foot-lbs.}$$

and,

$$\text{One kilowatt-hr.} = 3413 \text{ Btu} = 2.655 \times 10^{6} \text{ foot-lbs.}$$

Power is frequently defined as the rate of doing work and has units of energy per unit of time. For example, in a building the amount of energy required to produce a temperature suitable for occupancy is not constant over fixed period of time. In winter, the colder the day the more energy is required per hour to warm the building. Likewise, in summer, the hotter the day the more energy is used per hour to produce refrigeration for cooling. Therefore, power is a measure of the rate of which energy is used, expended, or transfered.

Power Measurement Units. In the English system of units, Btu per hour and horsepower are the most basic units. Power is expressed in watts in the metric system. The relationships between these units are

$$\text{One Btu} = 3.929 \times 10^{-4} \text{ horsepower-hr.} = 0.2931 \times 10^{-3} \text{ kilowatt-hr.} = 0.2931 \text{ watt-hr.}$$

$$\text{One watt} = 1 \text{ joule per second}$$

$$\text{One kilowatt} = 1.33 \text{ horsepower}$$

As stated in the definition, heat is energy transferred from one body to another due to differences in temperature. Thus, temperature and its scale of units, must also be established.

Temperature is the level of the thermal state of a body as indicated on a designated scale. The most frequently used temperature scales are the Fahrenheit Scale (°F) in the English System and the Celsius Scale (°C) in the Metric System.

> The *Fahrenheit (F) Scale* subdivides the temperature interval between the ice point and steam point into 180 degrees. The ice point is assigned the value of 32 degrees Fahrenheit and the steam point a temperature of 212 degrees Fahrenheit.
>
> The *Celsius (C) Scale** subdivides the temperature interval between the ice point and steam point into 100 degrees. The ice point is assigned the value of 0 degrees Celsius and the steam point has a temperature of 100 degrees Celsius.

Temperature is also measured on the Rankine and the Kelvin Scales. The Rankine (R) scale, which has the same temperature intervals as the Fahrenheit scale, uses the lowest attainable temperature (absolute zero) as its zero point (0°R). This gives the ice point a temperature of 491.7°R and the steam point a temperature of 671.7°R. The Kelvin (K) scale** also places its reference point at absolute zero, but uses the same temperature intervals as the Celsius scale, thus placing the ice point at 273.2°K and the steam point at 373.2°K. While the Rankine and Kelvin temperature scales are not used often in this book, temperature conversion tables for all scales are given in Appendix B for easy reference.

HEAT TRANSFER BY RADIATION

Having defined the major aspects of heat and its measurement, it is desirable to illustrate these concepts and establish the basic equations of heat transfer. This is necessary in order to specify the type and determine the quantity of thermal insulation which is to be installed in a particular building.

The visible portion of the electromagnetic spectrum extends from blue wavelengths at approximately 0.3 microns to the longest red wavelengths at approximately 0.74 microns. The invisible infrared wavelengths extend from 0.74 microns to 2.7 microns and are associated with the direct transfer of heat energy by radiation. The fundamental factors of heat transfer by radiation are illustrated in Figure 1 and include:

Surface emittance of a given object is the ratio of the total heat radiated per unit time per unit area to the total energy per unit time per unit area which would be emitted by an ideal radiator.

Thermal reflectance is the ratio of the radiant energy reflected by a body to the radiant energy that is incident upon it.

The representation of heat a path by radiation is shown as  in Figure 1. Equations 1-a and 1-b, which allow one to calculate the heat transferred by radiation between two surfaces at different temperatures, are given as follows:

*The Celsius scale was formerly known as the centigrade scale.

**On the Kelvin Scale a temperature interval of 1 degree kelvin is written 1 kelvin; hence temperatures can be expressed in kelvins without the use of the degree (°) sign. Thus 273°K can be expressed as 273 kelvins or simply 273K. In this book, we will continue to use the earlier notation (273°K).

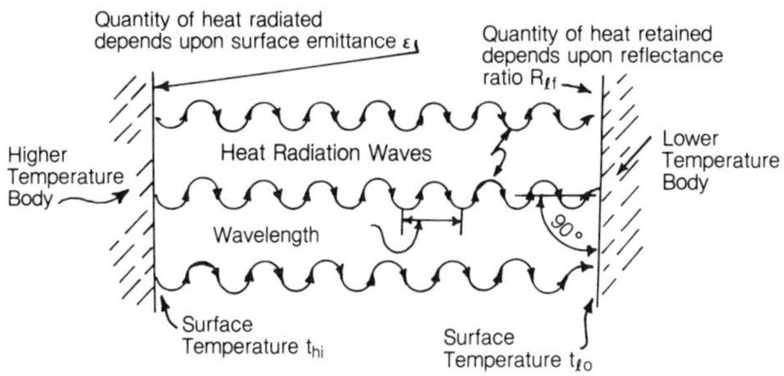

HEAT TRANSFER BY RADIATION — θ = 90°

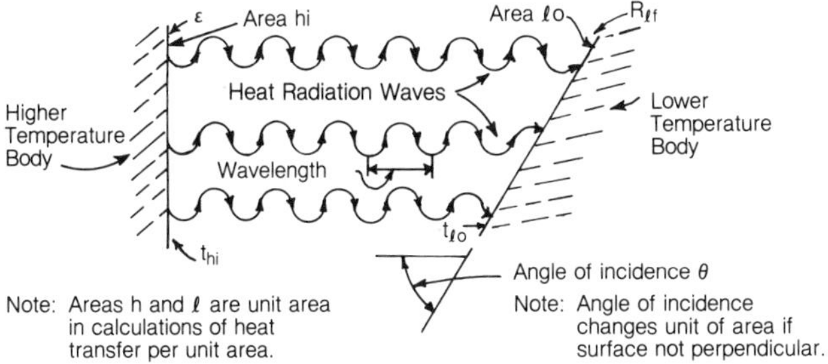

HEAT TRANSFER BY RADIATION BETWEEN PARALLEL SURFACES

HEAT TRANSFER BY RADIATION TO NON PARALLEL SURFACES

FACTORS OF HEAT TRANSFER BY RADIATION

Figure 1

In the English system of units, where

Q_r = heat transferred by radiation Btu per ft.^{2} per hour
ε = surface emittance (dependent on surface, material, and wavelength)
$R_{\ell f}$ = surface reflectance (dependent on surface, material and wavelength)
t_{hi} = temperature of hotter surface, °F
$t_{\ell o}$ = temperature of cooler surface, °F

then:

$$Q_r = 0.1714\varepsilon \left[\left(\frac{t_{hi} + 459.67}{100} \right)^4 - \left(\frac{t_{\ell o} + 459.67}{100} \right)^4 \right] \times R_{\ell f} \quad \text{Btu/ft.}^2\text{, hr.} \qquad (1)$$

In Metric units, when

Q_{mr} = heat transfer by radiation in W/m^2
t_{mhi} = temperature of hotter surface, °C
$t_{m\ell o}$ = temperature of cooler surface, °C

then:

$$Q_{mr} = 0.548\varepsilon \left[\left(\frac{t_{mhi} + 273}{55.55} \right)^4 - \left(\frac{t_{m\ell o} + 273}{55.55} \right)^4 \right] \text{ in W/m}^2. \qquad (2)$$

Equations, 1a and 1b, provide only the most basic factors for the calculation of heat transfer by radiation. Their use is limited to heat transfer from one flat surface to another at relatively close distances and under conditions of thermal equilibrium.

The formulas presented thus far are somewhat ideal and are, therefore, not directly applicable to applied problems. Figure 2 illustrates a practical situation for which the formulas do not take into consideration intensity variation due to absorption of dust, smoke or moisture between the sun and the surface. In addition, the incident angle of the radiation is always changing. Due to these complications, a method was developed to determine the heat gain through walls by adding degrees of temperature to the outer surface of the wall or roof to approximate the heat transfer through that particular construction.

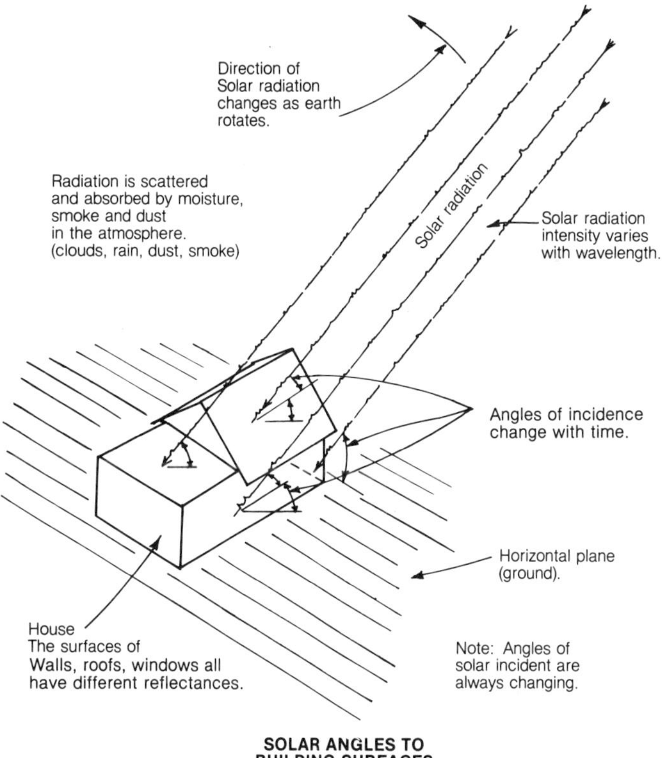

SOLAR ANGLES TO BUILDING SURFACES

Figure 2

The number of variables makes it necessary to generalize a system for practical design considerations. Tables of air temperatures and the temperature increases which must be added, due to radiation, in order to obtain design temperatures for calculations of heat transfer through walls and roofs of buildings are given in Chapter 11.

HEAT TRANSFER BY CONDUCTION

The heat transfer by conduction is a function of the thermal *conductivity* of the body, the temperature difference across the body, and cross sectional area of the body. Heat flow by conduction is illustrated in Figure 3 where the heat path is represented as
→→→→

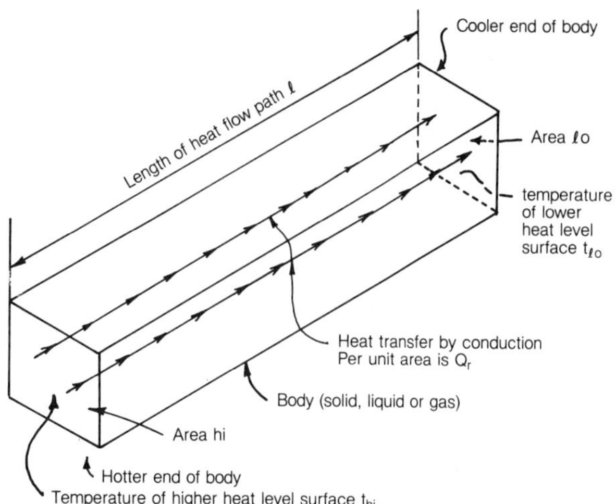

Note: Illustration is for unit surface areas at temperatures t_{hi} and $t_{\ell o}$. It assumes no heat transfer other than from area hi to area ℓo. It also assumes surfaces are parallel.

FACTORS OF HEAT TRANSFER BY CONDUCTION

Figure 3

The thermal *conductivity* k of a homogeneous body can be defined as the time rate of heat flow per unit area per unit of temperature difference per unit length of flow path. Conductivity values are a function of temperature and, therefore, may change in a particular material as the mean temperature of the material changes. The basic mathematical relationships for heat transfer by conduction are presented as follows.

For A SINGLE MATERIAL—INSTALLED FLAT—*English Units*

When:

Q_{co} = heat transfer by conduction, in Btu per ft.², per hr.
t_{hi} = temperature °F of hotter surface
$t_{\ell o}$ = temperature °F of lower temperature surface
ℓ = length between the two surfaces in inches
k = conductivity of the body at mean temperature

$$\frac{t_{hi} + t_{\ell o}}{2}\text{, in Btu in./ft.}^2\text{, hr., °F.}$$

Then:

$$Q_{co} = \frac{t_{hi} - t_{\ell o}}{\frac{\ell}{k}}\text{, in Btu/ft.}^2\text{, hr.} \qquad (3)$$

Note: For simplification and comparisons, frequently

$\frac{\ell}{k}$ is called Thermal Resistance R. Thus,

$$Q_{co} = \frac{t_{hi} - t_{\ell o}}{R}\text{, in Btu/ft.}^2\text{, hr.} \qquad (4)$$

In Metric units, when

Q_{mco} = heat transfer by conduction, in W/m²
t_{mhi} = temperature of hotter surface, °C
$t_{m\ell o}$ = temperature of cooler surface, °C
ℓ_m = length between the two surfaces in meters
k_m = conductivity of the body at mean temperature

$$\frac{t_{mhi} + t_{m\ell o}}{2}\text{, in W/m, °C}$$

then:

$$Q_{mco} = \frac{t_{mhi} - t_{m\ell o}}{\frac{\ell_m}{k_m}}\text{, in W/m}^2\text{.} \qquad (3a)$$

If $\frac{\ell_m}{k_m} = R_m$, then

$$Q_{mco} = \frac{t_{mhi} - t_{m\ell o}}{R_m}\text{, in W/m}^2\text{.} \qquad (4a)$$

It should be noted that R and R_m have no direct units of measure. When manufacturers advertise R values the usual meaning is insulation thickness divided by conductivity. Thus, for any given insulation material, if the thickness is doubled so is the "R value." Also, conductivity itself varies somewhat with temperature and may be affected by the orientation of the insulation and the direction of heat flow. Unless these variables are given, R values should only be considered rough approximations which are often used only for advertising purposes. It should be noted that R_m (thermal resistance in Metric units) has completely different values than does R in the English system of units. That is,

$$R = \frac{\text{inches}}{\text{Btu., inches/°F, ft.}^2\text{, hr.}} = \frac{\text{°F, ft.}^2\text{, hr.}}{\text{Btu}}$$

and

$$R_m = \frac{m^2\ °C}{W}$$

Note: $R_m = 0.1761$ R or R = $5.6786R_m$.

Note that the length of heat flow through a material directly affects the resistance to that heat flow; however, as this length appears in both the numerator and the denominator of the R equations, it is cancelled out. Thus, unless thickness (length of heat flow path) of an insulation is given, the R and R_m values are meaningless in the design of insulation systems.

For TWO LAYERED MATERIALS—INSTALLED FLAT—*English Units*

when:

Q_{co} = heat transfer by conduction in Btu per ft.², per hr.
t_{hi} = temperature of hotter surface °F
t_{bi} = temperature of surface of first material in contact with second material °F
$t_{\ell o}$ = temperature of cooler surface of second material °F
k_1 = conductivity of first material at a mean temperature of

$$\frac{t_{hi} + t_{bi}}{2}$$, in Btu, in/ft², hr., °F

k_2 = conductivity of second material at a mean temperature of

$$\frac{t_{bi} + t_{\ell o}}{2}$$, in Btu, in/ft.², hr., °F

ℓ_1 = length of heat path between hotter surface and contact surfaces of the two materials, inches

ℓ_2 = length of heat path from contact surface to lower temperature surface, inches

then:

$$Q_{co} = \frac{t_{hi} - t_{\ell o}}{\dfrac{\ell_1}{k_1} + \dfrac{\ell_2}{k_2}} \quad \text{Btu/ft.}^2\text{, hr.} \qquad (5)$$

Again, for simplification, when R_1 is used for $\frac{\ell_1}{k_1}$ and R_2 is used for $\frac{\ell_2}{k_2}$ then the above equation becomes

$$Q_{co} = \frac{t_{hi} - t_{\ell o}}{R_1 + R_2} \quad \text{Btu/ft.}^2\text{, hr.} \qquad (6)$$

This is illustrated in Figure 4.

In Metric units, when

Q_{mco} = heat transfer in W/m²

t_{mhi} = temperature of hotter surface °C

t_{mbi} = temperature of first material in contact with second material °C

$t_{m\ell o}$ = temperature of cooler surface, second material °C

k_{m1} = conductivity of first material at a mean temperature of

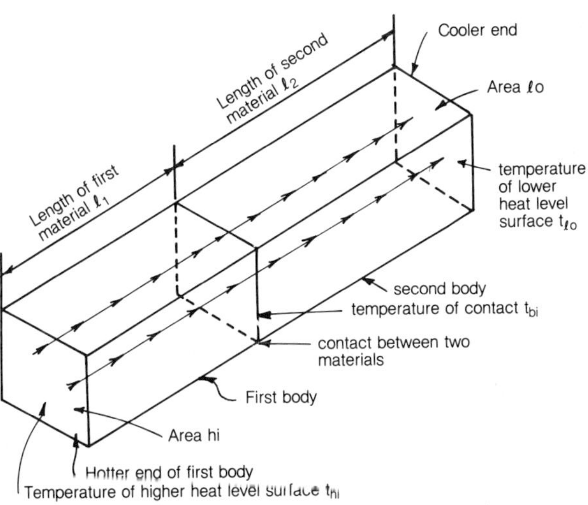

FACTORS OF HEAT TRANSFER BY CONDUCTION THROUGH TWO MATERIALS

Figure 4

$$\frac{t_{mhi} + t_{mbi}}{2} \text{ in W/m, °C}$$

k_{m2} = conductivity of second material at a mean temperature of

$$\frac{t_{mbi} + t_{m\ell o}}{2} \text{ in W/m, °C}$$

ℓ_{m1} = length of heat path between hotter surface contact surface of the two materials, in meters

ℓ_{m2} = length of heat path between contact surfaces of the two materials to lower temperature surface, in meters

then:

$$Q_{mco} = \frac{t_{mhi} - t_{m\ell o}}{\dfrac{\ell_{m1}}{k_{m1}} + \dfrac{\ell_{m2}}{k_{m2}}} \quad \text{W/m}^2 \qquad (5a)$$

Again, for simplification when R_{m1} is used for $\frac{\ell_{m1}}{k_{m1}}$ and R_{m2} is used for $\frac{\ell_{m2}}{k_{m2}}$ we have

$$Q_{mco} = \frac{t_{mhi} - t_{m\ell o}}{R_{m1} + R_{m2}} \quad \text{W/m}^2. \qquad (6a)$$

For A SINGLE MATERIAL— INSTALLED CYLINDRICALLY

In the previous equations, the heat was transferred from one flat surface to another flat surface, both parallel to each other. When installed in a cylindrical shape, the inner surface has less area than does the outer surface. This is shown in Figure 5.

In English units, when

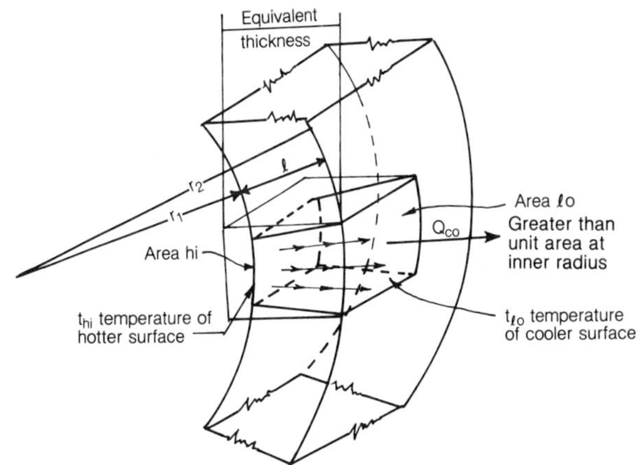

When Q_{co} is to be determined, based on outer radius r_2, calculations for resistance must be based on an "equivalent thickness" which is given as

$$\text{Equivalent Thickness} = r_2 \log_e \frac{r_2}{r_1} \quad \text{(in inches)}$$

FACTORS OF HEAT TRANSFER BY CONDUCTION THROUGH A SINGLE CYLINDRICAL MATERIAL

Figure 5

Q_{co} = heat transfer by conduction, in Btu per ft.², per hr. at outer surface

t_{hi} = temperature of hotter surface °F

t_{lo} = temperature of cooler surface °F
r_1 = radius of inner cylindrical shaped insulation, inches
r_2 = radius of outer cylindrical shaped insulation, inches
k = conductivity of the insulation at mean temperature

$$\frac{t_{hi} + t_{lo}}{2}, \text{ in Btu, in./ft.}^2\text{, hr., °F}$$

then: For one cylindrical material

$$Q_{co} = \frac{t_{hi} - t_{lo}}{\dfrac{r_2 \log_e \dfrac{r_2}{r_1}}{k}} \text{ Btu/ft.}^2\text{, hr.} \tag{7}$$

The previous equation can be written as

$$Q_{co} = \frac{t_{hi} - t_{lo}}{R},$$

where

$$R = \frac{r_2 \log_e \dfrac{r_2}{r_1}}{k}.$$

Note: The following example illustrates why "R" and "R_m" must be accurately defined in order to be significant in thermal design:

If an insulation at 50 °F mean temperature has a k (conductivity) of 0.3 is installed 2" in thickness (ℓ) on a flat surface with one side at t_{hi} = 90 °F and the other side at t_{lo} = 10 °F, then

$$Q_{co} = \frac{t_{hi} - t_{lo}}{\dfrac{\ell}{k}} = \frac{90 - 10}{\dfrac{2}{0.3}} = \frac{80}{6.67} = 12 \text{ Btu/ft.}^2\text{, hr.}$$

where $$R = \frac{\ell}{k} = \frac{2}{0.3} = 6.67.$$

If the same 2" thick insulation is now installed around a 4" outside diameter duct, then r_1 = 2" and $r_2 = r_1 + 2" = 4"$. Therefore,

$$Q_{co} = \frac{t_{hi} - t_{lo}}{\dfrac{r_2 \log_e \dfrac{r_2}{r_1}}{k}} = \frac{90 - 10}{\dfrac{4 \log_e \dfrac{4}{2}}{0.3}} = \frac{80}{\dfrac{4 \times 0.693}{0.3}} = \frac{80}{\dfrac{2.772}{0.3}}$$

$$= 8.65 \text{ Btu/ft.}^2\text{, hr.}$$

Note: $R = \dfrac{2.772}{0.3} = 9.24$ or $Q_{co} = \dfrac{t_{hi} - t_{lo}}{R} = \dfrac{80}{9.24}$.

Conversely, if the heat transfer were based on a heat flow of 12 Btu/ft^2, hr. the R value would be approximately 4.62 with heat flowing inward.

Thus, using the same insulation and thickness, at the same mean temperature, with the same temperature difference, the R value on a flat surface is 6.666 and on a 4" cylindrical surface the R value is 9.24 or 4.62, depending on the direction of heat flow. (See Figure 5).

In Metric units, when

Q_{mco} = heat transfer by conduction in W/m^2 at outer surface
t_{mhi} = temperature °C of hotter surface
t_{mlo} = temperature °C of cooler surface
r_{m1} = radius of inner cylindrical shaped insulation in meters
r_{m2} = radius of outer cylindrical shaped insulation in meters
k_m = conductivity of the insulation in W/m, °C evaluated at the mean temperature

$$\frac{t_{mhi} + t_{mlo}}{2}, \text{ in °C}$$

then: For one cylindrical material

$$Q_{mco} = \frac{t_{mhi} - t_{mlo}}{\dfrac{r_{m2} \log_e \dfrac{r_{m2}}{r_{m1}}}{k_m}} \text{ W/m}^2. \tag{7a}$$

Therefore, in Metric units, an insulation at 10 °C mean temperature having a conductivity of 0.04325 and installed in 0.0508 meter thickness on a flat surface at 32.2 °C on the hot side and −12.2 °C on the other side will transfer heat by conduction at the following rate:

$$Q_{mco} = \frac{t_{mhi} - t_{mlo}}{\dfrac{\ell_m}{k_m}} = \frac{32.2 - (-12.2)}{\dfrac{0.0508}{0.04325}} = \frac{44.4}{1.17} = 37.9 \text{ W/m}^2.$$

Note: $R_m = \dfrac{\ell_m}{k_m} = \dfrac{0.0508}{0.04325} = 1.17.$

However, if the same insulation is now installed around a 0.1016 meter diameter duct at the same thickness (0.0508 meter) r_{m2} would be 0.1016 meter and r_{m1} would be 0.0508 meter, then

$$Q_{mco} = \frac{t_{mhi} - t_{mlo}}{\dfrac{r_{m2} \log_e \dfrac{r_{m2}}{r_{m1}}}{k_m}} = \frac{32.2 - (-12.2)}{\dfrac{0.1016 \log_e \dfrac{0.1016}{0.0508}}{0.04325}}$$

$$= \frac{44.4}{\dfrac{0.1016 \times 0.693}{0.04325}} = \frac{44.4}{1.627} = 27.27 \text{ W/m}^2.$$

Note: Here R_m = 1.627.

Thus, using the same insulation and thickness, at the same temperature difference, the R_m value on a flat surface is 1.17 and on a cylindrical surface is 1.627 or 0.8135 depending on the direction of flow of heat.

It must be remembered that "thermal resistance" is a factor in a heat transfer equation—it is not a material property. Under identical conditions, the R value of thermal resistance can be converted from English to Metric units by multiplying the English units (°F, ft.^{2}, hr./Btu) by 0.1761 to obtain the Metric units (°C, m^2/W).

HEAT TRANSFER INFLUENCED BY CONVECTION

Due to mass movement of gases or liquids, heat transferred by conduction, is influenced by this movement. For ease of calculation (and understanding), we refer to this influence on heat transfer by mass movement as heat transfer by convection.

For purposes of calculation, the heat moved by natural convection currents of air across a hot surface, the following equation and constants were developed by Rice and Heilman.* In English units, when

Q_{cv} = heat flow from the surface by natural air convection, in Btu/ft.², hr.
c = convection constant—see the following table
d = cylinder diameter in inches. Note: if d is equal to or greater than 24″, then let d=24; also use d=24 for a flat surface.
T_{ave} = average temperature of a cylinder and the surrounding air in °R (°R = °F + 459.6)
Δt = temperature difference in °F between the surface of the cylinder and air

c Values for Free Convection (*English & Metric Units*)

Shape and Condition	Value of c
Horizontal cylinders	1.016
Long vertical cylinders	1.235
Vertical plates	1.394
Horizontal plates, warmer than air, facing upward	1.79
Horizontal plates, warmer than air, facing downward	0.89
Horizontal plates, cooler than air, facing upward	0.89
Horizontal plates, cooler than air, facing downward	1.79

then:

$$Q_{cv} = c \times \left(\frac{1}{d}\right)^{0.2} \times \left(\frac{1}{T_{ave}}\right)^{0.181} \times \Delta t^{1.266} \text{ in Btu/ft.}^2\text{, hr.} \tag{8}$$

In Metric units, when

Q_{mcv} = heat moved from surface by natural air convection, in W/m²
c = constant (same as for English units—use previous table)
d_m = cylinder diameter in meters with d_m to be taken as 0.6 meters for flat surfaces and for diameters greater than 0.6 meters
T_{mave} = average temperature in degrees Kelvin of cylinder and surrounding air (°K = °C + 273.2)
Δt_m = temperature difference in °C between cylinder surface and air

then:

$$Q_{mcv} = c \times \left(\frac{39.37}{d_m}\right)^{0.2} \times \left(\frac{0.555}{T_{mave}}\right)^{0.181} \times \left(1.8\Delta t_m\right)^{1.266} \text{ in W, hr./m}^2 \tag{8a}$$

These equations have been simplified in order that they apply to materials which have fairly low conductivities; such materials tend to have surface temperatures which approach that of the ambient air temperature. For this reason, they are frequently used as a part of equations used for calculation of heat transfer through insulation systems, walls, roofs, etc. This balancing effect reduces the need for a variable constant for different locations and directions of heat flow.

In English units, when

Q_{cv} = heat moved by natural air convection, in Btu/ft.², hr.
t_s = temperature of surface, °F
t_a = temperature of ambient air, °F

then, by Langmuir's Equation:

$$Q_{cv} = 0.296\ (t_s - t_a)^{5/4}. \tag{9*}$$

When air movement is by forced convection

V = velocity of the air in ft./min.
Q_{cvf} = heat moved by forced air movement, in Btu/ft.², hr.

then:

$$Q_{cvf} = 0.296\ (t_s - t_a)^{5/4} \sqrt{\frac{V + 68.9}{68.9}} \tag{10*}$$

In Metric units, when

Q_{mcv} = heat moved by natural air convection in Btu/ft.², hr.
t_{ms} = temperature of surface, °C
t_{ma} = temperature of surface, °C

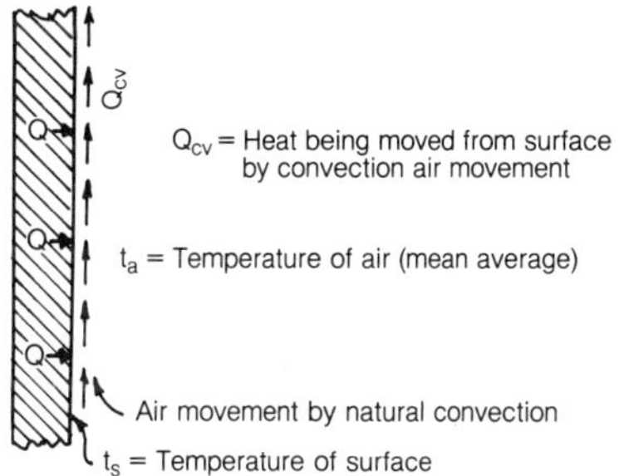

HEAT MOVED FROM SURFACE BY NATURAL CONVECTION

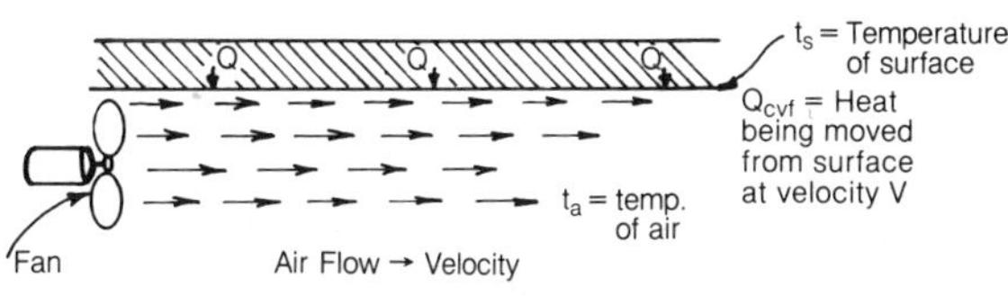

HEAT MOVED FROM SURFACE BY FORCED CONVECTION

Note: Position of surfaces may be any direction.

Figure 6

*R. H. Heilman, "Transactions, American Society of Mechanical Engineers," Vol. 1, Part 1 (1929).

R. H. Heilman, "Industrial and Engineering Chemistry," Vol. 28, No. 7 (1936).

*Langmuir equations on heat transfer by convection.

then:

$$Q_{mcv} = 1.957\,(t_{ms} - t_{ma})^{5/4}. \qquad (9a)$$

When air movement is by forced convection,

V_m = velocity of the air in meters per second
Q_{mcvf} = heat moved by forced air movement in W/m²

$$Q_{mcvf} = 1.957\,(t_{ms} - t_{ma})^{5/4} \sqrt{\frac{196.85V_m + 68.9}{68.9}}$$

$$= \text{W/m}^2 \qquad (10a)$$

The heat moved from surfaces by both natural and forced convection is shown in Figure 6.

HEAT TRANSFER THROUGH MATERIALS

When a temperature difference exists, heat (energy) will flow from regions of higher temperature to regions of lower temperature. No material can completely stop this transfer of energy which may be by one or several of the mechanisms of heat transfer previously discussed.

Thermal insulation is any material, or combination of materials, which provides resistance to the flow of heat energy. There are three major forms of thermal insulation which are commercially available today. These are: (1) vacuum insulations, (2) reflective insulations, and (3) mass or bulk insulations. Some insulation assemblies use two or more of these types.

Other than where partial vacuum is provided between two panes of glass or inside a glass block, vacuum types of insulation are seldom used in buildings. This is because it is very expensive to construct enclosures of metal or glass to resist the outside pressures which attempt to fill the vacuum space. Therefore, the insulation systems to be described next will be of the reflective type, the mass type, or a combination of the two.

Reflective Insulation consists of one or more sheets which have surfaces which reflect heat and which, on the other side, have low emittances. These surfaces must not be butted against solids since such contact transfers heat by conduction and the reflectance of the surface then has little or no influence on heat transfer. The heat transfer from a solid surface across an air space to a reflective surface is shown in Figure 7. To be effective, the movement of air must be restricted. It has been found that an air space in a vertical position should not be over 3/4″ (19 mm) thick. Thicker or wider air barriers will allow vertical air movement by natural convection thus lowering the insulating property of the barrier. Likewise, all spaces should have low conductivity and small cross sections in order to minimize conductive heat flow through the solid surfaces. Dry air with no movement has a low conductivity of 0.15 Btu in./ft.², hr., °F. (0.0216 W/m, °C).

Next we will develop the expression for the total heat transferred by convection, conduction and radiation across the air space illustrated in Figure 7. In English units, when

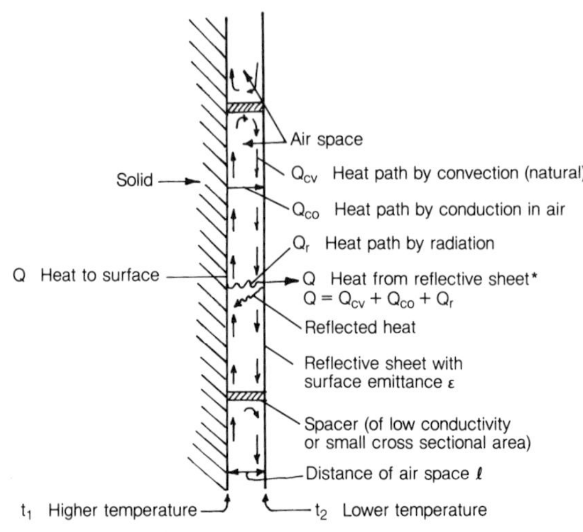

Note: To be effective reflective sheets must be so located to cause resistance to air movement by convection

*Q Heat from reflective sheet is dependent on surface emittance if sheet faces another air space on cooler side. If attached to solid emittance equals 1.

HEAT MOVED FROM SURFACE ACROSS AIR SPACE TO REFLECTIVE SURFACE

Figure 7

Q = total heat transfer by convection, conduction, and radiation in Btu/ft.², hr.
$Q = Q_{cv} + Q_{co} + Q_r$
Q_{cv} = heat transfer from one surface to another by convection of air in Btu/ft.², hr.
Q_{co} = heat transfer from one surface to another by conduction through air in Btu/ft.², hr.
Q_r = heat transfer from one surface to other by radiation in Btu/ft.², hr.
k = conductivity of air in Btu in./ft.², hr. °F
ℓ = distance across air space in inches
t_1 = higher temperature of solid, °F
t_2 = lower temperature of reflective sheet, °F
$R_{f\ell}$ = reflectance of sheet (dimensionless)
ε = surface emittance $(1.0 - R_{f\ell})$

Then (assuming that negligible conduction of heat occurs through the solid spacers):

$$Q_{cv} = 0.296\,(t_1 - t_2)^{5/4} \text{ in Btu/ft.}^2\text{, hr. (for natural convection)} \qquad (11\text{-a})$$

$$Q_{co} = \frac{t_1 - t_2}{\frac{\ell}{k}} \text{ in Btu/ft.}^2\text{, hr. (for conduction)} \qquad (12\text{-a})$$

$$Q_r = 0.174\varepsilon \left[\left(\frac{t_1 + 459.6}{100}\right)^4 - \left(\frac{t_2 + 459.6}{100}\right)^4\right] \text{ in Btu/ft.}^2\text{, hr. (for radiation)} \qquad (13\text{-a})$$

Therefore, the total heat transfer Q can be written as

$$Q = 0.296\,(t_1 - t_2)^{5/4} + \frac{t_1 - t_2}{\frac{\ell}{k}} + 0.174\varepsilon \left[\left(\frac{t_1 + 459.6}{100}\right)^4 - \left(\frac{t_2 + 459.6}{100}\right)^4\right] \text{ in Btu/ft.}^2\text{, hr.} \quad (14\text{-a})$$

In Metric units, when

Q_m = total heat transfer in W/m^2
Q_{mcv} = heat transfer from one surface to another by convection of air in W/m^2
Q_{mco} = heat transfer from one surface to another by conduction through air in W/m^2
Q_{mr} = heat transfer from one surface to another by radiation in W/m^2
k_m = conductivity of air in W/mk
ℓ_m = distance across air space in meters
t_{m1} = higher temperature of solid in °C
t_{m2} = lower temperatuer of reflective sheet in °C
T_{m1} = higher temperature of solid in °K
T_{m2} = lower temperature of solid in °K
$R_{mf\ell}$ = reflectance of sheet (dimensionless)
ε_μ = surface emittance $(1.0 - R_{mf\ell})$

Again, assuming negligible heat conduction occurs through the spacers, then

$$Q_{mcv} = 1.957\,(t_{m1} - t_{m2})^{5/4} \text{ in } W/m^2 \quad (11\text{-b})$$

$$Q_{mco} = \frac{t_{m1} - t_{m2}}{\frac{\ell_m}{k_m}} \text{ in } W/m^2 \quad (12\text{-b})$$

$$Q_{mr} = 0.548\varepsilon \left[\left(\frac{T_{m1}}{55.55}\right)^4 - \left(\frac{T_{m2}}{55.55}\right)^4\right] \text{ in } W/m^2. \quad (13\text{-b})$$

Thus, the total heat transfer Q_m is the sum of Q_{mcv}, Q_{mco}, and Q_{mr}; namely,

$$Q_m = 1.957\,(t_{m1} - t_{m2})^{5/4} + \frac{t_{m1} - t_{m2}}{\frac{\ell_m}{k_m}} + 0.548\varepsilon \left[\left(\frac{T_{m1}}{55.55}\right)^4 - \left(\frac{T_{m2}}{55.55}\right)^4\right] \text{ in } W/m^2. \quad (14\text{-b})$$

These equations give the fundamentals of heat transfer through one air space and one reflective shield. If more reflective sheets and spaces are added, they are added to the equation in series. Since, intermediate temperatures are unknown, problems have to be solved by reiteration using estimated intermediate values of temperature. Therefore, to avoid this complication, assemblies of reflective sheets are usually tested and their total conductance or resistance is determined for the calculation of heat transfer purposes. R is determined by heat transfer tests for the particular assembly and direction of heat flow. The resulting simplification is shown below. If R is the total thermal resistance in English units, then

$$Q = \frac{(t_1 - t_2)}{R} \text{ in Btu/ft.}^2\text{, hr.}$$

In Metric units,

$$Q_m = \frac{t_{m1} - t_{m2}}{R_m} \text{ in } W/m^2.$$

Note: R (English Units) × 0.1761 = R_m (Metric Units)

Mass Insulation consists of solids in the form of fibers, granules, or cells which contain air or gas filled pockets and voids in such arrangement as to retard the passage of heat. Similar to heat transfer for reflective-space insulation, the heat transferred from one side to the other side of mass insulation is the sum of heat transferred by convection in the spaces, conduction of the gas in the spaces, conduction of heat through the solids, and by radiation from one surface across the space to another surface. It makes no difference if the solid is made of flakes, granules, fibers, or cells, all modes of heat transfer can occur. This is illustrated in Figure 8.

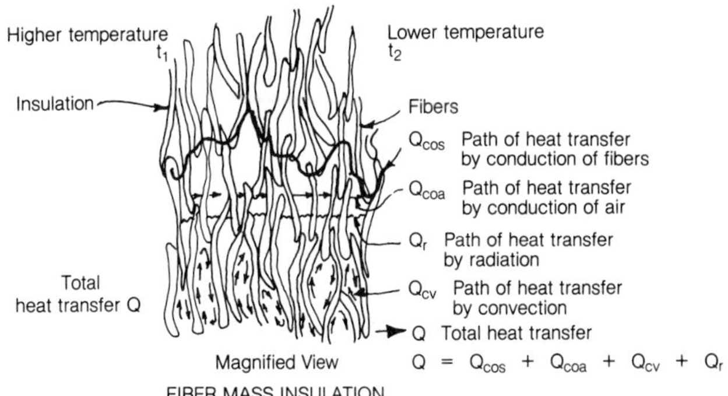

FIBER MASS INSULATION

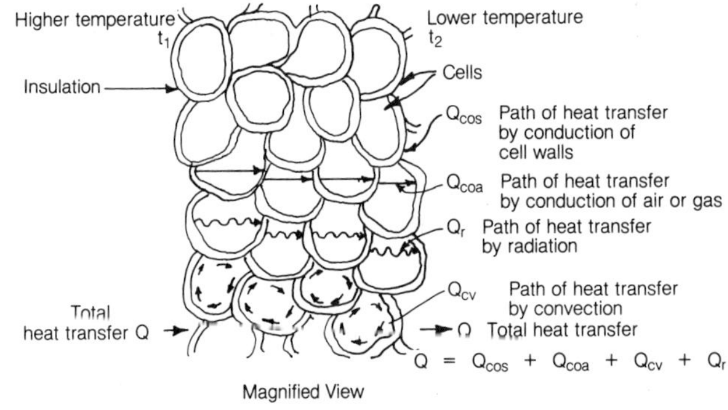

CELLULAR MASS INSULATION

SCHEMATIC HEAT FLOW THROUGH MASS INSULATION

Figure 8

To be effective, the fibers, flakes, granules, or cells of mass insulation must separate their occupied space into very small individual spaces with a minimum of solid contact between surfaces. Also, the lower and more opaque the surface emittance of the solids, the greater will be the reduction of heat transfer by radiation. Examination of Figure 8 illustrates that the total heat flow Q from one surface to the other through mass insulation is the total of all contributions due to conduction, convection, and radiation; namely,

$$Q = Q_{cos} + Q_{coa} + Q_{cv} + Q_r.$$

Mass insulation is not tested to determine the individual modes of heat transfer but instead, the test determines Q for a given set of conditions. As with reflective insulation, by knowing Q, the thermal resistance R can be calculated. Also thermal conductivity k can be calculated for the set of conditions under which it was tested. For simple parallel surfaces, when:

In English units

t_1 = higher temperature surface of insulation °F
t_2 = lower temperature surface of insulation °F
R = thermal resistance
Q = total heat transfer between two surfaces in Btu/ft.², hr.

then

$$Q = \frac{(t_1 - t_2)}{R} \text{ Btu/ft.}^2\text{, hr.}$$

As shown, no length of heat path from one surface to the other was given. For flat parallel surfaces, when

k = conductivity of the mass insulation Btu in./ft.², hr., °F
ℓ = heat path, from one surface to the other, in inches

then:

$$R = \frac{\ell}{k} \text{ °F, ft.}^2\text{, hr./Btu}$$

$$\text{or } Q = \frac{t_1 - t_2}{\frac{\ell}{k}} \text{ Btu/ft.}^2\text{, hr.}$$

For this reason, it must be remembered that thermal resistance is a function in an equation. R is not a thermal property of material. To illustrate: If a material has a thermal conductivity of 0.3 Btu, in./ft.², hr. °F, then 1 inch thickness has a resistance R of 3.33; 2 inches of thickness has a thermal resistance of 6.66; 3 inches of thickness has a thermal resistance of 9.99. Also, the conductivity of a material changes some with mean temperature, temperature difference, and direction of heat flow; all these factors must be given for the particular condition for which the R value of thermal resistance was determined and calculated.

The values of total heat transfer in Metric Q_m and thermal resistance R_m have different values than English values:

In Metric units, when

t_{m1} = higher temperature of surface of insulation, °C (Celsius)
t_{m2} = lower temperature of surface of insulation, °C (Celsius)
R_m = thermal resistance
Q_m = total heat transfer, from one surface to the other in W/m²

then

$$Q_m = \frac{(t_{m1} - t_{m2})}{R_m} \text{ W/m}^2$$

and when

k_m = conductivity of the mass insulation in W/mK
ℓ_m = heat path, from one surface to the other, in meters

then

$$R_m = \frac{\ell_m}{k_m}$$

The conversion factors between English and Metric units are:

Insert A

Quantity	***English Units***	***conversion***	***Metric Units***	***conversion***	***English Units***
Temperature	t (°F)	[t (°F) − 32] ÷ 1.8	= t_m (°C)	[t_m (°C) × 1.8] + 32	= t (°F)
	t (°F)	[t (°F) + 459.67] ÷ 1.8	= t_m (°K)	[t_m (°K) × 1.8] − 459.67	= t (°F)
	t (°R)	(°R) ÷ 1.8	= t_m (°K)	× 1.8	= t (°R)
	t (°R)	× [t (°R) ÷ 1.8] − 273.16	= t_m (°C)	(t °C × 1.8) + 491.67	= t (°R)
Temp. interval	Δt (°F)	× Δt (°F) ÷ 1.8	= Δt_m (°C) or Δt_m (°K)	× 1.8	= Δt (°F) or Δt (°R)
Length	ℓ″ (inches)	× 0.0254	= ℓ_m in meters	× 39.37	= ℓ″ (inches)
	ℓ′ (feet)	× 0.3048	= ℓ_m in meters	× 3.281	= ℓ′ (feet)
Area	ft² (sq ft)	× 0.0929	= m² (sq meters)	× 10.764	= ft² (sq ft)
Heat	Btu	× 1055.06	= J (joules)	× 0.000948	= Btu
Heat/time or Power	Btu/hr	× 0.293	= W (watts)	× 3.412	= Btu/hr
Heat/area, time	Q (Btu/ft², hr)	× 3.153	= Q_m (W/m²)	× 0.317	= Q (Btu/ft², hr)
Thermal conductivity	K (Btu in/ft², hr, °F)	× 0.1442	= k_m (W/mK)	× 6.935	= k (Btu in/ft², hr, °F)
Thermal resistance	R (°F, ft², hr, Btu)	× 0.176	= R_m (Km²W)	× 5.678	= R (°F, ft², hr, Btu)
Velocity	V (ft/min)	× 0.00508	= V_m (m/s)	× 196.85	= V (ft/min)

(Note: All energy conversions are based on International Table Units.)

Although the individual heat paths of thermal insulation are seldom measured, it is important that their relationship with insulation in service is understood.

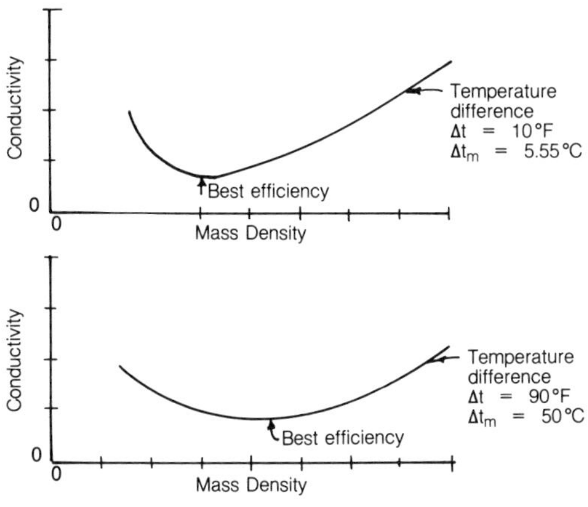

Identical fibers — Difference of density due to compaction. At the greater temperature differences the best efficiency is at higher mass density.

Note: Density variation is due to compaction. At higher temperature differences the best efficiency occurs at higher mass densities. The scale is arbitrary and only intended to show qualitative trends.

THE EFFECT OF MASS DENSITY ON THERMAL CONDUCTIVITY FOR TWO VALUES OF ΔT

Figure 9

The thermal conductivity of mass insulators consisting of flakes, fibers, etc., changes with density. Heat can be more easily transferred across the very light density insulator containing fewer, larger air spaces than when a higher density decreases the size but increases the number of air spaces. Conversely, the conductivity of the solid matter increases as density increases. Thus, for a given mean temperature and temperature difference from surface to surface, there is an optimum density which is the most efficient in reducing heat flow. This is illustrated in Figure 9. These facts are significant in that according to standard ASTM Test Methods C-177, C-518, and C-687. Density, Mean Temperature, Temperature Differences, and Direction of Heat Flow is to be reported with measured Thermal Resistance and Thermal Conductivity. Unfortunately, the manufacturers do not provide this information in most of their advertising so it is difficult to compare products or to know if the values which are given are appropriate for the intended installation.

The basic principle of all mass thermal insulation is that the solid particles will divide volume into a large number of very small air or gas spaces. With many small air spaces air convection currents are unable to form. This type of insulation also takes advantage of the fact that the conductivity of air at 70°F (21°C) is very low; namely, 0.15 Btu in./ft.², hr., °F (0.022 W/mK). As seen in Figure 8, the magnified view of mass insulation shows that these small spaces form the greatest part of the volume occupied by the insulation. If the insulation becomes wet, either due to being subjected to liquid water or by condensation of water vapor, the air spaces are all or partially filled with liquid water. The conductivity of water (with no convection movement) at 70°F (21°C) is 4.3 Btu in./ft.², hr., °F (0.62 W/mK). This conductivity is 28 times greater than that of air.

The thermal conductivities of the gases in cellular products differ with the material, ranging from 0.09 Btu in./ft.², hr., °F (0.013 W/mK) upward. Like fiberous insulation, if these cells become filled with liquid moisture (due to condensation) then the conductivity of the water in each cell is 4.3 Btu in./ft.², hr., °F (0.62 W/mK) and the wet insulation becomes a good conductor of heat.

All thermal insulations are advertised giving their thermal conductivity and/or thermal resistance in a dry state. The specimens tested are conditioned by drying in an oven to a constant weight. If the specimens are foamed plastics, unable to withstand drying temperatures above 212°F (100°C), they are dried in a sealed absorbent and absorbent system containing a desiccator.* The reason for such care is that any small amount of liquid moisture in the samples will greatly affect the thermal conductivity and hence the heat conducting properties of the insulation. This is illustrated in Figure 10, which was prepared from information published by Professor F. A. Joy and Mr. L. Adams, Fellow of ASHRAE. For comparative results, which are reproducible, it is necessary to use test dried insulations. As moisture content makes a great difference in thermal transmission of heat in insulations, accurate laboratory tests for insulations as used in service are not

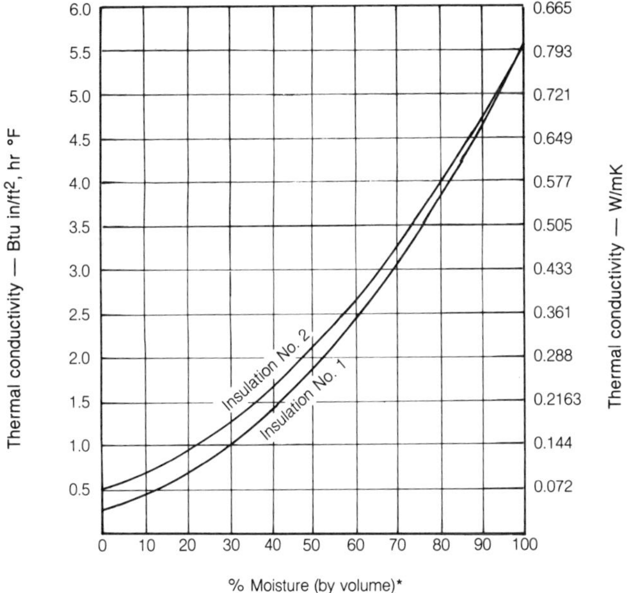

*Volume of air space replaced by liquid condensate

Insulation No. 1, in dry state, has conductivity of 0.25 Btu in/ft², hr, °F (0.036 W/mK)
Insulation No. 2, in dry state, has conductivity of 0.5 Btu in/ft², hr, °F (0.072 W/mK)

EFFECT OF MOISTURE CONDENSATION ON CONDUCTIVITY OF INSULATION

Figure 10

*See Specimen Preparation ASTM C-177 7.22 and ASTM C-518 6.2.2 and 6.2.3.

available. Thus, the advertised thermal conductivity and thermal resistance values must be modified to correctly design insulation for buildings and the results obtained by the use of a particular insulation system. The change in conductivity for two different conductivity materials is shown in Figure 10. It should be noted that regardless of the thermal conductivity of light weight insulations, as they gather water, their conductivity begins to approach the conductivity of that percentage of water in their volume.

Where thermal conductivity increases with water content, the thermal resistance "R" or "R_m" decreases. This is indicated in Tables 1 and 1a.

Not all thermal insulations are affected equally by moisture content, especially at low percentage levels. On glass fibers, the condensate tends to "bead", whereas with absorbent fibers, the condensate tends to "wet" the fiber itself. In cellular plastic insulations, the condensate will form in a plane at the temperature where the dew point occurs. Cells on each side of this plane remain dry until the change of thermal resistance causes a change in the temperature gradient across the insulation resulting in increased condensation in additional cells.

A logical question is why discuss the performance of wet insulation if it is to be installed inside a building where it will be kept dry. Unfortunately, just the opposite is often true. The manner in which insulation is installed in most buildings assures that it will collect condensate water. When it does, the insulation becomes thermally inefficient, and more heat may be lost than the heating system can produce. The control of water vapor is, therefore, essential in order to maintain thermally efficient buildings. Moisture control problems are discussed and design considerations to minimize these problems are presented in the next chapter.

EFFECT OF LIQUID WATER CONDENSATE IN INSULATION ON ITS THERMAL RESISTANCE (R) – ENGLISH UNITS

Insul. Thk. Inch	Btu, in. ft², hr, °F	Values of Thermal Resistance (R) at % Moisture Content												
		Dry	5%	10%	20%	30%	40%	50%	60%	70%	80%	90%	100%	
	k →	0.25	0.33	0.44	0.70	1.00	1.40	1.85	2.42	3.05	3.80	4.61	5.60	← k
1		4.0	3.0	2.2	1.4	1.0	0.7	0.5	0.4	0.3	0.3	0.2	0.2	
2		8.0	6.0	4.5	2.9	2.0	1.4	1.1	0.8	0.7	0.5	0.4	0.4	
3		12.0	9.0	6.8	4.3	3.0	2.1	1.6	1.2	1.0	0.8	0.7	0.5	
4		16.0	12.0	9.1	5.7	4.0	2.9	2.2	1.6	1.3	1.1	0.9	0.7	
5		20.0	15.0	11.4	7.1	5.0	3.6	2.7	2.0	1.6	1.3	1.1	0.9	
6		24.0	18.0	13.6	8.6	6.0	4.3	3.2	2.4	2.0	1.6	1.3	1.1	
7		28.0	21.0	15.9	10.0	7.0	5.0	3.8	2.8	2.3	1.8	1.5	1.3	
8		32.0	24.0	18.2	11.4	8.0	5.7	4.3	3.1	2.6	2.1	1.7	1.4	
9		36.0	27.0	20.6	12.9	9.0	6.4	4.9	3.5	3.0	2.4	2.0	1.6	
10		40.0	30.0	22.7	14.3	10.0	7.1	5.4	3.9	3.3	2.6	2.2	1.8	
	k →	0.50	0.60	0.70	0.92	1.25	1.66	2.10	2.65	3.25	4.00	4.70	5.60	← k
1		2.0	1.6	1.4	1.1	0.8	0.6	0.5	0.4	0.3	0.2	0.2	0.2	
2		4.0	3.3	2.9	2.2	1.6	1.2	1.0	0.8	0.6	0.5	0.4	0.4	
3		6.0	5.0	4.2	3.3	2.4	1.8	1.4	1.1	0.9	0.7	0.6	0.6	
4		8.0	6.7	5.7	4.3	3.2	2.4	1.9	1.5	1.2	1.0	0.8	0.7	
5		10.0	8.3	7.1	5.4	4.0	3.0	2.4	1.9	1.5	1.2	1.1	0.9	
6		12.0	10.0	8.6	6.5	4.8	3.6	2.9	2.3	1.8	1.5	1.3	1.1	
7		14.0	11.7	10.0	7.6	5.6	4.2	3.3	2.6	2.2	1.7	1.5	1.3	
8		16.0	13.3	11.4	8.7	6.4	4.8	3.8	3.0	2.5	2.0	1.7	1.4	
9		18.0	15.0	12.9	9.8	7.2	5.4	4.2	3.4	2.8	2.2	1.9	1.6	
10		20.0	16.6	14.3	10.9	8.0	6.0	4.8	3.8	3.1	2.5	2.1	1.8	

TABLE 1

EFFECT OF LIQUID WATER CONDENSATE IN INSULATION ON ITS THERMAL RESISTANCE (R_m) – METRIC UNITS

Insul. Thk. Meters	W/mK	Values of Thermal Resistance (R_m) at % Moisture Content												
		Dry	5%	10%	20%	30%	40%	50%	60%	70%	80%	90%	100%	
	k_m →	0.036	0.047	0.063	0.101	0.144	0.202	0.267	0.349	0.440	0.548	0.665	0.807	← k_m
0.0254		0.705	0.433	0.317	0.202	0.144	0.101	0.072	0.058	0.043	0.043	0.029	0.028	
0.0508		1.154	0.865	0.649	0.418	0.288	0.202	0.159	0.115	0.101	0.072	0.058	0.058	
0.0762		1.730	1.298	0.981	0.620	0.433	0.303	0.231	0.173	0.144	0.115	0.101	0.072	
0.1016		2.307	1.730	1.312	0.821	0.577	0.418	0.317	0.231	0.187	0.159	0.130	0.101	
0.127		2.884	2.163	1.644	1.023	0.721	0.519	0.389	0.288	0.231	0.187	0.159	0.130	
0.1524		3.461	2.596	1.961	1.240	0.865	0.620	0.461	0.346	0.288	0.231	0.187	0.159	
0.1778		4.037	3.028	2.292	1.442	1.009	0.721	0.548	0.404	0.332	0.260	0.216	0.187	
0.2032		4.614	3.461	2.624	1.644	1.158	0.822	0.620	0.447	0.374	0.302	0.245	0.202	
0.2286		5.191	3.893	2.971	1.860	1.298	0.923	0.706	0.505	0.433	0.346	0.288	0.231	
0.254		5.768	4.326	3.273	2.062	1.442	1.024	0.779	0.562	0.476	0.379	0.317	0.260	
	k_m →	0.072	0.086	0.101	0.133	0.180	0.239	0.302	0.382	0.469	0.577	0.678	0.807	← k_m
0.0254		0.288	0.231	0.202	0.159	0.115	0.087	0.072	0.058	0.043	0.029	0.029	0.028	
0.0508		0.577	0.476	0.418	0.317	0.231	0.173	0.144	0.115	0.087	0.072	0.058	0.058	
0.0762		0.865	0.721	0.606	0.475	0.346	0.260	0.159	0.159	0.130	0.101	0.087	0.087	
0.1010		1.158	0.966	0.822	0.620	0.461	0.346	0.274	0.216	0.173	0.144	0.115	0.101	
0.127		1.442	1.197	1.023	0.779	0.577	0.433	0.346	0.274	0.216	0.173	0.159	0.130	
0.1524		1.730	1.442	1.240	0.937	0.692	0.519	0.418	0.332	0.260	0.216	0.187	0.159	
0.1778		2.019	1.687	1.442	1.096	0.808	0.606	0.475	0.374	0.317	0.245	0.216	0.187	
0.2032		2.307	1.918	1.644	1.254	0.923	0.620	0.548	0.433	0.361	0.288	0.245	0.202	
0.2286		2.596	2.163	1.860	1.413	1.038	0.779	0.606	0.490	0.404	0.317	0.274	0.231	
0.254		2.884	2.393	2.062	1.572	1.154	0.865	0.692	0.598	0.447	0.332	0.302	0.260	

TABLE 1a

2 Fundamentals of Moisture Migration

Our world is soaking wet. Liquid water covers most of its surface. The ground is wet with water. The air contains water vapor and condensate. It rains liquid water. Following the same pattern, it is essential to control water moisture to control heat energy. Even the very definition of heat in the English units tie the two together, as it states that a British thermal unit (Btu) is the amount of heat required to heat one pound of water one degree Fahrenheit.

Water moisture is in two forms—liquid and vapor. Vapor has two basic forms; one is steam which is above the boiling point of water, and the other is water vapor. Water vapor is water in gaseous form but below the critical temperature where it can be liquefied by pressure alone without cooling to a lower temperature.

When a vapor pressure exists on either side of a solid, the vapor on one side will migrate through the solid in its attempt to equalize pressure. The amount of vapor per unit time that migrates through the solid depends upon the vapor permeability of that solid. If that material has any slight voids and if there exists a temperature gradient where the vapor temperature is lowered to liquefication temperature, then the vapor condenses into a liquid. This is the action which occurs in thermal insulation.

In the past, energy was plentiful and inexpensive; thus houses and buildings were not designed or built to conserve energy. Very little thermal insulation was used; windows and doors were not tight and sealed, and most construction masonry materials were not vapor tight. Thus, moisture in vapor form could escape to the atmosphere with little resistance. Now, to conserve energy, the walls and ceilings are filled with insulation; storm windows and doors are common, and cracks are sealed to prevent entry of cold air. With this development, it was overlooked that to accomplish thermal efficiency in walls, floors, and roofs, it was *first* necessary to control moisture. *If moisture is not controlled properly, insulation becomes ineffective and frame houses are likely to rapidly deteriorate due to decay and rot of the wooden members.* For these reasons, it is necessary to understand the properties of water vapor and its migration and condensation into liquid.

Water vapor is a gas which is invisible and which can travel through air at a great speed. It develops its own pressure. The amount of water vapor in the air as compared to the maximum amount that can be maintained in the air at that temperature is called relative humidity. This amount of vapor in the air causes a vapor pressure.

RELATIVE HUMIDITY

DIFFERENCE BETWEEN READING OF WET-BULB AND DRY-BULB TEMPERATURE °F — ENGLISH UNITS

PER CENT RELATIVE HUMIDITY

Eng. Units Dry-Bulb Temp °F	1	2	3	4	5	6	7	8	9	10	11	12	13	14	15	16	17	18	19	20	21	22	23	24	25	26	27	28	29	30	31	32	33	34	35	36	37	38	39	40	Metric Units Dry-Bulb Temp °C
0	70	40	10																																						− 17.8
5	75	50	26	4																																					− 15.0
10	80	60	40	22	4																																				− 12.2
15	84	68	52	34	20	5																																			− 9.4
20	86	72	59	46	33	20	7																																		− 6.7
21	86	73	60	48	34	22	11																																		− 6.1
22	87	74	61	50	36	25	13	2																																	− 5.6
23	87	75	62	51	38	27	16	5																																	− 5.0
24	88	76	64	53	41	30	19	8																																	− 4.4
25	88	77	65	54	42	32	22	11																																	− 3.9
26	88	77	66	55	44	34	24	14	4																																− 3.3
27	88	78	67	56	46	36	26	16	6																																− 2.8
28	89	78	68	57	48	37	28	18	9																																− 2.2
29	89	78	69	58	49	39	30	20	12	3																															− 1.7
30	89	80	70	60	51	41	32	23	14	6																															− 1.1
31	90	80	71	61	52	43	34	26	17	8																															− 0.6
32	90	81	72	62	53	44	36	27	19	10	3																														0
33	90	82	73	63	55	46	38	29	21	13	5																														0.6
34	90	82	74	65	56	48	40	32	24	16	8																														1.1
35	90	82	73	66	58	50	42	34	26	18	10	4																													1.7
36	91	82	74	65	59	51	44	36	28	20	13	6																													2.2
37	91	83	75	66	58	53	45	38	30	22	16	8																													2.8
38	91	83	75	67	59	51	47	39	32	24	17	10	4																												3.3
39	91	83	76	68	60	53	46	41	34	27	20	13	6																												3.9
40	92	84	77	68	61	53	47	39	36	28	22	15	8																												4.4
41	92	84	77	69	62	54	47	40	33	31	24	17	11	5																											5.0
42	92	84	77	70	62	55	48	41	34	28	26	19	13	6																											5.6
43	93	84	78	70	63	56	49	43	36	29	23	22	15	9	3																										6.1
44	93	85	78	71	64	57	50	44	37	31	24	18	17	12	5																										6.7
45	93	86	78	72	65	58	52	45	39	32	26	20	14	13	8	3																									7.2
46	93	86	79	72	65	58	53	46	40	33	27	22	16	11	11	5																									7.8
47	93	86	79	72	66	59	54	47	41	35	29	23	17	12	7	7	2																								8.3
48	93	86	80	73	67	60	54	48	42	36	30	24	19	14	8	4	4																								8.9
49	93	86	80	74	67	61	55	49	43	38	32	26	20	16	10	5	2																								9.4
50	93	87	81	75	68	62	56	50	45	39	33	28	23	18	13	7	3																								10.0
51	94	87	81	75	68	62	56	50	45	39	34	29	23	19	14	8	4																								10.6
52	94	87	81	75	69	63	57	51	46	40	35	30	24	20	15	10	5	2																							11.1
53	94	87	81	75	69	63	58	52	47	41	36	31	25	21	17	12	8	3																							11.7
54	94	88	82	76	70	64	58	53	47	42	37	32	27	23	18	14	9	4																							12.2
55	94	88	82	76	70	65	59	54	49	44	39	34	29	24	20	15	11	5	2																						12.8
56	94	88	82	76	71	66	60	55	50	45	40	35	30	25	21	17	13	8	4																						13.3
57	94	88	83	77	71	66	61	56	51	46	41	36	31	27	23	18	14	9	6	2																					13.9
58	94	88	83	77	72	67	62	57	52	47	42	37	32	28	24	20	15	11	7	4																					14.4
59	94	89	84	77	73	68	62	57	53	48	43	38	34	29	25	21	16	13	8	5	2																				15.0
60	94	89	84	78	73	68	63	58	53	48	44	39	35	31	27	22	18	14	10	6	3																				15.6
61	94	89	84	79	74	69	63	59	54	49	45	41	36	32	28	24	20	15	12	8	4																				16.1
62	94	89	84	79	74	69	64	59	54	50	46	42	37	33	29	25	21	17	13	9	6	3																			16.7
63	95	90	84	79	74	70	65	60	55	51	47	43	38	34	30	26	22	19	14	11	8	4																			17.2
64	95	90	85	80	75	70	65	61	56	52	48	44	39	34	31	27	23	20	16	12	9	6	2																		17.8
65	95	90	85	80	75	71	65	61	57	53	49	45	40	36	32	28	25	21	17	13	11	7	4																		18.3
66	95	90	85	80	75	71	66	62	57	53	49	46	41	37	33	30	26	22	19	15	12	8	5	2																	18.9
67	95	90	85	80	76	71	67	63	58	55	50	47	42	39	34	30	27	23	20	17	13	10	7	4																	19.4
68	95	90	86	81	76	72	67	63	58	55	51	48	43	39	35	31	28	24	21	18	14	11	8	5	2																20.0
69	95	90	86	81	76	72	68	64	59	56	52	48	44	40	36	32	29	25	22	19	15	12	9	6	3																20.6
70	95	90	86	81	76	72	68	64	60	56	52	48	45	41	37	33	30	27	23	20	17	14	10	7	5	2															21.1
71	95	90	86	81	76	72	69	64	60	56	53	49	45	41	38	34	31	28	24	21	18	15	12	8	6	3															21.7
72	95	91	86	82	77	73	69	65	61	57	53	50	46	42	39	35	32	28	25	22	19	16	13	10	7	4	2														22.2
73	95	91	86	82	77	73	69	65	61	58	54	50	46	43	39	36	32	29	26	23	20	17	14	11	8	5	3														22.8
74	95	91	86	82	78	74	70	65	62	58	54	51	47	43	40	36	33	30	27	24	21	18	15	12	9	6	4	2													23.3
75	95	91	87	82	78	74	70	66	62	58	55	52	48	45	41	37	34	31	28	25	22	19	16	13	11	8	6	3													23.9
76	96	91	87	83	78	75	70	66	63	59	55	52	49	45	42	38	35	32	29	26	23	20	17	15	12	9	7	4	2												24.4
77	96	91	87	83	78	75	71	67	64	59	56	53	49	46	42	39	36	33	30	27	24	21	18	16	13	11	8	5	3												25.0
78	96	91	87	83	79	75	71	67	64	60	57	53	50	46	43	40	37	34	31	27	25	22	19	17	14	12	9	6	4												25.6
79	96	91	87	83	79	75	71	67	64	60	57	54	50	47	44	40	37	34	32	28	26	23	20	18	16	13	10	7	5	3											26.1
80	96	91	87	83	79	75	72	68	64	61	57	54	50	47	44	41	38	35	32	29	27	24	21	19	17	14	11	8	6	5	2										26.7
82	96	92	88	84	80	76	72	69	65	62	58	55	51	48	45	42	39	37	34	31	28	25	23	21	18	16	13	11	9	7	4										27.8
84	96	92	88	84	80	76	73	69	66	63	59	57	53	50	47	44	41	38	35	32	30	28	25	22	20	17	15	12	11	8	5	4	2								28.9
86	96	92	88	84	80	77	73	70	66	63	60	57	54	51	48	45	42	39	37	34	32	29	26	24	22	19	17	14	13	11	8	6	4	2							30.0
88	96	92	88	85	81	78	74	70	67	64	61	58	55	52	49	46	43	41	38	35	33	31	28	26	24	21	18	17	15	13	11	8	6	4	2						31.1
90	96	92	89	85	81	78	74	71	68	65	61	59	55	53	50	47	44	42	39	37	34	32	29	27	25	22	20	18	16	14	12	10	8	6	4	2					32.2
92	96	92	89	85	82	79	75	72	68	65	62	59	56	54	51	48	46	43	41	38	36	33	31	28	26	24	22	20	18	16	14	12	10	8	6	4	2				33.3
94	96	93	89	85	82	79	75	72	69	66	63	60	57	55	52	49	47	44	42	39	37	34	32	30	28	26	24	21	19	17	15	13	12	10	8	7	5	3			34.4
96	96	93	89	86	82	79	76	73	70	67	64	61	58	56	53	50	48	45	43	40	38	35	33	31	29	27	25	23	21	19	17	15	14	12	10	8	6	4	3		35.6
98	96	93	89	86	82	79	77	73	70	67	64	62	59	57	54	51	49	46	44	41	39	36	34	32	30	28	26	24	22	20	18	16	15	13	11	9	8	6	4	3	36.7
100	96	93	89	86	83	80	77	74	71	68	65	62	60	57	54	51	49	47	44	42	40	37	35	33	31	29	27	25	23	22	19	18	16	14	13	11	9	8	6	5	37.8
Eng. Units Dry-Bulb Temp °F	0.5	1.1	1.6	2.2	2.7	3.3	3.8	4.4	5.0	5.5	6.1	6.6	7.2	7.7	8.3	8.8	9.4	10.0	10.5	11.1	11.6	12.3	12.7	13.3	13.5	14.4	15.0	15.5	16.1	16.6	17.2	17.7	18.3	18.8	19.4	20.0	20.5	21.1	21.6	22.2	Metric Units Dry-Bulb Temp °C

DIFFERENCE BETWEEN READING OF WET-BULB AND DRY-BULB CELCIUS TEMPERATURE °C — METRIC UNITS

NOTE: Values above heavy line are for ice, while those below line are for water.

TABLE 2

This vapor pressure affects the height of the mercury column of a barometer. In fact, the difference in the height of a mercury column produced by dry air and the height of that column when it contains moisture vapor is the pressure exerted by the vapor. In the British system of measurement, the height difference is measured in inches of mercury (" of Hg) and in the Metric system in centimeters of mercury (cm of Hg). These vapor pressures can be converted into English units, such as pounds per square foot, or metric (S.I.) units, such as pascals. (Note: 1 pascal or 1 Pa = 1 Newton/m^2).

The amount of moisture in the air varies from none to that of saturation. Saturated air occurs when it contains all the moisture in vapor form that it can retain. Any additional moisture or water vapor which might be put into the air would condense into a liquid. The relative amount of moisture that is contained in air at a given temperature to the maximum amount of moisture which could be held by the air at that same temperature is known as *relative humidity*.

Heat energy is required to change a liquid to a vapor. This is a fact that we all know and understand. For example; if we are uncomfortably hot, we get into a breeze of air flow which evaporates the moisture from the skin, and we are cooled due to the vaporization of that moisture. Using this principle gives an easy way for an accurate measurement of relative humidity by use of a sling psychrometer. This basic instrument for measuring wet and dry bulb temperatures is shown in Figure 11. The psychrometer is based on the law of evaporation. The whirling of the thermometer in air causes the air movement over the wet bulb to evaporate, causing a cooling effect. The rate of evaporation, thus cooling, is dependent upon the vapor content in the air. The evaporation and cooling continue until the water is cooled to a temperature at which its vapor pressure is just sufficient to keep driving off water molecules at a constant rate. The temperature indicated on that

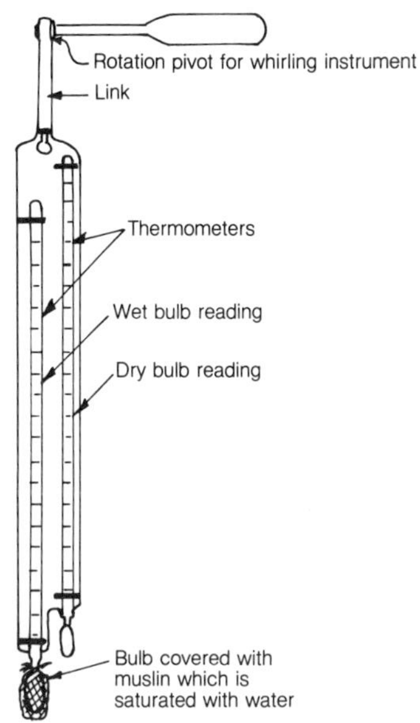

SLING PSYCHROMETER

Figure 11

thermometer is the *"wet bulb" temperature*. The other thermometer indicates the familiar *"dry bulb"* temperature.

Table 2 gives the *"dry bulb"* and *"wet bulb" temperatures* with respect to the relative humidity (RH) of the air. This is in both Celsius and Fahrenheit temperatures.

The *amount of moisture* contained in the air in a vapor state at specific combinations of air temperature and relative humidity is shown in English units in Table 3 and in Metric units in Table 4. This water vapor produces a *vapor pressure*. The amount of vapor in the air causes a change in this pressure. The vapor pressure of air containing the maximum amount of moisture (saturated) is shown in Table 5.

Air does not always contain the maximum amount of vapor. The percentages of the total it can contain are called % relative humidity. Dew point temperature is the temperature at which a cold surface will cause the air at that relative humidity of vapor to become saturated and condense as drops of water (dew). When a body or surface is at a lower temperature than the *dew point*, condensation of water will form on that surface. This act of condensation releases heat energy from the vapor and tends to increase the surface temperature up to the dew point temperature. The *dew point* temperatures in °F are given in Table 6 and in °C in Table 7.

"Sweating" occurs when moisture vapor condenses on a cool surface (at or below dew point). Sweating is a dew formation on a surface such as a cold glass of iced liquid or on the inner surfaces of window panes. This sweating (or condensation) can occur in walls, roofs, ceilings, and floors of buildings. When it does occur, it makes insulation inefficient, rots and decays wood, and causes deterioration of masonry materials. Unfortunately, as it is hidden from sight and the damage occurs slowly, it often goes unnoticed for long periods of time in most homes and buildings. Also, unfortunately with our tightening up of buildings and homes to conserve energy, many are being ruined by condensation in these structures. This is the reason for the graphs and tables of moisture in air. The basics must first be understood in order to properly design and build houses and other buildings which require heating in the winter for human comfort.

In the past, most buildings and homes were not built to conserve energy. Few storm windows and storm doors were installed, and for the most part, they leaked air and heat. Under these conditions, the moisture inside also leaked out with the air and through cracks and openings to the outside. Also, much less moisture was added to the interior. Occasionally, it might be observed that an owner of a wood frame house had difficulty keeping the paint from blistering. In these cases, the practice was to scrape off the blisters then repaint with oil base paint. Of course, the result was even more blisters because the paint resisted moisture vapor transmission greater than any other portion of the wall; thus, the vapor pressure pushed paint film from the wood in its effort to get to the lower pressure outdoors in the winter. The vapor also condensed into a liquid under the paint and in the wood if those were below the dew point.

With today's need for conservation of fuel, much of the literature advises that the outside of the house be caulked and sealed. The house should be sealed to prevent vapor migration on the inside or to a location where the temperature of the surface is

MOISTURE IN AIR *(English Units)*

% Rel. Hum.	GRAINS PER CUBIC FOOT AIR TEMPERATURE DEGREES FAHRENHEIT (F)												
	−20	−10	0	10	20	30	40	50	60	70	80	90	100
10	0.017	0.028	0.048	0.078	0.124	0.194	0.285	0.408	0.574	0.798	1.093	1.479	1.977
20	0.033	0.057	0.096	0.155	0.247	0.387	0.570	0.815	1.149	1.596	2.187	2.958	3.953
30	0.050	0.086	0.144	0.233	0.370	0.580	0.855	1.223	1.724	2.394	3.280	4.437	5.930
40	0.066	0.114	0.192	0.310	0.494	0.774	1.140	1.630	2.298	3.192	4.374	5.916	7.906
50	0.083	0.142	0.240	0.388	0.618	0.968	1.424	2.038	2.872	3.990	5.467	7.395	9.883
60	0.100	0.171	0.289	0.466	0.741	1.161	1.709	2.446	3.447	4.788	6.560	8.874	11.860
70	0.116	0.200	0.337	0.543	0.864	1.354	1.994	2.853	4.022	5.586	7.654	10.353	13.836
80	0.133	0.228	0.385	0.621	0.988	1.548	2.279	3.261	4.596	6.384	8.747	11.832	15.813
90	0.149	0.256	0.433	0.698	1.122	1.742	2.564	3.668	5.170	7.182	9.841	13.311	17.789
100	0.166	0.285	0.481	0.776	1.235	1.935	2.849	4.076	5.745	7.980	10.934	14.790	19.766

Moisture is in vapor state.

TABLE 3

MOISTURE IN AIR *(Metric Units)*

% Rel. Hum.	GRAMS PER CUBIC METERS AIR TEMPERATURE DEGREES CELSIUS (C)												
	−25	−20	−15	−10	−5	0	5	10	15	20	25	30	35
10	0.054	0.091	0.144	0.220	0.332	0.485	0.680	0.933	1.208	1.769	3.297	3.034	3.954
20	0.107	0.183	0.281	0.439	0.661	0.963	1.322	1.865	2.553	3.448	4.599	6.063	7.907
30	0.171	0.277	0.432	0.659	0.991	1.453	1.984	2.798	3.383	5.239	6.896	9.092	11.838
40	0.237	0.368	0.574	0.919	1.322	1.938	2.981	3.729	5.104	6.896	9.218	11.898	15.812
50	0.263	0.460	0.718	1.098	1.645	2.413	3.267	4.663	6.381	8.621	11.518	15.133	19.766
60	0.327	0.554	0.863	1.317	1.977	2.908	4.473	5.596	7.658	10.342	13.790	18.212	23.731
70	0.382	0.645	1.007	1.522	2.313	3.391	5.226	6.486	8.935	12.090	16.098	21.220	28.131
80	0.446	0.736	1.146	1.757	2.624	3.876	5.731	7.461	10.211	13.790	18.391	24.239	31.625
90	0.487	0.826	1.295	1.975	2.977	4.347	6.704	8.392	11.486	15.513	20.645	27.287	35.578
100	0.565	0.920	1.437	2.230	3.306	4.869	7.452	9.326	12.762	17.236	22.985	30.188	39.498

Moisture is in vapor state.

TABLE 4

VAPOR PRESSURES IN AIR AT 100% RELATIVE HUMIDITY

Above ice at temperature indicated					
Vapor Saturated Air (Dew Point) Temperature		Vapor Pressure			
		English Units		Metric Units	
°F	°C	In. of Hg	Lbs/sq ft	mm of Hg	Pa
−40	−40	.0039	.2758	0.099	13.2
−35	−37.2	.0052	.3678	0.132	17.6
−30	−34.4	.0070	.4951	0.178	23.7
−25	−31.7	.0094	.6648	0.239	31.8
−20	−28.9	.0126	.8912	0.320	42.7
−15	−26.1	.0167	1.181	0.424	56.6
−10	−23.3	.0220	1.556	0.559	74.5
− 5	−20.6	.0289	2.044	0.734	97.8
0	−17.8	.0377	2.667	0.957	127.7
1	−17.2	.0397	2.808	1.000	134.4
2	−16.7	.0419	2.964	1.064	141.8
3	−16.1	.0441	3.119	1.120	149.3
4	−15.6	.0464	3.282	1.178	157.1
5	−15	.0488	3.452	1.239	165.3
6	−14.4	.0514	3.636	1.306	174.1
7	−13.9	.0542	3.869	1.376	183.5
8	−13.3	.0570	4.032	1.448	193.0
9	−12.8	.0599	4.237	1.521	202.8
10	−12.2	.0629	4.449	1.598	213.0
11	−11.7	.0661	4.675	1.679	223.8

Above ice at temperature indicated					
Vapor Saturated Air (Dew Point) Temperature		Vapor Pressure			
		English Units		Metric Units	
°F	°C	In. of Hg	Lbs/sq ft	mm of Hg	Pa
12	−11.1	.0695	4.916	1.765	235.4
13	−10.6	.0730	5.163	1.854	247.2
14	−10.0	.0767	5.425	1.948	259.7
15	− 9.4	.0806	5.701	2.047	272.9
16	− 8.9	.0847	5.990	2.151	286.8
17	− 8.3	.0889	6.288	2.258	301.0
18	− 7.6	.0933	6.599	2.370	316.9
19	− 7.2	.0979	6.924	2.487	331.5
20	− 6.7	.1028	7.271	2.611	348.1
21	− 6.1	.1078	7.625	2.738	365.1
22	− 5.6	.1131	7.999	2.873	383.0
23	− 5.0	.1186	8.388	3.012	401.6
24	− 4.4	.1243	8.792	3.157	420.9
25	− 3.9	.1303	9.216	3.310	441.2
26	− 3.3	.1366	9.662	3.469	462.6
27	− 2.8	.1432	10.128	3.637	484.9
28	− 2.2	.1500	10.609	3.810	507.9
29	− 1.7	.1571	11.112	3.990	532.0
30	− 1.1	.1645	11.635	4.178	557.0
31	− 0.6	.1723	12.186	4.376	583.5

TABLE 5

VAPOR PRESSURES IN AIR AT 100% RELATIVE HUMIDITY

Above ice at temperature indicated					
Vapor Saturated Air (Dew Point) Temperature		Vapor Pressure			
		English Units		Metric Units	
°F	°C	In. of Hg	Lbs/sq ft	mm of Hg	Pa
32	0.0	.1803	12.752	4.580	610.6
33	0.6	.1878	13.283	4.770	636.0
34	1.1	.1955	13.827	4.966	662.0
35	1.7	.2035	14.393	5.170	689.1
36	2.2	.2118	14.981	5.380	717.2
37	2.8	.2203	15.582	5.596	746.0
38	3.3	.2292	16.211	5.822	776.2
39	3.9	.2383	16.855	6.053	807.0
40	4.4	.2478	17.527	6.294	839.1
41	5.0	.2576	18.220	6.551	872.3
42	5.6	.2677	18.934	6.799	906.5
43	6.1	.2782	19.677	7.066	942.1
44	6.7	.2891	20.448	7.343	979.0
45	7.2	.3004	21.247	7.630	1017.3
46	7.8	.3120	22.068	7.925	1056.6
47	8.3	.3240	22.916	8.230	1097.2
48	8.9	.3364	23.794	8.545	1139.2
49	9.4	.3493	24.706	8.872	1182.9
50	10.0	.3626	25.646	9.210	1227.9
51	10.6	.3764	26.623	9.561	1274.6
52	11.1	.3906	27.627	9.921	1322.7
53	11.7	.4052	28.659	10.292	1372.2
54	12.2	.4203	29.728	10.676	1423.3
55	12.8	.4359	30.831	11.072	1476.1
56	13.3	.4520	31.970	11.481	1530.6
57	13.9	.4686	33.144	11.902	1588.9
58	14.4	.4858	34.360	12.339	1645.1
59	15.0	.5035	35.612	12.789	1705.0
60	15.6	.5218	36.906	13.254	1767.0
61	16.1	.5407	38.244	13.734	1831.0
62	16.7	.5601	39.616	14.226	1896.7
63	17.2	.5802	41.037	14.737	1964.8
64	17.8	.6009	42.501	15.263	2034.9
65	18.3	.6222	44.008	15.804	2107.0
66	18.9	.6442	45.564	16.363	2181.5
67	19.4	.6669	47.169	16.939	2258.4
68	20.0	.6903	48.825	17.534	2337.6
69	20.6	.7144	50.529	18.146	2419.2
70	21.1	.7392	52.283	18.776	2503.2
71	21.7	.7648	54.044	19.426	2589.9
72	22.2	.7912	55.961	20.096	2679.3
73	22.8	.8183	57.878	20.785	2771.1
74	23.3	.8462	57.311	21.493	2865.6
75	23.9	.8750	61.889	22.225	2963.1
76	24.4	.9046	63.982	22.977	3063.3
77	25.0	.9352	66.147	23.754	3166.9
78	25.6	.9666	68.367	24.552	3273.3
79	26.1	.9989	70.652	25.372	3382.7
80	20.7	1.032	72.003	26.213	3494.7
81	27.2	1.066	75.398	27.076	3609.9
82	27.8	1.102	77.944	27.991	3731.8
83	28.3	1.138	80.490	28.905	3853.7
84	28.9	1.175	83.108	29.845	3979.0
85	29.4	1.213	85.795	30.810	4107.7
86	30.0	1.253	88.625	31.826	4243.1

Above ice at temperature indicated					
Vapor Saturated Air (Dew Point) Temperature		Vapor Pressure			
		English Units		Metric Units	
°F	°C	In. of Hg	Lbs/sq ft	mm of Hg	Pa
87	30.6	1.293	91.454	32.842	4378.6
88	31.1	1.335	92.425	33.909	4520.8
89	31.7	1.378	97.466	35.001	4666.4
90	32.2	1.422	99.578	36.119	4815.4
91	32.8	1.467	103.76	37.262	4967.8
92	33.3	1.513	107.01	38.430	5123.5
93	33.9	1.561	110.41	39.649	5286.1
94	34.4	1.610	113.87	40.894	5452.1
95	35.0	1.660	117.41	42.164	5621.4
96	35.6	1.712	121.09	43.485	5797.5
97	36.1	1.765	124.84	44.831	5977.0
98	36.7	1.819	128.65	46.202	6159.8
99	37.2	1.875	132.62	47.620	6349.0
100	37.8	1.933	136.72	49.090	6545.0
101	38.3	1.992	140.89	50.590	6746.0
102	38.9	2.052	145.14	52.120	6949.0
103	39.4	2.114	149.52	53.690	7158.0
104	40.0	2.178	154.05	55.320	7375.0
105	40.6	2.240	158.43	56.890	7585.0
106	41.1	2.310	163.39	58.670	7823.0
107	41.7	2.380	168.34	60.450	8060.0
108	42.2	2.450	173.29	62.230	8297.0
109	42.8	2.520	178.23	64.000	8533.0
110	43.3	2.600	183.89	66.040	8805.0
112	44.4	2.750	194.51	69.850	9313.0
114	45.6	2.910	205.82	73.910	9854.0
116	46.7	3.080	217.85	78.230	10430.0
118	47.8	3.260	230.58	82.800	11039.0
120	48.9	3.450	244.02	87.630	11683.0
122	50.0	3.640	257.45	92.460	12326.0
124	51.1	3.850	272.31	97.790	13038.0
126	52.2	4.060	287.16	103.120	18749.0
128	53.3	4.290	303.43	108.970	14528.0
130	54.4	4.530	320.41	115.060	15340.0
132	55.6	4.770	377.38	121.160	16153.0
134	56.7	5.030	355.77	127.760	17033.0
136	57.8	5.300	374.87	134.620	17948.0
138	58.9	5.590	395.38	141.990	18930.0
140	60.0	5.880	415.89	149.350	19912.0
142	61.1	6.190	437.82	157.230	20962.0
144	62.2	6.510	460.45	165.350	22045.0
146	63.3	6.850	484.50	173.990	23197.0
148	64.4	7.200	509.25	182.880	24381.0
150	65.6	7.570	535.42	192.280	25634.0
152	66.7	7.950	562.20	201.930	26922.0
154	67.8	8.350	590.59	212.090	28279.0
156	68.9	8.770	620.30	222.760	29669.0
158	70.0	9.200	650.71	233.680	31155.0
160	71.1	9.650	682.54	245.110	32679.0
162	72.2	10.120	715.78	257.050	34270.0

TABLE 5 (Continued)

DEW POINT TEMPERATURES *(English Units)*

% Rel. Hum.	DEGREES FAHRENHEIT (F) DRY BULB TEMPERATURE °F																	
	5	**10**	**15**	**20**	**25**	**30**	**35**	**40**	**45**	**50**	**55**	**60**	**65**	**70**	**75**	**80**	**85**	**90**
10	−35	−31	−28	−24	−20	−15	−12	−7	−4	−1	3	6	10	13	17	20	23	27
20	−25	−20	−16	−11	−8	−3	1	5	9	13	16	20	24	28	32	35	40	44
30	−17	−13	−8	−4	0	5	9	14	17	21	25	29	33	37	42	46	50	54
40	−12	−7	−3	2	6	11	15	19	23	27	32	35	40	45	49	54	58	62
50	−8	−3	1	6	10	15	20	24	28	32	37	42	46	50	55	60	64	69
60	−5	0	5	10	15	20	24	28	32	37	41	46	51	55	60	65	69	74
70	−2	3	8	13	18	23	27	31	36	41	45	50	55	60	64	69	74	79
80	1	5	10	15	20	25	30	35	39	44	49	54	59	64	69	74	78	83
90	3	8	13	18	23	28	33	38	43	47	52	57	62	67	72	77	82	87
100	5	10	15	20	25	30	35	40	45	50	55	60	65	70	75	80	85	90

Surface temperatures at which air at listed temperature and relative humidity will start condensing water vapor in the air to liquid on surface.

TABLE 6

DEW POINT TEMPERATURES *(Metric Units)*

% Rel. Hum.	DEGREES CELSIUS (C) DRY BULB TEMPERATURE °C											
	−15	**−10**	**−5**	**0**	**5**	**10**	**15**	**20**	**25**	**30**	**35**	**40**
10	−37	−38	−29	−25	−21	−18	−14	−11	−8	−5	−1	3
20	−32	−27	−22	−18	−15	−11	−7	−3	0	4	9	13
30	−27	−22	−18	−14	−10	−6	−2	1	5	10	15	19
40	−24	−19	−14	−11	−7	−3	2	5	9	14	19	24
50	−22	−17	−12	−9	−4	0	5	9	13	17	23	27
60	−20	−15	−9	−5	−2	2	8	12	15	20	26	31
70	−19	−13	−8	−3	0	4	10	14	17	23	29	34
80	−17	−12	−7	−2	2	6	12	16	19	25	31	36
90	−16	−11	−6	−1	3	8	14	18	21	28	33	38
100	−15	−10	−5	0	5	10	15	20	25	30	35	40

Surface temperatures at which air at listed temperature and relative humidity will start condensing water vapor in the air to liquid on surface.

TABLE 7

below the dew point. The outside of a building must be built to be water tight or to shed water. The inside must contain vapor retarders (barriers) to prevent the vapor pressure on the inside from forcing water vapor into the wall or roof where it drops to the dew point and condenses into liquid water. Table 8 provides the vapor pressures of air at various temperatures and relative humidities; it also gives the dew point temperature of the air.

In a report from U.S. Department of Commerce, National Bureau of Standards "Building Materials and Structures" Report BM 563 "Moisture Condensation in Building Walls" by Harold W. Woolley, the basic relationships of condensation in masonry and frame walls were established. This report showed that as the thermal resistance of the walls increased it was necessary to increase the vapor resistance so as to prevent condensation in the wall.

While methods by which condensation in insulated walls may be avoided are becoming fairly well known, a little consideration of the principles already explained here shows that possible remedies will include:

1. Lowering the indoor relative humidity either by lowering the rate at which water vapor is added to the air in the house or by increasing the ventilation of the house interior to the outdoors.
2. Increasing the vapor resistance on the warm side of the insulation.
3. Lowering the vapor resistance on the cold side of the insulation by venting to the outside air or by using less vapor-resistant outer wall materials without diminishing protection against driving rains.

Unfortunately, over the years these principles seem to have been forgotten; as it has become commmon practice to:

1. Raise the relative humidity in the inside air by use of humidifiers, clothes washers and driers, and shower baths. (The problem has been made worse by caulking and the installation of storm windows, etc.).

VAPOR PRESSURE OF AIR AT VARIOUS TEMPERATURES AND RELATIVE HUMIDITIES

Air Temp.		% Rel. Hum.	Dew Point Temp.		VAPOR PRESSURE			
					ENGLISH UNITS		METRIC UNITS	
°F	°C		°F	°C	In. of Hg	Lbs./ft^2	mm of Hg	Pa
		10	7	−13.9	0.054	3.87	1.37	183
		20	7	−13.6	0.055	3.94	1.39	188
		30	8	−13.3	0.057	4.03	1.44	193
		40	8	−13.1	0.058	4.10	1.47	196
10	−12.2	50	8	−13.0	0.059	4.17	1.49	199
		60	9	−12.8	0.060	4.24	1.52	202
		70	9	−12.6	0.061	4.33	1.54	206
		80	9	−12.4	0.062	4.41	1.57	207
		90	10	−12.2	0.063	4.50	1.60	213
		100	10	−12.0	0.066	4.59	1.67	220
		10	16	− 8.9	0.085	5.99	2.16	286
		20	16	− 8.6	0.087	6.14	2.21	293
		30	17	− 8.3	0.089	6.29	2.26	301
		40	17	− 8.1	0.091	6.44	2.31	309
20	−6.7	50	18	− 7.6	0.093	6.60	2.36	317
		60	18	− 7.4	0.095	6.70	2.41	323
		70	18	− 7.3	0.097	6.81	2.46	327
		80	19	− 7.2	0.098	6.92	2.49	331
		90	19	− 7.0	0.101	7.09	2.56	339
		100	20	− 6.7	0.103	7.27	2.62	348
		10	21	− 6.1	0.108	7.63	2.69	365
		20	22	− 5.6	0.113	8.00	2.86	383
		30	23	− 5.0	0.119	8.39	3.02	402
		40	24	− 4.4	0.124	8.79	3.15	420
30	−1.1	50	25	− 3.9	0.130	9.22	3.30	441
		60	26	− 3.3	0.137	9.66	3.47	463
		70	27	− 2.8	0.143	10.13	3.63	485
		80	28	− 2.2	0.150	10.61	3.81	507
		90	29	− 1.7	0.157	11.11	3.98	532
		100	30	− 1.1	0.165	11.63	4.19	557
		10	27	− 2.8	0.143	10.13	3.63	485
		20	28	− 2.2	0.150	10.61	3.81	507
		30	29	− 1.7	0.157	11.11	3.99	532
		40	31	− 0.6	0.172	12.19	4.37	583
40	4.4	50	33	0.6	0.188	13.28	4.78	636
		60	35	1.7	0.204	14.39	5.18	689
		70	37	2.8	0.220	15.58	5.59	746
		80	38	3.3	0.229	16.21	5.82	776
		90	39	3.9	0.238	16.86	6.04	807
		100	40	4.4	0.248	17.53	6.30	839

TABLE 8

VAPOR PRESSURE OF AIR AT VARIOUS TEMPERATURES AND RELATIVE HUMIDITIES

Air Temp.		% Rel. Hum.	Dew Point Temp.		VAPOR PRESSURE			
					ENGLISH UNITS		METRIC UNITS	
°F	°C		°F	°C	In. of Hg	Lbs./ft²	mm of Hg	Pa
		10	35	1.7	0.204	14.39	5.18	689
		20	37	2.8	0.220	15.58	5.59	746
		30	39	3.9	0.238	16.86	6.04	807
		40	40	4.4	0.248	17.53	6.30	839
50	10.0	50	42	5.6	0.268	18.93	6.81	906
		60	44	6.7	0.289	20.45	7.34	979
		70	45	7.2	0.300	21.25	7.62	1017
		80	47	8.3	0.324	22.92	8.23	1097
		90	49	9.4	0.349	24.70	8.86	1182
		100	50	10.0	0.363	25.65	9.22	1227
		10	41	5.0	0.258	18.22	6.55	872
		20	43	6.1	0.278	19.68	7.06	942
		30	46	7.8	0.312	22.07	7.92	1056
		40	48	8.9	0.336	23.79	8.53	1139
60	15.6	50	50	10.0	0.363	25.65	9.22	1227
		60	52	11.1	0.391	27.63	9.93	1323
		70	54	12.2	0.420	29.73	10.67	1423
		80	56	13.3	0.452	31.97	11.48	1530
		90	58	14.4	0.486	34.36	12.34	1645
		100	60	15.6	0.522	36.91	13.26	1767
		10	44	6.7	0.289	20.45	7.34	979
		20	47	8.3	0.324	22.92	8.23	1097
		30	49	9.4	0.349	24.70	8.86	1182
		40	52	11.1	0.391	27.63	9.93	1323
65	18.9	50	54	12.2	0.420	29.73	10.67	1423
		60	57	13.9	0.469	33.14	11.91	1588
		70	59	15.0	0.504	35.61	12.80	1705
		80	61	16.1	0.541	38.24	13.74	1831
		90	63	17.2	0.580	41.04	14.73	1964
		100	65	18.3	0.622	42.00	15.80	2107
		10	47	8.3	0.324	22.92	8.23	1097
		20	50	10.0	0.363	25.65	9.22	1227
		30	53	11.7	0.405	28.66	10.29	1372
		40	56	13.3	0.452	31.97	11.48	1530
70	21.1	50	59	15.0	0.504	35.61	12.80	1705
		60	61	16.1	0.541	38.24	13.74	1831
		70	63	17.2	0.580	41.04	14.73	1964
		80	66	18.9	0.644	45.56	16.36	2181
		90	68	20.0	0.690	48.82	17.53	2337
		100	70	21.1	0.739	52.28	18.77	2503

TABLE 8 (Continued)

VAPOR PRESSURE OF AIR AT VARIOUS TEMPERATURES AND RELATIVE HUMIDITIES

Air Temp.		% Rel. Hum.	Dew Point Temperature		VAPOR PRESSURE			
					ENGLISH UNITS		METRIC UNITS	
°F	°C		°F	°C	In. of Hg	Lbs./ft^2	mm of Hg	Pa
75	23.9	10	50	10.0	0.363	25.65	9.22	1227
		20	53	11.7	0.405	28.66	10.29	1327
		30	57	13.9	0.469	33.14	11.91	1588
		40	60	15.6	0.522	36.91	13.26	1767
		50	63	17.2	0.580	41.04	14.73	1964
		60	65	18.3	0.622	44.01	15.80	2107
		70	68	20.0	0.690	48.82	17.53	2337
		80	70	21.1	0.739	52.28	18.77	2503
		90	73	22.8	0.818	57.88	20.78	2771
		100	75	23.9	0.875	61.89	22.22	2963
80	26.7	10	53	11.7	0.405	28.66	10.29	1327
		20	57	13.9	0.469	33.14	11.91	1588
		30	60	15.6	0.522	36.91	13.26	1767
		40	64	17.8	0.601	42.50	15.26	2034
		50	67	19.4	0.667	47.17	16.94	2258
		60	70	21.1	0.739	52.28	18.77	2503
		70	72	22.2	0.791	55.96	20.09	2679
		80	75	23.9	0.875	61.89	22.22	2963
		90	78	25.6	0.966	68.37	24.54	3273
		100	80	26.7	1.032	72.99	26.21	3494
85	29.4	10	55	12.8	0.486	30.83	11.07	1476
		20	60	15.6	0.522	36.91	13.26	1767
		30	64	17.8	0.601	42.50	15.26	2039
		40	67	19.4	0.667	47.17	16.94	2258
		50	71	21.7	0.765	54.04	19.43	2589
		60	74	23.3	0.846	57.31	21.49	2865
		70	77	25.0	0.935	66.15	23.74	3166
		80	80	26.7	1.032	72.99	26.21	3494
		90	83	28.3	1.138	80.49	28.90	3853
		100	85	29.4	1.213	85.80	30.81	4107
90	32.2	10	58	14.4	0.486	34.36	12.34	1645
		20	63	17.2	0.580	41.04	14.73	1964
		30	68	20.0	0.690	48.82	17.53	2337
		40	71	21.7	0.765	54.04	19.43	2589
		50	75	23.9	0.875	61.89	22.22	2963
		60	79	26.1	0.999	70.65	25.37	3382
		70	82	27.8	1.102	77.94	27.99	3831
		80	86	30.0	1.253	88.63	31.82	4243
		90	88	31.1	1.335	92.43	33.91	4521
		100	90	32.2	1.422	99.58	36.12	4815

TABLE 8 (Continued)

2. Ignore the extreme importance of an inner vapor barrier, including the necessity for sealing around projections such as electrical outlet boxes.
3. Add vapor barriers on the outside in the form of low vapor permeability organic foams and make these even more damaging by covering them on the outside with aluminum foil.

This publication on "Moisture Condensation in Building Walls" illustrated the relationship in graphs based on a relatively homogeneous wall using a vapor resistance fraction which was plotted in reference to thermal resistance fraction.

Also, in this publication is an example which illustrates a method for the calculation of determining whether condensation will occur. This problem is illustrated using a very common type of brick wall.

The problem presents an 8″ thick brick wall with air space by furring strips with gypsum plaster board and plaster. The air temperature indoors is given as 70 °F at 30% RH, and outdoor air is 10 °F at 80% RH. By using vapor resistances and thermal resistances, it is possible to determine where the air at a particular vapor pressure will reach the dew point within the wall structure. The dew point temperatures can be determined from Table 6 or 7.

In this problem, the thermal resistances in English Units are:

Air to surface of plaster	0.61
Gypsum lath and plaster	0.36
Air space	0.78
Common Brick 4″	0.80
Face Brick 4″	0.44
Surface of face brick to outside air	0.17

The vapor resistances of the wall are:

Air to surface of plaster	0.001
Lath and plaster	0.045
Air space (between plaster and brick)	0.014
Brick 8″	0.90

The question is whether or not the vapor will condense on the inner surface of the brick, thus causing the air space to become saturated with water.

SOLUTION

The sum of the first three thermal resistances is 1.75 and the total is 3.16

$$1.75 \div 3.16 = 0.554$$

As total temperature difference is 70 − 10 = 60 °F, 60 °F × 0.55 = 33 °F; 70 − 33 = 37 °F on the brick surface facing the air space.

The vapor pressure change is determined in the same manner.

The sum of the first three is 0.060 and the total is 0.960; 0.060 ÷ 0.960 = 0.0625. As the vapor pressure for 70 °F, 30% RH is 15.6 lbs./ft.^{2} then (15.6 × 0.0625) = .975

$$15.6 - .975 = 14.625$$

From Table 5 the vapor pressure of saturated air at 37 °F is 15.582 lbs./ft^2.

Under these conditions, the surface of the inner surface of bricks would not condense moisture and the air space would stay dry. However, if the humidity of the interior were raised to 50% RH, then the indoor vapor pressure would be 25.6 lbs./ft^2 and thus 25.6 − (25.6 × 0.0625) = 25.6 − 1.6 = 24.0. The dew point is then approximately 48 °F; and under these conditions, condensate will collect on the inner surface of the brick. This is illustrated in Figure 12.*

Vapor retarder added in wall.

It can be observed that if a very vapor resistant film be installed in the air space so as to prevent the vapor pressure of the air from reaching the dew point then the wall would stay dry.

To illustrate this fact: If a vapor retarder having a vapor resistance of 5.5 were added over the furring strips, then the sum of air to plaster, lath and plaster, plus the vapor retarder would add up to 5.546 and the sum would be 6.46—then

$$5.546 \div 6.46 = 0.8585$$
$$(0.8585) \times 15.6 = 13.392$$
$$15.6 - 13.392 = 2.208 \text{ lbs./ft}^2$$

From Table 5, it is indicated that this vapor pressure on the outer side of the vapor retarder would have a dew point of approximately −3 °F (−20.6 °C) and that this vapor retarder would keep

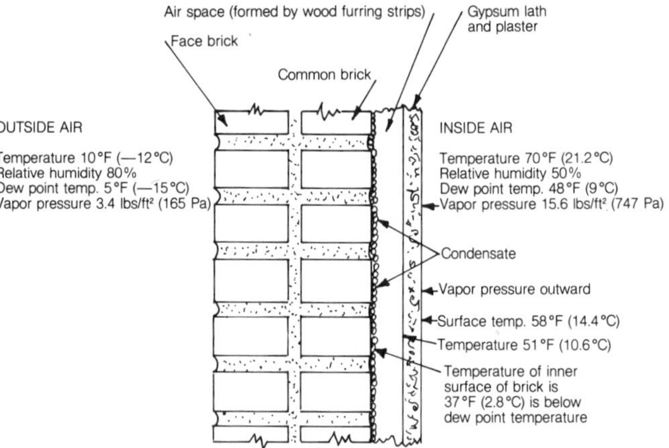

WALL COMPONENTS	THERMAL RESISTANCE		VAPOR RESISTANCE	
	English Units	Metric Units	English Units	Metric Units
Air to plaster surface	0.61	0.107	0.001	0.002
Gypsum lath and plaster	0.36	0.063	0.045	0.068
Air space	0.78	0.173	0.014	0.021
Common brick	0.80	0.141	0.900	1.368
Face brick	0.44	0.077		
Outside brick surface to air	0.17	0.029	0.001	0.002

TEMPERATURE GRADIENTS AND VAPOR PRESSURES IN MASONRY WALL

Figure 12

*Both English and Metric Units.

VAPOR PERMEANCE AND VAPOR RESISTANCE

BUILDING MATERIALS	THICKNESS		ENGLISH UNITS		METRIC UNITS (S.I.)	
	inches	cm	Permeance Perms	Resistance 1/Perms	Permeance Metric Perms	Resistance 1/Metric Perms
Brick – Facing	4	10.16	4.4	0.227	2.89	0.344
Brick – Common	4	10.16	5.0	0.20	3.29	0.303
Concrete block (cored,	8	20.32	2.6	0.385	1.65	0.607
limestone aggregate)	12	30.48	2.4	0.417	1.58	0.632
Fiberboard	0.5	1.27	120	0.008	79.08	0.013
Fiberboard – 1 surface coated with asphalt	0.5	1.27	16	0.063	10.54	0.095
Fiberboard – 2 layers cemented together	1.0	2.54	5.5	0.182	3.62	0.276
Fiberboard – compressed	0.19	0.48	20.3	0.049	13.38	0.075
Gypsum Lath (wall board)	0.50	1.27	105	0.009	69.09	0.014
Gypsum Lath with aluminum foil backing	0.50	1.27	0.35	2.857	0.23	4.340
Hardboard–Pressed wood–tempered	0.25	0.63	18.3	0.055	12.06	0.083
Hardboard–Pressed wood–	0.25	0.63	43.4	0.023	28.60	0.035
Plaster on wood lath	0.5	1.27	44.5	0.022	29.32	0.034
Plaster – gypsum	1.0	2.54	54.0	0.018	35.59	0.028
Plasterboard	0.375	0.95	75	0.013	49.43	0.020
Plywood, 3 ply Douglas Fir – Indoor	0.250	0.63	17.44	0.057	11.49	0.087
Plywood, 5 ply Douglas Fir, resin glue	0.500	1.27	10.60	0.094	8.99	0.143
Roof deck – concrete*	4.0	10.16	1.8	0.555	1.19	0.843
Roof deck – wood*	1.0	2.54	7.3	0.136	4.81	0.208
Roof deck– steel*			**	**	**	**
Siding – frame (siding board or shingles)	0.75		16.0	0.063	10.54	0.095
Siding – frame & stucco			16.0	0.063	10.54	0.095
Siding – brick veneer	4.0		15.5	0.065	10.21	0.098
Tile, wall with mortar	4.0	10.16	3.8	0.263	2.50	0.399
Wood – spruce	0.75	2.03	6.5	0.154	4.28	0.233
Wood – spruce	0.5	1.27	8.0	0.125	5.27	0.189
Wood – pine	0.75	2.03	3.7	0.270	2.43	0.410
Wood – pine	0.5	1.27	6.9	0.144	4.55	0.220

*Does not include insulation or roofing
**Depends completely on installation

TABLE 9

(Sheet 1 of 4)

VAPOR PERMEANCE AND VAPOR RESISTANCE

MATERIALS INSULATION MANUFACTURED ASSEMBLIES	THICKNESS		ENGLISH UNITS		METRIC UNITS (S.I.)	
	inches	cm	Permeance Perms	Resistance 1/Perms	Permeance Metric Perms	Resistance 1/Metric Perms
Fiberglas blanket with	1	2.54	0.03 to 0.5	33.33 to 2.00	0.019 to 0.329	50.58 to 3.03
attached aluminum foil	2	5.08	0.03 to 0.5	33.33 to 2.00	0.019 to 0.329	50.58 to 3.03
facing.	3	7.62	0.03 to 0.5	33.33 to 2.00	0.019 to 0.329	50.58 to 3.03
Note: The vapor resistance	4	10.16	0.03 to 0.5	33.33 to 2.00	0.019 to 0.329	50.58 to 3.03
is controlled by	5	12.7	0.03 to 0.5	33.33 to 2.00	0.019 to 0.329	50.58 to 3.03
thickness and quality of	6	15.24	0.03 to 0.5	33.33 to 2.00	0.019 to 0.329	50.58 to 3.03
the foil facing and its	7	17.78	0.03 to 0.5	33.33 to 2.00	0.019 to 0.329	50.58 to 3.03
application						
Fiberglas blanket	1	2.54	1.00	1.00	0.659	1.52
with kraft vapor	2	5.08	0.97	1.03	0.639	1.56
barrier paper	3	7.62	0.95	1.05	0.626	1.59
	4	10.16	0.93	1.07	0.613	1.63
	5	12.7	0.92	1.08	0.606	1.65
	6	15.24	0.91	1.10	0.599	1.66
	7	17.78	0.90	1.11	0.593	1.68
Fiberglas blanket or	1	2.54	0.007	142.8	0.0046	216.70
board with aluminum	1 1/2	3.81	0.007	142.8	0.0046	216.70
foil, skrim, and kraft	2	5.08	0.007	142.8	0.0046	216.70
paper	2 1/2	6.35	0.007	142.8	0.0046	216.70
	3	7.62	0.007	142.8	0.0046	216.70
Fiberglas board with	1	2.54	0.004	250.0	0.0026	379.36
skrim-vinyl film	1 1/2	3.81	0.004	250.0	0.0026	379.36
attached vapor barrier	2	5.08	0.004	250.0	0.0026	379.36
	2 1/2	6.35	0.004	250.0	0.0026	379.36
Foamed polystyrene	1/2	0.64	0.8 to 2.5	1.25 to 4.0	0.527 to 1.647	1.89 to 0.61
paper faced-both sides	3/4	1.91	0.5 to 1.25	2.0 to 0.8	0.329 to 0.527	3.03 to 1.89
	1	2.54	0.3 to 0.85	3.33 to 1.17	0.198 to 0.771	5.06 to 1.29
Foamed polystyrene	1/2	0.64	0.03 to 0.05	33.33 to 20.0	0.019 to 0.032	50.58 to 30.34
aluminum foil faced	3/4	1.91	0.03 to 0.05	33.33 to 20.0	0.019 to 0.032	50.58 to 30.34
both sides	1	2.54	0.03 to 0.05	33.33 to 20.0	0.019 to 0.032	50.58 to 30.34

TABLE 9

VAPOR PERMEANCE AND VAPOR RESISTANCE

MATERIALS INSULATION MANUFACTURED ASSEMBLIES	THICKNESS		ENGLISH UNITS		METRIC UNITS (S.I.)	
	inches	cm	Permeance Perms	Resistance 1/Perms	Permeance Metric Perms	Resistance 1/Metric Perms
Foamed polystrene bead	1/2	1.27	1.3 to 3.0	0.77 to 0.33	0.856 to 1.977	1.167 to 0.506
board paper faced both	3/4	1.91	0.9 to 2.0	1.11 to 0.50	0.593 to 1.313	1.686 to 0.758
sides	1	2.54	.5 to 1.5	2.0 to 0.66	0.329 to 0.988	3.035 to 1.011
Foamed polyurethane	1/2	1.27	0.03 to 0.05	33.33 to 20.00	0.019 to 0.329	50.58 to 3.03
board, aluminum foil	3/4	1.91	0.03 to 0.05	33.33 to 20.00	0.019 to 0.329	50.58 to 3.03
faced on one side	1	2.54	0.03 to 0.05	33.33 to 20.00	0.019 to 0.329	50.58 to 3.03
Foamed polyurethane	1/2	1.27	0.015 to 0.03	66.66 to 33.33	0.009 to 0.019	101.16 to 505.8
board—foil faced on	3/4	1.91	0.015 to 0.03	66.66 to 33.33	0.009 to 0.019	101.16 to 505.8
both sides	1	2.54	0.015 to 0.03	66.66 to 33.33	0.009 to 0.019	101.16 to 505.8
Expanded perlite	1 1/2	3.81	0.5 to 0.9	2.0 to 1.11	0.329 to 0.593	3.03 to 1.68
board chemically bonded	1 5/8	4.13	0.5 to 0.9	2.0 to 1.11	0.329 to 0.593	3.03 to 1.68
to top layer of polyurethane	2	5.08	0.5 to 0.9	2.0 to 1.11	0.329 to 0.593	3.03 to 1.68
foam, topped with	2 1/4	5.71	0.5 to 0.9	2.0 to 1.11	0.329 to 0.593	3.03 to 1.68
asphalt saturated	2 3/4	6.98	0.5 to 0.9	2.0 to 1.11	0.329 to 0.593	3.03 to 1.68
felt*	3 1/4	8.25	0.5 to 0.9	2.0 to 1.11	0.329 to 0.593	3.03 to 1.68
Mineral Wool-Batt and	1	2.54	0.03 to 0.5	33.33 to 2.00	0.019 to 0.329	50.58 to 3.03
Blanket-aluminum foil	2	5.08	0.03 to 0.5	33.33 to 2.00	0.019 to 0.329	50.58 to 3.03
faced. Note: The	3	7.62	0.03 to 0.5	33.33 to 2.00	0.019 to 0.329	50.58 to 3.03
vapor resistance is controlled	4	10.16	0.03 to 0.5	33.33 to 2.00	0.019 to 0.329	50.58 to 3.03
by the thickness and quality	5	12.7	0.03 to 0.5	33.33 to 2.00	0.019 to 0.329	50.58 to 3.03
of the foil facing and its application	6	15.24	0.03 to 0.5	33.33 to 2.00	0.019 to 0.329	50.58 to 3.03
Expanded perlite board	3/4	1.91	*32	0.031	21.05	0.047
with sealing coating on	1	2.54	25	0.040	16.45	0.061
top for attachment of	1 1/2	3.81	18	0.055	11.84	0.084
roofing**	2	5.08	13	0.078	8.55	0.116

*Board only
**As most vapor is on upper side of this roof insulation, a vapor retarder of less than 0.03 permeance must be installed underneath it.

TABLE 9

(Sheet 3 of 4)

VAPOR PERMEANCE AND VAPOR RESISTANCE

MATERIALS SHEETS, FILMS, PAPER FELTS	Thickness, plys or coverage	ENGLISH UNITS		METRIC UNITS (S.I.)	
		Permeance Perms	Resistance 1/Perms	Permeance Metric Perms	Resistance 1/Metric Perms
Aluminum foil (crinkled)	0.001" (0.025mm)	0.171	5.847	0.1127	8.87
Aluminum foil (smooth)	0.0015" (0.038mm)	0.0304	32.894	0.0200	49.91
Aluminum foil (smooth)	0.003" (0.076mm)	0.005	200.00	0.0033	303.49
Aluminum foil (0.005"), skrim, kraft paper	0.01" (0.254 mm)	0.0008	1250.00	0.0005	1896.81
Aluminum tape (0.005"THK) and adhesive	0.005" (0.127mm)	0.0057	175.438	0.0037	266.21
Asphalt coated felt	15 lb./500 sq. ft.	27.0	0.037	17.793	0.06
Asphalt coated felt	35 lb./500 sq. ft.	5.02	0.199	3.308	0.30
Asphalt coated felt	50 lb./500 sq. ft.	4.20	0.238	2.768	0.36
Asphalt impregnated roofing felt	35 lb./500 sq. ft.	2.10	0.476	1.384	0.72
Asphalt impregnated roofing felt	55 lb./500 sq. ft.	0.95	1.053	0.626	1.59
Friction tape (electrical) cloth	0.01 (025mm)	4.40	0.227	2.899	0.34
Kraft paper 1 sheet-plain	0.004 (0.1016mm)	336.0	0.003	221.00	0.004
Kraft paper asphalt saturated	45 lb./500 sq. ft.	0.36	2.777	0.237	4.215
Polyethylene sheet	0.006 (0.152mm)	0.900	1.111	0.593	1.68
Polyethylene sheet	0.03 (0.762mm)	0.279	3.579	0.183	5.44
Sand mastic (asphalt & sand)	0.125 (3.175mm)	0.75	1.333	0.494	2.02
Sheathing paper-std. dry	15 lb.	170.0	0.006	112.03	0.008
Sheathing paper-heavy dry	38 lb.	138.0	0.007	90.94	0.011
Vinylite sheet	0.006 (0.152 mm)	2.78	0.359	1.83	0.546
Vinylite sheet	0.03 (0.762mm)	0.266	3.759	0.175	5.704
Vinyl tape	0.01 (0.254mm)	0.361	2.770	0.243	4.123
Waxed paper (ordinary)	0.0015 (0.038mm)	7.2	0.138	4.745	0.211

TABLE 9

PERMEABILITY AND VAPOR RESISTIVITY

MATERIALS INSULATIONS, WOOD and CELLULOSE PRODUCTS	ENGLISH UNITS		METRIC UNITS (S.I.)	
	Permeability Perm-inch	Vapor Resistivity 1/Perm-inch	Permeability Perm-cm	Vapor Resistivity 1/Perm-cm
Calcium-silicate block	20 to 25	0.05 to 0.04	33.4 to 41.7	0.029 to 0.024
Cellulose fiber-loose & fill	90 to 100	0.011 to 0.01	150.3 to 167	0.006 to 0.006
Cellulose fiber insulating board	20 to 50	0.05 to 0.02	33.4 to 83.5	0.029 to 0.012
Cellular glass-block*	0.000+	∞	0.000+	∞
Corkboard-board	3 to 7	0.333 to 0.142	5.01 to 11.69	0.199 to 0.085
Diatomaceous silica-block	40 to 90	0.025 to 0.011	66.8 to 150.3	0.015 to 0.007
Expanded perlite-block	18 to 20	0.055 to 0.05	30.6 to 33.4	0.033 to 0.029
Fiberglas-board	35 to 80	0.028 to 0.013	58.4 to 133.6	0.017 to 0.007
Fiberglas-batt and blanket	60 to 90	0.017 to 0.011	100.2 to 150.3	0.010 to 0.007
Fiberglas-loose and fill	70 to 100	0.014 to 0.001	116.9 to 167	0.009 to 0.006
Foamed polystyrene, flexible	0.9 to 1.2	1.111 to 0.833	1.5 to 1.40	0.665 to 0.718
Foamed polystyrene-block board	0.3 to 0.85	3.333 to 1.176	0.50 to 1.40	2.000 to 0.704
Foamed polystryrene-bead board	1.5 to 2.0	0.666 to 0.5	2.05 to 3.34	0.399 to 0.299
Foamed polyurethane-board	1.2 to 1.7	0.833 to 0.588	2.00 to 2.84	0.499 to 0.352
Foamed rubber-flexible	0.3 to 0.5	3.333 to 2.00	0.5 to 0.84	2.0 to 1.197
Foamed urea-formaldehyde	20 to 25	0.05 to 0.04	33.4 to 41.7	0.029 to 0.024
Mineral wool-batt & blanket	50 to 85	0.02 to 0.012	83.5 to 141.9	0.012 to 0.007
Mineral wool-board & block	30 to 75	0.033 to 0.013	50.1 to 125.2	0.02 to 0.008
Mineral wool-loose & fill	50 to 80	0.02 to 0.125	83.5 to 133.6	0.012 to 0.007
Perlite expanded-board & block	18 to 20	0.055 to 0.05	30.1 to 33.4	0.033 to 0.030
Vegetable fiber insulating board	35 to 50	0.028 to 0.02	58.4 to 83.5	0.017 to 0.012
Vermiculite-loose & fill	60 to 75	0.016 to 0.013	100.2 to 125.3	0.010 to 0.008
Wood-Douglas Fir	5 to 5.5	0.2 to 0.182	8.35 to 9.18	0.119 to 0.109
Wood-pine	3.4 to 4.0	0.294 to 0.25	5.67 to 6.68	0.176 to 0.150
Wood-spruce	3.25 to 3.5	0.308 to 0.286	5.43 to 5.85	0.184 to 0.171

NOTE: Range in values are due to differences in manufactured products. In general, the higher the density the lower the permeability and higher resistivity.

*Lower permeability than can be accurately measured.

TABLE 10

THERMAL CONDUCTANCE AND RESISTANCE

BUILDING MATERIALS	ENGLISH UNITS			METRIC UNITS (S.I.)		
	Thickness	Conductance	Resistance	Thickness	Conductance	Resistance
	inches	$C = \frac{Btu}{hr.ft.^2 \, ^\circ F}$	$R = \frac{1}{C}$	cm	$C_m = \frac{W}{m^2 \, ^\circ K}$	$R_m = \frac{m^2 \, ^\circ K}{W}$
Brick-facing	4.0	1.25	0.80	10.16	7.094	0.1409
Brick-common	4.0	2.25	0.44	10.16	12.898	0.0775
Clay-tile 1 cell	4.0	0.90	1.11	10.16	5.112	0.1955
Concrete block-cinder aggregate	8.0	0.58	1.72	20.32	3.299	0.3031
Concrete block-cinder aggregate	12.0	0.53	1.88	30.48	3.018	0.3312
Concrete block-gravel aggregate	8.0	0.90	1.11	10.16	5.112	0.1955
Concrete block-gravel aggregate	12.0	0.78	1.28	30.48	4.433	0.2255
Fiberboard-1 sur. coated with asphalt	0.5	1.90	0.53	1.27	10.708	0.0933
Fiberboard-2 layers cemented with asphalt	1.0	0.88	1.13	2.54	5.022	0.1991
Fiberboard-compressed	0.19	4.3	0.23	0.48	24.675	0.0405
Gypsum lath wall board	0.5	3.10	0.32	1.27	17.735	0.0564
Gypsum lath wall board aluminum foil back	0.5	2.50	0.40	1.27	14.188	0.0705
Hardboard-Pressed wood-tempered	0.5	4.00	0.25	1.27	22.701	0.0441
Hardboard-Pressed wood	0.5	3.00	0.33	1.27	17.198	0.0581
Plaster on wood lath	0.5	2.05	0.49	1.27	11.582	0.0863
Plaster, gypsum	0.5	3.12	0.32	1.27	17.735	0.0564
Plasterboard	0.375	3.10	0.32	0.95	17.735	0.0564
Plywood 3 ply Douglas Fir	0.250	2.90	0.34	0.63	16.692	0.0599
Plywood 5 ply resin glue	0.500	1.5	0.66	1.27	8.599	0.1163
Roof deck-concrete	4.0	2.25	0.44	10.16	12.898	0.0775
Roof deck-wood	1.0	1.27	0.78	2.54	7.276	0.1374
Roof deck steel (no insulation)	Depends almost completely on thermal resistance of air filter.					
Siding-frame (siding board or shingles)	0.75	0.95	1.05	1.91	5.405	0.1850
Siding-frame and stucco	1.25	0.52	1.92	1.19	4.769	0.2096
Siding 4″ brick veneer	4.0	2.27	0.44	10.16	12.898	0.0775
Tile wall with mortar	4.0	1.25	0.80	10.16	7.094	0.1409
Wood (spruce, pine or fir)	0.75	0.94	1.06	1.91	5.354	0.1867
Wood (spruce, pine or fir)	0.5	1.41	0.71	1.27	7.993	0.1251

Conductance is per one degree for thickness as noted.

TABLE 11

(Sheet 1 of 4)

THERMAL CONDUCTANCE AND RESISTANCE

MATERIALS INSULATION MANUFACTURED ASSEMBLIES	ENGLISH UNITS			METRIC UNITS (S.I.)		
	Thickness	Conductance	Resistance	Thickness	Conductance	Resistance
	inches	$C = \frac{Btu}{hr.ft.^{2}°F}$	$R = \frac{1}{C}$	cm	$C_m = \frac{W}{m^{2}°K}$	$R_m = \frac{m^{2}°K}{W}$
Fiberglass blanket with attached aluminum foil facing (density 1.5 to 2.0 lbs. cubic ft.)	1	0.220	4.54	2.54	1.25	0.799
	2	0.125	8.54	5.08	0.66	1.504
	3	0.079	12.54	7.62	0.45	2.209
	4	0.060	16.55	10.16	0.34	2.916
	5	0.048	20.56	12.7	0.27	3.622
	6	0.041	24.56	15.24	0.23	4.327
	7	0.035	28.57	17.78	0.19	5.034
Fiberglass blanket with kraft paper vapor barrier	1	0.280	3.57	2.54	1.58	0.629
	2	0.133	7.51	5.08	0.75	1.323
	3	0.086	11.53	7.62	0.49	2.03
	4	0.064	15.53	10.16	0.36	2.73
	5	0.051	19.53	12.7	0.29	3.44
	6	0.043	23.52	15.24	0.24	4.14
	7	0.036	27.53	17.78	0.21	4.85
Fiberglass blanket or board with aluminum foil, skrim and kraft paper	1	0.24	4.16	2.54	1.36	0.733
	1 1/2	0.16	6.25	3.81	0.91	1.101
	2	0.12	8.33	5.08	0.68	1.467
	2 1/2	0.096	10.41	6.35	0.55	1.834
	3	0.080	12.5	7.62	0.45	2.202
Fiberglass board with skrim vinyl film attached vapor barrier	1	0.24	4.16	2.54	1.36	0.733
	1 1/2	0.16	6.25	3.81	0.91	1.101
	2	0.12	8.33	5.08	0.68	1.467
	2 1/2	0.096	10.41	6.35	0.55	2.202
Foamed polystrene, paper faced on both sides	1/2	0.50	2.00	1.27	2.83	0.352
	3/4	0.38	3.00	1.91	1.89	0.528
	1	0.25	4.00	2.54	1.42	0.705
Foamed polystrene aluminum foil faced on both sides	1/2	0.50	2.00	1.27	2.83	0.352
	3/4	0.38	3.00	1.91	1.89	0.528
	1	0.25	4.00	2.54	1.42	0.705

TABLE 11

(Sheet 2 of 4)

THERMAL CONDUCTANCE AND RESISTANCE

MATERIALS INSULATION MANUFACTURED ASSEMBLIES	ENGLISH UNITS			METRIC UNITS (S.I.)		
	Thickness inches	Conductance $C = \frac{Btu}{hr.ft.^2 {}^\circ F}$	Resistance $R = \frac{1}{C}$	Thickness cm	Conductance $C_m = \frac{W}{m^2 {}^\circ K}$	Resistance $R_m = \frac{m^2 {}^\circ K}{W}$
Foamed polystrene bead board, paper faced on both sides	1/2 3/4 1	0.50 0.38 0.25	2.0 2.63 4.0	1.27 1.91 2.54	2.84 2.15 1.42	0.352 0.463 0.704
Foamed polyurethane board, aluminum foil faced on one side	1/2 3/4 1	0.36 0.27 0.18	2.77 3.70 5.55	1.27 1.91 2.54	2.04 1.53 1.02	0.489 0.652 0.978
Foamed polyurethane board, aluminum foil faced on both sides	1/2 3/4 1	0.36 0.27 0.18	2.77 3.70 5.55	1.27 1.91 2.54	2.04 1.53 1.02	0.489 0.652 0.978
Expanded perlite board chemically bonded to top layer of polyurethane foam topped with asphalt saturated felt	1 1/2 1 5/8 2 2 1/4 2 3/4 3 1/4	0.15 0.13 0.10 0.08 0.07 0.05	6.67 7.69 10.00 12.50 14.29 20.00	3.81 4.13 5.08 5.71 6.98 8.25	0.85 0.74 0.57 0.45 0.40 0.28	1.174 1.354 1.761 2.201 2.516 3.522
Mineral wool-batt and blanket; aluminum foil faced. Note: Available in various qualities and thickness of foil facing	1 2 3 4 5 6	0.27 0.135 0.09 0.07 0.05 0.04	3.70 7.40 11.11 14.81 18.51 22.22	2.54 5.08 7.62 10.16 12.7 15.24	1.53 0.77 0.51 0.40 0.28 0.23	0.652 1.304 1.957 2.500 3.522 4.403
Expanded perlite board with sealing coating on top side for attachment of roofing	3/4 1 1 1/2 2	0.48 0.36 0.24 0.19	2.08 2.78 4.17 5.26	1.90 2.54 3.81 5.08	2.72 2.04 1.36 1.07	0.367 0.489 0.734 0.927

TABLE 11

(Sheet 3 of 4)

THERMAL CONDUCTANCE AND RESISTANCE

ENCLOSED AIR SPACES	Mean Temp. °F	Mean Temp. °C	ENGLISH UNITS Thickness inches	Conductance $C = \frac{Btu}{hr.ft.^2 °F}$	Resistance $R = \frac{1}{C}$	METRIC UNITS (S.I.) Thk cm	Conductance $C_m = \frac{W}{m^2 °K}$	Resistance $R_m = \frac{m^2 °K}{W}$
Horizontal: High emittance surfaces both sides. Heat Flow Up	40	4.4	1 1/2	1.06	0.94	3.81	6.038	0.1656
	40	4.4	3 1/2	1.04	0.96	8.89	5.912	0.1692
	90	32.2	1 1/2	1.25	0.80	3.81	7.094	0.1409
	90	32.2	3 1/2	1.21	0.82	8.89	6.921	0.1444
Horizontal: High emittance surfaces both sides. Heat Flow Down	40	4.4	1 1/2	0.87	1.15	3.81	4.935	0.2026
	40	4.4	3 1/2	0.81	1.24	8.89	4.577	0.2185
	90	32.2	1 1/2	1.06	0.94	3.81	6.037	0.1656
	90	32.2	3 1/2	1.00	1.00	8.89	5.675	0.1762
Horizontal: One facing surface with aluminum foil facing. Heat Flow Up	40	4.4	1 1/2	0.42	2.40	3.81	2.365	0.4229
	40	4.4	3 1/2	0.38	2.66	8.89	2.133	0.4687
	90	32.2	1 1/2	0.41	2.41	3.81	2.355	0.4246
	90	32.2	3 1/2	0.38	2.66	8.89	2.133	0.4687
Horizontal: One facing surface with aluminum foil facing. Heat Flow Up	40	4.4	1 1/2	0.21	4.66	3.81	1.218	0.8211
	40	4.4	3 1/2	0.23	4.36	8.89	1.301	0.7682
	90	32.2	1 1/2	0.22	4.55	3.81	1.247	0.8017
	90	32.2	3 1/2	0.23	4.33	8.89	1.311	0.7629
Vertical: High emittance surfaces both sides. Heat Flow Horizontal	40	4.4	1	0.98	1.02	2.54	5.564	0.1797
	40	4.4	3 1/2	0.99	1.01	8.89	5.619	0.1779
	90	32.2	1	1.15	0.87	2.54	6.523	0.1533
	90	32.2	3 1/2	1.18	0.85	8.89	6.677	0.1498
Vertical: One facing surface with aluminum foil facing. Heat Flow Horizontal	40	4.4	1	0.42	2.39	2.54	2.374	0.4212
	40	4.4	3 1/2	0.43	2.32	8.89	2.446	0.4088
	90	32.2	1	0.44	2.25	2.54	2.522	0.3964
	90	32.2	3 1/2	0.46	2.15	8.89	2.639	0.3788
Vertical: Both facing surfaces with aluminum foil facing. Heat Flow Horizontal	40	4.4	1	0.26	3.79	2.54	1.497	0.6677
	40	4.4	3 1/2	0.28	3.63	8.89	1.563	0.6396
	90	32.2	1	0.25	3.99	2.54	1.422	0.7030
	90	32.2	3 1/2	0.27	3.69	8.89	1.538	0.6502

Note: The reflective surfaces are aluminum foil mounted on component material facing the space. The reflective sheets *do not* divide the space.

TABLE 11

(Sheet 4 of 4)

FUNDAMENTALS OF MOISTURE MIGRATION – CONDUCTION AND THERMAL RESISTANCES

MATERIALS INSULATIONS, WOOD AND CELLULOSE PRODUCTS	ENGLISH UNITS							METRIC UNITS (S.I.)						
	Conductivity k $\frac{\text{Btu in.}}{\text{ft}^2\text{hr}^\circ\text{F}}$	*Thermal Resistance $\frac{\text{hr-ft}^2{}^\circ\text{F}}{\text{Btu}}$ Insulation Thickness - Inches						Conductivity k_m W/mK	*Metric Resistance $\frac{\text{m}^2\text{K}}{\text{W}}$ Insulation Thickness - cm					
		1	2	3	4	5	6		2.54	5.08	7.62	10.16	12.7	15.24
Calcium Silicate Block	0.35	2.85	5.71	8.57	11.42	14.28	17.14	0.0505	0.502	1.004	1.510	2.008	2.516	3.020
Cellulose Fiber, Loose & Fill	0.43	2.32	4.65	6.97	9.30	11.62	13.95	0.0620	0.408	0.816	1.228	1.632	2.047	2.456
Cellulose Fiber Board	0.41	2.43	4.86	7.31	9.75	12.19	14.62	0.0591	0.428	0.856	1.288	1.712	2.147	2.576
Cellular Glass	0.37	2.70	5.40	8.10	10.80	13.51	16.21	0.0533	0.475	0.950	1.427	1.900	2.380	2.854
Cork Board	0.33	3.03	6.06	9.09	12.12	15.15	18.18	0.0476	0.583	1.067	1.601	2.135	2.669	3.202
Diatomaceous Silica Block	0.50	2.00	4.00	6.00	8.00	10.00	12.00	0.0721	0.352	0.704	1.057	1.408	1.762	2.114
Expanded Perlite Block	0.39	2.56	5.12	7.69	10.24	12.82	15.38	0.0562	0.451	0.900	1.354	1.804	2.259	2.708
Fiber Glass – Board (3 lb. dens.)	0.28	3.57	7.14	10.71	14.28	17.85	21.42	0.0404	0.629	1.258	1.887	2.516	3.145	3.774
Fiber Glass – Batt & Blanket	0.28	3.57	7.14	10.71	14.28	17.85	21.42	0.0404	0.629	1.258	1.887	2.514	3.145	3.774
Fiber Glass – Loose and Fill	0.26	3.84	7.69	11.53	15.38	19.83	22.07	0.0375	0.676	1.353	2.031	2.706	3.494	4.065
Foamed Polystyrene – Flexible	0.35	2.85	5.71	8.57	11.42	14.28	17.14	0.0505	0.502	1.004	1.510	3.008	2.516	3.020
Foamed Polystyrene–Block, Board	0.23	4.34	8.69	13.04	17.33	21.74	26.08	0.0332	0.764	1.529	2.297	3.058	3.830	4.592
Foamed Urethane – Board	0.24	4.17	8.34	12.50	16.68	20.83	25.00	0.0346	0.734	1.469	2.202	2.939	3.670	4.404
Foamed Rubber – Flexible	0.29	3.44	6.88	10.34	13.76	17.24	20.69	0.0418	0.606	1.212	1.821	2.424	3.037	3.643
Foamed Urea Formaldehyde	0.27	3.70	7.40	11.11	14.81	18.52	22.22	0.0389	0.652	1.304	1.957	2.607	3.263	3.915
Mineral Wool – Batt & Blanket	0.29	3.44	6.88	10.34	13.76	17.24	20.68	0.0418	0.606	1.212	1.821	2.414	3.037	3.643
Mineral Wool – Board & Block	0.28	3.57	7.14	10.71	14.28	17.85	21.42	0.0404	0.629	1.258	1.887	2.516	3.145	3.774
Mineral Wool – Loose & Fill	0.29	3.44	6.88	10.34	13.76	17.24	20.68	0.0418	0.606	1.212	1.861	2.924	3.037	3.693
Perlite, Expanded – Board & Block	0.35	2.85	5.71	8.57	11.42	14.28	17.14	0.0505	0.502	1.004	1.510	2.008	2.516	3.020
Vegetable Fiber Insulating Board	0.39	2.56	5.12	7.69	10.24	12.82	15.38	0.0562	0.451	0.902	1.354	1.804	2.258	2.708
Vermiculite – Loose & Fill	0.97	2.12	4.24	6.38	8.48	10.63	12.76	0.0678	0.371	0.747	1.124	1.494	1.873	2.248
Wood – Douglas Fir	0.75	1.33	2.66	4.00	5.33	6.66	8.00	0.1032	0.234	0.468	0.705	0.936	1.173	1.410
Wood – Pine	0.71	1.40	2.30	4.22	5.60	7.04	8.44	0.1024	0.247	0.493	0.743	0.987	1.840	1.486
Wood – Spruce	0.74	1.35	2.70	4.05	5.40	6.75	8.10	0.1067	0.235	0.470	0.713	0.990	1.189	1.427

*Based on materials in dry state and at 70°F (21.1°C) mean temperature.

TABLE 12

the wall dry from condensation due to warm, moist air on the inside of the building. This is shown in Figure 13.

It should be noted that the units in measuring thermal resistance and vapor resistance were not stated. Any convenient unit could be used, provided that all quantities are expressed in that same unit of measurement. In Table 9, the vapor permeance and resistance of various materials and assemblies are given both in English and Metric units and the permeability and resistivity of insulation and building materials per unit of thickness are also given in English and Metric units. Table 10 provides the permeability and vapor resistance for various materials per unit of thickness in English and Metric units.

The thermal conductance and resistances are given in Table 11. Table 12 gives the thermal conductivity and thermal resistances of materials for various thicknesses.

The conductivities and resistances given for the insulation and construction materials are those which would be expected if the materials were dry. When used for roof or wall design these materials must be properly protected by vapor retarders (barriers). *However, should they become wet due to condensation, the thermal resistances given are no longer valid.*

It is necessary to know the thermal and moisture resistance properties of the individual parts of a wall, ceiling, or roof to properly design the insulation system for the structure. Methods of using these tables in the design will be given.

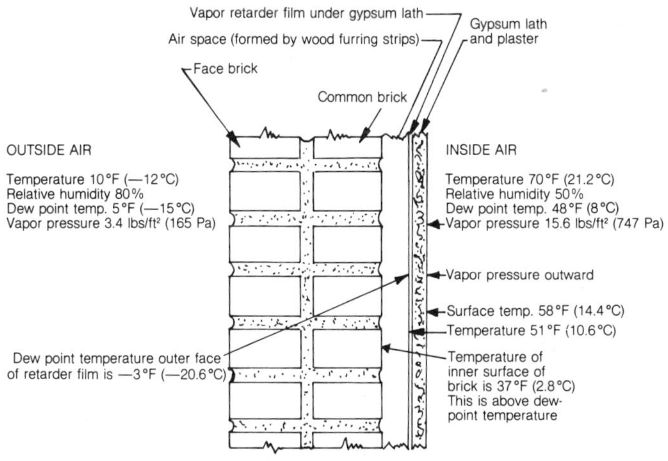

WALL COMPONENTS	THERMAL RESISTANCE		VAPOR RESISTANCE	
	English Units	Metric Units	English Units	Metric Units
Air to plaster surface	0.61	0.107	0.001	0.002
Gypsum lath and plaster	0.36	0.063	0.045	0.068
Vapor retarder film	—	—		
Air space	0.78	0.173	5.500	8.360
Common and face brick	1.24	0.218	0.900	1.368
Outside brick surface to air	0.17	0.029	0.001	0.002

TEMPERATURE GRADIENTS AND VAPOR PRESSURES IN MASONRY WALL WITH VAPOR RETARDER FILM

Figure 13

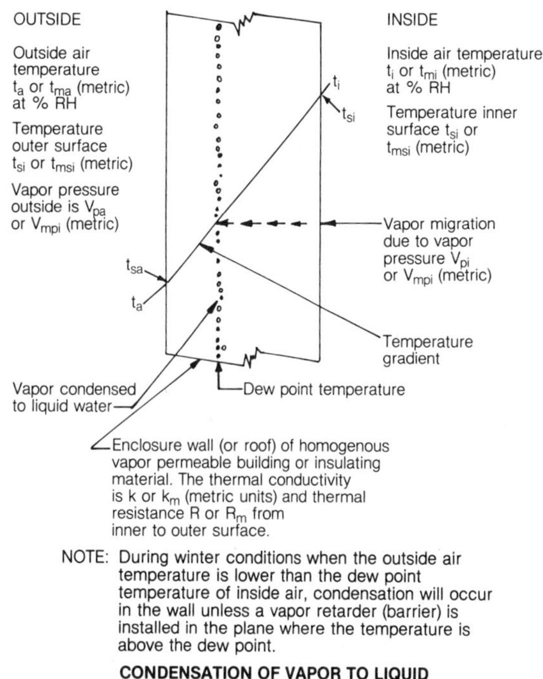

CONDENSATION OF VAPOR TO LIQUID IN ENCLOSURE WALL OR ROOF HAVING NO VAPOR RETARDER (BARRIER)

Figure 14

WALLS

Figure 14 illustrates the need for a vapor retarder or barrier to prevent condensation in a wall which consists of a homogeneous building material with high permeability of vapor migration. Vapor migrates through the wall in its effort to get from a higher to a lower pressure at some point in the enclosure. At some point, the water vapor will condense into liquid water. Further into the wall (or roof) section, this water may freeze into ice. Wet and iced up walls are inefficient thermal barriers. The variation of the thermal conductivities of bricks, concrete, cinder block, and wood, when they contain moisture, is shown in Figure 15.

If the building material were concrete, brick or cinder block, the porosity would have a tendency to pass the liquid outward where evaporation to the outside air would occur. Except in cases of severe freezing, most of these materials would not be physically damaged and the major result would be a loss of thermal efficiency during the winter. However, it is of utmost importance that the moisture (liquid or vapor) flow not be restricted beyond this dew point plane.

In order to keep any enclosure (walls, roofs, ceilings, or floors) dry in cold climates a vapor retarder should be installed on the warm (inner) side. This is illustrated in the following example.

Example 1—*English and Metric Units*
Illustrated in Figure 16 "A"

Consider an 8″ thick concrete block wall whose interior is covered with plaster board with vapor retarder aluminum foil back-

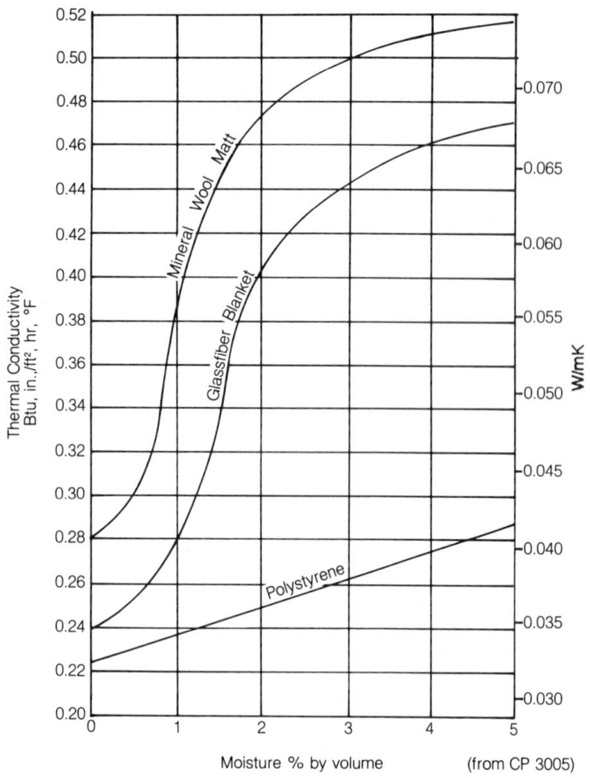

VARIATION OF THERMAL CONDUCTIVITY WITH MOISTURE CONTENT

Figure 15

ing. The exterior is frame siding 3/4″ (1.9 cm) thick. When the interior temperature is 70 °F (21.1 °C), what are the conditions of moisture condensation in the wall at various outside temperatures and interior relative humidities?

Solution: Thermal resistances are first obtained from Tables 11 and 12 as follows:

English Units		***Metric Units***
Thermal resistances are		
R_p = 0.40 = Thermal resistance of plaster		= R_{mp} = 0.0705
R_c = 1.72 = Thermal resistance of block		= R_{mc} = 0.3031
R_s = 1.05 = Thermal resistance of siding		= R_{ms} = 0.1805
3.17 = total resistance		0.5541 = total resistance
% resistance of plaster	= 0.4/3.17 = 12%	= 0.0705/0.5541
% resistance of concrete block	= 1.72/3.17 = 55%	= 0.3031/0.5541
% resistance of siding	= 1.05/3.17 = 33%	= 0.1805/0.5541

The temperature drop through the wall assembly, when the outside surface has a temperature of 0 °F (−17.8 °C), is:

$$0\,°F - 70\,°F = -70\,°F$$
$$-17.8\,°C - (21.1\,°C) = -38.9\,°C$$

The temperature at the plane between the plaster board and block is equal to the inside temperature minus the gradient; that is,

$$70 - (0.12 \times 70) = 70 - 8.4 = 61.6\,°F$$
$$21.1 - (0.12 \times 38.9) = 21.1 - 4.6 = 16.5\ °C.$$

The temperature at the plane between the outside siding and block is equal to the outside temperature plus the gradient; hence,

$$0 + (0.33 \times 70) = 23.1\,°F$$
$$-17.8 + (38.9 \times 0.33) = -17.8 + 12.8 = -5\,°C.$$

The inside temperature at 70 °F (21.1 °C) at 50% RH has a vapor pressure of 35.61 lbs./ft.² (1705 Pa). (See Table 8.) If the outdoor air at 0 °F (−17.88 °C) were completely saturated, its vapor pressure would be 2.66 lbs./ft.² (128 Pa). (See Table 5.) The vapor pressure difference would be 35.61 − 2.66 = 32.95 lbs./ft.² or 1705 Pa − 128 Pa = 1577 Pa.

The vapor resistances given to these building components are given in Table 9.

	English Units	***Metric Units***
Plaster board & aluminum foil, VR_i	= 2.857 (1 ÷ Perms) = VR_{mi}	= 4.340 (1 ÷ Metric Perms)
8″ concrete block VR_w	= 0.385 (1 ÷ Perms) = VR_{mw}	= 0.607 (1 ÷ Metric Perms)
Frame siding VR_s	= 0.063 (1 ÷ Perms) = VR_{ms}	= 0.095 (1 ÷ Metric Perms)
	3.305	5.042

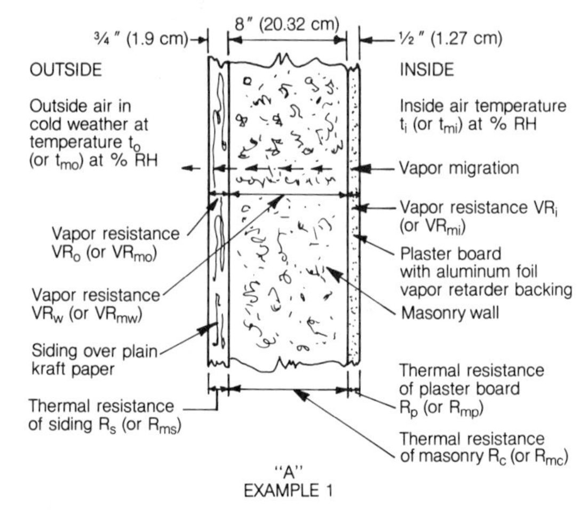

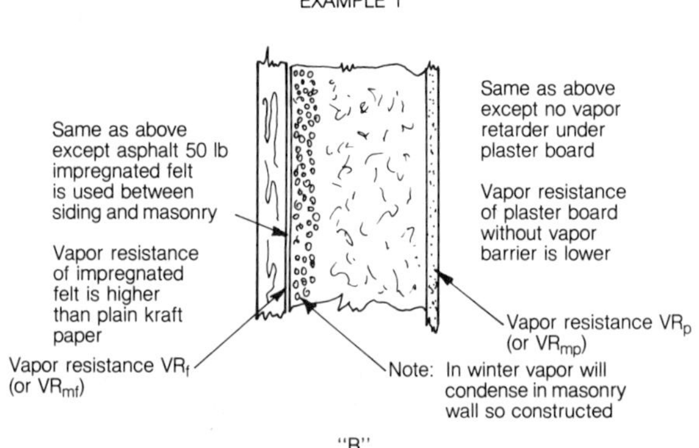

EFFECT ON MOISTURE CONDENSATION DUE TO LOCATION OF VAPOR RETARDERS (BARRIERS)

Figure 16

% Resistance to vapor migration is calculated as follows:

For plaster board and aluminum foil: 2.857 ÷ 3.305 = 86% = 4.34 ÷ 5.042.

For 8″ concrete block: 0.385 ÷ 3.305 = 11% = 0.607 ÷ 5.042.

English Units

The vapor pressure at the junction of these materials is: 35.61 − (32.95 × .86) = 35.61 − 28.31 = 7.30 lbs./ft.^{2} (English units) directly under the aluminum foil over the concrete block. It was previously determined that the temperature is 61.6°F at this plane. This would indicate (from Table 8) that the RH of the air would be approximately 5%; and from Table 5, the dew point temperature is approximately 20°F.

At the junction of the concrete block with the frame siding, the resistance to vapor migration is 86% + 11% = 97%; thus, the vapor pressure is found to be

$$35.61 - (32.93 \times .97) = 35.61 - 31.94 = 3.67 \text{ lbs./ft.}^2$$

The dew point at this vapor pressure is approximately 6°F; and as the temperature previously determined was 23.1°F, no condensation will occur where the plain kraft paper separates the wood siding from the concrete block.

Metric Units

The vapor pressure at the junction of the aluminum foil backed plaster board is:

$$1705 - (1577 \times .86) = 1705 - 1356 = 349 \text{ Pa.}$$

(From Table 5 the temperature at saturation is −6.7°C.)

As was previously determined, the temperature at this interface is 16.5°C, which is well above the dew point temperature of −6.7°C.

At the junction of the concrete block with the frame siding, the vapor pressure is 1705 − (1577 × .97) = 1705 − 1530 = 175 Pa. The dew point at this vapor pressure is −14.4°C; therefore, since the temperature at the interface of the siding-kraft paper with the concrete block is −5°C, no condensation will occur.

The wall, as illustrated in Figure 16 "A", is suitable for climates of low winter temperatures. As illustrated, the vapor retarders (barriers) were installed correctly so as to prevent vapor condensing to liquid water within the wall structure.

Unfortunately, the knowledge of correctly designing enclosures for buildings is not widespread; and as vapor is invisible, there is a tendency to construct buildings incorrectly. Since many do not understand the problem, the vapor retarder (barrier) is frequently not installed on the inner side of the enclosure. Since its omission causes condensation, as illustrated in Figure 7, any wood siding has a tendency to rot and decay. To prevent this, it has become the practice to install asphalt impregnated kraft paper or aluminum foil paper over the outer surface of the block, under the siding. This is completely in error. Assuming that this common practice is used on the wall shown in Figure 16 "B", determine the results.

Example 2—*English and Metric Units*
Illustrated in Figure 16 "B"

An 8″ (20.32 cm) thick concrete block wall is covered with plain plaster board on its inner side. On the outside, the block is covered with asphalt impregnated kraft paper over which the frame siding is installed. When the interior temperature is 70°F (21.1°C), what are the conditions of moisture condensation within the wall? First, determine the temperature gradients. When: (From Tables 11 and 12)*

English Units		*Metric Units*
R_p = 0.40	= Thermal resistance of plaster board	= R_{mp} = 0.0705
R_c = 1.72	= Thermal resistance of block	= R_{mc} = 0.3031
R_s = 1.05	= Thermal resistance of siding	= R_{ms} = 0.1805
3.17	= total resistance	0.5541 = total resistance

The % of thermal resistance:

Through plaster	12%
Through concrete block	55%
Through siding	33%

The temperature drop through the wall assembly, when the outside surface has a temperature of 0 °F (−17.8°C), is

$$0°F - (70°F) = -70°F$$
$$-17.8°C - (21.1°C) = -38.9°C$$

English Units	*Metric Units*

The temperature at the plane between plaster and block (inside temperature minus gradient)

70 − (0.12 × 70) = 70 − 8.4 = 61.6°F	21.1 − (0.12 × 38.9) = 21.1 − 4.6 = 16.5°C

Temperature at the plane of the asphalt impregnated kraft paper (outside temperature plus gradient)

0 + (0.33 × 70) = 23.1°F	−17.8 + (38.9 × 0.33) = −17.8 + 12.8 = −5°C

The inside temperature at 70°F (21.1°C) at 50% RH has a vapor pressure of 35.61 lbs./ft.^{2} (1705 Pa)—From Table 8. If the outside air at 0°F (−17.8°C) were completely saturated, its vapor pressure would be 2.66 lbs./ft.^{2} (128 Pa)—From Table 5. Then the vapor pressure difference would be 35.61 lbs./ft.^{2} − 2.66 lbs./ft.^{2} = 32.95 lbs./ft.^{2} or 1705 Pa − 128 Pa = 1577 Pa.

The vapor resistances given for this wall construction are given in Table 9.

*Based on the materials being relatively dry.

	English Units		Metric Units	
Plaster board	VR_p =	0.013	VR_{mp} =	0.020
8″ concrete block	VR_w =	0.385	VR_{mw} =	0.604
Impregnated paper	VR_f =	2.777	VR_{mf} =	4.215
Siding	VR_o =	0.063	VR_{mo} =	0.098
		3.238		4.937

As can be observed, the major vapor retarder is the impregnated paper.

% of Resistance to vapor barrier is 0.013 + 0.385 = 0.398 (English units). Resistance in Metric units is 0.020 + 0.604 = 0.624. The percentage of resistance = 0.398 ÷ 3.238 = 12% = 0.624 ÷ 4.937.

The vapor pressure at this junction then is:

35.61 − (32.93 × .12) = 35.61 − 3.95 = 31.16 lbs./ft.^{2} (English units) or 1705 − (1577 × .12) = 1705 − 189 = 1516 Pa (Metric units).

From Table 5, at these vapor pressures, the dew point temperature is 55.5 °F or 13 °C; and the temperature within the wall at this location is 23.1 °F or −5 °C. Thus, this wall will condense liquid water in the block behind the impregnated building paper. The amount of water condensed is controlled only by the amount of water entering the vapor stage on the inside.

The maximum relative humidity which can be tolerated inside without condensation occurring in the wall can also be determined.

As the temperature at this location in the wall was 23.1 °F (−5 °C), the maximum vapor pressure which could be tolerated is 8.38 lbs./ft.^{2} (401 Pa). As indicated, the vapor resistance to the vapor retarder was 12% of the total resistance. Thus, 12% more vapor pressure can be added to these in order to determine the allowable inside vapor pressure. That is,

8.38 × 1.12 = 9.38 lbs./ft.^{2} or 401 Pa × 1.12 = 449 Pa.

From interpretation of vapor pressures given in Table 5, the maximum allowable relative humidity inside would be approximately 3%. Any relative humidity above this would cause condensation within the wall structure. This condensation will cause the concrete block to become saturated with water and ice so that its thermal conductivity rises, resulting in greater heat loss. This indicates that the incorrect location of vapor retarders can turn relatively good insulation materials—with regard to thermal efficiency—into very poor insulation materials.

In the past two examples, the major portion of the wall structure was of masonry materials which do not rot and decay. The same examination shall be given for frame construction to evaluate the need for correct moisture control.

The insulation of frame walls must be designed and installed correctly as excessive moisture will cause wood to decay and rot. As mentioned before, the vapor retarder (barrier) of major resistance should be on the interior of walls (or ceilings) where the temperature is above the dew point.

With the emphasis on energy conservation, the organic foams have been recommended because of their low thermal conductivity (high thermal resistance). In recent years, there has been a trend to use this light weight material as wall panels installed under the siding or shingles.

The first major consideration is where such organic foams have aluminum foil sealing membranes on *both* sides of the panel. As shown in Example 2 "B" where vapor resistant felt is *under* the siding, condensation would occur in the masonry materials if the RH inside is greater than 3% at an outdoor temperature of 0 °F.

Aluminum foil has at least three times greater vapor resistance than asphalt saturated felts. *When aluminum foil is on the outer surface of organic siding board, this combination should never be used in cold climates.*

If the interior of the building is at 70 °F (21.1 °C) with 50% RH, the dew point temperature is approximately 59 °F (15 °C). Thus, water vapor will condense on this outer vapor barrier—in the insulation—whenever this barrier is at a lower temperature than 59 °F (15 °C). Since the only thermal protection outside this foil barrier is the siding or shingles, the foil would always be within 3° to 5 °F (1.5 to 3 °C) of the outdoor temperature. This type of siding board is unsuitable in any climate for which temperatures stay below 52 °F (11.1 °C) for any extended period of time.

For these reasons, the organic foam panel board selected for consideration in Example 3 will be a panel board having a vapor retarder (barrier) on its surface facing the interior.

Example 3*—English and Metric Units*

Consider a frame wall (2 × 4 studs) with 1/2″ (1.27 cm) plaster board on the interior (the outer surface consists of a 1 inch organic foam insulation panel board with 0.0015″ (0.038 mm) thick vapor retarder aluminum foil on the inner surface); overlaying the foam insulation is an exterior siding board as shown in Figure 17 "A"). The inside wall temperature is 70 °F (21.1 °C). In order to determine the moisture conditions through the wall, we must first establish the temperature in the wall by using the thermal resistances which are obtained from Tables 11 and 12. This example is illustrated in Figure 17 "A".

English Units			Metric Units	
R_p =	0.40	= Thermal resistance of plaster board	= R_{mp} =	0.0705
R_a =	1.01	= Thermal resistance of vertical 3 1/2″ (8.89 cm) air space	= R_{ma} =	0.1779
R_b =	5.55	= Thermal resistance of 1″ (2.54 cm) board	= R_{mb} =	0.9780
R_s =	1.05	= Thermal resistance of siding	= R_{ms} =	0.1805
	8.01			1.4069

Using the above values, the percent of resistance through each material is:

Through the interior plaster board	=	5%
Through the air space	=	13%
Through the insulation board	=	69%
Through the exterior siding	=	13%

The temperature drop across the assembly when the outside surface is 0 °F (−17.8 °C) is:

0 °F − (70 °F) = −70 °F
−17.8 °C − (21.1 °C) = −38.9 °C

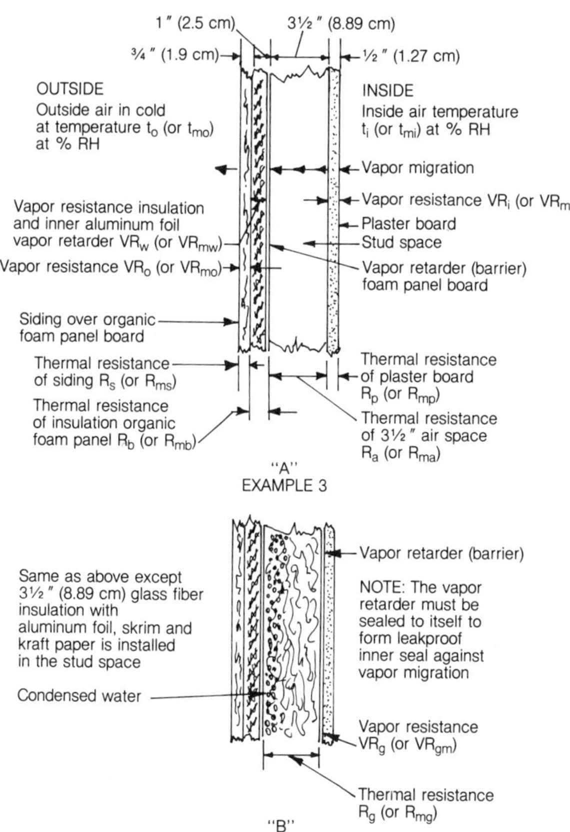

EFFECT OF MOISTURE CONDENSATION IN STUD WALLS DUE TO INSULATION BEING INSTALLED BETWEEN ORGANIC FOAM BOARD AND INTERIOR PLASTER BOARD

Figure 17

The thermal resistance percentage up to the vapor retarder is 5% + 13% = 18%.

Thus, the temperature on inner surface of the vapor barrier would be 70 °F − (70 °F × .18) = 70 − 12.6 = 57.4 °F or 21.1 °C − 7 °C = 14.1 °C. The dew point at 70 °F and 50% RH is 58.5 °F or 14.7 °C.

This wall construction will function down to 0 °F (−17.8 °C) outside wall temperature with no condensation of water vapor in the wall, as almost all the vapor retarders (barriers) are located above the dew point. Because of this, it is not even necessary to evaluate the pressure drops to the dew point plane.

In order to reduce energy losses, it is frequently recommended that the air space be filled with thermal insulation. The next example indicates what will occur to moisture control under this construction.

Example 4 "A" *English and Metric Units*

All construction and materials are the same except for the addition of 3 1/2″ (8.89 cm) of glass fiber insulation with an excellent vapor barrier seal below the plaster board. The addition of the insulation will, of course, change the temperature at various planes in the wall section. The moisture conditions within the wall structure, again, are established by using the thermal resistances and vapor resistances provided in Tables 9, 10, 11 and 12. This example is illustrated in Figure 17 "B."

English Units				***Metric Units***
R_p =	0.40	= Thermal resistance of plaster board	= R_{mp} =	0.0705
R_g =	13.80	= Thermal resistance of 3 1/2″ (8.89 cm) fiber glass	= R_{mg} =	2.4288
R_b =	5.55	= Thermal resistance of 1″ (2.54 cm) board	= R_{mb} =	0.9780
R_s =	1.05	= Thermal resistance of siding	= R_{ms} =	0.1805
	20.80			3.6578

The percentages of thermal resistances now become:

Through the interior plaster board	=	2%
Through the glass fiber insulation	=	66%
Through the organic foam board	=	27%
Through the exterior siding	=	5%

The temperature drop across the assembly when the outside surface is 0 °F (−17.8 °C) is:

$$0\,°F - (70\,°F) = -70\,°F$$
$$-17.8\,°C - (-21.1\,°C) = 38.9\,°C$$

The temperature at the plane between the plaster board and the glass fiber insulation is:

$$70 - (70 \times .02) = 70 - 1.4 = 68.6\,°F$$
$$21.1 - (38.9 \times .02) = 21.1 - 0.8 = 20.3\,°C$$

The temperature at the plane between the glass fiber insulation and the organic foam board is:

$$70 - (70 \times .68) = 70 - 46.2 = 23.8\,°F$$
$$21.1 - (38.9 \times .66) = 21.1 - 25.6 = -4.5\,°C$$

The temperature at the plane between the organic foam board and the exterior siding is:

$$70 - (70 \times 0.95) = 70 - 66.5 = 3.5\,°F$$
$$21.1 - (38.9 \times 0.95) = 21.1 - 37 = -15.9\,°C$$

It was previously determined that the dew point temperature at 70 °F (21.1 °C) and 50% RH is 58.5 °F (14.7 °C). As all the above determined temperatures are below the dew point temperature of the inside air, the dew point for each plane must be determined.

From Table 8, air at 70 °F (21.1 °C) at 50% RH has a vapor pressure of 35.61 lbs./ft.^{2} (1705 Pa).

As the vapor pressure of the outside air at 0 °F (−17.8 °C) at saturation is 2.66 lbs./ft.^{2} (128 Pa). The vapor difference from inside to outside is 35.61 − 2.66 = 32.95 lbs./ft.^{2} or 1705 − 128 = 1577 Pa.

From Tables 9 and 10, the vapor resistances can be determined for these components of wall construction. However, almost the entire vapor resistance of the glass fiber insulation is the foil-skrim-kraft barrier which must be taken with the vapor resistance at the plaster board.

English Units			***Metric Units***	
$VR_l + VR_g$	= 0.013 + 142.8 = vapor resistance of plaster board + film barrier on glass fiber insulation			
	= 142.813		$= VR_{ml} + VR_{mg}$	
			$= 0.02 + 216.8$	= 216.82
VR_w	= 33.3	= vapor resistance of organic board plus one foil face	$= VR_{mw}$	= 50.48
VR_o	= 0.063		$= VR_{mo}$	= 0.098
	176.176	= total resistance		267.398

The percentage of resistance:

$$VR_l + VR_g = 81\% \qquad = VR_{ml} + VR_{mg}$$
$$VR_w + VR_l + VR_g = 99\% \qquad = VR_{mw} + VR_{ml} + VR_{mg}$$

The vapor pressure in the glass fiber insulation is:

English Units	***Metric Units***
$35.61 - (32.95 \times .81) =$	$1705 - (1577 \times 0.81) =$
$35.61 - 26.68 = 8.93$ lbs./ft.^{2}	$1705 - 1277 = 428$ Pa

Air at this vapor pressure has a dew point of 24.5 °F (−4.1 °C). Thus even having a vapor barrier of foil, skrim, and kraft paper—assumed to be perfectly sealed—vapor will condense on the surface of the organic-aluminum faced (inward only) board. If the organic foam board had aluminum foil on both sides, the resistances would then be:

$$142.81 \div 209.47 \text{ or } 68\%$$

and the vapor pressure in the glass fiber insulation would be:

English Units	***Metric Units***
$35.61 - (32.95 \times .68) =$	$1705 - (1577 \times 0.68) =$
$35.61 - 22.46 = 13.15$ lbs./ft.^{2}	$1705 - 1062 = 643$ Pa

In this case, the dew point is 32.5 °F (0.3 °C) and vapor condensation would take place further inward in the glass fiber insulation. The additional layer of aluminum foil on the outer side of the organic foil makes it even more impractical to find materials suitable for sealing at the on inner surface to prevent condensation within the insulated stud space.

Example 4 "B"—*English and Metric Units*

It is common practice to recommend polyethylene film sheets to be installed over mineral fiber insulation installed in stud spaces. With this type of vapor retarder film it is very difficult to seal around projections such as electric conduit boxes or pipes. However, if it were possible, what would be the results? We assume all factors are identical to those listed in Example 4 "A", with the exception of the difference in the vapor migration properties of the two vapor retarders. As before, the vapor resistances are listed first as follows:

English Units			***Metric Units***	
$VR_l + VR_g$	= 0.013 + 3.579 = vapor resistance of plaster board + 0.03″ thick polyethylene film		$= VR_{ml} + VR_{mg}$	
	= 3.592		$= 0.02 + 5.44$	= 5.46
VR_w	= 33.3		$= VR_{mw}$	= 50.48
VR_o	= 0.063		$= VR_{mo}$	= 0.098
	36.955	= total vapor resistance		56.038

The percentage of vapor resistance supplied by the plaster board and the polyethylene film is 10%.

The vapor pressure in the glass fiber insulation is:

English Units	***Metric Units***
$35.61 - (39.95 \times 0.10) =$	$1705 - (1577 \times 0.10) =$
$35.61 - 3.99 = 31.62$ lbs./ft.^{2}	$1705 - 157.7 = 1547.3$ Pa

Air at this vapor pressure has a dew point of 56 °F (13.3 °C). Thus, when this construction is used—even if the vapor retarder sheet is perfectly sealed—vapor will condense in the fiber insulation.

To provide an escape for vapor getting into the stud space, it has been suggested that a corrugated strip be installed between the organic foam board and the horizontal 2 x 4 at the top of the stud space. This is illustrated in Figure 18.

As installed, the temperature of the strip will be very close to the temperature of the outdoor air. Even if completely sealed, the calculations for Example 4 "A" show the temperature of the inner surface of the organic foam board would be 21.1 °F (−5.5 °C). Thus, the exposed surface of the corrugated strip would average [when outdoor temperature is 0 °F (−17.8 °C)]

$$\frac{21.1° + 0°}{2} = 10.5\,°\text{F}$$

or −12.5 °C. At this temperature, the surface would condense vapor to ice in a few minutes. Since the strip is long and narrow with a high ratio of surface to vent area, the ice would effectively block further passage of vapor through this opening.

Vent of vapor by air ventilation:

If the outdoors is at 100% RH 0 °F (−17.8 °C), it has a vapor moisture content of 0.026 grains/ft.^{3} or 0.059 grams per cubic meter. At 10 °F (−12.2 °C) air at 100% RH has a moisture content of 0.776 grains/ft.^{3} or 1.77 grams per cubic meter. Thus, outside air at 10 °F (−12.2 °C), which is obtained for ventilation purposes, can absorb the following amount of moisture:

English Units	***Metric Units***
$0.776 - 0.026 = 0.75$ grains/ft^3 or	$1.77 - 0.059 = 1.611$ grams/m^3.

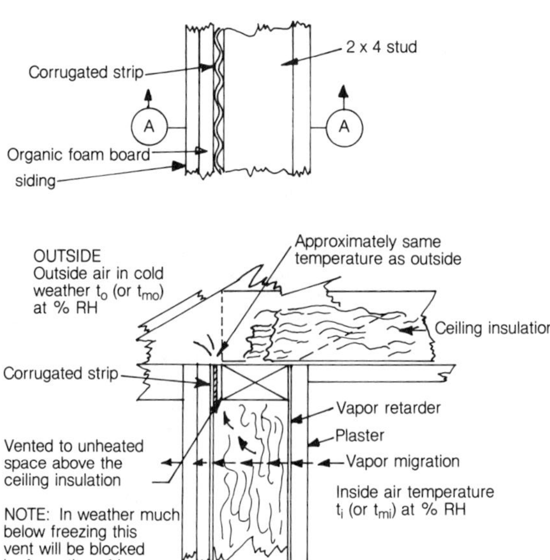

EFFECTIVENESS OF CORRUGATED STRIP AS VAPOR VENT

Figure 18

Under the conditions given in Example 4 "B", the vapor pressure difference between the inside enclosure and the glass fiber insulation is 39.95 × .10 = 3.99 lbs./ft.².

From Table 9, the permeance is 0.279 grains/lb. ft.² or 3.99 × 0.279 = 1.11 grains/hr. ft.² (0.812 grams/hr. m²). Therefore, under these conditions, the following amount of ventilation would be required:

English Units

$\frac{1.11}{0.75}$ = 1.48 cubic ft. of air/hr for each sq. ft. of wall surface. Each stud space 18″ × 8′0″ has an area of 12 ft² and would require 1.48 × 12 = 17.76 cubic feet of air per hour to prevent vapor from condensing in the insulation.

Metric Units

$\frac{0.812}{1.611}$ = 0.5 cubic meters of air/hr. for each m² of wall surface. Each 0.45 m × 2.44 m stud space has an area 2.44 × 0.45 = 1.09 m², and would require 109 × 0.5 = 0.54 cubic meters of air per hour to prevent vapor from condensing in the insulation.

If it is necessary to ventilate behind the organic foam insulation (or any other siding)* in order to compensate for a high vapor resistance on the wrong side of the major insulation, then the heat transfer must be calculated based on the temperature difference to the ventilation air. It is apparent that the thermal resistance of the siding and panel would have little effect on the heat transfer from the inside to the air being vented to prevent condensation.

To accomplish this ventilation to prevent condensation, the stud space must be provided with vent openings of sufficient cross sectional area so that they are not easily plugged with condensate and ice. Of course, a screen must be provided in these vents to prevent entry of bugs and insects. A typical installation is shown in Figure 19. *This practice is not recommended for new construction; it should only be used to correct the incorrect installation of organic foam sheathing or vapor resistance barriers under siding (shingles, etc.).*

To illustrate the principal involved, it is suggested that the reader try the following:

1. On a cold winter day (below freezing) fill up a glass with ice and water and set it outdoors. It will not condense water vapor on its surface.
2. Bring this cold glass of water inside to the heated indoor air. Water vapor will condense on its surface immediately because its surface temperature is below the dew point temperature of the air. Then place this glass directly under (or near) the kitchen vent fan. The air movement caused by the fan will not cause the condensate to evaporate from the glass surface, as the air has too high a dew point. This is illustrated in Figure 20.

This also illustrates that the use of air vents as shown in Figure 19 should only be used when the cavity space has not less than 2″ (5.08 cm) of insulation between the vents and the inner surface of

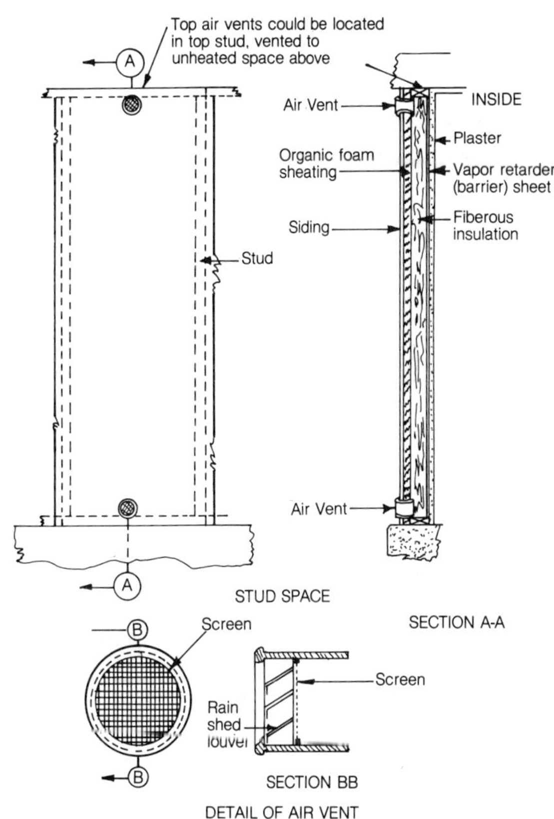

AIR VENTED STUD WALL TO RETARD VAPOR FROM CONDENSING IN STUD ENCLOSURE

Figure 19

*Where vapor resistant barriers—such as asphalt, impregnated felts, aluminum foil paper or aluminum sheets—have been installed under outer siding or sheathing board.

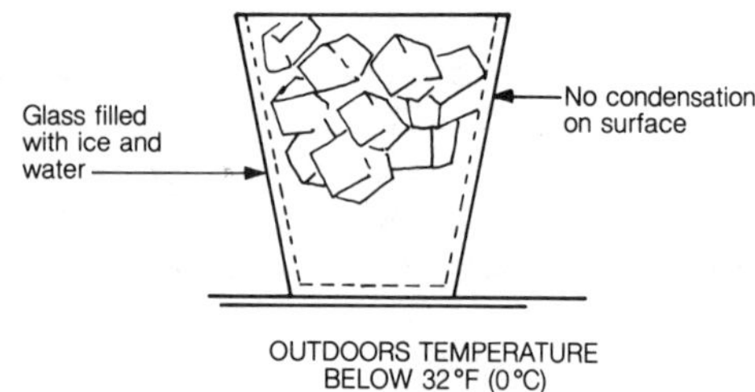

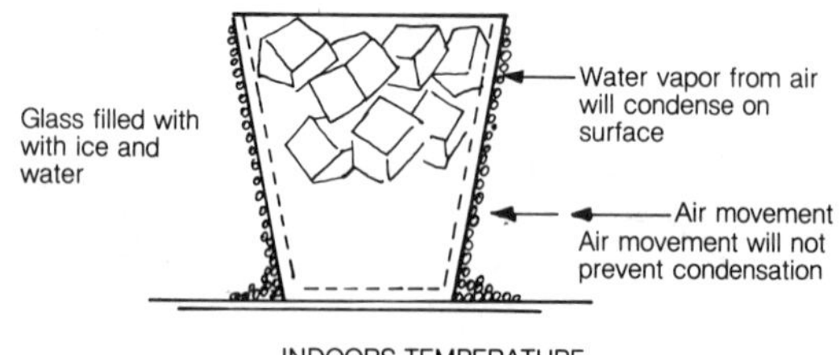

INDOORS TEMPERATURE
70°F (21.1°C)

WATER VAPOR CONDENSES OR DOES NOT CONDENSE ON COLD SURFACE DEPENDING ON TEMPERATURE AND RELATIVE HUMIDITY OF SURROUNDING AIR

Figure 20

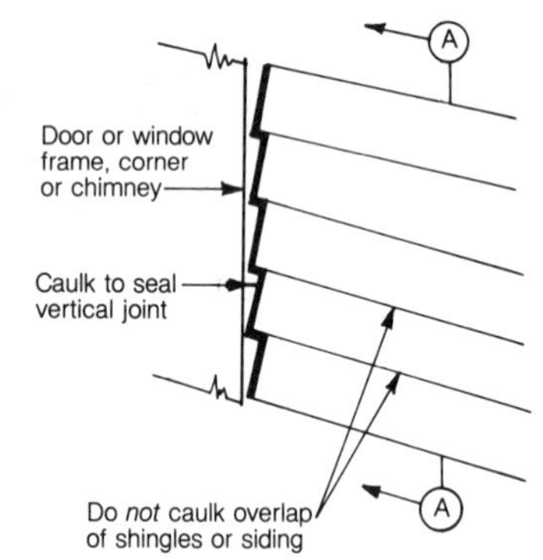

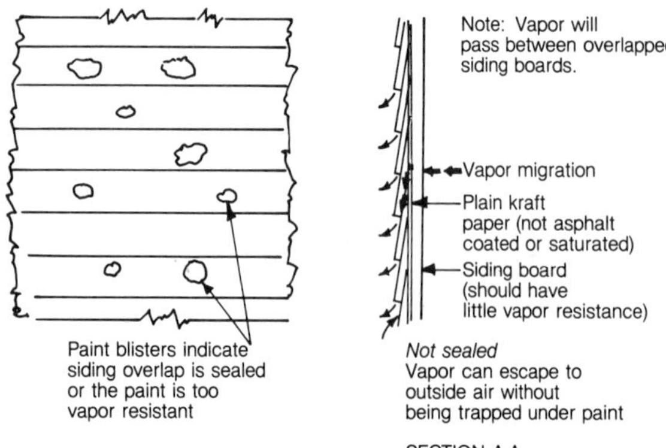

WHERE TO CAULK

Figure 21

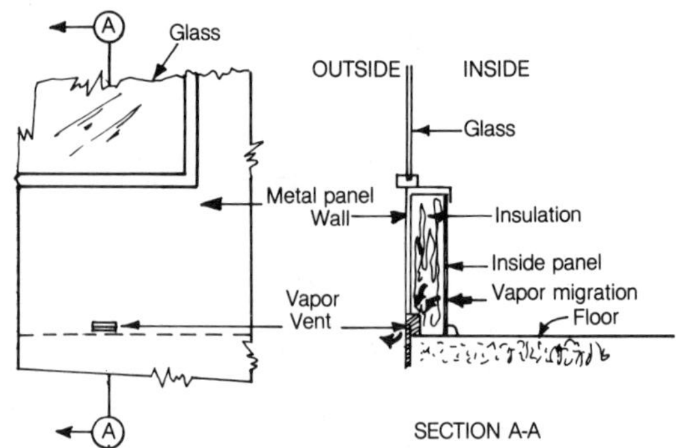

VAPOR VENTS IN METAL OUTSIDE PANEL WALLS

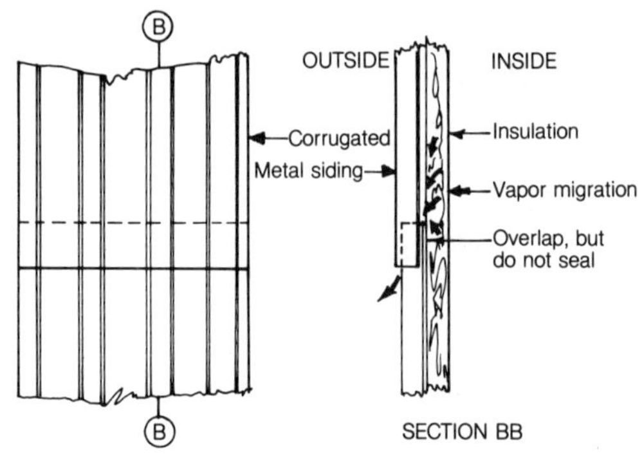

OVERLAP OF CORRUGATED METAL PANELS SERVE AS VAPOR VENT

Figure 22

the plaster board. Like the glass of water, if the air was ventilated into empty stud space, the low temperature outdoor air could cool the plaster wall to below the dew point temperature of the inside air.

As reported in USDA Forest Service Research Paper FPL 290, 1977, it is important that condensation be prevented in the wall spaces in order to keep the moisture content of the wood from exceeding 20.5%. It states that above this moisture content level wood is likely to decay (rot) thus endangering the entire structure.

In all cases, building walls must be constructed tightly enough to prevent the entry of rain water. However, the wall on the outside should not be vapor tight. Figure 21 illustrates where caulking should be used on outer shingle or siding walls and indicates that it should not be used where it prevents vapor migration. Aluminum metal siding should be provided with a small hole in its bottom extension every 18″ (45.7 cm) to allow vapor migration.

Also illustrated in Figure 21 is what happens when wood siding is coated with vapor resistant paint and too tightly sealed by caulking or paint at the overlap joints. The vapor *does go through* the wood; when it gets to the paint outside, it is trapped and in winter it is condensed to a liquid. When the temperature rises in summer, surfaces facing the sun can get up to 140°F to 150°F (60°C to 65.6°C). This entrapped liquid water is vaporized and develops a pressure of 415 lbs./ft.^{2} (19912 Pa) to 535 lbs./ft.^{2} (25634 Pa) and pushes the paint film off the wood into a blister.

It is equally important that metal siding is provided with vapor vents. Figure 22 illustrates a typical insulated panel in metal-glass office building. Section "AA" shows a proper vapor vent in such construction. Without proper vapor venting, trapped liquid water can cause serious rusting of steel members. Also, Figure 22 shows how corrugated metal panels can be overlapped to serve as vapor vents where such is used on buildings that require winter heating and humidification for human comfort.

ROOFS

Identical to walls, roof or ceiling insulation should be installed in such a manner that condensation does not occur within it. (As shown in Figure 23). The easiest way to prevent condensation in the insulation is to ventilate between the insulation and the roofing.

For a flat roof, ventilators should be installed in the walls above the insulation. These should be with sloped partitions to shed water and be provided with screens to prevent entry of insects. On large roofs, air vents that project through slab and roofing are required to obtain the needed air movement. This type of vapor migration removal is illustrated in Figure 23 "A".

A sloped roof, with an occupied attic space, is illustrated in Figure 23 "B". Like the flat roof in "A", the space between the insulation and the roof should be ventilated. When the attic space is not occupied, then the attic space itself may be used to ventilate above the insulation.

One of the most common errors in the installation of attic insulation is that it is installed upside down. The tendency for individuals putting it in existing homes is to place it with the foil, or film backing, on the top. This is completely wrong, as moisture is trapped and condenses in the insulation.

The ventilators installed to provide the air movement should have an open area—both inlet and outlet—of a minimum of 1 sq. ft. to 300 sq. ft. of insulation area (0.1 sq. meter to 30 sq. meter).

Flat roofs, where insulation is on top of the deck and on which the built up roof is installed, present a very difficult problem to prevent vapor condensation in the insulation. Where the insulation consists of low density fibers or particles with very high vapor permeability ordinary vapor barrier papers or films can be attached to the deck surface on which the insulation is installed. Air vents secured to a flat metal base can then be placed on top of the insulation. The built up roofing can then be installed over the insulation to prevent the entry of water, but the air vents will allow vapor escape. Of course, the vents must be above the height of the water drains and be equipped with screens to prevent the entry of insects. This is illustrated in Figure 24.

Oddly, a major problem exists with insulations having a low vapor permeability, but which is just sufficient to allow vapor entry. Specifically, these are organic foams, in either block application or sprayed in place. Both urethane and styrene foams are too vapor resistant for vents, as shown in Figure 24 "A", to be effective. Unfortunately, their vapor resistance, as compared to multiple layer built up roofing, is not sufficient to keep them free of vapor condensation. From Table 9, it can be determined that a single layer of 55 lb. roofing felt has a vapor resistance of 1.053 per layer (3 × 1.053 ³ 3.159) plus the resistance of the asphalt coat which would bring the total resistance to approximately 5.0. [Metric resistance (1.59 × 3) + 3.18 = 7.95]

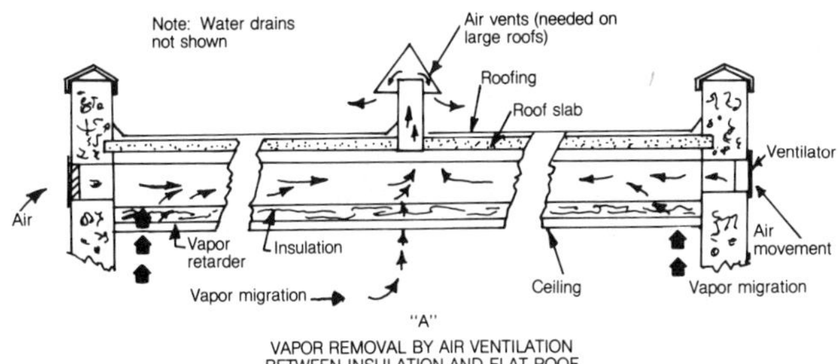

VAPOR REMOVAL BY AIR VENTILATION BETWEEN INSULATION AND FLAT ROOF

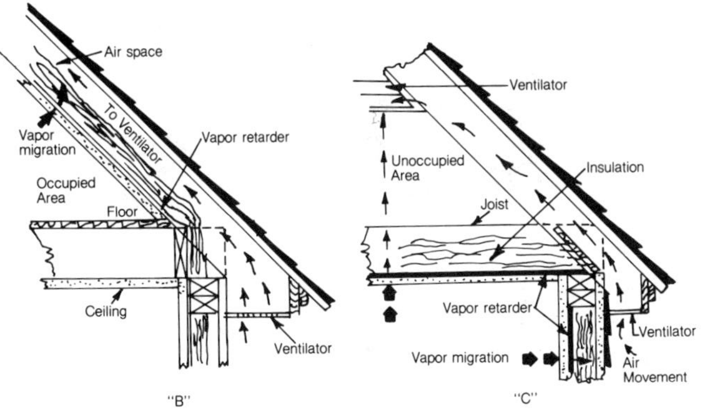

VAPOR REMOVAL BY AIR VENTILATION BETWEEN INSULATION AND SLOPE ROOF

Figure 23

Figure 24 "B" illustrates organic foam installed directly on top of roof deck and 24 "C" illustrates a combination of inorganic insulation over which the organic foam is installed. Unless the insulations have vapor resistance well over 5 (English units) or 3.18 (Metric units), the resistance of vapor retarder needed must be calculated. A suitable retarder of this resistance must be applied under the foam.

In a building that has 70 °F (21 °C) temperature at level of occupancy, the temperature directly under the ceiling will be not less than 80 °F (26.7 °C). This temperature at 50% relative humidity has a vapor pressure of 47.2 lbs./ft.^{2} (2258 Pa). The temperature directly under the roofing when outdoor temperature is 0 °F (−17.8 °C) would be no higher than 10 °F (−12 °C). At 100% relative humidity, the vapor pressure at this temperature is 4.6 lbs./ft.^{2} (220 Pa). Thus, the vapor pressure trying to drive vapor into the insulation is 47.2 ÷ 4.6 or (2258 ÷ 220) = 10.26 times greater than that of the vapor pressure at the surface where the vapor can escape into the atmosphere. To prevent build up of vapor moisture (and its condensation into liquid or ice), it is necessary that the vapor retarder be at least 10.26 times more resistant than the built up roofing. Therefore, under the conditions stated, the vapor resistance of the roof deck and vapor retarder membrane under the insulation must be no less than 10.26 × 5.0 = 51.3 English Vapor Resistance Units or 10.26 × 3.18 = 32.63 Metric Vapor Resistance Units.

On a metal deck roof this resistance can be obtained if all joints are caulked with vapor seal mastic over which is applied heavy aluminum adhesive tape of not less than 0.005″ thickness. Another way would be to seal weld all metal seams; however, this is difficult as these weld areas would all have to be cleaned and

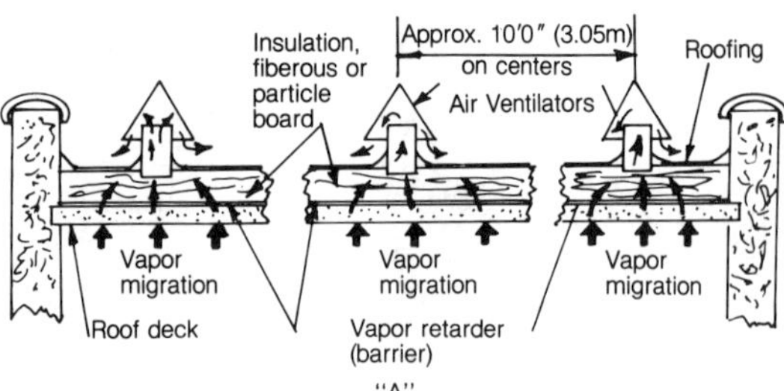

VAPOR REMOVAL BY AIR VENTILATION OF FIBEROUS (OR PARTICLE) INSULATION BOARD

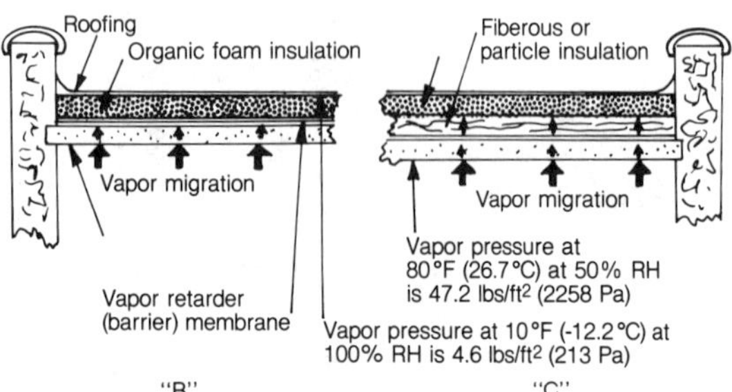

To prevent excessive moisture entrapment in the organic foam insulation, under stated conditions the vapor resistance must be at least (47.2 ÷ 4.5) = 10.5 times greater than the vapor resistance of insulation plus roofing.

VAPOR MIGRATION CONTROLLED BY VAPOR RETARDER (BARRIER) UNDER THE INSULATION

Figure 24

then recoated with primer and finish coating to prevent rusting. In using this type of metal deck as a vapor retarder (barrier), every overlap, hole, or joint must be vapor sealed.

On concrete slabs or wooden decks a vapor retarder must be installed over the deck before insulation is applied. The vapor retarder (barrier) must not have less than 51.13 English Vapor Units, or 32.63 Metric Units, of resistant to vapor passage. The values of resistance include joints. In the case of concrete or wooden decks, the vapor migrates through the concrete or between the wooden planks. These provide almost no vapor resistance, thus all the resistance to vapor migration must be obtained by the vapor retarder (barrier).

Note: To reduce the fire hazard, over wooden roof decks it is best to use either a non-combustible insulation—as shown in Figure 26 "B"—or a minimum of 1/2" thick gypsum board, this will also reduce the fire hazard caused by asphalt roofing or organic foam used as insulation.

The vapor retarder (barrier), to have this resistance to vapor migration, must be metallic type. The vapor barrier should be heavy aluminum foil of not less than 0.005" in thickness on skrim reinforcing or be backed with heavy kraft paper or plastic film. The vapor retarder and its backing shall be such that it can tolerate the temperature of melted asphalt without damage.

The aluminum foil vapor retarder must be adhered to the deck (or other substrate surface—such as gypsum board) with all joints lapped a minimum of 2 inches (50.8 mm) with sealant between the overlapped sheets. Care must be taken to prevent any puncture during application. Also, where installed over concrete decks, the aluminum foil must be protected with a coating over its surface facing the concrete so as to protect the metal from alkaline vapors and liquid. Such an application is shown in Figure 25 "A."

Thin galvanized steel or aluminum foils can be used as a vapor retarder membrane. If there is a risk that fire might contact the underside of the deck, then galvanized steel should be used. In applications where fire from inside is not a factor, aluminum 0.010" (0.254 mm) or 0.016" (0.406 mm) in thickness is an excellent, almost perfect, vapor retarder membrane. This application is shown in Figure 25 "B."

Some may consider that the cost is excessive for installing these types of vapor retarder membranes under organic foams or permeable fiber (or partial) forms of insulations. Note that basic, less expensive membranes and or less costly installations might be used. However, inefficient vapor retarder membranes will cause excessive damage to roofing, the insulation will become thermally inefficient and after a period of time may cause complete collapse of the entire roofing system.

The major problem of failure of built up roofing on flat roofs is water condensation getting into the roofing *from below*. During winter when the major vapor pressure difference exists, the vapor trying to escape into the atmosphere condenses in the seams and voids in the roofing. It is then subjected to freeze-thaw cycles

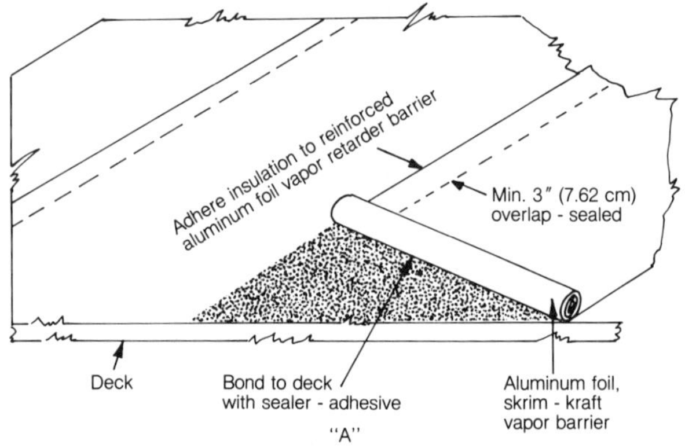

APPLICATION OF ALUMINUM FOIL VAPOR RETARDER MEMBRANE TO ROOF DECK

Adhere insulation to metal
Deck
Joint
Joint Detail
Bond to deck with sealer - adhesive
Thin gauge sheet metal
"B"

APPLICATION OF SHEET METAL VAPOR RETARDER MEMBRANE TO ROOF DECK

Figure 25

which stresses and breaks the roofing membrane. In the summer, this entrapped liquid in the roofing is vaporized by the high temperatures. It is not unusual for surface temperatures on top of black roofs to reach 160 °F (72.2 °C). At this temperature the vapor pressure is 68.20 lbs./ft.^{2} (or 32079 Pa). This causes the roofing to blister. To combat this, a common practice is to put 3 to 4 inches (7.6 to 10.1 cm) of stone on roofs to provide weight to push the blisters down. Also, as the asphaltic materials are quite soft at these high temperatures, the stones tend to cause small cracks in the surface which releases the vapor.

Therefore, on the top side the temperature causes very high vapor pressure as the entrapped liquid turns into a vapor. Why then does the vapor not escape downward in the manner that it entered? The reason is that the entrapped vapor has a 100% relative humidity and as this vapor moves downward to lower temperatures it again reaches the dew point and condenses back to liquid water. This is why permeable insulation must be provided with exterior vents as shown in Figure 24.

If not properly vented, the insulation becomes wet and its thermal conductivity increases. This reduces the thermal efficiency of the insulation and energy losses increase. In addition to the increase to heat transfer, the weight of the insulation increases. A light-weight foam insulation that weighs 1 1/2 lbs./cu.ft. (40.05 kgs/m^3) when completely filled with water or ice weighs 65.0 lbs./cu.ft.) (1041 kgs/m^3). Fiberous or partial inorganic insulations would weigh slightly more. This added weight, which for 9″ (10.5 cm) thickness could be as much as 20 lbs. (4.6 kilograms) per square foot, in winter with snow load could contribute to roof collapse. Another factor which contributes to roof failures is that the vapor after condensing into liquid does flow downward—this wets the surface of the supporting beams. Starting at the top, and out of sight, means that steel could be in a very weakened condition due to rust and corrosion without anyone knowing of the danger. Poorly designed and installed insulation on flat roofs may cause roof failures and possibly complete collapse.

To prevent these failures of flat roof insulation, the vapor retarder, as stated, must be installed correctly, or insulation, which is its own vapor barrier, must be used. Only cellular glass insulation meets this requirement. This insulation has a vapor permeability which is much lower than any obtainable plastic, films, foils or combination of these commonly used vapor retarders. The major concern is that the insulation be located in combination with other building materials or insulations, so that its surface facing the air of highest vapor pressure be above the dew point of that air.

The method in which this insulation should be installed on a flat roof with concrete slab is shown in Figure 26 "A." It is available in several forms: (1) Blocks 1 1/2″ thickness (3.8 cm) to 5″ (12.7 cm) in thickness in 1/2″ increments. (2) In boards 24″ × 48″ (0.61 m × 1.21 m) in 1 1/2″ thickness (3.8 cm) to 4″ thickness (10.2 cm); and (3) In tapered sheets to provide a 1/2″ to 1′0″ slope (1.3 cm to 30.4 cm). As shown in this figure, the insulation is bonded to the concrete slab with a suitable adhesive. All butt joints should be vapor sealed with vapor resistant, resilient mastic.

NOTE: Fire hazards as related to thermal insulations and their application will be discussed in Chapter 6—Safety.

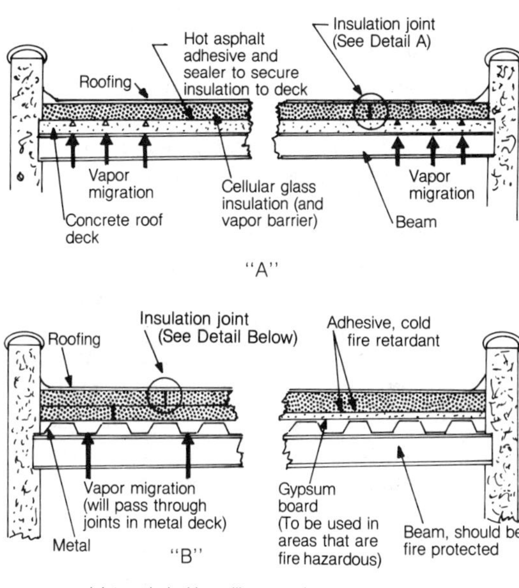

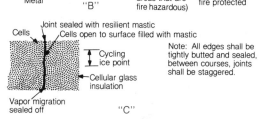

VAPOR MIGRATION CONTROLLED BY VAPOR RESISTANT INSULATION WITH SEALED JOINTS

Figure 26

Figure 26 "B" illustrates cellular glass installed on a corrugated metal deck. In this case the adhesive should be a two-part adhesive applied to the top of the steel member onto which the insulation is installed before the adhesive sets. (Ordinary solvent type adhesive between cellular glass and steel would require weeks for the solvent to escape and setting of the adhesive would occur.) A double layer of insulation is shown.

Figure 26 "C" illustrates cellular glass insulation over gypsum board which was applied to a metal roof deck before the insulation was installed. This type of construction is used where there may be an accidental fire in an area below the deck, and where flames could impinge on the corrugated metal. Although cellular glass is inorganic, noncombustible glass will crack when subjected to thermal shock. These cracks would then allow the heated bituminous roofing materials to flow and drip down into the fire. Where such construction is needed the beam must also be of fire-resistant construction.

FLOORS

Frame type floors over unheated or crawl spaces follow the same requirements for control of moisture as do frame walls. In this case the vapor retarder (barrier) must be on the top side of the insulation to prevent entry of moisture. Likewise, the underside should be properly ventilated to prevent build up of vapor or condensate. The same methods used to determine if moisture vapor will condense in walls are equally suitable for floors.

Over crawl spaces or moist unused areas there does exist the potential for moisture to enter from below. This would raise the dew point and condensation would occur at a much higher surface temperature. For this reason a vapor-water resistant cover should be installed over the soil or ground cover slab. This vapor-water resistant cover may be of plastic, reinforced aluminum foil, covered kraft or impregnated felt. The asphalt impregnated felt should only be used where the floor above is concrete or some other noncombustible flooring. This space should also be ventilated, with a minimum of 1 square foot of ventilation opening to 300 square feet of floor space (or 1 square meter to every 300 square meters).

In Figure 27 "A" a typical wood floor-joist over crawl space is shown. It is extremely important that the vapor barrier above the insulation be sealed as carefully as possible. If the vapor barrier is installed after the flooring is installed, its junction with joists should be sealed with aluminum adhesive tape. All joints of ground cover should be overlapped and sealed.

Slab floor construction over crawl spaces is shown in Figure 27 "B." The major difference of this application as compared to 27 "A" is that the vapor barrier under the slab may be a vapor seal coating which is applied after construction. Block insulation can

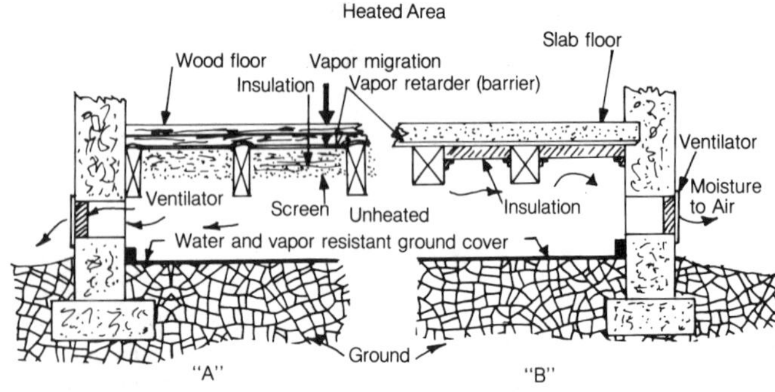

Note: The vapor barrier should be installed, if possible, to be over top of joists or beams. If not possible, then it should be sealed up to intersection of floor and joists or beams.

VAPOR MIGRATION CONTROLLED BY VENTILATION

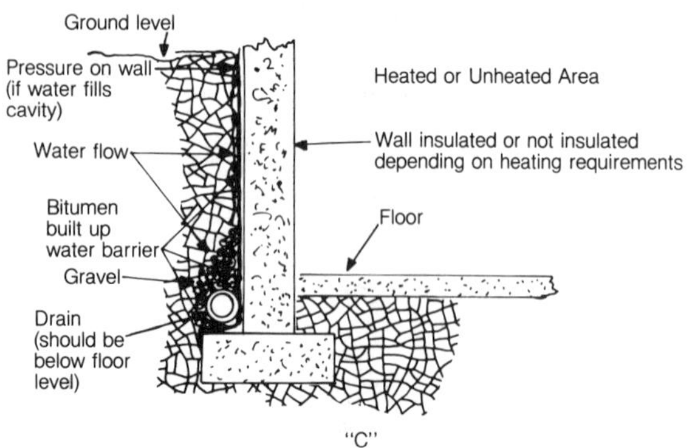

LIQUID MIGRATION CONTROLLED BY WALL, PROTECTED BY WATER BARRIER AND DRAIN

Figure 27

be used in this type of application by being installed in such a way that the blocks are held in position by angles or plates secured to the beams. Again, if the block insulation is an organic foam type, and a danger of fire exists below the floor, then the insulation should be protected by at least 1/2″ gypsum board (without paper) installed directly beneath it.

Another source of moisture is liquid water which seeps through walls. If walls are not water tight, the liquid which enters will evaporate within the space causing the relative humidity to get up to 100%. Figure 27 "C" shows the manner in which the wall below grade is protected by a water barrier on the outside. It is essential that water be provided a drain so as not to build a water head against the wall. The reason it is so important to prevent excessive water height is that water of any thickness between ground and wall will build up pressure per each linear foot of wall, per foot of depth as shown below:

TABLE 13

Total Horizontal Pressure on Wall Per Foot of Height (Depth) Per Linear Foot of Wall - (Horizontal)

Height of Water	Total Horizontal Pressure-Lbs.	Height of Water Ft.	Total Horizontal Pressure
1′ (0.3m)	31 lbs. (14 kg)	6′ (1.8m)	1124 lbs. (509 kg)
2′ (0.6m)	165 lbs. (56 kg)	7′ (2.1m)	1530 lbs. (604 kg)
3′ (0.9m)	281 lbs. (127 kg)	8′ (2.4m)	1997 lbs. (905 kg)
4′ (1.2m)	500 lbs. (226 kg)	9′ (2.7m)	2527 lbs. (1146 kg)
5′ (1.5m)	781 lbs. (354 kg)	10′ (3.0m)	3120 lbs. (1415 kg)

As indicated, the pressure build-up against a wall by water can be very high. It must be remembered 1/32″ thickness of water between the ground and a wall exerts the same pressure as if that water were a lake and the distance behind the wall were measured in miles.

To take an example: If the basement wall of a home (or business building) extended 10′0″ below grade and this wall were 30′0″ in length, what would be the pressure of the water against that wall if the drainage stopped up and the water from rain filled the space between the ground and the wall? The answer is 30′ × 3120 or 93,600 lbs. (42456 kg) pressure.

With this amount of pressure exerted against walls, when improperly drained, cracks are likely to occur. If cracks do develop, the water has sufficient pressure per unit of area to push through the cracks and into the inside of the building.

The pressure at the base of a column of water (height) is:

TABLE 14

Pressure of Water at Base - For Depth of Water

Depth of Water	Pressure Per Unit Area	Depth of Water	Pressure Per Unit Area
1′(0.3m)	0.43lbs/in² (302kgs/m²)	6′(1.8m)	2.60lbs/in² (1828kgs/m²)
2′(0.6m)	0.85lbs/in² (604kgs/m²)	7′(2.1m)	3.03lbs/in² (2130kgs/m²)
3′(0.9m)	1.30lbs/in² (914kgs/m²)	8′(2.4m)	3.46lbs/in² (2433kgs/m²)
4′(1.2m)	1.73lbs/in² (1216kgs/m²)	9′(2.7m)	3.90lbs/in² (2742kgs/m²)
5′(1.5m)	2.17lbs/in² (1525kgs/m²)	10′(3.0m)	4.35lbs/in² (3058kgs/m²)

The importance of keeping this liquid from entering the building can be better understood when it is recognized that all such water changes to vapor and this vapor must escape through the walls (above grade), windows, doors and roof of the building. In the process, it may be recondensed in the walls or roof with severe damage to the building.

It is necessary to point out that the essentials of preventing water build up outside of walls also applies to floor slabs. If the drainage of water from around the base of walls is not sufficient to remove it, then it will seep under the slab floors and exert the same type of hydraulic pressure under the slab causing it to move or fracture. To be effective, insulation under a slab must be dry, thus water must be controlled before it gets under the slab. If the slab floor is located above, or close to, ground level, frequently it is insulated around the perimeter where the greatest temperature difference, thus the greatest heat loss, occurs.

Perimeter insulation using organic foam insulation is shown in Figure 28 "A." Although dimensions are not given, in most instances the insulation extends in from the wall 18″ to 24″ (0.45 to 0.6 meters). These materials have little compressive strength being 15 lbs/in.^{2} (1.1 kg/cm^2) at 10% deformation. Thus, the concrete slab should be provided with reinforcing rods to provide cantilever strength so that heavy loads placed near the wall will not crack the floor. Also, these organic foams are not termite or rodent proof, hence a metal shield of galvanized or stainless steel should be installed.

With continued increases in energy costs, it will become much more important to insulate concrete slab floors of heated and air-conditioned areas. Among other advantages, an insulated ground floor slab reduces the human need for even higher room temperatures to compensate for the sensation of otherwise cold feet. Under an entire slab the insulation must be sufficiently strong to withstand the compressive loads imposed on it by the concrete slab and all the other loads imposed on the slab. Cellular glass has an ultimate compressive strength of 100 lbs./square inch (7.0 kg/cm^2) thus can be used under a concrete slab with the ordinary loads placed on such a floor. Such an installation is shown in Figure 28 "C." A water barrier sheet should be installed over the leveling sand so as to prevent liquid water from reaching the butt joints of the blocks. Over the cellular glass a mopped coating of asphalt coating should be applied to the surface of the insulation. This provides a cushion between the concrete and the insulation and prevents concrete bonding and shrinkage from cracking the insulation. Columns should be provided with their own footers so as not to cause excessive area loading of the insulation slab system.

As the insulation is almost completely impervious to vapor-migration and water-migration, it serves as a membrane to control moisture migration.

This chapter has established the basic properties of water vapor, its pressure, and the necessity of keeping walls, roofs, and floors relatively dry for maximum thermal efficiency. Although the subject of this text is thermal insulation, it is of importance to look at how all of the moisture in a building is generated. It is also of interest to determine what can be done, in an enclosure, to regulate the liquid vapor content of the air. For these reasons the next section presents the source of moisture in a home or building and the factors which control that moisture.

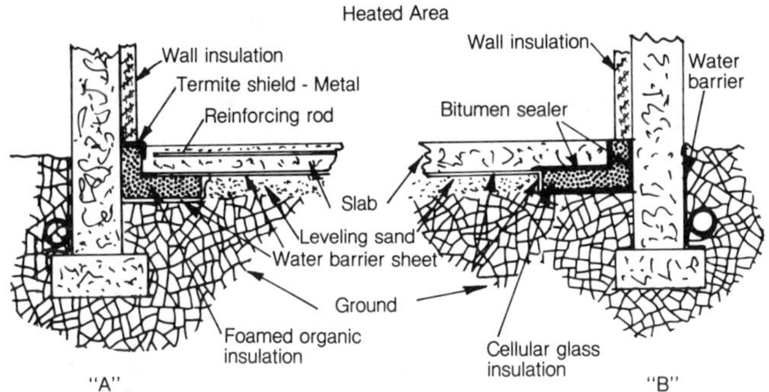

MOISTURE CONTROL OF SLAB AND PERIMETER INSULATION

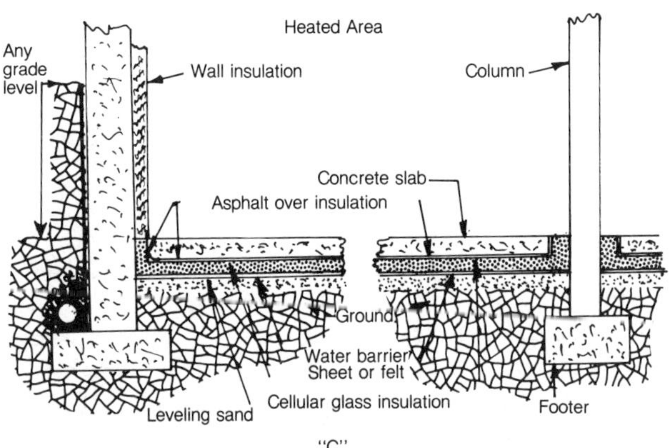

MOISTURE CONTROL OF SLAB INSULATION

Figure 28

SOURCES OF MOISTURE

It is fundamental that houses and buildings be constructed to keep out liquid water or frozen water (snow or sleet). The most difficult area to make water tight is the basement, or any area below grade. However, the previous section described how to construct areas below grade to be water tight. Where then does all the moisture come from?

In new construction, considerable water is used in the mixing of concrete, mortar, and plaster walls. A new house of medium size may require 300 gallons of water to be mixed in the concrete, 600 gallons in the concrete walls and 250 gallons in the plaster. For a number of months after construction, a house or new building requires ventilation and heat to remove this moisture. To illustrate the effect this will have on the air inside a house one must determine how many cubic feet of air would have this amount of water vapor.

The total amount of water is 300 + 600 + 250 = 1150 gallons; at 8.336 lbs. per gallon, this represents 9584 lbs. of water or 9584 × 7000 = 67,104,800 grains of water. At 70 °F and 50% RH each cubic foot of air contains 3.99 grains of water. Thus, this amounts to 16,818,246 cubic feet of air at 70 °F and 50% RH or 476,292 cubic meters of air at 21 °C and 50% RH.

If air for ventilation is originally at 40 °F (4.4 °C) and 50% RH, then its vapor content would be 1.42 grains. When heated to 70 °F (21.1 °C) at 50% RH, it will absorb 2.57 grains.

From the above it can be understood that it will take considerable time for the ventilated air to dry out the concrete and plaster. During this drying period do *not* turn on the humidifier. It may also be necessary (even in winter) to open one window at least 1/4" (.63 cm) to allow a direct moisture escape.

Moisture input is still a factor after houses or buildings have dried out from their construction. In industrial buildings it is necessary to evaluate the moisture input based on processes, use, and occupancy. In residences, the occupants use water and some of it goes into the air as vapor. Some activities that produce water vapor, and the approximate amounts of water that are vaporized into the air are:

Activity	***Vaporized Water Per Day***	
	Pints	***Liters***
Cooking—use of kettles, coffee pots, and pans of boiling liquids	5 to 7	2.4 to 3.3
	Water Vaporized Per Activity	
Clothes washing and drying	25 to 35	11.8 to 16.6
Hand and face washing	0.02	0.009
Bath	0.1	0.05
Shower	0.5	0.24

The dissipation of this moisture was not a problem in the older houses which were not built tightly and seldom were equipped with storm doors or windows. The infiltration air as it was heated had sufficient capacity to absorb this moisture. Houses and buildings now being constructed, however, are tighter and better insulated to reduce energy losses and cannot dissipate this moisure by infiltration alone. For this reason it is excellent practice to have an exhaust fan at each of the areas where moisture vapor is being produced.

Figure 29 "A" illustrates use of an exhaust fan in the bathroom. Figure 29 "B" shows the advantage of a vent hood over the stove and 29 "C" indicates that all clothes dryers should be directly ventilated to the outside.

One tends to think of air in a home or business as being at a certain temperature and relative humidity, and that when it encounters a sufficiently cold surface moisture will condense on that surface. However, a more correct analysis is that the lowest temperature surface (if it is sufficiently large) in an enclosure controls the relative humidity within that enclosure.

To illustrate, if a window, open to the rest of the house, has an inside inside surface temperature of 50°F (10.0°C) and the inside air temperature is 70°F (21.1°C) the maximum relative humidity that can be maintained is 20% (See Table 2) before condensate begins to collect on the windows. This points to the following recommendations:

1. In winter, to maintain desirable humidity, storm windows with double-pane glass are essential.
2. To prevent condensation in the walls, roofs and ceilings and in order to maintain human comfort over a range of relative humidities, it is essential to have an excellent vapor retarder (barrier) properly sealed.

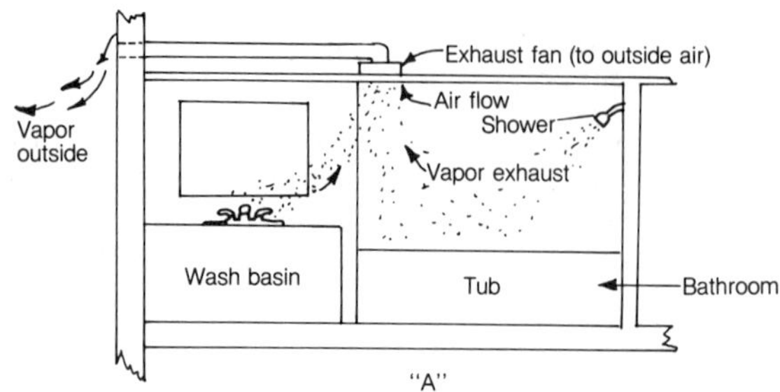

MOISTURE CONTROL BY USE OF EXHAUST FAN IN BATHROOM

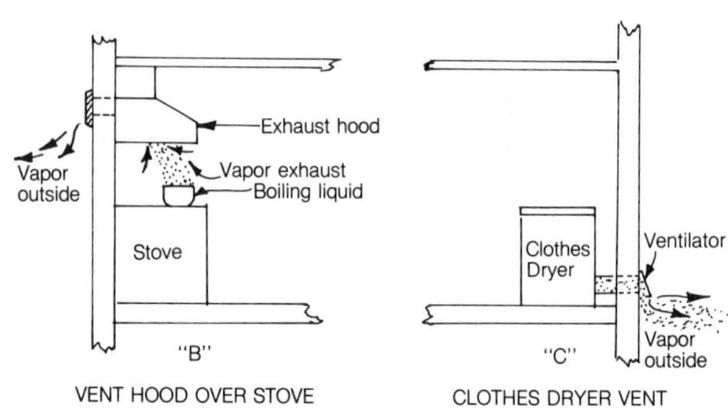

VENT HOOD OVER STOVE

CLOTHES DRYER VENT

MOISTURE CONTROL BY DIRECT EXHAUST OF VAPOR TO OUTSIDE AIR

Figure 29

3. Unless a building has an almost perfect vapor seal inside, where the temperature is 55°F or greater, the use of vapor resistant insulation such as organic urethane or styrene on the outside of frame walls, will cause condensation within the wall structure.
4. Never use asphalt impregnated felts, or foils over the stud panel boards (under siding or shingles) in cold climates.

Another most important item is that when condensation shows up on the windows, make sure the humidifier is turned off. When condensation is forming on the glass, the only thing that the humidifier accomplishes is to furnish more vapor for more condensation. Figure 30 "A" illustrates that the relative humidity in an enclosed space is controlled by the temperature of a large surface* if that surface is below the dew point of the air in the enclosure. Water added to the air stream will not raise the relative humidity in the space but only add to the amount of water vapor being condensed into a liquid.

In summer when the enclosure is cooled by mechanical refrigeration, the relative humidity is controlled by the temperature of the cooling coil, the amount of air passing over that coil and the time the coil is refrigerated. This moisture vapor migration from an enclosed space to the cooling coil in an air conditioning system is shown in Figure 30 "B." Much of the advantage of mechanical air conditioning is its ability to lower the relative humidity of the air within the building.

*The surface of a window, etc., as compared to a drinking glass.

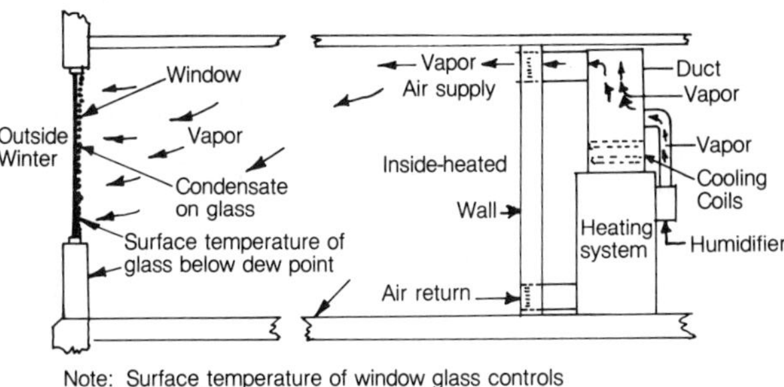

Note: Surface temperature of window glass controls relative humidity in room. When window condenses vapor of room air, additional vapor from humidifier provides more moisture for condensation on glass.

"A"

MOISTURE VAPOR MIGRATION TO WINDOW — WINTER

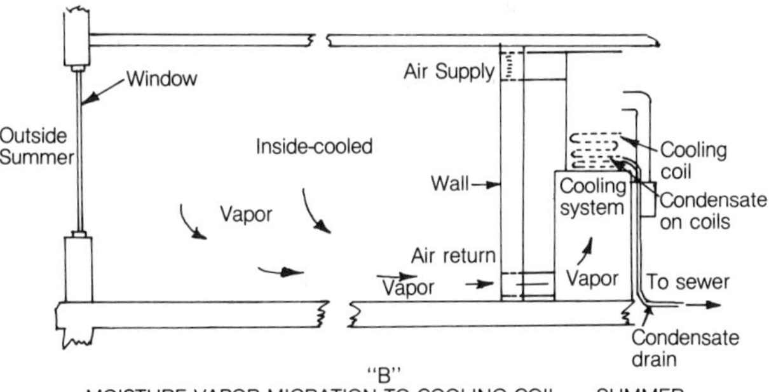

"B"

MOISTURE VAPOR MIGRATION TO COOLING COIL — SUMMER

Note: Cooling coil of air conditioning controls relative humidity in room. Vapor in air stream is condensed and liquid is drained to sewer.

Figure 30

In a manner similar to a cooling coil in an air-conditioning system, any duct work or pipe that is operating at a temperature lower than the dew point of the air surrounding it, will also condense moisture. This is illustrated in Figure 31 "A" for ducts and 31 "B" for pipes. This condensation will occur on ducts or pipes located indoors or outdoors whenever the temperature of either goes below the dew point temperature of the surrounding air. Likewise, cold water lines in heated areas, or exposed to outdoor summer air, will condense moisture on their surfaces.

Liquid condensate and water drip can be very damaging. As demonstrated in Chapter 1, if it saturates thermal insulation, the insulation becomes ineffective. It will rust steel and rot wood. Also, it is a hazard where it may cause damage to electrical switches and terminals and it can contribute to short circuits in the electrical system. Condensate can also cause discoloration of paint and interior decoration. For these reasons it is necessary to control condensation on pipe and duct surfaces.

Figure 31 "C" and "D" illustrate that if insulation is used over the duct or pipe, the surface temperature of the vapor retarder protecting the insulation will be above dew point and condensation will not occur.

The insulation illustrated in Figure 31 "C" was block insulation enclosed in a vapor retarder covering and in "D" it was pipe covering insulation also with vapor retarder covering. However, in many instances where ducts and cold water piping are located in a building, surface condensation can be prevented by the application of a heavy coating of cork-filled polyvinyl acetate. Where suitable, this type of mastic has the advantage that it can be applied, by trowel or spray, to the surfaces on which it is to retard condensation. The limitations of this mastic for the control of condensation are given in Table 15.

The thickness necessary to be used for such mass insulation of ducts, pipes, and equipment will be presented in the Thermal Insulation Design chapter of this text. The vapor resistance of the protective vapor retarder (barrier) must be determined by the temperature cycles and relative humidity cycles of the ambient air.

Where vapor permeable insulations (organic foams, fiberous or granular materials) are used on pipes or ducts subjected to temperature cycles, the drying cycle must be greater than the wetting cycle. During winter when the water supply is at 50° to 55°F (10° to 13°C), the vapor pressure of the pipe surface is 0.36 to 0.43" of Hg (9.2 to 11 mm of Hg) and the surrounding air at 75°F (23.9°C) and 50% RH has a vapor pressure of 0.58" of Hg (14.73 mm of Hg). This vapor pressure drives moisture into the insulation where it condenses and water is absorbed into the insulation. As this happens the insulation becomes less thermally efficient. During the summer months in air-conditioned areas the air at 70°F (23.9°C) at 40% relative humidity has a vapor pressure of 0.52" of Hg (13.26 mm of Hg) and the water which will be approximately 70°F will cause the saturated insulation to have a vapor pressure of 0.74" of Hg (18.78 mm of Hg). In this case the vapor migration is from the insulation to the air and the moisture flows from the insulation causing it to dry out.

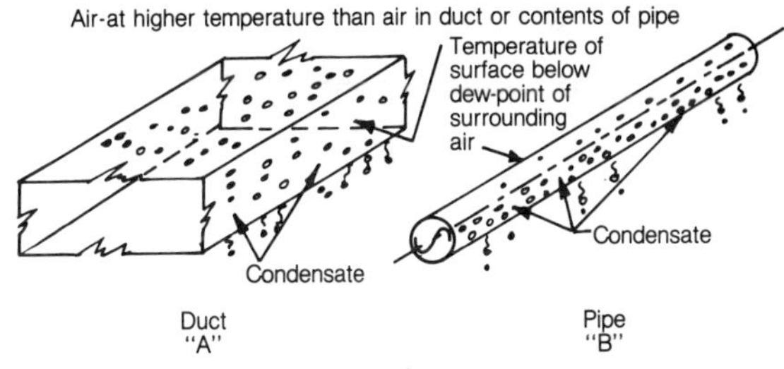

MOISTURE VAPOR MIGRATION TO SURFACE OF DUCT OR PIPE OPERATING BELOW AIR DEW POINT TEMPERATURE

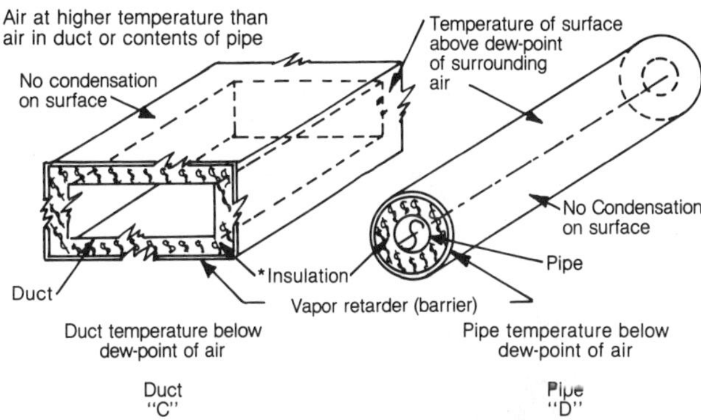

*Note: Where surface temperatures are only slightly below dew point, cork filled mastic may be used as insulation and vapor retarder.

CONTROL OF MOISTURE MIGRATION BY THE USE OF THERMAL INSULATION AND VAPOR RETARDER (BARRIER)

Figure 31

CONTROL OF CONDENSATION ON SURFACES OF DUCTS, PIPES AND EQUIPMENT BY COATING OF CORK FILLED PVA MASTIC

Duct, Pipe, or Equip. Temp. °F	CONDENSATION WILL OCCUR ON SURFACE WHEN RELATIVE HUMIDITY IS GREATER THAN AT THE AIR TEMPERATURE LISTED												Duct, Pipe, or Equip. °C Temp.
	1/4" (0.635 mm) of dried thickness of mastic												
	Ambient Air Temperature °F												
	45	50	55	60	65	70	75	80	85	90	95	100	
40	86	81	74	68	63	59	55	52	49	47	43	41	4.4
45		91	82	75	69	64	60	57	54	50	48	45	7.2
50			91	82	75	69	65	61	58	55	51	48	10.0
55				91	83	77	71	66	62	59	54	52	12.8
60					91	83	78	72	67	63	59	56	15.6
65						92	84	78	73	68	64	60	18.3
70							92	84	79	73	68	64	21.1
75								92	86	79	74	69	23.9
80									92	86	79	77	26.7
85										92	86	80	29.4
90											92	87	32.2
95												92	35.0
	7.2	10.0	12.8	15.6	18.3	21.1	23.9	26.7	29.4	32.2	35.0	37.8	
	Ambient Air Temperature °C												
	3/8" (0.95 mm) of dried thickness of mastic												
	Ambient Air Temperature °F												
°F	45	50	55	60	65	70	75	80	85	90	95	100	°C
40	92	84	79	76	71	66	63	59	57	55	53	50	4.4
45		92	85	80	75	70	67	64	61	58	56	53	7.2
50			92	85	80	76	71	68	64	62	59	57	10.0
55				92	86	80	76	72	69	66	63	60	12.8
60					92	86	81	77	73	69	67	63	15.6
65						92	87	82	78	73	70	68	18.3
70							93	88	82	78	74	71	21.1
75								93	90	82	79	74	23.9
80									93	88	83	79	26.7
85										93	88	84	29.4
90											93	88	82.2
95												93	35.0
	7.2	10.0	12.8	15.6	18.3	21.1	23.9	26.7	29.4	32.2	35.0	37.8	
	Ambient Air Temperature °C												

TABLE 15

Under these cyclic conditions, the organic foams or the fiberous or granular insulation can function relatively efficiently as long as the drying cycle exceeds the wetting cycle.

The ability of thermal insulation to function efficiently under the conditions described above leads many to recommend and use these types of insulation on surfaces of pipes, ducts, or equipment which operates continually below the dew point of the ambient air. If there is no drying cycle as would be true of cold brine lines or cold equipment, then the vapor always migrates to the lower temperature surface. It will condense at the location in the insulation where it reaches the dew point, then the liquid and super-saturated vapor will migrate to the freeze-thaw zone, and appear as frost on the pipe, duct, or equipment surfaces. It should be noted that until an insulated unit is removed for inspection, it

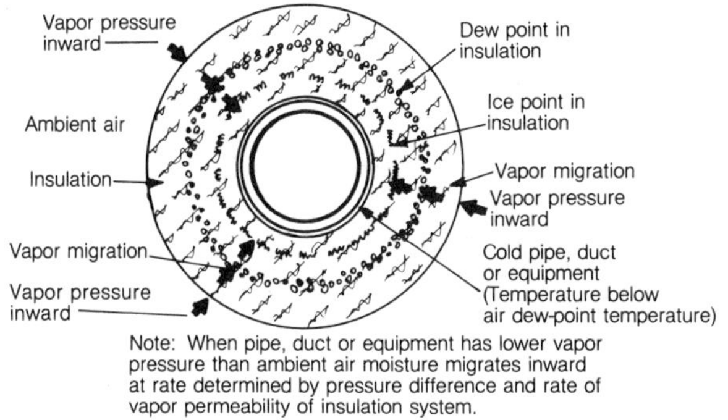

"A"

VAPOR MIGRATION INTO INSULATION

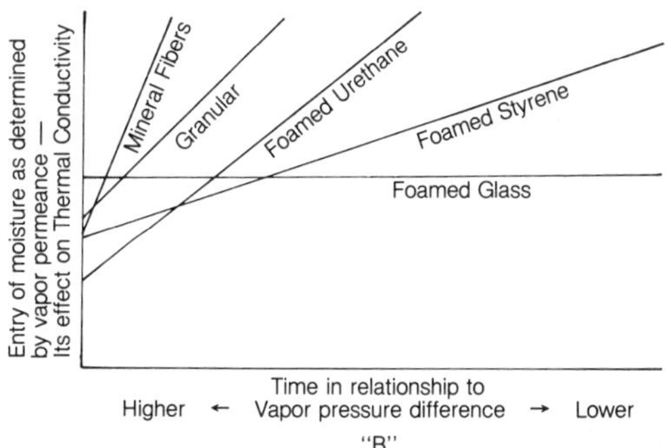

"B"

EFFECT OF VAPOR PRESSURE DIFFERENCE AND TIME ON THE THERMAL CONDUCTIVITY OF INSULATIONS

Figure 32

may look dry, as the dew point and freeze point is inside the insulation and the last surface to become wet is the outside when the insulation exceeds its ability to absorb more moisture. The moisture migration into insulation is shown in Figure 32 "A." The rate at which this migration occurs is a function of the pressure difference and vapor resistance of the insulation. The vapor permeance is a measure of this rate of migration. As shown, vapor permeance can be used as an indicator to estimate the length of time for insulation saturation or the recovery from saturation to a dry state. The conductivity of insulation changes with moisture content. Figure 32 "B" indicates the relative change in thermal conductivity with respect to the time and vapor pressure difference. These effects are cumulative if the moisture has no means of escape. The conductivity of thermal insulation, completely saturated with liquid water, is approximately 5.5 Btu in./ft.^{2}, hr., °F (0.976 W/m °C). This is why a very low thermal conductivity in dry state has little meaning in this type of installation. Of utmost importance is that the insulation, or its vapor retarder (barrier) system, has a permeability below 0.001 perms (English) or 0.0006 metric perms. The service life of insulation applied directly over metal which remains below ambient air dew point temperature is completely determined by its ability to restrict vapor migration into the system.

When insulation is used to retard heat flow into refrigerated spaces, a completely different set of moisture migration conditions exist. The basic difference is a proper design must allow some moisture migration and still remain efficient. This is because moisture which passes through the wall gets to the enclosed air space which in turn allows the moisture to collect on the refrigeration coil as frost. During the defrosting of the coil, the ice is melted and drained off as water. In this case the vapor barriers (which must be on the warm side) must restrict the amount of vapor passage so as not to exceed the ability of the defrosting cycle to remove the ice from the coils. This is illustrated in Figure 33. In the case of insulation enclosing refrigerated areas the vapor pressure is inward. Thus, the vapor retarder must be located on the outer surface as illustrated in Figure 33. An example, similar to the figure drawn, is given to illustrate the location and vapor resistance of the retarder.

Example 5A—*English and Metric Units*

The wall and ceiling are constructed as follows: The outer 1/2" (1.3 cm) is gypsum board to provide fire protection to the organic foam insulation. The insulation is a 4" (10.16 cm) thickness of foamed urethane board. The inner surface is also 1/2" (1.3 cm) gypsum board to provide fire protection to the inner surface of the refrigerated area. This refrigerated enclosure is located in a warehouse. The summer temperature (for design) is 85 °F (29.4 °C) with relative humidity of 60%. The refrigerated space maintains 20 °F (−6.7 °C).

First, from the thermal resistances, determine the temperature drops across the various components. From Table 11 the

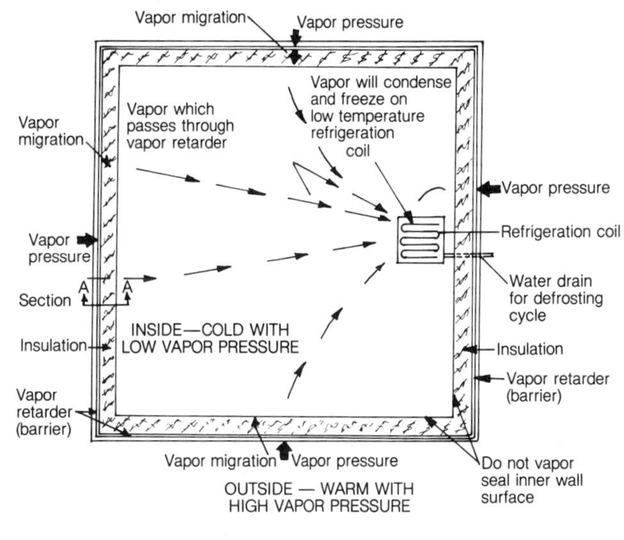

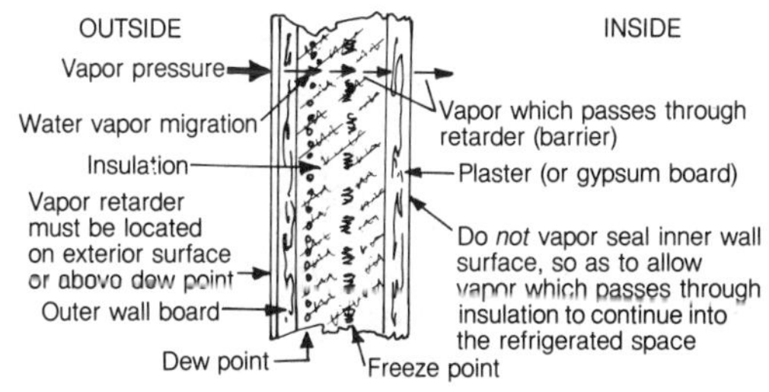

SECTION AA

MOISTURE MIGRATION INTO REFRIGERATED SPACE

Figure 33

thermal resistances of each component can be established. Here, R_1 = thermal resistance of outer gypsum (R_{m1}), R_2 = thermal resistance of urethane board insulation (R_{m2}) and R_3 = thermal resistance of inner gypsum plaster board (R_{m3}). These thermal resistances are tabulated below:

English Units	***Wall Component***	***Metric Units***
R_1 = 0.32	1/2″ outer gypsum wall board	R_{m1} = 0.056
R_2 = 16.68	4″ urethane insulation board	R_{m2} = 2.94
R_3 = 0.32	1/2″ inner gypsum plaster board	R_{m3} = 0.056
Total Resistance = 17.32		3.052

% Thermal resistance		
0.32/17.32 = 1.8%	1/2″ outer gypsum wall board	0.056/3.052 = 1.8%
16.68/17.32 = 96.3%	4″ urethane insulation board	2.94/3.052 = 96.3%
0.32/17.32 = 1.8%	1/2″ inner gypsum plaster board	0.056/3.052 = 1.8%

The temperature drop through the wall assembly, with outside temperature at 85 °F (29.4 °C) and the operating temperature of the refrigerated space at 20 °F (−6.7 °C) is:

$$85 - 20 = 65\,°\text{F} \quad \text{or} \quad 29.4 - (-6.7) = 36.1\,°\text{C}$$

From temperature gradients, the junction temperatures between components are found to be:

English Units	***Wall Location***	***Metric Units***
85 − (65 × .018) = 83.2 °F	Junction outer gypsum and urethane	29.4 − (36.1 × 0.018) = 28.8 °C
83.2 − (65 × 0.963) = 20.7 °F	Junction urethane and gypsum plaster	28.8 − (36.1 × 0.963) = −5.3 °C

The dew point of air at 85 °F (29.4 °C) is 74 °F (23.3 °C) and this is within the urethane insulation that has an outer temperature of 83.2 °F (28.8 °C) and an inner junction temperature of 20.7 °F (−5.3 °C). The vapor retarder (barrier) must be located above the dew point. Thus it must be located between the outer layer gypsum and the urethane insulation.

With a relatively even resistance to vapor migration, as is the case for urethane insulation, to retard vapor entry into the insulation through the vapor retarder located above, the dew point should be at least equal to the vapor resistance from that plane to the air side. The vapor resistance of the urethane and inner gypsum plaster then determines the necessary resistance of the barrier between the outer gypsum board and the urethane insulation.

The vapor resistance is (from Table 10):

English Units			***Metric Units***
Using 0.8 per inch	Urethane Insulation	Using 0.464 per cm	
4 × 0.8 = 3.2	Vapor Resistance	10.16 × 0.464 = 4.71	Vapor Resistance
= 0.009	1/2″ (1.27 cm) Gypsum	= 0.014	
3.209	Total	4.724	Vapor Resistance

From Table 9 it can be determined that the required vapor barrier between the outer gypsum board and the urethane could be one of several materials such as:

Aluminum Foil [(over 0.001″ (0.025 mm)]	at	5.84	R (English)	8.87	R (Metric)
Aluminum Foil, Scrim, Kraft	at	1250.	R (English)	1896.	R (Metric)
Viglite Sheet (over 0.03″ thick)	at	3.7	R (English)	5.7	R (Metric)
Polyethylene Sheet (over 0.03″ thick)	at	3.6	R (English)	5.4	R (Metric)

The selected vapor retarder (barrier) should be installed between the outer layer gypsum board and urethane insulation with all joints overlapped at least 3″ (7.6 cm) and sealed with sealing compound.

Example 5B

Consider conditions identical to those in Example 5A, except that aluminum foil is now on the *interior** of the urethane insulation.

From Table 9 it can be determined that urethane with one foil face has a vapor resistance of approximately 25 R (English) or 37.9 (Metric). Without the aluminum foil the resistance was 3.2 R (English), 4.71 (Metric). Thus, the aluminum foil provides a resistance of 25 − 3.2 = 22.8 R_{vi} (English) or 37.9 − 4.71 = 33.19 R_{mvi} (Metric).

The resistances must be in the correct ratio of the vapor pressure differences if moisture entrapment is to be prevented. For this reason with an additional vapor retarder which is below dew point it becomes necessary to determine the vapor pressure differences at each vapor retarder. The vapor resistance of vapor retarders located below dew point, divided by vapor pressure difference, must be less than the vapor resistance of retarders located above dew point, divided by the vapor pressure difference. In other words, the vapor retarders below the dew point must not "dam up" the moisture passed by the vapor retarder located above the dew point.

The temperatures at these junctions of materials were determined in Example 5A. They were:

At the junction of outer gypsum and urethane insulation
83.2 °F (28.8 °C)

At the junction of urethane insulation and inner gypsum
20.7 °F (−5.3 °C)

The vapor pressure against the outer vapor retarder was set by the outside vapor pressure of ambient air. As stated, at 85 °F (29.4 °C) at 60% RH the vapor pressure Vpo (from Table 8) is 0.846″ of Hg (21.49 mm of Hg = Vmpo).

The vapor pressure against the inner vapor retarder is the vapor pressure of the saturated air pressing against it minus the vapor pressure in the refrigerated room. The temperature at this inner vapor retarder is 20.7 °F (−5.3 °C) as previously calculated in Example 5. The vapor pressure at saturated conditions is 0.106″ of Hg (2.69 mm of Hg).

The vapor pressure inside the refrigerator is set by the temperature of the coil—it being the dew point. If the coil temperature is 0 °F (−17.8 °C) then the vapor pressure is 0.037″ of Hg (0.957 mm of Hg).

* It is mistakenly believed by some that the interior of refrigerated spaces should be vapor sealed.

The vapor pressure difference at this inner barrier is V_{pi} = 0.106 − 0.037 = 0.069″ of Hg (or 2.69 − 0.957 = 1.733 mm of Hg = V_{mpi}).

As stated, the ratio

$$\frac{R_{vi}\text{ (inner vapor retarder)}}{V_{pi}\text{ (vapor pressure-inner)}} < \frac{R_{vo}\text{ (outer vapor retarder)}}{V_{po}\text{ (outer vapor pressure)}}$$

Knowing three of the four terms, the fourth can be calculated. For example, determine the minimum value of the vapor resistance of the outer vapor retarder:

English Units		***Metric Units***	
R_{vi} = 22.8	R_{vo} = ?	R_{mv} = 33.19	R_{mvo} = ?
V_{pi} = 0.069	V_{po} = 0.846	V_{mpi} = 1.733	V_{mpo} = 21.49

$$\frac{22.8}{0.069} < \frac{R_{vo}}{0.846} \qquad \frac{33.19}{1.783} < \frac{R_{mvo}}{21.49}$$

Solving the above inequality for R_{vo} and R_{mvo} yields the following condition which they must obey:

$$R_{vo} > 279 \qquad R_{mvo} > 400$$

This result indicates that the outer barrier should have a minimum vapor resistance of R_{vo} = 279 (English Units) or R_{mvo} = 400 (Metric Units).

From Table 9 it can be determined that between the outer gypsum board and the outer surface the required vapor barrier could be an aluminum foil 0.005″ thick mounted on skrim and kraft paper, having R_{vo} = 1250 (English Units) which is greater than the 279 required minimum value. Also it is essential that this be installed with vapor sealed overlapping joints.

SUMMARY

All thermal insulation used in buildings contains dry air spaces. When these spaces become filled or saturated with liquid water or ice, the insulation becomes ineffective. Design of insulation systems is fundamentally one of designing a system that will stay dry. In brief, the simple basics can be summarized as follows:

1. All insulation must be protected from liquid water.
2. All insulation may be subjected to water vapor.
   a. Vapor retarders should always be installed on the warm or hot side of the insulation.
   b. *Do not* vapor seal the cold side of insulation as it will trap vapor which may be condensed into a liquid.

Remember, no matter how low a thermal conductivity an insulation may have, if it becomes saturated with water, its conductivitiy will be approximately 5.5 Btu. in./ft.^{2} hr., °F (0.796 W/m °C) and if this water becomes frozen, then the conductivity will be approximately 15.5 Btu. in./ft.^{2}, hr., °F (2.242 W/m °C).

3 Ventilation and Air Requirements

AIR—OXYGEN AND CARBON DIOXIDE REQUIREMENTS

Proper ventilation is necessary for every building of human occupancy. The architect and air-conditioning engineer must design the ventilation and air-conditioning system based on the building's occupancy, utilization, location, and heating and cooling requirements. This very complex subject cannot be covered in detail in this book. However, as the ventilation does affect the efficiency of the insulation system, and poor or inadequate ventilation can present a serious hazard, some of the basics will be pointed out.

The more efficiently buildings are constructed to prevent energy loss, the tighter they become in preventing air from entering their interiors by natural infiltration. The more tightly a building or home is sealed by interior vapor retarders, sealed windows and storm windows, the more consideration must be given to obtaining suitable air ventilation.

In the *ASHRAE Handbook* the following recommended minimum requirements for air are stated:

> The oxygen and carbon dioxide requirement can be met with outdoor-air ventilation of less than 4 cfm* per person. This may include infiltration. An outdoor-air ventilation rate is required to dilute odors, irritants, smoke density (cigar, cigarette) or other nontoxic conditions. The amount of ventilation required per person to remove human odors to an acceptable value, can be found in Figure 34.

In some manner, through the doors, or the air-conditioning (or heating) system, air from the outside must be introduced into the building or home. In summer, if the area is air conditioned, this air must be cooled and in winter it must be heated. This fundamental requirement cannot be changed by thermal insulation. In most homes and buildings more air is introduced into the enclosure than is required for human comfort. Excessive ventilation has the effect of by-passing the thermal insulation and storm windows and doors, resulting in greater energy consumption.

The air exhausted by bathroom or basement fans in vent hoods must be replaced with the same quantity of air that the fans

*4 cubic feet per minute (English Units) or 0.113 cubic meters per minute (Metric Units).

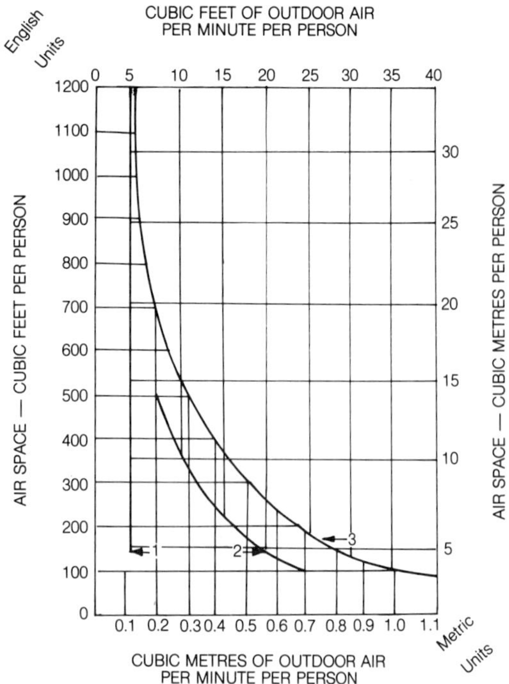

MINIMUM VENTILATION REQUIREMENTS

Figure 34

removed. Again, in the process, this air to supply the exhaust fans may use energy to be heated or cooled. For this reason, the exhaust fans should only be used when they are serving their function of removing vapor or odors.

AIR — FOR COMBUSTION

Other than electrical heating, all gas, oil, or coal heating systems require air for combustion. Depending upon the fuel, the air required is 6 to 12 times the weight of the fuel consumed. Years ago the old coal furnaces were constructed to have a pipe to the outdoors to provide air for combustion. As these were replaced by gas or oil-fired furnaces, the duct to the outside for combustion was eliminated. Houses were not very tightly built and fuel was cheap, so the combustion air was obtained from the air in the house, basement or utility room. However, now as fuel is more expensive, it is poor economics to use already heated air for combustion air. Even more important, with houses being constructed very tight with windows carefully sealed and storm doors being installed, the use of air from the home becomes hazardous. In very cold weather after constant use, the lack of air inside could prevent the flue from removing the products of combustion sufficiently to provide safe air inside.

A duct can be provided to supply outside air to the combustion chamber. A typical arrangement for such a duct is shown in Figure 35. It should be remembered that unless an outside source of air is provided, all the air exhausting from the chimney is air from the house or building.

Combustion of any fuel causes carbon dioxide, carbon monoxide, nitrogen, oxygen, sulfur dioxide and sulfur trioxide to be present in the flue gases. When sufficient moisture is present, the sulfur dioxide and sulfur trioxide form sulfurous or sulfuric acid. These are very corrosive. This is the reason that brick chimneys must be lined to retard corrosion and that metal pipes are used only to reach the chimney. While the flue gases are passing through the metal pipe, they should be above the dew point of these acids. The average dew-point temperatures of flue gas with natural gas as fuel is 127 °F (52.8 °C) and with oil as fuel is 111 °F (43.9 °C). Since these gases cool as they travel up the chimney, a general recommendation is that the flue gas in the metal flue pipe, from heater to chimney, not fall below 200 °F (93.3 °C).

Recently, because of the energy shortage, small exchanges and fans forcing air around the outer surface of flue gas metal pipes were recommended and sold as a means of heat conservation. These assist in the condensation process and are likely to cause corrosion of the flue pipe, and if sufficient condensation occurs, it will drain back into the heater and cause its corrosion. If a flue pipe has to be greater than 6′0″ in length to go from heater to chimney, it should be insulated. The insulation must be suitable for temperatures up to 600 °F (316 °C).

In order to conserve oil and gas, many people have installed coal or wood-burning stoves or have begun to use their fireplaces as a source of heat.

Stoves, like any furnace converting fuel to heat, require a flue to remove the products of combustion from the interior space. It is necessary to make sure that the stove pipe is well insulated when it is close to a flammable material. It should also be remembered

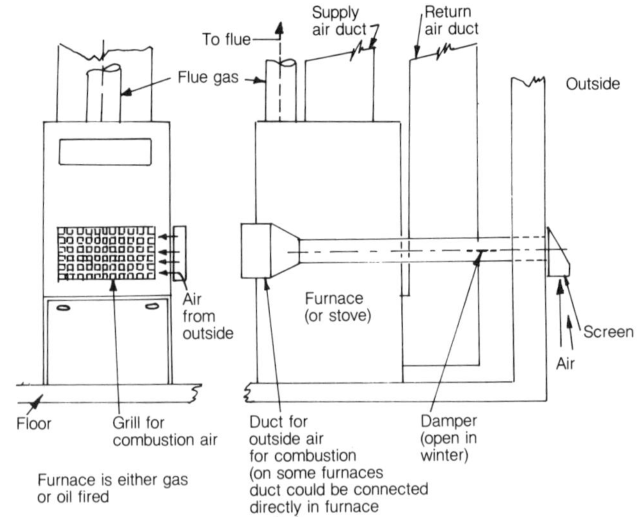

Note: Duct for combustion air from outside to be 2/3 to 3/4 the size of the flue gas duct.
Duct from outside must be insulated with 1″ thickness of glass wool, mineral wool, or foamed glass to prevent condensation of moisture on its surface.

CONSERVATION OF ENERGY BY USE OF DUCT TO OUTSIDE TO PROVIDE AIR FOR COMBUSTION

Figure 35

that flues and stove-piping for gas-burning equipment may be unsafe for use as flues for oil, coal, or wood-burning stoves or furnaces.

Stoves and fireplaces require large quantities of air for combustion. Depending on the fuel being burned, the air needed for combustion is 90 to 150 cubic ft. per lb. of fuel (5.6 to 9.4 cubic meters per kilogram). When stoves or fireplaces are not supplied with outside air, this amount of air is drawn from the room and possibly the rest of the house. The negative air pressure on the interior pulls cold air from the outside, causing a cooling effect except in the close area around the stove or fireplace. In many instances where stoves and fireplaces have been used to reduce oil and gas usage in a home, they have actually increased the use of the fuel to the central heating system. The flow of combustion air to these is illustrated in Figure 36.

The only way to prevent this loss under these conditions is to close the door of the room in which the stove or fireplace is located to retard the air flow from the outer parts of the house. The window nearest the stove or fireplace should be opened to allow approximately one inch of air space for entry of outdoor air. Of course, a better way is to have a direct supply of outside air to the stove or combustion area of the fireplace. Such an arrangement for a duct to supply outside air for combustion for stove or furnace is shown in Figure 37. A similar arrangement may be used for a duct to supply outside air for combustion to a fireplace. In this case, the grill should be located at one end of the hearth.

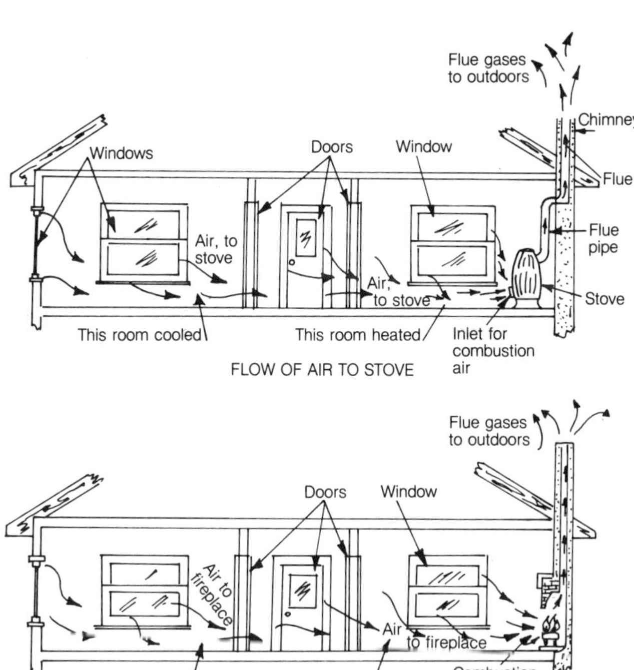

COOLING EFFECT OF STOVE OR FIREPLACE ON OTHER AREAS OF HOME DUE TO AIR IN HOUSE BEING USED AS AIR FOR COMBUSTION

Figure 36

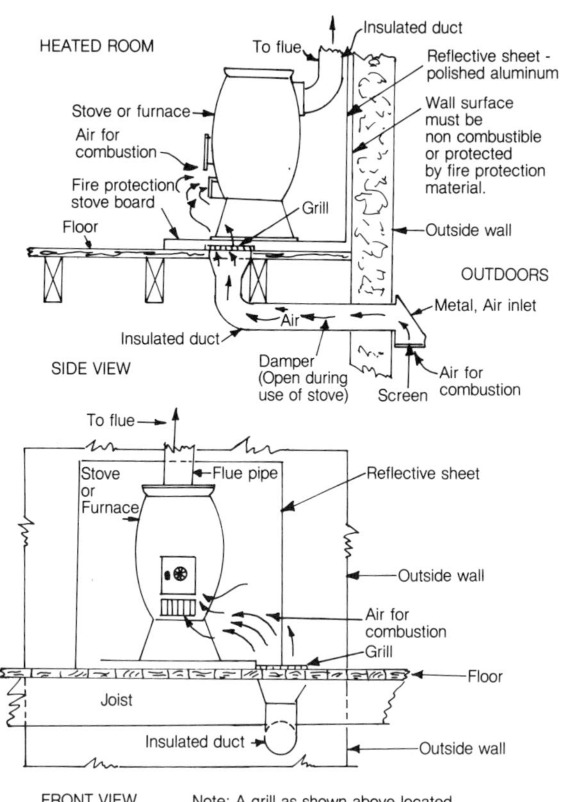

DUCT TO SUPPLY OUTSIDE AIR TO STOVE OR FURNACE FOR COMBUSTION AIR REQUIREMENTS

Figure 37

Where there is no crawl space or basement under the stove or furnace, it may be necessary to bring in the air at the side as shown for the furnace in Figure 35.

Fireplaces are probably the most inefficient means of converting energy in fuel to heat. They draw air from the other parts of the house for combustion air. They are located so that only about one-third of the heat of the flame is radiated inward to the room. If fireplaces are going to be used to reduce the energy shortage of oil and gas, it is essential that their efficiency in producing usable heat from wood or coal be improved.

The most important factor to improve fireplace performance is to provide a means for fresh air to be the primary source of air for combustion. One method is to install a duct for outside air to enter by a grill located at the end of the hearth, in a similar manner as recommended for the stove illustrated in Figure 37.

If the fireplace is equipped with an outdoor ash removal system as shown in Figure 38 "A," then a metal wedge should be used to hold the ash dump lid open at least 1" while the fireplace is in operation. Similarly, the ash pit door should also be kept open to the outside. This will allow outside air to flow in, under, and around the fuel for improved combustion. Of course, the fireplace damper must be open to allow the products of combustion to be exhausted out of the flue.

For fireplaces that have no outside ash pit and where it is difficult to install ducts, then (for fireplaces above outside grade) it is suggested that 6 to 8 holes be drilled, sloping downward from the inside to the outside from the back of the fireplace. These holes should be approximately 1" in diameter suitable for installing a 1" (25.4 mm) outside diameter piece of stainless steel tubing. Where these project outdoors, they should be provided with a metal jacket cover or damper that can be closed off when not in use. Also, the cover should be provided with a screen to prevent the entry of insects. This arrangement appears in Figure 38 "B."

Two additional mechanisms may be used to improve the thermal efficiency of fireplaces. One is to install a reflective shield of polished stainless, preferably having a 1/4" (0.6 cm) space between wall and sheet. The other is to install bent pipes that will take colder air at floor level, heat the air as a heat exchanger and exhaust it back to the room at a higher level. These are shown in Figure 38 "C." The reason that a book on thermal insulation discusses stoves and fireplaces is that all combustion heating devices must have air for the combustion process. If fresh air is not introduced to the combustion process efficiently and in sufficient quantity, it must come from the interior of the house. This same amount must enter the house by infiltration through doors or windows, when opened—in order to equalize the pressure. All this air effectively by-passes the thermal insulation in the walls, ceilings, or floors. This can reduce the thermal efficiency of the building considerably.

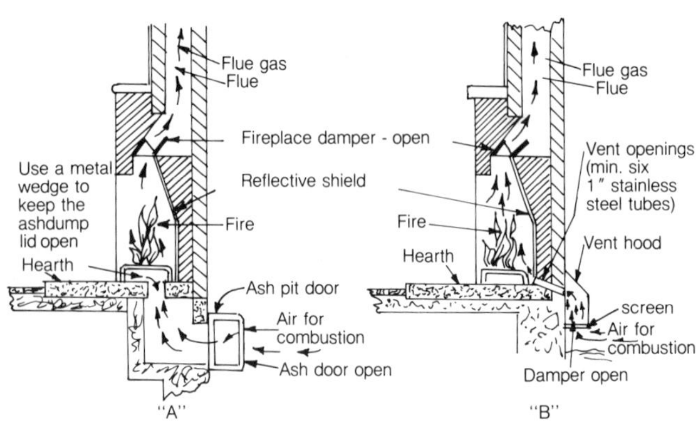

SECTIONS
FIREPLACES USING OUTSIDE AIR FOR COMBUSTION

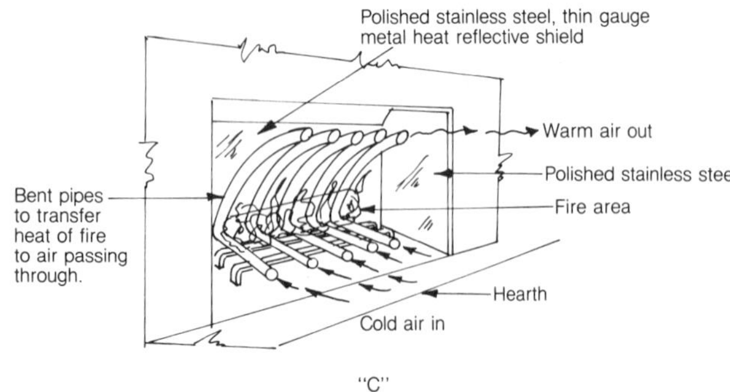

HEAT EXCHANGER PIPES AND REFLECTIVE SHIELD
TO IMPROVE FIREPLACE THERMAL EFFICIENCY

Figure 38

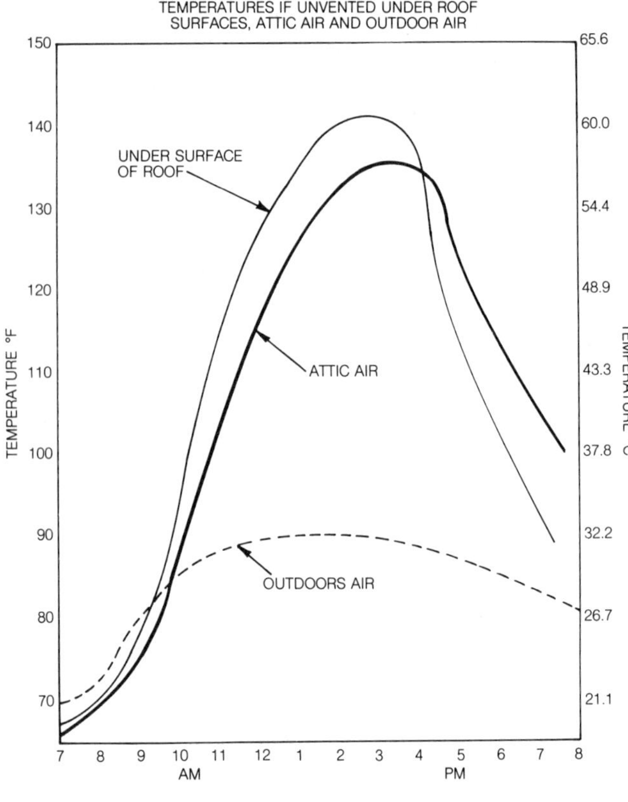

EASTERN DAYLIGHT SAVINGS TIME (AUGUST)

Figure 39

VENTILATION OF SPACE BETWEEN CEILING AND ROOF

Anyone who has even been in a poorly ventilated attic space under an unshaded roof on a hot day has experienced first hand the extremely high temperatures which can occur under these conditions.

In fact, on a sunny day the temperature of an unshaded roof of an unvented attic air space beneath it can rise as much as 40 to 50 degrees F (27.7°C) higher than the outdoor air temperature. This higher temperature produces in the room below, ceiling temperatures which are also high enough to be a source of discomfort.

On Figure 39 a chart shows the typical temperatures which would be expected on a clear day in August of the roof temperature, and attic temperature, above the ceiling, of an unvented roof dwelling. The temperatures will change under cloudy or rainy conditions; however, the solar heating will cause the attic space to become much hotter than ambient outdoor air unless the attic is well ventilated.

The chart on Figure 39 was based on observations of a structure at 39° latitude. In more tropical climates the measured temperatures would be higher. However, even at this relatively mild climate trapped air in unvented spaces under a roof can reach nearly 140°F (60°C). If an air space were maintained at 75°F (23.9°C) the temperature difference across the insulation would be 140°F − 75°F = 65°F (36°C) and if properly vented would be approximately 95 °F − 70 °F = 25 °F (13.9 °C). This means that much more energy would be required to remove the heat coming in from an unventilated roof space compared to the ventilated case.

Various types of roofs require various methods of ventilation. In Figure 40 "A" a gabled roof house is shown with ventilation

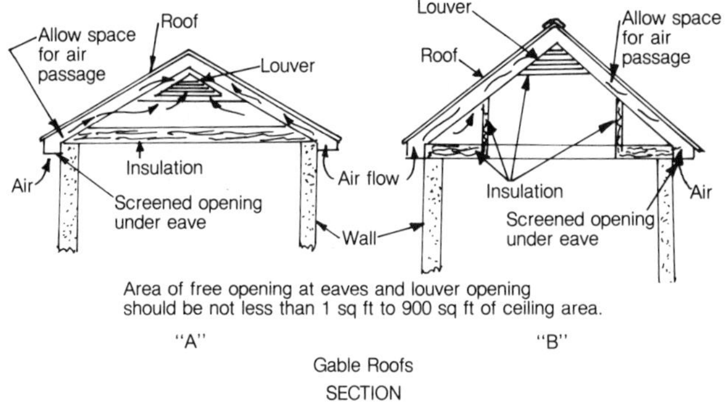

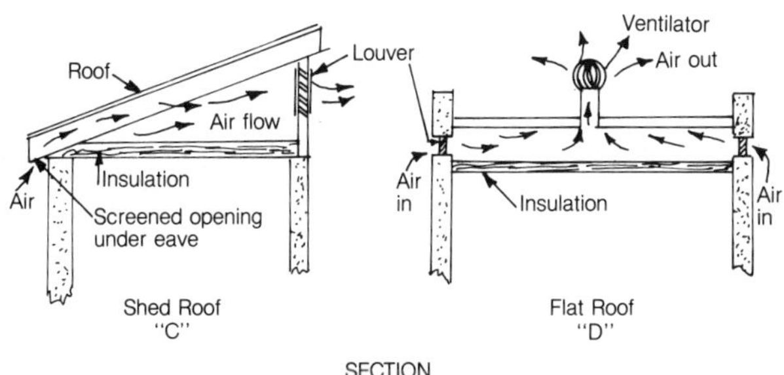

VENTILATION BETWEEN THERMAL INSULATION AND VARIOUS TYPES OF ROOFS

Figure 40

coming up from under the eaves into the attic space and out of the louver. In cases where ventilation can not be obtained, the louver area should be multiplied by three. Figure 40 "B" illustrates a gabled house with an attic room. Care must be taken not to block air passage into the attic area.

A shed-type roof is shown in Figure 40 "C." In this case air vents are not possible in the roof overhang on the lower side. Louvers should be placed at each end of the slope wall and be located near the lower end of the roof.

Flat roofs, as shown in Figure 40 "D," are the type which are most often incorrectly ventilated and insulated. On a flat roof, the insulation is frequently installed directly under the roofing materials and above the supporting members. On a hot sunny day the black roofing asphalt can reach 180 °F (82.2 °C). In this case the temperature difference from the inside to the outside surface temperature is 105 °F (58.3 °C) or more than 4 times the difference if the proper ventilation had been provided. Stated another way, when insulations are installed in roofs without ventilation, and they are needed to reduce the thermal load on the air conditioning, they lose most of their effectiveness at the time they are most needed.

This same basic error is made in the design and installation of insulation on the underside of roofs on metal buildings. Many metallic surfaces exposed directly to the sun will reach 160 °F (71.1 °C) and, if air is to be maintained at 75 °F (23.9 °C), the difference between this surface and air temperature will determine the heat transfer. Also, the value of the thermal conductivity will increase because of the higher mean temperature.

In the last chapter, it was shown that insulation will become less effective in winter if there is moisture condensation. In conclusion, this chapter has shown that without proper attic ventilation the insulation is also less effective in the summer because of higher temperature differences.

4 Climatic Considerations

To have a home or building comfortable under the worst climatic conditions it is necessary to size the heating system based on the maximum heat loss which could occur. This requires that an estimate be made of the lowest temperature to which the building might be exposed. Likewise, the cooling system must be based on the expected maximum temperature which could occur at that particular location. It should be pointed out that these "expected" maximum and minimum temperatures may be those anticipated every few years—not the "record" maximum and minimum experienced in that climatic location. The "expected" temperatures determine the "Design Temperature".

In this chapter, data as to expected maximum and minimum temperatures and temperature differences multiplied by time are given as a guide to insulation, selection and design. Many governmental agencies and insulation manufacturers have taken a simplified solution and based their recommended thermal resistance on the temperature zones. This, of course, can lead to error as this is only one factor in the determination of thermal insulation design for thermal efficiency. To illustrate this, in one area two subdivisions were built. These were less than one mile apart. However, in one, reasonably priced natural gas was available for heating and in the other only much more expensive electrical energy was available. In order to keep the energy costs the same, electrically heated homes needed a thermal resistance three times that of the gas heated homes under identical climatic conditions.

Another factor frequently overlooked is the fact that effective insulation of a building reduces the capacity requirement of both the heating and cooling systems. In uninsulated homes, the total cost of heating and cooling equipment required would be greater than the cost of heating/cooling systems and the insulation of a properly insulated house. In this case, the cost of the insulation might be nothing or it might provide a capital investment savings. In most instances, the question is not should the house or building be insulated, but how much insulation should be used so as to obtain no less than a 20% return on the investment on the last ½″ of material installed.

The maximum and minimum temperatures are the controlling factors as to the capital investment necessary for the heating and cooling systems and the thermal insulation.

The elapsed time and the difference between the indoor and outdoor temperatures determine the amount of energy which must be supplied to meet the heating or cooling needs of the home or building. Thus the amount of energy to supply heat in the winter and cooling in the summer are the controlling factors of the yearly cost. These are the costs which must be considered in determining the return on the investment in thermal insulation.

Other than air temperature outside and inside, the heat loss or gain from the inside of a building is influenced by solar radiation, wind, rain and snow. Thermal insulation in walls, roofs, ceilings, floors and under slabs is the means to reduce the heat transferred. With the exception of the fresh air which must be heated or cooled and which is induced into the building to supply needed oxygen, all other capacity of the heating or cooling equipment *is used to replace wasted energy* which escapes from the building or house through the roofs, walls, floors, doors and windows. Thermal insulation properly used in buildings and homes can reduce this energy loss by very significant amounts.

In most instances the question is not should the house or building be insulated but how much insulation should be used.

WINTER TEMPERATURE, EFFECT OF INSULATION ON HEAT LOSS

Example 6—*English and Metric Units*

To illustrate the importance of insulation, we will compare the heat loss through an uninsulated ceiling of 3/8″ plaster board installed on joists with the loss through a ceiling covered with 1/2″ sand plaster.

From the *ASHRAE Guide and Handbook* the thermal resistance for 3/8″ plaster and 1/2″ sand plaster is 0.41 (English Units) and 0.0722 (Metric Units).

If the temperature in the attic is taken to be 30 °F (−1 °C) and the temperature in the room is 75 °F (24 °C) then the heat loss per sq ft per hour can be found as follows:

$$\text{Heat Loss} = \frac{\text{Temperature Difference}}{\text{Thermal Resistance}}$$

English Units

$$\text{Heat Loss} = \frac{75-30}{0.41} = 109 \text{ Btu/ft}^2\text{, hr}$$

Metric Units

$$\text{Heat Loss} = \frac{24-(-1)}{0.0722} = 346 \text{ W/m}^2$$

Example 7—*English and Metric Units*

If an insulating batt, with suitable vapor retarder, was installed above plaster board and plaster, what heat losses would be the result of various thicknesses if the conductivity of the insulation was 0.28 Btu in/ft², hr °F (0.0404 w/mK). The thicknesses under consideration are 3″ (0.076 meter), 6″ (0.1524 meter) and 12″ (0.3048 meter). The thermal resistances of the insulations at the various thicknesses are:

English Units

$$3'' \text{ Thickness } \frac{3}{0.28} = 10.7 = R$$

$$6'' \text{ Thickness } \frac{6}{0.28} = 21.4 = R$$

$$12'' \text{ Thickness } \frac{12}{0.28} = 42 = R$$

Metric Units

$$0.076 \text{ m thickness } \frac{0.076}{0.0404} = 1.88 = R_m$$

$$0.1525 \text{ m thickness } \frac{0.1525}{0.0404} = 3.77 = R_m$$

$$0.3048 \text{ m thickness } \frac{0.3048}{0.0404} = 7.54 = R_m$$

The loss of the insulated ceiling is:

$$\text{Loss} = \frac{\text{Temperature Difference}}{\text{Resistance of Plaster Board and Plaster + Resistance of Insulation}}$$

For 3″ (0.076 m) thickness insulation the loss is:

English Units

$$\text{Loss} = \frac{75-30}{0.41+10.7} = \frac{45}{11.11} = 4.05 \text{ Btu/ft}^2\text{, hr}$$

Metric Units

$$\text{Loss} = \frac{24-(-1)}{0.072+1.88} = \frac{25}{1.952} = 12.85 \text{ W/m}^2$$

For 6″ (0.1525 m) thickness of insulation the loss is:

$$\text{Loss} = \frac{75-30}{0.41+21.4} = \frac{45}{21.81} = 2.06 \text{ Btu/ft}^2\text{, hr}$$

$$\text{Loss} = \frac{24-(-1)}{0.072+3.77} = \frac{25}{3.852} = 6.5 \text{ W/m}^2$$

For 12″ (0.3048 m) thickness of insulation the loss is:

$$\text{Loss} = \frac{75-30}{0.41+42} = \frac{45}{42.41} = 1.06 \text{ Btu/ft}^2\text{, hr}$$

$$\text{Loss} = \frac{24-(-1)}{0.072+7.54} = \frac{25}{7.612} = 3.28 \text{ W/m}^2$$

Thus the percentage of heat saved from escaping into the attic is:

For 3″ (0.076 m) thickness of insulation is 96%.
For 6″ (0.1525 m) thickness of insulation is 98%.
For 12″ (0.3048 m) thickness of insulation is 99%.

Of course, this illustrates the need in ceilings where only plaster board and plaster acts as a thermal barrier to the flow of heat from outside to inside. These percentages of savings would not be possible in walls where the air space does provide some thermal resistance. However, it does indicate that where homes and buildings require heating and/or cooling, thermal insulation is essential for efficient conservation of energy and money.

Example 8—*English and Metric Units*

Calculate the heat loss to an attic space at various temperatures and insulation thicknesses, with conditions as stated in Examples 6 and 7.

For ease of calculation, assume that the ceiling area is 1000 sq ft (92.9 sq meters); then the heat loss to the attic at various insulation thicknesses occurs at 30 °F (−1 °C) outside temperature.

English Units		Metric Units	
Insulation Thickness	**Heat Loss**	**Thickness**	**Heat Loss**
0	109,000 Btu/hr	0.0	31,944 w
3″	4,050 Btu/hr	0.076 m	1,186 w
6″	2,060 Btu/hr	0.1525 m	603 w
12″	1,060 Btu/hr	0.3048 m	310 w

From this, it is apparent that the cost of 3″ thickness for 1000 sq ft would be less than the cost of an additional 109,000 Btu/hr output in heat capacity. In any climate where heating is required for any period of time, it costs less for the first 3″ of insulation in a ceiling than it would cost for the additional cost of heating equipment to supply the loss.

As stated, the minimum atmospheric temperature for the particular locale controls the capacity requirement of the heating system. Examples 6 and 7 were taken at 30 °F (−1 °C) in attic space (same as outside air). Of course, this low is a moderate winter temperature for most areas and would only apply to the most southern U.S. states. It was selected to indicate that even for the most moderate conditions, insulation above ceilings and roofs is essential for economic design of buildings which require heating.

The effect of lower outside ambient air temperature is shown in Table 16 giving the heat loss under the same conditions as given in Example 7 except at lower ambient temperatures.

Examination of this table shows that at savings in heat capacity achieved by the use of insulation at lower ambient are even more signficant than that of the savings calculated at 30 °F (−1 °C) outside ambient temperature. This, of course, illustrates that the lower the outside temperature, the greater the need for insulation.

Based on the 3″ (0.076 m) thickness, additional insulation does not affect the heating capacity required as significantly as the first 3″ (0.076 m) of insulation. The additional thickness of insulation is desirable in the savings it provides in the cost of energy over the service life of the structure. To illustrate, if the ambient temperature was 0 °F, 6″ (0.1525 m) thickness of insulation would save over 1.0 kW of power per hour. At $0.03 per kWh this would amount to a savings of $0.72 per day in heat cost when the outside temperature was 0 °F (−18 °C) for the 1000 sq ft (92.9 m^2) area roof. If 12″ (0.3048 m) insulation had been provided, the savings per day over that provided by the 3″ (0.076 m) would be approximately $1.13 per day.

These same potential savings are true for the walls of a structure. This is illustrated in the next examples of frame house walls insulated and not insulated.

Example 9—*English and Metric Units*

Determine the savings that can be obtained in energy and money by insulation in a frame house wall at various temperatures. Heat losses for an insulated and uninsulated wall (at various outside temperatures) must be first determined in order to evaluate the effectiveness of insulation.

The wall under consideration is the most common of all frame houses, consisting of 1/2″ by 8″ lapped siding, over building paper over 25/32″ wood sheathing on 2″ x 4″ studs with 1/2″ gypsum board interior. The conductance of this type of construction is approximately 0.25 Btu/ft^2, hr °F (1.4135 W/mK) and the thermal resistance $R = \frac{1}{0.25} = 4$ (English Units) and $\frac{1}{1.4135} = 0.704$ (Metric Units).

Under these conditions, the heat transfer through the wall is based on the temperature difference between inside and outside air temperatures. Assuming the outside temperatures to be the same as those used in Example 7, then the losses would be:

$$\text{Loss} = \frac{\text{Inside temperature minus outside temperature}}{\text{Thermal resistance of uninsulated wall}}$$

English Units	Metric Units
At 30 °F outside temperature	At −1 °C outside temperature
$\text{Loss} = \frac{75-30}{4} = \frac{45}{4}$	$\text{Loss} = \frac{24-(-1)}{0.704} = \frac{25}{0.704}$
$= 11.25$ Btu/ft^2, hr	$= 35.5$ W/m^2
At 15 °F	At −9 °C
$\text{Loss} = \frac{75-15}{4} = \frac{60}{4}$	$\text{Loss} = \frac{24-(-9)}{0.704} = \frac{33}{0.704}$
$= 15$ Btu/ft^2, hr	$= 46.9$ W/m^2
At 0 °F	At −18 °C
$\text{Loss} = \frac{75-0}{4} = \frac{75}{4}$	$\text{Loss} = \frac{24-(-18)}{0.704} = \frac{42}{0.704}$
$= 18.75$ Btu/ft^2, hr	$= 59.7$ W/m^2
At −15 °F	At −26 °C
$\text{Loss} = \frac{75-(-15)}{4} = \frac{90}{4}$	$\text{Loss} = \frac{24-(-26)}{0.704} = \frac{50}{0.704}$
$= 22.5$ Btu/ft^2, hr	$= 71.0$ W/m^2

When the wall is insulated with 3 1/2″ (0.0889 meters) thickness of insulation in the cavity between the sheeting and gypsum board, the resistance added to the wall would be approximately 10 in English Units (1.761 in Metric Units).

The heat loss would then be:

$$\text{Loss} = \frac{\text{Inside temperature minus outside temperature}}{\text{Thermal insulated wall resistance}}$$

The insulated wall resistance is:

English Units	Metric Units
$R = 4 + 10 = 14$	$R_m = 0.704 + 1.761 = 2.465$

Then the heat transfer through the wall, based on the temperature difference between inside and outside air temperatures, for the various outside temperatures would be:

EFFECT OF AMBIENT TEMPERATURE ON HEAT LOSS FROM BUILDING TO UNHEATED ATTIC NOT INSULATED AND INSULATED

Based on 1000 sq. ft. (92.9 m²) Area
and Inside Temperature 75°F (24°C)

ENGLISH UNITS

AMBIENT AIR TEMP. °F	INSULATION ABOVE CEILING	R	HEAT LOSS Btu/hr	HEAT SAVINGS BY INSUL Btu/hr	HEAT SAVED BY ADD. INSUL Btu/hr	TOTAL % SAVED
30	No Insulation	0.41	109,000			
30	3″ Insul	11.11	4,050	104,950		96
30	6″ Insul	21.81	2,060	106,940	1,990	98
30	12″ Insul	42.41	1,060	107,940	2,990	99
15	No Insulation	0.41	145,333			
15	3″ Insul	11.11	5,399	139,934		96
15	6″ Insul	21.81	2,747	142,238	2,652	98
15	12″ Insul	42.41	1,413	143,920	3,986	99
0	No Insulation	0.41	181,600			
0	3″ Insul	11.11	6,750	174,917		96
0	6″ Insul	21.81	3,433	178,234	3,317	98
0	12″ Insul	42.41	1,766	179,901	4,984	99
−15	No Insulation	0.41	218,600			
−15	3″ Insul	11.11	8,100	209,900		96
−15	6″ Insul	21.81	4,120	213,880	3,980	98
−15	12″ Insul	42.41	1,766	179,901	4,984	99

METRIC UNITS

AMBIENT AIR TEMP. °C	INSULATION ABOVE CEILING	R	HEAT LOSS Watts	HEAT SAVINGS BY INSUL Watts	HEAT SAVED BY ADD. INSUL Watts	TOTAL % SAVED
− 1	No Insulation	0.0722	34,574			
− 1	0.076m Insul	1.9522	1,284	33,280		96
− 1	0.1525m Insul	3.842	653	33,921	631	98
− 1	0.3048m Insul	7.616	336	34,138	948	99
− 9	No Insulation	0.0722	46,099			
− 9	0.076m Insul	1.9522	1,712	44,387		96
− 9	0.1525m Insul	3.842	871	45,228	841	98
− 9	0.3048m Insul	7.616	336	45,651	1,264	99
−18	No Insulation	0.0722	57,624			
−18	0.076m Insul	1.9522	2,141	55,483		96
−18	0.1525m Insul	3.842	1,088	56,536	1,053	98
−18	0.3048m Insul	7.616	560	57,064	1,581	99
−26	No Insulation	0.0722	69,149			
−26	0.076m Insul	1.9522	2,569	66,580		96
−26	0.1525m Insul	3.842	1,309	67,840	1,260	98
−26	0.3048m Insul	7.616	672	68,477	1,897	99

TABLE 16

(Examples 8 and 8A)

EFFECT OF AMBIENT TEMPERATURE
ON HEAT LOSS FROM BUILDING TO AMBIENT AIR
THROUGH UNINSULATED AND INSULATED FRAME WALL

ENGLISH UNITS

AMBIENT AIR TEMPERATURE	FRAME WALL*	R	HEAT LOSS Btu/hr.	HEAT SAVINGS BY INSULATION Btu per hour	%
30°F	No Insul.	4	11,250		
	Insulated	14	3,210	8,041	71
15°F	No Insul.	4	15,000		
	Insulated	14	4,280	10,720	71
0°F	No Insul.	4	18,750		
	Insulated	14	5,360	13,390	71
−15°F	No Insul.	4	22,500		
	Insulated	14	6,430	16,070	71

METRIC UNITS

AMBIENT AIR TEMPERATURE	FRAME WALL*	Rm	HEAT LOSS Watts	HEAT SAVINGS BY INSULATION Watts	%
−1°C	No Insul.	0.704	3,295		
	Insulated	2.465	940	2,338	71
−9°C	No Insul.	0.704	4,397		
	Insulated	2.465	1,254	3,143	71
−10°C	No Insul.	0.704	5,556		
	Insulated	2.465	1,570	4,026	71
−26°C	No Insul.	0.704	6,595		
	Insulated	2.465	1,884	4,711	71

*The frame wall consisted of ½″ (12.7mm) X 8″ (203 mm) lapped siding over building paper which is over 25/32″ (19.5mm) wood sheathing attached to 2″ X 4″ (50.8mm X 101.6mm) studs with ½″ (12.7mm) gypsum board interior. The insulation used was mineral wool with thermal conductivity of 0.30 Btu, in/ft^2., hr., °F (0.0528 W/mK)

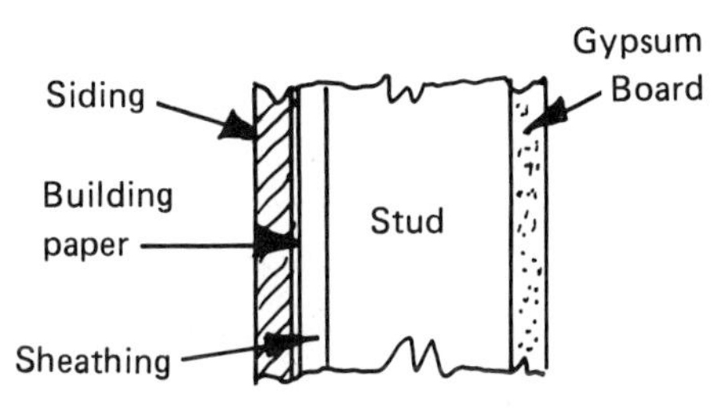

Wall – Not Insulated

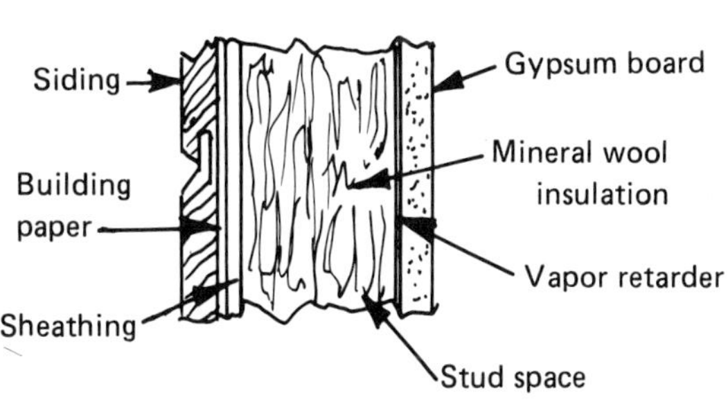

Wall – Insulated

TABLE 17

(Examples 9 and 9A)

English Units	***Metric Units***
At 30 °F outside temperature	At −1 °C outside temperature
$\text{Btu/ft}^2\text{, hr} = \dfrac{75-30}{14} = \dfrac{45}{14} = 3.21$	$\text{W/m}^2 = \dfrac{24-(-1)}{2.465} = \dfrac{25}{2.465} = 10.18$
At 15 °F outside temperature	At −9 °C outside temperature
$\text{Btu/ft}^2\text{, hr} = \dfrac{75-15}{14} = \dfrac{60}{14} = 4.28$	$\text{W/m}^2 = \dfrac{24-(-9)}{2.465} = \dfrac{33}{2.465} = 13.44$
At 0 °F outside temperature	At −18 °C outside temperature
$\text{Btu/ft}^2\text{, hr} = \dfrac{75-0}{14} = \dfrac{75}{14} = 5.36$	$\text{W/m}^2 = \dfrac{24-(-18)}{2.465} = \dfrac{42}{2.465} = 17.10$
At −15 °F outside temperature	At −26 °C outside temperature
$\text{Btu/ft}^2\text{, hr} = \dfrac{75-(-15)}{14} = \dfrac{90}{14} = 6.43$	$\text{W/m}^2 = \dfrac{24-(-26)}{2.465} = \dfrac{50}{2.465} = 20.35$

These results are tabulated in Table 17.

It is interesting to note that although it is impossible to obtain the 96 to 98% savings by insulation as possible for ceilings, the savings obtained by filling the cavity with mineral wool properly protected with a vapor retarder is over 70%.

Based on the fact that the colder the climate and the longer the time that heat is needed, various organizations have set up minimum recommended values of thermal resistance for the walls, roofs, ceilings and floors of buildings for numerous uses.

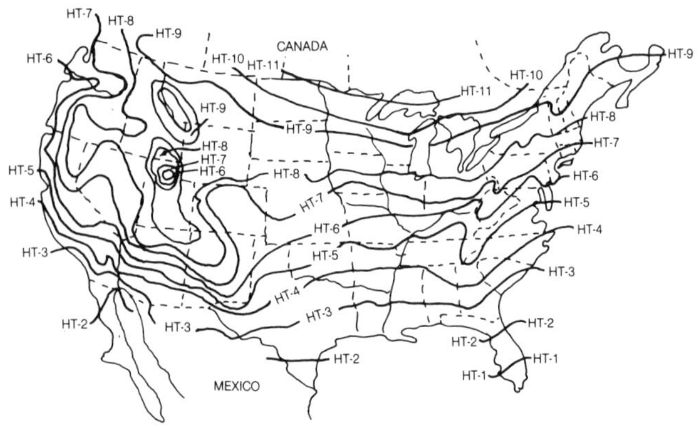

TEMPERATURE — TIME ZONES
Low Temperatures — Heating Season

	TEMPERATURE DIFFERENCE, OUTDOORS TO INDOORS MULTIPLIED BY TIME			
	ENGLISH UNITS (°F X TIME)		METRIC UNITS (°C X TIME)	
	DEGREE–DAYS*	DEGREE–HOURS	DEGREE–DAYS**	DEGREE–HOURS
HT-1	500	12,000	278	6,666
HT-2	1,000	24,000	556	13,333
HT-3	2,000	48,000	1,012	26,666
HT-4	3,000	72,000	1,666	40,000
HT-5	4,000	96,000	2,024	53,333
HT-6	5,000	120,000	2,777	66,666
HT-7	6,000	144,000	3,333	80,000
HT-8	7,000	168,000	3,889	93,333
HT-9	8,000	192,000	4,444	106,667
HT-10	9,000	216,000	5,000	120,000
HT-11	10,000	240,000	5,555	133,333

*The number of days times the difference of mean ambient air temperature below 65°F.
**The number of days times the difference of mean ambient air temperature below 18.3°C.

Figure 41

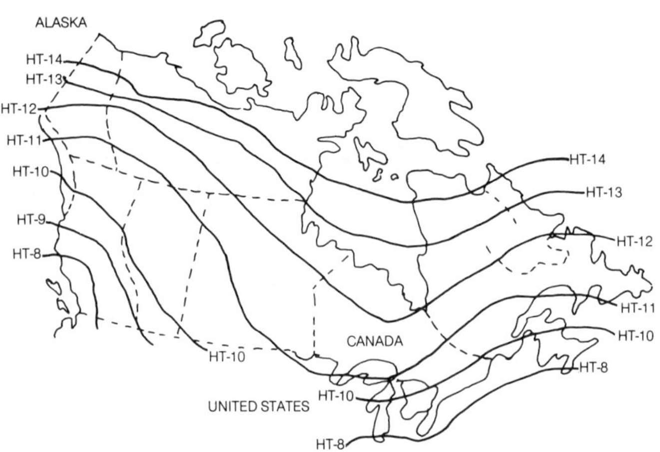

TEMPERATURE — TIME ZONES
Low Temperatures — Heating Season

HEATING TEMP. TIME ZONE	TEMPERATURE DIFFERENCE, OUTDOORS TO INDOORS MULTIPLIED BY TIME			
	ENGLISH UNITS (°F X TIME)		METRIC UNITS (°C X TIME)	
	DEGREE–DAYS*	DEGREE–HOURS	DEGREE–DAYS**	DEGREE–HOURS
HT-8	7,000	168,000	3,889	93,333
HT-9	8,000	192,000	4,444	106,667
HT-10	9,000	216,000	5,000	120,000
HT-11	10,000	240,000	5,555	133,333
HT-12	11,000	264,000	6,116	146,784
HT-13	12,000	288,000	6,672	160,128
HT-14	13,000	312,000	7,228	173,472

*The number of days times the difference of mean ambient air temperature below 65°F.
**The number of days times the difference of mean ambient air temperature below 18.3°C.

Figure 42

To keep energy loss within reasonable economic limits, the climatic conditions to which the home or building is to be subjected must be determined. To assist the designer of the building (or the owner) in proper selection of insulation systems, a Temperature-Time Zone Map of the United States and Canada is given in Figures 41 and 42. These are divided into 14 Temperature-Time Zones based on the number of degree-days (degree-hours) the ambient air temperature is below 65 °F (18 °C).

As shown on Table 17, the greater the temperature difference between inside and outside air, the greater the heat loss *per hour*. This loss per hour multiplied by number of hours the differences existed, determines the heat (or energy) loss. Based on Temperature-Time Zones, the American Society of Heating, Refrigerating and Air-Conditioning Engineers tabulated the minimum thermal resistance requirements for buildings in "Energy Conservation in New Building Design 90-75."

The maximum average heat transfer values per unit area are given in Table 18. It should be noted that these tabulated values represent upper limits which should not be exceeded on the amount of heat transfered per unit area. These maximum heat transfer values are not economical by todays standards and are not recommended by the author.

The minimum "R" values listed were based on gross areas of the structure. For walls, the gross area includes all surfaces such as opaque wall areas, window areas and door areas. Thus the values listed are the thermal average of the entire wall not just a thermal characteristic of the wall section by itself. The roof assembly includes all the components of the roof such as skylights.

MAXIMUM HEAT TRANSFER and MINIMUM THERMAL RESISTANCE OF WALLS

GROSS AREA HEAT TRANSFER
AVERAGE OF WALLS – INCLUDING DOORS and WINDOWS

HEATING TEMP. TIME ZONE	MAXIMUM AVERAGE THERMAL HEAT TRANSFER PER UNIT AREA								MINIMUM AVERAGE THERMAL RESISTANCE PER UNIT AREA							
	ENGLISH UNITS U in Btu/ft.^{2},hr°F				METRIC UNITS U_m in W/m^2 K				ENGLISH UNITS R = 1/U				METRIC UNITS $R_m = 1/U_m$			
	Types of Buildings				Types of Buildings				Types of Buildings				Types of Buildings			
	A	B	C	D	A	B	C	D	A	B	C	D	A	B	C	D
HT-1	0.32	0.39	0.38	0.49	1.81	2.21	2.15	2.78	3.12	2.56	2.63	2.04	0.55	0.45	0.46	0.36
HT-2	0.30	0.38	0.37	0.46	1.70	2.15	2.10	2.61	3.33	2.63	2.70	2.17	0.58	0.46	0.47	0.38
HT-3	0.28	0.37	0.35	0.43	1.58	2.10	1.98	2.44	3.57	2.70	2.85	2.32	0.63	0.47	0.50	0.41
HT-4	0.27	0.35	0.33	0.40	1.53	1.98	1.87	2.27	3.70	2.85	3.03	2.50	0.65	0.50	0.53	0.44
HT-5	0.25	0.33	0.31	0.38	1.42	1.87	1.76	2.15	4.00	3.03	3.22	2.63	0.70	0.53	0.57	0.46
HT-6	0.23	0.30	0.29	0.36	1.30	1.70	1.64	2.04	4.35	3.33	3.44	2.77	0.77	0.58	0.61	0.48
HT-7	0.22	0.28	0.27	0.33	1.24	1.58	1.53	1.87	4.54	3.57	3.70	3.03	0.80	0.63	0.65	0.53
HT-8	0.20	0.26	0.26	0.31	1.14	1.47	1.47	1.76	5.00	3.84	3.84	3.22	0.88	0.68	0.68	0.57
HT-9	0.18	0.24	0.24	0.30	1.02	1.36	1.36	1.70	5.55	4.16	4.16	3.33	0.98	0.73	0.73	0.58
HT-10	0.16	0.22	0.22	0.30	0.91	1.24	1.24	1.70	6.25	4.54	4.54	3.33	1.10	0.80	0.80	0.58
HT-11	0.15	0.20	0.20	0.29	0.85	1.14	1.14	1.64	6.66	5.00	5.00	3.44	1.17	0.88	0.88	0.61
HT-12	0.14	0.19	0.19	0.29	0.79	1.08	1.08	1.64	7.14	5.26	5.26	3.44	1.26	0.93	0.93	0.61
HT-14	0.13	0.18	0.18	0.28	0.74	1.02	1.02	1.58	7.69	5.55	5.55	3.57	1.35	0.98	0.98	0.63

Type A Buildings: Are detached one or two family dwellings.

Type B Buildings: Are multi-family dwellings, hotels and motels three stories and under.

Type C Buildings: Are hotels, offices, stores and commercial buildings three stories and under.

Type D Buildings: Are hotels, offices, stores and commercial buildings over three stories.

NOTE: The commercial buildings are those mainly used for human occupancy and do not include such buildings as processing buildings, cold storage buildings or warehouses.

TABLE 18

Although the major factor for the need for slab insulation is a function of depth, ground temperature, air temperature zones were used to indicate the maximum requirements of heat transfer per unit area and unit temperature difference. In the case of the unheated slab below heated interior air, the slab could be considered as part of the thermal resistance. For the heated slab, the thermal resistance should be calculated as material between the slab and the existing ground temperature.

The taller a building, the more area of walls in relation to the roof area. Also, the taller a building, the more it is directly exposed to wind, rain, and sun on its walls. It would be assumed correctly that tall buildings should be constructed with walls of lower thermal heat transfer than one or two-story buildings. Thus, it could be assumed that the minimum R values listed in Table 18 had the columns mistakenly reversed, but this is not so. The lesser "R" values for taller buildings are caused by the thermal inefficiency associated with their mostly glass walls. Of course, this is poor economic design and a waste of energy.

Tables 18 and 19 are not to be taken as the authors recommendations for the thermal insulation design of future buildings. These tables are provided to indicate what was acceptable in the past.

HUD and FHA published "Minimum Property Standards for One and Two Family Dwellings" in 1974. These standards listed maximum "U" values (heat transfer per unit area per degree temperature difference) for these dwellings based on "Heat Time-Temperature Zones". Based on these standards, Tables 20 and 21 are furnished to indicate the minimum requirements of dwellings.

MAXIMUM HEAT TRANSFER and MINIMUM THERMAL RESISTANCE
ROOFS, CEILINGS, FLOORS (OVER UNHEATED AREAS) and SLABS

GROSS AREA HEAT TRANSFER
AVERAGE OF WALLS AND CEILINGS – INCLUDING SKYLIGHTS, ETC.
AVERAGE OF FLOORS – INCLUDING ALL STRUCTURAL MEMBERS
AVERAGE OF SLAB – INCLUDES PERIMETER LOSSES

HEATING TEMP. TIME ZONE	MAXIMUM AVERAGE HEAT TRANSFER PER UNIT AREA								MAX. AVERAGE THERMAL RESISTANCE PER UNIT AREA							
	ENGLISH UNITS U in $Btu/ft^2,hr^\circ F$				METRIC UNITS U_m in $W/m^2 K$				ENGLISH UNITS in 1/U				METRIC UNITS in $1/U_m$			
	Roofs & Ceilings	Floors	Slab Heated	Slab Unheated	Roofs & Ceilings	Floors	Slab Heated	Slab Unheated	Roofs & Ceilings	Floors	Slab Heated	Slab Unheated	Roofs Ceilings	Floors	Slab Heated	Slab Unheated
HT-1	0.10	0.36	0.34	0.50	0.57	2.04	1.93	2.83	10.00	2.77	2.90	2.00	1.76	0.49	0.52	0.35
HT-2	0.10	0.33	0.31	0.50	0.57	1.87	1.76	2.83	10.00	3.03	3.25	2.00	1.76	0.33	0.56	0.35
HT-3	0.10	0.26	0.25	0.43	0.57	1.47	1.42	2.49	10.00	3.87	4.00	2.32	1.76	0.68	0.70	0.40
HT-4	0.10	0.19	0.21	0.35	0.57	1.07	1.19	1.98	10.00	5.26	4.75	2.85	1.76	0.98	0.84	0.50
HT-5	0.09	0.12	0.18	0.29	0.51	0.68	1.02	1.64	11.11	8.33	5.50	3.44	1.95	1.46	0.98	0.61
HT-6	0.08	0.08	0.16	0.24	0.45	0.45	0.91	1.36	12.50	12.50	6.25	4.15	2.20	2.20	1.09	0.73
HT-7	0.07	0.08	0.14	0.21	0.40	0.45	0.79	1.19	14.25	12.50	7.00	4.76	2.50	2.20	1.25	0.84
HT-8	0.06	0.08	0.13	0.18	0.34	0.45	0.74	1.02	16.66	12.50	7.69	5.55	2.93	2.20	1.35	0.98
HT-9	0.06	0.08	0.12	0.16	0.34	0.45	0.68	0.91	16.66	12.50	8.33	6.15	2.93	2.20	1.47	1.10
HT-10	0.06	0.08	0.10	0.15	0.34	0.45	0.57	0.85	16.66	12.50	10.00	6.85	2.93	2.20	1.75	1.17
HT-11	0.06	0.08	0.10	0.13	0.34	0.45	0.57	0.74	16.66	12.50	10.00	7.69	2.93	2.20	1.75	1.35
HT-12	0.06	0.08	0.09	0.12	0.34	0.45	0.34	0.68	16.66	12.50	10.25	8.33	2.93	2.20	2.94	1.47
HT-14	0.06	0.08	0.09	0.11	0.34	0.45	0.34	0.62	16.66	12.50	10.25	9.09	2.93	2.20	2.94	1.61

ROOFS and CEILINGS: The gross area is considered as all the components of the roof or ceiling envelope through which heat flows.

FLOORS: The gross area is considered as all the components of the floor over an unheated area.

SLAB, HEATED: The area under the heating elements in the slab which is on the ground.

SLAB, UNHEATED: The gross area of the slab between the ground and the heated area above.

TABLE 19

Table 20 provides the *Maximum Heat Transfer "U" Values and the Minimum Thermal Resistance "R" Values for Walls and Ceilings*. Table 21 provides *Maximum Heat Transfer "U" Values and the Minimum Thermal Resistance "R" Values for Flat Roofs, Floors and Slabs*. These are based on "Heat Time-Temperature Zones".

Because of basic differences in construction, the criteria are necessarily different for foundation walls surrounding heated basements and crawl spaces as compared to the walls above. Table 22 provides *Max. Heat Transfer "U" Values and the Minimum Thermal Resistance "R" Values for Walls Surrounding Heated Basements or Crawl Spaces*. This table also provides the criteria for maximum heat transfer (or minimum resistance) for windows and door areas.

The slab floor on, or below grade, must also be considered as to its heat loss when the area above is heated. Likewise, unheated areas must also be considered because of potential freezing of water piping. The minimum standards to reduce heat transfer of

STANDARD FOR: MAXIMUM HEAT TRANSFER "U" VALUES AND MINIMUM RESISTANCE "R" VALUES FOR DWELLING FLAT ROOFS, FLOORS and SLABS
ONE AND TWO FAMILY HOMES

BASED ON TEMPERATURE – TIME ZONES FOR HEATING
RECOMMENDATIONS FOR WALL AREA SEPARATE FROM
RECOMMENDATIONS FOR WINDOWS AND DOORS

HEATING TEMP. TIME ZONE	RECOMMENDED MAXIMUM HEAT TRANSFER								RECOMMENDED MINIMUM THERMAL RESISTANCE							
	ENGLISH UNITS U in Btu/ft², hr °F				METRIC UNITS U_m in W/m² K				ENGLISH UNITS R in 1/U				METRIC UNITS R_m in $1/U_m$			
	Masonry		Frame		Masonry		Frame		Masonry		Frame		Masonry		Frame	
	Walls	Ceilings	Walls	Ceilings	Walls	Ceilings	Walls	Ceilings	Walls	Ceilings	Walls	Ceilings	Walls	Ceilings	Walls	Ceilings
HT-1	0.17	0.08	0.10	0.08	0.96	0.45	0.57	0.45	6.00	12.50	10.00	12.50	1.03	2.20	1.76	2.20
HT-2	0.17	0.08	0.10	0.08	0.96	0.45	0.57	0.45	6.00	1.250	10.00	12.50	1.03	2.20	1.76	2.20
HT-3	0.17	0.08	0.10	0.08	0.96	0.45	0.57	0.45	6.00	12.50	10.00	12.50	1.03	2.20	1.76	2.20
HT-4	0.16	0.06	0.08	0.06	0.91	0.34	0.45	0.34	6.25	16.50	12.50	16.50	1.10	2.93	2.20	2.93
HT-5	0.14	0.05	0.07	0.05	0.79	0.28	0.39	0.28	7.00	20.00	14.00	20.00	1.25	3.52	2.51	3.52
HT-6	0.12	0.05	0.07	0.05	0.68	0.28	0.39	0.28	8.00	20.00	14.00	20.00	1.46	3.52	2.51	3.52
HT-7	0.12	0.05	0.07	0.05	0.68	0.28	0.39	0.28	8.00	20.00	14.00	20.00	1.46	3.52	2.51	3.52
HT-8	0.11	0.05	0.07	0.05	0.62	0.28	0.39	0.28	10.00	20.00	14.00	20.00	1.60	3.52	2.51	3.52
HT-9	0.10	0.05	0.07	0.05	0.57	0.28	0.39	0.28	10.00	20.00	14.00	20.00	1.76	3.52	2.51	3.52
HT-10	0.10	0.04	0.07	0.04	0.57	6.23	0.39	0.23	10.00	25.00	14.00	25.00	1.76	4.40	2.51	4.40
HT-11	0.10	0.04	0.07	0.04	0.57	0.23	0.39	0.23	10.00	25.00	14.00	25.00	1.76	4.40	2.51	4.40
HT-12	0.10	0.03	0.06	0.03	0.57	0.17	0.34	0.17	10.00	33.00	16.50	33.00	1.76	5.87	2.93	5.87
HT-14	0.10	0.03	0.05	0.03	0.57	0.17	0.28	0.17	10.00	33.00	20.00	33.00	1.76	5.87	3.52	5.87

The above table provides minimum thermal property standards for walls and ceilings of one and two family dwellings. These are *not* optimum thermal resistances for walls and ceilings. Optimum thermal resistances must be determined by economics in relation to time-temperature zones.

TABLE 20

slabs are presently based on perimeter insulation. Perimeter insulation is illustrated in Figure 28 "A" and "B". The standard presents information as to the distance the insulation should extend under the concrete and the thermal resistance "R" minimum. These standards are based the winter design temperatures rather than Time-Temperature Zones and is given in Table 23.

In addition to the temperature difference multiplied by the time that different influences act on the insulation, the minimum temperature that the insulation must withstand also influences the thermal insulation design. This is because the maximum heat losses are used to select and design the heating system. Table 24 provides the minimum (and maximum) design temperatures by cities in states (or provinces) arranged in alphabetical order. This table also provides the degree-day, degree-hours below 65°F (18°C) and above 80°F (27°C) in both English and Metric Units.

Using these tables (for winter conditions) it is then possible to determine the maximum temperature differences to be used in insulation systems for walls and roofs of heated buildings and homes. These tables also provide the degree-days (or degree-hours) for the determination of heat loss.

STANDARD FOR: MAXIMUM HEAT TRANSFER "U" VALUE and MINIMUM THERMAL RESISTANCE "R" VALUES FOR DWELLINGS FLAT ROOFS, FLOORS and SLABS
ONE and TWO FAMILY DWELLINGS
BASED ON TEMPERATURE – TIME ZONES FOR HEATING

RECOMMENDATIONS FOR: ROOF and INSULATION INSTALLED DIRECTLY OVER HEATED AREA – NO ROOF CAVITY ABOVE CEILING.
FLOORS, SLAB or WOOD, OF HEATED AREAS OVER UNHEATED AREA

HEATING TEMP. TIME ZONE	RECOMMENDED MAX. HEAT TRANSFER						RECOMMENDED MIN. THERM. RESISTANCE					
	ENGLISH UNITS U in Btu/ft^2,hr°F			METRIC UNITS U_m in W/m^2 K			ENGLISH UNITS R in 1/U			METRIC UNITS R_m in 1/U_m		
	Flat Roof & Insul.	Floors		Flat Roof & Insul.	Floors		Flat Roof & Insul.	Floors		Flat Roof & Insul.	Floors	
		Struc. Slab	Wood		Struc. Slab	Wood		Struc. Slab	Wood		Struc. Slab	Wood
HT-1	0.14	No Reg	No Reg	0.79	No Reg	No Reg	7.00	No Reg	No Reg	1.25	No Reg	No Reg
HT-2	0.14	No Reg	No Reg	0.79	No Reg	No Reg	7.00	No Reg	No Reg	1.25	No Reg	No Reg
HT-3	0.14	No Reg	No Reg	0.79	No Reg	No Reg	7.00	No Reg	No Reg	1.25	No Reg	No Reg
HT-4	0.14	0.15	0.10	0.79	0.85	0.57	7.00	6.50	10.00	1.25	1.17	1.76
HT-5	0.14	0.15	0.10	0.79	0.85	0.57	7.00	6.50	10.00	1.25	1.17	1.76
HT-6	0.09	0.12	0.08	0.51	0.68	0.45	11.00	8.25	12.50	1.96	1.46	2.20
HT-7	0.09	0.12	0.08	0.51	0.68	0.45	11.00	8.25	12.50	1.96	1.46	2.20
HT-8	0.09	0.12	0.08	0.51	0.68	0.45	11.00	8.25	12.50	1.96	1.46	2.20
HT-9	0.09	0.12	0.08	0.51	0.68	0.45	11.00	8.25	12.50	1.96	1.46	2.20
HT-10	0.09	0.12	0.08	0.51	0.68	0.45	11.00	8.25	12.50	1.96	1.46	2.20
HT-11	0.09	0.12	0.08	0.51	0.68	0.45	11.00	8.25	12.50	1.96	1.46	2.20
HT-12	0.07	0.10	0.07	0.39	0.57	0.39	14.00	10.00	14.00	2.52	1.76	2.52
HT-14	0.06	0.08	0.06	0.34	0.45	0.34	16.00	12.50	16.00	2.93	2.20	2.93

The above table provides minimum thermal property standards for flat roofs for one and two family dwellings. These are *not* optimum thermal resistances for flat roofs and floors. Optimum thermal resistances must be determined by economics in relation to time-temperature zones.

TABLE 21

STANDARD FOR: MAXIMUM HEAT TRANSFER "U" VALUE and MINIMUM THERMAL RESISTANCE "R" VALUES for DWELLINGS FOUNDATION WALL and WINDOWS ONE and TWO FAMILY DWELLINGS BASED ON TEMPERATURE – TIME ZONES for HEATING

RECOMMENDATIONS for FOUNDATION WALLS SURROUNDING HEATED BASEMENT or CRAWL SPACES.
RECOMMENDATIONS FOR 100% OF WINDOW or DOOR AREAS

HEATING TEMP. TIME ZONE	RECOMMENDED MAX. HEAT TRANSFER						RECOMMENDED MIN. THERM. RESISTANCE					
	ENGLISH UNITS U in Btu/ft^2,hr°F			METRIC UNITS U_m in W/m^2 K			ENGLISH UNITS R in I/U			METRIC UNITS R_m in I/U_m		
	Windows	Doors	Basement Foundation Walls	Windows	Doors	Basement Foundation Walls	Windows	Doors	Basement Foundation Walls	Windows	Doors	Basement Foundation Walls
HT-1	1.13	1.13	No Reg	6.41	6.41	No Reg	0.88	0.88	No Reg	0.15	0.15	No Reg
HT-2	1.13	1.13	No Reg	6.41	6.41	No Reg	0.88	0.88	No Reg	0.15	0.15	No Reg
HT-3	1.13	1.13	0.24	6.41	6.41	1.36	0.88	0.88	4.16	0.15	0.15	0.73
HT-4	1.13	1.13	0.24	6.41	6.41	1.36	0.88	0.88	4.16	0.15	0.15	0.73
HT-5	1.13	1.13	0.24	6.41	6.41	1.36	0.88	0.88	4.16	0.15	0.15	0.73
HT-6	0.65	0.65	0.17	3.69	3.69	0.96	1.53	1.53	5.88	0.27	0.27	1.00
HT-7	0.65	0.65	0.17	3.69	3.69	0.96	1.53	1.53	5.88	0.27	0.27	1.00
HT-8	0.65	0.65	0.17	3.69	3.69	0.96	1.53	1.53	5.88	0.27	0.27	1.00
HT-9	0.65	0.65	0.17	3.69	3.69	0.96	1.53	1.53	5.88	0.27	0.27	1.00
HT-10	0.65	0.65	0.17	3.69	3.69	0.96	1.53	1.53	5.88	0.27	0.27	1.00
HT-11	0.65	0.65	0.17	3.69	3.69	0.96	1.53	1.53	5.88	0.27	0.27	1.00
HT-12	0.47	0.65	0.17	2.67	3.69	0.96	2.12	1.53	5.88	0.37	0.27	1.00
HT-14	0.47	0.65	0.17	2.67	3.69	0.96	2.12	1.53	5.88	0.37	0.27	1.00

NOTE: Windows of U = 1.13 are single glass, U = 0.65 are double glass (regular sash plus storm windows), U = 0.47 is triple glass.

The above table provides minimum thermal properties. Optimum thermal resistance for basement or crawl space walls surrounding heated areas must be determined by economics in relation to time-temperature zones.

TABLE 22

STANDARD FOR MAXIMUM HEAT TRANSFER PER LINEAR FOOT OF EXPOSED EDGES OF SLAB FLOORS ON GROUND BASED ON WINTER DESIGN TEMPERATURE.

HEAT TRANSFER PER UNIT OF TEMPERATURE DIFFERENCE PER UNIT EXPOSED PERIMETER LENGTH

WINTER DESIGN TEMP.		English Units "F" = Btu/ft. hr						Metric Units "F_m" = W/m					
		Unheated Slab			Heated Slab**			Unheated Slab			Heated Slab**		
		Total Width of Insul,	Insul "R" ℓ/k		Total Width of Insul,	Insul "R" ℓ/k		Total Width of Insul	Insul "R_m" ℓ_m/k_m		Total Width of Insul	Insul "R_m" ℓ_m/k_m	
°F	°C	Inches	5.0	3.75	Inches	5.0	3.75	m	0.88	0.66	m	0.88	0.66
−30	−35	24	34	*	24	46	*	0.61	30	*	0.61	41	*
−25 to −29	−32 to −34	24	32	*	24	44	*	0.61	29	*	0.61	39	*
−20 to −24	−29 to −31	24	30	*	24	41	*	0.61	27	*	0.61	37	*
−15 to −19	−26 to −28	24	28	*	24	39	*	0.61	25	*	0.61	35	*
−10 to −14	−23 to −25	24	27	40	24	37	*	0.61	24	36	0.61	33	*
−5 to −9	−21 to −22	24	25	38	24	35	*	0.61	22	34	0.61	31	*
0 to −4	−18 to −20	24	24	36	24	32	48	0.61	21	32	0.61	29	43
5 to 1	−15 to −17	24	22	33	24	30	45	0.61	20	29	0.61	27	40
10 to 6	−12 to −14	18	21	31	18	25	38	0.46	19	28	0.46	22	34
15 to 11	−9 to −11	12	21	31	12	25	38	0.30	19	28	0.30	22	34

*Use insulation of greater thickness or lower conductivity to obtain required thermal resistance.

**Slab heated by heating pipes; hot ducts under the slab heated by electric cable has cable installed in the slab for better efficiency.

The above table provides the heat loss of slab perimeter exposed to outer surface air and/or ground temperatures. This standard is a minimum requirement and does not provide optimum design efficiency.

As energy cost increases, slab completely insulated underneath will be necessary for efficient economic installation.

TABLE 23

DESIGN TEMPERATURES, DEGREE DAYS AND DEGREE HOURS
HEATING AND COOLING – UNITED STATES CITIES – ENGLISH AND METRIC UNITS

CITY	STATE	DESIGN TEMPERATURES				WINTER HEATING DEG × TIME				SUMMER COOLING DEG × TIME			
		Winter		Summer		Δ ° × Days		Δ ° × Hours		Δ ° × Days		Δ ° × Hours	
		°F	°C	°F	°C	English	Metric	English	Metric	English	Metric	English	Metric
Anniston	Ala	17	−8	96	36	2,820	1,567	67,680	37,608	1,401	778	33,624	18,672
Birmingham	Ala	14	−10	97	36	2,780	1,544	67,720	37,056	1,371	761	32,904	18,264
Dothan	Ala	23	−5	97	36	1,700	944	40,800	22,656	1,800	1,000	43,200	24,000
Huntsville	Ala	8	−13	97	36	3,200	1,778	76,800	42,672	1,240	689	16,536	16,536
Mobile	Ala	26	−3	96	36	1,612	896	38,688	21,504	1,844	1,024	44,256	24,576
Montgomery	Ala	18	−8	98	37	2,137	1,187	51,288	28,488	1,743	968	41,832	23,232
Tuscaloosa	Ala	19	−7	98	37	2,529	1,405	60,696	33,720	1,533	851	36,792	20,424
Anchorage	Ak	−29	−34	73	23	10,789	5,994	258,936	143,856	6	3	140	72
Fairbanks	Ak	−59	−51	82	28	14,158	7,866	339,792	188,784	53	29	1,272	696
Juneau	Ak	−11	−24	75	24	8,888	4,938	213,312	118,512	6	3	144	72
Flagstaff	Arz	−10	−23	84	29	7,525	4,181	180,600	100,344	176	98	4,224	2,352
Kingman	Arz	18	−8	103	39	3,000	1,667	72,000	40,008	1,662	923	39,888	22,152
Phoenix	Arz	25	−4	108	42	1,698	943	40,752	22,632	2,845	1,581	68,280	37,944
Tucson	Arz	23	−5	105	41	1,776	987	42,624	23,688	2,445	1,358	58,680	32,592
Fayetteville	Ark	9	−13	97	36	3,800	2,111	91,200	50,664	1,207	670	28,968	16,080
Fort Smith	Ark	15	−9	101	38	3,188	1,771	76,512	42,504	1,535	852	36,816	20,448
Little Rock	Ark	13	−11	99	37	2,982	1,657	71,568	39,768	1,545	858	37,080	20,592
Pine Bluff	Ark	14	−10	99	37	2,800	1,556	67,200	37,344	1,641	912	39,384	21,888
Bakersfield	Calif	26	−3	103	39	2,115	1,175	50,760	28,200	1,648	915	39,552	21,960
Burbank	Calif	30	−1	97	36	1,808	1,004	43,392	24,096	775	430	18,600	10,320
Eureka	Calif	35	2	65	18	4,632	2,573	111,168	61,752	500	278	12,000	6,672
Fresno	Calif	25	−4	99	37	2,532	1,407	60,768	33,768	1,364	758	32,736	18,192
Long Beach	Calif	31	−1	84	29	2,000	1,111	48,000	26,664	250	139	6,000	3,336
Los Angeles	Calif	38	3	89	32	2,015	1,119	48,360	26,856	576	320	13,824	7,680
Oakland	Calif	30	−1	85	29	3,163	1,757	75,912	42,168	89	49	2,136	1,176
Redding	Calif	31	−1	103	39	4,000	2,222	96,000	53,328	700	389	16,800	9,336
Riverside	Calif	32	0	99	37	3,000	1,666	72,000	39,984	1,209	672	29,016	16,128
Sacramento	Calif	24	−4	100	38	2,600	1,444	62,400	34,666	1,006	559	24,144	13,416
Salinas	Calif	32	0	87	31	2,000	1,111	48,000	26,664	350	194	8,400	4,656
San Bernardino	Calif	31	−1	101	38	1,879	1,044	45,096	25,056	1,188	660	28,512	15,840
San Diego	Calif	38	3	86	30	1,574	874	37,776	29,976	130	72	3,120	1,728
San Francisco	Calif	35	2	81	27	3,421	1,900	82,104	45,600	42	23	1,008	552
San Jose	Calif	34	1	85	29	2,410	1,339	57,840	32,136	157	87	3,768	2,088
Santa Ana	Calif	33	1	92	33	2,500	1,389	60,000	33,336	900	500	21,600	12,000
Santa Barbara	Calif	34	1	87	31	1,995	1,108	47,880	26,592	44	24	1,056	576
Stockton	Calif	30	−1	99	37	2,675	1,486	64,200	35,664	1,017	565	24,408	13,560
Colorado Springs	Colo	−9	−23	90	32	6,254	3,474	150,096	83,376	508	282	12,192	6,768
Denver	Colo	−9	−23	92	33	6,132	3,407	147,168	81,768	647	359	15,528	8,616
Grand Junction	Colo	−2	−19	94	34	5,796	3,220	139,104	77,280	988	549	23,712	13,175
Pueblo	Colo	−14	−26	94	34	5,709	3,172	137,016	76,128	500	278	12,000	6,672
Hartford	Conn	1	−17	90	32	6,139	3,410	147,336	81,840	476	264	11,424	6,336
New Haven	Conn	0	−18	88	31	6,026	3,348	144,624	80,352	193	107	4,632	2,568
Dover	Del	8	−13	93	34	4,591	2,551	110,184	61,224	683	379	16,392	9,096
Wilmington	Del	6	−14	93	34	4,910	2,728	117,840	65,472	525	291	12,600	6,984
Washington	DC	14	−10	94	34	4,333	2,407	103,992	57,768	1,002	556	24,048	13,344
Fort Myers	Fla	34	−1	94	34	405	225	9,720	5,400	1,863	1,035	44,712	24,840
Jacksonville	Fla	26	−3	96	36	1,243	690	29,832	16,560	1,725	958	41,400	22,992
Miami	Fla	39	4	92	33	178	99	4,272	2,376	2,388	1,327	57,312	31,848
Orlando	Fla	29	−2	96	36	650	361	15,600	8,664	1,609	894	38,616	21,456
Panama City	Fla	28	−2	92	33	1,500	833	36,000	19,992	2,089	1,161	50,136	27,864
Pensacola	Fla	25	−4	92	33	1,435	797	34,440	19,128	1,884	1,046	45,216	25,104
Tallahassee	Fla	21	−6	96	36	1,519	844	36,456	20,256	1,538	854	36,912	20,496
Tampa	Fla	32	0	92	33	674	374	16,176	8,976	2,154	1,197	51,696	28,728
W. Palm Beach	Fla	36	2	92	33	248	138	5,952	3,312	2,000	1,111	48,000	26,664
Albany	Ga	26	−3	98	37	1,763	979	42,312	23,496	1,747	971	41,928	23,304
Atlanta	Ga	14	−10	95	35	2,826	1,570	67,824	37,680	1,151	639	27,624	15,336
Augusta	Ga	17	−8	98	37	2,138	1,188	51,312	28,512	1,431	795	34,344	19,080
Columbus	Ga	23	−5	98	37	2,396	1,331	57,504	31,944	1,611	839	36,264	20,136
Macon	Ga	21	−6	98	37	2,049	1,138	49,176	27,312	1,549	861	37,176	20,664
Savannah	Ga	24	−4	96	36	1,710	950	41,040	22,800	1,515	842	36,360	20,208
Valdosta	Ga	28	−2	96	36	1,525	847	36,600	20,328	1,539	855	36,936	20,520
Boise	Ida	0	−18	96	36	5,800	3,222	139,200	77,328	706	392	16,944	9,408
Coeur D'Alene	Ida	2	−17	94	34	7,000	3,889	168,000	93,336	500	278	12,000	6,672
Idaho Falls	Ida	−12	−24	91	33	7,500	4,167	180,000	100,008	700	389	16,800	9,336

TABLE 24

(Sheet 1 of 5)

DESIGN TEMPERATURES, DEGREE DAYS AND DEGREE HOURS
HEATING AND COOLING – UNITED STATES CITIES – ENGLISH AND METRIC UNITS

CITY	STATE	DESIGN TEMPERATURES Winter		DESIGN TEMPERATURES Summer		WINTER HEATING DEG × TIME Δ ° × Days		WINTER HEATING DEG × TIME Δ ° × Hours		SUMMER COOLING DEG × TIME Δ ° × Days		SUMMER COOLING DEG × TIME Δ ° × Hours	
		°F	°C	°F	°C	English	Metric	English	Metric	English	Metric	English	Metric
Lewiston	Ida	6	−14	98	37	5,483	3,046	131,592	73,104	878	488	21,072	11,712
Pocatello	Ida	−12	−24	94	34	6,976	3,876	167,424	93,024	601	334	14,424	8,016
Twin Falls	Ida	4	−16	96	36	8,000	4,444	192,000	106,656	700	389	16,800	9,336
Belleville	Ill	6	−14	97	36	4,670	2,594	112,080	62,256	1,150	639	27,600	15,336
Chicago	Ill	− 4	−20	95	35	6,310	3,505	151,440	84,120	727	404	17,448	9,696
Joliet	Ill	− 5	−21	94	34	6,578	3,654	157,872	87,696	676	376	16,224	9,024
Peoria	Ill	− 8	−22	94	34	6,087	3,381	146,088	81,144	612	340	14,688	8,160
Quincy	Ill	− 1	−18	97	36	5,302	2,945	127,248	70,680	794	441	19,056	10,584
Springfield	Ill	− 7	−22	95	35	5,693	3,162	136,632	75,888	867	482	20,808	11,568
Columbus	Ind	3	−16	95	35	5,255	2,919	126,120	70,056	848	471	20,352	11,304
Evansville	Ind	6	−14	96	36	4,360	2,422	104,640	58,128	1,195	664	28,680	15,936
Fort Wayne	Ind	− 5	−21	93	34	6,287	3,493	150,888	83,832	607	337	14,568	8,088
Indianapolis	Ind	− 5	−21	93	34	5,611	3,117	134,664	74,808	763	424	18,312	10,176
South Bend	Ind	− 6	−21	92	33	6,524	3,624	156,576	86,976	563	313	13,512	7,512
Terre Haute	Ind	3	−16	95	35	5,366	2,981	128,784	71,544	834	463	20,016	11,112
Burlington	Iowa	−10	−23	95	35	6,100	3,389	146,400	81,336	620	344	14,880	8,256
Cedar Rapids	Iowa	− 8	−22	92	33	6,800	3,778	163,200	90,672	540	300	12,960	7,200
Des Moines	Iowa	−13	−25	95	35	6,444	3,580	154,656	85,920	740	411	17,760	9,864
Mason City	Iowa	−13	−25	91	33	7,565	4,202	181,560	100,848	470	261	11,280	6,264
Ottumwa	Iowa	− 6	−21	95	35	6,300	3,500	151,200	84,000	910	506	21,840	12,144
Sioux City	Iowa	−17	−27	96	36	7,012	3,896	168,288	93,504	795	442	19,060	10,608
Dodge City	Kan	− 5	−21	99	37	5,059	2,810	121,416	67,440	1,135	631	27,240	15,144
Hutchinson	Kan	2	−17	99	37	4,676	2,598	112,224	62,352	1,275	708	30,600	16,992
Topeka	Kan	− 4	−20	99	37	5,210	2,894	125,040	69,456	1,155	641	27,720	15,384
Wichita	Kan	− 1	−18	102	39	4,571	2,539	109,704	60,936	1,280	711	30,720	17,064
Bowling Green	Ky	7	− 4	97	36	4,295	2,386	103,080	57,266	800	444	19,200	10,667
Corbin	Ky	5	−15	93	34	4,100	2,278	98,400	54,667	950	528	22,800	12,667
Lexington	Ky	0	−18	94	34	4,980	2,767	119,520	66,400	955	530	22,920	12,733
Louisville	Ky	1	−17	96	36	4,440	2,467	106,560	59,200	1,040	578	24,960	13,867
Paducah	Ky	10	−12	97	36	3,654	2,030	87,696	48,720	1,105	613	26,520	14,733
Alexandria	La	20	− 7	97	36	1,745	969	41,880	23,267	1,725	958	41,400	23,000
Baton Rouge	La	23	− 5	96	36	1,595	886	38,280	21,267	1,722	957	41,328	22,960
Lafayette	La	28	− 2	95	35	1,200	667	28,800	16,000	1,810	1,005	43,440	24,133
Lake Charles	La	25	− 4	95	35	1,545	858	37,080	20,600	1,805	1,003	43,320	24,067
Monroe	La	23	− 5	98	37	2,100	1,167	50,400	28,000	1,845	1,025	44,280	24,600
New Orleans	La	29	− 2	93	34	1,315	730	31,560	17,533	1,735	964	41,640	23,133
Shreveport	La	18	− 8	99	37	2,116	1,175	50,784	28,213	1,847	1,026	44,328	24,627
Augusta	Ma	− 7	−22	88	31	7,800	4,333	187,200	104,000	300	167	7,200	4,000
Bangor	Ma	−18	−28	88	31	7,500	4,167	180,000	100,000	220	162	5,280	2,933
Portland	Ma	−14	−26	87	31	7,680	4,267	184,230	102,350	235	130	5,640	3,133
Baltimore	Md	8	−13	91	33	4,611	2,561	110,664	61,480	835	464	20,040	11,133
Boston	Mass	− 1	−18	88	31	5,791	3,217	138,980	77,213	400	222	9,600	5,333
Springfield	Mass	− 3	−19	88	31	6,700	3,722	160,800	89,333	455	253	10,920	6,067
Worchester	Mass	− 8	−22	87	31	6,600	3,667	158,400	88,000	300	167	7,200	4,000
Detroit	Mich	0	−18	92	33	6,470	3,594	155,280	86,267	495	275	11,880	6,600
Flint	Mich	− 1	−18	89	32	7,180	3,989	172,320	95,733	510	283	12,240	6,800
Grand Rapids	Mich	− 3	−19	91	33	7,705	3,930	169,800	94,333	420	233	10,080	5,600
Lansing	Mich	− 4	−20	89	32	6,980	3,878	167,520	93,067	325	180	7,800	4,333
Saginaw	Mich	− 1	−18	88	31	7,065	3,925	169,560	94,200	390	217	9,360	5,200
Traverse City	Mich	0	−18	89	32	7,900	4,389	189,600	105,333	310	172	7,440	4,133
Duluth	Minn	−25	−32	85	29	9,935	5,519	238,440	132,467	133	74	3,192	1,773
Minneapolis	Minn	−19	−28	92	33	7,855	4,363	188,520	104,733	495	275	11,880	6,600
Rochester	Minn	−23	−31	90	32	8,096	4,494	194,304	107,946	335	186	8,040	4,466
St. Cloud	Minn	−20	−29	90	32	8,890	4,938	213,360	118,533	400	222	9,600	5,333
St. Paul	Minn	−14	−26	92	33	7,855	4,363	188,520	104,733	495	275	11,880	6,600
Columbia	Mo	− 4	−20	97	36	4,870	2,705	116,880	64,933	1,080	600	25,920	14,400
Jefferson City	Mo	3	−16	97	36	4,870	2,705	116,880	64,933	1,080	600	25,920	14,400
Joplin	Mo	5	−15	97	36	3,905	2,169	93,720	52,067	1,060	589	25,440	14,133
Kansas City	Mo	− 2	−19	100	38	4,890	2,717	117,360	65,200	1,095	608	26,280	14,600
St. Joseph	Mo	− 1	−18	97	36	5,333	2,962	127,992	71,107	1,125	625	27,000	15,000
St. Louis	Mo	− 2	−19	98	37	4,700	2,611	112,800	62,667	1,150	639	27,600	15,333
Springfield	Mo	0	−18	97	36	4,695	2,608	112,680	62,600	965	536	23,160	12,867

TABLE 24 (Continued) (Sheet 2 of 5)

DESIGN TEMPERATURES, DEGREE DAYS AND DEGREE HOURS
HEATING AND COOLING – UNITED STATES CITIES – ENGLISH AND METRIC UNITS

CITY	STATE	DESIGN TEMPERATURES Winter °F	Winter °C	Summer °F	Summer °C	WINTER HEATING DEG × TIME Δ °× Days English	Δ °× Days Metric	Δ °× Hours English	Δ °× Hours Metric	SUMMER COOLING DEG × TIME Δ °× Days English	Δ °× Days Metric	Δ °× Hours English	Δ °× Hours Metric
Billings	Mont	−19	−28	94	34	7,105	3,947	170,520	94,733	515	286	12,360	6,867
Butte	Mont	−24	−31	86	30	9,760	5,422	234,240	130,133	175	97	4,200	2,333
Great Falls	Mont	−29	−34	91	33	7,555	4,197	181,320	100,177	286	159	6,864	3,813
Helena	Mont	−27	−33	90	32	8,125	4,513	195,000	108,333	240	133	5,760	33,200
Miles City	Mont	−19	−28	97	36	7,820	4,344	187,680	104,267	595	330	1,480	822
Missoula	Mont	−16	−27	92	33	7,875	4,375	189,000	105,000	300	167	7,200	4,000
Grand Island	Neb	−13	−25	98	37	6,310	3,505	151,440	84,133	865	480	20,760	11,533
Lincoln	Neb	−10	−23	100	38	6,105	3,391	146,520	81,400	1,000	555	24,000	13,333
North Platte	Neb	− 6	−21	97	36	6,545	3,636	157,080	87,267	725	402	17,400	9,444
Omaha	Neb	−12	−21	97	36	6,160	3,422	147,840	82,133	900	500	21,600	12,000
Elko	Nev	−13	−25	94	34	7,335	4,075	176,040	97,800	661	367	15,864	8,813
Ely	Nev	− 6	−21	90	32	7,445	4,136	178,680	99,266	520	289	12,480	6,933
Las Vegas	Nev	18	− 8	108	42	2,709	1,504	65,016	36,016	2,360	1,311	56,640	31,467
Reno	Nev	− 2	−19	95	35	6,035	3,353	144,840	80,466	650	361	15,600	8,667
Concord	NH	−17	−27	91	33	7,610	4,228	182,640	101,466	400	222	9,600	5,333
Manchester	NH	− 5	−21	92	33	6,800	3,778	163,200	90,667	505	280	12,120	6,733
Atlantic City	NJ	14	−10	91	33	4,740	2,633	113,760	63,200	475	264	11,400	6,333
Newark	NJ	6	−14	94	34	5,252	2,917	126,048	70,027	595	330	14,280	7,933
Trenton	NJ	7	−14	92	33	5,070	2,817	121,680	67,500	450	250	10,800	6,000
Albuquerque	NM	6	−14	96	36	4,390	2,439	105,360	58,533	1,130	628	27,120	15,066
Clovis	NM	14	−10	99	37	4,200	2,333	100,800	56,000	1,200	667	28,000	15,555
Farmington	NM	6	−14	95	35	5,800	3,222	139,200	77,333	940	522	22,560	12,533
Roswell	NM	5	−15	101	38	3,425	1,902	82,200	45,667	1,615	897	38,760	21,533
Sante Fe	NM	7	−14	90	32	6,125	3,403	147,000	81,667	685	380	16,440	9,133
Tucumcari	NM	9	−13	99	37	4,000	2,222	96,000	53,333	1,000	555	24,000	13,333
Albany	NY	− 5	−21	91	33	6,962	3,868	167,088	92,826	420	233	10,080	5,600
Binghamton	NY	− 8	−22	91	33	7,535	4,186	180,840	100,467	255	142	6,120	3,400
Buffalo	NY	3	−16	88	31	6,840	3,800	164,160	91,200	347	193	8,326	4,625
Elmira	NY	1	−17	92	33	6,412	3,562	153,888	85,493	420	233	10,080	5,600
Massena	NY	−16	−27	86	30	7,832	4,351	187,968	104,426	280	155	6,720	3,733
New York	NY	12	−11	93	34	5,050	2,805	121,200	67,333	648	360	15,552	8,640
Rochester	NY	− 5	−21	91	33	6,865	3,813	164,760	91,533	420	233	10,080	5,600
Syracuse	NY	−10	−23	90	32	6,520	3,622	156,480	86,933	433	240	10,392	5,773
Asheville	NC	8	−13	91	33	4,072	2,262	97,728	54,293	610	339	14,640	8,133
Charlotte	NC	13	−11	96	36	3,205	1,780	76,920	42,773	1,140	633	27,360	15,200
Elizabeth City	NC	18	− 8	93	34	3,000	1,667	72,000	40,000	1,100	611	26,400	14,667
Fayettsville	NC	17	− 8	97	36	2,800	1,555	67,200	37,333	1,260	700	30,240	16,800
Greensboro	NC	9	−13	94	34	3,810	2,116	91,440	50,800	916	509	21,984	12,213
Raleigh/Durham	NC	13	−11	95	35	3,370	1,872	80,880	44,933	1,030	572	24,720	13,733
Wilmington	NC	19	− 7	93	34	2,325	1,291	55,800	31,000	1,245	692	29,880	16,600
Winston-Salem	NC	9	−13	94	34	3,720	2,067	89,280	49,600	805	447	19,320	10,733
Bismark	ND	−31	−35	95	35	9,033	5,018	216,792	120,440	471	262	11,304	6,280
Dickenson	ND	−23	−31	96	36	9,000	5,000	216,000	120,000	245	136	5,880	3,267
Fargo	ND	−28	−33	92	33	9,275	5,152	222,600	123,667	410	228	9,840	5,466
Grand Forks	ND	−26	−32	91	33	9,870	5,483	236,880	131,600	352	195	8,448	4,693
Jamestown	ND	−29	−34	95	35	9,406	5,225	225,744	125,413	450	250	10,800	6,000
Akron	OH	− 5	−21	89	32	6,203	3,446	148,872	82,707	415	230	9,960	5,533
Cincinnati	OH	2	−17	94	34	5,195	2,886	124,680	69,267	850	472	20,400	11,333
Cleveland	OH	− 2	−19	91	33	6,006	3,337	144,144	80,080	525	292	12,600	7,000
Columbus	OH	− 1	−18	92	33	5,615	3,119	134,760	74,867	760	422	18,240	10,133
Dayton	OH	− 2	−19	92	33	5,600	3,111	134,400	74,667	680	378	16,320	9,067
Mansfield	OH	− 5	−21	91	33	6,000	3,333	144,000	80,000	425	236	10,200	5,667
Toledo	OH	− 5	−21	92	33	6,395	3,552	153,480	85,267	590	328	14,160	7,867
Youngstown	OH	− 5	−21	89	32	6,175	3,340	148,200	82,333	390	217	9,360	5,200
Zanesville	OH	− 1	−18	92	33	5,484	3,046	131,616	73,120	615	342	14,760	8,200
Ardmore	Okla	15	− 9	103	39	2,900	1,611	69,600	38,667	1,596	887	38,304	21,280
Enid	Okla	10	−12	103	39	3,882	2,157	93,168	51,760	1,545	858	37,080	20,600
Lawton	Okla	13	−11	103	39	3,200	1,778	76,800	42,667	2,000	1,111	48,000	26,667
Muskogee	Okla	12	−11	102	39	3,616	2,009	86,784	48,213	1,400	778	33,600	18,667
Oklahoma City	Okla	4	−17	100	38	3,645	2,025	87,480	48,600	1,440	800	34,560	19,200
Tulsa	Okla	4	−17	102	39	3,585	1,992	86,040	47,800	1,420	788	34,080	18,933
Eugene	Ore	16	− 9	91	33	4,780	2,655	114,720	63,733	440	244	10,560	5,867
Klamath Falls	Ore	1	−17	89	32	6,465	3,592	155,160	86,200	339	188	8,136	4,520
Medford	Ore	21	− 6	98	37	4,547	2,526	106,128	58,960	630	350	15,120	8,400

TABLE 24 (Continued)

(Sheet 3 of 5)

DESIGN TEMPERATURES, DEGREE DAYS AND DEGREE HOURS
HEATING AND COOLING – UNITED STATES CITIES – ENGLISH AND METRIC UNITS

CITY	STATE	DESIGN TEMPERATURES				WINTER HEATING DEG × TIME				SUMMER COOLING DEG × TIME			
		Winter		Summer		Δ ° × Days		Δ ° × Hours		Δ ° × Days		Δ ° × Hours	
		°F	°C	°F	°C	English	Metric	English	Metric	English	Metric	English	Metric
Pendleton	Ore	− 2	−19	97	33	5,205	2,892	124,920	69,400	635	353	15,240	8,467
Portland	Ore	23	− 5	90	32	4,632	2,573	111,168	61,760	208	115	4,992	2,773
Salem	Ore	15	− 9	92	33	4,575	2,542	109,800	61,000	295	163	7,080	3,933
Allentown	Pa	− 2	−19	92	33	5,880	3,267	141,120	78,400	510	283	12,240	6,800
Altoona	Pa	1	−17	89	32	6,120	3,400	146,880	81,600	350	194	8,400	4,667
Erie	Pa	1	−17	88	31	6,115	3,397	146,760	81,533	270	150	6,480	3,600
Harrisburg	Pa	4	−16	92	33	5,260	2,922	126,240	70,133	745	414	17,880	9,933
Philadelphia	Pa	7	−14	93	34	4,865	2,703	116,760	64,867	702	390	16,848	9,360
Pittsburgh	Pa	1	−17	90	32	5,905	3,280	141,720	78,733	471	262	11,304	6,280
Wilkes-Barre	Pa	− 3	−19	89	32	6,500	3,611	156,000	86,667	400	222	9,600	5,333
Providence	RI	6	−14	89	32	6,125	3,403	147,000	81,667	315	175	7,560	4,200
Charleston	SC	23	− 5	94	34	1,973	1,096	47,352	26,307	1,250	694	30,000	16,667
Columbia	SC	16	− 9	98	37	2,435	1,353	58,440	32,467	1,359	755	32,616	18,120
Florence	SC	16	− 9	96	36	2,505	1,392	60,120	33,400	1,400	778	33,600	18,667
Greenville	SC	14	−10	95	35	3,200	1,778	76,800	42,667	1,093	608	26,280	14,600
Spartenburg	SC	18	− 8	95	35	3,044	1,691	73,056	40,587	1,026	570	24,624	13,680
Aberdeen	SD	−22	−30	95	35	8,600	4,778	206,400	114,667	550	305	13,200	7,333
Pierre	SD	−13	−25	98	37	7,285	4,047	174,840	97,133	550	305	13,200	7,333
Rapid City	SD	−17	− 8	96	36	7,535	4,186	180,840	100,466	570	316	13,680	7,600
Sioux Falls	SD	−21	−29	95	35	7,850	4,361	188,400	104,667	500	278	12,000	6,667
Watertown	SD	−20	−29	93	34	8,500	4,722	193,200	107,333	550	305	13,200	7,333
Chattanooga	Tenn	11	−12	97	36	3,383	1,879	81,192	45,107	1,250	694	30,000	16,667
Jackson	Tenn	14	−10	97	36	3,400	1,889	81,600	45,333	1,510	839	36,240	20,133
Knoxville	Tenn	9	−13	95	35	3,590	1,994	86,160	47,867	1,095	608	26,280	14,600
Memphis	Tenn	11	−12	98	37	3,137	1,743	75,288	41,827	1,509	838	36,216	20,120
Nashville	Tenn	6	−14	97	36	3,515	1,953	84,360	46,867	1,295	719	31,080	17,267
Abilene	Texas	17	− 8	101	38	2,655	1,475	63,720	35,400	2,005	1,114	48,120	26,733
Amarillo	Texas	2	−17	98	37	4,345	2,414	104,040	57,800	1,176	653	28,224	15,680
Austin	Texas	19	− 7	101	38	1,713	951	41,112	22,840	2,245	1,247	53,880	29,933
Bryan	Texas	27	− 3	100	38	1,700	944	40,800	22,667	2,091	1,162	50,184	27,880
Dallas	Texas	14	−10	101	38	2,270	1,261	54,480	30,267	2,304	1,280	55,296	30,720
El Paso	Texas	16	− 9	100	38	2,600	1,444	62,400	34,667	1,860	1,033	44,640	24,800
Fort Worth	Texas	14	−10	102	39	2,360	1,311	56,640	31,467	2,120	1,228	53,040	29,467
Galveston	Texas	28	− 2	91	33	1,233	685	29,592	16,440	2,639	1,466	63,336	35,186
Houston	Texas	24	− 4	96	36	1,390	772	33,360	18,533	2,080	1,155	49,920	27,733
Laredo	Texas	29	− 2	103	39	780	433	18,720	10,400	2,755	1,530	66,120	36,733
Lufkin	Texas	24	− 4	98	37	1,800	1,000	43,200	24,000	2,060	1,144	49,440	27,467
San Angelo	Texas	20	− 7	103	38	2,400	1,333	57,600	32,000	2,082	1,156	49,968	27,760
San Antonio	Texas	22	− 6	99	37	1,580	878	37,920	21,067	2,005	1,114	48,120	26,733
Tyler	Texas	17	− 8	98	37	2,200	1,222	58,800	32,667	1,855	1,030	44,520	24,733
Victoria	Texas	28	− 2	98	37	1,125	625	27,000	15,000	2,500	1,389	60,000	33,333
Waco	Texas	16	− 9	101	38	2,025	1,125	48,600	27,000	2,195	1,215	52,680	29,267
Wichita Falls	Texas	9	−13	103	39	3,025	1,680	72,600	40,333	2,050	1,138	48,200	26,778
Ogden	Utah	7	−14	92	33	5,538	3,077	132,912	73,840	817	453	19,608	10,893
Salt Lake City	Utah	− 2	−19	94	34	5,866	3,259	140,784	78,213	877	487	21,048	11,693
Burlington	Vt	−18	−28	85	29	7,865	4,369	188,760	104,867	284	158	6,816	3,787
Lynchburg	Va	10	−12	94	34	4,153	2,307	99,672	55,373	661	1,367	15,864	8,813
Norfolk	Va	18	− 8	94	34	3,454	1,919	82,896	46,053	990	550	23,760	13,200
Richmond	Va	10	−12	96	36	3,955	2,197	94,920	52,733	1,006	559	24,144	13,413
Roanoke	Va	9	−13	94	34	4,152	2,307	99,648	55,360	800	444	19,200	10,667
Everette	Wash	19	− 7	82	28	5,941	3,300	142,584	79,213	54	30	1,296	720
Olympia	Wash	15	− 9	85	29	5,501	3,056	132,024	73,346	125	69	3,000	1,667
Seattle	Wash	22	− 6	81	27	4,438	2,465	106,512	59,173	91	50	2,184	1,213
Spokane	Wash	− 5	−21	93	34	6,852	3,807	164,448	91,360	373	207	6,952	3,862
Tacoma	Wash	0	−18	85	29	4,866	2,703	116,784	64,880	122	67	2,928	1,627
Yakima	Wash	− 1	−18	94	34	5,845	3,247	140,280	77,933	551	306	13,224	7,397
Charleston	WVa	1	−17	92	33	4,417	2,453	106,008	58,893	779	432	18,698	10,388
Huntington	WVa	4	−16	95	35	4,073	2,485	97,752	54,307	979	543	23,496	13,053
Martinsburg	WVa	6	−14	96	37	4,200	2,333	100,800	56,000	810	450	19,440	10,800
Morgantown	WVa	0	−18	90	32	5,095	2,830	122,280	67,933	482	267	11,568	6,427
Parkersburg	WVa	2	−17	93	34	4,750	2,639	114,000	63,333	632	351	15,168	8,427
Eau Clair	Wis	−15	−26	90	32	7,970	4,428	191,280	106,267	300	167	7,200	4,000
Green Bay	Wis	−16	−27	88	31	8,259	4,588	198,216	110,120	264	147	6,336	3,220

TABLE 24 (Continued) (Sheet 4 of 5)

DESIGN TEMPERATURES, DEGREE DAYS AND DEGREE HOURS
HEATING AND COOLING – UNITED STATES CITIES – ENGLISH AND METRIC UNITS

CITY	STATE	DESIGN TEMPERATURES				WINTER HEATING DEG × TIME				SUMMER COOLING DEG × TIME			
		Winter		Summer		Δ ° × Days		Δ ° × Hours		Δ ° × Days		Δ ° × Hours	
		°F	°C	°F	°C	English	Metric	English	Metric	English	Metric	English	Metric
La Crosse	Wis	−12	−24	90	32	7,650	4,250	183,600	102,000	401	223	9,624	5,347
Madison	Wis	−13	−25	92	33	7,417	4,120	178,008	98,893	485	269	11,640	6,467
Milwaukee	Wis	−11	−24	90	32	7,205	4,002	172,920	96,067	358	199	8,592	4,773
Wausau	Wis	−18	−28	89	32	8,494	4,719	203,858	113,254	300	167	7,200	4,000
Casper	Wyo	−20	−29	92	33	7,638	4,243	183,312	101,840	546	303	14,304	7,947
Cheyenne	Wyo	−15	−27	89	32	7,562	4,201	181,488	100,287	370	206	8,880	4,933
Rock Springs	Wyo	−10	−23	86	30	8,473	4,707	203,352	112,973	300	167	7,200	4,000
Sheridan	Wyo	−21	−30	95	35	7,903	4,390	189,672	105,373	589	327	14,136	7,853

TABLE 24 (Continued) (Sheet 5 of 5)

DESIGN TEMPERATURES, DEGREE DAYS AND DEGREE HOURS
HEATING AND COOLING – CANADIAN CITIES – ENGLISH AND METRIC UNITS

CITY	STATE	DESIGN TEMPERATURES				WINTER HEATING DEG × TIME				SUMMER COOLING DEG × TIME			
		Winter		Summer		Δ ° × Days		Δ ° × Hours		Δ ° × Days		Δ ° × Hours	
		°F	°C	°F	°C	English	Metric	English	Metric	English	Metric	English	Metric
Calgary	Alta	−28	−33	82	23	12,060	6,700	289,440	160,800	225	125	5,400	3,000
Edmington	Alta	−29	−34	82	23	15,480	8,600	371,520	206,400	110	61	2,640	1,464
Vancover	BC	15	− 9	77	25	5,600	3,111	134,400	74,664				
Victoria	BC	20	− 7	76	24	5,500	3,055	132,000	73,320				
Churchill	Man	−43	−42	75	24	17,150	9,528	411,600	228,672				
Winnipeg	Man	−31	−35	86	30	11,000	6,111	264,000	146,664	130	72	3,120	1,728
Moncton	NB	−16	−27	81	27	8,500	4,722	204,000	113,328	190	105	4,560	2,520
St. John	NB	−15	−26	77	25	7,600	4,222	182,400	101,328	60	33	1,440	792
Gander	Nfld	− 5	−21	81	27	8,700	4,833	208,800	115,992	75	41	1,800	984
St. Johns	Nfld	2	−17	76	24	8,300	4,611	199,200	110,664				
Halifax	NS	0	−18	78		7,610	4,227	182,640	101,448				
Ft. Smith	NWT	−49	−45	75	24	19,000	10,555	456,000	253,320				
Ft. William	Ont	−27	−33	81	27	10,490	5,928	251,760	142,272	70	39	1,680	936
Hamilton	Ont	0	−18	87	31	7,120	3,955	170,880	94,920	425	236	10,200	5,664
London	Ont	− 9	−23	85	29	7,420	4,122	178,080	98,928	430	238	10,320	5,712
Ottawa	Ont	−11	−24	86	30	8,800	4,889	211,200	117,336	390	216	9,360	5,184
Toronto	Ont	−10	−23	87	31	7,375	4,097	177,000	98,328	325	180	7,800	4,320
Windsor	Ont	− 1	−18	88	31	6,800	3,777	163,200	90,648	490	272	11,760	6,528
Charlottetown	PEI	− 6	−21	78	26	8,550	4,750	205,200	114,000				
Montreal	PQ	−16	−27	85	29	8,400	4,667	201,600	112,008	280	155	6,720	3,720
Quebec	PQ	−19	−28	81	27	9,450	5,250	226,800	126,000	220	122	5,280	2,928
Saskatoon	Sask	−34	−37	86	30	10,700	5,944	256,800	142,656	380	211	9,120	5,064
Dawson	YT	−47	−44	73	23	15,300	8,500	367,200	204,000				

TABLE 25

HEAT TRANSFER FROM HIGHER TEMPERATURE SURFACE
TO LOWER TEMPERATURE AIR AT NATURAL CONVECTION
(0 Velocity) AND WITH WIND AT LISTED VELOCITIES
TEMPERATURE DIFFERENCE BETWEEN SURFACE AND
THE OUTSIDE AIR AS LISTED

ENGLISH UNITS

Temp. Diff. Surf. to Air °F	HEAT TRANSFER FROM SURFACE TO AIR IN BTU/FT², HR — WIND IN MILES PER HOUR																
	0	1	2	3	4	5	6	7	8	9	10	12	14	16	18	20	25	30
1	0.3	0.4	0.5	0.6	0.7	0.8	0.9	0.9	1.0	1.0	1.1	1.2	1.3	1.3	1.4	1.5	1.7	1.8
2	0.7	1.1	1.3	1.5	1.7	1.9	2.0	2.2	2.3	2.5	2.6	2.8	3.0	3.2	3.4	3.6	4.0	4.4
3	1.2	1.8	2.2	2.6	2.9	3.2	3.4	3.7	3.9	4.1	4.3	4.7	5.1	5.4	5.7	6.0	6.7	7.3
4	1.7	2.5	3.1	3.7	4.1	4.5	4.9	5.3	5.6	5.9	6.2	6.7	7.3	7.7	8.3	8.6	9.6	10.5
5	2.2	3.3	4.2	4.8	5.4	6.0	6.3	7.0	7.4	7.8	8.2	8.9	9.6	10.2	10.8	11.4	12.7	13.9
6	2.8	4.2	5.2	6.1	6.9	7.5	8.2	8.8	9.3	9.8	10.3	11.2	12.1	12.9	13.6	14.3	15.9	17.2
7	3.4	5.1	6.4	7.4	8.3	9.1	9.9	10.6	11.3	11.9	12.5	13.6	14.6	15.6	16.5	17.4	19.3	21.1
8	4.0	6.0	7.5	8.7	9.8	10.8	11.7	12.6	13.3	14.1	14.8	16.1	17.3	18.4	19.4	20.5	22.8	25.0
9	4.6	7.0	8.7	10.1	11.4	12.4	13.6	14.5	15.4	16.3	17.1	18.6	20.0	21.3	22.6	23.7	26.4	28.9
10	5.3	8.0	9.9	11.5	13.0	14.3	15.5	16.6	17.6	18.6	19.5	21.2	22.9	24.3	25.8	27.1	30.1	33.0
11	6.0	9.0	11.2	13.0	14.6	16.1	17.4	18.7	19.9	21.0	22.0	24.0	25.8	27.5	29.0	30.5	34.0	37.2
12	6.6	10.0	12.4	14.5	16.3	17.9	19.5	20.8	22.1	23.3	24.5	26.7	28.7	30.6	32.4	34.0	37.9	41.4
13	7.3	11.0	13.8	16.0	18.0	19.8	21.5	23.0	24.4	25.8	27.1	29.5	31.7	33.8	35.8	37.6	41.8	45.8
14	8.0	12.1	15.1	17.6	19.8	21.7	23.6	25.3	26.9	28.4	29.8	32.4	34.9	37.1	39.3	41.3	46.0	50.3
15	8.7	13.2	16.5	19.1	21.6	23.7	25.7	27.5	29.2	30.8	32.2	35.3	37.9	40.4	42.8	44.9	50.0	54.7

METRIC UNITS

Temp. Diff. Surf. to Air °C	HEAT TRANSFER FROM SURFACE TO AIR IN WATTS/m² — WIND IN KILOMETERS PER HOUR																
	0	1	2	3	4	5	6	7	8	9	10	12	14	16	18	20	25	30
1	2.0	2.6	3.1	3.6	4.0	4.4	4.7	5.0	5.3	5.6	5.9	6.4	6.8	7.2	7.7	7.8	8.9	9.8
2	4.7	6.2	7.5	8.6	9.5	10.3	11.2	11.9	12.6	13.2	13.9	15.1	16.2	17.2	18.1	18.6	21.2	23.1
3	7.7	10.3	12.4	14.2	15.8	17.2	18.6	19.8	21.1	22.1	23.1	25.1	26.9	28.6	30.2	30.9	35.3	38.4
4	11.1	14.8	17.8	20.4	22.6	24.6	26.6	28.3	30.0	31.6	33.1	35.9	38.5	41.0	43.3	44.3	50.5	55.0
5	14.6	19.6	23.5	26.9	29.9	32.5	35.1	37.5	39.7	41.7	43.7	47.5	51.0	54.1	57.2	58.5	66.7	72.7
6	18.4	24.6	29.6	33.8	37.6	40.8	44.1	47.1	49.8	52.4	55.0	59.6	64.0	66.0	71.9	73.5	83.8	91.4
7	22.3	29.9	35.8	41.0	45.5	50.0	53.5	57.0	60.4	63.6	66.6	72.3	77.5	82.4	87.1	89.1	101.6	110.7
8	26.2	35.3	42.2	48.2	53.6	58.5	63.0	67.1	71.1	74.9	78.7	85.1	91.3	97.1	102.5	105.3	120.1	130.9
9	30.5	40.9	49.1	56.1	62.3	67.7	73.2	78.1	82.7	87.1	91.2	99.0	106.2	112.9	119.3	122.0	139.1	151.6
10	35.8	46.6	56.0	64.0	71.1	77.3	83.5	89.1	94.3	99.3	104.1	112.9	121.1	128.8	136.0	139.2	158.7	173.0

TABLE 26

HEAT TRANSFER FROM HIGHER TEMPERATURE SURFACE TO LOWER TEMPERATURE AMBIENT AIR AND SURFACES BY RADIATION BASED ON 0°F (−18°C) AMBIENT SURROUNDINGS

ENGLISH UNITS

TEMP DIFF SURFACE TO AMB °F	HEAT TRANSFER FROM SURFACE SURROUNDINGS IN BTU/FT², HR SURFACE EMITTANCE									
	0.1	0.2	0.3	0.4	0.5	0.6	0.7	0.8	0.9	1.0
1	0.07	0.14	0.21	0.28	0.35	0.41	0.48	0.55	0.62	0.69
2	0.14	0.28	0.41	0.55	0.69	0.82	0.96	1.10	1.23	1.37
3	0.21	0.41	0.62	0.82	1.03	1.24	1.44	1.65	1.85	2.06
4	0.28	0.55	0.83	1.10	1.38	1.65	1.93	2.20	2.48	2.75
5	0.39	0.69	1.04	1.38	1.73	2.07	2.42	2.76	3.11	3.45
6	0.42	0.83	1.25	1.66	2.08	2.49	2.91	3.32	3.74	4.15
7	0.49	0.97	1.46	1.94	2.43	2.92	3.40	3.89	4.37	4.86
8	0.56	1.11	1.67	2.23	2.79	3.34	3.90	4.46	5.01	5.57
9	0.63	1.26	1.89	2.52	3.15	3.77	4.40	5.03	5.66	6.29
10	0.70	1.40	2.10	2.80	3.50	4.21	4.91	5.61	6.31	7.01
11	0.77	1.54	2.31	3.09	3.87	4.64	5.41	6.18	6.96	7.77
12	0.85	1.69	2.54	3.38	4.27	5.07	5.92	6.77	7.61	8.46
13	0.92	1.84	2.76	3.68	4.60	5.52	6.44	7.36	8.28	9.20
14	0.99	1.99	2.98	3.97	4.97	5.96	6.95	6.94	8.94	9.93
15	1.07	2.13	3.20	4.26	5.33	6.40	7.46	8.53	9.59	10.66

METRIC UNITS

TEMP DIFF SURFACE TO AMB °C	HEAT TRANSFER FROM SURFACE TO SURROUNDINGS IN WATTS/m² SURFACE EMITTANCE									
	0.1	0.2	0.3	0.4	0.5	0.6	0.7	0.8	0.9	1.0
1	0.42	0.82	1.26	1.68	2.10	2.51	2.93	3.53	3.77	4.19
2	0.84	1.68	2.52	3.37	4.22	5.06	5.90	6.74	7.59	8.43
3	1.27	2.54	3.81	5.08	6.36	7.63	8.90	10.18	11.45	12.72
4	1.71	3.42	5.11	6.82	8.53	10.23	11.94	13.64	15.35	17.05
5	2.14	4.28	6.43	8.58	10.72	12.86	15.00	17.15	19.30	21.44
6	2.58	5.16	7.74	10.33	12.92	15.50	18.08	20.66	23.24	25.83
7	3.04	6.08	9.11	12.15	15.13	18.23	21.27	24.30	27.34	30.38
8	3.49	6.98	10.47	13.97	17.45	20.95	24.44	27.94	31.43	34.92
9	3.95	7.90	11.86	15.81	19.75	23.71	27.66	31.62	35.57	39.52
10	4.44	8.88	13.25	17.67	22.09	26.50	30.92	35.34	39.75	44.17

TABLE 27

WINTER TEMPERATURE, EFFECT OF WIND ON HEAT LOSS

Wind effects the heat loss of a heated building by change of the outer surface thermal resistance and by infiltration.

The basic formula for heat transfer from a surface under natural convection (no forced air movement—such as wind), in English Units is:

$$Q_c = 0.296\,(t_s - t_a)^{5/4} \qquad (9)$$

where Q_c = Heat transfer by natural convection in Btu/ft², hr
t_s = The temperature of the surface in °F
t_a = The temperature of the outside air in °F

In Metric Units

$$Q_{cm} = 1.957\,(t_{sm} - t_{am})^{5/4} \qquad (9a)$$

where Q_{cm} = Heat transfer by natural convection in W/m²
t_{sm} = The temperature of the surface in °C
t_{am} = The temperature of the surface in °C

In English Units:

Under the effect of wind at velocity V, the rate of heat loss per unit area becomes

$$Q_{cvf} = 0.296\,(t_s - t_a)^{5/4}\sqrt{\frac{V + 68.9}{68.9}} \qquad (10)$$

where V = Velocity of forced convection (wind) in ft/min
Q_{cvf} = Heat moved from surface by forced convection (wind) in Btu/ft², hr

In Metric Units:

$$Q_{mcvf} = 1.957\,(t_{ms} - t_{ma})^{5/4}\sqrt{\frac{196.85V + 68.90}{68.9}} \qquad (10a)$$

where V_m = Velocity of air in meters per second
Q_{mcvf} = Heat moved from surface by forced convection (wind) in Watts/m²

These equations indicate that as the velocity of the air across a surface increases, the heat transfer from that surface to air at a lower temperature also increases. In the case of building walls and roofs, this wind (or air across the surface) tends to reduce the temperature of the higher temperature. For example, a wall under natural convection conditions may have a temperature of 9 °F above the ambient air temperature but with a 10 mph wind, its surface may only be 6 °F above the ambient air temperature.

Table 26 gives heat transfer from surfaces to air at lower temperatures under natural convection and at various wind velocities. This table gives the heat transfer in English and Metric Units.

The wind affects the quantity of the heat transferred to ambient surroundings by radiation. This is due to the effect of convection heat transfer causing the surface temperature to become closer to ambient temperature. For this reason, the radiation heat transfer must be combined with the convection heat transfer to obtain outer surface total heat transfer.

Table 27 is presented to provide the radiation from a surface to lower temperature surroundings at 0 °F (−18 °C). This table provides radiation heat transfer at various temperature differences and various surface emittances ε, in both English and Metric Units. This table was calculated using Equations 1 and 1a for radiation from a surface to its surroundings.

The ambient air temperature used to calculate the previous tables was 0 °F (−18 °C). This temperature was selected as being the one most frequently used in the United States for design of heating systems. This temperature affects only the radiation and not the convection heat transfer tables. A variation from this selected temperature makes little effect on combined radiation and convection heat transfer. Convection heat transfer is affected directly by the temperature difference and air velocity across the surface so these must be considered in all calculations.

Radiation from surface to ambient air is affected by temperature level, temperature difference, and surface emittance. However, the temperature level does not affect the heat transfer as directly as temperature difference and emittance. For this study the temperature level of the surrounding air was chosen as 0 °F (−18 °C). The difference, if a level of 20 °F (−6 °C) was chosen, is only 1.03 Btu/ft², hr at $\varepsilon = 1.0$ and $\Delta t = 15$ °F (8 °C), based on radiation only. When combined with the convection heat transfer, the error would be less than 3% for most heat loss transfer calculations, based on surface condition only.

Surface emittance ε does have a direct effect on the heat transfer by radiation. For this reason, selection of value of ε for outer surfaces of heated buildings must be considered for the calculation of the radiation part of heat loss. Table 28 lists the value of surfaces most common to residences and buildings. From this the outer surface emittance ε was selected as 0.8 for the calculations of combined convection and radiation heat transfer from a surface. This heat transfer by convection and radiation at various temperature differences and wind velocities is shown in Table 29.

The preceding tables are based on the wind velocity and the temperature difference between surface and ambient air and surroundings. In the case of wind velocity this factor is generally established by the design conditions for the individual area. Tables 31 and 32 provide suggested design wind velocities for various North American cities.

The temperature difference between outer surface and the ambient air and surroundings is an unknown factor. However, it can be determined by estimating the heat transfer from inner air to outer air. This is done by dividing the temperature difference by the total estimated thermal resistance. For this reason, it is necessary to establish surface thermal resistance by use of the information established in Table 29.

The total heat transfer from surface to air is by convection and radiation. The heat transferred by convection used in these tables

EMITTANCE (∈) OF VARIOUS SURFACES
(AVERAGED OVER INFRARED WAVE LENGTHS,
EXCEPT WHERE NOTED)

SURFACE	Emittance (∈)	Reflectance (r)
Aluminum—(24 ST)—Polished	0.08 to 0.09*	0.91 to 0.97*
Aluminum—(24 ST)—Oxidized	0.09 to 0.26*	0.11 to 0.74*
Asphalt	0.90 to 0.98	0.10 to 0.02
Brick—Red, Yellow & Buff	0.85 to 0.95	0.15 to 0.05
Cinder Block	0.90 to 0.98	0.10 to 0.02
Concrete & Concrete Block	0.85 to 0.95	0.15 to 0.05
Copper—Polished	0.03 to 0.07	0.97 to 0.93
Copper—Oxidized	0.70 to 0.80	0.30 to 0.20
Galvanized Steel—Bright	0.23 to 0.30	0.77 to 0.70
Glass, Window	0.90 to 0.95	0.10 to 0.05
Paint—Black, Dark Gray	0.85 to 0.95	0.15 to 0.05
Paint—Medium shades	0.80 to 0.91	0.20 to 0.09
Paint—White-Dirty	0.70 to 0.85	0.30 to 0.15
Paint—Aluminum, White-Clean	0.45 to 0.70*	0.55 to 0.30*
Paper	0.85 to 0.98	0.15 to 0.02
Plaster	0.85 to 0.95	0.15 to 0.05
Slate	0.90 to 0.98	0.10 to 0.02
Stone	0.85 to 0.93	0.15 to 0.05
Tar	0.93 to 0.98	0.07 to 0.02
Tile	0.85 to 0.93	0.15 to 0.07
Whitewash	0.85 to 0.95	0.15 to 0.05
Wood, Stained	0.90 to 0.97	0.10 to 0.03
Wood, Painted	As noted for paint	As noted for paint

*These values are not correct for wavelengths from 0.3 to 0.8 microns.

All values are based on ambient outside air temperatures and (except where noted by *) over entire total range of visible and infrared heat wavelengths.

TABLE 28

was calculated by Equations 10 and 10a. The heat transferred by convection was designated by Q_{cvt} (English Units) or Q_{mcvt} (Metric Units). The radiation was calculated by Equations 1 and 1a. The heat transferred by radiation was designated as Q_r (English Units) and Q_{mr} (Metric Units). The total heat transferred the sum of these.

In English Units:

$$Q_t = Q_{cvt} + Q_r$$

also, when t_s = Surface temperature °F
t_a = Air and surrounding temperature °F
R_o = Surface thermal resistance

then:

$$Q_t = \frac{t_s - t_a}{R_o} = Q_{cvt} + Q_r$$

and

$$R_o = \frac{t_s - t_a}{Q_{cvt} + Q_r} \quad (15)$$

In Metric Units:

Q_{mcvt} = Surface temperature °C
Q_{ma} = Air and surrounding temperature °C
R_{mo} = Surface thermal resistance

then:

$$Q_{mt} = \frac{t_{ms} - t_{ma}}{R_{mo}} = Q_{mcvt} + Q_{mr}$$

and

$$R_{mo} = \frac{t_{ms} - t_{ma}}{Q_{mcvt} + Q_{mr}} \quad (15a)$$

Using these equations and values given in Table 29, the thermal resistances of outer surfaces of buildings were calculated in reference to wind velocity and surface temperature differences. The results of calculated surface resistances are tabulated in Table 30. This information will be used to calculate the heat transfer through walls and roofs as shown in the following examples.

The heat transfer through a wall or ceiling and roof is equal to the total temperature difference divided by the total sum of the heat transfer resistances. This sum consists of the resistance from the inside air to the wall, plus the resistance of the wall plus the resistance of the surface.

In English Units:

If R_o = Outside surface thermal resistance (English Units)
R_w = Wall thermal resistance (English Units)
R_i = Inside surface thermal resistance (English Units) inside air to wall
t_o = Temperature of outside air in °F
t_i = Temperature of inside air in °F
Q = Heat transfer in Btu/ft², hr

then:

$$Q = \frac{t_i - t_o}{R_o + R_w + R_i} \text{ Btu/ft}^2\text{, hr} \quad (16)$$

In Metric Units:

If R_{mo} = Outside surface thermal resistance (Metric Units)
R_{mw} = Wall thermal resistance (Metric Units)
R_{mi} = Inside surface thermal resistance (Metric Units)
t_{mo} = Temperature of outside air in °C
t_{mi} = Temperature of inside air in °C
Q_m = Heat transfer in W/m²

then:

$$Q_m = \frac{t_{mi} - t_{mo}}{R_{mo} + R_{mw} + R_{mi}} \quad W/m^2 . \tag{16a}$$

If the outside surface temperature is put into the equations, then,

In English Units:

$$Q = \frac{t_i - t_s}{R_i + R_w} = \frac{t_s - t_a}{R_o} \tag{17}$$

when t_s = Outer surface temperature in °F.

In Metric Units:

$$Q_m = \frac{t_{mi} - t_{ms}}{R_{mi} + R_{mw}} = \frac{t_{ms} - t_{ma}}{R_{mo}} \tag{17a}$$

These equations show that the heat flow to the outer surface must equal the heat flow from that surface under static conditions.

Using these equations, it is then possible to graph

$$\frac{t_i - t_s}{R_i + R_w} = \frac{t_s - t_a}{R_o} \quad \text{or} \quad \frac{t_{mi} - t_{ms}}{R_{mi} + R_{mw}} = \frac{t_{ms} - t_{ma}}{R_{mo}}$$

for various values of

$$\frac{t_i - t_s}{R_i + R_w} \quad \text{or} \quad \frac{t_{mi} - t_{ms}}{R_{mi} - R_{mw}}$$

and

$$\frac{t_s - t_a}{R_o} \quad \text{or} \quad \frac{t_{ms} - t_a}{R_{mo}} .$$

For values of t_i = 70 °F (t_{mi} = 21 °C) and t_a = 0 °F (t_{ma} = −18 °C) with values $R_i + R_w$ or $R_{mi} + R_{mw}$ being assumed and wind velocity also assumed, it is possible to plot the relationship of these factors so as to determine $\Delta t_{sa} = t_s - t_a$ or $\Delta t_{msa} = t_{ms} - t_{ma}$. This graph is shown in Figure 43. After temperature difference Δt (or Δt_m) has been established, based on given wall resistances, wind factors R_o or R_{mo} can be determined using Figure 44. As heat transfer changes with wind it is necessary to establish design wind velocities. The wind velocities for design are given in Table 31 and 32. The design wind velocities are *not maximum* wind velocities which may occur at the listed city during winter. The listed velocities are average winds by which the heat loss can be calculated so that the heating system design will be practical.

Examination of Tables 29 and 30 shows that as wind velocity increases, it has less effect on the overall heat transfer through the building, wall or roof. This is because its maximum effect would be to cause the building surface temperature to be identical to the outside ambient air temperature. Also, it can be observed that as the thermal resistance of the structural partition increases, the percentage effect on heat loss is less. Because of these factors, it is possible to tabulate outside surface resistances of building structures based on wind velocities and thermal resistance through the building, wall or roof. Although there will be slight differences of these resistance values due to the level of temperature and surface emittance, the listed resistance values are suitable for most building and residential designs. Thermal resistances for the exterior surfaces of buildings facing winds up to 12 miles per hour are given in Table 33.

Although it was stated that increased velocity of wind would have less and less effect on the heat transfer through tight structural walls and roofs, this would not be true on heat loss due to air leakage around windows, window frames, doors, etc. The heat transfer from air passing through cracks and openings increases in almost direct proportion to wind velocity. For this reason, the use of weatherstripping, double windows and doors is very important in reduction of heat loss.

To illustrate the effect of wind loss on walls of various thermal resistance the following examples are presented. The first example will be one of a wall of frame construction consisting of siding, building paper, wood sheathing on 2″ × 4″'s with 3/8″ gypsum lath (no insulation). The first part will present the heat loss through this wall under still air conditions and the second part with wind velocity of 12 mph, to illustrate the effect of wind on heat loss.

SURFACE TEMPERATURE, TEMPERATURE DIFFERENCE AND HEAT TRANSFER BASED ON 0°F (−17.8°C) AMBIENT AIR TEMP. AND INSIDE AIR TEMP. 70°F (21.1°C) ENGLISH AND METRIC UNITS

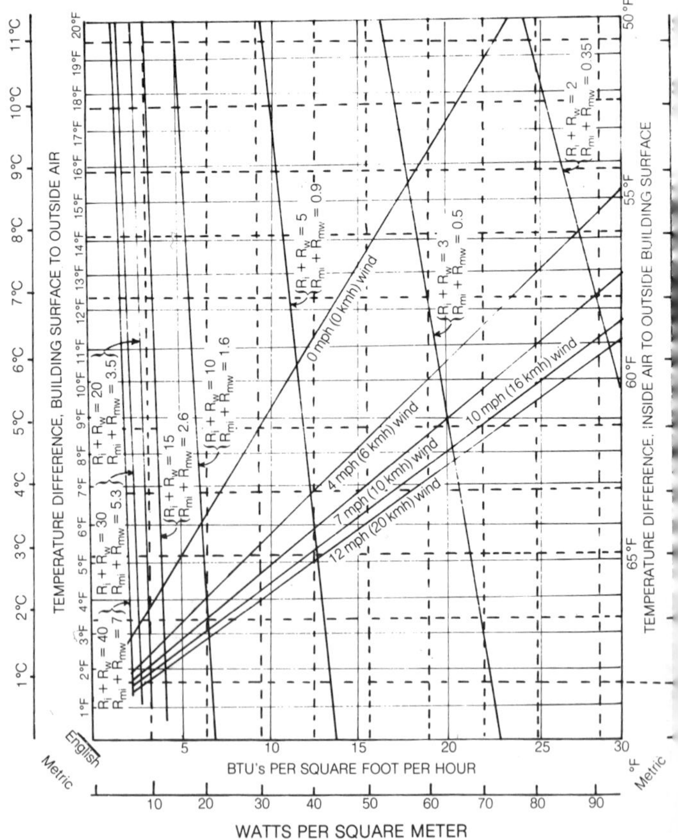

Figure 43

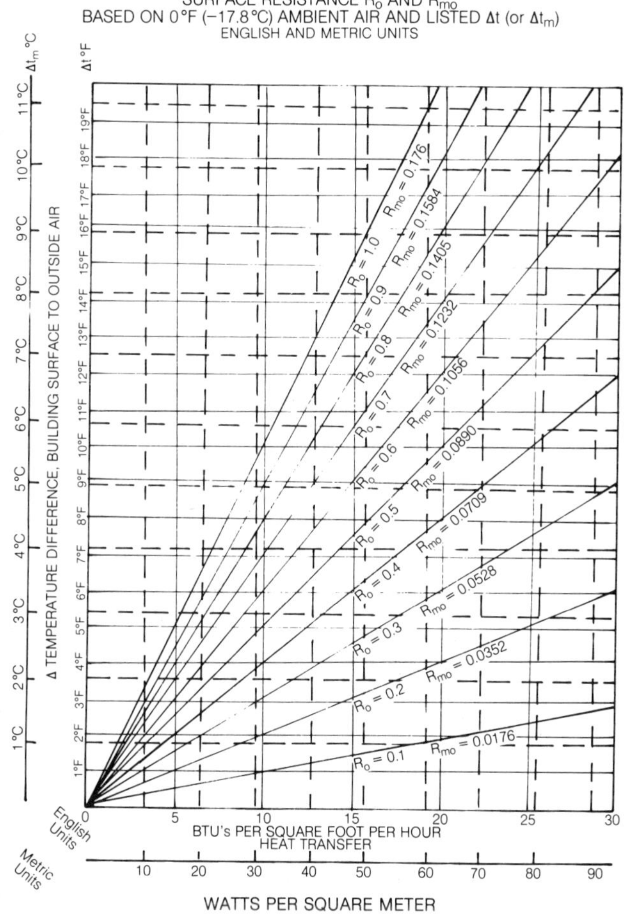

Figure 44

Example 10—*English Units*

The thermal resistance of the wall is $R_w = 3.2$
The thermal resistance of the inner surface is $R_i = 0.9$
Total resistance of wall, $R_w + R_i = 3.2 + 0.9 = 4.1$
Inside Air Temperature is $t_i = 70\,°F$, outside air temperature $t_o = 0\,°F$

From Table 33 using $R_w + R_i = 4.1$ at 0 mph wind velocity, one finds the thermal resistance to the outside surface to be $R_o = 0.89$ (by interpolation between 3 and 5); at 12 mph, R_o is 0.40.

At 0 mph wind, – ($R_o = 0.89$)

$$Q = \frac{t_i - t_a}{R_o + R_w + R_i} = \frac{70 - 0}{0.89 + 3.2 + 0.9} = \frac{70}{4.89} = 14.0 \text{ Btu/ft}^2\text{, hr}$$

At 12 mph, ($R_o = 0.4$)

$$Q = \frac{t_i - t_a}{R_o + R_w + R_i} = \frac{70 - 0}{0.4 + 3.2 + 0.9} = \frac{70}{4.5} = 15.56 \text{ Btu/ft}^2\text{, hr}$$

Thus a 12 mph wind on outer surface of the wall increases the heat loss by approximately 1.6 Btu/ft² per hour.

If this same wall is insulated with 3 1/2″ glass fiber insulation with aluminum foil vapor retarder under gypsum wall board, its thermal resistance would be increased by 14. The R_w would then be 17.2.

At 0 mph wind, $R_o = 1.02$ (From Table 33)

$$Q = \frac{t_i - t_a}{R_o + R_w + R_i} = \frac{70 - 0}{1.02 + 17.2 + .9} = \frac{70}{19.12} = 3.7 \text{ Btu/ft}^2\text{, hr}$$

At 12 mph, $R_o = 0.50$ (From Table 33)

$$Q = \frac{t_i - t_a}{R_o + R_w + R_i} = \frac{70 - 0}{0.50 + 17.2 + .9} = \frac{70}{18.6} = 3.8 \text{ Btu/ft}^2\text{, hr}$$

Thus a 12 mph wind on outer surface of this well insulated wall only increases the heat loss 3.8 − 3.7 = 0.1 Btu/ft² per hour.

Example 10a—*Metric Units*

The thermal resistance of the uninsulated wall is $R_{mw} = 0.563$. The thermal resistance of inner surface is $R_{mi} = 0.152$. The total resistance of the wall $R_{mw} + R_{mi} = 0.563 + 0.152 = 0.715$. Inside air temperature, t_{mi} is 21 °C and outside air temperature $t_{ma} = -18\,°C$.

From Table 33, using total wall resistance 0.715 at 0 km/hr one finds that R_{mo} is 0.157; at 20 km/hr, $R_{mo} = 0.071$.

At 0 km/hr wind, $R_{mo} = 0.157$

$$Q_m = \frac{t_{mi} - (t_{ma})}{R_{mo} + R_{mw} + R_{mi}} = \frac{21 - (-18)}{0.157 + 0.563 + 0.152} = \frac{39}{0.872}$$

$$= 44.72 \text{ W/m}^2$$

At 20 km/hr, $R_{mo} = 0.071$

$$Q_m = \frac{t_{mi} - (t_{mo})}{R_{mv} + R_{mw} + R_{mi}} = \frac{21 - (-18)}{0.071 + 0.563 + 0.152} = \frac{39}{0.786}$$

$$= 49.62 \text{ W/m}^2$$

Thus a 20 km/hr wind on the outer surface of the wall increases heat loss 49.62 − 44.72 = 4.9 W/m².

If this same wall is insulated with 8.9 cm of glass fiber insulation, with aluminum foil vapor retarder under the gypsum wall board, its thermal resistance would be increased by 2.465. Thus total resistance of wall is 2.465 + 0.563 = 3.028.

From Table 33, the thermal resistance of surface with $R_{mw} + R_{mi} = 3.18$ and wind of 0 km/hr, $R_{mo} = 0.183$.

At 0 km/hr wind, $R_{mo} = 0.179$

$$Q_m = \frac{t_{mi} - t_{ma}}{R_{mo} + R_{mw} + R_{mi}} = \frac{21 - (-18)}{0.183 + 2.465 + 0.563} = \frac{39}{3.21}$$

$$= 12.15 \text{ W/m}^2$$

HEAT TRANSFER FROM HIGHER TEMPERATURE SURFACE
TO LOWER TEMPERATURE AIR AND SURROUNDINGS
BY CONVECTION AND RADIATION, BASED ON 0°F (−18°C) AMBIENT
AIR AND SURROUNDINGS AND SURFACE EMITTANCE = 0.8

BRITISH UNITS $Q_{cvf} + Q_r = Q$

Temp. Diff. Surf. to Air °F	HEAT TRANSFER FROM SURFACE TO SURROUNDINGS IN Btu per ft² per hr — WIND IN MILES PER HOUR																
	0	1	2	3	4	5	6	7	8	9	10	12	14	16	18	20	25	30
1	0.8.	1.0	1.1	1.2	1.2	1.3	1.4	1.4	1.5	1.6	1.6	1.7	1.8	1.8	1.9	2.0	2.2	2.4
2	1.8	2.2	2.4	2.6	2.8	3.0	3.1	3.3	3.4	3.6	3.7	3.9	4.1	4.3	4.5	4.7	5.1	5.5
3	2.8	3.4	3.9	4.3	4.6	4.8	4.9	5.1	5.4	5.6	6.0	6.4	6.8	7.1	7.4	7.7	8.4	9.0
4	3.9	4.7	5.3	5.9	6.3	6.7	7.1	7.5	7.8	8.1	8.4	8.9	9.5	9.9	10.5	10.8	11.8	12.7
5	5.0	6.1	7.0	7.6	8.6	8.8	9.1	9.7	10.2	10.6	11.0	11.7	12.4	13.0	13.6	14.1	15.4	16.6
6	6.1	7.5	8.5	9.4	10.2	10.8	11.5	12.1	12.8	13.1	13.7	14.5	15.4	16.2	16.9	17.6	19.3	20.8
7	7.3	9.0	10.1	11.1	12.2	13.0	13.8	14.5	15.2	15.8	16.4	17.5	18.5	19.5	20.4	21.3	23.2	25.0
8	8.5	10.5	12.0	13.2	14.3	15.3	16.2	17.1	17.8	18.6	19.3	20.6	21.8	22.9	23.9	25.0	27.3	29.4
9	9.6	12.0	13.7	15.1	16.4	17.4	18.6	19.5	20.4	21.3	22.1	23.6	25.0	26.3	27.6	28.8	31.4	33.9
10	10.9	13.6	15.5	17.1	18.6	19.9	21.1	22.2	23.2	24.2	25.1	26.8	28.5	29.9	31.4	32.7	35.8	38.6
11	12.2	15.2	17.4	19.2	20.8	22.3	23.6	24.9	26.1	27.2	28.1	30.2	32.0	33.7	35.2	36.7	40.2	43.4
12	13.4	16.8	19.2	21.3	23.1	24.7	26.3	27.6	28.9	30.2	31.3	33.5	35.5	37.4	39.2	40.8	44.6	48.2
13	14.7	18.4	21.2	23.4	25.4	27.2	28.9	30.4	31.8	33.2	34.5	36.9	39.1	41.2	43.2	45.0	49.2	53.1
14	15.9	20.0	23.0	25.5	27.7	29.6	31.5	33.2	34.8	36.3	37.6	40.3	42.6	45.0	47.2	49.2	53.9	58.2
15	17.2	21.7	25.0	27.6	30.1	32.2	34.2	36.0	37.7	39.3	40.9	43.8	46.4	48.9	51.3	53.4	58.6	63.3

METRIC UNITS $Q_{mcvf} + Q_{mr} = Q_m$

Temp. Diff. Surf. to Air °C	HEAT TRANSFER FROM SURFACE TO SURROUNDINGS IN Watts per m² — WIND IN KILOMETRES PER HOUR																
	0	1	2	3	4	5	6	7	8	9	10	12	14	16	18	20	25	30
1	5.5	6.1	6.6	7.1	7.5	7.9	8.2	8.5	8.8	9.1	9.4	9.9	10.3	10.7	11.2	11.4	12.5	13.3
2	11.4	12.9	14.2	15.3	16.2	17.0	17.9	18.6	19.3	19.9	20.6	21.8	22.9	23.9	24.8	25.3	27.9	29.9
3	17.9	20.5	22.6	24.2	26.0	27.4	28.8	30.0	31.3	32.3	33.3	36.3	37.1	38.8	40.4	41.1	45.5	48.6
4	24.7	28.4	31.4	34.0	36.2	38.2	40.2	41.9	43.6	45.2	46.7	49.5	52.1	54.6	56.9	57.9	64.1	68.6
5	31.7	36.7	40.6	44.0	47.0	49.6	52.2	54.6	56.8	58.8	60.8	64.6	68.1	71.2	74.2	75.6	83.8	89.8
6	39.1	45.3	50.3	54.5	58.3	61.5	64.8	67.8	70.5	73.1	75.7	80.3	84.7	86.7	92.6	95.2	104.5	112.1
7	46.6	54.2	60.1	65.3	69.8	74.3	77.8	81.3	84.7	87.9	90.9	96.6	101.8	106.7	111.4	113.4	125.3	135.0
8	54.1	63.2	70.1	76.1	81.5	86.4	90.9	95.0	99.0	102.8	106.6	113.0	119.2	125.0	130.4	133.3	148.0	158.8
9	62.1	72.5	80.7	87.7	93.9	99.3	104.8	109.7	114.3	118.7	122.8	130.6	137.8	144.5	150.9	153.7	170.7	183.2
10	71.1	81.9	91.3	99.3	106.4	112.6	118.8	124.4	129.6	134.6	139.4	148.2	156.4	164.1	171.3	174.5	194.0	208.3

TABLE 29

At 20 km/hr, $R_{mo} = 0.090$

$$Q_m = \frac{t_{mi} - t_{ma}}{R_{mo} + R_{mw} + R_{mi}} = \frac{21 - (-18)}{0.09 + 2.465 + 0.563} = \frac{39}{3.118}$$

$$= 12.50 \text{ W/m}^2$$

Thus a 20 km/hr wind on the outer surface of the insulated wall increases the heat loss only 0.35 W/m². This, of course, is based on the wall being air tight and free of cracks or voids.

The temperature level of the outside air has very little effect on the thermal resistance of the outer surface wind factor. For this reason, the surface resistances listed in Table 33 can be used for most common levels of air temperature used in the calculation of winter heat losses. It might be noted that ASHRAE uses an English surface thermal resistance of 0.17 at 15 mph wind (Metric resistance of 0.03 at 27.8 km/hr wind) for their general calculations of thermal heat transfer through walls and ceilings regardless of temperature level, ε value, or wall thermal resistance. This value can be used for most heat loss calculations and provides a safety factor for most wall and roof constructions (other than uninsulated metal buildings).

In some construction where differential expansion and contraction may occur, then it becomes very important to determine

OUTSIDE SURFACE THERMAL RESISTANCE R_o or R_{mo}
WINTER CONDITIONS: BASED ON AIR AND SURROUNDINGS AT 0°F (−18°C) AND EMITTANCE OF OUTER SURFACE OF BUILDING MATERIAL = 0.8 $R_o = \frac{t_s - t_a}{Q}$ or $R_{mo} = \frac{t_{ms} - t_{ma}}{Q}$

BRITISH UNITS

Temp. Diff. Surf. to Air °F	THERMAL RESISTANCE R_o OF OUTER BUILDING SURFACE																	
	WIND IN MILES PER HOUR																	
	0	1	2	3	4	5	6	7	8	9	10	12	14	16	18	20	25	30
1	1.25	1.00	0.91	0.84	0.83	0.77	0.71	0.70	0.67	0.63	0.62	0.59	0.56	0.55	0.53	0.50	0.46	0.42
2	1.11	0.91	0.83	0.77	0.71	0.66	0.65	0.61	0.59	0.56	0.54	0.51	0.49	0.46	0.44	0.43	0.39	0.36
3	1.07	0.88	0.77	0.70	0.65	0.63	0.61	0.59	0.56	0.54	0.50	0.47	0.44	0.42	0.40	0.39	0.36	0.33
4	1.02	0.85	0.75	0.68	0.63	0.60	0.56	0.53	0.51	0.49	0.47	0.45	0.42	0.40	0.38	0.37	0.36	0.33
5	1.00	0.82	0.71	0.66	0.60	0.57	0.55	0.52	0.49	0.47	0.45	0.43	0.40	0.38	0.37	0.35	0.32	0.30
6	0.98	0.80	0.70	0.64	0.58	0.56	0.52	0.49	0.47	0.46	0.44	0.41	0.39	0.37	0.36	0.34	0.31	0.29
7	0.96	0.78	0.69	0.63	0.57	0.54	0.51	0.48	0.46	0.44	0.43	0.40	0.38	0.36	0.34	0.32	0.30	0.28
8	0.95	0.76	0.67	0.61	0.56	0.53	0.49	0.47	0.45	0.43	0.42	0.39	0.37	0.35	0.33	0.32	0.29	0.27
9	0.94	0.75	0.66	0.60	0.55	0.52	0.48	0.46	0.44	0.42	0.41	0.38	0.36	0.34	0.32	0.31	0.28	0.26
10	0.92	0.73	0.64	0.58	0.54	0.50	0.47	0.45	0.43	0.41	0.40	0.37	0.35	0.33	0.32	0.30	0.28	0.26
11	0.90	0.72	0.63	0.57	0.53	0.49	0.46	0.44	0.42	0.40	0.39	0.36	0.34	0.32	0.31	0.29	0.27	0.25
12	0.89	0.71	0.62	0.56	0.52	0.48	0.45	0.43	0.41	0.39	0.38	0.35	0.33	0.32	0.30	0.29	0.26	0.24
13	0.88	0.70	0.61	0.55	0.51	0.47	0.44	0.42	0.40	0.39	0.37	0.35	0.33	0.31	0.30	0.28	0.26	0.24
14	0.88	0.70	0.60	0.54	0.50	0.47	0.44	0.42	0.40	0.38	0.37	0.34	0.32	0.31	0.29	0.28	0.25	0.24
15	0.87	0.69	0.60	0.54	0.50	0.46	0.43	0.41	0.39	0.38	0.36	0.34	0.32	0.30	0.29	0.28	0.25	0.23

METRIC UNITS

Temp. Diff. Surf. to Air °F	THERMAL RESISTANCE R_{mo} OF OUTER BUILDING SURFACE																	
	WIND IN KILOMETRES PER HOUR																	
	0	1	2	3	4	5	6	7	8	9	10	12	14	16	18	20	25	30
1	0.182	0.164	0.152	0.141	0.133	0.127	0.122	0.118	0.114	0.110	0.106	0.101	0.097	0.093	0.089	0.087	0.080	0.075
2	0.175	0.155	0.141	0.131	0.123	0.118	0.112	0.108	0.103	0.101	0.097	0.092	0.087	0.084	0.081	0.079	0.072	0.067
3	0.168	0.146	0.133	0.124	0.115	0.109	0.104	0.100	0.096	0.093	0.090	0.083	0.081	0.077	0.074	0.073	0.066	0.062
4	0.162	0.140	0.127	0.118	0.110	0.105	0.099	0.095	0.092	0.088	0.086	0.081	0.077	0.073	0.070	0.069	0.062	0.058
5	0.158	0.136	0.123	0.114	0.106	0.100	0.096	0.092	0.088	0.085	0.082	0.077	0.073	0.070	0.067	0.066	0.059	0.056
6	0.153	0.132	0.119	0.110	0.103	0.097	0.093	0.088	0.085	0.082	0.079	0.075	0.071	0.069	0.065	0.063	0.057	0.054
7	0.150	0.129	0.116	0.107	0.100	0.094	0.090	0.086	0.083	0.079	0.077	0.072	0.069	0.066	0.063	0.062	0.056	0.052
8	0.147	0.127	0.114	0.105	0.098	0.093	0.089	0.084	0.081	0.078	0.075	0.071	0.067	0.064	0.061	0.060	0.054	0.050
9	0.144	0.124	0.112	0.103	0.096	0.091	0.086	0.082	0.079	0.076	0.073	0.069	0.065	0.062	0.060	0.059	0.053	0.049
10	0.141	0.122	0.110	0.100	0.094	0.089	0.084	0.080	0.077	0.074	0.072	0.067	0.064	0.061	0.058	0.057	0.051	0.048

TABLE 30

surface temperatures of the inner and outer surface materials. Table 33 will assist in the determination of outer surface temperatures and Equation 12 can be used to obtain heat transfer of inner surfaces.

SOLAR EFFECTS

In addition to the effect of outside air temperature, solar heat also affects the heat transfer from or to the inside of a residence or building. Solar energy to a surface is independent of the air temperature. However, air temperature of the surface is also influenced by convection and conduction to or from the ambient air.

DESIGN WIND VELOCITIES – MILES PER HOUR AND KILOMETRES PER HOUR
UNITED STATES CITIES – WINTER

CITY	STATE	WIND Miles per hr	km per hr
Anniston	Ala	7	10
Birmingham	Ala	7	10
Dothan	Ala	7	10
Huntsville	Ala	7	10
Mobile	Ala	10	16
Montgomery	Ala	7	10
Tuscaloosa	Ala	7	10
Anchorage	Alk	4	6
Fairbanks	Alk	4	6
Juneau	Alk	7	10
Flagstaff	Arz	4	6
Kingman	Arz	4	6
Phoenix	Arz	4	6
Tucson	Arz	4	6
Fayettsville	Ark	10	16
Ft. Smith	Ark	10	16
Little Rock	Ark	10	16
Pine Buff	Ark	7	10
Bakersfield	Calif	4	6
Burbank	Calif	4	6
Eureka	Calif	7	11
Fresno	Calif	4	6
Long Beach	Calif	4	6
Los Angeles	Calif	4	6
Oakland	Calif	4	6
Reading	Calif	4	6
Riverside	Calif	4	6
Sacramento	Calif	4	6
Salinas	Calif	4	6
San Bernardino	Calif	4	6
San Diego	Calif	4	6
San Francisco	Calif	4	6
San Jose	Calif	4	6
Santa Ana	Calif	4	6
Santa Barbara	Calif	4	6
Stockton	Calif	4	6
Colorado Springs	Colo	7	10
Denver	Colo	7	10
Grand Junction	Colo	4	6
Pueblo	Colo	4	6
Hartford	Conn	10	16
New Haven	Conn	12	20
Dover	Del	10	16
Wilmington	Del	10	16
Washington	DC	10	16
Fort Myers	Fla	10	16
Jacksonville	Fla	7	10
Miami	Fla	10	16
Orlando	Fla	7	10
Panama City	Fla	10	16
Penscaola	Fla	10	16
Tallahassee	Fla	7	10
Tampa	Fla	10	16
W. Palm Beach	Fla	10	16
Albany	Ga	7	10
Atlanta	Ga	12	16
Augusta	Ga	7	10
Columbus	Ga	7	10
Macon	Ga	7	10
Savannah	Ga	7	10
Valdosta	Ga	7	10
Boise	Ida	7	11
Coeur D'Alene	Ida	4	7
Idaho Falls	Ida	4	7
Lewiston	Ida	4	7
Pocatello	Ida	7	10
Twin Falls	Ida	7	10
Belleville	Ill	10	16
Chicago	Ill	10	16
Joliet	Ill	10	16
Peoria	Ill	10	16
Quincy	Ill	10	16
Springfield	Ill	10	16
Columbus	Ind	10	16
Evansville	Ind	10	16
Ft. Wayne	Ind	10	16
Indianapolis	Ind	10	16
So. Bend	Ind	10	16
Terre Haute	Ind	10	16
Burlington	Iowa	10	16
Cedar Rapids	Iowa	10	16
Des Moines	Iowa	10	16
Mason City	Iowa	10	16
Ottumwa	Iowa	10	16
Sioux City	Iowa	10	16
Dodge City	Kan	10	16
Hutchinson	Kan	12	20
Salina	Kan	12	20
Topeka	Kan	12	20
Wichita	Kan	12	20
Bowling Green	Ky	7	10
Corbin	Ky	7	10
Lexington	Ky	10	16
Louisville	Ky	7	10
Paducah	Ky	7	10
Alexandria	La	7	10
Baton Rouge	La	7	10
Lafayette	La	7	10
Lake Charles	La	10	16
Monroe	La	4	6
New Orleans	La	7	10
Shreveport	La	7	10
Augusta	Ma	10	16
Bangor	Ma	10	16
Portland	Ma	7	10
Baltimore	Md	10	16
Boston	Mass	12	20
Springfield	Mass	10	16
Worchester	Mass	10	16
Detroit	Mich	10	16
Flint	Mich	10	16
Grand Rapids	Mich	10	16
Lansing	Mich	10	16
Saginaw	Mich	10	16
Traverse City	Mich	10	16
Duluth	Minn	10	16
Minneapolis	Minn	7	10
Rochester	Minn	7	16
St. Cloud	Minn	7	10
St. Paul	Minn	7	10
Columbia	Mo	10	16
Jefferson City	Mo	10	16
Joplin	Mo	10	16
Kansas City	Mo	10	16
St. Joseph	Mo	10	16
St. Louis	Mo	10	16
Springfield	Mo	10	16
Billings	Mont	7	10
Butte	Mont	4	6
Great Falls	Mont	7	10
Helena	Mont	7	10
Miles City	Mont	7	10
Missoula	Mont	4	6
Ashville	NC	7	10
Charlotte	NC	7	10
Elizabeth City	NC	10	16
Fayetteville	NC	7	10
Greensboro	NC	7	10
Raleigh/Durham	NC	7	10
Wilmington	NC	7	10
Winston-Salem	NC	7	10
Bismark	ND	4	7
Dickenson	ND	7	10
Fargo	ND	7	10
Grand Forks	ND	7	10
Jamestown	ND	7	10
Grand Island	Neb	10	16
Lincoln	Neb	10	16
North Platte	Neb	10	16
Omaha	Neb	10	16
Elko	Nev	4	6
Ely	Nev	4	6
Las Vegas	Nev	4	6
Reno	Nev	4	6
Concord	NH	10	16
Manchester	NH	10	16
Atlantic City	NJ	12	20
Newark	NJ	10	16
Trenton	NJ	10	16
Albuquerque	NM	7	10
Clovis	NM	7	10
Farmington	NM	4	7
Roswell	NM	7	10
Santa Fe	NM	7	10
Turcumcari	NM	7	10
Albany	NY	7	10
Binghamton	NY	7	10
Buffalo	NY	10	16
Elmira	NY	7	10
Massena	NY	10	16
New York	NY	12	20
Rochester	NY	10	16
Syracuse	NY	10	16
Akron	Ohio	10	16
Cincinnati	Ohio	7	10
Cleveland	Ohio	10	16
Columbus	Ohio	10	16
Dayton	Ohio	10	16
Mansville	Ohio	10	16
Toledo	Ohio	10	16
Youngstown	Ohio	10	16
Ardmore	Okla	12	20
Enid-Vance	Okla	12	20
Lawton	Okla	12	20
Muskogee	Okla	10	16
Oklahoma City	Okla	12	20
Tulsa	Okla	12	20
Eugene	Ore	4	6
Klamath Falls	Ore	4	6
Medford	Ore	4	6
Pendleton	Ore	7	10

TABLE 31

DESIGN WIND VELOCITIES – MILES PER HOUR AND KILOMETRES PER HOUR
UNITED STATES CITIES – WINTER

CITY	STATE	WIND Miles per hr	WIND km per hr
Portland	Ore	7	10
Salem	Ore	4	6
Allentown	Pa	10	16
Altoona	Pa	7	10
Erie	Pa	10	16
Harrisburg	7	10	
Philadelphia	Pa	10	16
Pittsburgh	Pa	10	16
Wilkes-Barre	Pa	7	10
Providence	RI	10	16
Charleston	SC	7	10
Columbia	SC	7	10
Florence	SC	7	10
Greenville	SC	7	10
Spartanburg	SC	7	10
Aberdeen	SD	7	10
Pierre	SD	10	16
Rapid City	SD	10	16
Sioux Falls	SD	10	16
Watertown	SD	7	10
Chatanooga	Tenn	7	10
Jackson	Tenn	7	10
Knoxville	Tenn	7	10
Memphis	Tenn	7	10
Nashville	Tenn	7	10
Abilene	Tex	10	16
Amarillo	Tex	10	16
Austin	Tex	10	16
Bryan	Tex	10	16
Dallas	Tex	10	16
El Paso	Tex	7	11
Ft. Worth	Tex	12	20
Galveston	Tex	10	16
Houston	Tex	10	16
Loredo	Tex	10	16
Lufkin	Tex	10	16
San Angelo	Tex	10	16
San Antonio	Tex	7	16
Tyler	Tex	10	16
Victoria	Tex	10	16
Waco	Tex	10	16
Wichita Falls	Tex	12	20
Ogden	Utah	7	6
Salt Lake City	Utah	7	10
Burlington	VT	10	16
Lynchburg	Va	7	10
Norfolk	Va	7	10
Richmond	Va	7	10
Roanoke	Va	7	10
Everette	Wash	7	10
Olympia	Wash	7	10
Seattle	Wash	7	10
Spokane	Wash	4	6
Tacoma	Wash	7	10
Yakima	Wash	4	6
Eau Clair	Wis	7	10
Green Bay	Wis	10	16
LaCrosse	Wis	10	16
Madison	Wis	10	16
Milwaukee	Wis	10	16
Wausau	Wis	10	16
Charleston	WVa	7	10
Huntington	WVa	7	10
Martinsburg	WVa	7	10
Morgantown	WVa	7	10
Parkersburg	WVa	7	10
Casper	Wyo	7	10
Cheyenne	Wyo	10	16
Rock Springs	Wyo	4	6
Sheridan	Wyo	7	10

TABLE 31 (Continued)

DESIGN WIND VELOCITIES – MILES PER HOUR AND KILOMETRES PER HOUR
CANADIAN CITIES – WINTER

CITY	PROV	WIND Miles per hr	WIND km per hr
Calgary	Alta	10	16
Edmonton	Alta	4	6
Vancover	BC	7	10
Victoria	BC	10	16
Churchill	Mon	12	20
Winnipeg	Mon	10	16
Moncton	NB	12	20
St. John	NB	10	16
Gander	Nfld	12	20
St. John	Nfld	10	16
Halifax	NS	10	16
Ft. Smith	NWT	4	6
Ft. William	Ont	7	10
Hamilton	Ont	10	16
London	Ont	10	16
Ottawa	Ont	10	16
Toronta	Ont	10	16
Windsor	Ont	10	16
Charlottetown	PEI	12	20
Montreal	PQ	10	16
Quebec	PQ	10	16
Saskatoon	Sask	10	16
Dawson	YT	7	10

TABLE 32

SURFACE RESISTANCE R_o (English Units) and R_{mo} (Metric Units)
FOR VARIOUS VALUES OF $R_w + R_i$ ($R_{mw} + R_{mi}$) AND
WINTER WIND VELOCITIES – ALSO t_s - t_a (t_{ms} - t_{ma})

ENGLISH UNITS

Thermal Resistance Inside Surface and wall* Ri + Rw	WIND VELOCITIES Miles per hour: 0 — Therm Res Out Surf R_o	0 — Surf to Air t °F	4 — Therm Res Out Surf R_o	4 — Surf to Air t °F	7 — Therm Res Out Surf R_o	7 — Surf to Air t °F	10 — Therm Res Out Surf R_o	10 — Surf to Air t °F	12 — Therm Res Out Surf R_o	12 — Surf to Air t °F
2	0.84	20.5	0.50	14.5	0.43	12.4	0.39	11.2	0.36	10.9
3	0.87	15.7	0.53	10.5	0.46	9.2	0.42	8.3	0.39	7.9
5	0.91	10.8	0.57	7.2	0.49	6.2	0.45	5.6	0.42	5.4
10	0.97	6.3	0.62	4.2	0.55	3.7	0.50	3.3	0.47	3.2
15	1.02	4.2	0.65	2.9	0.60	2.5	0.52	2.2	0.50	2.2
20	1.02	4.2	0.65	2.9	0.60	2.5	0.52	2.2	0.50	1.8
30	1.09	2.80	0.71	2.0	0.65	1.7	0.56	1.6	0.54	1.5
40	1.12	2.30	0.73	1.8	0.66	1.6	0.57	1.5	0.55	1.4

METRIC UNITS

Thermal Resistance Inside Surface and Wall* Rmi + Rmw	WIND VELOCITIES km per hour: 0 — Therm Res Out Surf R_o	0 — Surf to Air t °C	6 — Therm Res Out Surf R_o	6 — Surf to Air t °C	10 — Therm Res Out Surf R_o	10 — Surf to Air t °C	16 — Therm Res Out Surf R_o	16 — Surf to Air t °C	20 — Therm Res Out Surf R_o	20 — Surf to Air t °C
0.35	0.147	11.3	0.088	8.0	0.075	6.9	0.069	6.2	0.963	6.1
0.50	0.153	8.7	0.093	5.8	0.081	5.1	0.073	4.6	0.069	4.4
0.90	0.160	6.0	0.100	4.0	0.086	3.4	0.079	3.1	0.073	3.0
1.80	0.171	3.5	0.109	2.3	0.097	2.0	0.088	1.8	0.083	1.8
2.60	0.179	2.3	0.114	1.6	0.106	1.4	0.092	1.2	0.088	1.2
3.50	0.187	1.8	0.119	1.3	0.111	1.2	0.095	1.1	0.092	1.0
5.30	0.191	1.6	0.125	1.1	0.114	0.9	0.099	0.9	0.095	0.8
7.00	0.197	1.3	0.128	1.0	0.116	0.8	0.100	0.8	0.096	0.7

TABLE 33

SURFACE TEMPERATURES

When designing air cooling systems, the solar heat influences the sizing of the system and the cost of maintaining the desired indoor temperature. Correct evaluation of solar radiation on the total heat load from outdoors to a cooled inside area is very complex. The solar heat increases the outside temperature of the walls or roofs of the building. This increases the temperature differential between the inside and outside surface.

The basic formula for obtaining this surface temperature under still air condition is:

$$t_s = t_a + \left\{ \frac{\alpha I_D + \alpha I_d + \varepsilon (I_s - I_\ell)}{f_{co}} \right\} \tag{18}$$

where, in English Units:

α_D = Absorptivity of outside surface of wall or roof for radiation over visible wavelengths (0.3 to 0.71 microns) under direct exposure

α_d = Absorptivity of outside surface of wall or roof for diffuse radiation (wavelengths 0.8 to 2.8 microns)

f_{co} = Unit conduction of outer surface in Btu/ft^2, hr

I_D = Incident direct solar radiation in Btu/ft^2, hr
I_d = Incident diffused radiation in Btu/ft^2, hr
I_s = Radiant energy falling on surface from outdoor surroundings in Btu/ft^2, hr
I_ℓ = Radiant energy emitted by black body at temperature t_s
t_a = Ambient outside air and surrounding temperature in °F
t_s = Surface temperature of outer surface of wall or roof in °F
ε = Emittance of surface at temperature t_s

As all these conditions change continually, t also changes continually. For this reason, the temperature to add to t has been estimated and is shown in Table 34. This table presents the temperature to be added in mid-summer with respect to time, location and surface emittance.

It should be noted that oxidized aluminum is *not* a good reflector of *solar* heat. All aluminum oxidizes when exposed to air, thus the use of aluminum as a reflector of solar heat is not effective. Tests runs by United Aircraft Corporation indicate that the reflectance over the visible portion of the spectrum of oxidized aluminum alloy 24ST averaged 0.31. Conversely, the emittance (ε) averaged 0.69. Over the non-visible thermal wavelengths the reflectance of oxidized aluminum ranged between 0.55 to 0.96.

A common error made by industry is to paint storage tanks and other equipment with aluminum paint to reduce solar radiation heat pick up. This is ineffective. A glossy white paint would be more effective.

Most materials are not so affected by the wave lengths of heat as aluminum and oxidized aluminum and their reflectance — although slightly affected — is more constant over all the wave lengths of radiant heat. Table 35 provides a guide to surface emittances to be used with Table 34.

Example 11—*English and Metric Units*

To illustrate how this method operates assume a building with outside siding of 0.85 surface emittance is so located, that one wall faces NE, another SE, another SW and the fourth faces NW. What would be the temperature of the outer surface on a clear day when the ambient air temperature is as listed below during the day?

Adding the temperature will give the surface temperature of each wall in the direction that it faces, as shown below. Rounding the given surface emittance up to 0.9 and using the last chart in Table 34 yields $\Delta T = t_s - t_a$, the differential surface temperature of each wall above ambient. Adding Δt to the summer design high temperatures yields the surface temperature of each wall for each hour of the day. These results are found in the accompanying table.

SOLAR STANDARD TIME	SURFACE TEMPERATURE OF WALLS							
	Facing NE		Facing SE		Facing SW		Facing NW	
	°F	°C	°F	°C	°F	°C	°F	°C
1 AM	72	22	72	22	72	22	72	22
2 AM	71	22	71	22	71	22	71	22
3 AM	71	22	71	22	71	22	71	22
4 AM	70	21	70	21	70	21	70	21
5 AM	71	22	71	22	71	22	71	22
6 AM	116	47	98	37	76	24	76	24
7 AM	131	55	120	49	82	28	82	28
8 AM	130	54	134	57	86	30	86	30
9 AM	121	49	140	60	90	32	90	32
10 AM	106	42	138	59	93	34	92	33
11 AM	96	36	131	55	104	40	95	35
12 Noon	96	36	118	48	122	50	96	36
1 PM	98	36	103	39	134	56	96	36
2 PM	99	37	100	38	145	63	113	45
3 PM	101	38	101	38	151	66	132	56
4 PM	103	39	103	39	140	60	146	63
5 PM	102	39	102	39	138	59	151	66
6 PM	98	37	102	39	120	49	138	59
7 PM	97	36	97	36	97	36	97	36
8 PM	90	32	90	32	90	32	90	32
9 PM	85	29	85	29	85	29	85	29
10 PM	80	27	80	27	80	27	80	27
11 PM	77	25	77	25	77	25	77	25
12 Midnight	75	24	75	24	75	24	75	24

DESIGN HIGH TEMPERATURE – SUMMER					
Time (Std)	Temperature		Time (Std)	Temperature	
AM	°F	°C	PM	°F	°C
1	72	22	1	85	29
2	71	21	2	87	28
3	71	21	3	90	32
4	70	21	4	93	34
5	71	21	5	95	35
6	73	23	6	95	35
7	75	24	7	94	34
8	77	25	8	90	32
9	79	26	9	85	29
10	80	27	10	80	27
11	82	28	11	77	25
12 Noon	83	28	12	75	24

DEGREES TEMPERATURE ABOVE SHADED WALL AND ROOF SURFACES WILL DIFFER FROM AMBIENT AIR TEMPERATURE BASED ON 40 deg. NORTH LATITUDE SOLAR CONDITIONS – July 21
Clear Weather
EMITTANCE of OUTSIDE SURFACE = 0.6

TIME	DIRECTION IN WHICH WALLS FACE																HOR.	
	N		NE		E		SE		S		SW		W		NW		ROOFS	
	°F	°C	°F	°C	°F	°C	°F	°C	°F	°C	°F	°C	°F	°C	°F	°C		
1 AM	0	0	0	0	0	0	0	0	0	0	0	0	0	0	0	0	-7	-4
2 AM	0	0	0	0	0	0	0	0	0	0	0	0	0	0	0	0	-7	-4
3 AM	0	0	0	0	0	0	0	0	0	0	0	0	0	0	0	0	-7	-4
4 AM	0	0	0	0	0	0	0	0	0	0	0	0	0	0	0	0	-7	-4
5 AM	0	0	0	0	0	0	0	0	0	0	0	0	0	0	0	0	-7	-4
6 AM	+7	+4	+11	+6	+13	+7	+12	+7	+1	+1	+1	+1	+1	+1	+1	+1	0	0
7 AM	+7	+4	+28	+15	+34	+19	+22	+12	+5	+3	+3	+2	+3	+2	+3	+2	+8	+4
8 AM	+7	+4	+28	+15	+37	+21	+28	+15	+10	+5	+3	+2	+3	+2	+3	+2	+18	+10
9 AM	+6	+3	+22	+12	+34	+19	+30	+17	+20	+11	+5	+3	+5	+3	+5	+3	+26	+14
10 AM	+6	+3	+15	+8	+27	+15	+29	+16	+23	+13	+6	+3	+6	+3	+6	+3	+33	+18
11 AM	+6	+3	+7	+4	+17	+9	+24	+13	+28	+15	+9	+5	+6	+3	+6	+3	+36	+20
12 NOON	+6	+3	+6	+3	+7	+4	+17	+9	+23	+13	+17	+9	+7	+4	+6	+3	+37	+21
1 PM	+6	+3	+6	+3	+6	+3	+9	+5	+22	+12	+24	+13	+17	+9	+7	+4	+37	+21
2 PM	+6	+3	+6	+3	+6	+3	+6	+3	+19	+10	+29	+16	+27	+15	+13	+7	+32	+18
3 PM	+5	+3	+5	+3	+5	+3	+5	+3	+18	+10	+30	+17	+34	+19	+21	+12	+26	+14
4 PM	+4	+2	+4	+2	+4	+2	+4	+2	+10	+5	+28	+15	+37	+21	+26	+14	+19	+10
5 PM	+4	+2	+3	+2	+3	+2	+3	+2	+7	+4	+22	+12	+28	+19	+28	+15	+10	+5
6 PM	+4	+2	+2	+1	+2	+1	+2	+1	+3	+2	+11	+6	+23	+13	+21	+12	+5	+3
7 PM	+2	+1	+2	+1	+2	+1	+2	+1	+2	+1	+2	+1	+2	+1	+2	+1	0	0
8 PM	0	0	0	0	0	0	0	0	0	0	0	0	0	0	0	0	-2	-1
9 PM	0	0	0	0	0	0	0	0	0	0	0	0	0	0	0	0	-4	-2
10 PM	0	0	0	0	0	0	0	0	0	0	0	0	0	0	0	0	-5	-3
11 PM	0	0	0	0	0	0	0	0	0	0	0	0	0	0	0	0	-6	-3
12 PM	0	0	0	0	0	0	0	0	0	0	0	0	0	0	0	0	-7	-4
AVG	+5	+3	+6	+3	+8	+4	+8	+4	+8	+4	+8	+4	+8	+6	+6	+3	10	6

For 30 degree latitude multiply plus numbers by 1.05
For 50 degree latitude multiply plus numbers by 0.96

(1 of 3)

DEGREES TEMPERATURE ABOVE SHADED WALL AND ROOF SURFACES WILL DIFFER FROM AMBIENT AIR TEMPERATURE BASED ON 40 deg. NORTH LATITUDE SOLAR CONDITIONS – July 21
Clear Weather
EMITTANCE of OUTSIDE SURFACE = 0.75

TIME	DIRECTION IN WHICH WALLS FACE																HOR. ROOFS	
	N		NE		E		SE		S		SW		W		NW			
	°F	°C	°F	°C	°F	°C	°F	°C	°F	°C	°F	°C	°F	°C	°F	°C	°F	°C
1 AM	0	0	0	0	0	0	0	0	0	0	0	0	0	0	0	0	-7	-4
2 AM	0	0	0	0	0	0	0	0	0	0	0	0	0	0	0	0	-7	-4
3 AM	0	0	0	0	0	0	0	0	0	0	0	0	0	0	0	0	-7	-4
4 AM	0	0	0	0	0	0	0	0	0	0	0	0	0	0	0	0	-7	-4
5 AM	0	0	0	0	0	0	0	0	0	0	0	0	0	0	0	0	-7	-4
6 AM	+11	+6	+27	+15	+30	+17	+19	+10	+2	+1	+2	+1	+2	+1	+2	+1	0	0
7 AM	+11	+6	+42	+23	+45	+25	+34	+19	+6	+3	+5	+3	+5	+3	+5	+3	+18	+10
8 AM	+10	+6	+41	+23	+55	+31	+43	+24	+16	+9	+6	+3	+6	+3	+6	+3	+32	+18
9 AM	+10	+6	+32	+18	+52	+19	+45	+25	+35	+19	+8	+4	+8	+4	+8	+4	+43	+24
10 AM	+9	+5	+21	+12	+41	+23	+43	+24	+36	+20	+10	+6	+9	+5	+9	+5	+53	+29
11 AM	+9	+5	+11	+6	+25	+14	+37	+21	+44	+24	+16	+9	+10	+6	+10	+6	+56	+31
12 NOON	+9	+5	+10	+6	+11	+6	+21	+12	+40	+22	+28	+15	+11	+6	+10	+6	+59	+33
1 PM	+9	+5	+10	+6	+10	+6	+14	+8	+31	+17	+37	+21	+26	+14	+10	+6	+59	+33
2 PM	+9	+5	+9	+5	+9	+5	+10	+6	+26	+14	+44	+24	+41	+23	+20	+12	+52	+29
3 PM	+8	+4	+8	+4	+8	+4	+8	+4	+23	+13	+45	+25	+46	+25	+32	+18	+43	+24
4 PM	+7	+4	+7	+4	+7	+4	+7	+4	+11	+6	+43	+24	+30	+28	+39	+22	+32	+18
5 PM	+6	+3	+5	+3	+5	+3	+5	+3	+7	+4	+33	+18	+42	+23	+42	+23	+19	+10
6 PM	+6	+3	+3	+2	+3	+2	+3	+2	+3	+2	+18	+10	+35	+19	+32	+18	+8	+4
7 PM	+4	+2	+3	+2	+3	+2	+3	+2	+3	+2	+3	+2	+3	+2	+3	+2	0	0
8 PM	0	0	0	0	0	0	0	0	0	0	0	0	0	0	0	0	-3	-2
9 PM	0	0	0	0	0	0	0	0	0	0	0	0	0	0	0	0	-5	-3
10 PM	0	0	0	0	0	0	0	0	0	0	0	0	0	0	0	0	-6	-3
11 PM	0	0	0	0	0	0	0	0	0	0	0	0	0	0	0	0	-7	-4
12 PM	0	0	0	0	0	0	0	0	0	0	0	0	0	0	0	0	-7	-4
AVG	+5	+3	+10	+6	+13	+7	+12	+7	+12	+7	+12	+7	+12	+7	+10	+6	+17	+9

For 30 degree latitude multiply plus numbers by 1.05
For 50 degree latitude multiply plus numbers by 0.96

(2 of 3)

TABLE 34

DEGREES TEMPERATURE ABOVE WALLS AND ROOF SURFACES WILL DIFFER FROM AMBIENT AIR TEMPERATURE BASED ON 40 deg. NORTH LATITUDE SOLAR CONDITIONS – July 21 EMITTANCE of OUTSIDE SURFACE = 0.9

TIME	DIRECTION IN WHICH WALLS FACE																HOR. ROOFS	
	N		NE		E		SE		S		SW		W		NW			
	°F	°C	°F	°C	°F	°C	°F	°C	°F	°C	°F	°C	°F	°C	°F	°C	°F	°C
1 AM	0	0	0	0	0	0	0	0	0	0	0	0	0	0	0	0	-7	-4
2 AM	0	0	0	0	0	0	0	0	0	0	0	0	0	0	0	0	-7	-4
3 AM	0	0	0	0	0	0	0	0	0	0	0	0	0	0	0	0	-7	-4
4 AM	0	0	0	0	0	0	0	0	0	0	0	0	0	0	0	0	-7	-4
5 AM	0	0	0	0	0	0	0	0	0	0	0	0	0	0	0	0	-7	-4
6 AM	+16	+9	+43	+24	+47	+26	+25	+14	+3	+2	+3	+2	+3	+2	+3	+2	+6	+3
7 AM	+16	+9	+56	+31	+60	+33	+45	+25	+7	+4	+7	+4	+7	+4	+7	+4	+27	+15
8 AM	+15	+8	+53	+29	+73	+41	+57	+32	+12	+7	+9	+5	+9	+5	+9	+5	+45	+25
9 AM	+14	+8	+42	+24	+68	+38	+61	+34	+50	+28	+11	+6	+11	+6	+11	+6	+60	+33
10 AM	+13	+7	+26	+14	+54	+30	+58	+32	+58	+32	+13	+7	+12	+7	+12	+7	+72	+40
11 AM	+13	+7	+14	+8	+37	+21	+49	+27	+63	+35	+22	+12	+13	+7	+13	+7	+79	+44
12 NOON	+13	+7	+13	+7	+14	+8	+35	+19	+58	+32	+39	+22	+14	+8	+13	+7	+82	+46
1 PM	+13	+7	+13	+7	+13	+7	+18	+10	+42	+24	+49	+27	+35	+19	+13	+7	+79	+44
2 PM	+12	+7	+12	+7	+12	+7	+13	+7	+33	+18	+58	+32	+54	+30	+26	+14	+72	+40
3 PM	+11	+6	+11	+6	+11	+6	+11	+6	+25	+14	+61	+34	+68	+38	+42	+23	+60	+33
4 PM	+10	+5	+10	+6	+10	+6	+10	+6	+12	+7	+57	+32	+74	+41	+53	+29	+45	+25
5 PM	+7	+4	+7	+4	+7	+4	+7	+4	+7	+4	+43	+24	+56	+31	+56	+31	+27	+15
6 PM	+3	+2	+3	+2	+3	+2	+3	+2	+3	+2	+25	+14	+47	+26	+43	+24	+10	+6
7 PM	+3	+2	+3	+2	+3	+2	+3	+2	+3	+2	+3	+2	+3	+2	+3	+2	0	0
8 PM	0	0	0	0	0	0	0	0	0	0	0	0	0	0	0	0	-4	-2
9 PM	0	0	0	0	0	0	0	0	0	0	0	0	0	0	0	0	-5	-3
10 PM	0	0	0	0	0	0	0	0	0	0	0	0	0	0	0	0	-7	-4
11 PM	0	0	0	0	0	0	0	0	0	0	0	0	0	0	0	0	-7	-4
12 PM	0	0	0	0	0	0	0	0	0	0	0	0	0	0	0	0	-7	-4
AVG	+7	+4	+13	+7	+16	+9	+16	+9	+16	+9	+16	+9	+16	+9	+16	+9	+24	+13

For 30 degree latitude multiply plus numbers by 1.05
For 50 degree latitude multiply plus numbers by 0.96

(3 of 3)

TABLE 34 (Continued)

EMITTANCE and REFLECTANCES R_ℓ of SURFACES*

MATERIAL SURFACE	TEMPERATURE		RATIO TO THERMAL BLACK BODY	
	°F	°C	EMITTANCE ∈	REFLECTANCE R_ℓ
Aluminum - Oxidized Surface	0 to 100	-18 to 38	0.32	0.68
Asphalt (Black unpainted)	70 to 150	21 to 66	0.95 to 0.98	0.02 to 0.05
Brick - Red	50 to 100	10 to 38	0.85 to 0.95	0.05 to 0.15
Chromium	0 to 100	-18 to 38	0.22	0.78
Concrete	0 to 100	-18 to 38	0.90 to 0.97	0.1 to 0.03
Copper - Oxidized Surface	0 to 100	-18 to 38	0.44	0.56
Fireclay Brick	300	149	0.75 to 0.8	0.20 to 0.25
Fiberglass - Insulation	50 to 100	10 to 38	0.76 to 0.82	0.18 to 0.24
Glass	50 to 150	10 to 38	0.82 to 0.96	0.04 to 0.18
Gypsum - Board	0 to 100	-18 to 38	0.90 to 0.93	0.10 to 0.07
Marble	0 to 100	-18 to 38	0.92 to 0.96	0.04 to 0.08
Paint - Aluminum, Bright	50 to 150	10 to 66	0.43 to 0.49	0.51 to 0.57
Paint - Black	50 to 150	10 to 66	0.93 to 0.97	0.03 to 0.07
Paint - Colors	50 to 150	10 to 66	0.73 to 0.90	0.10 to 0.27
Paint - White-clean, bright	50 to 150	10 to 66	0.68 to 0.83	0.17 to 0.32
Plastic - (Glass Fiber Reinforced)	0 to 100	-18 to 38	0.70 to 0.80	0.20 to 0.30
Siding, Painted - See Paint				
Shingles - Asphaltic or Slate	0 to 150	-18 to 66	0.87 to 0.98	0.02 to 0.17
Wood, Beech-Planed	70	21	0.91 to 0.94	0.06 to 0.09
Wood, Oak-Planed, Sanded	70	21	0.87 to 0.91	0.09 to 0.13
Wood, Pine-Sanded	70	21	0.86 to 0.90	0.10 to 0.14
Wood, Spruce-Sanded	70	21	0.82 to 0.88	0.12 to 0.18

*Based on Radiation of nonvisible wave lengths of 0.8 to 2.6 microns.

TABLE 35

In many instances, information for design high temperature summer conditions by the hour is not available. However, *Summer Design Temperature*, as given in Table 24, may be used for design purposes. When *Summer Design Temperature* is used to determine surface temperatures of walls or roofs, it should be noted that differences are slightly averaged and that some estimated surface temperatures may differ slightly from actual surface temperature. Where surface temperatures are estimated by adding temperature differentials (Table 34) to *Summer Design Temperatures* (Table 24) in early morning and late afternoon, they will be slightly higher than actual; but temperatures obtained for late morning and early afternoon will be slightly lower than actual. However, in most instances the surface temperatures obtained are sufficiently accurate for determining cooling requirements.

Example 12—*English and Metric Units*

Building: Brick Walls facing: N, E, S, W ($\varepsilon = 0.90$)
Roof — Flat, painted
Gray, surface emittance $\varepsilon = 0.92$

Location: New York, NY — Summer design temperature is 93 °F (34 °C)

Using Table 34 ($\varepsilon = 0.9$) to obtain ΔT above ambient which is then added to the 93 °F design temperature yields:

SOLAR STANDARD TIME	SURFACE TEMPERATURE OF WALLS — t_s								HORIZONTAL	
	Facing N		Facing E		Facing S		Facing W			
	°F	°C	°F	°C	°F	°C	°F	°C	°F	°C
1 to 5 AM	93	34	93	34	93	34	93	34	93	34
6 AM	109	43	140	60	96	36	96	36	96	36
7 AM	109	43	153	56	100	38	100	38	120	49
8 AM	108	42	166	74	105	41	102	39	138	59
9 AM	107	42	161	72	143	62	104	40	153	67
10 AM	106	41	147	64	151	66	105	41	165	74
11 AM	106	41	130	54	156	69	106	41	172	78
12 Noon	106	41	107	42	151	66	107	42	175	79
1 PM	106	41	106	41	135	57	128	53	172	78
2 PM	105	41	105	41	126	52	147	64	165	74
3 PM	104	40	104	40	119	48	161	72	153	67
4 PM	103	39	103	39	105	41	167	75	138	59
5 PM	100	38	100	38	100	38	149	65	120	49
6 PM	96	36	96	36	96	36	140	60	103	39
7 PM	96	36	96	36	96	36	96	36	93	34
8 to 12 MN	93	34	93	34	93	34	93	34	93	34

Once the outside surface temperature t_s of the wall or roof has been established, it is then possible to calculate the heat transfer to the interior for the time as stated. The temperature difference is the surface temperature t_s minus the interior air temperature t_i. The thermal resistance is the total resistance of the wall plus the resistance of the inner film R_i.

$$Q = \text{heat gain per hour} = \frac{t}{R_t} = \frac{t_s - t_i}{R_t} \qquad (13)$$

However, it can be observed that the surface temperature on the outside has direct effect on heat gain to the inside and this changes with time and ambient air temperature. These surface temperatures are necessary to enable the calculation of heat gain to inside severe conditions so as to obtain maximum cooling requirements.

NOTE: Units in equation must be all English Units or all Metric Units.

DIRECT RADIATION OF HEAT

Whereas in the case of roofs and walls of opaque materials, solar radiation incident on the surface will be converted to heat, raising the temperature and causing inward heat flow. In the case of glass or other transparent or translucent materials, a major part of the solar radiation can pass through into the inner area as heat energy. Therefore, it is essential to be able to determine radiation heat gains in a given situation.

For this reason, solar radiation levels to surfaces facing in various directions for design summer, fall, spring and winter are presented as a means to estimate the heat transfer through transparent or translucent materials. These tables are given for 24, 32, 40, 48 and 56 degree latitudes with surfaces facing directly nominal to the sun's rays and also North, Northeast, East, Southeast, South, Southwest, West, Northwest and Horizontal.

Table 36 was calculated so as to give the "design average" of radiation received for each of the four seasons, summer, autumn, winter and spring. The "design average" was presented heat as a means to calculate the expected radiation for the 21st day of each month (see *ASHRAE "Handbook of Fundamentals"*). In the design of heat gain for a building or home, the absolute maximum heat gain per hour is seldom required as it happens so infrequently. Also, the very mass of a building and its contents requires time for temperature change. Thus, these design average radiation tables should be sufficiently adequate for design of insulation systems and air cooling equipment.

When solar radiation is incident on windows, the clear glass provides little resistance to its entry into the building or home. Ordinary single glass will allow 80 to 87% of solar radiation to pass directly into the interior. This may be a significant heat gain for buildings with unshaded glass doors and windows. To illustrate the amount of heat that can be transmitted into a home on a clear summer day the following example is presented.

SPRING SOLAR HEAT GAIN AT STATED SOLAR TIME (DESIGN AVE.)
24 DEGREE NORTH LATITUDE

STD. SOLAR TIME	SURFACE POSITION																			
	DIRECT		VERTICAL SURFACE – FACING																HORIZONTAL	
	NOMINAL		NORTH		NORTHEAST		EAST		SOUTHEAST		SOUTH		SOUTHWEST		WEST		NORTHWEST		SURFACE	
	Btu/ft² hr	W/m²	Btu/ft² hr	W/m²	Btu/ft² hr	W/m²	Btu/ft² hr	W/m²	Btu/ft² hr	W/m²	Btu/ft² hr	W/m²	Btu/ft² hr	W/m²	Btu/ft² hr	W/m²	Btu/ft² hr	W/m²	Btu/ft² hr	W/m²
6 AM	97	305.79	25	78.81	68	214.37	72	226.98	33	104.03	5	15.76	4	12.61	5	15.76	4	12.61	16	50.44
7 AM	201	633.65	28	88.27	165	520.19	196	617.89	111	349.93	16	50.44	16	50.44	16	50.44	16	50.44	69	217.52
8 AM	251	791.27	35	110.84	173	545.38	218	687.24	136	428.74	25	78.81	24	75.66	25	78.81	24	75.66	136	428.72
9 AM	265	833.21	37	116.64	135	425.58	199	627.34	139	438.19	35	110.34	30	94.97	31	97.73	30	94.97	198	624.19
10 AM	287	904.76	38	119.79	88	277.42	152	479.18	118	371.99	43	135.56	34	107.18	35	110.34	37	116.64	245	769.21
11 AM	288	907.01	38	119.79	53	167.08	83	261.66	82	258.50	51	160.78	49	154.57	37	116.64	39	122.95	273	860.63
12 noon	289	911.07	39	122.95	41	129.25	40	126.10	58	182.84	53	167.08	58	182.84	40	126.10	41	129.25	281	885.84
1 PM	288	907.01	38	119.79	39	122.95	37	116.64	49	154.47	51	160.78	82	258.50	83	261.66	53	167.08	273	864.63
2 PM	287	904.76	38	119.79	37	116.64	35	110.34	34	107.18	41	129.25	118	371.99	152	479.18	88	277.42	245	769.21
3 PM	265	833.21	37	116.64	30	94.97	31	97.73	30	94.97	35	110.34	139	438.19	199	627.34	135	425.58	198	624.19
4 PM	251	791.27	35	110.34	24	75.66	25	78.81	24	75.66	25	78.81	136	428.74	218	687.24	173	545.38	136	428.74
5 PM	201	633.65	28	88.27	16	50.44	16	50.44	16	50.44	16	50.44	111	349.93	196	617.89	165	520.19	69	217.52
6 PM	97	305.79	25	78.81	4	12.61	5	15.76	4	12.61	5	15.76	33	104.03	72	226.98	68	214.37	16	50.44

32 DEGREE NORTH LATITUDE

STD. SOLAR TIME	SURFACE POSITION																			
	DIRECT		VERTICAL SURFACE – FACING																HORIZONTAL	
	NOMINAL		NORTH		NORTHEAST		EAST		SOUTHEAST		SOUTH		SOUTHWEST		WEST		NORTHWEST		SURFACE	
	Btu/ft² hr	W/m²	Btu/ft² hr	W/m²	Btu/ft² hr	W/m²	Btu/ft² hr	W/m²	Btu/ft² hr	W/m²	Btu/ft² hr	W/m²	Btu/ft² hr	W/m²	Btu/ft² hr	W/m²	Btu/ft² hr	W/m²	Btu/ft² hr	W/m²
6 AM	102	321.55	26	81.97	82	258.50	85	267.96	49	154.47	7	22.07	6	18.91	6	18.91	7	22.07	18	56.74
7 AM	209	658.02	28	88.27	151	476.02	203	639.95	124	390.91	17	53.59	16	50.44	16	50.44	17	53.59	77	242.74
8 AM	244	769.21	30	94.97	158	498.09	218	687.24	155	488.63	28	88.27	22	69.35	24	75.66	24	75.66	143	454.80
9 AM	264	835.25	33	104.03	121	381.45	199	627.34	160	504.40	58	182.84	30	94.97	30	94.97	30	94.97	197	621.04
10 AM	285	898.46	35	110.34	71	223.83	151	476.02	144	453.96	62	195.45	34	107.18	34	107.18	34	107.18	237	747.14
11 AM	287	904.76	37	116.64	47	148.17	82	258.50	98	308.94	77	242.74	39	122.95	36	113.49	36	113.49	265	833.41
12 noon	288	907.01	38	119.79	38	119.79	40	126.10	64	201.76	83	261.66	64	201.76	40	126.10	38	119.79	274	863.78
1 PM	287	904.76	37	116.64	36	113.49	36	113.49	39	122.95	77	242.74	98	308.94	82	258.50	47	148.17	265	833.41
2 PM	285	898.46	35	110.34	34	107.18	34	107.18	34	107.18	62	195.45	144	453.96	151	476.02	71	223.83	237	747.14
3 PM	264	835.25	33	104.03	30	94.97	30	94.97	30	94.97	58	182.84	160	504.40	199	627.34	121	381.45	197	621.04
4 PM	244	769.21	30	94.97	24	75.66	24	75.66	22	69.35	28	88.27	155	488.63	218	687.24	158	498.09	143	450.80
5 PM	209	658.82	28	88.27	17	53.59	16	50.44	16	50.44	17	53.59	124	390.91	203	639.95	151	476.02	77	242.74
6 PM	102	321.55	26	81.96	7	22.07	6	18.91	6	18.91	7	22.07	49	154.47	85	267.96	82	258.50	18	56.74

Heat gain per hour in position listed under clear, unclouded weather conditions.

(Sheet 1 of 12)

TABLE 36

SPRING SOLAR HEAT GAIN AT STATED SOLAR TIME (DESIGN AVE.)
40 DEGREE NORTH LATITUDE

STD. SOLAR TIME	SURFACE POSITION																			
	DIRECT		VERTICAL SURFACE – FACING																HORIZONTAL	
	NOMINAL		NORTH		NORTHEAST		EAST		SOUTHEAST		SOUTH		SOUTHWEST		WEST		NORTHWEST		SURFACE	
	Btu/ft² hr	W/m²	Btu/ft² hr	W/m²	Btu/ft² hr	W/m²	Btu/ft² hr	W/m²	Btu/ft² hr	W/m²	Btu/ft² hr	W/m²	Btu/ft² hr	W/m²	Btu/ft² hr	W/m²	Btu/ft² hr	W/m²	Btu/ft² hr	W/m²
6 AM	104	327.86	22	69.35	84	264.60	87	274.26	61	192.30	7	22.07	6	18.91	9	18.91	8	25.22	18	56.74
7 AM	180	567.44	24	72.51	162	510.70	205	646.26	133	419.30	19	59.90	18	56.74	18	56.74	18	56.74	81	255.35
8 AM	238	750.29	25	75.66	143	450.80	217	684.09	166	523.31	32	100.88	24	72.51	24	72.51	24	72.51	126	397.21
9 AM	272	857.47	31	97.73	102	321.55	197	621.04	175	551.68	55	173.39	29	91.42	29	91.42	29	91.42	182	573.75
10 AM	285	898.46	33	104.03	42	132.40	149	469.72	162	510.70	93	293.18	34	107.18	34	107.18	33	104.03	229	721.92
11 AM	292	920.52	35	110.34	38	119.79	81	255.35	116	365.69	120	378.30	43	135.56	37	116.64	34	107.18	256	807.04
12 noon	298	939.44	36	113.49	35	110.34	38	119.79	83	261.66	132	416.13	83	261.66	38	119.79	35	110.34	262	825.94
1 PM	292	920.52	35	110.34	34	107.18	37	116.64	43	135.56	120	378.30	116	365.69	81	255.35	38	119.79	256	807.04
2 PM	285	898.46	33	104.02	33	104.03	34	107.18	34	107.18	93	219.18	162	510.70	149	469.72	42	132.40	229	721.92
3 PM	275	857.47	31	97.73	29	91.51	29	91.42	29	91.42	55	173.39	175	551.68	197	621.04	102	321.53	182	573.75
4 PM	238	750.29	25	75.66	24	72.51	24	72.51	24	72.51	32	100.88	166	523.31	217	684.09	143	450.30	126	397.21
5 PM	180	567.44	24	72.51	18	56.74	18	56.74	18	56.74	19	59.70	133	419.30	205	646.26	162	510.70	81	255.35
6 PM	104	325.86	22	69.35	7	22.07	7	22.07	7	22.07	7	22.07	82	195.45	89	280.57	81	255.35	18	56.74

48 DEGREE NORTH LATITUDE

STD. SOLAR TIME	SURFACE POSITION																			
	DIRECT		VERTICAL SURFACE – FACING																HORIZONTAL	
	NOMINAL		NORTH		NORTHEAST		EAST		SOUTHEAST		SOUTH		SOUTHWEST		WEST		NORTHWEST		SURFACE	
	Btu/ft² hr	W/m²	Btu/ft² hr	W/m²	Btu/ft² hr	W/m²	Btu/ft² hr	W/m²	Btu/ft² hr	W/m²	Btu/ft² hr	W/m²	Btu/ft² hr	W/m²	Btu/ft² hr	W/m²	Btu/ft² hr	W/m²	Btu/ft² hr	W/m²
6 AM	106	334.16	22	69.35	83	261.65	93	293.18	72	226.98	7	22.07	6	18.91	7	22.07	8	25.22	18	56.74
7 AM	183	576.90	24	72.51	142	447.65	206	649.41	145	451.11	19	59.90	19	59.70	19	59.90	19	59.90	83	271.11
8 AM	243	766.05	25	75.66	128	403.32	217	684.09	178	561.14	37	116.64	25	75.66	25	75.66	25	75.66	110	346.77
9 AM	263	829.10	29	91.42	84	264.81	197	621.06	193	608.43	68	214.37	30	94.57	30	94.57	30	94.57	173	545.38
10 AM	281	885.84	31	97.73	39	122.95	146	462.26	182	573.75	110	346.77	34	107.18	34	102.18	34	107.10	216	680.96
11 AM	286	901.61	34	107.19	36	113.49	78	248.89	155	488.63	133	419.30	48	151.32	35	110.34	35	110.34	238	750.29
12 noon	288	907.01	34	107.19	34	107.19	36	113.49	112	353.08	156	491.79	112	353.08	36	113.43	37	116.60	241	759.75
1 PM	286	901.61	34	107.19	34	107.19	34	107.18	48	151.32	133	419.30	155	488.63	78	248.89	38	119.79	238	750.29
2 PM	281	885.04	31	97.73	34	107.18	34	107.18	35	110.34	110	346.77	182	573.75	146	462.26	39	122.95	216	680.96
3 PM	263	829.10	29	91.42	30	94.57	30	94.57	30	94.57	68	214.37	193	608.43	197	621.06	84	264.81	173	545.38
4 PM	243	766.05	25	75.66	25	75.66	25	75.66	25	75.66	37	116.64	178	561.14	217	684.09	128	453.02	110	346.77
5 PM	183	576.90	24	72.51	19	59.90	19	59.90	19	59.90	19	59.90	145	457.11	206	649.41	142	447.65	83	271.11
6 PM	106	334.16	22	69.35	8	25.22	7	22.07	7	22.07	7	22.07	75	236.44	93	293.18	81	255.35	18	56.74

Heat gain per hour in position listed under clear, unclouded weather conditions.

(Sheet 2 of 12)

TABLE 36 (Continued)

SPRING SOLAR HEAT GAIN AT LISTED SOLAR TIME (DESIGN AVE.)
56 DEGREE NORTH LATITUDE

STD. SOLAR TIME	SURFACE POSITION																			
	DIRECT NOMINAL		VERTICAL SURFACE – FACING																HORIZONTAL SURFACE	
			NORTH		NORTHEAST		EAST		SOUTHEAST		SOUTH		SOUTHWEST		WEST		NORTHWEST			
	Btu/ft² hr	W/m²	Btu/ft² hr	W/m²	Btu/ft² hr	W/m²	Btu/ft² hr	W/m²	Btu/ft² hr	W/m²	Btu/ft² hr	W/m²	Btu/ft² hr	W/m²	Btu/ft² hr	W/m²	Btu/ft² hr	W/m²	Btu/ft² hr	W/m²
6 AM	108	340.47	16	50.44	81	255.35	98	308.94	92	290.03	7	22.07	8	25.22	8	25.22	8	25.22	19	59.90
7 AM	182	573.75	16	50.44	112	353.08	176	554.84	136	428.74	21	66.20	12	37.83	12	37.83	12	37.83	54	170.23
8 AM	241	259.75	18	56.74	90	283.92	204	643.11	196	617.89	67	211.22	18	56.74	18	56.74	18	56.74	95	299.49
9 AM	264	832.25	23	72.51	43	135.56	170	535.32	217	684.09	125	394.06	22	69.35	22	69.35	22	69.35	133	419.30
10 AM	271	854.32	25	78.81	26	81.96	138	435.04	219	690.39	129	406.67	36	113.49	25	78.81	25	78.81	104	517.01
11 AM	277	873.24	26	81.96	26	81.96	72	226.98	193	614.73	201	633.64	87	274.27	26	81.96	26	81.96	182	573.75
12 noon	279	879.54	27	85.12	26	81.96	29	91.42	149	469.72	211	665.17	149	469.72	29	91.42	26	81.96	189	595.82
1 PM	277	873.24	26	81.96	26	81.96	26	81.96	87	274.27	201	633.65	195	614.73	72	226.98	26	81.96	182	573.75
2 PM	271	854.32	25	78.81	25	78.81	25	78.81	36	113.49	129	406.67	219	690.39	138	435.04	26	81.96	164	517.01
3 PM	264	832.25	23	72.51	22	69.35	22	69.35	22	69.35	125	394.06	217	684.09	170	535.32	43	135.56	133	419.30
4 PM	241	259.75	18	56.74	18	56.74	18	56.74	18	56.74	67	211.22	196	617.89	204	643.11	90	283.92	95	299.49
5 PM	182	573.75	16	50.44	12	37.83	12	37.83	12	37.83	21	66.20	136	428.74	176	554.84	112	353.08	54	170.23
6 PM	108	340.47	16	50.44	8	25.22	8	25.22	8	25.22	7	22.07	92	290.03	98	308.94	81	255.35	19	59.90

Heat gain per hour in position listed under clear, unclouded weather conditions.

(Sheet 3 of 12)

TABLE 36 (Continued)

SUMMER SOLAR HEAT GAIN AT LISTED SOLAR TIME (DESIGN AVE.)
24 DEGREE NORTH LATITUDE

STD. SOLAR TIME	SURFACE POSITION																			
	DIRECT		VERTICAL SURFACES – FACING																HORIZONTAL	
	NOMINAL		NORTH		NORTHEAST		EAST		SOUTHEAST		SOUTH		SOUTHWEST		WEST		NORTHWEST		SURFACE	
	Btu/ft² hr	W/m²	Btu/ft² hr	W/m²	Btu/ft² hr	W/m²	Btu/ft² hr	W/m²	Btu/ft² hr	W/m²	Btu/ft² hr	W/m²	Btu/ft² hr	W/m²	Btu/ft² hr	W/m²	Btu/ft² hr	W/m²	Btu/ft² hr	W/m²
6 AM	88	277.42	33	104.00	120	378.30	74	201.76	34	107.18	7	22.07	6	18.91	5	15.76	5	15.76	18	56.74
7 AM	197	621.04	35	110.34	165	520.16	196	617.89	120	378.20	20	63.05	18	56.74	17	53.59	17	53.59	79	249.05
8 AM	240	756.60	36	113.49	167	526.16	215	677.78	141	444.50	29	88.27	28	81.96	25	78.81	25	78.81	153	482.40
9 AM	263	829.10	37	116.64	148	466.57	220	693.95	145	457.11	34	107.18	31	91.73	31	97.73	31	97.73	203	639.95
10 AM	275	866.93	39	122.95	97	305.79	192	479.18	143	450.80	52	163.92	40	126.10	37	116.64	36	113.49	235	740.83
11 AM	280	882.69	41	129.25	80	252.20	84	264.81	132	416.13	62	195.45	53	162.08	39	122.95	38	119.79	268	844.86
12 noon	282	889.00	42	132.40	58	182.84	40	126.95	78	245.89	68	211.22	72	222.98	40	126.10	58	182.84	272	857.47
1 PM	280	882.69	41	129.25	39	122.95	39	122.95	53	167.08	66	195.45	132	416.13	84	264.81	80	252.20	268	844.86
2 PM	275	866.93	39	122.95	36	113.49	37	116.64	40	126.10	52	163.92	143	450.80	192	479.10	97	305.57	235	740.83
3 PM	263	829.10	37	116.64	31	97.73	31	97.73	31	97.73	34	107.18	145	457.11	220	693.55	148	466.57	203	639.95
4 PM	240	756.60	36	113.49	25	78.81	25	78.81	25	78.81	28	88.27	141	444.50	215	677.78	167	526.26	153	482.40
5 PM	197	621.04	35	110.34	16	50.44	11	34.68	18	56.74	20	63.05	120	378.30	196	617.89	165	520.16	79	249.05
6 PM	88	277.42	33	104.00	5	15.76	5	15.76	6	18.91	T	22.07	34	107.18	74	201.76	120	378.30	18	56.74

32 DEGREE NORTH LATITUDE

STD. SOLAR TIME	SURFACE POSITION																			
	DIRECT		VERTICAL SURFACES – FACING																HORIZONTAL	
	NOMINAL		NORTH		NORTHEAST		EAST		SOUTHEAST		SOUTH		SOUTHWEST		WEST		NORTHWEST		SURFACE	
	Btu/ft² hr	W/m²	Btu/ft² hr	W/m²	Btu/ft² hr	W/m²	Btu/ft² hr	W/m²	Btu/ft² hr	W/m²	Btu/ft² hr	W/m²	Btu/ft² hr	W/m²	Btu/ft² hr	W/m²	Btu/ft² hr	W/m²	Btu/ft² hr	W/m²
6 AM	87	274.27	26	81.96	58	182.84	83	261.66	54	170.23	9	28.37	5	15.76	7	22.07	6	18.91	18	56.74
7 AM	195	614.73	28	88.27	158	498.09	205	646.22	130	409.86	22	69.35	17	53.59	19	59.90	17	58.59	78	245.89
8 AM	236	742.99	30	95.57	163	513.85	217	684.09	150	472.87	30	94.57	25	78.81	26	81.86	25	78.81	147	463.41
9 AM	269	848.08	31	97.73	142	447.62	197	621.00	153	482.40	39	122.95	30	95.57	31	97.73	30	95.57	198	624.19
10 AM	276	870.08	34	107.18	93	293.19	160	504.40	161	507.55	54	170.23	42	132.40	36	113.49	35	110.34	238	750.29
11 AM	279	879.54	36	113.49	75	236.44	82	258.50	140	441.35	64	201.76	65	204.91	38	119.79	37	116.64	267	841.71
12 noon	281	885.84	39	122.95	39	122.95	39	122.95	80	252.20	70	220.67	80	252.20	39	122.95	39	122.95	275	866.93
1 PM	279	879.54	36	113.49	37	116.64	38	119.79	65	204.91	64	201.76	140	441.35	82	258.50	75	236.44	267	841.71
2 PM	276	870.08	34	107.18	35	110.34	36	118.49	42	132.40	54	170.23	161	507.55	160	504.40	93	293.18	238	750.29
3 PM	269	848.08	31	97.73	30	95.57	31	97.73	30	95.57	39	122.95	153	482.40	197	621.04	142	447.62	198	624.19
4 PM	236	743.99	30	95.97	25	78.81	26	81.96	25	78.81	30	94.57	150	472.87	217	684.09	163	513.35	147	463.41
5 PM	195	614.73	28	88.27	17	53.59	19	59.90	17	53.59	22	69.35	130	409.82	205	646.24	158	498.01	78	245.89
6 PM	87	274.27	26	81.96	6	18.91	7	22.07	5	15.76	9	28.37	54	170.23	83	261.06	58	182.84	18	56.74

Heat gain per hour in position listed under clear unclouded weather conditions.

(Sheet 4 of 12)

TABLE 36 (Continued)

SUMMER SOLAR HEAT GAIN AT LISTED SOLAR TIME (DESIGN AVE.)
40 DEGREE NORTH LATITUDE

STD. SOLAR TIME	SURFACE POSITION																			
	DIRECT NOMINAL		VERTICAL SURFACES – FACING																HORIZONTAL SURFACE	
			NORTH		NORTHEAST		EAST		SOUTHEAST		SOUTH		SOUTHWEST		WEST		NORTHWEST			
	Btu/ft² hr	W/m²	Btu/ft² hr	W/m²	Btu/ft² hr	W/m²	Btu/ft² hr	W/m²	Btu/ft² hr	W/m²	Btu/ft² hr	W/m²	Btu/ft² hr	W/m²	Btu/ft² hr	W/m²	Btu/ft² hr	W/m²	Btu/ft² hr	W/m²
6 AM	86	271.11	24	75.66	110	346.77	102	321.05	60	189.85	10	31.52	7	22.07	8	25.22	6	18.91	17	53.59
7 AM	196	617.80	27	85.12	155	491.79	206	649.41	130	409.82	24	75.66	23	72.51	20	63.05	17	53.59	75	236.44
8 AM	232	731.36	28	88.27	157	494.94	216	680.96	155	488.63	31	97.73	24	75.66	26	81.96	19	59.90	145	457.11
9 AM	252	794.42	30	94.57	140	441.35	193	608.43	163	513.85	48	151.32	30	94.57	31	97.73	28	88.27	190	598.31
10 AM	270	851.16	33	104.03	90	283.72	148	466.57	180	567.45	79	249.05	52	163.92	37	116.64	30	94.57	221	696.70
11 AM	276	870.08	35	110.34	62	195.45	79	249.05	142	447.65	108	340.47	80	252.20	40	126.10	33	104.00	246	775.51
12 noon	280	882.69	37	116.64	34	107.18	42	132.40	92	290.03	114	359.38	92	290.03	42	132.40	34	107.18	260	819.65
1 PM	276	870.08	35	110.34	33	104.03	40	126.10	80	252.20	108	340.47	142	447.65	79	249.05	62	195.45	246	775.51
2 PM	270	851.16	33	104.03	30	94.57	37	116.04	52	163.92	79	249.05	180	567.45	148	466.57	90	283.72	221	696.70
3 PM	252	794.42	30	94.57	28	88.27	31	97.73	30	94.57	48	151.32	163	513.85	193	608.43	140	441.35	190	598.31
4 PM	232	731.36	28	88.27	19	59.90	26	81.96	24	75.66	31	97.73	155	488.63	216	680.96	157	494.94	145	457.11
5 PM	196	617.80	27	85.12	17	53.59	20	63.05	23	72.51	24	75.66	130	409.82	206	649.41	155	491.79	75	236.44
6 PM	86	271.11	24	75.66	6	18.91	8	25.22	7	22.07	10	31.52	60	189.85	102	321.55	120	378.30	17	53.59

48 DEGREE NORTH LATITUDE

STD. SOLAR TIME	SURFACE POSITION																			
	DIRECT NOMINAL		VERTICAL SURFACES – FACING																HORIZONTAL SURFACE	
			NORTH		NORTHEAST		EAST		SOUTHEAST		SOUTH		SOUTHWEST		WEST		NORTHWEST			
	Btu/ft² hr	W/m²	Btu/ft² hr	W/m²	Btu/ft² hr	W/m²	Btu/ft² hr	W/m²	Btu/ft² hr	W/m²	Btu/ft² hr	W/m²	Btu/ft² hr	W/m²	Btu/ft² hr	W/m²	Btu/ft² hr	W/m²	Btu/ft² hr	W/m²
6 AM	85	267.96	22	69.33	114	359.38	115	362.53	98	308.94	11	34.68	12	37.83	8	25.22	7	22.07	9	28.37
7 AM	197	261.04	25	78.81	143	450.80	208	655.72	131	412.98	25	78.81	19	59.90	20	63.05	17	53.59	73	230.13
8 AM	230	725.07	27	85.12	152	479.18	211	665.17	160	504.40	36	113.49	22	69.35	25	78.81	19	59.90	130	409.82
9 AM	250	788.12	29	91.42	82	258.50	192	605.28	181	570.60	57	179.69	23	72.51	30	94.57	29	91.42	186	586.36
10 AM	264	844.86	32	100.88	35	110.34	143	450.80	190	598.31	98	308.94	34	107.18	37	116.64	30	94.57	216	680.96
11 AM	272	857.47	34	107.18	32	100.88	78	245.89	155	488.31	120	378.30	48	151.32	39	122.95	30	94.57	242	762.90
12 noon	276	870.08	36	113.49	31	97.73	41	129.25	101	318.40	148	450.80	101	318.40	41	129.25	31	97.73	258	813.34
1 PM	272	857.47	34	107.18	30	94.57	39	122.95	48	151.32	120	378.30	155	488.63	78	245.86	32	100.88	242	762.90
2 PM	264	844.86	32	100.88	30	94.57	37	116.64	34	107.18	98	308.94	190	598.31	143	450.80	35	110.34	216	680.96
3 PM	250	788.12	29	91.42	29	91.42	30	94.57	23	72.51	57	179.69	181	570.60	192	605.28	82	258.50	186	586.36
4 PM	230	725.07	27	85.12	19	59.90	25	78.81	22	69.35	36	113.49	160	504.40	211	665.19	152	479.18	130	409.82
5 PM	197	261.04	25	78.81	17	53.59	20	63.05	19	59.90	25	78.81	131	412.98	208	655.72	143	450.80	73	230.13
6 PM	85	267.96	22	69.33	7	22.07	8	25.22	12	37.83	11	34.68	98	308.94	115	362.53	114	359.38	9	28.31

Heat gain per hour in position listed under clear unclouded weather conditions.

(Sheet 5 of 12)

TABLE 36 (Continued)

SUMMER SOLAR HEAT GAIN AT LISTED SOLAR TIME (DESIGN AVE.)
56 DEGREE NORTH LATITUDE

STD. SOLAR TIME	SURFACE POSITION																			
	DIRECT		VERTICAL SURFACES – FACING																HORIZONTAL	
	NOMINAL		NORTH		NORTHEAST		EAST		SOUTHEAST		SOUTH		SOUTHWEST		WEST		NORTHWEST		SURFACE	
	Btu/ft² hr	W/m²	Btu/ft² hr	W/m²	Btu/ft² hr	W/m²	Btu/ft² hr	W/m²	Btu/ft² hr	W/m²	Btu/ft² hr	W/m²	Btu/ft² hr	W/m²	Btu/ft² hr	W/m²	Btu/ft² hr	W/m²	Btu/ft² hr	W/m²
6 AM	84	264.80	18	56.74	111	349.93	125	394.06	111	349.93	11	34.68	10	31.52	8	25.21	7	22.07	8	25.21
7 AM	196	617.89	21	60.20	140	441.35	202	636.80	145	457.11	22	69.35	18	56.74	18	56.74	17	53.59	55	173.39
8 AM	227	715.61	24	75.66	110	346.77	210	662.02	185	583.21	51	160.78	23	72.51	24	75.66	19	59.00	99	312.09
9 AM	243	766.05	26	81.96	60	189.85	185	588.21	204	643.11	106	334.16	27	85.12	28	88.27	28	88.27	139	438.19
10 AM	257	810.18	29	91.42	44	138.71	139	438.19	199	627.34	146	460.26	34	104.03	30	94.37	30	94.37	191	602.12
11 AM	263	829.10	32	100.88	34	107.18	75	236.44	179	564.29	173	545.58	130	409.82	31	97.73	31	97.73	209	658.87
12 noon	266	838.56	32	100.88	33	104.03	35	110.34	145	457.11	182	573.75	145	457.11	35	110.34	33	104.03	214	674.63
1 PM	263	829.10	32	100.88	31	97.73	31	97.72	130	409.82	173	545.58	179	564.29	75	236.44	34	107.18	209	658.87
2 PM	257	810.18	29	91.42	30	94.37	30	94.37	34	107.18	146	460.26	199	627.34	139	438.19	44	138.71	191	602.12
3 PM	243	766.05	26	81.96	28	88.27	28	88.27	27	85.12	106	334.16	204	643.11	185	583.21	60	189.85	139	438.19
4 PM	227	715.61	24	75.66	19	59.00	24	75.66	23	72.51	51	160.78	185	583.21	210	662.02	110	346.77	99	312.09
5 PM	196	617.88	21	60.20	17	53.59	18	56.74	18	56.74	22	69.35	145	457.11	202	636.80	140	441.35	55	173.39
6 PM	84	264.80	18	56.74	7	22.07	8	25.22	10	31.52	11	34.68	111	349.93	155	488.63	114	359.38	8	25.21

Heat gain per hour in position listed under clear unclouded weather conditions.

(Sheet 6 of 12)

TABLE 36 (Continued)

AUTUMN SOLAR HEAT GAIN AT LISTED SOLAR TIME (DESIGN AVE.)
24 DEGREE NORTH LATITUDE

STD. SOLAR TIME	SURFACE POSITION																			
	DIRECT		VERTICAL SURFACES – FACING																HORIZONTAL	
	NOMINAL		NORTH		NORTHEAST		EAST		SOUTHEAST		SOUTH		SOUTHWEST		WEST		NORTHWEST		SURFACE	
	Btu/ft² hr	W/m²	Btu/ft² hr	W/m²	Btu/ft² hr	W/m²	Btu/ft² hr	W/m²	Btu/ft² hr	W/m²	Btu/ft² hr	W/m²	Btu/ft² hr	W/m²	Btu/ft² hr	W/m²	Btu/ft² hr	W/m²	Btu/ft² hr	W/m²
7 AM	66	208.06	3	9.46	30	94.37	66	208.06	63	198.61	24	75.66	3	9.46	2	6.30	3	9.46	14	44.13
8 AM	230	725.07	13	40.98	43	135.56	190	598.31	210	662.02	108	340.47	13	40.96	10	31.52	13	40.98	53	167.88
9 AM	282	889.00	21	66.20	29	91.42	192	605.28	243	766.05	152	479.18	20	63.05	17	53.59	18	56.74	120	378.30
10 AM	301	943.90	24	75.66	24	75.66	142	447.65	244	769.21	199	627.34	39	122.95	24	75.66	22	69.35	173	545.38
11 AM	311	980.42	26	81.96	24	75.66	75	236.44	196	617.89	215	674.63	87	274.27	26	81.96	24	75.66	204	643.11
12 noon	314	989.88	27	85.12	24	75.66	31	97.73	158	498.09	220	662.02	158	498.06	31	97.73	24	75.66	217	684.09
1 PM	311	980.42	26	81.96	24	75.66	26	81.96	87	274.27	215	674.63	196	617.89	75	236.94	24	75.66	204	643.11
2 PM	301	943.90	24	75.66	22	69.35	24	75.66	39	122.95	199	627.34	244	769.21	142	447.63	24	75.66	173	545.38
3 PM	282	889.00	21	66.20	18	56.74	17	53.59	20	63.05	152	479.18	243	766.05	192	605.28	29	91.42	120	378.30
4 PM	230	725.07	13	40.98	13	40.98	10	31.52	13	40.96	108	340.37	210	662.02	190	598.31	43	135.56	53	167.80
5 PM	66	208.06	3	9.46	3	9.46	2	6.30	3	9.46	24	75.66	63	198.61	66	208.06	30	94.57	14	44.13

32 DEGREE NORTH LATITUDE

STD. SOLAR TIME	SURFACE POSITION																			
	DIRECT		VERTICAL SURFACES – FACING																HORIZONTAL	
	NOMINAL		NORTH		NORTHEAST		EAST		SOUTHEAST		SOUTH		SOUTHWEST		WEST		NORTHWEST		SURFACE	
	Btu/ft² hr	W/m²	Btu/ft² hr	W/m²	Btu/ft² hr	W/m²	Btu/ft² hr	W/m²	Btu/ft² hr	W/m²	Btu/ft² hr	W/m²	Btu/ft² hr	W/m²	Btu/ft² hr	W/m²	Btu/ft² hr	W/m²	Btu/ft² hr	W/m²
7 AM	35	110.34	1	3.15	13	40.98	53	167.08	57	179.69	7	22.07	1	3.15	2	6.30	1	3.15	8	25.21
8 AM	215	677.78	12	37.83	38	119.79	157	494.93	178	561.14	96	302.64	7	22.07	10	31.52	7	22.07	38	119.79
9 AM	263	829.10	15	47.29	25	78.81	175	551.68	240	756.60	170	535.32	10	31.52	17	53.59	17	53.59	116	365.69
10 AM	295	929.98	17	53.59	25	78.81	131	412.98	242	762.90	208	655.72	76	239.59	22	69.35	23	72.51	168	529.66
11 AM	300	945.74	23	72.51	24	75.66	66	208.06	214	674.63	222	699.85	108	340.47	23	72.51	24	75.66	198	624.19
12 noon	303	955.20	24	75.66	24	75.66	26	81.96	168	529.66	238	750.29	168	529.66	26	81.96	24	75.66	204	643.11
1 PM	300	945.74	23	72.51	24	75.66	23	72.51	108	340.47	222	699.85	214	674.63	66	208.06	24	75.66	198	624.19
2 PM	295	929.98	17	53.59	23	72.51	22	69.35	76	239.59	208	655.72	242	762.90	131	412.98	25	78.81	168	529.66
3 PM	268	829.10	15	47.29	17	53.59	17	53.59	10	31.52	170	535.32	240	756.60	175	551.68	25	78.81	116	365.69
4 PM	215	677.78	12	37.83	7	22.07	10	31.52	7	22.07	96	302.64	178	561.14	157	494.93	38	119.79	38	119.79
5 PM	35	110.34	1	3.15	1	3.15	2	6.30	1	3.15	7	22.07	57	179.62	53	167.08	13	40.98	8	25.21

Heat gain per hour in position listed under clear unclouded weather conditions.

(Sheet 7 of 12)

TABLE 36 (Continued)

AUTUMN SOLAR HEAT GAIN AT LISTED SOLAR TIME (DESIGN AVE.)
40 DEGREE NORTH LATITUDE

STD. SOLAR TIME	SURFACE POSITION																			
	DIRECT		VERTICAL SURFACES – FACING																HORIZONTAL	
	NOMINAL		NORTH		NORTHEAST		EAST		SOUTHEAST		SOUTH		SOUTHWEST		WEST		NORTHWEST		SURFACE	
	Btu/ft² hr	W/m²	Btu/ft² hr	W/m²	Btu/ft² hr	W/m²	Btu/ft² hr	W/m²	Btu/ft² hr	W/m²	Btu/ft² hr	W/m²	Btu/ft² hr	W/m²	Btu/ft² hr	W/m²	Btu/ft² hr	W/m²	Btu/ft² hr	W/m²
7 AM	28	88.27	1	3.15	9	28.37	47	148.17	52	163.92	11	34.68	5	15.76	3	9.46	3	9.46	8	25.22
8 AM	212	668.32	11	34.68	13	40.98	121	381.45	170	532.32	74	233.28	13	40.98	11	34.68	6	18.91	37	116.64
9 AM	252	794.42	15	47.29	14	44.13	162	510.70	230	725.07	156	491.79	27	85.12	18	56.74	16	50.44	108	340.47
10 AM	275	866.93	17	53.59	15	47.29	125	394.06	236	743.99	184	580.06	36	113.49	23	72.51	22	69.35	149	469.72
11 AM	285	898.46	20	63.05	16	50.44	63	198.61	207	652.56	214	674.63	85	267.96	24	75.66	24	75.66	172	642.23
12 noon	292	920.52	21	66.20	18	56.74	24	75.66	168	529.66	230	725.07	168	529.66	24	75.66	18	56.44	183	576.90
1 PM	285	898.46	20	63.05	24	75.66	24	75.66	85	267.96	214	674.63	207	652.56	63	198.61	16	50.44	172	642.23
2 PM	275	866.93	17	53.59	22	69.35	23	72.51	36	113.49	184	580.06	236	743.99	125	394.06	15	47.29	149	469.72
3 PM	252	794.42	15	47.29	16	50.44	18	56.74	27	85.12	156	491.79	230	725.07	162	510.70	14	44.13	108	340.47
4 PM	212	668.32	11	34.68	6	18.91	11	34.68	13	40.98	74	233.28	170	532.32	121	381.45	13	40.98	37	116.64
5 PM	28	88.27	1	3.15	3	9.46	3	9.46	5	15.76	11	34.68	52	163.92	47	148.17	9	28.37	8	25.22

48 DEGREE NORTH LATITUDE

STD. SOLAR TIME	SURFACE POSITION																			
	DIRECT		VERTICAL SURFACES – FACING																HORIZONTAL	
	NOMINAL		NORTH		NORTHEAST		EAST		SOUTHEAST		SOUTH		SOUTHWEST		WEST		NORTHWEST		SURFACE	
	Btu/ft² hr	W/m²	Btu/ft² hr	W/m²	Btu/ft² hr	W/m²	Btu/ft² hr	W/m²	Btu/ft² hr	W/m²	Btu/ft² hr	W/m²	Btu/ft² hr	W/m²	Btu/ft² hr	W/m²	Btu/ft² hr	W/m²	Btu/ft² hr	W/m²
7 AM	25	78.81	0	0	8	25.22	43	135.56	49	154.47	11	34.68	3	9.46	3	9.46	3	9.46	6	18.91
8 AM	150	472.37	5	15.76	12	37.83	112	353.08	161	507.53	73	230.13	13	40.98	11	34.68	10	31.52	32	100.88
9 AM	205	646.25	7	22.07	13	40.98	156	491.79	222	699.85	150	472.87	26	81.96	17	53.59	21	66.20	98	308.99
10 AM	230	725.07	11	34.68	14	44.13	118	371.99	228	718.77	178	561.14	35	110.34	22	69.35	25	78.81	136	428.74
11 AM	262	825.95	13	40.98	15	47.29	61	192.30	198	624.19	202	636.80	83	261.36	24	75.66	28	88.27	153	482.40
12 noon	264	832.25	16	50.44	17	53.59	33	104.03	160	504.40	222	699.85	160	504.40	33	104.03	17	53.59	162	510.70
1 PM	262	825.95	13	40.98	23	72.51	24	75.66	83	261.36	202	636.80	198	624.19	61	192.30	15	47.29	153	482.40
2 PM	230	725.07	11	34.68	22	69.35	22	69.35	35	110.34	178	561.14	228	718.77	118	371.99	14	44.13	136	428.74
3 PM	205	646.25	7	22.07	16	50.44	17	53.59	26	81.96	150	472.87	222	699.85	156	491.79	13	40.98	98	308.99
4 PM	150	472.37	5	15.76	6	18.91	11	34.68	13	40.98	73	230.13	161	507.55	112	353.08	12	37.38	32	100.88
5 PM	25	78.81	0	0	3	9.46	3	9.46	3	9.46	11	34.68	49	154.47	43	135.56	8	25.22	6	18.91

Heat gain per hour in position listed under clear, unclouded weather conditions.

(Sheet 8 of 12)

TABLE 36 (Continued)

AUTUMN SOLAR HEAT GAIN AT LISTED SOLAR TIME (DESIGN AVE.)
56 DEGREE NORTH LATITUDE

STD. SOLAR TIME	SURFACE POSITION																			
	DIRECT NOMINAL		VERTICAL SURFACES – FACING																HORIZONTAL SURFACE	
			NORTH		NORTHEAST		EAST		SOUTHEAST		SOUTH		SOUTHWEST		WEST		NORTHWEST			
	Btu/ft² hr	W/m²	Btu/ft² hr	W/m²	Btu/ft² hr	W/m²	Btu/ft² hr	W/m²	Btu/ft² hr	W/m²	Btu/ft² hr	W/m²	Btu/ft² hr	W/m²	Btu/ft² hr	W/m²	Btu/ft² hr	W/m²	Btu/ft² hr	W/m²
7 AM	23	72 51	0	0.00	7	22.07	34	107.18	30	94.37	6	18.91	2	6.30	2	6.30	2	6.30	4	12.61
8 AM	99	312.10	3	9.46	15	47.29	85	267.96	92	29.03	42	132.40	5	15.76	5	15.76	5	15.76	19	59.90
9 AM	179	564.29	7	22.07	16	50.44	116	365.69	159	501.24	107	337.32	10	31.52	8	25.22	8	25.22	42	132.40
10 AM	215	677.78	11	34.68	16	50.44	104	329.85	197	621.04	171	539.07	39	122.95	13	40.98	13	40.98	67	211.22
11 AM	236	743.99	13	40.98	16	50.44	53	167.88	190	598.31	209	655.72	101	318.40	15	47.29	15	47.29	84	264.81
12 noon	242	762.90	15	47.29	16	50.44	18	56.74	155	488.65	221	696.70	155	488.63	18	56.74	16	50.44	91	286.88
1 PM	236	743.99	13	40.98	15	47.29	15	47.29	101	318.40	209	655.72	190	598.31	53	167.88	16	50.44	84	264.81
2 PM	215	677.78	11	34.68	13	40.98	13	40.29	39	122.95	171	539.07	197	621.04	104	328.85	16	50.44	67	211.22
3 PM	179	564.29	7	22.07	8	25.22	8	25.22	10	31.52	107	337.32	159	501.24	116	365.69	16	50.44	42	132.40
4 PM	99	312.10	3	9.46	5	15.76	5	15.76	5	15.76	42	132.40	92	290.03	85	267.96	15	47.24	19	50.90
5 PM	23	72.51	0	0.00	2	6.30	2	6.30	2	6.30	6	18.91	30	94.37	34	107.18	7	22.07	4	12.61

Heat gain per hour in position listed under clear, unclouded weather conditions.

(Sheet 9 of 12)

TABLE 36 (Continued)

WINTER SOLAR HEAT GAIN AT LISTED SOLAR TIME (DESIGN AVE.)
24 DEGREE NORTH LATITUDE

STD. SOLAR TIME	SURFACE POSITION																			
	DIRECT NOMINAL		VERTICAL SURFACES – FACING																HORIZONTAL SURFACE	
			NORTH		NORTHEAST		EAST		SOUTHEAST		SOUTH		SOUTHWEST		WEST		NORTHWEST			
	Btu/ft² hr	W/m²	Btu/ft² hr	W/m²	Btu/ft² hr	W/m²	Btu/ft² hr	W/m²	Btu/ft² hr	W/m²	Btu/ft² hr	W/m²	Btu/ft² hr	W/m²	Btu/ft² hr	W/m²	Btu/ft² hr	W/m²	Btu/ft² hr	W/m²
7 AM	123	356.23	3	9.46	31	97.73	75	236.44	73	230.13	23	72.57	2	6.30	3	9.46	3	9.46	7	22.07
8 AM	232	731.36	12	37.83	50	157.62	185	583.21	216	680.96	104	329.85	12	37.83	12	37.83	10	31.52	56	176.54
9 AM	285	898.46	19	59.90	27	85.12	182	573.75	250	288.12	155	488.63	20	63.05	19	59.90	16	50.44	120	378.30
10 AM	308	970.96	24	75.66	24	75.66	145	457.11	243	766.05	198	624.19	35	110.34	24	75.66	19	59.90	178	561.14
11 AM	317	999.34	27	85.12	24	75.66	74	233.28	206	649.41	206	649.41	88	277.42	27	85.12	22	69.35	200	630.50
12 noon	320	1000.79	28	88.27	23	72.51	29	91.42	152	479.18	225	709.31	152	479.18	29	91.42	23	72.51	204	643.11
1 PM	317	999.34	27	85.12	22	69.35	27	85.12	88	277.42	206	649.41	206	649.41	74	233.28	24	75.66	200	630.50
2 PM	308	970.96	24	75.66	19	59.90	24	75.66	35	110.34	198	624.19	243	766.05	145	457.11	24	75.66	178	561.14
3 PM	285	898.46	19	59.90	16	50.44	19	59.90	20	63.05	155	488.62	250	788.12	182	573.75	27	85.12	120	378.30
4 PM	232	731.36	12	37.83	10	31.52	12	37.83	12	37.83	104	329.85	216	680.96	185	583.21	50	157.62	56	176.54
5 PM	123	356.23	3	9.46	3	9.46	3	9.46	2	6.30	23	72.57	73	230.13	75	236.44	31	97.73	7	22.07

32 DEGREE NORTH LATITUDE

STD. SOLAR TIME	SURFACE POSITION																			
	DIRECT NOMINAL		VERTICAL SURFACES – FACING																HORIZONTAL SURFACE	
			NORTH		NORTHEAST		EAST		SOUTHEAST		SOUTH		SOUTHWEST		WEST		NORTHWEST			
	Btu/ft² hr	W/m²	Btu/ft² hr	W/m²	Btu/ft² hr	W/m²	Btu/ft² hr	W/m²	Btu/ft² hr	W/m²	Btu/ft² hr	W/m²	Btu/ft² hr	W/m²	Btu/ft² hr	W/m²	Btu/ft² hr	W/m²	Btu/ft² hr	W/m²
7 AM	63	198.61	1	3.15	23	72.51	55	173.39	48	151.32	13	40.98	1	3.15	1	3.15	T	3.15	3	9.46
8 AM	196	617.89	6	18.91	36	113.49	166	523.31	193	608.43	97	305.79	9	28.37	9	28.37	9	28.37	39	122.95
9 AM	255	803.88	14	44.13	22	69.35	171	539.07	233	734.53	152	479.18	16	50.44	11	34.68	11	34.68	95	299.49
10 AM	289	911.07	20	63.05	20	63.05	128	403.52	247	778.66	204	643.11	42	132.40	17	53.59	17	53.59	143	450.80
11 AM	308	970.96	23	72.51	23	72.51	69	249.05	218	687.24	226	712.46	104	327.85	24	75.66	24	75.66	173	545.38
12 noon	311	980.42	24	75.66	24	75.66	27	85.12	162	510.90	238	750.29	162	510.70	27	85.12	24	75.66	183	576.90
1 PM	308	970.96	23	72.51	24	75.66	24	75.66	104	327.85	226	712.46	218	687.24	69	249.04	23	72.51	173	545.38
2 PM	289	911.07	20	63.05	17	53.59	17	53.59	42	132.40	204	643.11	247	778.66	128	403.52	20	63.05	143	450.80
3 PM	255	803.88	14	44.13	11	34.68	11	34.68	16	50.44	152	479.18	233	734.53	171	539.07	22	69.35	95	299.48
4 PM	196	617.89	6	18.91	9	28.37	9	28.37	9	28.37	97	305.79	193	608.43	166	523.31	36	113.46	39	122.95
5 PM	63	198.61	1	3.15	1	3.15	1	3.15	1	3.15	13	40.98	48	151.32	55	173.39	23	72.51	3	9.46

Heat gain per hour in position listed under clear, unclouded weather conditions.

(Sheet 10 of 12)

TABLE 36 (Continued)

WINTER SOLAR HEAT GAIN AT LISTED SOLAR TIME (DESIGN AVE.)
40 DEGREE NORTH LATITUDE

STD. SOLAR TIME	DIRECT NOMINAL		NORTH		NORTHEAST		EAST		SOUTHEAST		SOUTH		SOUTHWEST		WEST		NORTHWEST		HORIZONTAL SURFACE	
	Btu/ft² hr	W/m²	Btu/ft² hr	W/m²	Btu/ft² hr	W/m²	Btu/ft² hr	W/m²	Btu/ft² hr	W/m²	Btu/ft² hr	W/m²	Btu/ft² hr	W/m²	Btu/ft² hr	W/m²	Btu/ft² hr	W/m²	Btu/ft² hr	W/m²
7 AM	17	53.59	0	0	7	22.07	16	50.44	16	50.44	4	12.61	0	0	0	0	0	0	1	3.15
8 AM	149	469.72	5	15.76	24	75.66	120	378.30	138	435.04	72	226.98	6	18.91	6	18.91	6	18.91	20	63.05
9 AM	242	762.90	12	37.83	13	40.98	158	498.09	225	709.31	155	488.63	15	47.29	12	37.83	12	37.83	63	198.61
10 AM	281	885.84	18	56.74	17	53.59	126	397.21	241	759.75	178	561.14	48	151.32	17	53.59	17	53.59	105	331.01
11 AM	289	911.07	19	59.50	19	59.90	63	198.61	220	693.54	208	655.72	113	356.23	18	56.74	18	56.74	132	416.13
12 noon	294	926.83	20	63.05	20	63.05	22	69.35	175	551.68	226	712.46	175	551.68	22	69.35	20	63.05	142	447.65
1 PM	289	911.07	19	59.50	18	56.74	18	56.74	113	356.23	208	655.72	220	693.54	63	198.61	19	59.90	132	416.13
2 PM	281	885.84	18	56.74	17	53.59	17	53.59	48	151.32	178	561.14	241	759.75	126	397.21	17	53.59	105	331.01
3 PM	242	762.90	12	37.83	12	37.83	12	37.83	15	47.29	155	488.63	225	709.31	158	498.09	13	40.98	63	198.61
4 PM	149	469.72	5	15.76	6	18.91	6	18.91	6	18.91	72	226.98	138	435.04	120	378.30	24	75.66	20	63.05
5 PM	17	53.59	0	0	0	0	0	0	0	0	4	12.61	16	50.44	16	50.44	7	22.07	1	3.15

48 DEGREE NORTH LATITUDE

STD. SOLAR TIME	DIRECT NOMINAL		NORTH		NORTHEAST		EAST		SOUTHEAST		SOUTH		SOUTHWEST		WEST		NORTHWEST		HORIZONTAL SURFACE	
	Btu/ft² hr	W/m²	Btu/ft² hr	W/m²	Btu/ft² hr	W/m²	Btu/ft² hr	W/m²	Btu/ft² hr	W/m²	Btu/ft² hr	W/m²	Btu/ft² hr	W/m²	Btu/ft² hr	W/m²	Btu/ft² hr	W/m²	Btu/ft² hr	W/m²
7 AM	1	3.15	0	0	0	0	1	3.15	1	3.15	0	0	0	0	0	0	0	0	0	0
8 AM	121	381.45	3	9.46	13	40.98	59	186.10	66	208.06	33	104.03	2	6.30	3	9.46	3	9.46	8	25.91
9 AM	190	598.31	8	25.11	11	34.68	88	217.42	179	564.29	103	324.71	10	31.52	8	25.22	8	25.22	34	107.18
10 AM	242	762.90	11	34.68	16	50.44	109	340.47	217	684.09	193	604.43	45	141.86	12	37.83	12	37.83	65	204.91
11 AM	263	829.10	14	44.13	16	50.44	54	170.23	206	649.41	230	725.07	111	349.95	15	47.29	15	47.29	87	274.27
12 noon	269	848.02	15	15.76	16	50.44	17	53.59	170	535.32	244	769.21	170	535.32	17	53.59	16	50.44	95	299.49
1 PM	263	829.10	14	44.13	15	47.29	15	47.29	111	349.95	230	725.07	206	649.41	54	170.23	16	50.44	87	274.27
2 PM	242	762.90	11	34.68	12	37.83	12	37.83	45	141.86	193	608.43	217	684.09	109	340.47	16	50.44	65	204.91
3 PM	190	598.31	8	25.22	8	25.22	8	25.22	10	31.52	103	324.71	179	564.29	88	217.42	11	34.68	34	107.18
4 PM	121	381.45	3	9.46	3	9.46	3	9.46	2	6.30	33	104.03	66	208.06	59	186.10	13	40.98	8	25.91
5 PM	1	3.15	0	0	0	0	0	0	0	0	0	0	1	3.15	1	3.15	0	0	0	0

Heat gain per hour in position listed under clear, unclouded weather conditions.

(Sheet 11 of 12)

TABLE 36 (Continued)

WINTER SOLAR HEAT GAIN AT LISTED SOLAR TIME (DESIGN AVE.)
56 DEGREE NORTH LATITUDE

STD. SOLAR TIME	SURFACE POSITION																			
	DIRECT NOMINAL		VERTICAL SURFACES – FACING																HORIZONTAL SURFACE	
			NORTH		NORTHEAST		EAST		SOUTHEAST		SOUTH		SOUTHWEST		WEST		NORTHWEST			
	Btu/ft² hr	W/m²	Btu/ft² hr	W/m²	Btu/ft² hr	W/m²	Btu/ft² hr	W/m²	Btu/ft² hr	W/m²	Btu/ft² hr	W/m²	Btu/ft² hr	W/m²	Btu/ft² hr	W/m²	Btu/ft² hr	W/m²	Btu/ft² hr	W/m²
8 AM	38	119.79	1	3.15	6	18.91	31	97.73	38	119.79	18	56.24	1	3.15	1	3.15	1	3.15	3	9.46
9 AM	96	302.64	3	9.46	6	18.91	63	198.61	91	286.88	64	201.76	3	9.46	3	9.46	3	9.46	14	44.13
10 AM	176	534.84	7	22.07	7	22.07	78	245.89	160	540.40	145	457.11	15	47.29	9	28.37	9	28.37	31	97.73
11 AM	211	665.17	10	31.52	10	31.52	41	129.25	178	561.14	189	595.82	95	299.49	10	31.52	10	31.52	45	141.86
12 noon	221	696.70	10	31.52	10	31.52	12	37.83	144	453.96	206	649.41	144	453.96	12	37.83	10	31.52	51	160.78
1 PM	211	665.17	10	31.52	10	31.52	10	31.52	95	299.49	189	595.82	178	561.14	41	129.25	10	31.52	45	141.86
2 PM	176	534.84	7	22.07	9	28.37	9	28.37	15	47.29	145	457.11	160	504.40	78	245.89	7	22.07	31	97.73
3 PM	96	302.64	3	9.46	3	9.46	3	9.46	3	9.46	64	201.76	91	286.88	63	198.61	6	18.91	14	44.13
4 PM	38	119.79	1	3.15	1	3.15	1	3.15	1	3.15	18	56.74	38	119.79	31	97.73	6	18.91	3	9.46

(Sheet 12 of 12)

Heat gain per hour in position listed under clear, unclouded weather conditions.

TABLE 36 (Continued)

HEAT TRANSFER OF SOLAR RADIATION THROUGH GLASS (Summer)

Example 13—*English Units*

Determine heat gain into house due to direct solar radiation through glass during the clear summer day. The house is located approximately 32 degrees North latitude. Its wall positions are Northeast, Southeast, Southwest and Northwest. The surface area of glass in windows and doors are 62 sq ft in the Northeast wall, 86 sq ft in the Southeast wall, 108 sq ft in the Southwest wall and 86 sq ft in the Northwest wall. The direct transmittance of solar radiation through the glass is 85%.

From Table 36, Summer, 32 degrees North latitude, determine the solar heat gain to surfaces in the Northeast, Southeast, Southwest and Northwest for each hour. Multiply these values by 0.85 — the transmission of solar radiation of glass. These are also multiplied by the number or square feet of glass in each wall so as to obtain the total heat gain through glass for each wall. For any hour the total for all walls are added together. For total per day, all totals of gains per hour are added together. For a clear day the tabulation is as shown in the following chart.

Example 13a—*Metric Units*

Determine the heat gain into the house due to direct solar radiation during a clear summer day. The house is located approximately 32 degrees North latitude. Its wall positions are Northeast, Southeast, Southwest and Northwest.

The square meters of glass and doors in each of the walls is: 5.75m^2 in the Northeast wall, 8m^2 in the Southeast wall, 10m^2 in the Southwest wall and 8m^2 in the Northwest wall. The direct transmittance of solar radiation through the glass is 85%.

From Table 36, Summer, 32 degress North latitude determine the solar heat gain to surfaces in the Northeast, Southeast, Southwest and Northwest for each hour. Multiply these values by 0.85 — the transmission of solar radiation through the glass. This transmission is also multiplied by the square meters of glass in each wall so as to obtain the total heat gain through the glass for each wall. For any hour, the total gain into the house is all the glass heat transmittances added together. Likewise, the total gain into the house per day is the totals of the glass in all the walls added together. For a clear day, the tabulation is as shown on the following chart.

SOLAR HEAT THROUGH GLASS

STD. SOLAR TIME	Heat Gain of Building Glass as Described in Example 11 — Glass Btu Heat Gain — Summer — 32° N Latitude — Walls in Position as Noted — Sq ft Glass as Shown								TOTAL ALL WALLS BTU'S PER HR
	NORTHEAST		SOUTHEAST		SOUTHWEST		NORTHWEST		
	Sol. Rad. Btu/ft^2 hr (Table 36)	0.85 x 62 ft^2 glass = 52.7 ft^2	Sol. Rad. Btu/ft^2 hr (Table 36)	0.85 x 86 ft^2 glass = 73.1 ft^2	Sol. Rad. Btu/ft^2 hr (Table 36)	0.85 x 108 ft^2 glass = 91.8 ft^2	Sol. Rad. Btu/ft^2 hr (Table 36)	0.85 x 86 ft^2 glass = 73.1 ft^2	
6 AM	58	3,056	54	3,947	5	459	6	439	7,901
7 AM	158	8,326	130	9,503	17	1,561	17	1,243	20,633
8 AM	163	8,590	150	10.965	25	2,295	25	1,828	23,678
9 AM	192	7,483	153	11,184	30	2,754	30	2,193	23,614
10 AM	93	4,901	161	11,769	42	3,856	35	2,559	23,085
11 AM	75	3,952	140	10,234	65	5,967	37	2,705	22,858
12 noon	39	2,055	80	5,848	80	7,344	39	2,851	18,098
1 PM	37	1,949	65	4,751	140	12,852	75	5,483	25,035
2 PM	35	1,844	42	3,070	161	14,780	93	6,798	26,492
3 PM	30	1,581	30	2,193	153	14,045	142	10,380	26,199
4 PM	25	1,317	25	1,827	150	13,770	163	11,915	26,829
5 PM	17	896	17	1,243	130	11,934	158	11,550	25,623
6 PM	6	316	5	365	54	4,957	58	4,240	9,866
TOTALS PER DAY	878	46,266	1,052	76,899	1,052	96,574	878	62,184	281,923

* This total heat gain through the glass is the heat transferred by radiation only. Through this same area of glass, additional heat gain or lose will occur due to temperature differences between inside and outside.

Total Per Day All Glass*

EXAMPLE 13

SOLAR HEAT THROUGH GLASS

STD. SOLAR TIME	Heat Gain of Building as Described in Example 11A — Glass Watts Heat Gain – Summer — Walls in Position as Noted – Sq Meters of Glass as Shown								TOTAL ALL WALLS WATTS/ HOUR
	NORTHEAST		SOUTHEAST		SOUTHWEST		NORTHWEST		
	Solar Rad. w/m² (Table 36)	5.75m² x 0.85 = 4.89	Solar Rad. w/m² (Table 36)	8m² x 0.85 = 6.8	Solar Rad. w/m² (Table 36)	10m² x 0.85 = 8.5	Solar Rad. w/m² (Table 36)	8m² x 0.85 = 6.8	
6 AM	182.84	894.08	170.23	1,157.56	15.76	133.96	18.91	128.59	2,314.19
7 AM	498.09	2,435.66	409.86	2,787.05	53.59	455.52	58.59	398.41	6,076.64
8 AM	513.85	2,512.73	472.87	3,215.52	78.81	669.68	78.81	535.23	6,933.16
9 AM	447.62	2,188.86	482.40	3,280.32	95.57	812.35	95.57	649.88	6,931.41
10 AM	293.19	1,433.70	507.55	3,451.34	132.40	1,125.40	110.34	750.31	6,760.75
11 AM	236.44	1,156.19	441.35	3,001.18	204.91	1,741.74	116.64	793.15	6,692.26
12 noon	122.95	601.23	252.20	1,714.96	252.20	2,143.70	122.95	836.06	5,295.95
1 PM	116.64	570.37	204.91	1,393.39	441.35	3,751.48	236.44	1,607.79	7,323.03
2 PM	110.34	539.56	132.40	900.32	507.55	4,314.18	293.18	1,993.62	7.747.68
3 PM	95.57	467.34	95.57	647.88	482.40	4,100.40	447.62	3,043.82	8,259.44
4 PM	78.81	385.38	78.81	535.91	472.87	4,555.24	513.35	3,490.78	8,967.31
5 PM	53.59	262.05	53.59	364.41	409.82	3,483.47	498.01	3,386.47	7,496.40
6 PM	18.91	92.47	15.76	107.17	170.23	1,446.95	182.01	1,237.67	2,884.26
TOTALS PER DAY	2,769.04	13,533.55	3,317.49	22,557.01	3,317.49	28,734.07	2,772.42	18,821.78	83,682.48

* This 83,682.48 Watts of energy is the heat gain through the glass by solar radiation only. Additional heat will be transferred because of conduction and convection effects.

Total Per Day All Glass*

EXAMPLE 13a

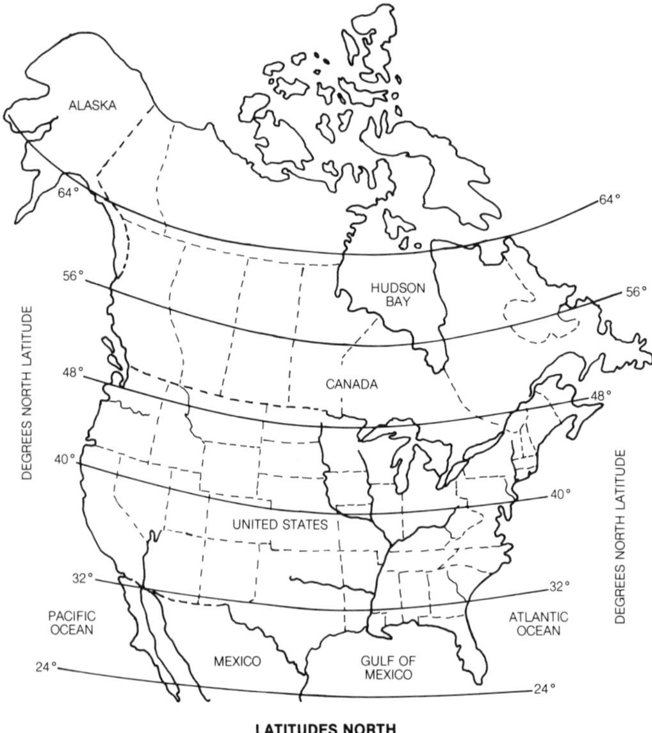

LATITUDES NORTH
USED TO ESTABLISH SOLAR RADIATION

Figure 45

The preceding example was selected to be at 32 degrees North latitude as a great number of buildings and homes requiring cooling during summer would be in this approximate zone. In more northern climates, the units of solar heat per unit area may be less, but they are still a very significant consideration in designing structures to be energy efficient.

Figure 45 provides a map of the United States and Canada to assist in the selection of the correct Solar Heat Gain tables to use for a specific installation.

To illustrate the difference in heat gain in summer by solar radiation, assume that the house described in Example 11 and 11A was located farther North near the 56 degree North latitude. At this location, find the direct transmittance of solar radiation through the glass.

EFFECT ON LOCATION, AT LATITUDE

Example 14—*English Units*

The house is located approximately 56 degrees North latitude. Its wall positions are Northeast, Southeast, Southwest and Northwest. The square feet of glass in windows and doors are 62 sq ft in the Northeast wall, 86 sq ft in the Southeast wall, 108 sq ft in the Southwest wall and 86 sq ft in the Northwest wall. Transmittance of solar radiation through glass is 85%. Based on information in Table 36, Summer, 56 degrees North latitude, and multiplying the values per unit area by 0.85 times the glass area, the tabulation will be as shown in the following chart.

SOLAR HEAT THROUGH GLASS

STD. SOLAR TIME	Heat Gain of Building Glass as Described in Example 12 — Glass Btu Heat Gain – Summer – 56° N. Latitude — Walls in Position as Noted – Sq meters of Glass as Shown								TOTAL GLASS ALL WALLS BTU'S PER HR
	NORTHEAST WALL		SOUTHEAST WALL		SOUTHWEST WALL		NORTHWEST WALL		
	Solar Rad. Btu/ft² hr (Table 36)	0.85 x 62 ft² = 52.7 Btu/hr	Solar Rad. Btu/ft² hr (Table 36)	0.85 x 86 ft² = 73.1 Btu/hr	Solar Rad. Btu/ft² hr (Table 36)	0.85 x 108 ft² = 91.8 Btu/hr	Solar Rad. Btu/ft² hr (Table 36)	0.85 x 86 ft² = 73.1 Btu/hr	
5 AM	69	3,636	47	3,436	5	459	5	365	7,896
6 AM	132	6,956	111	8,114	10	918	10	731	16,719
7 AM	140	7,378	145	10,600	18	1,652	18	1,316	20,946
8 AM	110	5,797	185	13,524	23	2,111	24	1,754	23,186
9 AM	60	3,162	204	14,912	27	2,479	28	2,046	22,599
10 AM	44	2,319	199	14,547	34	3,121	30	2,193	22,180
11 AM	34	1,792	179	13,085	130	11,934	31	2,266	29,044
12 noon	33	1,739	145	10,600	145	13,311	33	2,412	28,062
1 PM	31	1,634	130	9,503	179	16,432	34	2,485	30,054
2 PM	30	1,581	34	2,485	199	18,268	44	3,216	25,550
3 PM	28	1,473	27	1,974	204	18,727	60	4,386	26,560
4 PM	24	1,265	23	1,681	185	16,983	110	8,041	27,970
5 PM	18	949	18	1,316	145	13,311	140	10,234	25,810
6 PM	10	527	10	731	111	10,190	132	9,649	21,097
7 PM	5	263	5	365	47	4,314	69	5,044	9,986
TOTALS PER DAY	768	40,471	1,462	106,877	1,462	134,109	768	56,138	337,659

EXAMPLE 14

Example 14a—*Metric Units*

Using the same house and location as in Example 14, the surface area of windows and doors is 5.75m² in the Northeast wall, 10m² in the Southwest wall and 8m² in the Northwest wall. Calculated in the same manner as Example 14, using values obtained in Table 36, the tabulation is shown on the chart at the top of the following page.

It is apparent that the amount of solar radiation which can be transmitted through glass into an interior is very high regardless of the location by latitude. However, it must be remembered that calculations given in Examples 12 and 14 were based on sunlight per hour. Climatic conditions and locations effect the number of hours of direct sunlight. To illustrate this, the average number of hours of sunshine per average year is tabulated in Table 37.

In most cases where heat gain is being calculated, two things are of interest. The first is maximum heat gain per hour, for some relatively short period of time so as to determine maximum cooling requirements. The other is the amount of heat gain over a period of a month so as to calculate energy costs on a monthly basis. In addition, the solar heat radiated to a surface is influenced by weather conditions, thus there are no duplicate conditions for any one day or one month or one year. This forces most calculations of solar heating to be based on averages.

For this reason it is necessary to provide Table 38 which gives the mean heat gain total per day, averaged by season. These figures are based on solar radiation to surfaces in various positions on a clear, cloudless day. Unfortunately, solar hours are recorded as hours per month, without regard to solar time of day, thus it is necessary to establish average heat per hour mean radiation to surface. The *Average Heat Radiated per Sunshine Hour* for various degrees of North latitude and various surface positions is given in Table 39.

The average heat per hour and hours of sunshine per month are necessary in heat transfer calculations since weather conditions are so varied that particular conditions for a specific date in a year may not be duplicated for many, many years — if ever. Also, energy requirements are generally measured by the month. To illustrate the use of these tables the following problem is presented. This problem requires the determining of heat gain through glass of a home located in Savannah, Georgia.

SOLAR HEAT THROUGH GLASS

STD. SOLAR TIME	Heat Gain of Building Glass as Described in Example 12A — Glass Watt Heat Gain – Summer 56° N. Latitude — Walls in Position as Noted – Square Meters of Glass as Shown								TOTAL GLASS ALL WALLS WATTS PER HR
	NORTHEAST WALL		SOUTHEAST WALL		SOUTHWEST WALL		NORTHWEST WALL		
	Solar Rad. Watts/hr (Table 36)	0.85 x 5.75m² = 4.9 Watts/hr	Solar Rad. Watts/hr (Table 36)	0.85 x 8m² = 6.8 Watts/hr	Solar Rad. Watts/hr (Table 36)	0.85 x 10m² = 8.5 Watts/hr	Solar Rad. Watts/hr (Table 36)	0.85 x 8m² = 6.8 Watts/hr	
5 AM	217.52	1,066	148.17	1,007	15.76	134	15.76	107	2,314
6 AM	416.13	2,039	349.93	2,379	31.52	268	31.52	214	4,900
7 AM	441.35	2,163	456.11	3,108	56.74	482	56.74	386	6,139
8 AM	346.77	1,699	583.21	3,966	72.51	616	75.66	514	6,795
9 AM	189.85	930	643.11	4,373	85.12	723	88.27	600	6,626
10 AM	138.71	678	627.34	4,266	104.03	884	94.37	642	6,470
11 AM	107.18	525	564.29	3,837	409.82	3,483	97.73	664	8,509
12 noon	104.03	510	457.11	3,108	457.11	3,885	104.02	707	8,210
1 PM	97.73	479	409.82	2,787	564.29	4,796	107.18	729	8,791
2 PM	94.37	462	107.18	729	627.34	5,332	138.71	943	7,466
3 PM	88.27	433	85.12	579	643.11	5,466	187.85	1,277	7,755
4 PM	75.66	371	72.51	493	583.21	5,000	346.77	2,358	8,222
5 PM	56.74	278	56.74	386	457.11	3,885	441.34	3,001	7,550
6 PM	31.56	155	31.52	214	349.93	2,974	416.13	2,830	6,173
7 PM	15.76	77	15.76	107	148.19	1,260	217.52	1,479	2.923
TOTALS PER DAY	2,421.67	11,865	4,608.91	31,339	4,605.79	39,186	2,419.57	16,451	98,843

EXAMPLE 14a

HEAT GAIN OF SOLAR RADIATION, THROUGH PLAIN GLASS PER DAY (Summer)

Example 15—*English Units*

The house is located in Savannah, Georgia. The house is situated so that its walls face North, East, South and West. The North wall has 90 sq ft of glass windows and doors. The East wall has 120 sq ft of glass windows and doors. The South wall has 106 sq ft of glass windows and doors and the West has 96 sq ft of glass windows and doors.

The glass in windows and doors are single-pane with 85% transmittance of radiant solar heat.

From Table 37 it can be determined that the latitude N of Savannah is 32° 05′.

Using Table 39, the average heat for various seasons for North, East, South and West is given in Btu/sq ft per hour. These figures must be multiplied by the square feet of area of glass times 0.85 (the transmittance of radiant solar heat). For the various seasons this would be in Btu's per sun hour through the glass, as shown on the following chart.

SEASON	North Wall Btu/hr	East Wall Btu/hr	South Wall Btu/hr	West Wall Btu/hr
SPRING	90 x 0.85 x 32 = 2,448	120 x 0.85 x 86 = 8,772	106 x 0.85 x 45 = 4,054	96 x 0.85 x 86 = 7,017
SUMMER	90 x 0.85 x 31 = 2,371	120 x 0.85 x 89 = 9,078	106 x 0.85 x 39 = 3,514	96 x 0.85 x 89 = 7,262
AUTUMN	90 x 0.85 x 14 = 1,071	120 x 0.85 x 62 = 6,324	106 x 0.85 x 149 = 13,424	96 x 0.85 x 62 = 5,059
WINTER	90 x 0.85 x 17 = 1,300	120 x 0.85 x 75 = 7,650	106 x 0.85 x 180 = 16,218	96 x 0.85 x 75 = 6,120

LATITUDE, ELEVATION AND HOURS OF SUNSHINE – UNITED STATES CITIES

CITY	STATE	LATITUDE	ELEVATION		HOURS SUNSHINE FOR MONTH AS LISTED											
					MONTH											
			Feet	Meters	Jan	Feb	Mar	Apr	May	Jun	Jul	Aug	Sep	Oct	Nov	Dec
Anniston	Ala	33°40′	599	182	140	155	205	250	295	295	270	275	250	235	180	140
Birmingham	Ala	33°40′	610	185	140	155	210	250	295	290	270	275	250	235	180	135
Dothan	Ala	31°10′	321	97	175	180	225	265	310	310	280	285	245	260	205	165
Huntsville	Ala	34°35′	619	188	140	155	200	240	290	295	280	270	255	230	170	130
Mobile	Ala	30°45′	119	36	160	160	220	255	305	295	260	275	240	255	200	155
Montgomery	Ala	32°20′	195	59	160	160	220	305	305	280	285	250	250	200	150	
Tuscaloosa	Ala	33°10′	170	52	140	155	210	250	295	290	275	270	250	245	180	135
Anchorage	Alk	61°15′	99	30	10	20	70	110	140	150	160	130	115	60	20	5
Fairbanks	Alk	64°55′	436	132	10	15	60	90	130	140	150	120	110	50	15	5
Juneau	Alk	58°20′	17	15	15	20	75	120	150	170	165	140	130	70	30	10
Flagstaff	Arz	35°10′	6973	2125	200	195	255	295	335	360	315	295	290	255	215	200
Kingman	Arz	35°20′	3446	1050	235	240	300	330	385	400	375	345	330	300	255	230
Phoenix	Arz	33°30′	1117	340	240	250	310	350	400	405	365	340	330	310	270	235
Tucson	Arz	32°15′	2584	788	245	260	310	350	390	395	340	335	330	310	280	250
Fayetteville	Ark	36°05′	1253	382	140	160	200	235	275	300	320	300	260	235	180	140
Ft. Smith	Ark	35°25′	449	137	145	160	205	240	275	310	320	310	270	235	180	155
Little Rock	Ark	34°40′	257	76	140	155	205	245	285	310	320	310	265	245	180	140
Pine Bluff	Ark	34°15′	204	62	140	155	210	245	285	310	325	310	265	245	185	145
Bakersfield	Calif	35°20′	495	150	155	205	280	320	380	400	420	400	340	300	225	160
Burbank	Calif	34°05′	699	213	200	215	275	260	280	300	340	335	310	270	240	220
Eureka	Calif	40°55′	217	66	120	130	190	220	255	260	260	240	200	175	130	110
Fresno	Calif	36°50′	326	99	145	165	280	310	360	390	420	390	340	300	230	145
Long Beach	Calif	33°55′	34	10	200	215	275	260	280	300	340	335	310	270	240	220
Los Angeles	Calif	34°00′	125	38	200	215	275	260	280	300	340	335	310	270	240	220
Oakland	Calif	37°45′	3	1	160	165	255	280	320	340	340	320	290	250	200	150
Redding	Calif	40°30′	495	150	140	160	240	300	340	360	410	360	330	260	185	130
Riverside	Calif	33°55′	1511	460	200	215	275	230	280	300	340	335	310	270	240	220
Sacramento	Calif	38°35′	17	5	135	170	255	300	370	400	420	400	340	290	190	135
Salinas	Calif	36°40′	74	22	175	170	260	285	320	345	340	310	290	260	210	170
San Bernardino	Calif	34°10′	1125	342	190	220	280	270	330	340	360	350	330	290	260	220
San Diego	Calif	32°40′	19	6	210	215	275	255	280	280	320	300	275	260	250	215
San Francisco	Calif	37°50′	32	10	160	165	255	280	320	340	340	320	290	250	200	150
San Jose	Calif	37°20′	70	21	160	165	255	280	330	350	340	320	290	260	205	160
Santa Ana	Calif	33°40′	115	35	200	215	275	230	280	300	340	335	310	270	240	220
Santa Barbara	Calif	34°40′	100	30	185	210	275	285	300	315	350	330	300	260	230	225
Stockton	Calif	37°55′	28	9	140	170	260	305	370	400	420	400	340	295	205	135
Colorado Springs	Colo	38°50′	6173	1881	200	195	240	240	275	315	315	300	275	235	195	190
Denver	Colo	39°50′	5283	1610	200	200	240	240	270	300	320	290	265	240	200	180
Grand Junction	Colo	39°10′	4849	1477	160	180	240	265	305	340	340	300	295	255	195	165
Pueblo	Colo	38°20′	4639	1413	220	200	260	260	285	325	340	305	285	260	220	200
Hartford	Conn	41°50′	15	5	140	160	200	220	260	280	295	260	220	190	140	140
New Haven	Conn	41°20′	7	2	140	160	200	220	260	280	300	260	220	200	150	140
Dover	Del	39°00′	38	12	145	165	205	235	270	290	295	265	225	210	165	145
Wilmington	Del	39°40′	78	23	140	160	205	230	265	285	290	265	225	205	160	140
Washington	DC	38°50′	27	8	145	165	210	235	270	290	295	270	225	210	165	145
Ft. Myers	Fla	26°35′	13	4	225	220	260	280	300	265	260	255	220	235	220	205
Jacksonville	Fla	30°30′	24	7	190	190	240	170	300	260	260	255	215	215	200	175
Miami	Fla	25°50′	7	2	225	220	260	280	280	255	360	265	210	215	210	215
Orlando	Fla	28°30′	106	32	200	195	250	275	290	255	250	250	215	220	205	185
Panama City	Fla	30°05′	22	7	185	185	230	265	310	305	275	275	235	260	210	165
Pensacola	Fla	30°25′	13	4	170	170	225	265	305	305	270	285	235	260	205	160
Tallahassee	Fla	30°25′	58	18	190	190	235	270	210	295	280	265	235	260	210	170
Tampa	Fla	28°00′	19	6	220	210	260	280	320	265	250	250	220	240	220	190
W. Palm Beach	Fla	26°40′	15	5	215	210	255	275	280	250	360	255	210	215	210	210
Albany	GA	31°30′	224	68	180	180	230	265	320	305	290	285	245	250	210	170
Atlanta	GA	33°40′	1005	306	150	165	210	260	300	300	280	280	250	240	185	150
Augusta	GA	33°25′	143	44	170	180	240	280	320	310	290	290	255	245	205	165
Columbus	GA	32°30′	242	74	165	170	225	265	315	310	290	290	245	250	205	165

(Sheet 1 of 5)

TABLE 37

LATITUDE, ELEVATION AND HOURS OF SUNSHINE – UNITED STATES CITIES

CITY	STATE	LATITUDE	ELEVATION		HOURS SUNSHINE FOR MONTH AS LISTED											
					MONTH											
			Feet	Meters	Jan	Feb	Mar	Apr	May	Jun	Jul	Aug	Sep	Oct	Nov	Dec
Macon	GA	32°40′	356	108	170	175	225	265	320	310	290	290	255	245	205	165
Savannah	GA	32°05′	52	16	180	175	230	270	315	280	275	260	240	220	200	170
Valdosta	GA	30°40′	239	73	185	185	235	270	310	290	280	260	230	245	210	170
Boise	Idaho	43°30′	2842	866	110	145	205	265	320	340	400	360	300	225	140	105
Coeur D'Alene	Idaho	47°45′	2973	906	70	110	180	230	260	370	360	325	240	165	80	50
Idaho Falls	Idaho	43°30′	4730	1441	110	145	205	235	300	325	375	335	280	220	140	105
Lewiston	Idaho	46°25′	1413	431	75	110	190	260	305	315	400	345	270	190	90	60
Pocatello	Idaho	45°55′	4444	1355	115	150	210	235	300	330	380	340	285	230	145	110
Twin Falls	Idaho	42°30′	4148	1264	120	150	215	260	300	340	395	360	295	230	150	115
Belleville	Ill	38°30′	447	136	135	160	205	235	285	315	335	330	265	235	170	135
Chicago	Ill	41°50′	594	161	120	140	195	220	270	300	330	300	245	205	125	115
Joliet	Ill	41°30′	588	179	120	140	195	220	270	300	330	300	245	210	130	115
Peoria	Ill	40°40′	652	199	135	155	195	225	275	305	340	305	260	220	145	125
Quincy	Ill	40°00′	762	232	140	160	205	230	285	310	345	305	265	225	165	130
Springfield	Ill	39°50′	587	178	130	155	200	230	285	310	340	310	260	255	155	125
Columbus	Ind	39°15′	661	201	120	140	190	225	280	310	340	300	260	220	145	115
Evansville	Ind	38°00′	381	116	125	150	195	235	285	320	340	305	260	235	160	125
Ft. Wayne	Ind	41°05′	791	241	105	135	185	215	270	305	345	300	245	210	125	105
Indianapolis	Ind	39°40′	793	242	115	140	190	225	275	310	345	300	260	225	135	110
South Bend	Ind	41°40′	773	235	100	120	180	215	270	305	345	300	235	210	120	95
Terre Haute	Ind	39°30′	601	183	120	145	195	230	275	310	340	305	255	225	145	120
Burlington	Iowa	40°50′	692	210	140	160	205	230	280	305	345	310	265	225	155	130
Cedar Rapids	Iowa	42°00′	863	263	140	160	205	225	270	305	345	305	260	215	140	125
Des Moines	Iowa	41°35′	948	288	150	170	210	235	275	310	355	305	265	225	155	135
Mason City	Iowa	43°10′	1194	363	145	165	210	230	275	310	350	275	255	205	130	125
Ottumwa	Iowa	41°05′	842	257	145	165	210	230	275	305	350	305	265	225	160	130
Sioux City	Iowa	42°25′	1095	334	160	175	215	240	285	320	365	320	265	225	155	145
Dodge City	Kan	37°45′	2594	791	205	195	250	265	300	325	350	325	285	260	215	195
Hutchinson	Kan	38°00′	1524	464	185	185	230	250	290	320	350	320	280	250	205	185
Liberal	Kan	37°00′	2838	865	210	200	250	270	305	330	350	325	285	260	220	200
Topeka	Kan	39°00′	877	267	160	175	215	240	280	315	345	310	270	235	180	155
Wichita	Kan	37°40′	1321	402	180	180	225	245	285	320	345	320	280	250	205	180
Bowling Green	Ky	37°00′	535	163	120	145	190	240	285	305	310	285	255	225	165	120
Corbin	Ky	37°00′	1175	358	110	130	175	215	255	275	285	255	230	195	150	115
Lexington	Ky	38°00′	978	298	110	130	175	215	260	290	305	270	245	210	145	110
Louisville	Ky	38°15′	474	144	120	140	185	225	280	305	330	290	255	220	150	115
Paducah	Ky	37°05′	398	121	130	155	200	245	290	320	340	320	260	240	170	135
Alexandria	La	31°15′	92	28	150	150	210	235	290	305	390	290	255	255	190	170
Baton Rouge	La	30°30′	64	26	155	150	210	235	295	295	270	275	250	255	190	170
Lafayette	La	30°15′	38	12	155	150	210	235	290	305	270	275	250	255	185	170
Lake Charles	La	30°15′	14	4	155	150	210	235	290	305	285	285	255	255	185	170
Monroe	La	32°30′	78	24	145	150	210	235	290	305	310	305	260	250	190	170
New Orleans	La	30°00′	3	1	160	150	210	245	295	290	260	270	240	255	190	165
Shreveport	La	32°25′	252	76	150	150	210	235	285	315	330	310	265	250	190	165
Boston	Mass	42°20′	15	5	140	160	200	220	260	280	300	260	220	200	140	140
Springfield	Mass	42°10′	247	75	130	160	200	220	260	280	295	260	220	185	130	130
Worchester	Mass	42°15′	986	300	140	160	200	215	255	280	295	260	220	190	140	130
Baltimore	Md	39°10′	146	45	140	165	205	230	270	285	300	265	225	205	160	140
Augusta	Me	44°20′	350	107	140	160	200	220	260	280	300	260	220	200	130	130
Bangor	Me	44°50′	162	49	140	160	200	220	255	275	390	260	220	190	125	125
Portland	Me	43°40′	61	19	140	160	200	220	260	280	300	260	220	200	140	140
Detroit	Mich	42°40′	633	193	95	125	185	210	270	295	335	295	235	190	100	90
Flint	Mich	43°00′	766	233	90	125	185	215	265	300	335	295	225	185	95	80
Grand Rapids	Mich	42°50′	681	208	80	115	175	210	265	300	340	300	225	190	95	75
Lansing	Mich	42°45′	852	260	95	120	180	215	265	305	340	295	230	190	95	80
Saginaw	Mich	43°30′	662	201	90	125	185	220	265	300	335	290	220	180	90	80
Traverse City	Mich	44°40′	618	188	80	115	175	215	260	300	335	285	210	165	80	70

(Sheet 2 of 5)

TABLE 37 (Continued)

LATITUDE, ELEVATION AND HOURS OF SUNSHINE — UNITED STATES CITIES

CITY	STATE	LATITUDE	ELEVATION		HOURS SUNSHINE FOR MONTH AS LISTED											
					MONTH											
			Feet	Meters	Jan	Feb	Mar	Apr	May	Jun	Jul	Aug	Sep	Oct	Nov	Dec
Duluth	Minn	46°45′	1426	435	130	160	220	225	270	285	330	275	210	175	100	105
Minneapolis	Minn	44°55′	822	251	140	165	210	230	275	300	340	290	230	190	115	115
Rochester	Minn	44°00′	1297	395	140	165	205	230	270	305	340	290	240	195	120	115
St. Cloud	Minn	45°35′	1034	315	145	170	210	230	275	300	345	290	225	190	115	115
Willmar	Minn	45°10′	1133	345	145	170	210	235	280	305	350	295	235	195	120	125
Greenville	Miss	33°30′	139	42	140	150	205	250	290	305	310	305	260	245	180	140
Jackson	Miss	32°20′	330	100	130	150	200	245	290	320	315	300	260	245	175	135
Laurel	Miss	31°40′	264	80	150	150	205	245	295	290	270	275	250	250	185	145
Meridian	Miss	32°25′	294	90	140	150	205	250	295	290	275	275	250	250	185	140
Tupelo	Miss	34°20′	289	88	135	150	205	250	290	305	290	290	255	245	180	135
Columbia	Mo	39°00′	778	237	145	170	210	230	285	310	345	295	265	230	170	135
Jefferson City	Mo	38°40′	640	195	145	170	210	230	285	315	345	300	270	230	170	135
Joplin	Mo	37°05′	982	299	145	165	215	235	275	305	330	310	275	235	190	150
Kansas City	Mo	39°10′	742	226	155	170	215	235	275	315	350	305	270	235	175	150
St. Louis	Mo	38°45′	535	163	135	160	205	235	285	315	335	330	265	235	170	135
Springfield	Mo	37°10′	1265	386	130	155	200	230	285	310	340	310	260	225	155	125
Billings	Mont	45°45′	3567	1087	140	155	215	235	280	305	360	325	255	210	135	135
Butte	Mont	46°00′	5526	1684	110	140	195	225	280	275	345	325	250	190	120	105
Great Falls	Mont	47°30′	3664	1117	140	160	220	245	280	295	370	335	245	205	125	120
Helena	Mont	46°40′	3893	1187	125	160	210	230	280	285	365	330	245	200	125	110
Miles City	Mont	46°40′	2629	801	150	170	225	250	290	310	370	325	260	215	140	135
Missoula	Mont	46°50′	3200	975	90	110	170	215	265	270	360	330	245	160	90	80
Grand Island	Neb	41°00′	1841	561	180	180	220	250	290	310	350	315	270	245	185	170
Lincoln	Neb	40°50′	1150	350	170	175	215	245	285	315	355	310	270	235	175	160
North Platte	Neb	41°10′	2779	847	190	190	230	255	290	310	350	315	275	250	190	175
Omaha	Neb	41°20′	978	298	170	175	215	240	285	320	360	310	270	230	170	150
Carson City	Nev	39°10′	4675	1424	180	180	255	280	330	360	390	380	320	260	210	170
Elko	Nev	39°50′	5075	1547	150	175	235	255	290	350	385	355	310	250	180	155
Ely	Nev	39°15′	6257	1907	185	190	260	270	305	365	350	350	310	255	200	180
Las Vegas	Nev	36°10′	2162	659	225	240	285	315	380	400	375	360	340	290	250	240
Reno	Nev	39°30′	4490	1369	180	180	260	280	330	370	390	380	320	260	200	170
Concord	NH	43°15′	339	103	140	155	195	200	230	255	290	260	210	165	130	120
Keene	NH	42°55′	490	149	130	155	195	200	245	265	290	260	210	180	125	120
Manchester	NH	43°00′	253	77	140	160	200	210	240	270	295	260	215	175	135	130
Atlantic City	NJ	39°30′	11	3	140	160	205	230	270	285	300	270	220	205	160	140
Newark	NJ	40°40′	11	3	140	160	205	225	260	280	300	265	220	200	160	140
New Brunswick	NJ	40°30′	86	26	140	160	205	225	260	280	300	265	220	200	160	140
Trenton	NJ	40°15′	144	44	140	160	205	225	260	280	300	265	220	200	160	140
Albuquerque	NM	35°00′	5310	1618	220	210	260	290	340	355	320	305	300	270	240	210
Clovis	NM	34°25′	4279	1304	215	220	275	295	325	350	345	325	280	265	235	210
Farmington	NM	36°50′	5495	1675	185	195	250	285	320	350	325	300	290	260	205	190
Gallup	NM	35°30′	6465	1971	195	190	250	275	330	355	315	295	290	250	215	190
Las Cruces	NM	32°20′	3900	1189	240	240	300	330	375	375	330	330	305	285	260	235
Roswell	NM	33°25′	3643	1110	215	225	285	305	330	355	340	315	280	265	245	215
Sante Fe	NM	35°35′	7410	2259	200	205	245	275	310	340	315	295	280	260	210	200
Tucumcari	NM	35°10′	4053	1235	215	220	265	290	325	365	340	315	290	270	230	210
Albany	NY	42°45′	277	84	120	140	180	200	260	280	300	260	220	180	105	105
Binghamton	NY	42°10′	858	261	100	120	160	180	240	260	260	245	200	160	100	100
Buffalo	NY	42°55′	705	214	90	120	180	210	260	300	320	280	25	180	90	85
Elmira	NY	42°05′	860	262	95	120	155	175	235	255	160	240	195	165	95	80
New York	NY	40°50′	132	40	140	160	205	225	260	280	300	265	220	200	160	140
Niagara Falls	NY	43°10′	596	182	90	120	180	210	260	300	320	280	225	180	90	85
Rochester	NY	43°10′	543	166	90	120	175	210	260	300	320	280	220	175	90	85
Syracuse	NY	43°05′	424	129	90	110	170	195	260	280	300	260	195	160	90	80
Asheville	NC	35°30′	2170	661	145	160	190	230	280	265	260	240	220	210	180	140
Charlotte	NC	35°15′	735	239	165	175	220	265	300	300	285	275	240	240	190	165
Elizabeth City	NC	36°20′	10	3	155	170	220	250	300	300	290	280	230	225	185	160
Fayetteville	NC	35°05′	95	29	165	180	220	265	300	305	290	275	240	240	200	165

(Sheet 3 of 5)

TABLE 37 (Continued)

LATITUDE, ELEVATION AND HOURS OF SUNSHINE – UNITED STATES CITIES

CITY	STATE	LATITUDE	ELEVATION		HOURS SUNSHINE FOR MONTH AS LISTED											
					MONTH											
			Feet	Meters	Jan	Feb	Mar	Apr	May	Jun	Jul	Aug	Sep	Oct	Nov	Dec
Greensboro	NC	36°05′	879	268	155	170	215	255	290	300	285	275	240	230	185	160
Raleigh	NC	35°50′	433	132	155	170	215	255	290	295	285	275	235	230	185	160
Wilmington	NC	34°20′	30	9	180	180	225	265	305	310	285	275	235	235	205	165
Winston-Salem	NC	36°10′	967	294	155	170	210	250	290	290	285	275	240	230	185	160
Bismark	ND	46°45′	1647	502	150	175	210	240	290	300	355	310	245	200	135	130
Dickinson	ND	46°50′	2595	790	150	175	215	250	295	305	360	320	250	205	135	130
Fargo	ND	46°50′	900	274	140	170	215	235	285	295	350	295	220	190	120	115
Grand Forks	ND	48°00′	832	254	140	170	220	235	285	290	345	295	215	185	115	115
Jamestown	ND	47°00′	1492	455	145	170	215	235	290	295	355	305	230	195	130	125
Minot	ND	48°25′	1713	522	150	170	215	245	290	300	355	310	240	205	130	125
Akron	Ohio	41°05′	1210	369	95	115	170	205	265	290	320	280	235	195	105	90
Cincinnati	Ohio	39°10′	761	231	110	130	180	220	270	300	330	290	250	210	135	110
Cleveland	Ohio	41°35′	777	237	95	115	170	205	265	300	325	285	235	195	100	80
Columbus	Ohio	40°00′	812	247	100	120	175	210	265	295	320	280	240	205	125	100
Dayton	Ohio	39°45′	997	304	110	130	180	220	270	305	330	290	250	210	130	105
Mansfield	Ohio	40°45′	1297	395	100	125	175	210	265	295	325	285	245	200	115	90
Toledo	Ohio	41°40′	676	206	100	125	180	210	270	305	335	295	245	200	110	95
Zanesville	Ohio	40°00′	881	269	95	115	170	205	260	280	305	270	235	195	115	95
Ardmore	Okla	34°15′	880	268	170	180	230	245	290	330	345	325	275	245	205	175
Enid	Okla	36°25′	1287	392	185	190	235	250	290	325	345	325	285	250	210	185
Lawton	Okla	34°30′	1108	338	190	190	245	250	295	330	350	330	280	250	215	190
Muskogee	Okla	35°35′	610	186	155	165	215	230	275	315	335	315	275	240	295	160
Oklahoma City	Okla	35°25′	1280	390	180	185	235	250	290	325	345	325	285	250	210	180
Tulsa	Okla	36°10′	650	198	160	175	215	235	275	315	335	315	280	245	200	170
Eugene	Ore	44°00′	364	110	70	95	150	205	245	265	340	300	230	140	85	55
Klamath Falls	Ore	42°10′	4091	1247	110	110	190	245	300	330	380	350	285	220	140	95
Medford	Ore	42°20′	1298	396	90	100	175	220	260	260	350	330	270	180	120	85
Pendelton	Ore	45°35′	1492	455	75	125	205	255	300	330	395	355	280	195	95	85
Portland	Ore	45°30′	57	17	70	95	150	205	245	350	320	280	210	130	85	55
Salem	Ore	44°55′	195	59	70	95	150	205	245	355	320	280	210	140	85	60
Allentown	Pa	40°35′	376	114	130	160	200	220	260	280	295	260	220	195	145	130
Altoona	Pa	40°30′	1468	447	100	130	160	200	240	260	260	245	215	180	145	100
Erie	Pa	42°10′	732	223	80	115	170	210	260	285	300	280	230	185	95	75
Harrisburgh	Pa	40°15′	335	102	130	150	200	220	260	280	300	260	220	200	145	125
Johnston	Pa	40°20′	1214	370	100	120	160	195	240	255	260	240	215	175	100	80
Philadelphia	Pa	40°00′	7	2	140	160	205	225	265	285	300	360	220	205	160	140
Pittsburgh	Pa	40°25′	1137	347	90	115	150	195	240	260	280	250	225	185	105	80
Reading	Pa	40°20′	226	69	130	155	200	220	260	280	295	255	220	195	150	130
Wilkes-Barre	Pa	41°20′	905	275	120	140	180	200	245	265	285	245	205	185	125	110
Pawtucket	RI	41°30′	20	6	140	160	200	220	260	280	295	260	220	200	145	140
Providence	RI	41°40′	55	17	140	160	200	220	260	280	295	260	220	200	145	140
Charleston	SC	32°45′	30	9	180	185	240	265	320	300	280	280	240	240	200	180
Columbia	SC	34°00′	217	66	170	180	225	265	310	310	290	285	230	250	205	165
Florence	SC	34°10′	146	45	175	180	240	275	310	310	290	280	230	245	205	165
Greenville	SC	34°50′	957	291	160	170	220	260	300	295	280	280	225	240	190	160
Spartenburg	SC	34°55′	816	249	160	170	220	260	300	300	285	280	225	240	190	160
Aberdeen	SD	45°30′	1296	395	150	175	210	240	285	305	360	310	250	200	135	130
Pierre	SD	44°20′	1718	524	160	180	215	245	290	315	360	320	260	215	155	145
Rapid City	SD	44°00′	3165	964	160	180	230	250	285	305	340	315	265	225	165	150
Sioux Falls	SD	43°30′	1420	432	160	175	215	245	285	320	355	310	260	215	150	140
Watertown	SD	44°55′	1746	532	150	175	210	240	285	310	360	305	245	200	130	130
Chattanooga	Tenn	35°05′	670	204	135	145	185	230	285	290	275	265	245	225	165	125
Jackson	Tenn	35°40′	413	126	130	150	200	245	290	320	315	300	260	245	175	135
Knoxville	Tenn	35°55′	980	298	120	145	185	225	280	285	270	250	230	220	155	120
Memphis	Tenn	35°05′	263	80	135	150	205	250	250	320	320	300	260	245	180	140
Nashville	Tenn	36°10′	577	176	125	150	200	240	285	300	295	280	250	230	170	125
Abilene	Tex	32°30′	1759	536	200	200	255	255	295	340	350	330	275	250	225	200
Amarillo	Tex	35°10′	3607	1099	210	210	265	280	310	340	345	325	285	260	230	210

(Sheet 4 of 5)

TABLE 37 (Continued)

LATITUDE, ELEVATION AND HOURS OF SUNSHINE – UNITED STATES CITIES

CITY	STATE	LATITUDE	ELEVATION		HOURS SUNSHINE FOR MONTH AS LISTED											
					MONTH											
			Feet	Meters	Jan	Feb	Mar	Apr	May	Jun	Jul	Aug	Sep	Oct	Nov	Dec
Austin	Tex	30°25′	597	181	160	160	220	235	265	310	335	310	265	255	195	160
Bryan	Tex	30°40′	275	84	155	155	215	230	270	335	330	315	265	255	190	155
Dallas	Tex	32°50′	481	147	160	160	220	235	280	330	345	325	270	250	195	170
El Paso	Tex	31°50′	3918	1194	235	240	300	330	385	370	335	325	300	290	260	235
Ft. Worth	Tex	32°50′	544	166	160	170	230	240	280	330	345	320	270	250	200	175
Galveston	Tex	29°15′	5	2	150	155	215	230	280	330	295	295	255	255	190	150
Houston	Tex	29°45′	158	48	160	160	215	235	260	300	330	320	270	255	195	160
Laredo	Tex	27°30′	503	153	160	160	215	235	260	300	330	320	270	255	195	160
Lufkin	Tex	31°15′	286	87	150	155	215	230	280	315	325	320	265	255	190	155
San Angelo	Tex	31°20′	1878	572	195	200	235	255	290	335	345	330	270	250	220	195
San Antonio	Tex	29°30′	792	241	160	160	215	235	260	300	335	320	270	255	195	160
Tyler	Tex	32°20′	526	160	155	155	215	230	280	320	345	325	270	250	190	165
Victoria	Tex	28°45′	104	31	155	155	215	230	275	310	340	325	260	260	190	170
Waco	Tex	31°40′	500	152	160	160	220	235	275	320	340	325	270	250	195	170
Wichita Falls	Tex	34°00′	994	304	190	190	250	250	295	335	350	320	275	250	215	190
Ogden	Utah	41°15′	4400	1341	140	150	220	260	320	340	365	340	300	245	160	125
Price	Utah	39°40′	5580	1701	160	170	220	250	300	340	340	315	280	240	170	160
Salt Lake City	Utah	40°40′	4220	1286	140	150	225	260	320	340	360	340	300	245	160	125
Burlington	VT	44°30′	331	101	105	125	180	190	240	260	285	260	200	150	80	80
Lynchburg	VA	37°20′	947	289	145	165	205	240	280	275	280	275	235	210	165	140
Norfolk	VA	36°50′	26	8	150	170	220	250	300	300	295	280	230	220	185	160
Richmond	VA	37°30′	162	49	150	170	215	245	285	295	290	270	230	215	175	145
Roanoke	VA	37°20′	1172	357	140	165	200	240	280	270	270	260	220	205	160	140
Everette	Wash	47°50′	598	182	65	95	150	200	240	230	290	250	200	120	65	55
Olympia	Wash	47°00′	190	58	65	95	150	200	240	230	290	250	200	120	65	55
Seattle	Wash	47°40′	14	4	65	95	150	200	240	230	290	250	200	120	65	55
Spokane	Wash	47°40′	2357	718	70	120	190	245	300	310	385	350	265	175	85	55
Tacoma	Wash	47°15′	350	107	65	95	150	200	240	230	290	250	200	120	65	55
Yakima	Wash	46°35′	1061	323	80	130	220	250	320	340	390	350	270	190	100	60
Charleston	WVa	38°25′	639	195	95	115	155	195	240	265	280	255	225	190	130	110
Huntington	WVa	38°25′	565	172	95	115	160	200	250	270	290	255	230	195	130	110
Martinsburg	WVa	39°20′	537	163	125	115	200	215	260	280	280	250	225	195	145	125
Morgantown	WVa	39°40′	1245	379	95	115	155	195	235	250	265	240	220	185	115	100
Parkersburg	WVa	39°25′	615	187	95	115	160	200	245	270	290	260	230	190	120	100
Eau Clair	Wis	44°50′	888	271	135	160	205	225	265	295	330	285	230	185	110	110
Green Bay	Wix	44°30′	683	208	120	140	195	210	260	285	320	275	215	175	105	100
LaCrosse	Wis	43°50′	652	199	135	160	205	225	265	295	335	290	240	195	115	115
Madison	Wis	43°10′	858	261	130	150	195	215	260	295	330	290	235	195	120	115
Milwaukee	Wis	43°00′	672	204	120	140	190	215	265	295	330	290	235	195	120	115
Wausau	Wis	45°00′	1196	365	125	150	205	215	265	285	320	275	220	175	105	105
Casper	Wyo	43°50′	5319	1621	170	175	230	240	270	300	320	310	260	230	150	160
Cheyenne	Wyo	41°10′	6116	1867	185	180	240	235	255	300	315	290	265	240	170	170
Rock Springs	Wyo	41°40′	6741	2054	165	165	230	240	270	310	330	310	265	230	150	160
Sheridan	Wyo	44°50′	3942	1201	155	160	220	240	270	300	340	320	260	220	165	155

(Sheet 5 of 5)

TABLE 37 (Continued)

DAILY TOTAL HEAT GAIN AT LISTED LATITUDE – CLEAR CLOUDLESS DAY
SEASON MEAN HEAT GAIN TOTAL PER DAY – SOLAR RADIATION

LAT. DEGREE NORTH	SURFACE POSITION																			
	DIRECT NOMINAL		VERTICAL SURFACE FACING																HORIZONTAL SURFACE	
			NORTH		NORTHEAST		EAST		SOUTHEAST		SOUTH		SOUTHWEST		WEST		NORTHWEST			
	Btu/ft² hr	W/m²	Btu/ft² hr	W/m²	Btu/ft² hr	W/m²	Btu/ft² hr	W/m²	Btu/ft² hr	W/m²	Btu/ft² hr	W/m²	Btu/ft² hr	W/m²	Btu/ft² hr	W/m²	Btu/ft² hr	W/m²	Btu/ft² hr	W/m²
SPRING																				
24	3067	9668	441	1390	873	2752	1109	3496	834	2629	411	1295	834	2629	1109	3496	873	2752	2155	6793
32	3066	9665	416	1311	816	2572	1124	3543	943	2972	581	1831	943	2972	1124	3543	816	2572	2148	6771
40	3044	9596	376	1185	751	2367	1124	3543	1051	3319	784	2471	1051	3319	1124	3543	751	2367	2046	6449
48	3012	9495	362	1141	696	2194	1122	3537	1202	3789	904	2849	1202	3789	1122	3537	696	2194	1917	6043
56	2965	9347	275	856	523	1648	998	3146	1387	4372	1317	4132	1387	4372	998	3146	523	1648	1483	4575
SUMMER																				
24	2968	9356	484	1525	987	3111	1166	3675	966	3045	482	1519	966	3045	1166	3675	987	3111	2184	6885
32	2965	9347	409	1289	878	2767	1153	3634	1052	3316	506	1595	1052	3316	1153	3634	878	2767	2167	6831
40	2904	9154	391	1232	876	2761	1148	3619	1068	3366	734	2313	1068	3366	1148	3619	876	2761	2048	6456
48	2872	9053	374	1179	721	2272	1132	3568	1074	3385	842	2654	1074	3385	1132	3568	721	2272	1970	6210
56	2806	8845	332	1046	665	2096	1110	3499	1410	4444	1200	3782	1410	4444	1140	3499	665	2096	1616	5094
AUTUMN																				
24	2694	8492	201	633	254	800	775	2443	1276	4022	1616	5094	1276	4022	775	2443	254	800	1345	4240
32	2519	7941	160	504	221	696	682	2149	1301	4101	1644	5182	1301	4101	682	2149	221	696	1250	3940
40	2396	7553	149	469	156	491	621	1957	1229	3874	1508	4753	1229	3874	621	1957	156	491	1131	3565
48	2008	6330	88	277	149	469	600	1891	1178	3719	1450	4571	1178	3719	600	1891	149	469	1012	3190
56	1746	5504	83	261	129	406	453	1428	980	3089	1291	4069	980	3089	453	1428	129	406	523	1648
WINTER																				
24	2850	8984	160	504	249	784	773	2436	1297	4088	1597	5034	1297	4088	773	2436	249	784	1326	4180
32	2527	7966	152	479	210	662	678	2137	1271	4006	1622	5113	1271	4006	678	2137	210	662	1089	3433
40	2250	7093	128	403	153	482	558	1759	1197	3779	1460	4602	1197	3779	558	1759	153	482	784	2471
48	1903	5999	87	274	110	346	366	1153	1007	3174	1362	4293	1007	3174	366	1153	110	346	482	1519
56	1323	4170	52	163	62	195	248	781	725	2285	1038	3272	725	2285	248	781	62	195	237	747

TABLE 38

AVERAGE HEAT RADIATED PER SUNSHINE HOUR AT LISTED LATITUDE
HOURLY MEAN HEAT RADIATED TO SURFACE – AVERAGED OVER ONE DAY

LAT. DEGREE NORTH	SURFACE POSITION																			
	DIRECT NORMAL		VERTICAL SURFACE FACING																HORIZONTAL SURFACE	
			NORTH		NORTHEAST		EAST		SOUTHEAST		SOUTH		SOUTHWEST		WEST		NORTHWEST			
	Btu/ft² hr	W/m²	Btu/ft² hr	W/m²	Btu/ft² hr	W/m²	Btu/ft² hr	W/m²	Btu/ft² hr	W/m²	Btu/ft² hr	W/m²	Btu/ft² hr	W/m²	Btu/ft² hr	W/m²	Btu/ft² hr	W/m²	Btu/ft² hr	W/m²
SPRING																				
24	236	745	34	107	67	212	85	269	64	202	32	100	64	202	85	269	67	212	166	523
32	236	743	32	101	63	198	86	273	73	229	45	141	73	229	86	273	63	198	165	521
40	234	738	29	91	58	182	86	273	81	255	60	190	81	255	86	273	54	182	157	496
48	232	730	28	88	54	169	86	272	92	291	70	219	92	291	86	272	21	169	147	465
56	227	719	21	66	21	127	79	242	107	336	101	318	107	336	79	242	21	127	114	352
SUMMER																				
24	228	720	37	117	76	239	90	283	74	234	37	117	74	234	90	283	76	239	168	530
32	227	719	31	99	68	213	89	280	81	255	39	123	81	255	89	280	68	213	167	529
40	223	704	30	95	67	212	88	278	82	258	56	178	82	258	88	278	67	212	158	497
48	221	696	29	91	55	175	87	274	83	260	65	204	83	260	87	274	55	175	152	478
56	216	680	26	80	51	161	85	269	108	342	92	291	108	342	85	269	51	161	124	392
AUTUMN																				
24	245	772	18	58	23	73	70	222	116	366	146	463	116	366	70	222	23	73	122	385
32	229	722	15	46	20	63	62	195	118	373	149	471	118	373	62	195	20	63	114	358
40	218	685	14	43	14	45	56	178	112	352	137	432	112	352	56	178	14	45	103	324
48	182	575	8	25	13	43	55	172	107	338	132	415	107	338	55	172	13	43	92	290
56	159	500	7	24	12	37	41	130	89	281	117	370	89	281	41	130	12	37	48	150
WINTER																				
24	316	998	17	60	28	87	86	271	144	454	177	559	144	454	86	271	28	87	147	464
32	281	885	16	53	23	74	75	237	141	445	180	568	141	445	75	237	23	74	121	381
40	250	788	14	45	17	53	62	195	133	420	162	511	133	420	62	195	17	53	87	275
48	211	666	9	30	12	36	41	128	112	353	151	477	112	353	41	128	12	36	53	169
56	147	463	6	18	7	22	27	87	80	254	115	364	80	254	27	87	7	22	26	83

Based on average heat radiated to surface over sun hours per clear cloudless days. Sun hours per day taken for Spring and Summer taken at 13 hrs.; for Autumn 11 hours; for Winter 9 hours. (See Table 36)

TABLE 39

Table 37 gives the hours of sunshine per month as influenced by weather conditions for selected United States cities. To obtain the Btu per month for the heat gain of solar radiation through the glass area in each wall, the Btu's per hour established previously must be multiplied by the hours of sunshine for each month. In the case of Example 15 this is:

HEAT GAIN OF TOTAL GLASS IN WALL FACING (Glass in Wall in Vertical Position)										
	Sunshine Hours	North Wall		East Wall		South Wall		West Wall		
Month	Per Month	Btu/hr	Btu/Month	Btu/hr	Btu/Month	Btu/hr	Btu/Month	Btu/hr	Btu/Month
Jan	180	1,300	234,000	7,650	1,377,000	16,218	2,919,240	6,120	1,101,600
Feb	175	1,300	227,500	7,650	1,338,750	16,218	2,838,150	6,120	1,071,000
Mar;	230	1,300	299,000	7,650	1,759,500	16,218	3,730,140	6,120	1,407,600
Apr	270	2,448	660,960	8,772	2,368,440	4,054	1,094,580	7,017	1,894,590
May	315	2,448	771,120	8,772	2,763,180	4,054	1,277,010	7,017	2,210,355
June	280	2,448	685,440	8,772	2,456,160	4,054	1,135,120	7,017	1,964,760
July	275	2,371	652,025	9,078	2,496,450	3,514	966,350	7,262	1,997,050
Aug	260	2,371	616,460	9,078	2,360,280	3,514	913,640	7,262	1,888,120
Sept	240	2,371	569,040	9,078	2,178,720	3,514	843,360	7,262	1,742,880
Oct	230	1,071	246,100	6,324	1,454,520	13,424	3,087,520	5,059	1,163,570
Nov	200	1,071	214,000	6,324	1,264,800	13,424	2,684,400	5,059	1,011,800
Dec	170	1,071	181,900	6,324	1,075,080	13,424	2,282,080	5,059	860,030

Note: Btu's per hour from preceding calculations based on season.

The heat gain for the entire glass area in the house due to solar radiation on all sides by month would be as follows:

Month	Heat Gain/Btu's per Month
Jan.	5,631,840
Feb.	5,475,400
March	7,196,240
April	6,018,570
May	7,021,665
June	6,241,480
July	6,111,875
Aug.	5,778,550
Sept.	5,334,000
Oct.	5,951,710
Nov.	5,175,400
Dec.	4,399,000
TOTAL Btu's per Year	70,335,820

Example 15a—*Metric Units*

The house is located in Savannah, Georgia. The house is situated so that its walls face North, East, South and West. The North wall has 8.36 square meters of glass windows and doors, the East wall has 11.15 square meters of glass windows and doors, the South wall has 9.85 square meters of glass windows and doors, and the West has 8.9 square meters of glass windows and doors.

The glass in windows and doors is single-pane with 85% transmittance of solar radiant heat.

Using Table 39, the average heat for various seasons for North, East, South, and West walls in W/m² can be obtained. These figures must be multiplied by 0.85 (the transmittance of radiant heat).

For various seasons, this would be for sun hour watts through glass:

SEASON	North Wall Watts/hr	East Wall Watts/hr	South Wall Watts/hr	West Wall Watts/hr
SPRING	8.36 x 0.85 x 101 = 717	11.15 x 0.85 x 273 = 2,587	9.85 x 0.85 x 141 = 1,180	8.9 x 0.85 x 273 = 2,065
SUMMER	8.36 x 0.85 x 99 = 703	11.15 x 0.85 x 280 = 2,653	9.85 x 0.85 x 123 = 1,029	8.9 x 0.85 x 280 = 2,218
AUTUMN	8.36 x 0.85 x 44 = 312	11.15 x 0.85 x 195 = 1,848	9.85 x 0.85 x 471 = 3,943	8.9 x 0.85 x 195 = 1,475
WINTER	8.36 x 0.85 xx 53 = 376	11.15 x 0.85 x 237 = 2,246	9.85 x 0.85 x 568 = 4,755	8.9 x 0.85 x 237 = 1,792

Table 37 gives the hours of sunshine per month as influenced by weather conditions for selected United States cities. To obtain the Watt-Hrs per Month for heat gain of solar radiation through the glass area in each wall, the Watts previously established must be multiplied by the Hours of Sunshine for each month. This is given in the following chart.

Month	Heat Gain, Total Watts Per Month	Month	Heat Gain, Total Watts Per Month
Jan	1,650,420	Jul	1,788,325
Feb	1,604,575	Aug	1,690,780
Mar	2,090,870	Sep	1,560,720
Apr	1,768,230	Oct	1,742,940
May	2,062,935	Nov	1,515,600
Jun	1,833,720	Dec	1,288,260
Watts per year heat entering through glass: 20,597,375			

The heat gain for the entire glass area in the house due to solar radiation, on all sides, by the month would be:

HEAT GAIN OF TOTAL GLASS IN WALL FACING
(Glass in Wall in Vertical Position)

Month	Sunshine Hours Per Month	North Wall		East Wall		South Wall		West Wall	
		W/hr	Watts/Month	W/hr	Watts/Month	W/hr	Watts/Month	W/hr	Watts/Month
Jan	180	376	67,680	2,246	404,280	4,755	855,900	1,792	322,560
Feb	175	376	65,800	2,246	393,050	4,755	832,125	1,792	313,600
Mar	230	376	86,480	2,246	516,580	4,755	1,093,650	1,792	412,160
Apr	270	717	193,590	2,587	698,490	1,180	318,600	2,065	557,550
May	315	717	225,855	2,587	814,905	1,180	371,700	2,065	650,475
Jun	280	717	200,760	2,587	724,360	1,180	330,400	2,065	578,200
Jul	275	703	193,325	2,653	729,575	1,029	282,975	2,118	582,450
Aug	260	703	182,780	2,653	789,780	1,029	267,540	2,118	550,680
Sep	240	703	168,720	2,653	636,720	1,029	246,960	2,118	508,320
Oct	230	312	71,760	1,848	425,040	3,943	906,890	1,475	339,250
Nov	200	312	62,400	1,848	369,600	3,943	788,600	1,475	295,000
Dec	170	312	53,040	1,848	314,160	3,943	670,310	1,475	250,750

Note: Watts per hour from preceding calculation based on season.

The location of home or building makes considerable difference as to heat gain through glass due to solar radiation. To illustrate this, we calculate the heat gain of a house identical to the one in the previous example, but located in Seattle, Washington.

From Table 37, the latitude of Seattle, Washington is 47° 40′. Thus the solar heat transmittance obtained from Table 38 will be based on 48° North Latitude.

Example 16—*English Units*

The house has the same glass area as given in Example 15. Glass transmittance is also the same as in Example 15.

For various seasons the heat gain due to solar radiation is (Through the glass per average sunshine hours):

SEASON	North Wall Btu/hr	East Wall Btu/hr	South Wall Btu/hr	West Wall Btu/hr
SPRING	90 x 0.85 x 28 = 2,142	120 x 0.85 x 87 = 8,772	106 x 0.85 x 70 = 6,307	96 x 0.85 x 86 = 7,017
SUMMER	90 x 0.85 x 29 = 2,218	120 x 0.85 x 87 = 8,874	106 x 0.85 x 65 = 5,856	96 x 0.85 x 97 = 7,099
AUTUMN	90 x 0.85 x 8 = 612	120 x 0.85 x 55 = 5,610	106 x 0.85 x 132 = 11,893	96 x 0.85 x 55 = 4,488
WINTER	90 x 0.85 x 9 = 688	120 x 0.85 x 41 = 4,182	106 x 0.85 x 151 = 13,605	96 x 0.85 x 41 = 3,345

From Table 37, the hours of sunshine per month in the city of Seattle is obtained. Using Btu's per Hour for each area of glass, in each direction, multiplied by hours of sunshine per month, the heat gain per month is obtained. The Heat Gain per Month is:

HEAT GAIN OF TOTAL GLASS IN WALL FACING
(Glass in Wall in Vertical Position)

Month	Sunshine Hours Per Month	North Wall		East Wall		South Wall		West Wall	
		Btu/hr	Btu/Month	Btu/hr	Btu/Month	Btu/hr	Btu/Month	Btu/hr	Btu/Month
Jan	65	688	44,720	4,182	271,830	13,605	884,325	3,345	217,425
Feb	95	688	65,360	4,182	397,290	13,605	1,292,475	3,345	317,775
Mar	150	688	103,200	4,182	627,300	13,605	2,040,750	3,345	501,750
Apr	200	2,142	428,400	8,772	1,754,400	6,307	1,261,400	7,017	1,403,400
May	240	2,142	514,080	8,772	2,105,280	6,307	1,513,680	7,017	1,684,080
Jun	230	2,142	492,660	8,772	2,017,560	6,307	1,450,610	7,017	1,613,910
Jul	290	2,218	643,220	8,874	2,573,460	5,856	1,698,240	7,090	2,056,100
Aug	250	2,218	554,500	8,874	2,218,500	5,856	1,464,000	7,090	1,772,500
Sep	265	2,218	587,770	8,874	2,351,670	5,856	1,551,840	7,090	1,878,850
Oct	175	612	107,100	5,610	981,750	11,893	2,081,275	4,488	785,400
Nov	85	612	52,020	5,610	476,850	11,893	1,010,960	4,488	381,480
Dec	55	612	33,660	5,610	308,550	11,893	654,115	4,488	246,840

Note: Btu's per hour from preceding calculation based on season.

The heat gain for the entire glass area of the house, due to solar radiation, on all sides per month, would be:

Month	Heat Gain Btu's Per Month	Month	Heat Gain Btu's Per Month
Jan	1,418,300	Jul	6,971,020
Feb	2,072,900	Aug	6,009,500
Mar	3,273,000	Sep	6,370,130
Apr	4,847,600	Oct	3,955,525
May	5,817,100	Nov	1,921,310
Jun	5,554,740	Dec	1,243,165
Btu's per year heat entering through glass in the house: $49,454,310			

Example 16a—*Metric Units*

The house has the same glass area as given in Example 15a. Glass transmittance is also the same as 15a. The difference is that this house is located in Seattle, Washington, at 47° 40′ North Latitude.

Using Table 39, the average for the various seasons for North, East, South and West can be obtained. In this case, the heat gain is that which is listed for 48° North Latitude.

For the various seasons, the heat gain average due to solar radiation through glass at 0.85 solar transmittance is:

SEASON	North Wall Watts/hr	East Wall Watts/hr	South Wall Watts/hr	West Wall Watts/hr
SPRING	8.36 x 0.85 x 88 = 625	11.15 x 0.85 x 272 = 2,577	9.85 x 0.85 x 221 = 1,847	8.9 x 0.85 x 272 = 2,057
SUMMER	8.36 x 0.85 x 91 = 646	11.15 x 0.85 x 274 = 2,596	9.85 x 0.85 x 204 = 1,715	8.9 x 0.85 x 274 = 2,072
AUTUMN	8.36 x 0.85 x 25 = 178	11.15 x 0.85 x 173 = 1,639	9.85 x 0.85 x 416 = 3,483	8.9 x 0.85 x 173 = 1,308
WINTER	8.36 x 0.85 x 28 = 199	11.15 x 0.85 x 129 = 1,222	9.85 x 0.85 x 477 = 3,993	8.9 x 0.85 x 129 = 976

From Table 37, the hours of sunshine per month in the city of Seattle is obtained. Using the Watts for each area of glass, in each direction multiply hours of Sunshine/Month for heat gain of that season to obtain heat gain in Watt-Hrs per month as follows:

HEAT GAIN THROUGH GLASS, IN WALLS FACING
(Glass in Walls in Vertical Position)

Month	Sunshine Hours Per Month	North Wall W/hr	North Wall Watts/Month	East Wall W/hr	East Wall Watts/Month	South Wall W/hr	South Wall Watts/Month	West Wall W/hr	West Wall Watts/Month
Jan	65	199	12,935	1,222	79,430	3,993	259,545	976	63,440
Feb	95	199	18,905	1,222	116,090	3,993	379,335	976	92,720
Mar	150	199	29,850	1,222	183,300	3,993	598,950	976	146,400
Apr	200	625	125,000	2,577	515,400	1,847	369,400	2,057	411,400
May	240	625	150,000	2,577	616,480	1,847	443,280	2,057	493,680
Jun	230	625	143,750	2,577	591,330	1,847	424,810	2,057	473,110
Jul	290	646	187,340	2,596	752,840	1,715	497,350	2,072	600,880
Aug	250	646	161,500	2,596	649,000	1,715	428,750	2,072	518,000
Sep	265	646	171,190	2,596	687,970	1,715	454,475	2,072	549,080
Oct	175	178	31,150	1,639	287,825	3,483	609,525	1,308	228,900
Nov	85	178	15,130	1,639	139,315	3,483	296,055	1,308	111,180
Dec	55	178	9,790	1,639	90,145	3,483	191,565	1,308	71,940

Note: Watts per hour from preceding calculations based on season.

The heat gain for the entire glass area of the house, due to solar radiation, on all sides per month, is shown in the following:

Month	Heat Gain Watt-Hr Per Month	Month	Heat Gain Watt-Hr Per Month
Jan	415,350	Jul	2,038,410
Feb	607,050	Aug	1,757,250
Mar	958,500	Sep	1,862,715
Apr	1,421,200	Oct	1,156,400
May	1,703,440	Nov	561,680
Jun	1,633,000	Dec	363,440
Watt hours per year entering through the glass in the house: 14,478,435			

COMPARISON OF EFFECT OF BUILDING OF HOUSE OR BUILDING

It is apparent by comparison of the results that geographic location and prevailing weather conditions can make a considerable difference in the amount of solar energy incident on a house or building. As calculated, heat gain for the house located in Savannah, Georgia, was 70,335,820 Btu's (20,597,375 Watt-hrs) per year; this same house, if located in Seattle, Washington, would gain 49,454,310 Btu's (14,478,435 Watt-hrs) per year. It should also be observed that the major difference in heat gain was in the period from October through March. At the more southern location, the heat gain through this six month period was 33,829,680 Btu's (9,892,700 Watt-hrs) and at the more northern location only 13,884,200 Btu's (4,062,420 Watt-hrs). Thus a difference of 19,945,480 Btu's (5,830,280 Watt-hrs) occurred in heat gain during these months for an identical house in a different geographical location.

One examining the figures of heat gain may wonder why the values fluctuate rather than show a constant rise in summer and reduction to winter. The reason, of course, is that the patterns of heat energy from the sun are related to the position of the earth in relation to the sun. This is illustrated in Figure 46.

A typical pattern of heat transfer to glass in vertical position facing in various directions is shown in Figure 47. As previously shown in Tables 37, 38 and 39, the heat transmittance to glass (or any vertical surface) changes in respect to season, latitude and directional position.

The transmittance of heat energy from solar radiation through the glass to the inside of the house in problems 15, 15a, 16 and 16a was taken to be 85%. This transmittance can be reduced by several methods. Two glass panes, as used in most storm windows, will lower solar transmittance to 0.85 × 0.85 = 0.72. Other means of reducing solar heat gain include awnings, reflective blinds, curtains, solar reflective glass panels and reflective films attached to the inner surface of the glass. Figure 48 gives the heat transmission of solar radiation through single and two pane windows of clear glass, darkened glass and glass with metallic film attached.

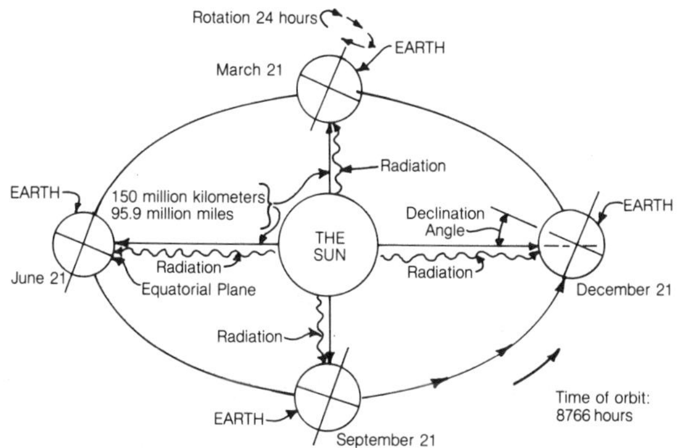

THE EARTH'S MOTION AROUND THE SUN AS IT AFFECTS SOLAR RADIATION

Figure 46

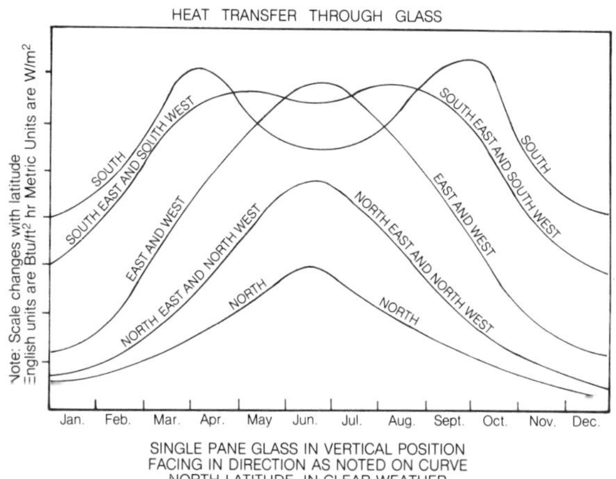

TYPICAL DAILY RATE OF HEAT TRANSMISSION THROUGH GLASS

Figure 47

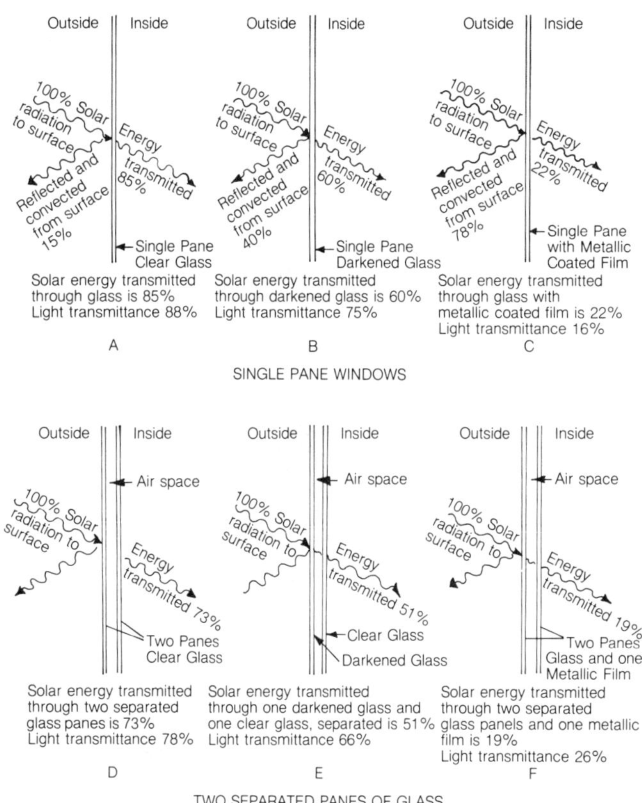

Figure 48

To illustrate the difference in heat transmittance that can be obtained by the use of two panes of glass, one of which has a metallic film installed on its inner surface, the following example is presented:

Example 17—*English Units*

This example will be identical to the house as described in Example 13, with the exception that instead of plain, clear, single glass in the windows and doors, the glass in this house will be two paned with one having a metallic film attached for reflection of solar radiation. The transmission of solar energy through the two panes of glass, one of which was covered with a reflective metallic-coated film, as shown in "F" Figure 48, is 19% of the energy incident on its outer surface.

This house is located approximately 32° North latitude. Its wall positions are Northeast, Southeast, Southwest and Northwest. The total area of glass in windows and doors is 62 sq ft in the Northeast wall, 86 sq ft in the Southeast wall, 108 sq ft in the Southwest wall and 86 sq ft in the Northwest wall. The transmittance of the double pane reflective metallic coated film glass is 19%.

From Table 36, Summer, 32° North latitude determine the solar heat gain to surfaces for each hour in the Northeast, Southeast, Southwest and Northwest positions. In this case multiply the listed energy to these glass areas by 0.19 to determine the heat transmittance of solar energy to the interior per unit of area. These values must be multiplied by the number of square feet of glass in each wall position to obtain the total gain for each wall. The total gain of heat energy per hour for the entire house is shown below.

The tabulation for heat gain to the house for a clear day is:

STD. SOLAR TIME	Heat Gain into House as Described in Example 17 — Glass Btu Heat Gain – Summer, 32° N Latitude — Walls in Position as Noted – Sq ft of Glass as Shown								TOTAL ALL WALLS BTU'S/ HOUR
	NORTHEAST WALL		SOUTHEAST WALL		SOUTHWEST WALL		NORTHWEST WALL		
	Solar Rad. (Table 36) Btu/ft²	0.19 x ft² (ft² = 62) 11.78	Solar Rad. (Table 36) Btu/ft²	0.19 x ft² (ft² = 86) 16.34	Solar Rad. (Table 36) Btu/ft²	0.19 x ft² (ft² = 108)	Solar Rad. (Table 36) Btu/ft²	0.19 x ft² (ft² = 86) 16.34	
6 AM	58	683	54	882	5	102	6	98	1,766
7 AM	158	1,861	130	2,124	17	348	17	278	4,612
8 AM	163	1,920	150	2,451	25	513	25	409	5,293
9 AM	142	1,673	153	2,500	30	616	30	490	5,279
10 AM	93	1,096	161	2,631	42	862	33	539	5,160
11 AM	75	884	140	2,288	65	1,334	37	605	5,109
12 noon	39	459	80	1,307	80	1,642	39	637	4,045
1 PM	37	435	65	1,062	140	2,873	75	1,225	5,596
2 PM	35	412	42	686	161	3,303	93	1,519	5,920
3 PM	30	353	30	490	153	3,139	142	2,320	6,302
4 PM	25	295	25	408	150	3,078	163	2,663	6,444
5 PM	17	200	17	278	130	2,667	158	2,582	5,727
6 PM	6	71	5	82	54	1,108	58	948	2,205
TOTAL	878	10.341	1,052	17,187	1,052	21,587	878	13,900	63,458

Example 17a—*Metric Units*

Identical to Example 17 with units of area of glass in square meters instead of square feet. The total surface of glass in windows and doors is 5.75m² in the Northeast wall, 8m² in the Southeast wall, 10m² in the Southwest wall and 8m² in the Northwest wall.

From Table 36, under summer conditions at 32° North latitude, determine the heat gain to surface in Watts/m² to surfaces in the Northeast, Southeast, Southwest and Northwest positions. In this case, multiply the listed energy to these glass areas by 0.19 to determine the heat transmittance to the interior. These values must be multiplied by the number of square meters of glass in each wall position to obtain solar heat gain for each wall. The total heat gain through the glass area and the total of all glass in the house is calculated and tabulated in the chart which follows.

The tabulation for heat gain to the house for a clear day is:

STD. SOLAR TIME	Heat Gain into House as Described in Example 17a — Glass Watt Heat Gain – Summer, 32° N Latitude — Walls in Position as Noted – Sq Metres of Glass as Shown								TOTAL ALL WINDOW WATTS/ HOUR
	NORTHEAST WALL		SOUTHEAST WALL		SOUTHWEST WALL		NORTHWEST WALL		
	Solar Rad. w/m² (Table 36)	5.75m² x 0.19 = 1.09 1.09 x w/m²	Solar Rad. w/m² (Table 36)	8m² x 0.19 = 1.52 1.52 x w/m²	Solar Rad. w/m² (Table 36)	10m² x 0.19 = 1.90 1.9 x w/m²	Solar Rad. w/m² (Table 36)	8m² x 0.19 = 1.52 1.52 x w/m²	
6 AM	182.84	199.30	170.23	258.75	15.76	29.94	18.91	18.91	516.73
7 AM	498.09	542.92	409.86	622.99	53.59	101.82	58.59	89.06	1,356.79
8 AM	513.85	560.10	472.87	718.76	78.81	149.74	78.81	119.79	1,548.39
9 AM	447.62	487.90	482.40	733.25	95.57	181.58	95.57	145.27	1,548.00
10 AM	293.19	317.57	507.55	771.47	132.40	251.56	110.34	167.72	1,508.32
11 AM	236.44	257.72	441.35	670.85	204.50	388.55	116.64	175.77	1,492.89
12 noon	122.95	134.02	252.20	383.34	252.20	419.18	122.95	186.88	1,183.62
1 PM	116.64	127.14	204.51	310.85	441.35	838.56	236.44	359.39	1,635.97
2 PM	110.34	120.27	132.40	201.25	507.55	964.34	293.19	445.65	1,731.41
3 PM	95.57	104.17	95.57	145.26	482.40	916.56	447.62	680.38	1,846.37
4 PM	78.81	85.90	78.81	119.79	472.87	898.45	513.85	781.05	1,885.19
5 PM	53.59	58.41	53.59	81.46	409.86	778.73	498.04	757.02	1,675.62
6 PM	18.91	20.61	15.76	23.95	170.23	323.44	182.84	277.92	645.92
TOTAL	2,769.04	3,018.25	3,317.49	5.042.58	3,317.49	6,303.23	2,769.04	4,208.90	18,530.22

From the results obtained in Examples 16 and 17, it is apparent that the use of metallic film on the interior of the windows can reduce the solar heat flow into the house. Plain glass, without the reflective film allowed the transmission of heat, in the summer, to the interior of the house of 281,923 Btu's (83,682 Watt-Hrs) per day.* Glass with metallic film transmitted 63,458 Btu's (18,530 Watts) per day. The heat therefore blocked by the metallic film was 218,465 Btu's (64,152 Watt-Hrs) per day. It is apparent that reflective metallic film is very effective in retarding the transmission of solar heat through glass. Unfortunately, it also reduces the visibility through the glass. For this reason, it is seldom used by homeowners.

As illustrated in Examples 13 and 17, solar radiation into the house through glass during the summer months affects the cooling load of the air-conditioning system and its cost of operation. The more energy entering an air-conditioned house, the more energy that must be used to remove this energy from the house. Conversely, energy from solar radiation through glass into the house will reduce the energy load of the heating system.

When calculated on a monthly average similar to what was done in Example 15 and 16, the heat transfer through glass (plain and metallic film-treated) can be determined so as to obtain the total heat gain per average month. Using the same dimensions, location and latitude of the house in Example 13, calculate the monthly average of heat gain due to solar radiation.

HEAT GAIN DUE TO SOLAR RADIATION, PER AVERAGE MONTH, EACH MONTH OF THE YEAR

Example 18—*English Units*

Determine the heat gain into the house due to solar radiation through plain glass and glass with metallic film on the interior per average month over the year. The house is located in Savannah, Georgia, approximately 32° North latitude. Its walls are positioned Northeast, Southeast, Southwest and Northwest. The total area of glass in the windows and doors is 62 sq ft in the Northeast wall, 86 sq ft in the Southeast wall, 108 sq ft in the Southwest wall and 86 sq ft in the Northwest wall.

Determine the heat transfer through plain glass having a transmittance of solar heat of 85% and glass having a metallic film on its interior with transmittance of solar heat of 19%.

From Table 39 obtain the direct solar heat transmitted to surfaces in various positions. To obtain the answers in this problem, these values must be multiplied by transmittance through the glass, the area of the glass and the total average sunshine hours (given in Table 37).

For the various seasons, the heat gain per average sunshine hour due to solar radiation through plain glass and glass with metallic film is given in the table which follows:

HEAT GAIN — Btu's

SEASON	Northeast Wall	Southeast Wall	Southwest Wall	Northwest Wall
PLAIN GLASS – SOLAR TRANSMITTANCE = 0.85				
SPRING	62 x 0.85 x 63 = 3,320	86 x 0.85 x 73 = 5,336	108 x 0.85 x 73 = 6,701	86 x 0.85 x 63 = 4,605
SUMMER	62 x 0.85 x 68 = 3,584	86 x 0.85 x 81 = 5,921	108 x 0.85 x 81 = 7,436	86 x 0.85 x 68 = 4,970
AUTUMN	62 x 0.85 x 20 = 1,054	86 x 0.85 x 118 = 8,625	108 x 0.85 x 118 = 10,832	86 x 0.85 x 20 = 1,462
WINTER	62 x 0.85 x 23 = 1,212	86 x 0.85 x 141 = 10,307	108 x 0.85 x 141 = 12,944	86 x 0.85 x 23 = 1,681
GLASS WITH METALLIC FILM – SOLAR TRANSMITTANCE = 0.19				
SPRING	62 x 0.19 x 63 = 742	86 x 0.19 x 73 = 1,193	108 x 0.19 x 73 = 1,498	86 x 0.19 x 63 = 1,029
SUMMER	62 x 0.19 x 68 = 801	86 x 0.19 x 81 = 1,323	108 x 0.19 x 81 = 1,662	86 x 0.19 x 68 = 1,111
AUTUMN	62 x 0.19 x 20 = 164	86 x 0.19 x 112 = 1,830	108 x 0.19 x 112 = 2,298	86 x 0.19 x 14 = 229
WINTER	62 x 0.19 x 23 = 270	86 x 0.19 x 141 = 2,304	108 x 0.19 x 141 = 2,893	86 x 0.19 x 23 = 375

Table 37 gives the hours of sunshine per month, as influenced by weather conditions for selected United States cities. To obtain the Btu per month for heat gain of solar radiation through the glass area in each wall, the Btu's per hour established must be multiplied by hours of sunshine for each month. For Example 18, this is:

FOR PLAIN GLASS — HEAT GAIN in Btu's (British Units) ON WALLS FACING IN DIRECTION NOTED (Example 18)

Month	Sunshine Hours Per Month	Northeast Wall		Southeast Wall		Southwest Wall		Northwest Wall		Total Btu All Walls Per Month
		Btu/hr	Btu/Month	Btu/hr	Btu/Month	Btu/hr	Btu/Month	Btu/hr	Btu/Month	
Jan	180	1,212	218,160	10,307	1,855,260	12,944	2,329,920	1,681	302,580	4,705,920
Feb	175	1,212	212,100	10,307	1,803,725	12,944	2,265,200	1,681	294,175	4,575,200
Mar	230	1,212	278,760	10,307	2,370,610	12,944	2,977,120	1,681	386,630	6,013,120
Apr	270	3,320	896,400	5,336	1,440,720	6,701	1,809,270	4,605	1,243,350	5,389,740
May	315	3,320	1,045,800	5,336	1,680,840	6,701	2,110,815	4,605	1,450,575	6,288,030
Jun	280	3,320	929,600	5,336	1,494,080	6,701	1,876,280	4,605	1,289,400	5,589,360
Jul	275	3,584	985,600	5,921	1,639,275	7,436	2,044,900	4,970	1,366,750	6,036,525
Aug	260	3,584	931,840	5,921	1,549,860	7,436	1,933,360	4,970	1,292,200	5,707,260
Sep	240	3,584	860,160	5,921	1,430,640	7,436	1,784,640	4,970	1,192,800	5,268,240
Oct	230	1,054	242,420	8,625	1,983,750	10,832	2,491,360	1,463	336,490	5,054,020
Nov	200	1,054	210,800	8,625	1,725,000	10,832	2,166,400	1,463	292,600	4,394,600
Dec	170	1,054	179,160	8,625	1,466,250	10,832	1,841,440	1,463	284,710	3,735,560

*Example 13

FOR GLASS WITH METALLIC FILM – HEAT GAIN in Btu's ON WALLS FACING IN DIRECTION NOTED										
Month	Sunshine Hours Per Month	Northeast Wall		Southeast Wall		Southwest Wall		Northwest Wall		Total Btu All Walls Per Month
		Btu/hr	Btu/Month	Btu/hr	Btu/Month	Btu/hr	Btu/Month	Btu/hr	Btu/Month	
Jan	180	270	48,600	2,304	414,720	2,893	520,740	375	67,500	1,051,560
Feb	175	270	47,250	2,304	403,200	2,893	506,275	375	65,625	1,022,350
Mar	230	270	62,100	2,304	529,920	2,893	665,390	375	86,250	1,343,660
Apr	270	742	200,340	1,193	322,100	1,498	404,460	1,029	277,830	1,204,730
May	315	742	233,730	1,193	375,795	1,498	471,870	1,029	324,135	1,405,530
Jun	280	742	207,760	1,193	334,040	1,498	419,440	1,029	286,440	1,247,680
Jul	275	801	220,275	1,323	363,825	1,662	457,050	1,111	305,525	1,346,675
Aug	260	801	208,260	1,323	343,980	1,662	432,120	1,111	288,860	1,273,220
Sep	240	801	192,240	1,323	317,520	1,662	398,880	1,111	266,640	1,175,280
Oct	230	164	37,720	1,830	420,900	2,298	528,540	229	52,670	1,039,830
Nov	200	164	32,800	1,830	366,000	2,298	459,600	229	45,800	904,200
Dec	170	164	27,880	1,830	311,100	2,298	390,660	229	38,930	768,570

During the months of July, August and September the heat gain through plain glass is 17,333,145 Btu's and the heat gain through the glass with metallic film is 3,795,175 Btu's. This is a savings of 13,537,970 Btu's in heat load on the air-conditioning system. However, in the months of November, December, January, February and March the heat gain through the plain glass is 22,779,840 Btu's and only 5,090,340 Btu's for glass with metallic film or 17,689,500 Btu's which must be added to the heating load.

Example 18a—*Metric Units*

Determine the heat gain into the house due to solar radiation through plain glass and glass with metallic film on the interior, per average month over the year. The house is located in Savannah, Georgia, approximately 32° North latitude and positioned Northeast, Southeast, Southwest and Northwest. The surface area of glass in windows and doors is 5.75m² in the Northeast wall, 8m² in the Southeast wall, 10m² in the Southwest wall and 8m² in the Northwest wall. The transmittance of solar heat through plain glass is 85% and glass with metallic film on the interior is 19%.

From Table 39, obtain the direct solar heat transmitted to surfaces in the Northeast, Southeast, Southwest and Northwest. To obtain the answers in this problem, these values must be multiplied by the transmittance through the glass (or glass and metallic film), the area and the total average of sunshine hours. (Given in Table 37).

For various seasons, the heat gain due to solar radiation as described in Example 18a, through glass and glass with metallic film per average sunshine hour is:

HEAT GAIN IN WATTS PLAIN GLASS SOLAR TRANSMITTANCE = 0.85				
SEASON	Northeast Wall	Southeast Wall	Southwest Wall	Northwest Wall
SPRING	5.75 x 0.85 x 198 = 968	8 x 0.85 x 229 = 1,557	10 x 0.85 x 229 = 1,947	8 x 0.85 x 198 = 1,346
SUMMER	5.75 x 0.85 x 213 = 1,041	8 x 0.85 x 255 = 1,734	10 x 0.85 x 255 = 2,168	8 x 0.85 x 213 = 1,482
AUTUMN	5.75 x 0.85 x 63 = 308	8 x 0.85 x 373 = 2,536	10 x 0.85 x 373 = 3,171	8 x 0.85 x 63 = 428
WINTER	5.75 x 0.85 x 74 = 362	8 x 0.85 x 445 = 3,026	10 x 0.85 x 445 = 3,783	8 x 0.85 x 74 = 503
GLASS WITH METALLIC FILM – SOLAR TRANSMITTANCE = 0.19				
SPRING	5.75 x 0.19 x 198 = 216	8 x 0.19 x 229 = 348	10 x 0.19 x 229 = 435	8 x 0.19 x 198 = 301
SUMMER	5.75 x 0.12 x 213 = 233	8 x 0.19 x 255 = 388	10 x 0.19 x 255 = 485	8 x 0.19 x 213 = 324
AUTUMN	5.75 x 0.19 x 63 = 69	8 x 0.19 x 373 = 567	10 x 0.19 x 373 = 709	8 x 0.19 x 63 = 96
WINTER	5.75 x 0.19 x 74 = 81	8 x 0.19 x 445 = 676	10 x 0.19 x 445 = 845	8 x 0.19 x 74 = 112

Table 37 gives the hours per month as influenced by weather conditions. Multiply the heat gain transmitted through the plain glass and glass with metallic film times the hours of sunshine per month to obtain the heat gain per month of each. For Example 18a this is shown below:

FOR PLAIN GLASS – HEAT GAIN in WATTS (Metric Units) ON WALLS FACING in DIRECTION NOTED (Example 18a)										
Month	Sunshine Hours Per Month	Northeast Wall		Southeast Wall		Southwest Wall		Northwest Wall		Total Watts Per Month
		W/hr	W/Month	W/hr	W/Month	W/hr	W/Month	W/hr	W/Month	
Jan	180	362	65,160	3,026	544,680	3,783	680,940	503	90,540	1,381,320
Feb	175	362	63,350	3,026	529,550	3,783	662,025	503	88,025	1,342,950
Mar	230	362	83,260	3,026	695,980	3,783	870,090	503	115,690	1,765,020
Apr	270	968	261,360	1,557	420,390	1,947	525,690	1,346	363,420	1,570,860
May	315	968	304,920	1,557	490,455	1,947	613,305	1,346	423,990	1,832,670
Jun	280	968	271,040	1,557	435,960	1,947	545,160	1,346	376,880	1,629,040
Jul	275	1,041	286,275	1,734	476,850	2,168	596,200	1,482	407,550	1,766,875
Aug	260	1,041	270,660	1,734	450,840	2,168	563,680	1,482	385,320	1,670,500
Sep	240	1,041	249,840	1,734	416,160	2,168	498,640	1,482	355,680	1,520,320
Oct	230	308	70,840	2,536	583,280	3,171	729,330	428	98,440	1,481,890
Nov	200	308	61,600	2,536	507,200	3,171	634,200	428	85,600	1,288,600
Dec.	170	308	52,360	2,536	431,120	3,171	539,070	428	72,760	1,095,310

FOR GLASS WITH METALLIC FILM – HEAT GAIN IN WATTS (Metric Units) ON WALLS FACING IN DIRECTION NOTED (Example 18a)										
	Sunshine Hours	Northeast Wall		Southeast Wall		Southwest Wall		Northwest Wall		Total Watts
Month	Per Month	W/hr	W/Month	W/hr	W/Month	W/hr	W/Month	W/hr	W/Month	Per Month
Jan	180	81	14,580	676	121,680	845	152,100	112	20,160	308,520
Feb	175	81	14,175	676	118,300	845	147,875	112	19,600	299,950
Mar	230	81	18,630	676	155,480	845	194,350	112	25,760	394,220
Apr	270	216	58,320	348	93,960	435	117,450	301	81,270	351,000
May	315	216	68,040	348	109,620	435	137,025	301	94,815	409,500
Jun	280	216	60,480	348	97,440	435	121,800	301	84,280	364,000
Jul	275	233	64,075	388	106,700	485	133,375	324	89,100	393,250
Aug	260	233	60,580	388	100,880	485	126,100	324	84,240	371,600
Sep	240	233	55,920	388	93,120	485	111,550	324	77,760	338,350
Oct	230	69	15,870	567	130,410	709	163,070	96	22,080	331,430
Nov	200	69	13,800	567	113,400	709	141,800	96	19,200	288,200
Dec	170	69	11,730	567	96,390	709	120,530	96	16,320	244,970

During the months of July, August and September, the heat gain through plain glass is 4,957,695 watts and the heat gain through glass with metallic film is 1,103,200 watts. This is a savings of 3,854,495 watts heat load on the air-conditioning system. However, the heat gain through plain glass during the months of November, December, January, February and March is 6,873,200 watts and only 1,535,860 watts for glass with metallic film. This 5,337,340 watts must be added to the heating load.

HEAT TRANSFER THROUGH GLASS WITH LOUVERED SCREENS TO SHADE GLASS

Screens with angle louvers can be used to reduce solar radiation into a building either when windows are opened or closed.

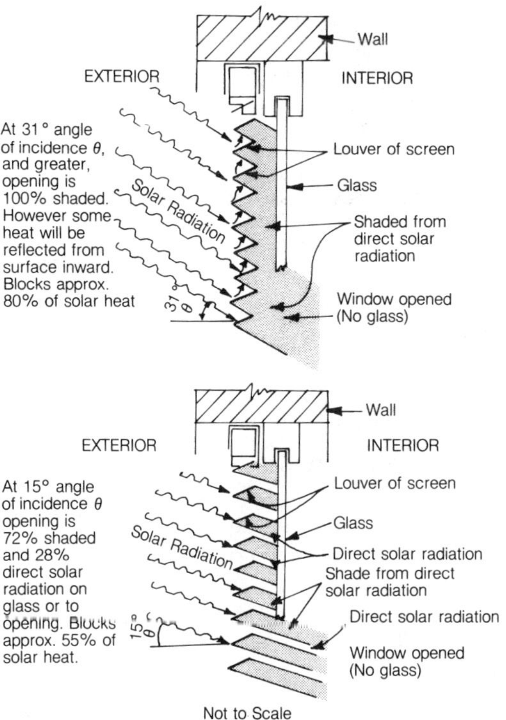

HOW LOUVERED SCREENS REDUCE SOLAR RADIATION TRANSMISSION THROUGH GLASS OR OPENING

Figure 49

These screens use louvers instead of wire and the louvers are so formed to shade the glass (or interior) during periods of the year when interior may required cooling. These screens are most frequently formed of aluminum alloy, the louvers being approximately 0.06″ (15mm) wide by 0.9″ (23mm) long and spaced 17 to 18 per inch (1.5 per mm). After they are formed, they are finished with coating for appearance and to extend service life. When the sun is at or above 31° elevation, these screens shade 100% of the window glass or open area. This effectively reduces the heat gain to the interior through this area by approximately 80%. Such a screen is illustrated in Figure 49.

Although it might be thought that the screens would be more effective as a heat reflector if they were not painted (coated), such is not the case. Oxidized aluminum is not an effective solar heat reflector in general, even though its effectiveness increases as the infrared wavelengths decrease.

These aluminum louvered screens reduce the entry of solar radiation through a window in cold weather as well as hot weather. For this reason, if solar heat energy can be used in winter to reduce fuel costs, it is recommended that the screens be removed from windows during the heating seasons.

HEAT TRANSFER THROUGH GLASS WITH VARIOUS TYPES OF VENETIAN BLINDS

Venetian blinds can be used to reduce solar radiation through the glass openings of a building. By adjusting the shades so that the radiation is reflected outward, the heat of solar radiation to the interior of a building can be reduced. The effectiveness of the venetian blinds when adjusted for 100% shading is completely dependent upon the reflectance of solar heat of surface facing outward. With good reflectance, 0.6 to 0.9, that amount of solar heat will be reflected back toward the glass and through to the exterior. All heat absorbed by the surface heats the venetian blinds themselves which is then transmitted to the interior by conduction and convection to the surrounding air. This illustrated in Figure 50.

In the case where venetian blinds are used to reduce solar radiation entry into a window, their efficiency depends upon the reflectance of the surface facing the sun. In the case of the surface being highly reflective and the blind adjusted for 100% shading so as to reflect the sun's radiation outward, the solar heat entry into the building can be reduced by approximately one-half.

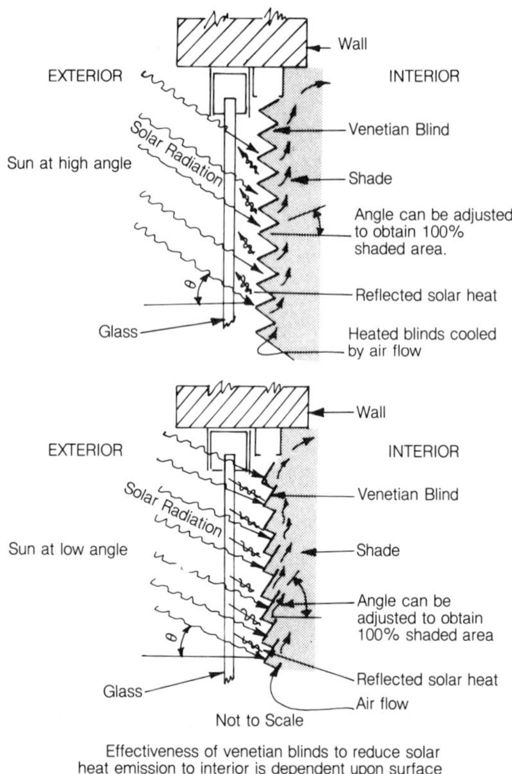

Effectiveness of venetian blinds to reduce solar heat emission to interior is dependent upon surface heat reflectance of surface facing outward at wavelengths of visable heat. (0.38 to 0.76 microns).

HOW VENETIAN BLINDS CAN BE USED TO REDUCE SOLAR RADIATION THROUGH WINDOWS

Figure 50

However, if the blind is of a dark color, although it may shade the interior, the blind itself will be heated by the solar radiation. This heat is then transmitted by conduction and convection to the air surrounding the venetian blind and transferred to the interior of the room.

The effectiveness of the venetian blind in reducing solar radiation into a room as a function of colors is as follows:

Based on Blind Being Adjusted for 100% Shading	Percent of Solar Radiation Entering Interior - Approx.
Aluminum	90
White-bright	50
Pastel colors	75
Dark colors	98

*Not considering reflectance of glass.

The previous discussion was directed toward the function of louver screens and blinds to reduce solar heat into the interior during the cooling season. During the winter when heat is required, the venetian blinds can be adjusted to allow maximum amount of transmission of solar heat energy to the interior. This is illustrated in Figure 51.

*Adapted from "Reflective Insulating Blinds" report by Hanna B. Shapira and Paul R. Barnes, Solar and Special Studies Section Energy Division, Oak Ridge National Laboratory, Oak Ridge, Tennessee 37830.

Roller shades and drapes can also be used to shade windows and doors to retard the heat gain due to solar radiation. The effectiveness of these depend upon the ability of the surface facing outward to reflect the sun's rays. Although a roller shade or drape may be able to provide 100% shade if the surface is dark with high surface emittance factor, the rays onto the surface will heat the surface to a higher temperature than surrounding air. By conduction and convection, the air passing over this surface will be heated to balance the amount of heat absorbed by the surface.

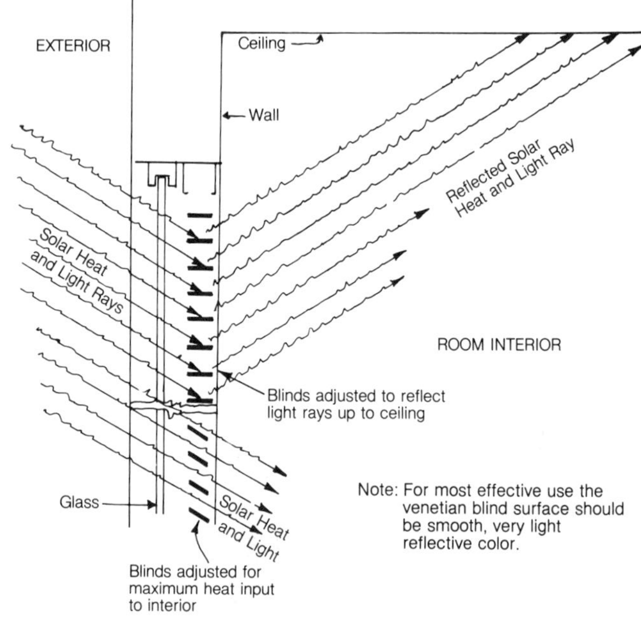

Venetian blind adjusted to allow solar heat and energy into interior so as to reduce heating load.

HOW VENETIAN BLINDS CAN BE USED TO UTILIZE SOLAR HEAT IN WINTER

Figure 51

CONTROL OF SOLAR RADIATION BY USE OF AWNINGS AND OVERHANGING PROJECTIONS

Because the solar altitude angle changes with the time of year, awnings or overhang projections can be designed as part of the building to help control solar transmission to window or door openings.

Solar radiation can be achieved by the use of overhangs, awnings or other opaque surfaces to regulate shadow areas on windows so that sunlight can be shut out during the summer and enter during the winter. This difference in shading, depending on the angle of the sun's rays, is shown in Figure 52.

It can also be observed from this figure that the solar altitude angle changes in respect to both time of day and time of year. The angle of incidence which the sun's rays are to a vertical wall is influenced by the wall's orientation with respect to North and South and latitude location. The relationship of these angles are shown in Figure 53. It is essential to relate these angles to design projec-

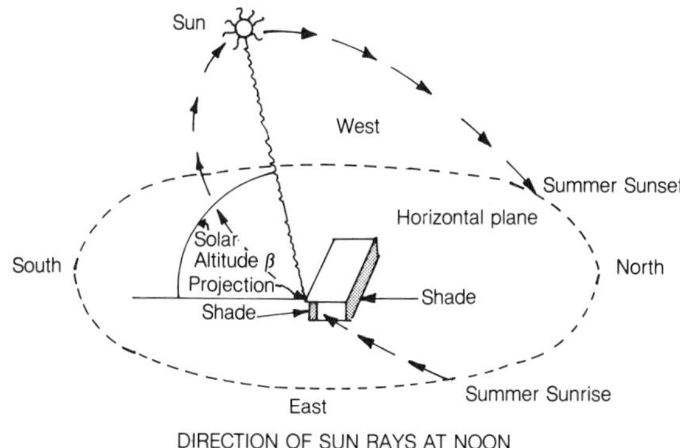

DIRECTION OF SUN RAYS AT NOON
NORTH LATITUDE — SUMMER

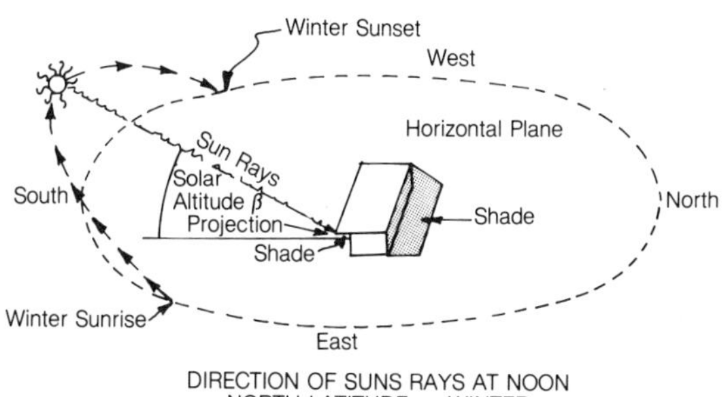

DIRECTION OF SUNS RAYS AT NOON
NORTH LATITUDE — WINTER

SOLAR ALTITUDE ANGLE β

Figure 52

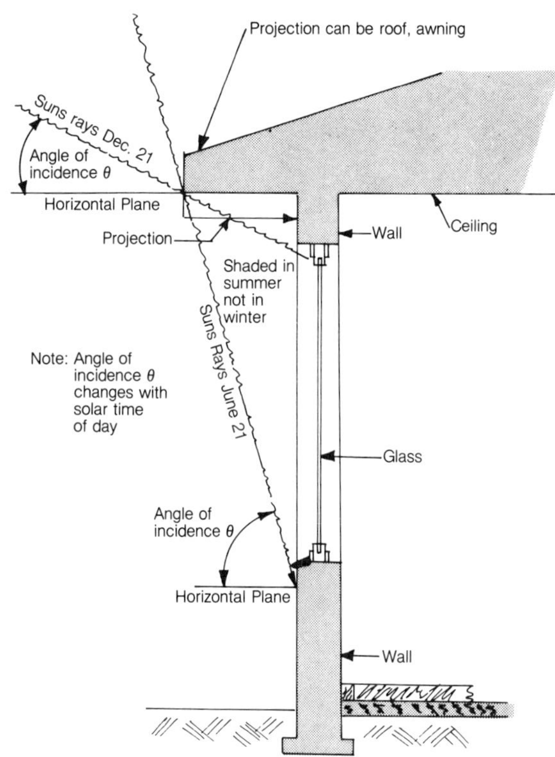

HOW OVERHANGS SHADE SOLAR HEAT RAYS IN SUMMER AND ADMIT SOLAR HEAT IN WINTER

Figure 54

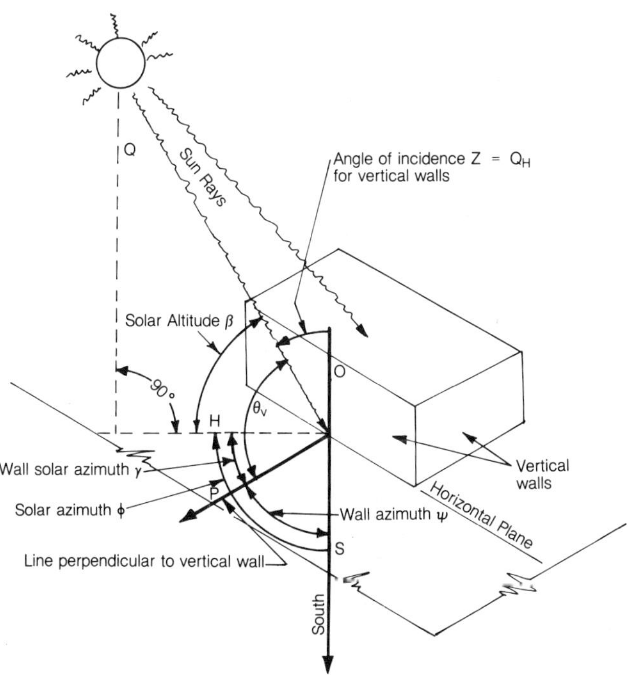

RELATIONAHIP OF SOLAR ANGLE θ TO HORIZONTAL AND VERTICAL SURFACES

Figure 53

tions that serve to shade windows in summer and not shade these windows in winter so as to take advantage of solar energy for heating. This is illustrated in Figure 54.

These overhangs, projections or awnings providing the shade or partly shaded openings must be done as part of the architectural design, thus it is not strictly thermal insulation. However, the shading effect does influence heat gain or loss. Shading depends upon the angle of incidence on vertical surfaces. The following provides information for assistance in the architectural design of projections, overhangs and awnings.

The incident sun angle θ as shown in Figure 53 has the following relationship with other solar ray angles which are determined by date, solar time and latitude.

β = altitude angle of sun
γ = azimuth angle of sun with respect to wall
θ = angle between sun and perpendicular to wall

$$\cos\theta = \cos\beta\cos\gamma \qquad (19)$$

Angle β is an angle determined by date, solar time and latitude and can be determined from Table 40. This table provides solar position angles, altitude and azimuth based on date, solar time, and Northern latitude.

Angle γ is dependent upon position of the sun azimuth angle ϕ, and ψ, the angle measured from south to the wall perpendicular. This is shown in Figure 53.

SOLAR POSITION ALTITUDE and AZIMUTH
IN REFERENCE TO DATE, TIME and LATITUDE

DATE	SOLAR TIME AM	SOLAR TIME PM	24° North Lat Altitude ∠β	24° North Lat Azimuth ∠φ	32° North Lat Altitude ∠β	32° North Lat Azimuth ∠φ	40° North Lat Altitude ∠β	40° North Lat Azimuth ∠φ	48° North Lat Altitude ∠β	48° North Lat Azimuth ∠φ	56° North Lat Altitude ∠β	56° North Lat Azimuth ∠φ
Jan 21	7	5	5	66	1	65	0	63	0	62	0	59
	8	4	17	58	13	57	8	56	4	55	0	53
	9	3	28	48	23	46	17	44	11	43	5	42
	10	2	37	36	31	33	24	31	17	29	10	29
	11	1	44	20	36	18	28	16	21	15	13	15
	12	12	46	0	38	0	30	0	22	0	14	0
Feb 21	7	5	9	74	7	73	4	72	2	72	1	63
	8	4	22	66	19	64	14	62	11	60	7	59
	9	3	34	57	29	53	24	50	19	47	14	46
	10	2	45	44	39	39	32	35	26	33	19	31
	11	1	52	25	45	21	37	19	30	17	22	16
	12	12	55	0	47	0	39	0	31	0	23	0
Mar 21	7	5	14	84	13	82	11	80	10	79	8	78
	8	4	27	77	25	73	23	70	20	67	16	64
	9	3	40	68	37	62	33	57	28	53	23	50
	10	2	52	55	47	48	42	42	35	38	29	35
	11	1	62	33	55	27	48	23	40	20	33	18
	12	12	66	0	58	0	50	0	42	0	34	0
Apr 21	6	6	5	101	6	100	7	99	7	98	9	96
	7	5	18	95	19	92	19	90	19	87	18	84
	8	4	32	89	32	84	30	79	29	75	26	71
	9	3	46	82	44	74	41	67	38	61	34	56
	10	2	59	72	56	60	51	51	46	45	40	40
	11	1	71	52	65	38	59	29	52	24	44	21
	12	12	78	0	70	0	62	0	54	0	46	0
May 21	5	7	0	116	0	114	2	115	5	114	9	113
	6	6	8	108	10	107	13	106	15	104	16	101
	7	5	21	103	23	100	24	97	25	93	25	89
	8	4	35	99	35	93	35	87	35	82	33	76
	9	3	48	94	48	85	47	76	44	68	41	62
	10	2	62	88	61	73	58	61	53	51	48	44
	11	1	76	77	72	52	66	37	60	29	52	23
	12	12	86	0	78	0	70	0	62	0	54	0
Jun 21	5	7	0	119	0	118	4	117	8	117	11	115
	6	6	9	112	12	110	15	108	17	106	19	104
	7	5	22	107	24	103	26	100	27	96	28	92
	8	4	36	103	37	97	37	91	37	85	36	79
	9	3	49	99	50	89	49	80	47	72	44	64
	10	2	63	95	62	80	60	66	56	55	51	46
	11	1	76	91	74	61	69	42	63	31	56	25
	12	12	89	0	81	0	74	0	65	0	57	0

(Sheet 1 of 2)

TABLE 40

SOLAR POSITION ALTITUDE and AZIMUTH
IN REFERENCE TO DATE, TIME and LATITUDE

DATE	SOLAR TIME AM	SOLAR TIME PM	24° North Lat Altitude ∠β	24° North Lat Azimuth ∠ϕ	32° North Lat Altitude ∠β	32° North Lat Azimuth ∠ϕ	40° North Lat Altitude ∠β	40° North Lat Azimuth ∠ϕ	48° North Lat Altitude ∠β	48° North Lat Azimuth ∠ϕ	56° North Lat Altitude ∠β	56° North Lat Azimuth ∠ϕ
Jul 21	5	7	0	119	0	117	2	115	6	115	9	114
	6	6	8	109	11	108	13	106	15	104	17	102
	7	5	21	104	23	101	24	97	25	94	25	90
	8	4	35	99	36	94	36	88	35	82	34	77
	9	3	48	95	48	86	47	77	45	69	41	62
	10	2	62	89	61	74	58	62	54	52	48	45
	11	1	76	79	72	53	67	38	60	29	53	24
	12	12	87	0	79	0	71	0	63	0	55	0
Aug 21	5	7	0	106	0	105	0	107	1	108	2	109
	6	6	5	101	7	101	8	100	9	98	10	97
	7	5	19	96	19	93	19	90	19	87	19	83
	8	4	32	90	32	85	31	80	29	75	27	71
	9	3	46	83	44	75	42	68	38	62	34	57
	10	2	59	73	56	61	52	52	46	45	41	40
	11	1	72	53	66	38	59	30	52	24	45	21
	12	12	78	0	70	0	62	0	54	0	46	0
Sep 21	7	5	14	84	13	82	11	80	10	79	8	78
	8	4	27	77	25	73	23	70	20	67	16	64
	9	3	40	68	37	62	33	58	28	53	23	50
	10	2	52	55	47	48	42	42	35	38	29	35
	11	1	62	33	55	27	48	23	40	20	33	18
	12	12	66	0	58	0	50	0	42	0	34	0
Oct 21	7	5	9	74	7	73	5	72	2	72	0	68
	8	4	22	67	19	64	15	62	11	60	7	59
	9	3	34	57	30	53	25	50	19	47	14	46
	10	2	45	44	39	39	32	36	26	33	19	31
	11	1	53	25	45	21	38	19	30	17	22	16
	12	12	56	0	48	0	40	0	32	0	24	0
Nov 21	7	5	5	66	2	65	0	62	0	67	0	68
	8	4	17	58	13	57	8	55	4	55	0	54
	9	3	28	49	23	46	17	44	11	43	5	42
	10	2	37	36	31	33	24	31	17	30	10	29
	11	1	44	20	36	18	29	16	21	15	13	15
	12	12	46	0	38	0	30	0	22	0	14	0
Dec 21	7	5	3	63	0	62	0	60	0	53	0	52
	8	4	15	55	10	54	6	53	0	47	0	46
	9	3	26	46	20	44	14	42	8	41	2	41
	10	2	34	34	28	31	21	29	14	28	7	28
	11	1	40	18	33	16	25	15	17	14	10	14
	12	12	43	0	35	0	27	0	19	0	11	0

(Sheet 2 of 2)

TABLE 40 (Continued)

COSINES AND SINES

Degree Angle θ	Natural COSINE θ		Degree Angle θ	Natural COSINE θ		Degree Angle θ	Natural COSINE θ	
1	0.9998	89	31	0.8572	59	61	0.4848	29
2	0.9994	88	32	0.8480	58	62	0.4695	28
3	0.9986	87	33	0.8387	57	63	0.4540	27
4	0.9976	86	34	0.8290	56	64	0.4384	26
5	0.9962	85	35	0.8192	55	65	0.4226	25
6	0.9945	84	36	0.8090	54	66	0.4067	24
7	0.9925	83	37	0.7986	53	67	0.3907	23
8	0.9903	82	38	0.7880	52	68	0.3746	22
9	0.9877	81	39	0.7771	51	69	0.3584	21
10	0.9848	80	40	0.7660	50	70	0.3420	20
11	0.9816	79	41	0.7547	49	71	0.3256	19
12	0.9781	78	42	0.7431	48	72	0.3090	18
13	0.9744	77	43	0.7314	47	73	0.2924	17
14	0.9703	76	44	0.7193	46	74	0.2756	16
15	0.9659	75	45	0.7071	45	75	0.2588	15
16	0.9613	74	46	0.6947	44	76	0.2419	14
17	0.9536	73	47	0.6820	43	77	0.2250	13
18	0.9511	72	48	0.6691	42	78	0.2076	12
19	0.9455	71	49	0.6561	41	79	0.1908	11
20	0.9397	70	50	0.6428	40	80	0.1736	10
21	0.9336	69	51	0.6293	39	81	0.1564	9
22	0.9272	68	52	0.6157	38	82	0.1392	8
23	0.9205	67	53	0.6018	37	83	0.1219	7
24	0.9135	66	54	0.5878	36	84	0.1045	6
25	0.9063	65	55	0.5736	35	85	0.0872	5
26	0.8988	64	56	0.5592	34	86	0.0698	4
27	0.8910	63	57	0.5446	33	87	0.0523	3
28	0.8829	62	58	0.5299	32	88	0.0349	2
29	0.8746	61	59	0.5150	31	89	0.0175	1
30	0.8660	60	60	0.5000	30	90	0.0000	0
	Natural SINE θ	Degree Angle θ		Natural SINE θ	Degree Angle θ		Natural SINE θ	Degree Angle θ

$$\text{Cosine } \theta = \frac{b}{c}$$

$$\text{Sine } \theta = \frac{a}{c}$$

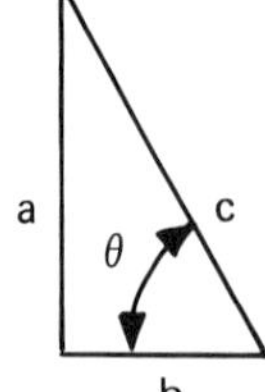

TABLE 41

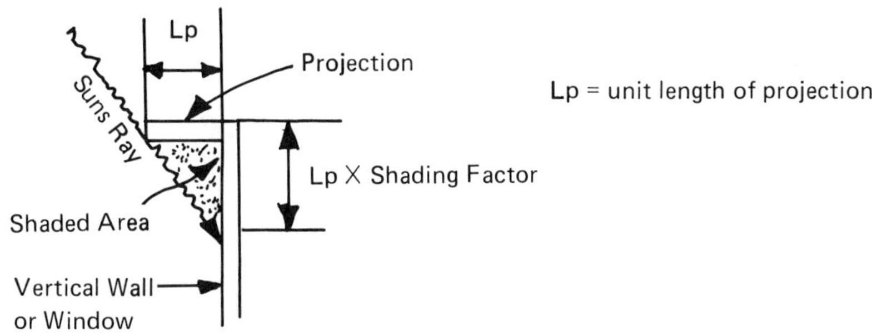

VERTICAL SURFACE AT LATITUDE POSITION AND SOLAR TIME
SHADING FACTOR BASED ON PROVIDING SHADE APRIL 11 TO SEPTEMBER 11

Latitude North	SOLAR TIME A.M.	SHADING FACTOR — WINDOW OR WALLS FACING					SOLAR TIME P.M.	SHADING FACTOR — WINDOWS OR WALL FACING				
		NE	E	SE	S	SW		NW	W	SW	S	SE
	8	0.92	0.61	0.63	6.25	S	1	S	3.57	2.63	3.70	S
	9	1.88	1.00	0.93	4.35	S	2	4.35	1.72	1.33	3.70	S
24°	10	4.35	1.72	1.33	3.70	S	3	1.88	1.00	0.93	4.35	S
	11	S	3.57	2.63	3.70	S	4	0.92	0.61	0.63	6.25	S
	12	S	S	4.54	3.57	4.54	5	0.60	0.19	0.20	S	S
	8	1.00	0.57	0.70	3.33	S	1	S	3.57	1.96	2.38	8.33
	9	2.16	0.97	1.00	2.63	S	2	7.14	1.66	1.36	2.38	S
32°	10	7.14	1.66	1.36	2.38	S	3	2.16	0.97	1.00	2.63	S
	11	S	3.57	1.96	2.38	8.33	4	1.00	0.57	0.70	3.33	S
	12	S	S	3.33	2.38	3.33	5	0.62	0.18	0.24	5.88	S
	8	1.03	0.53	0.62	2.33	S	1	S	3.22	1.53	1.66	4.54
	9	2.12	0.89	0.86	1.85	S	2	7.69	1.54	1.09	1.72	S
40°	10	7.69	1.54	1.09	1.72	S	3	2.12	0.89	0.86	1.85	S
	11	S	3.22	1.53	1.64	4.54	4	1.03	0.53	0.62	2.33	S
	12	S	S	2.32	1.58	2.33	5	0.64	0.16	0.20	5.55	S
	8	1.07	0.51	0.53	1.33	S	1	S	3.13	1.19	1.19	3.13
	9	3.03	0.83	0.73	1.33	S	2	S	1.36	0.83	1.19	S
48°	10	S	1.36	0.83	1.19	S	3	3.03	0.83	0.73	1.33	S
	11	S	3.13	1.19	1.19	3.13	4	1.07	0.51	0.53	1.53	S
	12	S	S	1.66	2.78	1.66	5	0.66	0.14	0.18	2.77	S
	8	1.04	0.36	0.28	1.28	S	1	S	2.56	0.98	0.94	2.12
	9	3.45	0.74	0.61	1.07	S	2	S	1.26	0.77	0.96	7.14
56°	10	S	1.26	0.77	0.96	7.14	3	3.45	0.74	0.61	1.07	S
	11	S	2.56	0.98	0.94	2.12	4	1.04	0.36	0.28	1.28	S
	12	S	S	1.33	0.93	1.33	5	0.60	0.08	0.10	1.83	S

S indicates in shade due to position. North considered shaded by position.

Note: The shading presented represents the minimum length of shade for a particular projection in the time period of April 11 to Sept. 11, for given solar time. In winter months, the shading factor is much less for walls facing in southernly direction.

TABLE 42

The solar azimuth γ can be found for vertical walls as follows:

For Walls Facing East of South

Solar morning hours AM	$\gamma = \phi - \psi$	
Noon	$\gamma = \phi$	
Solar afternoon hours PM	$\gamma = \phi + \psi$	(20)

For Walls Facing West of South

Solar morning hours AM	$\gamma = \phi + \psi$	
Noon	$\gamma = \phi$	
Solar afternoon hours PM	$\gamma = \phi - \psi$	(21)

When γ is in negative degrees, the cosine shall be used the same as if it is in positive degrees. When γ is greater than 90°, this indicates the facing surface is in shade without any projection above.

DETERMINATION OF SOLAR AZIMUTH OF A WALL

Example 19

Find the sun angle θ on a wall at 40° North latitude at 2:00 p.m., Jan. 21, with the wall facing 15° West of South. Also find shade per unit of projection in the direction of the sun's rays.

From Table 40 altitude angle $\beta = 24°$
From Table 40 azimuth angle $\phi = 31°$

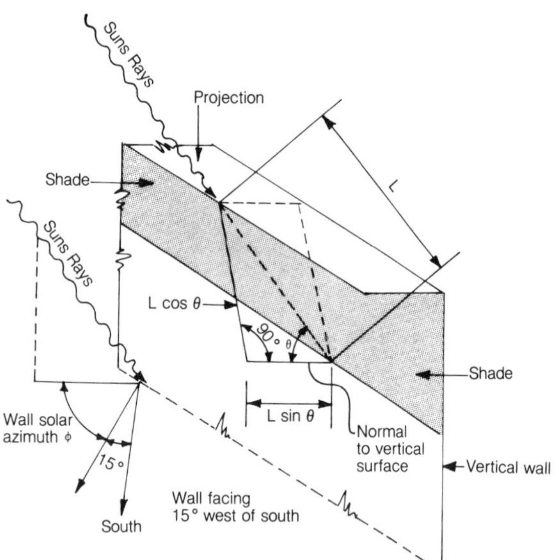

At 2:00 pm, Latitude 40° North, $\beta = 24°$ and $\phi = 31°$.
Angle θ incident to vertical surface calculated to be 29°.
L is the length of suns rays from projection to vertical wall.
Length of projection is L sin θ.
The length of shade, in direction of suns rays, on vertical wall equals L cos θ.

ILLUSTRATION OF EXAMPLE 19

Figure 55

From Equation (21) for a wall facing 15° West of South (ψ = 15°)

$$\gamma = 31° - 15° = 16°$$

$$\begin{aligned} \cos\theta &= \cos\beta\cos\gamma \\ \cos\theta &= \cos 24° \cos 16° \\ \cos\theta &= 0.9135 \times 0.9613 = 0.8781 \\ \theta &= 28.6° \end{aligned}$$

From Table 41 θ is approx 29° (See Figure 55).

The solar angles to any surface are under continuous change. Sufficient information for determining angles in respect to time, date and location has been provided for those who need this specific design. However, in many cases of residences, the major concern is to provide shading for windows during the summer months to reduce heat load on the air-conditioning equipment. For this reason, a table for guidance in projection shading for the period of April 11 to September 11 is provided. This "Shading Factor for Projections" is presented in Table 42.

To illustrate how this table functions consider the problem below.

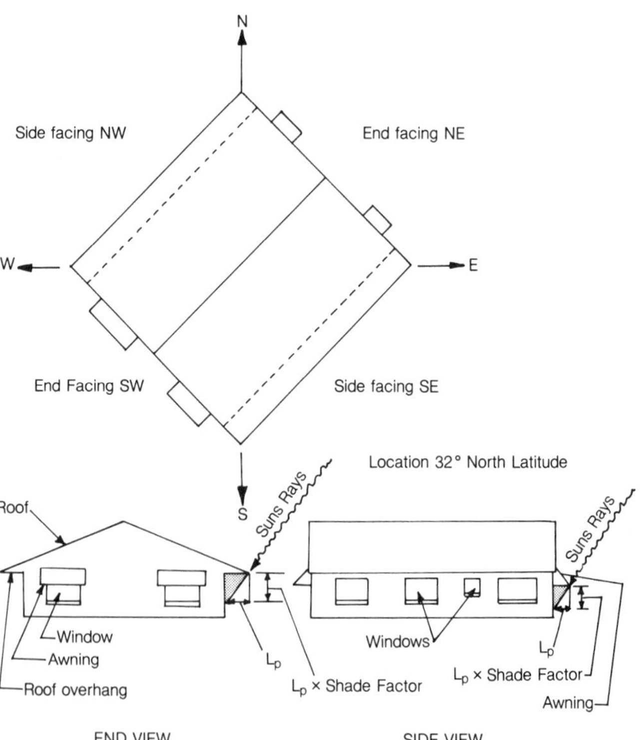

Example 20 L_p of Projection of roof is 3 feet
L_p of Projection of awning is 2 feet
Example 20A L_p of Projection of roof is 0.9 metres
L_p of Projection of awning is 0.6 metres

SHADING ON VERTICAL WALLS BY PROJECTIONS

Figure 56

DETERMINATION OF LENGTH OF SHADING ON VERTICAL WALLS BY PROJECTIONS

Example 20—*English Units*

A house is located so that its sides are facing Northwest and Southeast. These sides have a 3′0″ overhang roof. The ends facing Northeast and Southwest have awnings which project 2′0″ from the wall. What is the length of the shade patterns downward from these projections for various times of the day, during the time period of April 11 to September 11? These are minimum shadings during this period. The house is located 32° North latitude. See Figure 56.

From Table 42 determine the shading factor to multiply the projections so as to determine the length of shade path downward from projections of awnings. The solution to Example 20 is shown in the following:

Example 20a—*Metric Units*

A house is located so that its sides are facing Northwest and Southeast. These sides have a 0.9 meter overhang roof. The ends facing Northeast and Southwest have awnings which project 0.6 meter from the wall. What is the length of the shade patterns downward from these projections for various times of the day during the time period of April 11 to September 11? These are minimum shadings during this period. The house is located 32° North latitude. See Figure 56.

As done in Example 17, determine the shading factor from Table 42.

By the method illustrated in Examples 20 and 20a, the area of window or door openings shaded by projections can be determined for the summer season. Reduction of solar radiation to windows and doors affects the capacity needed by the cooling systems in residences and buildings and reduces energy use requirements.

SOLAR RADIATION TO FLAT HORIZONTAL ROOF SURFACES

On flat, horizontal surfaces the effect of solar heat gain to the interior is generally greater than on vertical surfaces. As shown in Figure 53, the solar angle to flat roofs in horizontal positions is the same as solar altitude, which can be determined directly from Table 40.

The direct design average* heat gain for horizontal surfaces at stated solar times at various North latitudes is given in Table 36. This table is based on the surface having a radiation absorbance = 1.0. For a surface having a value less than 1.0, the heat which is transmitted to the surface is this value of absorbance multiplied by the heat gain as listed. This is illustrated in the following example.

Example 21—*English and Metric Units*

Find the heat radiation to the surface of a flat roof during the summer months for various times of the day for a building located in Dallas, Texas. The roof is painted white and has a surface radiation absorbance of 0.68.

For Dallas use Table 36 - Summer Solar Heat Gain at 32°N latitude. The values given for horizontal surfaces in this table shall be multiplied by 0.68.

The solution of Example 21 is as follows:

HORIZONTAL SURFACE					
		British Units		Metric Units	
Solar Time	Surface Emittance	Gain Btu/ft^{2}hr At = 1.0	Gain W/m^2 At = 0.68	Gain Btu/ft^{2}hr At = 1.0	Gain W/m^2 At = 0.68
6	0.68	18	12.2	56.74	38.58
7	0.68	78	53.0	245.89	167.20
8	0.68	147	100.0	463.41	315.12
9	0.68	198	134.6	624.19	424.45
10	0.68	238	161.8	750.29	510.20
11	0.68	267	181.6	841.71	572.36
12	0.68	275	187.0	864.93	588.15
1	0.68	267	181.6	841.71	572.36
2	0.68	238	161.8	750.29	510.20
3	0.68	198	134.0	624.19	424.45
4	0.68	147	100.0	463.41	315.12
5	0.68	78	53.0	245.89	167.20
6	0.68	18	12.2	56.74	30.58

These heat gains are for clear days. This method must be used to find design capacity for residence or building cooling.

The solar heat load is affected by weather conditions. For this reason, it is necessary to determine daily average solar energy received for various cities, as a means of estimating energy gain to roofs as influenced by weather. This is found in Table 43, "Daily Average of Solar Energy Received on Horizontal Surfaces".

The information as to direct normal solar heat as given in Table 36 and average solar heat received on vertical and horizontal surfaces provide information as to solar heat transmitted to this surface. When solar radiation raises the temperature of a surface above ambient air temperature some of the heat will be transferred to the air. At thermal equilibrium the rate at which heat is transferred to a surface must equal the rate at which heat leaves that surface.

*Note: This is design average for clear days of the season noted.

SOLUTION TO EXAMPLE 20

Solar Time	Shading Length Due to Roof Projection = 3 ft.						Shading Length Due to Awning Projection = 2 ft.					
	Wall Facing SE			Wall Facing NW			End Facing SW			End Facing NE		
	Shading Factor	Proj. Length Lp	Ver. Shade Length	Shading Factor	Proj. Length Lp	Ver. Shade Length	Shading Factor	Proj. Length Lp	Shade Length	Shading Factor	Proj. Length	Shade Length
8	0.70	3	2.10′	In Shade	3	Total Shade	In Shade	2	Total Shade	0.92	2	1.84
9	1.00	3	3.00′	In Shade	3	Total Shade	In Shade	2	Total Shade	1.88	2	3.76
10	1.36	3	4.08′	In Shade	3	Total Shade	In Shade	2	Total Shade	4.35	2	
11	1.96	3	5.88′	In Shade	3	Total Shade	In Shade	2	Total Shade	In Shade	2	Total Shade
12	3.33	3	9.99′	In Shade	3	Total Shade	4.54	2	9.08′	In Shade	2	Total Shade
1	In Shade	3	Total Shade	In Shade	3	Total Shade	2.63	2	5.26′	In Shade	2	Total Shade
2	In Shade	3	Total Shade	7.14	3	21.42′	1.33	2	2.66′	In Shade	2	Total Shade
3	In Shade	3	Total Shade	2.16	3	7.48′	0.93	2	1.86′	In Shade	2	Total Shade
4	In Shade	3	Total Shade	1.00	3	3.00′	0.63	2	1.26′	In Shade	2	Total Shade
5	In Shade	3	Total Shade	0.62	3	1.86′	0.20	2	0.40′	In Shade	2	Total Shade

Answer Example 20
Total shade indicates shade by wall position.

SOLUTION TO EXAMPLE 20a

Solar Time	Shading Length Due to Roof Projection is 0.9 Metres						Shading Length Due to Awning Projection is 0.6 Metres					
	Wall Facing SE			Wall Facing NW			End Facing SW			End Facing NE		
	Shading Factor	Proj. Length Lp	Ver. Shade Length	Shading Factor	Proj. Length Lp	Ver. Shade Length	Shading Factor	Proj. Length Lp	Shade Length	Shading Factor	Proj. Length	Shade Length
8	0.70	0.9	0.63m	In Shade	0.9	Total Shade	In Shade	0.6	Total Shade	0.92	0.6	0.55m
9	1.00	0.9	0.90m	In Shade	0.9	Total Shade	In Shade	0.6	Total Shade	1.88	0.6	1.13m
10	1.36	0.9	1.22m	In Shade	0.9	Total Shade	In Shade	0.6	Total Shade	4.35	0.6	2.61m
11	1.96	0.9	1.76m	In Shade	0.9	Total Shade	In Shade	0.6	Total Shade	In Shade	0.6	Total Shade
12	3.33	0.9	3.00m	In Shade	0.9	Total Shade	4.54	0.6	2.72m	In Shade	0.6	Total Shade
1	In Shade	0.9	Total Shade	In Shade	0.9	Total Shade	2.63	0.6	1.58m	In Shade	0.6	Total Shade
2	In Shade	0.9	Total Shade	7.14	0.9	6.42m	1.33	0.6	0.80m	In Shade	0.6	Total Shade
3	In Shade	0.9	Total Shade	2.16	0.9	1.94m	0.93	0.6	0.56m	In Shade	0.6	Total Shade
4	In Shade	0.9	Total Shade	1.00	0.9	0.90m	0.63	0.6	0.38m	In Shade	0.6	Total Shade
5	In Shade	0.9	Total Shade	0.68	0.9	0.61m	0.20	0.6	0.12m	In Shade	0.6	Total Shade

Answer Example 20a
Total shade indicates shade by wall position.

In English Units when:

Q_s = Solar heat to the surface in Btu/ft², hr
t_s = temperature of surface, facing outward in °F
t_a = temperature of ambient air in °F
t_i = temperature of inside surface of wall or roof
V = Velocity of air in ft/min
R_t = total thermal resistance of wall or roof
ε = emittance of surface facing the sun

then the rate of heat transfer per unit area is

$$\varepsilon\, Q_s - 0.296\,(t_s - t_a)^{5/4} \sqrt{\frac{V + 68.9}{68.9}} = \frac{t_s - t_i}{R_t} . \quad (22)$$

The unknown factor is the surface temperature t_s of wall of roof facing outward.

In Metric Units when:

Q_{sm} = solar heat to the surface in W/m²
T_{sm} = temperature of surface facing outward in °K
T_{am} = temperature of air in °K
t_{sm} = temperature of surface facing outward in °C
t_{am} = temperature of ambient air in °C
t_{im} = temperature of inner surface of wall or roof
R_{mf} = total metric resistance of wall or roof
ε = radiation emittance of surface facing sun
V_m = velocity of air, m/s

then the rate of heat transfer per unit area is

$$\varepsilon\, Q_{sm} - 1.957\,(t_{im} - t_{sm})^{5/4} \sqrt{\frac{196.85V_m + 68.9}{68.9}} = \frac{t_{sm} - t_{im}}{R_{mf}} \quad (22a)$$

Solutions to these equations are difficult in that the unknown factor to obtain the balance is t_s (or t_{sm}) and appears on both sides of the equation. Values for this temperature must be assumed and calculations are performed by iteration.

The heat transferred to the interior of the building is based on the temperature difference between inside and outside surfaces of the wall divided by the thermal resistance. As Equations 22 and 22a are quite difficult to solve, Table 44 is provided as a means to estimate what the outside surface temperature would be when subjected to solar radiation. This table is an average over summer months and geographical locations above 32° North latitude.

To illustrate the manner in which Table 44 is to be used to estimate the outside surface temperatures t_s of buildings, the following examples are given.

Example 21—Determination of Wall Surface Temperatures
English Units

Find the surface temperatures of a building with walls facing North, East, South and West. These are light-colored walls with surface emittance ε = 0.75. When the temperature of a summer day is as listed in this example, what would the surface temperature of these walls be at a particular solar time? The solution from Table 44 is shown below:

Solar Time	t_a Ambient Air Temp °F	t_s = Surface Temperature of Walls (°F) North t_s-t_a °F	North t_s Wall °F	East t_s-t_a °F	East Wall °F	South t_s-t_a °F	South t_s Wall °F	West t_s-t_a °F	West t_s Wall °F
6 AM	71.0	0.0	71.0	27.0	98.0	0.0	71.0	0.0	71.0
7 AM	73.0	0.0	73.0	42.0	115.0	0.0	73.0	0.0	73.0
8 AM	75.0	0.0	75.0	45.0	120.0	8.3	83.3	0.0	75.0
9 AM	78.0	0.0	78.0	43.5	121.5	18.0	96.0	0.0	78.0
10 AM	81.0	0.0	81.0	28.5	109.5	26.3	107.3	0.0	81.0
11 AM	84.0	0.0	84.0	16.5	100.5	33.0	117.0	0.0	84.0
12 Noon	86.0	0.0	86.0	0.0	86.0	36.8	122.8	0.0	86.0
1 PM	87.0	0.0	87.0	0.0	87.0	37.5	124.5	14.3	101.3
2 PM	88.0	0.0	88.0	0.0	88.0	34.5	122.5	27.3	115.3
3 PM	90.0	0.0	90.0	0.0	90.0	29.3	119.3	38.3	128.3
4 PM	88.0	3.8	91.8	0.0	88.0	21.0	109.0	42.0	130.0
5 PM	86.0	6.0	92.0	0.0	86.0	16.5	102.5	39.0	125.0
6 PM	83.0	9.0	92.0	0.0	83.0	13.5	96.5	23.3	116.3
Answers Example 21			↑		↑		↑		↑

*From Table 44

Example 21a—Determination of Roof Surface Temperatures
Metric Units

Find the surface temperatures of a building with walls facing North, East, South and West. These are light-colored walls with surface emittance ε = 0.75. When the temperatures of a summer day are listed below, what would the surface temperature of the wall be at a particular solar time? Solution shown on the top of the following page.

Solar Time	t_a Ambient Air Temp °C	t_s = Surface Temperature of Walls: North $t_s - t_a$ °C	North t_s Wall °C	East $t_s - t_a$ °C	East t_s Wall °C	South $t_s - t_a$ °C	South t_s Wall °C	West $t_s - t_a$ °C	West t_s Wall °C
6 AM	21.7	0.0	21.7	15.0	36.7	0.0	0.0	0.0	0.0
7 AM	22.8	0.0	22.8	23.3	46.1	0.0	0.0	0.0	0.0
8 AM	23.9	0.0	23.9	25.4	49.3	4.6	28.5	0.0	0.0
9 AM	25.6	0.0	25.6	24.1	49.7	10.0	35.6	0.0	0.0
10 AM	27.2	0.0	27.2	15.8	53.0	14.6	41.8	0.0	0.0
11 AM	28.9	0.0	28.9	9.2	38.1	18.3	47.2	0.0	0.0
12 Noon	30.0	0.0	30.0	0.0	30.0	20.4	50.4	0.0	0.0
1 PM	30.6	0.0	30.6	0.0	30.6	20.8	51.4	7.9	38.5
2 PM	32.2	0.0	32.2	0.0	32.2	19.2	51.4	15.2	47.9
3 PM	32.2	0.0	32.2	0.0	32.2	16.3	48.5	21.3	53.5
4 PM	31.1	2.1	33.2	0.0	31.1	11.7	42.8	23.3	54.4
5 PM	30.0	3.3	33.3	0.0	30.0	9.2	39.2	21.7	51.7
6 PM	28.3	5.5	33.5	0.0	28.3	7.5	35.8	12.9	41.2
Answers Example 21a			↑		↑		↑		↑

Example 22—Determination of Wall Surface Temperature
English Units

A building has a flat roof, coated with a black asphalt material having a surface emittance $\varepsilon = 0.98$. When the temperatures of a summer day are as shown t_s in the following table, what would be the surface temperature of the roof facing the sun during a typical clear summer day?

Solar Time	Ambient Air Temp Summer Day Temp °F	Degree F Surface (Table 44) Will be Above Ambient Temp Horizontal for = 1.0*	Surface Temp °F of Roof
6 AM	71	13	84
7 AM	73	28	101
8 AM	75	30	105
9 AM	78	50	128
10 AM	81	64	145
11 AM	84	76	160
12 Noon	86	88	174
1 PM	87	92	179
2 PM	88	86	174
3 PM	90	75	165
4 PM	88	60	148
5 PM	86	34	120
6 PM	83	22	105

*From Table 44 use values when = 1.0 Answer Example 21 ↑

Example 22a—Determination of Roof Surface Temperatures
Metric Units

A building has a flat roof, coated with black asphalt with a surface emittance of $\varepsilon = 0.98$. When temperatures are listed, as below, what would be the surface temperature of the asphalt roof in the sun during a typical clear, summer day?

Solar Time	Ambient Air Temp Summer Day Temp °C	Degree C Surface (Table 44) Will be Above Ambient Temp Horizontal Surface = 1.0*	Surface Temp °C of Roof
6 AM	21.7	7.2	28.9
7 AM	22.8	15.5	36.3
8 AM	23.9	21.1	45.0
9 AM	25.6	27.8	53.4
10 AM	27.2	35.6	62.8
11 AM	28.9	42.2	71.1
12 Noon	30.0	48.9	78.9
1 PM	30.6	51.1	81.7
2 PM	31.1	47.8	78.9
3 PM	32.2	41.7	73.9
4 PM	31.1	33.3	64.4
5 PM	30.0	18.9	48.9
6 PM	28.3	12.2	40.5

*From Table 44 use values when = 1.0 Answer Example 22a ↑

DAILY AVERAGE OF SOLAR ENERGY RECEIVED ON HORIZONTAL SURFACE UNITED STATES CITIES

CITY	STATE	DAILY AVERAGE OF SOLAR ENERGY TO HORIZONTAL SURFACE												ANNUAL AVERAGE		
		SPRING			SUMMER			AUTUMN			WINTER					
		Btu/ft^2	*$gcal/cm^2$	W/m^2	Btu/ft^2	*$gcal/cm^2$	W/m^2	Btu/ft^2	*$gcal/cm^2$	W/m^2	Btu/ft^2	*$gcal/cm^2$	W/m^2	Btu/ft^2	*$gcal/cm^2$	W/m^2
Anniston	Ala	1733	470	5463	1917	520	6044	1290	350	4068	756	205	2383	1438	390	4533
Birmingham	Ala	1733	470	5463	1917	520	6044	1290	350	4068	756	205	1383	1438	390	4533
Dothan	Ala	1862	505	5870	1936	525	6102	1309	355	4126	903	245	2848	1549	420	4881
Huntsville	Ala	1622	440	5114	1917	520	6044	1364	370	4300	701	190	2208	1401	380	4417
Mobile	Ala	1805	490	5695	1954	530	6160	1549	420	4581	885	240	2790	1549	420	4881
Montgomery	Ala	1843	500	5812	1954	530	6160	1364	370	4300	848	230	2673	1493	405	4707
Tuscaloosa	Ala	1733	470	5463	1954	530	6160	1364	370	4300	774	210	2441	1456	395	4591
Anchorage	Alk	848	220	2557	1770	480	5579	664	180	2092	147	40	465	848	230	2673
Fairbanks	Alk	737	200	2324	1807	490	5695	627	170	1916	18	5	58	193	215	2499
Juneau	Alk	959	260	3080	1843	500	5812	848	230	2673	166	45	523	302	260	3022
Flagstaff	Arz	2360	640	7497	2470	670	7788	1512	410	4765	1143	310	3603	1862	505	5870
Kingman	Arz	2323	630	7380	2470	670	7788	1512	410	4765	1143	310	3603	1862	505	5870
Phoenix	Arz	2360	640	7497	2470	670	7788	1549	420	4881	1180	320	3719	1880	510	5928
Tucson	Arz	2433	660	7671	2470	670	7788	1549	420	4881	1180	320	3719	1899	515	5986
Fayetteville	Ark	1604	435	5056	2157	585	6800	1309	355	4126	682	185	2150	1438	390	4533
Ft. Smith	Ark	1622	440	5114	2175	590	6858	1327	360	4184	737	200	2324	1456	395	4591
Little Rock	Ark	1604	435	4056	2065	560	6509	1364	370	4300	682	185	2150	1347	365	4242
Pine Bluff	Ark	1622	440	5114	2102	570	6625	1364	370	4300	737	200	2324	1419	385	4475
Bakersfield	Calif	2175	590	6858	2507	680	7904	1475	400	4649	995	270	3138	1788	485	5637
Burbank	Calif	1991	540	6277	2397	650	7555	1475	400	4649	922	250	2906	1696	460	5347
Eureka	Calif	1530	415	4824	1696	460	5347	995	270	3138	516	140	1627	1180	320	3719
Fresno	Calif	2138	580	6741	2507	680	4904	1475	400	4649	737	200	2324	1714	465	5405
Longbeach	Calif	1991	540	6277	2397	650	7555	1475	400	4649	922	250	2906	1696	460	5347
Los Angeles	Calif	1991	540	6277	2397	650	7555	1475	400	4649	922	250	2906	1696	460	5347
Oakland	Calif	1954	530	6160	2397	650	7555	1217	330	3835	553	150	1743	1530	415	4824
Redding	Calif	1917	520	6044	2581	700	8136	1106	300	3487	627	170	1976	1549	420	4881
Riverside	Calif	1991	540	6277	2397	650	7555	1475	400	4649	995	270	3138	1714	465	5405
Sacremento	Calif	2028	550	6393	2507	680	7904	1290	350	4068	664	180	2092	1585	440	5114
Salinas	Calif	1954	530	6160	2397	650	7555	1290	350	4068	586	160	1860	1549	420	4881
San Bernadino	Calif	2102	570	6625	2507	680	7904	1475	400	4649	1106	300	3487	1807	490	5695
San Diego	Calif	1954	530	6160	2360	640	7439	1475	400	4649	922	250	2906	1678	455	5289
San Francisco	Calif	1954	530	6160	2397	650	7555	1217	330	3835	553	150	1743	1530	415	4824
San Jose	Calif	1954	530	6160	2433	660	7671	1254	340	3952	553	150	1743	1549	420	4881
Santa Ana	Calif	1991	540	6277	2397	650	7555	1475	400	4649	995	270	3133	1714	465	5405
Santa Barbara	Calif	1991	540	6277	2397	650	7555	1475	400	4649	848	230	2673	1678	455	5298
Stockton	Calif	1991	540	6277	2470	670	7788	1327	360	4184	664	180	2092	1604	435	5056
Colorado Springs	Colo	1990	540	6277	2397	650	7555	1364	370	4300	922	250	2906	1659	450	5230
Denver	Colo	1954	530	6160	2360	640	7439	1290	350	4068	848	230	2673	1604	435	5056
Grand Junction	Colo	1991	540	6277	2433	660	7671	1401	380	4417	885	240	2790	1678	455	5289
Pueblo	Colo	1991	540	6277	2397	650	7555	1364	370	4300	953	260	3022	1678	455	5289
Hartford	Conn	1475	400	4649	1880	510	5928	885	240	2790	553	150	1743	1198	325	3778
New Haven	Conn	1475	400	4649	1880	510	5928	922	250	2906	553	155	1802	1217	330	3835
Dover	Del	1549	420	4881	1954	530	6160	995	270	3138	627	170	1976	1235	335	3894
Wilmington	Del	1549	420	4881	1954	530	6160	995	270	3138	627	170	1976	1235	335	3894
Washington	DC	1585	430	4998	1954	530	6160	1032	280	3254	627	170	1976	1254	340	3952
Ft. Myers	Fla	1991	540	6297	2028	550	6393	1364	370	4300	1254	340	3952	1659	450	5230
Jacksonville	Fla	1880	510	5928	1917	520	6044	1735	335	3894	1014	275	3196	1530	415	4824
Miami	Fla	1991	540	6277	2028	550	6393	1433	390	4533	1254	340	3952	1696	460	5347
Orlando	Fla	2028	550	6393	2028	550	6393	1290	350	4068	1106	300	3787	1622	440	5114
Panama City	Fla	1917	520	6044	1954	530	6160	1364	370	4300	953	260	3022	1549	420	4881
Pensacola	Fla	1862	505	5870	1954	530	6160	1383	375	4359	940	255	2964	1530	415	4824
Tallahassee	Fla	1899	515	5986	1936	525	6102	2046	355	4126	977	265	3080	1530	415	4824
Tampa	Fla	1991	540	6277	2028	550	6393	1401	380	4417	1217	330	3885	1659	450	5230
W. Palm Beach	Fla	1991	540	6277	2028	550	6393	1439	390	4533	1254	340	3952	1673	455	5289
Albany	Ga	1917	520	6044	1991	540	6277	1327	360	4184	953	260	3022	1549	420	4881
Atlanta	Ga	1807	490	5695	1954	530	6160	2046	355	4126	756	205	2383	1456	395	4591
Augusta	Ga	1862	505	5870	1936	525	6102	1735	335	3894	866	235	2731	1475	400	4649
Columbus	Ga	1880	510	5928	1954	530	6160	1327	360	4184	885	240	2790	1512	410	4765
Macon	Ga	1862	505	5870	1954	530	6160	2190	350	4068	866	235	2731	1493	405	4707
Savannah	Ga	1917	520	6044	1973	535	6218	1272	345	4010	953	260	3022	1530	415	4824
Valdosta	Ga	1917	520	6044	1973	535	6218	1290	350	4068	995	270	3138	1549	420	4881
Honolulu	Haw	1991	540	6277	2249	610	7090	1696	460	5347	1585	430	4998	1880	510	5928
Boise	Ida	1770	480	5579	2416	655	7613	1143	310	3603	516	140	1627	1456	395	4591
Coeur D'Alene	Ida	1696	460	5347	2397	560	7555	848	230	2673	442	120	1395	1346	365	4242
Idaho Falls	Ida	1751	475	5521	2268	615	7148	1180	320	3719	553	150	1743	1438	390	4533
Lewiston	Ida	1696	460	5347	2212	600	6974	922	250	2906	442	120	1395	1364	370	4300
Pocatello	Ida	1733	470	5463	2286	620	7206	1475	400	4649	627	170	1976	1580	415	4824
Twin Falls	Ida	1806	490	5695	2360	640	7439	1180	320	3719	627	170	1976	1493	405	4707

*Langley Unit = gram calories per square centimeter.

(Sheet 1 of 4)

TABLE 43

DAILY AVERAGE OF SOLAR ENERGY RECEIVED ON HORIZONTAL SURFACE
UNITED STATES CITIES

CITY	STATE	DAILY AVERAGE OF SOLAR ENERGY TO HORIZONTAL SURFACE												ANNUAL AVERAGE		
		SPRING			SUMMER			AUTUMN			WINTER					
		Btu/ft^2	*gcal/cm²	W/m^2	Btu/ft^2	*gcal/cm²	W/m^2	Btu/ft^2	*gcal/cm²	W/m^2	Btu/ft^2	*gcal/cm²	W/m^2	Btu/ft^2	*gcal/cm²	W/m^2
Bellville	Ill	1549	420	4881	2065	560	6509	1143	310	3603	627	170	1976	1346	365	4242
Chicago	Ill	1475	400	4649	1991	540	6277	922	250	2906	553	150	1743	1235	335	3894
Joliet	Ill	1475	400	4649	1991	540	6277	922	250	2906	553	150	1743	1235	335	3894
Peoria	Ill	1512	410	4765	1991	540	6277	995	270	3138	596	160	1860	1272	345	4010
Quincy	Ill	1549	420	4881	2065	560	6509	1063	290	3371	627	170	1976	1327	360	4184
Springfield	Ill	1512	410	4765	2028	550	6393	1063	290	3371	627	170	1976	1309	355	4126
Columbus	Ind	1475	400	4649	1991	540	6277	1063	290	3371	553	150	1743	1272	345	4010
Evansville	Ind	1622	440	5114	1991	540	6277	1180	320	3719	586	160	1860	1327	360	4184
Ft. Wayne	Ind	1383	375	4359	1973	535	6218	1180	250	2906	516	140	1627	1235	325	3798
Indianapolis	Ind	1419	385	4475	1991	540	6277	1014	275	3196	516	140	1627	1235	335	3894
South Bend	Ind	1383	375	4353	1973	535	6218	922	250	2906	516	140	1627	1198	325	3778
Terre Haute	Ind	1493	405	4707	1901	540	6277	1063	285	3313	553	150	1743	1272	345	4010
Burlington	Iowa	1549	420	4881	2028	550	6393	1032	280	3254	627	170	1976	1309	355	4126
Cedar Rapids	Iowa	1493	405	4707	1973	535	6218	922	250	2906	627	170	1976	1254	340	3952
Des Moines	Iowa	1530	415	4824	2009	545	6334	995	270	3138	627	170	1976	1290	350	4068
Mason City	Iowa	1549	420	4881	1991	540	6277	922	250	2906	627	170	1976	1272	345	4100
Ottumwa	Iowa	1549	420	4881	2009	545	6334	977	265	3080	627	170	1976	1290	350	4068
Sioux City	Iowa	1567	425	4939	2120	575	6682	1032	280	3254	664	180	2092	1346	365	4242
Dodge City	Kan	1917	520	6044	2286	620	7206	1327	360	4184	953	260	3022	1622	440	5114
Hutchinson	Kan	1751	475	5521	2194	595	6916	1254	340	3952	848	230	2673	1512	410	4765
Liberal	Kan	1936	525	6102	2304	625	7265	1364	370	4300	953	260	3022	1641	445	5289
Topeka	Kan	1622	440	5114	2183	580	6741	1217	330	3885	701	190	2208	1438	390	4533
Wichita	Kan	1751	475	5521	2194	585	6800	1254	340	3958	811	220	2557	1493	405	4707
Bowling Green	Ky	1622	440	5114	1991	540	6277	1217	330	3835	627	170	1976	1364	370	4300
Corbin	Ky	1641	445	5172	1936	525	6102	1180	320	3719	627	170	1976	1346	365	4242
Lexington	Ky	1696	460	5374	2249	610	7090	1364	370	4300	627	170	1976	1512	410	4765
Louisville	Ky	1659	450	5230	1991	540	6277	1143	310	3603	586	160	1860	1346	365	4242
Paducah	Ky	1622	440	5114	1991	540	6277	1217	330	3835	627	170	1976	1365	370	4300
Alexandria	La	1714	465	5405	2023	550	6393	1456	395	4591	848	230	2673	1512	410	4765
Baton Rouge	La	1751	475	5521	2009	545	6334	1475	400	4649	885	240	2790	1530	415	4824
Lafayette	La	1714	465	5405	1973	535	6218	1475	400	4649	885	240	2790	1512	410	4765
Lake Charles	La	1696	460	5374	1991	540	6277	1475	400	4649	885	240	2790	1512	410	4765
Monroe	La	1659	450	5230	2065	560	6509	1401	380	4417	811	220	2557	1475	400	4649
New Orleans	La	1549	420	4881	1991	540	6277	1475	400	4649	885	240	2790	1475	400	4649
Shreveport	La	1512	410	4765	2065	560	6509	1549	420	4881	848	230	2673	1493	405	4707
Boston	Mass	1401	380	4417	1880	510	5928	922	250	2906	516	140	1627	1180	320	3719
Springfield	Mass	1290	350	4068	1880	510	5928	885	240	2790	516	140	1627	1143	310	3603
Worchester	Mass	1309	355	4126	1880	510	5928	866	235	2731	516	140	1627	1143	310	3603
Baltimore	Md	1549	420	4881	1954	530	6160	1032	280	3254	627	170	1976	1290	350	4068
Augusta	Me	1364	370	4300	1880	510	5928	922	250	2906	479	130	1511	1161	315	3661
Bangor	Me	1401	380	4417	1917	520	6044	922	250	2906	479	130	1511	1180	320	3719
Portland	Me	1438	390	4533	1954	530	6160	995	270	3138	553	150	1743	1235	335	3894
Detroit	Mich	1346	365	4242	1936	525	6102	953	260	3022	479	130	1511	1180	320	3719
Flint	Mich	1346	365	4242	1936	525	6102	953	260	3022	479	130	1511	1180	320	3719
Grand Rapids	Mich	1327	360	4184	1917	520	6044	885	240	2790	516	140	1627	1161	315	3661
Lansing	Mich	1309	355	4126	1899	515	5986	922	250	2906	442	120	1395	1143	310	3603
Saginaw	Mich	1198	325	3778	1936	525	6102	953	260	3022	479	130	1511	1180	320	3719
Traverse City	Mich	1383	375	4359	1936	525	6102	848	230	2673	479	130	1511	1161	315	3661
Duluth	Minn	1493	405	4707	2009	545	6334	848	230	2673	553	150	1743	1235	335	3894
Minneapolis	Minn	1512	410	4765	2023	550	6393	885	240	2790	586	160	1860	1254	340	3952
Rochester	Minn	1493	405	4707	1973	535	6218	885	240	2790	586	160	1860	1235	335	3894
St. Cloud	Minn	1549	420	4881	2023	550	6393	848	230	2673	586	160	1860	1234	340	3952
Willmar	Minn	1549	420	4881	2138	580	6741	922	250	2908	627	170	1976	1309	355	4126
Greenville	Miss	1604	435	5056	1973	535	6218	1364	370	4300	737	200	2324	1419	385	4475
Jackson	Miss	1696	460	5370	1991	540	6277	1401	380	4417	811	220	2557	1475	400	4649
Laurel	Miss	1733	470	5463	1991	540	6277	1401	380	4417	848	230	2673	1493	405	4707
Meridian	Miss	1714	465	5405	1973	535	6218	1401	380	4417	811	230	2557	1475	400	4649
Tupelo	Miss	1604	435	5056	1973	535	6218	1327	360	4184	701	190	2208	1401	380	4417
Columbia	Mo	1585	430	4998	2065	560	6509	1143	310	3603	664	180	2092	1364	370	4300
Jefferson City	Mo	1585	430	4998	2102	570	6625	1180	320	3719	664	180	2092	1383	375	4359
Joplin	Mo	1622	440	5114	2138	580	6741	1290	350	4068	701	190	2208	1438	390	4533
Kansas City	Mo	1622	440	5114	2102	570	6625	1063	290	3371	664	180	2092	1364	370	4300
St Louis	Mo	1549	420	4881	2065	560	6509	1143	310	3603	627	170	1976	1346	365	4242
Springfield	Mo	1585	430	4998	2102	570	6625	1254	340	3952	664	180	2092	1401	380	4417
Billings	Mont	1751	475	5521	2304	625	7264	1106	300	3487	586	160	1860	1438	390	4533
Butte	Mont	1696	460	5374	2286	620	7206	1032	280	3254	516	140	1627	1383	375	4359
Grand Falls	Mont	1622	440	5114	2249	610	7090	995	270	3138	516	140	1627	1346	365	4242
Helena	Mont	1641	445	5172	2268	615	7148	1032	280	3254	516	140	1627	1364	370	4300
Miles City	Mont	1770	480	5579	2286	620	7206	1063	290	3371	627	170	1976	1438	390	4533
Missoula	Mont	1659	450	5230	2286	620	7206	953	260	3022	479	130	1511	1346	365	4242

* Langley Unit = gram calories per square centimeter.

TABLE 43 (Continued)

(Sheet 2 of

DAILY AVERAGE OF SOLAR ENERGY RECEIVED ON HORIZONTAL SURFACE
UNITED STATES CITIES

CITY	STATE	DAILY AVERAGE OF SOLAR ENERGY TO HORIZONTAL SURFACE												ANNUAL AVERAGE		
		SPRING			SUMMER			AUTUMN			WINTER					
		Btu/ft²	*gcal/cm²	W/m²	Btu/ft²	*gcal/cm²	W/m²	Btu/ft²	*gcal/cm²	W/m²	Btu/ft²	*gcal/cm²	W/m²	Btu/ft²	*gcal/cm²	W/m²
Grand Island	Neb	1658	450	5230	2212	600	6974	1106	300	3487	701	190	2208	1419	385	4475
Lincoln	Neb	1585	430	4998	2102	570	6625	1032	280	3254	664	180	2092	1346	365	4242
North Platte	Neb	1806	490	5695	2175	590	6858	1143	310	3603	774	210	2441	1475	400	4649
Omaha	Neb	1622	440	5114	2065	560	6509	1032	280	3254	664	180	2092	1346	365	4242
Carson City	Nev	2138	580	6741	2765	750	8716	1364	370	4300	885	240	2790	1788	485	5637
Elko	Nev	1843	500	5812	2323	630	7323	1254	340	3952	701	190	2208	1530	415	4824
Ely	Nev	2065	560	6509	2360	640	7439	1401	380	4417	843	230	2673	1659	450	5230
Las Vegas	Nev	2360	640	7439	2344	690	8020	1512	410	4765	1063	290	3371	1862	505	5870
Reno	Nev	2138	580	6741	2728	740	8601	1364	370	4300	843	230	2673	1770	480	5579
Concord	NH	1475	400	4649	1880	510	5928	848	230	2673	516	140	1627	1180	320	3719
Keene	NH	1327	360	4184	1880	510	5928	848	230	2673	516	140	1627	1143	310	3603
Manchester	NH	1475	400	4649	1880	510	5928	848	230	2673	516	140	1627	1180	320	3719
Atlantic City	NJ	1549	420	4881	1917	520	6044	995	270	3138	627	170	1976	1272	345	4010
Newark	NJ	1475	400	4649	1917	520	6044	922	250	2906	553	150	1743	1217	330	3835
New Brunswick	NJ	1512	410	4765	1917	520	6044	995	270	3138	586	160	1860	1254	340	3952
Trenton	NJ	1475	400	4649	1954	530	6160	953	260	3022	627	170	1976	1254	340	3952
Albuquerque	NM	2286	620	7206	2433	660	7671	1512	410	4765	1143	310	3603	1843	500	5812
Clovis	NM	2138	580	6741	2433	660	7671	1438	390	4533	1063	290	3371	1770	480	5579
Farmington	NM	2212	600	6974	2433	660	7671	1512	410	4765	1063	290	3371	1807	490	5695
Gallup	NM	2286	620	7206	2433	660	7671	1512	410	4765	1143	310	3603	1843	500	5812
Las Cruces	NM	2397	650	7555	2433	660	7671	1512	410	4765	1180	320	3719	1180	510	5928
Roswell	NM	2286	620	7206	2433	660	7671	1475	400	4649	1106	300	3487	1825	495	5753
Santa Fe	NM	2212	600	6974	2433	660	7671	1512	410	4765	1063	290	3371	1807	490	5695
Tucumcari	NM	2138	580	6741	2433	660	7671	1438	390	4533	1063	290	3371	1770	480	5579
Albany	NY	1254	340	3952	1807	490	5695	885	240	2790	479	130	1511	1106	300	3487
Binghamton	NY	1290	350	4068	1843	500	5812	885	240	2790	479	130	1511	1125	305	3545
Buffalo	NY	1254	340	3952	1843	500	5812	885	230	2673	479	130	1511	1106	300	3487
Elmira	NY	1254	340	3952	1880	510	5928	885	240	2790	479	130	1511	1125	305	3545
New York	NY	1475	400	4649	1917	520	6044	922	250	2906	553	150	1743	1217	330	3835
Niagara Falls	NY	1254	340	3952	1843	500	5812	848	230	2673	479	130	1511	1106	300	3487
Rochester	NY	1254	340	3952	1843	500	5812	848	230	2673	479	130	1511	1106	300	3487
Syracuse	NY	1254	340	3952	1843	500	5812	848	230	2673	479	130	1511	1106	300	3487
Ashville	NC	1696	460	5347	1917	520	6044	1217	330	3835	701	190	2208	1383	375	4359
Charlotte	NC	1843	500	5812	1991	540	6277	1217	330	3835	774	210	2441	1456	395	4591
Elizabeth City	NC	1843	500	5812	1991	540	6277	1143	310	3603	774	210	2441	1438	390	4533
Fayetteville	NC	1862	505	5870	1973	535	6218	1180	320	3719	811	220	2557	1456	395	4591
Greensboro	NC	1788	485	5637	1973	535	6218	1180	320	3719	737	200	2324	1419	385	4475
Raleigh	NC	1807	490	5695	1991	540	6277	1180	320	3719	774	210	2441	1438	390	4533
Wilmington	NC	1880	510	5928	2023	550	6383	1217	330	3835	885	240	2790	1512	410	4765
Winston-Salem	NC	1788	485	5637	1973	535	6218	1180	320	3719	737	200	2324	1419	385	4475
Bismark	ND	1585	430	4998	2212	600	6974	1032	280	3254	627	170	1976	1364	370	4300
Dickinson	ND	1658	450	5230	2212	600	6974	1032	280	3254	627	170	1976	1383	375	4359
Fargo	ND	1585	430	4998	2175	590	6858	922	250	2906	627	170	1976	1327	360	4184
Grand Forks	ND	1512	410	4765	2138	580	6741	922	250	2906	586	160	1860	1290	350	4068
Jamestown	ND	1585	430	4998	2212	600	6974	953	260	3022	627	170	1976	1346	365	4242
Minot	ND	1622	440	5114	2249	610	7090	995	270	3138	586	160	1860	1364	370	4300
Akron	OH	1254	340	3952	1954	530	6160	953	260	3022	479	130	1511	1161	315	3661
Cincinnati	OH	1438	390	4533	1954	530	6160	1063	290	3371	553	150	1743	1254	340	3952
Cleveland	OH	1254	340	3952	1954	530	6160	953	260	3022	479	130	1511	1161	315	3661
Columbus	OH	1364	370	4300	1954	530	6160	1032	280	3254	516	140	1627	1217	330	3835
Dayton	OH	1438	390	4533	1954	530	6160	1032	280	3254	516	140	1627	1735	335	3894
Mansfield	OH	1327	360	4184	1954	530	6160	995	270	3138	516	140	1627	1198	325	3778
Toledo	OH	1290	350	4068	1954	530	6160	995	270	3138	479	130	1511	1180	320	3719
Zanesville	OH	1364	370	4300	1954	530	6160	1032	280	3254	516	140	1627	1217	320	3835
Ardmore	Okla	1733	470	5463	2249	610	7090	1364	370	4300	922	250	2906	1567	425	4939
Enid	Okla	1770	480	5579	2249	610	7090	1327	360	4184	922	250	2906	1567	425	4939
Lawton	Okla	1807	490	5695	2286	620	7206	1364	370	4300	953	260	3022	1604	435	5056
Muskogee	Okla	1678	455	5289	2157	585	6800	1327	360	4184	737	200	2324	1475	400	4649
Oklahoma City	Okla	1751	475	5521	2231	605	7032	1364	370	4300	922	250	2906	1567	425	4939
Tulsa	Okla	1696	460	5347	2212	600	6974	1327	360	4184	811	220	2557	1512	410	4765
Eugene	Ore	1622	440	5114	2286	620	7206	774	210	2441	332	90	1046	1254	340	3952
Klamath Falls	Ore	1770	480	5579	2544	690	8020	1063	290	3371	516	140	1627	1475	400	4649
Medford	Ore	1217	330	3835	1843	500	5812	774	210	2441	295	80	930	1032	280	3254
Pendelton	Ore	1711	465	5405	2415	655	7603	995	270	3138	406	110	1279	1383	375	4359
Portland	Ore	1622	440	5114	2286	620	7200	848	230	2673	332	90	1046	1254	340	3952
Salem	Ore	1604	435	5056	2378	645	7497	848	230	2673	332	90	1046	1290	350	4068
Allentown	PA	1419	385	4475	1899	515	5986	922	250	2906	553	150	1743	1198	325	3778
Altoona	PA	1401	380	4417	1917	520	6044	922	250	2906	553	150	1743	1198	325	3778
Erie	PA	1254	340	3952	1917	520	6044	922	250	2906	479	130	1511	1143	310	3603
Harrisburgh	PA	1438	390	4533	1917	520	6044	953	260	3022	553	150	1743	1217	330	3835

*Langley Unit = gram calories per square centimeter.

(Sheet 3 of 4)

TABLE 43 (Continued)

DAILY AVERAGE OF SOLAR ENERGY RECEIVED ON HORIZONTAL SURFACE
UNITED STATES CITIES

CITY	STATE	DAILY AVERAGE OF SOLAR ENERGY TO HORIZONTAL SURFACE												ANNUAL AVERAGE		
		SPRING			SUMMER			AUTUMN			WINTER					
		Btu/ft^2	*gcal/cm^2	W/m^2	Btu/ft^2	*gcal/cm^2	W/m^2	Btu/ft^2	*gcal/cm^2	W/m^2	Btu/ft^2	*gcal/cm^2	W/m^2	Btu/ft^2	*gcal/cm^2	W/m^2
Harrisburgh	PA	1438	390	4533	1917	520	6044	953	260	3022	553	150	1743	1217	330	3835
Johnston	PA	1364	370	4300	1917	520	6044	922	250	2906	516	140	1627	1180	320	3719
Philadelphia	PA	1512	410	4765	1917	520	6044	1032	280	3254	627	170	1976	1272	345	4010
Pittsburgh	PA	1401	380	4417	1954	530	6160	922	250	2906	516	140	1627	1198	325	3778
Reading	PA	1438	390	4533	1917	520	6044	953	260	3022	553	150	1743	1217	330	3835
Wilkes-Barre	PA	1327	360	4184	1917	520	6044	885	240	2790	516	140	1627	1161	315	3661
Pawtucket	RI	1512	410	4765	1880	510	5928	922	250	2906	553	150	1743	1217	330	3835
Providence	RI	1512	410	4765	1880	510	5928	922	250	2906	553	150	1743	1217	330	3835
Charleston	SC	1917	520	6044	1973	535	6218	1198	325	3778	953	260	3022	1512	410	4765
Columbia	SC	1862	505	5870	1991	540	6277	1198	325	3778	848	230	2673	1475	400	4649
Florence	SC	1862	505	5870	1991	540	6277	1198	325	3778	848	230	2673	1475	400	4649
Greenville	SC	1843	500	5812	1991	540	6277	1217	330	3835	774	210	2441	1456	395	4591
Spartanburg	SC	1807	490	5695	1991	540	6277	1217	330	3835	737	200	2324	1438	390	4533
Aberdeen	SD	1807	490	5695	2138	580	6741	1106	300	3487	701	190	2208	1438	390	4533
Pierre	SD	1659	450	5230	2212	600	6974	1063	290	3371	664	180	2092	1401	380	4417
Rapid City	SD	1807	490	5695	2212	600	6974	1106	300	3487	701	190	2208	1456	395	4591
Sioux Falls	SD	1585	430	4998	2138	580	6741	995	270	3138	664	180	2092	1346	365	4242
Watertown	SD	1585	430	4998	2175	590	6858	995	270	3138	627	170	1976	1346	365	4242
Chattanooga	Tenn	1622	440	5114	2065	560	6509	1327	360	4184	737	200	2324	1438	390	4533
Jackson	Tenn	1622	440	5114	1991	540	6277	1290	350	4068	701	190	2208	1401	380	4417
Knoxville	Tenn	1622	440	5114	1954	530	6160	1217	330	3835	664	180	2092	1364	370	4300
Memphis	Tenn	1622	440	5114	2028	550	6383	1327	360	4184	701	190	2208	1419	385	4495
Nashville	Tenn	1622	440	5114	1991	540	6277	1254	340	3952	664	180	2092	1383	375	4359
Abilene	Tex	1825	495	5753	2341	635	7380	1401	380	4417	995	270	3138	1641	445	5172
Amarillo	Tex	2083	565	6567	2378	645	7497	1401	380	4417	995	270	3138	1714	465	5405
Austin	Tex	1659	450	5230	2286	620	7206	1401	380	4417	995	270	3138	1585	430	4998
Bryan	Tex	1622	440	5114	2249	610	7090	1438	390	4533	953	260	3022	1567	525	2615
Dallas	Tex	1622	440	5114	2249	610	7090	1401	380	4417	922	250	2906	1549	420	4881
El Paso	Tex	2433	660	7671	2470	670	7788	1567	425	4939	1198	325	3778	1917	520	6044
Ft. Worth	Tex	1770	480	5579	2360	640	7439	1438	390	4533	995	270	3138	1641	445	5172
Galveston	Tex	1696	460	5347	2138	580	6741	1475	400	4649	953	260	3022	1567	425	2615
Houston	Tex	1696	460	5347	2138	580	6741	1475	400	4649	953	260	3022	1567	425	2615
Laredo	Tex	1622	440	5114	2286	620	7206	1475	400	4649	1032	280	3254	1604	435	5056
Lufkin	Tex	1622	440	5114	2175	590	6858	1438	390	4533	885	240	2790	1530	415	4824
San Angelo	Tex	1807	490	5695	2323	630	7323	1456	395	4591	1051	285	3313	1659	450	5236
San Antonio	Tex	1622	440	5114	2286	620	7206	1475	400	4649	1032	280	3254	1604	435	5056
Tyler	Tex	1622	440	5114	2212	600	6974	1401	380	4417	885	240	2790	1530	415	4824
Victoria	Tex	1641	445	5172	2231	605	7032	1475	400	4649	995	270	3138	1585	430	4998
Waco	Tex	1512	410	4765	2249	610	7090	1401	380	4417	953	260	3022	1530	415	4834
Wichita Falls	Tex	1807	490	5695	2286	620	7206	1364	370	4300	953	260	3022	1604	435	5056
Ogden	Utah	1770	480	5579	2323	630	7323	1217	330	3835	664	180	2092	1493	405	4707
Price	Utah	1917	520	6044	2397	650	7555	1364	370	4300	811	220	2557	1622	440	5114
Salt Lake City	Utah	1770	480	5579	2323	630	7323	1217	330	3835	664	180	2092	1493	405	4707
Burlington	Vt	1438	390	4533	1696	460	5347	885	240	2790	479	130	1511	1125	305	3545
Lynchburg	Va	1733	470	5463	1991	540	6277	1143	310	3603	664	180	2092	1383	375	4359
Norfolk	Va	1770	480	5579	1991	540	6277	1106	300	3487	234	200	2324	1401	380	4417
Richmond	Va	1696	460	5347	1991	540	6277	1063	290	3371	701	190	2208	1364	370	4300
Roanoke	Va	1733	470	5463	1991	540	6277	1143	310	3603	664	180	2092	1383	375	4359
Everette	Wash	1549	420	4881	1843	500	5812	701	190	2208	382	90	1046	1106	300	3487
Olympia	Wash	1585	430	4998	1880	510	5928	701	190	2208	382	90	1046	1125	305	3545
Seattle	Wash	1549	420	4881	1770	480	5579	701	190	2208	382	90	1046	1088	295	3429
Spokane	Wash	1696	460	5347	2397	650	7555	811	220	2557	406	110	1279	1327	360	4184
Tacoma	Wash	1549	420	4881	1843	500	5812	701	190	2208	382	90	1046	1106	300	3487
Yakima	Wash	1659	450	5230	2397	650	7555	885	240	2790	368	100	1162	1327	360	4184
Charleston	WV	1659	450	5230	1954	530	6160	1106	300	3487	586	160	1860	1327	360	4184
Huntington	WV	1659	450	5230	1954	530	6160	1106	300	3487	586	160	1860	1327	360	4184
Martinsburg	WV	1512	410	4765	1954	530	6160	995	270	3138	553	150	1743	1254	340	3952
Morgantown	WV	1475	400	4649	1317	520	6044	995	270	3138	553	150	1743	1235	335	3894
Parkersburg	WV	1512	410	4765	1317	520	6044	1032	280	3254	553	150	1743	1254	340	3952
Eau Clair	Wis	1475	400	4649	1991	540	6277	848	230	2673	553	150	1743	1217	330	3835
Green Bay	Wis	1401	380	4417	1991	540	6277	848	230	2673	479	130	1511	1180	320	3719
LaCrosse	Wis	1512	410	4765	1991	540	6277	885	240	2790	553	150	1743	1235	335	3894
Madison	Wis	1438	390	4533	1991	540	6277	885	240	2790	553	150	1743	1217	330	3835
Milwaukee	Wis	1475	400	4649	1991	550	6383	922	250	2906	516	140	1685	1235	335	3894
Wausau	Wis	1438	390	4533	1991	540	6277	848	230	2673	516	140	1685	1198	325	3778
Casper	Wyo	1917	520	6044	2323	630	7323	1254	340	3952	774	210	2441	1567	425	4939
Cheyenne	Wyo	1917	520	6044	2323	630	6323	1254	340	3952	843	230	2673	1585	430	4998
Rock Springs	Wyo	1880	510	5928	2323	630	7323	1327	360	4184	811	220	2557	1585	430	4998
Sheridan	Wyo	1880	510	5928	2323	630	7232	1180	320	3719	664	180	2092	1512	410	4765

*Langley Unit = gram calories per square centimeter.

(Sheet 4 of 4)

TABLE 43 (Continued)

SURFACE TEMPERATURE ABOVE AMBIENT AIR TEMPERATURE ($t_s - t_a$)
DUE TO EXPOSURE TO SOLAR RADIATION, AVERAGED OVER
SUMMER MONTHS AND ABOVE 32°N. LATITUDE.

SOLAR TIME	SURFACE EMITTANCE = 1.0									
	British Units—Degree F					Metric Units—Degree C				
	Hor.	Walls Facing				Hor.	Walls Facing			
	Sur.	N	E	S	W	Sur.	N	E	S	W
6 AM	13.0	0.0	36.0	0.0	0.0	7.2	0.0	20.0	0.0	0.0
7 AM	28.0	0.0	56.0	0.0	0.0	15.5	0.0	31.0	0.0	0.0
8 AM	38.0	0.0	61.0	11.0	0.0	21.1	0.0	33.9	6.1	0.0
9 AM	50.0	0.0	58.0	24.0	0.0	27.8	0.0	32.2	13.3	0.0
10 AM	64.0	0.0	38.0	35.0	0.0	35.6	0.0	21.1	19.4	0.0
11 AM	76.0	0.0	22.0	44.0	0.0	42.2	0.0	12.2	24.4	0.0
12 NOON	88.0	0.0	0.0	49.0	0.0	48.9	0.0	0.0	27.2	0.0
1 PM	92.0	0.0	0.0	50.0	19.0	51.1	0.0	0.0	27.8	10.5
2 PM	86.0	0.0	0.0	46.0	37.0	47.8	0.0	0.0	25.6	20.6
3 PM	75.0	0.0	0.0	39.0	51.0	41.7	0.0	0.0	21.7	28.3
4 PM	60.0	5.0	0.0	28.0	56.0	33.3	2.8	0.0	15.6	31.1
5 PM	34.0	8.0	0.0	22.0	52.0	18.9	4.4	0.0	12.2	28.9
6 PM	22.0	12.0	0.0	18.0	31.0	12.2	6.7	0.0	10.0	17.2
SURFACE EMITTANCE = 0.75										
6 AM	9.8	0.0	27.0	0.0	0.0	5.4	0.0	15.0	0.0	0.0
7 AM	21.0	0.0	42.0	0.0	0.0	11.7	0.0	23.3	0.0	0.0
8 AM	28.5	0.0	45.8	8.3	0.0	15.8	0.0	25.4	4.6	0.0
9 AM	37.5	0.0	43.5	18.0	0.0	20.8	0.0	24.1	10.0	0.0
10 AM	48.0	0.0	28.5	26.3	0.0	26.7	0.0	15.8	14.6	0.0
11 AM	57.0	0.0	16.5	33.0	0.0	31.7	0.0	9.2	18.3	0.0
12 NOON	66.0	0.0	0.0	36.8	0.0	36.7	0.0	0.0	20.4	0.0
1 PM	69.0	0.0	0.0	37.5	14.3	38.3	0.0	0.0	20.8	7.9
2 PM	64.5	0.0	0.0	34.5	27.3	35.8	0.0	0.0	19.2	15.2
3 PM	56.2	0.0	0.0	29.3	38.3	31.2	0.0	0.0	16.3	21.3
4 PM	45.0	3.8	0.0	21.0	42.0	25.0	2.1	0.0	11.7	23.3
5 PM	25.5	6.0	0.0	16.5	39.0	14.2	3.3	0.0	9.2	21.7
6 PM	16.5	9.0	0.0	13.5	23.3	9.2	5.0	0.0	7.5	12.9
SURFACE EMITTANCE = 0.50										
6 AM	6.5	0.0	18.0	0.0	0.0	3.6	0.0	10.0	0.0	0.0
7 AM	14.0	0.0	28.0	0.0	0.0	7.8	0.0	15.6	0.0	0.0
8 AM	19.0	0.0	30.5	5.5	0.0	10.6	0.0	16.9	3.1	0.0
9 AM	25.0	0.0	29.0	12.0	0.0	13.9	0.0	16.1	6.7	0.0
10 AM	32.0	0.0	19.0	17.5	0.0	17.8	0.0	10.6	9.7	0.0
11 AM	38.0	0.0	11.0	22.0	0.0	21.1	0.0	6.11	12.2	0.0
12 NOON	44.0	0.0	0.0	24.5	0.0	24.4	0.0	0.0	13.6	0.0
1 PM	46.0	0.0	0.0	25.0	9.5	25.6	0.0	0.0	13.9	5.3
2 PM	43.0	0.0	0.0	23.0	18.5	23.9	0.0	0.0	12.8	10.3
3 PM	37.5	0.0	0.0	19.5	25.5	20.8	0.0	0.0	10.8	14.2
4 PM	30.0	2.5	0.0	14.0	28.0	16.7	1.4	0.0	7.8	15.6
5 PM	17.0	4.0	0.0	11.0	26.0	9.4	2.2	0.0	6.1	14.4
6 PM	11.0	6.0	0.0	9.0	15.5	6.1	3.3	0.0	5.0	8.6

The above are estimated design temperatures based on averages of many variables.

TABLE 44

As shown in Examples 22 and 22a, roof temperature can be quite high due to solar heat transmitted to the surface. The temperature difference between surface temperature on the top to surface temperature of the interior of a building divided by the thermal resistance of the roof, insulation and ceiling (if these are single-integrated mass) determines the heat transmitted to the interior. Due to the high temperature of the surface facing upward, the heat transmitted can be multiplied 4 to 5 times that if only ambient air temperature is the factor controlling surface temperature of the insulation. This causes thermal insulation directly under the roofing to be ineffective. Wherever possible, a ventilated space should be provided between the flat roof and the insulation above the interior ceiling.

Figure 57 illustrates that when a ventilated space is provided between a roof and the insulation, the insulation surface temperature, t_a, is very close to the same temperature as the ambient air. In this manner, the roof acts as a shading mechanism to reduce sun solar heat load to the insulation, the same as projections and awnings reduce the solar sun heat load to walls, windows and doors. For more efficient operation in winter, ventilators provided with closures can be used to change this space from a ventilated space to a dead air space.

Control and use of solar heat is very important for efficient conservation of energy.

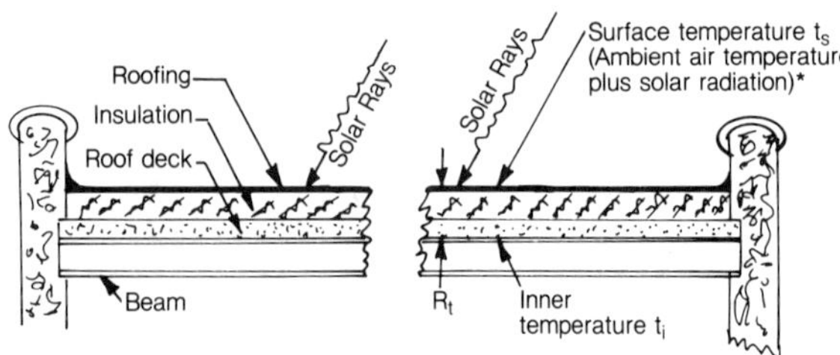

R_t = total thermal resistance of roofing, insulation and roof deck

Q_s = heat transfer = $\frac{t_s - t_i}{R_s}$

INSULATION DIRECTLY UNDER ROOFING

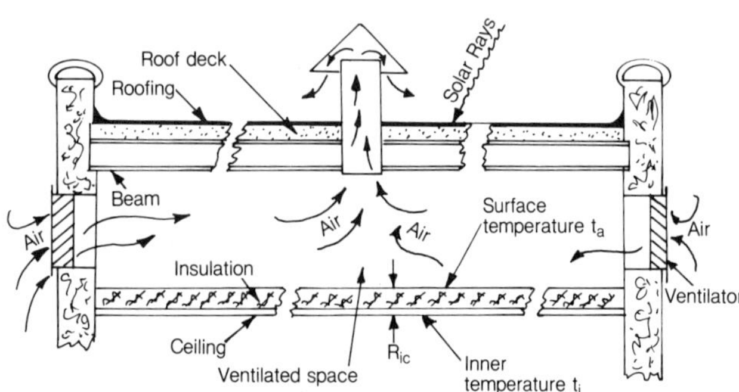

R_{ic} = total thermal resistance of insulation and ceiling

Q_s = heat transfer $\frac{t_a - t_i}{R_{ic}}$

t_a in a ventilated space will be 30°F (16.6°C) to 70°F (38°C) less than t_s surface temperature of roofing surface in summer conditions with solar heat gain.

VENTILATED SPACE BETWEEN ROOF AND INSULATION

ADVANTAGE OF VENTILATION BETWEEN ROOF AND THERMAL INSULATION

Figure 57

EFFECT OF RAIN ON SURFACE TEMPERATURES OF ROOFS AND WALLS

Just the opposite of solar radiation, rain will reduce the surface temperature of walls and roofs. In the case of rain falling on surfaces, the temperature of the surface will probably be slightly less than that of the rain. This is because as the rain changes from a liquid to a vapor, approximately 1100 Btu's (1,160,500 joules) for each pound (0.4536 kilograms) evaporates from that surface. This explains the cooling effect of rain storms any time of the year.

This rapid change in temperature causes very high stresses in building structures because of the change of dimensions of materials due to their expansion - contraction coefficients.

Another effect of rain is the possibility of the building material becoming wet (see Table 45). This wetting causes a major change of thermal conductivity of any mass insulation or building materials such as brick, or cinder block. Again, it must be remembered that a minimum of 1100 Btu's (1,160,500 joules) is required to dry the insulation, block or brick. From this it is apparent that residences or buildings should be constructed so as to shed water without absorption of the same. However, the outside should not be so highly water vapor resistant that moisture build-up inside the building is unable to escape.

ABSORPTION OF WATER
PARTS BY WEIGHT
OF STONE AND MASONRY MATERIALS

MATERIAL	SPECIFIC GRAVITY DRY	POTENTIAL ABSORPTION OF WATER PARTS BY WEIGHT
Granite	2.67	0.0013
Limestone	2.53	0.0263
Limestone (oolithic)	2.48	0.0434
Marble	2.72	0.0033
Sandstone	2.22	0.0417
Slate	2.77	0.0333
Brick, Common	2.00	0.333
Brick, Face	2.10	0.166
Brick, Sand-Lime	1.85	0.100
Cinder Block	1.50	0.650
Concrete Block	1.90	0.500

TABLE 45

5 Expansion and Contraction Considerations

TEMPERATURE EFFECTS ON STRUCTURES

As shown in the previous chapter, atmospheric conditions cause major changes in the outer temperature of residences and buildings, while the inner temperatures, in most cases, stay approximately the same. These changes in temperature and temperature differentials cause high tensile or compressive forces in or between materials of construction. The forces can cause crushing, fracture or deformation of materials of construction.

The major reason that this factor is overlooked is that in the past buildings were masonry and stone, which had relatively low thermal expansion coefficients. Also, the sizes of the buildings were limited and temperature differences between the inside and outside were relatively minor. Now buildings are constructed of many different materials, which have different expansion coefficients; they may be very large in size and by the use of air conditioning, heating, etc., major differences between inside and outside temperatures occur.

This book is written to provide information on the thermal insulations used in their construction. It is not a book on structural design. Therefore, it will be limited to those cases where the thermal insulation is affected or affects the design of the structure.

It is essential to be able to calculate changes in the size of structures of large dimension. For this reason, Table 46 is provided which presents the coefficient of expansion of many building materials. To illustrate the magnitude of dimensional changes the following example is presented.

Example 23—Expansion and Contraction of Building Columns in Vertical Walls—*English Units*

A building is constructed of metal siding supported by steel columns. The height of the beams is 60′0″. The walls facing the sun can rise to 150 °F in summer. The winter temperature goes down to −10 °F. Inside, the interior panels are separated from the columns by thermal insulation. Find the difference in length of columns between summer and winter.

SOLUTION: As the column is insulated from the interior temperatures, the temperature of the columns may be as low as −10 °F in winter and as high as 150 °F in summer. $\Delta T = 150\,°F - (-10\,°F) = 160\,°F$ temperature difference. From Table 46, the expansion coefficient is 0.00000670 per °F times length. Length of column is 60′0″, thus difference in length from low to high temperature, in feet is 60 × 160 × 0.00000670 = 0.0643 feet or .0643 × 12 = 0.7718 inches (Answer).

A building is constructed of metal siding supported by steel columns. The height of the beams is 20 meters. The walls facing the sun heat to 66 °C in summer. The winter temperature goes down to −23 °C. Inside the interior panels are separated from the columns by thermal insulation. Find the difference in length of the columns between summer and winter.

COEFFICIENT OF LINEAR EXPANSION OF BUILDING MATERIALS

MATERIAL	Average Coefficient of Expansion*	
	English Units Per Degree F	Metric Units Per Degree C
Aluminum	0.00001233	0.00002219
Asphalt Felt Roofing	0.00004560	0.00008208
Brass (Cast)	0.00001042	0.00001875
Bronze	0.00001024	0.00001843
Concrete Roof Deck	0.00000552	0.00000994
Copper	0.00000926	0.00001666
Glass	0.00000495	0.00000891
Glass Fiber Felt-Asphalt Membrane	0.00000575	0.00001035
Granite	0.00000401	0.00000721
Gypsum Roof Deck	0.00000861	0.00001550
Ice	0.00002833**	0.00005099**
Iron-Soft	0.00000672	0.00001209
Iron-Cast	0.00000589	0.00001060
Lead	0.00001516	0.00002728
Limestone	0.00000451	0.00000811
Marble	0.00000452	0.00000813
Masonry (Average)	0.00000420	0.00000756
Perlite Board	0.00001000	0.00001800
Sandstone	0.00000553	0.00000995
Slate	0.00000582	0.00001048
Steel, Cast	0.00000613	0.00001103
Steel, Galvanized	0.00000670	0.00001206
Steel, Hard	0.00000731	0.00001315
Steel, Pipe and Structural	0.00000670	0.00001206
Steel, Soft	0.00000615	0.00001107
Steel, Stainless	0.00000994	0.00001789
Wood Decking	0.00000310	0.00000558
Wood-Parallel to Fiber, Beech	0.00000143	0.00000257
Wood-Parallel to Fiber, Chestnut	0.00000361	0.00000630
Wood-Parallel to Fiber, Elm	0.00000314	0.00000563
Wood-Parallel to Fiber, Maple	0.00000354	0.00000637
Wood-Parallel to Fiber, Oak	0.00000273	0.00000491
Wood-Parallel to Fiber, Pine	0.00000301	0.00000541
Wood-Across Fiber, Beech	0.00003411	0.00006139
Wood-Across Fiber, Chestnut	0.00001806	0.00003251
Wood-Across Fiber, Elm	0.00002926	0.00004367
Wood-Across Fiber, Maple	0.00002234	0.00004039
Wood-Across Fiber, Oak	0.00003022	0.00005440
Wood-Across Fiber, Pine	0.00001894	0.00003409
Zinc	0.00001736	0.00003125

*Based on average coefficient of expansion at temperatures −30° F (−34° C) to 150° F (65° C)
**Based on mean temperatures −40° F (−20° C) to 32° F (°C)

TABLE 46

SOLUTION: As the beam is insulated from the interior temperature, the temperature of the beams may go as low as −23 °C in winter and as high as 66 °C in summer. The temperature difference between summer and winter is 66 °C − (−23 °C) = 89 °C. From Table 46 the expansion coefficient is 0.00001206 per °C. As the length of the columns is 20 meters, the difference in length from low to high temperature in meters is: 20 × 0.00001206 × 89 = 0.0214 meters or 21.4 millimeters (Answer).

The change in dimensions of columns and beams necessitates means of providing for differential expansion joints between these and exterior siding and internal paneling. In Examples 23 and 23a, the beam was stated to be insulated from the interior temperature thus its temperature was approximately identical as to the surface temperature of the siding. However, in most instances, the beam is influenced by inside air temperature. This then means that differential expansion occurs between structural members and the outside siding. For this reason, metal buildings are designed and constructed so that beam expansion and contraction can occur. This beam expansion is illustrated in Figure 58-B. Where siding is of single sections it is secured by "Z" supports which will bend slightly to compensate for differential movement. This type of securement is shown in Figure 58-C.

When metal building walls are insulated between the structure and the internal facing, the differential movement between the two

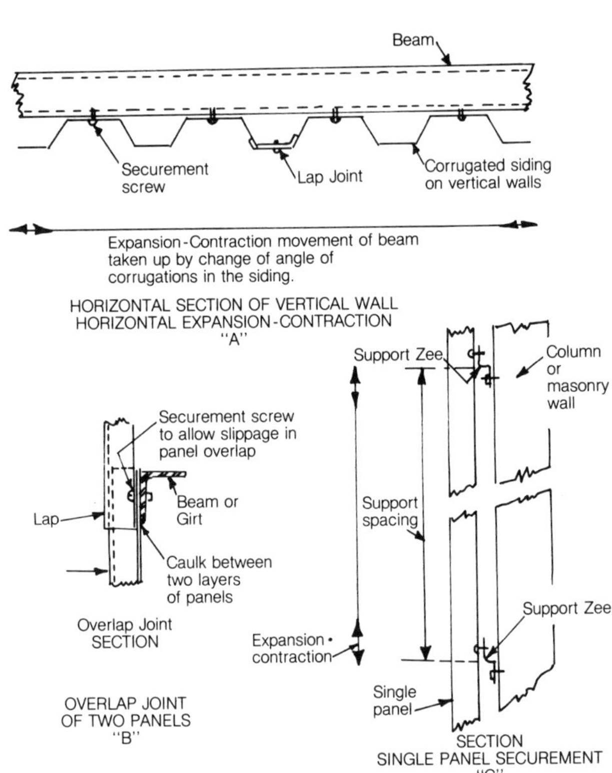

EXTERNAL METAL PANELS SECUREMENT

Figure 58

must be compensated for by some design mechanism. Although the temperature may be constant on the interior of the building thus causing little change in dimensions on the inner panels, the panels must be secured to the metal column girts or beams that do change in dimensions. This difference in dimension change must be compenstated by the use of "Z" supports, slip joints and corrugations in the metal panels. Plain, flat sheets of thin metal secured on all four sides will buckle (fish mouth) between joints. The metal will bulge where compression is exerted in a plane or the fastening will be loosened when tension is exerted. Recommended construction is shown in Figure 59.

EXPANSION AND CONTRACTION OF BEAMS IN FLAT ROOFS

When thermal insulation is installed on top of flat roofs, the beam being below the insulation stays at a much more constant temperature than the insulation. If the insulation has a coefficient of expansion, or changes dimension for other reasons, then differential movement between the beams and roof insulation systems will occur.

A listing of commonly used thermal insulations used is given in Table 47. The properties listed are combustibility, coefficient of expansion, density, compressive strength and vapor migration. These are listed as they are of utmost importance in the design of the insulation.

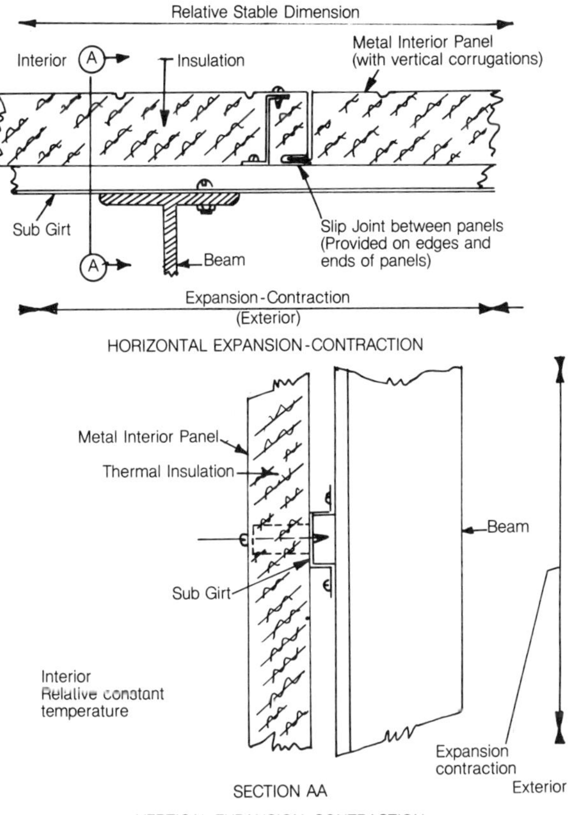

INTERNAL METAL PANELS SECUREMENT

Figure 59

The combustibility is listed first as this may be the most important property for selection of an insulation suitable for the safety of the structure as determined by usage and the fire resistivity of the other parts of the roof system.

The coefficient of expansion of the materials is given so that differential movement can be determined.

The compressive strength of materials is provided so that the ability of a material to compress under a load of expansion and contraction can be evaluated. Fiberous glass wool or mineral wool panels in most instances will compress in the plane of length and width and compensate for the compression forces. However, the facing materials, having a different coefficient of expansion, can cause buckling and distortion of the panels.

Cellular glass insulation having a relatively low thermal expansion coefficient, only about 67% that of structural steel, helps to compensate for some of the difference in movement of the structural steel on the interior, which stays at a more constant temperature than the insulation above it. For this reason, the dimension change can be, under ordinary building conditions, compensated by the adhesive cement or asphalt in the butt joints. However, where there is an expansion—contraction joint in the structural roof system, a corresponding joint must be put into the cellular glass roof insulation. Such an expansion joint is necessary with all forms of insulation. A typical insulation joint for roof insulation at the structural expansion joints is shown in Figure 60.

Cellular organic insulations have very large coefficients of expansion. To illustrate this, consider the following example:

LINEAR EXPANSION OF RIGID URETHANE INSULATION

Example 24—*English Units*

If polyurethane insulation is placed in position on a roof and its linear movement is unrestricted, what would be its increase per 100 ft in length (or width) when the temperature to which it is exposed changes from 70 °F to 150 °F?

Temperature difference is 150° − 70 °F = 80 °F. From Table 47 the coefficient of expansion of polyurethane is given as 0.000060 (English Units). Thus, the expansion in feet is: 100 × 80 × 0.00006 = 0.48 feet or 0.48 × 12 = 5.76 inches (Answer).

Example 24a—*Metric Units*

If polyurethane insulation is placed in position on a roof and its linear movement is unrestricted, what would be its increase per 30.5 meters in length (or width) when the temperature to which it is exposed changes from 21.1 °C to 65.6 °C?

Temperature difference is 65.6° −21.1 °C = 44.5 °C. From Table 47 the coefficient of expansion of polyurethane is given as 0.000108 (Metric Units). Thus, the expansion in meters is: 30.5 × 44.5 × 0.000108 = 0.1465 meters (Answer).

PROPERTIES OF ROOF BOARD AND PANEL INSULATIONS

MATERIAL	FORM	COMBUSTIBILITY	COEFFICIENT OF EXPANSION		DENSITY		COMPRESSIVE STRENGTH at 10% DEFORM.		WATER VAPOR MIGRATION	
			°F	°C	Lbs per ft³	kgs per m³	lbs/in²	kPa	perm inch	perm cm
Cellular Glass	Board	Noncombustible	0.000046	0.0000082	7.0 to 9.5	112 to 152	* 100 (min)	* 690 (min)	0.00001	0.00002
Cellular Glass Between Layers of Felt	Panels	Insulation Noncombustible Felt will burn	0.0000046	0.0000082	7.0 to 9.5	112 to 152	* 100 (min)	* 690 (min)	0.00001	0.00002
Cellulosic Fiber Roof Insulating Board Panels	Panels	Combustible	0.000050	0.000090	15 to 18	240 to 288	35 to 60	241 to 413	** 0.5 to 1.2	** 0.8 to 2.0
Glass, fiber with organic binder, Asphalt - foil facing	Board and Panel	Insul. will char; Asphalt will burn	0.000005	0.000009	7 to 9	112 to 144	10 to 15	69 to 103	** 0.1 to 0.5	** 0.16 to 0.8
Glass fiber with organic binder, Alumn. facing	Panel	Insulation will char	0.000005	0.000009	5 to 8	80 to 128	5 to 15	34 to 103	** 0.1 to 0.5	** 0.16 to 0.8
Mineral fiber with organic binder	Panel	Insulation will smolder	0.000006	0.000011	15 to 20	240 to 320	3 to 8	21 to 55	***	***
Polystyrene foam (with or without facings)	Panel	Combustible	0.000035	0.000063	1.5 to 3.3	24 to 5.2	8 to 30	55 to 207	1.5 to 5.0	2.5 to 8.3
Polyurethane foam (with or without facings)	Board and Panel	Combustible	0.000060	0.000108	1.7 to 4.0	27 to 64	15 to 30	103 to 207	2.5 to 5.0	4.1 to 8.3
Polyurethane foam - sprayed in place	Foamed in Position	Combustible	0.000060	0.000108	1.0 to 4.0	16 to 64	15 to 30	103 to 207	2.5 to 8.0	4.1 to 13.4
Polyurethane bonded to treated wood fiber	Panels	Combustible	0.000060	0.000108	33 to 36	528 to 576	15 to 30	103 to 207	4.0 to 10.0	6.7 to 16.7
Expanded perlite and fibers with or without facing	Panels	Insulation Combustible	0.000010	0.000018	10 to 12	160 to 192	30 to 50	207 to 345	***	***
Siliceous fiber with binders and aluminum facing	Panels	Noncombustible	0.000006	0.000011	4.0 to 5.0	64 to 80	30 to 40	207 to 276	***	***

*Crushing strength

**Total permeance one surface to other - including facings

***Depends upon vapor retarder installed on inner surface

TABLE 47

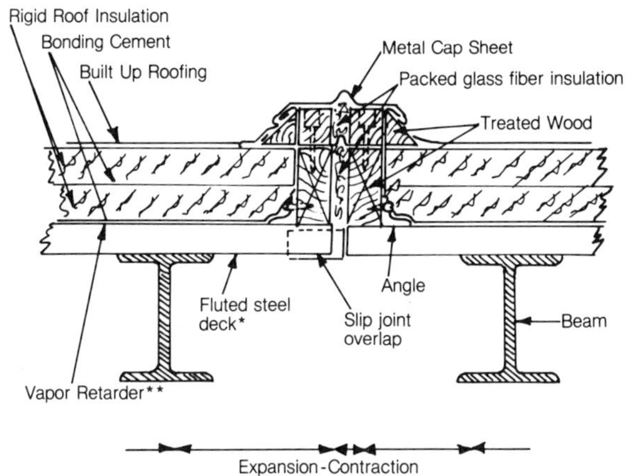

SECTION — ROOF

* Shown is fluted steel roof deck. The deck may be steel tees or channels. Although the slip joint of these will be different to that shown, the expansion-contraction joint of the insulation will be approximately as shown.

** Vapor retarder is located under insulation on heated buildings. It would be located above the insulation if the space below was refrigerated.

TYPICAL ROOF INSULATION EXPANSION-CONTRACTION JOINT

Figure 60

It is very evident that organic foams are dimensionally unstable materials with respect to temperature change and that they must be restrained to function as building insulation.

If these foams were restrained between two strong, stable materials, the expansion force would equal the compressive strength and the material would, in effect, be compressed at the high temperature. Then having been compressed, when the temperature is reduced, the materials would shrink and may eventually crack.

In actual practice, these insulations are bonded to the wood, masonry or metal building slab. Where they are over a heated area, a vapor retarder must also be included in between the slab and the insulation. In summer, when the outer surface of insulation reaches a higher temperature than the inner surface, the outer surface tries to expand more than the inner surface. This causes uneven stresses within panels or sprayed organic foam insulation. If the stresses in the insulation exceed the bonding strength of the adhesive between the roof slab and the insulation, bowing or blistering of the insulation will occur. This is shown in Figure 61.

Where insulation is placed above a slab or metal roof and becomes the base for the roofing membrane, the insulation becomes only one part of the total system. The success of the roofing-insulation system depends upon the ability of all the materials to function together and their ability to resist deforming stresses imposed by internal and external forces and loads.

LINEAR EXPANSION OF POLYSTYRENE INSULATION

Using the same units as were used in Example 24 and 24a, what would the lateral expansion of polystyrene insulation be as compared to polyurethane insulation?

Example 25—*English Units*

If polystyrene insulation is placed in position on a roof and its linear movement is unrestricted, what would be its increase per 100 ft in length (or width) when the temperature to which it is exposed changes from 70 °F to 150 °F?

Temperature difference is 150 °F − 70 °F = 80 F. From Table 47, the coefficient of expansion of polystyrene is given as 0.000036 (English Units). Thus, the expansion in feet is: 100 × 80 × 0.000036 = 0.288 feet or 0.288 × 12 = 3.465 inches (Answer).

Example 25a—*Metric Units*

If polystyrene insulation is placed in position on a roof and its linear movement is unrestricted, what would be its increase per 30.5 meters of length (or width) when the temperature to which it is exposed changes from 21.1 °C to 65.5 °C?

Temperature difference is 65.6 °C − 21.1 °C = 44.5 °C. From Table 47 the coefficient of expansion of polystyrene is 0.000063 (Metric Units). Thus, the expansion in meters is 30.5 × 44.5 × 000063 = 0.0855 meters (Answer).

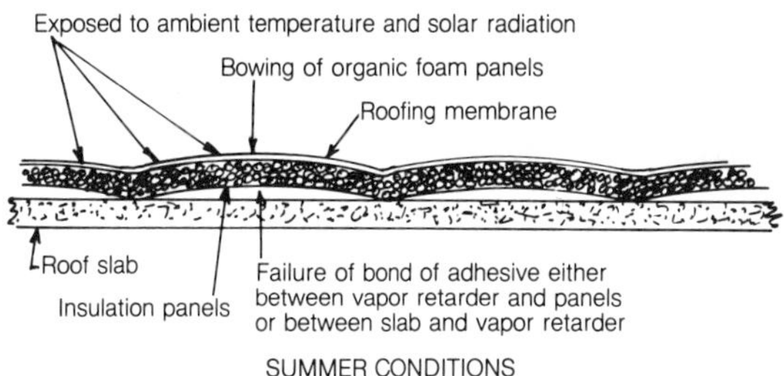

SUMMER CONDITIONS

EFFECT OF HIGH TEMPERATURE EXPOSURE ON PANELS DUE TO BOND FAILURE OR LACK OF SUFFICIENT EXPANSION-CONTRACTION JOINTS

"A"

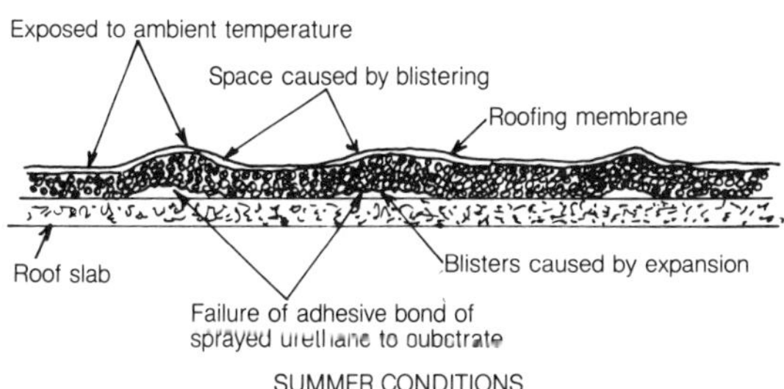

SUMMER CONDITIONS

EFFECT OF HIGH TEMPERATURE EXPOSURE ON SPRAYED URETHANE INSULATION IF ADHESION BOND FAILS

"B"

Figure 61

Internal forces occur when two materials are bonded together but have different coefficients of expansion. As temperature changes at this bonding plane, shearing forces will produce fractures between the two different materials. The larger the ratio of the coefficient of expansion-contraction between two materials, the more likely shearing is to take place at the bonding plane. As materials have different expansion properties at different mean temperatures, the ratio may change with mean ambient temperature of the materials. Table 48 gives the ratios of expansion between insulations and roofing membrane in "A" part and between insulations and roof materials in "B" part.

Where thermal insulation is part of the roof system, it is necessary to know the properties of each segment of the system so as to design a satisfactory system. For convenience of design, the basic properties of roof membranes and coatings are provided in Tables 49 and 50.

These properties illustrate why the roofing and the insulation systems are highly sensitive to changes in temperature. These changes in temperature cause the insulation's outer surface and roofing membrane to expand and contract and change in physical properties.

Another factor is the change of moisture within the system. Water increases volume approximately 1500 times when changed from a liquid to vapor at 112°F (44.4°C). If confined to a constant volume, the change of 100°F (37.8°C), it can develop 600 lbs per sq in. (4,137,000 Pa) pressure. All of these cause stress beween the materials in the insulation-roofing system and blisters between

RATIOS OF EXPANSION BETWEEN INSULATIONS AND ROOF MEMBRANE MATERIALS

ROOF MEMBRANE MATERIAL	MEAN TEMPERATURE			ROOF INSULATION MATERIALS				
	°F	°C	Organic Fiberboard	Fiberous Glass	Cellular Glass	Polystyrene	Polyurethane	Perlite Block
Organic Felt Asphalt Membrane	−20 100	−28.9 37.8	1:1.3 2.3:1	1:3.6 1.6:1	1:3.9 1.5:1	2.2:1 11:1	3.2:1 20:1	1:1.3 2.3:1.0
Organic Felt Coal Tar Membrane	−20 100	−28.9 37.8	1:1.1 7:1	1:3 5:1	1:3.3 4.6:1	2.7:1 33:1	3.9:1 60:1	1.1:1 7:1
Inorganic Felt, Asphalt Membrane	−20 100	−28.9 37.8	1.3:1 3.5:1	1:2 2.5:1	1:2.2 2.3:1	4:1 16.5:1	5.8:1 30:1	1.3:1 3.5:1
Glass Fiber Felt, Asphalt Membrane	−20 100	−28.9 37.8	1:2.3 2.3:1	1:6.1 1.6:1	1:6.6 1.5:1	1.3:1 11:1	1.9:1 20:1	1:2.3 2.3:1

"A"

RATIOS OF EXPANSION BETWEEN INSULATIONS AND ROOF DECK MATERIALS

ROOF MEMBRANE MATERIAL	MEAN TEMPERATURE			ROOF INSULATION MATERIALS				
	°F	°C	Organic Fiberboard	Fiberous Glass	Cellular Glass	Polystyrene	Polyurethane	Perlite Block
Fluted Steel Deck	−20 100	−28.9 37.8	2.2:1 1:1	1:1.3 1:1.3	1:1.4 1:1.4	6:1 5:1	9:1 9:1	2.2:1 1:1
Concrete Deck	−20 100	−28.9 37.8	2.6:1 1.4:1	1:1 1:1	1:1.1 1:1.1	8:1 6.7:1	11:1 12:1	2.6:1 1.4:1
Wood Deck	−20 100	−28.9 37.8	4.4:1 2.1:1	1.5:1 1:1.5	1.5:1 1:1.3	12:1 10:1	17:1 18:1	4.4:1 2.1:1
Gypsum Deck	−20 100	−28.9 37.8	1.8:1 1.3:1	1:2 1:1	1:1.6 1:1.1	4:1 6.5:1	6:1 11.5:1	1.8:1 1.3:1

"B"

TABLE 48

PROPERTIES OF ROOF MEMBRANE MATERIAL
SATURATED ROOFING FELTS

FELT PLYS MEMBRANE MATERIAL	MEAN TEMPERATURE		EXPANSION COEFFICIENT		TENSILE STRENGTH		MODULUS OF ELASTICITY		COMPRESSIVE STRENGTH		FLEXURAL STRENGTH		VAPOR PERMEABILITY	
	°F	°C	English	Metric	PSI	Pa	PSI	Pa	PSI	Pa	PSI	Pa	English	Metric
Organic Felt Asphalt Sat. Membrane Thk.0.125''(3.17mm)	−10	−23.3	18 X10^{-6}	124 X10^{-6}	480	330 X10^{4}	43 X10^{3}	296 X10^{6}						
	0	−17.8	5 X10^{-6}	34 X10^{-6}	400	275 X10^{4}	28 X10^{3}	193 X10^{6}	100	689 X10^{3}	40	276 X10^{3}	25	41.75
	70	21.1	3 X10^{-6}	21 X10^{-6}	190	130 X10^{4}	19 X10^{3}	131 X10^{6}						
Organic Felt Coal Tar Membrane Thk.0.125''(3.17mm)	−10	−23.3	14 X10^{-6}	96 X10^{-6}	380	261 X10^{4}	40 X10^{3}	276 X10^{6}						
	0	−17.8	13 X10^{-6}	90 X10^{-6}	300	206 X10^{4}	17 X10^{3}	117 X10^{6}	100	689 X10^{3}	50	345 X10^{3}	27	45.09
	70	21.1	12 X10^{-6}	83 X10^{-6}	160	110 X10^{4}	13 X10^{3}	90 X10^{6}						
Inorganic Felt, Asphalt Membrane Thk.0.125''(3.17mm)	−10	−23.3	10 X10^{-6}	68 X10^{-6}	480	330 X10^{4}	43 X10^{3}	296 X10^{6}						
	0	−17.8	6 X10^{-6}	41 X10^{-6}	400	275 X10^{4}	28 X10^{3}	193 X10^{6}	100	689 X10^{3}	35	241 X10^{3}	20	33.4
	70	21.1	2 X10^{-6}	13 X10^{-6}	190	130 X10^{4}	19 X10^{3}	131 X10^{6}						
Glass Fiber Felt, Asphalt Membrane Thk.0.125''(3.17mm)	−10	−23.3	31 X10^{-6}	213 X10^{-6}	230	158 X10^{4}	40 X10^{3}	276 X10^{6}						
	0	−17.8	6 X10^{-6}	41 X10^{-6}	200	138 X10^{4}	17 X10^{3}	117 X10^{6}	100	689 X10^{3}	50	345 X10^{3}	75	125.25
	70	21.1	3 X10^{-6}	20 X10^{-6}	180	124 X10^{4}	13 X10^{3}	90 X10^{6}						

PHYSICAL PROPERTIES OF ROOFING FELTS

TABLE 49

COATING–ADHESIVE	Flash Point Temperature		Softening Temperature		Liquidification Temperature		Combustibility of Material in Dried, Solid State
	°F	°C	°F	°C	°F	°C	
Asphalt Coating Regular Softening Temperature	100 to 125	37.8 to 51.7	150 to 175	65.6 to 79.4	400 to 500	204 to 254	Combustible
Asphalt Coating Higher Temp. Softening	115 to 140	46.1 to 60.0	200 to 225	93.3 to 107	500 to 550	254 to 288	Combustible
Tar Coating and Adhesive	75 to 100	23.9 to 37.8	125 to 150	51.7 to 65.6	300 to 350	149 to 171	Combustible

EFFECT OF TEMPERATURE ON HOT APPLIED BITUMINOUS COATINGS

TABLE 50

the components. These blisters occur where adhesion of the roofing membrane to the insulation is the weakest.

The increase of temperatures causing the expansion and blisters from trapped moisture is always followed by periods of decreasing temperature, such as night temperatures being cooler than day temperatures plus solar radiation. As the temperature goes down, the bitumen coating becomes rigid and brittle with resultant loss of adhesion between membrane and the insulation (or slab). Of course, aging causes the felts and coatings to harden and lose elasticity.

When the roofing membrane is not adhered to the insulation, it will not be able to resist the lift of wind caused by the air-foil effect. A wind lifting effect of only 0.0217 lbs per sq in. (150 Pa) will lift a roofing membrane up from the surface on which it is resting. To prevent roofing membrane from being lifted off the insulation (or slab), it is standard practice to put not less than 3″ (7.62 cm) thickness of gravel on the roof so that its weight will hold the roof membrane down to the insulation or slab. This also reduces the fire hazard of exposed asphalt or tar materials which might be ignited by only a very small spark or fire.

The gravel also serves to press the membrane and adhesive coating to the insulation or slab where it may rebound if temperatures go above 150°F (65.5°C). However, this does not eliminate the need for correct design of the roof slab, insulation and roofing membrane expansion joints.

Physical and combustion properties of roofing felts are given in Tables 49 and 50.

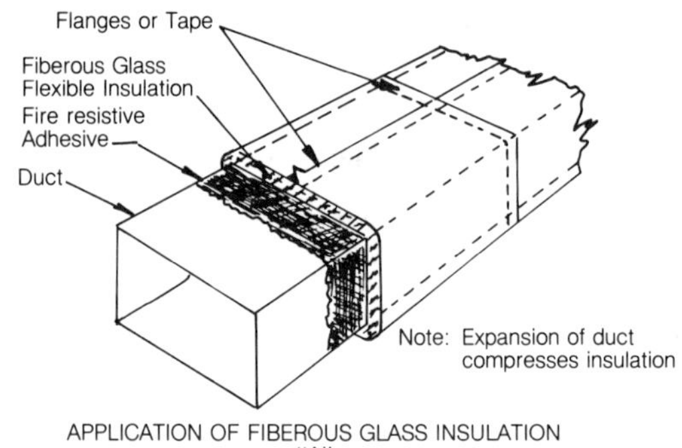

APPLICATION OF FIBEROUS GLASS INSULATION
"A"

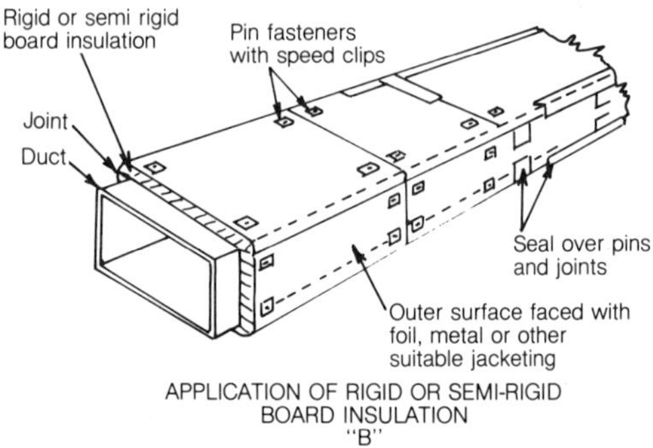

APPLICATION OF RIGID OR SEMI-RIGID BOARD INSULATION
"B"

INSTALLATION OF INSULATION ON DUCTS

Figure 62

EXPANSION AND CONTRACTION OF DUCTS

On ducts of relatively little temperature rise or fall, such as air-conditioning ducts, the insulation is most frequently fiberglass insulation with metal foil or similar type backing. In this instance the compressibility and recovery of the insulation are generally sufficient to take care of dimensional differences caused by temperature changes. Such applications should be limited to a maximum temperature of approximately 212°F (100°C). This is shown in Figure 62-A.

Where ducts are larger and possibly have maximum temperatures up to 250°F (121°C), semi-rigid or rigid board insulations should be used to obtain better strength. These are secured to the surface by welding pins and speed clips and installed so overlapping 90° joints are installed at the corner. This is shown in Figure 62-B.

Where it is necessary to provide slip expansion joints or metal expansion joints in the duct work (or other flat surfaces) it is necessary to put thermal insulation expansion joints also. Where the metal expansion joint is a slip joint, then a ship-lap insulation slip joint at the metal expansion joint should be installed. This is shown in Figure 63-A. When the expansion-contraction is taken up by a curved, metal expansion joint, the insulation over this curved metal should be made of mineral wool flexible insulation secured in position by a metal mesh blanket. This is shown in Figure 63-B.

EXPANSION AND CONTRACTION OF PIPING SYSTEMS

This discussion will be limited to hot piping as refrigerated pi ing is more frequently considered as industrial insulation rathe than building insulation. The ordinary pipe for air-conditionin systems is at temperatures which are so moderate that expansic is not a major problem.

The common hot-water piping is most frequently insulated wi light-weight fiberous insulation which has sufficient flexibility that compensates for pipe expansion—especially when it is installe with each section pushing against the next to be compressed durir installation.

Steam piping and equipment do expand to the point that pr per expansion joints are required.

When rigid insulation, such as calcium silicate or expande silica is used on high temperature piping, it should be installed staggered positions to form a secured cylinder of insulation. Eac cylinder should extend 18′ to 21′ along a horizontal pipe and that junction with another cylinder around the pipe, an expansic joint should be installed. The manner of installing and securing th single-layer insulation to a pipe is shown in Figure 64-A. Where th temperature of the pipe is sufficiently high to require multip layers of insulation, they should be applied as shown in Figu 64-B.

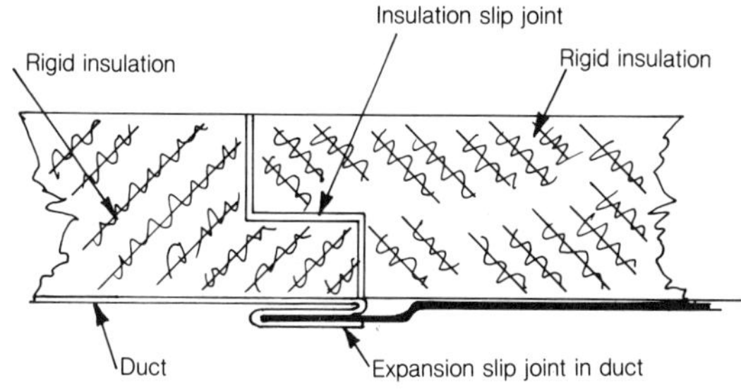

No scale

INSULATION SLIP EXPANSION JOINT "A"

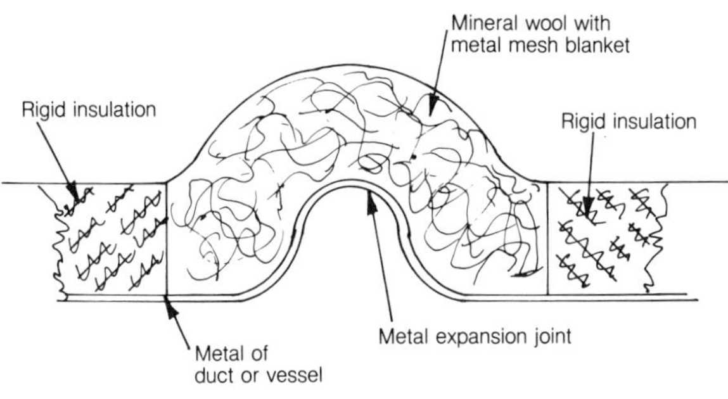

No scale

FLEXIBLE INSULATION EXPANSION JOINT "B"

INSULATION EXPANSION-CONTRACTION JOINTS

Figure 63

All round flue gas ducts should be insulated with multi-layer high temperature insulation as shown in Figure 64-B so as to prevent any direct leak of heat or gas from the inside of the duct to the ambient air.

INSULATION OF HOT VESSELS OR EQUIPMENT

Because most hot vessels and equipment are purchased with the insulation as part of the unit, their insulation application is not generally considered as a part of the building insulation. However, should it become necessary to insulate a hot tank or vessel, the same principles of expansion-contraction presented for pipe and ducts must be observed.

To assist in such insulation, Figure 65 should be used for guidance. Shown in this is a vertical vessel with rigid insulation on the vertical cylindrical walls. Shown are the base insulation support and expansion joints for top and middle supports if the vessel is greater than 18′0″ (5.5 meters) in length on the cylindrical surface. This is shown in Figure 65A. In Figure 65B the method of securing the rigid insulation to flat vertical sections is illustrated.

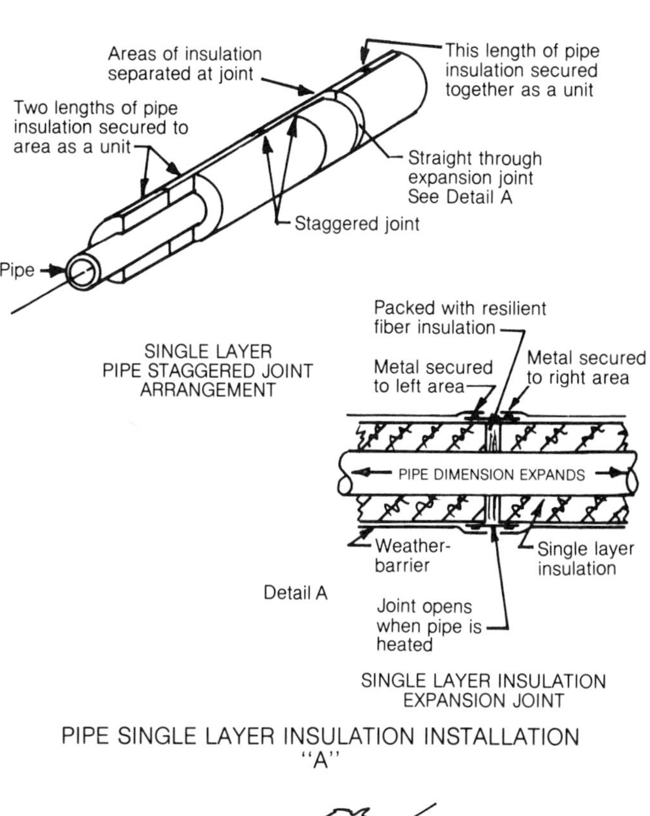

PIPE SINGLE LAYER INSULATION INSTALLATION "A"

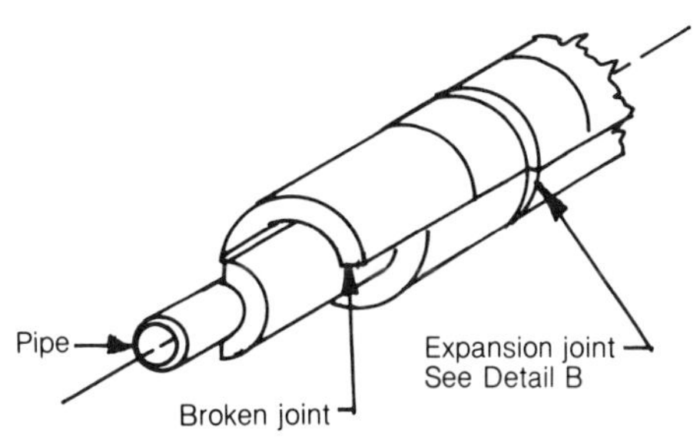

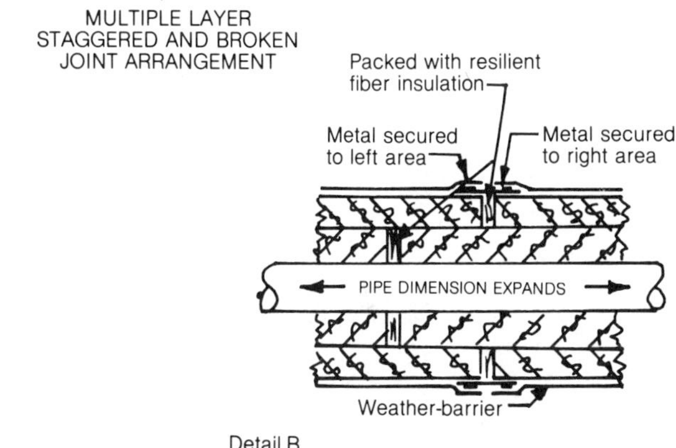

PIPE MULTIPLE LAYER INSULATION INSTALLATION "B"

APPLICATION OF RIGID INSULATION TO HOT PIPE

Figure 64

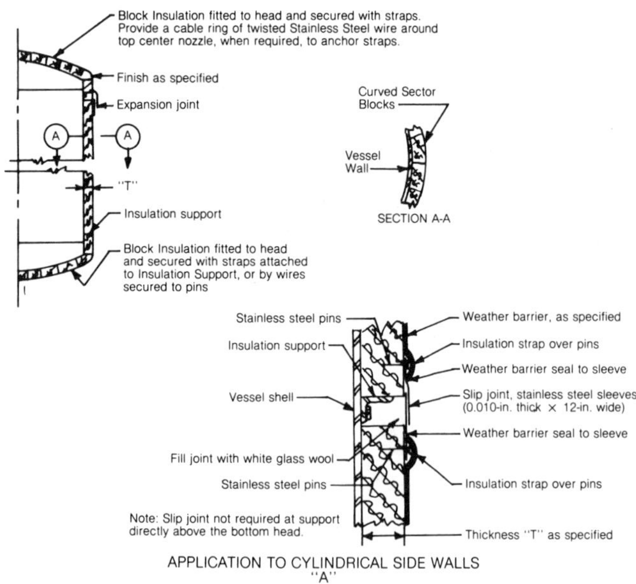

APPLICATION TO CYLINDRICAL SIDE WALLS
"A"

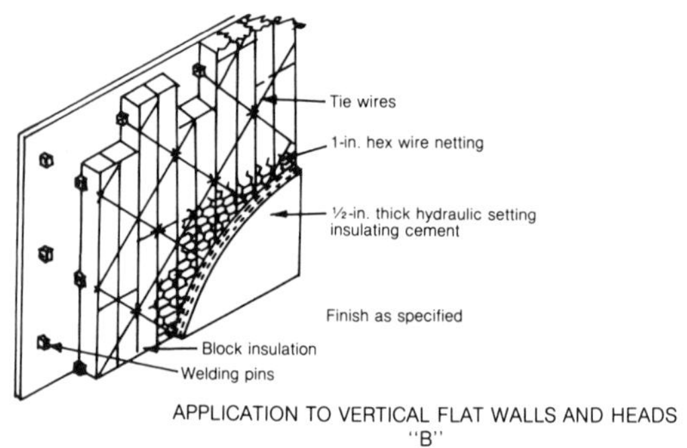

APPLICATION TO VERTICAL FLAT WALLS AND HEADS
"B"

VESSEL INSULATION

Figure 65

It should be noted that most high temperature rigid insulation *shrinks* as temperature increases. This further explains why expansion and contraction joints are so vital to satisfactory thermal insulation performance.

6 Safety Considerations

Summary from *"Health Aspects of Fibrous Glass In Air Duct Systems"* published by Thermal Insulation Manufacturers Association (TIMA)

1. Fibrous glass is an extremely valuable basic material which benefits everyone every day.
2. In the construction industry, fibrous glass is broadly accepted and specified for thermal and acoustical insulation requirements.
3. Fiber glass duct insulation materials meet the requirements of nationally recognized standards-making bodies and are approved for use by key governmental agencies.
4. Extensive industrial hygiene engineering research studies have been carried out by highly qualified, university-affiliated occupational health and environmental engineers. Studies both in the laboratory and in actual installations demonstrate no measurable increase in airborne fibrous glass due to the passage of the air stream over fibrous glass duct linings.
5. The preponderance of medical and engineering scientific research to date shows that fibrous glass insulation can be used safely.
6. Fibrous glass manufacturers have supported health and safety research for nearly 50 years. The Thermal Insulation Manufacturers Association will continue to dedicate time, energy and resources in the interests of the health and safety of the men and women who manufacture, install, and use fibrous glass products.

From *"Health and Safety Aspects of Fiber Glass"* published by TIMA, October 1987, Appendix A.

NIOSH Criteria Document Fiber Glass

In April 1977, The U.S. National Institute for Occupational Safety and Health (NIOSH) published a criteria document on occupational exposure to fiber glass.

The criteria document recommended that the U.S. Occupational Safety and Health Administration (OSHA) adopt an occupational exposure standard limiting exposure to fiber glass to 3.0 f/cc, composed of fibers less than 3.5 microns in diameter and greater than 10 microns in length.

OSHA classifies fiber glass as a nuisance dust, with a workplace permissible exposure limit of 5.0 mg/m3 for respirable dust and 15.0 mg/m3 for total dust. OSHA defines the permissible exposure limit as the time weighted average level that a worker can be exposed to in any 8-hour work shift of a 40-hour work week.

Safety is an important consideration in a study of thermal insulation. Under certain conditions some insulations have hazardous properties such as dust, high combustibility, and toxic fumes. A knowledge of these properties as well as of correct installation design greatly reduces any risk. Since there are hazards inherent in installation of the materials, a review of common safety procedures follows.

Protection from Falls

The most common accidents in construction of residences are falls from ladders or scaffolds. For this reason, the following fundamental suggestions for use of ladders are given:

Use of Ladders

1. The ladder must be suitable for the intended use and in good, safe condition.
2. The foot of the ladder should be placed approximately 1/4 of its length from the vertical plane of its top support.
3. Ladder feet should be placed on substantially level base.
4. Both top and bottom of ladder should be secured to prevent movement.
5. Ladders leading to platforms or walkways should extend at least 36″ (1 meter) above that level.
6. Long ladders must be braced at intermediate points to prevent springing.
7. Objects should not be carried in hands while climbing or descending a ladder.
8. Metal ladders should not be used around electrical circuits because they may accidently come in contact with a "hot" wire and become a conductor of electricity.

Use of Scaffolds

Construction of safe scaffolds depends on the following:

1. The scaffolding must be on firm footing. Footing should be secured against movement by recessing, stacking or other means.
2. All uprights must be plumb. Two-inch diameter tubing in scaffolding must be limited to 75 ft (23 m) in height.
3. Platforms of wood must be firm and steady. All joints should be at ledges; minimum 6″ drop lap on each side of the

ledge. All planks should be secured by wire and/or nails to provide solid footing and prevent them from being blown off by the wind.
4. Guardrails and toe-boards should be installed.
5. Overhead danger signs should be installed.

The use of scaffolds for safe working conditions depends upon the following:

1. Scaffold working platform should be free of ice, snow, oil, slick dust, or other debris while being used.
2. No fires should be permitted upon or near the wooden platforms or scaffolds, nor near any canvas or flammable sheeting used for rain or wind protection.
3. No materials should be stockpiled on scaffolds.

Prevention of Accidental Fires

Precaution to prevent fires during installation is of utmost importance. Many may think that all thermal insulations and accessories are relatively fire safe. However, many insulations and accessories are combustible. Also, plastic sheeting and cut-back coatings and sealers are combustible. All organic foam insulations, cellulose fiber insulation and insulation with organic binders are combustible. The degree of combustibility of these materials will be presented later in this chapter.

In the handling of combustible mastics, coatings and solvents, care must be taken not to damage container identification on the cans. Care should be taken to prevent contamination. All drums should be grounded to make them safe from static build-up when material is poured or pumped into small containers or on to surfaces. Adequate fire protection should be provided at storage locations. When any combustible insulation coating, sealer or adhesive is used, care must be taken that none is spilled on to hot surfaces which might cause it to ignite.

Where installation is near any welding, the following precautions should be taken:

1. The frame of all welding machines must be grounded.
2. All combustible materials should be removed from the area where welding is being performed.
3. All wooden scaffolding should be shielded with metal.
4. Proper protective clothing such as eye shields, flame-resistant gauntlet gloves and flame-resistant aprons and hats should be worn.

Prevention of Injury to Eyes, Skin, or Lungs Due to Dust

Where insulation containing mineral fibers or fine dust are cut, installed, removed or sprayed, such dust can cause eye injury, skin infection or lung problems. Where this problem occurs, individuals in the area shall be protected as follows:

1. Safety glasses, goggles or face shields should be worn.
2. Suitable gloves should be worn.
3. Long sleeve shirts, long trousers and caps should be worn to avoid contact with the mineral fibers or dust.
4. Good, personal hygiene practices are essential. Thoroughly wash exposed skin areas periodically and shower at the end of the work day. Change and launder work clothing frequently, (separate from other articles).
5. Barrier creams applied to exposed skin areas will provide protection against skin irritation. Medical doctors' recommendations should be followed.

Prevention of Injury or Poisoning During Spraying

Any compressed air pressure pump or spray gun can be dangerous. Loose jets or tips can be ejected by the pressure; hose or containers can be split by pressure. Thus it is important that all equipment be in good condition and operated in accordance with manufacturer's instructions. Some basic precautions are:

1. Never exceed the recommended air pressure.
2. Even when unit is stopped and air to pump is turned off, there can still be dangerous pressure in the system. This pressure must be relieved to prevent accidents while servicing or handling.
3. Handle gun with care. Liquid or gas velocity is dangerous. At close range the pressure stream can penetrate the skin and cause serious injury.
4. Always use extreme care when removing hose from gun. A plugged line contains liquid under pressure. Always release pressure by closing pump air valve and opening dump valve on spray gun.

If the material being sprayed is combustible, care must be taken that no sparks or open flames are in the area.

Many organic foams contain toxic and hazardous components. For example, in polyurethane foams are toluene (TPI) and diphenylmethane diisocyanate (MDI). The National Institute for Occupational Safety and Health recommends that for eight-hour exposure the vapor particles in the air should not be greater than five parts per billion. As it is impossible to spray within these limits, all workers applying materials *and all other persons within 10′* should be equipped with Type C continuous flow supplied air-positive pressure hoods. Others in the area should wear a full-face mask with an organic vapor cartridge and a particular filter (30 C FRI3). All workers within 10 ft of spraying and in close areas should wear impervious clothing, gloves and footwear.

It must be remembered that in the spraying of polyurethane insulation, the fumes and droplets in the surrounding air are toxic, combustible and hazardous; thus strict safety precautions are imperative for safe application.

Installation of formaldehyde foam insulation is also hazardous. Where this material is poured into walls or ceilings, the material should be separated from the inside of the structure by a completely sealed vapor barrier—preferably metallic, such as aluminum foil. It should be installed so that the fumes are vented to the exterior of the building. All applicators should wear impervious clothing, gloves and footwear. If the applicators are located in confined areas, they must wear suitable masks as for spraying polyurethane.

All organic foam boards, sheets, panels of polyurethane, polyisocyanurates, and styrene are combustible. Likewise, so are cellulose insulations either in batt, loose or panel forms. All must be protected from accidental spark or flame during storage and installation.

SAFETY CONSIDERATIONS—IN SERVICE DUST PROBLEM—WALL AND CEILING INSULATION

Mineral fibers and glass fiber insulation can give off small particles (or dust). The dust, if it gets into the area of use by individuals, can cause emphysema and skin irritations. When used in walls, these materials should not be left exposed to the interior of the home or building. Where used in ceilings or roofs of attics and left exposed, the floor below should be completely sealed off. The doors or ceiling entrance should be properly sealed to prevent dust getting through the cracks into the living quarters.

DUST PROBLEM—DUCT INSULATION

Air conditioning and heating ducts are frequently lined with fiberous thermal insulation. Also, many ducts are made of rigid glass fiber insulation or soft insulation for flexible ducts. In all cases the inner surface of the fiberous insulation is subjected to the velocity of the air flowing through the ducts.

When inner layers of glass fiber insulation are fabricated to fit the inner surface of the duct, or fabricated of the glass fiber insulation, all the cut fabricated edges should be vacuum cleaned and then coated with a suitable adhesive sealer to prevent glass particles from getting into the air stream when the system is in operation.

Many individuals have suffered from breathing problems and itching skin and did not know that such was caused by dust blowing out of the duct systems in their home or office. Suitable tests have not been devised to test for airborne concentrations of fiberous dust; however, it is suggested that manufacturers be requested to submit test results of dust count per cu ft when 100 lin ft of their material is subjected to four times the recommended velocity.

A report concerning the "Health Aspects of Fiberous Glass", Bulletin No. 13-68, was issued by the Industrial Hygiene Foundation of America, Inc., 4400 Fifth Avenue, Pittsburgh, Pennsylvania 15213, in 1968. Additional references to this subject are listed below:

1. Wright, George W.: Airborne Fiberous Glass Particles, AMA Arch. Environ. Health *16*:175-181, February 1968.
2. Gross, Paul, M. L. Westrick, and J. M. McNerney: Glass Dust: A Study of Its Biologic Effects, AMA Arch. Industrial Health *21*:10-23, January 1960.
3. Hatch, T. F. and Paul Gross: *Pulmonary Deposition and Retention of Inhaled Aerosols*, Academic Press, New York 1964.
4. McCord, Carey P.: Fiber Glass: Chemistry and Technology, J. Occ. Med. *9*:339-344, July 1967.
5. Nasr, Ahmed, N. M.: Pulmonary Hazards from Exposure to Glass Fibers, J. Occ. Med. *9*:345-348, July 1967.
6. Heisel, Eldred R. and John H. Mitchell: Cutaneous Reaction to Fiberglas, Ind. Med. & Surg. *26*:547-550, December 1957.
7. The Saranac Laboratory for the Study of Tuberculosis of the Edward L. Trudeau Foundation, Annual Report, Saranac Lake, New York, 1940.
8. Siebert, W. J.: Fiberglass Health Hazard Investigation, Industrial Medicine, January 1942.
9. Sulzberger, Marian B. and Randolph L. Baer, Dept. of Medicine, New York University College of Medicine: The Effects of Fiber Glass on Animal and Human Skin, Industrial Medicine, October 1942.
10. Irwin, J. R., Boeing Aircraft Company, Seattle, Washington: Fiberglas-Plastics—Industrial Medical Aspects and Experiences, Industrial Medicine, September 1947.
11. Schepers, G. W. H., T. M. Durkan, A. B. Delahant, A. J. Redlin, J. G. Schmidt, F. T. Creedon, J. W. Jacobsen, and D. A. Bailey: The Biological Action of Fiberglas-Plastic Dust, Arch. of Industrial Health *18*:134, 1958.
12. Bjure, J., B. Soderholm and J. Widminsky: Cardiopulmonary Function Studies in Workers Dealing with Asbestos and Glasswool. Thorax *19*:22, 1964.
13. Keane, W. T. and M. R. Zavon: Occupational Hazards of Pipe Insulators, Arch. of Environ. Health *13*:171, 1966.

Dust from asbestos-containing products such as insulation, asbestos board or asbestos felts is much more hazardous than mineral fiber dust.

Individuals exposed to asbestos dust developed asbestoses of the lungs and lung cancer. Because of the problem of asbestos fibers being such a health hazard, insulation manufacturers have removed these fibers from their products. However, if work is being done with existing insulations, wall boards or felts, every one in the area should wear fresh air masks and protective clothing.

Reports as to potential hazards of Chrysotile, Amosite, and Crocidolite asbestos hazards have been published by the U. S. Public Health Service, Industrial Hygiene Foundation, Mt. Sinai Environmental Sciences Laboratories and the National Insulation Manufacturers Association.

In all cases, any materials containing asbestos removed from service should be disposed of as hazardous waste.

GAS AND ODOR PROBLEMS

Carbon Monoxide

One major problem has been intensified by the use of stoves, open heaters, and fireplaces used to conserve energy—the build up of potentially lethal carbon monoxide in the living areas.

Carbon monoxide is produced when carbon-containing fuels such as natural or manufactured gas, coal, wood or gasoline are burned. Thus, unless proper ventilation and sufficient outside air are available for combustion, all fuel-burning furnaces, stoves, ovens, water heaters, clothes dryers, fireplaces, gas refrigerators and automobiles are potential threats to health and life.

The tightening up of air leakage around windows and doors to conserve energy can increase the hazard of carbon monoxide poisoning, unless all heat-producing equipment from the fuel is properly vented.

The hazard of carbon monoxide is underestimated because individuals are not familiar with the danger. Hundreds are killed by carbon monoxide poisoning each year. Thousands become so sick from carbon monoxide that they must have medical help, but they do not know the cause of their illness. This is because carbon monoxide is an odorless, colorless gas. It kills, not by poisoning, but by asphyxiation. It suffocates its victim by displacing oxygen in the blood.

Carbon monoxide is more dangerous than other pollutants because the red blood cells that carry oxygen from the lungs to the body combine 210 times more readily with CO than with oxygen. A very small amount of carbon monoxide is a threat to human safety. Air containing one percent of carbon monoxide will kill a person in five minutes. It is essential that ventilation principles and recommendations for combustion air and its venting to outdoor given in Chapter 3 be followed for safety of the occupants of buildings and residences.

Formaldehyde Vapor

Urea formaldehyde foam insulation releases formaldehyde vapor or gas for a long period of time after it is foamed into place. Likewise, particle board, made of wood fragments held together with urea formaldehyde glue can release formaldehyde vapor and gas. The release of the formaldehyde gas has occurred immediately after installation, and in other instances the problem was not detectible for as long as a year after installation.

Homeowners with formaldehyde insulation or particle board have complained about the odors and have stated that it has caused breathing difficulties, serious eye and skin irritation, headaches, dizziness, nausea, vomiting and severe nose bleeds. Because of the potential difficulties, these insulations must be installed so that they are sealed from the inner area of the residences and buildings and are allowed to breathe to the outside air.

The inner area of a building should be vapor sealed from formaldehyde insulation by heavy gauge reinforced aluminum foil vapor retarder. All edges, joints and locations where such goes around wiring must be sealed by tape or vapor resistant adhesive. The installation of the vapor retarder installation for stud walls with urea-formaldehyde particle board on the outside and cinder block walls with voids in blocks filled with urea-formaldehyde foam is shown in Figure 66.

Where urea-formaldehyde foam is used above the ceiling—in attic space—the aluminum foil vapor retarder must be installed directly above the plaster board—below the insulation foam. This is illustrated in Figure 67.

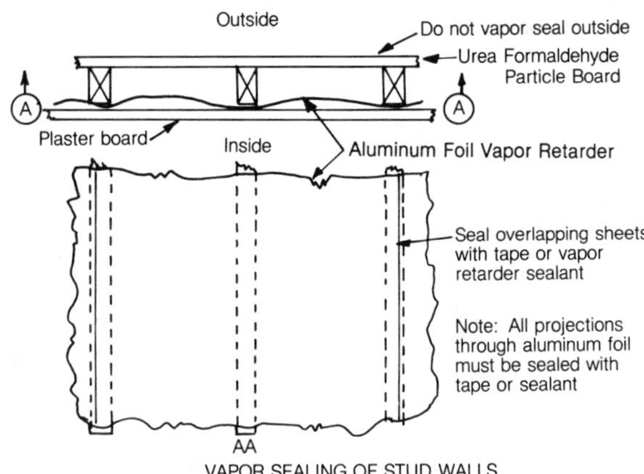

VAPOR SEALING OF STUD WALLS INSULATED WITH UREA FORMALDEHYDE PARTICLE BOARD

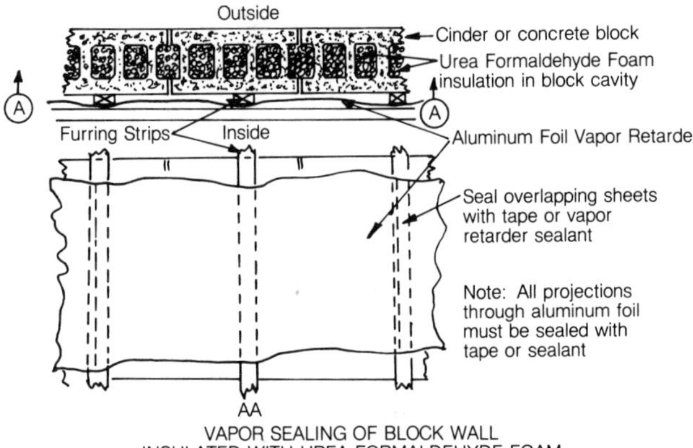

VAPOR SEALING OF BLOCK WALL INSULATED WITH UREA FORMALDEHYDE FOAM

SEALING OF WALLS TO PREVENT ODORS INSIDE

Figure 66

It must also be remembered that the attic space must be vented to the outside and that all doors or entries into the attic must be weather stripped to prevent passage of air into the living area.

Some references as to the serious nature of incorrect use of urea-formaldehyde products have been presented in:

1. *Business Week*, December 24, 1979 — Environment
2. *Engineering Times*, September 1980 — Formaldehyde Vapors Causes Headaches
3. *The Washington Post*; Real Estate Section, Saturday, December 22, 1979

Fire Hazard General

At present, the buildings being constructed for human occupancy are much more hazardous than those built before 1940. Thermal insulation contributes to some of these added hazards but some are due to change of construction methods and ignorance as to fire safety. Some of the basic changes that are only partly related to thermal insulation will be presented first.

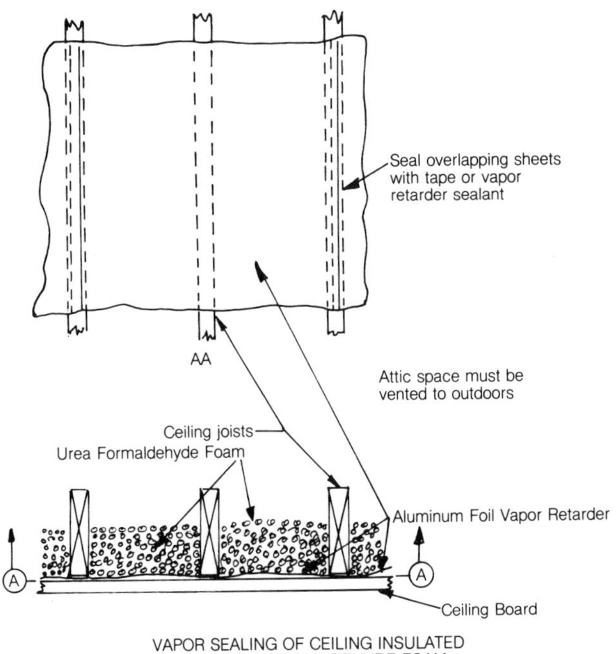

SEALING OF ATTIC INSULATION TO PREVENT ODORS IN LIVING AREA BELOW

Figure 67

The all glass sealed-in buildings are a fire hazard to the occupants. This can be reduced by extensive and complete water sprinkler systems, but there is no way of obtaining outside air. There are no outside fire escapes, so they still remain a fire hazard.

Over the years, the emphasis has been placed on constructing buildings of glass, metal and concrete or so called "fire proof" construction. Unfortunately, this is only the shell of the building and once in use it becomes filled with combustible items such as paper, wooden furniture, cloth drapes, plastic foam rug mats and cushions for furniture (including mattresses in hotels), plastic fiber drapes and carpeting and plastic floor tile. All are very combustible and when they burn, they produce very toxic gases. When these materials catch fire and burn, although the building does not collapse, many of the occupants may be injured or killed.

To improve architectural appearance, the outdoor fire escapes were put inside with fire resistant concrete walls and with fire resistant doors to the stairs. In most cases the exit doors at lower levels were inside the building. These were no longer fire escapes, they were only indoor stairs enclosed with a non-burning material. As a matter of fact, when a fire would start on lower floors, individuals trying to escape from these floors had to open the doors to the stairway and in a major fire these fire escapes turned into a chimney to carry off the smoke and fire due to natural draft upward.

In some cases the enclosed indoor stairways are equipped with blowers to increase the air pressure inside the stairways so that fire and gases would not flow inward. The major difficulties in such a system are: (1) Where can the air intake be located to make sure its inlet always obtains fresh, clean air? (2) What are the assurances that such blowers will remain operative in a fire?

Insulation materials and non-combustible materials are used to make residences and buildings more fire safe. Yet, it must be remembered that although a furnace, boiler, chimney and smoke stack are made of non-combustible materials, when fuel is burned inside the furnace or boiler, a human being would also be burned. Therefore, in a building constructed of fire resistant materials, when its furnishings and contents are combustible, provisions must be taken to put out the fire with water-sprinkler systems and provide a safe means of exiting for the occupants. Buildings and residences must be designed on the basis that fires will occur inside and that safety measures are provided for the occupants. To illustrate the necessity of this thinking, in the United States hotel fires now average over 100 per month. It should also be pointed out that the number of persons killed per number of population in the United States is four times that of other industrialized nations.

One of the major problems causing the construction of unsafe residences is that there is no concentrated effort in the education of architects or engineers in fire safety procedures. There has been very little research as to potential hazards of materials. Mr. Carlos J. Hilado, of Product Safety Corporation, has done most of the research as to the toxic hazards of organic plastic materials.

Three laboratories which test materials for combustibility are: Factory Mutual Systems, Fire Research Section of Southwest Research Institute and the Underwriters Laboratory of Chicago. These laboratories test materials under laboratory conditions with fixed limits. There is no other way to test materials, but such tests can only provide the particular property of combustion under certain given conditions.

Tests which have been used by a number of laboratories and companies are submitted to the American Society for Test and Materials for consideration to become "standard" test methods. These standard ASTM test methods are as follows:

D-568	Rate of Burning and/or Extent and Time of Burning of Flexible Plastics in Vertical Position
D-635	Rate of Burning and/or Extent and Time of Burning of Self-supporting Plastics in Horizontal Position
D-1230	Clothing Textiles, Flammability of
D-1360	Paints, Fire Retardancy of
D-1433	Rate of Burning and/or Extent and Time of Burning of Flexible Thin Plastic Sheeting Supported on a 45 degree Incline
E-69	Treated Wood, Combustible Properties of by the Fire Tube Apparatus
E-84	Surface Burning Characteristics of Building Materials
E-108	Roof Coverings, Fire Tests of
E-119	Building Construction and Materials, Fire Tests of
E-136	Behavior of Materials in Vertical Tube Furnace at 750 °C
E-152	Door Assemblies, Fire Tests of
E-160	Treated Wood, Combustible Properties of, by the Fire Tube Apparatus
E-162	Surface Flammability of Materials, Using a Radiant Energy Source

E-163 Window Assemblies, Fire Tests of
E-286 Surface Flammability of Building Materials Using an 8 ft (2.44 m) Tunnel Furnace
E-648 Critical Radiant Flux of Floor Covering Systems Using a Radiant Heat Energy Source
E-662 Specific Optical Density of Smoke Generated by Solid Materials
E-736 Cohesion/Adhesion of Sprayed Fire-Resistive Materials Applied to Structural Members

Laboratory tests must be used with caution, as in many instances, what is a suitable test for one material is not suitable for another. To illustrate the care that must be taken the following was published to warn the public of improper use of some of the tests listed previously.

Important Notice Regarding The Flammability of Cellular Plastics Used in Building Construction, and Low Density Cellular Plastics Used in Furniture

The flammability characteristics of cellular plastics used in building construction, and low density cellular plastics used in furniture are tested under numerous test methods and standards. Included among these are ASTM D-568, 635, 757, 1433, E-84, 162 and 286; UL 94 and 723; and NFPA 255. The Federal Trade Commission considers that these standards are not accurate indicators of the performance of the tested materials under actual fire conditions, and that they are only valid as a measurement of the performance of such materials under specific, controlled test conditions. The terminology associated with the above tests or standards, such as "non-burning", "self-extinguishing", "noncombustible" or "25 (or any other) flame spread" is not intended to reflect hazards presented by such products under actual fire conditions. Moreover, some hazards associated with numerical flame spread ratings for such products derived from test methods and standards may be significantly greater than those which would be expected of other products with the same numerical rating.

The Commission considers that under actual fire conditions, such products, if allowed to remain exposed or unprotected, will under some circumstances, produce rapid flame spread, quick flashover, toxic or flammable gases, dense smoke and intense and immediate heat and may present a serious fire hazard. The manufacturer of the particular product or The Society of the Plastics Industry, Inc., should be consulted for instructions for use to minimize the risks that may be involved in the use of these products.

The Federal Trade Commission, Washington, D.C. 20580, requested that any representation that is inconsistent with the terms of this Notice be brought to its attention.

This Notice is distributed by The Society of the Plastics Industry, Inc., 250 Park Avenue, New York, NY 10017.

Not many standard tests exist for determining the potential fire hazards to occupants of residences, offices, hotels, factories and places of entertainment.

Few Municipal or State Fire Commissions are provided with monies for research or testing. In most instances they are the most *under-staffed* department in government. Because of this, the codes they prepare are based on laboratory tests of one or more laboratories. As they are not staffed to make a safety engineering study of every building constructed in their jurisdiction, building codes are limited to very minimal restrictions. The fact that a building passes existing building codes does not mean that it is free of fire hazards. It is the responsibility of architects and engineers to design residences and buildings safely for the use and occupancy intended. It is the responsibility of the owners to maintain safe conditions in a building.

To assist the users of insulation, this book will list combustible properties of materials as can be determined by standard or nonstandard tests at the time of its writing. However, use of this information must be applied and evaluated with occupancy of the building. In many instances hazardous practices are noted but it was not intended that this publication would be used exclusively to design safe residences and buildings.

To further illustrate the poor position we are in regarding safety of occupants of buildings, just a few years ago a very deceptive test method was used to evaluate foamed plastics used as carpet padding, cushions and insulation. Based on this test, these highly combustible materials which produce toxic gas when burned were given evaluation ratings as "non-burning" and "self-extinguishing". Labels with these terms are on furniture and will be for many years, giving a false sense of security.

As to terminology, the words used to describe fire and combustion properties have been misused by manufacturers to mislead the public. Mr. C. H. Yuill presented an article in "Materials Research and Standards", October, 1970. In this article it is pointed out that the term "Fireproof" has been officially dropped from NFPA publication as misleading. Also "noncombustible" has been distorted through application to apply to materials having limited degrees of flammability. Likewise, after fire retardants are added to combustible materials to slow down their burning rate, many manufacturers then describe these slower burning products as "fire retardant" or "flame retardant", which they are not.

Because of the lack of understanding terms related to thermal insulation properties and properties of combustion, a Glossary is located in Appendix A.

The properties of materials as related to fire and use of materials for specific purposes follow.

COMBUSTIBILITY OF MATERIALS

It is very difficult to provide direct comparison of the relative combustibility of materials because of wide differences of relationships of similar properties. For example, E-84 test is used to measure the flame spread, fuel contribution and smoke density of materials being burned on the underside in horizontal positions as compared to dry, red oak. In the case of red oak, the ignition point of the gases is relatively close to gasification temperature. When this test is used to test a number of foamed plastics, such as foamed polyurethane, the gasification temperature is so much lower that the combustibles are blown out from the test before they reach ignition point. In the case of foamed styrene, its melting point is so low that it melts and flows down the tunnel before under-slab ignition occurs. The statement from the Federal Trade Commission that this test was unsuitable to test cellular plastics used in furniture is also true of foamed-plastic insulations.

In the case of plastics which gasify at low temperature before they ignite, these combustible gases can collect in the top of a room or building and ignition of the same is called "flash over combustion". It is a small order explosion and spreads fire over a wide area in a small instant of time.

The faster a material burns the greater its potential hazard. This is why flame-spread characteristics are so important. Unfortunately, this is one of the most difficult properties to measure when comparing one material to another because temperature and direction all affect results and cause inconsistancies.

The amount of heat released from a burning and the rate the heat is released is relative to fire safety. Heat release, amount of heat and spread of burning influence the severity of potential injuries and deaths and the extent of property damage.

Smoke is another important consideration of fire safety. Smoke affects visibility, which is a factor in the ability of occupants to find means of exiting a burning building or residence.

The toxicity of the gases being released from burning materials is, of course, of utmost importance. Almost all individuals that lose their lives in a fire are killed by the toxic poisonous gases before the heat of the fire can cause death. In addition, these gases cause occupants to lose their ability to try to escape.

The reason that these factors are being put into this book is that if the construction of a building is of non-combustible materials, including the thermal insulation, it is considered fire safe. This is not true if its contents are combustible, and means for controlling the fire and suitable escape exits are not provided. However, proper thermal insulation can be used to assist in containment of fire and to protect structural members in a building.

To establish relationships of combustibility, tables giving the heat release of very combustible materials such as flammable gases and liquids and commonly used materials such as wood, plastics and cotton are provided. Other tables giving the properties of thermal insulation are presented so as to guide the designer in choosing materials for the individual residence or building.

It should also be pointed out that the values given in the tables are average values obtained from a number of references which use various test methods. Likewise, it was impossible to determine the exact formulation of each material tested—thus each value is that for typical material of the type listed.

Table 51 lists various solids, liquids and gases used as fuels and their heat release when burned. Also listed are their self-ignition temperatures. This is the temperature at which ignition will occur without spark or flame impingement.

Table 52 is a condensation of references as to the combustibility of materials used in large quantities in the interiors of residences, office buildings and hotels. Comparison to Table 51 shows that materials used in furniture, rugs, cushions, rug pads, etc., are combustible and contain considerable energy for heat release during combustion.

Because of the thousands of fabrics, rugs, furniture and different formulas used in the production of foams and fabrics, Table 52, by necessity, is averaged out for the generic type of material listed. For this reason, the particular products used in a building should be evaluated as to their fire hazard so that safety provisions can be determined. These two tables are furnished in this book to show that the fire hazards in buildings are determined as much, or more, by their contents as by their construction. Even though the shell of a building is fire resistant, it should be designed so that accidental fire is confined and extinguished.

Unfortunately, few residences or buildings are fire safe. In many cases thermal insulations which are improperly used con-

COMBUSTIBILITY PROPERTIES OF MATERIALS

FUELS				
FUEL	HEAT RELEASED BY COMBUSTION		SELF IGNITION TEMP.	
	Btu per Pound	Joule per Kilogram	°F	°C
Acetylene C_2H_2	21,500	56,459,000	580 to 825	307 to 441
Anthrocite Coal	14,500	38,077,000	840 to 1100	449 to 599
Bituminous Coal	13,500	35,451,000	760 to 850	404 to 454
Charcoal	9,500	24,947,000	600 to 800	316 to 427
Ethane C_2H_6	22,300	58,559,800	880 to 1100	471 to 593
Gasoline Vapor	18,000	47,268,000	500 to 800	260 to 427
Kerosene Vapor	17,500	45,955,000	490 to 560	254 to 293
Methane CH_2	21,500	56,459,000	1170 to 1380	632 to 749
Wood	8,000	21,008,000	500 to 550	260 to 288

HEAT RELEASE AND SELF IGNITION TEMPERATURE OF FUELS

TABLE 51

MATERIALS USED IN FURNITURE, FURNISHINGS OR PANELS INSIDE RESIDENCES, OFFICE BUILDINGS AND HOTELS

MATERIAL	IGNITION TIME* SECONDS	HEAT RELEASED BY COMBUSTION		TOXICITY Minutes to Death**
		Btu per Pound	Joule per Kilogram	
Cellulose Fiberboard	16	9,000	20,916,000	19.9
Chipboard	23	9,500	24,947,000	18.2
Cotton	12	8,300	19,289,200	14.2
Hardboard	29	9,000	20,916,000	15.9
Nylon Carpeting	42	11,000	25,564,000	12.5
Nylon Fabric	26	11,000	25,564,000	13.5
Paper	10	8,500	19,754,000	16.0
Polyiscyanutate Foam	72	16,000	37,184,000	22.5
Polyurethane Cushion Foam	48	16,000	37,184,000	9.2
Polyvinyl Flooring Tile	95	7,300	16,965,200	7.3
Rayon	16	11,000	25,564,000	8.1
Rubber Cushion Foam	18	12,500	29,050,000	9.6
Silk	16	8,600	20,218,800	9.4
Wood-Cedar	30	8,000	21,008,000	11.7
Wood-Fir	40	8,300	19,056,800	13.6
Wood-Oak	24	8,000	21,008,000	14.5
Wood-Pine	30	8,600	19,986,400	15.4
Wool Carpeting	19	7,900	18,359,600	6.8
Wool Fabric	52	7,900	18,359,600	6.5

* Based on subjecting material to heat flux of 10.5 W/cm²

** Based on pyrolysis tube, pyrolysis boat and rat exposure chamber and described in article by Carlos J. Hilado and Heather J. Cumming, *Fire and Materials* Vol. 2, No. 2, 1978.

IGNITION TIME, HEAT RELEASE AND TOXICITY OF GASES OF MATERIALS USED INSIDE RESIDENCES, BUILDINGS AND HOTELS

TABLE 52

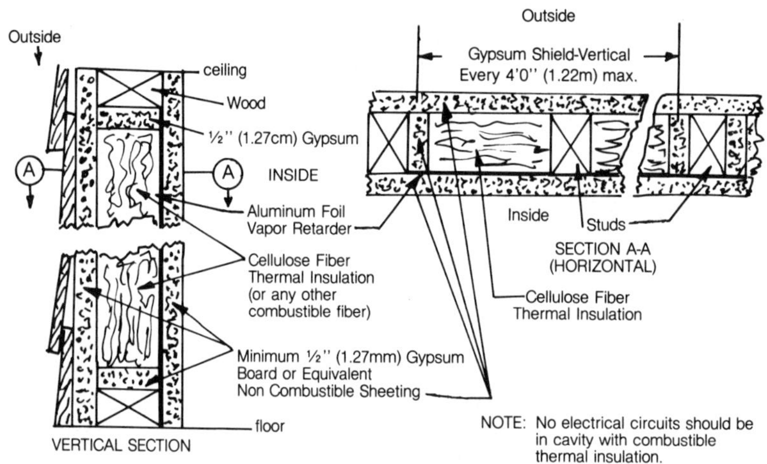

INSTALLATION OF FIRE BARRIER AROUND COMBUSTIBLE FIBER INSULATION IN FRAME WALL

Figure 68

COMBUSTION PROPERTIES OF COMBUSTIBLE THERMAL INSULATIONS

THERMAL INSULATION	FORM	FLASH POINT TEMP.*		FLAME POINT TEMP.**		HEAT RELEASED BY COMBUSTION	
		°F	°C	°F	°C	Btu per lb.	Joule per kg
Cork, Vegetable Molded	Semi-Rigid	575	303	575	303	8,500	19,771,000
Glass Fiber (Binder is combustible)	Semi-Rigid	750	399	750	390	Smolders	Smolders
Mineral Fibers (Binders will oxidize)	Semi-Rigid	***	***	***	***	Can Smolder	Can Smolder
Phenolic Foam	Semi-Rigid	780	416	850	454	13,000	30,238,000
Polystyrene (Melts before burning)	Semi-Rigid	730	388	750	399	16,000	37,216,000
Polyurethane Foam	Semi-Rigid	590	310	600	316	16,500	38,379,000
Wood fibers with binders	Semi-Rigid	500	260	500	260	8,000	18,608,080
Elastameric sheets & pipe cov.	Flexible	560	293	590	293	15,000	34,890,000
Glass Fiber (Binder is combustible)	Flexible	750	399	750	399	Smolders	Smolders
Polyurethane-Air Sprayed	Sprayed	590	310	600	316	16,000	37,216,000
Urea-Formaldehyde	Foamed	***	***	1280	654	Slow burning	Slow burning
Cork Granulated	Loose	570	299	620	327	8,000	18,608,000
Cellulose Fibers	Loose	500	260	500	260	8,000	18,608,000
Gilsonite		600	316	600	316	17,000	39,542,000

*Flash point is temperature at which materials will emit gases that can flash into flame when ignited by spark or flame above.

**Flame point is temperature of materials that flash flame will continue to burn.

***When ignited, binders will smolder.

FLASH POINT, FLAME POINT AND HEAT RELEASE OF COMBUSTIBLE INSULATIONS

TABLE 53

tribute to the hazard of a residence, office building, hotel or commercial building. For this reason, it is necessary to evaluate the combustibility characteristics of thermal insulations, as well as other materials. Because combustible insulations are subject to being heated to temperatures above ambient temperature, the flash-point temperatures and fire-point temperatures are given in Table 53.

Fire losses in the United States of residences and buildings are billions of dollars each year. Over 6,000 persons a year are killed by fires in homes, buildings and hotels. It is imperative that when combustible insulations are used in homes and buildings, that they be used in a safe manner. The following are given examples of dangerous ways in which combustible insulations are used. Also, recommendations are given as to installation methods to reduce fire hazards.

USE OF COMBUSTIBLE INSULATIONS IN RESIDENCES AND BUILDINGS

WALLS

When fiber cellulose insulation is used in walls, it should be enclosed on both sides, top and bottom with a non-combustible material, such as, 1/2″ thick fire-resistant gypsum sheet or board. This would be equally true of any other organic fibers or inorganic fibers which contain combustible binders or oil lubricants that are combustible.

When installed in a frame wall, both sides can be fire protected by gypsum board, minimum 1/2″ (1.27mm) thick. This will keep the temperature of the combustible below the self-ignition point for approximately 15 minutes. Installation of fire barriers around combustible fiber insulation in frame walls is shown in Figure 68.

Likewise, rigid organic foam must be protected from fire with minimum 1/2″ (1.27mm) gypsum board from inside or outside a house. The installation of fire barriers on both sides of the organic foam on the outside frame wall is shown in Figure 69. This is confirmed by the Urethane Foam Contractors Association, Position Statement, shown below.

Thermal Barriers
for
Rigid Polyurethane Foams*

It is the UFCA stated policy that all rigid foams for interiors of buildings where life safety is a factor are required to comply with applicable building codes.

*NOTE: UFCA has developed several standardized foams which can be utilized to help assure compliance with the applicable building codes.

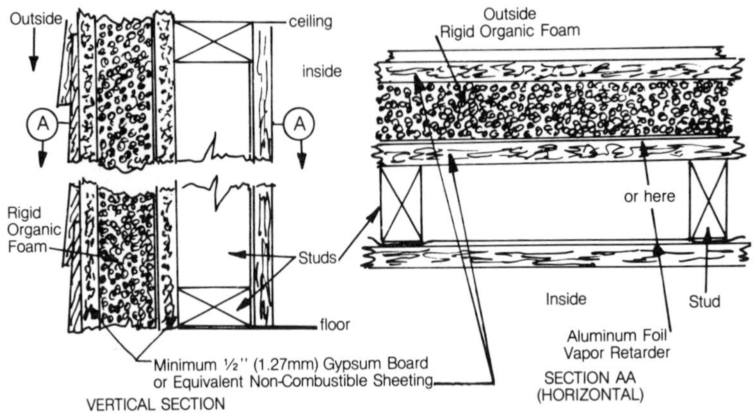

INSTALLATION OF FIRE BARRIER ON BOTH SIDES OF ORGANIC FOAM ON OUTSIDE OF FRAME WALL

Figure 69

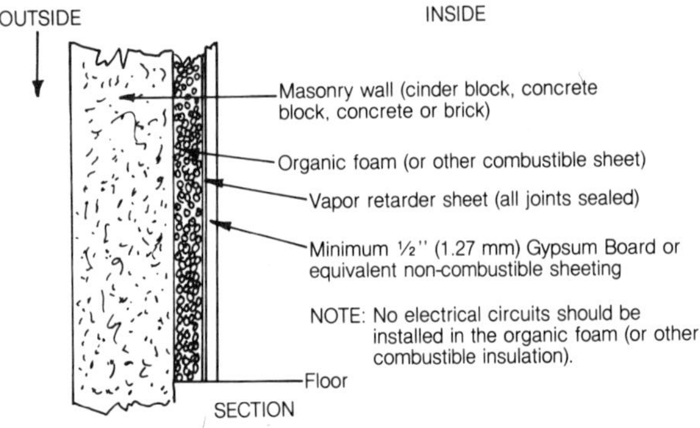

INSTALLATION OF FIRE BARRIER OVER COMBUSTIBLE INSULATION INSIDE MASONRY WALL

Figure 70

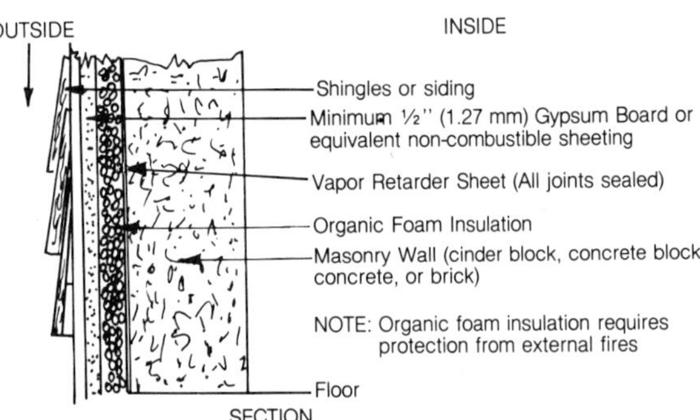

INSTALLATION OF FIRE BARRIER OVER ORGANIC FOAM INSULATION OUTSIDE MASONRY WALL

Figure 71

All model building codes require that rigid foam plastic insula tion be protected by a thermal barrier equal in fire resistance t 1/2" gypsum board. Such thermal barriers are those that limit th temperature rise of the unexposed surface to not more than 25 degrees above ambient after 15 minutes of fire exposure comply ing with the ASTM E-119 standard time-temperature curve and re main in place during these test conditions. Thermal barriers tha have been evaluated by diversified tests may also be acceptabl to the local building code official.

Where building codes are applicable, foam contractors shoul get specific approval from the local building code official showin compliance with the local building code. Where no building code exist, it is still UFCA policy to apply thermal barriers for your ow protection and the safety of others.

Where organic foam, combustible sheets or boards are use on the interior or exterior of masonry constructed walls, they als should be protected from fire exposure.

In the interior the need for fire protection of highly combustib materials is obvious. The heat and toxic gas release of burnin combustible insulations are very hazardous.

It is necessary to provide fire protection for organic foams i stalled on the outer surface of a masonry wall. This is becaus even simple things, such as a trash fire, will ignite organic foa very quickly and as its flame spreads very rapidly, the fire cou spread upward to the roof in a very short period of time. Figure 7 shows a method of installing the fire barrier over combustible i sulations on the inside of a masonry wall. Figure 71 shows method of installing the fire barrier over organic foam insulati on the outside of a masonry wall.

WALLS — METAL

One of the misconceptions regarding organic foam is that it can be sprayed on a sheet of steel applied to metal building walls and roof and the metal will protect it from external fire. Metal is very poor thermal insulation and will transmit heat from an external fire impingement very rapidly. Because fires can start in this manner, the following burn test is given.

Test to Establish Combustibility from Outside Trash Fire of Organic Foam Installed Inside a Metal Wall

A burn test of simulated wall panel was performed on 2′0″ x 3′0″ sections of corrugated steel exterior wall panels which had been insulated with polyurethane foam on the interior surface. The exposed surface of foam was protected with an intumescent coating.

The 2′0″ x 3′0″ section of corrugated metal insulated with 2″ thick sprayed-on polyurethane was installed in a fire-resistant panel so that approximately 2′6″ x 1′9″ was exposed to the fire impingement from the outside. The fire was produced by burning 50 pieces of spruce arranged in 19″ cubic geometry and 7 pieces were placed leaning against the metal wall panel. The assembly was on solid concrete blocks representing the building foundation. The test section was held in position over the opening of the fire shield by heavy steel straps. All potential openings were sealed with high-temperature mineral fiber blankets which were compressed to prevent passage of sparks, flame or heated gases.

Thermocouples were installed in the rib and flat sections of the corrugated siding to measure the temperature of the steel. Thermocouples were located under the insulation on the side away from the fire.

TEST RESULTS: The temperature of the ambient air was 44°F (6.6°C). After the ignition of the wooden crib, the temperature of the steel on which the flame impingement occurred began to increase. The temperature of the steel exposed to the fire was:

Time Exposed to Fire, Min-Sec	Temperature of Steel Rib Section	Temperature of Steel Flat Section	Observations/Conditions Side Away from Fire
0:00	44°F(6.7°C)	44°F(6.7°C)	
1:10	726°F(385°C)	408°F(208°C)	
2:00	1052°F(567°C)	1048°F(563°C)	Organic foam begins to burn at base
3:00	1185°F(641°C)	1220°F(000°C)	Smoke emitted
4:00	1259°F(676°C)	1472°F(800°C)	
5:00	1380°F(749°C)	1647°F(897°C)	Voluminous smoke and flame
6:00	1353°F(845°C)	1510°F(821°C)	All polyurethane insulation burned
10:00	1253°F(676°C)	724°F(385°C)	Some smoking of charred material

As can be noticed, the temperature of tne steel exceeded that at which, when subjected to load, it would have failed.

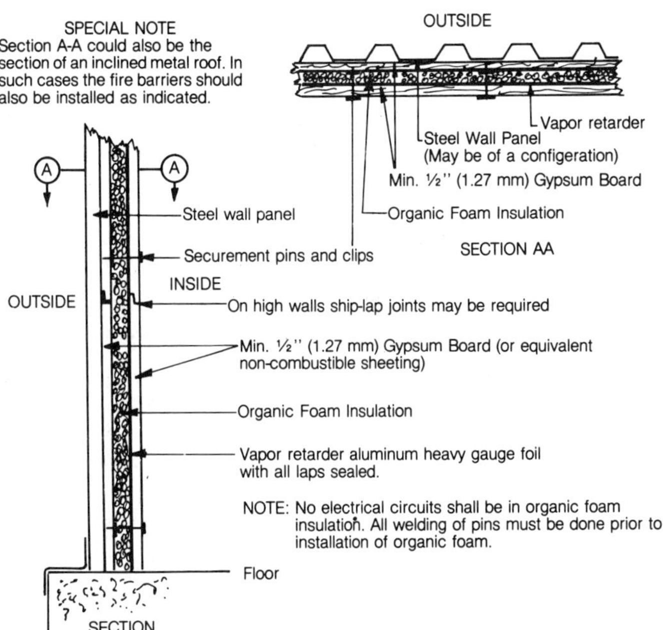

INSTALLATION OF FIRE BARRIERS OVER ORGANIC FOAM INSULATION USED IN METAL WALL PANELS

Figure 72

A fire on the inside of a building would also ignite organic foam (or other combustible sheet insulation) even if the interior were faced with metal. Thus, it is necessary, for safety reasons, to provide fire barriers of non-combustible material on both sides of the combustible insulation when installed in metal panel walls or sloped roofs. Metals have such a high thermal conductivity that, when backed with an insulation, flame impingement raises their temperature very rapidly. For this reason, thin metals are ineffective as a fire barrier, especially when they are in contact with a combustible material. Methods of installing fire barriers in metal panel walls and inclined metal roofs are shown in Figure 72.

FLOORS

Floors constructed of wood or other combustible materials also contribute to fire spread in a residence or building. Such floors over crawl space or other unheated areas now are being insulated. Where the insulation is combustible, cellulose boards, fiber or organic foam sheets contribute to the fire hazard. It should be noted that although the test conditions, as given in ASTM-84 (UL-723) "Test for Surface Burning Characteristics of Building Materials" are similar to expected fire exposure, this test is *unsuitable* for evaluation of foamed organic insulations.*

To reduce the hazard of fire spreading upward when combustible fiber blankets or batts are installed between the floor joists, a 1/2" (1.27 mm) minimum thick gypsum board should be installed under the joists. This is shown in Figure 73.

*REFERENCE: Federal Trade Commission's Cellular Plastics Rule of July 22, 1975.

If organic foam insulation is used under the floor, it should be protected from fire below with a 1/2" (1.27 mm) minimum thick gypsum board. This is illustrated in Figure 74. If there are any ducts or wiring in the joist spaces, then the fire protective gypsum boards must be installed on both sides of the organic foam insulation for adequate fire protection.

ROOFS — INCLINED

Frame-constructed roofs, including those with insulation, should be fire protected in a similar manner as described above for floors and as illustrated in Figures 73 and 74.

ROOFS — FLAT

Flat roofs constitute a very difficult fire-safety problem, as roofing felts, asphalts and other bituminous coatings used as the roofing membranes are very combustible.

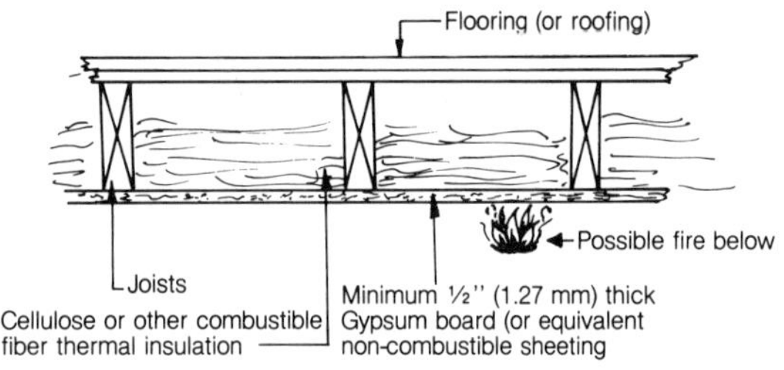

NOTE: No electrical circuits shall be run in joist area containing combustible fiber insulation.

FLOOR SECTION* OVER CRAWL SPACES OR OTHER OPEN AREAS

INSTALLATION OF FIRE BARRIER BELOW COMBUSTIBLE FIBEROUS INSULATION IN JOIST SPACE

Figure 73

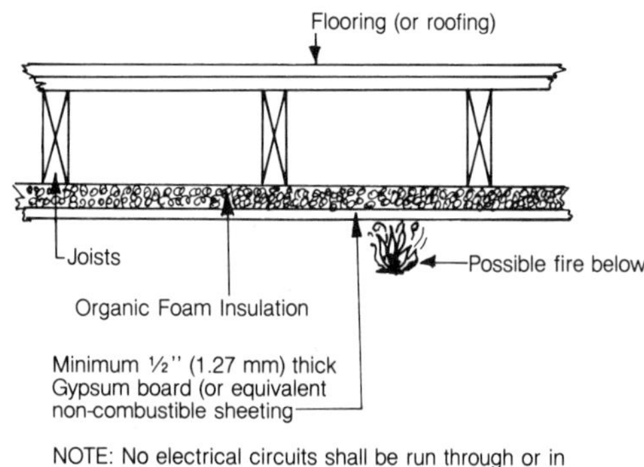

NOTE: No electrical circuits shall be run through or in area above organic foam insulation.

FLOOR SECTION* OVER CRAWL SPACES OR OTHER OPEN AREAS

INSTALLATION OF FIRE BARRIER BELOW ORGANIC FOAM INSULATION ATTACHED TO JOISTS

Figure 74

*NOTE: Also represents section of inclined frame roofs

When combustible insulations are used in flat roofs, they increase the potential fire hazard unless some provision is made to retard their ignition by fire from above and below. The most common method of reducing the fire hazard from fires above and to restrict the burning rate of built-up roofing is to cover the roof with a minimum 3" (7.62 mm) thickness of stone or gravel. However, this will not protect combustible insulation from ignition from below.

If the insulation is applied over a concrete deck, the concrete will protect the insulation from fires underneath it. However, combustible insulation installed over wood decking or metal panel decking requires additional fire protection. A sheeting of 1/2" (1.27 mm) minimum thickness of gypsum board or equivalent fire-resistant sheeting should be installed over the wood decking or metal panel decking prior to installing the combustible insulation on the flat roof. Installation of fire barrier on a frame flat roof is shown in Figure 75 and on a metal panel deck in Figure 76. In addition to protection of the insulation, the wood joists and decking should also be protected with fire-resistant panels attached under the joists.

The illustrations depict installation of a fire barrier to retard the ignition of combustible insulation and to give minimum protection. To illustrate, the 1/2" (1.27 mm) gypsum board with ordinary fire impingement on one side will keep the organic foam below its ignition temperature for 10 to 15 minutes. This provides time for the fire to be extinguished or burn out.

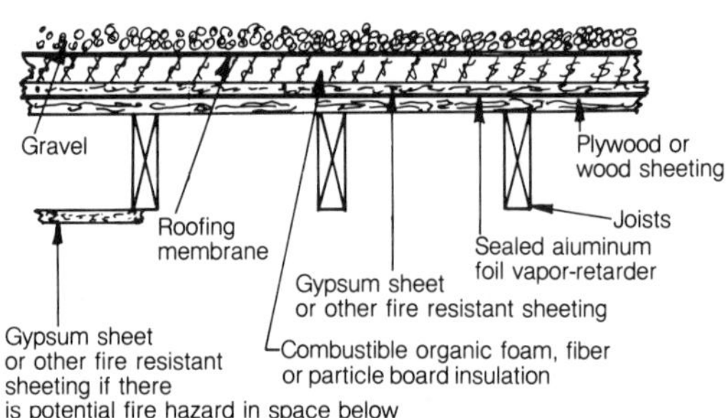

INSTALLATION OF FIRE BARRIERS — FRAME FLAT ROOF WITH COMBUSTIBLE INSULATION

Figure 75

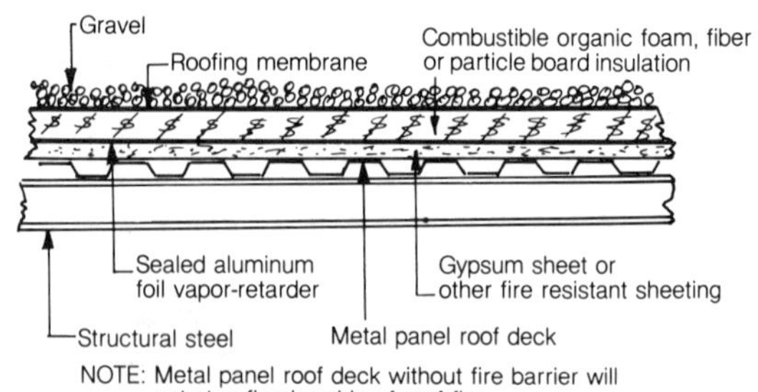

NOTE: Metal panel roof deck without fire barrier will not stop flaming drip of roof fire.

INSTALLATION OF FIRE BARRIERS — METAL PANEL ROOF WITH COMBUSTIBLE INSULATION

Figure 76

FIRE PROTECTION OF STRUCTURAL MEMBERS

Steel structural members are subject to failure when exposed to flame. Ordinary flame temperatures of burning wood, trash, etc., exceed 2000 °F (1093 °C). Above 600 °F (316 °C) steel begins to lose its strength rapidly. Over 1000 °F (538 °C) it loses its load-bearing capacity. When exposed to ordinary fire exposure, depending on the size, flame impingement, and flame temperature, unprotected steel will reach this critical 1000 °F (538 °C) in four to six minutes. The effect of temperature on strength of steel is shown in Figure 77.

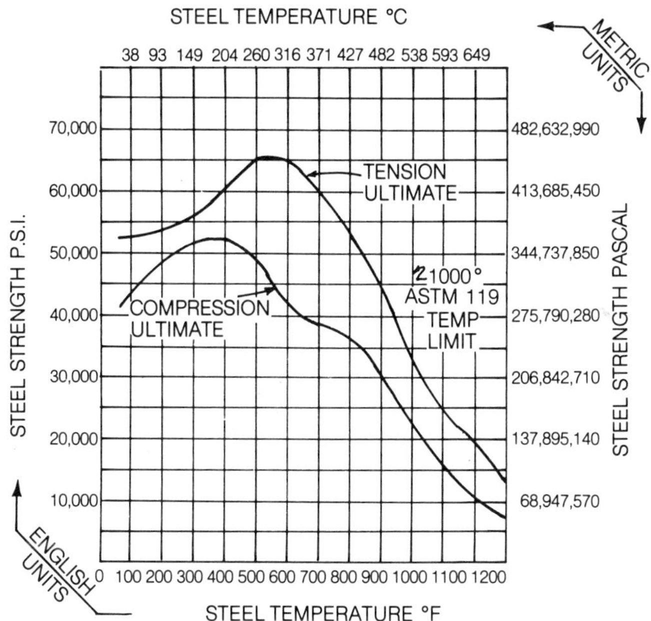

STRENGTH OF STRUCTURAL STEEL UNDER FIRE EXPOSURE

Figure 77

The materials that provide fire protection require almost completely different physical properties to function effectively than does lightweight thermal insulation. Although low thermal conductivity is important, in order for a material to be effective as a fire barrier it needs a low thermal diffusivity, high mass, and a relatively high specific heat. Another helpful property is the ability to hold quantities of moisture by being hydroscopic. Since a minimum of 1150 Btu's is required to evaporate a pound of water, a material containing even a little water requires a lot of heat to dry it before its temperature rises above 212 °F (100 °C).

Because of shrinkage of the protective materials when block or boards or sheets are used over structural members, broken joint construction is recommended. Where poured materials, such as cement are used, then reinforcing metal mesh or wire is recommended. There are so many shapes, configurations and combinations of structural members as columns, beams, bents, etc., that it is impossible to present detailed recommendations in this book.

The most commonly used materials to retard the heat flow into structural members are:

1. Portland cement concrete
2. Lumnite cement concrete
3. Gypsum plaster
4. Gypsum plaster board
5. Perlite plaster
6. Vermiculite plaster
7. High temperature, heavy density, mineral wool fiber sprayed on
8. Concrete block
9. Concrete slabs
10. Cinder blocks
11. Brick

Using these materials to a minimum 3″ (6.45 mm) thickness over the outer surface of steel members will slow down the rise in temperature, in ordinary fire, so that the steel retains its strength for two to four hours. Of course, if surfaces are wet down with water from sprinkler systems or hose, the outside surface temperature will not exceed 212 °F (100 °C). However, even with sprinkler systems, structural metal should be fire protected with fire-resistant, thermal-resistant materials, as there is no way to make sure all surfaces are wet and there will be water available at every sprinkler head.

CONTAINMENT OF FIRE TO SMALL AREAS

The design of office buildings, stores and hotels has been that of partition type walls with ducts, electrical cables, etc., running above the ceiling and the floor above. In case of fire, flames could travel over large areas with no restriction.

In buildings of relatively high occupancy with combustible furnishings and contents, fire containment walls should be provided to restrict spread of fire. These walls should be constructed of one or more of the materials listed above to protect steel members.

Air-conditioning ducts can also spread a fire. In buildings of large occupancy, when these ducts are insulated, they should be insulated with non-combustible insulations. In addition, where they pass through fire containment walls, they should be equipped with automatic interior fire doors.

ELECTRICAL HAZARDS

Electrical wire cables must dissipate to the surrounding ambient air the heat generated by the current passing through them. Too much current or too great a thermal resistance to the heat flow from the cable surface to the air will cause the temperature to rise to the point of failure (either the wire or the electrical insulation). Electrical cables and wires are rated on their ability to carry a certain current without overheating. For these reasons, thermal insulation should never be packed around wire and cables unless the current capacity is reduced—in most cases to 1/4 or 1/3—at the fuse box or circuit breaker. In construction of many buildings, especially mobile homes, this basic electrical hazard is forgotten. Insulation cut back from electrical appliance and switch box is shown in Figures 78 and 79.

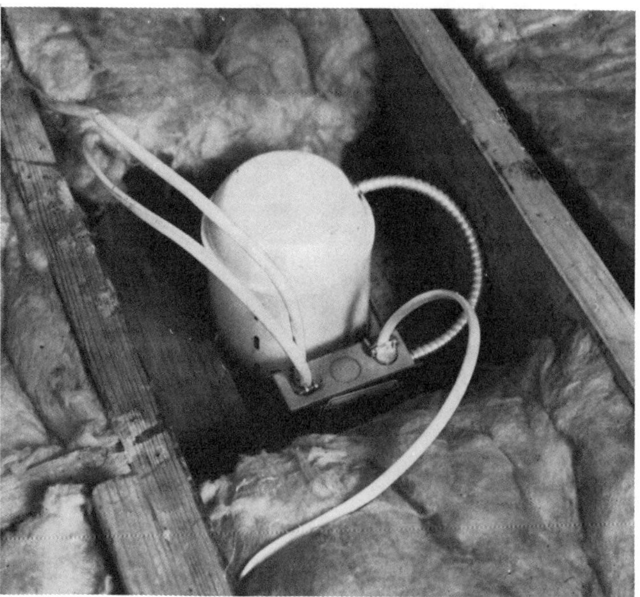

Courtesy of Owens-Corning Fiberglass Corp., Toledo, Ohio

GLASS FIBER INSULATION CUT BACK FROM ELECTRICAL APPLIANCE

Figure 78

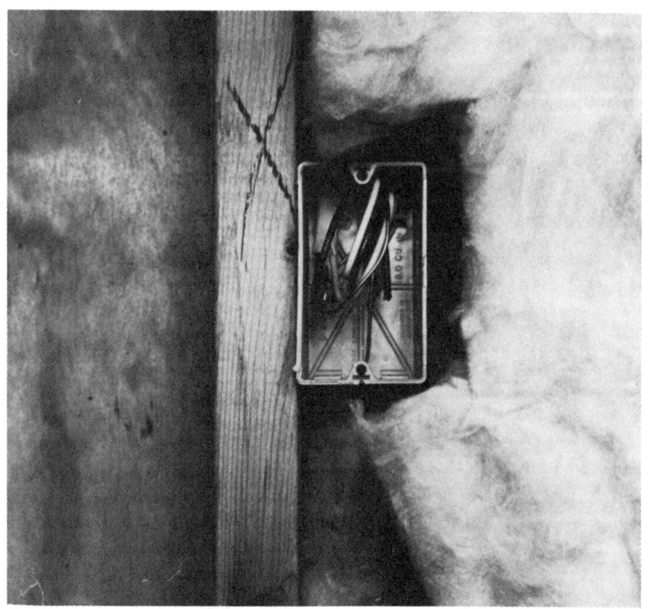

Courtesy of Owens-Corning Fiberglass Corp., Toledo, Ohio

GLASS FIBER INSULATION CUT BACK FROM SWITCH BOX

Figure 79

ELECTRICAL WIRES OR CABLE RUN THROUGH NON-COMBUSTIBLE INSULATION

When insulation is installed in joists or stud spaces where wiring is present, a space should be formed around the wires with stove-pipe or sheet metal so as to allow air circulation around the wiring. The air space around the cable should be 2 1/2″ to 3″ (6.35 to 7.62 mm) to allow air movement and provide sufficient external surface area for heat transfer through the insulation. This is basically true for ceiling areas, areas below floors and insulated exterior walls. Illustration of a metal jacket used around wire or cable to provide air space through an insulated area is shown in Figure 80.

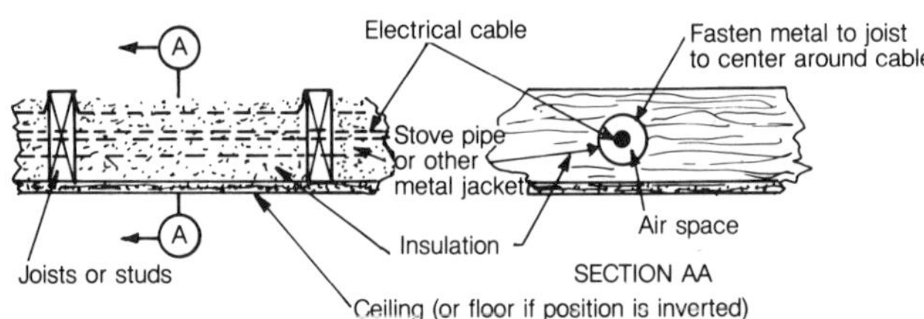

Cable or Wire Run Through Holes in Joists

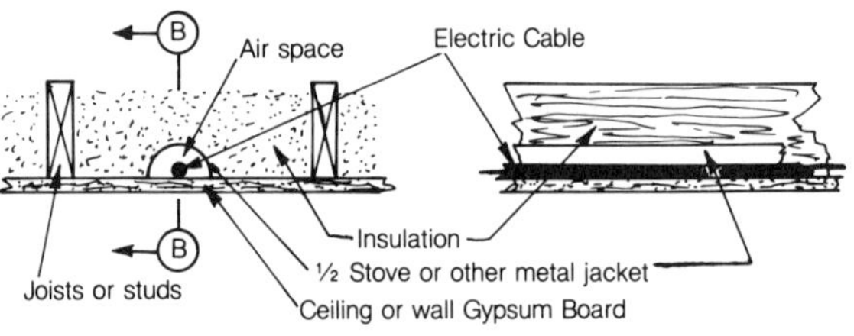

Cable or Wire Run Against Gypsum Board

METAL JACKET USED AROUND WIRE OR CABLE TO PROVIDE AIR SPACE IN INSULATION

Figure 80

ELECTRICAL WIRES OR CABLES RUN THROUGH COMBUSTIBLE INSULATION

All thermal insulations that are effective in reducing heat transfer will cause wires or cables embedded or encased in them to heat to higher temperatures than if installed in air. The tests for current capacity of wires are based on the outside surface area being able to transfer the heat by radiation, convection and conduction to surrounding air and objects at atmospheric temperature. In many instances, these tests are based on 80°F (26.7°C) ambient air. When enclosed with low thermal conductivity insulation, wires which are subjected to their "rated" load will rise in temperature above what the electrical insulation will tolerate. When the electrical insulation fails, a short or grounding of the electrical wires will occur, causing sparks and fire.

When encased or embedded in combustible fiber or foam insulation, the electrical heating and sparks will cause the insulation to ignite. At this time, organic styrene foam may melt into a combustible liquid and organic polyurethane will burn as a solid. *For this reason, no electrical circuits should be run where they are en-*

cased, surrounded, or in direct contact with combustible insulation.

In many instances the walls of homes have been filled with combustible insulation blown in around the wiring. In other instances, such as mobile homes, wire has been installed and then polyurethane foam or urea-formaldehyde foam poured or filled around it. In other instances, foam panels are grooved out to fit around the wiring. Where such a hazard exists, the wiring should be disconnected and installed inside, away from the hazardous insulation.

Where it is necessary to put circuits inside the walls of residences and mobile homes to reduce the hazard of electrical circuits in the wall interior, wire-conduit systems are now available. Such a conduit box for mounting on the wall, including the receptacles, is shown in Figure 79. Also shown are typical installations inside a room at the floor level.

SURFACE METAL RACEWAY WITH RECEPTACLES

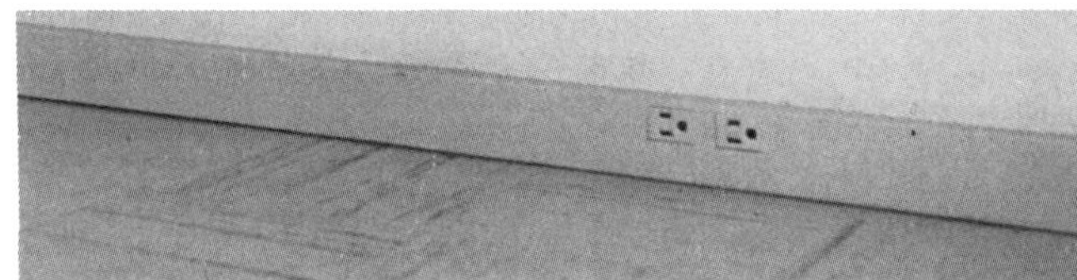

Courtesy of The Wiremold Co., West Hartford, CT

BASEBOARD INSTALLATIONS OF SURFACE METAL RACEWAYS

Figure 81

CABLE TRAYS

Inside large buildings and industrial plants, cables are grouped and supported on trays. These grouped electrical cables present a dangerous fire hazard. The so called "flame-retardant" plastic electrical insulation burns and fails rapidly when subjected to exterior fire. This causes the cables to short circuit and cease functioning, increasing the fire. When the cables no longer deliver power, all pumps, equipment and computers cease operation. This may cause secondary fires or explosions.

In one major company such a fire caused a loss of millions of dollars in property damage, materials and time out of production.

Based on this experience, it was decided to fire protect the trays of cables by enclosing them with 1" (2.54 mm) thick high temperature silica-asbestos board. Unfortunately, due to the fact that the fire-protection board was thermally efficient, the current carrying capacity of the cables was reduced to 50% of the rated load because of overheating. For this reason, the fire protective board had to be removed.

The only solution then remaining was a water-sprinkling system to extinguish the fires and cool the cables to keep the electrical insulation below its failure temperature. Unfortunately, another fire proved that this system failed. The failure temperature of the cable insulation was reached before water-spray systems were activated by the temperature of the fire. To solve this problem, a coating was developed with a sufficient time lag to minimize the temperature rise of the electrical insulation until the water systems could function. It did not lower the current capacity of the cables.

The coating was made of refractory solids and emulsion-type binders that required considerable heat for decomposition. When this coating is installed on cables it absorbs heat during decomposition, thus holding the temperature of the cable below its failure for a minimum of 15 minutes under average fire conditions.

The outside temperature of the electrical insulation was measured by thermocouples as the cables were heated while carrying their rated amperage. Tests were run until temperature equilibrium was established. In the following test, temperatures of the 12 locations were recorded before and after applying the non-flammable latex emulsion cable coating.

CABLES WITHOUT COATING													
	THERMOCOUPLE NUMBERS												
AMPERES	1	2	3	4	5	6	7	8	9	10	11	12	UNITS
87	105°F	106°F	102°F	115°F	106°F	112°F	118°F	106°F	105°F	108°F	115°F	116°F	English
87	40.6°C	37.8°C	38.9°C	46.1°C	41.1°C	44.4°C	47.8°C	41.1°C	40.6°C	40.6°C	46.1°C	46.7°C	Metric

CABLES AFTER FIRE PROTECTIVE COATING WAS APPLIED													
	THERMOCOUPLES NUMBER												
AMPERES	1	2	3	4	5	6	7	8	9	10	11	12	UNITS
87	108°F	102°F	108°F	116°F	109°F	114°F	120°F	109°F	109°F	110°F	115°F	118°F	English
87	42.2°C	38.9°C	42.2°C	46.7°C	42.8°C	45.6°C	48.9°C	42.8°C	42.8°C	43.3°C	46.1°C	47.8°C	Metric

The coating was applied so as to obtain 1/16″ (0.159 mm) dried thickness. As indicated, the temperature rise on the outside of electrical insulation was relatively small, thus the current carry ing capacity of the cable was not reduced. An illustration of this cable coating installed on cables in the tray is shown in Figure 82.

FIRE EXTINGUISHMENT

Residences. Most residences are filled with combustible furnishings, such as wooden furniture, cushions and rug pads of very flammable polyurethane, drapes, clothes, etc. Few homes are built with automatic fire extinguishers. However, in most instances, the number of exits and the building heights are such that in case of fire the people can escape, if properly warned. All residences should be equipped with smoke and fire detectors to warn (or wake from sleep) the occupants in case of fire so that they can escape.

It is also advisable to have dry chemical fire extinguishers in the home so that small cooking fires, etc. can be rapidly extinguished. Where fire occurs in foamed organic cushions, mattresses, rugs, etc., it is almost impossible to put out these fires before being overcome by the toxic gases of combustion. For this reason, where such fires occur, the occupants should leave the residence immediately.

COMMERCIAL BUILDINGS

High-rise buildings, such as offices or hotels, may be constructed completely of metal and fire-resistant masonry material. All piping may also be protected with fire-resistant insulation and electrical circuits in metal conduit. These buildings are built to withstand fires without collapse. However, when used as offices, hotels, etc., these buildings become full of combustible materials, such as furniture, carpets, drapes and possibly tons of paper. In case of ignition, unless some means is provided for watering or chemically extinguishing the fire before it spreads, these buildings are fire traps for the occupants. All high-rise buildings, especially those with inside, enclosed so called fire escapes, must be equipped with correctly designed, installed and operative water-sprinkler systems to contain and extinguish fires before they spread and kill the occupants. To assist in the confinement of accidental fires, the space should be divided into relatively small areas, enclosed by fire-resistant building materials and fire-resistant thermal insulation.

In industrial buildings where processes may occur using combustible liquids and/or electrical circuits and transformers, water-sprinkler systems are unsuitable. In such cases, carbon-dioxide systems, properly designed, may be used to extinguish fires of these types.

Courtesy of Vimasco Corp., Nitro, W. Va.

NOTE: Cables with combustible electrical insulation so treated to retard flame spread and for protection until surfaces can be wetted by sprinkler system.

CABLES IN TRAY PROTECTED WITH FIRE RETARDANT COATING

Figure 82

7 Basic Properties of Thermal Insulation

There are two major classes of insulation which are used to insulate residences and buildings today. One type is *reflective* insulation and the other is *mass* or bulk insulation whose insulating properties are largely due to the air or gas spaces present in a solid or semi-solid material.

Reflective insulation is composed of parallel thin sheets, or foil, of high-thermal reflectance and so spaced as to reflect radiant heat back toward its source. The spacing is designed to provide a restrictive air space. The spacing should be of such dimensions that it causes resistance to air movement so that little heat is transferred from one surface to the other by *convection*. The overall heat transfer of reflective insulation is dependent upon its spacing, heat flow direction and method of installation. It is a system, rather than a single homogenous material. For this reason, its resistance to heat flow and thermal efficiences must be presented as a system installed with other components of construction.

Mass insulation consists of flake, fiberous, granular or cellular solids so composed or installed to form many air spaces. Although the space occupied is mostly air, the resultant product is considered to be a homogenous material having definite thermal properties which can be established. Many mass insulations are made with a combination of flake, fibers, granules, and binders. The insulation is produced in many forms to fulfill installation requirements. For example, mass insulation may be loose so that it can be poured into cavities or it may be rigid and relatively strong so as to resist loading, such as would be necessary for insulation under concrete slabs. In other installations it may be in blanket form so it can be applied between studs. Because of the numerous installation requirements, mass insulation is produced in the following various forms:

Forms of Mass Insulation

1. Rigid — in pre-formed shapes such as boards, blocks and pipe covering
2. Semi-rigid boards, sheets and pre-formed shapes
3. Flexible boards, sheets and pre-formed shapes
4. Blankets — faced or unfaced
5. Batts — faced or unfaced

6. Felts
7. Tapes
8. Cements
9. Loose — flakes or fibers
10. Fill — granules or flakes
11. Mastics

The following tables include many insulations that are not used in residences or building walls, roofs or floors. Such insulations as Aluminum-Silica fiber, Diatomaceus Silic, etc., are used in high temperature applications involving high-pressure steam piping or boilers. However, they may also be used to provide heat protection to steel beams located near boilers in steam plants. For these reasons, they have been included in the listing of properties and conductivities of insulations. The properties are given in Tables 54-59 and the conductivities in Tables 60-68.

Eventually each of these forms of insulation is installed and becomes a part of an insulation *system*. Its properties then become important to ensure that it will function properly in the system. Thus the insulation's properties become of utmost importance to its economic installation, service life, and thermal efficiency.

The large number of types of insulation and the forms in which it is produced ensure that their thermal properties will vary widely. This is both expected and desirable as the applications for the insulation will also vary widely.

Properties of Mass Insulations

Knowledge of the properties of insulations is necessary for selection of the proper insulation for specific uses. Many times, in a given installation, a certain property may be of little importance, whereas that same property may be of utmost important in a different application. One of the problems in the insulation industry is that some of the manufacturers resist providing information as to their material's properties. They feel that too much information may limit the use of their material; this is a mistake. Without sufficient information their insulation may be misused, resulting in failures and a loss of confidence in the manufacturer and his product by the contractor and the user.

Many of the following listed properties are not freely available from the manufacturers nor do they appear in their published literature; therefore they have not been included in the Tables of Properties, which are presented later in this chapter. Occasionally, however, some of the omitted properties may assume extreme importance. In this case effort should be made to obtain them on an individual basis from the appropriate manufacturers.

Properties (and Their Significance) of Rigid and Semi-Rigid Insulations

The basic properties of mass insulations and the significance of these properties in choosing an insulation are as follows:

Abrasion resistance. The property of a material measuring its ability to withstand abrasion without wearing away. Abrasion is caused by the rubbing together of two abutting objects or by an external force or object rubbing against the surface of one of the objects.

Alkalinity (pH) or acidity. The tendency of a material to have a basic (alkaline) or acidic reaction. It is measured on the pH scale, with all readings above 7.0, alkaline; below 7.0, acidic. A reading of 7.0 is chemically neutral. The pH value indicates how corrosive or caustic a particular material will be to a particular metal.

Breaking strength. The property of a material which indicates its maximum strength in flexure when loaded as a simply supported beam with a concentrated load at its center.

This property is of significance in installations where the material must "bridge" or span an unsupported space.

Capillarity. The property of a material which enables it to draw liquid into or through itself.

Should a water leak occur, insulation having high capillarity will spread the damage.

Chemical reaction. The property of a material which measures its tendency to chemically combine (or react) with other materials which may come into contact with or be absorbed by it.

This information is very important since a chemical reaction may be a fire hazard.

Coefficient of thermal expansion (contraction). The property of a material which measures its dimensional change resulting from a temperature change.

This information is necessary in order to solve problems related to the spacing and design of expansion and contraction joints in the insulation system.

Combustibility. The property of a material which measures (on an arbitrary scale referred to some common material) its tendency to burn.

This information is vital in determining any possible fire hazard. If a material is combustible, its potential contribution to a fire hazard is determined by the combination of the following:

1. Flash point temperature
2. Flame point temperature
3. Self-ignition point temperature (if any)
4. Burning rate
5. Rate of flame travel
6. Heat contribution
7. Explosion index
8. Auto-internal heating
9. Toxicity of products of combustion
10. Smoke density
11. Melting point temperature

Table 54 Properties of Rigid and Semi-Rigid Insulation

Material Identification Number				1	2	3
Composition				Alumina-Silica fiber & binder-Block	Alumina-silica fiber & binder-Block	Alumina-silica ceramic fiber & binders-Board
Abrasion Resistance % of Weight Loss			1st Run	30		
			2nd Run	60		
Alkalinity pH						
Capillarity				Will Wick	Will Wick	Will Wick
Coefficient of Expansion			English	Will Shrink	Will Shrink	Will Shrink
			Metric	Will Shrink	Will Shrink	Will Shrink
Combustibility				Non-combustible	Non-combustible	Non-combustible
If Combustible	Flash Point Temp.		°F	"	"	"
			°C	"	"	"
	Flame Point Temp.		°F	"	"	"
			°C	"	"	"
	Heat Release		Btu/lb	"	"	"
			J/kg	"	"	"
	Self Internal Heating			"	"	"
	Smoldering – Afterglow			"	"	"
Corrosion – Rusting						
Corrosion – Stress						
Density			lb/ft^3	15 to 17	18 to 22	10 to 16
			kg/m^2	240 to 272	288 to 352	160 to 256
Hardness – mm penetration						
Hygroscopicity % Volume				High	High	High
Resistance to Acids				Good, except HF & H_3PO_4	Good, except HF & H_3 PO_4	Good, except HF & H_3PO_4
Resistance to Caustics				Fair to Poor	Fair to Poor	Fair to Poor
Resistance to Solvents				Excellent	Excellent	Excellent
Shrinkage %			Linear	2.5 at 1800°F (982°C)	2 to 5 at 2300°F (1260°C)	3.0 at 1800°F (982°C)
			Volumetric			
Specific Gravity			Real			
			Apparent	0.25	0.32	0.24
Specific Heat			Btu/lb	0.27	0.27	0.27
			w/kg	0.15	0.15	0.15
Strength	Compressive		psi			
			kPa			
	Flexural		psi/m			
			kPa			
	Modulus of Rupture		psi		125 after 24 hrs 2300°F	31 after 24hrs. 1500°F
			kPa		858 after 24hrs 1260°C	213 after 24hrs. 816°C
	Tensile		psi			
			kPa			
Temperature Limits, Service	Short Time		°F	2300	2300	2300
			°C	1260	1260	1260
	Continuous		°F	2300	2300	2300
			°C	1260	1260	1260
Melting Temperature			°F	3200	3200	3200
			°C	1980	1980	1980
Thermal Diffusivity			ft^2/hr			
			m^2/hr			
Thermal Shock Resistance				Excellent	Excellent	Excellent
Vibration Resistance				Good	Good	Good
Water Absorption			% by Volume	91	90	90 to 92
Water Vapor Transmission			perm-inch	Very High	Very High	Very High
			perm-cm	Very High	Very High	Very High

Table 54 Properties of Rigid and Semi-Rigid Insulation

Material Identification Number			4	5	6
Composition			Alumina-silica ceramic fiber & binder-Board	Calcium Silicate-Type 1: bulk & Pipe Covering	Calcium Silicate-Type 2: Block & Pipe Cover
Abrasion Resistance % of Weight Loss		1st Run		(max.) 30	(max.) 20
		2nd Run		(Max.) 50	(max.) 45
Alkalinity pH				8 to 10.5	8 to 10.5
Capillarity			Will Wick	Will Wick	Will Wick
Coefficient of Expansion		English	Will Shrink	0	0
		Metric	Will Shrink	0	0
Combustibility			Non-combustible	Non-combustible	Non-combustible
If Combustible	Flash Point Temp.	°F	"	"	"
		°C	"	"	"
	Flame Point Temp.	°F	"	"	"
		°C	"	"	"
	Heat Release	Btu/lb	"	"	"
		J/kg	"	"	"
	Self Internal Heating		"	"	"
	Smoldering – Afterglow		"	"	"
Corrosion – Rusting				Will not contribute	Will not contribute
Corrosion – Stress				Will contribute	Will contribute
Density		lb/ft^3	21 to 90	(max.) 14.5	(max.) 15
		kg/m^2	336 to 640	(max.) 230	(max.) 240
Hardness – mm penetration				40	60
Hygroscopicity % Volume			High	10 to 15	10 to 14
Resistance to Acids			Good, except HF & H_8 Po_4		
Resistance to Caustics			Fair to Poor		
Resistance to Solvents			Excellent		
Shrinkage %		Linear		2.0 at 1200° F (650° C)	2.5 at 1500° F (982° F)
		Volumetric			
Specific Gravity		Real			
		Apparent	0.48	0.22	0.24
Specific Heat		Btu/lb	0.27	0.28	0.28
		w/kg	0.15	0.16	0.16
Strength	Compressive	psi		60 to 100 at 5% defor	60 to 250 at 5% defor
		kPa		414 to 600 at 5% defor	412 to 1725 at 5% defor
	Flexural	psi/m		35 to 50	38
		kPa		241 to 300	262
	Modulus of Rupture	psi	125 after 24hrs. 2300° F		
		kPa	858 after 24hrs. 1260° C		
	Tensile	psi			
		kPa			
Temperature Limits, Service	Short Time	°F	2300	1200	1500
		°C	1260	649	816
	Continuous	°F	2300	1200	1500
		°C	1260	649	816
Melting Temperature		°F	3200		
		°C	1980		
Thermal Diffusivity		ft^2/hr		0.016	0.016
		m^2/hr		4.12×10^{-7}	4.12×10^{-7}
Thermal Shock Resistance			Excellent	Excellent	Excellent
Vibration Resistance			Good	Fair	Fair
Water Absorption		% by Volume	84	90	90
Water Vapor Transmission		perm-inch	Very High	Very High	Very High
		perm-cm	Very High	Very High	Very High

Table 54 Properties of Rigid and Semi-Rigid Insulation

Material Identification Number			7	8	9
Composition			Diatomaceous-Silica Type 1: Block & Pipe Covering	Diatomaceous-Silica Type 2: Block & Pipe Covering	Cork Vegetable Board
Abrasion Resistance % of Weight Loss		1st Run	(max.) 55	(max.) 55	
		2nd Run	(max.) 80	(max.) 80	
Alkalinity pH			7 to 9	7 to 9	7
Capillarity			Will Wick	Will Wick	0
Coefficient of Expansion		English	0	0	
		Metric	0	0	
Combustibility			Non-combustible	Non-combustible	Combustible
If Combustible	Flash Point Temp.	°F	"	"	575
		°C	"	"	303
	Flame Point Temp.	°F	"	"	575
		°C	"	"	303
	Heat Release	Btu/lb	"	"	8000 to 9000
		J/kg	"	"	1.8 to 2.0 × 10^7
	Self Internal Heating		"	"	
	Smoldering – Afterglow		"	"	
Corrosion – Rusting			Will not contribute	Will not contribute	Steel should be protected
Corrosion – Stress					Will not contribute
Density		lb/ft^3	(max.) 24	(max.) 26	6 to 8
		kg/m^2	(max.) 384	(max.) 416	96 to 128
Hardness – mm penetration			50 to 60	50 to 60	
Hygroscopicity % Volume			High	High	1
Resistance to Acids					Poor
Resistance to Caustics					Good
Resistance to Solvents					Good
Shrinkage %		Linear	2.5 at 1600°F (871°C)	2.5 at 1900°F (1038°C)	
		Volumetric	5 to 7 at 1600°F (871°C)	5 to 7 at 1900°C (1038°C)	
Specific Gravity		Real	2.2 to 2.3	2.2 to 2.3	
		Apparent	0.34 to 0.40	0.34 to 0.40	0.096 to 0.128
Specific Heat		Btu/lb	0.22 to 0.28	0.22 to 0.28	.43
		w/kg	0.12 to 0.16	0.12 to 0.16	.25
Strength	Compressive	psi	50 at 5% defor	65 at 5% defor	5 at 5% defor
		kPa	345 at 5% defor	448 at 5% defor	34 at 5% defor
	Flexural	psi/m	60	65	15
		kPa	414	448	103
	Modulus of Rupture	psi			
		kPa			
	Tensile	psi			
		kPa			
Temperature Limits, Service	Short Time	°F	1600	1900	200
		°C	871	1038	93
	Continuous	°F	1600	1900	180
		°C	871	1038	80
Melting Temperature		°F			Combustible
		°C			"
Thermal Diffusivity		ft^2/hr	0.008 to 0.01 at 500°F	0.008 to 0.01 at 500°F	0.007 at 70°F
		m^2/hr	2.06 to 2.58 × 10 at 260°C	2.06 × 2.58 × 10 at 260°C	1.08 × 10^{-7} at 21°C
Thermal Shock Resistance			Excellent	Excellent	
Vibration Resistance			Fair	Fair	Good
Water Absorption		% by Volume	85	83	5% (surface only)
Water Vapor Transmission		perm-inch	Very High	Very High	3 to 7
		perm-cm	Very High	Very High	5.4 to 12.6

Table 54 Properties of Rigid and Semi-Rigid Insulation

Material Identification Number	10	11	12
Composition	Cork Vegetable Pipe Covering	Cellular glass Block & Pipe Covering	Cellular glass between layers of felt-Board
Abrasion Resistance % of Weight Loss — 1st Run		Poor	Poor
Abrasion Resistance % of Weight Loss — 2nd Run		Poor	Poor
Alkalinity pH	7	7 to 8	7 to 8
Capillarity	0	0	0
Coefficient of Expansion — English		46×10^{-7}	46×10^{-7}
Coefficient of Expansion — Metric		83×10^{-7}	83×10^{-7}
Combustibility	Combustible	Non-combustible	Felt cov. is combustible
If Combustible — Flash Point Temp. °F	575	"	Felt-550°F Ins. non-combustible
If Combustible — Flash Point Temp. °C	303	"	Felt-288°C Ins. non-combustible
If Combustible — Flame Point Temp. °F	575	"	Felt-575°F Ins. non-combustible
If Combustible — Flame Point Temp. °C	303	"	Felt-302°C Ins. non-combustible
If Combustible — Heat Release Btu/lb	8000 to 9000	"	Ins. non-combustible
If Combustible — Heat Release J/kg	$1.8 \text{ to } 2.0 \times 10^7$	"	Ins. non-combustible
If Combustible — Self Internal Heating		"	
If Combustible — Smoldering – Afterglow		"	
Corrosion – Rusting	Steel should be protected	Will not contribute	Will not contribute
Corrosion – Stress	Will not contribute	Will not contribute	Will not contribute
Density — lb/ft^3	7 to 14	7.5 to 9.5	7.5 to 9.5
Density — kg/m^2	112 to 224	120 to 152	120 to 152
Hardness – mm penetration			
Hygroscopicity % Volume	1	0	0
Resistance to Acids	Poor	Excellent except HF & H_3PO_4	Excellent except HF & H_3PO_4
Resistance to Caustics	Good	Excellent	Excellent
Resistance to Solvents	Good	Excellent	Excellent
Shrinkage % — Linear		0 at 800°F (427°C)	0 at 800°F (427°C)
Shrinkage % — Volumetric		0 at 800°F (427°C)	0 at 800°F (427°C)
Specific Gravity — Real		2.5	2.5
Specific Gravity — Apparent	0.096 to 0.128	0.012 to 0.152	0.012 to 0.152
Specific Heat — Btu/lb	.43	0.18	0.18
Specific Heat — w/kg	.25	0.10	0.10
Strength — Compressive psi	5 at 5% defor.	75 to 100 (max.)	75 to 100 (max.)
Strength — Compressive kPa	34 at 5% defor.	517 to 690 (max.)	517 to 690 (max.)
Strength — Flexural psi/m	18 to 21	60 to 80	60 to 80
Strength — Flexural kPa	123 to 144	414 to 552	414 to 552
Strength — Modulus of Rupture psi			
Strength — Modulus of Rupture kPa			
Strength — Tensile psi		84	84
Strength — Tensile kPa		580	580
Temperature Limits, Service — Short Time °F	200	940	350
Temperature Limits, Service — Short Time °C	93	504	427
Temperature Limits, Service — Continuous °F	180	800	350
Temperature Limits, Service — Continuous °C	80	427	427
Melting Temperature °F	Combustible	1500	(insul.) 1500
Melting Temperature °C	"	816	816
Thermal Diffusivity — ft^2/hr	0.007 at 70°F	0.018 at 70°F	0.018 at 70°F
Thermal Diffusivity — m^2/hr	1.08×10^{-7} at 21°C	1.08×10^{-7} at 21°C	1.08×10^{-7} at 21°C
Thermal Shock Resistance		Poor	Surface protected by felt
Vibration Resistance	Good	Poor	Surface protected by felt
Water Absorption — % by Volume	3% (surface only)	5% (surface only)	2% (surface only)
Water Vapor Transmission — perm-inch	3 to 7	0.002 at 80°F	0.002 at 80°F
Water Vapor Transmission — perm-cm	5.4 to 12.6	0.0036 at 27°C	0.0036 at 27°C

Table 54 Properties of Rigid and Semi-Rigid Insulation

Material Identification Number			13	14	15
Composition			Diatomaceous earth-Type 1-Block & pipe Cov.	Diatomaceous earth-Type 2-Block & pipe Cov.	Glass fiber with organic binder-Boards
Abrasion Resistance % of Weight Loss		1st Run	30 to 37		
		2nd Run			
Alkalinity pH			7 to 9	7 to 9	8 to 10
Capillarity			Will Wick	Will Wick	Negligable
Coefficient of Expansion		English	12 X 10^{-8} on reheat	28 X 10^{-7} on reheat	
		Metric			
Combustibility			Non-combustible	Non-combustible	Binder is combustible
If Combustible	Flash Point Temp.	°F	"	"	(Binder) 750° F
		°C	"	"	(Binder) 399° C
	Flame Point Temp.	°F	"	"	Smolders
		°C	"	"	"
	Heat Release	Btu/lb	"	"	Fibers Non-combustible
		J/kg	"	"	"
	Self Internal Heating		"	"	
	Smoldering – Afterglow		"	"	Binder will smolder
Corrosion – Rusting			Will not contribute	Will not contribute	Will not contribute
Corrosion – Stress					Will not contribute
Density		lb/ft^3	17 to 24	26	1.4 to 1.6
		kg/m^2	217 to 384	416	22 to 26
Hardness – mm penetration			0.6 to 0.8	0.5 to 0.6	
Hygroscopicity % Volume			2 to 8	2 to 8	2
Resistance to Acids			Good	Good	Good, except HF & H_3PO_4
Resistance to Caustics			Good	Good	Good from pH 7 to 10
Resistance to Solvents			Resistant	Resistant	Good
Shrinkage %		Linear	2% at 1600° F (870° C)	2% at 1900° F (1040° C)	0 up to 450° F (232° C)
		Volumetric	6 to 18% at 1600° F (870° C)	5 to 7% at 1900° F (1040° C)	0 up to 450° F (232° C)
Specific Gravity		Real	2.2 to 2.3	2.2 to 2.3	2.5
		Apparent	0.38	0.41	0.02
Specific Heat		Btu/lb	.22 to .28	.22 to .28	0.20
		w/kg	.12 to .16	.12 to .16	0.11
Strength	Compressive	psi	50 at 5% deform.	65 at 5% deform.	0.03 at 10% deform.
		kPa	344 at 5% deform.	448 at 5% deform.	0.21 at 10% deform.
	Flexural	psi/m	60	65	
		kPa	413	448	
	Modulus of Rupture	psi			
		kPa			
	Tensile	psi	10 to 20	15 to 20	
		kPa	69 to 137	103 to 137	
Temperature Limits, Service	Short Time	°F	1600	1900	max. 450
		°C	870	1040	max. 232
	Continuous	°F	1600	1900	max. 450
		°C	870	1040	max. 232
Melting Temperature		°F			1200
		°C			649
Thermal Diffusivity		ft^2/hr	0.008 to 0.01	0.009 to 0.012	0.04 at 75°F
		m^2/hr	2.06 X 10^{-7} to 2.58 X 10^{-7}	2.32 X 10^{-7} to 3.1 X 10^{-7}	1.03 X 10^{-6} at 24°C
Thermal Shock Resistance			Excellent	Excellent	Excellent
Vibration Resistance					
Water Absorption		% by Volume	High	High	93
Water Vapor Transmission		perm-inch	150 to 250	High	Very High
		perm-cm	270 to 450		Very High

Table 54 Properties of Rigid and Semi-Rigid Insulation

Material Identification Number		16	17	18
Composition		Glass fiber with organic binder-Boards	Glass fiber with organic binder-Boards	Glass fiber w/organic binder-Duct board foil reinforced
Abrasion Resistance % of Weight Loss	1st Run		40	
	2nd Run			
Alkalinity pH		8 to 10	8 to 10	8 to 10
Capillarity		Negligable	Negligable	Negligable
Coefficient of Expansion	English			
	Metric			
Combustibility		Binder is combustible	Binder is combustible	Binder is combustible
If Combustible: Flash Point Temp.	°F	(Binder) 750°F	(Binder) 750°F	(Binder) 750°F
	°C	(Binder) 399°C	(Binder) 399°C	(Binder) 399°C
If Combustible: Flame Point Temp.	°F	Smolders	Smolders	Smolders
	°C	"	"	"
If Combustible: Heat Release	Btu/lb	Fibers Non-combustible	Fibers Non-combustible	Fibers Non-combustible
	J/kg	"	"	"
If Combustible: Self Internal Heating				
If Combustible: Smoldering – Afterglow		Binder will smolder	Binder will smolder	Binder will smolder
Corrosion – Rusting		Will not contribute	Will not contribute	Will not contribute
Corrosion – Stress		Will not contribute	Will not contribute	Will not contribute
Density	lb/ft^3	2.5 to 3.5	5 to 7	2.5 to 3.5
	kg/m^2	40 to 56	80 to 112	40 to 56
Hardness – mm penetration				
Hygroscopicity % Volume		2	2	2
Resistance to Acids		Good, except HF & H_3PO_4	Good, except HF & H_3PO_4	Good, except HF & H_3PO_4
Resistance to Caustics		Good from pH 7 to 10	Good from pH 7 to 10	Good from pH 7 to 10
Resistance to Solvents		Good	Good	Good
Shrinkage %	Linear	0 up to 450°F (232°C)	0 up to 450°F (232°C)	0 up to 250°F (121°C)
	Volumetric	0 up to 450°F (232°C)	0 up to 450°F (232°C)	0 up to 250°F (121°C)
Specific Gravity	Real	2.5	2.5	2.5
	Apparent	o.5	1.0	0.5
Specific Heat	Btu/lb	0.20	0.20	0.20
	w/kg	0.11	0.11	0.11
Strength: Compressive	psi	0.7 at 10% defor.	2.4 at 10% defor.	
	kPa	4.8 at 10% defor.	16.6 at 10% defor.	
Strength: Flexural	psi/m			
	kPa			
Strength: Modulus of Rupture	psi			
	kPa			
Strength: Tensile	psi			
	kPa			
Temperature Limits, Service: Short Time	°F	max. 450	max. 450	max. 250
	°C	max. 232	max. 232	max. 121
Temperature Limits, Service: Continuous	°F	max. 450	max. 450	max. 250
	°C	max. 232	max. 232	max. 121
Melting Temperature	°F	1200	1200	1200
	°C	649	649	649
Thermal Diffusivity	ft^2/hr	0.02 at 75°F	0.015 at 75°F	0.015 at 75°F
	m^2/hr	5.16×10^{-7} at 24°C	3.8×10^{-7} at 24°C	3.87×10^{-7} at 24°C
Thermal Shock Resistance		Excellent	Excellent	Excellent
Vibration Resistance				
Water Absorption	% by Volume	92	93	93
Water Vapor Transmission	perm-inch	Very High	Very High	Depends on covering
	perm-cm	Very High	Very High	Depends on covering

Table 54 Properties of Rigid and Semi-Rigid Insulation

Material Identification Number			19	20	21
Composition			Glass fiber-organic binder semi-rigid light weight pipe cov.	Glass fiber with organic binder-heaver density pipe cov.	Mineral fiber (rock slab or glass) Class I block or Board
Abrasion Resistance % of Weight Loss		1st Run	50	40	
		2nd Run	95	90	
Alkalinity pH			8 to 10	8 to 10	7 to 9
Capillarity			Will Wick	Negligable	Will Wick
Coefficient of Expansion		English			Will Shrink
		Metric			Will Shrink
Combustibility			Binder is combustible	Binder is combustible	Binder is combustible
If Combustible	Flash Point Temp.	°F	(Binders) 750°F	(Binders) 750°F	Fibers non-combustible
		°C	(Binders) 399°C	(Binders) 399°C	"
	Flame Point Temp.	°F	Smolders	Smolders	"
		°C	"	"	"
	Heat Release	Btu/lb	Fibers non-combustible	Fibers non-combustible	
		J/kg	"	"	
	Self Internal Heating		Can self heat	Can self heat	Can self heat
	Smoldering – Afterglow		Will smolder	Will smolder	Will smolder
Corrosion – Rusting			Will not contribute	Will not contribute	Will not contribute
Corrosion – Stress			Depends on binder	Depends on binder	Depends on binder
Density		lb/ft^3	3 to 5	5 to 10	9 to 11
		kg/m^2	54 to 80	80 to 160	144 to 176
Hardness – mm penetration					
Hygroscopicity % Volume			2	2	2.5
Resistance to Acids			Good, except HF & H_3PO_4	Good, except HF & H_3PO_4	Good, except HF & H_3PO_4
Resistance to Caustics			Good, pH 7 to 10	Good, pH 7 to 10	Good, pH 7 to 10
Resistance to Solvents			Good	Good	Good
Shrinkage %		Linear	0 up to 450°F (232°C)	0 up to 450°F (232°C)	(max.) 2 up to 400°F (204°C)
		Volumetric	0 up to 450°F (232°C)	0 up to 450°F (232°C)	
Specific Gravity		Real	2.5	2.5	2.5
		Apparent	0.08 (max.)	0.16 (max.)	0.14 (max.)
Specific Heat		Btu/lb	0.20	0.20	0.20
		w/kg	0.12	0.12	0.12
Strength	Compressive	psi			
		kPa			
	Flexural	psi/m			
		kPa			
	Modulus of Rupture	psi			
		kPa			
	Tensile	psi			
		kPa			
Temperature Limits, Service	Short Time	°F	(max.) 500	(max.) 500 to 650	(max.) 400 to 500
		°C	(max.) 260	(max.) 260 to 343	(max.) 204 to 260
	Continuous	°F	(max.) 450 to 500	(max.) 450 to 500	(max.) 400 to 500
		°C	(max.) 232 to 260	(max.) 232 to 260	(max.) 204 to 200
Melting Temperature		°F	1200	1200	1200
		°C	649	649	649
Thermal Diffusivity		ft^2/hr	0.018 at 70°F	0.020 at 70°F	0.030 at 70°F
		m^2/hr	4.6×10^{-7} at 21°C	5.16×10^{-7} at 21°C	7.7×10^{-7} at 21°C
Thermal Shock Resistance			Good	Good	Good
Vibration Resistance			Fair	Fair	Fair
Water Absorption		% by Volume		0.2	
Water Vapor Transmission		perm-inch	Depends on Jacket	Depends on Jacket	Very High
		perm-cm	Depends on Jacket	Depends on Jacket	Very High

Table 54 Properties of Rigid and Semi-Rigid Insulation

Material Identification Number			22	23	24
Composition			Mineral fiber (rock slab or glass) Class 2 Block or Board	Mineral fiber Class 3 Block or Board	Mineral fiber Class 4 Block or Board
Abrasion Resistance % of Weight Loss		1st Run			
		2nd Run			
Alkalinity pH			7 to 9	7 to 9	7 to 9
Capillarity			Will Wick	Will Wick	Will Wick
Coefficient of Expansion		English	Will Shrink	Will Shrink	Will Shrink
		Metric	Will Shrink	Will Shrink	Will Shrink
Combustibility			Binder is combustible	Some binders will burn	Some binders oxidize
If Combustible	Flash Point Temp.	°F	Fibers non-combustible	Fibers non-combustible	Fibers non-combustible
		°C	"	"	"
	Flame Point Temp.	°F	"	"	"
		°C	"	"	"
	Heat Release	Btu/lb			
		J/kg			
	Self Internal Heating		Can self heat	Can self heat	Can self heat
	Smoldering – Afterglow		Will smolder	Can smolder	Can smolder
Corrosion – Rusting			Will not contribute	Will not contribute	Will not contribute
Corrosion – Stress			Depends on binder	Depends on binder	Will not contribute
Density		lb/ft^3	11 to 13	7 to 12	8 to 13
		kg/m^2	176 to 192	112 to 192	128 to 208
Hardness – mm penetration					
Hygroscopicity % Volume			2.5	2.5	2.5
Resistance to Acids			Good, except HF & H_3PO_4	Good, except HF & H_3PO_4	Good, except, HF & H_3PO_4
Resistance to Caustics			Good pH 7 to 10	Good	Good
Resistance to Solvents			Good	Good	Good
Shrinkage %		Linear	(max.) 2 up to 400°F (204°C)	2.5% at 350°F (434°C)	2% at 1000°F (649°C)
		Volumetric			
Specific Gravity		Real	2.5	2.5	2.5
		Apparent	0.20 (max.)	0.11 to 0.19	0.12 to 0.20
Specific Heat		Btu/lb	0.20	0.22	0.22
		w/kg	0.12	0.13	0.13
Strength	Compressive	psi	25	0.15 to 0.25 at 10% defor.	1.0 to 2.0 at 10% defor.
		kPa	172	2.4 to 4.4 at 10% defor.	33.5 to 67 at 10% defor.
	Flexural	psi/m			25
		kPa			172
	Modulus of Rupture	psi			
		kPa			
	Tensile	psi			
		kPa			
Temperature Limits, Service	Short Time	°F	(max.) 400 to 500	850 to 1000	1000 to 1200
		°C	(max.) 204 to 260	454 to 540	540 to 649
	Continuous	°F	(max.) 400 to 500	850 to 1000	1000 to 649
		°C	(max.) 204 to 260	454 to 590	540 to 649
Melting Temperature		°F	1200		
		°C	649		
Thermal Diffusivity		ft^2/hr	0.035 at 70°F	0.01 to 0.07 at 70°F	0.005 to 0.01 at 70°F
		m^2/hr	9.03×10^{-7} at 21°C	2.5 to 5.1×10^{-7} at 21°C	1.25 to 2.6×10^{-7} at 21°C
Thermal Shock Resistance			Good	Good	Good
Vibration Resistance			Fair		
Water Absorption		% by Volume		85 to 93	85 to 93
Water Vapor Transmission		perm-inch	Very High	Very High	Very High
		perm-cm	Very High	Very High	Very High

Table 54 Properties of Rigid and Semi-Rigid Insulation

Material Identification Number				25	26	27
Composition				Mineral fiber Class 5 Board & Block	Mineral fiber Class 1 Pipe Cov.	Mineral fiber Class 2 Pipe Cov.
Abrasion Resistance % of Weight Loss			1st Run			
			2nd Run			
Alkalinity pH				7 to 9	7 to 9	7 to 9
Capillarity				Will Wick	Will Wick	Will Wick
Coefficient of Expansion			English	Will Shrink	Will Shrink	Will Shrink
			Metric	Will Shrink	Will Shrink	Will Shrink
Combustibility				Non-combustible	Some binders will burn	Some binders oxidize
If Combustible	Flash Point Temp.		°F	"	Fibers non-combustible	Fibers non-combustible
			°C	"	"	"
	Flame Point Temp.		°F	"	"	"
			°C	"	"	"
	Heat Release		Btu/lb	"	"	"
			J/kg	"	"	"
	Self Internal Heating			"	Can self heat	Can self heat
	Smoldering – Afterglow			"	Can smolder	Can smolder
Corrosion – Rusting				Will not contribute	Will not contribute	Will not contribute
Corrosion – Stress				Will not contribute	Depends on binder	Depends on binder
Density			lb/ft^3	12 to 25	9 to 11	11 to 14
			kg/m^2	192 to 400	144 to 176	176 to 224
Hardness – mm penetration						
Hygroscopicity % Volume				2	2	2
Resistance to Acids				Good, except HF & H_3PO_4	Good, except HF & H_3PO_4	Good, except HF & H_3PO_4
Resistance to Caustics				Good	Good	Good
Resistance to Solvents				Good	Good	Good
Shrinkage %			Linear	5% at 1900°F (1038°C)	2% at 450°F (232°C)	2% at 650°F (345°C)
			Volumetric			
Specific Gravity			Real	2.5	2.5	2.5
			Apparent	0.19 to 0.40	0.11 to 0.17	0.17 to 0.22
Specific Heat			Btu/lb	0.22	0.22	0.22
			w/kg	0.13	0.13	0.13
Strength	Compressive		psi	10 to 15 at 10% defor.		
			kPa	69 to 104 at 10% defor.		
	Flexural		psi/m	40		
			kPa	276		
	Modulus of Rupture		psi			
			kPa			
	Tensile		psi			
			kPa			
Temperature Limits, Service	Short Time		°F	1800 to 1900	450	650
			°C	982 to 1038	232	345
	Continuous		°F	1800 to 1900	450	650
			°C	982 to 1038	232	345
Melting Temperature			°F			
			°C			
Thermal Diffusivity			ft^2/hr	0.004 to 0.005 at 70°F	0.005 to 0.01 at 70°F	0.004 to 0.008 at 70°F
			m^2/hr	1.03 to 2.06×10^{-7} at 21°C	1.25 to 2.6×10^{-7} at 21°C	1.03 to 2.06×10^{-7} at 21°C
Thermal Shock Resistance				Good	Good	Good
Vibration Resistance						Good
Water Absorption			% by Volume	85 to 90	85 to 90	90
Water Vapor Transmission			perm-inch	Very High	Very High	Very High
			perm-cm	Very High	Very High	Very High

Table 54 Properties of Rigid and Semi-Rigid Insulation

Material Identification Number			28	29	30
Composition			Mineral Fiber Class 3 Pipe Cov.	Mineral Fiber Class 2 Pipe Cov.	Mineral Fiber Class 3 Pipe Cov.
Abrasion Resistance % of Weight Loss		1st Run		35 to 40	30 to 40
		2nd Run		60 to 70	60 to 70
Alkalinity pH			7 to 9	7 to 8	7 to 8
Capillarity			Will Wick	Negligible	Will Wick
Coefficient of Expansion		English	Will Shrink	Will Shrink	Will Shrink
		Metric	Will Shrink	Will Shrink	Will Shrink
Combustibility			Non-combustible	Non-combustible	Non-combustible
If Combustible	Flash Point Temp.	°F	"	"	"
		°C	"	"	"
	Flame Point Temp.	°F	"	"	"
		°C	"	"	"
	Heat Release	Btu/lb	"	"	"
		J/kg	"	"	"
	Self Internal Heating		"	"	"
	Smoldering – Afterglow		"	"	"
Corrosion – Rusting			Will not contribute	Will not contribute	Will not contribute
Corrosion – Stress			Will not contribute	Will not contribute	Will not contribute
Density		lb/ft^3	16 to 20	13 to 15	12 to 15
		kg/m^2	256 to 320	208 to 240	240
Hardness – mm penetration					
Hygroscopicity % Volume			2	4	16
Resistance to Acids			Good, except HF & M_3PO_4	Excellent	Excellent
Resistance to Caustics			Good	Excellent	Excellent
Resistance to Solvents			Good	Good	Good
Shrinkage %		Linear	2% at 1200°F (650°C)	2% at 1500°F (816°C)	2% at 1500°F (816°C)
		Volumetric			
Specific Gravity		Real	2.5	1.8	1.8
		Apparent	0.25 to 0.32	0.14 to 0.16	0.14 to 0.16
Specific Heat		Btu/lb	0.22	0.216	0.216
		w/kg	0.13	0.13	0.13
Strength	Compressive	psi		60 at 5% deform.	55 at 5% deform.
		kPa		415 at 5% deform.	379 at 5% deform.
	Flexural	psi/m		35	
		kPa		241	
	Modulus of Rupture	psi			
		kPa			
	Tensile	psi		32	
		kPa		221	
Temperature Limits, Service	Short Time	°F	1200	1500	1500
		°C	650	816	816
	Continuous	°F	1200	1500	1500
		°C	650	816	816
Melting Temperature		°F			
		°C			
Thermal Diffusivity		ft^2/hr	0.007 to 0.008 at70°F	0.014 at 70°F	0.014 at 70°F
		m^2/hr	1.8 to 2.06 $\times 10^{-7}$ at 21°C	3.6 $\times 10^{-7}$ at 21°C	3.6 $\times 10^{-7}$ at 21°C
Thermal Shock Resistance			Good	Good	Good
Vibration Resistance			Good	Fair	Fair
Water Absorption		% by Volume	90	9	180
Water Vapor Transmission		perm-inch	Very High	18	40
		perm-cm	Very High	32	72

Table 54 Properties of Rigid and Semi-Rigid Insulation

Material Identification Number			31	32	33
Composition			Perlite expanded with binders and fibers Roof board	Phenolic foam Board and pipe Cov.	Polystyrene expanded beads molded to shape block, board pipe Cov.
Abrasion Resistance % of Weight Loss		1st Run	30 to 40		
		2nd Run	65 to 70		
Alkalinity pH			7 to 8	6.5 to 7.5	6.5 to 7.5
Capillarity			Negligable	Negligable	0.5
Coefficient of Expansion		English	0.2 to 0.9% at 50% RH	Shrinks	35×10^{-6}
		Metric	0.2 to 0.9% at 50% RH	Shrinks	63×10^{-6}
Combustibility			Resists combustion	Slow burning	Combustible
If Combustible	Flash Point Temp.	°F		780	645
		°C		416	345
	Flame Point Temp.	°F		850	756
		°C		454	357
	Heat Release	Btu/lb			16,000
		J/kg			3.71×10^{7}
	Self Internal Heating				Melts and burns
	Smoldering – Afterglow				Melst and burns
Corrosion – Rusting			Will not contribute	Will not contribute	Does not contribute
Corrosion – Stress			Will not contribute	Will not contribute	Does not contribute
Density		lb/ft^3	10	2 to 3	1.0 to 1.5
		kg/m^2	160	32 to 48	16.01 to 24.01
Hardness – mm penetration			(Brinnel) 60 (X 1000)		
Hygroscopicity % Volume					0.5 to 1.5
Resistance to Acids			Good	Good	Good to most acids
Resistance to Caustics			Good	Good	Good to most caustics
Resistance to Solvents			Good	Fair	Poor to many solvents
Shrinkage %		Linear			Expands to temp. limits
		Volumetric			"
Specific Gravity		Real	2.2	2.3	
		Apparent	.16	0.3 to 0.5	0.016 to 0.024
Specific Heat		Btu/lb	.25		0.27 at 40°F
		w/kg	.14		0.15 at 4°C
Strength	Compressive	psi	35 at 10% defor.	13 to 22 at 10% defor.	12 at 10% defor.
		kPa	240 at 10% defor.	70 to 151 at 10% defor.	82 at 10% defor.
	Flexural	psi/m	40		
		kPa	276×10^{3}		
	Modulus of Rupture	psi			
		kPa			
	Tensile	psi	4		
		kPa	28		
Temperature Limits, Service	Short Time	°F	200	250	180
		°C	93	121	82
	Continuous	°F	200	250	180
		°C	93	121	82
Melting Temperature		°F			200 to 230
		°C			93 to 110
Thermal Diffusivity		ft^2/hr			0.063 at 40°F
		m^2/hr			1.6×10^{-6} at 4°C
Thermal Shock Resistance					
Vibration Resistance					Excellent
Water Absorption		% by Volume	1.5 max.	0.4 to 0.8	3.8 to 4.0
Water Vapor Transmission		perm-inch	25	High	2.0
		perm-cm	45	High	3.6

Table 54 Properties of Rigid and Semi-Rigid Insulation

Material Identification Number			34	35	36
Composition			Polystyrene expanded foam-cut Shapes Board & Pipe Cov.	Polystyrene expanded foam-molded in shape Board & Pipe Cov.	Polystyrene expanded foam-molded in shape board
Abrasion Resistance % of Weight Loss		1st Run			
		2nd Run			
Alkalinity pH			6.5 to 7.5	6.5 to 7.5	6.5 to 7.5
Capillarity			0	0	0
Coefficient of Expansion		English	35 X 10^{-6}	35 X 10^{-6}	35 X 10^{-6}
		Metric	63 X 10^{-6}	63 X 10^{-6}	63 X 10^{-6}
Combustibility			Combustible	Combustible	Combustible
If Combustible	Flash Point Temp.	°F	730	730	730
		°C	388	388	388
	Flame Point Temp.	°F	750	750	750
		°C	399	399	399
	Heat Release	Btu/lb	16,000	16,000	16,000
		J/kg	3.71 X 10^{7}	3.71 X 10^{7}	3.71 X 10^{7}
	Self Internal Heating		Melts and burns	Melts and burns	Melts and burns
	Smoldering – Afterglow		Melts and burns	Melts and burns	Melts and burns
Corrosion – Rusting			Does not contribute	Does not contribute	Does not contribute
Corrosion – Stress			Does not contribute	Does not contribute	Does not contribute
Density		lb/ft^3	1.75 to 2.0	2.0 to 2.2	2.25 at 2.5
		kg/m^2	28.02 to 32.2	32.02 to 35.22	36.02 to 40.03
Hardness – mm penetration					
Hygroscopicity % Volume			0.5 to 1.0	0.6 to 1.1	0.6 to 1.0
Resistance to Acids			Good to most acids	Good to most acids	Good to most acids
Resistance to Caustics			Good to most caustics	Good to most caustics	Good to most caustics
Resistance to Solvents			Poor to many solvents	Poor to many solvents	Poor to many solvents
Shrinkage %		Linear	Expands to temp. limit	Expands to temp. limit	Expands to temp. limit
		Volumetric	"	"	"
Specific Gravity		Real			
		Apparent	0.028 to 0.032	0.032 to 0.035	0.036 to 0.04
Specific Heat		Btu/lb	0.27 at 40° F	0.27 at 40° F	0.27 at 40° F
		w/kg	0.15 at 4° C	0.15 at 4° C	0.15 at 4° C
Strength	Compressive	psi	40 at 10% deform.	40 at 10% deform.	45 at 10% deform.
		kPa	275 at 10% deform.	275 at 10% deform.	310 at 10% deform.
	Flexural	psi/m	40	60	100
		kPa	275	413	689
	Modulus of Rupture	psi			
		kPa			
	Tensile	psi	70	70	85
		kPa	482	482	586
Temperature Limits, Service	Short Time	°F	165	165	165
		°C	74	74	74
	Continuous	°F	165	165	165
		°C	74	74	74
Melting Temperature		°F	200 to 230	200 to 230	200 to 230
		°C	93 to 110	93 to 110	93 to 110
Thermal Diffusivity		ft^2/hr	0.017 at 40° F	0.017 at 40° F	0.017 at 40° F
		m^2/hr	4.4 X 10^{-7} at 4° C	4.4 X 10^{-7} at 4° C	4.4 X 10^{-7} at 4° C
Thermal Shock Resistance					
Vibration Resistance			Excellent	Excellent	Excellent
Water Absorption		% by Volume	0.5 to 0.7	0.6 to 0.8	0.5 to 0.7
Water Vapor Transmission		perm-inch	1.0 to 1.4	0.4 to 1.0	0.4 to 0.8
		perm-cm	1.8 to 2.5	0.7 to 1.8	0.7 to 1.4

Table 54 Properties of Rigid and Semi-Rigid Insulation

Material Identification Number				37	38	39
Composition				Polystrene expanded foam molded to shape roof boards	Polyurethane expanded foam Block, Board & Pipe Cov.	Polyurethane expanded foam Block, Board & Pipe Cov.
Abrasion Resistance % of Weight Loss			1st Run		6 to 20	6 to 20
			2nd Run			
Alkalinity pH				6.5 to 7.5	6.5 to 7.5	6.5 to 7.5
Capillarity				0	0	0
Coefficient of Expansion			English	35×10^{-6}	5×10^{-5}	5×10^{-5}
			Metric	63×10^{-6}	9×10^{-5}	9×10^{-5}
Combustibility				Combustible	Combustible	Combustible
If Combustible	Flash Point Temp.		°F	730	590 to 750	590 to 750
			°C	388	310 to 399	310 to 399
	Flame Point Temp.		°F	750	600 to 750	600 to 750
			°C	399	316 to 399	316 to 399
	Heat Release		Btu/lb	16,000	16,000 to 17,000	16,000 to 17,000
			J/kg	3.71×10^7	3.71×10^7 to 3.9×10^7	3.71×10^7 to 3.9×10^7
	Self Internal Heating			Melts and burns	Will react with some chemicals	Will react with some chemicals
	Smoldering – Afterglow			Melts and burns	Chars	Chars
Corrosion – Rusting				Will not contribute	Does not contribute	Will contribute
Corrosion – Stress				Does not contribute	Does not contribute	Does not contribute
Density			lb/ft^3	2.5 to 2.7	Up to 1.7	1.7 to 2.5
			kg/m^2	40.0 to 43.2	Up to 27.2	27.2 to 40.0
Hardness – mm penetration						
Hygroscopicity % Volume				0.6 to 0.7	1.0 to 1.8	0.9 to 1.6
Resistance to Acids				Good to most acids	Resistant, to most dilute acids	Resistant to dilute acids
Resistance to Caustics				Good to most caustics	Resistant to most dilute caustics	Resistant to dilute caustics
Resistance to Solvents				Poor to many solvents	Resistant to most solvents	Resistant to most solvents
Shrinkage %			Linear	Expands to temp. limit	Expands to temp. limit	Expands to temp. limit
			Volumetric	Exp Expands to temp. limit	Expands to temp. limit	Expands to temp. limit
Specific Gravity			Real			
			Apparent	0.04 to 0.043	0.027	0.027 to 0.04
Specific Heat			Btu/lb	0.27 at 40°F	0.4 at 40°F	0.4 at 40°F
			w/kg	0.15 at 4°C	0.22 at 4°C	0.22 at 4°C
Strength	Compressive		psi	60 at 10% deform.	15 at 10% deform.	17 at 10% deform.
			kPa	413 at 10% deform.	103 at 10% deform.	117 at 10% deform.
	Flexural		psi/m	125	15	20
			kPa	707	105	140
	Modulus of Rupture		psi			
			kPa			
	Tensile		psi	200	10 to 30	15 to 50
			kPa	1378	69 to 205	130 to 344
Temperature Limits, Service	Short Time		°F	165	230	220
			°C	74	110	104
	Continuous		°F	165	200	200
			°C	74	94	94
Melting Temperature			°F	200 to 230	Chars	Chars
			°C	93 to 110	Chars	Chars
Thermal Diffusivity			ft^2/hr	0.017 at 40°F	0.011 to 0.15 at 40°F	0.015 to 0.02 at 40°F
			m^2/hr	4.4×10^{-7} at 4°C	2.8×10^{-7} to 3.6×10^{-7} at 4°C	3.6×10^{-7} to 5.1×10^{-7} at 4°C
Thermal Shock Resistance						
Vibration Resistance				Excellent	Excellent	Excellent
Water Absorption			% by Volume	0.5 to 0.7	1.5	1.3
Water Vapor Transmission			perm-inch	0.5 to 0.8	4 to 6	3 to 5
			perm-cm	0.9 to 1.4	6.7 to 10	5 to 8.4

Table 54 Properties of Rigid and Semi-Rigid Insulation

Material Identification Number			40	41	42
Composition			Polyurethane expanded foam Block, Board & Pipe Cov.	Expanded rubber Rigid Board	Expanded silica with binders-Block & Pipe Cov.
Abrasion Resistance % of Weight Loss		1st Run	6 to 20		35 to 40
		2nd Run			65 to 70
Alkalinity pH			6.5 to 7.5		8 to 9.5
Capillarity			0	10%	
Coefficient of Expansion		English	5×10^{-5}	Shrinks	Shrinks
		Metric	9×10^{-5}	Shrinks	Shrinks
Combustibility			Combustible	Combustible	Non-combustible
If Combustible	Flash Point Temp.	°F	590 to 750	560	"
		°C	310 to 399	293	"
	Flame Point Temp.	°F	600 to 750	590	"
		°C	316 to 399	310	"
	Heat Release	Btu/lb	16,000 to 17,000	15,000	"
		J/kg	3.71×10^7 to 3.9×10^7	3.486×10^7	"
	Self Internal Heating		Will react with some chemicals		"
	Smoldering – Afterglow		Chars		"
Corrosion – Rusting			Will contribute	Does not contribute	Does not contribute
Corrosion – Stress			Does not contribute	Does not contribute	Does not contribute
Density		lb/ft^3	2.5 to 5.0	4 to 6	13
		kg/m^2	40 to 80	64 to 96	208
Hardness – mm penetration					
Hygroscopicity % Volume			0.8 to 1.4		0.5
Resistance to Acids			Resistant to dilute acid	Good	Excellent
Resistance to Caustics			Resistant to dilute caustic	Good	Excellent
Resistance to Solvents			Resistant to most solvents	Poor	Good
Shrinkage %		Linear	Expands to temp. limit	10% at 200°F	1% at 1500°F
		Volumetric	Expands to temp. limit		
Specific Gravity		Real			2.2
		Apparent	0.04 to 0.08	0.07	0.22
Specific Heat		Btu/lb	0.4 at 40°F	0.20	0.14 to 0.18
		w/kg	0.22 at 4°C		0.08 to 0.10
Strength	Compressive	psi	30 at 10% deform.	40 at 10% deform.	82 at 5% deform.
		kPa	207 at 10% deform.	275 at 10% deform.	565 at 5% deform.
	Flexural	psi/m	40		46
		kPa	275		179
	Modulus of Rupture	psi			
		kPa			
	Tensile	psi	20 to 108	60	
		kPa	137 to 670	413	
Temperature Limits, Service	Short Time	°F	220	200	1500
		°C	104	93	815
	Continuous	°F	200	200	1500
		°C	94	93	815
Melting Temperature		°F	Chars		
		°C	Chars		
Thermal Diffusivity		ft^2/hr	0.02 to 0.03 at 40°F		0.015
		m^2/hr	5.1×10^7 to 7.7×10^7 at 4°C		3.8×10^7
Thermal Shock Resistance					Excellent
Vibration Resistance			Excellent		Good
Water Absorption		% by Volume	1.1	1	
Water Vapor Transmission		perm-inch	2.5 to 5	0.1	High
		perm-cm	4.2 to 8.4	0.2	High

Table 54 Properties of Rigid and Semi-Rigid Insulation

Material Identification Number			43	44
Composition			Wood fibers with binders-Board	Borosilica closed cell inorganic foam-Block
Abrasion Resistance % of Weight Loss		1st Run		Poor abrasion
		2nd Run		Resistance
Alkalinity pH			4.5 to 6	6 to 7
Capillarity				None
Coefficient of Expansion		English		1.6 X 10^{-6} / °F
		Metric		2.8 X 10^{-8} / °C
Combustibility			Combustible	Non-combustible
If Combustible	Flash Point Temp.	°F	500	"
		°C	260	"
	Flame Point Temp.	°F	500	"
		°C	260	"
	Heat Release	Btu/lb	8,000	"
		J/kg	1.859 X 10^7	"
	Self Internal Heating			"
	Smoldering – Afterglow			"
Corrosion – Rusting				Does not contribute
Corrosion – Stress				Does not contribute
Density		lb/ft³	15 to 18	12
		kg/m²	240 to 288	190
Hardness – mm penetration				
Hygroscopicity % Volume			14	0
Resistance to Acids				Impervious
Resistance to Caustics				Impervious
Resistance to Solvents				Impervious
Shrinkage %		Linear	5.4% at 200° F	None
		Volumetric		None
Specific Gravity		Real	0.95	2.3
		Apparent	0.24	0.19
Specific Heat		Btu/lb		0.2
		w/kg		0.11
Strength	Compressive	psi		200 to 210
		kPa		1400 to 1970
	Flexural	psi/m		90 to 120
		kPa		580 to 770
	Modulus of Rupture	psi	50	
		kPa	344	
	Tensile	psi		
		kPa		
Temperature Limits, Service	Short Time	°F	200	960 (unloaded)
		°C	93	517 (unloaded)
	Continuous	°F	200	800 (with applied load)
		°C	93	425 (with applied load)
Melting Temperature		°F		3100
		°C		1704
Thermal Diffusivity		ft²/hr		
		m²/hr		
Thermal Shock Resistance				Excellent
Vibration Resistance			Fair	Fair
Water Absorption		% by Volume	10	2% (surface only)
Water Vapor Transmission		perm-inch	5	0.00000+
		perm-cm	9	0.00000+

Table 55 Properties of Flexible and Faced Insulation

Material Identification Number			45	46
Composition			Elastomeric sheet and pipe covering	Glass fiber, flexible with various facings
Alkalinity pH				7 to 8
Capillarity			Negligable	Will Wick
Combustibility			Combustible	Binder is combustible
If Combustible	Flash Point Temp.	°F	560	(Binders) 750
		°C	293	(Binders) 399
	Flame Point Temp.	°F	590	Smolders
		°C	310	Smolders
	Heat Release	Btu/lb	15000	Fibers non-combustible
		J/kg	3.486 X 10^7	Fibers non-combustible
	Self Internal Heating			
	Smoldering or Afterglow			Can Smolder
Corrosion – Rusting			Does not contribute	Does not contribute
Corrosion – Stress			Does not contribute	Does not contribute
Density of Insulation (only)		lb/ft³	4.5 to 8.5	0.6 to 3.0
		kg/m³	36 to 72	10.4 to 48.0
Hygroscopicity % Weight				
Maximum Temperature Limits	Short	°F	220	400
		°C	104	204
	Continuous	°F	220	400
		°C	104	204
Resistance to Acids			Resistant to dilute acids	Good from pH 5 to 7 except HF & H_3PO_4
Resistance to Caustics			Resistant to dilute acids	Good from pH 7 to pH 10
Specific Heat		Btu/lb		.20
		w/kg		.11
Strength Compression		psi		902 to 0.09 at 10% deform.
		Pa		
Thermal Diffusivity		ft²/hr		0.13 to 0.62 at 10% deform.
		m²/hr		0.06 at 70°F
Water Absorption % Weight			4.6 to 8.4	9.126 X 10^{-6} at 21°C
Water Vapor Transmission		perm-inches	0.15 to 0.30	Depends on facing
		perm-cm	0.27 to 0.54	Depends on facing

Material Identification Number			47	48
Composition			Glass fiber flexible with various facings**	Glass fiber, flexible semi-rigid w/various facings**
Alkalinity pH			7 to 8	7 to 8
Capillarity			Will Wick	Will Wick
Combustibility			Binder is combustible	Binder is combustible
If Combustible	Flash Point Temp.	°F	(Binder) 750°F	(Binder) 750°F
		°C	(Binder) 399	(Binder)
	Flame Point Temp.	°F	Smolders	Smolders
		°C	Smolders	Smolders
	Heat Release	Btu/lb	Fibers non-combustible	Fibers non-combustible
		J/kg	Fibers non-combustible	Fibers non-combustible
	Self Internal Heating			
	Smoldering or Afterglow		Can Smolder	Can Smolder
Corrosion – Rusting			Does not contribute	Does not contribute
Corrosion – Stress			Depends on binder	Depends on binder
Density of Insulation (only)		lb/ft³	3 to 4.5	4.5 (min.)
		kg/m³	48 to 72	72 (min.)
Hygroscopicity % Weight				
Maximum Temperature Limits	Short	°F	400	400
		°C	204	204
	Continuous	°F	400	400
		°C	204	204
Resistance to Acids			Good from pH 5 to 7 except HF & H_3PO_4	Good from pH 5 to 7 except HF & H_3PO_4
Resistance to Caustics			Good from pH 7 to pH 10	Good from pH 7 to pH 10
Specific Heat		Btu/lb	.20	.20
		w/kg	.11	.11
Strength Compression		psi	0.5 at 10% deform.	0.5 to 1.5 at 10% deform.
		Pa	3.45 at 10% deform.	3.45 to 10.35 at 10% deform.
Thermal Diffusivity		ft²/hr	0.16 at 70°F	0.05 to 0.15 at 70°F
		m²/hr	4.126 X 10^{-6} at 21°C	1.25 to 3.75 X 10^{-6} at 21°C
Water Absorption % Weight				
Water Vapor Transmission		perm-inches	Depends on facing	Depends on facing
		perm-cm	Depends on facing	Depends on facing

**Facings may be woven wire mesh, expanded metal lath, (copper bearing) on one or both sides.

Table 55 Properties of Flexible and Faced Insulation

Material Identification Number			49	50
Composition			Kaolin ceramic fiber in ss* metal mesh	Kaolin ceramic fiber in ss* metal mesh
Alkalinity pH				
Capillarity			Will Wick	Will Wick
Combustibility			Non-combustible	non-combustible
If Combustible	Flash Point Temp.	°F	"	"
		°C	"	"
	Flame Point Temp.	°F	"	"
		°C	"	"
	Heat Release	Btu/lb	"	"
		J/kg	"	"
	Self Internal Heating		"	"
	Smoldering or Afterglow		"	"
Corrosion – Rusting			Does not contribute	Does not contribute
Corrosion – Stress			Does not contribute	Does not contribute
Density of Insulation (only)		lb/ft³	3 to 4	6 to 8
		kg/m³	48 to 64	96 to 128
Hygroscopicity % Weight				
Maximum Temperature Limits	Short	°F	3000	3000
		°C	1649	1649
	Continuous	°F	2300	2300
		°C	1260	1260
Resistance to Acids			Fair except HF & H_3PO_4	Fair except HF & H_3PO_4
Resistance to Caustics			Fair to Poor	Fair to Poor
Specific Heat		Btu/lb	0.285 at 1800°F	0.285 at 1800°F
		w/kg	0.158 at 982°C	0.158 at 982°C
Strength Compression		psi		
		Pa		
Thermal Diffusivity		ft²/hr		
		m²/hr		
Water Absorption % Weight				
Water Vapor Transmission		perm-inches	High	High
		perm-cm	High	High

Material Identification Number			53	54
Composition			Mineral fibers w/metal mesh facing** Class II	Urethane foam, block, pipe cov.
Alkalinity pH			7 to 9	6.5 to 7.5
Capillarity			Will Wick	Negligable
Combustibility			Non-combustible	Combustible
If Combustible	Flash Point Temp.	°F	"	590 to 750
		°C	"	310 to 399
	Flame Point Temp.	°F	"	600 to 750
		°C	"	316 to 300
	Heat Release	Btu/lb	"	15,000
		J/kg	"	3.71×10^7
	Self Internal Heating		"	Will react with various liquids and gases
	Smoldering or Afterglow		"	Chars
Corrosion – Rusting			Will not contribute	Will not contribute
Corrosion – Stress			Will not contribute	Will not contribute
Density of Insulation (only)		lb/ft³	12 (ave. max.)	Up to 1.5
		kg/m³	192 (ave. max.)	Up to 24
Hygroscopicity % Weight			1.5	
Maximum Temperature Limits	Short	°F	1200	210
		°C	650	99
	Continuous	°F	1200	185
		°C	650	85
Resistance to Acids			Poor	Resistant to dilute acids
Resistance to Caustics			Poor	Resistant to dilute caustics
Specific Heat		Btu/lb	.22	.25
		w/kg	,12	.14
Strength Compression		psi	25 at 10% deform.	8 at 5% deform.
		Pa	172 at 10% deform.	5.5 at 5% deform.
Thermal Diffusivity		ft²/hr	0.22 at 70°F	0.011 to 0.027 at 70°F
		m²/hr	5.67×10^{-6} at 21°C	2.83×10^{-7} to 6.93×10^{-7} at 21°C
Water Absorption % Weight			96	5
Water Vapor Transmission		perm-inches	High	5
		perm-cm	High	8.3

* ss = Stainless steel metal mesh.

** Facings may be woven wire mesh, expanded metal lath (copper bearing) on one, or both sides.

Table 55 Properties of Flexible and Faced Insulation

Material Identification Number			51	52
Composition			Mineral fibers-spun glass fibers metal mesh**	Mineral fibers w/metal mesh facing-** Class I
Alkalinity pH			7 to 8	7 to 9
Capillarity			Will Wick	Will Wick
Combustibility			Non-combustible	Non-combustible
If Combustible	Flash Point Temp.	°F	"	"
		°C	"	"
	Flame Point Temp.	°F	"	"
		°C	"	"
	Heat Release	Btu/lb	"	"
		J/kg	"	"
	Self Internal Heating		"	"
	Smoldering or Afterglow		"	"
Corrosion – Rusting			Does not contribute	Does not contribute
Corrosion – Stress			Does not contribute	Does not contribute
Density of Insulation (only)		lb/ft^3	10 to 14	6 (ave. max.)
		kg/m^3	160 to 224	96 (ave. max.)
Hygroscopicity % Weight			1.5	2.0
Maximum Temperature Limits	Short	°F	1200	1000
		°C	649	540°C
	Continuous	°F	1200	1000
		°C	649	540°C
Resistance to Acids			Fair to Poor	Fair to Poor
Resistance to Caustics			Fair to Poor	Fair to Poor
Specific Heat		Btu/lb	.20	.22
		w/kg	.11	.12
Strength Compression		psi		10 at 10% deform.
		Pa		69 at 10% deform.
Thermal Diffusivity		ft^2/hr	0.02 to 0.04 at 70°F	0.02 to 0.04 at 70°F
		m^2/hr	5.16×10^{-7} to 1.03×10^{-6} at 21°C	5.16×10^{-7} to 1.03×10^{-7} at 21°C
Water Absorption % Weight				96
Water Vapor Transmission		perm-inches	High	High
		perm-cm	High	High

Material Identification Number			55	56
Composition			Urethane foam, block, pipe cov.	Alumina-silica fiber, blanket and strip
Alkalinity pH			6.5 to 7.5	
Capillarity			Negligable	Will Wick
Combustibility			Combustible	Non-combustible
If Combustible	Flash Point Temp.	°F	590 to 750	"
		°C	310 to 750	"
	Flame Point Temp.	°F	600 to 750	"
		°C	316 to 399	"
	Heat Release	Btu/lb	16,000	"
		J/kg	3.71×10^7	"
	Self Internal Heating		Will react with various liquids and gases	"
	Smoldering or Afterglow		Chars	"
Corrosion – Rusting			Will contribute	Does not contribute
Corrosion – Stress			Will contribute	Does not contribute
Density of Insulation (only)		lb/ft^3	Over 1.5	3 to 4
		kg/m^3	Over 24	48 to 64
Hygroscopicity % Weight			210	3
Maximum Temperature Limits	Short	°F	99	2300
		°C	185	1260
	Continuous	°F	85	2300
		°C	.24	1260
Resistance to Acids			Resistant to dilute acids	Good, except HF & H_3PO_4
Resistance to Caustics			Resistant to dilute caustics	Good to poor
Specific Heat		Btu/lb	.25	0.25
		w/kg	.14	0.14
Strength Compression		psi	30 at 5% deform.	
		Pa	205 at 5% deform.	
Thermal Diffusivity		ft^2/hr	0.011 to 0.027	
		m^2/hr	2.83×10^{-7} to 6.99×10^{-7}	
Water Absorption % Weight			2	Saturates
Water Vapor Transmission		perm-inches	1.5	High
		perm-cm	2.5	High

* ss = Stainless steel metal mesh.

** Facings may be woven wire mesh, expanded metal lath (copper bearing) on one, or both sides.

Table 55 Properties of Flexible and Faced Insulation

Material Identification Number			57	58
Composition			Alumina-silica fiber, blanket and strip	Alumina silica fiber, blanket and strip
Alkalinity pH				
Capillarity			Will Wick	Will Wick
Combustibility			Non-combustible	Non-combustible
If Combustible	Flash Point Temp.	°F	"	"
		°C	"	"
	Flame Point Temp.	°F	"	"
		°C	"	"
	Heat Release	Btu/lb	"	"
		J/kg	"	"
	Self Internal Heating		"	"
	Smoldering or Afterglow		"	"
Corrosion – Rusting			Does not contribute	Does not contribute
Corrosion – Stress			Does not contribute	Does not contribute
Density of Insulation (only)		lb/ft³	6 to 8	12
		kg/m³	96 to 128	192
Hygroscopicity % Weight			3	3
Maximum Temperature Limits	Short	°F	2300	2300
		°C	1260	1260
	Continuous	°F	2300	2300
		°C	1260	1093
Resistance to Acids			Good, except HF & H_3PO_4	Good, except HF & H_3PO_4
Resistance to Caustics			Fair to poor	Fair to poor
Specific Heat		Btu/lb	0.25	0.25
		w/kg	0.14	0.14
Strength Compression		psi		
		Pa		
Thermal Diffusivity		ft²/hr		
		m²/hr		
Water Absorption % Weight			Saturates	Saturates
Water Vapor Transmission		perm-inches	High	High
		perm-cm	High	High

Material Identification Number			61	62
Composition			Glass fine-fiber with binder, Blanket	Glass fiber, with binder
Alkalinity pH			6 to 10	7 to 9
Capillarity			None	Negligable
Combustibility			Non-combustible	Binder-Combustible
If Combustible	Flash Point Temp.	°F	650 to 700	650 to 700
		°C	343 to 371	343 to 371
	Flame Point Temp.	°F	1100	1100
		°C	593	593
	Heat Release	Btu/lb	Only binder burns	Only binder burns
		J/kg	Only binder burns	Only binder burns
	Self Internal Heating			
	Smoldering or Afterglow		Smolders- (Binder only)	Smolders- (Binder only)
Corrosion – Rusting			Will not contribute	Will not contribute
Corrosion – Stress			s.s. should be protected	s.s. should be protected
Density of Insulation (only)		lb/ft³	1/2 to 2	1 to 2
		kg/m³	8 to 32	16 to 32
Hygroscopicity % Weight			0.2	0.2
Maximum Temperature Limits	Short	°F	370	400
		°C	188	204
	Continuous	°F	370	400
		°C	188	204
Resistance to Acids			Good, except HF & H_3PO_4	Good, except HF & H_3PO_4
Resistance to Caustics			Good from pH 7 to pH 10	Good from pH 7 to pH 10
Specific Heat		Btu/lb	0.20	0.20
		w/kg	0.11	0.11
Strength Compression		psi	*0.02 at 10% deform.	4.5 at 10% deform.
		Pa	**0.13 at 10% deform.	31 at 10% deform.
Thermal Diffusivity		ft²/hr	0.10 at 100°F (1 lb)	6.10 at 100°F
		m²/hr	2.58×10^{-6} at 38°C	2.58×10^{-6} at 38°C
Water Absorption % Weight			98	98
Water Vapor Transmission		perm-inches	High	High
		perm-cm	High	High

* 2 pounds per cubic foot density.
** 32 kg/m³ density.

Table 55 Properties of Flexible and Faced Insulation

Material Identification Number			59	60
Composition			Alumina silica fiber, rope, cord, yarn	Glass, fine fiber (no binder) Blanket and Matt
Alkalinity pH				6 to 8
Capillarity			Will Wick	None
Combustibility			Non-combustible	Non-combustible
If Combustible	Flash Point Temp.	°F	"	"
		°C	"	"
	Flame Point Temp.	°F	"	"
		°C	"	"
	Heat Release	Btu/lb	"	"
		J/kg	"	"
	Self Internal Heating		"	"
	Smoldering or Afterglow		"	"
Corrosion – Rusting			Does not contribute	Will not contribute
Corrosion – Stress			Does not contribute	Will not contribute
Density of Insulation (only)		lb/ft^3	25	11 to 12
		kg/m^3	400	176 to 192
Hygroscopicity % Weight			3	5
Maximum Temperature Limits	Short	°F	2300	1200
		°C	1260	649
	Continuous	°F	2000	1200
		°C	1093	649
Resistance to Acids			Good except HF & H_3PO_4	Excellent except, HF & H_3PO_4
Resistance to Caustics			Fair to Poor	Good from pH 7 to pH 10
Specific Heat		Btu/lb	0.25	0.20
		w/kg	0.14	0.11
Strength Compression		psi		20 at 63% deform.
		Pa		137 at 63% deform.
Thermal Diffusivity		ft^2/hr		0.08 at 150°F
		m^2/hr		2.06×10^{-6} at 66°C
Water Absorption % Weight			Saturates	93
Water Vapor Transmission		perm-inches	High	High
		perm-cm	High	High

Material Identification Number			63	64
Composition			Glass fiber with binder	Mineral fiber, **Blanket & Batt**
Alkalinity pH			7 to 9	7 to 9
Capillarity			Negligable	Will Wick
Combustibility			Binder-Combustible	*Non-combustible
If Combustible	Flash Point Temp.	°F	650 to 700	Fibers-non-combustible
		°C	343 to 371	"
	Flame Point Temp.	°F	1100	"
		°C	593	"
	Heat Release	Btu/lb	Only binder burns	"
		J/kg	Only binder burns	"
	Self Internal Heating			*Depends on manufacturer
	Smoldering or Afterglow		Smolders- (Binder only)	*Depends on manufacturer
Corrosion – Rusting			Will not contribute	Will not contribute
Corrosion – Stress			s.s. should be protected	Will not contribute
Density of Insulation (only)		lb/ft^3	2.5 to 3.5	8 to 12
		kg/m^3	40 to 56	128 to 192
Hygroscopicity % Weight			0.2	0.5 to 1.0
Maximum Temperature Limits	Short	°F	400	1200
		°C	204	649
	Continuous	°F	400	1200
		°C	204	649
Resistance to Acids			Good, except HF & H_3PO_4	Poor to fair
Resistance to Caustics			Good from pH 7 to pH 10	Poor to good
Specific Heat		Btu/lb	0.20	0.20 to 0.22
		w/kg	0.11	0.11 to 0.12
Strength Compression		psi	100 at 10% deform.	4 to 6 at 10% deform.
		Pa	689 at 10% deform.	25 to 41 at 10% deform.
Thermal Diffusivity		ft^2/hr	0.35 at 100°F	0.012 at 200°F
		m^2/hr	9.03×10^{-6} at 38°C	3.09×10^{-7} at 93°C
Water Absorption % Weight			96	90
Water Vapor Transmission		perm-inches	High	No resistance
		perm-cm	High	No resistance

* Some oil is used to lubricate fibers. Slow oxidation can cause internal heating and smoldering.

Table 55 Properties of Flexible and Faced Insulation

Material Identification Number				65	66
Composition				Mineral fibers with binders, **Blanket**	Silica fibers, Blankets & Felts
Alkalinity pH				7 to 9	5 to 7
Capillarity				Will Wick	Will Wick
Combustibility				*Non-combustible	Non-combustible
If Combustible	Flash Point Temp.		°F	Fibers- non combustible	"
			°C	"	"
	Flame Point Temp.		°F	"	"
			°C	"	"
	Heat Release		Btu/lb	"	"
			J/kg	"	"
	Self Internal Heating			*Depends on binders	"
	Smoldering or Afterglow			*Depends on binders	"
Corrosion — Rusting				Will not contribute	
Corrosion — Stress				Will not contribute	
Density of Insulation (only)			lb/ft³	7 to 9	6 to 9.5
			kg/m³	112 to 144	96 to 152
Hygroscopicity % Weight				2	10
Maximum Temperature Limits	Short		°F	1000	1800
			°C	540	982
	Continuous		°F	1000	1800
			°C	540	982
Resistance to Acids				Poor to Fair	Good except HF & H_3PO_4
Resistance to Caustics				Poor to Good	Poor
Specific Heat			Btu/lb	0.24	0.19
			w/kg	0.13	0.10
Strength Compression			psi	120 at 10% deform.	10 at 50% deform.
			Pa	703 at 10% deform.	69 at 50% deform.
Thermal Diffusivity			ft^2/hr	0.01 to 0.026 at 200° F	0.039 to 0.054 at 1000° F
			m^2/hr	2.6×10^{-7} to 6.7×10^{-7} at 93° C	1.0×10^{-6} to 1.39×10^{-6} at 540° C
Water Absorption % Weight				93	High
Water Vapor Transmission			perm-inches	No resistance	No resistance
			perm-cm	No resistance	No resistance

* Some oil is used to lubricate fibers. Slow oxidation can cause internal heating and smoldering.

Table 56 Properties of Insulating and Finishing Cements

Material Identification Number			67	68	69
Composition			Alumina-silica semi-refractory hydraulic setting	Alumina-silica ceramic fiber, air setting mold mix cement	Alumina-silica ceramic-spray mix cement
Abrasion Resistance % of Weight Loss		1st Run			
		2nd Run			
Alkalinity pH			9 to 10	9.7	9 to 10
Capillarity			Will Wick	Will Wick	Will Wick
Coefficient of Expansion		English	Shrinks 2% at 1800°F	Shrinks 2.1% at 1800°F	Shrinks 1.4% at 1800°F
		Metric	Shrinks 2% at 987°C	Shrinks 2.1% at 987°C	Shrinks 1.4% at 982°C
Combustibility			Non-combustible	Non-combustible	Non-combustible
If Combustible	Flash Point Temp.	°F	"	"	"
		°C	"	"	"
	Flame Point Temp.	°F	"	"	"
		°C	"	"	"
	Heat Release	Btu/lb	"	"	"
		w/kg	"	"	"
	Self Internal Heating		"	"	"
	Smoldering – Afterglow		"	"	"
Corrosion – Rusting					
Corrosion – Stress					
Coverage Dried or Set		board ft/100 lbs	15 to 25		7 to 8
		m^2 cm/100 kgs	7.8 to 13		6.2 to 6.8
Density Dried or Set		lbs/ft^3	48 to 70	75	12 to 13
		kgs/m^3	768 to 1120	1200	192 to 208
Hardness – mm penetration					
Hygroscopicity % Weight					
Maximum Temperature Limits	Short	°F	2000	2300	2300
		°C	1093	1257	1257
	Continuous	°F	2000	2300	1800
		°C	1093	1257	982
Shrinkage Wet to Dry		Volumetric			
Specific Gravity		Real			
		Apparent	.77 to 1.12	1.2	0.21
Specific Heat		Btu/lb	.27	.27	.27
		w/kg	.15	.15	.15
Strength	Compressive	psi		6500	0.6 at 5% deform.
		kPa		451,500	4.1 at 5% deform.
	Shear	psi			
		kPa			
	Tensile	psi			
		kPa			
Thermal Diffusivity		ft^2/hr			
		m^2/hr			
Thermal Shock Resistance			Excellent	Excellent	Excellent
Vibration Resistance					
Water Absorption % Weight			High	High	High
Water Vapor Transmission		perm-inch	High	High	High
		perm-cm	High	High	High

Table 56 Properties of Insulating and Finishing Cements

Material Identification Number			70	71	72
Composition			Diatomateous silica & binders air setting cement	Kaowool ceramic fiber & inorganic binders-spray cement	Kaowool and mineral fiber blend-spray cement
Abrasion Resistance % of Weight Loss		1st Run	10		
		2nd Run	20		
Alkalinity pH			9		
Capillarity			Will Wick		
Coefficient of Expansion		English	Shrinks 5% at 1900° F	Shrinks 1.5% at 2000° F	
		Metric	Shrinks 5% at 1040° C	Shrinks 1.5% at 1093° C	
Combustibility			Non-combustible	Non-combustible	Non-combustible
If Combustible	Flash Point Temp.	°F	"	"	"
		°C	"	"	"
	Flame Point Temp.	°F	"	"	"
		°C	"	"	"
	Heat Release	Btu/lb	"	"	"
		w/kg	"	"	"
	Self Internal Heating		"	"	"
	Smoldering – Afterglow		"	"	"
Corrosion – Rusting					
Corrosion – Stress			s.s. should be protected		
Coverage Dried or Set		board ft/100 lbs	35	100	170 to 180
		m^2 cm/100 kgs	18		
Density Dried or Set		lbs/ft^3	23 to 32	12 to 13	7 to 15
		kgs/m^3	368 to 512	192 to 206	112 to 288
Hardness – mm penetration			0.95 to 1.20		
Hygroscopicity % Weight					
Maximum Temperature Limits	Short	°F	1900	2000	800 to 1600
		°C	1040	1093	427 to 871
	Continuous	°F	1900	1000	800 to 1600
		°C	1040	982	427 to 871
Shrinkage Wet to Dry		Volumetric	30%		
Specific Gravity		Real	2.3		
		Apparent	0.37 to 51		
Specific Heat		Btu/lb	0.22 to 0.28		
		w/kg	0.12 to 0.16		
Strength	Compressive	psi	25	0.6 at 5% deform.	55 at 5% deform.
		kPa	173		2630 at 5% deform.
	Shear	psi			
		kPa			
	Tensile	psi			
		kPa			
Thermal Diffusivity		ft^2/hr	0.01		
		m^2/hr	2.58×10^{-7}		
Thermal Shock Resistance			Excellent	Excellent	Excellent
Vibration Resistance			Poor		
Water Absorption % Weight			40 to 50		
Water Vapor Transmission		perm-inch	High	High	High
		perm-cm	High	High	High

Table 56 Properties of Insulating and Finishing Cements

Material Identification Number			73	74	75
Composition			Mineral fiber water mix insulating cement	Mineral fiber binders hydraulic setting cement	Vermiculte expanded with binders water mix-air setting
Abrasion Resistance % of Weight Loss		1st Run	25 to 45	15 to 35	35
		2nd Run	50 to 75	25 to 65	60
Alkalinity pH			8.5 to 11	8½ to 12½	6.9
Capillarity			High	High	High
Coefficient of Expansion		English	Shrinks 3% to 1800° F	Shrinks 2% to 1200° F	Shrinks 3% at 1800° F
		Metric	Shrinks 3% to 980° C	Shrinks 2% to 689° C	Shrinks 3% at 982° C
Combustibility			Non-combustible	Non-combustible	Non-combustible
If Combustible	Flash Point Temp.	°F	"	"	"
		°C	"	"	"
	Flame Point Temp.	°F	"	"	"
		°C	"	"	"
	Heat Release	Btu/lb	"	"	"
		w/kg	"	"	"
	Self Internal Heating		"	"	"
	Smoldering – Afterglow		"	"	"
Corrosion – Rusting			Holds Water	Holds Water	Holds Water
Corrosion – Stress			s.s. must be protected	s.s. must be protected	
Coverage Dried or Set		board ft/100 lbs	40	30	65
		m^2 cm/100 kgs	21	15.3	34
Density Dried or Set		lbs/ft^3	22 to 30	27 to 44	18 to 19
		kgs/m^3	352 to 512	432 to 704	288 to 304
Hardness – mm penetration			1 to 2.5		
Hygroscopicity % Weight			2.5 to 5	0.8 to 2.0	
Maximum Temperature Limits	Short	°F	1800	1200	1800
		°C	980	650	980
	Continuous	°F	1800	1200	1800
		°C	980	650	980
Shrinkage Wet to Dry		Volumetric	25%	10%	20%
Specific Gravity		Real	2.3	2.5	2.4
		Apparent	0.35 to .48	0.43 to 0.70	0.28 to 0.30
Specific Heat		Btu/lb	0.22 to 0.23	0.22 to 0.23	0.22
		w/kg	0.12 to 0.13	0.12 to 0.13	0.12
Strength	Compressive	psi	10 at 5% deform.	100 at 5% deform.	15 at 5% deform.
		kPa	69 at 5% deform.	689 at 5% deform.	103 at 5% deform.
	Shear	psi			
		kPa			
	Tensile	psi			
		kPa			
Thermal Diffusivity		ft^2/hr	0.006 to 0.01 at 200° F	0.007 at 200° F	0.015 at 200° F
		m^2/hr	1.54×10^{-7} to 2.58×10^{-7} at 93° C	1.8×10^{-7} at 93° C	3.87×10^{-7} at 93° C
Thermal Shock Resistance			Good	Good	Good
Vibration Resistance					
Water Absorption % Weight			High	High	High
Water Vapor Transmission		perm-inch	High	High	High
		perm-cm	High	High	High

Table 57 Properties of Foamed or Sprayed Insulations

Material Identification Number			76	77	78
Composition			Polyurethune – 2 part mix foamed air-sprayed in place	Urea-formaldehyde foamed in place	Poly Vinyl-acetate cork filled mastic sprayed or trowelled
Abrasion Resistance % of Weight Loss		1st Run			
		2nd Run			
Alkalinity pH			6.5 to 7.5		7 to 7.5
Capillarity					
Coefficient of Expansion		English	Expands	1.8% to 3% at 70°F	
		Metric	Expands	1.8% to 3% at 21°C	
Combustibility			Combustible	Slow burning	Slow burning
If Combustible	Flash Point Temp.	°F	590		
		°C	310		
	Flame Point Temp.	°F	630	1280°F	
		°C	332	654	
	Heat Release	Btu/lb	16,000		
		w/kg	3.71×10^7		
	Self Internal Heating		Will react with some chemicals		Will not self internal heat
	Smoldering – Afterglow		Will smolder	Will char	Will char
Corrosion – Rusting			Will contribute	Will contribute	Pre coat steel
Corrosion – Stress			Must be protected		Must be pre-coated
Coverage Dried or Set		board ft/100 lbs	20 to 30		¼" thk. 100 ft^2 – 24 gal.
		m^2 cm/100 kgs	11 to 16		6.35 mm thk. 10 m^2 – 100 litre
Density Dried or Set		lbs/ft^3	1.6 to 3.0	0.6 to 1.0	53
		kgs/m^3	25 to 48	9.6 to 16	848
Hardness – mm penetration					
Hygroscopicity % Weight			0.8 to 1.4	2	
Maximum Temperature Limits	Short	°F	220	212	180
		°C	104	100	82
	Continuous	°F	200	180	180
		°C	94	82	82
Shrinkage Wet to Dry		Volumetric		3	30
Specific Gravity		Real			
		Apparent		0.009 to 0.016	25 to 35
Specific Heat		Btu/lb	.4 at 40°F		
		w/kg	0.22 at 4°C		
Strength	Compressive	psi	15 at yield		
		kPa	103 at yield		
	Shear	psi			
		kPa			
	Tensile	psi			
		kPa			
Thermal Diffusivity		ft^2/hr	0.02 to 0.03 at 40°F		
		m^2/hr	5.1×10^{-7} to 7.7×10^{-7} at 4°C		
Thermal Shock Resistance					Excellent
Vibration Resistance			Excellent		Excellent
Water Absorption % Weight					
Water Vapor Transmission		perm-inch	2.5 to 5.0	15 to 38	
		perm-cm	4.2 to 8.4	24 to 61	

Table 58 Properties of Loose and Fill Insulations

Material Identification Number			79	80	81	82
Composition			Alumina-silica fibers-Bulk	Alumina-silica fibers-Chopped and milled	Alumina-silica fibers-long fibers-bulk	Cork, vegetable Granulated
Alkalinity pH						6.5 to 7.5
Combustibility			Non-combustible	Non-combustible	Non-combustible	Combustible
If Combustible	Flash Point Temp.	°F	"	"	"	570
		°C	"	"	"	299
	Flame Point Temp.	°F	"	"	"	620
		°C	"	"	"	327
	Heat Release	Btu/lb	"	"	"	8000
		w/kg	"	"	"	1.85×10^7
	Self Internal Heating		"	"	"	
	Smoldering or Afterglow		"	"	"	
Density As Received		lb/ft^3	3 to 4	7.5	3	8 to 12
		kg/m^3	48 to 64	120	48	128 to 192
Packed		lb/ft^3	6	7.5	6	12 to 14
		kg/m^3	96	120	96	192 to 224
Hydroscopicity % Weight						
Maximum Temperature Limits	Short Time	°F	2300	2300	2300	200
		°C	1256	1256	1256	93
	Continuous	°F	2300	2300	2300	200
		°C	1256	1256	1256	93
Specific Heat		Btu/lb	0.20	0.20	0.20	0.27
		w/kg	0.11	0.11	0.11	0.15
Thermal Diffusivity		ft^2/hr				
		m^2/hr				
Water Absorption % Weight			High	High	High	Low

Material Identification Number			83	84	85	86
Composition			Cellulose fiber Loose Fill	Diatomaceous silica fine powder	Diatomaceous silica-Caloined powder	Diatomaceous silica-Coarse powder
Alkalinity pH			4.5 to 6.0	5 to 7	6 to 7	5 to 7
Combustibility			Combustible	Non-combustible	Non-combustible	Non-combustible
If Combustible	Flash Point Temp.	°F	500	"	"	"
		°C	260	"	"	"
	Flame Point Temp.	°F	500	"	"	"
		°C	260	"	"	"
	Heat Release	Btu/lb	8000	"	"	"
		w/kg	1.859×10^7	"	"	"
	Self Internal Heating			"	"	"
	Smoldering or Afterglow		Can smolder	"	"	"
Density As Received		lb/ft^3		10 to 12	22	25 to 27
		kg/m^3		160 to 192	352	400 to 432
Packed		lb/ft^3	10 to 15	13 to 17	32	25 to 31
		kg/m^3	160 to 240	208 to 272	496	400 to 496
Hydroscopicity % Weight						
Maximum Temperature Limits	Short Time	°F	200	1600	2000	1600
		°C	93	871	1093	871
	Continuous	°F	200	1600	2000	1600
		°C	93	871	1093	871
Specific Heat		Btu/lb		0.25	0.25	0.25
		w/kg		0.14	0.14	0.14
Thermal Diffusivity		ft^2/hr		0.01	0.01	0.009
		m^2/hr		2.58×10^{-7}	2.58×10^{-7}	2.32×10^{-7}
Water Absorption % Weight			High	High	High	High

Table 58 Properties of Loose and Fill Insulations

Material Identification Number			87	88	89	90
Composition			Gilsonite granules, for packing around underground pipe	Glass fiber, unbonded no binders loose fibers	Glass fibers Loose & Blowing Wool	Gypsum pellets Bulk
Alkalinity pH			7	8 to 10	8 to 10	
Combustibility			Combustible	Non-combustible	Fiber coating or lubi. maybe combustible	Non-combustible
If Combustible	Flash Point Temp.	°F	600	"	Fibers non-combustible	"
		°C	316	"	"	"
	Flame Point Temp.	°F	600	"	"	"
		°C	316	"	"	"
	Heat Release	Btu/lb	17000	"		"
		w/kg	3.95 × 10^7	"		"
	Self Internal Heating			"		"
	Smoldering or Afterglow		Will smolder	"	May smolder	"
Density As Received		lb/ft³	40	2 to 12		12 to 20
		kg/m³	640	32 to 192		
Packed		lb/ft³	44 to 50	2 to 12	2 to 6	20 to 30
		kg/m³	704 to 800	32 to 192	32 to 96	320 to 480
Hydroscopicity % Weight			Very Low	0	0	
Maximum Temperature Limits	Short Time	°F	300 to 460	1000	350	800
		°C	149 to 238	538	177	427
	Continuous	°F	300 to 460	1000	350	800
		°C	149 to 238	538	177	427
Specific Heat		Btu/lb	0.5 at 200°F	0.2 at 200°F	0.2 at 100°F	
		w/kg	0.27 at 93°C	0.11 at 93°C	0.11 at 38°C	
Thermal Diffusivity		ft²/hr	0.0023 at 200°F	0.0167 at 100°F		
		m²/hr	5.93 × 10^{-8} at 93°C	4.3 × 10^{-7} at 38°C		
Water Absorption % Weight			Low	800	1000	High

Material Identification Number			91	92	93	94
Composition			Mineral fiber with small amounts of oil, Loose	Mineral fiber, oil free, Loose	Mineral fiber granulated	Perlite, expanded Loose powder
Alkalinity pH			7 to 9	7 to 9	7 to 9	
Combustibility			Lub. oil is combustible	Non-combustible	Non-combustible	Non-combustible
If Combustible	Flash Point Temp.	°F	Fibers are non-combus.	"	"	"
		°C	"	"	"	"
	Flame Point Temp.	°F	"	"	"	"
		°C	"	"	"	"
	Heat Release	Btu/lb	"	"	"	"
		w/kg	"	"	"	"
	Self Internal Heating		Can develop self internal heating	"	"	"
	Smoldering or Afterglow		Will smolder	"	"	"
Density As Received		lb/ft³	4 to 11	4 to 11	7 to 10	
		kg/m³	64 to 176	64 to 176	112 to 160	
Packed		lb/ft³	4 to 12	10 to 15	6 to 12	5 to 8
		kg/m³	64 to 192	160 to 240	96 to 192	80 to 128
Hydroscopicity % Weight						
Maximum Temperature Limits	Short Time	°F	1000	1200	1000 to 1200	1800
		°C	538	649	538 to 649	982
	Continuous	°F	1000	1200	1000 to 1200	1800
		°C	538	649	.22	982
Specific Heat		Btu/lb	.22	.22	.12	
		w/kg	.12	.12		
Thermal Diffusivity		ft²/hr				
		m²/hr				
Water Absorption % Weight			High	High	High	High
Capillarity			Negligable	Negligable	Negligable	Negligable

Table 58 Properties of Loose and Fill Insulations

Material Identification Number			95	96	97	98
Composition			Quartz fibers Loose	Silica fibers Loose	Silica aerogel Granules	Vermiculite Flakes
Alkalinity pH			5 to 7	5 to 7	3.5 to 4.0	6 to 10
Combustibility			Non-combustible	Non-combustible	Non-combustible	Non-combustible
If Combustible	Flash Point Temp.	°F	"	"	"	"
		°C	"	"	"	"
	Flame Point Temp.	°F	"	"	"	"
		°C	"	"	"	"
	Heat Release	Btu/lb	"	"	"	"
		w/kg	"	"	"	"
	Self Internal Heating		"	"	"	"
	Smoldering or Afterglow		"	"	"	"
Density As Received		lb/ft^3		6 to 12	3.5 to 5.5	4 to 10
		kg/m^3		96 to 192	56 to 88	64 to 160
Packed		lb/ft^3	6 to 12		4 to 5.5	4 to 10
		kg/m^3	96 to 192		64 to 88	64 to 160
Hydroscopicity % Weight				2 to 10	10 to 15	2
Maximum Temperature Limits	Short Time	°F	2500	1800	1300	1400
		°C	1371	982	704	760
	Continuous	°F	2500	1800	1300	1400
		°C	1371	982	704	760
Specific Heat		Btu/lb		0.19 at 300°F	0.2 at 150°F	0.24 at 300°F
		w/kg		0.11 at 149°F	0.11 at 66°C	0.13 at 149°C
Thermal Diffusivity		ft^2/hr		0.054 at 1000°F*	0.02 at 150°F	0.025 at 150°F
		m^2/hr		1.39×10^{-6} 538°C**	5.16×10^{-7} at 66°C	6.45×10^{-7} at 66°C
Water Absorption % Weight			High	High	100	300
Capillarity			Will Wick	Will Wick	Will Wick	Will Wick

*At 6 lb/ft^3 density

**At 96 kg/m^3 density

Table 59 Properties of Reflective Insulation — foils and sheets

Material Identification Number			99	100
Composition			Aluminum sheet	Aluminum foil, various thickness
Form			Any thickness sheet	Rolls
Capilarity			Negligable	Negligable
Combustibility			Non-combustible	Non-combustible
If Combustible	Flash Point Temp.	°F	"	"
		°C	"	"
	Flame Point Temp.	°F	"	"
		°C	"	"
	Heat Release	Btu/lb	"	"
		w/kg	"	"
	Self Internal Heating		"	"
Melting Temp.		°F	1200	1200
		°C	650	650
Corrosion — Rusting			Will not contribute to	Will not contribute to
Corrosion — Stress			Will not contribute to	Will not contribute to
Emissivity — Heat			0.05	0.05
Maximum Temperature Limits	Short Time	°F	1000	600
		°C	538	316
	Continuous	°F	1000	600
		°C	538	316
Resistance to Acids			Resistant to acids	Resistant to Acids
Resistance to Caustics			Not resistant	Not resistant to Caustics
Resistance to Solvents			Resistant	Resistant
Reflectance — Heat			0.95	0.95
Thermal Shock Resistance			Excellent	Excellent
Vibration Resistance			Excellent	Excellent
Water Absorption % Weight			0	0
Construction			Installed to form air spaces which are enclosed and/or separated by sheets of aluminum or aluminum surfaced paper	

Material Identification Number			101	102
Composition			Aluminum foil or membrane paper or other reinforcement	Aluminum-kraft paper—Accordian formed
Form			Foil attached to kraft paper or scrim	Reflective sheet separated to form spaces
Capilarity			Negligable	Negligable
Combustibility			Paper-combustible	Paper is combustible
If Combustible	Flash Point Temp.	°F	500	500
		°C	260	260
	Flame Point Temp.	°F	500	500
		°C	260	260
	Heat Release	Btu/lb	Paper — 8000	Paper — 8000
		w/kg	Paper — 1.859×10^7	Paper — 1.854×10
	Self Internal Heating		Not self heating	Not self heating
Melting Temp.		°F	Aluminum — 1200	Aluminum — 1200
		°C	Aluminum — 650	Aluminum — 650
Corrosion — Rusting			Will not contribute to	Will not contribute to
Corrosion — Stress			Will not contribute to	Will not contribute to
Emissivity — Heat			0.05	0.05
Maximum Temperature Limits	Short Time	°F	200	200
		°C	93	93
	Continuous	°F	200	200
		°C	93	93
Resistance to Acids			Paper-not resistant	Paper-not resistant
Resistance to Caustics			Paper-not resistant	Paper-not resistant
Resistance to Solvents			Resistant to	Resistant to
Reflectance — Heat			0.95	0.95
Thermal Shock Resistance			Excellent	Excellent
Vibration Resistance			Excellent	Excellent
Water Absorption % Weight			0	0
Construction			Installed to form air spaces which are enclosed and/or separated by sheets of aluminum or aluminum surfaced paper	To be installed into cavities to form spaces separated by sheets

Table 59 Properties of Reflective Insulation – Performed to Fit Pipe or Equipment

Material Identification Number			103	104	105
Composition			Aluminum casing & aluminum reflective sheets	Stainless steel casing & aluminum reflective sheets	Stainless steel casing & ss reflective sheets
Form			Factory fabricated to fit equipment and pipe	Factory fabricated to fit equipment and pipe	Factory fabricated to fit equipment and pipe
Capillarity % by Weight			Negligable	Negligable	Negligable
Coefficient of Expansion		English	0.142 X 10^4	Casing 0.961 X 10^5 Sheets 0.142 X 10^4	0.961 X 10^5
		Metric	0.257 X 10^4	Casing 1.72 X 10^5 Sheets 0.257 X 10^4	1.72 X 10^5
Combustibility			Non-combustible	Non-combustible	Non-combustible
If Combustible	Flash Point Temp.	°F	"	"	
		°C	"	"	
	Flame Point Temp.	°F	"	"	
		°C	"	"	
	Heat Release	Btu/lb	"	"	
		w/kg	"	"	
	Self Internal Heating		"	"	
Melting Temp.		°F	1200	Casing 2600 Sheets 1200	2600
		°C	649	Casing 1423 Sheets 649	1423
Corrosion – Rusting			Will not contribute	Will not contribute	Will not contribute to
Corrosion – Stress			Will not contribute	Will not contribute	Will not contribute to
Density – Apparent			5 to 7 lbs/ft^3, 80 to 112 kgs/m^3	6 to 8 lbs/ft^3, 96 to 123 kgs/m^3	8 to 10 lbs/ft^3, 128 to 160 kgs/m^3
Hygroscopicity % Weight			0	0	0
Maximum Temperature Limits	Short Time	°F	1000	1000	1500
		°C	538	538	816
	Continuous	°F	1000	1000	1500
		°C	538	538	816
Resistance to Acids			Excellent	Excellent	Excellent
Resistance to Caustics			Poor	Fair	Excellent
Resistance to Solvents			Excellent	Excellent	Excellent
Specific Gravity		Real	2.7	3.0 to 5.2	5.76
		Apparent	.17	.16	.12
Shrinkage			No shrinkage	No shrinkage	No shrinkage
Thermal Shock Resistance			Excellent	Excellent	Excellent
Vibration Resistance			Excellent	Excellent	Excellent
Water Absorption % Weight			0	0	0

Because of the emphasis on energy conservation, more insulation is being used in plants, buildings and homes, including large quantities of combustible organic foams and cellulose fibers. Where these are used their fire hazard should be evaluated. Unfortunately, the properties of combustibility as listed above are difficult to obtain for individual products. It should be noted that such classifications as flame spread index, smoke index, determined from a single tunnel test method are of little value in the determination of the fire hazard as related to a particular installation. Such indexes only relate as a comparison of combustibles, and can be very misleading. It should be noted, tests where such indexes are determined, such as ASTM E84 and E286, contain the following statement in their scope:

> "This standard should be used to measure and describe the properties of materials, products or systems in response to heat and flame under controlled laboratory conditions and should not be used for the description or appraisal of the fire hazard of materials, products, or systems under actual fire condition."

Compressive strength. The property of a material which neasures its ability to resist a load tending to squeeze or shorten t.

It is one of the properties of insulation which determines its erformance in service. The desirable compressive strength may e high or low. Where the insulation should support a load without rushing, high compression strength is needed. Conversely when nsulation is used to take up dimensional change, a low compressive strength may be needed. In the latter case the percentge of recovery to original size upon the relief of stress becomes nportant.

orrosion to substrates. The property of a material which inicates its chemical effect on various metals upon which it may be nstalled.

This is of upmost importance. If an insulation may cause corroion to a substrate metal, either another insulation should be elected or the substrate metal should be protected by a proper oating. The fact that certain combinations of insulations and etals may cause a corrosion of the metal makes it necessary to etermine the corrosive effect of each insulation on, for example, arbon steel, stainless steel, monel, copper, and aluminum. Of articular importance is the stress-corrosion effect on austenitic tainless steels.

ensity. The weight of a unit volume of insulation, such as lb/ft^3 g/m^3).

It is necessary to know the density to calculate loadings and e heating rate when mass is one of the functions.

imensional stability. The property of a material which indicates s ability to retain its size or shape after aging, cutting, or being ubjected to temperature or moisture.

As every insulation must be installed to given dimensions, this roperty affects the ease of installation and also its service life.

Emittance. The ability of a body to emit radiant energy as a consequence of temperature only to the corresponding emittance of a perfect emitter or black body at same temperature.

Flexural strength. The property of a material which measures its ability to resist bending (flexing) without breaking.

It is one of the mechanical properties which determines the suitability of an insulation during application and in service.

Hardness. The property of a material which measures its ability to resist penetration.

In some applications a hard surface is needed, while in others a soft surface is desired. It is, again, one of the mechanical properties to be considered in the selection of a material for a given application.

Hygroscopicity. The property of a material which measures its ability to absorb and retain water from the ambient air.

This property points up the fact that water in insulation, however it got there, must be allowed to escape without damage to the outer insulation protection. In other words, the higher the hygroscopicity, the more water the insulation will initially absorb and eventually have to reject when subjected to heat. This emphasizes the need for near-perfect moisture protection in the first place. If moisture does get in, however, then the outer barrier must be of the "breathing" type to let it back out when heated.

Incidence of cracking. The property of a material which indicates its tendency to crack when applied to a hot surface at its recommended maximum temperature.

This cracking affects the strength of the insulation and its thermal conductivity.

Reflectance. The ratio of radiant flux reflected by a body to that incident upon it.

Resistance to acids. The property of a material which indicates its ability to resist decomposition by various acids to which it may be subjected.

As insulation is used in all kinds of atmospheres and conditions of chemical contamination, its resistance to the acids to which it may be subjected affects its service life.

Resistance to caustics. Information on the definition and significance of resistance to strong alkaline solutions is the same as that given for "resistance to acids," but it is paraphrased to read "caustics" instead of "acids."

Resistance to solvents. Information on the definition and significance of resistance to solvents is the same as that given for "resistance to acids," but it is paraphrased to read "solvents" instead of "acids."

Shear strength. The property of a material which indicates its ability to resist cleavage.

It is another of the mechanical properties which indicates the suitability of an insulation during application and in service.

Shrinkage. The property of a material which indicates its proportionate loss in dimensions or volume when its temperature is changed.

It is a major cause of cracks in high-temperature applications. It will occur both as linear (length and width) and volumetric loss. Insulation's shrinkage in relation to temperature should be determinded in order to properly design insulation expansion-contraction joints for thermal vessels and piping.

Specific gravity or relative density. The ratio of the specific weight of a material, including all its voids, to the weight of an equal volume of water. This can be obtained by the simple method of dividing the density of the material (lbs per cubic foot) by the weight of a cubic foot of water.

Specific heat. The ratio of the amount of heat required to raise a unit mass of material 1 °F to that required to raise a unit mass of water 1 °F at some specified temperature. It is essential to know this property in order to solve problems related to heat storage and temperature-time lag.

Temperature limits. The range of temperatures, as determined by its properties, up to and down to which a material will experience no essential change in its properties.

Within its limits a material should not fail mechanically, chemically, or thermally. These limits are determined by the effect of the temperature on the properties listed in this chapter. Time exposure to temperature and the number of changes in temperature have an effect on some materials. For this reason, in most instances, it is necessary to determine the following:

Max. temp.—continuous operation
Max. temp.—intermittent operation
Max. temp.—cyclic operation
Min. temp.—cyclic operation
Min. temp.—continuous cyclic operation

Temperature rise—self-internal heating. A property of some materials indicating that when heated above a certain temperature an internal reaction will occur, causing an internal temperature rise in excess of the temperatures to which they are subjected.

This internal temperature rise may not only cause damage to the material itself, but can be a very serious fire hazard.

Tensile strength. The property of a material which measures its ability to resist a stress tending to pull it apart.

It is another of the mechanical properties which evaluate the service performance of the insulation.

Thermal conductivity. The property of a homogeneous body measured by the ratio of steady state heat flux (time rate of heat flow per unit area) to the temperature gradient (temperature difference per unit length of heat flow path) in the direction perpendicular to the area.

As the property varies with mean temperature it is essential to know the conductivity curve of an insulation to calculate heat transfer. However, the prime value of insulation is its *resistance* to heat transfer.

Thermal diffusivity. The property of a material which measures the rate of temperature rise through it. It is *not* a measure of the amount of heat or heat transfer.

In many cases thermal diffusivity is an important factor. For example, in cyclic operations where rapid dissipation of temperature is desired a high rate of thermal diffusivity is important. Conversely, when insulation is used as fire protection a slow rate of thermal diffusivity is most important.

Thermal resistance. The "thermal resistivity" of a material is the reciprocal of the thermal conductivity of that material.

In Metric Units the conductivity of one insulation given in English Units as 0.25 Btu, in./ft^2, hr °F is 0.03605 W/mk and the one given as 0.45 Btu, in./ft^2, hr °F is 0.06489 W/mk. However, the first still has 1.8 more thermal resistivity than the second. The use of thermal resistance values is now being used extensively and as such includes the thickness of the insulation. For example, a 4″ thick insulation with a conductivity of 0.25 is advertised as an insulation of 16.0. Thermal resistance (4/0.25 = 16.00) — English Units. In Metric Units the thermal resistance would be 0.0254 × 4/0.03605 = 2.82. For this reason it is essential that the system in which "thermal resistance" values are quoted must be identified, as no terms have yet been assigned to differentiate between thermal resistance in English and Metric Units.

Thermal shock resistance. The property of a material which indicates its ability to be subjected to rapid temperature changes without physical failure.

This property is important when insulations are used for fire protection or on cyclic operations which involve fast heat ups.

Vapor (water) migration. The property of a material which measures the rate at which water vapor will penetrate it due to vapor pressure differences between its surfaces.

The water vapor which enters insulation has serious effects on its thermal conductivity and on some of its physical and chemical properties. It is well known that water vapor entering low temperature insulation will at some point condense into liquid or freeze into ice and will eventually destroy the thermal efficiency of the insulation. However, vapor migration into insulation on hot pipes may cause swelling of the insulation and contribute to rusting of steel or stress corrosion of stainless steel.

Vapor (water) permeability. The water vapor transmission of a homogeneous body under vapor pressure differences between two specific surfaces.

Vibration resistance. The property of a material which indicates its ability to resist mechanical vibration, without wearing away, settling, or dusting off.

Almost any insulation used in an industrial application will be subjected to some vibration, such as compressor vibration, fan pulsations, or vibrations caused by the fluids or gases passing through the lines or vessels.

Warpage. The change in dimension of one surface of insulation as compared to that of another surface due to a difference in temperature of the two surfaces.

Warpage will cause cracks in the finished insulation and damage to the weather and vapor barriers.

Water absorption. The property of a material which measures the amount of water it will soak up when submerged in water.

It is a measure of the amount of water which may be taken into an insulation due to water leaks in the weather barrier or during construction. It is also important as an indicator of the amount of combustible or toxic liquids which an insulation could absorb in case of spillage. However, a material may be water repellent and still be quite absorptive when subjected to hydrocarbon solvents or other penetrating liquids. In special cases like these, the absorptivity of the insulation must be tested with the chemical to which it may be subjected.

Properties of Insulating Cements, Foams, and Insulating Mastics

After they are applied and dried, insulating cements, insulating mastics, foamed in place insulation, sprayed foam, and sprayed inorganic insulations all become rigid or semi-rigid and their properties are the same as those listed for the rigid or semi-rigid material. However, from the wet state to the dried (or cured) state, additional properties must be considered, as follows:

Adhesion. The property of a material which indicates its ability to stick to the surface to which it has been applied without sliding or falling off.

As the temperature at the time of application and the temperature of the surface to which the material is applied affect the adhesion, either the limits of these two temperatures must be established or the adhesion for various temperatures determined. This property has direct bearing on the durability and cost of an installation.

Shrinkage—wet to dry. The property of a material which measures the volumetric and linear changes which occur in the drying of insulating cements and mastics.

This change affects not only the quantity of material required, but it can also affect the finished installation if cracks or breaks develop because of excessive shrinkage.

Expansion—wet state to cured state. The property of a material which measures the volumetric change in poured, sprayed, or foamed-in-place organic insulations.

This information is necessary in the calculation of quantities of material necessary to obtain a given amount of mass insulation.

Coverage—wet. The property of a material which measures the amount of material necessary to cover a given area to obtain a specific dried or cured thickness.

Of course this information is necessary for calculation of quantities necessary for a particular application.

Properties of Blanket and Batt Insulation

Since blankets and batt insulations are essentially non-rigid, such physical properties as flexural strength and shear, listed as properties of rigid materials, are not applicable. However, many of those properties given for rigid materials are also properties of non-rigid thermal materials. These include:

- Alkalinity (pH)
- Capillarity
- Combustibility (or non-combustibility)
- Chemical reaction
- Compressive strength (although generally of relatively low value as compared to rigid material)
- Corrosion to substrates
- Density
- Hygroscopicity
- Resistance to acids
- Resistance to caustics
- Resistance to solvents
- Specific gravity
- Specific heat
- Tensile strength
- Thermal conductivity
- Thermal diffusivity
- Thermal resistance
- Temperature limits
- Temperature rise (self internal heating)
- Vapor migration
- Vibration resistance
- Water absorption

In addition, properties other than those listed for rigid materials do become important in this form of insulation; namely:

Compaction or setting. The property of the blankets or batts which measures their change in density and thickness resulting from loading or vibration thus producing change of thermal efficiency.

Some materials tend to fluff up when subjected to vibration whereas others tends to compact. This factor has a definite relation to suitability of the insulation for certain installations.

Recovery of thickness after compression of blankets or batts is a vital factor in the use of material as cushion blankets, or in expansion joints.

Resistance to air movement. The property which indicates the ability of a blanket-type material to resist erosion by air currents over its surface.

The importance of a blanket being able to withstand high air velocities across its surface is demonstrated most frequently when blankets are used to line the inside of ducts.

Table 60 Thermal Conductivities of Insulations – *English Units* Btu, in./ft., hr., °F

Mean Temp.		Identification No. 1	Identification No. 2	Identification No. 3	Identification No. 4	Identification No. 5	Identification No. 6	Identification No. 7
		Generic Type: Alumina-silica fiber and binder Form: Block	Generic Type: Alumina-silica fiber and binder Form: Block	Generic Type: Alumina-silica fiber and binder Form: Board	Generic Type: Alumina-silica fiber and binder Form: Board	Generic Type: Calcium-Silicate Type 1 Form: Block and Pipe Cov.	Generic Type: Calcium-Silicate Type 2 Form: Block and Pipe Cov.	Generic Type: Diatomaceous-silica Type 4 Form: Block and Pipe Cov.
°R	°F	Density lbs/ft.^{3} 15 to 17	Density lbs/ft.^{3} 18 to 22	Density lbs/ft.^{3} 10 to 16	Density lbs/ft.^{3} 21 to 40	Density lbs/ft.^{3} 14 (max.)	Density lbs/ft.^{3} 15 (max.)	Density lbs/ft.^{3} 24 (max.)
260	−200							
360	−100							
460	0							
560	100					0.35		
660	200	0.42	0.47	0.33	0.57	0.43	0.58	0.62
760	300	0.44	0.51	0.36	0.61	0.50	0.62	0.64
860	400	0.45	0.55	0.40	0.65	0.55	0.66	0.67
960	500	0.50	0.60	0.46	0.70	0.60	0.70	0.70
1060	600	0.56	0.65	0.52	0.75	0.66	0.76	0.72
1160	700	0.62	0.70	0.58	0.80	0.72	0.82	0.75
1260	800	0.68	0.75	0.64	0.85		0.94	0.77
1360	900	0.74	0.80	0.70	0.90		0.01	0.80
1460	1000	0.81	0.85	0.76	0.95		1.08	0.83
1560	1100	0.88	0.91	0.83	1.00		1.16	0.87
1660	1200	0.96	0.97	0.90	1.05		1.25	0.91

Metric Units w/mK

°K	°C	Density kgs/m^3 240 to 272	Density kgs/m^3 288 to 352	Density kgs/m^3 160 to 256	Density kgs/m^3 336 to 640	Density kgs/m^3 224 (max.)	Density kgs/m^3 240 (max.)	Density kgs/m^3 384 (max.)
173	−100							
223	−50							
273	0							
323	50							
373	100	0.060	0.067	0.047	0.082	0.066	0.083	0.089
423	150	0.063	0.073	0.052	0.088	0.072	0.089	0.092
273	200	0.065	0.079	0.058	0.091	0.079	0.095	0.096
523	250	0.073	0.085	0.066	0.098	0.085	0.100	0.100
573	300	0.081	0.091	0.074	0.106	0.093	0.109	0.103
623	350	0.089	0.098	0.082	0.114	0.100	0.118	0.106
673	400	0.097	0.105	0.090	0.120		0.128	0.109
723	450	0.107	0.112	0.098	0.126		0.136	0.112
773	500	0.116	0.119	0.106	0.132		0.144	0.115
823	550	0.126	0.126	0.114	0.138		0.152	0.118
873	600	0.137	0.133	0.122	0.144		0.161	0.122
923	650	0.138	0.140	0.130	0.151		0.170	0.126

Table 60 Thermal Conductivities of Insulations — *English Units* Btu, in./ft., hr., °F

Mean Temp.		Identification No. 8	Identification No. 9	Identification No. 10	Identification No. 11	Identification No. 12	Identification No. 13	Identification No. 14
		Generic Type: Diatomaceous-silica Type 2 Form: Block and Pipe Cov.	Generic Type: Cork, vegetable Form: Board	Generic Type: Cork, vegetable Form: Pipe Cov.	Generic Type Cellular Glass Form:	Generic Type: Cellular Glass between layers of felt Roof Form: Board	Generic Type: Diatomaceous-silica Type 1 Class I Form: Block and Pipe Cov.	Generic Type: Diatomaceous-silica Type 2 Class II Form: Block and Pipe Cov.
°R	°F	Density lbs/ft.^{3} 26 (max.)	Density lbs/ft.^{3} 6 to 8	Density lbs/ft.^{3} 7 to 14	Density lbs/ft.^{3} 7.5 to 9.5	Density lbs/ft.^{3} 7.5 to 9.5	Density lbs/ft.^{3} 17 to 24	Density lbs/ft.^{3} 26
260	−200				0.22			
360	−100				0.27			
460	0		0.31	0.35	0.31	0.35		
560	100		0.35	0.39	0.36	0.43		
660	200	0.71			0.42	0.51	0.68	0.75
760	300	0.74			0.49		0.69	0.76
860	400	0.76			0.57		0.70	0.78
960	500	0.78			0.70		0.72	0.80
1060	600	0.80			0.82		0.74	0.82
1160	700	0.82			1.03		0.76	0.84
1260	800	0.84					0.78	0.87
1360	900	0.87					0.80	0.90
1460	1000	0.90					0.83	0.93
1560	1100	0.93					0.86	0.96
1660	1200	0.96					0.90	0.99

Metric Units **w/mK**

°K	°C	Density kgs/m^3 416 (max.)	Density kgs/m^3 96 to 128	Density kgs/m^3 112 to 224	Density kgs/m^3 120 to 152	Density kgs/m^3 120 to 152	Density kgs/m^3 272 to 584	Density kgs/m^3 416
173	−100				0.031			
223	−50				0.039	0.039		
273	0		0.046	0.051	0.047	0.047		
323	50		0.051	0.056	0.055	0.055		
373	100	0.102			0.063	0.063	0.098	0.108
423	150	0.106			0.073		0.099	0.109
273	200	0.109			0.085		0.101	0.112
523	250	0.112			0.098		0.103	0.115
573	300	0.115					0.105	0.117
623	350	0.118					0.108	0.119
673	400	0.121					0.111	0.123
723	450	0.124					0.114	0.127
773	500	0.127					0.116	0.131
823	550	0.130					0.121	0.135
873	600	0.133					0.125	0.140
923	650	0.136					0.130	0.143

Table 60 Thermal Conductivities of Insulations — *English Units* Btu, in./ft., hr., °F

Mean Temp.		Identification No. 15	Identification No. 16	Identification No. 17	Identification No. 18	Identification No. 19	Identification No. 20	Identification No. 21
		Generic Type: Glass fiber with organic binder Form: Board	Generic Type: Glass fiber with organic binder Form: Board	Generic Type: Glass fiber with organic binder Form: Board	Generic Type: Glass fiber with organic binder foil reinforced Form: Duct Board	Generic Type: Glass fiber with organic binder Form: Pipe Cov.	Generic Type: Glass fiber with organic binder Form: Pipe Cov.	Generic Type: Mineral fiber (rock, slag, or glass) Form: Block or Board
°R	°F	Density lbs/ft.^{3} 1.4 to 1.6	Density lbs/ft.^{3} 2.5 to 3.5	Density lbs/ft.^{3} 5 to 7	Density lbs/ft.^{3} 2.5 to 3.5	Density lbs/ft.^{3} 3 to 5	Density lbs/ft.^{3} 5 to 10	Density lbs/ft.^{3} 9 to 11
260	−200							
360	−100							
460	0	0.22						
560	100	0.29	0.28	0.27	0.30	0.28	0.28	0.28
660	200	0.36	0.32	0.30	0.34	0.32	0.32	0.35
760	300	0.47	0.39	0.35		0.39	0.42	0.43
860	400	0.62	0.50	0.46		0.48	0.50	0.54
960	500							
1060	600							
1160	700							
1260	800							
1360	900							
1460	1000							
1560	1100							
1660	1200							

Metric Units w/mK

°K	°C	Density kgs/m^3 22 to 26	Density kgs/m^3 40 to 56	Density kgs/m^3 80 to 112	Density kgs/m^3 40 to 56	Density kgs/m^3 54 to 80	Density kgs/m^3 80 to 160	Density kgs/m^3 144 to 176
173	−100							
223	−50							
273	0	0.033			0.030			
323	50	0.040	0.039	0.038	0.043	0.040	0.040	0.040
373	100	0.053	0.047	0.045		0.047	0.047	0.052
423	150	0.068	0.056	0.051		0.056	0.059	0.062
273	200	0.089	0.072	0.066		0.069	0.071	0.078
523	250							
573	300							
623	350							
673	400							
723	450							
773	500							
823	550							
873	600							
923	650							

Table 60 Thermal Conductivities of Insulations — *English Units* Btu, in./ft., hr., °F

Mean Temp.		Identification No. 22	Identification No. 23	Identification No. 24	Identification No. 25	Identification No. 26	Identification No. 27	Identification No. 28
		Generic Type: Mineral fiber (rock, slag or glass) Form: Block or Board	Generic Type: Mineral fiber Class 3 Form: Block and Board	Generic Type: Mineral fiber Class 4 Form: Block and Board	Generic Type: Mineral fiber Class 5 Form: Block and Board	Generic Type: Mineral fiber Class 1 Form: Pipe Cov.	Generic Type: Mineral fiber Class 2 Form: Pipe Cov.	Generic Type: Mineral fiber Class 3 Form: Pipe Cov.
°R	°F	Density lbs/ft.^{3} 11 to 13	Density lbs/ft.^{3} 7 to 12	Density lbs/ft.^{3} 8 to 13	Density lbs/ft.^{3} 12 to 25	Density lbs/ft.^{3} 9 to 11	Density lbs/ft.^{3} 11 to 14	Density lbs/ft.^{3} 16 to 20
260	−200							
360	−100							
460	0					0.24		
560	100	0.28	0.30	0.36		0.27	0.33	0.36
660	200	0.35	0.36	0.42	0.40	0.33	0.39	0.42
760	300	0.43	0.42	0.50	0.46	0.42	0.46	0.49
860	400	0.54	0.49	0.57	0.51		0.53	0.56
960	500			0.64	0.57			0.64
1060	600				0.62			
1160	700				0.68			
1260	800				0.75			
1360	900				0.82			
1460	1000				0.90			
1560	1100				0.98			
1660	1200				1.07			

Metric Units w/mK

°K	°C	Density kgs/m^3 176 to 208	Density kgs/m^3 112 to 192	Density kgs/m^3 128 to 208	Density kgs/m^3 192 to 400	Density kgs/m^3 144 to 176	Density kgs/m^3 176 to 224	Density kgs/m^3 256 to 320
173	−100							
223	−50							
273	0							
323	50	0.040	0.046	0.054		0.036	0.046	
373	100	0.052	0.053	0.062		0.049	0.056	0.061
423	150	0.062	0.060	0.072	0.066	0.060	0.066	0.071
273	200	0.078	0.072	0.082	0.073		0.076	0.081
523	250			0.091	0.082			0.099
573	300				0.087			
623	350				0.095			
673	400				0.105			
723	450				0.111			
773	500				0.121			
823	550				0.132			
873	600				0.140			
923	650				0.154			

Table 60 Thermal Conductivities of Insulations — *English Units* Btu, in./ft., hr., °F

Mean Temp.		Identification No. 29	Identification No. 30	Identification No. 31	Identification No. 32	Identification No. 33	Identification No. 34	Identification No. 35
		Generic Type: Perlite expanded and water repellent binder Form: Block and Pipe Cov.	Generic Type: Perlite expanded with binders Form: Block and Pipe Cov.	Generic Type: Perlite expanded with binder and fibers Form: Roof Board	Generic Type: Expanded Phenolic foam Form: Block and Pipe Cov.	Generic Type: Poly-strene expanded beads, molded to shape Form: Block, Board Pipe Cov.	Generic Type: Poly-styrone expanded foam-cut shapes Form: Board and Pipe Cov.	Generic Type: Poly-styrone expanded foam-molded shapes Form: Board and Pipe Cov.
°R	°F	Density lbs/ft.^{3} 13 to 15	Density lbs/ft.^{3} 13 to 15	Density lbs/ft.^{3} 10	Density lbs/ft.^{3} 2 to 3	Density lbs/ft.^{3} 1.0 to 1.5	Density lbs/ft.^{3} 1.75 to 2.0	Density lbs/ft.^{3} 2.0 to 2.2
260	−200							
360	−100							
460	0			0.36	0.22	0.24	0.20	0.20
560	100	0.40	0.40	0.40	0.26	0.28	0.24	0.24
660	200	0.45	0.45					
760	300	0.50	0.50					
860	400	0.55	0.55					
960	500	0.60	0.60					
1060	600	0.65	0.65					
1160	700	0.71	0.71					
1260	800	0.77	0.77					
1360	900	0.83	0.83					
1460	1000	0.90	0.90					
1560	1100							
1660	1200							

Metric Units w/mK

°K	°C	Density kgs/m^3 208 to 240	Density kgs/m^3 208 to 240	Density kgs/m^3 160	Density kgs/m^3 32 to 48	Density kgs/m^3 16.01 to 24.01	Density kgs/m^3 28.01 to 32.01	Density kgs/m^3 32.02 to 35.22
173	−100							
223	−50							
273	0			0.053	0.032	0.035	0.026	0.026
323	50			0.057	0.058	0.041	0.032	0.032
373	100	0.068	0.068					
423	150	0.072	0.072					
273	200	0.078	0.078					
523	250	0.085	0.085					
573	300	0.091	0.091					
623	350	0.098	0.098					
673	400	0.105	0.105					
723	450	0.112	0.112					
773	500							
823	550							
873	600							
923	650							

Table 60 Thermal Conductivities of Insulations —

English Units Btu, in./ft., hr., °F

Mean Temp.		Identification No. 36	Identification No. 37	Identification No. 38	Identification No. 39	Identification No. 40	Identification No. 41	Identification No. 42
		Generic Type: Polystyrone expanded foam-molded into shape Form: Board	Generic Type: Polystyrone foam molded into shapes Form: Roof Board	Generic Type: Polyurethane expanded foam Form: Block, Board Pipe Cov.	Generic Type: Polyurethane expanded foam Form: Block, Board Pipe Cov.	Generic Type: Polyurethane expanded foam Form: Block, Board Pipe Cov.	Generic Type: Expanded rubber Form: Rigid Board	Generic Type: Expanded silica and binders Form: Block and Pipe Cov.
°R	°F	Density lbs/ft.^{3} 2.25 to 2.5	Density lbs/ft.^{3} 2.5 to 2.7	Density lbs/ft.^{3} Up to 1.7	Density lbs/ft.^{3} 1.7 to 2.5	Density lbs/ft.^{3} 2.5 to 5.0	Density lbs/ft.^{3} 4 to 6	Density lbs/ft.^{3} 13
260	−200							
360	−100							
460	0	0.20	0.20	0.23	0.22	0.20	.26	
560	100	0.24	0.24	0.27	0.26	0.24	.28	.40
660	200							.42
760	300							.45
860	400							.50
960	500							.55
1060	600							.60
1160	700							.65
1260	800							.70
1360	900							.76
1460	1000							.82
1560	1100							
1660	1200							

Metric Units w/mK

°K	°C	Density kgs/m^3 36.02 to 40.03	Density kgs/m^3 90.03 to 43.22	Density kgs/m^3 27.3	Density kgs/m^3 27.2 to 40.0	Density kgs/m^3 40 to 80	Density kgs/m^3 64 to 96	Density kgs/m^3 208
173	−100							
223	−50							
273	0	0.026	0.026	0.033	0.032	0.028	0.037	
323	50	0.032	0.032	0.039	0.040	0.034	0.041	0.057
373	100							0.061
423	150							0.065
273	200							0.072
523	250							0.078
573	300							0.085
623	350							0.091
673	400							0.098
723	450							0.105
773	500							0.113
823	550							
873	600							
923	650							

Table 60 Thermal Conductivities of Insulations — *English Units* Btu, in./ft., hr., °F

Mean Temp.		Identification No. 43	Identification No. 44	Identification No. 45	Identification No. 46	Identification No. 47	Identification No. 48	Identification No. 49
		Generic Type: Wood fibers and binders Form: Board	Generic Type: Borosilica closed cell inorganic foam Form: Block	Generic Type: Elastomeric Flexible Form: Sheet and Pipe Cov.	Generic Type: Glass fiber, flexible with various facings Form: Roll or Sheets	Generic Type: Glass fiber, flexible with various facings** Form: Sheets	Generic Type: Glass fiber, semi-rigid with various facings** Form: Sheets	Generic Type: Kaolin ceramic fiber in stainless steel metal mesh Form: Strips & Shaped
°R	°F	Density lbs/ft.^{3} 15 to 18	Density lbs/ft.^{3} 12	Density lbs/ft.^{3} 9.5 to 8.5	Density lbs/ft.^{3} 0.6 to 3.0	Density lbs/ft.^{3} 3 to 4.5	Density lbs/ft.^{3} 4.5 (min.)	Density lbs/ft.^{3} 3 to 4
260	−200							
360	−100							
460	0	0.40		0.29	0.30	0.28	0.27	
560	100	0.42	0.58	0.32	0.40	0.36	0.33	
660	200		0.64		0.54	0.46	0.40	0.37
760	300		0.69		0.68	0.60	0.50	0.40
860	400		0.75					0.45
960	500		0.82					0.57
1060	600							0.70
1160	700							0.82
1260	800							0.96
1360	900							1.11
1460	1000							1.30
1560	1100							1.47
1660	1200							1.65

Metric Units w/mK

°K	°C	Density kgs/m^3 240 to 288	Density kgs/m^3 190	Density kgs/m^3 72	Density kgs/m^3 10.4 to 8	Density kgs/m^3 98 to 72	Density kgs/m^3 72 (min.)	Density kgs/m^3 48 to 64
173	−100							
223	−50							
273	0			0.0425	0.044	0.042	0.040	
323	50	0.059	0.0865	0.0476	0.059	0.053	0.049	
373	100	0.062	0.0937		0.079	0.067	0.059	0.053
423	150		0.0995		0.101	0.082	0.071	0.058
273	200		0.1082					0.065
523	250		0.1168					0.084
573	300							0.099
623	350							0.114
673	400							0.128
723	450							0.146
773	500							0.170
823	550							0.194
873	600							0.211
923	650							0.237

**Facing may be woven wire mesh, expanded metal lath (copper bearing) on one or both sides.

Table 60 Thermal Conductivities of Insulations – English Units Btu, in./ft., hr., °F

Mean Temp.		Identification No. 50	Identification No. 51	Identification No. 52	Identification No. 53	Identification No. 54	Identification No. 55	Identification No. 56
		Generic Type: Kaolin ceramic fiber in stainless steel metal mesh Form: Strips & Shaped	Generic Type: Mineral wool-spun glass-stainless steel metal mesh Form: Various Shapes	Generic Type: Mineral fibers with metal mesh facing** Class I Form: Sheets	Generic Type: Mineral fibers with metal mesh facing** Class II Form: Sheets	Generic Type: Urethane foam Form: Block, Pipe Cov.	Generic Type: Urethane foam Form: Block, Pipe Cov.	Generic Type Alumina fibers Form: Blanket and Felts
°R	°F	Density lbs/ft.^{3} 6 to 8	Density lbs/ft.^{3} 10 to 14	Density lbs/ft.^{3} 6 (max. ave.)	Density lbs/ft.^{3} 12 (ave. max.)	Density lbs/ft.^{3} Up to 1.5	Density lbs/ft.^{3} Over 1.5	Density lbs/ft.^{3} 3 to 4
260	−200							
360	−100							
460	0		0.26			0.29	0.27	
560	100		0.30		0.31	0.31	0.29	
660	200		0.34	0.29	0.37			0.36
760	300		0.40	0.36	0.44			0.38
860	400	0.35	0.47	0.46	0.51			0.40
960	500	0.40	0.55	0.56	0.59			0.49
1060	600	0.47	0.62	0.67				0.60
1160	700	0.56	0.76	0.67				0.74
1260	800	0.68	0.87					0.90
1360	900	0.81	0.98					1.05
1460	1000	0.95	1.10					1.25
1560	1100	1.07						1.42
1660	1200	1.20						1.65

Metric Units w/mK

°K	°C	Density kgs/m^3 96 to 128	Density kgs/m^3 160 to 224	Density kgs/m^3 96 (max. ave.)	Density kgs/m^3 96 (ave. max.)	Density kgs/m^3 Up to 24	Density kgs/m^3 Over 24	Density kgs/m^3 40 to 64
173	−100							
223	−50							
273	0		0.039			0.043	0.040	
323	50		0.044	0.036	0.039	0.046	0.043	
373	100		0.050	0.044	0.046			0.052
423	150		0.057	0.057	0.054			0.054
273	200	0.051	0.067	0.069	0.063			0.058
523	250	0.059	0.078	0.082	0.072			0.069
573	300	0.066	0.088					0.084
623	350	0.077	0.103					0.097
673	400	0.089	0.118					0.118
723	450	0.106	0.132					0.140
773	500	0.125	0.150					0.163
823	550	0.142	0.162					0.187
873	600	0.157						0.209
923	650	0.173						0.238

**Facing may be woven wire mesh, expanded metal lath (copper bearing) on one or both sides.

Table 60 Thermal Conductivities of Insulations — *English Units* Btu, in./ft., hr., °F

Mean Temp.		Identification No. 57	Identification No. 58	Identification No. 59	Identification No. 60	Identification No. 61	Identification No. 62	Identification No. 63
		Generic Type: Alumina fibers Form: Blanket and Felts	Generic Type: Alumina fibers Form: Blanket and Felts	Generic Type: Alumina fibers Form: Rope, cord, yarn	Generic Type: Glass fiber, fine fiber (no binder) Form: Blanket and Batt	Generic Type: Glass fiber, fine fiber with binder Form: Blanket	Generic Type: Glass fiber with binder Form: Blanket	Generic Type: Glass fiber with binder Form: Blanket
°R	°F	Density lbs/ft.^{3} 6 to 8	Density lbs/ft.^{3} 12	Density lbs/ft.^{3} 25	Density lbs/ft.^{3} 11 to 12	Density lbs/ft.^{3} ½ to 2	Density lbs/ft.^{3} 1 to 2	Density lbs/ft.^{3} 2.5 to 3.5
260	−200							
360	−100							
460	0				0.26	0.23		
560	100				0.30	0.24	0.38	0.35
660	200	0.28	0.27	0.32	0.34	0.30	0.54	0.48
760	300	0.30	0.28	0.35	0.40	0.38		
860	400	0.33	0.30	0.40	0.47			
960	500	0.38	0.36	0.49	0.55			
1060	600	0.44	0.42	0.60	0.62			
1160	700	0.53	0.49	0.74	0.76			
1260	800	0.62	0.57	0.90	0.87			
1360	900	0.72	0.65	1.05	0.98			
1460	1000	0.83	0.72	1.25	1.10			
1560	1100	0.94	0.82	1.42				
1660	1200	1.06	0.92	1.65				

Metric Units w/mK

°K	°C	Density kgs/m^3 96 to 128	Density kgs/m^3 192	Density kgs/m^3 400	Density kgs/m^3 176 to 192	Density kgs/m^3 8 to 32	Density kgs/m^3 16 to 32	Density kgs/m^3 40 to 56
173	−100							
223	−50							
273	0				0.039	0.033		
323	50				0.044	0.037	0.060	0.055
373	100	0.040	0.039	0.046	0.050	0.044	0.079	0.071
423	150	0.043	0.040	0.050	0.057	0.054		
273	200	0.047	0.430	0.058	0.067			
523	250	0.053	0.050	0.069	0.078			
573	300	0.061	0.058	0.084	0.088			
623	350	0.072	0.066	0.097	0.103			
673	400	0.083	0.077	0.118	0.118			
723	450	0.095	0.086	0.140	0.132			
773	500	0.104	0.099	0.163	0.150			
823	550	0.123	0.104	0.187	0.162			
873	600	0.140	0.121	0.209				
923	650	0.153	0.133	0.238				

Table 60 Thermal Conductivities of Insulations — *English Units* Btu, in./ft., hr., °F

Mean Temp.		Identification No. 64	Identification No. 65	Identification No. 66	Identification No. 67	Identification No. 68	Identification No. 69	Identification No. 70
		Generic Type: Mineral Fibers Form: Blanket and Batt	Generic Type: Mineral Fibers with binders Form: Blankets	Generic Type: Silica fibers Form: Blanket, Felt	Generic Type: Alumina-silica, semi refractory hydraulic setting Form: Cement	Generic Type: Alumina silica-ceramic fiber air setting cement Form: Ready-mix	Generic Type: Alumina silica-ceramic Form: Spray mix cement	Generic Type: Diatomaceous silica and binder Form: Water mix cement
°R	°F	Density lbs/ft.^{3} 8 to 12	Density lbs/ft.^{3} 7 to 9	Density lbs/ft.^{3} 6 to 9.5	Density lbs/ft.^{3} 48 to 70	Density lbs/ft.^{3} 115 to 122	Density lbs/ft.^{3} 12 to 13	Density lbs/ft.^{3} 23 to 32
260	−200							
360	−100							
460	0							
560	100	0.40	0.30	0.40				
660	200	0.43	0.34	0.41	1.31			0.75
760	300	0.49	0.38	0.43	1.33			0.80
860	400	0.57	0.44	0.45	1.35		0.40	0.85
960	500	0.68	0.52	0.47	1.38	1.80	0.46	0.90
1060	600	0.82		0.50	1.42	2.00	0.52	0.95
1160	700			0.56	1.47	2.20	0.58	1.00
1260	800			0.67	1.52	2.60	0.65	1.06
1360	900			0.77	1.58	3.00	0.71	1.12
1460	1000			0.90	1.64	3.50	0.78	1.19
1560	1100			1.02	1.70	4.20	0.86	
1660	1200			1.20	1.76	5.00	0.94	

Metric Units w/mK

°K	°C	Density kgs/m^3 128 to 192	Density kgs/m^3 112 to 140	Density kgs/m^3 96 to 152	Density kgs/m^3 768 to 1120	Density kgs/m^3 1840 to 1952	Density kgs/m^3 192 to 208	Density kgs/m^3 368 to 512
173	−100							
223	−50							
273	0							
323	50	0.058	0.045	0.058				
373	100	0.062	0.051	0.059	0.188			0.109
423	150	0.071	0.055	0.062	0.191			0.115
273	200	0.082	0.063	0.065	0.194		0.058	0.123
523	250	0.097	0.076	0.067	0.198	0.250	0.004	0.129
573	300	0.121		0.071	0.203	0.259	0.073	0.135
623	350			0.078	0.210	0.288	0.081	0.140
673	400			0.086	0.216	0.316	0.088	0.146
723	450			0.102	0.223	0.374	0.097	0.153
773	500			0.120	0.231	0.431	0.107	0.162
823	550			0.133	0.239	0.505	0.115	0.174
873	600			0.152	0.248	0.610	0.125	
923	650			0.173	0.254	0.721	0.136	

Table 60 Thermal Conductivities of Insulations — *English Units* Btu, in./ft., hr., °F

Mean Temp.		Identification No. 71	Identification No. 72	Identification No. 73	Identification No. 74	Identification No. 75	Identification No. 76	Identification No. 77
		Generic Type: Kaowool ceramic fiber and inorganic binders Form: Spray cement	Generic Type: Kaowool fiber and mineral fiber blend Form: Spray cement	Generic Type: Mineral fiber, binder and hydraulic setting cement Form: Cement	Generic Type: Mineral fiber Form: Cement	Generic Type: Vermiculite expands, with binders Form: Water mix-air setting	Generic Type: Polyurethane, 2 part mix Form: Foamed or sprayed	Generic Type: Ureaformaldehyde Form: Foamed in place
°R	°F	Density lbs/ft.^{3} 12 to 13	Density lbs/ft.^{3} 7 to 15	Density lbs/ft.^{3} 22 to 30	Density lbs/ft.^{3} 27 to 44	Density lbs/ft.^{3} 18 to 19	Density lbs/ft.^{3} 1.6 to 3.0	Density lbs/ft.^{3} 0.6 to 1.0
260	−200							
360	−100							
460	0						0.22	0.25
560	100						0.26	0.29
660	200	0.48	0.48	0.93	0.73	0.98		
760	300	0.51	0.52	0.98	0.78	1.03		
860	400	0.54	0.57	1.03	0.84	1.08		
960	500	0.58	0.65	1.08	0.90	1.15		
1060	600	0.61	0.70	1.10	0.96	1.22		
1160	700	0.67	0.76		1.02	1.29		
1260	800	0.73	0.83					
1360	900	0.80	0.91					
1460	1000	0.86						
1560	1100	0.93						
1660	1200	0.99						

Metric Units w/mK

°K	°C	Density kgs/m^3 192 to 208	Density kgs/m^3 112 to 240	Density kgs/m^3 352 to 512	Density kgs/m^3 482 to 704	Density kgs/m^3 288 to 304	Density kgs/m^3 25 to 48	Density kgs/m^3 9.6 to 16.0
173	−100							
223	−50							
273	0						0.032	0.035
323	50						0.038	0.042
373	100	0.069	0.071	0.134	0.103	0.140		
423	150	0.073	0.075	0.139	0.109	0.147		
273	200	0.077	0.082	0.144	0.115	0.154		
523	250	0.083	0.092	0.148	0.123	0.162		
573	300	0.086	0.099	0.153	0.129	0.170		
623	350	0.094	0.108		0.136	0.182		
673	400	0.101	0.115					
723	450	0.109	0.124					
773	500	0.118	0.134					
823	550	0.125						
873	600	0.136						
923	650	0.143						

Table 60 Thermal Conductivities of Insulations – *English Units* Btu, in./ft., hr., °F

Mean Temp.		Identification No. 78	Identification No. 79	Identification No. 80	Identification No. 81	Identification No. 82	Identification No. 83	Identification No. 84
		Generic Type: Poly-vinyl cork filled mastic Form: Sprayed or trowelled	Generic Type: Alumina-silica fibers Form: Bulk	Generic Type: Alumina silica fiber Form: Chopped and milled	Generic Type: Alumina silica fiber-long fiber Form: Bulk	Generic Type: Cork, vegetable granulated Form: Suitable for pouring	Generic Type: Cellulose fiber Form: Loose-Fill	Generic Type: Diatomaceous silica Form: Fine powder
°R	°F	Density lbs/ft.^{3} 53	Density lbs/ft.^{3} 6	Density lbs/ft.^{3} 7.5	Density lbs/ft.^{3} 6	Density lbs/ft.^{3} 12 to 14	Density lbs/ft.^{3} 10 to 15	Density lbs/ft.^{3} 13 to 17
260	−200							
360	−100							
460	0					0.31	0.42	
560	100	0.96				0.34	0.46	
660	200		0.63	0.36	0.37			0.42
760	300		0.66	0.38	0.39			0.45
860	400		0.69	0.40	0.43			0.48
960	500		0.73	0.43	0.50			0.51
1060	600		0.78	0.47	0.61			0.55
1160	700		0.85	0.55	0.75			0.58
1260	800		0.97	0.65	0.90			0.62
1360	900		1.09	0.75	1.05			0.65
1460	1000		1.22	0.85	1.20			0.69
1560	1100		1.35	0.95	1.35			0.74
1660	1200		1.55	1.05	1.52			0.80

Metric Units w/mK

°K	°C	Density kgs/m^3 848	Density kgs/m^3 96	Density kgs/m^3 120	Density kgs/m^3 96	Density kgs/m^3 124 to 224	Density kgs/m^3 160 to 240	Density kgs/m^3 208 to 272
173	−100							
223	−50							
273	0					0.046		
323	50	0.138				0.047	0.059	
373	100		0.091	0.052	0.053			0.062
423	150		0.095	0.055	0.057			0.064
273	200		0.099	0.058	0.062			0.069
523	250		0.105	0.062	0.071			0.073
573	300		0.112	0.068	0.085			0.078
623	350		0.121	0.078	0.107			0.082
673	400		0.133	0.088	0.120			0.087
723	450		0.147	0.099	0.139			0.091
773	500		0.166	0.114	0.161			0.097
823	550		0.180	0.125	0.177			0.101
873	600		0.200	0.139	0.200			0.108
923	650		0.224	0.151	0.219			0.115

Table 60 Thermal Conductivities of Insulations — *English Units* Btu, in./ft., hr., °F

Mean Temp.		Identification No. 85	Identification No. 86	Identification No. 87	Identification No. 88	Identification No. 89	Identification No. 90	Identification No. 91
		Generic Type: Diatomaceous silica calcined Form: Powder	Generic Type: Diatomaceous silica Form: Course Powder	Generic Type: Gilsonite granules-for packing around underground pipe Form:	Generic Type: Glass fiber, unbonded Form: Loose fibers	Generic Type: Glass fibers Form: Loose and Blowing Wool	Generic Type: Gypsum pellits Form: Bulk	Generic Type: Mineral fiber, with small amounts of oil Form: Loose
°R	°F	Density lbs/ft.^{3} 32 (max.)	Density lbs/ft.^{3} 25 to 31	Density lbs/ft.^{3} 44 to 50	Density lbs/ft.^{3} 6 (ave.)	Density lbs/ft.^{3} 2 to 3	Density lbs/ft.^{3} 20 to 30	Density lbs/ft.^{3} 4 to 12
260	−200							
360	−100							
460	0				0.25			
560	100			0.75	0.26	0.35	0.45 to 0.95	0.30
660	200	0.92	0.63	0.85	0.31			0.35
760	300	0.97	0.68	1.00	0.36			0.44
860	400	1.03	0.77		0.45			0.54
960	500	1.08	0.84					0.65
1060	600	1.13	0.92					0.75
1160	700	1.18	0.99					0.86
1260	800	1.25	1.06					0.97
1360	900	1.31	1.14					
1460	1000	1.37	1.22					
1560	1100	1.43	1.32					
1660	1200	1.49	1.42					

Metric Units w/mK

°K	°C	Density kgs/m^3 512 (max.)	Density kgs/m^3 400 to 496	Density kgs/m^3 704 to 800	Density kgs/m^3 86 (ave.)	Density kgs/m^3 32 to 43	Density kgs/m^3 320 to 480	Density kgs/m^3 64 to 192
173	−100							
223	−50							
273	0							
323	50			0.111	0.038	0.052	0.067 to 0.143	0.045
373	100	0.134	0.092	0.124	0.046			0.052
423	150	0.140	0.098	0.144	0.052			0.063
273	200	0.148	0.111		0.064			0.078
523	250	0.155	0.120					0.092
573	300	0.162	0.131					0.105
623	350	0.169	0.141					0.120
673	400	0.177	0.150					0.134
723	450	0.184	0.158					
773	500	0.192	0.168					
823	550	0.200	0.183					
873	600	0.207	0.193					
923	650	0.215	0.204					

Table 60 Thermal Conductivities of Insulations — *English Units* Btu, in./ft., hr., °F

Mean Temp.		Identification No. 92	Identification No. 93	Identification No. 94	Identification No. 95	Identification No. 96	Identification No. 97	Identification No. 98
		Generic Type: Mineral fiber, oil free Form: Loose	Generic Type: Mineral fiber, granulated Form: Loose	Generic Type: Perlite expanded Form: Loose powder	Generic Type: Quartz fibers Form: Loose	Generic Type: Silica fibers Form: Loose	Generic Type: Silica aerogel granuals Form: Loose	Generic Type: Vermiculite flakes Form: Loose
°R	°F	Density lbs/ft.^{3} 10 to 15	Density lbs/ft.^{3} 6 to 12	Density lbs/ft.^{3} 5 to 8	Density lbs/ft.^{3} 6 to 12	Density lbs/ft.^{3} 9	Density lbs/ft.^{3} 4 to 5.5	Density lbs/ft.^{3} 8 to 10
260	−200							
360	−100							
460	0							
560	100		0.30	0.29			0.21	0.47
660	200	0.53	0.35	0.35	0.35	0.34	0.23	0.57
760	300	0.58	0.44		0.41	0.36	0.24	0.63
860	400	0.64	0.54		0.47	0.39		0.74
960	500	0.70	0.65		0.53	0.42		0.84
1060	600	0.76	0.75		0.59	0.45		0.94
1160	700	0.83	0.86		0.66	0.52		1.05
1260	800	0.91	0.97		0.73	0.60		1.15
1360	900				0.80	0.71		1.25
1460	1000				0.87	0.81		1.36
1560	1100				0.95	0.92		
1660	1200				1.04	1.02		

Metric Units w/mK

°K	°C	Density kgs/m^3 160 to 240	Density kgs/m^3 96 to 192	Density kgs/m^3 80 to 120	Density kgs/m^3 96 to 192	Density kgs/m^3 144	Density kgs/m^3 64 to 88	Density kgs/m^3 128 to 160
173	−100							
223	−50							
273	0							
323	50		0.045	0.042			0.032	0.069
373	100	0.078	0.052	0.049	0.049	0.049	0.034	0.079
423	150	0.084	0.063		0.059	0.052	0.036	0.091
273	200	0.092	0.078		0.067	0.056		0.106
523	250	0.099	0.092		0.073	0.060		0.119
573	300	0.108	0.105		0.083	0.063		0.133
623	350	0.117	0.120		0.093	0.072		0.146
673	400	0.131	0.134		0.103	0.082		0.160
723	450				0.113	0.094		0.173
773	500				0.123	0.106		0.186
823	550				0.133	0.119		0.200
873	600				0.142	0.133		
923	650				0.150	0.148		

Table 61 Thermal Conductances of Reflective Insulations —

Material Identification number	Generic type and class or description of insulating material	Form
99	Aluminum Sheet	Various Thicknessess
100	Aluminum Foil	Various Thicknessess
101	Aluminum foil, both sides, or membrane paper or other reinforcement	Laminate
102	Aluminum foil accordian formed roll insulation-Opened in place	Expandable roll for cavities

English Units

Position	Direction of Heat Flow	Conductance in Btu/ft^2 hr °F 75° Mean Temperature Air spaces 3/4" $\epsilon = 0.05$ (Aluminum) Air spaces enclosed* and separated by reflective shield									
		Number of spaces									
		1	2	3	4	5	6	7	8	9	10
Vertical	Across	0.35	0.18	0.12	0.09	0.07	0.06	0.05	0.045	0.04	0.035
Horizontal	Upward	0.48	0.25	0.17	0.12	0.10	0.08	0.07	0.06	0.055	0.05
Horizontal	Downward	0.35	0.18	0.12	0.09	0.08	0.07	0.055	0.055	0.04	0.035

Metric Units

Position	Direction of Heat Flow	Conductance in w/m^{2}k 24°C Mean Temperature Air spaces 19mm $\epsilon = 0.05$ (Aluminum) Air spaces enclosed* and separated by reflective shield									
		Number of spaces									
		1	2	3	4	5	6	7	8	9	10
Vertical	Across	1.987	1.022	0.681	0.511	0.397	0.341	0.284	0.256	0.227	0.159
Horizontal	Upward	2.725	1.419	0.965	0.681	0.568	0.454	0.397	0.341	0.312	0.284
Horizontal	Downward	1.997	1.022	0.681	0.511	0.454	0.397	0.312	0.256	0.277	0.189

*All spaces constructed to prevent air movement from one space to another.

Note: Heat transfer is given in conductance not conductivity.

Table 62 Thermal Conductivities of Preformed Reflective Insulations

English Units Btu, in/ft, hr, °F

		Identification Number 103	Identification Number 104	Identification Number 105
Mean Temp.		Generic Type: Aluminum casing & aluminum reflective sheets Form: Factory formed for equipment & pipe	Generic Type: Stainless steel casing & aluminum reflective sheets Form: Factory formed for equipment & pipe	Generic Type: Stainless steel casing & reflective sheets Form: Factory formed for equipment & pipe
°R	°F	Density Lbs/ft³ 5 to 7	Density Lbs/ft³ 6 to 8	Density Lbs/ft³ 8 to 10
260	−200			
360	−100			
460	0			
560	100	0.250	0.265	Varied to suit
660	200	0.307	0.328	need
760	300	0.378	0.400	
860	400	0.444	0.466	
960	500	0.507	0.537	
1060	600	0.570	0.600	
1160	700			
1260	800			
1360	900			
1460	1000			
1560	1100			
1660	1200			

Metric Units w/mK

°K	°C	Density kgs/m³	Density kgs/m³	Density kgs/m³
173	−100			
223	−50			
273	0			
323	50	0.037	0.040	
373	100	0.046	0.049	Varied to suit
423	150	0.055	0.058	need
273	200	0.064	0.067	
523	250	0.073	0.076	
573	300	0.082	0.085	
623	350	0.091	0.094	
673	400			
723	450			
773	500			
823	550			
873	600			
923	650			

Based on spacing of one reflective sheet per 1/2" or 1.27 cm enclosed space.

Properties of Fill Insulations

Although the particular set of properties which are important to fill material are slightly different from the forms of insulation previously discussed, each of the individual properties have already been mentioned. The important properties of fill insulations follow:

Absorptivity
Alkalinity
Capillarity
Combustibility (or non-combustibility)
Chemical reaction
Compaction or setting
Corrosion to substrates
Hygroscopicity
Resistance to acids
Resistance to caustics
Resistance to solvents
Specific gravity
Specific heat
Thermal conductivity
Thermal diffusivity
Thermal resistance
Temperature limits
Temperature rise—self internal heating
Vapor migration

Measurement of Properties of Thermal Insulation

Determining what insulation properties are important only indicates the information needed and provides the basis of investigation. The next questions then arise, how can these properties be measured and under what conditions should they be measured? Another question which follows is whether or not measurements made by laboratory tests are truly indicative of the properties of the material in service. Does aging affect the properties? If so, how much and at what rate? What effect does temperature have on test results? What effect does moisture content have on the properties?

Unfortunately, testing methods and studies of properties of material are not yet able to answer all these questions. Nor is there a laboratory available which is accepted as the authority on testing of properties of insulation materials. This statement does not imply that there are no standard test methods and laboratories to perform these standard test methods, but it does mean that the insulation industry lacks even the minimum requirements to provide reliable information. In many instances the manufacturers of insulation have developed their own test methods and can furnish some information regarding their materials. However, the test method used by one manufacturer is not always identical with that of another, so that the values of one cannot be compared with the results obtained by another. This makes direct comparison of material properties quite difficult. Another problem is that most of the mechanical properties are measured at atmospheric conditions which are not at the temperatures to which the material is subjected in service. Inadequate information leads to poor design, the use of improper materials, and the installation failures to which the insulation industry has become accustomed.

In recent years progress has been made in correcting this confusion and presently there exists a number of American Society of Testing and Materials Standard Test Methods. However, even with standard test methods, sometimes it is necessary, because of the wide range in characteristics, to perform more than one test for a single property, as one test may be incapable of measuring all the characteristics for the wide variety of insulation materials. Thus, duplication of test methods for a single property of insulations does exist and the proper selection must be made for the insulation being tested. A list of American Society of Testing and Materials Test Methods for various properties follows:

ASTM Test Methods No.	For
	ADHESION OF THERMAL INSULATING CEMENTS
C-353	Adhesion of Dried Thermal Insulation or Finishing Cement
C-383	Adhesion, Wet, of Thermal Insulating Cements to Metal
	BREAKING LOAD
C-203	Test for Breaking Load and Calculated Flexural Strength of Block Type Thermal Insulation
C-446	Test for Breaking Load and Calculated Modulus of Rupture of Preformed Insulation for Pipes
	COMBUSTIBILITY
D-92	Flash and Fire Points by Cleveland Open Cup Tester
D-93	Flash Point by Pensky-Martens Closed Tester
D-568	Test for Flammability of Flexible Plastics
D-635	Test for Flammability of Self-Supporting Plastics
D-1525	Vicat Softening Temperature of Plastics
D-2015	Gross Calorific Value of Solid Fuel by the Adiabatic Bomb Calorimeter Tester
D-2582	Test for Heat of Combustion of Hydrocarbon Fuels by Bomb Calorimeter
D-2843	Measure of Density of Smoke From Burning or Decomposition of Plastics
D-2863	Test for Flammability of Plastics Using Oxygen Index Method
D-3014	Flammability of Rigid Cellular Plastics (Vertical Position)
D-3211	Test for Relative Density of Black Smoke (Ringlemann Method)
E-84	Test for Surface Burning Characteristics of Building Materials
E-136	Test for Noncombustibility of Elementary Materials
E-162	Test for Surface Flammability of Materials Using Radiant Heat Energy Source
E-286	Test for Surface Flammability of Building Materials Using 8′ (244 m) Tunnel Furnace
	COMPRESSIVE STRENGTH
C-165	Recommended Practice for Measuring Compressive Properties of Thermal Insulation
C-354	Tests for Compression Strength of Thermal Insulating or Finishing Cements
C-495	Test for Compressive Strength of Lightweight Insulating Concrete
D-1621	Test for Compressive Properties of Rigid Cellular Plastics

ASTM Test Methods No.	For
	CORROSION
C-464	Corrosion Effect of Thermal Insulating Cements on Base Metal
C-692	Stress Corrosion Effect of Wicking-Type Thermal Insulation on Stainless Steel
	COVERING CAPACITY
C-166	Covering Capacity and Volume Change upon Drying of Thermal Insulating Cements
	DENSITY
C-519	Density of Fibrous Loose Fill Building Insulations
C-520	Density of Granular Loose Fill Insulations
C-303	Density of Preformed Block-Type Insulation
C-302	Density of Preformed Pipe-Covering Type Thermal Insulation
D-1622	Apparent Density of Rigid Cellular Plastics
	DIMENSIONAL STABILITY
C-548	Dimensional Stability of Low Temperature Thermal Block and Pipe Insulation
	DROPPING (RESISTANCE TO)
C-487	Resistance to Dropping of Preformed Block Type Thermal Insulation
	EMITTANCE
C-445	Normal Total Emittance of Surfaces of Materials 0.01" or Less in Thickness at Approx. Room Temperature
C-835	Test for Total Hemispherical Emittances of Surfaces from 20 to 1400° C (68 to 2550° F)
	HARDNESS
C-569	Test for Indention Hardness of Preformed Thermal Insulation
	HEAT FLUX
C-745	Heat Flux Through Evacuated Insulations Using a Flat Plate Boiloff Calorimeter
	HOT SURFACE PERFORMANCE
C-411	Hot-surface Performance of High-Temperature Thermal Insulation
	MAXIMUM USE TEMPERATURE
C-447	Recommended Practice for Estimating Maximum Use Temperature of Preformed Homogeneous Thermal Insulation
	MECHANICAL STABILITY
C-421	Mechanical Stability of Preformed Thermal Insulation, Tumbling Test
	RESISTANCE TO EXTERNAL LOADS
C-854	Resistance to External Loads on Metal Reflective Insulation
	SHRINKAGE DUE TO HEAT
C-356	Shrinkage, Linear of Preformed High-Temperature Thermal Insulation Subjected to Soaking Heat
	SPECIFIC HEAT
C-351	Mean Specific Heat of Thermal Insulation
	THERMAL HEAT TRANSFER
C-177	Steady State Thermal Conductivity Properties by Means of the Guarded Hot Plate
C-518	Steady State Thermal Conductivity Properties by Means of the Heat Flow Meter
C-236	Thermal Conductance and Transmittance of Built Up Sections by Means of the Guarded Hot Plate
C-335	Thermal Conductivity of Pipe Insulation
C-691	Thermal Transference of Nonhomogeneous Pipe Insulation Above Ambient Temperatures
	WATER ABSORPTION
C-209	Insulating Board (part 13)
	WATER VAPOR TRANSMISSION
E-96	Water Vapor Transmission of Materials in Sheet Form
C-355	Water Vapor Transmission of Thick Materials

As can be seen from the above, standard test methods do not exist for all the properties of insulation listed previously. One reason is that some of these properties are difficult to determine. In other cases test equipment is quite expensive, few manufacturers or laboratories have the required equipment, and an industry standard has never been devised.

It should be remembered that any laboratory test is a measure of a small sample. For certain characteristics, a small sample will provide an accurate appraisal, whereas for other materials the answer may only *indicate* the characteristics of that insulation. Other tests may attempt to predict the service life of insulation. Such service life tests are quite difficult to devise and evaluate. Yet this type of test is essential, and as the demand for it increases, eventually such tests will be devised by the industry.

Property Tables—Not Including Conductivity

The properties in Tables 54 through 59 are not all of the relevant properties of the materials listed previously in this chapter. The tables do list, however, the most important properties and those most commonly needed to make an intelligent selection of an insulation for a specific purpose.

Note that many of the materials have only a few properties listed and that none of them are complete. The properties were compiled from information obtained from many sources, but most were secured directly from the manufacturers of the insulations, or from their published literature. As can easily be understood, further testing and checking by the author of each individual value obtained involved a task much beyond the scope of this book. However, based upon the author's knowledge of insulation materials, the values presented were reviewed to establish that each one was reasonable for that particular insulation and that particular property.

All properties listed are not significant for all materials. For example, the water vapor transmission rate of an insulation used for high temperature service is unimportant because the high temperature pipe or vessel causes a high vapor pressure which drives moisture outward and the insulation remains dry and effective. For this reason, the vapor transmission rate on high temperature materials is seldom measured.

For certain properties, such as alkalinity, there has been no test method agreed upon by the insulation industry. Some manufacturers provide the information in terms of pH and others with respect to percentage of Na_2O. Again because of these limitations, despite the fact that this may cause confusion, the only recourse was to list the values obtainable.

Another factor which should be remembered in the use of these tables is that a particular generic material may be made by several manufacturers. As there are differences in the properties of a generic material produced by different manufacturers, the properties are a mean of the values listed for the products of the various manufacturers. Thus, for those products made by several manufacturers, no single manufacturer's product will have exactly the values as listed. It follows then, that these tables are presented to serve as a *guide* for selection of the correct *generic material* for an installation. However, the final choice of the particular manufactured product must be made by the individual by comparing competitive products.

The property "Maximum Temperature Limits" is given first as this is the first requirement that must be filled. "Density" is second, as it frequently identifies or classifies a particular material as being in a given generic class, and within a generic class it affects the thermal conductivity.

All other properties are listed in alphabetical order.

Property Tables—Conductivity

The thermal conductivities given in Tables 60 through 62 have been gathered from various sources and while deemed accurate, are not the result of tests by the author. In most cases, they are the results of tests performed upon laboratory-dried samples by the manufacturers. They are based upon representative values for generic material within their ranges of densities and are neither the highest nor the lowest of a particular listing of competitive materials. They represent averages which we expect will provide relatively accurate information for use in heat transfer calculations.

As all values listed are taken as representative of the means of generic materials available and are provided for engineering calculation only, they are not deemed suitable for use as material specification limits. Any one particular manufacturer's product is not likely to have the mean values of all the materials manufactured in that particular generic classification.

For a particular installation, the material specification must be written by deciding what property values are essential and, with knowledge of materials available, the specification must be written to obtain those essential properties within the ranges of the materials available.

Hopefully, this book will indicate the need for more complete information on properties of materials and for any missing information to be made available by the manufacturers. When insulation consumers understand the need for designing insulation systems on a scientific basis the manufacturers will quickly respond by providing the necessary technical information.

The abbreviations in the Tables have the following meanings:

amts.	amounts
Exp. Co.	Coefficient of expansion
Comb.	Combustible
hex.	hexagon
Insul.	Insulation
Neg.	Negligible
NC	Noncombustible
PC	Pipe Covering
RH	Relative Humidity
Temp.	Temperature
thk.	thickness
SS	Stainless steel

Listing of Thermal Insulation and Their Properties

An identification number is given in the listing of individual generic materials. This same number is used to identify the properties of that material in Tables 54 through 59. That same number is used to identify the material in the Conductivity Tables 60 through 62.

Wherever possible the physical properties listed were properties established by standard ASTM Test Methods. Unfortunately standard test methods have not yet been established for a number of properties needed for good engineering design. It should also be pointed out that all the strength properties listed were obtained at atmospheric temperatures and not at the temperature at which the material may be expected to operate. Many of the important properties for effective insulation design were not obtainable from the manufacturers of the insulation materials. In some instances, to guide the user, descriptive wording was entered on the property sheets where exact test results were not obtainable. In other instances where information was not obtainable the space for entry of that property was left blank. Where such information is essential the consumer may be able to obtain the information from the manufacturer, or he may have the tests for these properties done by others.

It should also be noted that the heat transfer test results in accordance with the listed ASTM Test Methods are based on laboratory dried samples. Therefore, for absorbent fibrous, cellular, or granular insulation the actual heat transfer will be greater than that obtained from dried test samples. The amount of conductivity increase depends upon the degree of absorbance of the insulation in question and also on the particular service conditions involved.

Properties related to fire hazards are difficult to obtain. Few manufacturers of combustible insulations list or publish the basic combustible properties (as listed under combustibility) of their materials. Instead they list flame spread index, smoke index, etc., which do not provide a guide for good engineering design. Where combustible insulations are used in buildings, spray systems and other fire protection measures should be provided. Likewise combustible insulations should be used with caution in industrial installations.

It must be remembered that the following listing of properties and conductivities was the most accurate that could be collected at the time of the preparation of this book and that manufacturers change their products. Thus where a property is critical to design it is well to check with the manufacturer or manufacturers of that product as to characteristics at that time.

Rigid and Semi-Rigid Insulations

Material identification number	Insulation material, generic type	Description and general characteristics
1	Alumina-silica fibers, combined with binders.	Block insulation for exceptionally high temperature service, resists thermal shock.

Rigid and Semi-Rigid Insulations

Material identification number	Insulation material, generic type	Description and general characteristics
2	Alumina-silica fibers, combined with binders.	Block insulation—stronger than (1) above. For exceptionally high temperature services. Resists thermal shock.
3	Refractory-fiber molded, predominantly alumina and silica.	High temperature lightweight board.
4	Alumina-silica ceramic fibers, combined with binders.	High temperature board, resists thermal shock.
5	Calcium-silicate, hydrous, reinforced with small amounts of mineral fiber. Type 1.	Molded into rigid block and pipe covering. Strong, for its weight. Good cutting characteristics. Water absorbent, but retains most of its strength when wet.
6	Calcium, silicate, hydrous, reinforced with small amounts of mineral fiber, extra high temperatures. Type 2.	Molded into rigid block and pipe covering. Strong, for its weight. Good cutting characteristics. Water absorbent.
7	Diatomaceous silica, blended and bonded. Class I.	Formed into rigid block and pipe covering. Good for high temperatures. Excellent thermal shock resistance.
8	Diatomaceous silica, blended and bonded. Extra high temperature. Class II.	Formed into rigid block and pipe covering. Good for higher temperatures than material (7). Higher density. Excellent thermal shock resistance.
9	Cork, vegetable, compressed. Board.	Molded into rigid board or pipe covering. Strong, for its weight. Good cutting characteristics. Will resist liquid water, but has relatively high rate of vapor transmission.
10	Cork, vegetable, compressed. Pipe covering.	Molded into rigid board or pipe covering. Strong for its weight. Good cutting characteristics. Will resist liquid water, but has relatively high rate of vapor transmission.
11	Glass, cellular	Rigid block and pipe covering. Good strength characteristics. Poor abrasion resistance. Can be formed and fabricated rapidly. Water resistant. Has almost perfect resistance to the migration of moisture vapor.
12	Glass, cellular, between layers of felt	Same a material (11) except it is fabricated into sheets between layers of felt for roof insulation.
13	Diatomaceous earth, Type 1	Light density rigid block and pipe covering.
14	Diatomaceous earth, Type 2	Moderate density rigid block and pipe covering.
15 16 17	Glass, fiber, with organic binder	Semi-rigid formed board or sheets of various densities. The higher the density the stronger the material. The higher densities also provide a lesser conductivity. Available with various types of factory attached outer facings.
18	Glass, fiber, with organic binder, faced one side with vapor resistant facing	Formed into rigid sheets for constructing air conditioning and heating ducts.
19	Glass, fiber, with organic binder	Semi-rigid formed pipe covering. Light density pipe insulation. Available with various types of factory attached jacketing.
20	Glass, fiber, with organic binder	Formed into rigid pipe covering. Heavy density pipe insulation. Available with various types of factory attached jacketing.
21	Mineral fiber, Class 1, lower temperature limit	Semi-rigid formed block or board, moderate density. Available with various types of factory attached jacketing.
22	Mineral fiber, Class 2, inorganic binder	Formed semi-rigid board and block.
23	Mineral fiber, Class 3, with inorganic binders	Formed into semi-rigid block and board.
24	Mineral fiber, Class 4, with inorganic binder	Semi-rigid formed block and board.
25	Mineral fiber, with organic binder, Class 5	Formed into semi-rigid board and block.
26	Mineral fiber, with inorganic binders, Class 1	Formed into pipe covering.
27 29	Mineral fiber, with inorganic binders, Class 2	Formed into pipe covering.
28 30	Mineral fiber, with inorganic binders, Class 3	Formed into pipe covering.
31	Perlite, expanded, blended and bonded with binders	Roof board. Low shrinkage at high temperatures. Water repellent.
32	Phenolic foam	Formed into board and pipe covering.
33	Polystyrene, expanded beads, molded to shape	Formed into rigid block, board and pipe coverings. Light weight, excellent cutting characteristics.
34	Polystyrene, cellular foam, molded	Molded rigid board and pipe covering. Light weight.
35	Polystyrene, cellular foam	Rigid board, sheets and pipe covering cut from slab stock Light weight. Good cutting characteristics. Combustible.
36	Polystyrene, cellular foam	Rigid boards. Light weight. Good cutting characteristics.
37	Polystyrene, cellular	Molded to shape for roof boards.

Rigid and Semi-Rigid Insulations

Material identification number	Insulation material, generic type	Description and general characteristics
38	Polyurethane, cellular foam	Formed into rigid boards, sheets, and pipe covering. Light weight. Good cutting characteristics.
39	Polyurethane, cellular foam	Formed into rigid boards, sheets, and pipe covering. Light weight. Good cutting characteristics.
40	Polyurethane, cellular foam	Formed into block, board and pipe covering.
41	Rubber resin, cellular foam	Rigid board. Good cutting characteristics.
42	Expanded silica and binders	4'x8' (approx.) semi-rigid panels for insulation of large diameter and flat surfaces.
43	Wood fiber and binder	Formed into rigid panels for roof deck insulation.
44	Borosilicate closed cell inorganic foam	Formed into rigid board.
45	Elastomeric sheet	Seamless flexible pipe covering.
46 47 48	Glass, fiber and inorganic binder, available with various facings	Flexible rolls or sheets. All are of similar material except for density. As the density affects both strength and flexibility and also conductivity, divisions in density were provided to establish the relationships.
49	Kaolin cermaic fiber in S.S. metal mesh	Flexible rolls or sheets.
50	Kaolin ceramic fiber in S.S. metal mesh	Flexible rolls or sheets.
51	Mineral and spun glass fibers-metal mesh	Formed into blankets.
52	Mineral fibers-metal mesh facing, Class 1	Formed into blankets with metal mesh facing.
53	Mineral fibers-metal mesh facing, Class 2	Formed into blankets with metal mesh facing.
54	Urethane cellular foam	Formed into block and pipe covering.
55	Urethane cellular foam	Formed into block and pipe covering.

Blanket, Felt and Batt Insulations

Material identification number	Insulation material, generic type	Description and general characteristics
56 57 58	Alumina-silica fiber	Formed into blanket and strip for high temperature service.
59	Alumina-silica fiber	Formed into rope, cord and yarn for high temperature service.
60	Glass, fine fiber, (no binder)	Formed into blanket and matt.
61	Glass, fine fiber with binder	Formed into blanket.

Blanket, Felt and Batt Insulations

Material identification number	Insulation material, generic type	Description and general characteristics
62	Glass fiber, batts with various facings, with binder	Formed into batts and encased in various facings. Mainly used for building ceiling and wall cavity insulation.
63	Glass fiber, bonded with thermosetting organic binders. Available in various facings	Formed into blankets. Numerous densities available. General purpose blanket.
64	Mineral fiber	Formed into blankets and batts.
65	Mineral fiber with binders, coated one side	Blanket specially coated on one side to resist air erosion. For lining heating or air conditioning ducts.
66	Silica fibers	Formed into woven blankets and felts for high temperature installations.

Insulating and Finishing Cements

Material identification number	Insulation material, generic type	Description and general characteristics
67	Alumina-silica, semi-refractory hydraulic setting cement	Semi-refractory and insulating cement. Water mix—hydraulic setting.
68	Alumina-silica, ceramic fibers, air setting mold cement	High temperature—water mix—cement.
69	Alumina-silica ceramic spray mix, cement	Water mix, insulating and finishing cement.
70	Diatomaceous silica and binders, air setting cement	Water mix, finishing cement.
71	Kaowool ceramic fiber and inorganic binders-spray cement	Water mix, hard surface finishing cement
72	Kaowool and mineral fiber blend-spray cement	Water mix, finishing cement.
73	Mineral fiber insulating cement	Water mix, insulating and finishing cement.
74	Mineral fiber, binders, hydraulic setting cement	Water mix—hydraulic setting cement.
75	Vermiculite-expanded with binders, water mix-air setting	Water mix insulating cement. High shrinkage.

Formed or Foamed Sprayed-in-Place Insulations

Material identification number	Insulation material, generic type	Description and general characteristics
76	Urethane, two part liquid mix, rigid foam. Self-extinguishing	Liquids mixed by spray gun. Liquid on surface foams to form rigid insulation. Formulas obtainable for insulation will not continue burning after fire is removed.

Formed or Foamed Sprayed-in-Place Insulations

Material identification number	Insulation material, generic type	Description and general characteristics
77	Urea-formaldehyde foamed in place	Formulated to be used as fire protection and high temperature insulation.
78	Polyvinyl acetate cork filled mastic	Premixed mastic. Can be applied by spray, trowel or palm. Used to control condensation.

Fill and Loose Insulation

Material identification number	Insulation material, generic type	Description and general characteristics
79	Alumina-silica fibers, bulk	For fill applications such as packing furnace joints.
80	Alumina-silica, chopped and milled	Same as (79)
81 83	Alumina-silica, long fibers, bulk cellulose fiber—loose fill	Same as (79)
82	Cork, granulated	Available in various grades and particle sizes.
84	Diatomaceous silica, fine powder	Milled to as fine powder, for filling cavities.
85	Diatomaceous silica, calcined	Calcined to withstand higher temperatures. Used as fill, or with concrete to form insulating concrete.
86	Diatomaceous silica, coarse powder	Milled to a coarse powder, for filling cavities.
87	Gilsonite granules	Available in various grades for various service temperatures. Used as fill around underground hot piping.
88	Glass fibers, loose or shredded	Various grades available.
89	Glass fibers, loose and blowing wool	For pouring or blowing into cavity spaces.
90	Gypsum pellets—bulk	For pouring into cavity spaces.
91	Mineral fibers, with small amount of oil lubricant	Fibers in loose form for filling cavity spaces.
92	Mineral fibers, oil free, loose	Loose fiber for filling cavity spaces.
93	Mineral fiber, granulated	For pouring or blowing in cavity spaces.
94	Perlite, expanded, loose	Loose for filling cavity spaces. Available in various size spheres.
95	Quartz fibers-loose	For filling cavity spaces.
96	Silica fibers	Soft fibers for high temperature service.
97	Silica-aerogel spheres	Small spheres. Very low conductivity. Will flow through a very small opening.

Fill and Loose Insulation

Material identification number	Insulation material, generic type	Description and general characteristics
98	Vermiculite flakes	Expanded mica. Available in various sizes. For filling cavity spaces.

Reflective Insulation/Foils and Sheets

Material identification number	Insulation material, generic type	Description and general characteristics
99	Aluminum Sheet	Polished sheet, thickness as required for strength. Installed to enclose and separate air spaces.
100	Aluminum Foil	Thickness as required for strength. Installed to enclose or separate air spaces.
101	Aluminum Foil on membrane paper, or other reinforcements such as glass fiber	Foil with various reinforcements. Installed to enclose or separate air spaces.
102	Preformed aluminum with kraft paper, accordian formed. Wall and ceiling insulation	Preformed into strips or rolls, then when opened and installed, layers of foil will form enclosed spaces in cavity installation.

Reflective Insulation/Preformed Pipe and Equipment Insulation

Material identification number	Insulation material, generic type	Description and general characteristics
103	Preformed aluminum case with aluminum reflective sheets	Custom formed to fit piping or equipment. Non absorbent. Can be removed and replaced rapidly. Light weight.
104	Preformed stainless steel case with aluminum reflective sheets	Same as (103) above. Slightly more efficient, but limited to lesser service temperatures.
105	Preformed stainless steel case with stainless steel reflective sheets	Custom formed to fit piping or equipment. Non absorbent. Can be removed and replaced rapidly. Light weight.

Thermal conductivity tables were not provided for reflective insulations, as the conductance across a given space depends upon the number of reflective surfaces, size and shapes of air space and the direction of the heat flow. For this reason it is necessary to provide "Surface Emissivity and Reflectance" tables for aluminum. Notice in Table 63 that the reflectance of highly polished aluminum is fairly high; however, aluminum exposed to ambient air soon becomes oxidized and its reflectance to solar radiation decreases significantly.

SURFACE EMITTANCE AND REFLECTANCE OF ALUMINUM FOILS AND SHEETS

			99 and 100 (ALUMINUM FOILS AND SHEETS)	
Condition of Surface	**Radiation**	**Wavelengths**	**Emittance (At ambient temp.)**	**Reflectance R_r (At ambient temp.)**
Polished	Visible	0.36 to 0.75 microns	0.40 to 0.25	0.60 to 0.75
Polished	Infrared	0.75 to 2.8 microns	0.03 to 0.06	0.97 to 0.94
Dull	Visible	0.36 to 0.75 microns	0.60 to 0.48	0.40 to 0.52
Dull	Infrared	0.75 to 2.8 microns	0.06 to 0.09	0.94 to 0.91
Oxidized	Visible	0.36 to 0.75 microns	0.90 to 0.68	0.10 to 0.32
Oxidized	Infrared	0.75 to 2.8 microns	0.10 to 0.12	0.90 to 0.88

TABLE 63

CONDUCTANCE (U) VALUES OF AIR SPACES WITH ALUMINUM FOIL SPACER SHEETS

Air Spaces Divided by Aluminum Foil	**Mean Temp. °F**	**Conductance U Btu/ft² hr° F**	**Mean Temp. °C**	**Conductance U W/m² K**
Air spaces and aluminum foil spacers—vertical				
1 1/2" space divided by				
Aluminum foil (bright both sides)	50	0.23 (U)	10.0	1.3060 (U)
3/4" space divided by				
Aluminum foil (bright both sides)	50	0.31 (U)	10.0	1.7603 (U)
2 1/4" space divided by two curtains				
Aluminum foil (bright both sides)	50	0.15 (U)	10.0	0.8517 (U)
3" space divided by three curtains				
Aluminum foil (bright both sides)	50	0.11 (U)	10.0	0.6246 (U)
3 3/4" space divided by four curtains				
Aluminum foil (bright both sides)	50	0.09 (U)	10.0	0.5110 (U)
Air spaces and aluminum foil spacers—horizontal				
3 5/8" faced both sides with aluminum foil				
Vertical (heat flow across)	50	0.56 (U)	10.0	3.1798 (U)
Horizontal (heat flow up)	50	0.94 (U)	10.0	5.3376 (U)
Horizontal (heat flow down)	50	0.41 (U)	10.0	2.3281 (U)
Air spaces with ordinary building materials				
3 5/8" face with material of emissivities = .83				
Vertical (heat flow across)	50	0.17 (U)	10.0	6.6436 (U)
Horizontal (heat flow up)	50	1.32 (U)	10.0	7.4953 (U)
Horizontal (heat flow down)	50	0.94 (U)	10.0	5.3376 (U)

TABLE 64

Air spaces divided by aluminum foil can be very effective in providing resistance to heat flow. In an ordinary stud space (with no insulation or reflective sheets) cavity the heat flow from one inner surface to the other is approximately 65% to 80% by radiation, 15% to 28% by conduction and 7% by convection. From this, it is evident that control of heat transfer by radiation by use of reflective sheets can be an efficient means to reduce heat transfer across the space. In addition to reducing the radiation heat transfer, when sheets are spaced 1/2″ (1.27 cm) or closer, they also reduce the heat transfer by convection.

To illustrate the effectiveness of reflective sheets facing or separating an air space, see Table 64. It must be remembered, however, that this table does not take into consideration the heat transfer by conduction of the studs, joists or other supporting members. For a wall or roof, etc., the heat transfer caused by these must be part of the total heat transfer.

In the calculation of heat transfer through the floors, walls, roofs, etc., of a residence or building, it is necessary to know the thermal conductivities of all of the construction materials.

In the case of thermal insulations, the physical properties tables were provided to assist in the selection of the insulation to fulfill installation requirements. Strength and corrosion information is not provided for building materials due to space limitations. However, necessary thermal properties of selected structual and construction materials are provided later in this chapter.

The conductivity tables of construction materials are as follows:

Table 65 — Conductivities of Masonry and Construction Materials
Table 66 — Conductivities of Metals
Table 67 — Conductivities of Miscellaneous Materials
Table 68 — Conductivities of Wood

Information as to windows is not included in view of the fact that the overall conductance is dependent upon the number of panes, type of glass, spacing, etc. The conductivities for windows are presented elsewhere.

In the calculations of heating or cooling requirements, it is often necessary to determine heat capacity of the construction materials. Also, it is necessary to know the surface emittance or reflectance R, as these values influence heat transfer from surrounding surfaces when exposed to direct solar radiation.

Table 69 provides the radiation emittance and reflectance of the surfaces of miscellaneous construction materials and painted surfaces.

Table 70 provides the radiation emittance and reflectance of oxidized surfaces of metals.

Table 71 provides the radiation emittance and reflectance of weather barrier materials.

When it is necessary to determine the quantity of heat necessary to change the temperature of a material, it is necessary to know the specific heat and the density of that material.

Table 72 provides specific heats and weights of materials commonly used in the construction of buildings and residences.

Thermal insulations are materials having air or gas-filled pockets, void spaces or heat reflecting surfaces which retard the flow of heat. Used in residences and buildings, they serve to control energy loss, surface temperatures, rate of temperature change and energy costs. The need for insulation may be based on one or more requirements of the residence or building. These needs for insulation are presented in the following chapter.

Table 65 Conductivities of Masonry and Construction Materials – English and Metric Units

Material			Mean Temp. °F	Conductivity k Btu in/ft²hr°F	Density kg/m³	Mean Temp. °C	Conductivity W/m K
BUILDING CONSTRUCTION MATERIALS							
Asphalt, street	132#	Density	68	5.28	2116	20.0	0.7614
Beaver board							
Cane fiber	13.8#	Density	75	0.33	221	23.9	0.0476
Spruce fiber	31#	Density	75	1.97	497	23.9	0.2541
Cane fiber board	(Celotex)						
	13.8#	Density	75	0.33	221	23.9	0.0476
			90	0.34			0.0490
Ebonite	74#	Density	68	0.41	1185	20.0	0.0591
Glass							
Flint			59	4.16		15.0	0.5999
Plate			68	5.55		20.0	0.6003
Soda	161#	Density	68	4.94	2579	20.0	0.7123
Quartz			212	13.27		100.0	1.9135
Gypsum block	42.7#	Density	32	1.69	684	0.0	0.2437
			68	1.86		20.0	0.2682
			86	1.94		30.0	0.2797
Gypsum board							
Gypsum board							
covered with paper	62#	Density	70	1.44	993	21.1	0.2076
1/2" Thick	53.5#	Density	68	2.60 (U)	857	20.0	14.7634 (U)
Linoleum			32	1.21		0.0	0.1745
			68	1.29		20.0	0.1860
			75	1.36		23.9	0.1961
Masonite	20#	Density	75	0.33	320	23.9	0.0476
Plaster and lath							
Metal lath and plaster			70	4.4 (U)		21.1	24.984 (U)
Wood lath and plaster			70	2.6 (U)		21.1	14.763 (U)
Plaster board							
Covered with paper	61#	Density			977		
3/8" Thick			70	3.73 (U)		21.1	21.1799 (U)
1/2" Thick			70	2.83 (U)		21.1	16.0695 (U)
Plaster, gypsum	52.4#	Density	68	1.77	839	20.0	0.2552
	46.2#	Density	86	2.32	740	30.0	0.3343
Porcelain			329	11.30		165.0	1.6295
Quartz			212	13.27		100.0	1.9135
Rubber	68.6#	Density	86	1.22	1099	30.0	0.1759
Shingles							
Asbestos	65#	Density	75	6.0 (U)	1041	23.9	34.0696 (U)
Asphalt	70#	Density	75	6.5 (U)	1121	23.9	36.9087 (U)
Slate	201#	Density	75	10.37 (U)	3220	23.9	58.8835 (U)
Wood			75	1.28 (U)		23.9	7.2682 (U)
Slate							
Across cleavage			50	9.15 to 10.45		10.0	1.319 to 1.506
Along cleavage			50	10.00 to 18.9		10.0	1.442 to 2.725
Strawboard	43#	Density	86	0.50	689	30.0	0.721
Textan							
Rubber composition	81#	Density	86	1.17	1297	30.0	0.1687
Wood pulp board	43#	Density	86	0.49	689	30.0	0.0707
Wood felt	20.6#	Density	86	0.37	330	30.0	0.0534
Wood fiber board	19.8#	Density	75	0.33	317	23.9	0.0476
	11.9#	Density	86	0.30	191	20.0	0.0433
	28.5#	Density	75	0.50	457	23.9	0.0721

Note: Where values under the conductivity K column are marked (U) they are conductance values (not conductivity) and are given in the following units:
English: Btu/ft²hr°F
Metric: W/m²K

Table 65 Conductivities of Masonry and Construction Materials — English and Metric Units

Material			Mean Temp. °F	Conductivity k Btu in/ft² hr° F	Density kg/m³	Mean Temp. °C	Conductivity W/m K
MASONRY MATERIALS							
Brick							
Low density			70	5.0		21.1	0.7214
High density			70	9.2		21.1	1.3266
Red building, soft burned			600	4.3		316.0	0.6201
			800	4.6		427.0	0.6633
			1000	5.9		538.0	0.7066
Red building, hard burned			600	7.4		316.0	1.0671
			800	8.2		427.0	1.1824
			1000	9.0		538.0	1.2973
Brick—slag	87.4#	Density	59	87.4	1400	15.0	12.6031
Cement plaster			70	8.0		21.1	1.1536
Cinder block							
4 x 8 x 16—Solid			40	1.00 (U)		4.4	5.6783 (U)
8 x 8 x 16—With standard hollow spaces			40	0.58 (U)		4.4	3.2934 (U)
12 x 8 x 16—With standard hollow spaces			40	0.53 (U)		4.4	3.0094 (U)
Concrete							
Typical			40	12.00		4.4	1.7304
Concrete Block							
8 x 8 x 16—Sand and gravel aggregate (hollow)			40	0.9 (U)		4.4	5.1104 (U)
8 x 8 x 16—Limestone aggregate (hollow)			40	0.86 (U)		4.4	4.8833 (U)
12 x 8 x 16—Sand and gravel aggregate (hollow)			40	0.78 (U)		4.4	4.4290 (U)
Solid			88	8.2		31.1	1.1824
Concrete, cellulated							
	40#	Density	75	1.06	640	23.9	0.1528
	50#	Density	75	1.44	800	23.9	0.2076
	60#	Density	75	1.80	561	23.9	0.2596
	70#	Density	75	2.18	1121	23.9	0.3144
Concrete, cinder							
1:2:2.75 Ratio	104#	Density	75	4.63	1666	23.9	0.6676
1:2.75:4.5 Ratio	99#	Density	75	4.30	1585	23.9	0.6201
1:3.5:5.5 Ratio	92#	Density	75	3.73	1474	23.9	0.5379
Concrete—cork filled							
1 Portland: 2 Sand: 3 Granulated cork	79#	Density	185	1.79	1265	85.0	0.2581
Concrete gypsum							
87.5% Gypsum, 12.5% Wood chips	51#	Density	74	1.66	816	23.3	0.2394
Concrete, Haydite							
1:2:2.75 Ratio	80#	Density	75	4.15	1291	23.9	0.5984
1:2.75:4.5 Ratio	75#	Density	75	3.78	1201	23.9	0.5451
1:3.5:5.5 Ratio	72#	Density	75	3.67	1153	23.9	0.5292
1:8 Ratio	67#	Density	75	2.90	1073	23.9	0.4182
Concrete, limestone							
1:2:2.7 Ratio	135#	Density	75	11.2	2162	23.9	1.6150
1:2.75:4.5 Ratio	138#	Density	75	12.0	2210	23.9	1.7304
1:3.5:5.5 Ratio	136#	Density	75	11.5	2178	23.9	1.6583
Concrete, sand and gravel							
1:2:2.75 Ratio	145#	Density	75	13.1	2322	23.9	1.8890
1:2.75:4.5 Ratio	146#	Density	75	12.9	2338	23.9	1.8602
1:3.5:5.5 Ratio	145#	Density	75	13.2	2322	23.9	1.9034
Dolomite, compact			70	13.6 to 16.3		21.1	1.96 to 2.35
Domont brick (Terracotta)	113#	Density	196	4.62	1610	91.1	0.6662
Glagstone							
Across cleavage			70	12.8		21.1	1.8458
Along cleavage			70	18.4		21.1	2.6533
Freestone, sandstone			70	6.1		21.1	0.8796
Glass block			100	0.46 (U)		37.8	2.6120 (U)
			200	0.49 (U)		93.3	2.7823 (U)
			300	0.53 (U)		149.0	3.0095 (U)
			400	0.56 (U)		204.0	2.1798 (U)
			500	0.60 (U)		260.0	3.4070 (U)
Granite			70	15.0 to 22.0		21.1	2.16 to 3.17
Gravel							
Fine (0.16" to 0.35")	91#	Density	185	1.83	1467	85.0	0.2350
Dry Stone (1" to 3")	115#	Density	32	2.34	1642	0.0	0.3374
			68	2.58		20.0	0.3720
			104	2.83		40.0	0.4081

Note: Where values under the conductivity K column are marked (U) they are conductance values (not conductivity) and are given in the following units:
English: Btu/ft² hr° F
Metric: W/m² K

Table 65 Conductivities of Masonry and Construction Materials – English and Metric Units

Material		Mean Temp. °F	Conductivity k Btu in/ft²hr°F	Density kg/m³	Mean Temp. °C	Conductivity W/m K
MASONRY MATERIALS (continued)						
Gypsum Board						
Gypsum board						
covered with paper	62# Density	70	1.44	993	21.1	0.2076
	1/2" Thick 53.5# Density	68	2.6	856	20.0	14.7634 (U)
Gypsum plaster						
	52.4# Density	68	1.77	839	20.0	0.2552
	46.2# Density	86	2.32	740	20.0	0.3345
Gypsum tile		50	.46		10.0	0.0663
3 x 3 x 16	67# Density	40	.50 (U)	1073	4.4	0.088 (U)
Gypsum tile		50	.46		10.0	0.0663
Haydite block						
8 x 8 x 16	67# Density	40	.50 (U)	1073	4.4	0.088 (U)
8 x 12	77# Density	40	.46 (U)	1233	4.4	0.081 (U)
Insulux Glass Block		100	.46 (U)		37.8	0.081 (U)
		200	.49 (U)		93.3	0.086 (U)
		300	.53 (U)		149.0	0.093 (U)
		400	.57 (U)		204.0	0.100 (U)
		500	.50 (U)		26.0	0.088 (U)
Lime						
Hard		50	25.57		10.0	3.687
Limestone						
		32	4.8		0.0	0.692
		59	4.9		15.0	0.707
		68	5.1		20.0	0.735
		77	5.2		25.0	0.750
Marble		86	14.5 to 19.9		80.0	2.09 to 2.74
Millstone	78.5# Density	177	3.36	1257	80.6	0.485
Mortar	107# Density	191	2.24	1714	88.3	0.323
	117# Density	191	3.71	1874	88.3	0.535
Onyx		86	16.14		30.0	2.327
Red Brick		200	5.4		98.3	0.779
		400	5.7		204.0	0.822
		600	6.1		316.0	0.880
		800	6.5		427.0	0.937
		1000	8.6		538.0	1.240
Sand						
Fine (less than .08"–Dry)	96# Density	32	2.10	1538	0	0.303
		68	2.26	1538	20.0	0.326
Fine–common moisture	98# Density	68	8.60	1570	20.0	1.240
Sandstone						
Fresh cut–natural gray	141# Density	50	10.72	2259	10.0	1.546
		68	11.62		30.0	1.676
		104	12.75		40.0	1.839
Slate						
Across cleavage		50	9.15 to 10.45		10.0	1.31 to 1.51
Along cleavage		50	16.00 to 18.9		10.0	2.31 to 2.73
Soapstone	171# Density	158	23.22	2739	70.0	3.398
Stucco		50	12.00		10.0	1.730
Terra Cotta	112# Density	196	4.62	1794	91.1	0.666
Terrazzo		50	12.00		10.0	1.730
Tile, clay hollow						
4"		50	1.00 (U)		10.0	0.176 (U)
6"		50	0.64 (U)		10.0	0.113 (U)
8"		50	0.60 (U)		10.0	0.106 (U)
10"		50	0.58 (U)		10.0	0.102 (U)
12"		50	0.40 (U)		10.0	0.070 (U)
Tile, gypsum						
4" hollow		50	0.46 (U)		10.0	0.081 (U)

Note: Where values under the conductivity K column are marked (U) they are conductance values (not conductivity) and are given in the following units:
English: Btu/ft²hr°F
Metric: W/m²K

Table 66 Conductivities of Metals — English and Metric Units

Material			Mean Temp. °F	Conductivity k Btu in/ft²hr°F	Density kg/m³	Mean Temp. °C	Conductivity W/m K
			METALS				
Aluminum		168 to 170# Density	−290	1396.0	2690 to 2723	−178.9	201.30
			32	1396.0		0.0	201.30
			210	1430.0		100.0	206.21
			570	1597.0		300.0	230.29
			930	1855.0		500.0	267.49
Antimony		413# Density	32	128.35	6615	0.0	18.51
			212	115.0		100.0	16.58
Brass	Yellow Brass		32	592.5		0.0	85.44
			212	738.0		100.0	106.42
	Red Brass		32	714.0		0.0	102.95
			212	820.0		100.0	118.24
Bronze			68	410.0		20.0	59.12
			210	492.0		98.9	70.94
Copper			32	2190.0		0.0	315.80
			212	2324.0		100.0	335.12
			390	2574.0		200.0	371.17
Gold			−420	1048.0		−251.1	151.12
			32	2160.0		0.0	311.47
			390	2145.0		200.0	309.31
Iron							
Cast			86	432.5		20.0	62.37
Iron wrought		492# Density	65	417.0	7881	18.3	60.13
			212	412.0		100.0	59.41
Lead			−297	313.8		−182.8	45.25
			10.4	276.3		− 12.0	39.84
			32	244.5		0.0	35.26
			64.4	241.0		18.0	34.75
			210	233.0		98.8	33.60
Magnesium		108# Density	210	1089.0	1729	98.8	157.03
Mercury			32	50.2		0.0	7.24
			212	62.8		100.0	9.06
Nickel (99%)			−256	374.5		−160.0	54.00
			50	403.0		10.0	58.18
			930	331.0		500.0	47.73
			1650	306.0		900.0	44.12
Platinum			212	485.0		100.0	69.94
Silver			−256	2900.0		−160.0	418.18
			32	3135.0		0.0	452.07
			212	2880.0		100.0	415.30
			644	2920.0		340.0	421.06
Steel							
Less 0.1% carbon			210	379.0		100.0	54.65
			570	347.0		300.0	50.03
			1110	258.0		600.0	37.20
			1650	234.0		900.0	33.74
Less than 0.6% carbon			210	290.0		100.0	41.82
			1110	234.0		600.0	33.74
			1650	202.0		900.0	29.13
Approximately 1.5% carbon			210	258.0		100.0	37.20
			570	249.0		300.0	35.91
			1110	234.0		600.0	33.74
			1650	202.0		900.0	29.13
Steel chromium			86	213.0 to 291.0		30.0	30.71 to 41.96
Steel, puddled			59	319		15.0	46.0
Steel Wool							
No. 2 Size Fiber		4.74# Density	132	0.63	76	55.6	0.0908
		6.3# Density	132	0.61	101	55.6	0.0897
		9.48# Density	132	0.55	152	55.6	0.0793
Tin			32	443.0		0.0	63.88
			212	413.0		100.0	59.55

Note: Where values under the conductivity K column are marked (U) they are conductance values (not conductivity) and are given in the following units:
English: Btu/ft²hr°F
Metric: W/m²K

Table 67 Conductivities of Miscellaneous Materials — English and Metric Units

Material				Mean Temp. °F	Conductivity k Btu in/ft^2 hr°F	Density kg/m^3	Mean Temp. °C	Conductivity W/m K
				MISCELLANEOUS				
Celluloid, white				86	1.46		30.0	0.2105
Chalk				70	6.48		21.1	0.9344
Charcoal		11.85#	Density	32	0.41	189	0.0	0.0591
				104	0.46		40.0	0.0663
				176	0.51		80.0	0.0735
Clay	Dried			50	3.60		10.0	0.5192
	Wet			50	16.09		10.0	2.3202
Clinkers, from boilers		46.8#	Density	32	1.05	750	0.0	0.1151
				68	1.13		10.0	0.1629
Coal dust	Dry	62.4#	Density	32	0.97	999	0.0	0.1399
				68	1.05		10.0	0.1151
Lamp black		12.05#	Density	132	0.22	193	55.6	0.0317
				316	0.27		156.0	0.0389
				441	0.32		230.0	0.0461
Leather		62#	Density	50	1.10	993	10.0	0.1586
Linen				50	0.61		10.0	0.0879
Paraffin		55#	Density	86	1.60	861	30.0	0.2307
Peat Moss	Dry	11.8#	Density	32	0.33	189	0.0	0.0475
				68	0.34		20.0	0.0490
	Damp	12.17#	Density	68	0.57	195	20.0	0.0822
Plaster of Paris								
Powder				50	7.55		10.0	1.0887
Set				50	2.04		10.0	0.2942
Rubber	Hard	74.3#	Density	99	1.11	1190	37.2	0.1600
	Soft	68.6#	Density	86	1.22	1098	30.0	0.1759
	Sponge	14#	Density	50	1.38	224	10.0	0.1989
Sawdust								
Various, dry		12#	Density	90	0.41	192	32.2	0.0591
Pine, loose, dry		3.6#	Density	166	0.57	57	74.4	0.0822
Silk Fibers		9.2#	Density	32	0.32	147	0.0	0.0189
				122	0.38		50.0	0.0548
				212	0.42		100.0	0.0606
Soil	Dry			50	0.96		10.0	0.1384
	Including stones— Normal dampness			32	3.47		0.0	0.5004
				68	3.63		20.0	0.5234
				158	4.03		70.0	0.5811
	Wet			50	4.64		10.0	0.6691
Vacuum								
Silvered vacuum jacket								
Residual air pressure 0.001 MM of Hg				77	0.0042 (U)		25.0	0.0238 (U)

Note: Where values under the conductivity K column are marked (U) they are conductance values (not conductivity) and are given in the following units:
English: Btu/ft^2 hr°F
Metric: W/m^2 K

Table 68 Conductivities of Wood – English and Metric Units

Material			Mean Temp. °F	Conductivity k Btu in/ft²hr°F	Density kg/m³	Mean Temp. °C	Conductivity W/m K
			WOOD				
Balsa							
Across grain	20.6#	Density	86	0.59	330	30.0	0.0851
	7.05#	Density	86	0.32	113	30.0	0.0461
California Redwood (across grain)							
0% Moisture	22#	Density	75	0.66	352	23.9	0.0952
8% Moisture	22#	Density	75	0.70	352	23.9	0.1009
16% Moisture	22#	Density	75	0.74	352	23.9	0.1067
0% Moisture	28#	Density	75	0.70	448	23.9	0.1009
8% Moisture	28#	Density	75	0.75	448	23.9	0.1008
16% Moisture	28#	Density	75	0.80	448	23.9	0.1154
Cypress (across grain)							
0% Moisture	22#	Density	75	0.67	352	23.9	0.0966
8% Moisture	22#	Density	75	0.71	352	23.9	0.1024
16% Moisture	22#	Density	75	0.79	352	23.9	0.1139
0% Moisture	32#	Density	75	0.79	512	23.9	0.1139
8% Moisture	32#	Density	75	0.84	512	23.9	0.1211
16% Moisture	32#	Density	75	0.90	512	23.9	0.1298
Elm–Soft (across grain)							
0% Moisture	28#	Density	75	0.73	449	23.9	0.1052
8% Moisture	28#	Density	75	0.77	449	23.9	0.1110
16% Moisture	28#	Density	75	0.81	449	23.9	0.1168
0% Moisture	34#	Density	75	0.88	544	23.9	0.1269
8% Moisture	34#	Density	75	0.93	544	23.9	0.1341
16% Moisture	34#	Density	75	0.97	544	23.9	0.1399
Fir (across grain)							
0% Moisture	26#	Density	75	0.61	416	23.9	0.0880
8% Moisture	26#	Density	75	0.66	416	23.9	0.0952
16% Moisture	26#	Density	75	0.76	416	23.9	0.1096
0% Moisture	34#	Density	75	0.67	544	23.9	0.0966
8% Moisture	34#	Density	75	0.75	544	23.9	0.1082
16% Moisture	34#	Density	75	0.82	544	23.9	0.1182
Hemlock, Eastern (across grain)							
0% Moisture	22#	Density	75	0.60	352	23.9	0.0865
8% Moisture	22#	Density	75	0.63	352	23.9	0.0908
16% Moisture	22#	Density	75	0.67	352	23.9	0.0966
0% Moisture	30#	Density	75	0.76	480	23.9	0.1096
8% Moisture	30#	Density	75	0.81	480	23.9	0.1108
16% Moisture	30#	Density	75	0.85	480	23.9	0.1226
Hemlock, West Coast (across grain)							
0% Moisture	22#	Density	75	0.68	352	23.9	0.0980
8% Moisture	22#	Density	75	0.73	352	23.9	0.1053
16% Moisture	22#	Density	75	0.78	352	23.9	0.1125
0% Moisture	30#	Density	75	0.79	480	23.9	0.1139
8% Moisture	30#	Density	75	0.85	480	23.9	0.1226
16% Moisture	30#	Density	75	0.91	480	23.9	0.1312
Mahogany (across grain)	34#	Density	86	0.90	544	30.0	0.1298
Maple, Hard							
Across Grain	45#	Density	127	1.26	720	52.8	0.1817
Along Grain	45#	Density	127	3.02	720	52.8	0.4355
0% Moisture	40#	Density (Across Grain)	75	1.01	640	23.9	0.1456
8% Moisture	40#	Density (Across Grain)	75	1.08	640	23.9	0.1557
16% Moisture	40#	Density (Across Grain)	75	1.15	640	23.9	0.1658
0% Moisture	46#	Density (Across Grain)	75	1.05	736	23.9	0.1514
8% Moisture	46#	Density (Across Grain)	75	1.13	736	23.9	0.1629
16% Moisture	46#	Density (Across Grain)	75	1.21	736	23.9	0.1745
Maple, Soft (Across grain)							
0% Moisture	36#	Density	75	0.89	576	23.9	0.1283
8% Moisture	36#	Density	75	0.96	576	23.9	0.1384
16% Moisture	36#	Density	75	1.01	576	23.9	0.1456
0% Moisture	42#	Density	75	0.95	672	23.9	0.1370
8% Moisture	42#	Density	75	1.02	672	23.9	0.1408
16% Moisture	42#	Density	75	1.09	672	23.9	0.1572

Note: Where values under the conductivity K column are marked (U) they are conductance values (not conductivity) and are given in the following units:
English: Btu/ft²hr°F
Metric: W/m²K

Table 68 Conductivities of Wood – English and Metric Units

Material		Mean Temp. °F	Conductivity k Btu in/ft² hr°F	Density kg/m³	Mean Temp. °C	Conductivity W/m K
		WOOD (continued)				
Oak						
Across grain	51# Density	32	1.38	816	0.0	0.1990
		59	1.46		15.0	0.2105
Along grain	51# Density	54	2.42	816	12.2	0.3490
		60	2.50		15.6	0.3605
		120	2.99		48.9	0.4312
0% Moisture	38# Density (Across Grain)	75	0.98	608	23.9	0.1413
8% Moisture	38# Density (Across Grain)	75	1.03	608	23.9	0.1485
16% Moisture	38# Density (Across Grain)	75	1.07	608	23.9	0.1543
0% Moisture	48# Density (Across Grain)	75	1.18	769	23.9	0.1702
8% Moisture	48# Density (Across Grain)	75	1.24	769	23.9	0.1783
16% Moisture	48# Density (Across Grain)	75	1.29	769	23.9	0.1860
Pine, Norway (across grain)						
0% Moisture	22# Density	75	0.62	352	23.9	0.0894
8% Moisture	22# Density	75	0.68	352	23.9	0.0981
16% Moisture	22# Density	75	0.74	352	23.9	0.1067
0% Moisture	32# Density	75	0.74	512	23.9	0.1067
8% Moisture	32# Density	75	0.83	512	23.9	0.1197
16% Moisture	32# Density	75	0.92	512	23.9	0.1327
Pine, Sugar (across grain)						
0% Moisture	22# Density	75	0.54	352	23.9	0.0779
8% Moisture	22# Density	75	0.59	352	23.9	0.0851
16% Moisture	22# Density	75	0.65	352	23.9	0.0937
0% Moisture	30# Density	75	0.64	480	23.9	0.0923
8% Moisture	30# Density	75	0.71	480	23.9	0.1024
16% Moisture	30# Density	75	0.78	480	23.9	0.1125
Pine, White						
Across Grain	28# Density	167	0.74	448	75.0	0.1067
Along Grain	28# Density	133	1.78	448	56.1	0.2567
Across Grain	34# Density	86	0.80	544	30.0	0.1154
Pine, Yellow, Long Leaf (across grain)						
0% Moisture	30# Density	75	0.76	480	23.9	0.1096
8% Moisture	30# Density	75	0.83	480	23.9	0.1197
16% Moisture	30# Density	75	0.89	480	23.9	0.1283
0% Moisture	40# Density	75	0.86	640	23.9	0.1240
8% Moisture	40# Density	75	0.95	640	23.9	0.1370
16% Moisture	40# Density	75	1.03	640	23.9	0.1485
Pine, Yellow, Short Leaf (across grain)						
0% Moisture	26# Density	75	0.74	416	23.9	0.1067
8% Moisture	26# Density	75	0.79	416	23.9	0.1139
16% Moisture	26# Density	75	0.84	416	23.9	0.1211
0% Moisture	30# Density	75	0.91	480	23.9	0.1312
8% Moisture	30# Density	75	0.97	480	23.9	0.1399
16% Moisture	30# Density	75	1.04	480	23.9	0.1500
Sawdust						
Various, dry	12# Density	90	0.41	192	32.2	0.0591
Pine, loose, dry	3.6# Density	166	0.57	58	74.4	0.0822
Shavings–Planer						
Red Wood Bark	3# Density	90	0.31	48	32.2	0.0447
Red Wood Bark	5# Density	75	0.26	80	23.9	0.0375
Various	8.75# Density	86	0.41	140	30.0	0.0591
Beech and Birch	13.2# Density	90	0.36	211	32.2	0.0519
Teak Wood						
Across grain	40.5# Density	32	1.13	649	0.0	0.1629
Across grain	40.5# Density	59	1.21	649	15.0	0.1745
Across grain	40.5# Density	122	1.38	649	50.0	0.1990
Along grain	40.5# Density	32	2.59	649	0.0	0.3735
Along grain	40.5# Density	59	2.67	649	15.0	0.3850
Along grain	40.5# Density	122	2.75	649	50.0	0.3966

Note: Where values under the conductivity K column are marked (U) they are conductance values (not conductivity) and are given in the following units:
English: Btu/ft² hr°F
Metric: W/m² K

EMITTANCE AND REFLECTANCE OF CONSTRUCTION MATERIALS AND PAINTED SURFACES

English and Metric Units

Miscellaneous materials	Surface Temp. °F	°C	Total Normal emittance ∈	Total Normal reflectance R
Asbestos board	100	38	0.96	0.04
Asbestos paper	100-700	38-371	0.93-0.95	0.07-0.05
Asbestos cloth	200	93	0.90	0.10
Asphalt pavement	Solar	Solar	0.93	0.07
Brick				
Glazed			0.75	0.25
Red, rough	Solar	Solar	0.70	0.30
Silica	2500	1371	0.84	0.16
Refractory	200-1000	93-538	0.92-0.97	0.08-0.03
Carbon black	70-700	21-371	0.95	0.05
Concrete			0.63	0.37
Concrete	Solar	Solar	0.65	0.35
Glass	72	22	0.93	0.07
Gypsum	70	21	0.90	0.10
Ice	32	0	0.96-0.99	0.04-0.01
Marble, polished	70	21	0.93	0.07
Mica	200	93	0.84	0.16
Oak, planed	70	21	0.89	0.11
Oil film	68	20	0.27-0.82	0.73-0.18
Paint, black	200-600	93-316	0.92-0.95	0.08-0.05
Black	100	38	0.97	0.03
Black	Solar	Solar	0.90	0.10
Green	200-600	93-316	0.90-0.93	0.10-0.07
Green	100	38	0.80	0.20
Green	Solar	Solar	0.50	0.50
White	200-600	93-316	0.92-0.84	0.08-0.16
White	100	38	0.68	0.32
White	Solar	Solar	0.30	0.70
Aluminum	212	100	0.27-0.67	0.73-0.33
Oil paint			0.92-0.94	0.08-0.06
Paper, white	70	21	0.92-0.94	0.08-0.06
Paper, black roofing	100	38	0.95	0.05
Porcelain, glazed	70	21	0.92	0.08
Refractories	1500	816	0.65-0.90	0.35-0.10
Roofing paper	100	38	0.91	0.09
Rubber				
Hard, gloss	70	21	0.94	0.06
Soft, rough	76	24	0.85	0.15
Water	70	21	0.85-0.96	0.15-0.04
Wood	100	38	0.83-0.92	0.17-0.08

TABLE 69

RADIATION EMITTANCE AND REFLECTANCE
OF OXIDIZED METALS

Oxide surfaces	Probable value ∈ for oxide forms on smooth metal	Reflectance R_r
Alumel (Oxidized)	0.87	0.13
Cast Iron (Oxidized)	0.70	0.30
80 Ni20 Cr (Oxidized)	0.90	0.10
60 Ni 24 Fe 16 Cr (Oxidized)	0.83	0.17
55 Fe 37:5 Cr 7.5 Al (Oxidized)	0.78	0.22
70 Fe 23 Cr 5 Al 2 Co (Oxidized)	0.75	0.25
Constantan (SS Co. 45 Ni) Oxidized)	0.84	0.16
Carbon Steel (Oxidized)	0.80	0.20
Stainless Steel (18-8) (Oxidized)	0.85	0.15

TABLE 70

RADIATION EMITTANCE AND REFLECTANCE TABLE,
WEATHER-BARRIERS AND FINISHES

Weather-Barrier or Surface Finish	Conditions	Emittance ∈ (At Surface Temp. of Approx 100 F)	Reflectance R_r
Aluminum Paint	New	0.20 to 0.30	0.80 to 0.70
	After weathering	0.40 to 0.70	0.60 to 0.30
Asbestos Paper	Clean	0.90 to 0.94	0.10 to 0.06
Asphalt Asbestos Felts		0.93 to 0.96	0.07 to 0.06
Asphalt Mastics		0.90 to 0.95	0.10 to 0.05
Galvanized Steel	New-bright	0.06 to 0.10	0.94 to 0.90
	Dull	0.20 to 0.30	0.80 to 0.70
Paints	White-clean	0.55 to 0.70	0.45 to 0.30
	Green-clean	0.65 to 0.80	0.35 to 0.20
	Gray-clean	0.80 to 0.90	0.20 to 0.10
	Black-clean	0.90 to 0.95	0.10 to 0.05
Painted Canvas	Color as painted	Will be approx. the same as ∈ for color of paint used	
PVA Mastics	White-clean	0.60 to 0.70	0.40 to 0.30
	Green-clean	0.70 to 0.80	0.30 to 0.20
	Gray-medium-clean	0.85 to 0.90	0.15 to 0.10
	Black	0.85 to 0.95	0.15 to 0.05
Roofing Felts		0.90 to 0.95	0.10 to 0.05
Stainless Steel	Polished	0.22 to 0.26	0.78 to 0.74
	No. 4 mill finish	0.35 to 0.40	0.65 to 0.60
	Oxidized	0.80 to 0.85	0.20 to 0.15

TABLE 71

SPECIFIC HEATS AND WEIGHTS OF MATERIALS

English and Metric Units

Material	Temperature		Density		Mean Specific Heat
	°F	°C	lbs/cu ft	kgs/m³	
Aluminum	32-212	0-100	168	2691	0.215
Asbestos	32-212	0-100	150	2403	0.20
Bakelite	32-212	0-100	80	1282	0.3-0.4
Brass, yellow	32-212	0-100	534	8554	0.088
Brass, red	32-212	0-100	534	8554	0.09
Bronze	32-212	0-100	509-554	8154-8875	0.014
Brick	32-212	0-100	125-143	2002-2201	0.20-0.22
Carbon	32-212	0-100	139	2227	0.165
Chalk	32-212	0-100	143	2290	0.215
Charcoal	32-212	0-100	25	400	0.20
Cinders	32-212	0-100	70-80	1121-1281	0.18
Coal	32-212	0-100	81-94	1297-1505	0.24-0.3
Concrete	32-212	0-100	137	2194	0.156
Cork	32-212	0-100	12	192	0.485
Coke	32-212	0-100	75	1201	0.203
Copper	32-212	0-100	556	8907	0.094
Glass	32-212	0-100	162	2599	0.12-0.19
Graphite	32-212	0-100	135	2162	0.201
Gold	32-212	0-100	1205	19304	0.031
Granite	32-212	0-100	168	2691	0.195
Gypsum	32-212	0-100	155	2483	0.259
Humus (soil)	32-212	0-100	76-100	1217-1602	0.44
Ice	0	minus 18	56	897	0.465
Ice	32	0	64	1025	0.487
Iron, cast	32-212	0-100	442	7080	0.130
Iron, wrought	32-212	0-100	485	7770	0.110
Iron, wrought	32-572	0-300	485	7770	0.122
Iron, at high temperatures	1382-1832	759-982	485	7770	0.213
Lead	32-212	0-100	710	11374	0.031
Limestone	70	21	155-162	2483-2595	0.217
Marble	32-212	0-100	170	2723	0.210
Mercury	32-212	0-100	850	13617	0.033
Masonry, brick	32-212	0-100	125-150	2002-2403	0.20-0.22
Nickel	32-212	0-100	537	8602	0.109
Oil, machine	0	minus 18	60	961	0.400
Porcelain	32-212	0-100	160	2263	0.22
Quartz	32-212	0-100	165	2643	0.17-0.28
Sand	32-212	0-100	100-125	1602-2002	0.195
Sandstone	32-212	0-100	143	2290	0.22
Silver	32-212	0-100	655	10493	0.056
Silica	32-212	0-100	180	2883	0.191
Steel, mild	32-212	0-100	485	7769	0.116
Steel, high carbon	32-212	0-100	485	7769	0.117
Stone, average	70	21	150	2403	0.200
Tin	32-212	0-100	459	7353	0.056
Water	70	21	62.4	1000	1.000
Wood, fir	70	21	25-32	400-512	0.650
Wood, oak	70	21	42-54	676-865	0.570
Wood, pine	70	21	27-42	432-673	0.67
Zinc	32-212	0-100	440	7048	0.095

TABLE 72

8 The Function of Thermal Insulation

The Function of Thermal Insulation to Provide a Heat Transfer Retarder to Maintain Temperature Within A Structure

To provide for proper processing or storage of materials, some buildings must be maintained at specific inside air temperatures.

For example, many warehouses are designed so that the inside winter temperature does not drop below the tolerance of the materials in storage. On the other hand, thread manufacturing and weaving plants require a moderate air temperature with high relative humidity to reduce the static charge collecting in the thread or fabric being manufactured. In these manufacturing plants, where high relative humidity in the air is maintained, the

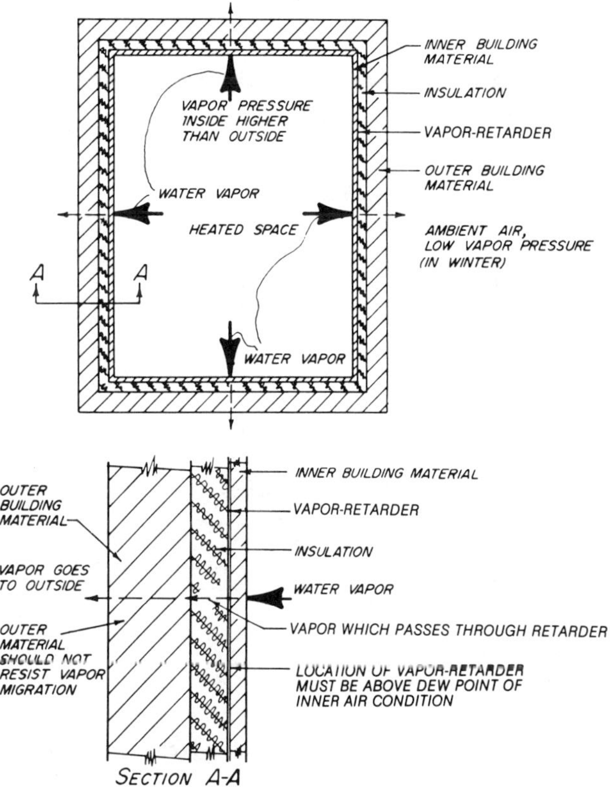

Figure 84 Schematic diagram of insulated wall in heated building.

wall and roof insulation must be protected from this internal moisture by a very efficient, low perm vapor retarder. This retarder must be located in the structural section of the wall or roof where it is always above the dew-point temperature of the air at that relative humidity and temperature. This is illustrated in Figure 84.

The Function of Thermal Insulation to Provide Heat Transfer Resistance to Maintain Low Temperature Within a Structure

The moisture vapor pressure on the outside of refrigerated spaces is higher than the vapor pressure in cold spaces. For this reason, the vapor retarder must be installed on the outer part of the structure to prevent moisture condensation in the wall, ceiling or floor. This vapor retarder system must be installed to keep excessive amounts of water vapor from entering the thermal insulation. However, as perfect vapor barriers are impossible, the innermost surfaces of the walls and ceilings should be constructed to have little vapor resistance. With this construction, small amounts of moisture penetrating the vapor retarder can continue through the insulation and will condense on the cold refrigeration coils or other cold surfaces within the room. This is illustrated in Figure 85.

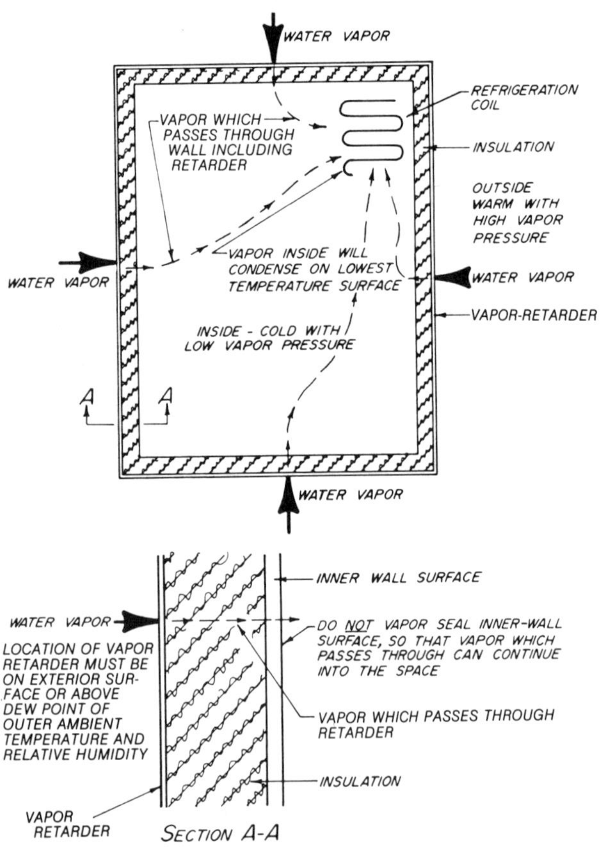

Figure 85 Schematic diagram of insulated refrigerated space.

As shown, the inner area, due to its low temperature, has a low vapor pressure which is below the ambient vapor pressure. Thus, the higher pressure outer vapor tries to migrate to the low pressure region inside. This migration is resisted by the vapor retarder, which should be located within the construction that is above dew-point temperature of the outer ambient air. Otherwise condensation will occur on its outer surface.

Even the best vapor retarders do allow some moisture to pass through. Any moisture which does pass this vapor retarder should not be impeded in its attempt to condense on the coldest surface within the room. In most cases, this is the refrigeration coil.

Although many factors can affect the overall design, a rule of thumb, which has proved to be satisfactory for most applications is that the outer vapor retarder should be ten times more resistant to the passage of vapor than the total of all other resistances in the path between the vapor retarder and the interior of the room. *In no case* should the interior wall, ceiling or floor be vapor sealed, as the resultant build-up of water and ice will rapidly ruin the structure of the building.

The Function of Thermal Insulation to Provide Heat Transfer Resistance to Maintain Temperature Requirements in Heated and Cooled Buildings

Residences and buildings which are heated in the winter and cooled in the summer require thermal insulation to maintain a comfortable interior temperature. To keep the insulation dry

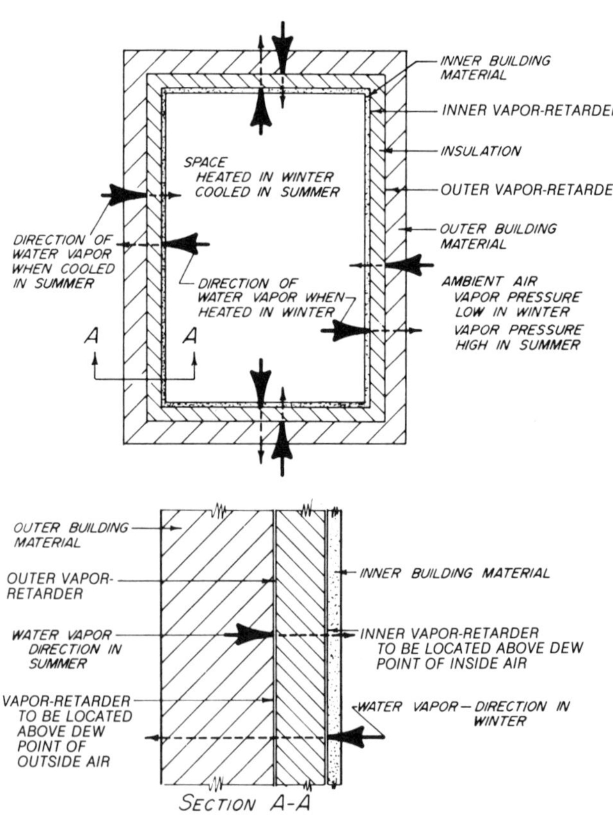

Figure 86 Schematic diagram of insulated wall heated and cooled building.

vapor retarders are required on both sides of the insulation. When non-waterproof construction materials are used, the placement of vapor retarders is a very important factor in determining the service life of the structure and how satisfactorily it will function during that time.

Vapor retarders should be located on both sides of the thermal insulation. The vapor retarder should always be placed at a location which is above the dew-point temperature which exists in the wall or roof structure to be insulated. This is illustrated in Figure 86.

The importance of vapor retarders as part of the function of thermal insulation to provide thermal resistance so as to maintain temperature cannot be overemphasized. Most dry thermal insulations used in building construction have a thermal conductivity of .25 to 0.35 Btu in./ft², hr. °F, (0.036 to 0.050 W/mK). When saturated with condensate, the conductivity can be above 5.0 Btu in./ft², hr, (0.721 W/mK). Compared to its dry state, the conductivity of wet insulation increases by a factor of 15. That is, wet insulation is approximately 7% as efficient as when dry. Thus, to be effective, insulation to control heat transfer must be installed with correctly designed and properly installed vapor retarders. *The lack or improper use of vapor retarders in buildings and residences is probably the most common architectural design and construction error.*

The Function of Thermal Insulation to Provide Thermal Resistance to Maintain Temperature Within Pipes

Thermal insulation on high temperature pipes within a building is used to conserve energy and to provide safe surface temperatures; on low-temperature pipe it is used to prevent condensation. In other instances, piping through unheated areas, or outdoors, must be heat traced and insulated to prevent their contents from freezing.

The need for and location of vapor retarders depends largely upon the temperature of the pipe which is to be insulated.

High Temperature Piping and Equipment

Thermal insulation is used on high-temperature piping mainly to reduce the heat loss from that piping. The design of the insulation system (including its type, placement, and thickness) is based on economic principles which are presented in "Thermal Insulation Design Economics for Pipes and Equipment" by W. C. Turner and J. F. Malloy. The manner in which this book is used to design the insulation system will be further presented in Chapter 9, Installation of Thermal Insulation Systems. However, it should be pointed out that these tables and methods of calculating thickness of insulation are not correct for hot pipes passing through air-conditioned spaces. In this situation, the cost of the energy lost from the pipe is not the only factor. An even more important cost factor is the cost of energy needed to remove this heat from the space.

Another function of thermal insulation is to provide sufficient thermal resistance between the outer surface and the hot piping so that the outer surface will be at a safe temperature. A safe surface temperature is one where a surface can be touched by the bare skin and not cause a burn. This safe surface temperature is not the same for all surfaces. It depends upon the type of surface and the thermal conductivity of the material of that surface. When the exposed surface is metal, any temperature above 140°F (60°C) is considered hazardous; whereas, mastics or fiberous surfaces can be up to 150°F (63.6°C) without being considered hazardous.

Piping which operates above the dew-point temperature of the ambient air needs no vapor retarder. The higher inside temperature causes a higher pressure so that moisture vapor flow is outward. In some instances, the insulation on the hot pipe will give the appearance of being wet when condensate occurs under the cooler surface jacket. This and the flow of vapor on insulated high-temperature pipe and equipment is shown in Figure 87.

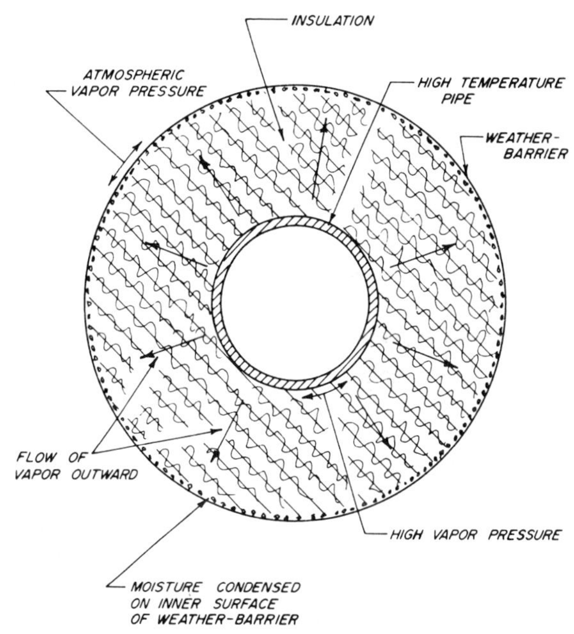

Figure 87 Schematic diagram of vapor flow insulated high temperature pipe.

Low Temperature Piping and Equipment

Piping and equipment which is operated constantly at temperatures below the ambient air dew point are, from a thermal standpoint, very difficult to insulate for long, efficient service. Water vapor will pass through almost any material in an attempt to equalize pressure from one side to the other. Thus there is no absolute barrier to vapor migration for the protection of insulation on low-temperature piping or equipment. When the vapor reaches the dew point within an insulation, it condenses into liquid water. This liquid condensate saturates the insulation and becomes ice when it is in an area where the temperature is below 32 °F (0 °C). This is illustrated in Figure 88.

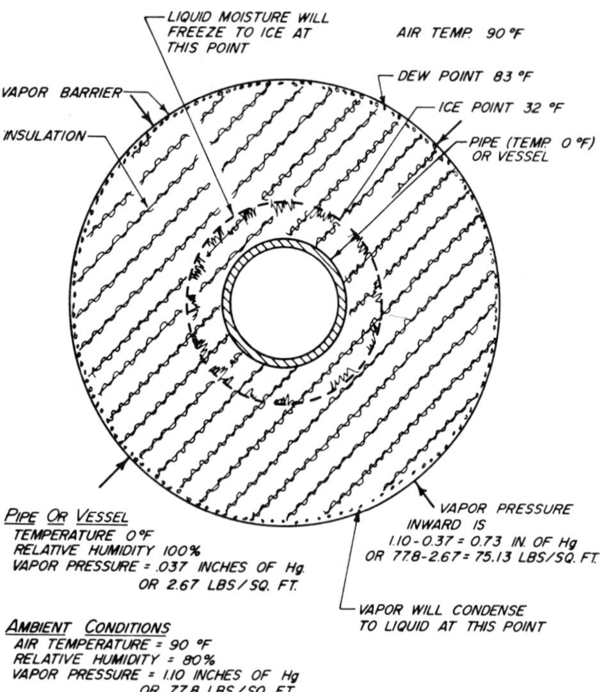

Figure 88 Vapor-retarder diagram.

When the vapor migration is from the outside to the surface of a colder pipe the effective life of the insulation is completely dependent upon the vapor resistance of the insulation and its outer vapor retarder. Since insulation saturated with liquid water has a thermal conductivity of 3.5 Btu in./ft^2, hr °F (0.5 W/mK) and iced up insulation has a conductivity of 15.5 Btu in./ft^2, hr °F (2.23 W/mK), the importance of dry insulation is evident. For this reason, the vapor resistance of the insulation is a much more important factor than a very low reported value thermal conductivity, as these thermal conductivities are always quoted for material in a dry state.

For example, one organic foam has a vapor permeability of 2.5 perm-inches (4.2 metric perms), whereas a cellular glass insulation has a vapor permeability of 0.005 perm-inch (0.0084 meter perm). Installed with no additional vapor barriers on a low temperature surface below the ambient dew point, the expected efficient life of the cellular glass would be 2.5/0.005 or 500 times that of the organic foam under the same conditions.

In the case of a low temperature pipe, the primary function of insulation is to prevent heat gain to the pipe. Therefore, the insulation (or insulation system) must provide adequate resistance to liquid and vapor moisture migration from the atmosphere to the pipe. To illustrate the magnitude of the forces which drive vapor toward the pipe, determine the vapor pressure difference when at 100% relative humidity. If the pipe temperature is 40 °F (4.4 °C), the vapor pressure at its surface would be 17.5 lbs/sq ft (839 Pa) and on the outer surface; if it is moist and at 75 °F, the vapor pressure would be 619 lbs/sq ft (2963 Pa). Thus, the vapor pressure difference (which is directly proportional to the force driving moisture from the outside surface inward to the pipe) is 61.9 − 17.5 = 44.4 lbs/sq ft or 2963 Pa − 839 Pa = 2024 Pa.

In inches of mercury pressure units, the pressure on the outer surface is 0.875 and on the inner side of insulation (next to the pipe) the pressure is 0.248. The vapor pressure difference in these units is 0.875 − 0.248 = 0.627 inches of mercury. At this vapor pressure difference, one inch thickness of insulation having a permeability of 2.5 perm-inches would allow moisture penetration of 2.5 × 0.627 or 1.57 grains of moisture per square foot per hour. In one year, this would amount to $\frac{1.57 \times 8760}{7000} = 2.1274$ pounds of condensed moisture in each square foot of the insulation (or 10.38 kilograms per square meter).*

This example indicates that insulation systems installed on constant low temperature piping and equipment must have a very low moisture vapor migration rate to be effective in controlling heat transfer.

CYCLIC TEMPERATURE PIPING AND EQUIPMENT

In a number of installations of heating and air conditioning systems the piping is used to supply cold water during the cooling season and hot water when heating is needed. In such instances the vapor migration is inward when the pipe temperature is lower than ambient air, but it is outward when the pipe is at a higher temperature than the ambient air. In those instances where the cooling season is of moderate length, insulation, such as organic foams, can be used satisfactorily. This is particularly true if the moisture entering is a relatively small amount since the drying effect which occurs during the heating season will restore the insulation to a satisfactory thermal barrier state.

It is also true that the higher the temperature, the greater the vapor pressure difference for identical temperature differences. For example, for low-temperature piping, when the pipe is at 50 °F (10 °C) and the ambient air is 75 °F (23.9 °C) for a temperature difference of 25 °F (14 °C), the vapor difference is 44.4 lbs/sq ft (212 Pa). Now consider a higher pipe temperature of 100 °F (37.8 °C) with the same temperature difference of 25 °F. That is, at an air temperature of 75 °F (23.9 °C) and 100% humidity the vapor pressure is 61.9 lbs/sq ft (2963 Pa). For the pipe at 100 °F (37.8 °C) the surface vapor pressure would be 136.6 lbs/sq ft (6545 Pa). Under these conditions, the vapor pressure difference is 74. lbs/sq ft (3577 Pa).

Thus, for a 25 °F difference in temperature above ambient, the vapor pressure difference is approximately 1.7 times greater than when the pipe temperature is 25 °F below ambient air temperature.

The correct design of thermal insulation on pipes saves energy when the pipe is operating at both low and high temperatures. The control of heat on low temperature piping and equipment can also be designed so as to prevent condensation and drip.

*From vapor pressure tables.

The Function of Thermal Insulation to Prevent Freezing of Pipe and Equipment Located in Unheated Areas

Where water pipes are located in unheated areas that reach low temperatures, some means must be taken to prevent these pipes from freezing. If a sufficient flow of water at a temperature above freezing is maintained, then this input of heat will prevent the freezing of uninsulated pipes. However, if it is not practical to run water through the pipes, then insulation will assist in the prevention of freezing. Table 73A shows the flow of inlet water at 40 °F (4.4 °C) necessary to prevent freezing in bare and insulated pipe in ambient air at −20 °F (−23 °C).

It should be noted that at no flow, insulated water lines will eventually freeze. If no flow is likely to occur, then heat must be added by an external method.

PREVENTION OF FREEZING OF WATER LINES
BASED ON AMBIENT AIR TEMPERATURE OF -20° F (-23° C)
AND MAXIMUM AIR MOVEMENT OF 15 mph (2.4 km per hr.)

FLOW OF WATER REQUIRED TO PREVENT FREEZING

NPS PIPE SIZE – "	***FLOW IN GALLON PER MINUTE PER 100 FEET OF PIPE		PIPE SIZE mm	****FLOW IN LITRES PER MINUTE PER 10 METRES OF PIPE	
	No Insulation	*2" Thk. Insulation		No Insulation	**53mm Thk. Insulation
1/4	1	1/4	16	1.15	0.29
1/2	1 1/4	1/4	21	1.44	0.29
3/4	1 1/2	1/4	27	1.72	0.29
1	1 5/8	3/8	33	1.93	0.43
1 1/4	3 3/4	1/2	42	2.01	0.58
1 1/2	1 7/8	1/2	48	2.16	0.58
2	2	1/2	60	2.31	0.58
2 1/2	2 1/4	1/2	73	2.60	0.58
3	2 1/2	1/2	89	2.88	0.58
4	3	1/2	114	3.45	0.58
6	4	1/2	168	4.61	0.58
8	5	5/8	219	5.76	0.72
10	6	3/4	244	6.92	0.86
12	7	7/8	273	8.07	1.00
14	8	7/8	324	9.23	1.00
16	9	1	406	10.37	1.15

*Based on thermal conductivity not exceeding 0.4 Btu., in/ft^2, hr°F

**Based on thermal conductivity not exceeding 0.577 W/mK

***Based on inlet water temperature of 40° F

***Based on inlet water temperature of 4.4° C

FLOW REQUIRED TO PREVENT FREEZING OF WATER LINES

TABLE 73A

ELECTRIC TRACED PIPE TO PREVENT WATER PIPES FROM FREEZING

The term "traced" indicates that an external source of heat was added on the outside of the pipe. This external "tracer" may contain heat furnished by hot liquid, vapor, or electricity. In the case of water lines that are heated to prevent freezing, electrical heat tracers are most commonly used. A typical type of installation of electric heat tracers on pipe is shown in Figure 89.

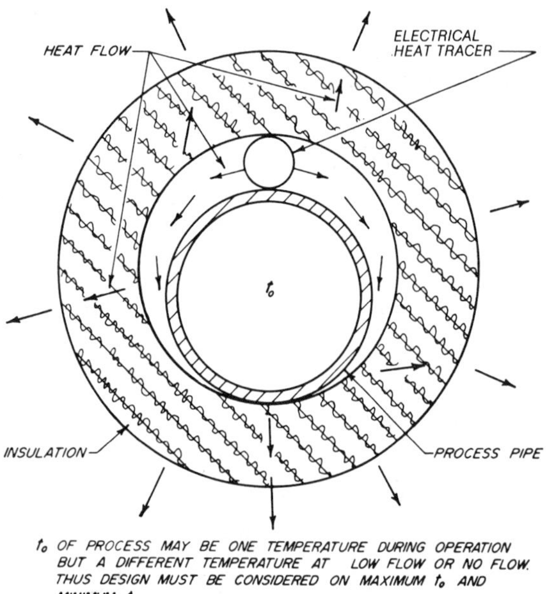

t_o OF PROCESS MAY BE ONE TEMPERATURE DURING OPERATION BUT A DIFFERENT TEMPERATURE AT LOW FLOW OR NO FLOW. THUS DESIGN MUST BE CONSIDERED ON MAXIMUM t_o AND MINIMUM t_o

IN CASES OF NO FLOW, ALL HEAT SUPPLIED BY TRACER EQUALS HEAT LOSS FROM INSULATION

Figure 89 Schematic diagram of heat traced process pipe.

The electrical tracer heating cables may be either parallel or series resistances. They may be furnished with protective steel wire braid or with pre-molded heat transfer plastic. These cables can also be obtained for various output capacities and voltage ratings. Figure 90 shows the types of heating cables available and their outputs and rating. Figure 91 shows the electrical connection to the tracer. The type of cable and the output required are determined by the size of the pipe, inlet temperature of water flow (if any), the ambient air temperature, the rate of air movement, and the thickness and effective thermal resistance of the insulation. As the ambient air temperature changes, the heat requirements to maintain a given temperature also change. For this reason, most electric-traced systems are equipped with thermostatic controls. The heat required for various conditions is presented in Table 73B.

The Function of Thermal Insulation to Control Heat Transfer to or From Ducts

The same thermal insulation used on piping is used on ducts to conserve energy or to prevent condensation of moisture on low-

ECONOTRACE®* CUT-TO-LENGTH HEATING CABLES

PARALLEL RESISTANCE HEATING CABLES

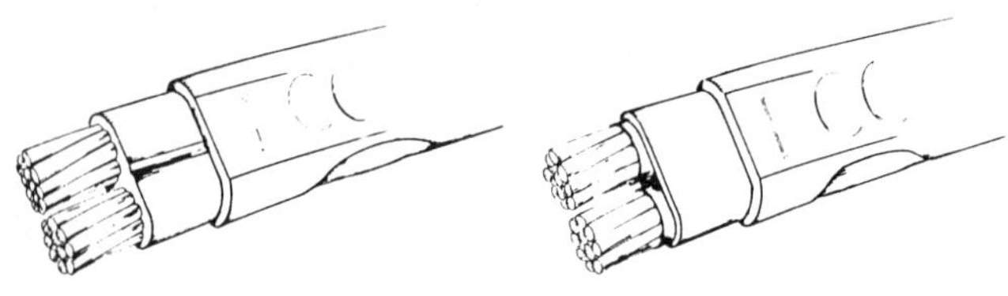

ECONOTRACE* EL
Watt densities from 2.5–8.7 W/Ft.
Voltage ratings from 110V–277V
Insulation– Elexar T.M.**

ECONOTRACE* FP
Watt densities from 2.25–14.5 W/Ft.
Voltage ratings from 110V–480V
Insulation– FEP***

SERIES RESISTANCE HEATING CABLES

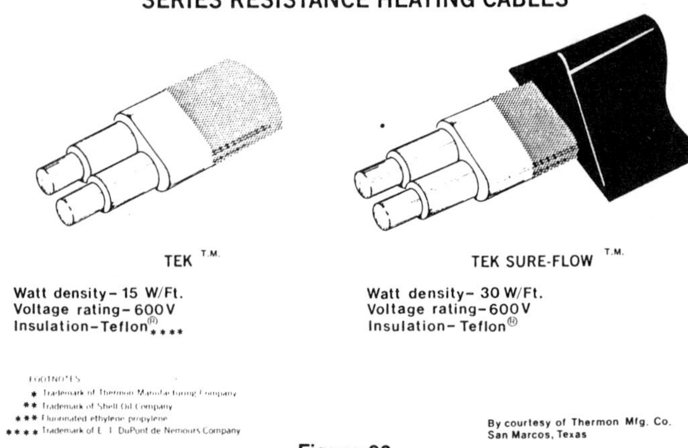

TEK T.M.
Watt density– 15 W/Ft.
Voltage rating– 600V
Insulation–Teflon®****

TEK SURE-FLOW T.M.
Watt density– 30 W/Ft.
Voltage rating–600V
Insulation– Teflon®

FOOTNOTES
* Trademark of Thermon Manufacturing Company
** Trademark of Shell Oil Company
*** Fluorinated ethylene propylene
**** Trademark of E. I. DuPont de Nemours Company

By courtesy of Thermon Mfg. Co. San Marcos, Texas

Figure 90

temperature duct surfaces. On very hot ducts, such as flue gas ducts, insulation also serves to provide a safe surface temperature.

High Temperature Ducts

Thermal insulation on high-temperature ducts is used to provide thermal resistance to reduce heat loss to ambient air and, in some instances, to control the temperature of the exposed surface. Like piping, if the duct work is always above the condensation temperature of ambient air, then no vapor retarder in the system is necessary.

Low Temperature Ducts

Where low temperature air flows through ducts, causing the outer insulation surface temperature to be below the dew point of the ambient air, a vapor retarder must be installed in the insulation system. The vapor retarder specifications necessary to maintain insulation in a dry condition are identical to those for low temperature piping and equipment (See Figure 86). However, the problem of vapor sealing is not as critical if the temperature of the duct does not drop below the dew-point temperature of the ambient air. This is because the vapor in the air will not condense into a liquid within the insulation. When duct temperatures do not rise

ENERGY REQUIREMENTS OF ELECTRIC TRACING TO MAINTAIN WATER CONTENTS OF PIPE AT TEMPERATURES ABOVE FREEZING

AMBIENT AIR TEMPERATURE AT 20° F (-6.7° C)
BASED ON AIR MOVEMENT 15 MILES (25 Kilometres) PER HOUR

TO MAINTAIN WATER TEMPERATURE AT 40° F (4.4° C)

PIPE SIZE		INSULATION SIZE		WATTS PER LINEAR FOOT — INSULATION THICKNESS			WATTS PER LINEAR METRE — INSULATION THICKNESS		
NPS	mm	NPS	mm	1" (25mm)	1½" (38mm)	2" (51mm)	1" (25mm)	1½" (38mm)	2" (51mm)
½	21.3	1 ¼	42.2	1.31	0.99	0.90	3.99	3.01	2.74
¾	26.7	1 ¼	42.2	1.31	0.99	0.90	3.99	3.01	2.74
1	33.4	1 ½	48.3	1.33	1.09	0.98	4.05	3.32	2.99
1 ½	48.3	2	60.3	1.55	1.24	1.06	4.72	3.78	3.23
2	60.3	2 ½	73.0	1.78	1.28	1.12	5.42	4.25	3.41
2 ½	73.0	3	88.9	2.10	1.62	1.40	6.40	4.94	4.26
3	88.9	3 ½	101.6	2.29	1.77	1.50	6.98	5.39	4.57
3 ½	101.6	4	114.3	2.47	1.95	1.64	7.53	5.94	5.00
4	114.3	4 ½	127.0	3.16	2.52	2.15	9.63	7.68	6.55
6	168.3	7	193.7	3.95	2.98	2.47	12.04	9.08	7.53
8	219.1	9	244.5	4.83	3.60	2.83	14.72	10.97	8.63
10	273.0	11	298.4	5.74	4.16	3.40	17.50	12.68	10.36
12	323.8	14	355.6	6.76	5.25	4.21	20.60	16.00	12.83

TO MAINTAIN WATER TEMPERATURE AT 60° F (15.6° C)

PIPE SIZE		INSULATION SIZE		WATTS PER LINEAR FOOT — INSULATION THICKNESS			WATTS PER LINEAR METRE — INSULATION THICKNESS		
NPS	mm	NPS	mm	1" (25mm)	1½" (38mm)	2" (51mm)	1" (25mm)	1½" (38mm)	2" (51mm)
½	21.3	1 ¼	42.2	2.69	2.03	1.85	8.20	6.19	5.64
¾	26.7	1 ¼	42.2	2.69	2.03	1.85	8.20	6.19	5.64
1	33.4	1 ½	48.3	2.75	2.25	1.93	8.38	6.86	5.88
1 ½	48.3	2	60.3	3.19	2.55	2.18	9.72	7.77	6.64
2	60.3	2 ½	73.0	3.66	2.63	2.31	11.16	8.01	7.04
2 ½	74.0	3	88.9	4.33	3.34	2.88	13.20	10.18	8.77
3	88.9	3 ½	101.6	4.59	3.46	2.96	13.99	10.55	9.02
3 ½	101.6	4	114.3	5.11	4.01	3.37	15.58	12.22	10.27
4	114.3	4 ½	127.0	5.56	4.50	3.88	16.95	13.72	11.83
6	168.3	7	193.7	8.16	6.16	5.06	24.87	18.78	15.42
8	219.1	9	244.5	10.00	7.44	5.82	30.48	22.68	17.74
10	273.0	11	298.4	11.88	8.58	7.00	36.21	26.15	21.34
12	323.8	14	355.6	13.98	10.87	8.69	42.61	33.13	26.49

TABLE 73B

ENERGY REQUIREMENTS OF ELECTRIC TRACING TO MAINTAIN WATER CONTENTS OF PIPE AT TEMPERATURES ABOVE FREEZING

AMBIENT AIR TEMPERATURE AT 0°F (-17.8°C)
BASED ON AIR MOVEMENT 15 MILES (24 Kilometres) PER HOUR

TO MAINTAIN WATER TEMPERATURE AT 40°F (4.4°C)

PIPE SIZE		INSULATION SIZE		WATTS PER LINEAR FOOT			WATTS PER LINEAR METRE		
				INSULATION THICKNESS			INSULATION THICKNESS		
NPS	mm	NPS	mm	1" (25mm)	1½" (38mm)	2" (51mm)	1" (25mm)	1½" (38mm)	2" (51mm)
½	21.3	1 ¼	42.2	2.60	1.96	1.78	7.92	5.97	5.42
¾	26.7	1 ¼	42.2	2.60	1.96	1.78	7.92	5.97	5.42
1	33.4	1 ½	48.3	2.66	2.17	1.95	8.11	6.61	5.94
1 ½	48.3	2	60.3	3.09	2.47	2.10	9.41	7.53	6.40
2	60.3	2 ½	73.0	3.54	2.54	2.22	10.79	7.74	6.76
2 ½	73.0	3	88.9	4.19	3.23	2.77	12.77	9.84	8.44
3	88.9	3 ½	101.6	4.36	3.34	2.80	13.29	10.18	8.53
3 ½	101.6	4	114.3	4.95	3.87	3.25	15.09	11.80	9.91
4	114.3	4 ½	127.0	5.70	4.81	4.06	17.37	14.66	12.37
6	168.3	7	193.7	7.90	5.95	4.88	24.08	18.13	14.87
8	219.1	9	244.5	9.68	7.19	5.62	29.50	21.92	17.13
10	273.0	11	298.4	11.52	8.31	6.76	35.11	25.33	20.60
12	323.8	14	355.6	13.54	10.51	8.39	41.27	32.03	25.57

TO MAINTAIN WATER TEMPERATURE AT 60°F (15.6°C)

PIPE SIZE		INSULATION SIZE		WATTS PER LINEAR FOOT			WATTS PER LINEAR METRE		
				INSULATION THICKNESS			INSULATION THICKNESS		
NPS	mm	NPS	mm	1" (25mm)	1½" (38mm)	2" (51mm)	1" (25mm)	1½" (38mm)	2" (51mm)
½	21.3	1 ¼	42.2	4.00	3.01	2.74	12.19	9.17	8.35
¾	26.7	1 ¼	42.2	4.00	3.01	2.74	12.19	9.17	8.35
1	33.4	1 ½	48.3	4.09	3.33	2.88	12.47	10.15	8.77
1 ½	48.3	2	60.3	4.75	3.79	3.29	14.48	11.59	10.03
2	60.3	2 ½	73.0	5.44	3.89	3.41	16.58	11.86	10.39
2 ½	73.0	3	88.9	6.45	4.96	4.26	19.66	15.15	12.98
3	88.9	3 ½	101.6	6.65	5.18	4.29	20.27	15.79	13.07
3 ½	101.6	4	114.3	7.61	5.96	5.00	23.20	18.17	15.24
4	114.3	4 ½	127.0	8.22	6.86	5.81	25.05	20.91	17.70
6	168.3	7	193.7	12.17	9.14	7.50	37.09	27.86	22.86
8	219.1	9	244.5	14.91	11.05	8.64	45.45	33.68	26.33
10	273.0	11	298.4	17.74	12.76	10.39	54.07	38.89	31.67
12	323.8	14	355.6	20.84	16.17	12.90	63.52	49.29	39.32

TABLE 73B

ENERGY REQUIREMENTS OF ELECTRIC TRACING TO MAINTAIN WATER CONTENTS OF PIPE AT TEMPERATURES ABOVE FREEZING

AMBIENT AIR TEMPERATURE AT -20° F (-28.9°C)
BASED ON AIR MOVEMENT 15 MILES (24 Kilometres) PER HOUR

TO MAINTAIN WATER TEMPERATURE AT 40° F (4.4°C)

PIPE SIZE		INSULATION SIZE		WATTS PER LINEAR FOOT INSULATION THICKNESS			WATTS PER LINEAR METRE INSULATION THICKNESS		
NPS	mm	NPS	mm	1" (25mm)	1½" (38mm)	2" (51mm)	1" (25mm)	1½" (38mm)	2" (51mm)
½	21.3	1 ¼	42.2	3.87	2.90	2.64	11.80	8.84	8.05
¾	26.7	1 ¼	42.2	3.87	2.90	2.64	11.80	8.84	8.05
1	33.4	1 ½	48.3	3.95	3.21	2.88	12.04	9.78	8.77
1 ½	48.3	2	60.3	4.59	3.65	3.11	13.99	11.12	9.48
2	60.3	2 ½	73.0	5.26	3.75	3.29	16.03	11.43	10.03
2 ½	73.0	3	88.9	6.23	4.78	4.11	18.99	14.60	12.53
3	88.9	3 ½	101.6	6.88	4.80	4.13	20.97	14.64	12.59
3 ½	101.6	4	114.3	7.36	5.75	4.81	22.43	17.53	14.66
4	114.3	4 ½	127.0	7.98	6.65	4.83	24.32	20.27	14.72
6	168.3	7	193.7	11.79	8.82	7.24	35.94	26.88	22.07
8	219.1	9	244.5	14.44	10.68	8.32	44.01	32.55	25.36
10	273.0	11	298.4	17.18	12.23	10.02	52.36	37.28	30.54
12	323.8	14	355.6	20.21	15.62	12.44	61.60	47.61	37.92

TO MAINTAIN WATER TEMPERATURE AT 60° F (15.6°C)

PIPE SIZE		INSULATION SIZE		WATTS PER LINEAR FOOT INSULATION THICKNESS			WATTS PER LINEAR METRE INSULATION THICKNESS		
NPS	mm	NPS	mm	1" (25mm)	1½" (38mm)	2" (51mm)	1" (25mm)	1½" (38mm)	2" (51mm)
½	21.3	1 ¼	42.2	5.28	3.95	3.59	16.09	12.04	10.94
¾	26.7	1 ¼	42.2	5.28	3.95	3.59	16.09	12.04	10.94
1	33.4	1 ½	48.3	5.39	4.38	3.82	16.43	13.35	11.64
1 ½	48.3	2	60.3	6.26	4.98	4.23	19.08	15.18	12.89
2	60.3	2 ½	73.0	7.18	5.18	4.48	21.88	15.60	13.66
2 ½	73.0	3	88.9	8.51	6.52	5.60	25.94	19.87	17.07
3	88.9	3 ½	101.6	9.01	6.55	5.64	27.46	19.66	17.19
3 ½	101.6	4	114.3	10.04	7.84	6.56	30.60	23.90	19.99
4	114.3	4 ½	127.0	10.52	8.71	6.59	32.06	26.55	20.09
6	168.3	7	193.7	16.09	12.04	9.88	49.04	36.70	30.11
8	219.1	9	244.5	19.71	14.57	11.36	60.08	44.41	34.63
10	273.0	11	298.4	23.45	16.83	13.67	71.48	51.30	41.67
12	323.8	14	355.6	27.59	21.32	16.98	84.09	64.98	51.76

TABLE 73B

ENERGY REQUIREMENTS OF ELECTRIC TRACING TO MAINTAIN WATER CONTENTS OF PIPE AT TEMPERATURES ABOVE FREEZING

AMBIENT AIR TEMPERATURE AT -40°F (-40°C)
BASED ON AIR MOVEMENT 15 MILES (24 Kilometres) PER HOUR

TO MAINTAIN WATER TEMPERATURE AT 40°F (4.4°C)

PIPE SIZE		INSULATION SIZE		WATTS PER LINEAR FOOT — INSULATION THICKNESS			WATTS PER LINEAR METRE — INSULATION THICKNESS		
NPS	mm	NPS	mm	1" (25mm)	1½" (38mm)	2" (51mm)	1" (25mm)	1½" (38mm)	2" (51mm)
½	21.3	1 ¼	42.2	5.10	3.81	3.46	15.54	11.61	10.54
¾	26.7	1 ¼	42.2	5.10	3.81	3.46	15.54	11.61	10.54
1	33.4	1 ½	48.3	5.20	4.22	3.58	15.84	12.86	10.91
1 ½	48.3	2	60.3	6.04	4.80	4.08	18.41	14.63	12.44
2	60.3	2 ½	73.0	6.93	4.92	4.31	21.12	14.99	13.14
2 ½	73.0	3	88.9	8.22	6.29	5.39	25.05	19.17	16.43
3	88.9	3 ½	101.6	8.73	6.31	5.42	26.61	19.23	16.52
3 ½	101.6	4	114.3	9.71	7.55	6.32	29.60	23.01	19.26
4	114.3	4 ½	127.0	10.19	8.42	6.35	31.06	25.66	19.35
6	168.3	7	193.7	15.56	11.62	9.51	47.43	35.42	28.99
8	219.1	9	244.5	19.08	14.06	10.94	58.15	42.85	33.35
10	273.0	11	298.4	22.68	16.23	13.17	69.13	49.47	40.14
12	323.8	14	355.6	26.70	20.59	16.37	81.38	62.76	49.90

TO MAINTAIN WATER TEMPERATURE AT 60°F (15.6°C)

PIPE SIZE		INSULATION SIZE		WATTS PER LINEAR FOOT — INSULATION THICKNESS			WATTS PER LINEAR METRE — INSULATION THICKNESS		
NPS	mm	NPS	mm	1" (25mm)	1½" (38mm)	2" (51mm)	1" (25mm)	1½" (38mm)	2" (51mm)
½	21.3	1 ¼	42.2	6.52	4.86	4.42	19.87	14.81	13.47
¾	26.7	1 ¼	42.2	6.52	4.86	4.42	19.87	14.81	13.47
1	33.4	1 ½	48.3	6.65	5.39	4.52	20.27	16.43	13.77
1 ½	48.3	2	60.3	7.72	6.13	5.21	23.53	18.68	15.88
2	60.3	2 ½	73.0	8.86	6.29	5.50	27.00	19.17	16.76
2 ½	73.0	3	88.9	10.50	8.03	6.89	32.00	24.48	21.00
3	88.9	3 ½	101.6	10.88	8.06	6.93	33.16	24.57	21.12
3 ½	101.6	4	114.3	12.41	9.66	8.08	37.83	29.44	24.63
4	114.3	4 ½	127.0	12.74	10.49	8.10	38.83	31.97	24.69
6	168.3	7	193.7	19.89	14.85	12.16	60.62	45.26	37.06
8	219.1	9	244.5	24.38	17.97	13.98	74.31	54.77	42.61
10	273.0	11	298.4	29.00	20.75	16.83	88.39	63.25	51.30
12	323.8	14	355.6	34.13	26.32	20.93	104.02	80.22	63.79

TABLE 73B

Figure 91 Electrical connection to electric heat tracer to prevent freezing of water pipe

Courtesy of
Thermon Manufacturing Co.
San Marcos, Texas

above the dew point, then vapor sealing or the use of vapor resistance insulation, such as cellular glass, is needed to ensure efficient insulation. Properly insulated ducts reduce the need for refrigeration producing capacity and are therefore, cost effective from the standpoint of improved energy efficiency.

Cyclic Temperature Ducts

In many instances, air-conditioning ducts in residences and buildings do not go above room temperature in winter or below room temperature in summer. Thus, direction of heat transfer is dependent upon which side the thermal insulation is at the higher temperature. The air in the duct is often slightly below the dew point of the ambient air during the cooling cycle. If there is a fair resistance to vapor migration, the insulation will dry out when the duct is serving the heating cycle. However, if the ducts are constructed of steel, it is of major importance that the steel surfaces be protected from the condensed water to prevent rusting. This is particularly true of low-temperature duct work.

Insulation of Ducts — General

Insulation can be located on the interior of ducts. On low temperature or cyclic temperature air conditioning ducts, placing insulation on the interior of metal ducts is advantageous, since the metal then functions as the vapor retarder. On the warm side of the thermal insulation where the air velocity is moderate, rigid fiberous insulation can be secured to the interior of the duct by adhesive. On the other hand, when the air velocity inside the duct is high, the rigid fiberous insulation should be secured by mechanical fasteners in addition to duct adhesive. Also, when the air velocity is high, the insulation used must be such that particles or fibers are not released into the air stream, as mineral or glass fibers are considered to be a serious health hazard if blown into occupied areas. If the air velocity is above 3000 FPM (15.2 meters per second), a liner over the insulation exposed to the moving air should be used. It is also important that all adhesives used to secure the insulation and/or liner be fire-resistant.

Ducts insulated on the inside can obtain noise attenuation from the insulation in addition to heat control. This reduces the noise of the air conditioning and heating equipment and also serves to reduce noise transfer from one room, or office, to another. In most instances these internally insulated ducts are equipped with an internal perforated metal liner.

The fan noise reduction depends upon the efficiency of the sound absorber fiber lining, diameter of duct and length from fan to outlet. "Cross talk" from one room to another can also be reduced by a sound absorber inner lining. In contrast, an unlined metal duct is an effective "speaking tube" from outlet to outlet. Since accoustical and noise control in buildings is a very complex subject, those design considerations are beyond the scope of this book. Likewise, the air friction of insulated lined ducts is different from smooth, metal ducts, and correct air friction tables must be used to design air conditioning and heating ducts.

Typical applications of insulation applied to ducts are presented in Chapter 9.

The Function of Thermal Insulation as Fire Protection of Pipe and Structural Members

Thermal insulation can be used effectively to retard failure of pipe or structural members in a fire. In providing protection to pipe and structural members, the lower the thermal diffusivity of the fire protective material, the longer the protection for a given thickness. Although the thermal diffusivity is very important to reduce heat transfer, specific heat and specific unit weight directly affect the thermal diffusivity as shown by the following basic formula:

when

α = Thermal diffusivity
k = Thermal conductivity
c = Specific heat per unit weight
p = Weight per unit volume,

then, the basic equation for thermal diffusivity is

$$\alpha = \frac{k}{cp} \text{ in units of area per unit time.} \qquad (30)$$

It should be noted that thermal diffusivity, like thermal conductivity, changes as the mean temperature of the material changes. To illustrate why relatively heavy materials having moderately low values of conductivity are more effective than light weight materials of low conductivity in retarding temperature changes, consider the diffusivities of concrete and mineral wool insulation.

In English Units — Concrete, when

k = Conductivity = 1.35 Btu/ft, hr, °F
p = Weight density = 145 lbs per cubic ft
c = Specific heat = 0.16 Btu/lb, °F

then:

$$\alpha = \frac{k}{cp} = \frac{1.35}{145 \times 0.16} = \frac{1.35}{23.2} = 0.0582 \text{ ft}^2/\text{hr}$$

(English Units)

Mineral wool, when

k = Conductivity = 0.033 Btu, ft, hr, °F
p = Weight density = 12 lbs per cubic foot
c = Specific heat = 0.18 Btu/lb, °F

then, diffusivity is:

$$\alpha = \frac{0.033}{12 \times 0.18} = \frac{0.033}{2.16} = 0.0152 \text{ ft}^2/\text{hr.}$$

(English Units)

Even though the mineral wool has a thermal conductivity of only 2.4% that of concrete due mainly to the weight difference, its thermal diffusivity is 26% that of concrete.

In Metric Units — Concrete, when

k_m = Conductivity = 2.334 w/m, °C
p_m = Weight density = 2322 kg/m^3
c_m = Specific heat = 0.16 k cal/kg, °C
Conversion factor = 4186 joules/k cal.

then:

$$\alpha_m = \frac{2.334}{2322 \times 0.16 \times 4186} = 1.5 \times 10^{-6} \text{ m}^2/\text{sec}$$

(Metric Units)

Mineral wool, when

Conductivity k_m = 0.057 w/m, °C
Density p_m = 192 kg/m^3
Specific heat = 0.18 k cal/kg, °C
Conversion factor = 4186 joules/k cal.

then:

$$\alpha_m = \frac{0.057}{192 \times 0.18 \times 4186} = 3.94 \times 10^{-7} \text{ m}^2/\text{sec}$$

(Metric Units)

Thermal diffusivity is an extremely important factor in the protection of pipes and structural members during a fire because the high temperature of the fire may affect a large area. The total structural mass is heated by the fire with no means to cool the interior. This is illustrated in Figure 92 which shows a steel column protected by concrete in order to extend the time before the failure temperature of the steel is reached when it is exposed to fire.

Non-combustible insulation serves as a fire protective medium, as well as a material to conserve energy under normal conditions. This is shown in Figure 93.

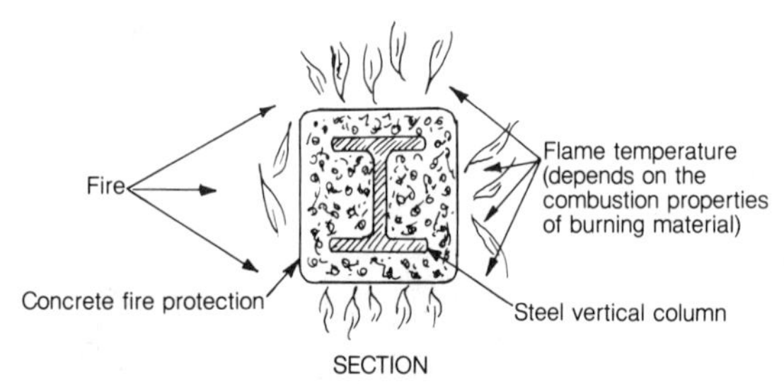

FIRE EXPOSURE OF CONCRETE PROTECTED COLUMN

Figure 92

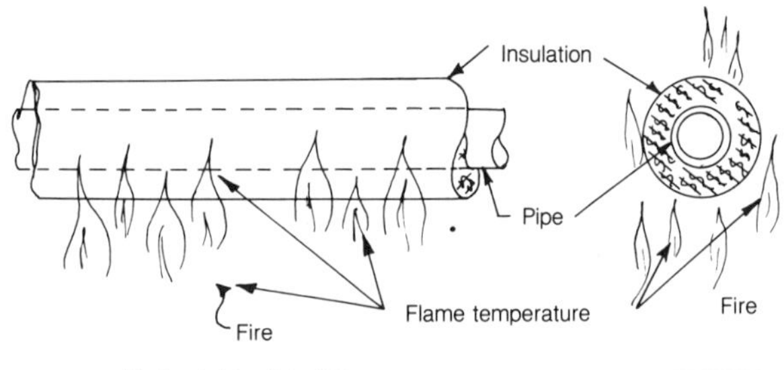

FIRE EXPOSURE OF THERMALLY INSULATED PIPE

Figure 93

The effects of fire impingement on structural members and pipes are so complex that it is impractical to attempt to set up formulas for calculating time required to reach failure temperature. In general, when fire protection is needed, non-combustible materials which have low thermal diffusivity, and which can maintain strength and dimensional stability at high temperatures will provide good fire protection. Also, fire protection of structural members should be part of the architectural design of the building. Although this book does not cover structural design, a table listing the thermal characteristics of commonly used building materials is given in Table 74. It should be noted that this table is based on atmospheric temperatures and both conductivity and diffusivity values increase as the mean temperature increases.

In most cases, it is assumed that the thermal insulation and its jacketing will provide fire protection for the pipe itself. Unfortunately, this is not always correct and in many cases insulated pipes and equipment add to the fire hazard.

The exposed surface of insulated pipe and equipment is the jacketing or coating. Certain types of jacketing or coating can contribute to the fire hazard or cause rapid failure of the thermal insulation on the pipe or equipment. For this reason it is necessary

THERMAL CHARACTERISTICS OF MATERIALS USED FOR FIRE PROTECTION
BASED ON MEAN AMBIENT AIR TEMPERATURE 70°F (21.1°C)

MATERIAL	Conductivity Btu/hr.ft.°F	Density Lbs/ft^3	Specific Heat % H_2O	Diffusivity	Conductivity w/mk	Density kg/m^3	Specific Heat % H_2O	Diffusivity
BRICK	0.50	128	0.22	0.0178	0.865	2050	0.22	0.0019
Clay-Hard	0.42	112	0.20	0.0187	0.727	1794	0.20	0.0020
Clay-Med	0.35	103	0.18	0.0188	0.606	1649	0.18	0.0020
CONCRETE								
Aerated	1.44	70	0.19	0.0109	2.491	1121	0.19	0.0012
Gravel	1.04	145	0.20	0.0358	1.799	2322	0.20	0.0039
Gypsum	0.142	51	0.30	0.0031	0.246	816	0.30	0.0003
Slag and Cinders	0.435	120	0.13	0.0200	0.753	1922	0.18	0.0022
Stone	1.35	145	0.16	0.0582	2.334	2322	0.16	0.0063
GYPSUM								
Blocks	0.25	78	0.26	0.0123	0.432	1249	0.26	0.0013
Molded	0.12	50	0.26	0.0092	0.208	800	0.26	0.0010
INSULATION								
Alumina-Silica	0.030	18	0.18	0.0093	0.052	288	0.18	0.0010
Calcium-Silicate	0.029	13	0.20	0.0112	0.050	208	0.20	0.0012
Diatomaceous-Silica	0.037	24	0.22	0.0070	0.064	384	0.22	0.0007
Mineral Fiber (HT)	0.033	12 to 25	0.18	0.015 to 0.007	0.057	192 to 400	0.18	0.0015 to 0.0007
Perlite (Expanded)	0.033	14	0.18	0.013	0.057	224	0.18	0.0014
Silica (Expanded)	0.033	13	0.18	0.0141	0.057	208	0.18	0.0015
Borosilica Foam	0.048	12	0.21	0.0190	0.083	192	0.21	0.0020
IRON and STEEL								
Steel-High Carb.	26.20	489	0.115	0.4659	45.326	7833	0.115	0.0503
Steel-Medium	26.20	489	0.115	0.4659	45.326	7833	0.115	0.0503
Steel-Soft	26.20	489	0.115	0.4659	45.326	7833	0.115	0.0503
Wrought Iron	34.90	485	0.12	0.5996	60.377	7768	0.12	0.0647
PLASTER								
Cement	0.67	138	0.21	0.0231	1.159	2210	0.21	0.0249
Gypsum	0.28	90	0.26	0.0119	0.484	1441	0.26	0.0013
Lime	0.42	103	0.20	0.0203	0.744	1649	0.20	0.0022
REFRACTORIES								
Fire Clay	0.50	147	0.19	0.0179	0.865	2355	0.19	0.0019
High Aluminia	0.50	130	0.23	0.0167	0.865	2082	0.23	0.0013
Kalolin	0.92	133	0.22	0.0314	1.592	2130	0.22	0.0034
STONE								
Granite	1.00	162	0.20	0.0308	1.730	2594	0.20	0.0033
Lime	0.77	154	0.22	0.0227	1.332	2467	0.22	0.0025
Marble	1.20	166	0.21	0.0344	2.075	2659	0.21	0.0037
Sand	0.75	143	0.22	0.0238	1.297	2290	0.22	0.0026
TERRA COTTA								
Block	0.35	118	0.15	0.0198	0.605	1890	0.15	0.0021

TABLE 74

to determine the combustible properties of commonly used jacketing or coating materials. The description of the test and results follow.

Fire Test of Insulation Jackets and Coatings

Test Procedure

Two thermocouples were located on the inside wall of sections of 3″ NPS Schedule 40 steel pipe. These were located on each side of the pipe 180° apart in the center of the area to be tested by fire exposure. Another was located in an exposed fire area.

In this case, 1″ thick calcium silicate was used as the pipe insulation over which the jackets or coatings to be tested were applied. The pipe section was located over a pan 4′0″ x 7′6″ x 1′6″ deep. The pan was charged with water so that the level was 1′8″ below the center line of the pipe. Fuel was fed into the pan and was ignited after the water surface had been covered with a thin layer of fuel.

The fuel flow to the pan was controlled at a rate which raised the temperature at the area of the exposed surface in accordance with the ASTM standard time temperature curve for fire exposure tests. [1000°F (538°C) in 5 min, 1300°F (704°C) in 10 min, 1399°F (760°C) in 15 min, 1500°F (816°C) in 30 min, and 1700° (927°C) in 69 mins]. These temperatures were also recorded on bare 3″ NPS pipe exposed to the fire.

Test Results

Aluminum Jacketing:
Began to melt and drip after 2 minutes of fire exposure. Note: Temperature of bare test pipe had only reached 600°F, but due to the insulation between the aluminum and the test pipe, the jacketing reached its melting point of 1220°F (660°C) in 2 minutes. The aluminum jacket provided no fire protection for the insulation and could contribute to fire hazard due to the hot molten drip.

Stainless Steel Jacketing:
Did not fail. Some distortion of shape and discoloration occurred after 60 minutes of fire exposure. This material is a safe and protective jacketing.

Polyvinyl Chloride Film Jacketing:
Started to melt and burn in 20 seconds. Provided no fire protection and contributed some fuel to the fire and also produced some dangerous fumes.

55 Pound Roofing Felt:
Started to burn after 1 minute and 30 seconds resulting in rapid flame spread and a flaming drip. Considered to be fire hazardous as a jacketing.

Emulsion Asphalt Mastic Insulation Coating:
Started to burn in 35 seconds. Rapid flame spread and flaming drip. Considered to be fire hazardous as a covering.

Emulsion Polyvinyl Acetate Fabric Reinforced Coat:
Slight burning noted in the area after 53 seconds. No flame spread observed outside of test area. Carbonized material stayed on pipe and no flaming drip was produced. This material is not fire hazardous as a covering.

This provides a guide for determining what jacketing or coating is suitable, depending upon the hardness of the insulation of the insulated pipe.

The stainless steel jacketing was the most fire resistant material tested. Asphaltic materials and plastic films would have contributed to the fire thus raising the surface temperature. Since the emulsion polyvinyl-acetate coating had a high surface emittance and very slow burning rate, it was selected to serve as the covering over the insulation being fire tested. Its properties would have the least effect on time-temperature results.

Fire Tests of Thermal Insulation

Test Procedure

Two thermocouples were installed on the inside wall of the sections of 3″ NPS Schedule 40 steel pipe. The thermocouples were located in a horizontal position 180° apart in the pipes

Figure 94

Figure 95 Fire test of insulation jackets and coatings

covered by the insulation to be tested and in the bare pipe used to control and record the temperature of the fire area.

All the insulations tested were 2 1/2″ nominal thickness, broken-joint construction. The inside insulation had butt joints in vertical position and the outer layer butt joints in horizontal position. Circumfertial joints were installed in staggered positions. Outer coating was 1/16″ dry fabric reinforced polyvinyl-acetate emulsion coating.

The test pipes and bare pipes were located in a horizontal plane above the fire pan. The thermocouples were located in the center of the fire. The pan was 4′0″ (1.219 m) x 7′6″ (2.286 m) x 1′6″ (0.456 m) deep.

The pan was filled with water so that water level was 1′8″ (0.508 m) below the center of the insulated pipe. Fuel was fed into the pan and was ignited after the water surface had been covered. The test section was subject to fire impingement as shown in Figures 94 and 95.

The fuel flow to the pan was controlled at a rate which raised the temperature of the bare pipe section in accordance with the ASTM Standard Time-Temperature Curve for fire exposure tests. This time-temperature standard is given below:

TEMPERATURE OF BARE PIPE
Time of Exposure — Minutes

	5	10	15	20	25	30	35	40	45	50	55	60
°F	1000	1300	1399	1492	1510	1550	1584	1613	1638	1661	1681	1700
°C	538	704	754	811	821	843	862	879	892	905	917	927

Steel pipe and structural members lose their physical strength very rapidly at temperatures above 1000 °F (538 °C). Thus, this temperature is considered the maximum safe working temperature of steel. Aluminum has much lower critical strength temperature factor, as it begins to lose its strength rapidly when its temperature exceeds 300 °F (149 °C).

Test Results

Material	Thick	Insulated Pipe Temp. Rise	Remarks
Cellular Urethane	2 1/2″	14 min-50 sec 1000 °F (538 °C)	Burned beyond test tank
Cellular Polystyrene	2 1/2″	11 min-30 sec 1000 °F (538 °C)	Burning drip started immediately
Glass Fiber 6-lb density	2 1/2″	31 min-10 sec 1000 °F (538 °C)	All melted in exposed area
Cellular Glass	2 1/2″	65 min - 760 °F (404 °C)	Cracked into pieces
Calcium Silicate	2 1/2″	65 min - 220 °F (104 °C)	Shrinkage - deformed
Borosilica Foam	2 1/2″	65 min - 800 °F (427 °C)	Some hairline cracks
Diatomaceous Silica	2 1/2″	65 min - 190 °F (88 °C)	Some slight shrinkage
Expanded Silica	2 1/2″	65 min - 211 °F (100 °C)	Some shrinkage & powdering

The plotted curves of the recorded time-temperature of the insulated pipe of these tests are given in Charts 1 to 8.

Conclusions

1. These fire tests confirm that organic foam insulation contributes fuel to a fire.
2. Light-weight insulations having high thermal diffusivity are not good materials for fire protection.
3. High density, non-combustible materials with low thermal conductivities provide the best fire protection.

THE FUNCTION OF THERMAL INSULATION TO PROVIDE SAFE SURFACE TEMPERATURES

High-temperature process, steam lines and flue gas ducts should be so designed that the surface temperature of the jacketing will be such as not to cause burns when touched.

The outer surface influences the temperature which can be touched without causing skin burns. Metallic jackets, due to their high thermal conductivity, will transfer heat to the skin very rapidly. For this reason, their safe surface temperature is lower than mastics or fiberous coverings having lower thermal conductivities.

The surface emittance influences the surface temperature. Metal has a very low thermal emittance and this causes the surface temperature to be higher than would occur with coatings or fabric coverings which have a much greater surface emittance.

It has been determined that a metallic surface can be touched briefly without causing immediate burns when it is not over 140 °F (66 °C). Pain starts when the skin temperature exceeds 112 °F (44.4 °C); therefore, 140 °F (66 °C) is above the temperature at which an object can be gripped and held tightly without discomfort.

In many instances metal covered insulation at too high a surface temperature can be made safe by the simple procedure of covering it with a coating of mastic.

The thicknesses of insulation needed to provide safe surface temperatures for aluminum jackets, stainless steel jacketings and mastic coatings are given in Table 75.

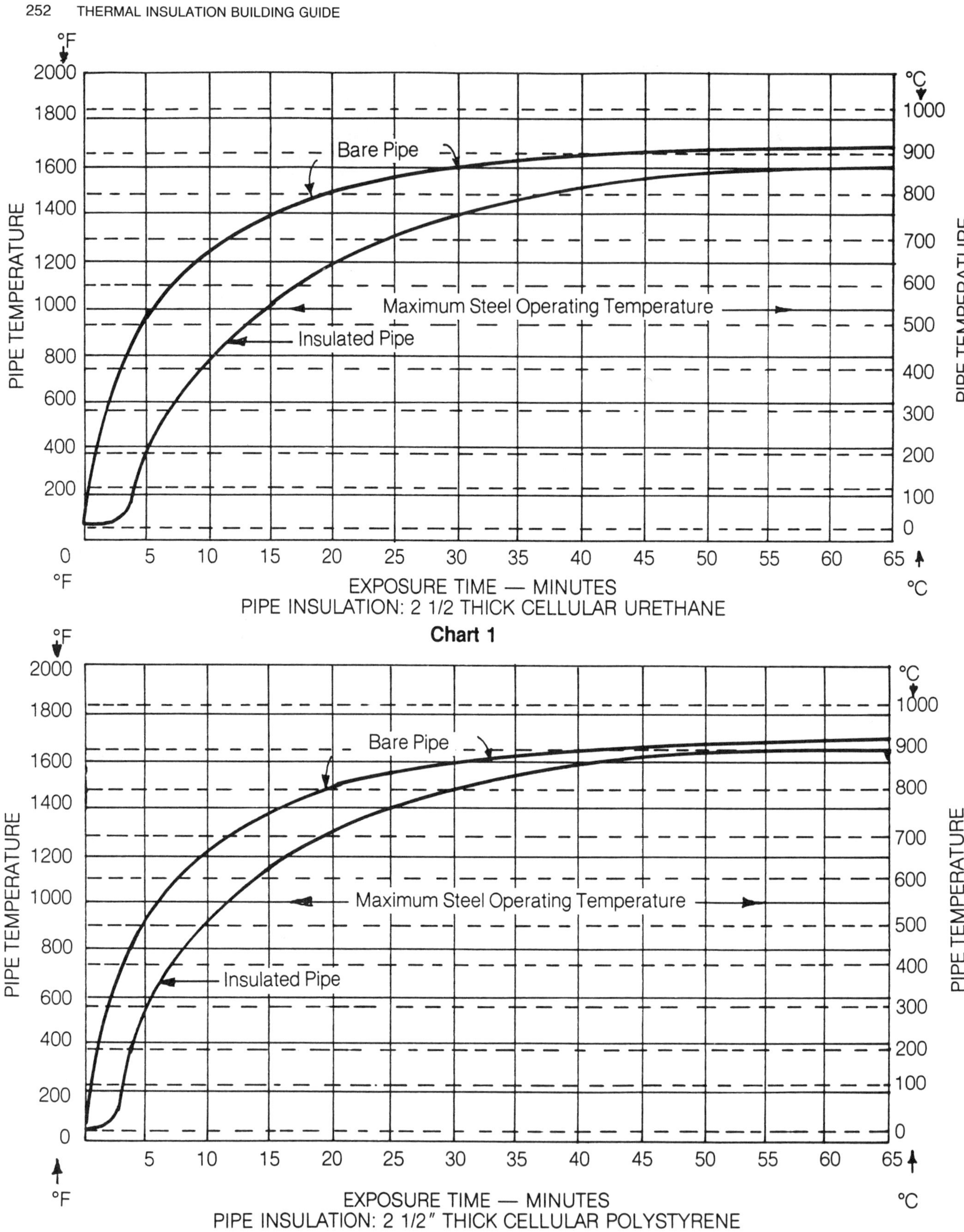

Chart 1

Chart 2

INSULATION FIRE EXPOSURE TESTS

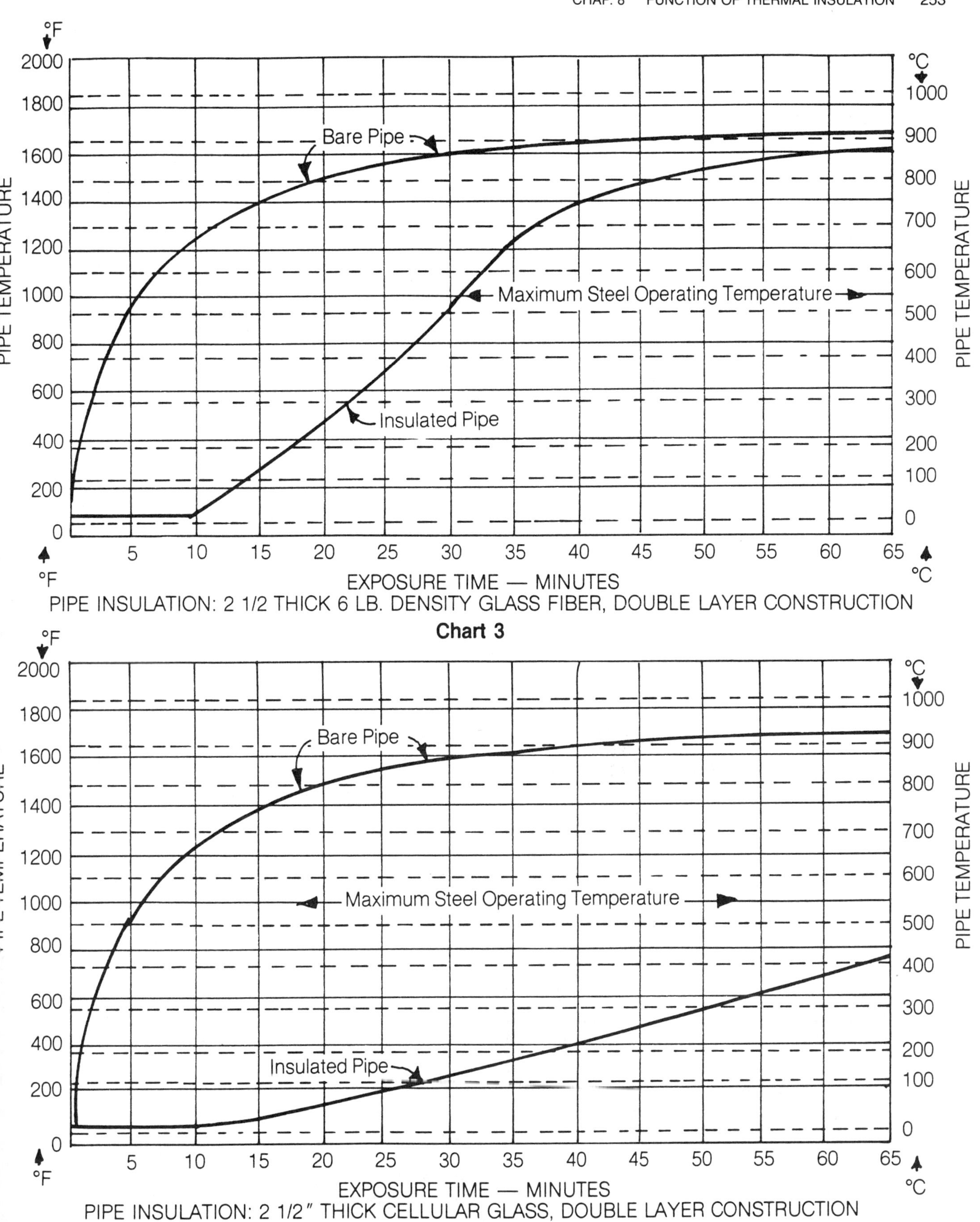

Chart 3

Chart 4

INSULATION FIRE EXPOSURE TESTS

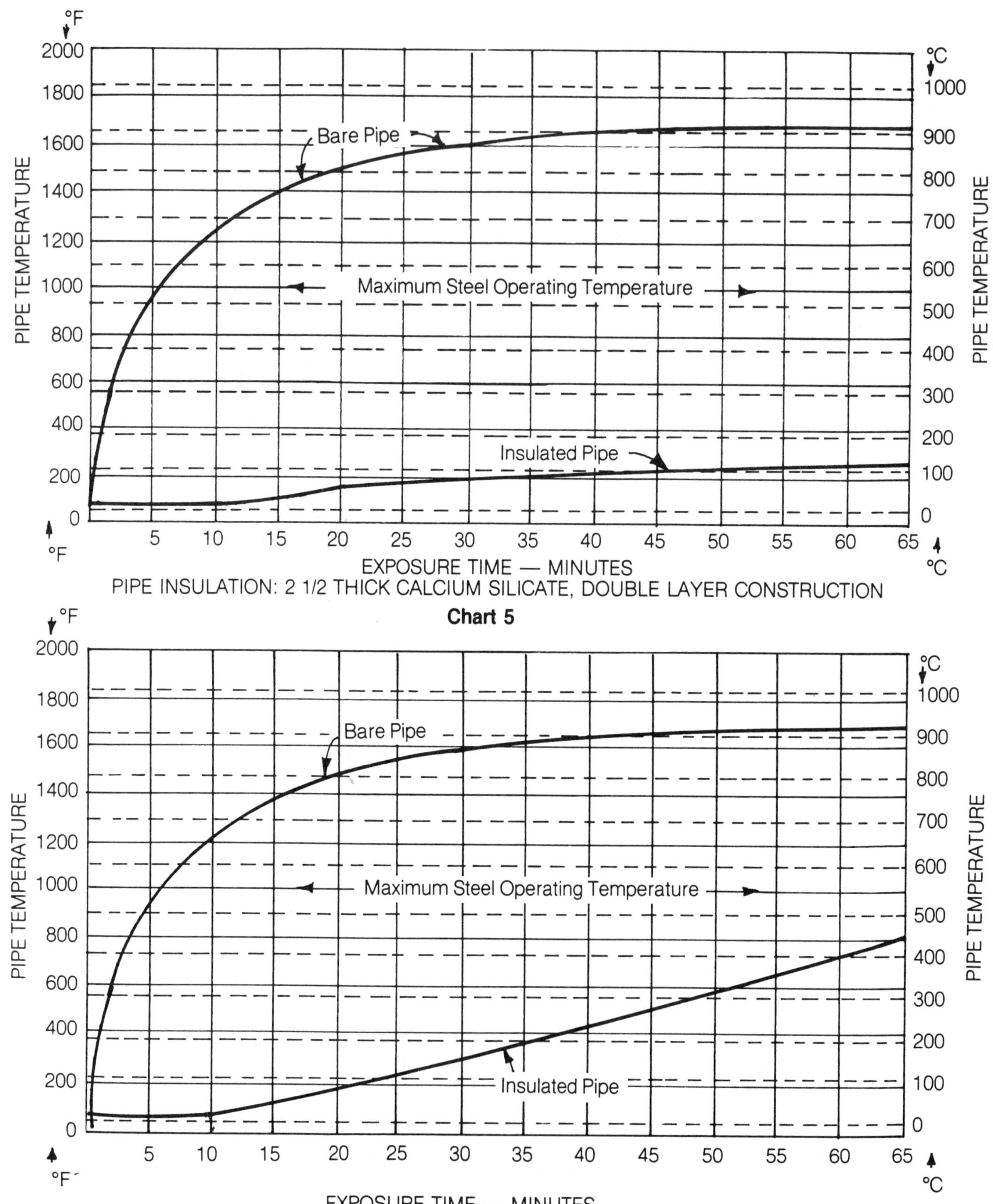

PIPE INSULATION: 2 1/2 THICK CALCIUM SILICATE, DOUBLE LAYER CONSTRUCTION

Chart 5

PIPE INSULATION: 2 1/2" THICK BOROSILICA FOAM, DOUBLE LAYER CONSTRUCTION

Chart 6

INSULATION FIRE EXPOSURE TESTS

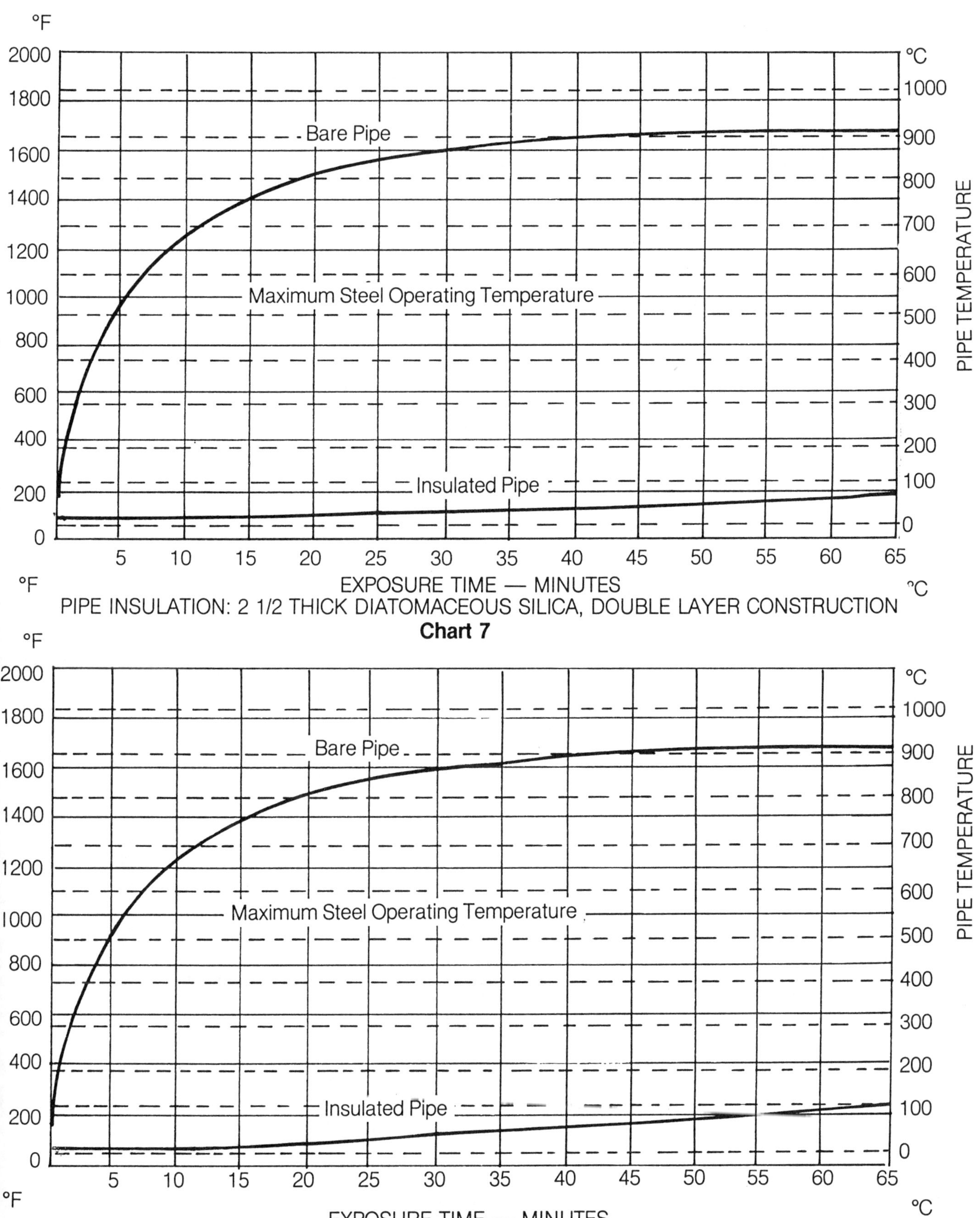

Chart 7

Chart 8

INSULATION FIRE EXPOSURE TESTS

Table 75 Safety insulation thicknesses

Outer Covering: Aluminum Jacket Emissivity 0.05

Nom. Pipe Size	Operating Temperature °C																				
	60 to 79	80 to 99	100 to 119	120 to 139	140 to 159	160 to 179	180 to 199	200 to 219	220 to 239	240 to 259	260 to 279	280 to 299	300 to 319	320 to 339	340 to 359	360 to 379	380 to 399	400 to 419	420 to 439	440 to 459	460 to 479
½	1	1	1	1	1	1	1	1½	1½	1½	2	2	2	2	2	2½	2½	2½	2½	2½	2½
¾	1	1	1	1	1	1	1	1½	1½	1½	2	2	2	2	2½	2½	2½	2½	2½	2½	3
1	1	1	1	1	1	1	1	1½	1½	1½	2	2	2	2	2½	2½	2½	2½	3	3	3
1¼	1	1	1	1	1	1	1½	1½	1½	1½	2	2	2	2	2½	2½	2½	3	3	3	3
1½	1	1	1	1	1	1	1½	1½	1½	2	2	2	2	2	2½	2½	2½	3	3	3	3½
2	1	1	1	1	1	1	1½	1½	1½	2	2	2	2½	2½	2½	3	3	3	3½	3½	3½
2½	1	1	1	1	1	1	1½	1½	1½	2	2	2	2½	2½	2½	3	3	3	3½	3½	3½
3	1	1	1	1	1	1	1½	1½	2	2	2	2½	2½	2½	3	3	3	3½	3½	3½	4
3½	1	1	1	1	1	1	1½	1½	2	2	2	2½	2½	2½	3	3	3	3½	3½	3½	4
4	1	1	1	1	1	1	1½	1½	2	2	2½	2½	2½	3	3	3	3½	3½	4	4	4
6	1½	1½	1½	1½	1½	1½	2	2	2	2½	2½	2½	3	3	3	3½	3½	4	4	4	4½
8	1½	1½	1½	1½	1½	1½	2	2	2	2½	2½	2½	3	3	3½	3½	3½	4	4	4½	4½
10	1½	1½	1½	1½	1½	1½	2	2	2	2½	2½	3	3	3	3½	3½	4	4	4½	4½	5
12	1½	1½	1½	1½	1½	1½	2	2	2	2½	2½	3	3	3½	3½	4	4	4½	4½	5	5
14	1½	1½	1½	1½	1½	1½	2	2	2½	2½	3	3	3	3½	3½	4	4	4½	4½	5	5
16	1½	1½	1½	1½	1½	1½	2	2	2½	2½	3	3	3½	3½	3½	4	4	4½	5	5	5½
18	1½	1½	1½	1½	1½	1½	2	2	2½	2½	3	3	3½	3½	4	4	4½	4½	5	5	5½
20	1½	1½	1½	1½	1½	1½	2	2	2½	2½	3	3	3½	3½	4	4	4½	4½	5	5	5½
22	1½	1½	1½	1½	1½	1½	2	2	2½	2½	3	3	3½	3½	4	4	4½	4½	5	5	5½
24	1½	1½	1½	1½	1½	1½	2	2	2½	2½	3	3	3½	3½	4	4	4½	4½	5	5½	5½
26	1½	1½	1½	1½	1½	1½	2	2	2½	2½	3	3	3½	3½	4	4	4½	4½	5	5½	5½
28	1½	1½	1½	1½	1½	1½	2	2	2½	2½	3	3	3½	3½	4	4	4½	4½	5	5½	5½
30	1½	1½	1½	1½	1½	1½	2	2	2½	2½	3	3½	3½	3½	4	4½	4½	5	5	5½	6
36	1½	1½	1½	1½	1½	1½	2	2	2½	2½	3	3½	3½	3½	4	4½	4½	5	5	5½	6
Equip. Flat	1½	1½	1½	1½	1½	1½	2	2	2½	3	3	3½	3½	4	4½	4½	5	5½	6	6	6½
	140 to 175	176 to 211	212 to 247	248 to 283	284 to 319	320 to 355	356 to 391	392 to 427	428 to 463	464 to 499	500 to 535	536 to 571	572 to 607	608 to 643	644 to 679	680 to 715	716 to 751	752 to 787	788 to 823	824 to 859	860 to 895
	Operating Temperature °F																				

Based on 90° F (32° C) ambient air temperature and conductivity curves of calcium silicate or expanded silica insulation.

Table 75 Safety insulation thicknesses

Outer Covering: Stainless Steel Jacket Emissivity 0.40

Nom. Pipe Size	Operating Temperature °C																				
	60 to 79	80 to 99	100 to 119	120 to 139	140 to 159	160 to 179	180 to 199	200 to 219	220 to 239	240 to 259	260 to 279	280 to 299	300 to 319	320 to 339	340 to 359	360 to 379	380 to 399	400 to 419	420 to 439	440 to 459	460 to 479
½	1	1	1	1	1	1	1	1	1	1	1½	1½	1½	1½	1½	1½	1½	2	2	2	2
¾	1	1	1	1	1	1	1	1	1	1½	1½	1½	1½	1½	1½	2	2	2	2	2	2
1	1	1	1	1	1	1	1	1	1	1½	1½	1½	1½	1½	1½	2	2	2	2	2	2
1¼	1	1	1	1	1	1	1	1	1	1½	1½	1½	1½	1½	1½	2	2	2	2	2	2
1½	1	1	1	1	1	1	1	1	1	1½	1½	1½	1½	1½	2	2	2	2	2	2	2
2	1	1	1	1	1	1	1	1	1	1½	1½	1½	1½	2	2	2	2	2	2½	2½	2½
2½	1	1	1	1	1	1	1	1	1½	1½	1½	1½	1½	2	2	2	2	2	2½	2½	2½
3	1	1	1	1	1	1	1	1	1½	1½	1½	1½	1½	2	2	2	2½	2½	2½	2½	3
3½	1	1	1	1	1	1	1	1	1½	1½	1½	1½	1½	2	2	2	2½	2½	2½	2½	3
4	1	1	1	1	1	1	1	1	1½	1½	1½	1½	2	2	2	2	2½	2½	2½	3	3
6	1½	1½	1½	1½	1½	1½	1½	1½	1½	1½	1½	2	2	2	2½	2½	2½	2½	3	3	3
8	1½	1½	1½	1½	1½	1½	1½	1½	1½	1½	1½	2	2	2	2½	2½	2½	2½	3	3	3
10	1½	1½	1½	1½	1½	1½	1½	1½	1½	1½	1½	2	2	2	2½	2½	2½	3	3	3	3½
12	1½	1½	1½	1½	1½	1½	1½	1½	1½	1½	1½	2	2	2	2½	2½	2½	3	3	3	3½
14	1½	1½	1½	1½	1½	1½	1½	1½	1½	2	2	2	2	2½	2½	2½	3	3	3	3½	3½
16	1½	1½	1½	1½	1½	1½	1½	1½	1½	2	2	2	2	2½	2½	3	3	3	3½	3½	3½
18	1½	1½	1½	1½	1½	1½	1½	1½	1½	2	2	2	2½	2½	2½	3	3	3	3½	3½	3½
20	1½	1½	1½	1½	1½	1½	1½	1½	1½	2	2	2	2½	2½	2½	3	3	3	3½	3½	3½
22	1½	1½	1½	1½	1½	1½	1½	1½	1½	2	2	2	2½	2½	2½	3	3	3	3½	3½	3½
24	1½	1½	1½	1½	1½	1½	1½	1½	1½	2	2	2	2½	2½	2½	3	3	3	3½	3½	4
26	1½	1½	1½	1½	1½	1½	1½	1½	1½	2	2	2	2½	2½	2½	3	3	3	3½	3½	4
28	1½	1½	1½	1½	1½	1½	1½	1½	1½	2	2	2	2½	2½	2½	3	3	3	3½	3½	4
30	1½	1½	1½	1½	1½	1½	1½	1½	1½	2	2	2	2½	2½	2½	3	3	3½	3½	3½	4
36	1½	1½	1½	1½	1½	1½	1½	1½	1½	2	2	2	2½	2½	2½	3	3	3½	3½	3½	4
Equip. Flat	1½	1½	1½	1½	1½	1½	1½	1½	1½	2	2	2	2½	2½	3	3	3½	3½	3½	4	4
	140 to 175	176 to 211	212 to 247	248 to 283	284 to 319	320 to 355	356 to 391	392 to 427	428 to 463	464 to 499	500 to 535	536 to 571	572 to 607	608 to 643	644 to 679	680 to 715	716 to 751	752 to 787	788 to 823	824 to 859	860 to 895
	Operating Temperature °F																				

Based on 90° F (32° C) ambient air temperature and conductivity curves of calcium silicate or expanded silica insulation.

Table 75 Safety insulation thicknesses

Outer Covering: Mastic or Jacket With Emissivity 0.9

Nom. Pipe Size	Operating Temperature °C																				
	60 to 79	80 to 99	100 to 119	120 to 139	140 to 159	160 to 179	180 to 199	200 to 219	220 to 239	240 to 259	260 to 279	280 to 299	300 to 319	320 to 339	340 to 359	360 to 379	380 to 399	400 to 419	420 to 439	440 to 459	460 to 479
½	1	1	1	1	1	1	1	1	1	1	1	1	1	1	1	1½	1½	1½	1½	1½	1½
¾	1	1	1	1	1	1	1	1	1	1	1	1	1	1	1	1½	1½	1½	1½	1½	1½
1	1	1	1	1	1	1	1	1	1	1	1	1	1	1	1	1½	1½	1½	1½	1½	1½
1¼	1	1	1	1	1	1	1	1	1	1	1	1	1	1½	1½	1½	1½	1½	1½	1½	1½
1½	1	1	1	1	1	1	1	1	1	1	1	1	1	1½	1½	1½	1½	1½	1½	2	2
2	1	1	1	1	1	1	1	1	1	1	1	1	1	1½	1½	1½	1½	1½	1½	2	2
2½	1	1	1	1	1	1	1	1	1	1	1	1	1½	1½	1½	1½	1½	1½	1½	2	2
3	1	1	1	1	1	1	1	1	1	1	1	1½	1½	1½	1½	1½	1½	2	2	2	2
3½	1	1	1	1	1	1	1	1	1	1	1	1½	1½	1½	1½	1½	1½	2	2	2	2
4	1	1	1	1	1	1	1	1	1	1	1	1½	1½	1½	1½	1½	2	2	2	2	2
6	1½	1½	1½	1½	1½	1½	1½	1½	1½	1½	1½	1½	1½	1½	1½	2	2	2	2	2	2½
8	1½	1½	1½	1½	1½	1½	1½	1½	1½	1½	1½	1½	1½	1½	1½	2	2	2	2	2	2½
10	1½	1½	1½	1½	1½	1½	1½	1½	1½	1½	1½	1½	1½	1½	1½	2	2	2	2	2	2½
12	1½	1½	1½	1½	1½	1½	1½	1½	1½	1½	1½	1½	1½	1½	1½	2	2	2	2	2½	2½
14	1½	1½	1½	1½	1½	1½	1½	1½	1½	1½	1½	1½	1½	1½	2	2	2	2	2½	2½	2½
16	1½	1½	1½	1½	1½	1½	1½	1½	1½	1½	1½	1½	1½	1½	2	2	2	2	2½	2½	2½
18	1½	1½	1½	1½	1½	1½	1½	1½	1½	1½	1½	1½	1½	1½	2	2	2	2	2½	2½	2½
20	1½	1½	1½	1½	1½	1½	1½	1½	1½	1½	1½	1½	1½	1½	2	2	2	2	2½	2½	2½
22	1½	1½	1½	1½	1½	1½	1½	1½	1½	1½	1½	1½	1½	1½	2	2	2	2	2½	2½	2½
24	1½	1½	1½	1½	1½	1½	1½	1½	1½	1½	1½	1½	1½	1½	2	2	2	2½	2½	2½	2½
26	1½	1½	1½	1½	1½	1½	1½	1½	1½	1½	1½	1½	1½	1½	2	2	2	2½	2½	2½	2½
28	1½	1½	1½	1½	1½	1½	1½	1½	1½	1½	1½	1½	1½	1½	2	2	2	2½	2½	2½	2½
30	1½	1½	1½	1½	1½	1½	1½	1½	1½	1½	1½	1½	1½	1½	2	2	2	2½	2½	2½	2½
36	1½	1½	1½	1½	1½	1½	1½	1½	1½	1½	1½	1½	1½	1½	2	2	2	2½	2½	2½	2½
Equip. Flat	1½	1½	1½	1½	1½	1½	1½	1½	1½	1½	1½	1½	1½	2	2	2	2	2½	2½	2½	2½
	140 to 175	176 to 211	212 to 247	248 to 283	284 to 319	320 to 355	356 to 391	392 to 427	428 to 463	464 to 499	500 to 535	536 to 571	572 to 607	608 to 643	644 to 679	680 to 715	716 to 751	752 to 787	788 to 823	824 to 859	860 to 895
	Operating Temperature °F																				

Based on 90°F (32°C) ambient air temperature and conductivity curves of calcium silicate or expanded silica insulation.

9 Installation of Thermal Insulation Systems

GENERAL

As there are numerous types of insulations with varied physical properties no single one would be appropriate for all situations. Insulation must be selected on the basis of its physical and thermal properties as dictated by the construction and use of the building in which it is to be installed.

Long life thermal efficiency depends upon insulation being installed correctly with the other components of the residence or building construction. The location within the structure, the construction, inside temperatures and humidities, outside weather conditions and potential hazards all influence the selection and installation of thermal insulation. These all determine the conditions to which the insulation may be subjected, such as, compression, compaction, tensile stress, vibration, temperature, temperature difference, expansion, contraction, differential vapor pressure, air movement and possibly fire. Proper selection of correct insulation is essential for thermally efficient systems.

LOCATION OF INSULATION IN BUILDING

Generally thermal insulation will be installed in the roof (ceiling), walls or floors of a building. Since the structural design is different in each of these locations, the insulation system must be designed and installed with these structural variations in mind.

Roofs and Ceilings

Flat Roof Installation of Insulation Systems — Basic Considerations

Vapor Migration. The indoor air temperature usually considered most comfortable for human occupancy is 75°F (23.9°C). The relative humidity is 50 percent, the dew-point temperature is 58°F (14.4°C), and the vapor pressure under these conditions is 0.4858 inches of Hg or 34.36 lb/ft^2 (1.2 cm of Hg or 167.7 kgs per square meter). When the ambient air temperature is lower than 58°F (14.4°C), vapor pressure inside attempts to drive the moisture to the outside where the vapor pressure is lower.

Built-up roofs with insulation directly below are the most difficult of all buildings components to keep insulation dry during the winter months. This is because the built-up roofing itself is resistant to vapor passage. Thus, to be effective, the vapor retarder below the insulation must provide more resistance to vapor flow into the insulation than does the restriction of the built-up roofing to the flow of moisture from the inside of the building to the outside air. Where the vapor retarder is insufficient to provide this resistance, vents must be provided. Vapor retarder and vents as used on concrete roofs (or slabs) are shown in Figures 24 and 26. Figure 26 also illustrates installation over metal decks.

Fire. When organic foams or cellulose insulations are used over a steel deck, the deck should be installed with all roof joints liquid tight. This is because a very small fire below the roof would set the urethane, styrene or cellulose insulation on fire. This in turn would melt the asphalt roofing materials and the flaming molten drip could cause additional fires in the area below.

Where organic insulations are used with asphaltic built-up roofs, the roof structure and insulation should be designed to prevent this potential burning mass from leaking or falling into the interior of the building if a fire should occur.

Although many advertise their products as being acceptable as tested by the ASTM E-84 Surface Burning Characteristics of Building Material, this test does not indicate the quantities of combustible gases released or the melting point of solids into combustible liquids. For this reason, ASTM states in the scope of these tests:

> This standard should be used to measure and describe the properties of materials, products, or assemblies in response to heat or a flame under laboratory conditions and *should not be used to describe or appraise the fire hazard or risk of the materials, products or assemblies under actual fire conditions. However, results of this* test may be used as elements of a *fire risk assessment which takes* into account all *the factors which are pertinent* to an assessment of the fire hazard of a particular use.

ASTM E-84 does not measure the highly combustible and poisonous gases produced by a material before it reaches ignition point. Also, it does not measure the quantity of material which melts and drips away as a combustible liquid prior to ignition. Because of these limitations, it is not a suitable test for evaluation of the fire hazards of foamed, fiberous or solid plastics, unless the melting point, gasification temperature, flash-point temperature, flame-point temperature, self-ignition temperature and heat contribution are also given.

Although the statements are true for insulation used any place in a structure, they are presented here because of the extreme hazard they cause when improperly used in a roof.

At the present time no cellular-organic foam has been developed which is fire safe. All foam deteriorates or melts rapidly in a fire, contributes fuel and produces very toxic gases. Although a heavy topping of gravel or crushed stone does provide some protection from fire on top of a roof, it provides no protection from fire below. Protection of combustible insulation from an interior fire must be provided by non-combustible boards, sheets of materials such as gypsum, concrete, expanded silica, alumina-silica or diatomaceous silica.

The general installation considerations follow.

ROOFS

A. *Flat Roof with Insulation Located on Top of the Deck*

All large roofs should be constructed with the expansion-contraction joints designed to include the expansion-contraction strength factors, not only of the columns, beams and deck, but also of the rigid insulation. A typical roof expansion-contraction joint is shown in Figure 96.

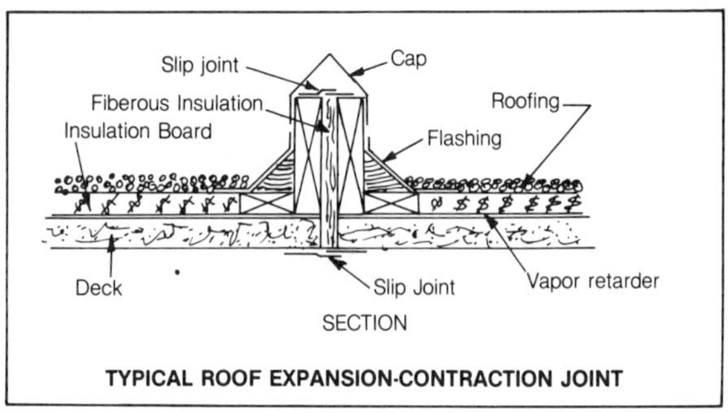

Figure 96

Where there is any fire hazard potential, decks of metal or wood supporting flammable type insulations, such as organic foam or fibers, should be covered with gypsum board or similar fire-resistant, non-combustible material prior to installation of insulation.

In some cases, fire-proofing boards and insulation are secured to the metal deck by welding pins and speed clips. All pins shall be welded in position before combustible insulations and roofing materials are moved to the deck area.

The specified vapor retarder film, foil or mastic shall be installed over the deck or fire-resistant boards, whichever is below the insulation. This shall be carefully sealed and free of any pin holes or vapor leaks. This vapor retarder must have excellent bond and adhesion to the surface to which it is installed. See Figure 25, Chapter 2. However, where cellular glass insulation is used, the vapor retarder film or coating may be eliminated and insulation can be adhered directly to the deck with hot asphalt or other suitable adhesive. Such an installation is shown in Figure 97.

When cellular plastic, fiberous boards, or other vapor permeable insulations are used, the vapor retarder installed below the insulation is the most important factor for long life thermal efficiency. In addition, vapor vents for the insulation should be provided as shown in Figure 24-A.

Insulation sheets or blocks shall be adhered to the substrate with hot asphalt or adhesive. This type of installation is shown in Figure 98.

The built-up roofing shall be applied over the installed insulation with all vent openings, drains and flashings properly sealed.

Where foot traffic or other loading occurs on the roof, suitable walkways or traffic pads should be installed.

INSTALLING FLAT CELLULAR GLASS BOARD TO ROOF.

Figure 97

Courtesy of
Pittsburgh Corning Corp.
Pittsburgh, PA

INSTALLING INSULATING ROOF BOARD TO METAL DECK ROOF

Figure 98

Courtesy of
Johns-Manville Corp.
Denver, CO

Although roofing design is not part of this book, the roofing must perform satisfactorily, otherwise the thermal insulation below it will be ruined. The roof must withstand fluctuation of pressure caused by wind movement across the roof. Parapets cause low and high pressures depending upon the direction and velocity of the wind. The rise and fall motion induces stress in the roofing membrane, which may cause cracks or rupture of the bond to the insulation.

Structural motion may cause the roofing to break by expansion-contraction, compression, tension or shear. This movement also contributes to the loosening of the roof bond with resultant ridges and blisters. In time, these ridges will crack and leak. Proper expansion joints and flashing on the walls and projections will relieve the effects of movement and direction differences.

Water which collects in puddles on a flat roof creates areas of cooler temperature than the dry hot areas around the water. The sun's rays cause the areas that are at the wet/dry interface to deteriorate much more rapidly than areas that drain well. In winter, water turning to ice is destructive to the roofing. This flat roof problem can be overcome by the use of tapered insulation blocks installed on the roof so as to provide slope to the water drains. Typical tapered rigid thermal roof insulation blocks and their arrangement on the roof are shown in Figures 99 and 100.

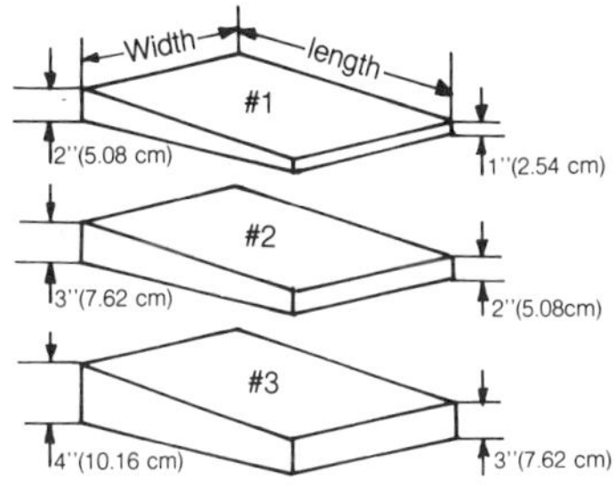

TAPERED RIGID BLOCK ROOF INSULATION
SHAPE OF BLOCKS

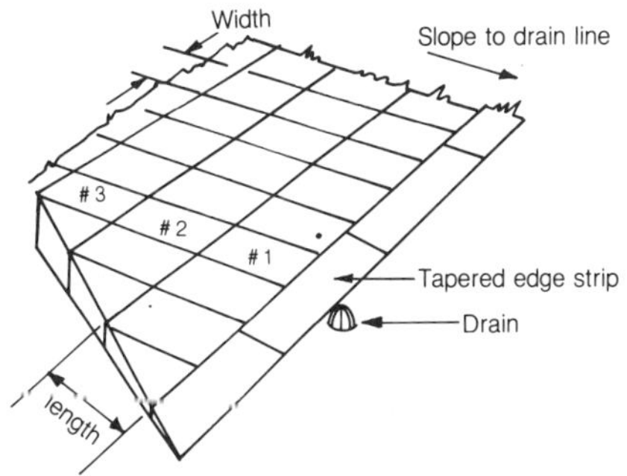

TYPICAL INSTALLATION ARRANGEMENT OF INSULATION

TAPERED INSULATION TO PROVIDE SLOPE TO FLAT ROOF

Figure 99

TAPERED CELLULAR GLASS BLOCKS INSTALLED ON FLAT METAL SO AS TO PROVIDE WATER DRAINAGE.

Figure 100

Courtesy of
Pittsburgh Corning Corp.
Pittsburgh, PA

B. *Flat Roof with Insulation Located Below a Metal Deck*

In any climate where the winter air temperatures go below 50°F for a moderate period of time, construction of a satisfactory, long-term insulation system is very difficult. The reason is that the metal deck is extremely effective as a vapor retarder. This makes it almost impossible to provide an interior vapor retarder film or metal shield sufficiently tight so as to prevent vapor from entering from below then condensing in the insulation as it reaches the dew-point. The only practical way to keep the insulation dry is to provide insulation roof vents. The insulation itself must have very little vapor migration resistance so that vapor can get to vents. This is true whether the roofs are horizontal or sloped. These roof vents are illustrated in Figure 101. The insulation can be secured to the roof with speed clips, and the studs fastened to the roof. However, very special care must be taken that the vapor retarder foil or sheet be sealed over, or to each stud. As shown in Figure 101, the insulation is non-combustible block or sheet type.

When combustible insulation is used directly under any roof, the insulation should be fire protected by a non-combustible material. In this case the protective fire-resistant, non-combustible block or sheets should be secured by metal fasteners to the roof deck or supporting members. In the case of a metal deck or wooden deck roofs, the insulation must be protected on the top side as well as on the bottom side. When attached to a concrete deck or slab, protection is necessary only below the insulation. This is illustrated in Figure 102.

Shown in Figure 102 are the attachments which are to be welded to the roof. All welding operations or other spark-producing construction should be completed before combustible insulation is moved into the area. Likewise, where sprayed polyurethane is used as the insulation, welding or any other ignition source must be removed from the area before the polyurethane is sprayed on. During spraying operations, the workers and all others in the area should be provided with protective clothes to prevent damage to skin by toxic vapors and gases. They must also be provided with Type C continuous-flue, supplied-air, positive-pressure impervious hoods.

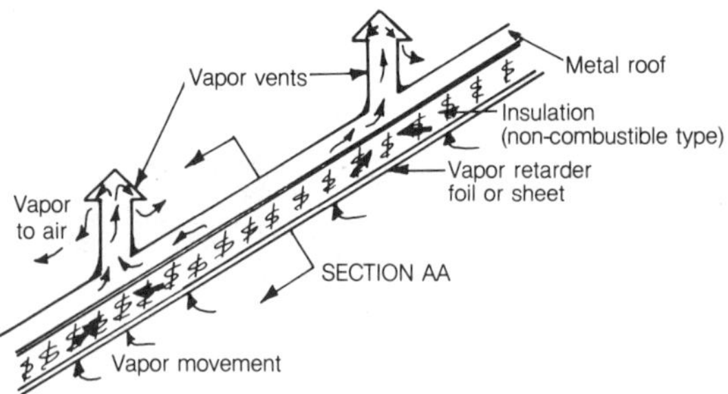

Note: Vapor retarder on inside surface should be more vapor resistant than all resistance from its inner surface to outside air. If vapor is not vented to outside air it will condense in the insulation.

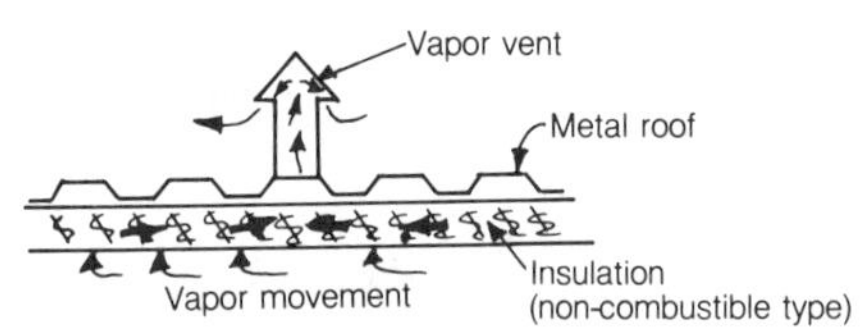

SECTION AA

VAPOR VENTS FOR METAL DECK ROOFS HORIZONTAL OR SLOPED

Figure 101

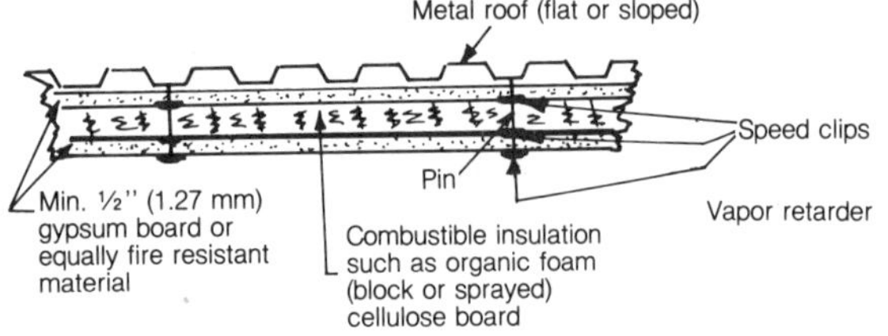

METAL ROOF DECK

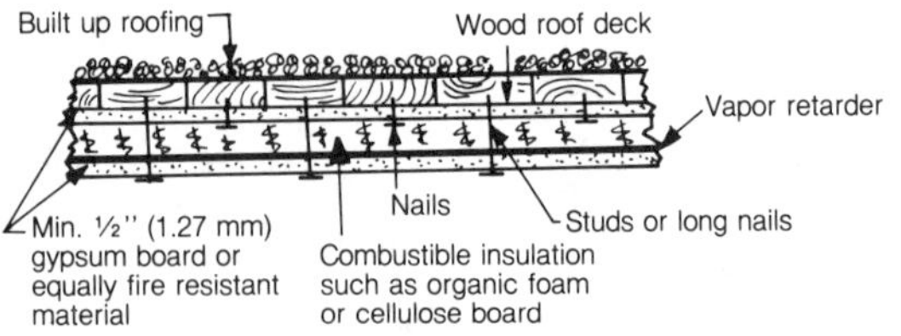

WOOD ROOF DECK

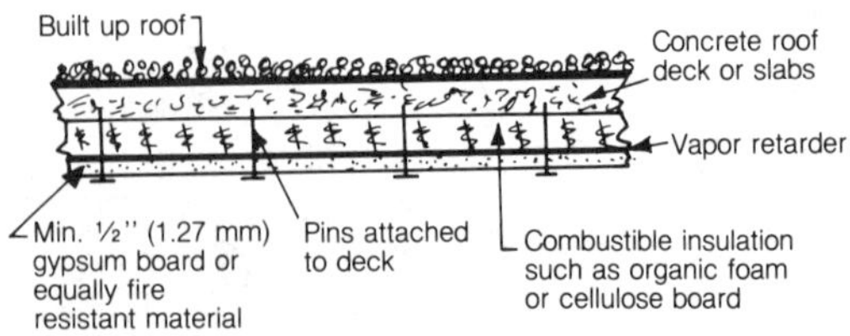

CONCRETE DECK OR CONCRETE SLAB ROOF

FIRE PROTECTION OF COMBUSTIBLE INSULATION INSTALLED ON UNDERSIDE OF ROOF

Figure 102

General Recommendations

All pins used for securement shall be attached to the concrete decks or slabs. Pins welded to metal should be installed prior to the application of the insulation system.

Where combustible insulation is used below metal and wooden roof decks, a fire-resistant board or sheet should be installed before application of insulation. These boards shall be secured in position by pins and speed clips or other mechanical attachments. All edges shall be tightly butted. Any voids shall be filled with mineral fiber, high-temperature insulation cement.

The thermal insulation should be installed below the fire protective material or directly to the concrete roof deck or pads. It can be secured in position by pins and speed clips or other mechanical attachments.

Vapor resistant cellular glass should have all butt joints sealed with vapor resistant sealant to provide a continuous vapor retarder system.

All other insulations should have vapor barrier aluminum foil or sheets with edges overlapped a minimum of two inches. These overlaps are sealed with vapor resistant adhesive or taped with pressure-sensitive aluminum tape. This vapor barrier must be carefully sealed around all securement pins and over speed clips with vapor resistant sealant.

All combustible insulation should be preceded by installation of a non-combustible board or sheet secured to the assembly by projecting pins and speed clips. Where smooth finish is desired, the non-combustible board or sheet shall be cut out sufficiently deep around the pin so that the speed clip and the pin are below the surface. This indentation shall then be filled with gypsum or mineral cement trowelled flush to the surface. This is shown in Figure 103.

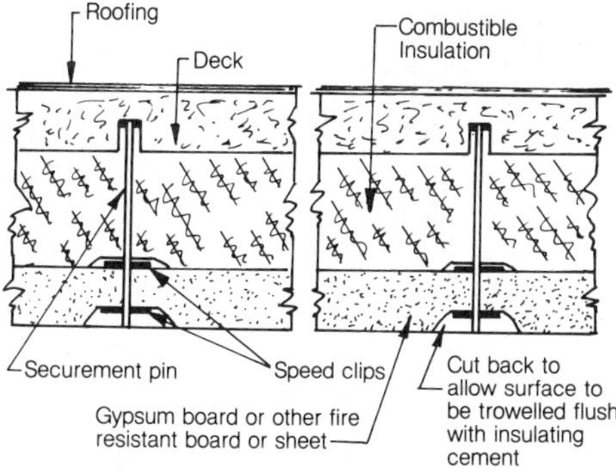

INSTALLATION OF FIRE RESISTANT BOARD OR SHEET BELOW INSULATION

Figure 103

NOTE: The thickness of the fire protective inner layer material must be determined by the fire hazard within the building and the critical time required for safe structural strength.

The insulation thickness should be determined by climatic conditions, use of the building, temperature and humidity inside the building and the cost of energy. For these reasons, insulation thickness is not mentioned in this chapter unless it is dictated by basic construction.

C. *Metal Roof with Insulation Secured by Structural Members Installed from Above*

Fiberous insulation can be installed by support or securement to structural members. In each method the insulation becomes a part of the metal assembly. Thus it is difficult to present insulation installation criteria apart from the major building construction installation instructions. A number of methods are appropriate for this type of construction. These methods will be illustrated in Figure 104 with comments regarding each type of construction.

One method of installing insulation below metal roofs is to install the mineral fiber (rock or glass) wool insulation and its vapor retarder film over the perlins before installation of the metal roof. This is shown in Figure 104 "A". Another method is to wrap the insulation and retarder film around each perlin before locating it across to the next perlin. This is shown in Figure 104 "B". The advantage of this method is that in addition to being more thermally efficient, it also prevents moisture condensation on the perlin.

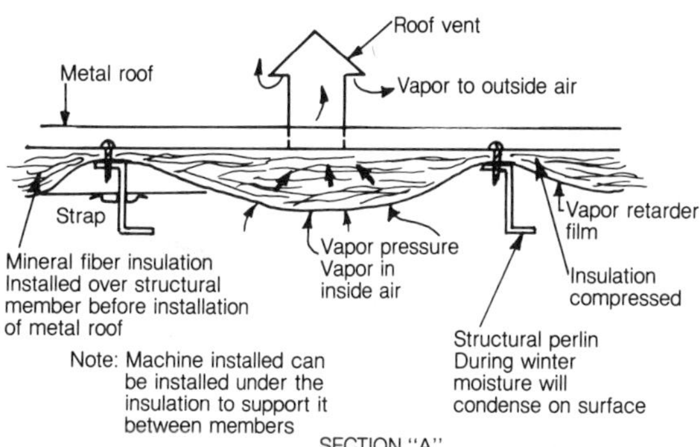

SECTION "A"

FIBROUS INSULATION SUPPORTED BY STRUCTURAL MEMBER

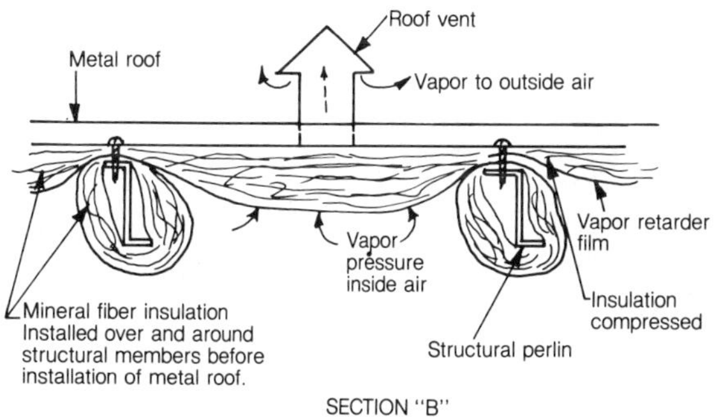

SECTION "B"

FIBROUS INSULATION AROUND STRUCTURAL MEMBERS AND UNDER METAL ROOF

INSTALLATIONS OF FIBROUS INSULATION UNDER METAL ROOF

Figure 104

The vapor retarder film may be plastic or reinforced aluminum foil. It should be located below the insulation. The vapor retarder film shall be sealed to the perlins with vapor-sealed cement or secured by pressure-sensitive aluminum tape (1/2 over the perlin and 1/2 over the film). No film or jacket should be on the top side of the insulation.

In many instances, such as in existing buildings, the insulation must be installed after the metal roof is in place. In this case, it is necessary to install the insulation from below.

D. *Metal Roof with Insulation Installed from Below*

The simplest application of blanket insulation, where no ceiling board is required, is to attach the vapor barrier film directly to the perlins. The insulation blanket must be the correct width so that the film extending beyond the blanket (or board insulation) can be installed to the perlins. Such an installation is shown in Figure 105 "A". Note that the structural strength to support the insulation is supplied by the vapor retarder film portion of the batt. For this reason, a strong film, such as plastic and fiber-reinforced aluminum foil vapor retarder is recommended. Battens should be of treated wood or corrosion-resistant metal, since in cold weather they can be below the dew-point temperature. Where the ends of two batts are butted together, the vapor retarder films should be joined and sealed to prevent vapor bypass at this junction.

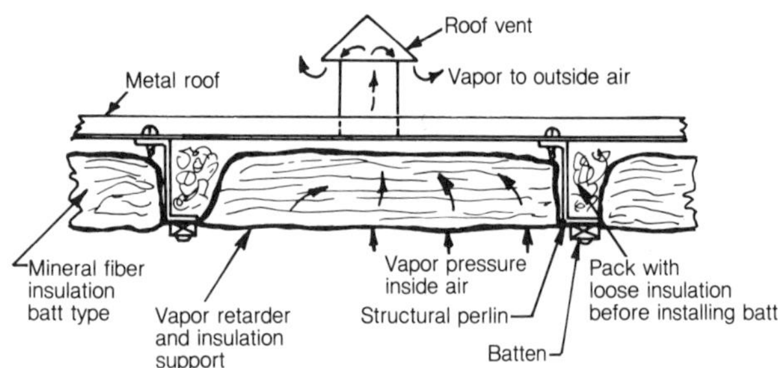

SECTION "A"

FIBROUS INSULATION SUPPORTED BY VAPOR RETARDER REINFORCED FILM FASTENED TO PERLINS BY BATTENS

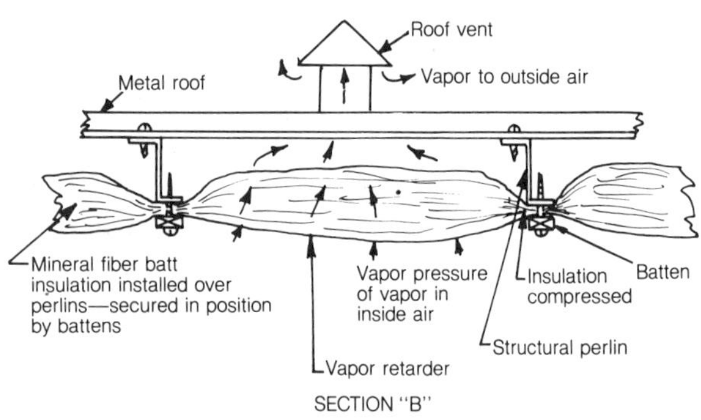

SECTION "B"

FIBROUS INSULATION SUPPORTED BY VAPOR RETARDER REINFORCED FILM HELD TO PERLINS BY BATTENS

Figure 105

Large batt mineral fiber insulation with vapor retarder on the underside can be secured by battens held in position by screws or studs and clips. In this installation the insulation is compressed between the perlins and the battens. The vapor retarder sheet must have sufficient strength to support the insulation. At the end of the batts the vapor retarder of one batt must be seated to the

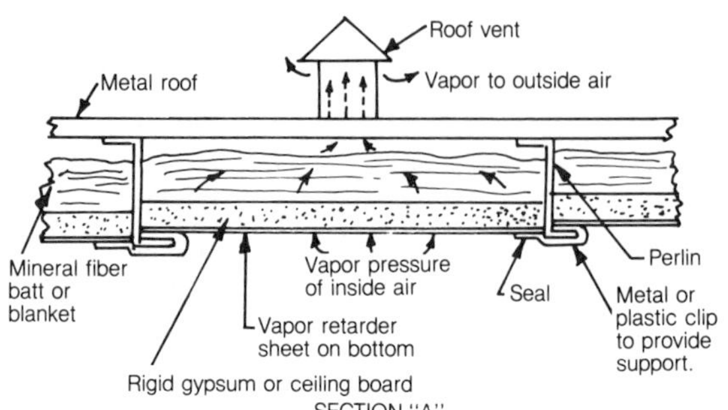

SECTION "A"

FIBROUS INSULATION SUPPORTED BY RIGID BOARD OR SHEET

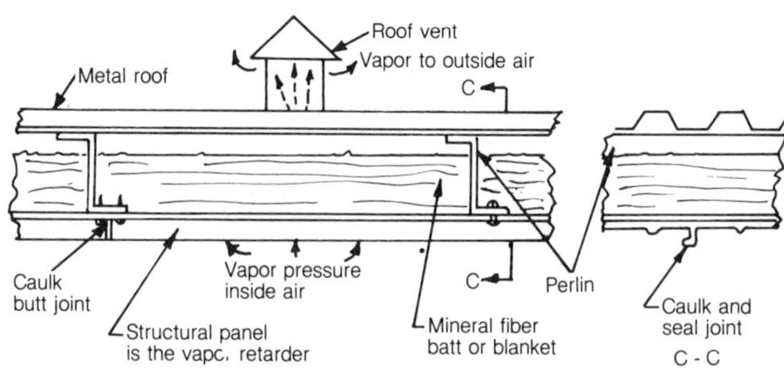

SECTION "B"

FIBROUS INSULATION SUPPORTED BY METAL PANEL

Figure 106

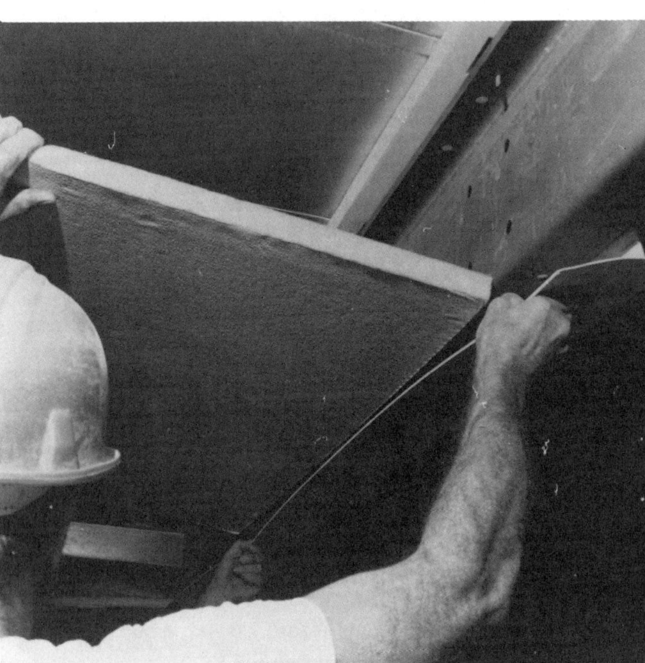

INSTALLING FACED INSULATING BOARD BETWEEN PERLINS

Figure 107

Courtesy of Johns-Manville Corp. Denver, CO

adjacent butt. This method of installing thermal insulation to the underside of perlins is shown in Figure 105 "B".

Another method of installing insulation below metal roofs is to support it by separate rigid ceiling boards or sheets. One of the major problems involved in such installations is placing and sealing the vapor retarder film. Gypsum board or rigid mineral fiber board is excellent for this purpose; however, the underside should be provided with vapor retarder covering of foil or plastic. This should be sealed at the junction with perlin and butt ends. A section of this construction is shown in Figure 106 "A". This type of installation is illustrated in Figure 107.

The blanket or batt insulation can also be supported by metal panels. Panels may have one of a number of corrugation profiles, but should be provided with an edge sealing system. Butt ends, where attached to perlins, should be caulked with vapor seal mastic. This construction is illustrated in Figure 106 "B". The expansion and contraction differential between the outer structure is taken care of by the corrugations and the caulked spaces of butt ends. Flat sheets fastened to the perlins with screws will pull loose or distort due to differential inside and outside expansion.

Organic foam insulation boards or sheets can be installed between perlins in a similar manner as other rigid ceiling boards, the difference being that the organic board or sheet is the primary insulation. Such an installation is shown in Figure 108 "A". Sprayed on organic foam insulation is shown in Figure 108 "B".

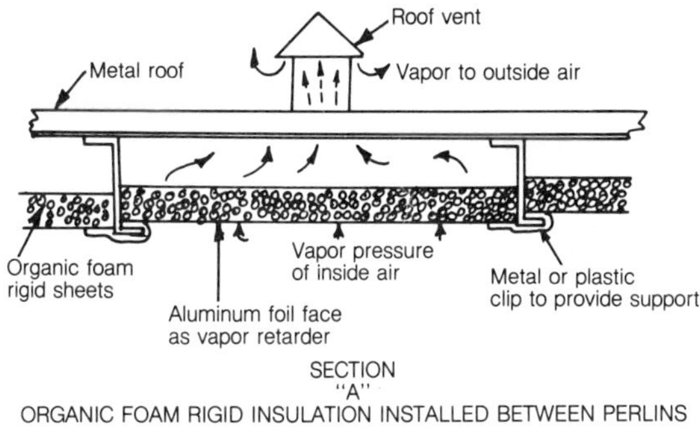

SECTION
"A"
ORGANIC FOAM RIGID INSULATION INSTALLED BETWEEN PERLINS

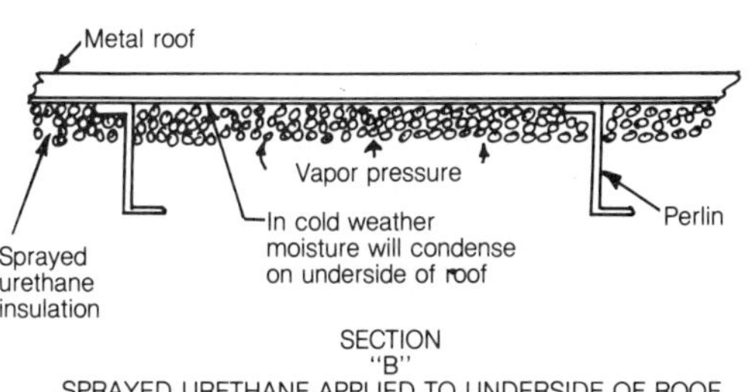

SECTION
"B"
SPRAYED URETHANE APPLIED TO UNDERSIDE OF ROOF

Note: Due to combustibility of organic foam the above installations should only be used in structures of low occupancy and minimum fire hazard potential.

ORGANIC FOAM INSTALLED UNDER METAL ROOF

Figure 108

Metal buildings are an attraction for lightning. When organic foam, or any other flammable insulation is used in the ceiling and/or sidewall, they should be protected by a lightning-rod system. Even though an unprotected building will carry the lightning current to the ground at the point of impact, the metal may reach temperatures that will ignite combustible insulation. As an additional safety precaution, no sparks, flame or hot metal should be allowed close to this insulation from inside the building.

E. *Insulation Located Below Flat Roof Beamed Ceiling*

In many residences the living room and dining area are designed with beamed ceilings. In many cases these are built with the wooden deck as the only thermal barrier. Where it is possible to install gypsum board below the wood, insulation can be installed between the gypsum (or other ceiling board) and the wooden deck. This is illustrated in Figure 109 "A".

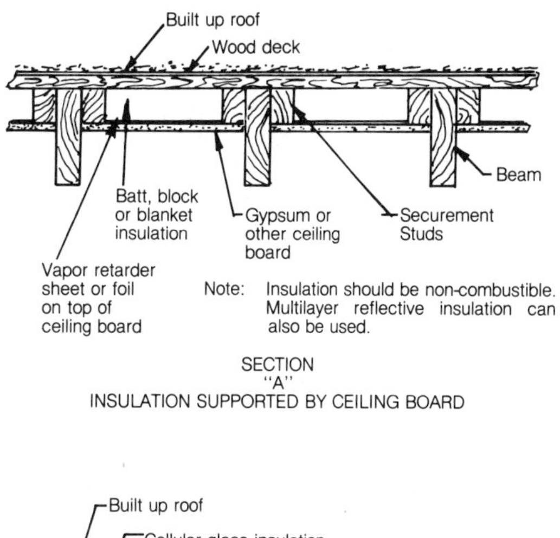

SECTION
"A"
INSULATION SUPPORTED BY CEILING BOARD

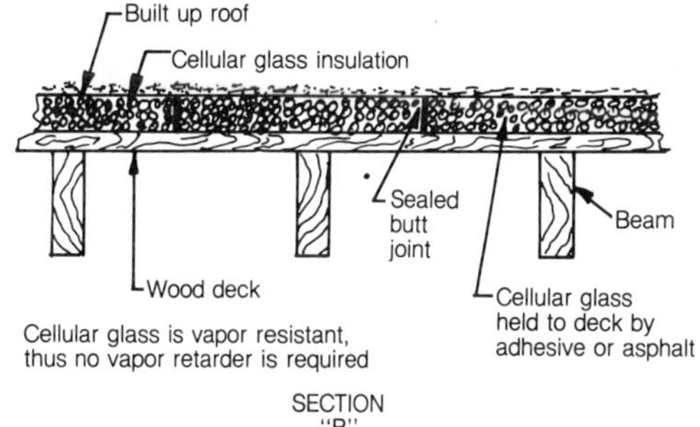

SECTION
"B"
CELLULAR GLASS ON WOOD DECK ROOF

INSULATION OF BEAMED CEILING

Figure 109

When the insulation must be on top of the wooden deck, it must support the built-up roof, provide some fire protection from a fire on the roof and be sufficiently resistant so that a vapor retarder is not required. Only cellular glass insulation has this combination of properties. Figure 109 "B" shows cellular glass installation on the wooden deck. When beams are reinforced concrete, and the deck is concrete slabs (and fire protection is not vital), other applications can be considered.

From Below with Reflective Insulation

Reflection insulation can be installed in multiple layers of single sheets secured to the stud or joists. It can also be obtained in pre-fabricated rolls, such that, when installed, kraft paper acts as a spacer causing it to open up into air spaces faced with the reflective foil surfaces.

Blanket insulation is an excellent material to use when light-weight insulation is needed under a roof. It also has an advantage in that it provides no adsorbent material in which moisture vapor can condense. Correctly installed, it provides its own vapor retarder.

Where open-web, rod-type joists are used to support the roof, the insulation can be installed by stapling the flange of one roll to the flange of the next roll around the rods. This is shown in Figure 110 "A".

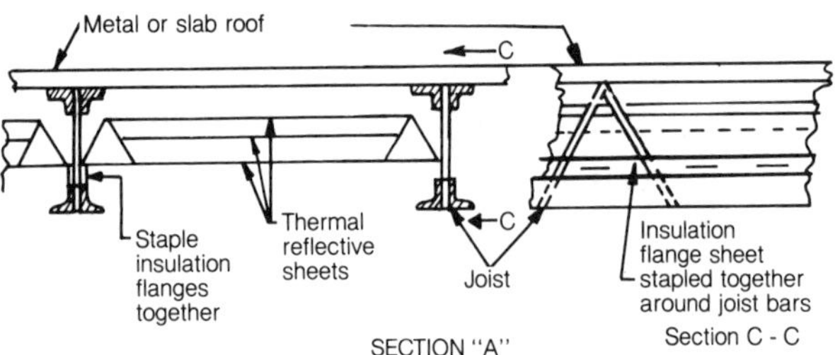

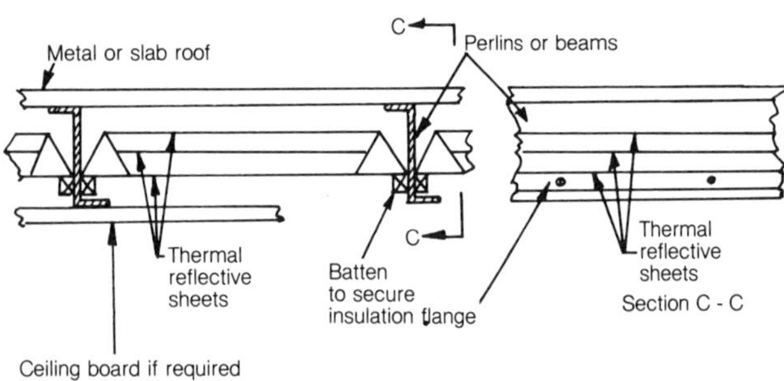

REFLECTIVE INSULATION INSTALLED BELOW ROOF

Figure 110

Where the roof is supported by perlins or beams, the insulation may be installed as shown in Figure 110 "B". As the reflective insulation is not fiberous, granular, etc., it is not absorbent to moisture, and, when flanges are tightly attached and sealed, the undermost sheets become an effective vapor retarder. Reflective insulation being installed is shown in Figure 111.

Where ceiling board or fire-retardant gypsum sheets are specified, they can be directly attached to the underside of the joists, perlins or beams.

F. *Insulation Installed in Ceiling Below Flat Roof*

Locating the insulation in the ceiling has both advantages, and disadvantages. In most instances it is much easier to provide vapor migration venting, but achieving proper installation around

INSTALLING REFLECTIVE INSULATION BELOW METAL BUILDING ROOF

Figure 111

Courtesy of
Foilpleat Insulation Inc.
Gardena, CA

wires or lighting fixtures may be more difficult. All ducts running through vented areas must be insulated. Water pipes may require electric tracing. In buildings of high occupancy, such as offices, hotels, stores, etc., if flammable insulations are used, then fire protection spray systems should also be used. In this case, dry pipe fire spray systems are required. As with roof insulation and depending on the type of construction, ceiling insulation may be installed from above or below.

From Above with Mass Insulation

Where insulation is installed to the ceiling from space above, the installation of the vapor retarder (film, foil or combination) is the most critical procedure. This vapor retarder should have all its joints sealed or taped together and should be sealed around all projections. The insulation can be in blanket, batt, board or loose form. Such an installation is shown in Figure 112 "A".

Where the ceiling is supported from below, the installation is identical except that there are no supports through the insulation. This is shown in Figure 112 "B".

Care must be taken that insulation does not enclose electrical conduits, nor should it be installed over and around flush-mounted light fixtures which must be allowed to dissipate heat.

Sloped Frame Roofs

The placement of insulation in this area depends upon the use of the space below the roof. The area may be used as just a space between the roof and the ceiling below. In other constructions it may be used as a heated or cooled space for occupancy, and in other designs, this space may be divided into room areas by vertical walls and ceilings. In each condition the placement of the insulation should differ.

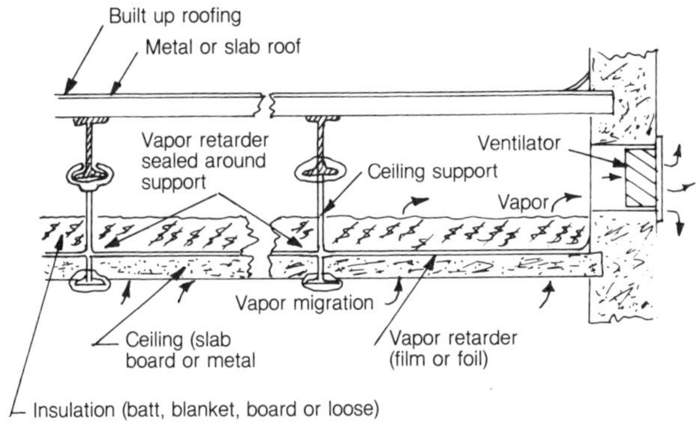

"A"
INSULATION INSTALLED TO SUSPENDED CEILING

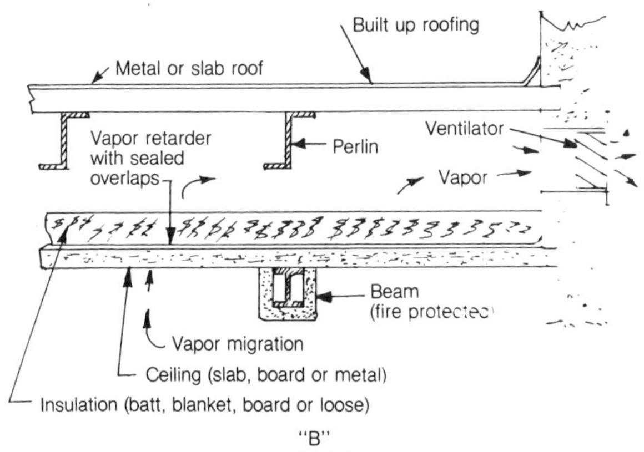

"B"
INSULATION INSTALLED TO SUPPORTED CEILING

INSULATION INSTALLED TO CEILING FROM ABOVE

Figure 112

G. *Sloped Frame Roof Over Unheated Attic Space*

In this type of installation the insulation should be placed in the ceiling of the occupied space below the attic. The vapor retarder must be placed facing the room below. Wherever possible, the insulation should be installed before ceiling sheets so that the vapor retarder flange can be secured to the bottom edge of ceiling joists and overlapped to obtain a sealed joint. Figure 113 illustrates glass fiber insulation being installed between joists.

The attic space above the insulation should be provided with vents to reduce heat build-up during the summer. Even with vents, this attic space can reach temperatures on hot sunny days from 130 °F (54 °C) to 150 °F (66 °C). Without vents, attic temperatures under these same conditions have been known to go above 180 °F (82 °C). In winter, the vents allow any water vapor which gets through the vapor retarder to be transported to the outside air. Typical installation for these conditions is shown in Figure 114.

H. *Sloped Frame Roof Over Heated Attic Space*

Where the space directly below the roof is used as heated and/or cooled space for occupancy, then the insulation must be put into the roof structure itself. This is shown in Figure 115. The insulation must be installed so as to leave space between the insulation and roof sheeting. This space is used so that air can move

INSTALLING GLASS FIBER INSULATION IN CEILING BELOW UNHEATED ATTIC.

Note: Ceiling panels to be installed after insulation and vapor retarder are completely in place.

Figure 113

Courtesy of
Owens-Corning Fiberglas Corp.
Toledo, OH

OUTSIDE
Shingles
Roof sheets
Roof joists
Attic ventilator
Air flow to vent
Ceiling joists
C
UNHEATED SPACE
Wall
Ceiling
C
Vapor pressure
Vapor retarder
Air flow into ventilator
Ventilator
HEATED SPACE (and/or COOLED)

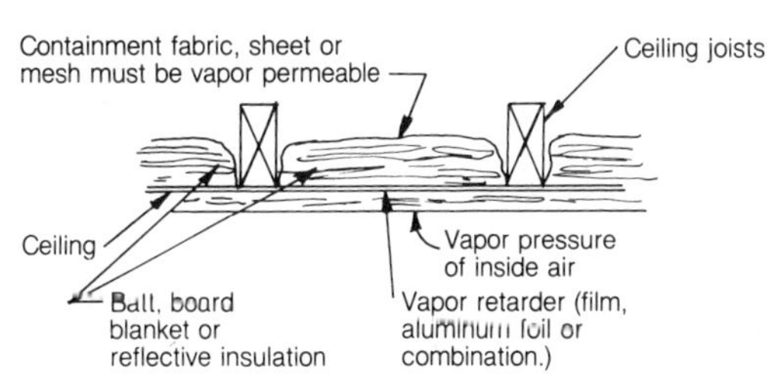

SECTION C-C

PLACEMENT OF INSULATION IN CEILING BELOW UNHEATED ATTIC

Figure 114

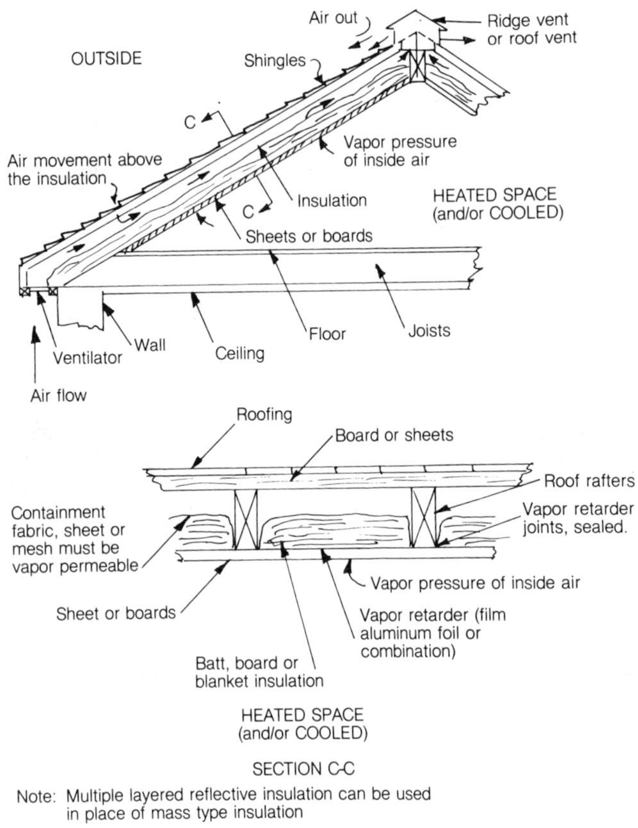

PLACEMENT OF INSULATION IN FRAME ROOF

Figure 115

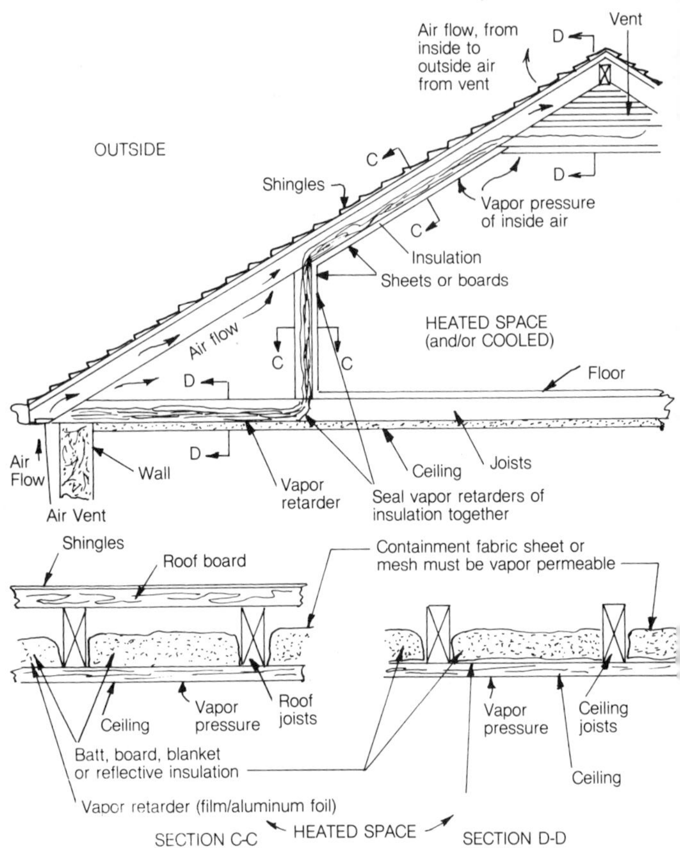

PLACEMENT OF INSULATION — ATTIC ROOMS

Figure 116

from a ventilator in the overhang passes upward to the roof vents at the top. Again, care must be taken to seal all joints of the vapor retarder facing.

I. *Sloped Frame Roof — Rooms in Attic Space*

Many larger homes have rooms in the attic space. In many instances the roof and vertical inner enclosure form a wall which is both vertical and sloped. The insulation should be installed so as to form an insulation space around the room. In spaces where the roof itself is not insulated, the sloped section of the wall must be ventilated. Such an arrangement is shown in Figure 116. Insulation being installed above ceiling panels is shown in Figure 117.

One of the most common errors in putting thermal insulation in existing homes and buildings is that, when installed from above, the insulation is installed with vapor retarder film on top. This, of course, is wrong, as the vapor retarder film holds back the moisture vapor which will then condense and thus remain in the insulation. This wet insulation is ineffective as a thermal barrier and can also cause the wood joists (or rafters) to rot. This is true of insulation made of mineral fiber, cellulose fiber, particle board and foam organic plastic. The only exception is cellular glass which is sufficiently vapor resistant to resist vapor penetration.

Insulation should not be packed around electric wires, cables or outlet boxes as it may cause overheating of the electrical circuit. A method of installation using shields is shown in Figure 80.

INSTALLING BATT INSULATION ABOVE SUSPENDED CEILING.

Figure 117

Courtesy
Johns-Manville Cor
Denver, C

WALLS — ABOVE GRADE

The major factors regarding vapor migration in ceiling insulation also apply to insulation in walls. The major difference is that it is generally easier to construct a wall which will shed water and still have suitable vapor migration release. Fire safety must also be considered in wall construction.

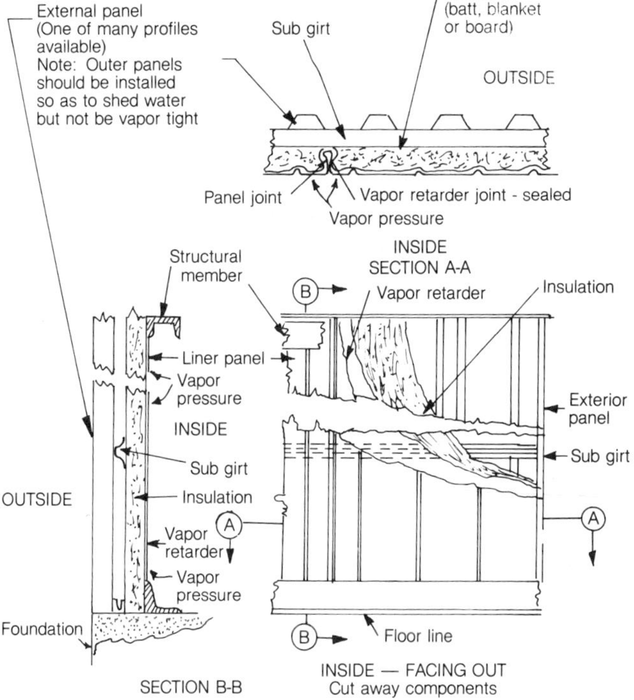

Note: Combustible insulations such as organic foams or cellulose fibers should not be used if there is potential fire exposure inside or outside

INSULATION SECURED BY INNER PANEL IN METAL WALL

Figure 118

A. *A Metal Wall with Insulation Secured by Inner Panel*

A typical metal building is shown in Figure 118. This construction is such that the insulation is installed as part of the inner panel system. The insulation is located between the panel joints thus the vapor retarder film must be sealed to the liner panels before erection. The major problem is the vapor sealing of the butt ends of the insulation and panel. The inner panels may be overlapped to provide a watershed joint, however, butt ends must be so constructed so as to allow for differential lateral movement caused by difference in temperatures of inner and outer panels. To illustrate the need for taking differential expansion into account the following example is given:

A 50′0″ outside metal panel and wall will be approximately 0.6″ longer in the summer, when exposed to the sun, than in the winter at 0 °F. If the interior temperature is maintained at 75 °F throughout the year, then the length of the inner steel will remain unchanged. Construction must be such that it does not cause buckling of inner panels.

One major misconception regarding the use of insulation in metal buildings is that, being faced by metal on two sides, combustible insulations are safe. Unfortunately, this is not true. When metal is heated (inside or out) by fire, impingement ignition of combustible insulation will occur when the metal reaches the insulation's self-ignition temperature.

B. *Metal Wall with Insulation Held in Position by Batten Secured to Girts*

In some metal buildings, such as those used for warehouses or fabrication plants, an inner panel is not installed. In these buildings batt or blanket insulation is held to the wall by battens secured to the girts or structural members. The vapor retarder backing is exposed to the interior of the building. For this reason the vapor retarder needs good tensile strength and puncture resistance. It is recommended under these conditions that vapor retarder sheeting be constructed of aluminum foil and plastic film reinforced with fiber mesh. This vapor retarder must have the joints sealed with pressure-sensitive tape or vapor-sealant adhesive. An installation of metal building walls insulated by blanket or batt insulation with vapor retarder film sheeting is shown in Figure 119.

One of the major disadvantages of this type of installation is that the vapor retarder sheeting has no protection and can be punctured because of its interior exposure.

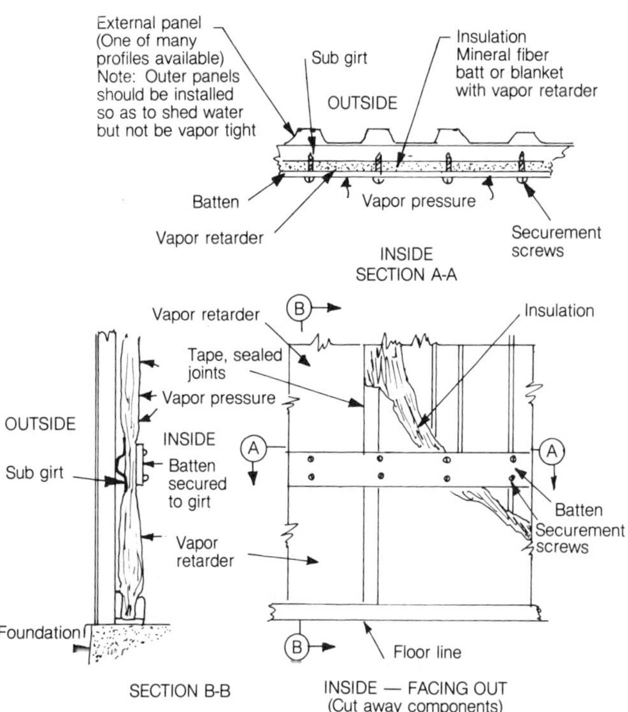

Note: Combustible fiber insulation should not be used if there is potential fire exposure inside or outside of the wall.

INSULATION SECURED BY BATTEN TO WALL GIRTS

Figure 119

C. *Metal Wall with Insulation Held in Position by Panel Liner*

When the interior of a metal building is finished with a liner panel, this liner panel can be used to secure the blanket or batt insulation in place. The biggest construction difficulty in this type of installation is the sealing of vapor retarder laps as the insulation is erected.

The corrugations in the liner panel will take care of the differential movement of inner and outer metal in the direction across the panel. The end joints of these panels should not be secured together but should be installed as a slip overlap joint or spaces mastic sealed butt joint. To illustrate the need for such joints consider the following example:

> The height of outer metal structure is 50′ (15.24 m) and its temperature in winter goes down to 0°F (−17.8°C) and up to 150°F (65.6°C) in the summer. The length in summer will be 0.6″ (1.5 cm) longer than in winter. If the interior is maintained at constant temperature, then the inner panel will undergo no dimensional change. This illustrates the need for the internal expansion joints.

The installation of insulation as secured to the wall in a metal building is shown in Figure 120.

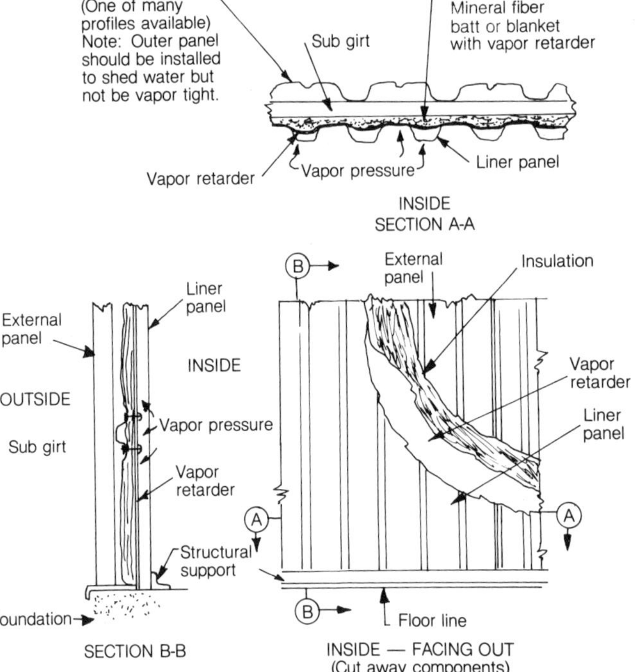

Note: Combustible fiber insulation should not be used if there is potential fire exposure to inside or outside of the wall.

INSULATION SECURED BY LINER PANEL OF WALL

Figure 120

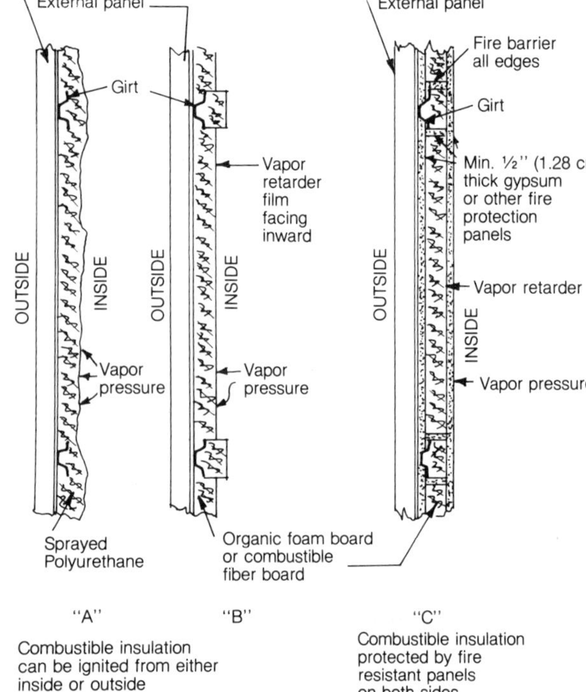

COMBUSTIBLE INSULATION OF METAL BUILDING WALL

Figure 121

D. *Metal Building with Combustible Insulation Inside Walls*

The use of unprotected combustible insulations on the interio of the metal buildings changes the building from a fire-resistar structure to a very flammable structure. This is especially true c organic foam materials. Figures 121 "A" and "B" illustrate suc an installation.

Shown in Figure 121 "A" is a metal wall insulated with spraye polyurethane on the inner surface of a wall. Where the structure i heated and provided with 50 percent relative humidity and the a temperature is below the dew point of inside air, water vapor wi condense in the insulation as there is no known spray vapor seale which has as little moisture permeance as the outer metal pane Of course, during the summer the heat of the outer panel wi vaporize any condensed moisture which will then migrate to the ir side air.

Figure 121 "B" illustrates a metal building wall insulated wit combustible insulation boards or sheets with a vapor retarder fil and/or foil on the inner face. When all the edges of this retarde film are sealed with aluminum pressure-sensitive tape, or othe overlap type sealing, the vapor retarder will be effective.

Figure 121 "C" shows how thermally efficient organic boar or sheets can be used safely in most metal building structure The combustible materials are encased in non-combustible pane of sufficient thickness and mass that even if the insulation did i nite, the fire would be contained. Also, these panels serve to pro vide a barrier to keep any accidental fire or spark (inside or out) from getting to the combustible mass.

A rapid flame spread upward followed by a top flash-over can cause a rapid temperature rise. In several accidental fires of interior combustible insulation in metal buildings the temperature of the fires caused the roofs to collapse within 8 minutes.

E. *Panel Walls of Multiple Glass-Metal Buildings*

There are numerous arrangements for the panel sections of walls of multiple story buildings depending upon the type of heating and air conditioning, office or room arrangements, electrical distribution, structural steel design and fire protection requirements. With all these variables it is impossible to present a typical illustration of insulation installation. Insulation being installed in a wall panel section is shown in Figure 122.

INSTALLING INSULATING BOARD IN EXTERIOR WALL

Figure 122

Courtesy of
Johns-Manville Corp.
Denver, CO

With the high-rise, glass-metal faced building there are a few important factors which influence the choice and design of insulation in these structures. These are as follows:

a. Combustible fiber or organic foam should never be used in these building walls as there are no practical fire spray systems for confined wall insulation.
b. As there is no practical means to vent the outside walls, insulation in winter will be subjected to vapor condensation where dew point occurs in its structure.
c. Insulation must not be installed around electrical cables, causing them to overheat.
d. Insulation where effective aluminum foil vapor retarder can be installed can be glass fiber, non-absorbent fire-resistant binder type.
e. Where possible to be installed, multiple layer reflective foil insulation can be used effectively.
f. Where wall venting is impossible and continuous vapor retarder films or foil are impractical then cellular glass insulation should be used.
g. In calculation of the heating or cooling load, the loss or gain through the structural members and floor that join the wall must be added to that of the wall itself.

In many such buildings the major wall area is glass. The glass may be single or double pane, clear or tinted. Unfortunately, the heat transfer and radiation factors are such that the thermal efficiency of these walls is very low. Heat transfer coefficients and

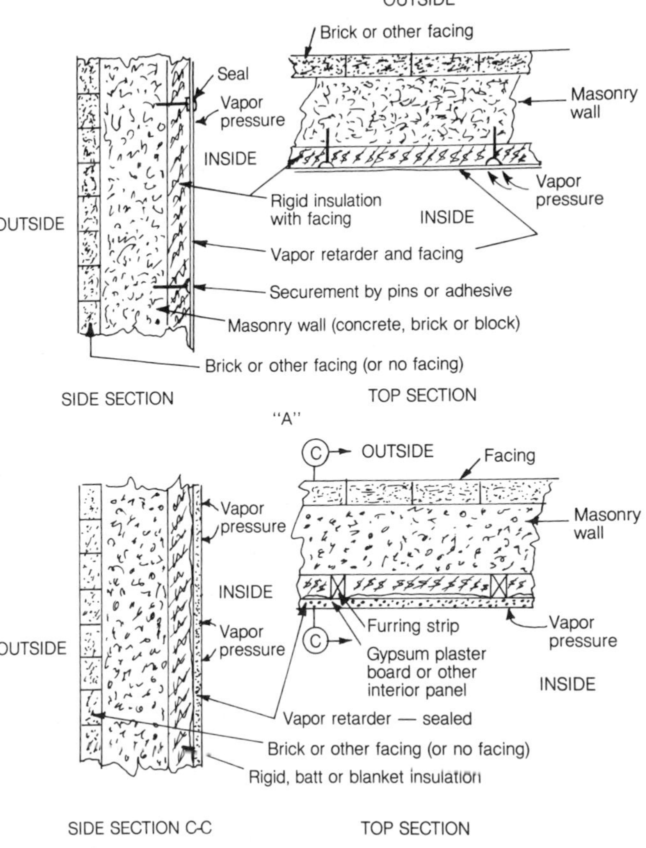

MASONRY WALLS WITH INSULATION ON INSIDE

Figure 123

radiation of various types of glass will be given in the tables on heat transfer through construction materials.

F. *Masonry Walls with Insulation on Inside*

Masonry walls may be of brick, tile, block or concrete. When insulation is installed on the interior, in most cases some type of protective inner panel is required. However, in some instances inner panels may provide sufficient thermal resistance to be considered the thermal insulation.

Some thermal insulation panels are made with a vapor-retarder facing. These can be secured to the masonry walls with suitable adhesives, securement pins or cement nails. The major problem when pins or nails are used is that they must be countersunk and then be cemented over to provide a smooth, inner surface. The vapor sealer must also be sealed around these securements. As shown in Figure 123 "A", the masonry wall may be of concrete blocks, cinder blocks, brick or concrete. A facing on the outside may be aluminum or wood siding, brick or shingles.

Another frequently used construction is to use furring strips on the inner side for securement of panels. This provides a space for the installation for the insulation. The insulation can be batt, blanket or rigid board. The inner surface of the insulation should be supplied with vapor retarder film or foil. This vapor retarder film or foil should be sealed at the furring strips, top and bottom. Such an installation is shown in Figure 123 "B".

It should be noted that if a flammable insulation or wall board is used in either of these installations, the masonry wall is no longer fire safe. However, when organic foam or cellulose board insulations are used in a manner as shown in Figure 123 "B", they can be made fire resistant by using gypsum board (or other fire-resistant panel) as interior panels and by using fire stops of similar fire-resistant material around all edges to complete the enclosure.

G. *Masonry Walls with Reflective Insulation*

Reflective insulation can be used very effectively with masonry construction. Correctly installed, it functions as a vapor retarder as well as an insulation. In many instances, where additional thermal resistance is needed for existing walls, the addition of reflective insulation can be very effective.

Where space is at a premium, furring strips 3/4" (1.9 cm) in thickness can be installed so that reflective aluminum sheets can be fastened to the wall. This construction will divide this air space into two spaces and provide three reflective sides of aluminum. Over the furring strips, interior panels of gypsum, etc. can be installed. Such an installation is shown in Figure 124 "A".

When more space is available for the insulation, the reflective insulation can be installed as shown in Figure 124 "B". This installation provides four air spaces and four reflective surfaces.

In addition to the manufactured multiple layer, as illustrated, sheets of regular aluminum foil can be installed in a similar manner.

H. *Block or Tile Wall with Cavities Filled with Poured or Foamed Organic Insulation*

Filling the cavities of block or tile masonry walls with appropriate insulation results in a much more thermally efficient

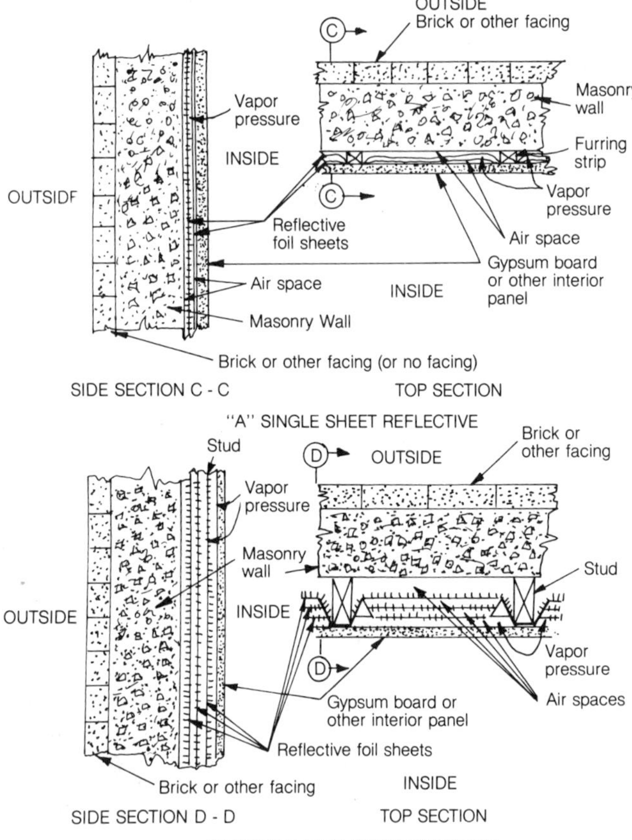

MASONRY WALL WITH REFLECTIVE INSULATION

Figure 124

wall. These cavities may be filled from the top, as the wall is erected, with granules, pellets or flakes. Figure 125 "A" shows such a cavity block wall filled with pour type insulation. In such an installation the vapor retarder must be placed on the surface facing the inside of the building. Typical of the pour type insulations are diatomaceous silica powder, gypsum pellets, perlite powder, silica aerogel powder and vermiculite flakes. All of these are noncombustible and add to the fire resistance of the wall.

Where there is need for additional thermal resistance, the gypsum or other type of wall panels can be furred away from the wall in order to provide space for additional insulation. Aluminum foil insulation installed in this furred space will provide additional thermal insulation, and, installed correctly with sealed joints, will serve as the interior vapor retarder. Such installation is shown in Figure 124 "A" or 124 "B".

The cavities can also be filled with foamed, organic insulations, such as polyurethane or urea-formaldehyde. In the case of polyurethane, the wall should be capped with solid masonry to act as a fire stop.

In both cases the vapor retarder must be installed on the inside wall surface, similar to that described above. However, when urea-formaldehyde is used, the interior must be vapor sealed with foil, vapor resistant sheets with all joints vapor sealed and taped. The outer surface must *not* be vapor sealed, as the vapors from

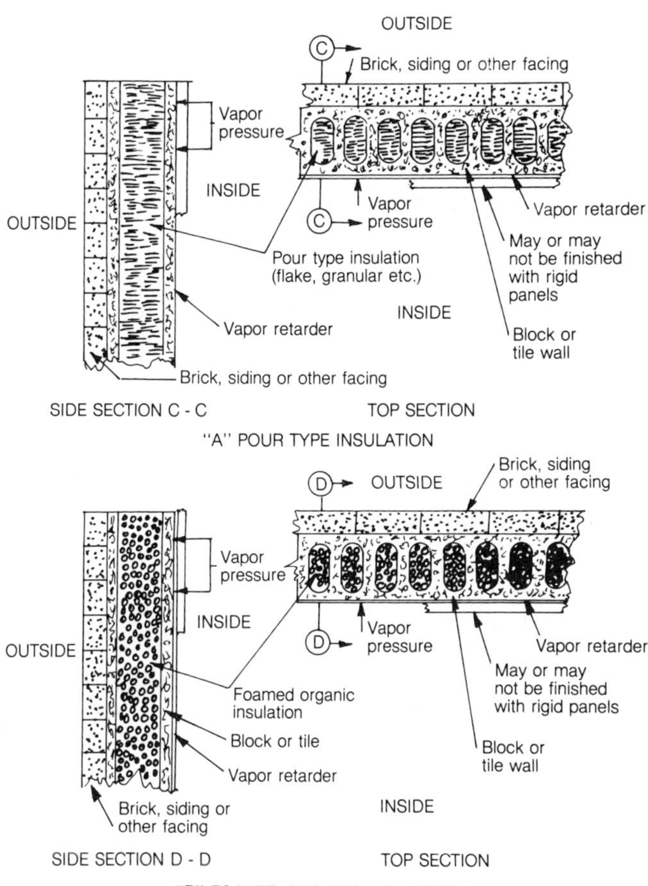

BLOCK OR TILE WALL WITH CAVITIES FILLED WITH POURED OR FOAMED ORGANIC INSULATION

Figure 125

this insulation must be allowed to escape outside. This material will give off fumes for a very long time after installation depending on the temperature, moisture and vapor. These fumes must be sealed out of the interior of a house carefully with a 100 percent sealed vapor retarder on the inner side of the wall and ceiling. These fumes cause nasal and respiratory discomfort and have been linked to cancer by studies conducted by the chemical industry.*

As stated previously, where additional thermal resistance is required, the walls can be furred out and reflective or mass insulation can be installed in the furred space, as shown in Figure 124 and 125.

I. *Masonry Wall with Insulation Between Wall and its Outer Facing*

The use of insulation on the outside of the basic wall structure of block or concrete has been increasing. Located on the outside, it has the advantage of stabilizing the temperature major mass of the masonry structure. Also, such insulation can be installed on the outside of the existing walls; then the exterior can be finished with brick or siding.

*Report of *Engineering Times*, September, 1980 and February, 1981 and May, 1981.

Where such walls are constructed, an outer cavity might be built in the wall and then filled with foamed-in-place polyurethane. This is possible (but impractical) as the walls and their facings must be able to contain a foam capable of putting 5 psi (34.5 kPa) expansion pressure. Also it would be very difficult to install a suitable vapor retarder.

Even preformed polyurethane rigid insulation sheeting must be installed with space for thickness expansion to take place. Otherwise, it too will exert pressure. The reason for this was explained in the building research report of the National Research Council of Canada, "Polyurethane Foam as a Thermal Insulation", Revised Edition, November, 1978. This report states:

> Polyurethane is basically a dimensionally unstable material. This is because the gas pressure in the cells increases to as much as 12 psi (83 kPa) above atmospheric pressure as air permeates into the cells. Thus, there is a potential for a release of large forces when restrained or a large expansion if unrestrained.
>
> As explained previously, at temperatures above 120°F (49°C) or at relative humidities approaching 90 to 95 percent and temperatures above normal ambient, the material begins to "grow" if unrestrained. The cell walls stretch and the material expands.
>
> High humidity has a detrimental effect on the material at high temperatures. Polyurethane is plasticized by the water vapor causing it to lose its strength and allowing the closed cells to return to a spherical form and then to expand. If expansion is restrained in one or two directions, larger expansions occur in

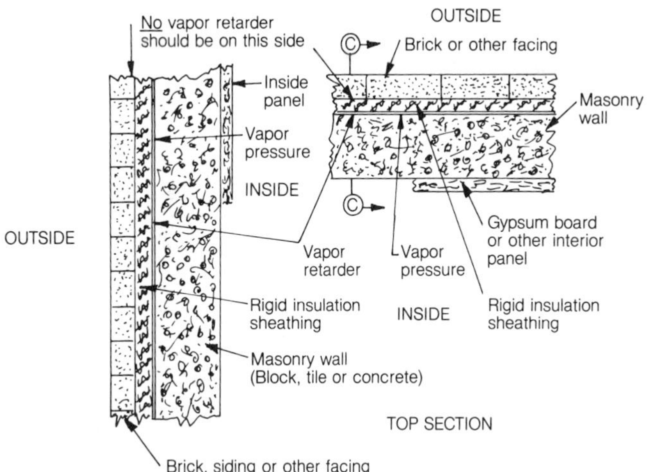

Note: Insulation may be rigid mineral fiber or particle board, foamed glass or organic foam. If polyurethane foam is used, space for increase in thickness should be provided.
In no case should cavity between wall and siding be poured with polyurethane foam as its expansion can cause distortion or failure of facing.
Vapor retarder facing inward is vital for protection of moisture vapor from inside.
Any facing on rigid insulation facing outward must allow vapor passage. Any aluminum foil used on outer surface must be perforated, so that vapor passes to outside air.

MASONRY WALL WITH INSULATION INSTALLED BETWEEN WALL AND ITS OUTER FACING

Figure 126

the unrestrained directions. If it is completely confined, the material applies a force to the surrounding structure. This can large enough to cause damage even to such strong structures as prefabricated concrete panels.

Polyurethane is best used in a cool, dry environment. A good design will ensure its protection from extremes in temperature and from becoming wet. It will also keep the foam dry by providing a vapor barrier that will keep excessive moisture from entering the material and condensing or freezing in the colder parts of the foam. Polyurethane does not have good freeze-thaw resistance. If the material is wet and is subjected to freeze-thaw cycles it may disintegrate after only a few dozen cycles.

Other rigid insulation sheathing can be used. However, in all cases the vapor retarder must be on the surface facing inward. The surface facing outward must be free to pass the vapor outward. Wherever these insulations are faced on the outer surface with film or aluminum foil, this facing must be perforated or be of a material of high vapor permeability. If not, this facing will allow a build-up of vapor which will then be condensed within the sheathing. The preceding is illustrated in Figure 126.

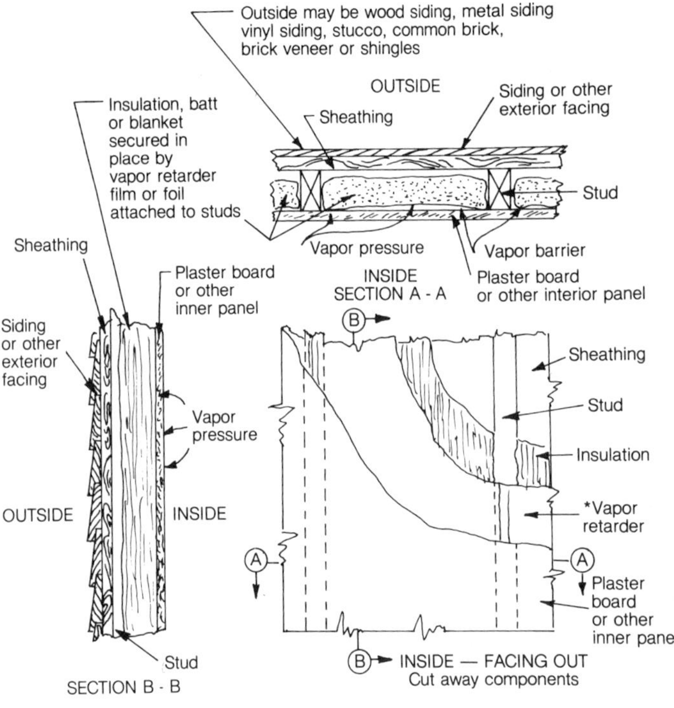

*Vapor retarder stapled and sealed to side of stud or over face of stud. It must be sealed with adhesive or pressure sensitive tape to be an effective vapor retarder. Also it must be sealed at bottom and top of stud space.

FRAME WALL WITH INSULATION IN STUD SPACE

Figure 127

J. *Frame Wall with Insulation in Stud Space*

A wood stud wall is shown in Figure 127. The studs could be 2 x 4's (approximate dimensions 1 5/8" x 3 1/2" or 4.1 cm x 8.9 cm) or 2 x 6's (approximate dimensions 1 5/8" x 5 1/2" or 4.1 cm x 14 cm). These studs support the inside sheeting and outer sheathing

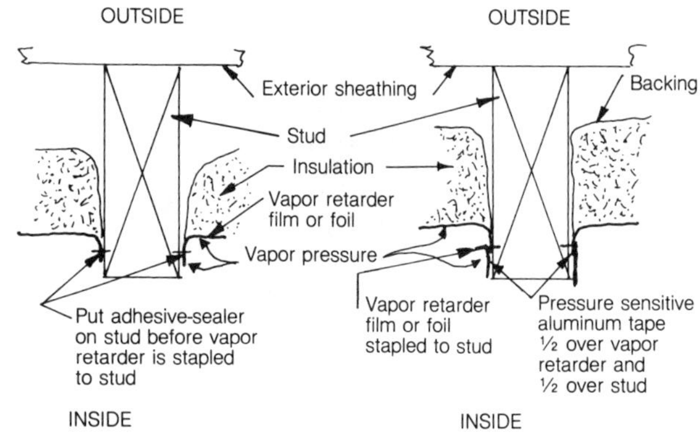

INSULATION RECESSED BETWEEN STUDS
"A"

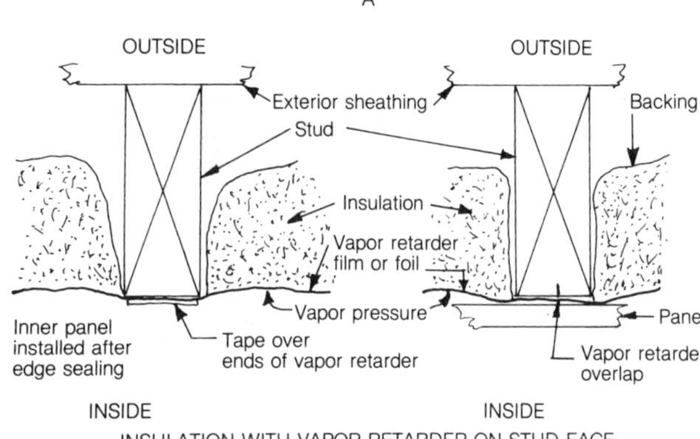

INSULATION WITH VAPOR RETARDER ON STUD FACE
"B"

Figure 128

and facing. The cavity between these can be used for the installation of thermal insulation.

This cavity can be poured full of loose fiber of granular insulation. However, the vapor retarder film or foil must be placed over the studs and sealed before the application of gypsum plaster board (or other interior paneling) and before the insulation is poured. *If a suitable vapor retarder is not provided, moisture vapor will condense to water in the insulation. In addition to making the insulation ineffective, it will also cause the wooden members to rot.*

The vapor retarder is equally important with batt, blanket or board, fiberous or particle insulations. Many of these are furnished with vapor retarder film or foil as part of their construction. To be effective, this vapor retarder film or foil must be sealed to complete a vapor tight inner enclosure. Methods of sealing the vapor retarder to studs are shown in Figure 127 and 128. Of course, the top and bottom must also be sealed to complete the enclosure. Figure 129 shows glass fiber insulation with aluminum foil vapor retarder being installed in stud space. Figure 130 shows plastic film vapor retarder being installed over glass fiber insulation already in the stud space.

Wherever pipes, electrical cables, conduit boxes, etc., extend through the vapor retarder, these projections must also be vapor sealed.

It should be noted that incomplete sealing of the vapor retarder film or foil is the most common insulation error in building construction. However, the correct installation of the vapor

INSTALLING GLASS FIBER INSULATION WITH ALUMINUM FOIL VAPOR RETARDER IN FRAME WALL.

Figure 129

Courtesy of
Owens-Corning Fiberglas Co.
Toledo, OH

NSTALLING VAPOR RETARDER PLASTIC FILM OVER GLASS FIBER NSULATION IN FRAME WORK.

ote: Wall panels to be installed after insulation and vapor retarder are completely in place.

Figure 130

Courtesy of
Owens-Corning Fiberglas Corp.
Toledo, OH

etarder is essential to the effectiveness of thermal insulation in ontrolling heat transfer.

Reflective Insulation in Stud Space

Reflective insulation is also used in stud spaces. This is available in several configurations, however the greater the number of reflective sheets and dead air spaces the more thermally efficient the insulation.

The innermost reflective sheet should be installed so that it provides a sealing vapor retarder to prevent condensation in the stud cavity. A typical application is shown in Figure 131. Installation is shown in Figure 132.

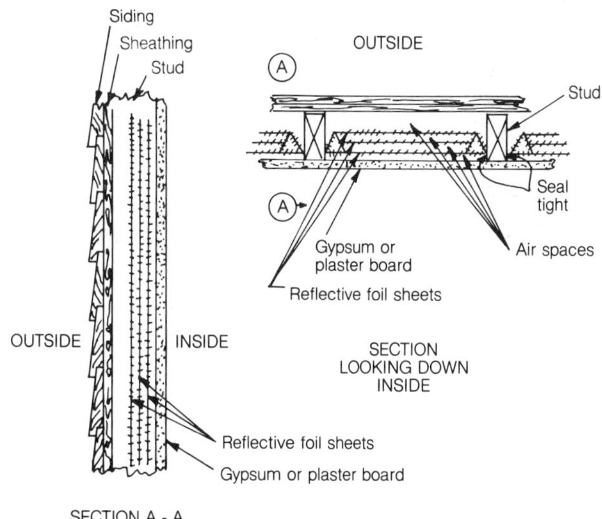

MULTIPLE REFLECTIVE SHEET INSULATION IN STUD WALL

Figure 131

Water Pipes or Drain Pipes in Outside Walls

Where water pipes or drain pipes are in the outside wall cavity, the insulation must be installed between the piping and the outside sheathing. *It should never be placed over the pipe on the inner side, as this will cause a serious reduction of pipe temperature in winter.* Under sufficiently cold outside temperatures, the water lines and rain lines may freeze. Also, thermal insulation should never be packed around electric wires, cables or boxes. An air space not less than 2 inches (5.00 cm) should be provided between the insulation and the wires and cables. *Electric wiring in service must be able to dissipate heat produced by the current flow. Thermally insulated wires will overheat, then short or ground out.* Even though the insulation is non-combustible, it might result in damaging the electrical circuit. Correct installation is shown in Figure 133.

K. *Frame Wall with Insulation Sheathing Only*

In an effort to conserve energy, outer sheathing of low thermal conductivity has been developed to replace wood and fiberboard as sheathing. The major insulation sheathings now marketed are (1) foamed styrene, (2) foamed polyurethane and (3) glass fiber.

INSTALLING REFLECTIVE INSULATION IN WALLS AND CEILING OF FRAME HOUSE.

Figure 132

Courtesy of
Foilpleat Insulation Co.
Gardena, CA

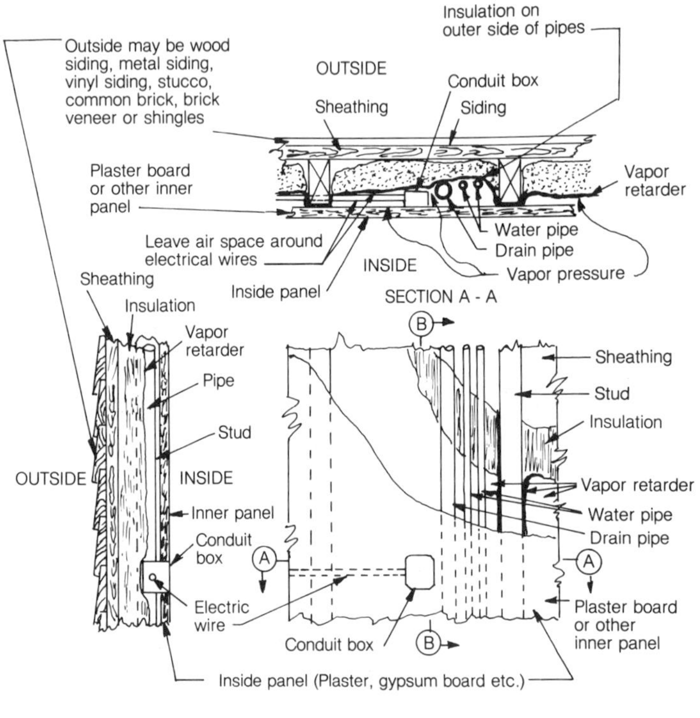

Note: Any necessary projections of pipes, cables or wires through vapor retarder film or foil must be vapor sealed with sealant or tape. Electric wires or cables overheat when enclosed in thermal insulation. Air spaces should be provided around the wires and cables for cooling.

PLACEMENT OF INSULATION WHERE WIRES, PIPES AND DRAINS ARE IN WALL CAVITY

Figure 133

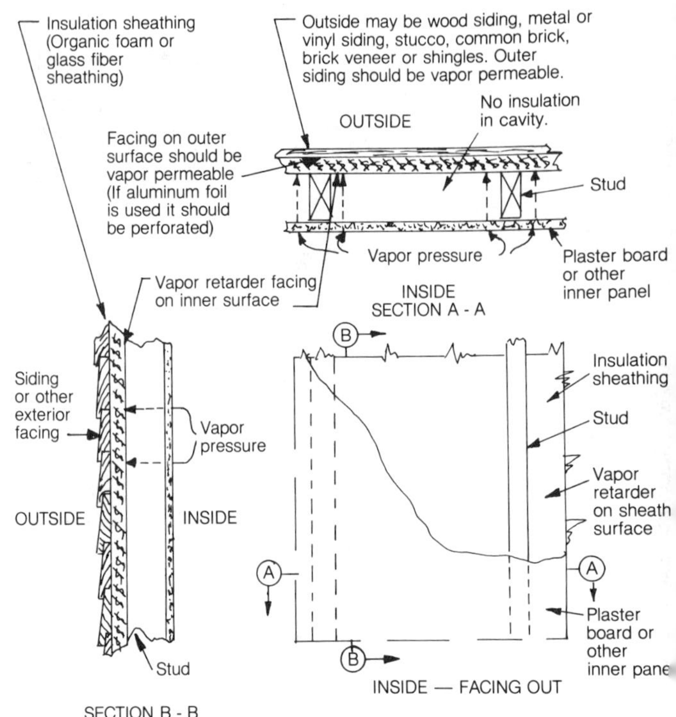

Note: Glass fiber sheathing is most frequently manufactured with permeable outer facing to allow moisture to escape to outside air. Organic foam panels are frequently faced on both sides with aluminum foil. The outer foil facing should be perforated to prevent moisture entrapment within the sheathing.

FRAME WALL WITH INSULATING SHEATHING

Figure 134

The organic foams are in most instances furnished with aluminum foil surfaces, the inner surface acting as a suitable vapor barrier. Some are furnished with full foil-faced vapor barriers on the outer surface and should be used only when outside ambient air temperatures seldom go below 55°F (12.8°C). When used in moderate or low temperature climates, the outer surface should be perforated aluminum foil to allow passage of moisture vapor. High vapor permeable facings may also be used over the outer surface of organic foam insulation. Its perm rating should be over 25 (English Units) or 16.5 (Metric Units).

Glass fiber insulation is furnished with a vapor permeable outer facing. The inner vapor retarder on the inside may be placed on its inner surface or under the inner panel. This is shown in Figure 134.

If there is any possibility of external fire, such as a brush fire or burning leaves, and organic foam sheathing is used in the construction, then the outer facing should be brick or cement stucco. If the facing is wood, metal vinyl siding, or shingles, a minimum of 1/2″ (1.35 cm) thick gypsum board should be placed over the organic sheathing before the siding or shingles are applied. This prevents rapid ignition.

L. Frame Wall with Insulation Sheathing and Cavity Insulation

Where glass or mineral fiber insulation sheathing is used on the outside of the frame wall, additional thermal resistance can be obtained by installing insulation in the cavity space.

The outer facing may be brick, siding or shingles. They must be installed so as to shed water, but they should not be vapor sealed. Vapor from the inside air must be allowed to escape through the sheathing and outer facing. The vapor retarder installed on the inner side of the insulation is illustrated in Figure 135 "B". The glass or mineral fiber insulation can be any thickness up to the limitation of the cavity. This construction is shown in Figure 135 "A".

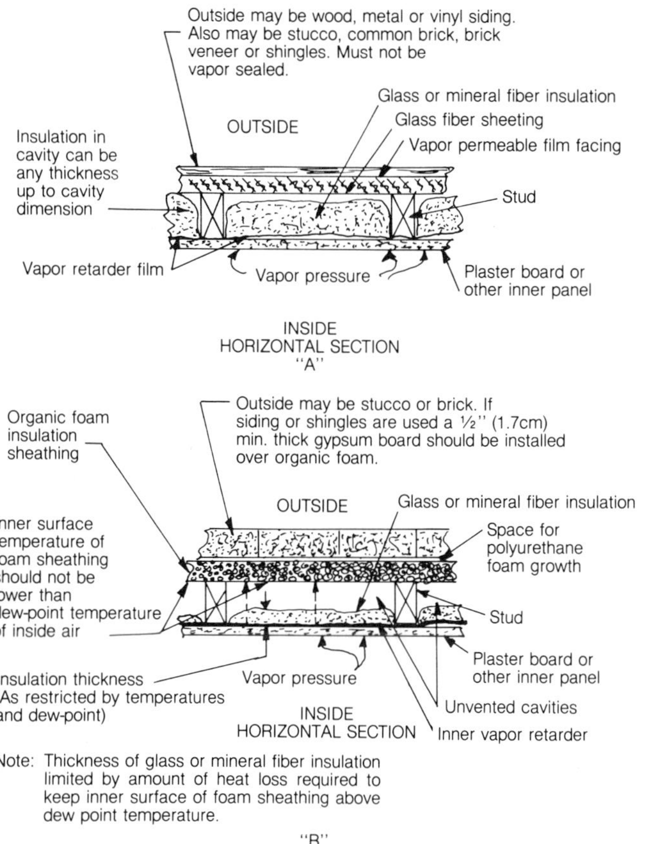

FRAME WALL WITH INSULATION SHEATHING AND CAVITY INSULATION

Figure 135

Where sheathing is organic foam and is installed to completely enclose the stud cavities, the cavity insulation is limited. The organic foam with its aluminum facing has a very low vapor permeability. These materials have a permeance of 0.03 (English Units) and 0.0197 (Metric Units). To prevent entrapment of moisture in the cavity, a vapor retarder with effective permeance of less than 0.005 (English Units) and 0.0033 (Metric Units) would be necessary on the inner side of the cavity. Such vapor resistance is practically impossible with standard building materials and installation procedures. For this reason, the air temperature in the cavity must remain above the dew point of the air inside the residence or building. The thickness of the cavity insulation must be limited to keep vapor from condensing in it and on the inner surface of the organic foam sheathing. This is shown in Figure 135 "B"

The temperature change due to heat transfer is directly determined by thermal resistance. Thus the allowable thickness of insulation in the cavity can be calculated for various conditions of ambient temperature and thickness of organic foam sheathing.

Example 32—*English Units*

A stud frame wall with wood siding over 1" thick organic foam with the foil facing on its inner surface is used. The inner panel of 1/2" gypsum wall board over 6 mil polyethylene film is to have glass fiber insulation installed in the cavity. At 20 °F what minimum thickness of glass fiber insulation is possible without causing the inner surface of sheathing to be below the dew point? As the inside air is at a temperature of 75 °F at 50% relative humidity, the dew point temperature is 62 °F. The design factors are as follows:

	R_e Values (*English Units*)
Outside surface of siding — film factor	0.15
Wood siding	1.1
1" thick organic foam sheathing	4.5
1" thick glass fiber insulation	4.0
1/2" gypsum wall board	0.5
Inside surface — film factor	0.6

The temperature difference from the inside air temperature to the dew-point temperature is 75 °F − 62 °F = 13 °F, and the difference from dew-point temperature to ambient air temperature is 62 °F − 20 °F = 42 °F.

To determine thermal resistances as allowed by these temperature differences, the percent of total thermal resistance to maintain at the inner surface can be established by

Total temperature difference is 75 °F − 20 °F = 55 °F

$$\% \text{ outside to inner surface of sheathing} = \frac{42}{55} = 76\%$$

$$\% \text{ inner surface of sheathing to inside air} = \frac{13}{55} = 24\%$$

The thermal resistance from the outside air to the inner sheathing surface is

$$R_e = 0.15 + 1.1 + 4.5 = 5.75$$

Allowable thermal resistance from this surface inward would then be

$$R_e = \frac{24\%}{76\%} \times 5.75 = 1.8$$

As the gypsum board has thermal resistance of 0.5 and inside thermal resistance of 0.6, the allowable additional insulation is

$$R_e = 1.81 - (0.5 + 0.6) = 0.71$$

Thus the allowable maximum thickness of glass fiber insulation, would, at resistance of 4.0 per inch, be $\frac{0.71}{4.00} = 0.177$ inches.

This, of course, is completely impractical and indicates that, as constructed, any additional insulation in the cavity would cause condensation to occur there and thus on the surface of the organic foam sheathing itself.

Example 32-A—*Metric Units*

A stud wall with wood siding over 2.54 cm thick organic foam with foil facing on its inner surface is used. The inner panel of 1.27 cm gypsum wall board over 0.15 mm polyethlene film is to have glass fiber in the cavity. At −6.7 °C what is the maximum thickness of glass fiber possible without causing inner surface of sheathing to be below dew-point temperature. As the inside air is at a temperature of 23.9°C at 50% relative humidity, the dew-point temperature is 16 °C.

The design factors are:

	R_m Values (*Metric Units*)
Outside surface of siding — film factor	0.026
Wood siding	0.194
2.54 cm organic foam sheathing	0.792
2.54 cm glass fiber insulation	0.704
1.27 cm Gypsum wall board	0.088
Inside surface — film factor	0.106

The temperature difference from inside air temperature to dew-point temperature is 23.9°C − 16 °C = 7.9 °C, and the difference from dew-point to ambient air temperature is 16 °C − (−6.7 °C) = 22.7 °C.

To determine the thermal resistance to maintain as allowed by these temperature differences the

Percent of the total difference is 23.9 °C − (−6.7 °C) = 30.6 °C

$$\% \text{ outside to inner surface of sheathing} = \frac{22.7}{30.6} = 75\%$$

$$\% \text{ inner surface of sheathing to inside air} = \frac{7.9}{30.6} = 25\%$$

The thermal resistance from the outside air to inner sheathing surface is

$$R_m = 0.026 + 0.194 + 0.792 = 1.012$$

Allowable thermal resistance from this surface inward would be

$$R_m = \frac{25\%}{75\%} \times 1.012 = 0.337$$

As the gypsum board has thermal resistance of 0.088 and inside film thermal resistance is 0.106, the allowable additional insulation resistance is

$$R_m = 0.337 - (0.088 + 0.106) = 0.143$$

The allowable maximum thickness of glass fiber insulation would, at a resistance of 0.704 per 2.54 cm, be $\frac{0.143}{0.704} \times 2.54 =$ 0.515 cm. As stated in Example 32, this is impractical.

Where the cavity is to be filled with mass insulation, the sheathing and exterior surfacing must have a relatively high vapor permeability or be vented in some manner to provide vapor release. This is particularly true in climates where ambient air temperatures go below the dew-point for any length of time.

M. *Frame Wall with Organic Foam Sheathing and Cavity Insulation (Vented Cavity)*

In areas where the ambient temperature goes below 32 °F in January, moisture entrapped in the cavity insulation and sheathing surface may be released by providing a vent strip. This vent strip is installed on the top stud beam so that moisture can migrate to the vented section of the attic. This is shown in Figure 136.

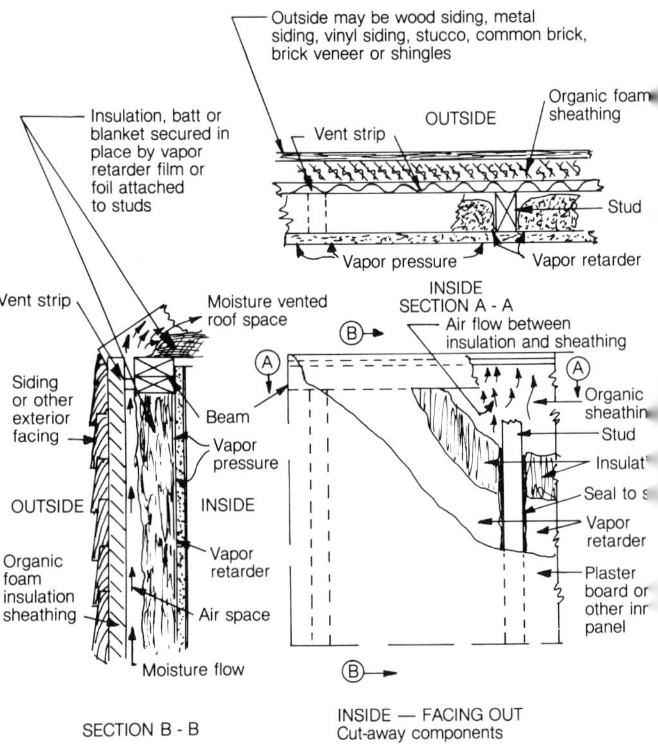

Note: Vapor retarder on inner surface must be completely vapor sealed at studs or joints. Also at top and bottom of stud space.

FRAME WALL WITH ORGANIC FOAM SHEATHING AND BATT, BLANKET OR RIGID INSULATION IN STUD SPACE

Figure 136

Although there is no opening at the bottom, the circulating a[ir] currents in a space will cause heat that escaped through the cavi[ty] insulation to continue up to the attic space. For this reason, th[e] thermal resistance of the organic sheathing and outer exterior fac[-] ing is much less than published values. When resistance of insula[-] tion batt or blanket and inner wall is approximately $R_e = 10$ (English Units), $R_m = 1.76$ (Metric Units) or over, and air temperature [is] below freezing, then effective thermal resistance of the organic foam sheathing is approximately 25% of the tested laborato[ry] results.

Note that the vent strip also provides a means for rapid fire spread to the roof if any fires occur in the insulation or sheathing or in the wall cavity. For this reason, no electrical circuits should be placed in walls of this construction. Also, care should be taken that dry leaves, paper or dry bushes are not close to the face of the exterior wall.

WALLS — BELOW GRADE

A. *Masonry Walls Below Grade — With Insulation on the Inside*

Masonry walls below grade may be brick, block or concrete. The major problem involved is to keep outside water from rain, snow or other sources from penetrating into the wall structure. For this reason, the exterior water sealing of these walls and the providing of base water drainage are of upmost importance. Where the water level is near the surface of the ground, it may be necessary to have a sump pump to keep the drain tile from filling with water. This is necessary to prevent build up of water head against the wall.

Drain tile must be installed below the level of the basement floor. This is necessary so that a water head does not get under the floor and cause a lifting pressure. Several feet of water head under the basement floor will exert over 60 lb/ft^2 (2.5 kg per sq meter). When a concrete floor is waterproofed and insulated properly, water drainage is essential.

The outside surface of the masonry wall, which is below grade, should be waterproofed with bitumen mastics and felts in a manner similar to that of built up flat roofing. This waterproofing should be carried up on the wall just a sufficient distance so that the top of water seal is just above potential water level during the heaviest rain. This waterproofing is also a very effective vapor retarder and should only be used where the soil keeps the temperature at a moderate level and not above grade where ambient air temperatures can be quite low.

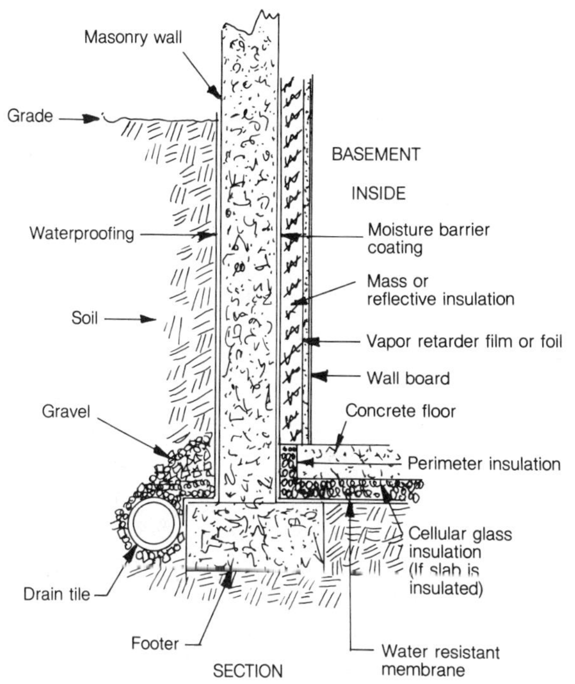

MASONRY WALL BELOW GRADE — WITH INSULATION ON THE INSIDE

Figure 137

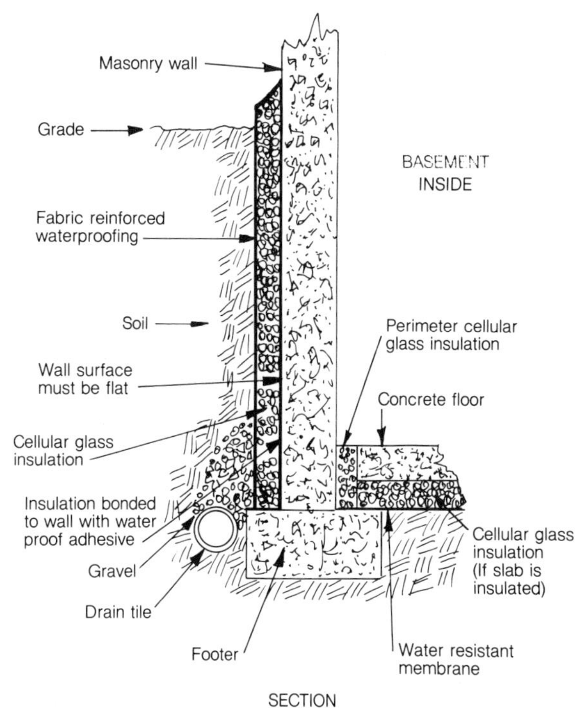

MASONRY WALL, BELOW GRADE — WITH INSULATION ON THE OUTSIDE

Figure 138

Figure 137 shows a typical masonry wall below grade.

The insulation on the wall is shown to be in a furred space between the masonry wall and the wall board. The insulation may be mass insulation installed as shown in Figure 123 or reflective insulation as shown in Figure 124. The mass insulation may be batts, blankets, rigid board or cellular glass. Where cellular glass insulation is used on the wall and its butt ends are sealed together, a separate vapor retarder of film or foil on inside is not necessary. In all cases, cellular insulation should be installed as perimeter insulation between wall and concrete floor. This perimeter insulation must be water sealed to the wall and to the concrete floor. As forced air heat is ineffective in heating concrete on grade floors, the placement of a hot-water pipe system of heating the slab is frequently used. This will raise the inside surface temperature to a comfortable level.

B. *Masonry Wall, Below Grade, with Insulation on its Outside Surface*

The only insulation which is both sufficiently strong and water resistant to insulate walls which are below grade, is cellular glass.

The cellular glass insulation is much lighter than water, thus it will tend to float. For this reason, care must be taken that the exterior of the masonry wall is smooth and flat so that the insulation block can be tightly bonded to the wall surface with a strong, waterproof adhesive. All butt joints of the block insulation must be sealed to be water tight. Properly installed and sealed, the cellular glass insulation is so vapor resistant that no inside vapor retarder is required.

The outer water barrier is installed on the outer surface of the insulation. Although the cellular glass insulation has sufficient strength to resist the compression of the back fill, the water barrier must be a strong fabric, reinforced bitumen sealer. A strong water seal is needed to prevent puncturing of the waterproofing by rocks or other sharp objects. The major advantage of installing the cellular glass on the outside of the masonry wall is that by being waterproof it assures that the masonry wall stays very dry. Installation is shown in Figure 138.

FLOORS — ABOVE GRADE

Frame doors over unheated spaces can be insulated with mass or reflective insulation.

Where mass insulation is used, the vapor retarder film or foil must be installed above the insulation. When possible, it is easier to install this vapor retarder before the flooring is installed. The vapor retarder can be installed with overlapping edges on joists so that when flooring is installed, a tight seal is accomplished. Where this is not possible, the edges of the vapor retarder film or foil should be carefully sealed with an adhesive or taped to the joists.

Blanket or batt insulation may be supported by a membrane attached to the vapor barrier or one-inch hexagonal galvanized wire netting attached to the joists. Rigid insulation may be secured in position by cleats nailed to the joists.

Wherever there is electrical wire or cables in these spaces, *do not pack insulation around them*, as it may cause overheating.

Reflective insulation can also be installed in the wood joist spaces. It may be installed either flush or recessed in the joist space.

If the installation is in a space over exposed soil or ground, it is suggested that the exposed soil be covered with a film of vapor retarder film such as 2 mil polyethylene, over which 1/2″ to 1″ (1.77 to 2.54 mm) of dry sand is placed. This area should also be vapor vented to the outside so as to prevent build up of water vapor.

Mass insulation is shown in Figure 139 "A" and reflective insulation in Figure 139 "B".

Mass type glass fiber insulation being installed between joists of floor, above an unheated area, is shown in Figure 140.

A. *Masonry Floors on Steel Beams Over Unheated Space — Insulation Installed Below Slab*

Masonry floors supported by beams may be insulated between the beams with rigid thermal insulation block, board or sheet. If the insulation is vapor permeable, then a vapor retarder film, foil or coating must be installed over the top of the insulation. In Figure

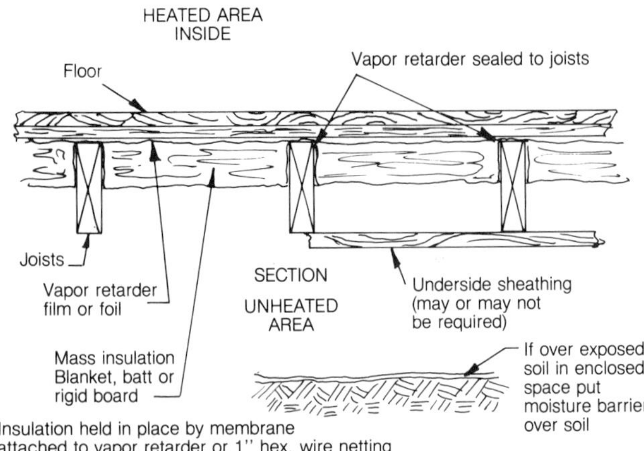

FRAME FLOOR INSULATED WITH MASS INSULATION
"A"

Floor
HEATED AREA
INSIDE
Floor
Joists
Reflective floor insulation
Flush Application
Recessed Application
SECTION
UNHEATED AREA

FRAME FLOOR INSULATED WITH REFLECTIVE INSULATION
INSULATION FOR FRAME FLOORS OVER UNHEATED SPACE
"B"

Figure 139

INSTALLING GLASS FIBER INSULATION BELOW FLOOR OVE[R] UNHEATED AREA.

Figure 140

Courtesy
Owens-Corning Fiberglas Cor[p.]
Toledo, O[hio]

141 "A", the vapor retarder is shown as an adhesive sealant which vapor seals the concrete and also acts as a bond to secure the insulation to the ceiling. The rigid insulation used in this type of application must have sufficient tensile strength from top to bottom so that it does not separate when supported from the top. The insulation must be shaped to fit in and around the structural steel beams.

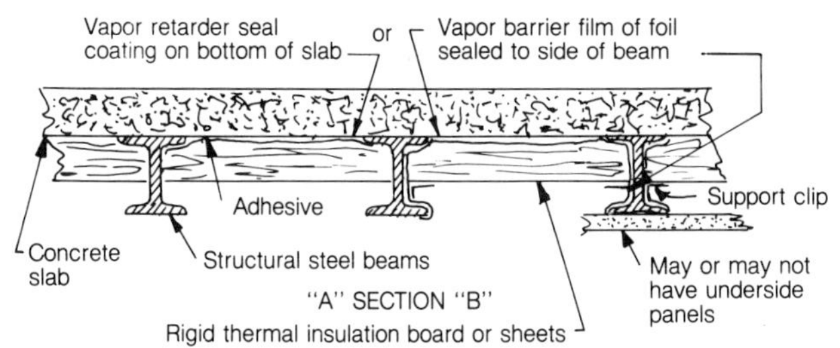

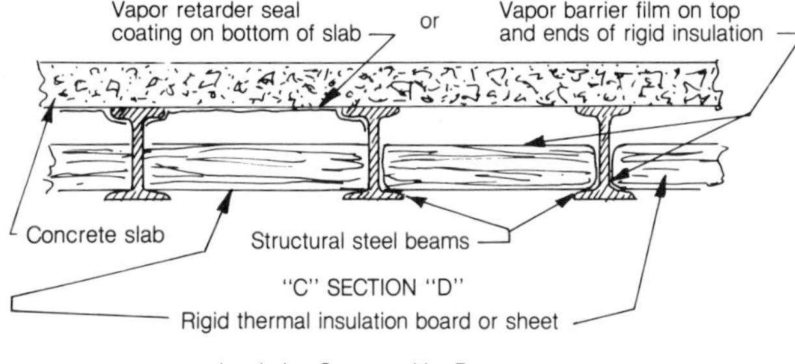

INSULATION FOR CONCRETE FLOORS OVER UNHEATED SPACE

Figure 141

Figure 141 "B" shows a rigid thermal insulation which has a vapor retarder film or foil as part of its construction. If the insulation has sufficient transverse tensile strength and the vapor retarder is securely bonded to the surface, it may be installed by adhesives, as shown in "A". Otherwise, it will have to be supported by clips, cleats, etc., as shown in Figure 141 "B".

The flanges of the structural steel beam can be used to support the insulation. Where the insulation does not have a vapor retarder film or foil on its higher temperature surface, then the bottom of the concrete floor must be sealed with a vapor-resistant sealer. This is shown in Figure 141 "C".

When the insulation does have a vapor retarder film or foil on higher temperature surfaces, it should be installed as shown in Figure 122 'D".

The one exception in the need for vapor retarder between the insulation and concrete slab is cellular glass, which has sufficient vapor resistance as a basic product.

B. *Masonry Floors on Concrete Beams and Block Over Unheated Space, Insulation Installed on Underside*

Rigid insulation boards and sheets can be installed by pins and speed clips as shown in Figure 142 "A". The clips will depress the surface of the insulation. This depression can be filled with insulating cement to level the surface.

Where cellular glass insulation block is used under the flat masonry floor assembly, it can be attached to the surface with fast-drying sealer adhesive. This adhesive should also be put in the butt joints. This is shown in Figure 142 "B".

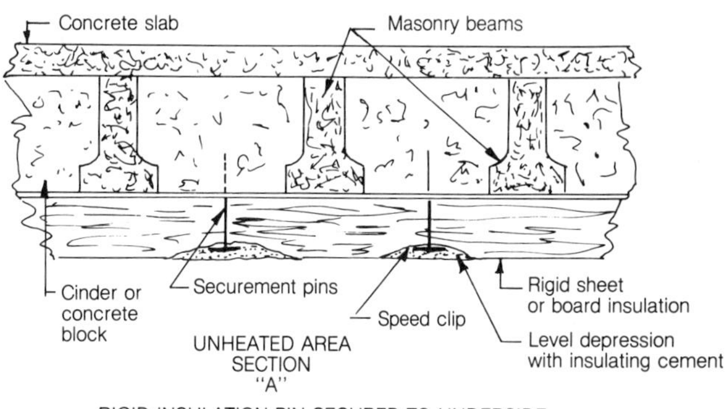

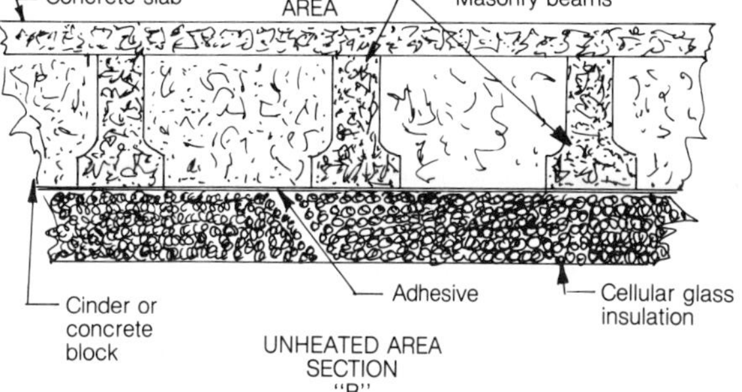

INSULATION FOR MASONRY FLOOR OVER UNHEATED SPACE

Figure 142

C. *Masonry Floors on Grade — Perimeter Insulation*

In Figure 143 "A" the slab is insulated by water-resistant organic foam or inorganic glass foam extending under the slab a minimum of 18″ (0.45 m). Where organic foams, such as foamed styrene, or polyurethane insulations are used, the slab over the insulation, and a minimum of 18″ (0.45 m) beyond, should be reinforced with reinforcing rods in 4″ (10 mm) centers. This is necessary to retard cracking of the slab at the line of juncture of insulation with soil, because of the dimensional unstability of organic foams. Also, the organic foams have little compressive strength, 15 to 30 psi (103 to 207 kPa) at 10% deformation. If the insulation was 3″ thick (7.6 mm), this compression would be 0.3″ or 0.7 mm, sufficient to cause the overhang section of the concrete to crack.

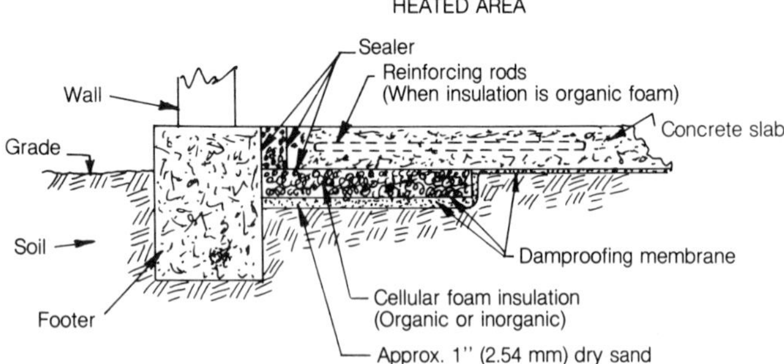

Note: When organic foam insulation is used as perimeter insulation under slab the edges must be reinforced by rods extending a minimum of 18" (0.45 m) from inner edge of the insulation

As shown the concrete slab is constructed on grade. Same construction is suitable below grade with proper drain tile installed.

ON GRADE (AND BELOW) PERIMETER INSULATION OF SLAB
"A"

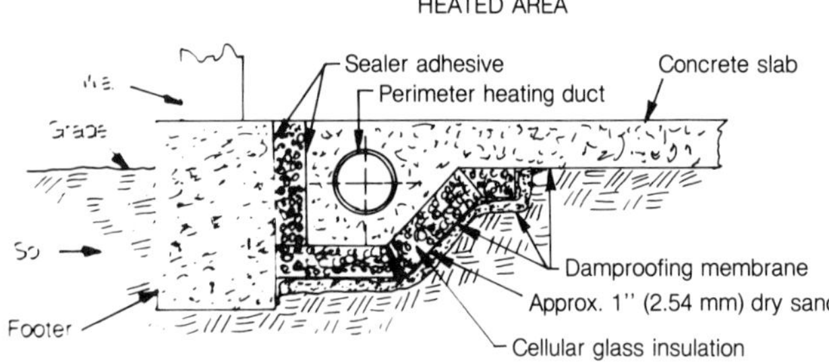

PERIMETER INSULATION OF SLAB WITH HEATING DUCT
"B"

PERIMETER INSULATION FOR SLABS

Figure 143

In all cases, the insulation and the slab should be installed over a dampproofing membrane, such as 2 mil thick polyethylene sheets. Where insulation is installed, a leveling layer of 1" (2.54 cm) of dry sand should be installed along with another layer of dampproofing membrane before the insulation is put in place. After the insulation is installed, it should be coated with a heavy coat of protective sealant to reduce the possibility of puncture when the concrete is poured in place.

Where a perimeter heating duct is installed in the concrete floor, slab cellular glass insulation should be installed as shown in Figure 124 "B". Cellular glass should be used because of the much greater weight of this construction at this location. The cellular glass has a compressive strength of 75 psi to 100 psi (517 to 690 kPa) at almost no dimensional compression.

D. *Thermal Insulation Under Concrete Floors*

Because of the need for dimensional stability and high compression strength, cellular glass insulation is used under concrete floors. Figure 144 'A" shows such an installation where the floor is located below grade and Figure 144 "B" shows the floor located on grade.

Over the leveled soil a 1" thick layer of dry sand should be placed and this covered with a dampproofing membrane of 2 mil polyethylene film or water-resistant asphalt and felts. This serves as a flat, protective surface on which to install the insulation, block or sheet. All butt ends should be bonded and sealed with suitable adhesive-sealer. Over the insulation a membrane of asphalt and felt is installed. This protects the insulation from damage as the concrete floor is poured into place.

Perimeter insulation should also be placed so that the concrete slab is insulated from the wall. Insulation on the interior of walls may be mass insulation or reflective insulation in furred space.

REFRIGERATED SPACES IN BUILDING

If a space in a building is refrigerated to a temperature below that of ambient air, then vapor migration will be inward instead of outward. When a building is simply air conditioned for human comfort, this vapor migration is seldom a problem, as in most cases the wall and roof (or ceiling) structures are still above the dew point of the atmospheric air. Also, the cooling coil is very effective in condensing moisture from the air supplied to the space. For this reason, the vapor barrier is located in the structural components of homes and buildings as determined by winter conditions, and assuming the interior of building will be heated. All the roof, ceiling, wall and floor drawings in this chapter up to this point are based on these conditions. Figure 84, Chapter 8, illustrates the location of vapor in heated residences and buildings.

When a building, or part of a building, is refrigerated for cold storage or used for other low temperature purposes, then the vapor retarder film or foil must be located on the outer side of the thermal insulation to prevent condensation and freezing in the wall or ceiling structure. A schematic diagram of these conditions is shown in Figure 85, Chapter 8. The inner surface should *not* be vapor sealed so that any moisture which passes through the

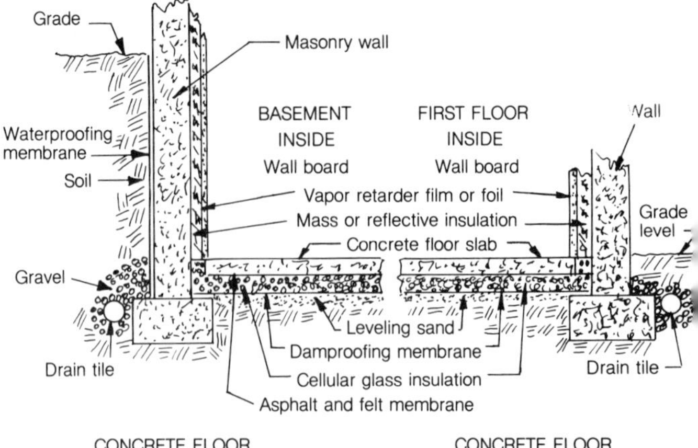

CONCRETE FLOOR
BELOW GRADE LEVEL
SECTION "A"

CONCRETE FLOOR
AT GRADE LEVEL
SECTION "B"

Note: Over the insulation a layer of 43 lb. (19.5 kg) felt should be applied over the asphalt coating. All seams should be sealed with flashing cement.

THERMAL INSULATION UNDER CONCRETE FLOOR BELOW OR ON GRADE

Figure 144

retarder can pass on into the space and can be condensed by the refrigeration coil.

When the space is refrigerated, the location of the vapor retarder as illustrated by the following figures, should be changed from inner surface of insulation to surface on insulation facing to the outside. These Figures are: 101, 102, 104, 105, 106, 118, 119, 120, 121, 125 and 126 in Chapter 9.

A special problem involving differential expansion occurs when insulation is to be installed under a concrete floor which contains refrigeration pipe or supports a frozen surface, such as an ice skating rink. Concrete has a coefficient of expansion of 8. × 10^{-6}, English Units (14.4 × 10^{-6}, Metric Units) and cellular glass, a coefficient of expansion of 4.6 × 10^{-6} British Units (8.3 × 10^{-6} Metric Units). The concrete is exposed to major temperature differences as compared to the cellular glass, causing the concrete to undergo a much greater dimensional change. To take care of this difference in dimensions between the concrete and the cellular glass insulation, slip sheets must be provided between them to allow movement between the two. Figure 145 "A" shows an insulated concrete floor installed over a concrete pad. Figure 145 "B" shows an insulated concrete floor installed over an earth base.

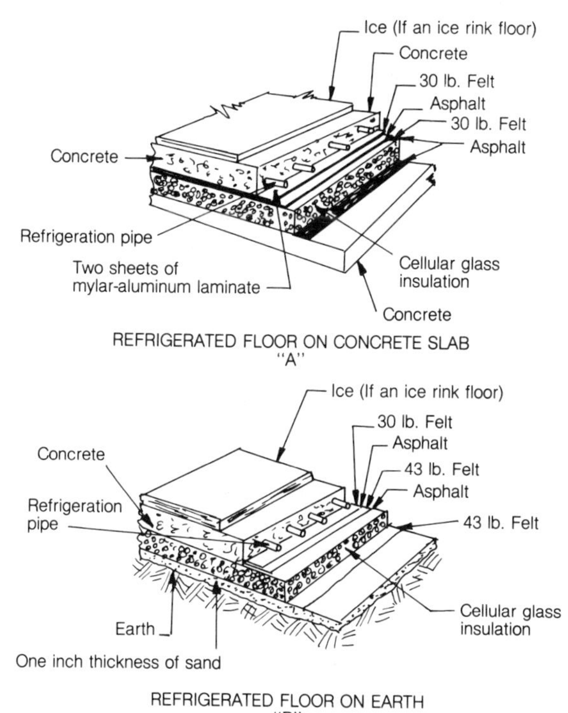

INSULATED REFRIGERATED CONCRETE SLAB FLOOR

Figure 145

It should be noticed that whereas cellular glass has a coefficient of expansion, most thermal insulations shrink when heated. In the case of refrigerated floors, the materials shrink when the temperature is reduced; however, in most instances, especially metal structures, the problem is that the metal expands as the insulation shrinks. Where long spans and areas are involved, the differential expansion between materials must be considered as a potential problem.

To assist in the evaluation of the differential expansion between various materials, Table 76 provides the coefficients of linear expansion.

TABLE OF AVERAGE* COEFFICIENT OF LINEAL EXPANSION

MATERIAL	Coefficient of Lineal Expansion**	
	English Units per °F	Metric Units per °C
Aluminum	0.000,012.3	0.000,022,1
Brass, cast	0.000,010,4	0.000,018,7
Brick	0.000,003,1	0.000,005,4
Bronze	0.000,010,4	0.000,018,7
Cellular, Glass Insulation	0.000,004,6	0.000,008,3
Concrete	0.000,008,0	0.000,014,0
Copper	0.000,009,3	0.000,016,7
Glass, plate	0.000,005,0	0.000,009,0
Iron, cast	0.000,005,9	0.000,010,6
Iron, soft forged	0.000,006,3	0.000,011,3
Limestone	0.000,001,4	0.000,002,5
Marble	0.000,006,5	0.000,011,7
Stainless Steel	0.000,008,1	0.000,014,5
Steel, hard	0.000,005,6	0.000,010,1
Steel, soft	0.000,006,3	0.000,011,3
Tin	0.000,012,7	0.000,022,8
Wood, Ash (parallel to fiber)	0.000,005,3	0.000,009,5
Wood, Oak (parallel to fiber)	0.000,002,7	0.000,004,8
Wood, Pine (parallel to fiber)	0.000,003,0	0.000,005,4
Zinc	0.000,016,5	0.000,029,7

*Based on average between temperatures 32 and 212 degrees F (0 and 100°C).

**Coefficient of contraction is identical when temperature is reduced.

TABLE 76

INSULATION OF DUCTS

This discussion is limited to application of insulation to air-conditions or heating ducts used in residences and buildings. Ducts may be rigid metal not insulated, insulated inside or insulated outside. They may also be made of rigid insulation materials and shapes may be rectangular or cylindrical. The choice of the duct system must be made by the air-conditioning engineer.

A. *Rigid Glass Fiber Insulation Duct*

The duct system can be made of rigid glass fiber insulation. The insulation can be grooved and formed into a multiple-sided duct or can be a rectangular duct. In this case, the insulation is supplied with an outer aluminum foil (or reinforced aluminum foil) face which becomes the exterior of the duct. All butt ends of joints must be protected by a continuous layer of foil or should be taped with strong pressure sensitive aluminum tape. End sections are

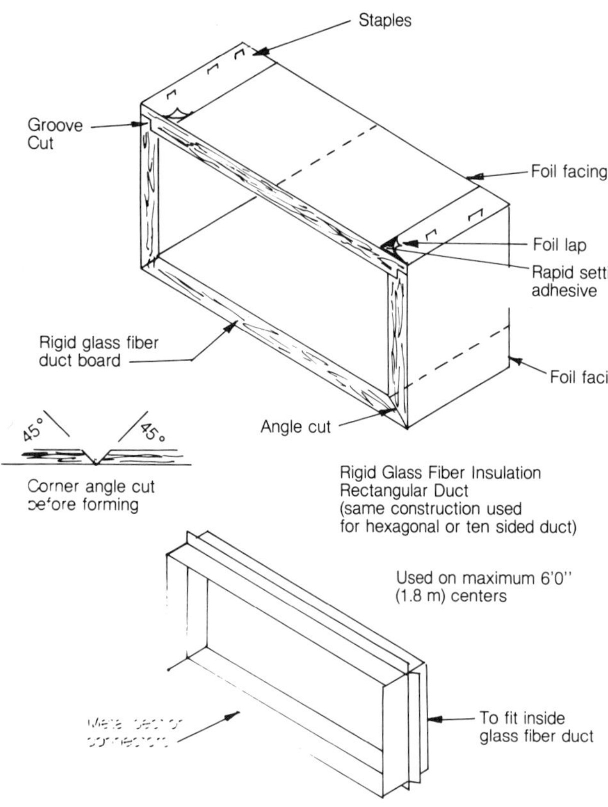

Note: Put hangers at ends of sections with metal connectors.

RIGID GLASS FIBER INSULATION DUCT

Figure 146

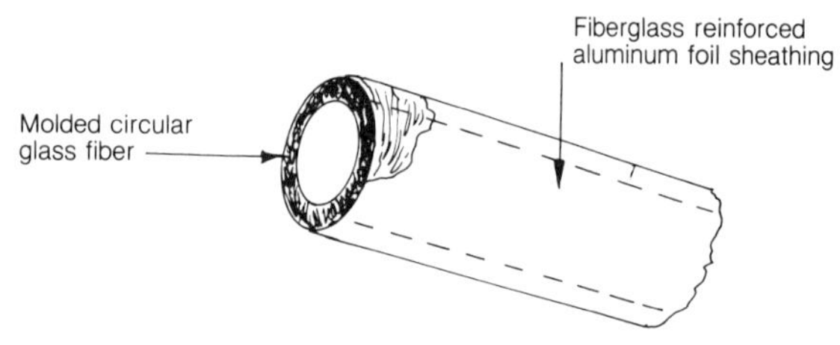

RIGID CYLINDRICAL GLASS FIBER INSULATION DUCT
"A"

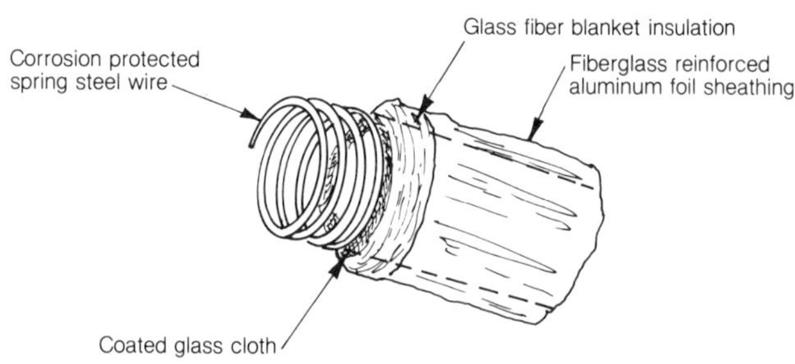

FLEXIBLE CYLINDRICAL GLASS FIBER INSULATION DUCT
"B"

CYLINDRICAL GLASS FIBER INSULATION DUCTS

Figure 147

joined by metal connections. This type of insulated duct is shown in Figure 146. Such ducts are limited to air temperatures below 212°F (100°C) and air velocities below 2000 feet per minute (610 meters per minute).

B. *Cylindrical Glass Fiber Insulation Ducts*

Glass fiber can be molded into cylindrical shapes so that the insulation itself is basically the air-conditioning or heating ducts. The outer surface is protected with a fiber glass reinforced aluminum sheathing, which also acts as a vapor retarder film. Such an insulation air-conditioning/heating duct is shown in Figure 147 "A".

By the use of spring steel wire covered with a coated glass cloth and fiber glass blanket, a flexible air duct may be constructed. This, too, must be protected with a fiber glass reinforced aluminum foil sheathing. The flexible cylindrical glass fiber duct system is shown in Figure 147 "B".

It should be noted that these ducts are limited to low and medium pressure systems and moderate air velocities. Also, both provide sound and noise reduction. This is advantageous in fan noise reduction.

These duct systems have relatively little time resistance to fire.

C. *Cylindrical Glass Fiber Insulated Metal Ducts*

Cylindrical ducts are manufactured with glass fiber insulation installed between the inner metal liner and the outer metal duct. Where very high air velocity is necessary or where air pressure is significantly higher then the insulation must be protected on its inner surface with sheet metal. Such a duct system is shown in Figure 148 "A".

When it is desired that the thermal insulation also act as a sound absorber material, the inner layer is perforated to obtain this noise-reduction property. Such an installation serves to reduce fan noises in the duct system and to reduce the potential of cross talk between rooms or offices. A cylindrical metal duct with such an application is shown in Figure 148 "B".

As the insulation is protected on both sides by steel sheet metal, it provides a good time delay before failure when exposed to fire.

D. *Flexible Fibrous Duct Insulation on Exterior of Duct*

Duct wrap flexible fiber insulation is used as an external insulation on heating and air-conditioning ducts. The duct wrap insulation is available in a number of facings, such as foil - skrim - kraft or heavy vinyl film vapor retarders. To be effective, the duct wrap should be cemented to the duct surface by a duct adhesive and the jacket laps sealed with rapid setting adhesive and/or pressure sensitive tape. This installation of insulation is shown in Figure 149. This type of installation is suitable for ducts operating at temperatures from 50°F (10°C) to 200°F (93°C).

E. *Vapor Retarder Over Rigid Fibrous Insulation on Exterior of Duct*

Rigid insulation can be installed on exterior ducts by the use of speed clips and adhesives. Where duct temperature gets below

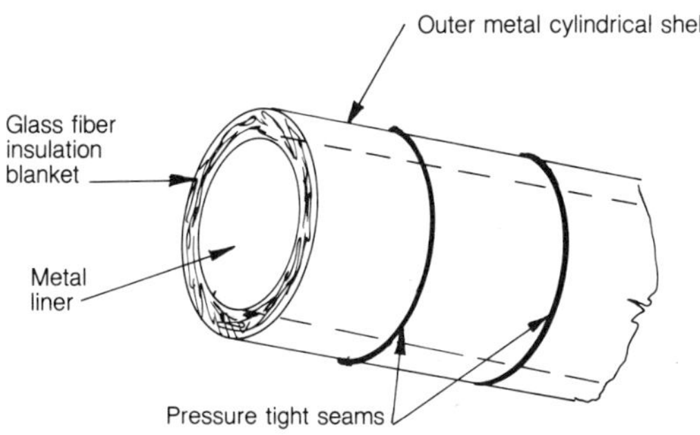

HIGH PRESSURE, HIGH VELOCITY INSULATED DUCT
"A"

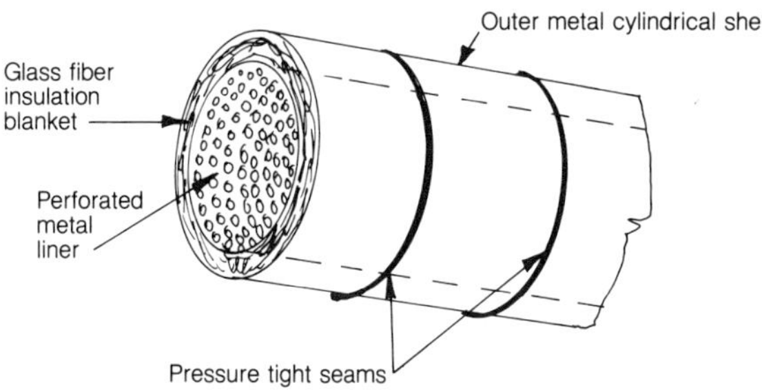

ACOUSTICAL — SOUND ABSORBING INSULATED DUCT
"B"

CYLINDRICAL GLASS FIBER INSULATED METAL DUCT

Figure 148

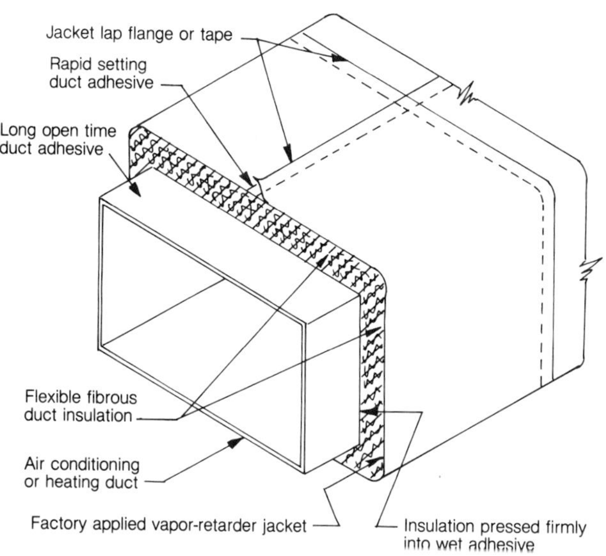

Note: Duct may be rectangular or cylindrical

METAL DUCT
APPLICATION OF ADHESIVE TO SECURE FLEXIBLE FIBROUS DUCT INSULATION TO EXTERIOR OF DUCT

Figure 149

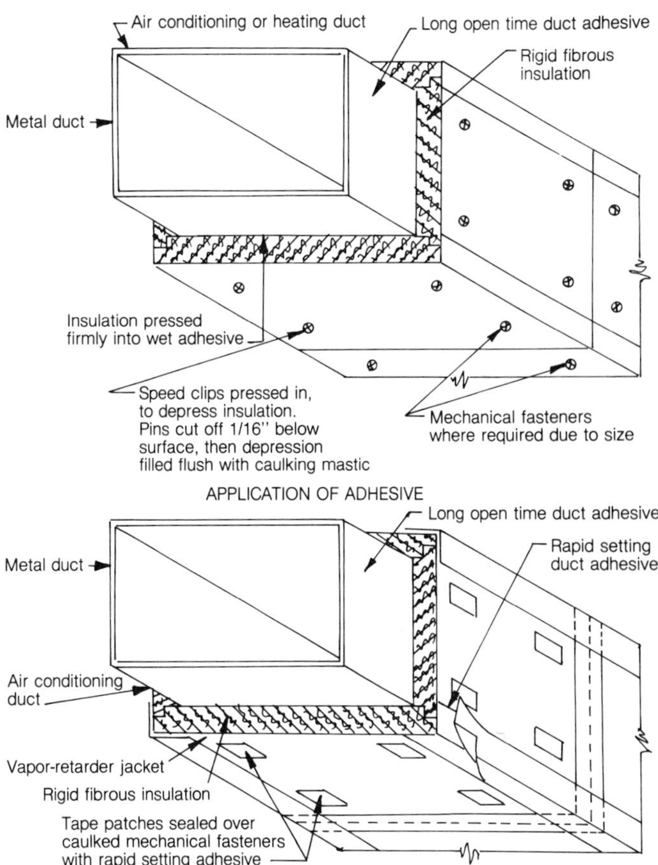

VAPOR-RETARDER JACKET OVER RIGID FIBROUS INSULATION ON EXTERIOR OF DUCT

Figure 150

room temperature, a vapor retarder jacket must be installed to reduce moisture entry into the insulation. The installation of this application is shown in Figure 150. This type of application is suitable for duct temperatures of 50°F (10°C) to 212°F (100°C).

F. *Installation of Cellular Glass on Exterior of Cold Duct*

Although the title states that this is an application for a cold duct, the temperature range for this type of installation is −40°F (−40°C) to 250°F (212°C). The emphasis here is on the cold temperature ranges because installations of rigid fiber insulations are not suitable for those cases where the duct operates most of the time at a temperature below the ambient air dew point. With cyclic temperatures, condensate/moisture can be vaporized and forced out of the insulation. For continuous low temperature operation, there is no practical way to seal and maintain a vapor retarder that is at all close to the vapor resistance of cellular glass. Cellular organic insulations will also become wet, as their water vapor transmissions are 0.4 to 5.0, as compared to cellular glass which is less than 0.001.

The outer surface may be protected with foil or scrim sheathing; however, a mastic emulsion coating of polyvinyl-acetate or acrylic with extension fabric reinforcement provides an excellent covering. This is illustrated in Figure 151.

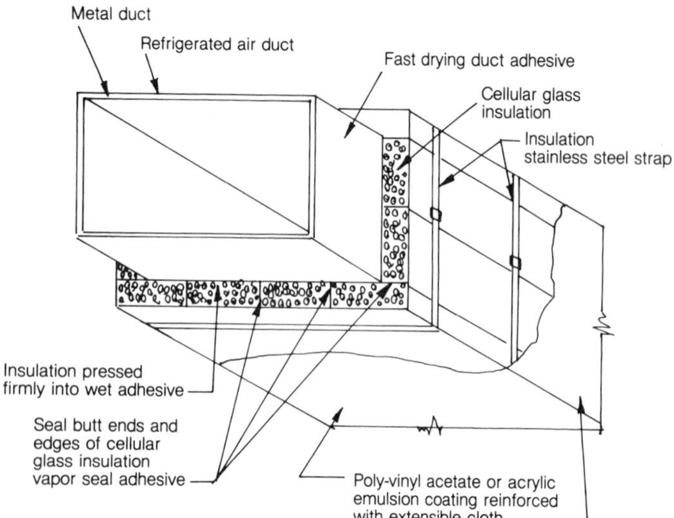

Note: Cellular glass may be mitered to fit curves and bends or cut to fit cylindrical duct.

LOW TEMPERATURE DUCT WHICH OPERATES AT TEMPERATURE BELOW DEW-POINT OF AMBIENT AIR IN THE SPACE AREA

INSTALLATION OF CELLULAR GLASS INSULATION ON EXTERIOR OF COLD DUCTS

Figure 151

G. *Installation of Rigid Fibrous Insulation to Interior of Air-Conditioning and Heating Duct*

In many instances where sound absorption is desired, the duct systems are insulated on the inside with rigid fibrous insulation. In Figure 152 "A" rigid fibrous insulation is shown installed inside a duct. This installation is limited to air velocity up to 2000 FPM (609 meters per minute). Above these velocities fibers may break off and get into the air stream.

For higher air velocities, the inside must be lined with duct liner, either supplied as part of the insulation or added by sticking a suitable adhesive to the insulation. Properly protected, this installation of rigid fiberous insulation may be used when air velocities in the duct are up to 6000 feet per minute (1829 meters per minute).

These installations are suitable for duct temperatures of 32 °F (0 °C) to 250 °F (121 °C).

INSULATION ON PIPES

The selection of insulation for pipes depends upon two major factors. The first is the operating temperature of the pipe and the second is the location of the pipe.

When operating temperature is always below the ambient air dew point, then the installation which has the lowest vapor migration permeability will give the longest thermally efficient service. For those cases that the operating temperature only occasionally dips below the dew point, then insulation with slight vapor migration permeability is tolerable. When operating temperature is always above the ambient air dew-point temperature, then it is permissible for the insulation to have a very high vapor permeability.

Besides the dew point of the ambient air, the location of pipes, such as indoors or outdoors, makes a difference in the selection of the outer jacket or covering. Also, if located where they can be touched by people, then those pipes must have a safe outside temperature (see Tables 73, 74 and 75, Chapter 8). The expected outside physical abuses that an insulated pipe might be subjected to can influence the selection of insulation—especially for those operating in the high-temperature range.

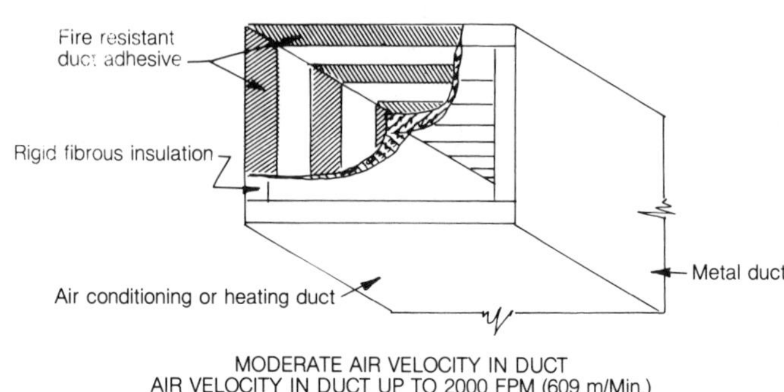

MODERATE AIR VELOCITY IN DUCT
AIR VELOCITY IN DUCT UP TO 2000 FPM (609 m/Min.)
"A"

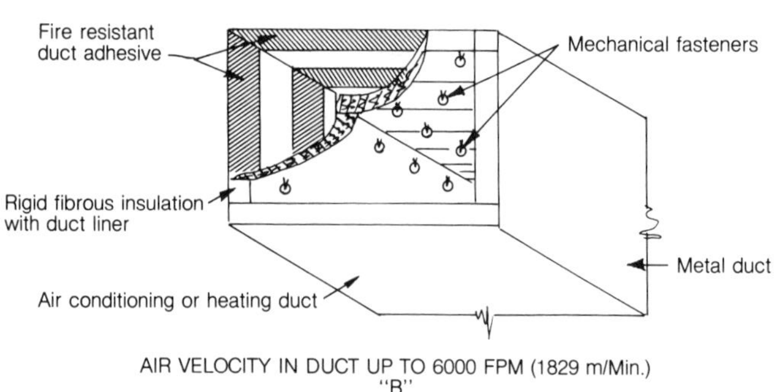

AIR VELOCITY IN DUCT UP TO 6000 FPM (1829 m/Min.)
"B"

INSTALLATION OF RIGID FIBROUS INSULATION TO INTERIOR OF AIR CONDITIONING AND HEATING DUCT

Figure 152

A. *Cellular Glass Insulation on Pipe*

On low-temperature piping, which is always below the dew point of ambient air, cellular glass will give the longest efficient service because of its very low vapor permeability. The metal pipe will shrink more than the cellular glass as the operating temperature goes down from ambient temperature. For this reason, insulation contraction joints should be provided. Installation of cellular glass on low-temperature pipe is shown in Figure 153.

Although the illustration indicates the installation for low temperatures down to −250 °F (−157 °C), if high temperature, anti-abrasive coating is used in the slip joints, the application is suitable up to 400 °F (204 °C). Where temperatures are very low or very high, the pipe covering should be installed in multiple layer, broken-joint construction.

The external surface may be protected with aluminum jacketing or extensible fabric-reinforced PVA or acrylic emulsion

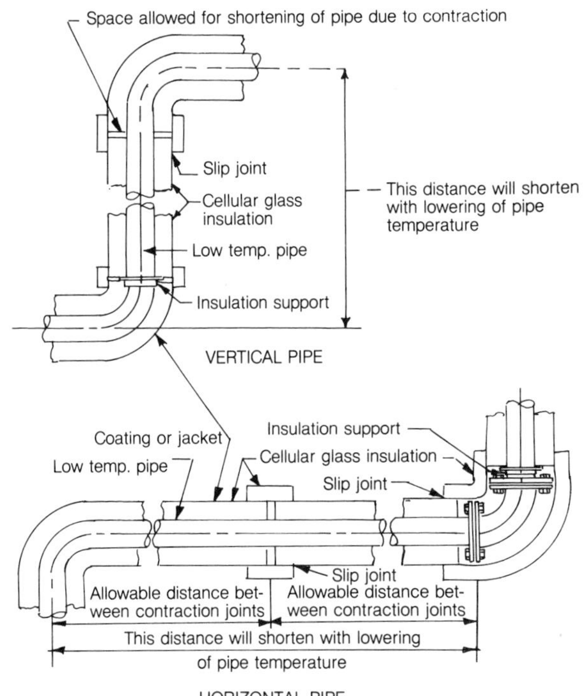

Note: Relative movement illustrates the need for contraction joints and anti-abrasive coatings on surfaces of large dimensional differentials.
For low temperature as shown, however same application is suitable for high temperature with high temperature anti-abrasive coating at slip joints.

CELLULAR GLASS INSULATION ON PIPES AND FITTINGS

Figure 153

mastic. On high-temperature installations, care must be taken that when metal jacket is used, it should have safe surface temperature.

B. *Organic Foam Insulation on Low and Medium Temperature Pipe and Fittings*

When the operating temperature of the pipe is less than ambient air temperature, a vapor retarder outer cover protecting the insulation must be used. This vapor retarder film, aluminum foil jacket or vapor retarder sealer is necessary to reduce the vapor migration into the insulation.

If the pipes operate most of the time above ambient air temperature, then no vapor retarder outer protection is required. However, the film and/or aluminum jackets may be used as an outer protection material. The extensible fabric reinforced emulsion PVA or acrylic mastics may also be used as a protective outer barrier, especially if there may be any chance of small flame or spark exposure. However, all organic foams are combustible.

These applications are suitable for operating temperatures of 50 °F (10 °C) to 190 °F (88 °C). This application is shown in Figure 154.

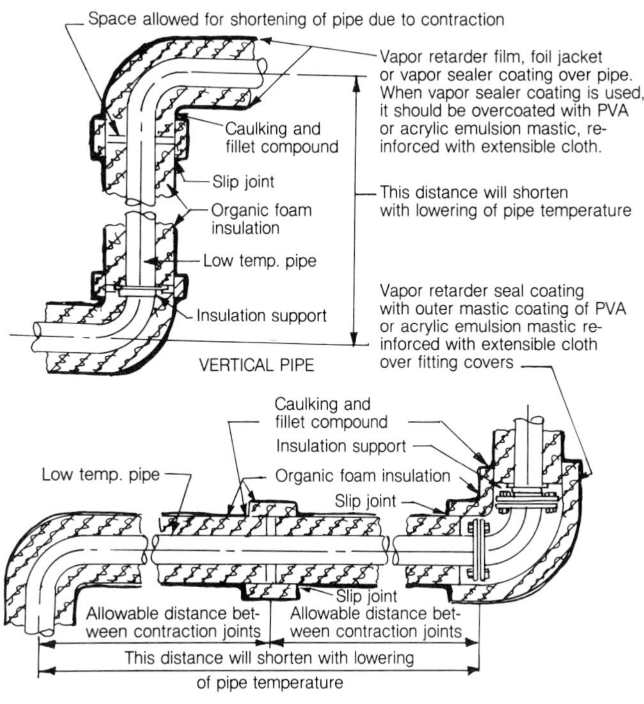

ORGANIC FOAM INSULATION ON LOW AND MODERATE TEMPERATURE PIPE AND FITTINGS

Figure 154

C. *Insulation on High-Temperature Pipes and Fittings*

The insulation on high-temperature pipes and fittings (150 °F or above) may be one of many types of insulation available for this service. In most instances the most practical is pre-formed insulation to fit the pipe or tubing. The compressive strength, tensile strength, weight mass and upper temperature limits must be suitable for expected service conditions. The upper temperature limits of the various pipe insulations are listed in Table 77.

Insulation Pipe Covering Generic Material	***Temperature °F***	***Limits °C***
Calcium Silicate	1200	649
Cellular Glass	580	350
Diatomaceous Silica	1600	871
Expanded Silica	1500	815
Glass Fiber with Organic Binder	450	232
Mineral (Rock) Fiber	850	454

TABLE 77

Where there is a potential fire hazard and pipes must be protected by the insulation, then high service temperature and high mass insulation should be used.

High temperature pipes and fittings should be installed with insulation-expansion joints. This is shown in Figure 155.

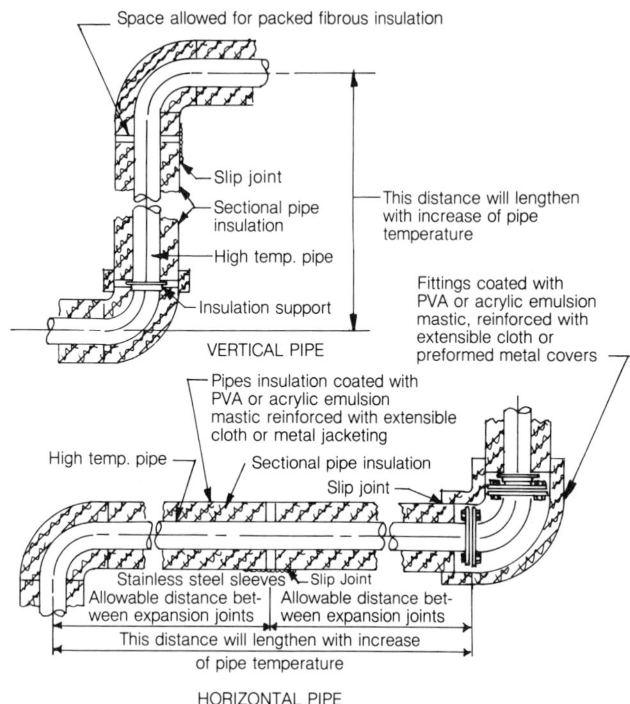

INSULATION ON HIGH TEMPERATURE PIPES AND FITTINGS

Figure 155

Prefabricated or pre-formed fitting covers are recommended. The dimensions for these are given in "Prefabrication and Field Fabrication of Thermal Insulation Fitting Covers", ASTM Recommended Standard C-450.

The insulation must be protected externally with jacketing or mastic.

10 Heat Transfer Equations for Building Components

One of the major problems related to correctly designing efficient residences and buildings is the difficulty of understanding all types of energy and its potential states. For this reason, it is necessary to define the following energy and related terms.

Energy is the result of a variety of forces and may occur in either *stored* or *transient* form.

> STORED ENERGY is commonly classified into two major forms of potential energy.
>
> MECHANICAL POTENTIAL ENERGY is energy possessed by a body by virtue of its vertical distance above a horizontal plane.
>
> INTERNAL POTENTIAL ENERGY is the energy within a body such as a gas, liquid or solid. Energy stored within any material and which can be released by chemical reaction comes under this classification.
>
> STORED ENERGY can be transformed into transient energy which has two classifications:
>
> WORK is energy in transient form and can be defined as a force acting through a distance.
>
> HEAT is also energy in transient form. It is energy in transition or transfer from one body to another by virtue of a temperature difference existing between the bodies and will flow from the body at the higher temperature to the body at the lower temperature. Conversion of mechanical energy to heat energy can be by friction or by chemical reaction. Thus, heat is the result of random molecular motion.
>
> FRICTION converts mechanical energy into heat energy thereby increasing the temperature of the body upon which the mechanical energy is acting.
>
> CHEMICAL REACTION, or combustion of a substance converts internal chemical potential energy into heat energy thus raising the temperature of the gas, liquid, or solid involved.

(Note: These definitions, as presented, would not be acceptable to a physicist, as they are too general and not sufficiently detailed or restrictive. They are given to establish the relationship of terms used to present the fundamentals of heat transfer.)

Not only can energy be converted from one form to another, but it can also be transmitted from one body to another. For example, electrical energy is conducted from one point to another by wire, or heat energy may be transmitted from one body to another by conduction through a material or by radiation through space.

The basic law of the flow energy is defined as follows:

> A steady flow of energy through any medium of transmission is directly proportional to the force causing the flow and inversely proportional to the resistance to that force.

In simple terms:

$$\text{Energy flow } \alpha \ \frac{\text{Force}}{\text{Resistance}}$$

This simple relationship is the basis upon which most insulation calculations are built. From this basic equation it is now possible to expand or restrict each factor pertinent to this branch of engineering.

For example, *heat* is a form of energy flow and *temperature difference* provides the driving force for this flow. Therefore,

$$\text{Heat flow } \alpha \ \frac{\text{Temperature difference}}{\text{Resistance to heat flow}}$$

In order for heat to flow it must have a means of transfer or transmission of which there are two. One heat flow mechanism is CONDUCTION and the other is RADIATION. They can be defined as follows:

> CONDUCTION is the transfer of energy within a body or between two bodies in physical contact. The transfer is from a higher temperature region to a lower temperature region by tangible contact.
>
> RADIATION is the transfer of energy from a higher temperature body, through space, to another lower temperature body or bodies some distance away. True radiation is the transfer of heat between these bodies which does not raise the temperature of the medium through which the heat passes.

A third natural phenomenon is often considered as a means of heat transfer. This is a movement of mass called CONVECTION.

> CONVECTION is the movement of a mass with its associated energy, from one location to another.

Liquids or gases in contact with a body of a higher temperature have energy transmitted to them by conduction and radiation. This energy increases the temperature of the liquid or gas, which in turn changes density. This change in density causes a movement of the gas or liquid. As the liquids or gases of this new density move away from the higher temperature body, lower temperature masses of the gas or liquid move in. Thus, the process of heating the close molecules by conduction and radiation and their subsequent movement away from the higher temperature body continues. This movement is called NATURAL CONVECTION. The rate of this movement may be increased by some outside influence such as wind or by a fan. This is called FORCED CONVECTION.

The process of convection is so closely related to heat transmission that it has been accepted as a means of heat transfer. Even if not strictly correct, this definition is convenient. For this reason it must be understood that from this point on reference will be made to "heat transfer by convection" as if it were a separate mechanism.

The concept of TEMPERATURE is one with which we are all familiar. Indeed, it is one of the most important concepts in thermodynamics.

TEMPERATURE is an indicated level by which temperature differences may be measured. It is the thermal state of a body in reference to its ability to impart heat to other bodies. Temperature is a measure of the thermal intensity level of a system.

When energy is used, or transmitted, at a certain rate, then the element of time becomes a factor. POWER is the time rate of doing work, or the rate of expending energy. Therefore power has dimensions of energy (or work) per unit time.

Units of Measurement

HEAT

Any form of energy may be expressed quantitatively in any unit of energy. These units are derived from the standards of mass, length, and time. The fundamental energy unit in the English system is the foot-pound. The fundamental unit of heat energy in the English system is the British thermal unit (Btu).

The BRITISH THERMAL UNIT (Btu) was originally defined as the quantity of energy (heat) required to raise the temperature of 1 lb of water 1 °F, from and at 32 °F. At a later time the Btu was redefined to be 1/180 of the quantity of energy (heat) required to change 1 lb of water from the ice point to the steam point at standard atmospheric pressure. The difficulty of setting up standards for heat measurements (taking into account the now known variations of specific heat with both pressure and temperature) was so great that no single standard was ever generally accepted. To relate the measurement of heat with other energies and to arrive at an acceptable standard Btu, the International Steam Table Conference in 1926 recommended that the International Calorie be set at 1/860 of an International Watt-hour. By calculation, this established the Btu as 778.26 ft-lb.

> A JOULE is the work done when a force of one Newton acts through a distance of one meter in the direction of the force.
>
> A WATT is the power which gives rise to the transformation of energy at the rate of one joule per second.
>
> A CALORIE (International Table) = 4.1878 Joule (J).

Conversion

Btu (British thermal unit) = 1055.06 J (Joule)
J (Joule) = 9.481×10^{-4} Btu

The purpose of a temperature scale is to assign a number to every level of thermal state. In the English system two scales are used. The most common scale is the FAHRENHEIT (F) scale and next the RANKINE (R) scale. In the Metric system two scales are also used. These are the CELSIUS (C) — in the past known as centigrade — and the KELVIN (K) scales.

Three important reference points are identified on all temperature scales. These are: (1) the theoretically lowest thermal state — sometimes referred to as absolute zero, (2) the ice point of water, and (3) the steam point of water.

The FAHRENHEIT (F) scale subdivides the temperature interval between the ice point and the steam point into 180 parts. The ice point is assigned the value of 32 degrees F so the steam point has a temperature of 212 degrees F.

The RANKINE (R) scale subdivides the temperature interval between the ice point and the steam point into 180 parts. The theoretically lowest thermal state was assigned a value of 0 degrees R so the ice point of water then has a value of 491.7 degrees R and the steam point of water has a value of 671.7 degrees R.

The CELSIUS (C) scale subdivides the temperature interval between the ice point of water and the steam point of water into 100 parts and assigns the value of 0 degrees C to the freezing point of water, and 100 degrees C to the steam point of water. Absolute zero then has a value of −273.16 degrees C.

The KELVIN (K) scale subdivides the temperature interval between the ice point of water and the steam point of water into 100 parts and assigns the value of lowest thermal state as 0 degrees K. The ice point then has a value of 273.2 degrees K and the steam point has a value of 373.2 degrees K.

THERMAL RESISTANCE

Thermal resistance is that property of a material that opposes the passage of heat energy through the material. Heat flow occurs as a consequence of a temperature difference between two bodies or between one region of a body and another region of the same body. If a homogeneous body is at a higher temperature on one side than it is on the other, then the amount of heat energy passing from the higher to the lower side is determined by the resistance of the body to heat flow. The thermal resistance of a homogeneous body of uniform cross section varies in direct proportion to its thickness.

Units for thermal resistance have neither been named nor standardized. However, thermal resistance is the reciprocal of conductance.

The conductance of a homogeneous material is the amount of heat transmitted through a unit area of the material per unit time, through its *total* thickness, and with a unit of temperature difference between the surfaces of the two opposite sides.

Thus, for a given thickness of a homogeneous body, the resistance is

$$\text{Thermal resistance} = \frac{\text{Thickness of homogeneous body}}{\text{Conductivity of that body}}$$

Heat Transfer Equation for a Single, Flat Homogeneous Body (Insulation)

From the basic relationship between energy, force, and resistance, the first equation of heat transfer was given as

$$\text{Heat flow} = \frac{\text{Temperature difference}}{\text{Thermal resistance}},$$

where,

$$\text{Thermal resistance} = \frac{\text{Thickness of homogeneous body}}{\text{Conductivity of that body}}.$$

Thus,

$$\text{Heat flow} = \frac{\text{Temperature difference}}{\left(\dfrac{\text{Thickness of homogeneous body}}{\text{Conductivity of that body}}\right)}.$$

In all applications covered in this book, the homogeneous body will be thermal insulation. Therefore,

$$\text{Heat flow} = \frac{\text{Temperature difference}}{\left(\dfrac{\text{Thickness of insulation}}{\text{Conductivity of insulation}}\right)}.$$

The units of measurement used have absolutely no effect on the fundamental equation of heat transfer. However, all equations must be solved using measurement units of a single system. The units of the English and Metric systems must not be intermingled.

In both systems there are many different units of measurement for length, area, volume, temperature, energy, power, etc. With these numerous units, conversion from the English system to the Metric system (or vice versa) becomes complex. To assist in handling this conversion, the most commonly used units in heat transfer problems are provided, and conversion factors between the English and Metric systems are given.

For simplification, the English units given in this book will have no special identification. For example temperature and lengths will be written as t, t_1, t_2 and ℓ, ℓ_1, ℓ_2 respectively. The Metric units will be given as t_m, t_{m1}, and t_{m2} for temperature and ℓ_m, ℓ_{m1}, ℓ_{m2} for length. Note that the Metric symbols have a small "m" subscript.

The relationship between various English units and Metric units is given in Table 78. This conversion table is limited to units of heat, temperatures, thermal conductivities, thermal resistances and air volocities which are commonly used in heat transfer equations.

HEAT TRANSFER THROUGH A SINGLE FLAT MATERIAL

The heat flow through a material is due to the temperature of one side being higher than that of the other side. The energy which passes from the hotter side to the cooler side is heat. The quantity of heat is determined by the thermal resistance of the material.

TEMPERATURE AND HEAT UNITS, ENGLISH AND METRIC CONVERSIONS

English Units	− *conversion*	= *Metric Units*	− *conversion*	= *English Units*
t °F	(t °F − 32) ÷ 1.8	= t_m °C	(t_m °C × 1.8) + 32	= t °F
t °F	(t °F + 459.67) ÷ 1.8	= t_m °K	(t_m °K × 1.8) − 459.67	= t °F
t °R	t °R ÷ 1.8	= t_m °K	× 1.8	= t °R
t °R	(t °R ÷ 1.8) − 273.16	= t_m °C	(t °C × 1.8) + 491.67	= t °R
Δt °F	t °F ÷ 1.8	= Δt_m °C or Δt_m °K	× 1.8	= Δt °F or Δt °R
ℓ" (inches)	× 0.0254	= ℓ_m in metres	× 39.37	= ℓ" (inches)
ℓ' (feet)	× 0.3048	= ℓ_m in metres	× 3.280839	= ℓ' (feet)
ft^2 (sq ft)	× 0.092903	= m^2 (sq metres)	× 10.76391	= ft^2 (sq ft)
Btu	× 1055.056	= J (joule)	× 0.0009478	= Btu
Btu/hr	× 0.2930711	= W (watt)	× 3.412141	= Btu/hr
Q (Btu/ft^2, hr)	× 3.152591	= Q_m (W/m^2)	× 0.317199	= Q (Btu/ft^2, hr)
U (Btu/ℓft, hr)	× 0.9609097	= $U_{m\ell m}$ (W/ln m)	× 1.04068	= $U_{\ell f}$ (Btu/ℓft, hr)
f (Btu/ft^2, hr, °F)	× 5.67826	= f_m (W/m^2 K)	× 0.17611	= f (Btu/ft^2, hr, °F)
k (Btu in./ft, hr, °F)	× 0.1442	= k_m (W/mK)	× 6.933	= k (Btu in./ft^2, hr, °F)
R (°F, ft, hr, Btu)	× 0.1761	= R_m (Km^2W)	× 5.6782	= R (°F, ft^2, hr, Btu)
V (ft/min)	× 0.00508	= V_m (m/s)	× 196.85039	= V (ft/min)

(*Note*: All energy conversions are based on International Table Units.)

TABLE 78

Thermal resistance is defined as the length of flow path across the material divided by the thermal conductivity (at that particular mean temperature). This is illustrated in Figure 156.

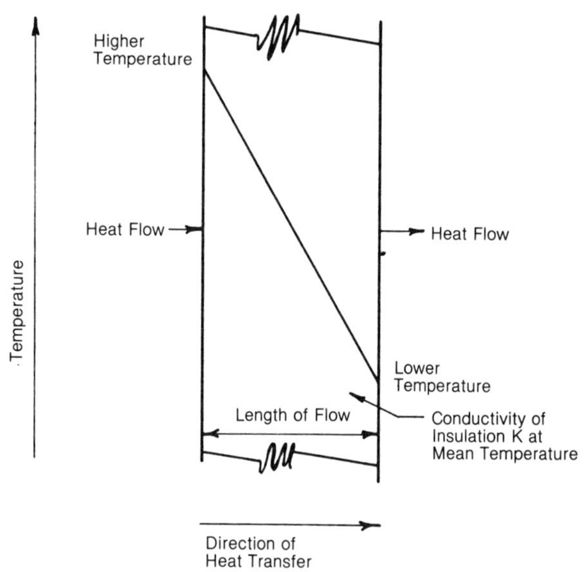

SCHEMATIC DIAGRAM OF HEAT TRANSFER THROUGH A SINGLE MATERIAL

Figure 156

In English Units when:

Q = Heat flow in Btu/ft², hr
Δt = Temperature difference between the two surfaces in degrees F
ℓ = Length of heat path (thickness of insulation) in inches
k = Conductivity in Btu's through 1-inch thickness of insulation, per one square ft of area, per one hour, per one degree of temperature difference in °F, (Btu in/ft², °F). Value of k is at the mean temperature of the insulation [higher temperature, (t_1) plus lower temperature, (t_2) divided by 2 or $(t_1 + t_2)/2$]

then:

$$Q = \frac{\Delta t}{\frac{\ell}{k}}$$

Here, $\Delta t = t_1 - t_2$

where, t_1 = °F of surface at higher temperature
t_2 = °F of surface at lower temperature

Thus:

$$Q = \frac{t_1 - t_2}{\frac{\ell}{k}} \quad (3)$$

In Metric Units when:

Q_m = Heat flow in W/m²
t_{m1} = °C or °K of surface at higher temperature
t_{m2} = °C or °K of surface at lower temperature
ℓ_m = Length of heat path in meters
k_m = Conductivity in W/mK. Value of k_m is at the mean temperature [higher temperature t_{m1} plus lower temperature, t_{m2} divided by 2, or $(t_{m1} + t_{m2})/2$]

then:

$$Q_m = \frac{t_{m1} - t_{m2}}{\frac{\ell_m}{k_m}} \quad (3a)$$

Example 33—*English Units*

An 8″-thick concrete wall has a temperature of 70 °F on the higher temperature side and 35 °F temperature on the lower temperature side. At a mean temperature of $\frac{70+35}{2} = 52.5\,°F$, the conductivity is 13.1 Btu in./ft², hr, °F. What is the heat transfer in Btu/ft², hr?

Solution:

$$Q = \frac{t_1 - t_2}{\frac{\ell}{k}}$$

where:

t_1 = 70 °F
t_2 = 35 °F
ℓ = 8″
k = 13.1 Btu in./ft², hr, °F

then:

$$Q = \frac{70 - 35}{\left(\frac{8}{13.1}\right)} = \frac{35}{0.6107} = 57.3 \text{ Btu/ft}^2\text{, hr}$$

If the outer surface temperature was 0 °F, then $Q = \frac{70 - 0}{0.6107} =$ 114 Btu/ft², hr.

Example 33a—*Metric Units*

A 0.2032 m thick concrete wall has a temperature of 21.2 °C on the higher side and 1.7° on the lower side. At a mean temperature of $\frac{21.2 + 1.7}{2} = 11.45\,°C$ the conductivity is 1.889 W/mK. What is the heat transfer in W/m²?

Solution:

t_{m1} = 21.2 °C
t_{m2} = 1.7 °C
ℓ_m = 0.2032 m
k_m = 1.889 W/mk

then:

$$Q_m = \frac{21.2 - 1.7}{\left(\frac{0.2032}{1.889}\right)} = \frac{19.5}{0.1075} = 181.27 \text{ W/m}^2$$

If outer surface temperature was −17.8 °C, then $Q = \frac{21.2 - (-17.8)}{0.1075}$ = 361 W/m².

Check: 57.3 Btu/ft² hr × 3.15291 = 181 W/m², also 114 Btu/ft² hr × 3.15291 = 360 W/m².

Thickness Required for Maximum Given Heat Transfer

In the case where the maximum heat transfer is known and the problem is to determine the thickness necessary to restrict the heat flow from the hotter to the cooler surface, this same basic equation can be used and solved for ℓ. In the English System

when:

t_1 = higher surface temperature (°F)
t_2 = lower surface temperature (°F)
k = conductivity in Btu in./ft², hr, °F
ℓ = thickness (length of heat flow path) in inches
Q = maximum heat transfer in Btu/ft², hr

Q is known — ℓ is unknown

then:

$$\ell = k\,\frac{t_1 - t_2}{Q} \text{ inches} \quad (31)$$

In metric units when:

t_{m1} = higher surface temperature °C
t_{m2} = lower surface temperature °C
k_m = conductivity in W/mK
ℓ_m = thickness (length of heat flow path) in meters
Q_m = maximum heat transfer W/m², hr

Q_m is known — ℓ_m is unknown

then:

$$\ell_m = k_m\,\frac{t_{m1} - t_{m2}}{Q_m} \text{ meters} \quad (31a)$$

Heat Transfer Through Two or More Flat Materials

All resistances in the path of heat flow are additive. A schematic diagram is shown in Figure 157.

In English Units when:

ℓ_1 = thickness in inches of first insulation material
ℓ_2 = thickness in inches of second insulation material
ℓ_3 = thickness in inches of third insulation material
k_1 = conductivity in Btu in./sq ft, hr, °F of first material
k_2 = conductivity in Btu in./sq ft, hr, °F of second material
k_3 = conductivity in Btu in./sq ft, hr, °F of third material

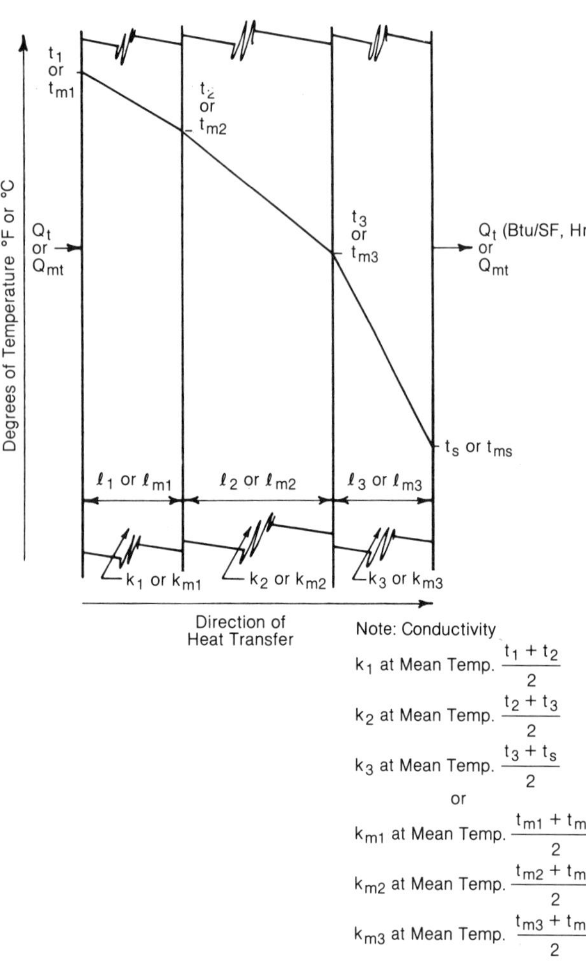

SCHEMATIC DIAGRAM OF HEAT TRANSFER THROUGH THREE MATERIALS EQUATIONS 32 and 32A

Figure 157

then:

Resistances R_1, R_2 and R_3 of each material are

$$\frac{\ell_1}{k_1}, \frac{\ell_2}{k_2}, \frac{\ell_3}{k_3}, \text{ respectively.}$$

$$Q = \frac{t_1 - t_s}{\frac{\ell_1}{k_1} + \frac{\ell_2}{k_2} + \frac{\ell_3}{k_3}} \quad \text{Btu/ft}^2\text{, hr} \tag{32}$$

In Metric Units when:

ℓ_{m1} = thickness in meters of first material
ℓ_{m2} = thickness in meters of second material
ℓ_{m3} = thickness in meters of third material
k_{m1} = conductivity in W/mK of first material
k_{m2} = conductivity in W/mK of second material
k_{m3} = conductivity in W/mK of third material

then:

Resistances R_{m1}, R_{m2} and R_{m3} of each material are

$$\frac{\ell_{m1}}{k_{m1}}, \frac{\ell_{m2}}{k_{m2}}, \frac{\ell_{m3}}{k_{m3}}, \text{ respectively.}$$

$$Q = \frac{(t_{1m} - t_{sm})}{\frac{\ell_{1m}}{k_{1m}} + \frac{\ell_{2m}}{k_{2m}} + \frac{\ell_{3m}}{k_{3m}}} \quad \text{W/m}^2. \tag{32a}$$

Example 34—*English Units*

An 8″-thick concrete wall with 2″ thickness of rigid insulation over which is installed 1/2″ gypsum wallboard. Calculate the heat transfer when:

t_1 = 70°F
t_2 = 0°F
ℓ_1 = 8″
ℓ_2 = 2″
ℓ_3 = 1/2″
k_1 = 13.1
k_2 = 0.3
k_3 = 1.4

then:

$$Q = \frac{(t_1 - t_2)}{\frac{\ell_1}{k_1} + \frac{\ell_2}{k_2} + \frac{\ell_3}{k_3}} = \frac{(70 - 0)}{\frac{8}{13.1} + \frac{2}{0.3} + \frac{0.5}{1.4}}$$

Note: $\frac{\ell_1}{k_1} = R_1 = 0.6107$, $\frac{\ell_2}{k_2} = R_2 = 6.6666$ and

$$\frac{\ell_3}{k_3} = R_3 = 0.3571$$

$$Q = \frac{(70 - 0)}{R_1 + R_2 + R_3} = \frac{70}{0.6107 + 6.6666 + 0.3571} = \frac{70}{7.63}$$

$$= 9.17 \text{ Btu/ft}^2\text{, hr.}$$

It should be noted that at 0°F to 70°F temperature difference, the addition of 2″ of insulation plus 1/2 gypsum wallboard changed the heat loss of the wall from 114 Btu/ft², hr to 9.17 Btu/ft², hr. This represents a 92 percent savings in energy.

Example 34a—*Metric Units*

A 0.2032 m thick concrete wall with 0.0508 m thickness of rigid insulation over which a 0.0127 m gypsum wallboard is installed. Determine the heat transfer when:

t_{m1} = 21.2°C
t_{m2} = −17.8°C
ℓ_{m1} = 0.2032 m
ℓ_{m2} = 0.0508 m
ℓ_{m3} = 0.0127 m
k_{m1} = 1.889 W/mK
k_{m2} = 0.0432 W/mK
k_{m3} = 0.2016 W/mK

then:

$$Q_m = \frac{(t_{m1} - t_{m2})}{\frac{\ell_{m1}}{k_{m1}} + \frac{\ell_{m2}}{k_{m2}} + \frac{\ell_{m3}}{k_{m3}}} = \frac{\{21.2 - (-17.8)\}}{\frac{0.2032}{1.889} + \frac{0.0508}{0.0432} + \frac{0.127}{0.2016}}$$

Note: $\frac{\ell_{m1}}{k_{m1}} = R_{m1} = 0.1075$; $\frac{\ell_{m2}}{k_{m2}} = R_{m2} = 1.1759$ and

$$\frac{\ell_{m3}}{k_{m3}} = R_{m3} = 0.0630$$

then:

$$Q_m = \frac{39.0}{(0.1075 + 1.1759 + 0.0630)} = \frac{39.0}{1.3464} = 28 \text{ W/m}^2$$

The addition of the rigid insulation and gypsum board provided 92 percent savings in energy.

CHECK: English to Metric Units; 9.17 Btu/ft², hr × 3.1529 = 28 W/m².

HEAT FLOW FROM, OR TO, A SURFACE

In the previous discussion of heat transfer the equations were all based upon the conductance of heat through a material. Even the air films on one or both sides were treated as a mass which had a given conductance to heat. Unfortunately, heat transfer from a surface to air and surrounding bodies is not so simple.

Heat is transferred from (or to) a surface (or from surrounding air and bodies) by radiation, convection, and conduction. In most instances the amount of heat transferred by conduction through air is negligible. This is due to the fact that air, without movement, has a conductivity of approximately 0.16 Btu in/ft², hr, °F. Thus, the resistance to heat flow from one surface to another just three feet away would be 36 (inches) divided by 0.16, or 225. This illustrates that absolutely still air would be a good insulator if it were not for radiation and for the fact that air never remains still when temperature differences occur. For most cases, then, the conductivity of air may be neglected when calculating heat transfer from surfaces, as still air never occurs in practice.

Radiation

For the bases of heat transfer from surfaces we must return to the work of Langmuir and Stefan-Boltzmann. According to the Stefan-Boltzmann law on heat transfer by radiation:

In English Units when:

Q_r = heat transfer by radiation in Btu/ft², hr, °F
ε = surface emittance (dimensionless)
T_s = absolute temperature of hot surface °R
T_a = absolute temperature of ambient air and bodies °R
T (°R) = t (°F) + 459.6

$$Q_r = 0.174\,\varepsilon \left[\left(\frac{T_s}{100}\right)^4 - \left(\frac{T_a}{100}\right)^4\right] \tag{1}$$

Stefan-Boltzmann Law. Surface emittance is the ability of an opaque material to emit radiant energy as a result of its temperature. It is measured by the ratio of the rate of radiant emission of the material to the corresponding emission of a thermally black body at the same temperature.

In Metric Units when:

Q_{mr} = heat transfer by radiation in W/m²
ε = surface emittance
T_{ms} = absolute temperature of hot surface in °K
T_{ma} = absolute temperature of air and bodies in °K

$$Q_{mr} = 0.548\,\varepsilon \left[\left(\frac{T_{ms}}{55.55}\right)^4 - \left(\frac{T_{ma}}{55.55}\right)^4\right] \tag{1a}$$

Example 35—*English Units*

What would be the heat loss by radiation from a wall if its surface temperature t_s was 35°F and ambient air and surrounding objects were at a temperature t_s of 10°F? The surface emittance ε is 0.9. Determine the heat loss Q_r in Btu/ft², hr.

$$T_s = t_s + 459.6 = 35.0 + 459.6 = 494.6\,°R$$
$$T_a = t_a = 10.0 + 459.6 = 469.6\,°R$$
$$Q_r = 0.174 \times 0.9 \left[\left(\frac{494.6}{100}\right)^4 - \left(\frac{469.6}{100}\right)^4\right]$$
$$Q_r = 0.174 \times 0.9\,(4.946^4 - 4.696^4) = 0.174 \times 0.9\,(598 - 486)$$
$$= 0.174 \times 0.9\,(112) = 17.6 \text{ Btu/ft}^2\text{, hr}$$

Example 35a—*Metric Units*

Metric Conversion: T_a in temperature °R ÷ 1.8 = T_{ma} in temperature °K.

What would be the heat loss by radiation from a wall if its surface temperature T_{ma} was 338.7°K and the ambient air and surrounding objects were at a temperature T_{ms} = 294.2°K? The surface emittance is 0.9. Determine the heat loss Q_{mr} in W/m².

$$Q_{mr} = 0.548 \times 0.9 \left[\left(\frac{338.7}{55.5}\right)^4 - \left(\frac{294.2}{55.5}\right)^4\right]$$
$$= 0.548 \times 0.9 \times (1387 - 789) = 0.548 \times 0.9 \times 598$$
$$= 294 \text{ W/m}^2$$

It must be pointed out that an ambient conduction where air and all surrounding objects are a single given temperature just does not exist in nature. For example, the calculations presented do not take into account solar radiation. In the previous example if the surface under consideration were located so that it faced the sun directly, it would not be emitting heat from its surface but would be receiving heat, even in 0°F ambient air. At night the opposite is true. On clear nights a horizontal surface pointing upward will radiate much greater quantities of heat than calculations based upon ambient air temperature would indicate. (This is the reason we must scrape frost off the modern car windshields, which face almost upward, at air temperatures up to 40°F, but we have little trouble with side windows which are in the vertical position.)

Convection

As stated in the definition, convection is a movement of a mass of gas or liquid due to temperature difference, and physical contact of the gas or liquid is the actual method of heat transfer. Thus, the speed at which this gas or liquid passes over a surface will change the rate of heat transfer. As convection is not a basic physical law but a combination of physical phenomena, it cannot be arranged into a nice, clean-cut mathematical formula. For this reason the equations used by some will differ from those used by others. Also, the equations must be applied within the particular limits and under the conditions set forth by the investigators.

Because of these facts, statements must be made as to what the conditions are that apply to each equation.

English Units

The formula for heat transfer from a surface, at relatively moderate temperatures, to air moved by "natural convection" was developed by Langmuir as being:

$$Q_c = 0.296\,(t_s - t_a)^{5/4}$$

(Langmuir's Equation) (9)*

when:

Q_c = heat transferred by natural convection in Btu/sq ft, hr
t_s = temperature of surface, °F
t_a = temperature of ambient air, °F

Note: Natural convection is the movement of air caused by variation in its density as its temperature changes. As air is heated, its density becomes less and it tends to rise. As the heated air rises, it is replaced by cooler air, which in its turn is heated, rises, and is replaced in a continuous, recurring process.

When air has movement (caused by an external force) other than the natural movement caused by any change of its temperature, this movement will increase the rate of heat transfer. The effect of this movement on heat transfer was determined by Langmuir to be:

*Five-Fourths Power of Numbers in Table 79.

$$Q_{cv} = Q_c \sqrt{\frac{V + 68.9}{68.9}}$$

where V is the velocity of the air in ft/min.

Thus, the equation for heat transfer by convection by air at a forced velocity is Q_{cv}:

English Units

$$Q_{cv} = 0.296\,(t_s - t_a)^{5/4} \sqrt{\left(\frac{V + 68.9}{68.9}\right)}$$

(Langmuir) (10)

where:

Q_{cv} = heat transfer by forced convection in Btu/sq ft, hr
V = air velocity, feet/min

Note: There are two different elements of time in the single equation. Although confusing, the equation is simpler to solve as shown than by having air velocity expressed in units per hour and having much larger compensating constants in the square factor.

Metric Units—Natural Convection

$$Q_{mc} = 1.957\,(T_{ms} - T_{ma})^{5/4}$$

(Langmuir) (9a)

when:

Q_{mc} = heat transferred by natural convection, in W/m²
T_{ms} = temperature of surface, °C (or °K)*
T_{ma} = temperature of ambient air, °C (or °K)*

Metric Units—Forced Convection

Q_{mc} = heat transfer by forced convection in W/m²
V_m = air velocity, meter per second, m/s

$$Q_{mcv} = (T_{ms} - T_{ma})^{5/4} \sqrt{\left(\frac{196.85V_m - 68.9}{68.9}\right)}$$

(10a)

Note: For convenience, a table for five fourths power of numbers is provided in Table 79.

Example 36—*English Units*

What would be the heat loss of natural convection (no wind) from a wall if its surface temperature t_s was 35°F and the ambient air t_a was 10°F? Determine Q_c the heat loss by natural convection in Btu/ft², hr.

*Note: T_{ms} and T_{ma} must be in the same units.

TABLE 79 Five-fourths power of numbers

N	0	1	2	3	4	5	6	7	8	9
0	0.000	1.000	2.3784	3.9482	5.6569	7.4767	9.3905	11.386	13.454	15.588
10	17.783	20.033	22.335	24.685	27.081	29.520	32.000	34.519	37.076	39.668
20	42.295	44.955	47.646	50.369	53.121	55.910	58.711	61.547	64.409	67.297
30	70.210	73.148	76.109	79.094	82.101	85.130	88.182	91.254	94.347	97.461
40	100.59	103.75	106.92	110.11	113.32	116.55	119.80	123.06	126.34	129.64
50	132.96	136.29	139.64	143.00	146.38	149.78	153.19	156.62	160.06	163.52
60	166.99	170.48	173.98	177.49	181.02	184.56	188.12	191.69	195.27	198.87
70	202.48	206.10	209.73	213.38	217.04	220.71	224.40	228.09	231.80	235.52
80	239.26	243.00	246.76	250.52	254.30	258.09	261.89	265.70	269.53	273.73
90	277.21	281.06	284.93	288.80	292.69	269.59	300.50	304.41	308.34	312.28
100	316.23	320.19	324.15	328.13	332.12	336.11	340.12	344.14	348.16	352.19
110	356.24	360.29	364.35	368.42	372.50	376.59	380.69	384.80	388.91	393.04
120	397.17	401.31	405.46	409.62	413.79	417.96	422.15	426.34	430.54	434.75
130	438.96	443.19	447.42	451.66	455.91	460.17	464.43	468.71	472.99	477.27
140	481.57	485.87	490.19	494.50	498.83	503.16	507.51	511.85	516.21	520.57
150	524.95	529.32	533.71	538.10	542.50	546.91	551.32	555.74	560.17	564.61
160	569.05	573.50	577.95	582.42	586.89	591.36	595.85	600.34	604.83	609.34
170	613.85	618.37	622.89	627.42	631.96	636.50	641.05	645.60	650.17	654.74
180	659.31	663.89	668.48	673.08	677.68	682.28	686.90	691.52	696.14	700.77
190	705.41	710.05	714.70	719.36	724.02	728.69	733.37	738.04	742.73	747.42
200	752.12	756.82	761.53	766.25	770.97	775.70	780.43	785.17	789.91	794.66
210	799.42	804.18	808.95	813.72	818.50	823.28	828.07	832.87	837.67	842.47
220	847.28	852.10	856.92	861.75	866.58	871.42	876.27	881.11	885.97	890.83
230	895.69	900.56	905.44	910.32	915.21	920.21	925.00	929.90	934.81	939.72
240	944.63	949.56	954.48	959.42	964.36	969.30	974.25	979.20	984.16	989.12
250	994.09	999.06	1004.0	1009.0	1014.0	1019.0	1024.0	1029.0	1034.0	1039.0
260	1044.0	1049.1	1054.1	1059.1	1064.2	1069.2	1074.2	1079.3	1084.3	1089.4
270	1094.5	1099.5	1104.5	1109.7	1114.8	1119.9	1125.0	1130.1	1135.2	1140.3
280	1145.4	1150.5	1155.6	1160.7	1165.9	1171.0	1176.1	1181.3	1186.4	1191.6
290	1196.7	1201.9	1207.1	1212.2	1217.4	1222.6	1227.8	1232.9	1238.1	1243.3
300	1248.5	1253.7	1259.0	1264.2	1269.4	1274.6	1279.8	1285.1	1290.3	1295.5
310	1300.8	1306.0	1311.3	1316.5	1321.8	1327.1	1332.3	1337.6	1342.9	1348.2
320	1353.4	1358.7	1364.0	1369.3	1374.6	1379.9	1385.2	1390.5	1395.9	1401.2
330	1406.5	1411.8	1417.2	1422.5	1427.9	1433.2	1438.5	1443.9	1449.3	1454.6
340	1460.0	1465.4	1470.7	1476.1	1481.5	1486.9	1492.3	1497.7	1503.1	1508.5
350	1513.9	1519.3	1524.7	1530.1	1535.5	1540.9	1546.4	1551.8	1557.2	1562.7
360	1568.1	1573.6	1579.0	1584.5	1589.9	1595.4	1600.9	1606.3	1611.8	1617.3
370	1622.8	1628.2	1633.7	1639.2	1644.7	1650.2	1655.7	1661.2	1666.7	1672.2
380	1677.8	1683.3	1688.8	1694.3	1699.9	1705.4	1710.9	1716.5	1722.0	1727.6
390	1733.1	1738.7	1744.1	1749.8	1755.4	1760.9	1766.5	1772.1	1777.7	1783.3
400	1788.9	1794.4	1800.0	1805.6	1811.2	1816.8	1822.5	1828.1	1833.7	1839.3
410	1844.9	1850.6	1856.2	1861.8	1867.5	1873.1	1878.7	1884.4	1890.0	1895.7
420	1901.3	1907.0	1912.7	1918.3	1924.0	1929.7	1935.4	1941.0	1946.7	1952.4
430	1958.1	1963.8	1969.5	1975.2	1980.9	1986.6	1992.3	1998.0	2003.7	2009.5
440	2015.2	2020.9	2026.6	2032.4	2038.1	2043.9	2049.6	2055.3	2061.1	2066.8
450	2072.6	2078.4	2084.1	2089.9	2095.7	2101.4	2107.2	2113.0	2118.8	2124.5
460	2130.3	2136.1	2141.9	2147.7	2153.5	2159.3	2165.1	2170.9	2176.7	2182.6
470	2188.4	2194.2	2200.0	2205.9	2211.7	2217.5	2223.4	2229.2	2235.0	2240.9
480	2246.7	2252.6	2258.4	2264.3	2270.2	2276.0	2281.9	2287.8	2293.6	2299.5
490	2305.4	2311.3	2317.2	2323.0	2328.9	2334.8	2340.7	2346.6	2352.5	2358.4

TABLE 79 Five-fourths power of numbers (continued)

N	0	1	2	3	4	5	6	7	8	9
500	2364.4	2370.3	2376.2	2382.1	2388.0	2393.9	2399.9	2405.8	2411.7	2417.7
510	2423.6	2429.5	2335.5	2441.4	2447.4	2453.3	2459.3	2465.3	2471.2	2477.2
520	2483.2	2489.1	2495.1	2501.1	2507.1	2513.0	2519.0	2525.0	2531.0	2537.0
530	2543.0	2549.0	2555.0	2561.0	2567.0	2573.0	2579.0	2585.0	2591.1	2597.1
540	2603.1	2609.1	2615.2	2621.2	2627.2	2633.3	2639.3	2645.4	2651.4	2657.5
550	2663.5	2669.6	2675.6	2681.7	2687.7	2693.8	2699.9	2705.9	2712.1	2718.1
560	2724.2	2730.3	2736.3	2742.4	2748.5	2754.6	2760.7	2766.8	2772.9	2779.0
570	2785.1	2791.2	2797.3	2803.5	2809.6	2815.7	2821.8	2827.9	2834.1	2840.2
580	2846.3	2852.5	2858.6	2864.7	2870.9	2877.0	2883.2	2889.3	2895.5	2901.6
590	2907.8	2914.0	2920.1	2926.3	2932.5	2938.6	2944.8	2951.0	2957.2	2963.4
600	2969.5	2975.7	2981.9	2988.1	2994.3	3000.5	3006.7	3012.9	3019.1	3025.3
610	3031.5	3037.7	3044.0	3050.2	3056.4	3062.6	3068.9	3075.1	3084.3	3087.5
620	3093.8	3100.0	3106.3	3112.5	3118.8	3125.0	3131.3	3137.5	3143.8	3150.0
630	3156.3	3162.5	3168.8	3175.1	3181.4	3187.6	3193.9	3200.2	3206.5	3212.7
640	3219.0	3225.3	3231.6	3237.9	3244.2	3250.5	3256.8	3263.1	3269.4	3275.7
650	3282.0	3288.3	3294.7	3301.0	3307.3	3313.6	3319.9	3326.3	3332.6	3338.9
660	3345.3	3351.6	3357.9	3364.3	3370.6	3377.0	3383.3	3389.7	3396.0	3402.4
670	3408.7	3415.1	3421.5	3427.8	3434.2	3440.6	3446.9	3453.3	3459.7	3466.1
680	3472.5	3478.8	3485.2	3491.6	3498.0	3504.4	3510.8	3517.2	3523.6	3530.0
690	3536.4	3542.8	3549.2	3555.6	3562.0	3568.5	3574.9	3581.3	3587.7	3594.2
700	3600.6	3607.0	3613.4	3919.9	3626.3	3632.8	3639.2	3645.6	3652.1	3658.5
710	3665.0	3671.4	3677.9	3684.4	3690.8	3697.3	3703.7	3710.2	3716.7	3723.2
720	3729.6	3736.1	3742.6	3749.1	3755.5	3762.0	3768.5	3775.0	3781.5	3788.0
730	3794.5	3801.0	3807.5	3814.0	3820.5	3827.0	3833.5	3840.0	3846.5	3853.1
740	3859.6	3866.1	3872.6	3879.1	3885.7	3892.2	3898.7	3905.3	3911.8	3918.3
750	3924.9	3914.4	3939.0	3944.5	3951.1	3957.6	3964.2	3970.7	3977.3	3983.8
760	3990.4	3997.0	4003.5	4010.1	4016.7	4023.2	4029.8	4036.4	4043.0	4049.6
770	4056.1	4062.7	4069.3	4075.9	4082.5	4089.1	4095.7	4102.3	4108.9	4115.5
780	4122.1	4128.7	4135.3	4141.9	4148.5	4155.2	4161.8	4168.4	4175.0	4181.6
790	4188.3	4194.9	4201.5	4208.1	4214.8	4221.4	4228.1	4234.7	4241.3	4248.0
800	4254.6	4261.3	4267.9	4274.6	4281.2	4287.9	4294.6	4301.2	4307.9	4314.5
810	4321.2	4327.9	4334.6	4341.2	4347.9	4354.6	4361.3	4367.9	4374.6	4381.3
820	4388.0	4394.7	4401.4	4408.1	4414.8	4421.5	4428.2	4434.9	4441.6	4448.3
830	4455.0	4461.7	4468.4	4475.1	4481.8	4488.6	4495.3	4502.0	4508.7	4515.5
840	4522.2	4528.9	4535.7	4542.4	4549.1	4555.9	4562.6	4569.3	4576.1	4582.8
850	4589.6	4596.3	4603.1	4609.8	4616.6	4623.4	4630.1	4636.9	4643.6	4650.4
860	4657.2	4663.9	4670.7	4677.5	4684.3	4691.0	4697.8	4704.6	4711.4	4718.2
870	4725.0	4731.8	4738.6	4745.3	4752.1	4758.9	4765.7	4772.5	4779.3	4786.1
880	4793.0	4799.8	4806.6	4813.4	4820.2	4827.0	4833.8	4840.7	4847.5	4854.3
890	4861.1	4868.0	4874.8	4881.6	4888.5	4895.3	4902.1	4909.0	4915.8	4922.7
900	4929.5	4936.4	4943.2	4950.1	4956.9	4963.8	4970.6	4977.5	4984.3	4991.2
910	4998.1	5004.9	5011.8	5018.7	5025.5	5032.4	5039.3	5046.2	5053.0	5059.9
920	5066.8	5073.7	5080.6	5087.5	5094.4	5101.3	5108.2	5115.0	5121.9	5128.0
930	5135.7	5142.7	5149.6	5156.5	5163.4	5170.3	5177.2	5184.1	5191.0	5198.0
940	5204.9	5211.8	5218.7	5225.6	5232.6	5239.5	5246.4	5253.4	5260.3	5267.2
950	5274.2	5281.1	5288.1	5295.0	5301.9	5308.9	5315.8	5322.8	5329.8	5336.7
960	5343.7	5350.6	5357.6	5364.5	5371.5	5378.5	5385.4	5392.4	5399.4	5406.4
970	5413.3	5420.3	5427.3	5434.3	5441.3	5448.2	5455.2	5462.2	5469.2	5476.2
980	5483.2	5490.2	5497.2	5504.2	5511.2	5518.2	5525.2	5532.2	5539.2	5546.2
990	5553.2	5560.2	5567.3	5574.2	5581.3	5588.3	5595.3	5602.3	5609.4	5616.4
1000	5623.4									

$$Q_c = 0.296\,(t_a - t_s)^{5/4} = 0.296\,(35 - 10)^{5/4} = 0.296\,(25)^{5/4}$$
$$= 0.296\,(55.9)$$
$$Q_c = 16.5\ \text{Btu/ft}^2\text{, hr}$$

Under identical conditions, except the wind causes 10 miles per hour air movement across the surface, what would be the heat loss in Btu/ft², hr?

$$Q_c = 0.296\,(t_a - t_s)^{5/4}\sqrt{\left(\frac{V + 68.9}{68.9}\right)}$$

10 miles per hour is $10 \times \frac{5280}{60} = 10 \times 88 = 888$ ft/minute

$$Q_c = 16.5\sqrt{\left(\frac{888 + 68.9}{68.9}\right)} = 16.5 \times \sqrt{\left(\frac{956.9}{68.9}\right)}$$
$$= 61.5\ \text{Btu/ft}^2\text{, hr.}$$

Example 36a—*Metric Units*

What would be the heat loss by natural convection (no wind) from a wall if its surface temperature t_{ms} was 1.7 °C and the ambient temperature t_{ma} was −12.2 °C? Determine Q_m the heat loss by natural convection in W/m².

$$Q_{mc} = 1.957\,(t_{ms} - t_{ma})^{5/4} = 1.957\,\{1.7 - (-12.2)\}^{5/4}$$
$$= 1.957\,(13.9)^{5/4} = 1.957 \times 27 = 52\ \text{W/m}^2$$

Under identical conditions except that wind causes a 4.51 m/sec air movement, what would be the heat loss in Btu/ft², hr?

$$Q_m = 1.957\,(t_{ms} - t_{ma})^{5/4}\sqrt{\left(\frac{196.85V_m + 68.9}{68.9}\right)}$$
$$= 52 \times \sqrt{\left(\frac{956.69}{68.9}\right)} = 52 \times \sqrt{13.9} = 193.8\ \text{W/m}^2$$

Conversion: 723 W/m × 0.317 = 229 Btu/ft², hr.

HEAT LOSS FROM AIR TO LOWER TEMPERATURE SURFACE

The previous examples illustrated the heat transfer from an outside surface at a temperature higher than ambient air. When the air is the higher temperature, then the heat transfer by radiation and convection would be to a colder surface. This is true of ceilings, walls, and floors.

In the case of walls, roofs, floors above grade, the heat transfer is, in most instances, from (or to) inside air, through the building segment to (or from) the outside air. However, in the case of concrete, or slab, on grade or below grade, the heat transferred goes to the ground. Even with heat being transferred to the soil or ground from inside a building, the temperature of the ground is mostly influenced by atmospheric conditions and the heat from the inside has an effect on its temperature.

Ground temperatures lower than 55 °F (13.3 °C) are common in winter in most parts of the United States. As an uninsulated concrete floor has little resistance to heat flow, its temperature will be close to the temperature of the soil on which it is placed.

Equations 34, 34a, 35, and 35a can be used directly to calculate heat loss from ambient air and surroundings to a colder floor.

Example 37—*English Units*

Calculate the heat loss to a concrete slab floor at temperature $t_s = 60$ °F from its surroundings which are in equilibrium with air at temperature $t_a = 75$ °F. The surface emittance ε of the floor is 0.9. There is no forced air movement. Determine the heat loss Q_{rc} in Btu/ft², hr.

$$Q_{rc} = 0.174 \times \varepsilon\left[\left(\frac{T_a}{100}\right)^4 - \left(\frac{T_s}{100}\right)^4\right] \times 0.296\,(t_a - t_s)^{5/4}$$
in Btu/ft², hr

$$T_a = 75\,°\text{F} + 459.6 = 534.6\,°\text{R},\quad T_s = 60\,°\text{F} + 459.6 = 519.9\,°\text{R}$$
$$Q_{cv} = 0.174 \times 0.9 \times \left[\left(\frac{534.6}{100}\right)^4 - \left(\frac{519.6}{100}\right)^4\right] \times 0.296\,(75 - 60)^{5/4}$$
$$Q_{cv} = 0.1566\,(816 - 728) \times 0.296\,(15)^{5/4} = 0.1566 \times 88 \times 0.296\,(29.5)$$
$$Q_c = 13.8 + 8.7 = 22.5\ \text{Btu/ft}^2\text{, hr.}$$

Example 37a—*Metric Units*

Calculate the heat loss to a concrete slab floor at a temperature of t_{ms} 15.6 °C from surroundings above a temperature of t_{ms}, 23.9 °C. The surface emittance ε of the floor is 0.9. There is no forced air movement. Determine the heat loss Q_{mcv} in W/m².

$$T_{ma} = t_{ma} + 273.2 = 23.9 + 273.2 = 297.1\,°\text{K}$$
$$t_{ms} = t_{ms} + 273.2 = 15.6 + 273.2 = 288.8\,°\text{K}$$
$$Q_{mcv} = 0.548\,\varepsilon\left[\left(\frac{T_{ms}}{55.5}\right)^4 - \left(\frac{T_{ma}}{55.5}\right)^4\right] + 1.957\,(T_{ms} - T_{ma})^{5/4}$$
$$Q_{mc} = 0.548 \times 0.9\left[\left(\frac{297.1}{55.5}\right)^4 - \left(\frac{288.8}{55.5}\right)^4\right] + 1.957 \times (297.1 - 288.8)^{5/4}$$
$$Q_{mc} = 0.4932\,(821 - 733) + 1.957\,(8.8)^{5/4} = 0.4932\ 88 \times 1.957\,(14.1)$$
$$= 1198\ \text{W/m}^2$$

It should be noted that at these very mild conditions the heat loss to the slab floor is two to three times more than is considered to be an acceptable loss for walls. In addition to having high heat loss, the uninsulated concrete slab with its low surface temperature causes the feet and lower legs of the occupants to be uncomfortably cold. Therefore, except in southern climates, concrete floors should be insulated.

SHADED WALLS AND ROOFS

When heat passes from air to a single material, through the material and from its surface into air at a lower temperature, then the previous equations are used in combination.

Under static conditions, the heat transfer to one surface equals the heat transfer through the material which also equals the heat transfer from that surface into ambient air and the surroundings.

EQUATION 37—*English Units when:*

t_i = temperature of warmer air (°F)
T_i = temperature of warmer air (°R)*
t_{s1} = temperature of first surface (°F)
T_{s1} = temperature of first surface (°R)
t_{s2} = temperature of second (lower temperature) surface (°F)
T_{s2} = temperature of second (lower temperature) surface (°R)
t_a = temperature of cooler air (°F)
T_a = temperature of cooler air (°R)
V_i = velocity of air over higher temperature surface in ft/min
V_a = velocity of air over lower temperature surface in ft/min
ℓ = thickness of material (length of heat path) in inches
k = conductivity of material in Btu, in/ft², hr, °F
ε_1 = surface emittance of higher temperature side
ε_2 = surface emittance of lower temperature side
Q_t = heat transfer in Btu/ft², hr

then:

$$Q_t = 0.174\,\varepsilon_1 \left[\left(\frac{T_i}{100}\right)^4 - \left(\frac{T_{s1}}{100}\right)^4\right] + 0.296\,(t_i - t_{s1}) \times \sqrt{\left(\frac{V_i + 68.9}{68.9}\right)} = \frac{(t_{s1} - t_{s2})}{\frac{\ell}{k}} =$$

$$0.174\,\varepsilon_2 \left[\left(\frac{T_{s2}}{100}\right)^4 - \left(\frac{T_a}{100}\right)^4\right] + 0.296\,(t_{s2} - t_{sa}) \times \sqrt{\left(\frac{V_a + 68.9}{68.9}\right)} \text{ Btu/ft}^2\text{, hr} \tag{33}$$

Illustrated in Figure 158.

EQUATION 33a—*Metric Units when:*

t_{mi} = temperature of warmer air (°C)
T_{mi} = temperature of warmer air (°K)**
t_{ms1} = temperature of first surface (°C)
T_{ms1} = temperature of first surface (°K)
t_{ms2} = temperature of second (lower temperature) surface (°C)
T_{ms2} = temperature of second (lower temperature) surface (°K)
t_{ma} = temperature of cooler air (°C)
T_{ma} = temperature of cooler air (°K)
V_{mi} = velocity of air over higher temperature surface in m/sec
V_{ma} = velocity of air over lower temperature surface in m/sec
ℓ_m = thickness of material (length of heat path) in meters
k_m = conductivity of material in W/mK
ε_1 = surface emittance higher temperature side
ε_2 = surface emittance lower temperature side
Q_{mt} = heat transfer in W/m²

*T (in °R) = t(in °F) + 459.69
**T (°K) = t (in °C) + 273.16

then:

$$Q_{mt} = 0.548\,\varepsilon_1 \left[\left(\frac{T_{mi}}{55.5}\right)^4 - \left(\frac{T_{ms1}}{55.5}\right)^4\right] + 1.957\,(t_{m1} - t_m)^{5/4} \times \sqrt{\left(\frac{196.85V_{m1} + 68.9}{68.9}\right)} = Q_{mt} = \frac{t_{m1} - t_{m2}}{\frac{\ell_m}{k_m}} =$$

$$0.548\,\varepsilon_2 \left[\left(\frac{T_{m2}}{55.5}\right)^4 - \left(\frac{T_{ma}}{55.5}\right)^4\right] + 1.967\,(t_{ms2} - t_{ma})^{5/4} \times \sqrt{\frac{196.85V_{ma} + 68.9}{68.9}} \tag{33a}$$

In most instances walls, ceilings or roofs have more than one component material in their construction, thus it is necessary to develop equations for two or more components. Heat transfers from higher temperature air to surface of three materials through the materials then from the surface of the outer material to lower temperature air.

In many building walls, ceilings and roofs the construction is of more than one material. Thus the heat transfer goes from one to another to yet another until the heat passes through the entire component. These multiple materials of one building component may be of two materials, three materials or more. The following schematic equation and diagram show three materials; however it could be only two or more than three, and the equation would have the same format with the subtraction or addition of materials.

EQUATION 34—*English Units when:*

t_i = temperature of higher temperature air °F
T_i = temperature of higher temperature air °R
t_{si} = temperature of first surface of component °F
T_{si} = temperature of first surface of component °R
t_{s2} = temperature of junction of 1st and 2nd materials °F
T_{s2} = temperature of junction of 1st and 2nd materials °R
t_{s3} = temperature of junction of 2nd and 3rd materials °F
T_{s3} = temperature of junction of 2nd and 3rd materials °R
t_{s4} = temperature of surface of 3rd material, facing air °F
T_{s4} = temperature of surface of 3rd material, facing air °R
t_a = temperature of lower temperature air °F
T_a = temperature of lower temperature air °R
V_i = velocity of air over higher temperature surface in ft/min
V_a = velocity of air over lower temperature surface in ft/min
ℓ_1 = thickness of 1st material in inches

Equation 33

English Units

$$Q_t = 0.174\,\varepsilon\left[\left(\frac{T_i}{100}\right)^4 - \left(\frac{T_{s1}}{100}\right)^4\right] + 0.296\,(t_i - t_{s1})^{5/4}\sqrt{\frac{V_i + 68.9}{68.9}} = \frac{t_{s1} - t_{s2}}{\frac{\ell}{k}} = Q_t \text{ Btu/ft}^2\text{, hr}$$

$$\text{also } \frac{t_{s1} - t_{s2}}{\frac{\ell}{k}} = Q_t = 0.174\,\varepsilon\left[\left(\frac{T_{s2}}{100}\right)^4 - \left(\frac{T_a}{100}\right)^4\right] + 0.296\,(t_2 - t_a)^{5/4}\sqrt{\frac{V_a + 68.9}{68.9}}$$

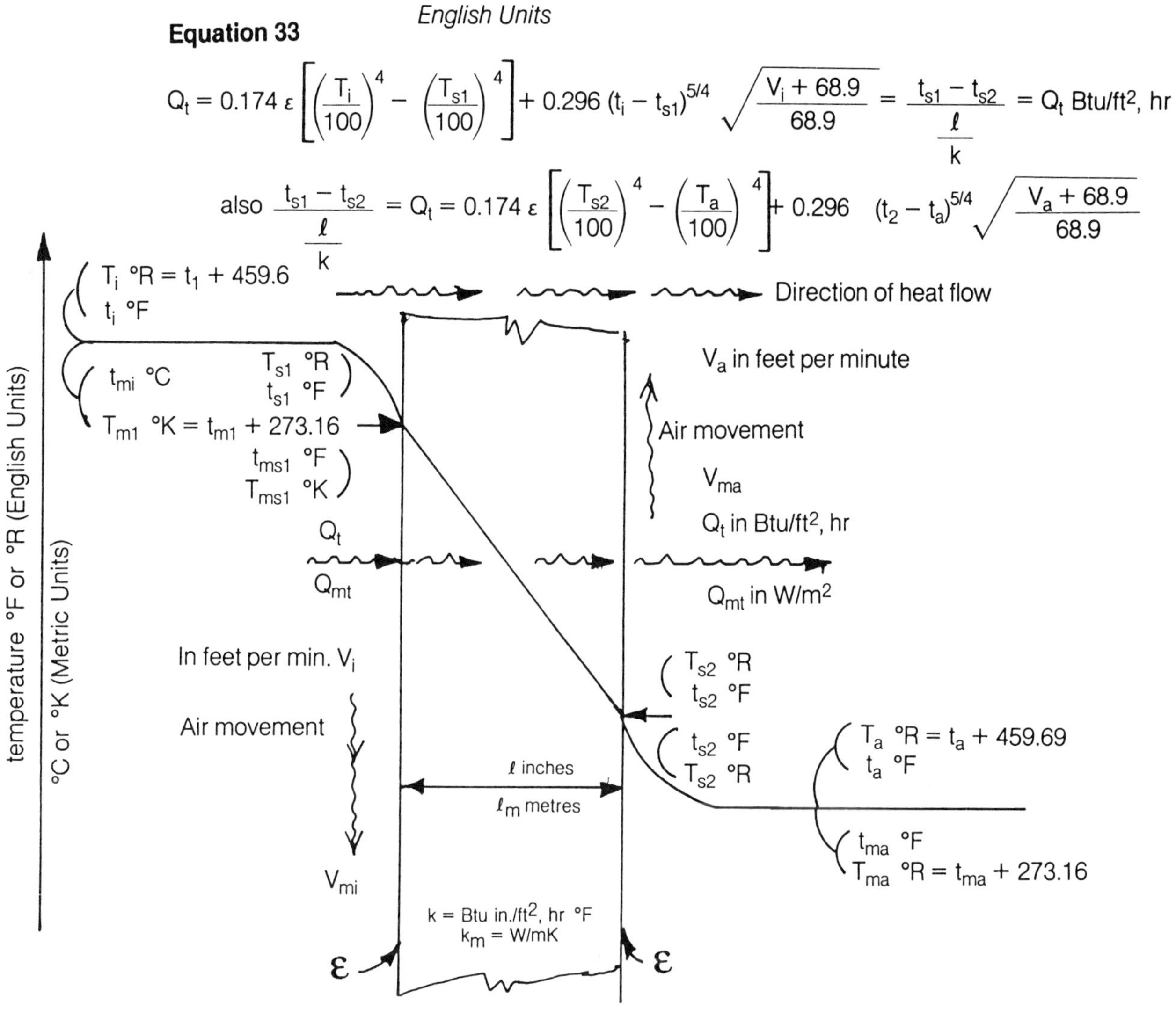

Equation 33a

Metric Units

$$Q_{mt} = 0.548\,\varepsilon\left[\left(\frac{T_{mi}}{55.5}\right)^4 - \left(\frac{T_{ms1}}{55.5}\right)^4\right] + 1.957\,(t_{mi} - t_{ms1})^{5/4}\sqrt{\frac{195.85\,V_m + 68.9}{68.9}} = \frac{t_{m1} - t_{m2}}{\frac{\ell}{k_m}} = Q \text{ W/m}^2$$

$$\frac{t_{mi} - t_{m2}}{\frac{\ell}{k_m}} = Q_m = 0.548\,\varepsilon\left[\left(\frac{T_{m2}}{55.5}\right)^4 - \left(\frac{T_{ma}}{55.5}\right)^4\right] + 1.957\,(t_{m2} - t_{ma})^{5/4}\sqrt{\frac{195.85\,V_{ma} + 68.9}{68.9}}$$

SCHEMATIC DIAGRAM OF HEAT TRANSFER FROM HIGHER TEMPERATURE AIR TO SURFACE OF MATERIAL, THROUGH MATERIAL AND FROM SURFACE OF MATERIAL TO LOWER TEMPERATURE AIR

Figure 158

Equation 34 *English Units*

$$Q_t = 0.174\,\varepsilon\left[\left(\frac{T_i}{100}\right)^4 - \left(\frac{T_{s1}}{100}\right)^4\right] + 0.296\,(t_i - t_{sa})^{5/4}\sqrt{\frac{V_i + 68.9}{68.9}} = Q_t = \frac{t_{s1} - t_{s4}}{\frac{\ell_1}{k_1} + \frac{\ell_2}{k_2} + \frac{\ell_3}{k_3}} = Q_t \text{ Btu/ft}^2\text{, hr} =$$

$$0.174\,\varepsilon_2\left[\left(\frac{T_{s4}}{100}\right)^4 - \left(\frac{T_a}{100}\right)^4\right] + 0.296\,(t_{s4} - t_a)^{5/4}\sqrt{\frac{V_a + 68.9}{68.9}} \text{ Btu/ft}^2\text{, hr}$$

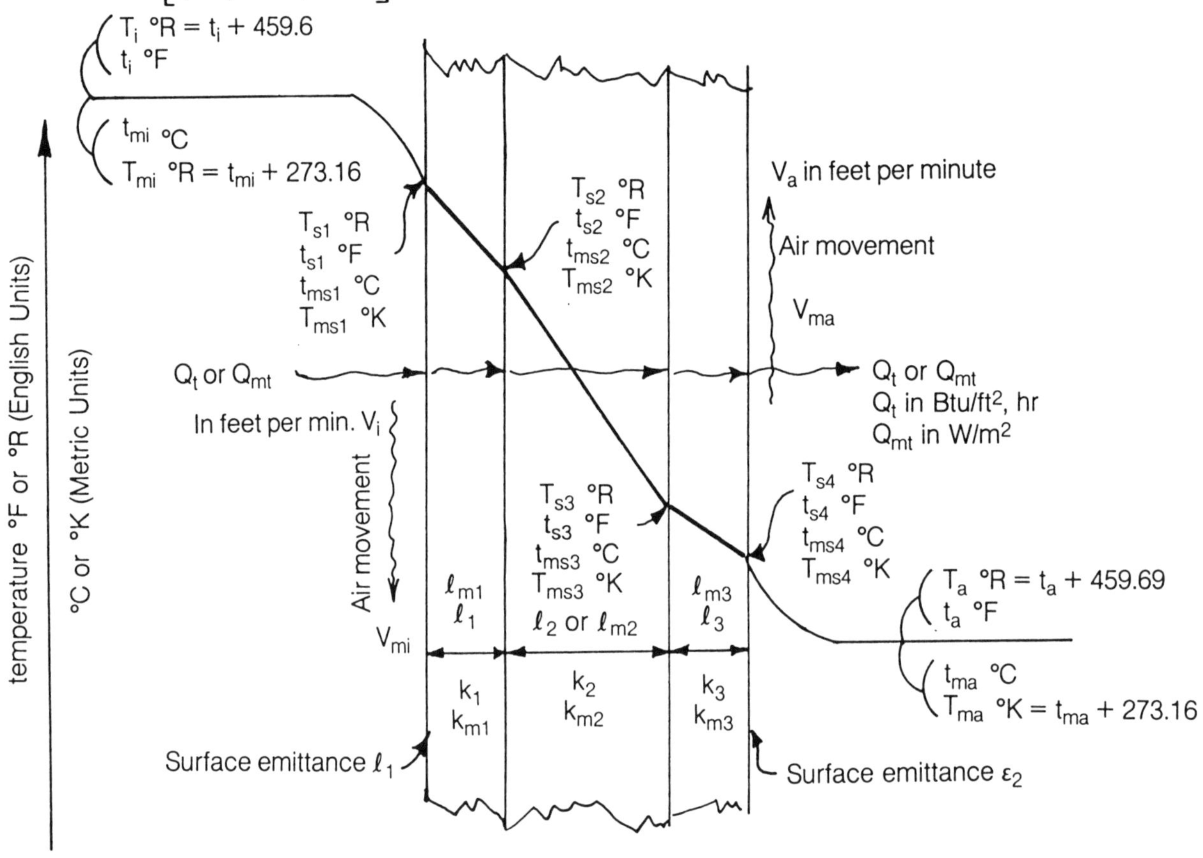

T_i °R = t_i + 459.6
t_i °F
t_{mi} °C
T_{mi} °R = t_{mi} + 273.16
temperature °F or °R (English Units)
°C or °K (Metric Units)
T_{s1} °R
t_{s1} °F
t_{ms1} °C
T_{ms1} °K
T_{s2} °R
t_{s2} °F
t_{ms2} °C
T_{ms2} °K
V_a in feet per minute
Air movement
V_{ma}
Q_t or Q_{mt}
In feet per min. V_i
Air movement
V_{mi}
Q_t or Q_{mt}
Q_t in Btu/ft², hr
Q_{mt} in W/m²
T_{s3} °R
t_{s3} °F
t_{ms3} °C
T_{ms3} °K
T_{s4} °R
t_{s4} °F
t_{ms4} °C
T_{ms4} °K
ℓ_{m1} ℓ_1
ℓ_2 or ℓ_{m2}
ℓ_{m3} ℓ_3
k_1 k_{m1}
k_2 k_{m2}
k_3 k_{m3}
T_a °R = t_a + 459.69
t_a °F
t_{ma} °C
T_{ma} °K = t_{ma} + 273.16
Surface emittance ℓ_1
Surface emittance ε_2

Equation 34a *Metric Units*

$$Q_{mt} = 0.548\,\varepsilon_1\left[\left(\frac{T_{mi}}{55.5}\right)^4 - \left(\frac{T_{ms1}}{55.5}\right)^4\right] + 1.957\,(t_{mi} - t_{ms1})^{5/4}\sqrt{\frac{195.85\,V_{mi} + 68.9}{68.9}} = Q_{mt} \text{ W/m}^2 =$$

$$\frac{t_{ms} - t_{ms4}}{\frac{\ell_{m1}}{k_{m1}} + \frac{\ell_{m2}}{k_{m2}} + \frac{\ell_{m3}}{k_{m3}}} = Q_{mt} = 0.548\,\varepsilon_2\left[\left(\frac{T_{ms4}}{55.5}\right)^4 - \left(\frac{T_{ma}}{55.5}\right)^4\right] + 1.957\,(t_{ms4} - t_{ma})^{5/4}\sqrt{\frac{195.85\,V_{ma} + 68.9}{68.9}} \text{ in W/m}$$

SCHEMATIC DIAGRAM OF HEAT TRANSFER FROM HIGHER AIR TEMPERATURE TO MASS OF THREE MATERIALS THEN FROM SURFACE OF OUTER MATERIAL TO LOWER TEMPERATURE AIR

Figure 159

In Metric Units

$$Q_{mc} = 1.957\,(t_{mi} - t_{msi})^{5/4}\sqrt{\frac{195.85\,V_{mi} + 68.9}{68.9}}$$

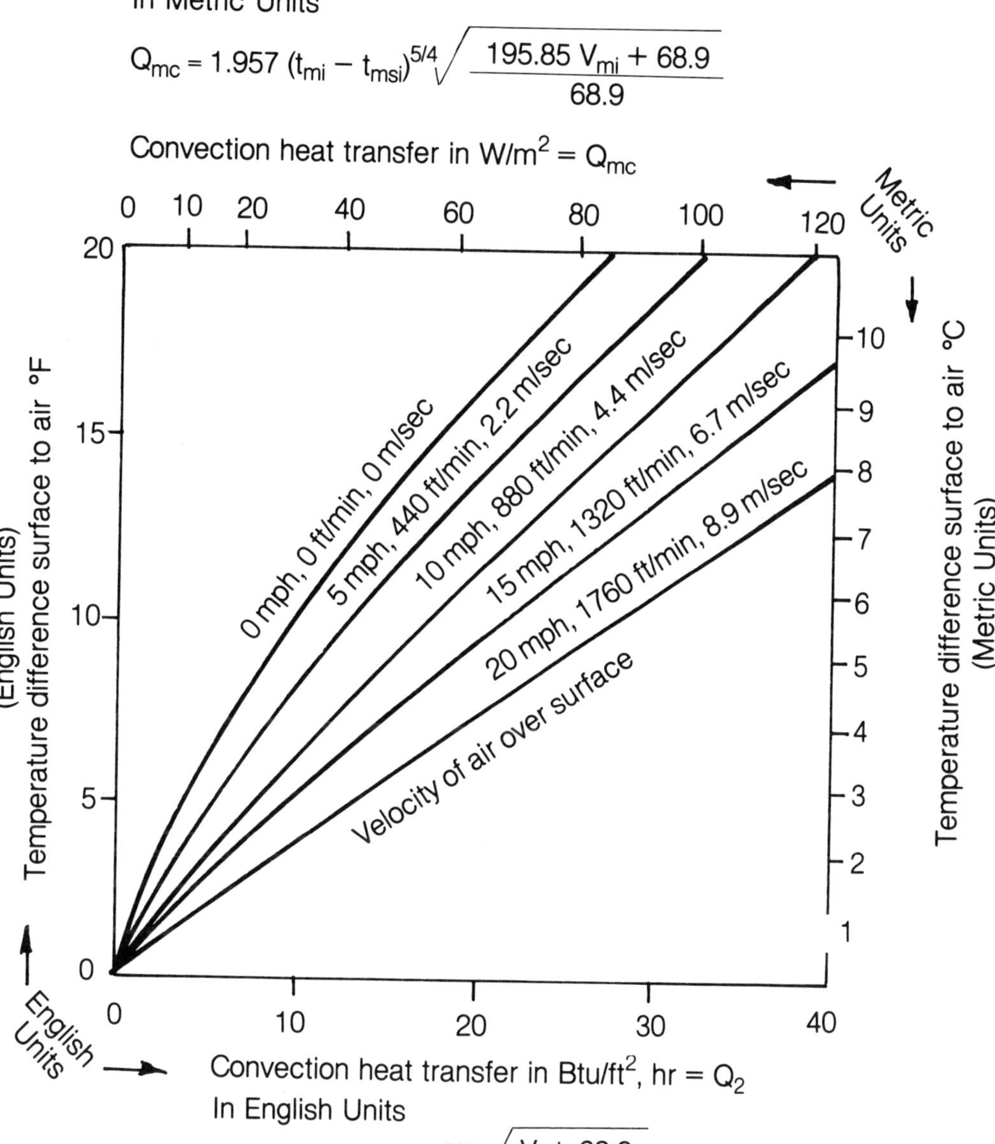

In English Units

$$Q_c = 0.296\,(t_i - t_{si})^{5/4}\sqrt{\frac{V_i + 68.9}{68.9}} = \text{Btu/ft}^2\text{, hr}$$

CONVECTION
HEAT TRANSFER TO, OR FROM, SURFACE
FROM (OR TO) AIR AT VARIOUS AIR VELOCITIES
OVER THE SURFACE

Graph 9

In Metric Units

$$Q_{mr} = 0.548\,\varepsilon \left[\left(\frac{T_{mi}}{55.5}\right)^4 - \left(\frac{T_{msi}}{55.5}\right)^4\right] \text{ in W/m}^2$$

Radiation heat transfer in W/m² = Q_{mr}

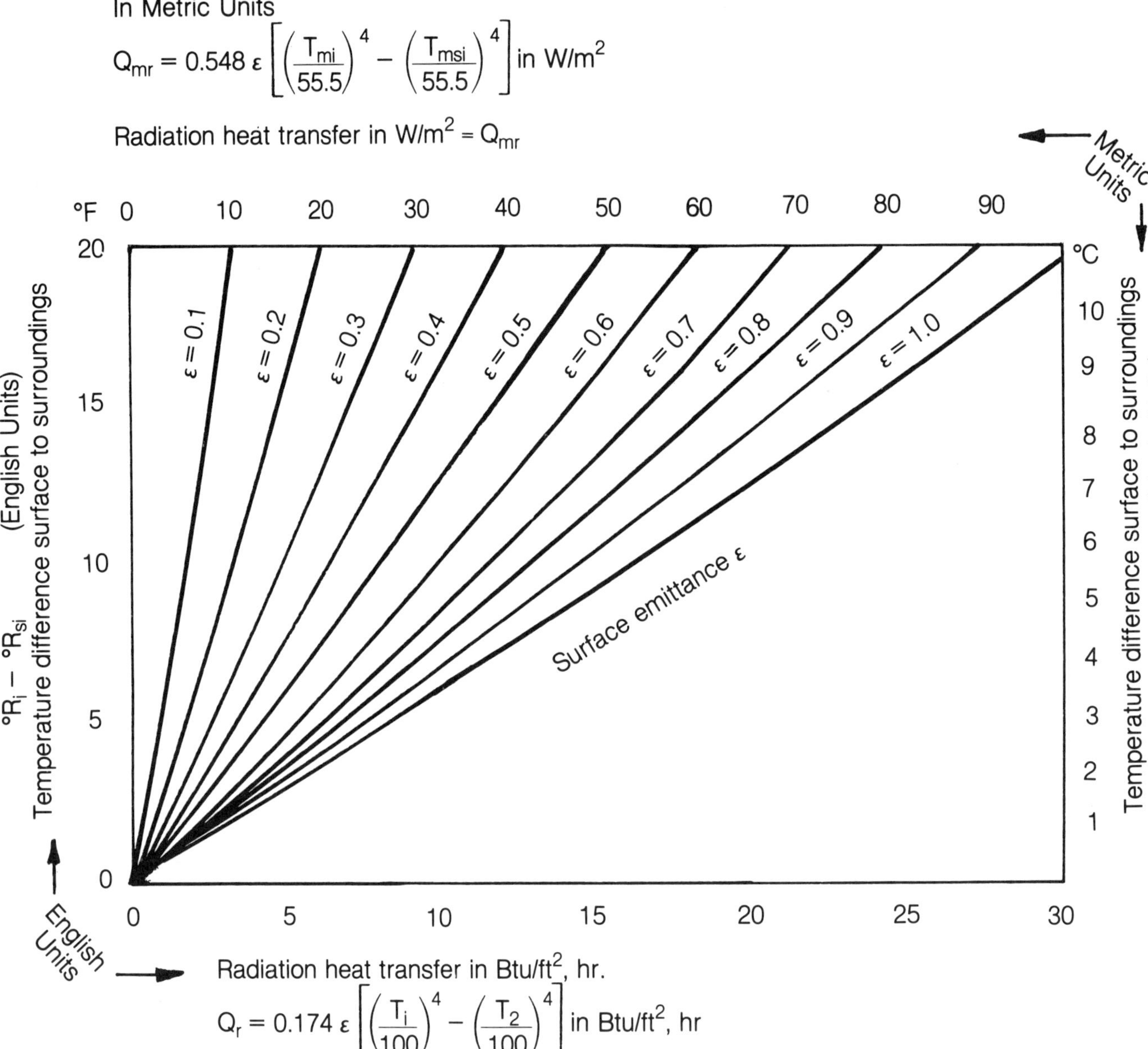

Radiation heat transfer in Btu/ft², hr.

$$Q_r = 0.174\,\varepsilon \left[\left(\frac{T_i}{100}\right)^4 - \left(\frac{T_2}{100}\right)^4\right] \text{ in Btu/ft}^2\text{, hr}$$

HEAT TRANSFER FROM SURFACE TO AIR AND SURROUNDINGS AT VARIOUS EMITTANCES ε

Graph 10

ℓ_2 = thickness of 2nd material in inches
ℓ_3 = thickness of 3rd material in inches
k_1 = conductivity of first material in Btu, in/ft², hr, °F
k_2 = conductivity of second material in Btu, in/ft², hr, °F
k_3 = conductivity of third material in Btu, in/ft², hr, °F
ε_1 = surface emittance higher temperature side ratio to black body
ε_2 = surface emittance lower temperature side ratio to black body
Q_t = heat transfer in Btu/ft², hr

then:

$$Q_t = 0.174\,\varepsilon_1 \left[\left(\frac{T_i}{100}\right)^4 - \left(\frac{T_2}{100}\right)^4\right] + 0.296\,(t_i - t_{sa})^{5/4} \times \sqrt{\left(\frac{V_a + 68.9}{68.9}\right)} = \frac{t_{si} - t_{s4}}{\frac{\ell_1}{k_1} + \frac{\ell_2}{k_2} + \frac{\ell_3}{k_3}} =$$

$$0.174\varepsilon \left[\left(\frac{T_{s4}}{100}\right)^4 - \left(\frac{T_a}{100}\right)^4\right] + 0.296\,(t_{sa} - t_a)^{5/4} \times \sqrt{\left(\frac{V_a + 68.9}{68.9}\right)} \text{ in Btu/ft}^2\text{, hr} \tag{34}$$

The convection part of this section is shown in Graph 9 and the radiation part of this equation is shown in Graph 10.

EQUATION 35—*English Units*

The previous equation can be modified by conversion of various units to thermal resistances. For materials of known conductivity and length of heat path (thickness).

$$R = \frac{\ell}{k} \text{ for more than one material}$$

$$R_1 = \frac{\ell_1}{k_1},\ R_2 = \frac{\ell_2}{k_2},\ R_3 = \frac{\ell_3}{k_3},\ \text{etc.}$$

The heat transfer from (or to) air to (or from) the surface is not quite as simple. Particularly, in most cases, the surface temperature t_{s1} or t_{s4} are unknown.

As the basic heat transfer equation is $Q = \frac{\Delta t}{R}$, then in Equation 38 on the lower side we have

$$Q_t = \frac{(t_{s4} - t_a)}{R_{s4}} \text{ or, } R_{s4} = \frac{(t_{s4} - t_a)}{Q_t}. \text{ Thus}$$

$$R_{s4} = \frac{(t_{si} - t_a)}{0.174\,\varepsilon_2 \left[\left(\frac{T_{s4}}{100}\right)^4 - \left(\frac{T_a}{100}\right)^4\right] + 0.296\,(t_{s4} - t_a)^{5/4}\sqrt{\left(\frac{V + 68.9}{68.9}\right)}}$$

Using resistance values for R_{s4} and R_{si}, Q_t could then be determined by

$$Q_t = \frac{t_i - t_a}{R_{si} + R_1 + R_2 + R_3 + R_{s4}} \text{ in Btu/ft}^2\text{ hr} \tag{35}$$

EQUATION 34a—*Metric Units when:*

t_{mi} = temperature of warmer air (°C)
T_{mi} = temperature of warmer air (°R)
t_{msi} = temperature of initial surface component (°C)
T_{msi} = temperature of initial surface component (°R)
t_{ms2} = temperature of junction of first and second materials (°C)
T_{ms2} = temperature of junction of first and second materials (°R)
t_{ms3} = temperature of junction of second and third materials (°C)
T_{ms4} = temperature of junction of second and third materials (°R)
t_{ma} = temperature of cooler air (°C)
T_{ma} = temperature of cooler air (°R)
V_{m1} = velocity of air over lower temperature surface in m/sec
V_{ma} = velocity of air over lower temperature surface in m/sec
ℓ_{m1} = thickness of first material in meters
ℓ_{m2} = thickness of second material in meters
ℓ_{m3} = thickness of third material in meters
k_{m1} = conductivity of first material in W/mK
k_{m2} = conductivity of second material in W/mK
k_{m3} = conductivity of third material in W/mK
ε_1 = surface emittance of higher temperature side
ε_2 = surface emittance of lower temperature side
Q_{m2} = heat transfer in W/m²

then:

$$Q_{mt} = 0.548\,\varepsilon_a \left[\left(\frac{T_{mi}}{55.5}\right)^4 - \left(\frac{T_{msi}}{55.5}\right)^4\right] + 1.957\,(t_{mi} - t_{msi})^{5/4} \times \sqrt{\left(\frac{196.85V_{mi} + 68.9}{68.9}\right)} = Q_{mt} =$$

$$\frac{t_{s1} - t_{s4}}{\frac{\ell_1}{k_1} + \frac{\ell_2}{k_2} + \frac{\ell_3}{k_3}} = Q_{mt} = 0.548\,\varepsilon_2 \left[\left(\frac{T_{ms4}}{55.5}\right)^4 - \left(\frac{T_{ma}}{55.5}\right)^4\right] + 1.957\,(t_{s4} - t_a)^{5/4}\sqrt{\left(\frac{196.85V_{m2} + 68.9}{68.9}\right)} \text{ in W/m}^2 \tag{34a}$$

The previous equation can be modified by conversion of various units to thermal resistance. For materials of known conductivity and length of heat path (thickness) we have

$$R_m = \frac{\ell_m}{k_m} \text{ and for more than one material } R_{m1} = \frac{\ell_{m1}}{k_{m2}},$$

$$R_{m2} = \frac{\ell_{m2}}{k_m},\ R_{m3} = \frac{\ell_{m3}}{k_{m3}},\ \text{etc.}$$

The heat transfer from (or to) air to (or from) the surface is not quite as simple, since, in most cases, the surface temperatures t_{m1} and t_{m4} are unknown.

From the basic equation we solve for R_{ms4} as follows:

$$Q_{mt} = \frac{(t_{m4} - t_{ma})}{R_{ms4}} \quad \text{or} \quad R_{ms4} = \frac{(t_{ms4} = t_{ma})}{Q_{mt}} \quad .$$

Substituting in the expression for Q_{mt} yields

$$R_{ms4} = \frac{(t_{ms1} - t_{ma})}{0.548\,\varepsilon_2\left[\left(\frac{T_{ms4}}{55.5}\right)^4 - \left(\frac{T_{ma}}{55.5}\right)^4\right] + 1.957\,(t_{ms4} - t_{ma})^{5/4}\sqrt{\left(\frac{196.85V_{m2} + 68.9}{68.9}\right)}}$$

Using resistance values for R_{ms4} and R_{msi}, Q_{mt} could then be determined by

$$Q_{mt} = \frac{(t_{mi} - t_{ma})}{(R_{msi} + R_1 + R_2 + R_3 + R_{ms4})} \quad \text{in W/m}^2 \qquad (35a)$$

The major problem in solving Equation 34 or 34a is that surface temperatures t_{s1} and t_{s4} (or t_{ms1} and t_{ms4}) are unknown. Solution of this equation could only be accomplished by assuming a value for these surface temperatures and solving by interaction until the heat transfer Q_t (or Q_{mt}) for each of the three parts is equal. This is a very complex mathematical procedure.

To simplify solution of these calculations it would be desirable that the heat transfer be directly obtained by dividing the total temperature difference from inside air to outside air by the sum of the resistances of all components of the wall structure and surfaces. Then the heat transfer equation used would be Equation 35 or 35a.

Knowing the conductivities of the components of the structure and the thicknesses of each component, the resistances R_1, R_2, R_3, etc., are easy to determine as was done earlier:

$$\text{Resistance} = \frac{\text{Length of Heat Flow Path (Thickness of Component)}}{\text{Thermal Conductivity of the Material}}$$

or $R = \frac{\ell}{k}$ (English Units); $R_m = \frac{m}{k_m}$ (Metric Units) and in Equation 39, $R_1 = \frac{\ell_1}{k_1}$, $R_2 = \frac{\ell_2}{k_2}$, $R_3 = \frac{\ell_3}{k_3}$, etc. (English Units), and in Equation 39a, $R_{m1} = \frac{\ell_{m1}}{k_{m1}}$, $R_{m2} = \frac{\ell_{m2}}{k_{m2}}$, $R_{m3} = \frac{\ell_{m3}}{k_{m4}}$ (Metric Units).

Using the resistances and approximating the temperature drop from higher to lower temperature, an approximate heat transfer Q_t can be obtained or

$$\text{English Units:} \quad Q_t = \frac{(t_{s1} - t_{s4})}{\frac{\ell_1}{k_1} + \frac{\ell_2}{k_2} = \frac{\ell_3}{k_3}} \quad \text{in Btu/ft}^2\text{, hr}$$

$$\text{Metric Units:} \quad Q_{mt} = \frac{(t_{ms1} - t_{ms4})}{\frac{\ell_{m1}}{k_{m1}} + \frac{\ell_2}{k_{m2}} + \frac{\ell_{m3}}{k_{m3}}} \quad \text{in W/m}^2$$

Once having established this approximate heat transfer Q_t or Q_{mt} and knowing the surface conditions, such as surface emittance and air flow across the surface, it is possible to determine the approximate temperature difference from surface to ambient air. These precalculations are given in English Units in Table 80 and in Metric Units in Table 80a.

Once the temperature difference between surface and air is known, the thermal resistance R_{si} or R_{msi} can be determined. For specific surface conditions and surface to air temperature differences, the thermal resistance is precalculated in English Units in Table 81 and in Metric Units in Table 81a.

In many instances the conductivity of a building material is not given because the material, such as concrete block, has open core spaces. In these instances the overall conductance C of the material is given. For an 8″ concrete block, the conductance is given as 0.90. In this case the resistance R through the 8″ block is

$$R = \frac{1}{C}$$

This R can then be used in Equation 35 and 35a.

To illustrate how heat transfer can be determined through a wall using the preceding equations and tables, the following example is presented.

Example 38—*English Units*

Determine the heat transfer from the inside air of a building through the wall to outside air.

when:

the *air inside* is at 70 °F and moves over all surfaces by natural convection;

the *inner component* of the wall is 1/2″ gypsum board which has a conductance of 2.25; surface emittance is 0.8;

the *middle layer component* is rigid cellular glass block, 2″ thick with conductivity of 0.36;

the *outer layer* is 8″ concrete block with a conductance of 0.9 and a surface emittance of 0.9;

the *outside air* is 0 °F and has a 10 mph wind over the outer surface of the block.

then:

$$Q_t = \frac{t_i - t_a}{R_{si} + R_1 + R_2 + R_3 + R_{s4}}$$

$$Q_t = \frac{70 - 0}{R_{si} + \frac{1}{2.25} + \frac{2}{0.36} + \frac{1}{0.9} + R_{s4}}$$

Q_t HEAT TRANSFER TO, OR FROM, A SURFACE TO AMBIENT AIR

EMIT-TANCE	AIR MOVEMENT ACROSS SURFACE	TEMPERATURE DIFFERENCE SURFACE TO AIR OR AIR TO SURFACE IN DEGREES F														
		1	2	3	4	5	6	7	8	9	10	12	14	16	18	20
1.0	By Radiation Only	0.9	1.7	2.6	3.4	4.3	5.2	6.0	6.7	7.6	8.5	10.1	11.8	13.4	14.9	16.5
	+															
	Natural Convection	1.2	2.4	3.8	5.1	6.5	7.9	9.4	10.7	12.2	14.0	16.7	19.8	22.9	25.9	29.0
	5 mph, 440 ft/min	1.7	3.6	5.7	8.0	10.3	12.5	15.2	17.5	20.1	22.9	28.0	33.5	39.2	44.8	50.5
	10 mph, 880 ft/min	2.0	4.3	6.9	9.6	12.4	14.4	18.4	21.5	24.6	28.0	34.7	41.5	48.6	55.6	63.0
	15 mph, 1320 ft/min	2.2	4.8	7.4	10.9	14.2	17.7	21.1	24.6	28.3	32.1	39.8	47.8	55.9	64.2	70.1
	20 mph, 1760 ft/min	2.4	5.3	8.6	12.0	15.7	19.5	23.4	27.2	31.4	35.6	44.1	53.1	62.2	71.4	81.0
0.9	By Radiation Only	0.8	1.5	2.3	3.1	3.9	4.7	5.4	6.1	6.9	7.7	9.1	10.6	12.1	13.4	14.9
	+															
	Natural Convection	1.1	2.2	3.5	4.8	6.1	7.4	8.8	10.0	11.5	13.1	15.7	18.6	21.6	24.4	27.4
	5 mph, 440 ft/min	1.6	3.4	5.4	7.7	9.9	12.0	14.6	16.8	19.4	22.1	27.0	32.3	37.9	43.3	48.9
	10 mph, 880 ft/min	1.9	4.1	6.6	9.3	12.0	13.9	17.8	20.9	23.9	27.2	33.7	40.3	47.3	54.0	61.4
	15 mph, 1320 ft/min	2.1	4.6	7.1	10.6	13.8	17.2	20.5	24.0	27.6	31.3	38.8	46.6	54.6	62.7	68.5
	20 mph, 1760 ft/min	2.3	5.1	8.3	11.7	15.3	19.0	23.2	26.6	30.7	34.8	43.1	51.9	60.9	69.9	79.4
0.8	By Radiation Only	0.7	1.4	2.1	2.7	3.4	4.2	4.8	5.4	6.1	6.8	8.1	9.4	10.7	11.9	13.2
	+															
	Natural Convection	1.0	2.1	3.3	4.5	5.6	6.9	8.2	9.4	10.7	12.1	14.7	17.4	20.2	22.9	25.7
	5 mph, 440 ft/min	1.5	3.3	5.2	7.3	9.4	11.5	14.0	16.2	18.6	21.2	26.0	31.1	36.5	41.8	47.2
	10 mph, 880 ft/min	1.8	4.0	6.4	8.9	11.5	13.4	17.2	20.2	23.1	26.3	32.7	39.1	45.9	52.6	59.7
	15 mph, 1320 ft/min	2.0	4.5	7.0	10.2	13.3	16.7	19.9	23.4	26.8	30.4	37.8	45.4	53.2	61.2	66.8
	20 mph, 1760 ft/min	2.2	5.0	8.1	11.3	14.8	18.5	22.2	25.9	29.9	33.9	42.1	50.7	59.7	68.4	77.7
0.7	By Radiation Only	0.6	1.2	1.8	2.4	3.0	3.6	4.2	4.7	5.4	6.0	7.1	8.3	9.4	10.4	11.6
	+															
	Natural Convection	0.9	1.9	3.0	4.1	5.2	6.3	7.6	8.5	10.0	11.4	13.7	16.3	18.9	21.4	24.1
	5 mph, 440 ft/min	1.4	3.1	4.9	7.0	9.0	10.9	13.4	15.5	17.9	20.4	25.0	30.0	35.2	40.3	45.6
	10 mph, 880 ft/min	1.7	3.8	6.1	8.6	11.1	13.6	16.6	19.5	22.4	25.5	31.7	38.0	44.6	51.1	58.1
	15 mph, 1320 ft/min	1.9	4.3	6.6	9.9	12.9	16.1	19.3	22.6	26.1	29.6	36.8	44.3	51.9	59.7	65.2
	20 mph, 1760 ft/min	2.1	4.8	7.8	11.0	14.4	17.9	21.6	25.2	29.0	33.1	41.1	49.6	58.2	66.9	76.1
0.6	By Radiation Only	0.5	1.0	1.5	2.0	2.5	3.0	3.5	4.0	4.5	5.0	6.0	7.0	8.0	8.9	9.9
	+															
	Natural Convection	0.8	1.7	2.7	3.7	4.7	5.8	6.9	8.0	9.1	10.4	12.7	15.1	17.5	19.9	22.4
	5 mph, 440 ft/min	1.3	2.9	4.6	6.6	8.5	10.4	12.7	14.8	17.0	19.5	24.0	28.8	33.8	38.8	43.9
	10 mph, 880 ft/min	1.6	3.6	5.8	8.2	10.6	13.2	15.9	18.8	21.5	24.5	30.6	36.7	43.2	49.6	56.4
	15 mph, 1320 ft/min	1.8	4.1	6.3	9.5	12.4	15.5	18.6	21.9	25.2	28.6	35.7	43.0	50.5	58.2	63.5
	20 mph, 1760 ft/min	2.0	4.6	7.5	10.6	13.9	17.3	20.9	24.5	28.3	32.1	40.0	48.0	56.8	65.4	74.3
0.5	By Radiation Only	0.4	0.9	1.3	1.7	2.1	2.6	3.0	3.5	3.9	4.3	5.1	5.9	6.7	7.5	8.3
	+															
	Natural Convection	0.7	1.6	2.5	3.4	4.3	5.3	6.4	7.4	8.5	9.6	11.7	13.9	16.2	18.5	20.6
	5 mph, 440 ft/min	1.2	2.8	4.4	6.3	8.1	9.9	12.2	14.2	16.4	18.7	23.0	27.6	32.6	37.4	42.3
	10 mph, 880 ft/min	1.5	3.5	5.6	7.9	10.2	12.8	15.4	18.3	20.9	23.8	29.7	35.6	41.9	48.2	54.8
	15 mph, 1320 ft/min	1.7	4.0	6.1	9.2	12.0	15.1	18.1	21.4	24.6	27.9	54.8	41.9	49.2	56.8	61.9
	20 mph, 1760 ft/min	1.9	4.5	7.3	10.3	13.5	16.9	20.4	24.0	27.7	31.4	39.1	47.2	55.5	64.0	72.8

ENGLISH UNITS

TABLE 80

Q_{mt} HEAT TRANSFER TO, OR FROM, A SURFACE TO AMBIENT AIR

EMITTANCE	SURFACE	TOTAL HEAT TRANSFER per sq. meter per hr. (W/m²)									
	AIR MOVEMENT ACROSS SURFACE	TEMPERATURE DIFFERENCE, SURFACE TO AIR OR AIR TO SURFACE IN DEGREES C									
	km/h m/sec	1	2	3	4	5	6	7	8	9	10
1.0	By Radiation Only	4.7	9.4	14.2	19.2	24.1	29.0	34.0	37.8	42.2	46.6
	+										
	Natural Convection	6.6	14.0	21.9	30.2	38.7	47.4	56.3	64.1	72.7	81.4
	5 km/h, 1.38 m/s	8.9	19.6	31.3	43.6	56.5	69.8	83.5	96.2	109.9	123.9
	10 km/h, 2.78 m/s	10.3	23.1	37.1	52.0	67.6	83.8	100.5	116.2	133.1	150.3
	15 km/h, 4.17 m/s	11.5	25.9	41.8	58.6	76.4	94.9	113.8	133.0	151.4	171.2
	20 km/h, 5.76 m/s	12.5	27.9	45.8	64.3	83.9	104.4	125.4	144.6	167.2	189.3
0.9	By Radiation Only	4.2	8.5	12.8	17.3	21.7	26.1	30.6	34.0	38.0	41.9
	+										
	Natural Convection	6.1	13.1	20.5	28.3	36.3	44.5	52.9	60.3	68.5	76.7
	5 km/h, 1.38 m/s	8.4	18.7	29.9	41.7	54.1	66.9	80.1	92.4	105.7	119.2
	10 km/h, 2.78 m/s	9.8	22.2	35.7	50.1	65.2	80.9	97.1	112.4	128.9	145.6
	15 km/h, 4.17 m/s	11.0	25.0	40.4	56.7	74.0	92.0	110.4	129.2	147.2	166.5
	20 km/h, 5.56 m/s	12.2	27.0	44.4	62.4	81.5	101.5	122.0	141.8	163.0	184.6
0.8	By Radiation Only	3.8	7.5	11.4	15.4	19.3	23.2	27.2	30.2	33.8	37.3
	+										
	Natural Convection	5.7	12.1	19.1	26.4	33.9	41.6	49.5	56.5	64.3	72.1
	5 km/h, 1.38 m/s	8.0	17.7	28.5	39.8	51.7	64.0	76.7	88.6	101.5	114.6
	10 km/h, 2.78 m/s	9.4	21.2	34.3	48.2	62.8	78.0	93.7	108.6	124.7	141.0
	15 km/h, 4.17 m/s	10.6	24.0	39.0	54.8	71.6	89.1	107.0	125.4	143.0	161.9
	20 km/h, 5.56 m/s	11.6	26.0	43.0	60.5	79.1	98.6	118.6	138.0	158.8	179.0
0.7	By Radiation Only	3.3	6.6	9.9	13.4	16.9	20.3	23.8	26.5	29.5	32.6
	+										
	Natural Convection	5.2	11.2	17.6	24.4	31.5	38.7	45.1	52.8	60.0	67.4
	5 km/h, 1.38 m/s	7.5	16.8	27.0	37.8	49.3	61.1	73.3	84.9	97.2	109.9
	10 km/h, 2.78 m/s	8.9	20.3	32.8	46.2	60.4	75.1	90.3	104.9	120.4	136.3
	15 km/h, 4.17 m/s	10.1	23.1	37.5	52.8	69.2	86.2	103.6	121.7	138.7	157.2
	20 km/h, 5.56 m/s	11.1	25.1	41.5	58.5	76.7	95.7	115.2	134.3	154.5	175.3
0.6	By Radiation Only	2.8	5.6	8.5	11.5	14.5	17.4	20.4	22.7	25.3	28.0
	+										
	Natural Convection	4.7	10.2	16.2	22.5	29.1	35.8	42.7	49.0	55.8	62.8
	5 km/h, 1.38 m/s	7.0	15.8	25.6	35.9	46.9	58.2	69.9	81.1	93.0	105.3
	10 km/h, 2.78 m/s	8.4	19.3	31.4	44.3	58.0	72.2	86.9	101.1	116.2	131.7
	15 km/h, 4.17 m/s	9.6	22.1	36.1	50.9	66.8	83.3	100.2	117.9	134.5	152.6
	20 km/h, 5.56 m/s	10.6	24.1	40.1	56.6	74.3	92.8	111.8	130.5	150.3	170.7
0.5	By Radiation Only	2.4	4.7	7.1	9.6	12.1	14.5	17.0	18.9	21.1	23.3
	+										
	Natural Convection	4.3	9.3	14.8	20.6	26.7	32.9	39.3	45.2	51.6	58.1
	5 km/h, 1.38 m/s	6.6	14.9	24.2	34.0	44.5	55.3	66.5	77.3	88.8	100.6
	10 km/h, 2.78 m/s	8.0	18.4	30.0	42.4	55.6	69.3	83.5	97.3	112.0	126.0
	15 km/h, 4.17 m/s	9.2	21.2	34.7	49.0	64.4	80.4	96.8	114.1	130.3	147.9
	20 km/h, 5.56 m/s	10.2	23.2	38.7	54.7	71.9	89.9	108.4	126.8	146.1	166.0

METRIC UNITS

TABLE 80a

SURFACE RESISTANCE TO HEAT TRANSFER R

SURFACE		THERMAL RESISTANCE OF SURFACE R (English Units)														
EMITTANCE	AIR MOVEMENT ACROSS SURFACE	TEMPERATURE DIFFERENCE, SURFACE TO AIR OR AIR TO SURFACE IN DEGREES F														
	mph ft/min	1	2	3	4	5	6	7	8	9	10	12	14	16	18	20
1.0	Natural Convection	0.83	0.83	0.79	0.78	0.77	0.76	0.75	0.74	0.73	0.71	0.71	0.70	0.69	0.68	0.68
	5 mph, 440 ft/min	0.59	0.56	0.53	0.50	0.48	0.47	0.46	0.45	0.45	0.44	0.44	0.43	0.42	0.41	0.40
	10 mph, 880 ft/min	0.50	0.47	0.44	0.42	0.41	0.40	0.38	0.37	0.36	0.35	0.34	0.34	0.33	0.33	0.32
	15 mph, 1320 ft/min	0.45	0.42	0.40	0.37	0.35	0.34	0.33	0.32	0.32	0.31	0.30	0.30	0.29	0.29	0.28
	20 mph, 1760 ft/min	0.42	0.38	0.35	0.33	0.32	0.31	0.30	0.29	0.29	0.28	0.28	0.27	0.26	0.24	0.24
0.9	Natural Convection	0.91	0.91	0.86	0.83	0.82	0.81	0.80	0.79	0.78	0.76	0.75	0.74	0.73	0.72	0.72
	5 mph, 440 ft/min	0.63	0.59	0.56	0.52	0.51	0.50	0.48	0.47	0.46	0.45	0.45	0.44	0.43	0.42	0.41
	10 mph, 880 ft/min	0.53	0.49	0.45	0.43	0.43	0.42	0.40	0.38	0.37	0.36	0.36	0.35	0.34	0.34	0.33
	15 mph, 1320 ft/min	0.48	0.43	0.42	0.38	0.36	0.35	0.34	0.33	0.32	0.32	0.32	0.31	0.31	0.30	0.29
	20 mph, 1760 ft/min	0.43	0.39	0.36	0.34	0.33	0.32	0.30	0.30	0.29	0.29	0.29	0.28	0.27	0.26	0.25
0.8	Natural Convection	1.00	0.95	0.91	0.89	0.88	0.87	0.85	0.85	0.84	0.83	0.82	0.80	0.79	0.78	0.77
	5 mph, 440 ft/min	0.67	0.61	0.58	0.55	0.53	0.52	0.50	0.49	0.48	0.47	0.46	0.45	0.44	0.43	0.42
	10 mph, 880 ft/min	0.56	0.50	0.47	0.45	0.44	0.43	0.41	0.40	0.39	0.38	0.37	0.36	0.35	0.35	0.34
	15 mph, 1320 ft/min	0.50	0.44	0.42	0.39	0.37	0.36	0.35	0.34	0.33	0.33	0.33	0.32	0.32	0.31	0.30
	20 mph, 1760 ft/min	0.45	0.40	0.37	0.35	0.34	0.33	0.32	0.31	0.30	0.29	0.29	0.28	0.27	0.26	0.25
0.7	Natural Convection	1.11	1.05	1.00	0.98	0.96	0.95	0.94	0.91	0.90	0.88	0.87	0.86	0.85	0.84	0.83
	5 mph, 440 ft/min	0.71	0.64	0.61	0.57	0.56	0.54	0.52	0.51	0.50	0.49	0.48	0.47	0.46	0.45	0.44
	10 mph, 880 ft/min	0.59	0.53	0.49	0.48	0.46	0.44	0.42	0.41	0.40	0.39	0.38	0.37	0.36	0.35	0.34
	15 mph, 1320 ft/min	0.53	0.47	0.44	0.41	0.39	0.37	0.36	0.35	0.34	0.33	0.33	0.32	0.32	0.32	0.31
	20 mph, 1760 ft/min	0.48	0.42	0.38	0.36	0.35	0.34	0.33	0.32	0.31	0.30	0.29	0.28	0.27	0.26	0.26
0.6	Natural Convection	1.25	1.18	1.11	1.08	1.06	1.03	1.01	1.00	0.98	0.96	0.94	0.92	0.91	0.90	0.88
	5 mph, 440 ft/min	0.77	0.69	0.65	0.61	0.59	0.58	0.56	0.55	0.53	0.51	0.50	0.49	0.48	0.47	0.46
	10 mph, 880 ft/min	0.63	0.56	0.52	0.49	0.47	0.46	0.44	0.42	0.41	0.41	0.40	0.39	0.37	0.36	0.35
	15 mph, 1320 ft/min	0.53	0.49	0.47	0.42	0.40	0.39	0.38	0.37	0.36	0.35	0.34	0.33	0.32	0.31	0.31
	20 mph, 1760 ft/min	0.50	0.43	0.40	0.38	0.36	0.35	0.33	0.33	0.32	0.31	0.30	0.29	0.28	0.28	0.27
0.5	Natural Convection	1.43	1.25	1.20	1.17	1.16	1.13	1.09	1.08	1.05	1.04	1.02	1.00	0.99	0.97	0.96
	5 mph, 440 ft/min	0.88	0.71	0.68	0.63	0.61	0.61	0.57	0.56	0.55	0.53	0.52	0.51	0.50	0.49	0.47
	10 mph, 880 ft/min	0.67	0.57	0.54	0.51	0.50	0.47	0.45	0.44	0.43	0.42	0.51	0.39	0.38	0.37	0.36
	15 mph, 1320 ft/min	0.59	0.50	0.49	0.43	0.42	0.40	0.39	0.37	0.36	0.36	0.35	0.34	0.33	0.32	0.32
	20 mph, 1760 ft/min	0.53	0.44	0.41	0.39	0.37	0.36	0.34	0.33	0.32	0.32	0.31	0.30	0.29	0.28	0.27

ENGLISH UNITS

TABLE 81

SURFACE		THERMAL RESISTANCE OF SURFACE R_{ms4} (Metric Units)									
EMITTANCE	AIR MOVEMENT ACROSS SURFACE	TEMPERATURE DIFFERENCE, SURFACE TO AIR OR AIR TO SURFACE, IN DEGREES C or K									
	Km/hr m/sec	1	2	3	4	5	6	7	8	9	10
1.0	Natural Convection	0.152	0.143	0.137	0.132	0.129	0.126	0.124	0.125	0.124	0.123
	5 km/h, 1.38 m/sec	0.112	0.102	0.096	0.092	0.089	0.086	0.084	0.083	0.082	0.081
	10 km/h, 2.78 m/sec	0.097	0.086	0.081	0.077	0.073	0.071	0.070	0.069	0.067	0.066
	15 km/h, 4.17 m/sec	0.087	0.077	0.072	0.068	0.065	0.063	0.062	0.060	0.059	0.058
	20 km/h, 5.56 m/sec	0.080	0.071	0.066	0.062	0.060	0.057	0.056	0.055	0.054	0.053
0.9	Natural Convection	0.164	0.153	0.146	0.141	0.138	0.135	0.134	0.133	0.132	0.130
	5 km/h, 1.38 m/sec	0.119	0.107	0.100	0.096	0.092	0.090	0.087	0.086	0.085	0.084
	10 km/h, 2.78 m/sec	0.102	0.090	0.084	0.080	0.077	0.074	0.072	0.071	0.070	0.069
	15 km/h, 4.17 m/sec	0.091	0.080	0.074	0.071	0.068	0.065	0.063	0.062	0.061	0.060
	20 km/h, 5.56 m/sec	0.081	0.074	0.068	0.064	0.061	0.059	0.057	0.056	0.054	0.054
0.8	Natural Convection	0.175	0.165	0.157	0.152	0.147	0.144	0.141	0.140	0.139	0.138
	5 km/h, 1.38 m/sec	0.125	0.113	0.105	0.101	0.097	0.094	0.091	0.090	0.089	0.087
	10 km/h, 2.78 m/sec	0.106	0.094	0.087	0.083	0.080	0.077	0.074	0.073	0.072	0.071
	15 km/h, 4.17 m/sec	0.094	0.083	0.077	0.073	0.070	0.067	0.065	0.064	0.063	0.062
	20 km/h, 5.56 m/sec	0.086	0.076	0.069	0.066	0.063	0.061	0.059	0.058	0.057	0.056
0.7	Natural Convection	0.192	0.178	0.170	0.163	0.159	0.157	0.155	0.152	0.150	0.148
	5 km/h, 1.38 m/sec	0.133	0.119	0.111	0.106	0.101	0.098	0.096	0.094	0.092	0.091
	10 km/h, 2.78 m/sec	0.112	0.096	0.091	0.087	0.083	0.080	0.078	0.076	0.075	0.073
	15 km/h, 4.17 m/sec	0.099	0.086	0.080	0.076	0.072	0.070	0.068	0.066	0.065	0.064
	20 km/h, 5.56 m/sec	0.090	0.079	0.072	0.068	0.065	0.063	0.061	0.059	0.058	0.057
0.6	Natural Convection	0.212	0.196	0.185	0.178	0.172	0.168	0.165	0.163	0.161	0.159
	5 km/h, 1.38 m/sec	0.143	0.127	0.117	0.111	0.107	0.103	0.101	0.099	0.097	0.094
	10 km/h, 2.78 m/sec	0.119	0.104	0.096	0.090	0.086	0.083	0.081	0.079	0.077	0.076
	15 km/h, 4.17 m/sec	0.104	0.090	0.083	0.079	0.075	0.072	0.070	0.068	0.067	0.066
	20 km/h, 5.56 m/sec	0.094	0.083	0.075	0.071	0.068	0.065	0.063	0.061	0.060	0.059
0.5	Natural Convection	0.232	0.215	0.203	0.194	0.187	0.184	0.178	0.176	0.174	0.172
	5 km/h, 1.38 m/sec	0.152	0.134	0.124	0.117	0.112	0.108	0.105	0.103	0.101	0.099
	10 km/h, 2.78 m/sec	0.125	0.108	0.100	0.094	0.090	0.087	0.084	0.082	0.080	0.079
	15 km/h, 4.17 m/sec	0.108	0.094	0.086	0.082	0.078	0.075	0.072	0.070	0.069	0.068
	20 km/h, 5.56 m/sec	0.098	0.086	0.077	0.073	0.070	0.067	0.066	0.063	0.061	0.060

METRIC UNITS

TABLE 81a

To find R_{si}, first assume that the temperature difference from surface to surface is 60°F. Then, the estimated heat transfer would be $Q_t = \frac{60}{0.44 + 5.55 + 1.11} = \frac{60}{7.10} = 8.44.$

From Table 81 $\varepsilon = 0.9$ for the outer surface and air movement at 10 mph we see that a temperature difference of three Fahrenheit degrees gives a resistance of 0.45; a Δt of four degrees yields a resistance of 0.43. Thus the surface resistance R_s is approximately 0.44.

The inner air is at $\varepsilon = 0.8$ and air movement is natural convection. From Table 81 the temperature difference ΔT is slightly higher than 7°F. Going to Table 81, the thermal resistance at $\Delta t = 8$°F and $\varepsilon = 0.8$ for natural convection is 0.85. Using these resistances, yields

$$Q_t = \frac{70 - 0}{0.85 + 0.44 + 5.55 + 1.11 + 0.44} = \frac{70}{8.4} = 8.3 \text{ Btu/ft}^2\text{, hr}$$

This is very close to the heat transfer used to determine R_{s1} and R_{s4}.

Example 38a—*Metric Units*

when:

the *air inside* is 21.1°C and moves over the wall by natural convection;

the *inner component* of the wall is 1.27 cm gypsum board and has a conductance of 12.77;

the *middle layer* is rigid cellular glass 5.04 cm thick with conductivity of 0.623 W/mK;

the *outer component* is 20.32 cm thick cinder block with a conductance of 5.118;

the *air outside* is −17.8°C and has a 4.47 m/sec velocity over the surface of the block.

$$Q_{mt} = \frac{t_{mi} - t_{ma}}{R_{msi} + R_{m1} + R_{m2} + R_m + R_{s4}}$$

$$= \frac{21.1 - (-17.8)}{R_{m1} + \frac{1}{12.77} + \frac{0.0504}{0.0518} + \frac{1}{5.118} + R_{msa}}$$

To find R_{ms1} start with the fact that the temperature difference from surface to surface is 33°C; then the estimated heat transfer becomes

$$Q_{mt} = \frac{33}{0.078 + 0.973 + 0.195} = \frac{33}{1.246} = 26 \text{ W/m}^2.$$

From Table 81a the outer surface at $\varepsilon = 0.9$ and air movement at 4.47 m/sec, the temperature differential is approximately 3°C using the thermal resistance value of 0.074.

From Table 80a at natural convection with $\varepsilon = 0.8$, the temperature is approximately 6° and from Table 81a, the thermal resistance is 0.144.

$$Q_{mc} = \frac{38.9}{0.144 + 0.078 + 0.973 + 0.195 + 0.078} = \frac{38.9}{1.47} = 26.4 \text{ W/m}^2$$

CHECK: 26.4 W/m² × 0.317 = 8.3 Btu/ft², hr.

It should be noted that 66 percent of the total resistance to heat flow is provided by the insulation. Without the insulation, the heat loss in English Units would be $\frac{70}{2.85} = 24.5$ Btu/ft², hr and in Metric Units it would have been $\frac{37.9}{0.497} = 78.2$ W/m².
This represents a savings of 16.2 Btu/ft², hr or 51.8 W/m².

HEAT TRANSFER FROM ONE SURFACE OF A STRUCTURAL COMPONENT, CONTAINING AN AIR SPACE CAVITY, TO SURFACE ON THE OUTER SIDE

A cavity in a building structural component would transfer heat from one surface to the enclosed air then to the surface on the other side. The heat transfer would be by conduction, convection and radiation. A schematic diagram of this heat transfer is shown in Figure 160.

Because of the complexity of the equations, the overall resistance to heat transfer it generally gives in R_{cav} the thermal resistance of the cavity is, in English Units

$$Q_t = \frac{t_{s1} - t_{s2}}{R_{cav}} \text{ or } R_{cav} = \frac{t_{si} = t_{s2}}{Q_t}$$

when:

t_1 = higher temperature surface of mass on high side (°F)
t_2 = lower temperature of surface of mass on low side (°F)
ℓ_1 = thickness of mass (high side) inches
ℓ_2 = thickness of mass (low side) inches
k_1 = conductivity of mass (high side) Btu, in/ft², hr, °F
k_2 = conductivity of mass (low side) Btu, in/ft², hr, °F
R_1 = thermal resistance (high side)
R_2 = thermal resistance (low side)
R_{cav} = thermal resistance of cavity
Q_t = heat transfer from surface S_1 to surface S_2, Btu/ft², hr

then:

$$Q_t = \frac{t_1 - t_2}{\frac{\ell_1}{k_1} + R_{oav} + \frac{\ell_2}{k_2}} \text{ in Btu/ft}^2\text{, hr}$$

or,

$$Q_t = \frac{t_1 - t_2}{R_1 + R_{cav} + R_2} \text{ in Btu/ft}^2\text{, hr}$$

If outer surface resistances are included (air film resistance)

CONVERSION TABLE — Btu/ft², hr to W/m²

Btu/ft², hr	W/m² (Per Hour) 0	1	2	3	4	5	6	7	8	9
0	0	3.15	6.30	9.46	12.61	15.76	18.91	22.07	25.22	28.37
10	31.52	34.68	37.83	40.98	44.13	47.29	50.44	53.59	56.74	59.90
20	63.05	66.20	69.35	72.51	75.66	78.81	81.96	85.12	88.27	91.42
30	94.57	97.73	100.88	104.03	107.18	110.34	113.49	116.64	119.79	122.95
40	126.10	129.25	132.40	135.56	138.71	141.86	145.01	148.17	151.32	154.47
50	157.62	160.78	163.98	167.08	170.23	173.39	176.54	179.69	182.84	186.00
60	189.85	192.30	195.45	198.61	201.76	204.91	208.06	211.22	214.37	217.52
70	220.67	223.83	226.98	230.13	233.28	236.44	239.59	242.74	245.89	249.05
80	252.20	255.35	258.50	261.66	264.81	267.96	271.11	274.77	277.42	280.57
90	283.72	286.88	290.03	293.18	296.33	299.49	302.79	305.79	308.94	312.10
100	315.24	318.40	321.55	324.71	327.85	331.01	334.16	337.32	340.47	343.62
110	346.77	349.93	353.08	356.23	359.38	362.54	365.69	368.84	371.99	379.15
120	378.30	381.45	384.60	387.76	390.91	394.06	397.21	400.37	403.52	406.67
130	409.82	412.98	416.13	419.30	422.43	425.58	428.74	431.89	435.04	438.19
140	441.35	444.50	447.65	450.80	453.96	457.11	460.26	463.41	466.57	469.72
150	472.87	476.02	479.18	482.40	485.48	488.63	491.79	494.94	498.09	501.24
160	504.40	507.55	510.70	513.85	517.01	520.16	523.31	526.46	529.66	532.77
170	535.32	539.07	542.23	545.38	548.53	551.68	554.84	557.99	561.14	564.29
180	567.45	570.60	573.75	576.90	580.06	583.21	586.36	589.59	592.67	595.82
190	598.31	602.12	605.28	608.43	611.58	614.73	617.89	621.04	624.19	627.54
200	630.49	633.65	636.80	639.95	643.11	646.26	649.41	652.56	655.72	658.87
210	662.02	665.17	668.32	671.48	674.63	677.78	680.96	684.09	687.24	690.39
220	693.54	696.70	699.85	703.00	706.15	709.31	712.46	715.61	718.77	721.92
230	725.07	728.22	731.36	734.53	737.68	740.83	743.99	747.14	750.29	753.44
240	756.60	759.75	762.90	766.05	769.21	772.36	775.51	778.66	781.82	784.97
250	788.12	791.27	794.42	797.58	800.73	803.88	807.04	810.19	813.34	816.49
260	819.65	822.80	825.95	829.10	832.25	833.41	838.56	841.71	844.86	848.02
270	851.16	854.32	857.47	860.63	863.78	866.93	870.08	873.24	876.39	879.54
280	882.69	885.84	889.00	892.15	895.30	898.46	901.61	904.76	907.91	911.07
290	914.21	917.37	920.52	923.68	926.83	929.98	933.13	936.27	939.44	942.59
300	945.74	948.98	952.05	955.20	958.35	961.66	964.66	967.81	970.96	974.12
310	977.27	980.42	983.57	986.73	989.88	993.03	996.18	999.34	1002.49	1005.64
320	1008.79	1011.95	1015.10	1018.25	1021.40	1024.56	1027.71	1030.86	1034.01	1037.17
330	1040.31	1043.47	1046.62	1049.78	1052.93	1056.08	1059.23	1062.39	1065.54	1066.69
340	1071.84	1075.00	1078.15	1081.30	1084.45	1087.61	1090.76	1093.01	1097.00	1100.22
350	1103.37	1106.52	1109.67	1112.83	1115.98	1119.13	1122.28	1125.43	1128.59	1131.74
360	1134.89	1138.05	1141.20	1144.35	1147.50	1150.86	1152.81	1156.96	1160.11	1163.27
370	1166.42	1169.57	1172.72	1175.88	1179.03	1182.18	1185.33	1188.49	1191.64	1194.79
380	1197.94	1201.09	1204.24	1207.40	1210.55	1213.71	1216.86	1220.01	1223.16	1226.32
390	1229.47	1232.62	1235.77	1238.93	1242.08	1245.23	1248.38	1251.53	1254.69	1257.84
400	1261.00	1264.14	1267.30	1270.45	1273.60	1276.75	1279.91	1283.06	1286.21	1289.36
410	1292.52	1295.67	1298.82	1301.97	1305.13	1308.27	1311.43	1314.58	1317.74	1320.89
420	1324.04	1327.19	1330.35	1333.50	1336.65	1339.60	1342.56	1346.11	1349.26	1352.41
430	1355.56	1358.72	1361.87	1365.02	1368.18	1371.33	1374.48	1377.63	1380.79	1383.94
440	1387.09	1390.24	1393.40	1396.55	1399.70	1402.85	1406.00	1409.16	1412.31	1415.46
450	1418.62	1421.77	1424.92	1428.07	1431.23	1434.37	1437.53	1440.68	1443.84	1446.99
460	1450.14	1453.29	1456.45	1459.60	1462.75	1465.90	1469.06	1472.21	1475.36	1478.51
470	1481.67	1484.81	1487.97	1491.12	1494.28	1497.43	1500.58	1503.73	1506.89	1510.04
480	1513.19	1516.34	1519.50	1522.65	1525.80	1528.93	1532.11	1535.26	1536.41	1541.56
490	1544.72	1547.87	1551.02	1554.17	1557.33	1560.48	1563.63	1566.78	1569.94	1573.09
500	1576.24	1579.39	1582.54	1585.70	1588.85	1592.00	1595.16	1598.31	1601.46	1604.61
510	1607.76	1610.92	1614.07	1617.22	1620.36	1623.53	1626.68	1629.83	1632.99	1636.14
520	1639.29	1642.44	1645.60	1648.75	1651.90	1655.05	1658.20	1661.36	1664.51	1667.66
530	1670.81	1673.97	1677.12	1680.27	1683.42	1686.58	1689.73	1692.88	1696.03	1699.19
540	1702.34	1705.49	1708.64	1711.80	1714.95	1718.10	1721.25	1724.41	1727.56	1730.71
550	1733.86	1737.02	1740.17	1743.32	1746.47	1749.63	1752.78	1755.93	1759.08	1762.24
560	1765.39	1768.54	1771.69	1774.85	1778.00	1781.15	1784.30	1787.46	1790.61	1793.76
570	1796.91	1800.06	1803.22	1806.37	1809.52	1812.68	1815.83	1818.98	1822.13	1825.29
580	1828.44	1831.59	1834.74	1837.89	1841.05	1844.20	1847.35	1850.51	1853.66	1856.81
590	1855.96	1863.12	1866.27	1869.42	1872.57	1875.73	1878.88	1882.03	1885.18	1888.34
600	1891.49	1894.64	1897.79	1900.95	1904.10	1907.25	1910.40	1913.56	1916.71	1919.86
610	1923.01	1926.17	1925.32	1932.47	1935.62	1938.78	1941.93	1945.08	1948.23	1951.38
620	1954.54	1957.69	1960.84	1964.00	1967.15	1970.30	1973.45	1976.61	1979.76	1982.91
630	1986.06	1989.22	1992.37	1995.52	1998.67	2001.83	2004.98	2008.13	2011.28	2014.44
640	2017.59	2020.74	2023.89	2027.05	2030.20	2033.35	2036.50	2039.66	2042.81	2045.96
650	2049.11	2052.26	2055.42	2058.57	2061.72	2064.88	2068.03	2071.18	2074.33	2077.48
660	2080.64	2083.79	2086.94	2090.09	2093.24	2096.40	2099.55	2102.70	2105.86	2109.01
670	2112.16	2115.31	2118.47	2121.62	2124.77	2127.92	2131.08	2134.23	2137.38	2140.53
680	2143.69	2146.84	2149.99	2153.14	2156.30	2159.45	2162.60	2165.75	2166.91	2172.06
690	2175.21	2178.36	2181.52	2184.67	2187.82	2190.97	2194.13	2197.28	2200.43	2203.58
700	2206.74	2209.89	2213.04	2216.19	2219.35	2222.50	2225.65	2226.80	2231.96	2235.11
710	2238.26	2241.50	2244.57	2247.72	2250.87	2254.02	2257.17	2260.33	2263.48	2266.64
720	2269.79	2272.94	2276.09	2279.24	2282.40	2285.55	2288.70	2291.85	2295.01	2298.16
730	2301.31	2304.46	2307.62	2310.77	2314.92	2317.07	2320.23	2323.38	2326.53	2329.69
740	2332.80	2335.99	2339.14	2342.29	2345.45	2348.60	2351.75	2354.00	2358.06	2361.21
750	2364.36	2367.51	2370.67	2373.82	2376.97	2380.12	2383.28	2386.43	2389.58	2392.70
760	2395.89	2399.04	2402.19	2405.34	2408.50	2411.65	2414.80	2417.95	2421.11	2424.26
770	2427.41	2430.56	2433.72	2436.87	2440.02	2443.17	2446.33	2449.48	2452.63	2455.78
780	2458.94	2462.09	2465.24	2468.39	2471.55	2474.70	2477.85	2481.00	2484.15	2487.31
790	2490.46	2493.61	2496.76	2499.92	2503.07	2506.22	2509.37	2512.53	2515.68	2518.83
800	2521.98	2525.14	2528.29	2531.44	2534.59	2537.75	2540.90	2544.05	2547.20	2550.36
810	2553.51	2556.66	2559.81	2562.97	2566.12	2569.27	2572.42	2575.58	2578.73	2581.88
820	2585.03	2588.19	2591.34	2594.49	2597.64	2600.80	2603.95	2607.10	2610.25	2613.41
830	2615.32	2619.71	2622.86	2626.02	2629.17	2632.32	2635.47	2638.63	2641.78	2644.93
840	2648.08	2651.24	2654.39	2657.54	2660.69	2663.85	2667.00	2670.15	2673.30	2676.46
850	2679.60	2682.76	2685.91	2689.07	2692.22	2695.37	2698.52	2701.68	2704.83	2707.98
860	2711.13	2714.29	2717.43	2720.59	2723.74	2726.90	2730.05	2733.20	2736.35	2739.51
870	2742.65	2745.81	2748.96	2752.12	2755.27	2758.42	2761.37	2764.73	2767.88	2771.03
880	2774.18	2777.33	2780.49	2783.64	2786.79	2789.95	2793.10	2796.25	2799.40	2802.55
890	2805.71	2808.86	2812.01	2815.17	2818.32	2821.47	2824.62	2827.78	2830.93	2834.08
900	2837.23	2840.39	2843.54	2846.60	2849.84	2852.99	2856.15	2859.30	2862.45	2865.61
910	2868.76	2871.91	2875.06	2878.22	2881.37	2884.52	2887.67	2890.83	2893.98	2897.13
920	2900.28	2903.44	2906.59	2909.74	2912.89	2916.04	2919.20	2922.35	2925.50	2928.65
930	2931.81	2934.96	2938.11	2941.26	2944.42	2947.57	2950.72	2953.87	2957.02	2960.18
940	2963.33	2966.48	2969.64	2972.79	2975.94	2979.09	2982.25	2985.40	2988.55	2991.70
950	2994.86	2998.00	3001.16	3004.31	3007.47	3010.62	3013.77	3016.92	3020.08	3023.23
960	3026.38	3029.53	3032.69	3035.84	3038.99	3042.14	3045.30	3048.45	3051.60	3054.75
970	3057.90	3061.06	3064.21	3067.36	3070.52	3073.67	3076.82	3079.97	2083.13	3086.28
980	3089.43	3092.58	3095.74	3098.89	3102.04	3105.19	3108.35	3111.50	3114.65	3117.80
990	3120.96	3124.11	3127.26	3130.41	3133.57	3136.72	3139.87	3143.02	3146.18	3149.33
1000	3152.48									

TABLE 82

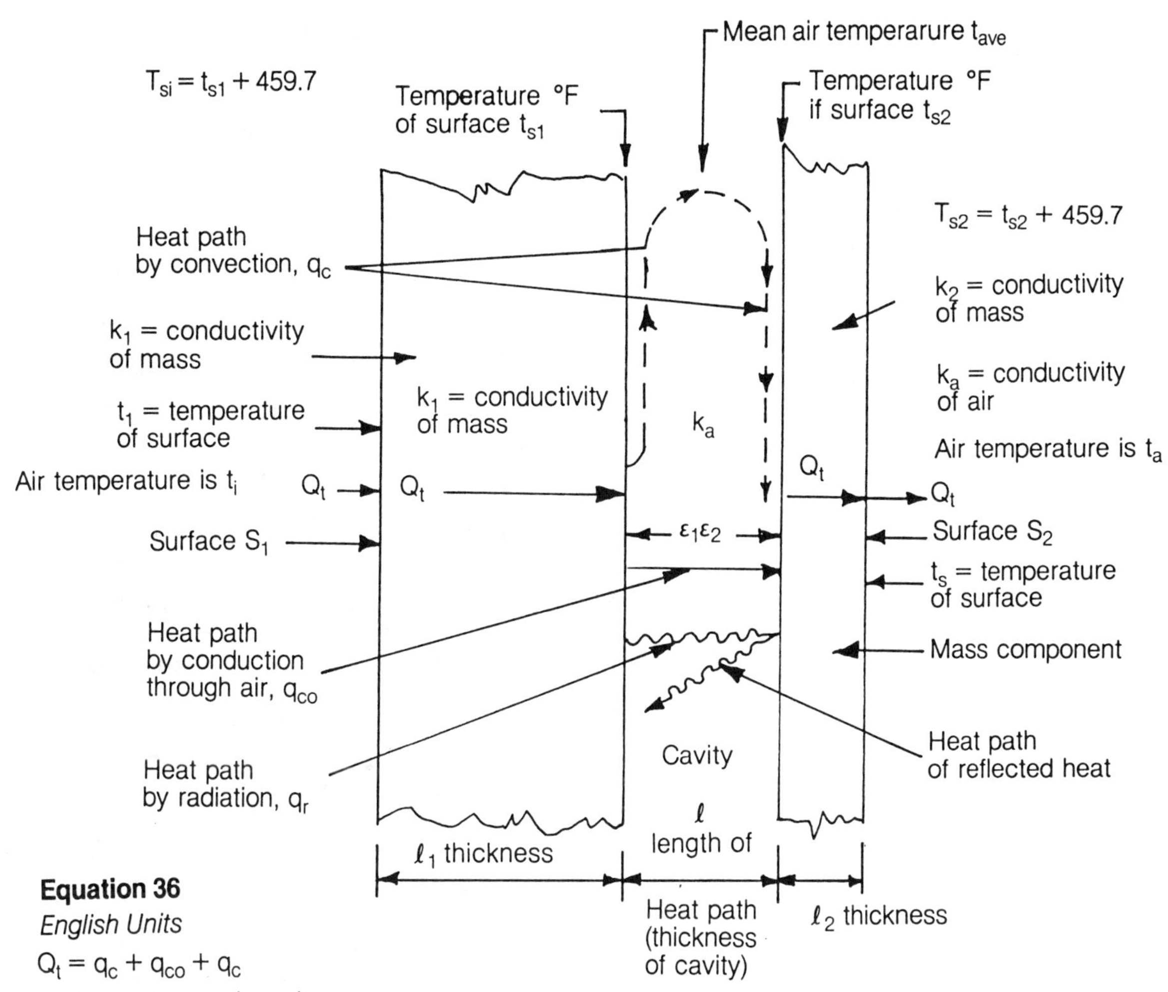

Equation 36

English Units

$Q_t = q_c + q_{co} + q_c$

Heat transfer across cavity only.

$$Q_t = 0.174\,\ell_1 \left[\left(\frac{T_{s1}}{100}\right)^4 - \left(\frac{T_{s2}}{100}\right)^4\right] - 0.174\,(1 - \ell_2)\left[\left(\frac{T_{s1}}{100}\right)^4 - \left(\frac{t_{s1}}{100}\right)^4\right] + \frac{\ell}{k_a} + 0.296\,(t_{s1} - t_{ave})^{5/4}$$

Illustration above is in English Units only

Equation 36a

The equation in Metric Units would be: $Q_{mt} = q_{mc} + q_{mco} + q_{mc}$ in W/m^2

$$Q_{mt} = 0.548\,\varepsilon_1 \left[\left(\frac{T_{ms1}}{55.5}\right)^4 - \left(\frac{T_{ms2}}{55.5}\right)^4\right] - 0.174\,(1 - \varepsilon_2)\left[\left(\frac{T_{ms1}}{55.5}\right)^4 - \left(\frac{T_{ms2}}{55.5}\right)^4\right] + \frac{\ell_m}{k_{ma}} + 1.957\,(t_{ms1} - t_{ms2})^{5/4}$$

SCHEMATIC DIAGRAM OF HEAT FLOW ACROSS ONE CAVITY SPACE

Figure 160

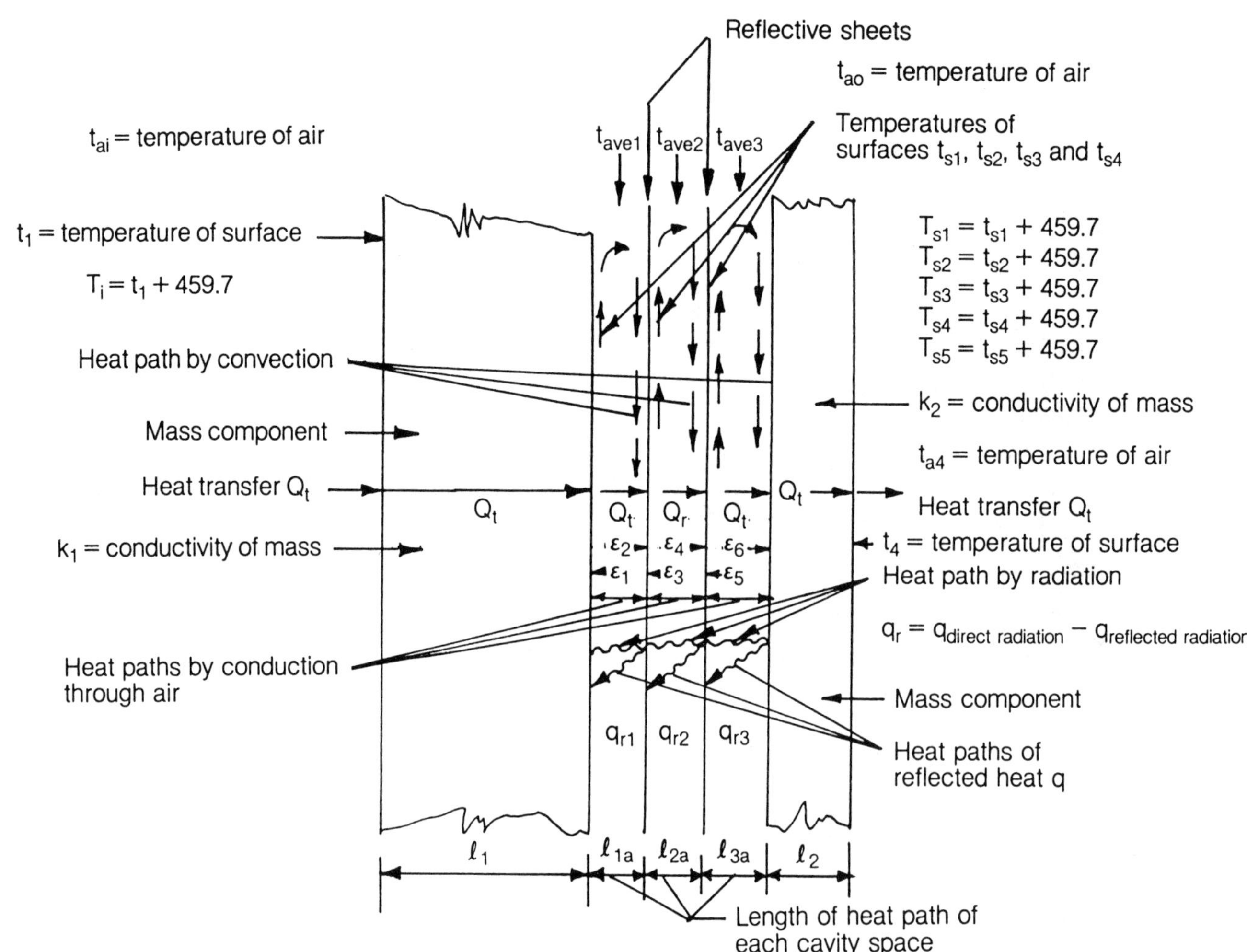

English Units:

When Q_t = its total heat transfer
q_u = heat transfer by convection, first cavity; q_{c2} = second cavity; q_{c3} = third cavity
q_{co1} = heat transfer by conduction, first cavity; q_{co2} = second cavity; q_{co3} = third cavity
q_{r1} = heat transfer by radiation, first cavity; q_{r2} = second cavity; q_{r3} = third cavity

Then, in English Units:

$Q_t = q_{c1} + q_{co1} + q_{r1} = q_{c2} + q_{co2} + q_{r2} = q_{c3} + q_{co3} + q_{r3}$ in Btu/ft^2, hr

Note: Each space has equation as shown for one cavity space in Figure 144.

SCHEMATIC HEAT FLOW DIAGRAM OF CAVITY SPACE DIVIDED BY REFLECTIVE SHEETS OR FOILS

Figure 161

R_i = thermal resistance air to surface (air film resistance)
R_a = thermal resistance surface to air
t_i = temperature of warmer air (°F)
t_a = temperature of cooler (°F)

then:

$$Q_t = \frac{(t_{ai} - t_{a4})}{(R_i + R_1 + R_{cav} + R_2 + R_a} \text{ in Btu/ft}^2\text{, hr}$$

In Metric Units when:

t_{mi} = higher temperature surface with mass on hot side (°C)
t_{m2} = lower temperature surface with mass on cooler side (°C)
ℓ_{m1} = thickness of mass (high side) in meters
ℓ_{m2} = thickness of mass (low side) in meters
k_{m1} = conductivity of mass (high side) in W/mK
k_{m2} = conductivity of mass (low side) in W/mK
R_{m1} = thermal resistance (high side)
R_{m2} = thermal resistance (low side)
R_{mcav} = thermal resistance of cavity
Q_{mt} = heat transfer in W/m²

then:

$$Q_{mt} = \frac{t_{m1} - t_{m2}}{\frac{\ell_{m1}}{k_{m1}} + R_{mcav} + \frac{\ell_{m2}}{k_{m2}}} \text{ in W/m}^2$$

or,

$$Q_{mt} = \frac{t_{mi} - t_{m2}}{R_{m1} + R_{mcav} + R_{m2}} \text{ in W/m}^2$$

If outer surface resistances are included

R_{mi} = thermal resistance air to surface (air film resistance)
R_{ma} = thermal resistance surface to air (air film resistance)
t_{mi} = temperature of warmer air (°C)
t_{m2} = temperature of cooler air (°C)

then:

$$Q_{mt} = \frac{t_{mi} - t_{ma}}{R_{m1} + R_{m1} + R_{mcav} + R_{m2} + R_{ma}} \text{ in W/m}^2$$

HEAT TRANSFER BETWEEN SURFACES OF A STRUCTURAL COMPONENT CONTAINING A SPACE WITH REFLECTIVE SHEETS

Any divisions of an enclosed cavity space by plastic sheets or felts will increase the overall thermal resistance of that space. Using reflective foil or thin sheets to divide the cavity space greatly increases the thermal resistance.

Figure 160 shows a cavity in a building structure divided by two reflective sheets or foils thus changing the space into three cavity spaces. Each of these spaces would have the heat transfer equation as shown in Figure 161. Since the various temperature gradients are now known solutions, the solution of the equation by iteration is not practical. The only practical method is to use the thermal resistance (or conductance) across the cavity. The thermal resistance or conductivity values are established by direct measurement testing.

GENERAL EQUATIONS FOR HEAT TRANSFER THROUGH BUILDING COMPONENTS

As shown previously, given a known thickness of flat material and a known conductivity, the thermal resistance can be determined by dividing thickness of the material by its conductivity; that is

$R = \frac{\ell}{k}$ English Units $\qquad R_m = \frac{\ell_m}{k_m}$ Metric Units.

Also the heat transfer Q is equal to the temperature difference divided by the thermal resistance as follows:

$Q = \frac{\Delta t}{R}$ English Units $\qquad Q_m = \frac{\Delta t_m}{R_m}$ Metric Units.

Although thermal resistance R (or R_m) is not a property of materials, cavity spaces, reflective spaces or surfaces, it can be determined for each as shown previously. This provides an easy method for determination of heat transfer through structural components, cavities, reflective shields and exposed surfaces.

The major variable which must always be determined for individual cases is the temperature, both inside and outside. For this reason, all the resistances are added to determine the total resistance of the component as determined by some selected mean temperature. The total temperature difference divided by the total thermal resistances will give the heat transfer.

In Chapter 11 the total resistance for various structural components will be given for several selected temperature conditions. Also, the conductance U values will be given. The thermal conductance U is the reciprocal of R. That is, U = 1/R.

For a single material we have $R_t = R_1$ and the conductance U across the material is

English Units	*Metric Units*
$R_t = R_1$	$R_{mt} = R_{m1}$
$U = \frac{1}{R_t} = \frac{1}{R_1}$	$U_m = \frac{1}{R_{mt}} = \frac{1}{R_{m1}}$

For two or more materials with a cavity between (R_{cav} or R_{mcav}) the conductance U is still given as $U = 1/R_t$ where R_t, the total thermal resistance is expressed as follows:

English Units	*Metric Units*
$R_t = R_1 + R_{cav} + R_2 + \text{etc.}$	$R_{mt} = R_{m1} + R_{mcav} + R_{m2} + \text{etc.}$

For two or more materials with a cavity separated by reflective sheets or foils (R_{ref} or R_{mref}) the total thermal resistance becomes

English Units	*Metric Units*
$R_t = R_1 + R_{ref} + R_2 + \text{etc.}$	$R_{mt} = R_{m1} + R_{mref} + R_{m2} + \text{etc.}$

HEAT TRANSFER THROUGH CYLINDRICAL INSULATION (Such as Pipe Insulation)

As heat flows through a cylindrical insulation, the difference in the inner and outer areas within a given length affects the amount of heat flow per unit area. Solutions to problems involving areas of insulation in cylindrical form can be accomplished by using actual thicknesses and solving the problems based on the mean area involved. Many engineers use this method of solution, but it does not lend itself to precalculated tables for easy reference.

A more common method of calculating heat transfer is to base all heat transfer calculations on the outer area of the insulation and to use "equivalent thicknesses" in the heat transfer formulas to arrive at resistances.

When:

r_1 = inner radius of a single layer of cylindrical insulation
r_2 = outer radius of a single layer of cylindrical insulation

then:

$$\text{Equivalent thickness} = r_2 \log_e \frac{r_2}{r_1}$$

A schematic diagram is shown in Figure 162.

Thus, the equation for heat transfer through one thickness of cylindrical insulation is in

English Units:

$$Q = \frac{(t_1 - t_2)}{\dfrac{r_2 \log_e \dfrac{r_2}{r_1}}{k}} \text{ in Btu/ft}^2\text{, hr (Outside Area)} \qquad (7)$$

Metric Units:

$$Q_m = \frac{(t_{m3} - t_{m2})}{\dfrac{r_{m2} \log_e \dfrac{r_{m2}}{r_{m1}}}{k_m}} \text{ in W/m}^2 \text{ (Outside Area)} \qquad (7a)$$

Example 39—*English Units*

Find the heat loss per sq ft of outer surface area of insulation enclosing a flue pipe.

when:

the flue pipe is 6″ in diameter ($r_1 = 3''$)

the insulation is 4″ thick ($r_2 = 3'' + 4'' = 7''$)

the temperature of the flue pipe (t_1) is 400 °F

the temperature of the surface of the insulation (t_2) is 100 °F

the thermal conductivity (k) of the insulation is 0.6 Btu/ft², hr, °F

What is the heat transfer Q from inner to outer surface of the insulation in Btu/ft², hr?

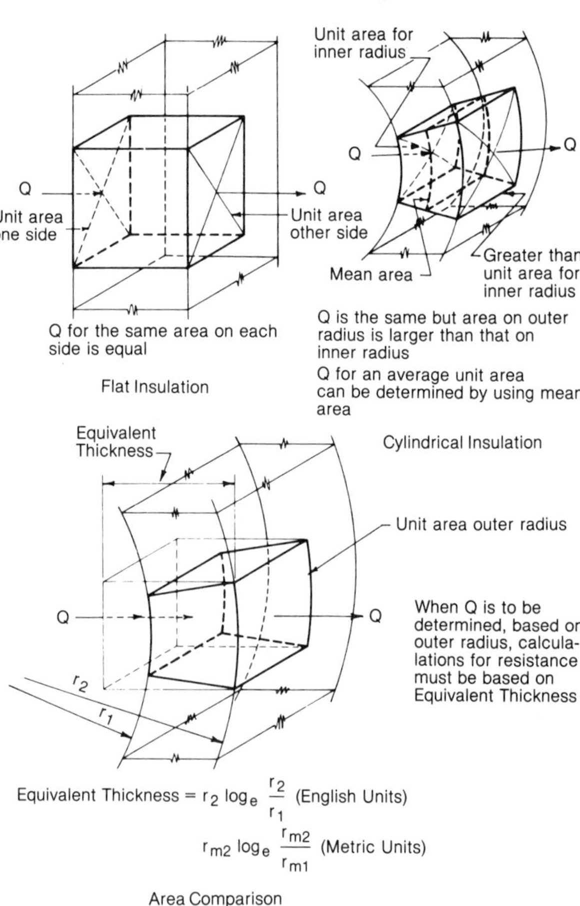

SCHEMATIC DIAGRAM OF EQUIVALENT THICKNESS

Figure 162

Then:

$$Q = \frac{400 - 100}{\dfrac{7 \log_e \dfrac{7}{3}}{0.6}} = \frac{300}{\dfrac{7 \log_e 2.33}{0.6}}$$

From the table of natural logarithms, $\log_e 2.33 = 0.8459$, thus

$$Q = \frac{300}{\frac{7 \times 0.8459}{0.6}} = \frac{300}{9.9} = 30.3 \text{ Btu/ft}^2\text{, hr (outer radius surface)}$$

U_f, the heat transfer per linear foot of the cylinder, can be calculated as follows:

$$U_f = Q \times \frac{2\pi r_2}{12}$$

$$U_f = 30.3 \times \frac{2\pi\, 7}{12} = 30.3 \times 3.65 = 110.6 \text{ Btu/lin ft, hr}$$

Example 39a—*Metric Units*

Find the heat loss per square meter of outer surface area of an insulation enclosing a flue pipe.

when:

the flue pipe is 0.1524m in diameter ($r_{m1} = 0.0762$m)

the insulation is 0.1016m thick ($r_{m2} = 0.0762 + 0.1016 = 0.1778$m)

the temperature of the flue pipe (t_{m1}) is 204 °C

the temperature of the surface of the insulation (t_{m2}) is 37.8 °C

the thermal conductivity (k_m) of the insulation is 0.08652 W/mK

What is the heat transfer Q_m from inner to outer surface of the insulation in W/m²?

then:

$$Q_m = \frac{t_{m1} - t_{m2}}{\frac{r_{m2} \log_e \left(\frac{r_{m2}}{r_{m1}}\right)}{k_m}} = \frac{204 - 378}{\frac{0.1778 \log_e \left(\frac{0.1778}{0.0762}\right)}{0.08652}}$$

$$= \frac{166.2}{\frac{0.1778 \log_e 2.333}{0.08652}} = \frac{166.2}{\frac{0.1778 \times 0.8459}{0.08552}}$$

$$= \frac{166.2}{1.7383} = 95.6 \text{ W/m}^2$$

CHECK: $Q_m = 95.6 \text{ W/m}^2 \times 0.317199 = 30.3 \text{ Btu/ft}^2 = Q$

HEAT TRANSFER THROUGH ONE CYLINDRICAL INSULATION AND ONE SURFACE RESISTANCE (Such as Pipe Insulation)

Surface Resistance Located on Outermost Radius

In an identical manner as shown for flat insulations, the surface resistance* can be added to a series of resistances in the basic equation of heat transfer:

In English Units when:

t_1 = temperature of inner radius surfaces (°F)
t_s = temperature of outer radius surface (°F)
t_a = temperature of the ambient air (°F)
r_1 = inner radius of insulation in inches
r_2 = outer radius of insulation in inches
k = conductivity of insulation in Btu, in/ft², hr, °F
f = film conductance in Btu/ft², hr, °F
q_f = total heat transfer in Btu/lin ft, hr
Q = heat transfer in Btu/ft², hr — (outer surface)

then:

$$Q = \frac{t_1 - t_a}{\frac{r_2 \log_e \left(\frac{r_2}{r_1}\right)}{k} + \frac{1}{f}} \qquad (37)$$

In Metric Units when:

t_{m1} = temperature of inner radius surface in °C (or °K)
t_{ms} = temperature of outer surface in °C (or °K)
t_{ma} = temperature of ambient air in °C (or °K)
r_{m1} = inner radius of insulation in meters
r_{m2} = outer radius of insulation in meters
k_m = conductivity of insulation in W/mK
f_m = film conductance in W/m²
q_m = heat transfer in W/lin m
Q_m = heat transfer in W/m² (outer surface)
R_{ms} = Surface resistance = $1/f_m$.

then:

$$Q_m = \frac{t_{m1} - t_{ma}}{\frac{r_{m2} \log_e \left(\frac{r_{m2}}{r_{m1}}\right)}{k_m} + \frac{1}{f_m}} \qquad (37a)$$

Note: Direction of flow will be indicated by plus or minus sign. Heat transfer to the air will have a plus sign, and heat transfer from surrounding ambient air will have a minus sign.

Example 40—*English Units*

Determine the heat transfer per square foot and per linear foot through insulation installed on a pipe. Also determine the surface temperaure t_s of the insulation.

*Surface resistance $R_s = \frac{1}{f}$

when:

the surface temperature of the inner radius of the insulation (t_1) is 300 °F (pipe temperature also)

the temperature of the ambient air (t_a) is 90 °F

the inner radius of the insulation is 2 1/4″ (outer radius of the pipe)

the outer radius of the insulation is 3 3/4″

the conductivity of the insulation is 0.4 Btu, in/ft^2, hr, °F

the film surface conductance (f_1) is 2.2

the film resistance R_s is $1 \div f_1 = 0.45$

then:

$$Q = \frac{300 - 90}{\dfrac{3.75 \log_e \dfrac{3.75}{2.25}}{0.4} + \dfrac{1}{2.2}}$$

$$= \frac{210}{\dfrac{3.75 \times \log_e 1.667}{0.4} + \dfrac{1}{2.2}}$$

$$= \frac{210}{\dfrac{(3.75 \times 0.5086)}{0.4} + 0.45}$$

$$= \frac{210}{\dfrac{1.9}{0.4} + 0.45} = \frac{210}{4.75 + 0.45}$$

$$= \frac{210}{5.2} = 40.3 \text{ Btu/sq ft, hr (outer surface)}$$

Temperature gradients can be determined starting from either direction from a known temperature:

$$t_s = t_a + Q \times \frac{1}{f} = 90\,°\text{F} + (40.3) \times 0.45 = 90 + 18 = 108\,°\text{F}$$

The heat transfer q_f per linear foot of insulation can be found as follows:

$$q_f = Q \times \frac{2\pi r_2}{12}$$

$$q_f = 40.3 \times \frac{2\pi \times 3.75}{12}$$

$$q_f = 40.3 \times 1.96 = 79 \text{ Btu/lin ft, hr}$$

Example 40a—*Metric Units*

when:

the surface temperature of the inner radius of the insulation (t_{m1}) is 149.9 °C (pipe temperature also)

the temperature of the ambient air (t_a) is 32.2 °C

the inner radius of the insulation is 0.0571m (which is also outer radius of the pipe)

the conductivity (k_m) of the insulation is 0.05768 W/mK

the film surface conductance (f_{m1}) is 12.492 W/m^{2}K

the film resistance $R_{ms} = 1/f_{m1} = 0.08$

The conversion factors used were:

English Values (as given)	Conversion		Metric Values	
t_1 = 300 °F	(300 − 32) ÷ 1.8	=	149.9 °C	= t_{m1}
t_a = 90 °F	(90 − 32) ÷ 1.8	=	32.2 °C	= t_{ma}
r_1 = 2.25″	× 0.0254	=	0.0571m	= r_{m1}
r_2 = 3.75″	× 0.0254	=	0.0952m	= r_{m2}
k = 0.4 Btu in. ft^2, hr, °F	× 0.1442	=	0.05768W/mK	= k_m
f_1 = 2.2 Btu/ft^2, hr, °F	× 5.6782	=	12.492W/m^{2}K	= f_{m1}

To determine t_{mas}, outer surface temperature, it is first necessary to determine Q_m, the heat transfer per square meter per hour:

$$Q_m = \frac{t_{m1} - t_{ma}}{\dfrac{r_{m2} \log_e \dfrac{r_{m2}}{r_{m1}}}{k_{m2}} + \dfrac{1}{f_m}}$$

$$= \frac{148.9 - 32.2}{\dfrac{0.0952 \log_e \dfrac{0.0952}{0.0571}}{0.05768} + \dfrac{1}{12.492}}$$

$$= \frac{116.7}{\dfrac{0.0952 \times 0.5068}{0.05768} + 0.08}$$

$$= \frac{116.7}{\dfrac{0.0482}{0.05768} + 0.08}$$

$$= \frac{116.7}{[0.8364 + 0.08]} = \frac{116.7}{0.9165} = 127.3 \text{ W/m}^2$$

CHECK: Q_m = 127.3 W/m^2 × 0.317199 = 40.3 Btu/ft^2, hr.

Determine outer surface temperature of insulation t_{ms}

$$t_{ms} = t_{ma} + Q_m \times \frac{1}{f_m} = 32.2 + (127.3 \times 0.08)$$

$$= 32.2 + 10.2 = 42.4\,°\text{C } (108\,°\text{F})$$

While the above example served to calculate heat loss from a high-temperature pipe to the atmosphere, the same equations can also be used to determine the heat transfer from ambient air through the thermal insulation to a low-temperature pipe. The following example illustrates this.

Example 41—*English Units*

With all factors the same except the inner surface temperature (previously given as 300 °F), assume that the inner surface is refrigerated and now maintained at a temperature of 30 °F. Then Q, the heat loss per square foot of outer surface, is

$$Q = \frac{30 - 90}{\dfrac{3.75 \log_e \dfrac{3.75}{2.25}}{0.4} + \dfrac{1}{2.2}} = \frac{-60}{5.2}$$

$$= -11.5 \text{ Btu/sq ft, hr}$$

(The minus sign indicates that the heat transfer is from the air to the insulation.)

$$t_s = t_a + Q \times \frac{1}{f} = 90 + (-11.5 \times 0.45)$$

$$= 90 + (-5.2) = 84.8\,°\text{F}$$

$$q_f = \text{heat loss per lin ft}$$

$$q_f = -11.5 \times (1.96) = -22.5 \text{ Btu lin/ft, hr}$$

Example 41a—*Metric Units*

All same as Examples 12 and 13 — except

English Values (as given)	Conversion	Metric Values
t_1 = 30 °F	(30 − 32) ÷ 1.8 =	−1.1 °C = t_{m1}

$$Q_m = \frac{-1.1 - 32.2}{\dfrac{0.0952 \times 0.5068}{0.05768} + 0.08} = \frac{-33.3}{0.9165} = -36.3 \text{ W/m}^2$$

CHECK: $Q_m = -36.3 \text{ W/m}^2 \times 0.317199 = -11.5 \text{ Btu/ft}^2\text{, hr.}$

(Minus sign indicates that the heat transfer is from air to insulation.)

$$t_{ms} = t_{ma} + Q_m \frac{1}{f_{m1}} = 32.2 + (-36.3 \times 0.08)$$

$$= 32.2 - 29.3\,°\text{C } (84.8\,°\text{F})$$

11 Calculated Thermal Conductances and Transmittances — Residence and Building Construction

This chapter will determine the thermal resistance and conductances of typical residences and building constructions.

As discussed in earlier chapters, the heat transfer through a unit area per unit of time is the temperature difference divided by the total thermal resistance of the barrier. Therefore, in order to calculate heat transfer it is first necessary to determine the thermal resistances of various building constructions. The thermal resistances are given in both English and Metric units.

Heat transfer can also be calculated by multiplying the temperature difference by the overall transmittance of the construction.

In most instances the thermal resistances will include the surface resistance of both sides of the particular construction, as the heat transfer is from the air on one side to the air on the other. However, when the heat transfer is from the outside inward, and the outside surface is exposed directly to the sun, then the temperature of the surface exposed to the sun, rather than outside air, determines the temperature differential. For this reason, only the resistance of the inner surface will be included as part of the total resistance.

Most frequently ignored is the fact that the conductivity of most materials changes their mean temperature changes. This means that the conductivity values lie along a curve with respect to temperature. Thus, it is necessary to establish the conditions under which the conductivities of various construction materials are determined.

Also when absorbant and adsorbant materials are used in construction, their conductivities are affected by moisture content. For this reason, their thermal efficiency is improved by proper water shedding and the use of vapor retarders. For example, at 70 °F (21 °C) dry brick has an approximate thermal conductivity of 5 Btu, in./ft^2, hr, °F (8.65 W/mK) while water-saturated brick has a thermal conductivity of 7.8 Btu, in./ft^2, hr, °F (13.5 W/mK). If this water freezes to ice, the conductivity increases to 17.3 Btu, in./ft^2, hr, °F (29.9 W/mK).

Thus the stated conductivities, conductances, and resistances are evaluated at the mean temperature under the given temperature conditions. As it is impractical to list these under all temperature conditions, the temperatures selected as being the

*Table 82 in Chapter 10 establishes the relationship between units of energy per unit of area in the English and Metric Systems.

most representative for the United States are 40°F (4.4°C) for winter conditions and 85°F (29.4°C) for summer conditions. However, when a wall or roof is directly exposed to the sun, an even higher mean temperature must be used. Directly solar exposed wall or roof surfaces can be 130°F to 170°F (54.4°C to 76.7°C), thus, the mean temperature selected was 110°F (43.3°C). In all cases it has been assumed that proper vapor retarders will be installed and that the insulation and other construction components will stay relatively dry.

The thermal conductivity k, conductance C, and thermal resistance R for building materials are given in Table 83 in English Units and in Table 85 in Metric Units. The coefficients of heat transmission, resistance, and light reflectance for various types of glass and glass block, are given in Table 84 in English Units and in Table 86 in Metric Units. For easier reference, all the tables mentioned in this chapter are placed at the end of the chapter.

CONSTRUCTION ASSEMBLIES

Tables 83 through 86 provide data on the conductivities and thermal resistances of individual materials. However, almost all residences and buildings are constructed with a combination of materials in the roofs, walls, and floors. The heat transfer therefore is through the various materials and the spaces or cavities within the construction assembly.

The fundamental heat transfer through a single flat material is given below:

when

ℓ = length of heat flow (thickness of material)
k = conductivity
Δt = temperature difference across the material
Q = heat transfer from high to lower temperature surface

then

$$Q = \frac{\Delta t}{\frac{\ell}{k}} \quad (3)$$

where, in English Units, Q is Btu/ft², hr, Δt is in °F, ℓ is in inches, and k is in Btu, in/ft², hr, °F

In Metric Units Q_m is in W/m², Δt_m is in °C, ℓ is in meters, and k is expressed in W/m,°C.

where

$$Q_m = \frac{\Delta t_m}{\frac{\ell_m}{k_m}} \quad (3a)$$

For a single material $R = \frac{\ell}{k}$ or $R_m = \frac{\ell_m}{k_m}$ for the English and Metric systems respectively. Thus

$$Q = \frac{\Delta t}{R} \text{ (English Units)} \quad (4)$$

and

$$Q_m = \frac{\Delta t_m}{R_m} \text{ (Metric Units)} \quad (4a)$$

For two or more materials

$$Q = \frac{\Delta t}{R_1 + R_2 \text{ etc.}} \text{ Btu/ft}^2\text{, hr} \quad (35)$$

$$Q_m = \frac{\Delta t_m}{R_{m1} + R_{m2} \text{ etc.}} \quad (35a)$$

If R_{si} is the surface resistance on one side and R_{so} is the surface resistance on the other side and assuming the construction had three materials and was exposed to ambient air on both sides, then:

$$Q = \frac{\Delta t}{R_{si} + R_1 + R_2 + R_3 + R_{so}} \text{ (English Units)} \quad (38)$$

$$Q_m = \frac{\Delta t_m}{R_{msi} + R_{m1} + R_{m2} + R_{m3} + R_{mso}} \text{ (Metric Units)} \quad (38a)$$

For convenience the total thermal resistance of a building construction is going to be designated as R (English Units) and R_m (Metric Units).

$$R = R_{si} + R_1 + R_2 + R_3 \text{ etc.} + R_{so} \text{ (English Units)}$$

$$R_m = R_{msi} + R_{m1} + R_{m2} + R_{m3} \text{ etc.} + R_{mso} \text{ (Metric Units)}$$

$$\text{and } Q = \frac{\Delta t}{R} \text{ (English Units)} \quad (39)$$

$$Q_m = \frac{\Delta t_m}{R_m} \text{ (Metric Units)} \quad (39a)$$

Calculated values of thermal resistances R and R_m for common construction assemblies are given in the following pages and tables.

To determine the heat transfer directly divide the temperature difference between inside and outside by the resistance per unit of area to obtain the heat loss per unit of area, as given in Equations 39 and 39a.

Another way to determine the heat transfer is to multiply the temperature difference between inside and outside by U or U_m; that is,

$Q = U \times \Delta t$ (English Units), or $Q_m = U_m \times \Delta t_m$ (Metric Units).

The values of $U = \frac{1}{R}$ and $U_m \frac{1}{R_m}$ are also given in the following tables. To illustrate how these tabular values can be used, consider the following examples.

Example 42—*English Units*

If the inside temperature is 70 °F and outside it is 10 °F and the thermal resistance through the wall without insulation is 2.5, then what is the heat loss in Btu/ft², hr?

$$Q = \frac{t_1 - t_o}{R} = \frac{\Delta t}{R} = \frac{70 - 10}{2.5} = \frac{60}{2.5} = 24 \text{ Btu/ft}^2\text{, hr}$$

If this same wall is insulated and its thermal resistance R is 13, then what is the heat loss in Btu/ft², hr?

$$Q = \frac{t_1 - t_o}{R} = \frac{70 - 10}{13} = \frac{60}{13} = 4.62 \text{ Btu/ft}^2\text{, hr}$$

Thus the insulation provides a saving of 24 − 4.62 = 19.38 Btu/ft², hr.

Example 42a—*Metric Units*

If the temperature inside is 21.1 °C and outside is −12.2 °C and the thermal resistance R_m is 0.441, what is the loss in W/m²

$$Q_m = \frac{t_{m1} - t_{mo}}{R_m} = \frac{21.1 - (-12.2)}{0.441} = \frac{33.3}{0.441} = 75.5 \text{ W/m}^2$$

If this same wall is insulated and its R_m is 2.291

$$Q = \frac{t_{m1} - t_{mo}}{R_m} = \frac{33.3}{2.291} = 14.5 \text{ W/m}^2$$

Here the insulation provides a savings of 61.0 W/m²

U and U_m values of transmittance can be used in similar manner to obtain heat transfer except that $Q = U \times \Delta t$ and $Q_m = U_m \times \Delta t_m$.

Example 43—*English Units*

If the temperature inside is 70 °F and outside is 10 °F, and the thermal transmittance of the wall is U = 0.40, then what is the heat loss in Btu/ft², hr?

$$Q - U \times \Delta t - 0.40 \times (70 - 10) = 0.40 \times (60) = 24 \text{ Btu/ft}^2\text{, hr}$$

If the same wall is insulated with transmittance U = 0.077, then calculate the heat loss in Btu/ft², hr.

$$Q = U \times \Delta t = 0.077 \times (60) = 4.62 \text{ Btu/ft}^2\text{, hr}$$

Thus the insulation provides a savings of 24 − 4.62 = 19.38 Btu/ft², hr.

Example 43a—*Metric Units*

If the temperature inside is 21.1 °C and outside is −12.2 °C, and the thermal transmittance of the wall U_m is 2.267, what is the loss in W/m²?

$$Q_m = U_m \times \Delta t = 2.267\,[21.1 - (-12.2)] = 2.267\,[33.3] = 75.5 \text{ W/m}^2$$

If the same wall is insulated with transmittance $U_m = 0.436$

$$Q_m = U_m \times \Delta t = 0.436\,[21.1 - (-12.2)] = 0.436\,[33.3] = 14.4 \text{ W/m}^2$$

Thus the insulation in this example provides a savings of 61 W/m²

If thermal resistance or transmittance is known for a particular construction assembly the heat transfer can be determined by simple division or multiplication. The following tables are the calculated values of R, U, R_m and U_m for the most common of construction assemblies of roofs, ceilings, walls and floors.

Since the resistances and transmittances of materials do change with mean temperature, the calculations were based on several mean temperatures. Because wind affects surface resistance, typical wind factors were included.

For heating, the basis of selecting the mean temperature is based on the fact that 10 °F (−12.2 °C) is a weather temperature common in winter. The inside temperature was taken to be 70 °F (21.1 °C) thus the tables for calculating heat loss during the winter season were based on an average temperature of $\frac{10 + 70}{2}$ = 40 °F (4.4 °C).

For cooling (not considering solar radiation effect) the mean temperature selected was determined using 100 °F (37 °C) as the high temperature giving a mean temperature of $\frac{100 + 70}{2}$ = 85 °F (29.4 °C).

On the outer surfaces directly exposed to solar radiation the mean temperature was established using a surface temperature of 150 °F (65.6 °C) with 70 °F indoor temperature, for an average of 110 °F (43.2 °C).

In this case the Δt must be determined from outside surface exposed to solar radiation and not the ambient air temperature. See Table 34.

For some unexplainable reason the Federal Register does not require manufacturers to publish the conductivity, or thermal resistance of their insulations at various mean temperatures so that a thermal conductivity or resistance curve can be established. The regulations state that the conductivities must be determined at a mean temperature of 75 °F. If the indoor temperature is 70 °F then the higher surface temperature would have to be 80 °F, an atmospheric state where no thermal insulation is needed.

In anothor part of the regulation it states that a temperature differential of 30 °F is required, thus the higher temperature must be 90 °F and lower temperature 60 °F, neither of which is suitable for indoor comfort conditions. For this reason, conductivities, resistances, and transmittances will differ in this book from manufacturers' published data.

Another factor which influences the conductivity, resistance or transmittance is the moisture content of the material.

To achieve reproducible test results all test methods require that insulation materials be dried either in a ventilated oven or by enclosure in a sealed container with a desiccant. The fact that this is necessary indicates that moisture content does affect the thermal properties of materials. Unfortunately, most test results are not accurate guides as to the thermal performance of insulations and building materials in the state they are purchased or as they occur in service.* The efficient operation of insulation materials requires that they be kept relatively dry by properly installed vapor retarders and rain shields.

Again, because of these variables, the factors of thermal resistance and transmittance have been modified to make the calculated results more accurate under service conditions. Therefore, values of thermal resistance R and R_m and thermal transmittance U and U_m, have been calculated using corrected values of conductivity for various construction assemblies and are presented in Tables 87 through 162.

These tables provide a fast easy way to determine both heat losses and the quantity of heat saved by insulation.

Example 44—*English Units*

Determine the heat loss through an uninsulated ceiling under a vented space with no insulation and through one with 3″ of glass fiber insulation, or with 6″ of fiber glass insulation. The interior temperature of the air below the ceiling is 75°F and the attic space is vented to the atmosphere and is at 20°F.

This situation is shown in Construction Assembly XII. The resistance R for an uninsulated ceiling is given as 1.40 and U is given as 0.714.

The temperature difference is 75°F − 20°F = 55°F.
Q, the heat loss, is:

Using R:

$$Q = \frac{t_1 - t_o}{R} = \frac{75 - 20}{1.40} = \frac{55}{1.40} = 39.3 \text{ Btu/ft}^2\text{, hr}$$

Using U:

$$Q = (t_1 - t_o)\,U = (75 - 20)\,0.714 = 55 \times 0.714 = 39.3 \text{ Btu/ft}^2\text{, hr}$$

In a similar way: From Table 113

For 3″ fiberglass insulation: R = 13.90, U = 0.072

Using R for 3″ insulation:

$$Q = \frac{t_1 - t_o}{R} = \frac{75 - 20}{13.90} = \frac{55}{13.90} = 3.9 \text{ Btu/ft}^2\text{, hr}$$

Using U for 3″ insulation:

$$Q = (t_1 - t_o)\,U = 55 \times 0.072 = 3.9 \text{ Btu/ft}^2\text{, hr}$$

*Refer to Figure 15 "Variation of Thermal Conductivitity with Moisture Content."

Thus the savings using 3″ insulation is:

$$39.3 - 3.9 = 35.4 \text{ Btu/ft}^2\text{, hr (90\%)}$$

For 6″ fiberglass: R = 26.4, U = 0.038

Using R for 6″ insulation:

$$Q = \frac{t_1 - t_o}{R} = \frac{55}{26.4} = 2.1 \text{ Btu/ft}^2\text{, hr}$$

Using U for 6″ insulation:

$$Q = (t_1 - t_o)\,U = 55 \times 0.038 = 2.1 \text{ Btu/ft}^2\text{, hr}$$

Thus the savings using 6″ insulation is

$$39.3 - 2.1 = 37.2 \text{ Btu/ft}^2\text{, hr (94\%)}$$

Comment: Although it might be concluded that 6″ of insulation may not be warranted as it only provides an additional 1.8 Btu/ft², hr savings, such is not true. In most instances the cost of installation for 6″ insulation is approximately the same as for 3″ insulation and the only major difference in cost is the additional cost of the insulation. With the higher and higher cost of energy, the additional cost of the insulation can be recovered in a very short pay back period.

Example 44a—*Metric Units*

Determine the heat loss through an uninsulated ceiling under a vented attic space with no insulation, and through one with 0.076 meters thickness of glass fiber insulation or with 0.152 meters thickness of glass fiber insulation. The interior temperature is −6.7°C and the space below the ceiling is 23.9°C.

This construction is shown in Construction Assembly XII. The resistance R_m for uninsulated ceiling is given as 0.247. The temperature difference is 23.9°C − (−6.7°C) = 30.6°C. Q_m, the W/m², hr heat loss for the uninsulated ceiling, is:

Using R_m:

$$Q_m = \frac{\Delta t_m}{R_m} = \frac{30.6}{0.247} = 123.7 \text{ W/m}^2\text{, hr}$$

Check 123.9 W/m², hr × 0.317199 = 39.3 Btu/ft², hr

Using U_m:

$$Q_m = \Delta t_m \times U_m = 30.6 \times 4.048 = 123.8 \text{ W/m}^2\text{, hr}$$

In a similar way from Table 113

Using R_m for 0.076 meters of glass fiber insulation, R_m = 2.449

$$Q_m = \frac{\Delta t_m}{R_m} = \frac{30.6}{2.449} = 12.5 \text{ W/m}^2\text{, hr}$$

Using U_m:

$$Q_m = \Delta t_m \times U_m = 12.5 \text{ W/m}^2\text{, hr}$$

Thus the savings using 0.076 meters of insulation is:

$$123.9 - 12.5 = 111.4 \text{ W/m}^2\text{, hr}$$

Check 111.4 W/m², hr × 0.317109 = 35.33 Btu/ft², hr (90%).

Using R_m for 0.152 meters of glass fiber insulation, from Table 113, $R_m = 4.652$

$$Q_m = \frac{\Delta t_m}{R_m} = \frac{30.6}{4.652} = 6.6 \text{ W/m}^2\text{, hr}$$

Using U_m, from Table 113, $U_m = 0.214$

$$Q_m = \Delta t_m \times U_m = 30.6 \times 0.214 = 6.5 \text{ W/m}^2\text{, hr}$$

The savings then, as related to an uninsulated ceiling, is 123.9 − 6.5 = 117.4 W/m², hr (94%).

Check 117.4 × 0.317109 = 37.2 Btu/ft², hr.

Example 45—*English Units*

Determine the heat loss through an uninsulated frame wall with no insulation and through one with 2 reflective sheets (reflective on both sides) located in the stud space. The interior temperature is 75 °F and the outside temperature is 20 °F.

This situation is shown in Construction Assembly XIX.

The resistance R for the uninsulated wall is given in Table 129 as 3.38. The temperature difference is 75 °F − 20 °F = 55 °F.

then:

$$Q = \frac{\Delta t}{R} = \frac{55}{3.38} = 16.27 \text{ Btu/ft}^2\text{, hr}$$

Using U which is given as 0.296:

$$Q = 55 \times 0.296 = 16.28 \text{ Btu/ft}^2\text{, hr}$$

For the same wall insulated with stud space divided by two reflective sheets (reflective on both sides), from Table 129, R = 10.04 and U = 0.099.

Using R:

$$Q = \frac{\Delta t}{R} = \frac{55}{10.04} = 5.47 \text{ Btu/ft}^2\text{, hr}$$

or using U:

$$Q = \Delta t \times U = 55 \times 0.099 = 5.45 \text{ Btu/ft}^2\text{, hr}$$

Thus the savings is 16.27 − 5.47 = 10.80 Btu/ft², hr (66%).

Example 45a—*Metric Units*

Determine the heat loss through an uninsulated frame wall with no insulation and through one with 2 reflective sheets located in the stud space. The interior temperature is 23.9 °C and the outside temperature is −6.7 °C.

The temperature difference is 23.9 °C − (−6.7 °C) = 30.6 °C. Uninsulated the $R_m = 0.596$ and $U_m = 1.678$, from Table 129.

Using R_m:

$$Q_m = \frac{\Delta t_m}{R_m} = \frac{30.6}{0.596} = 51.34 \text{ W/m}^2\text{, hr}$$

Using U_m:

$$Q_m = \Delta t_m \times U_m = 30.6 \times 0.565 = 17.30 \text{ W/m}^2\text{, hr}$$

Insulated the heat loss, using $R_m = 1.769$ and $U_m = 0.565$, from Table 129.

Using R_m:

$$Q_m = \frac{\Delta t_m}{R_m} = \frac{30.6}{1.769} = 17.30 \text{ W/m}^2\text{, hr}$$

Using U_m:

$$Q_m = \Delta t_m \times U_m = 30.6 \times 0.565 = 17.30 \text{ W/m}^2\text{. hr}$$

The savings in heat loss is 51.35 − 17.30 W/m22, hr (66%).

Check 34.05 × 0.317109 = 10.8 Btu/ft², hr.

Example 46—*English Unit*

Determine the heat loss through an uninsulated 4″ concrete slab on soil and through a 4″ slab insulated with 3″ cellular glass insulation on soil. The inside temperature is 75 °F and the soil temperature is 45 °F.

This construction is shown in Construction Assembly XXXII.

For this construction R = 0.83 and U = 1.205 (no insulation). The temperature difference is 75 °F − 45 °F = 30 °F.

Using R:

$$Q = \frac{\Delta t}{R} = \frac{30}{0.83} = 36.1 \text{ Btu/ft}^2\text{, hr}$$

Using U:

$$Q = \Delta t \times U = 30 \times 1.205 = 36.1 \text{ Btu/ft}^2\text{, hr}$$

Insulated with 3″ cellular glass, from Table 158, R = 9.21 and U = 0.109.

Using R:

$$Q = \frac{\Delta t}{R} = \frac{30}{9.21} = 3.3 \text{ Btu/ft}^2\text{, hr}$$

Using U:

$$Q = \Delta t \times U = 30 \times 0.109 = 3.3 \text{ Btu/ft}^2\text{, hr}$$

The savings in heat is $34.1 - 3.3 = 30.8$ Btu/ft², hr (90%).

This indicates the fallacy that thermal insulation under concrete could not be used for economic advantage.

Example 46a—*Metric Units*

Determine the heat loss through an uninsulated 10.1 cm concrete slab on soil and through a 10.1 cm slab with 0.076 m cellular glass on soil. The inside temperature is 23.9 °C and the soil temperature is 7.2 °C.

For this construction $R_m = 0.146$ and $U_m = 6.849$, (no insulation), Construction Assembly XXXII.

The temperature difference is 23.9 °C − 7.2 °C = 16.7 °C.

Using R_m:

$$Q_m = \frac{\Delta t_m}{R_m} = \frac{16.7}{0.146} = 114.4 \text{ W/m}^2\text{, hr}$$

Using U_m:

$$Q_m = \Delta t_m \times U_m = 16.7 \times 6.849 = 114.4 \text{ W/m}^2\text{, hr}$$

Check $114.4 \times 0.317109 = 36.2$ Btu/ft², hr

Insulated with 0.076 meters thick cellular glass, from Table 158, $R_m = 1.623$ and $U_m = 0.616$.

Using R_m:

$$Q_m = \frac{\Delta t_m}{R_m} = \frac{16.7}{1.623} = 10.3 \text{ W/m}^2\text{, hr}$$

Using U_m:

$$Q_m = \Delta t_m \times U_m = 16.7 \times 0.616 = 10.3 \text{ W/m}^2\text{, hr}$$

Check: 10.3 W/m², hr × 0.317109 = 3.3 Btu/ft², hr

The savings is $114.9 - 10.3 = 104.1$ W/m², hr (90%).

CONSTRUCTION ASSEMBLIES REFRIGERATED SPACE

Vapor migration is always toward regions of lower temperature. In the case of refrigerated spaces the vapor migration is from outside the space to the inside. This is illustrated in Figure 83. Also, the inner surfaces of walls, ceilings, and floors should be vapor permeable so that any vapor which passes through the outer vapor retarder can continue to the cold coil surface. It is a common error to seal the inside of refrigerator spaces since this usually causes the insulation to become wet and inefficient from a thermal standpoint.

CONSTRUCTION ASSEMBLIES DUCTS

Heat gains or losses will be caused by ducts which operate at a different temperature than the space surrounding them. In many instances the economic losses are twice that of the ducts themselves. To illustrate: If a duct of heated air or gases passes through an air conditioned space, its economic loss is its loss of heat to the space. However, the second loss will occur in the cost of energy used by the air conditioning equipment to remove that heat from the ambient air.

Another reason for insulating ducts is to prevent moisture condensation on the surface of ducts which contain cooled or refrigerated air. If the insulation is on the interior of the duct then fiberous insulation can be used satisfactorily as the duct itself acts as the vapor retarder.

When the duct is on the interior no vapor retarder on vapor permeable insulation can provide a vapor resistance sufficiently effective (as compared to the metal duct) to keep the insulation dry. For this reason cellular glass insulation is the only insulation (because of its own vapor resistance) suitable for application on cold metal ducts which operate below ambient air dew point.

No heat transfer information was provided for cellulose insulation or organic foam insulation as duct insulations. This is because these are combustible and combustible insulation should not be used on ducts because of fire hazards. It should be pointed out that, as a distribution system for conditioned air, ducts of combustible materials would also serve to distribute a fire through the structure.

Tables 167 through 170 provide thermal resistance and transmittances for flat ducts. Tables 171 through 174 provide this information for cylindrical ducts. The following are examples using Tables 167 through 174.

Example 47—*English Units*

Determine the heat loss per square foot of both a rectangular uninsulated duct and one insulated with 1″ thick glass fiber insulation. The duct runs through an unheated space, at 20 °F and contains 95 °F air for heating other areas in the building.

Uninsulated — From Construction Assembly XXXVII R = 0.51

$$Q = \frac{t_1 - t_2}{R} = \frac{95 - 20}{R}$$

$$Q = \frac{75}{0.51} = 147 \text{ Btu/ft}^2\text{, hr}$$

Insulated — From Table 167 R = 4.85

$$Q = \frac{t_1 - t_2}{R} = \frac{75}{4.85} = 15.5 \text{ Btu/ft}^2\text{, hr}$$

A savings of 147.0 – 15.5 = 131.5 Btu/ft², hr, (89%) is realized by using insulation.

Example 47a—Metric Units

Determine the heat loss per square meter of both a rectangular uninsulated duct and one insulated with 0.025 in. thick glass fiber insulation. The duct runs through an unheated space which has a temperature of –6.7 °C. The air temperature in the duct is 35 °C.

Uninsulated — From Construction Assembly XXXVII R = 0.090

$$Q_m = \frac{t_{m1} - t_{m2}}{R_m} = \frac{35 - (-6.7)}{R_m}$$

$$Q_m = \frac{41.7}{0.090} = 463.3 \text{ W/m}^2\text{, hr}$$

Check: 463.3 W/m² × 0.317109 = 147 Btu/ft², hr

Insulated — From Table 167 R = 0.855

$$Q_m = \frac{t_{m1} - t_{m2}}{R_m} = \frac{35 - (-6.7)}{R_m} = \frac{41.7}{0.855} = 48.8 \text{ W/m}^2\text{, hr}$$

This yields a savings of 463.3 – 48.8 = 414.5 W/m², hr (89%).

Example 48—*English Units*

Determine the heat loss per square foot across the outer surface area of a cylindrical duct which has a 12 in. OD with 2″ thick insulation over its outer surface. The air temperature inside the duct is 110 °F, and the duct is located in an area that is maintained at 70 °F. What is the heat loss to the ambient air?

Uninsulated — From Construction Assembly XL R = 0.53 — outer surface)

$$Q = \frac{t_1 - t_2}{R}$$

$$Q = \frac{110 - 70}{0.53} = \frac{40}{0.53} = 75.5 \text{ Btu/ft}^2\text{, hr}$$

Insulated — From Table 173 R = 7.94

$$Q = \frac{t_1 - t_2}{R} = \frac{110 - 70}{7.94} = \frac{40}{7.94} = 8.8 \text{ Btu/ft}^2\text{, hr}$$

This is a savings of 75.5 – 8.8 = 66.7 Btu/ft², hr (89%).

Example 48a—*Metric Units*

Determine the heat loss per square meter across the outer surface area of a cylindrical duct which has a 0.254 m OD with 0.051 thickness of insulation over its outer surface. The air temperature in the duct is 43.3 °C and the duct is located in an area that is maintained at 21.1 °C. What is the heat loss to ambient air?

Uninsulated — R_m for Construction Assembly XL = 0.093 — (outer surface)

$$Q_m = \frac{t_{m1} = t_{m2}}{R_m}$$

$$Q_m = \frac{43.3 - 21.1}{0.093} = \frac{22.2}{0.098} = 238.7 \text{ W/m}^2\text{, hr}$$

Check: 238.7 × 0.317109 = 75.6 Btu/ft², hr

Insulated — R_m from Table 173 = 1.399

$$Q_m = \frac{t_{m1} - t_{m2}}{R_m} = \frac{43.3 - 21.1}{1.399} = \frac{22.2}{1.399} = 15.9 \text{ W/m}^2\text{, hr}$$

Here the savings is 238.7 – 15.9 = 212.8 W/m², hr (89%).

Example 49—*English Units*

A duct containing cold air at 45 °F is located in an area where the temperature is 75 °F at 50% relative humidity. Under these conditions the duct will condense moisture on its outer surface as the dew point temperature is 55 °F (Table 175). For this reason 2″ thick cellular glass is installed. What is the heat gain in Btu/ft², hr for both a bare duct and for an insulated duct?

Uninsulated — R for Construction Assembly XXXVIII = 0.53

$$Q = \frac{t_1 - t_2}{R}$$

$$Q = \frac{75 - 45}{0.53} = \frac{30}{0.53} \text{ Btu/ft}^2\text{, hr}$$

Insulated — R from Table 170 = 6.78

$$Q = \frac{t_1 - t_2}{R} = \frac{75 - 45}{6.78} = \frac{30}{6.78} = 4.4 \text{ Btu/ft}^2\text{, hr}$$

This yields a savings of 56.6 – 4.4 = 52.2 Btu/ft², hr (92%).

Example 49a—*Metric Units*

A duct containing cold air at 7.2 °C is located in an area where the temperature is 23.9 °C at 50% relative humidity (Table 176). Under these conditions the duct will condense moisture on its outer surface as the dew point is 12.8 °C. For this reason 0.051 m thickness of cellular glass insulation is installed. What is the heat gain in W/m² for both a bare duct and an insulated duct?

Uninsulated — R_m for Construction Assembly XXXVIII = 0.093

$$Q_m = \frac{t_{m1} - t_{m2}}{R_m}$$

$$Q_m = \frac{23.9 - 7.2}{0.093} = \frac{16.7}{0.093} = 179.6 \text{ W/m}^2\text{, hr}$$

Check: 179.6 × 0.317109 = 36.7 Btu/ft², hr

Insulated — R_m from Table 170 = 1.195

$$Q_m = \frac{t_{m1} - t_{m2}}{R_m} = \frac{23.9 - 7.2}{1.195} = \frac{16.7}{1.195} = \text{ W/m}^2\text{, hr}$$

This yields a savings of 179.6 − 14 = 165.6 W/m², hr (92%).

CONTROL OF WATER DRIP FROM SURFACES

In many instances when ducts and pipes have a temperature only slightly below the dew point of ambient air, they drip water due to condensation on the surface. Often a full installation of insulation is not warranted and the control of water drip is by the application of cork filled poly vinyl acetate insulating mastic.

This mastic may be applied to any metal surface. However, it is a water mix mastic and steel surfaces must be galvanized or protected by a good grade of rust resistant paint.

The metal surface must be clean and dry and the mastic should only be installed when the surface temperature is 10 °F (5.55 °C) above dew point temperature, but not above 160 °F (71.1 °C). The surface must remain above dew point until the mastic has dried and cured. A minimum drying period of 24 hours should be allowed for each 1/8″ (0.317 cm) thickness of insulating mastic applied.

Table 177 provides information relating the dew point temperature to the proper thickness of mastic in order to control water drip.

PIPES

As heat is transmitted through a cylindrical insulation the difference between inner and outer areas affects the amount of heat transferred per unit of area. As the outer surface of insulation on a pipe is larger than the inner surface, the outer surface heat transfer per a unit of area is less than that of the pipe itself. In order to compensate for this difference in area we base heat transfer on the outer area of the insulation and use "equivalent thicknesses" in the heat transfer formulas to arrive at resistances R or R_m.

In English Units when:

r_1 = Inner radius of the cylindrical insulation
r_2 = Outer radius of the cylindrical insulation

then, the equivalent thickness is expressed as

$$\ell_e = r_2 \log_e \frac{r_2}{r_1} \qquad (47)$$

In Metric Units when:

r_{m1} = Inner radius of the cylindrical insulation
r_{m2} = Outer radius of the cylindrical insulation

then, the equivalent thickness is

$$\ell_{me} = r_{m2} \log_e \frac{r_{m2}}{r_{m1}} \qquad (47a)$$

Note: Logarithm is to base e

A schematic diagram for the cylindrical geometry discussed above is shown in Figure 162, Chapter 10.

The resistance R_e of the cylindrical insulation and given thermal conductivity is:

In English Units when r_1 is inner radius and r_2 is outer radius

$$R_e = \frac{r_2 \log_e \frac{r_2}{r_1}}{k} \qquad (48)$$

In Metric Units

$$R_{me} = \frac{r_{m2} \log_e \frac{r_{m2}}{r_{m1}}}{k_m} \qquad (48a)$$

Precalculated values of equivalent thickness, values of conductivities for various materials at various mean temperatures and typical surface resistances are presented in this text to enable fast calculation of thermal resistance. Equivalent thicknesses for standard thicknesses of insulation for NPS pipe and tubes are given in Tables 178, 179, 180 and 181. Typical thermal conductivities of pipe insulation are presented in Tables 182 and 183.

CALCULATING HEAT LOSS OR GAIN FROM INSULATED PIPES OR TUBES

Tables 178 through 191 are provided to simplify the determination of heat transfer to or from a bare or insulated pipe or tube

Where considerable piping exists and its heat loss is expensive, the insulation thickness should be determined to obtain the most economic thickness. However, this is complex and must be done by precalculated tables or computer. The precalculated tables are found in "Handbook of Thermal Insulation Economics," by W. C. Turner and J. F. Malloy which is published by Krieger Publishing Co.

Where the insulation thickness has already been established or is limited by space or investment considerations, the heat loss can be easily determined from the tables provided. Surface temperatures can also be calculated. The method is illustrated by the following examples.

Example 50—*English Units*

A 4″ ips steam pipe operating at 300°F (50 psi steam) is located in an area of 70°F ambient air temperature. What is its loss per linear foot when bare? What is its loss when insulated with 1″, with 1 1/2″, and with 2″ nominal insulation? The insulation is rigid expanded silica sectional pipe insulation, protected with fabric reinforced PVA mastic (gray color).

Heat loss when bare.

The *bare loss* from the 300°F pipe is 783 Btu/ft², hr (Table 191).

Heat loss when insulated with 1″ thickness of expanded silica insulation with mastic covering:

Factors

Equivalent Thickness = 1.19″ — (Table 178)
$k = 0.42$ — (Table 182)
$R_s = 0.58$ — (Table 184)
$t_1 = 300°F$
$t_2 = 70°F$

$$R = \frac{\text{Equivalent thickness}}{k} + R_s$$

$$R = \frac{1.19}{0.42} + 0.58 = 2.83 + 0.58 = 3.41$$

$$Q = \frac{t_1 - t_2}{R} = \frac{300 - 70}{3.41} = \frac{230}{3.41} = 67 \text{ Btu/ft}^2\text{, hr}$$

(Heat loss from outer surface, ft², hr)

To convert to linear feet (from Table 186 surface area/lin ft) for 4″ NPS pipe with 1″ insulation = 1.73 ft.

$$Q_{\ell f} = 67 \times 1.73 = 115.9 \text{ Btu/lin ft, hr}$$

This yields a savings of 783 − 115.9 = 667.1 Btu/lin ft, hr (85% savings).

The surface temperature of the outer jacket will be: temperature of air + (heat transfer per sq ft multipled by surface resistance). $t_s = t_1 + (Q \times R_s)$ in °F

In this case surface temperature: $t_s = 70 + (67 \times 0.58)$, $t_s = 70 + 38.9 = 108.9$°F.

Heat loss when insulated with 1 1/2″ thickness of expanded silica insulation with mastic covering:

Factors

Equivalent Thickness = 1.91 — (Table 178)
$k = 0.42$ — (Table 182)
$R_s = 0.68$ — (Table 184)

$$R = \frac{\text{Equivalent thickness}}{k} + R_s$$

$$R = \frac{1.91}{0.42} + 0.68 = 4.55 + 0.68 = 5.23$$

$$Q = \frac{t_1 - t_2}{R} = \frac{300 - 70}{5.23} = 43.98 \text{ Btu/ft}^2\text{, hr}$$

(Heat loss from outer surface, ft²)

To convert to linear feet (from Table 186) for 4″ NPS pipe 1 1/2″ insulation = 2.00.

$$Q_{\ell f} = 43.98 \times 2.00 = 87.96 \text{ Btu/lin ft, hr}$$

(Heat loss from outer surface, lin ft)

A savings of 783 − 87.96 = 695.04 (88%) is realized.

The surface temperature $t_s = t_1 + (Q \times R_s)$ °F, $t_s = 70 + (43.98 \times 0.68) = 70 + 29.9 = 99.99$°F.

Heat loss when insulated with 2″ thickness of expanded silica insulation with mastic covering:

Factors

Equivalent Thickness = 2.71 — (Table 178)
$k = 0.42$ — (Table 182)
$R_s = 0.73$ — (Table 184)

$$R = \frac{2.71}{0.42} + 0.73 = 6.45 + 0.73 = 7.18$$

$$Q = \frac{t_1 - t_2}{R} = \frac{300 - 70}{7.18} = 32.03 \text{ Btu/ft}^2\text{, hr}$$

To convert to linear feet (from Table 186) for 4″ NPS pipe having 2″ insulation = 2.26.

$$Q_{\ell f} = 32.03 \times 2.26 = 72.3 \text{ Btu/lin ft, hr}$$

This yields a savings of 783 − 72.3 = 710.7 Btu/lin ft, hr (91% savings).

The surface temperature = $t_s = t_r + (Q \times R_s)$ °F, $t_s = 70 + (32.02 \times 0.73) = 70 + 23.4 = 93.4$ °F.

The material used as an outside covering or jacket affects the heat transfer of insulated pipes because of surface emittance ε. Also the surface temperature can be changed considerably by the change of ε even if the insulation and its thickness remains the same. In general, the higher the ε value, the closer the surface temperature will be to that of the ambient air temperature but the higher the heat transfer will be. This appears to be a contradictory statement, as it is just assumed the higher the surface temperature the greater the heat transfer. This assumption is only correct if the emittance ε remains constant. For this reason the previous example using mastic as the outer coating will now be recalculated using aluminum jacket as the outer covering.

Heat loss when insulated with 1″ thickness of expanded silica insulation with aluminum jacket outer covering:

Factors

Equivalent Thickness = 1.19″ — (Table 178)
k = 0.42 — (Table 182)
R_s = 1.24 — (Table 184)

$$R = \frac{\text{Equivalent thickness}}{k} + R_s$$

$$R = \frac{1.19}{0.42} + 1.24 = 2.83 + 1.24 = 4.07$$

$$Q = \frac{t_1 - t_2}{R} = \frac{300 - 70}{4.07} = 56.5 \text{ Btu/ft}^2\text{, hr}$$

(Heat loss from outer surface, ft^2)

To convert to linear feet (from Table 186) for 4″ pipe 1″ insulation = 1.73.

$$Q_{lf} = 56.5 \times 1.73 = 97.7 \text{ Btu/lin ft, hr}$$

Here the savings is 783 − 97.7 = 685.3 Btu/lin ft, hr (87% savings).

The surface temperature of the aluminum jacket is: $t_s = t_1 + (Q \times R_s) = 70 + (56.5 \times 1.24) = 70 + 70.1 = 141.1$ °F.

Heat loss when insulated with 1 1/2″ thickness of expanded silica insulation with aluminum jacket outer covering:

Factors

Equivalent Thickness = 2.71 — (Table 178)
k = 0.42 — (Table 182)
R_s = 1.55 — (Table 184)

$$R = \frac{\text{Equivalent thickness}}{k} + R_s$$

$$R = \frac{2.71}{0.42} + 1.55 = 6.45 + 1.55 = 8.0$$

$$Q = \frac{300 - 70}{8.0} = \frac{230}{8.0} \text{ Btu/ft}^2\text{, hr}$$

(Heat loss from outer surface, ft^2)

To convert to linear feet (from Table 186) for 4″ pipe with 2″ insulation = 2.26.

$$Q_{lf} = 28.8 \times 2.26 = 65.0 \text{ Btu/lin ft, hr}$$

This represents a savings of 783 − 65.0 = 718 Btu/lin ft, hr (92% savings).

The surface temperature of the aluminum jacket is: $t_s = t_1 + (Q \times R_s) = 70 + (28.8 \times 1.55) = 70 + 44.6 = 114.6$ °F.

TABULATION OF RESULTS

Example 50—*English Units*

This example is presented to show the differences of heat transfer and surface temperatures possible using different thicknesses of insulation and coverings on a pipe operating at the same temperature.

Insulation Thickness	Covering	Heat Loss	Heat Loss	Surface Temperature °F	% Energy Savings
1″	Dark Mastic	66.1	114.3	113	85
1 1/2″	Dark Mastic	43.9	88.0	99.9	88
2″	Dark Mastic	32.03	72.0	93.4	91
1″	Aluminum Jacket	56.5	97.7	140.1	87
1 1/2″	Aluminum Jacket	40.2	80.4	125.0	90
2″	Aluminum Jacket	28.8	65.0	114.6	92

Example 50a—*Metric Units*

A 114 mm steam pipe operating at 149 °C is located in an area with a 21.1 °C ambient air temperature. What is the heat loss per linear meter for bare pipe? What is the heat loss when the pipe is insulated with 0.025 m nominal thickness, 0.037 m nominal thickness and 0.051 m nominal thickness of insulation? The insulation is rigid expanded silica sectional pipe insulation, protected with fabric reinforced PVA mastic (gray color).

Heat loss when bare.

The loss from the bare pipe at 149 °C is 752 Btu/lin m, hr.

Heat loss when insulated with 0.025 m thickness of expanded silica insulation with mastic covering:

Factors

Equivalent Thickness = 0.030 m — (Table 179)
k = 0.061 W/mK — (Table 183)
R_{ms} = 0.120 — (Table 185)

$$R_m = \frac{\text{Equivalent thickness}}{k_m} + R_{ms}$$

$$R_m = \frac{0.030}{0.061} + 0.120 = 0.492 + 0.120 = 0.612$$

$$Q_m = \frac{t_{m1} - t_{m2}}{R_m} = \frac{149 - 21.1}{0.612} = \frac{127.9}{0.612} = 208 \text{ W/m}^2\text{, hr}$$

Check (from Ex. 50) 66.1 Btu/ft², hr × 3.1525 = 208 W/m², hr.

To convert to linear meters (from Table 187) for 114 mm pipe with 0.025 m nominal thickness insulation 0.527 × m² = linear meters.

$$Q_{\ell mm} = 208 \text{ W/m}^2 \times 0.527 = 109.8 \text{ W/lin m}$$

Check (from Ex. 50) 109.8 W/lin m × 1.0406 = 114 Btu/lin ft, hr.

This yields a savings of 752 − 109.8 = 642 W/lin m, hr (85% savings).

Heat loss when insulated with 0.037 m thickness of expanded silica insulation with mastic covering:

Factors

Equivalent Thickness = 0.049 — (Table 179)
k_m = 0.061 W/mK — (Table 183)
R_{ms} = 0.129 — (Table 185)

$$R_m = \frac{\text{Equivalent thickness}}{k_m} + R_{ms}$$

$$R_m = \frac{0.049}{0.061} + 0.129 = 0.932$$

$$Q_m = \frac{t_{m1} - t_{m2}}{R_m} = \frac{149 - 21.1}{0.932} = 137 \text{ W/m}^2\text{, hr}$$

Check (from Ex. 50) 43.9 Btu/ft², hr × 3.1525 = 138 W/m².

To convert to linear meters (from Table 187) for 114 mm pipe with 0.037 nominal thickness 0.610.

$$Q_{\ell mm} = 137 \times 0.610 = 83.6 \text{ W/lin m}$$

Check (from Ex. 50) 83.6 W/lin m × 1.0406 = 87 Btu/lin ft

This yields a savings of 752 − 83.6 = 668.4 W/lin m (88% savings).

The surface temperature of outer surface of mastic is: $t_{ms} = t_{m1} + (Q_m \times R_{ms}) = 21.1 + (137 \times 0.129) = 21.1 + 17.7 = 38.8$ °C.

Heat loss when insulated with 0.051 m thickness of expanded silica insulation with mastic coating:

Factors

Equivalent Thickness = 0.069 — (Table 179)
k_m = 0.061 W/mK — (Table 183)
R_{ms} = 0.136 — (Table 185)

$$R_m = \frac{\text{Equivalent thickness}}{k_m} + R_{ms}$$

$$R_m = \frac{0.069}{0.061} + 0.136 = 1.131 + 0.136 = 1.267$$

$$Q_m = \frac{t_{m1} - t_{m2}}{R_m} = \frac{149 - 21.1}{1.267} = 100.9 \text{ W/m}^2\text{, hr}$$

Check (from Ex. 50) 32 Btu/ft², hr × 3.1525 = 100.9 W/m², hr.

To convert to linear meters (from Table 187) for 114 mm pipe 0.05 nominal thickness 0.689 × m² lin meters.

$$Q_{\ell m} = 100.9 \times 0.689 = 69.5 \text{ W/lin m}$$

Check (from Ex. 50) 69.5 W/lin m × 1.0406 = 72 Btu/lin ft

This yields a savings of 752 − 69.5 = 682.5 W/lin m (91% savings).

The surface temperature $t_{ms} = t_{mi} + (Q_m \times R_s)$ °C, $t_{ms} = 21.1 + (100.9 \times 0.136) = 21.1 + 13.7 = 34.8$ °C.

As previously stated, outside covering affects the heat transfer and surface temperature. For this reason this part of the example will be calculated for an aluminum jacketing outer covering.

Heat loss when insulated with 0.025 thickness of expanded silica insulation covered with aluminum jacket:

Factors

Equivalent Thickness = 0.030 — (Table 179)
k_m = 0.061 W/mK — (Table 183)
R_{ms} = 0.241 — (Table 185)

$$R_m \frac{\text{Equivalent thickness}}{k_m} + R_{ms}$$

$$R_m = \frac{0.030}{0.061} + 0.241 = 0.492 + 0.241 = 0.733$$

$$Q_m = \frac{t_{m1} - t_{m2}}{R_m} = \frac{149 - 21.1}{0.733} = 174.5 \text{ W/m}^2\text{, hr}$$

Check (from Ex. 50): 56.3 Btu/ft², hr × 3.1525 = 175 W/m².

To convert to linear meters (from Table 187) for 114 mm pipe with 0.025 meters thickness insulation 0.527 × m.

$$Q_{\ell m} = 174.5 \times 0.527 = 91.7 \text{ W/lin m, hr}$$

Check (from Ex. 50) 91.7 W/lin m × 1.0406 = 96 Btu/lin ft

Here the savings is 752.0 − 91.7 = 660.3 W/lin m (88% savings).

The surface temperature of the outer surface is: $t_{ms} = t_{mi} + (Q_m \times R_{ms}) = 21.1 + (174 \times 0.241) = 21.1 + 41.9 = 63.0\,°C$.

Heat loss when insulated with 0.037 m thickness of expanded silica with aluminum jacket covering:

Factors

Equivalent Thickness = 0.049 — (Table 179)
k_m = 0.061 W/mK — (Table 183)
R_{ms} = 0.273 — (Table 185)

$$R_m = \frac{\text{Equivalent thickness}}{k_m} + R_{ms}$$

$$R_m = \frac{0.049}{0.061} + 0.273 = 0.803 + 0.273 = 1.076$$

$$Q_m = \frac{t_{m1} - t_{m2}}{R_m} = \frac{149 - 21.1}{1.076} = \frac{127.9}{1.076} = 119 \text{ W/m}^2$$

Check: 119 W/m² × 0.317199 = 38.

To convert to linear meters (from Table 187) for 114 mm pipe with 0.037 m nominal thickness 0.610 × m² = linear meters.

$$Q_{\ell mm} = 119 \times 0.61 = 72.6 \text{ W/lin m}$$

This yields a savings of 752 − 72.6 = 679.4 (90% savings).

The surface temperature of the outer surface is: $t_{ms} = t_m + (Q_m + R_{ms}) = 21.1 = (119 \times 0.273) = 21.1 + 32.5 = 53.6\,°C$.

Heat loss when insulated with 0.051 thickness of expanded silica insulation with aluminum jacket covering:

Factors

Equivalent Thickness = 0.069 — (Table 179)
k_m = 0.061 W/mK — (Table 185)
R_{ms} = 0.296 — (Table 185)

$$R_m = \frac{\text{Equivalent thickness}}{k_m} + R_{ms}$$

$$R_m = \frac{0.062}{0.061} + 0.296 = 1.131 + 0.296 = 1.427$$

$$Q_m = \frac{t_{m1} - t_{m2}}{R_m} = \frac{149 - 21.1}{1.427} = 89.6 \text{ W/m}^2$$

Check: 89.6 W/m² × 0.317199 = 28.4.

To convert to linear meters (from Table 187) for 144 mm pipe with 0.051 m nominal thickness 0.689 × m² = lin meters.

$$Q_{\ell mm} = 89.6 \times 0.689 = 61.7 \text{ W/lin m}$$

Check: 61.7 W/lin m × 1.0406 = 64 Btu/lin ft

This yields a savings of 752 − 61.7 = 690.3 W/lin m (92%).

The surface temperature of the outer surface is: $t_{ms} = t_m + (Q_m \times R_{ms}) = 21.1 + (89.6 \times 0.296) = 21.1 + 26.5 = 47.6\,°C$.

TABULATION OF RESULTS

Example 49a—*Metric Units*

Insulation Thickness	Covering	Heat Loss	Heat Loss	Surface Temperature °C	% Energy Savings
0.025 m	Dark Mastic	208.0	109.8	46.1	85
0.037 m	Dark Mastic	137.0	83.6	38.8	88
0.051 m	Dark Mastic	100.9	69.5	34.8	91
0.025 m	Aluminum Jacket	174.5	91.7	63.0	88
0.037 m	Aluminum Jacket	119.0	72.6	53.6	90
0.051 m	Aluminum Jacket	189.6	61.7	47.6	92

Example 50—*English Units*

A 3″ NPS brine pipe operating at 10 °F passes through a heated and air conditioned area kept at 70 °F. Calculate the heat gain to pipe from air, natural convection for 1 1/2″, 2″, 2 1/2″, and 3″ thickness of cellular glass insulation with conductivity k at 0.33 Btu/ft², hr, °F. Calculate the surface temperature when the pipe is coated with green colored mastic reinforced with monocrylic fiber cloth. Also calculate the surface temperature assuming the insulation is jacketed with bright aluminum.

CALCULATIONS FOR HEAT GAINS, AND SURFACE TEMPERATURES OF OUTER SURFACES OF MASTIC OR ALUMINUM OVER VARIOUS THICKNESSES OF CELLULAR GLASS INSULATION.

Mastic Covering ε = 0.9, 1″ thickness of Cellular glass. — Insulation k = 0.33 Btu/ft², hr, °F.

Factors

Equivalent Thickness = 1.24 — (Table 178)
R_s = 0.87 — (Table 186)
t_a = 70 °F
t_p = 10 °F

$$R = \frac{\text{Equivalent thickness}}{k} = R_s = \frac{1.24}{0.33} + 0.87 = 3.76 + 0.86 = 4.63$$

$$Q = \frac{t_a - t_p}{R} = \frac{70 - 10}{4.63} = \frac{60}{4.63} = 12.96 \text{ Btu/ft}^2\text{, hr}$$

(Heat Gain)

To convert to linear foot (from Table 188) surface area ft²/lin ft for 3″ NPS pipe with 1″ thickness of insulation is 1.46.

$$Q_{lf} = 12.96 \times 1.46 = 18.92 \text{ Btu/lin ft, hr}$$

(Heat Gain)

The surface temperature of the outer coating surface will be: temperature of air minus (heat transfer per sq ft multiplied by surface resistance).

$$t_s = t_a - (Q \times R_s) = 70 - (12.96 \times 0.87) = 70 - 11.3 = 58.7°$$

Relative humidity above 68% will cause surface condensation (Table 175).

Mastic Covering ε = 0.9, 1 1/2″ thickness Cellular Glass — Insulation k = 0.33 Btu, in./ft², hr, °F

Factors

Equivalent Thickness = 2.02 — (Table 178)
$R_s = 0.88$ — (Table 186)
$t_a = 70°F$
$t_p = 10°F$

$$R = \frac{\text{Equivalent thickness}}{k} = R_s = \frac{2.02}{0.33} + 0.88 = 6.12 + 0.88 = 7.00$$

$$Q = \frac{t_a - t_p}{R} = \frac{70 - 10}{7.00} = \frac{60}{7.00} = 8.6 \text{ Btu/ft}^2\text{, hr}$$

(Heat Gain)

To convert to linear foot (from Table 188) surface area ft²/lin ft or 3″ NPS pipe with 1 1/2″ thickness of insulation is 1.73.

$$Q_{lf} = 8.6 \times 1.73 = 14.8 \text{ Btu/lin ft, hr}$$

(Heat Gain)

The surface temperature of the outer coating will be: emperature of air minus (heat transfer per sq ft multiplied by surace resistance).

$$t_s = t_a - (Q \times R_s) = 70 - (8.6 \times 0.88) = 70 - 7.5 = 62.5°F$$

Relative humidity above 76% will cause surface condensation 'able 175).

Mastic Covering ε = 0.9, 2″ thickness of Cellular Glass — Inulation k = 0.33 Btu, in./ft², hr, °F

Factors

Equivalent Thickness = 2.85 — (Table 178)
$R_s = 0.9$ — (Table 186)
$t_a = 70°F$
$t_p = 10°F$

$$= \frac{\text{Equivalent thickness}}{k} + R_s = \frac{2.85}{0.33} = 89 = 8.64 + 0.89 = 9.5$$

$$Q = \frac{t_a - t_p}{k} = \frac{70 - 10}{9.52} = \frac{60}{9.53} = 6.29 \text{ Btu/ft}^2\text{, hr}$$

(Heat Gain)

To convert to linear feet (from Table 188) surface area per linear foot for 3″ NPS pipe, and a 2″ thickness of insulation is 2.0.

$$Q_{lf} = 6.29 \times 2 = 12.6 \text{ Btu/lin ft, hr}$$

(Heat Gain)

The surface temperature of the outer coating will be:

$$t_s = t_a - (Q \times R_s) = 70 - (6.29 \times 0.88) = 70 - 5.5 = 64.5°F.$$

Relative humidity above 80% will cause surface condensation (Table 175).

Mastic Covering ε = 0.9, 2 1/2″ thickness of Cellular Glass — Insulation k = 0.33 Btu in./ft², hr, °F

Factors

Equivalent Thickness = 3.78 — (Table 178)
$R_s = 0.91$ — (Table 186)
$t_a = 70°F$
$t_p = 10°F$

$$R = \frac{\text{Equivalent thickness}}{k} + R_s = \frac{3.78}{0.33} + 0.91 = 11.45 + 0.91 = 12.36$$

$$Q = \frac{t_a - t_p}{k} = \frac{70 - 10}{12.36} = 4.85 \text{ Btu/ft}^2\text{, hr}$$

(Heat Gain)

To convert to linear feet (from Table 188) the surface area per linear foot for 3″ NPS pipe with a 2″ thickness of insulation is 2.26. Thus

$$Q_{lf} = 4.85 \times 2.26 = 10.97 \text{ Btu/lin ft, hr.}$$

(Heat Gain)

The surface temperature of the outer coating will be:

$$t_s = t_a - (Q \times R_s) = 70 - (4.85 \times 0.91) = 70 - 4.4 = 65.6°F.$$

Relative humidity above 86% will cause surface condensation (Table 175).

The previous calculations were based on fabric reinforced mastic covering which had a surface emittance of 0.09. When low temperature pipe insulation is covered with aluminum jacketing its surface emittance of 0.09 causes the surface resistance to be greater resulting in a lowering of the surface temperature, with corresponding lowering of the dew point temperature. This will cause condensation on the surface. The calculations for aluminum jacketing over the insulation follow:

Aluminum Jacket $\varepsilon = 0.5$, 1″ thickness of Cellular Glass — Insulation k = 0.33 Btu, in./ft², hr, °F

Factors

Equivalent Thickness = 1.24 — (Table 178)
$R_s = 1.82$ — (Table 186)
$t_s = 70°F$
$t_p = 10°F$

$$R = \frac{\text{Equivalent thickness}}{k} + R_s = \frac{1.24}{0.33} + 1.82 = 3.76 + 1.82 = 5.58$$

$$Q = \frac{t_a - t_p}{R} = \frac{70 - 10}{5.58} = 10.75 \text{ Btu/ft}^2\text{, hr}$$

(Heat Gain)

To convert to linear feet (from Table 188) surface area ft²/lin ft for 3″ NPS pipe with 1″ thickness of insulation is 1.46.

$$Q_{\ell f} = 10.75 \times 1.46 = 15.70 \text{ Btu/lin ft, hr}$$

(Heat Gain)

The surface temperature of the outer aluminum jacket will be: temperature of air minus (heat transfer sq ft multiplied by surface resistance).

$$t_s = t_a - (Q \times R_s) = 70 - (10.75 \times 1.82) = 70 - 19.6 = 51.4°F$$

Relative humidity above 52% will cause surface condensation (Table 175).

Aluminum Jacket $\varepsilon = 0.5$, 1 1/2″ thickness Cellular Glass — Insulation k = 0.33 Btu, in./ft², hr, °F

Factors

Equivalent Thickness = 2.02 — (Table 178)
$R_s = 1.90$ — (Table 186)
$t_a = 70°F$
$t_p = 10°F$

$$R = \frac{\text{Equivalent thickness}}{k} + R_s = \frac{2.02}{0.33} + 1.90 = 6.12 + 1.90 = 8.02$$

$$Q = \frac{t_a - t_s}{R} = \frac{70 - 10}{8.02} = \frac{60}{8.02} = 7.48$$

(Heat Gain)

To convert to linear feet (from Table 188) surface area ft²/ lin ft for 3″ NPS pipe, 1 1/2″ thickness of insulation is 1.73.

$$Q_{\ell f} = 7.48 \times 1.73 = 12.94 \text{ Btu/lin ft, hr.}$$

(Heat Gain)

The surface temperature of outer aluminum jacket will be:

$$t_s = t_a - (Q \times R_s) = 70 - (7.98 \times 1.9) = 70 - 14.2 = 55.8°F$$

Relative humidity above 63% will cause surface condensation (Table 175).

Aluminum Jacket $\varepsilon = 0.05$, 2″ thickness Cellular Glass — Insulation k = 0.33 Btu, in./ft², hr, °F

Factors

Equivalent Thickness = 2.85 — (Table 178)
$R_s = 2.02$ — (Table 186)
$t_a = 20°F$
$t_p = 10°F$

$$R = \frac{\text{Equivalent thickness}}{k} + R_s = \frac{2.85}{0.33} + 2.02 = 8.64 + 2.02 = 10.66$$

$$Q = \frac{t_a - t_p}{R} = \frac{70 - 10}{10.66} = \frac{60}{10.66} = 5.6 \text{ Btu/ft}^2\text{, hr}$$

(Heat Gain)

To convert to linear feet (from Table 188) surface area ft²/lin ft for 3″ NPS pipe, 2″ thickness of insulation is 2.0.

$$Q_{\ell f} = 5.6 \times 2.0 = 11.2 \text{ Btu/lin ft, hr}$$

(Heat Gain)

The surface temperature of the aluminum jacket will be:

$$t_s = t_a - (Q \times R_s) = 70 - (5.6 \times 2.02) = 70 - 11.3 = 58.7°F$$

Relative humidity above 68% will cause surface condensation (Table 175).

Aluminum Jacket $\varepsilon = 0.05$, 2 1/2″ thickness Cellular Glass — Insulation k = 0.33 Btu, in./ft², hr, °F

Factors

Equivalent Thickness = 3.78
$R_s = 2.09$ — (Table 186)
$t_a = 70°F$
$t_p = 10°F$

$$R = \frac{\text{Equivalent thickness}}{k} + R_s = \frac{3.78}{0.33} + 2.09 = 11.45 + 2.09 = 13.54$$

$$Q = \frac{t_a - t_p}{R} = \frac{70 - 10}{11.45} = \frac{60}{11.45} = 4.43 \text{ Btu/ft}^2\text{, hr}$$

(Heat Gain)

To convert to linear feet (from Table 188) surface area ft²/lin ft for 3″ pipe 2 1/2″ thickness of insulation is 2.26.

$$Q_{\ell f} = 4.43 \times 2.26 = 10.0 \text{ Btu/lin ft, hr}$$

(Heat Gain

The surface temperature of the aluminum jacket will be:

$$t_s = t_a - (Q \times R_s) = 70 - (4.43 \times 2.09) = 70 - 9.3 = 60.7°$$

Relative humidity above 71% will cause surface condensation (Table 175).

Example 50a—*Metric Units*

A 88.9 mm diameter brine pipe operating at −12.2 °C passes through a heated and air conditioned area kept at 21.1 °C. Calculate the heat gain to the pipe for 0.025 m, 0.039 m, 0.051 m, 0.064 m, and 0.076 m thickness of cellular glass insulation with conductance of 0.047 W/mK. Calculate the surface temperature when the pipe is coated with green colored mastic reinforced with monocrylic fiber cloth. Also, calculate the surface temperature assuming the insulation is jacketed with bright aluminum.

Note: The results in heat gain per square meter will be multiplied by 0.317199 to check with the English Units Btu/ft², hr and W/lin m will be multiplied by 1.0406 to check with Btu/lin ft, hr.

Mastic Jacket $\varepsilon = 0.9$, 0.025 m thickness of Cellular Glass — Insulation $k_m = 0.047$ W/mK.

Factors

Equivalent Thickness = 0.031 — (Table 179)
$R_{ms} = 0.152$ — (Table 187)
$t_{ms} = 21.1$ °C
$t_{mp} = -12.2$ °C

$$R_m = \frac{\text{Equivalent thickness}}{k_m} + R_m = \frac{0.031}{0.047} + 0.152 =$$
$$0.660 + 0.151 = 0.811$$

$$Q_m = \frac{t_{ma} - t_{mp}}{R_m} = \frac{21.1 - (-12.2)}{0.811} = \frac{33.3}{0.811} = 41.1 \text{ W/m}^2$$

(Heat Gain)

Check 41.1 × 0.317199 = 13.0 Btu/ft², hr.

To convert to linear meters (from Table 189) surface area m²/lin m for 88.9 diameter pipe with 0.025 m thickness of insulation is 0.445.

$$Q_{\ell m} = 41.1 \times 0.445 = 18.29 \text{ W/lin m}$$

(Heat Gain)

Check 18.29 W/lin m × 1.0406 = 19 Btu/lin ft, hr.

The surface temperature of the outer coating will be: temperature of air − (Heat transfer in m² multiplied by surface resistance).

$$t_{ms} = t_a - (Q_m \times R_s) = 21.1 - (41.1 \times 0.152) = 21.1 - 6.2 = 14.9°$$

Relative humidity above 69% will cause surface condensation.

Mastic Jacket $\varepsilon = 0.9$, 0.039 m thickness of Cellular Glass — Insulation k = W/mK.

Factors

Equivalent Thickness = 0.051
$t_{ma} = 21.1$ °C
$t_{mp} = -12.2$
$R_s = 0.155$ — (Table 187)

$$R_m = \frac{\text{Equivalent thickness}}{k_m} + R_{ms} = \frac{0.051}{0.047} + 1.240$$

$$Q_m = \frac{t_{ma} - t_{mp}}{R_m} = \frac{21.1 - (-12.2)}{R_m} = \frac{33.3}{1.24} = 26.8 \text{ W/m}^2$$

Check 26.8 W/m × 0.317199 = 8.5 Btu/ft², hr.

To convert to linear foot (from Table 189) surface area m²/lin m 88.9 mm diameter pipe with 0.039 m thickness of insulation is 0.527.

$$Q_{\ell m} = 26.8 \times 0.527 = 14.12 \text{ W/lin m}$$

(Heat Gain)

Check 14.12 W/lin ft × 1.0406 = 14.7 Btu/lin ft, hr.

The surface temperature of the outer coating will be: temperature of air − (Heat transfer per m² multiplied by surface resistance).

$$t_{ms} = t_a - (Q_m \times R_s) = 21.1 - (26.8 \times 0.155) = 21.1 - 4.2 = 16.9° \text{ C.}$$

Relative humidity above 76% will cause surface condensation (Table 176).

Mastic Covering $\varepsilon = 0.9$, 0.051 m thickness of Cellular Glass — Insulation k = 0.047 W/mK.

Factors

Equivalent Thickness = 0.073 — (Table 179)
$R_{ms} = 0.158$ — (Table 187)
$t_{ma} = 21.1$ °C
$t_{mp} = -12.2$ °C

$$R_m = \frac{\text{Equivalent thickness}}{k} + R_{ms} = \frac{0.073}{0.047} + 0.158 =$$
$$0.553 + 0.158 = 1.711$$

$$Q_m = \frac{t_{ma} - t_{mp}}{k_m} = \frac{21.1 - (-12.2)}{1.711} = \frac{33.3}{1.711} = 19.40 \text{ W/m}^2$$

(Heat Gain)

Check 19.46 W/m² × 0.317199 = 6.2 Btu/ft², hr.

To convert to linear meter (from Table 189) surface area m²/lin m for 88.9 mm diameter pipe with 0.051 m thickness of insulation is 0.610.

$$Q_{\ell m} = 19.46 \times 0.610 = 11.87 \text{ W/lin ft}$$

(Heat Gain)

Check 11.87 W/lin ft × 1.0406 = 12.4.

The surface temperature of the outer coating will be:

$$t_{ms} = t_{ma} - (Q_m \times R_{ms}) = 21.2 - (19.46 \times 0.158) = 21.2 - 31 = 18.1\,°C$$

Relative humidity above 80% will cause surface condensation (Table 176).

Mastic Covering ε = 0.9, 0.066 m thickness of Cellular Glass — Insulation k_m = 0.047 W/mK.

Factors

Equivalent Thickness = 0.096 — (Table 179)
R_s = 0.160 — (Table 187)
t_{ma} = 21.1 °C
t_{mp} = − 12.2 °C

$$R_m = \frac{\text{Equivalent thickness}}{k_m} + R_{ms} = \frac{0.096}{0.047} + 0.160 =$$

2.042 = 2.202

$$Q_m = \frac{t_{ma} - t_{mp}}{k_m} = \frac{21.1 - (-12.2)}{2.202} = \frac{33.3}{2.202} = 15.12 \text{ W/m}^2$$

(Heat Gain)

Check 15.12 × 0.317199 = 4.8 Btu/ft², hr.

To convert to linear meter (from Table 189) surface area m²/lin m for 88.9 mm diameter pipe with 0.064 m thickness of insulation is 0.689.

$$Q_{\ell m} = 15.12 \times 0.689 = 10.42 \text{ W/lin m}$$

(Heat Gain)

Check 10.42 W/lin m × 1.0406 = 10.8.

The surface temperature of the outer coating will be:

$$t_{ms} = t_{ma} - (Q_m \times R_{ms}) = 21.2 - (15.12 \times 0.160) = 21.2 - 2.4 = 18.8\,°C$$

Relative humidity above 86% will cause surface condensation (Table 176).

As noted in the English Units part of this example, a change of outer surface changes the heat gain and surface temperature. The following is for the brine pipe insulated with cellular glass using an aluminum outer jacket.

Aluminum Jacket ε = 0.05, 0.025 m thickness of Cellular Glass — Insulation k_m = 0.047 W/mK.

Factors

Equivalent Thickness = 0.031
R_s = 0.310 — (Table 187)
t_{ma} = 21.1 °C
t_{mp} = −12.2 °C

$$R_m = \frac{\text{Equivalent thickness}}{k} + R_{ms} = \frac{0.031}{0.047} + 0.299 =$$

0.660 + 0.310 = 0.970

$$Q_m = \frac{t_{ma} - t_{mp}}{R_m} = \frac{21.1 - (-12.2)}{0.970} = \frac{33.3}{0.970} = 34.3 \text{ W/m}^2$$

(Heat Gain)

Check 34.3 W/m² × 0.317199 = 10.8 Btu/ft², hr.

To convert to linear foot (from Table 189) surface are m/lin m for 88.9 mm diameter pipe with 0.025 m thickness of insulation is 0.445.

$$Q_{\ell m} = 34.3 \text{ W/m}^2 \times 0.445 = 15.26 \text{ W/lin m}$$

(Heat Gain)

Check 15.26 W/lin m × 1.0406 = 15.8 Btu/lin ft, hr.

The surface temperature of the aluminum jacket will be:

$$t_{ms} = t_{ma} - (Q_m \times R_m) = 21.1 - (34.3 \times 0.31) = 21.1 - 10.6 = 10.5\,°C$$

Relative humidity above 68% will cause surface condensation.

Aluminum Jacket ε = 0.05, 0.039 m thickness Cellular Glass — Insulation k_m = 0.047 W/mK.

Factors

Equivalent Thickness = 0.051
t_{ma} = 21.1 °C
t_{mp} = −12.2
R_{ms} = 0.321 — (Table 187)
t_{ma} = 21.1 °C
t_{mp} = −12.2 °C

$$R_m = \frac{\text{Equivalent thickness}}{km} + R_s = \frac{0.051}{0.047} + 0.321 =$$

1.085 + 0.321 = 1.406

$$Q_m = \frac{t_{ma} - t_{mp}}{R_m} = \frac{21.1 - (-12.2)}{1.406} = \frac{33.3}{1.406} = 23.7 \text{ W/m}^2$$

(Heat Gain)

Check 23.7 × 0.317199 = 7.5 Btu/ft², hr.

To convert to linear meters (from Table 189) surface area m²/lin m for 88.9 mm diameter pipe with 0.039 thickness of insulation is 0.527.

$$Q_{\ell m} = 23.7 \times 0.527 = 12.5 \text{ W/lin m}$$

(Heat Gain)

Check 12.5 W/lin m × 1.0406 = 12.9 Btu/ft², hr.

The surface temperature of the aluminum jacket will be:

$$t_{ms} - (Q_m \times R_{ms}) = 21.1 - (23.7 \times 0.321) = 21.1 - 7.6 = 13.5\,°C$$

Relative humidity over 63% will cause surface condensation (Table 176).

Aluminum Jacket ε = 0.05, 0.051 m thickness of Cellular Glass — Insulation km = 0.047 W/mK.

Factors

Equivalent Thickness = 0.073 — (Table 179)
R_{ms} = 0.355 — (Table 187)
t_{ms} = 21.1 °C
t_{mp} = −12.2 °C

$$R_m = \frac{\text{Equivalent thickness}}{\text{km}} + R_s + \frac{0.073}{0.047} + 0.355 =$$
$$1.553 + 0.355 = 1.908$$

$$Q_m = \frac{t_{ma} - t_{mp}}{R_m} = \frac{21.1 - (-12.2)}{1.908} = 17.45 \text{ W/m}^2$$

(Heat Gain)

Check 17.45 W/m² × 0.317199 = 5.5 Btu/ft², hr.

To convert to linear meters (from Table 189) surface area m²/lin m for 0.889 mm diameter pipe, 0.051 m thickness of insulation is 0.610.

$$Q_{\ell m} = 17.45 \times 0.610 = 10.64 \text{ W/lin m}$$

Check 10.64 W/lin m × 1.0406 = 11.1 Btu/lin ft, hr.

The surface temperature of the aluminum jacket will be:

$$t_{ms} = t_{ma} - (Q_m \times R_{ms}) = 21.1 - (17.45 \times 0.355) = 21.11 - 6.2 = 14.9\,°C$$

Relative humidity above 68% will cause surface condensation (Table 176).

Aluminum Jacket ε = 0.05, 0.064 m thickness of Cellular Glass — Insulation km = 0.047 W/mK.

Factors

Equivalent Thickness = 0.096 — (Table 179)
R_{ms} = 0.368 — (Table 187)
t_{ma} = 21.1 °C
t_{mp} = −12.2 °C

$$R_m = \frac{\text{Equivalent thickness}}{\text{km}} + R_s = \frac{0.096}{0.047} + 0.368 =$$
$$2.042 - 0.368 = 2.410$$

$$Q_m \; \frac{t_{ma} - t_{mp}}{R_m} = \frac{21.1 - (-12.2)}{2.41} = \frac{33.3}{2.41} = 13.81 \text{ W/m}^2$$

(Heat Gain)

Check 13.81 W/m² × 0.317199 = 4.38 Btu/ft², hr.

To convert to linear meters (from Table 189) surface area m²/lin m for 0.889 mm diameter pipe with 0.064 m thickness of insulation is 0.689.

$$Q_{\ell m} = 13.81 \times 0.689 = 9.515 \text{ W/lin m}$$

(Heat Gain)

Check 9.515 W/lin m × 1.0406 = 9.9 Btu/lin ft, hr.

The surface temperature of the aluminum jacket will be:

$$t_{ms} = t_{ma} - (Q_m \times R_m) = 21.1 - (13.81 \times 0.368) = 21.1 - 5.1 = 15\,°C$$

Relative humidity above 71% will cause surface condensation (Table 175).

As shown on the tabulation for Examples 50 and 50a, a greater insulation thickness is required to prevent condensation when the outer surface has a low surface emittance ε. Table 194 presents the thickness of cellular glass insulation required to prevent condensation on the outer surface having a surface emittance ε of 0.9 indoors where the air is conditioned to be 75 °F (23.9 °C) at 60% relative humidity. See page 424 for tabulation.

Table 195 provides the approximate wall thicknesses of standard nominal thickness insulation for NPS pipes in English Units and Table 196 provides this information in Metric Units. Table 197 provides the approximate wall thicknesses of standard nominal thicknesses of insulation for Tube Sizes in English Units and Table 198 provides this information in Metric Units.

The dimensions for "Inner and Outer Diameters of Rigid Thermal Insulation for Nominal Sizes of Pipe and Tubing" are given in the following tables. Table 199 "Dimensions of Insulation for NPS (Dec. & Fractions) English Units." Table 200 "Dimensions of Insulation for NPS Pipe — Metric Units." Table 201 "Dimension of Insulation for Tubes — Metric Units." These provide the dimensions needed for such designs as pipe racks.

THERMAL CONDUCTIVITY K, CONDUCTANCE C AND RESISTANCE R OF BUILDING MATERIALS

*English Units**

Material	Density lb/ft³	40°F Mean Temp.			85°F Mean Temp.			110°F Mean Temp.		
		K	C	R	K	C	R	K	C	R
Air										
(No heat transfer by radiation or convection)		0.17			0.18			0.20		
Air spaces and aluminum foil spacers — vertical										
1 1/2" space divided by										
Aluminum foil (bright both sides)			0.22	4.54		0.23	4.34		0.24	4.16
3/4" space divided by										
Aluminum foil (bright both sides)			0.31	3.22		0.32	3.12		0.33	3.03
2 1/4" space divided by two curtains										
Aluminum foil (bright both sides)			0.15	6.66		0.16	6.25		0.17	5.88
3" space divided by three curtains										
Aluminum foil (bright both sides)			0.11	9.09		0.12	8.33		0.13	7.69
3 3/4" space divided by four curtains										
Aluminum foil (bright both sides)			0.09	11.11		0.10	10.0		0.11	9.09
Air spaces and aluminum foil spacers										
3 5/8" faced both sides with aluminum foil										
Vertical (heat flow across)			0.56	1.78		0.57	1.75		0.58	1.72
Horizontal (heat flow up)			0.94	1.06		0.96	1.04		1.00	1.00
Horizontal (heat flow down)			0.41	2.43		0.43	2.32		0.49	2.04
Air spaces with ordinary building materials										
3 5/8" face with material of emissivities = .83										
Vertical (heat flow across)			1.70	0.56		0.80	0.55		1.90	0.53
Horizontal (heat flow up)			1.32	0.76		1.35	0.74		1.40	0.71
Horizontal (heat flow down)			0.94	1.06		0.97	1.03		1.01	0.99
Aluminum	166 to 170	1396.0			1400.0			1420.0		
Brass	534	1035.0			1044.0			1060.0		
Bronze	509	820.0			828.0			835.0		
Beaver board										
Cane fiber	18.8	0.33			0.35			0.36		
Spruce fiber	31.0	1.97			2.03			2.12		
Brick										
Low density	103.0	5.0			5.2			5.3		
High density	128.0	9.2			9.4			9.5		
Red building, soft burned	112.0	4.3			4.5			4.6		
Brick—slag	87.4	8.8			9.0			9.1		
Cement plaster	96.0	8.0			8.2			8.3		
Cinder block										
4 x 8 x 16—Solid	104.0		1.00	1.00		1.10	0.91		1.25	0.80
8 x 8 x 16—With standard hollow spaces			0.58	1.72		0.64	1.56		0.72	1.39
12 x 8 x 16—With standard hollow spaces			0.53	1.88		0.58	1.72		0.66	1.51
Concrete										
Typical	144.0	12.00			12.45			12.96		
Concrete Block										
8 x 8 x 16—Sand and gravel aggregate (hollow)			0.90	1.11		0.99	1.01		1.12	0.89
8 x 8 x 16—Limestone aggregate (hollow)			0.86	1.19		0.95	1.05		1.09	0.92
12 x 8 x 16—Sand and gravel aggregate (hollow)			0.78	1.28		0.86	1.16		0.99	1.01
Solid										
Concrete, cellulated	40.0	1.06			1.16			1.31		
Concrete, cellulated	50.0	1.44			1.58			1.31		
Concrete, cellulated	60.0	1.80			1.99			2.18		
Concrete, cellulated	70.0	2.18			2.29			2.42		
Concrete, cinder										
1:2:2.75 Ratio	104.0	4.63			4.98			5.40		
1:2.75:4.5 Ratio	99.0	4.30			4.75			5.25		
1:3.5:5.5 Ratio	92.0	3.78			3.99			4.30		
Concrete, Haydite										
1:2:2.75 Ratio	80.0	4.28			4.77			5.28		
1:2.75:4.5 Ratio	75.0	3.78			3.99			4.30		
1:3.5:5.5 Ratio	72.0	3.67			3.88			4.17		
1:8 Ratio	67.0	2.90			2.95			3.07		
Concrete, limestone										
1:2:2.7 Ratio	135.0	11.20			11.50			11.80		
1:2.75:4.5 Ratio	138.0	12.00			12.35			12.70		
1:3.5:5.5 Ratio	139.0	12.50			11.85			12.20		
Concrete, sand and gravel										
1:2:2.75 Ratio	145.0	13.1			13.40			13.70		
1:2.75:4.5 Ratio	146.0	12.9			13.20			13.50		
1:3.5:5.5 Ratio	148.0	13.2			13.50			13.80		
Basic Ambient Design Conditions		**Winter Season**			**Summer Season**			**Solar Exposed**		

*K = Btu, in./ft², hr, °F, C = Btu/ft², hr, °F, $R = \frac{1}{C}$ or when ℓ is in inches $R = \frac{\ell}{K}$

TABLE 83

(Sheet 1 of 4

THERMAL CONDUCTIVITY K, CONDUCTANCE C AND RESISTANCE R OF BUILDING MATERIALS
English Units

Material	Density lb/ft³	40°F Mean Temp.			85°F Mean Temp.			110°F Mean Temp.		
		K	C	R	K	C	R	K	C	R
Clay										
Dried	63.0	3.60			3.62					
Wet	71.0	4.20			4.28					
Copper	556.0	2190.00			2205.00			2218.00		
Dolomite, compact	108.0	14.50			14.61			14.80		
Domont brick (Terracotta)	113.0	4.62			4.69			4.78		
Earth (Dry and Packed)	95.0	3.80			3.83			0.49		
Ebonite	74.0	0.41			0.44			0.49		
Flagstone										
Across cleavage	140.0	12.80			12.80			13.00		
Along cleavage	149.0	17.40			17.40			17.30		
Freestone, sandstone	135.0	18.40			18.40			19.00		
Glass block (with cavity)	154.0		6.10	0.16		6.50	0.15		6.90	0.14
Glass										
Flint	267.0	4.16			4.66			5.20		
Plate	154.0	5.55			5.95			6.45		
Soda	161.0	4.94			5.52			6.02		
Quartz	163.0	12.67			13.50			14.05		
Granite	93.0	18.00			19.50			22.00		
Gravel										
Fine (0.16″ to 0.35″)	91.0	1.63			1.89			2.08		
Dry Stone (1″ to 3″)	115.0	2.35			2.52			2.73		
Gypsum block	42.7	1.69			1.75			1.83		
Gypsum board	72.0	3.00			3.06			3.11		
Gypsum board covered with paper 1/2″ thick	68.0		2.60	0.38		2.80	0.36		3.00	0.33
Gypsum plaster										
Heavy density	52.4	1.73			1.83			1.83		
Light density	46.2	2.32			2.40			2.52		
Gypsum tile										
3 x 3 x 16	67.0		0.50	2.00		0.55	1.81		0.60	1.67
Haydite block										
8 x 8 x 16	67.0		0.52	1.92		0.54	1.85		0.58	1.72
8 x 8 x 12	77.0		0.48	2.00		0.52	1.92		0.56	1.78
Insulation for Building										
Cements										
Mineral Fiber	8.7	0.35			0.37			0.40		
Loose										
Aluminum Silica Fibers	6.0 to 7.5	0.33			0.35			0.38		
Cellulose	10.0 to 15.0	0.31			0.33			0.36		
Glass Fiber	4.0 (Avg.)	0.25			0.27			0.31		
Gypsum Pellets	2.0 to 3.0	0.53			0.54			0.55		
Mineral Fiber	4.0 to 12.0	0.30			0.31			0.33		
Perlite	5.0 to 8.0	0.29			0.31			0.34		
Silica aerogel granules	4.0 to 5.5	0.21			0.24			0.29		
Vermiculite flakes	8.0 to 10.0	0.47			0.48			0.50		
Blankets and Flexible Faced										
Elastomeric	8.0 to 9.0	0.30			0.32			0.34		
Glass Fiber	1.0 to 3.0	0.30			0.32			0.34		
Glass Fiber	3.0 to 4.5	0.28			0.30			0.34		
Glass Fiber with Alum. Ref. Sheet	4.0 to 4.5	0.27			0.29			0.33		
Mineral Wool	6.0 to 7.5	0.27			0.29			0.33		
Rigid and Semi-Rigid										
Cellular Glass	7.5 to 9.0	0.36			0.37			0.38		
Glass Fiber Board and Sheets	3.0 to 4.5	0.26			0.30			0.36		
Mineral Fiber	7.0 to 12.0	0.32			0.34			0.38		
Polystyrene	1.5 to 2.0	0.25			0.29			0.31		
Polyurethane	1.7 to 4.5	0.24			0.30			0.33		
Rubber—expanded	4.0 to 6.0	0.28			0.29			0.41		
Silica—expanded	13.0	0.39			0.40			0.42		
Wood Fibers and Binders	15.0 to 18.0	0.42			0.47			0.47		
Insulating Boards—Structural										
Glass Fiber organic bonded	4.0 to 9.0	0.27			0.28			0.30		
Foamed rubber	4.0 to 4.5	0.25			0.26			0.27		
Foamed polystyrene	1.8	0.28			0.29			0.30		
Foamed polystyrene	2.2	0.24			0.26			0.28		
Foamed polystyrene	3.5	0.23			0.25			0.27		
Foamed polyurethane	1.5	0.18			0.20			0.29		
Basic Ambient Design Conditions		**Winter Season**			**Summer Season**			**Solar Exposed**		

K = Btu, in./ft², hr, °F, C = Btu/ft², hr, °F, $R = \frac{1}{C}$ or when ℓ is in inches $R = \frac{\ell}{K}$

TABLE 83 (Continued) (Sheet 2 of 4)

THERMAL CONDUCTIVITY K, CONDUCTANCE C AND RESISTANCE R OF BUILDING MATERIALS

English Units

Material	Density lb/ft^3	40°F Mean Temp.			85°F Mean Temp.			110°F Mean Temp.		
		K	C	R	K	C	R	K	C	R
Foamed polyurethane	2.5	0.17			0.19			0.30		
Mineral fiber with resin binder	15.0	0.29			0.30			0.32		
Mineral fiber core insulation	16.0 to 17.0	0.34			0.35			0.37		
Mineral fiber acoustical tile	18.0 to 21.0	0.37			0.38			0.40		
Plaster or gypsum board	50.0	1.12			1.15			1.20		
Particle board low density	37.0	0.54			0.60			0.68		
Particle board medium density	50.0	0.94			1.10			1.21		
Particle board high density	62.5	1.18			1.30			1.50		
Preformed roof insulation 1/2" thick			0.72	1.38		0.73	1.36		0.75	1.33
Preformed roof insulation 1" thick			0.36	2.78		0.38	2.63		0.40	2.50
Preformed roof insulation 1 1/2" thick			0.24	4.17		0.26	3.85		0.28	3.57
Preformed roof insulation 2" thick			0.19	5.26		0.20	5.00		0.21	4.76
Preformed roof insulation 2 1/2" thick			0.15	6.67		0.16	6.25		0.17	5.88
Preformed roof insulation 3" thick			0.12	8.33		0.13	7.69		0.14	7.14
Iron—wrought	485.00	417.00			415.00			412.00		
Masonry Materials										
Cement mortar	116.00	5.00			5.50			6.00		
Light weight concrete (Using light weight aggregates)	100.0	3.60			3.65			3.80		
Light weight concrete	80.00	2.50			2.55			2.63		
Light weight concrete	60.00	1.70			1.75			1.79		
Concrete (sand and gravel aggregates)	140.00	10.50			10.70			10.90		
Stucco	116.00	5.00			5.25			5.50		
Nickel	537.00	403.00			413.00			420.00		
Plastering Materials										
Plaster gypsum board 3/8" thick	68.00		3.73	0.27		3.75	0.26			
Plaster gypsum board 1/2" thick	61.00		2.83	0.35		2.85	0.35			
Cement plaster, sand aggregate	116.00	5.00			5.20					
Cement plaster, sand aggregate 3/8" thick	116.00		13.30	0.07		13.35	0.07			
Cement plaster, sand aggregate 1/2" thick	116.00		9.95	0.10		10.00	0.10			
Cement plaster, sand aggregate 3/4" thick	116.00		6.60	0.15		6.65	0.15			
Roofing Materials										
Asphalt shingles	170.00		6.50	0.15		7.00	0.14		7.70	0.13
Cement shingles	120.00		8.00	0.17		6.50	0.15		7.20	0.14
Slate shingles	201.00		19.50	0.05		20.00	0.05		21.00	0.05
Wood shingles	65.00		1.28	0.78		1.38	0.72		1.55	0.64
Built up roofing (felts and asphalts)	70.00		7.00	0.14		7.25	0.14		2.00	0.11
Siding Materials										
Aluminum siding (with backing)			1.60	0.625		1.66	0.602		1.70	0.583
Cement shingles			4.76	0.210		4.96	0.201		5.20	0.192
Insulating backing board 3/8" thick			1.61	0.621		1.70	0.588		1.90	0.526
Wood shingles 16" (7 1/2" exposure)			1.15	0.869		1.20	0.833		1.30	0.769
Wood, drop siding 1" x 8"			1.27	0.787		1.30	0.769		1.34	0.746
Wood bevel 1/2" x 8" lapped			1.23	0.813		1.27	0.787		1.31	0.763
Wood bevel 3/4" x 10" lapped			0.95	1.052		0.98	1.020		1.02	0.980
Wood—plywood 3/8" lapped			1.59	0.629		1.65	0.606		1.73	0.578
Steel—structural	487.00	245.0			260.0			275.0		
Stone—solid, marble	155.00	12.50			12.78			12.93		
Sand—fine, dry	97.00	2.10			2.15			2.30		
Sand—common moisture	99.00	2.26			2.35			2.50		
Sandstone—dry	141.00	10.72			11.80			12.80		
Slate across cleavage	160.00	10.45			11.05			12.28		
Slate along cleavage	160.00	18.90			19.55			21.10		
Soapstone	170.00	23.00			23.60			24.10		
Terrazzo	120.00	12.00			12.60			13.20		
Soil—dry	75.00	0.96			1.01					
Soil—normal dampness	80.00	3.47			3.63					
Structural building boards—sheathing and panels										
Fiber—cement board	120.00	4.10			4.30			4.60		
Fiber—cement board 1/8" thick	120.00		33.00	0.03		34.40	0.03		36.80	0.03
Fiber—cement board 1/4" thick	120.00		16.50	0.06		17.20	0.06		18.40	0.05
Hardboard—medium density 7/16" thick	40.00		1.49	0.67		1.63	0.61		1.70	0.58
Hardboard—high density	55.00	0.82			0.86			0.93		
Hardboard—high density tempered	63.00	1.10			1.15			1.20		
Particle board—low density	37.00	0.54			0.57			0.60		
Particle board—medium density	50.00	0.94			0.98			1.04		
Particle board—high density	62.50	1.18			1.24			1.30		
Basic Ambient Design Conditions		**Winter Season**			**Summer Season**			**Solar Exposed**		

K = Btu, in./ft^2, hr, °F, C = Btu/ft^2, hr, °F, $R = \frac{1}{C}$ or when ℓ is in inches $R = \frac{\ell}{K}$

TABLE 83 (Continued)

(Sheet 3 of

THERMAL CONDUCTIVITY K, CONDUCTANCE C AND RESISTANCE R OF BUILDING MATERIALS

English Units

Material	Density lb/ft^3	40°F Mean Temp.			85°F Mean Temp.			110°F Mean Temp.		
		K	C	R	K	C	R	K	C	R
Plywood	34.00	0.80			0.84			0.89		
Plywood 1/4″ thick	34.00		3.20	0.31		3.36	0.29		3.56	0.28
Plywood 3/8″ thick	34.00		2.13	0.47		2.24	0.45		2.37	0.42
Plywood 1/2″ thick	34.00		1.60	0.62		1.68	0.59		1.78	0.56
Plywood 3/4″ thick	34.00		1.06	0.94		1.12	0.89		1.18	0.85
Plywood 1″ thick	34.00		0.80	1.25		0.84	1.19		0.89	1.12
Wood subfloor 3/4″ thick	40.00		1.06	0.94		1.12	0.89		1.15	0.87
Tile										
Clay hollow—1 cell, 3″ wide			1.25	0.80		1.30	0.76		1.35	0.74
Clay hollow—1 cell, 4″ wide			0.90	1.11		0.95	1.05		1.00	1.00
Clay hollow—2 cells, 6″ wide			0.66	1.51		0.70	1.42		0.74	1.39
Clay hollow—2 cells, 8″ wide			0.54	1.85		0.58	1.72		0.62	1.61
Clay hollow—2 cells, 10″ wide			0.45	2.22		0.49	2.04		0.53	1.88
Clay hollow—3 cells, 12″ wide			0.40	2.50		0.43	2.23		0.46	2.17
Wood (across gain conductivities)										
California redwood, 0% moisture	22.00	0.65			0.66			0.69		
California redwood, 8% moisture	24.00	0.69			0.70			0.73		
California redwood, 16% moisture	28.00	0.73			0.74			0.77		
Cypress, 0% moisture	22.00	0.66			0.67			0.70		
Cypress, 8% moisture	27.00	0.80			0.81			0.84		
Cypress, 16% moisture	28.00	0.83			0.84			0.87		
Maple, 0% moisture	36.00	0.88			0.89			0.92		
Maple, 8% moisture	38.00	0.98			0.99			1.02		
Maple, 16% moisture	40.00	1.08			1.09			1.12		
Oak, 0% moisture	38.00	1.02			1.03			1.06		
Oak, 8% moisture	43.00	1.15			1.16			1.19		
Oak, 16% moisture	48.00	1.27			1.29			1.32		
Pine, 0% moisture	30.00	0.75			0.76			0.78		
Pine, 8% moisture	35.00	0.89			0.90			0.93		
Pine, 16% moisture	40.00	1.01			1.03			1.06		
Wood pulp board	20.60	0.37			0.39			0.92		
Wood fiber board	11.90	0.30			0.31			0.33		
Wood fiber board	19.80	0.33			0.39			0.41		
Wood fiber board	28.50	0.50			0.52			0.55		
Wood doors										
1″ thickness			0.64	1.56		0.65	1.53		0.67	1.49
With wood storm door			0.30	3.33		0.81	3.22		0.33	3.30
With metal storm door			0.39	2.56		0.40	2.50		0.42	2.35
1 1/2″ thickness			0.49	2.04		0.50	2.00		0.52	1.92
With wood storm door			0.27	3.70		0.28	3.57		0.30	3.33
With metal storm door			0.33	3.03		0.34	2.94		0.36	2.78
2″ thickness			0.43	2.32		0.44	2.27		0.46	2.17
With wood storm door			0.24	4.16		0.25	4.00		0.27	3.70
With metal storm door			0.29	3.45		0.30	3.33		0.32	3.12
Steel doors 1 3/4″ thick										
With mineral fiber core			0.59	1.69		0.61	1.63		0.65	1.53
With organic foam core			0.47	2.12		0.53	1.88		0.59	1.69
With gypsum core			0.90	1.11		0.92	1.08		0.93	1.07
Basic Ambient Design Conditions		**Winter Season**			**Summer Season**			**Solar Exposed**		

K = Btu, in./ft^2, hr, °F, C = Btu/ft^2, hr, °F, $R = \frac{1}{C}$ or when ℓ is in inches $R = \frac{\ell}{K}$

TABLE 83 (Continued) (Sheet 4 of 4)

COEFFICIENT OF HEAT TRANSMISSION U, RESISTANCE R AND LIGHT REFLECTANCES OF VARIOUS TYPES OF GLASS AND GLASS BLOCK

English Units

Position	Description	Thickness Inches	Surface Coating	Winter U	Winter R	Summer U	Summer R	% Solar Reflectance	% Light Reflectance Outdoors	% Light Reflectance Indoors
Vertical	Single Flat Glass	3/32	Clear	1.16	0.86	1.03	0.97	7	8	8
	Single Flat Glass	1/8	Clear	1.16	0.86	1.04	0.96	7	8	8
	Single Flat Glass	1/4	Clear	1.13	0.88	1.04	0.96	7	8	8
	Single Flat Glass	3/8	Clear	1.11	0.90	1.03	0.97	7	8	8
	Single Flat Glass	1/2	Clear	1.09	0.93	1.03	0.97	7	8	8
	Single Flat Glass	3/4	Clear	1.05	0.95	1.01	0.99	7	8	8
	Single Flat Glass	1	Clear	1.01	0.99	0.99	1.01	7	8	8
	Single Flat Glass—Coated	1/8	Bronze	1.16	0.86	1.10	0.91	6	35	18
	Single Flat Glass—Coated	1/4	Bronze	1.13	0.88	1.10	0.91	6	35	14
	Single Flat Glass—Coated	3/8	Bronze	1.11	0.90	1.09	0.92	6	35	11
	Single Flat Glass—Coated	1/8	Gray	1.16	0.86	1.09	0.92	5	6	6
	Single Flat Glass—Coated	1/4	Gray	1.14	0.88	1.10	0.91	5	6	6
	Single Flat Glass—Coated	3/8	Gray	1.13	0.88	1.10	0.91	5	6	6
	Single Flat Glass—Coated	1/8	Reflect.	1.13	0.86	1.12	0.89	26	6	6
	Single Flat Glass—Coated	1/4	Reflect.	1.13	0.86	1.12	0.89	26	6	6
	Single Flat Glass—Coated	3/8	Reflect.	1.13	0.86	1.12	0.89	30	5	5
	Double Flat Glass									
	1/4″ Air Space	1/8	Clear	0.58	1.72	0.62	1.61	11	15	15
	1/4″ Air Space	1/4	Clear	0.56	1.78	0.60	1.67	11	14	14
	1/2″ Air Space	1/2	Clear	0.49	2.04	0.58	1.72	11	14	14
	Double Flat Glass									
	1/4″ Air Space	1/4	Bronze	0.58	1.72	0.64	1.56	26	36	23
	1/2″ Air Space	1/2	Bronze	0.50	2.00	0.57	1.75	29	36	23
	Double Flat Glass									
	1/4″ Air Space	1/4	Gray	0.58	1.72	0.64	1.56	13	28	6
	1/2″ Air Space	1/2	Gray	0.49	2.04	0.58	1.72	13	23	5
	Double Flat Glass									
	1/4″ Air Space	1/4	Reflect.	0.49	2.04	0.57	1.75	45	22	35
	Triple Flat Insulating Glass									
	1/4″ Air Space	1/4	Clear	0.47	2.12	0.45	2.22	15	16	16
	1/2″ Air Space	1/4	Clear	0.36	2.77	0.35	2.86	15	16	16
	Storm Windows									
	1″ to 4″ Air Space	1/4	Clear	0.56	1.78	0.54	1.85	11	15	15
	Glass Block									
	6″ x 6″ x 4″ Thick		Clear	0.60	1.66	0.57	1.75	13		
	8″ x 8″ x 4″ Thick		Clear	0.88	1.72	0.54	1.85	13		
	With Cavity Divider		Clear	0.46	2.17	0.46	2.17	19		
	12″ x 12″ x 4″ Thick		Clear	0.52	1.92	0.50	2.00	13		
	With Cavity Divider		Clear	0.44	2.27	0.42	2.38	19		
Horizontal (Skylights)	Single Flat Glass	1/2	Clear	1.18	0.85	0.83	1.20	7		
	Single Flat Glass	3/4	Clear	1.14	0.88	0.81	1.23	7		
	Single Flat Glass	1	Clear	1.10	0.91	0.79	1.26	7		
	Single Flat Glass—Coated	1/2	Bronze	1.20	0.83	0.94	1.06	6		
	Single Flat Glass—Coated	1/2	Gray	1.22	0.82	0.95	1.05	5		
	Single Flat Glass—Coated	1/2	Reflect.	1.23	0.81	0.97	1.03	26		
	Double Flat Glass—1/2″ Air Space	1/2	Bronze	0.57	1.75	0.53	1.88	6		
	Double Flat Glass—1/2″ Air Space	1/2	Gray	0.43	2.53	0.58	1.72	5		
	Double Flat Glass—1/2″ Air Space	1/2	Reflect.	0.42	2.98	0.52	1.92	26		
	Glass Block									
	12″ x 12″ x 4″ Thick		Clear	0.51	1.96	0.34	2.94	19		
	With Cavity Divider		Clear	0.51	1.96	0.34	2.94	19		
	Plastic Bubbles									
	Single Walled			1.15	0.87	0.80	1.25	12		
	Doubled Walled			0.70	1.42	0.46	2.17	14		
Vertical	Windows									
	Wood Sash—80% Glass		Clear	1.02	0.98	1.10	0.91	7		
	Wood Sash—60% Glass		Clear	0.90	1.11	0.99	1.01	7		
	Double Glass—Wood Sash—80% Glass		Clear	0.45	2.22	0.49	2.04	11		
	Metal Sash—Single Glass		Clear	1.13	0.88	1.24	0.80	11		
	Metal Sash—Double Glass		Clear	0.56	1.78	0.70	1.42	11		
	Sliding Door									
	Wood Frame			1.08	0.92	1.04	0.96	11		
	Metal Frame			1.13	0.88	1.14	0.88	11		

$U = Btu/ft^2, hr, °F, \quad R = \frac{1}{U}$

TABLE 84

THERMAL CONDUCTIVITY K_m, CONDUCTANCE C_m AND RESISTANCE R_m OF BUILDING MATERIALS
Metric Units

Material	Density	4.4°C Mean Temp.			29.4°C Mean Temp.			43.3°C Mean Temp.		
	Kilograms per m^3	K_m	C_m	R_m	K_m	C_m	R_m	K_m	C_m	R_m
Air										
(No heat transfer by radiation or convection)										
Air spaces and aluminum foil spacers — vertical										
3.8 cm space divided by										
Aluminum foil (bright both sides)			1.24	0.600		1.30	0.766		1.36	0.734
1.9 cm space divided by										
Aluminum foil (bright both sides)			1.76	0.568		1.81	0.550		1.87	0.583
5.7 cm space divided by two curtains										
Aluminum foil (bright both sides)			0.85	1.174		0.91	1.101		0.96	1.036
7.6 cm space divided by three curtains										
Aluminum foil (bright both sides)			0.62	1.060		0.68	1.467		0.74	1.355
9.5 cm space divided by four curtains										
Aluminum foil (bright both sides)			0.51	1.957		0.57	1.761		0.62	0.624
Air spaces and aluminum foil spacers										
9.2 cm faced both sides with aluminum foil										
Vertical (heat flow across)			3.17	0.314		3.23	0.308		3.29	0.303
Horizontal (heat flow up)			5.33	0.187		5.45	0.183		5.68	0.176
Horizontal (heat flow down)			2.32	0.429		2.55	0.391		2.78	0.359
Air spaces with ordinary building materials										
9.2 cm face with material of emissivities = .83										
Vertical (heat flow across)			9.65	0.104		10.22	0.097		10.78	0.092
Horizontal (heat flow up)			7.49	0.133		7.68	0.130		7.94	0.126
Horizontal (heat flow down)			5.33	0.187		5.50	0.181		5.73	0.174
Aluminum	2659 to 2723	201.303			201.880			204.764		
Brass	8554	149.247			150.545			152.852		
Bronze	8154	118.244			119.397			120.407		
Beaver board										
Cane fiber	288	0.047			0.050			0.057		
Spruce fiber	497	0.284			0.293			0.306		
Brick										
Low density	1650	0.721			0.749			0.764		
High density	2051	1.327			1.355			1.369		
Red building, soft burned	1794	0.620			0.649			0.663		
Brick—slag	1393	1.269			1.298			1.312		
Cement plaster	1538	1.154			1.182			1.107		
Cinder block										
10 x 20 x 40 cm—Solid	1666		5.68	0.176		6.24	0.160		7.09	0.140
20 x 20 x 40 cm—With standard hollow spaces			3.29	0.304		3.63	0.275		4.09	0.244
30 x 20 x 40 cm—With standard hollow spaces			3.00	0.333		3.29	0.304		3.75	0.267
Concrete										
Typical	2306	1.730			1.795			1.869		
Concrete Block										
20 x 20 x 40 cm—Sand and gravel aggregate (hollow)			5.11	0.196		5.62	0.178		6.35	0.157
20 x 20 x 40 cm —Limestone aggregate (hollow)			4.88	0.204		5.39	0.185		6.18	0.162
30 x 20 x 40 cm—Sand and gravel aggregate (hollow)			4.42	0.226		4.88	0.204		5.62	0.177
Solid										
Concrete, cellulated	640	0.152			0.167			0.188		
Concrete, cellulated	801	0.208			0.228			0.257		
Concrete, cellulated	961	0.260			0.287			0.314		
Concrete, cellulated	1121	0.314			0.330			0.349		
Concrete, cinder										
1:2:2.75 Ratio	1666	0.668			0.718			0.779		
1:2.75:4.5 Ratio	1580	0.620			0.684			0.757		
1:3.5:5.5 Ratio	1473	0.545			0.575			0.620		
Concrete, Haydite										
1:2:2.75 Ratio	1281	0.617			0.688			0.761		
1:2.75:4.5 Ratio	1201	0.545			0.575			0.620		
1:3.5:5.5 Ratio	1153	0.529			0.559			0.601		
1:8 Ratio	1073	0.418			0.425			0.443		
Concrete, limestone										
1:2:2.7 Ratio	2162	1.615			1.658			1.702		
1:2.75:4.5 Ratio	2210	1.730			1.780			1.831		
1:3.5:5.5 Ratio	2226	1.638			1.708			1.750		
Concrete, sand and gravel										
1:2:2.75 Ratio	2322	1.889			1.932			1.976		
1:2.75:4.5 Ratio	2338	1.860			1.903			1.947		
1:3.5:5.5 Ratio	2398	1.903			1.947			1.989		
Basic Ambient Design Conditions		**Winter Season**			**Summer Season**			**Solar Exposed**		

K_m = W/mK, C_m = W/m^2, $R_m = \frac{1}{C_m}$ or when ℓ_m is in meters $R_m = \frac{\ell_m}{K_m}$

TABLE 85

(Sheet 1 of 4)

THERMAL CONDUCTIVITY K_m, CONDUCTANCE C_m AND RESISTANCE R_m OF BUILDING MATERIALS

Metric Units

Material	Density	4.4°C Mean Temp.			29.4°C Mean Temp.			43.3°C Mean Temp.		
	Kilograms per m³	K_m	C_m	R_m	K_m	C_m	R_m	K_m	C_m	R_m
Clay										
Dried	1009	0.519			0.522					
Wet	1137	0.606			0.617					
Copper	8907	315.798			317.361			319.836		
Dolomite, compact	1730	2.091			2.107			2.134		
Domont brick (Terracotta)	1810	0.666			0.676			0.682		
Earth (Dry and Packed)	1521	0.547			0.552					
Ebonite	1185	0.059			0.063			0.070		
Flagstone										
Across cleavage	2242	1.846			1.846			1.875		
Along cleavage	2242	2.509			2.509			2.495		
Freestone, sandstone	2162	2.653			2.653			2.733		
Glass block (with cavity)	2467		34.64	0.029		36.91	0.027		39.18	0.026
Glass										
Flint	4277	0.600			0.671			0.749		
Plate	2467	0.800			0.857			0.930		
Soda	2579	0.712			0.795			0.868		
Quartz	2611	1.816			1.947			2.026		
Granite	1490	2.596			2.812			3.172		
Gravel										
Fine	1457	0.235			0.273			0.299		
Dry Stone	1842	0.338			0.363			0.364		
Gypsum block	684	0.244			0.252			0.264		
Gypsum board	1153	0.433			0.441			0.448		
Gypsum board covered with paper 1.27 cm thick	1089		14.76	0.068		15.90	0.063		17.03	0.059
Gypsum plaster										
Heavy density	839	0.255			0.264			0.270		
Light density	740	0.334			0.346			0.363		
Gypsum tile										
7.6 cm x 7.6 cm x 40.6 cm	1073		2.83	0.352		3.122	0.320		3.41	0.294
Haydite block										
20.3 cm x 23.32 cm x 46.6 cm	1073		2.95	0.338		3.06	0.326		3.17	0.304
20.3 cm x 30.5 cm x 30.5 cm	1233		2.72	0.367		2.95	0.338		3.17	0.314
Insulation for Building										
Cements										
Mineral Fiber	139	0.051			0.053			0.057		
Loose										
Aluminum Silica Fibers	36 to 120	0.047			0.051			0.055		
Cellulose	160 to 240	0.045			0.047			0.052		
Glass Fiber	64 (Avg.)	0.036			0.038			0.045		
Gypsum Pellets	32 to 48	0.076			0.078			0.079		
Mineral Fiber	64 to 192	0.043			0.045			0.048		
Perlite	80 to 128	0.042			0.045			0.049		
Silica aerogel granules	64 to 88	0.030			0.035			0.042		
Vermiculite flakes	128 to 160	0.068			0.069			0.072		
Blankets and Flexible Faced										
Elastomeric	128 to 144	0.043			0.046			0.049		
Glass Fiber	16 to 48	0.043			0.046			0.049		
Glass Fiber	48 to 72	0.040			0.043			0.049		
Glass Fiber with Alum. Ref. Sheet	64 to 72	0.039			0.042			0.047		
Mineral Wool	96 to 120	0.039			0.042			0.047		
Rigid and Semi-Rigid										
Cellular Glass	96 t 122	0.052			0.053			0.055		
Glass Fiber Board and Sheets	32 to 48	0.037			0.043			0.052		
Mineral Fiber	112 to 192	0.046			0.049			0.055		
Polystyrene	24 to 32	0.036			0.042			0.044		
Polyurethane	27 to 72	0.036			0.043			0.048		
Rubber—expanded	64 to 96	0.040			0.042			0.044		
Silica—expanded	208	0.056			0.057			0.060		
Wood Fibers and Binders	240 to 288	0.060			0.063			0.068		
Insulating Boards—Structural										
Glass Fiber organic bonded	64 to 144	0.038			0.040			0.043		
Foamed rubber	64 to 72	0.036			0.037			0.038		
Foamed polystyrene	28	0.040			0.041			0.043		
Foamed polystyrene	35	0.035			0.037			0.040		
Foamed polystyrene	56	0.033			0.036			0.038		
Foamed polyurethane	24	0.026			0.029			0.038		
Basic Ambient Design Conditions		**Winter Season**			**Summer Season**			**Solar Exposed**		

$K_m = W/mK$, $C_m = W/m^2$, $R_m = \frac{1}{C_m}$ or when ℓ_m is in meters $R_m = \frac{\ell_m}{K_m}$

TABLE 85 (Continued)

(Sheet 2 of 4

THERMAL CONDUCTIVITY K_m, CONDUCTANCE C_m AND RESISTANCE R_m OF BUILDING MATERIALS
Metric Units

Material	Density	4.4°C Mean Temp.			29.4°C Mean Temp.			43.3°C Mean Temp.		
	Kilograms per m^3	K_m	C_m	R_m	K_m	C_m	R_m	K_m	C_m	R_m
Foamed polyurethane	40	0.024			0.027			0.043		
Mineral fiber with resin binder	240	0.042			0.043			0.046		
Mineral fiber core insulation	256 to 272	0.049			0.050			0.053		
Mineral fiber acoustical tile	288 to 336	0.053			0.054			0.057		
Plaster or gypsum board	801	0.161			0.165			0.173		
Particle board low density	592	0.078			0.086			0.098		
Particle board medium density	801	0.135			0.158			0.174		
Particle board high density	1001	0.170			0.187			0.216		
Preformed roof insulation 1.72 cm thick			4.08	0.244		4.14	0.241		4.25	0.234
Preformed roof insulation 2.54 cm thick			2.04	0.489		2.15	0.463		2.24	0.440
Preformed roof insulation 3.26 cm thick			1.36	0.733		1.47	0.677		1.58	0.629
Preformed roof insulation 5.08 cm thick			1.07	0.976		1.13	0.880		1.19	0.834
Preformed roof insulation 6.84 cm thick			0.85	1.174		0.99	1.101		0.96	1.036
Preformed roof insulation 7.62 cm thick			0.68	1.355		0.74	1.355		0.79	1.257
Iron—wrought	7769	60.13			59.84			59.41		
Masonry Materials										
Cement mortar	1850	0.721			0.723			0.865		
Light weight concrete										
(Using light weight aggregates)	1602	0.519			0.526			0.548		
Light weight concrete	1281	0.360			0.367			0.379		
Light weight concrete	961	0.245			0.252			0.258		
Concrete (sand and gravel aggregates)	2242	1.514			1.542			1.572		
Stucco	1858	0.721			0.757			0.793		
Nickel	8602	58.113			59.555			60.564		
Plastering Materials										
Plaster gypsum board 0.95 cm thick	977		21.17	0.047		21.29	0.047			
Plaster gypsum board 1.27 cm thick	977		16.06	0.062		16.18	0.061			
Cement plaster, sand aggregate 2.54 cm thick	1858									
Cement plaster, sand aggregate 0.95 cm thick	1858		75.51	0.013		75.80	0.013			
Cement plaster, sand aggregate 1.27 cm thick	185800		56.49	0.017		56.78	0.013			
Cement plaster, sand aggregate 1.90 cm thick	1858		37.47	0.027		37.75	0.026			
Roofing Materials										
Asphalt shingles	1121		36.91	0.027		39.74	0.025		43.72	0.023
Cement shingles	1922		34.07	0.029		36.91	0.027		40.88	0.024
Slate shingles	3220		110.72	0.009		113.56	0.008		119.23	0.008
Wood shingles	1041		7.26	0.137		7.89	0.127		8.80	0.113
Built up roofing (felts and asphalts)	1121		39.74	0.025		41.16	0.024		51.10	0.019
Siding Materials										
Aluminum siding (with backing)			9.08	0.110		9.42	0.106		9.62	0.103
Cement shingles			27.03	0.036		28.04	0.035		12.52	0.034
Insulating backing board 0.95 cm thick			9.14	0.109		9.65	0.103		10.79	0.093
Wood shingles 40.6 cm (19 cm exposure)			6.53	0.153		6.81	0.146		7.38	0.135
Wood, drop siding 2.54 cm x 20.3 cm			7.21	0.138		7.38	0.135		7.60	0.131
Wood bevel 1.72 cm x 20.3 cm lapped			6.98	0.143		7.21	0.138		7.44	0.134
Wood bevel 1.90 cm x 25.4 cm lapped			5.39	0.185		5.56	0.180		5.79	0.173
Wood—plywood 0.95 cm lapped			9.02	0.111		9.36	0.107		9.82	0.101
Steel—structural	7801	35.326			37.492			39.655		
Stone—solid, marble	2483	1.802			1.843			1.864		
Sand—fine, dry	1554	0.302			0.310			0.332		
Sand—common moisture	1585	0.325			0.339			0.350		
Sandstone—dry	2258	1.545			1.902			1.846		
Slate across cleavage	2563	1.506			1.593			1.759		
Slate along cleavage	2563	2.725			2.819			3.042		
Soapstone	2723	3.317			3.403			3.475		
Terrazzo	1970	1.730			1.816			1.903		
Soil—dry	1201	0.138			0.146					
Soil—normal dampness	1281	0.500			0.523					
Structural building boards—sheathing and panels										
Fiber—cement board	1922									
Fiber—cement board 0.31 cm thick	1922		187.37	0.005		195.32	0.005		208.95	0.004
Fiber—cement board 0.61 cm thick	1922		93.68	0.011		97.66	0.010		104.47	0.009
Hardboard—medium density	640		8.46	0.118		9.25	0.108		9.65	0.103
Hardboard—high density	881	0.118			0.124			0.134		
Hardboard—high density tempered	1009	0.158			0.166			0.173		
Particle board—low density	592	0.078			0.082			0.087		
Particle board—medium density	601	0.135			0.141			0.149		
Particle board—high density	1001	0.170			0.179			0.187		
Basic Ambient Design Conditions		**Winter Season**			**Summer Season**			**Solar Exposed**		

$K_m = W/mK$, $C_m = W/m^2$, $R_m = \frac{1}{C_m}$ or when ℓ_m is in meters $R_m = \frac{\ell_m}{K_m}$

TABLE 85 (Continued)

THERMAL CONDUCTIVITY K_m, CONDUCTANCE C_m AND RESISTANCE R_m OF BUILDING MATERIALS

Metric Units

Material	Density	4.4°C Mean Temp.			29.4°C Mean Temp.			43.3°C Mean Temp.		
	Kilograms per m³	K_m	C_m	R_m	K_m	C_m	R_m	K_m	C_m	R_m
Plywood	544	0.115			0.121			0.128		
Plywood 0.61 cm thick	544		18.16	0.055		19.07	0.052		20.21	0.049
Plywood 0.95 cm thick	544		12.09	0.083		12.71	0.079		13.45	0.074
Plywood 1.72 cm thick	544		9.08	0.110		9.54	0.105		10.11	0.099
Plywood 1.90 cm thick	544		6.01	0.166		6.35	0.157		6.52	0.153
Plywood 2.54 cm thick	544		4.54	0.220		4.76	0.210		5.05	0.197
Wood subfloor 1.9 cm thick	640		6.01	0.166		6.35	0.157		6.52	0.153
Tile										
Clay hollow—1 cell			7.09	0.141		7.38	0.135		7.66	0.130
Clay hollow—1 cell			5.11	0.195		5.39	0.185		5.68	0.176
Clay hollow—2 cells			3.74	0.266		3.97	0.251		4.20	0.238
Clay hollow—2 cells			3.06	0.326		3.29	0.304		3.52	0.284
Clay hollow—2 cells			2.55	0.391		2.78	0.359		3.00	0.338
Clay hollow—3 cells			2.27	0.440		2.44	0.409		2.61	0.388
Wood (across gain conductivities)										
California redwood, 0% moisture	352	0.093			0.095			0.099		
California redwood, 8% moisture	384	0.099			0.101			0.105		
California redwood, 16% moisture	448	0.105			0.106			0.111		
Cypress, 0% moisture	352	0.095			0.096			0.101		
Cypress, 8% moisture	432	0.115			0.116			0.121		
Cypress, 16% moisture	448	0.119			0.121			0.125		
Maple, 0% moisture	576	0.126			0.128			0.133		
Maple, 8% moisture	606	0.141			0.142			0.147		
Maple, 16% moisture	640	0.153			0.157			0.161		
Oak, 0% moisture	608	0.147			0.148			0.153		
Oak, 8% moisture	688	0.166			0.167			0.171		
Oak, 16% moisture	768	0.183			0.186			0.190		
Pine, 0% moisture	480	0.108			0.109			0.112		
Pine, 8% moisture	560	0.128			0.130			0.134		
Pine, 16% moisture	640	0.145			0.148			0.152		
Wood pulp board	330	0.053			0.056			0.060		
Wood fiber board	190	0.043			0.045			0.047		
Wood fiber board	317	0.047			0.056			0.059		
Wood fiber board	640	0.072			0.075			0.079		
Wood doors										
2.54 cm thickness			3.63	0.275		3.69	0.271		3.80	0.262
With wood storm door			1.70	0.578		1.76	0.568		1.87	0.533
With metal storm door			2.21	0.451		2.27	0.440		2.38	0.419
3.81 cm thickness			2.78	0.359		2.84	0.352		2.95	0.338
With wood storm door			1.53	0.652		1.58	0.628		1.70	0.578
With metal storm door			1.87	0.534		1.93	0.517		2.04	0.489
5.08 cm thickness			2.44	0.409		2.49	0.400		2.61	0.382
With wood storm door			1.36	0.734		1.41	0.704		1.53	0.652
With metal storm door			1.64	0.607		1.70	0.578		1.81	0.550
Steel doors 4.44 cm thick										
With mineral fiber core			3.35	0.298		3.46	0.288		3.69	0.271
With organic foam core			2.66	0.374		3.00	0.333		3.35	0.298
With gypsum core			5.11	0.195		5.22	0.191		5.28	0.189
Basic Ambient Design Conditions		**Winter Season**			**Summer Season**			**Solar Exposed**		

K_m = W/mK, C_m = W/m^2, $R_m = \frac{1}{C_m}$ or when ℓ_m is in meters $R_m = \frac{\ell_m}{K_m}$

TABLE 85 (Continued)

(Sheet 4 of 4)

COEFFICIENT OF HEAT TRANSMISSION U_m, RESISTANCE R_m AND LIGHT REFLECTANCES OF VARIOUS TYPES OF GLASS AND GLASS BLOCK

Metric Units

Position	Description	Thickness mm	Surface Coating	Winter U_m	Winter R_m	Summer U_m	Summer R_m	% Solar Reflectance	% Light Reflectance Outdoors	% Light Reflectance Indoors
Vertical	Single Flat Glass	2.318	Clear	6.58	0.152	5.84	0.170	7	8	8
	Single Flat Glass	3.175	Clear	6.58	0.152	5.90	0.169	7	8	8
	Single Flat Glass	6.35	Clear	6.42	0.156	5.90	0.169	7	8	8
	Single Flat Glass	9.525	Clear	6.30	0.158	5.84	0.170	7	8	8
	Single Flat Glass	12.7	Clear	6.18	0.161	5.84	0.170	7	8	8
	Single Flat Glass	19.05	Clear	5.96	0.168	5.73	0.174	7	8	8
	Single Flat Glass	25.4	Clear	5.73	0.174	5.62	0.177	7	8	8
	Single Flat Glass—Coated	3.175	Bronze	6.58	0.152	6.24	0.160	6	35	18
	Single Flat Glass—Coated	6.35	Bronze	6.42	0.156	6.24	0.160	6	35	14
	Single Flat Glass—Coated	9.525	Bronze	6.30	0.158	6.18	0.162	6	35	11
	Single Flat Glass—Coated	3.175	Gray	6.58	0.152	6.18	0.162	5	6	6
	Single Flat Glass—Coated	6.35	Gray	6.30	0.158	6.24	0.160	5	6	6
	Single Flat Glass—Coated	9.525	Gray	6.42	0.159	6.24	0.160	5	6	6
	Single Flat Glass—Coated	3.175	Reflect.	6.42	0.156	6.35	0.157	26	6	6
	Single Flat Glass—Coated	6.35	Reflect.	6.42	0.156	6.35	0.157	26	6	6
	Single Flat Glass—Coated	9.525	Reflect.	6.42	0.156	6.35	0.157	30	5	5
	Double Flat Glass									
	6.35 mm Air Space	3.175	Clear	3.29	0.303	3.52	0.284	11	15	15
	6.35 mm Air Space	6.35	Clear	3.17	0.314	3.40	0.293	11	14	14
	12.7 mm Air Space	12.7	Clear	2.78	0.359	3.29	0.303	11	14	14
	Double Flat Glass									
	6.35 mm Air Space	6.35	Bronze	3.29	0.303	3.63	0.275	26	36	23
	6.35 mm Air Space	12.7	Bronze	2.84	0.352	3.23	0.308	29	36	23
	Double Flat Glass									
	6.35 mm Air Space	6.35	Gray	3.29	0.303	3.63	0.275	13	28	6
	12.7 mm Air Space	12.7	Gray	2.78	0.359	3.29	0.303	13	23	5
	Double Flat Glass									
	12.7 mm Air Space	6.35	Reflect.	2.78	0.359	3.23	0.308	45	22	35
	Triple Flat Insulating Glass									
	6.35 mm Air Space	6.35	Clear	2.66	0.375	2.55	0.391	15	16	16
	12.7 mm Air Space	6.35	Clear	2.04	0.489	1.98	0.503	15	16	16
	Storm Windows									
	2.5 to 10 cm Air Space	6.35	Clear	3.17	0.314	3.06	0.326	11	15	15
	Glass Block									
	15.24 x 15.24 x 10.96 cm Thick		Clear	3.40	0.293	3.23	0.308	13		
	20.32 x 20.32 x 10.16 cm Thick		Clear	3.29	0.303	3.06	0.326	13		
	With Cavity Divider		Clear	2.61	0.382	2.61	0.382	19		
	30.93 x 30.93 x 10.16 cm Thick		Clear	2.95	0.339	2.84	0.352	13		
	With Cavity Divider		Clear	2.49	0.400	2.38	0.419	19		
Horizontal (Skylights)	Single Flat Glass	12.7	Clear	6.70	0.149	4.82	0.207	7		
	Single Flat Glass	19.05	Clear	6.47	0.154	4.98	0.200	7		
	Single Flat Glass	25.4	Clear	6.24	0.160	4.48	0.222	7		
	Single Flat Glass—Coated	12.7	Bronze	6.80	0.147	5.38	0.187	6		
	Single Flat Glass—Coated	12.7	Gray	6.92	0.144	5.39	0.185	5		
	Single Flat Glass—Coated	12.7	Reflect.	6.98	0.143	5.51	0.181	26		
	Double Flat Glass—Air Space	12.7	Bronze	3.29	0.308	3.00	0.333	6		
	Double Flat Glass—Air Space	12.7	Gray	2.44	0.409	3.29	0.309	5		
	Double Flat Glass—Air Space	12.7	Reflect.	2.38	0.419	2.93	0.339	26		
	Glass Block									
	30.48 x 30.48 x 10.16 cm Thick									
	With Cavity Divider		Clear	2.90	0.345	1.93	0.517	19		
	Plastic Bubbles									
	Single Walled			6.52	0.153	4.54	0.220	12		
	Doubled Walled			3.97	0.252	2.61	0.382	14		
Vertical	Windows									
	Wood Sash—80% Glass		Clear	5.79	0.172	6.24	0.160	7		
	Wood Sash—60% Glass		Clear	5.11	0.195	5.62	0.177	7		
	Double Glass—Wood Sash—80% Glass		Clear	2.55	0.391	2.78	0.359	11		
	Metal Sash—Single Glass		Clear	6.41	0.155	7.04	0.142	11		
	Metal Sash—Double Glass		Clear	3.17	0.314	3.97	0.251	11		
	Sliding Door									
	Wood Frame			6.13	0.163	5.90	0.169	11		
	Metal Frame			6.42	0.156	6.47	0.154	11		

$U_m = W/m^2$, $R_m = \frac{1}{U_m}$

TABLE 86

FLAT ROOFS

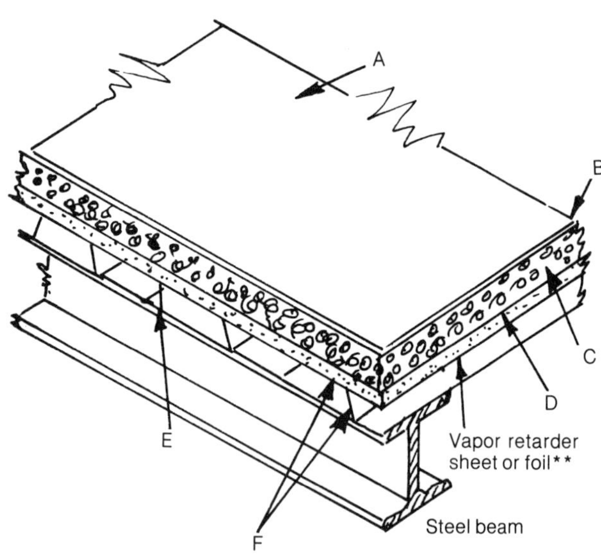

CONSTRUCTION ASSEMBLY I
INSULATION ABOVE METAL DECK

Note: Gypsum boards or other fire resistant sheets should be used over metal deck to prevent an inside fire from igniting roof or a roof fire from causing flaming drip to fall inside when combustible insulation is used on deck.

**When organic foams and other insulations which are vapor permeable are used in cold climates very vapor resistant retarder installed below is essential.

NO THERMAL INSULATION
THERMAL RESISTANCE R and R_m

Mean Temperature 40° (4.4°C), Wind 10 mph (16.1 km/hr)

	Components	Eng. Units		Metric Units
A	Outside surface	0.20		0.035
B	Built up roofing	0.33		0.058
C	Insulation	—		—
D	Gypsum sheets 1" (2.74 cm) thk.	0.76		0.133
E	Metal deck	0.01		0.002
F	Inside surface	0.85		0.149
	(TOTAL) RESISTANCE R =	2.15	R_m =	0.377
	TRANSMITTANCE U =	0.46	U_m =	2.652

Mean Temperature 85° (29.4°C), Wind 5 mph (8 km/hr)

	Components	Eng. Units		Metric Units
A	Outside surface	0.19		0.033
B	Built up roofing	0.31		0.054
C	Insulation	—		—
D	Gypsum sheets 1" (2.74 cm) thk.	0.76		0.133
E	Metal deck	0.01		0.002
F	Inside surface	0.83		0.146
	(TOTAL) RESISTANCE R =	2.10	R_m =	0.368
	TRANSMITTANCE U =	0.47	U_m =	2.718

Mean Temperature 110° (43.2°C), No Wind

	Components	Eng. Units		Metric Units
A	Outside surface	0.15		0.026
B	Built up roofing	0.22		0.039
C	Insulation	—		—
D	Gypsum sheets 1" (2.74 cm) thk.	0.76		0.133
E	Metal deck	0.01		0.002
F	Inside surface	0.80		0.141
	(TOTAL) RESISTANCE R =	1.94	R_m =	0.341
	TRANSMITTANCE U =	0.51	U_m =	2.936

1. POLYURETHANE OR POLYISOCYANURATE FOAM INSULATION
THERMAL RESISTANCE AND TRANSMITTANCE OF ASSEMBLY I WITH POLYURETHANE OR POLYISOCYANURATE FOAM INSULATION:

k = 0.19 Btu, in./ft^2, hr °F at 40°F, k_m = 0.027 W/mK at 4.4°C, K = 0.24 Btu, in./ft^2, hr °F at 85°F, k_m = 0.035 W/mK at 29.4°C, k = 0.28 Btu, in./ft^2, hr °F at 110°F, k_m = 0.040 W/mK at 43.2°C

Insulation Thickness		Mean Temp. 40°F (4.4°C)				Mean Temp. 85°F (29.4°C)				Mean Temp. 110°F (43.2°C)			
		English Units		Metric Units		English Units		Metric Units		English Units		Metric Units	
Inches	Meters	R	U	R_m	U_m	R	U	R_m	U_m	R	U	R_m	U_m
3/4	0.019	6.09	0.164	1.072	0.932	5.22	0.191	0.919	1.087	4.62	0.216	0.814	1.229
1	0.025	7.41	0.135	1.305	0.766	6.27	0.159	1.014	0.906	5.51	0.181	0.970	1.030
1 1/2	0.038	10.04	0.099	1.768	0.566	8.35	0.119	1.470	0.680	7.29	0.137	1.284	0.779
2	0.051	12.68	0.079	2.232	0.448	10.43	0.096	1.836	0.544	9.08	0.110	1.599	0.625
2 1/2	0.064	15.31	0.653	2.696	0.371	12.52	0.079	2.024	0.453	10.86	0.092	1.912	0.523
3	0.076	17.94	0.056	3.159	0.316	14.60	0.068	2.571	0.388	12.65	0.079	2.228	0.449
4	0.102	23.20	0.043	4.085	0.244	18.77	0.053	3.305	0.308	16.22	0.062	2.856	0.350
5	0.127	28.46	0.035	5.012	0.199	22.93	0.044	4.037	0.247	19.79	0.051	3.485	0.287

R = °F, ft^2, hr, Btu R_m = km^{2}W   U = Btu per ft^2 per hr U_m = W/m^{2}K

TABLE 87

FLAT ROOFS
CONSTRUCTION ASSEMBLY I

2. CELLULAR GLASS INSULATION

THERMAL RESISTANCE AND TRANSMITTANCE OF ASSEMBLY WITH INSULATION CELLULAR GLASS INSULATION: k = 0.33 Btu, in./ft^2, hr °F at 40 °F, k_m = 0.047 W/mK at 4.4 °C, k = 0.35 Btu, in./ft^2, hr °F at 85 °F, k_m = 0.050 W/mK at 29.4 °C, k = 0.36 Btu, in./ft^2, hr °F at 110 °F, k_m = 0.052 W/mK at 43.2 °C

Insulation Thickness		Mean Temp. 40 °F (4.4 °C)				Mean Temp. 85 °F (29.4 °C)				Mean Temp. 110 °F (43.2 °C)			
		English Units		Metric Units		English Units		Metric Units		English Units		Metric Units	
Inches	Meters	R	U	R_m	U_m	R	U	R_m	U_m	R	U	R_m	U_m
1 1/2	0.038	6.72	0.149	1.184	0.847	6.37	0.157	1.122	0.893	6.06	0.165	1.068	0.935
2	0.051	8.25	0.121	1.454	0.690	7.80	0.128	1.374	0.730	7.43	0.135	1.309	0.763
2 1/2	0.064	9.77	0.103	1.721	0.581	9.22	0.108	1.625	0.617	8.81	0.114	1.552	0.645
3	0.076	11.30	0.088	1.991	0.503	10.65	0.094	1.876	0.532	10.18	0.098	1.794	0.557
3 1/2	0.088	12.82	0.078	2.259	0.442	12.07	0.083	2.127	0.469	11.56	0.087	2.037	0.491
4	0.102	14.35	0.070	2.528	0.395	13.50	0.074	2.379	0.420	12.93	0.077	2.228	0.449
4 1/2	0.114	15.87	0.063	2.796	0.357	14.92	0.067	2.629	0.380	14.30	0.070	2.520	0.397
5	0.127	17.39	0.058	3.064	0.327	16.35	0.061	2.881	0.347	15.68	0.064	2.763	0.362

Note: Since glass is a non-combustible inorganic glass, it can be used in hazardous areas. Where installation of a very highly resistant vapor retarder is difficult, this insulation can be used as it is the most vapor resistant of all thermal insulations.

Tapered blocks are available so as to achieve positive drainage on a flat roof deck. Heat transfer for total area with this type of installation must be calculated the sum total of heat transfer for each area at each average thickness.

TABLE 88

3. PERLITE THERMAL INSULATION BOARD

THERMAL RESISTANCE AND TRANSMITTANCE OF ASSEMBLY WITH INSULATION PERLITE, BINDERS AND FIBERS INSULATION BOARD: k = 0.36 Btu, in./ft^2, hr °F at 40 °F, k_m = 0.052 W/mK at 4.4 °C, k = 0.39 Btu, in./ft^2, hr °F at 85 °F, k_m = 0.056 W/mK at 29.4 °C, k = 0.43 Btu, in./ft^2, hr °F at 110 °F, k_m = 0.062 W/mK at 43.2 °C

Insulation Thickness		Mean Temp. 40 °F (4.4 °C)				Mean Temp. 85 °F (29.4 °C)				Mean Temp. 110 °F (43.2 °C)			
		English Units		Metric Units		English Units		Metric Units		English Units		Metric Units	
Inches	Meters	R	U	R_m	U_m	R	U	R_m	U_m	R	U	R_m	U_m
3/4	0.019	4.23	0.236	0.745	1.342	4.02	0.249	0.708	1.412	3.68	0.271	0.688	1.542
1	0.025	4.92	0.203	0.867	1.154	4.66	0.214	0.821	1.217	4.26	0.234	0.751	1.332
1 1/2	0.038	6.51	0.158	1.147	0.872	5.94	0.168	1.047	0.955	5.42	0.184	0.955	1.047
2	0.051	7.70	0.129	1.356	0.737	7.22	0.138	1.272	0.786	6.59	0.152	1.161	0.861

TABLE 89

FLAT ROOFS
CONSTRUCTION ASSEMBLY I

4. SHEETS OF CHEMICALLY TREATED WOOD FIBER BONDED WITH PORTLAND CEMENT

THERMAL RESISTANCE AND TRANSMITTANCE OF ASSEMBLY WITH INSULATION WOOD FIBER AND PORTLAND CEMENT, Used as parts C & D in Assembly I: k = 0.4 Btu, in./ft^2, hr °F at 40 °F, k_m = 0.057 W/mK at 4.4 °C, k = 0.043 Btu, in./ft^2, hr °F at 85 °F, k_m = 0.062 W/mK at 29.4 °C, k = 0.47 Btu, in./ft^2, hr °F at 110 °F, k_m = 0.068 W/mK at 43.2 °C

Insulation Thickness		Mean Temp. 40°F (4.4°C)				Mean Temp. 85°F (29.4°C)				Mean Temp. 110°F (43.2°C)			
		English Units		Metric Units		English Units		Metric Units		English Units		Metric Units	
Inches	Meters	R	U	R_m	U_m	R	U	R_m	U_m	R	U	R_m	U_m
1	0.025	3.89	0.257	0.685	1.459	3.66	0.273	0.645	1.551	3.41	0.293	0.601	1.664
1 1/2	0.038	5.14	0.194	0.906	1.104	4.82	0.207	0.849	1.178	4.47	0.224	0.788	1.269
2	0.051	6.39	0.156	1.125	0.888	5.99	0.167	1.055	0.947	5.53	0.181	0.974	1.026
2 1/2	0.064	7.64	0.131	1.346	0.743	7.15	0.140	1.250	0.794	6.60	0.152	1.163	0.860
3	0.076	8.89	0.112	1.556	0.638	8.32	0.120	1.466	0.682	7.66	0.130	1.349	0.741
3 1/2	0.088	10.14	0.099	1.787	0.560	9.47	0.105	1.669	0.599	8.73	0.114	1.538	0.650
4	0.102	11.39	0.088	2.007	0.498	10.64	0.094	1.875	0.533	9.79	0.102	1.725	0.580

Note: Where the treated wood fiber portland cement binder insulation is used above metal deck the gypsum fire protective sheet is seldom used due to fact that most of these products have a flame spread index of 25 as per ASTM Test Method E-84.

TABLE 90

5. HEAVY DENSITY MINERAL WOOL INSULATING BOARD

THERMAL RESISTANCE AND TRANSMITTANCE OF ASSEMBLY WITH INSULATION HEAVY DENSITY, 15 lb/ft^3 (240 kg/m^3), MINERAL WOOL BOARDS (OR SHEETS): k = 0.36 Btu, in./ft^2, hr °F at 40 °F, k_m = 0.052 W/mK at 4.4 °C, k = 0.39 Btu, in./ft^2, hr °F at 85 °F, k_m = 0.056 W/mK at 29.4 °C, k = 0.43 Btu, in./ft^2, hr °F at 110 °F, k_m = 0.062 W/mK at 43.2 °C

Insulation Thickness		Mean Temp. 40°F (4.4°C)				Mean Temp. 85°F (29.4°C)				Mean Temp. 110°F (43.2°C)			
		English Units		Metric Units		English Units		Metric Units		English Units		Metric Units	
Inches	Meters	R	U	R_m	U_m	R	U	R_m	U_m	R	U	R_m	U_m
7/16	0.011	2.60	0.383	0.458	2.183	2.45	0.408	0.432	2.316	2.19	0.455	0.386	2.591
1	0.025	4.16	0.239	0.733	1.364	3.90	0.256	0.687	1.455	3.51	0.285	0.618	1.617
1 1/2	0.038	5.55	0.180	0.978	1.023	5.18	0.193	0.913	1.096	4.67	0.214	0.823	1.215
2	0.051	6.94	0.144	1.223	0.818	6.46	0.155	1.138	0.878	5.83	0.171	1.027	0.973
2 1/2	0.064	8.33	0.120	1.468	0.681	7.75	0.129	1.366	0.732	6.99	0.143	1.232	0.812

Note: Like above, gypsum board is seldom used between metal deck and mineral fiber resin binder board. Thus this also functions as C & D in Assembly I.

TABLE 91

FLAT ROOFS

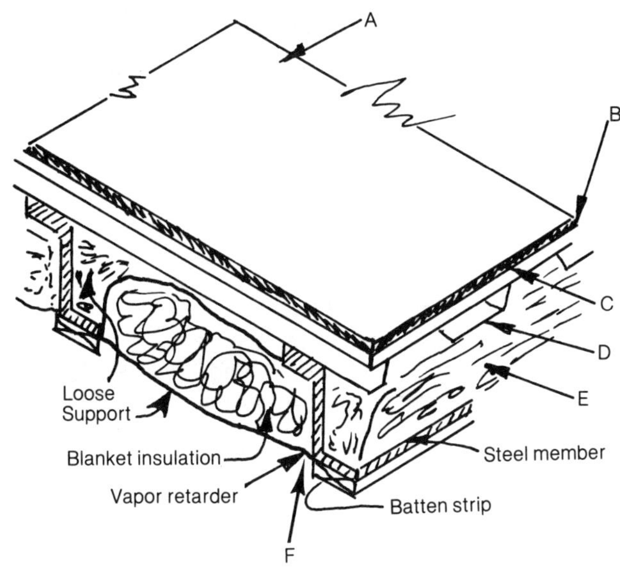

CONSTRUCTION ASSEMBLY II
INSULATION BELOW METAL DECK

Note: Because steel members act as a thermal bypass through the insulation a transmittance factor must be added to transmittance calculations of the assembly. This is based on unit area of conductivity of steel, length of heat flow path and percentage of total area. In this example total resistance R per sq ft of insulation was reduced 1.88 (R_m was reduced 0.33).

NO THERMAL INSULATION
THERMAL RESISTANCE R and R_m

Mean Temperature 40° (4.4°C), Wind 10 mph (16.1 km/hr)

	Components	Eng. Units		Metric Units
A	Outside surface	0.20		0.035
B	Built up roofing	0.33		0.058
C	Deck Board	0.76		0.133
D	Metal deck	0.01		0.002
E	Thermal Insulation	—		—
F	Inside surface	0.85		0.149
	(TOTAL) RESISTANCE R =	2.15	R_m =	0.244
	TRANSMITTANCE U =	0.46	U_m =	2.652

Mean Temperature 85° (29.4°C), Wind 5 mph (8 km/hr)

	Components	Eng. Units		Metric Units
A	Outside surface	0.19		0.033
B	Built up roofing	0.31		0.054
C	Deck Board	0.76		0.122
D	Metal deck	0.01		0.002
E	Thermal Insulation	—		—
F	Inside surface	0.83		0.146
	(TOTAL) RESISTANCE R =	2.10	R_m =	0.368
	TRANSMITTANCE U =	0.47	U_m =	2.718

Mean Temperature 110° (43.2°C), No Wind

	Components	Eng. Units		Metric Units
A	Outside surface	0.15		0.026
B	Built up roofing	0.22		0.039
C	Deck Board	0.76		0.122
D	Metal deck	0.01		0.002
E	Thermal Insulation	—		—
F	Inside surface	0.80		0.141
	(TOTAL) RESISTANCE R =	1.94	R_m =	0.341
	TRANSMITTANCE U =	0.51	U_m =	2.936

6. MINERAL OR GLASS WOOL BLANKET OR BATT INSULATION (HEAVY DENSITY) THERMAL RESISTANCE AND TRANSMITTANCE OF ASSEMBLY II WITH HEAVY DENSITY MINERAL OR GLASS WOOL INSULATION:

k = 0.34 Btu, in./ft^2, hr °F, k_m = 0.059 W/mK at 4.4°C, k = 0.38 Btu, in./ft^2, hr °F at 85°F, k_m = 0.067 W/mK at 29.4°C, k = 0.41 Btu, in./ft^2, hr °F at 110°F, k_m = 0.072 W/mK at 43.2°C

Insulation Thickness		Mean Temp. 40°F (4.4°C)				Mean Temp. 85°F (29.4°C)				Mean Temp. 110°F (43.2°C)			
		English Units		Metric Units		English Units		Metric Units		English Units		Metric Units	
Inches	Meters	R	U	R_m	U_m	R	U	R_m	U_m	R	U	R_m	U_m
1	0.025	2.90	0.345	0.511	1.957	2.85	0.351	0.502	1.991	2.49	0.400	0.439	2.279
2	0.054	5.53	0.181	1.974	1.026	5.48	0.182	1.966	1.036	4.93	0.203	0.896	1.151
3	0.076	7.16	0.140	1.262	0.793	8.11	0.123	1.429	0.699	7.37	0.136	1.299	0.770
4	0.102	10.79	0.093	1.901	0.526	10.75	0.093	1.894	0.528	9.81	0.102	1.729	0.589
5	0.127	13.44	0.075	2.368	0.423	13.78	0.075	2.428	0.411	12.26	0.082	2.160	0.463
6	0.152	16.06	0.062	2.829	0.353	16.01	0.062	2.821	0.354	14.69	0.068	2.588	0.386

R = °F, ft^2, hr, Btu R_m = km^{2}W U = Btu per ft^2 per hr U_m = W/m^{2}K

TABLE 92

FLAT ROOFS

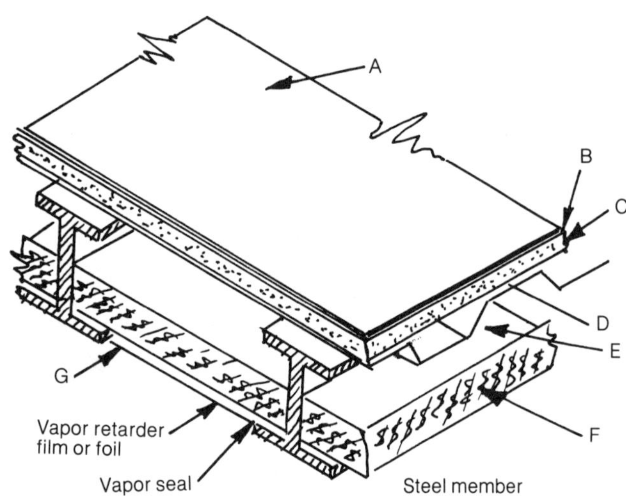

CONSTRUCTION ASSEMBLY III
INSULATION BETWEEN SUPPORTS

Note: To prevent condensation of moisture in the insulation or underside of metal deck a sealed vapor retarder film is necessary on underside of insulation, sealed to metal supports. Under low temperature weather conditions condensate may collect on bottom of metal support beams.

NO THERMAL INSULATION
THERMAL RESISTANCE R and R_m

Mean Temperature 40° (4.4°C), Wind 10 mph (16.1 km/hr)

	Components	Eng. Units		Metric Units
A	Outside surface	0.20		0.035
B	Built up roofing	0.33		0.058
C	Deck board	0.76		0.133
D	Metal deck	0.01		0.002
E	Air space	.88		0.155
F	Thermal insulation	—		—
G	Bottom surface	0.85		0.149
	(TOTAL) RESISTANCE R =	3.03	R_m =	0.532
	TRANSMITTANCE U =	0.330	U_m =	1.879

Mean Temperature 85° (29.4°C), Wind 5 mph (8 km/hr)

	Components	Eng. Units		Metric Units
A	Outside surface	0.19		0.033
B	Built up roofing	0.31		0.054
C	Deck board	0.76		0.133
D	Metal deck	0.01		0.002
E	Air space	0.85		0.150
F	Thermal insulation	—		—
G	Bottom surface	0.83		0.146
	(TOTAL) RESISTANCE R =	2.95	R_m =	0.519
	TRANSMITTANCE U =	0.339	U_m =	1.972

Mean Temperature 110° (43.2°C), No Wind

	Components	Eng. Units		Metric Units
A	Outside surface	0.15		0.026
B	Built up roofing	0.22		0.039
C	Deck board	0.76		0.133
D	Metal deck	0.01		0.002
E	Air space	0.99		0.174
F	Thermal insulation	—		—
G	Bottom surface	0.80		0.141
	(TOTAL) RESISTANCE R =	2.93	R_m =	0.516
	TRANSMITTANCE U =	0.341	U_m =	1.984

7. SHEET OF CHEMICALLY TREATED WOOD FIBER BONDED WITH PORTLAND CEMENT THERMAL RESISTANCE AND TRANSMITTANCE OF ASSEMBLY III WITH WOOD FIBER AND PORTLAND CEMENT BOARD INSULATION:

k = 0.4 Btu, in./ft^2, hr °F at 40°F, k_m = 0.057 W/mK at 4.4°C, k = 0.43 Btu, in./ft^2, hr °F at 85°F, k_m = 0.062 W/mK at 29.4°C, k = 0.47 Btu, in./ft^2, hr °F at 110°F, k_m = 0.068 W/mK at 43.2°C

Insulation Thickness		Mean Temp. 40°F (4.4°C)				Mean Temp. 85°F (29.4°C)				Mean Temp. 110°F (43.2°C)			
		English Units		Metric Units		English Units		Metric Units		English Units		Metric Units	
Inches	Meters	R	U	R_m	U_m	R	U	R_m	U_m	R	U	R_m	U_m
2	0.051	8.03	0.125	1.414	0.707	7.49	0.133	1.319	0.758	7.19	0.139	1.267	0.789
2 1/2	0.064	9.28	0.108	1.635	0.612	8.63	0.116	1.521	0.658	8.25	0.121	1.453	0.688
3	0.076	10.53	0.095	1.855	0.539	9.77	0.102	1.721	0.581	9.31	0.107	1.640	0.609
3 1/2	0.088	11.78	0.085	2.076	0.482	10.90	0.092	1.921	0.521	10.38	0.096	1.829	0.547
4	0.102	13.03	0.077	2.296	0.436	12.04	0.083	2.121	0.471	11.44	0.087	2.016	0.496

R = °F, ft^2, hr, Btu R_m = km^{2}W   U = Btu per ft^2 per hr U_m = W/m^{2}K

TABLE 93

FLAT ROOFS
CONSTRUCTION ASSEMBLY III

8. WOOD FIBER AND PORTLAND CEMENT BOARD WITH FOAMED URETHANE INSULATION ON TOP (Conductivities of board given in II-7 and insulation in I-1) THERMAL RESISTANCE AND TRANSMITTANCE OF BOARD, INSULATION AND ASSEMBLY III

Thickness				Mean Temp. 40°F (4.4°C)				Mean Temp. 85°F (29.4°C)				Mean Temp. 110°F (43.2°C)			
Fiber Bd.		Urethane		English Units		Metric Units		English Units		Metric Units		English Units		Metric Units	
in.	m	in.	m	R	U	R_m	U_m	R	U	R_m	U_m	R	U	R_m	U_m
2	0.051	1	0.025	14.00	0.071	2.467	0.405	13.51	0.074	2.380	0.420	13.31	0.075	2.345	0.426
2	0.051	1/2	0.038	17.35	0.058	3.057	0.327	16.85	0.059	2.969	0.337	16.45	0.061	2.898	0.345
2 1/2	0.064	1	0.025	14.95	0.067	2.634	0.380	14.35	0.070	2.528	0.395	14.02	0.071	2.470	0.404
2 1/1	0.064	1 1/2	0.038	18.29	0.055	3.223	0.310	17.13	0.058	3.018	0.331	16.63	0.060	2.930	0.341
3	0.088	1	0.025	15.89	0.063	2.799	0.357	15.32	0.064	2.699	0.370	14.80	0.067	2.481	0.403

Note: Because of combustible properties of urethane foam, electric circuits should not be installed in space above the space above insulated board. Bottom of board should be faced with vapor retarder.

R = °F, ft^2, hr, Btu R_m = km^{2}W U = Btu per ft^2 per hr U_m = W/m^{2}K

TABLE 94

9. HEAVY DENSITY MINERAL WOOL INSULATING BOARD OR SHEET (Conductivities given in I-5) THERMAL RESISTANCE AND TRANSMITTANCE OF ASSEMBLY I WITH INSULATION

Insulation Thickness		Mean Temp. 40°F (4.4°C)				Mean Temp. 85°F (29.4°C)				Mean Temp. 110°F (43.2°C)			
		English Units		Metric Units		English Units		Metric Units		English Units		Metric Units	
Inches	Meters	R	U	R_m	U_m	R	U	R_m	U_m	R	U	R_m	U_m
1	0.025	8.29	0.120	1.460	0.684	7.13	0.140	1.256	0.796	6.40	0.156	1.127	0.887
1 1/2	0.038	10.92	0.092	1.924	0.520	9.20	0.108	1.621	0.617	8.28	0.121	1.459	0.685
2	0.051	13.56	0.074	2.389	0.419	11.28	0.089	1.987	0.503	10.07	0.099	1.774	0.563
2 1/2	0.064	16.19	0.062	2.853	0.351	13.37	0.074	2.356	0.424	11.85	0.084	2.088	0.479
3	0.076	18.82	0.053	3.316	0.302	15.45	0.065	2.722	0.367	13.63	0.073	2.402	0.416
4	0.102	24.08	0.042	4.243	0.236	19.62	0.051	3.457	0.286	17.21	0.058	3.032	0.330
5	0.127	29.34	0.034	5.169	0.193	23.81	0.042	4.195	0.238	20.78	0.048	3.661	0.273

Note: Underside should be faced with vapor retarder film or foil.

R = °F, ft^2, hr, Btu R_m = km^{2}W U = Btu per ft^2 per hr U_m = W/m^{2}K

TABLE 95

10. SEMI-RIGID GLASS FIBER INSULATION BOARD — FACED

k = 0.25 Btu, in./ft^2, hr °F at 40°F, k_m = 0.036 W/mK at 4.4°C, k = 0.28 Btu, in./ft^2, hr °F at 85°F, k_m = 0.040 W/mK at 29.4°C, k = 0.32 Btu, in./ft^2, hr °F at 110°F, k_m = 0.046 W/mK at 43.2°C

THERMAL RESISTANCE AND TRANSMITTANCE OF ASSEMBLY WITH INSULATION

Insulation Thickness		Mean Temp. 40°F (4.4°C)				Mean Temp. 85°F (29.4°C)				Mean Temp. 110°F (43.2°C)			
		English Units		Metric Units		English Units		Metric Units		English Units		Metric Units	
Inches	Meters	R	U	R_m	U_m	R	U	R_m	U_m	R	U	R_m	U_m
1	0.025	6.15	0.163	1.084	0.923	5.67	0.176	0.999	1.000	5.38	0.185	0.940	1.054
1 1/2	0.038	8.15	0.123	1.436	0.696	7.45	0.134	1.313	0.762	6.62	0.151	1.166	0.857
2	0.051	10.15	0.098	1.788	0.559	9.24	0.108	1.628	0.614	8.19	0.122	1.443	0.693
2 1/2	0.064	12.15	0.082	2.141	0.467	11.03	0.091	1.943	0.514	9.75	0.102	1.630	0.614
3	0.076	14.15	0.071	2.493	0.401	12.81	0.078	2.257	0.443	11.32	0.088	1.995	0.501

Note: Underside should be faced with vapor retarder film or foil.

R = °F, ft^2, hr, Btu R_m = km^{2}W U = Btu per ft^2 per hr U_m = W/m^{2}K

TABLE 96

FLAT ROOFS

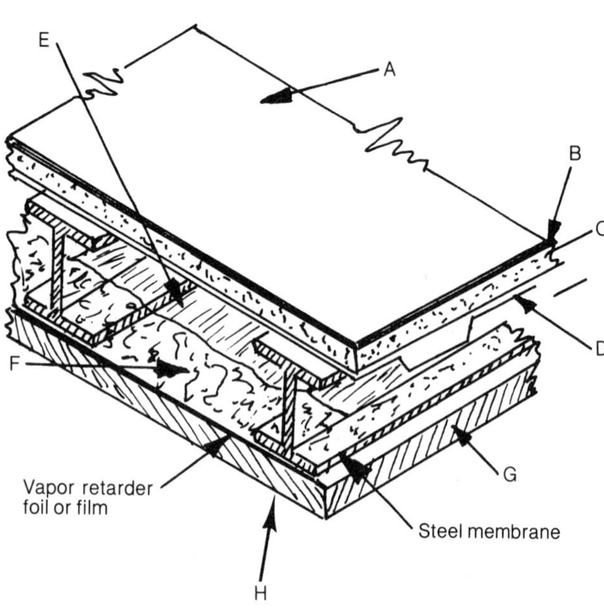

CONSTRUCTION ASSEMBLY IV
INSULATION BETWEEN SUPPORTS
GYPSUM AND OR PLASTER BELOW SUPPORTS

Note: The vapor retarder film can be placed above the gypsum board or plaster as during heating season. The inner facing stays above dew point temperature.

NO THERMAL INSULATION
THERMAL RESISTANCE R and R_m

Mean Temperature 40° (4.4°C), Wind 10 mph (16.1 km/hr)

	Components	Eng. Units		Metric Units
A	Outside surface	0.20		0.035
B	Built up roofing	0.33		0.058
C	Deck Board	0.76		0.133
D	Metal deck	0.01		0.002
E	Air space	0.88		0.155
F	Thermal insulation	—		—
G	Gypsum board or plaster 1″	0.87		0.155
H	Bottom surface	0.85		0.149
	(TOTAL) RESISTANCE R =	3.90	R_m =	0.687
	TRANSMITTANCE U =	0.00	U_m =	0.000

Mean Temperature 85° (29.4°C), Wind 5 mph (8 km/hr)

	Components	Eng. Units		Metric Units
A	Outside surface	0.19		0.033
B	Built up roofing	0.31		0.054
C	Deck board	0.76		0.134
D	Metal deck	0.01		0.002
E	Air space	0.85		0.150
F	Thermal insulation	—		—
G	Gypsum board or plaster 1″	0.82		0.144
H	Bottom surface	0.83		0.146
	(TOTAL) RESISTANCE R =	3.77	R_m =	0.651
	TRANSMITTANCE U =	0.00	U_m =	1.536

Mean Temperature 110° (43.2°C), No Wind

	Components	Eng. Units		Metric Units
A	Outside surface	0.15		0.026
B	Built up roofing	0.22		0.039
C	Deck board	0.76		0.134
D	Metal deck	0.01		0.002
E	Air space	0.99		0.174
F	Thermal insulation	—		—
G	Gypsum board or plaster 1″	0.82		0.144
H	Bottom surface	0.80		0.141
	(TOTAL) RESISTANCE R =	3.75	R_m =	0.645
	TRANSMITTANCE U =	0.00	U_m =	0.000

11. GLASS FIBER INSULATION
THERMAL RESISTANCE AND TRANSMITTANCE OF ASSEMBLY WITH INSULATION:

k = 0.25 Btu, in./ft², hr °F at 40°F, k_m = 0.036 W/mK at 4.4°C, k = 0.28 Btu, in./ft², hr °F at 85°F, k_m = 0.040 W/mK at 29.4°C, k = 0.32 Btu, in./ft², hr °F at 110°F, k_m = 0.046 W/mK at 43.2°C

Insulation Thickness		Mean Temp. 40°F (4.4°C)				Mean Temp. 85°F (29.4°C)				Mean Temp. 110°F (43.2°C)			
		English Units		Metric Units		English Units		Metric Units		English Units		Metric Units	
Inches	Meters	R	U	R_m	U_m	R	U	R_m	U_m	R	U	R_m	U_m
1	0.025	7.90	0.126	1.392	0.718	7.34	0.136	1.293	0.773	6.87	0.145	1.210	0.826
2	0.064	11.90	0.084	2.097	0.476	10.91	0.092	1.922	0.820	10.00	0.100	1.762	0.568
3	0.076	15.90	0.063	2.802	0.356	14.48	0.069	2.551	0.392	13.12	0.076	2.311	0.432
4	0.102	19.90	0.050	3.506	0.285	18.05	0.055	3.180	0.314	16.25	0.062	2.863	0.349
5	0.127	23.90	0.042	4.211	0.273	21.63	0.046	3.811	0.262	19.37	0.052	3.413	0.293
6	0.152	27.90	0.036	4.916	0.203	25.19	0.039	4.436	0.225	22.50	0.044	3.964	0.252

R = °F, ft², hr, Btu R_m = km²W U = Btu per ft² per hr U_m = W/m²K

TABLE 97

FLAT ROOFS OF CONSTRUCTION ASSEMBLIES I, II, III AND IV
CORRECTIONS OF RESISTANCES DUE TO SUBSTITUTIONS OF COMPONENTS

The construction of these assemblies previously given is with the metal deck being covered with a 1″ thick gypsum board or sheet. In some constructions the metal deck is covered by light weight concrete slabs. In other instances reinforced concrete slabs replace both the board and the metal deck. The following gives corrections to the calculated resistance tables.

LIGHT WEIGHT AGGREGATE INSTEAD OF 1″ (2.54 cm) DECK BOARD

Thickness of Light Wt. Cement		Units of Resistance to be Added to Tables 87 through 96					
		Mean Temp. 40°F (4.4°C)		Mean Temp. 85°F (29.4°C)		Mean Temp. 110°F (43.2°C)	
Inches	Meters	Eng. Res. R	Met. Res. R_m	Eng. Res. R	Met. Res. R_m	Eng. Res. R.	Met. Res. R_m
2	0.051	+1.46	+0.257	+1.41	+0.248	+1.35	+0.238
3	0.076	+2.57	+0.452	+2.52	+0.444	+2.48	+0.436
4	0.102	+3.68	+0.648	+3.63	+0.639	+3.53	+0.622

TABLE 98

Some constructions use gravel cement slabs directly over the steel beams. This replaces both the metal roof structure and the board. In this case due to the high thermal conductivity of gravel cement the total thermal resistance will be less than shown in Tables 87 through 97.

GRAVEL CEMENT SLABS INSTEAD OF METAL DECK AND DECK BOARD

Thickness of Concrete Slab		Units of Resistance to be Subtracted from Tables 87 through 96					
		Mean Temp. 40°F (4.4°C)		Mean Temp. 85°F (29.4°C)		Mean Temp. 110°F (43.2°C)	
Inches	Meters	Eng. Res. R	Met. Res. R_m	Eng. Res. R	Met. Res. R_m	Eng. Res. R	Met. Res. R_m
4	0.102	−0.45	−0.079	−0.44	−0.077	−0.42	−0.074
6	0.153	−0.29	−0.051	−0.28	−0.049	−0.27	−0.047
8	0.204	−0.13	−0.022	−0.13	−0.022	−0.12	−0.021

TABLE 99

FLAT ROOFS

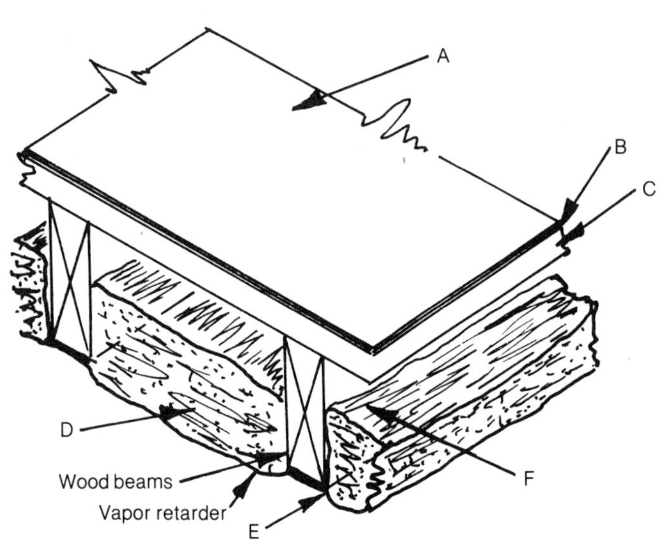

CONSTRUCTION ASSEMBLY V

Note: When insulation thickness is less then height of wood beam then resistance for air space (F) must also be added. Based on 2" (5.08 cm) on 16" (0.406 m) on center this added resistance factor was calculated to be for R = 0.8 and R_m = 0.141. This is included in Tables 12 and 13.

NO THERMAL INSULATION
THERMAL RESISTANCE R and R_m

Mean Temperature 40° (4.4°C), Wind 10 mph (16.1 km/hr)

	Components	Eng. Units		Metric Units
A	Outside surface	0.17		0.030
B	Built up roofing	0.33		0.058
C	1" (2.54 cm) Plywood deck	1.06		0.187
D	Insulation	—		—
E	Inside surface	0.61		0.107
	(TOTAL) RESISTANCE R =	2.17	R_m =	0.382
	TRANSMITTANCE U =	0.46	U_m =	2.617

Mean Temperature 85° (29.4°C), Wind 5 mph (8 km/hr)

	Components	Eng. Units		Metric Units
A	Outside surface	0.16		0.028
B	Built up roofing	0.33		0.058
C	1" (2.54 cm) Plywood deck	0.99		0.174
D	Insulation	—		—
E	Inside surface	0.59		0.104
	(TOTAL) RESISTANCE R =	2.07	R_m =	0.364
	TRANSMITTANCE U =	0.48	U_m =	2.747

Mean Temperature 110° (43.2°C), No Wind

	Components	Eng. Units		Metric Units
A	Outside surface	0.16		0.028
B	Built up roofing	0.33		0.058
C	1" (2.54 cm) Plywood deck	0.96		0.169
D	Insulation	—		—
E	Inside surface	0.59		0.104
	(TOTAL) RESISTANCE R =	2.04	R_m =	0.359
	TRANSMITTANCE U =	0.49	U_m =	2.786

12. GLASS FIBER LIGHT DENSITY BATT OR BLANKET INSULATION THERMAL RESISTANCE AND TRANSMITTANCE OF ASSEMBLY V WITH GLASS FIBER INSULATION: k = 0.24 Btu, in./ft^2, hr °F at 40°F, k_m = 0.036 W/mK at 4.4°C, k = 0.28 Btu, in./ft^2, hr °F at 85°F, k_m = 0.040 W/mK at 29.4°C, k = 0.34 Btu, in./ft^2, hr °F at 110°F, k_m = 0.049 W/mK at 43.2°C

Insulation Thickness		Mean Temp. 40°F (4.4°C)				Mean Temp. 85°F (29.4°C)				Mean Temp. 110°F (43.2°C)			
		English Units		Metric Units		English Units		Metric Units		English Units		Metric Units	
Inches	Meters	R	U	R_m	U_m	R	U	R_m	U_m	R	U	R_m	U_m
1	0.025	7.13	0.140	1.256	0.796	6.44	0.155	1.135	0.881	5.78	0.173	1.018	0.982
2	0.051	11.30	0.088	1.991	0.502	10.01	0.100	1.767	0.567	8.72	0.115	1.536	0.651
3	0.076	15.47	0.064	2.726	0.367	13.58	0.074	2.393	0.418	11.66	0.086	2.054	0.486
4	0.102	19.64	0.051	3.461	0.289	17.16	0.058	3.024	0.331	14.60	0.068	2.573	0.389
5	0.127	23.80	0.042	4.194	0.238	20.72	0.048	3.651	0.274	17.46	0.056	3.076	0.325
6	0.152	27.97	0.036	4.928	0.203	24.30	0.041	4.282	0.234	20.48	0.048	3.608	0.227
7	0.178	32.14	0.031	5.663	0.177	27.87	0.035	4.911	0.204	23.43	0.042	4.128	0.242
8	0.203	36.30	0.027	6.396	0.156	31.44	0.031	5.539	0.181	26.37	0.038	4.646	0.215
9	0.229	40.47	0.024	7.131	0.140	35.01	0.028	6.169	0.162	29.31	0.034	5.164	0.196
10	0.254	44.64	0.022	7.836	0.128	38.58	0.025	6.964	0.143	32.25	0.031	5.682	0.175
11	0.279	48.80	0.020	8.599	0.116	42.15	0.023	7.427	0.135	35.19	0.028	6.200	0.161
12	0.305	52.88	0.019	9.317	0.107	45.03	0.022	7.934	0.126	37.33	0.027	6.577	0.152

R = °F, ft^2, hr, Btu R_m = km^{2}W U = Btu per ft^2 per hr U_m = W/m^{2}K

TABLE 100

FLAT ROOF

13. MINERAL FIBER, HEAVY DENSITY, BATT OR BLANKET INSULATION THERMAL RESISTANCE AND TRANSMITTANCE OF ASSEMBLY V WITH HEAVY DENSITY MINERAL WOOL INSULATION:

k = 0.36 Btu, in./ft², hr °F at 40 °F, k_m = 0.052 W/mK at 4.4 °C, k = 0.39 Btu, in./ft², hr °F at 85 °F, k_m = 0.056 W/mK at 29.4 °C, k = 0.43 Btu, in./ft², hr °F at 110 °F, k_m = 0.062 W/mK at 43.2 °C

Insulation Thickness		Mean Temp. 40 °F (4.4 °C)				Mean Temp. 85 °F (29.4 °C)				Mean Temp. 110 °F (43.2 °C)			
		English Units		Metric Units		English Units		Metric Units		English Units		Metric Units	
Inches	Meters	R	U	R_m	U_m	R	U	R_m	U_m	R	U	R_m	U_m
1	0.025	5.75	0.174	1.013	0.987	5.43	0.184	0.957	1.045	5.16	0.194	0.909	1.100
2	0.051	8.53	0.117	1.503	0.665	8.00	0.125	1.410	0.709	7.49	0.133	1.320	0.758
3	0.076	11.30	0.088	1.991	0.502	10.56	0.095	1.861	0.537	9.81	0.102	1.728	0.579
4	0.102	14.08	0.071	2.481	0.404	13.13	0.076	2.313	0.432	12.41	0.082	2.187	0.457
5	0.127	15.86	0.059	2.795	0.358	15.63	0.064	2.754	0.363	14.47	0.069	2.550	0.392
6	0.152	19.64	0.050	3.461	0.288	18.25	0.055	3.216	0.311	16.79	0.059	2.958	0.338
7	0.178	22.41	0.045	3.910	0.256	20.82	0.048	3.668	0.273	19.11	0.052	3.367	0.297
8	0.203	25.19	0.040	4.438	0.225	23.33	0.043	4.111	0.243	21.44	0.047	3.778	0.265
9	0.229	27.97	0.036	4.923	0.203	25.95	0.039	4.572	0.219	23.77	0.042	4.188	0.239
10	0.254	30.75	0.033	5.418	0.185	28.51	0.035	5.023	0.199	26.10	0.038	4.599	0.217
11	0.279	33.53	0.030	5.908	0.169	31.07	0.032	5.475	0.183	28.42	0.035	5.008	0.200
12	0.305	35.50	0.028	6.255	0.160	32.84	0.030	5.786	0.173	29.95	0.033	5.277	0.198

R = °F, ft², hr, Btu R_m = km²W   U = Btu per ft² per hr U_m = W/m²K

TABLE 101

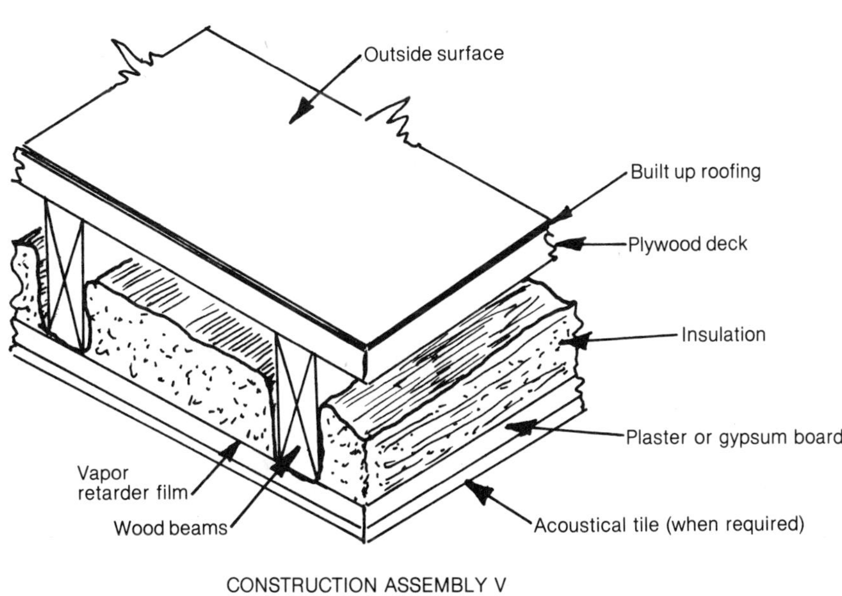

CONSTRUCTION ASSEMBLY V
WITH CEILING

FLAT ROOF CONSTRUCTION ASSEMBLY V CORRECTIONS OF RESISTANCE DUE TO ADDITION OF CEILING PLASTER OR BOARD

Added Ceiling Components	Add to Resistances Listed in Tables 100 and 101	
	R	R_m
1/2″ gypsum wallboard	0.45	0.079
Metal lath and 3/4″ (1.9 cm) light weight agg.	0.47	0.082
1/2″ gypsum board and 1/2″ acoustical tile	1.70	0.299

FLAT ROOFS

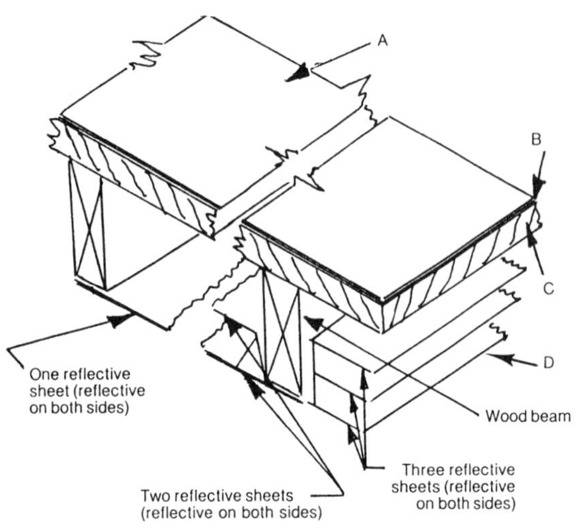

CONSTRUCTION ASSEMBLY V

Note: Seal inner sheet for vapor retarder.

NO THERMAL INSULATION
THERMAL RESISTANCE R and R_m

Mean Temperature 40° (4.4°C), Wind 10 mph (16.1 km/hr)

	Components	Eng. Units		Metric Units
A	Outside surface	0.17		0.030
B	Built up roofing	0.33		0.058
C	Deck 1″ plywood	1.06		0.189
D	Insulation, reflective (Number of sheets & spaces as shown below)	—		—
	(TOTAL) RESISTANCE R =	1.56	R_m =	0.275
	TRANSMITTANCE U =	0.641	U_m =	3.638

Mean Temperature 110° (43.2°C), No Wind

	Components	Eng. Units		Metric Units
A	Outside surface	0.16		0.028
B	Built up roofing	0.33		0.058
C	Deck 1″ plywood	0.96		0.169
D	Insulation, reflective	—		—
	(TOTAL) RESISTANCE R =	1.45	R_m =	0.255
	TRANSMITTANCE U =	0.670	U_m =	3.922

14. REFLECTIVE ALUMINUM FOIL INSULATION
THERMAL RESISTANCE AND TRANSMITTANCE OF ASSEMBLY V WITH REFLECTIVE INSULATION SHEETS. SURFACE EMITTANCE ε = 0.05

Horizontal Reflective Spaces		Mean Temp. 40°F (4.4°C) Heat Flow Up (Heating)				Mean Temp. 110°F (43.2°C) Heat Flow Down (Cooling)			
		English Units		Metric Units		English Units		Metric Units	
Number of Enclosed Air Spaces	Aluminum Reflective Sheets (Reflective on Both Sides)	R	U	R_m	U_m	R	U	R_m	U_m
1	1 (attached to bottom)	8.31	0.120	1.464	0.682	9.38	0.107	1.653	0.605
3	2 (between beams)	13.71	0.073	2.415	0.414	15.96	0.063	2.812	0.356
4	3 (between beams)	19.11	0.052	3.367	0.297	23.52	0.043	4.144	0.241

TABLE 102

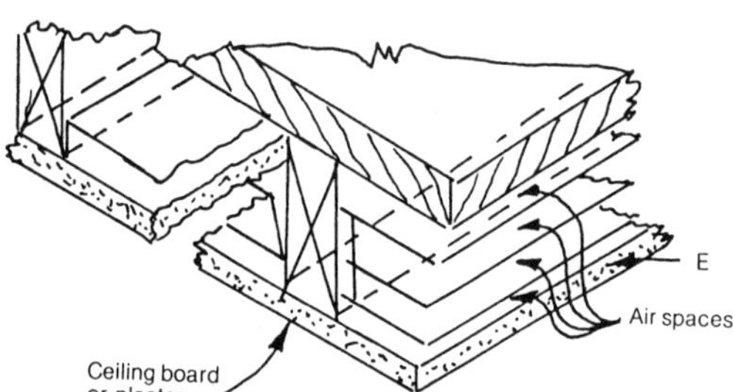

Both mean temperatures
E Plaster or Gypsum ceiling board plus one additional air space — R = 1.35, R_m = 0.238 (Add to values for Assembly V above)

CONSTRUCTION ASSEMBLY V WITH CEILING

THERMAL RESISTANCE AND TRANSMITTANCE OF ROOF ASSEMBLY V AND CEILING BOARD WITH REFLECTIVE INSULATION SHEETS. ε = 0.05

Horizontal Reflective Spaces		Mean Temp. 40°F (4.4°C) Heat Flow Up (Heating)				Mean Temp. 110°F (43.2°C) Heat Flow Down (Cooling)			
		English Units		Metric Units		English Units		Metric Units	
Number of Enclosed Air Spaces	Aluminum Reflective Sheets (Reflective on Both Sides)	R	U	R_m	U_m	R	U	R_m	U_m
2	1 (between beams)	8.31	0.120	1.464	0.682	9.38	0.107	1.653	0.605
3	2 (between beams)	13.71	0.073	2.415	0.414	15.96	0.063	2.812	0.356
4	3 (between beams)	19.11	0.052	3.367	0.297	23.52	0.043	4.144	0.241

TABLE 103

FLAT ROOFS

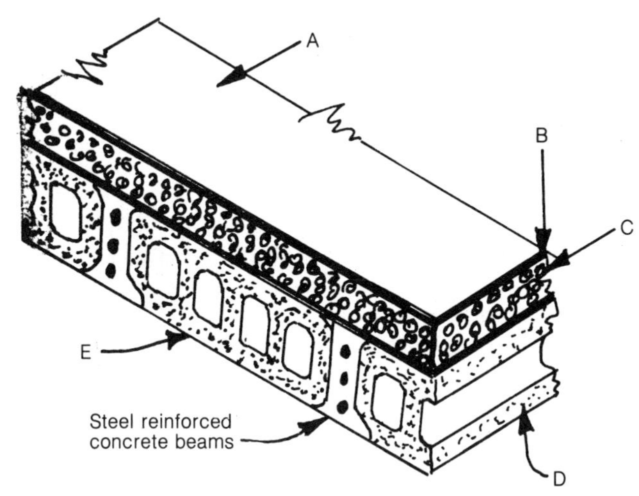

CONSTRUCTION ASSEMBLY VI

Note: Cellular glass insulation is available in boards, the insulation laminated between sheets of kraft paper. Also is available in flat and tapered blocks.

NO THERMAL INSULATION
THERMAL RESISTANCE R and R_m

Mean Temperature 40° (4.4°C), Wind 10 mph (16.1 km/hr)

	Components	Eng. Units		Metric Units
A	Outside surface	0.20		0.035
B	Built up roofing	0.33		0.058
C	Insulation	—		—
D	8″ Cinder blocks and beams	1.61		0.284
E	Inside surface	0.85		0.149
	(TOTAL) RESISTANCE R =	2.99	R_m =	0.524
	TRANSMITTANCE U =	0.334	U_m =	1.906

Mean Temperature 85° (29.4°C), Wind 5 mph (8 km/hr)

	Components	Eng. Units		Metric Units
A	Outside surface	0.19		0.033
B	Built up roofing	0.31		0.054
C	Insulation	—		—
D	8″ Cinder blocks and beams	1.57		0.276
E	Inside surface	0.85		0.149
	(TOTAL) RESISTANCE R =	2.92	R_m =	0.512
	TRANSMITTANCE U =	0.342	U_m =	1.953

Mean Temperature 110° (43.2°C), No Wind

	Components	Eng. Units		Metric Units
A	Outside surface	0.15		0.026
B	Built up roofing	0.22		0.039
C	Insulation	—		—
D	8″ Cinder blocks and beams	1.51		0.266
E	Inside surface	0.87		0.153
	(TOTAL) RESISTANCE R =	2.75	R_m =	0.484
	TRANSMITTANCE U =	0.364	U_m =	2.064

16. CELLULAR GLASS INSULATION
THERMAL RESISTANCE AND TRANSMITTANCE OF ASSEMBLY VI WITH CELLULAR GLASS INSULATION:

INSULATION: k = 0.33 Btu, in./ft^2, hr °F at 40°F, k_m = 0.047 W/mK at 4.4°C, k = 0.35 Btu, in./ft^2, hr °F at 85°F, k_m = 0.050 W/mK at 29.4°C, k = 0.36 Btu, in./ft^2, hr °F at 110°F, k_m = 0.052 W/mK at 43.2°C

Insulation Thickness		Mean Temp. 40°F (4.4°C)				Mean Temp. 85°F (29.4°C)				Mean Temp. 110°F (43.2°C)			
		English Units		Metric Units		English Units		Metric Units		English Units		Metric Units	
Inches	Meters	R	U	R_m	U_m	R	U	R_m	U_m	R	U	R_m	U_m
1 1/2	0.038	7.56	0.132	1.332	0.752	7.19	0.139	1.266	0.787	6.87	0.146	1.210	0.826
2	0.051	9.09	0.110	1.601	0.625	8.62	0.116	1.519	0.658	8.24	0.121	1.452	0.690
3	0.076	12.14	0.082	2.139	0.467	11.47	0.087	2.021	0.495	10.99	0.091	1.936	0.515
4	0.102	15.18	0.066	2.675	0.373	14.32	0.069	2.523	0.397	13.74	0.073	2.421	0.413
5	0.127	18.23	0.055	3.212	0.312	17.17	0.058	3.025	0.330	16.49	0.061	2.906	0.344
6	0.152	21.22	0.047	3.738	0.267	20.01	0.050	3.526	0.283	19.23	0.052	3.388	0.295

R = °F, ft^2, hr, Btu R_m = km^{2}W U = Btu per ft^2 per hr U_m = W/m^{2}K

TABLE 104

FLAT OR PITCHED ROOFS

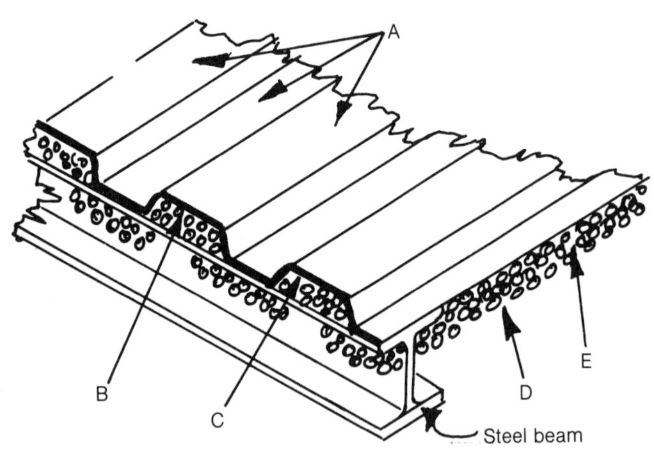

CONSTRUCTION ASSEMBLY VII
NO PLASTER BOARD OR TILE BELOW BEAM

Note: Unless some sealed metal expansion joints are provided for corrugated deck, a fire on the interior side could cause roofing to melt and drip through butt joints.

In cold weather the bare steel beam will condense water from vapor in the inside air.

NO THERMAL INSULATION
THERMAL RESISTANCE R and R_m

Mean Temperature 40° (4.4°C), Wind 10 mph (16.1 km/hr)

	Components	Eng. Units		Metric Units
A	Outside surface	0.20		0.035
B	Built up roofing	0.33		0.058
C	Metal deck	0.01		0.002
D	Inside surface	0.85		0.244
	(TOTAL) RESISTANCE R =	1.39	R_m =	0.244
	TRANSMITTANCE U =	0.72	U_m =	4.098

Mean Temperature 85° (29.4°C), Wind 5 mph (8 km/hr)

	Components	Eng. Units		Metric Units
A	Outside surface	0.19		0.033
B	Built up roofing	0.31		0.054
C	Metal deck	0.01		0.002
D	Inside surface	0.80		0.141
	(TOTAL) RESISTANCE R =	1.31	R_m =	0.230
	TRANSMITTANCE U =	0.76	U_m =	4.347

Mean Temperature 110° (43.2°C), No Wind

	Components	Eng. Units		Metric Units
A	Outside surface	0.15		0.026
B	Built up roofing	0.22		0.039
C	Metal deck	0.01		0.002
D	Inside surface	0.72		0.126
	(TOTAL) RESISTANCE R =	1.10	R_m =	0.193
	TRANSMITTANCE U =	0.91	U_m =	5.181

17. SPRAYED ON POLYURETHANE INSULATION
THERMAL RESISTANCE AND TRANSMITTANCE OF ROOF ASSEMBLY VII INSULATED BELOW WITH SPRAYED ON POLYURETHANE: k = 0.24 Btu, in./ft^2, hr °F at 40°F, k_m = 0.035 W/mK at 4.4°C, k = 0.25 Btu, in./ft^2, hr °F at 85°F, k_m = 0.036 W/mK at 29.4°C, k = 0.28 Btu, in./ft^2, hr °F at 110°F, k_m = 0.040 W/mK at 43.2°C

Insulation Thickness		Mean Temp. 40°F (4.4°C)				Mean Temp. 85°F (29.4°C)				Mean Temp. 110°F (43.2°C)			
		English Units		Metric Units		English Units		Metric Units		English Units		Metric Units	
Inches	Meters	R	U	R_m	U_m	R	U	R_m	U_m	R	U	R_m	U_m
1	0.025	5.55	0.180	0.977	1.023	5.31	0.188	0.935	1.069	4.67	0.214	0.822	1.216
1 1/2	0.038	7.64	0.131	1.345	0.743	7.31	0.137	1.287	0.777	6.45	0.155	1.136	0.880
2	0.051	9.72	0.103	1.711	0.584	9.31	0.107	1.639	0.609	8.24	0.121	1.451	0.689
2 1/2	0.064	11.81	0.084	2.079	0.481	11.31	0.088	1.991	0.502	10.02	0.099	1.764	0.566
3	0.076	13.89	0.072	2.446	0.409	13.31	0.075	2.343	0.426	11.81	0.084	2.079	0.481
3 1/2	0.088	15.97	0.063	2.812	0.356	15.31	0.065	2.696	0.371	13.60	0.074	2.395	0.417
4	0.102	18.06	0.055	3.180	0.314	17.31	0.058	3.048	0.328	15.38	0.065	2.708	0.369

Note: Due to the high combustibility of polyurethane foam all welding (or other sparks or fire) must be prohibited from insulated area. All electrical circuits should be kept away from the organic foam insulation. This method of insulation is not satisfactory in northern (cold) areas as there is no suitable vapor retarder to prevent moisture vapor from entering and condensing in the insulation.

R = °F, ft^2, hr, Btu R_m = km^{2}W   U = Btu per ft^2 per hr U_m = W/m^{2}K

TABLE 105

PITCHED ROOFS

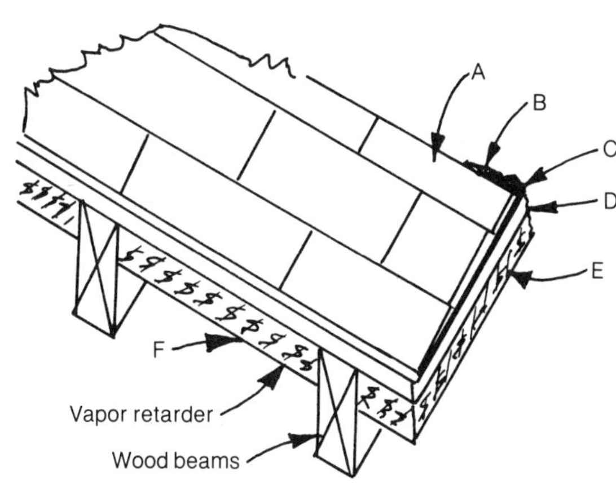

Heated and/or cooled area below

CONSTRUCTION ASSEMBLY VIII

Note: If some suitable vapor retarder film is not installed below insulation condensation will occur within the insulation during heating season. In this case the insulation becomes ineffective.

NO THERMAL INSULATION
THERMAL RESISTANCE R and R_m

Mean Temperature 40° (4.4°C), Wind 10 mph (16.1 km/hr)

	Components	Eng. Units		Metric Units
A	Outside surface	0.20		0.035
B	Shingles	0.44		0.078
C	Building paper	0.05		0.009
D	Plywood deck	0.80		0.141
E	Insulation	—		—
F	Inside surface	0.60		0.106
	(TOTAL) RESISTANCE R =	2.09	R_m =	0.369
	TRANSMITTANCE U =	0.478	U_m =	2.175

Mean Temperature 85° (29.4°C), Wind 5 mph (8 km/hr)

	Components	Eng. Units		Metric Units
A	Outside surface	0.19		0.033
B	Shingles	0.40		0.070
C	Building paper	0.04		0.007
D	Plywood deck	0.75		0.132
E	Insulation	—		—
F	Inside surface	0.65		0.114
	(TOTAL) RESISTANCE R =	2.03	R_m =	0.356
	TRANSMITTANCE U =	0.493	U_m =	2.795

Mean Temperature 110° (43.2°C), No Wind

	Components	Eng. Units		Metric Units
A	Outside surface	0.17		0.029
B	Shingles	0.36		0.063
C	Building paper	0.03		0.005
D	Plywood deck	0.70		0.123
E	Insulation	—		—
F	Inside surface	0.70		0.123
	(TOTAL) RESISTANCE R =	1.96	R_m =	0.343
	TRANSMITTANCE U =	0.510	U_m =	2.915

18. SHEET OF CHEMICALLY TREATED WOOD FIBERS BONDED WITH PORTLAND CEMENT THERMAL RESISTANCE AND TRANSMITTANCE OF ASSEMBLY VIII WITH WOOD FIBER CEMENT BOARD:

k = 0.4 Btu, in./ft², hr °F at 40°F, k_m = 0.057 W/mK at 4.4°C, k = 0.43 Btu, in./ft², hr °F at 85°F, k_m = 0.062 W/mK at 29.4°C, k = 0.47 Btu, in./ft², hr °F at 110°F, k_m = 0.068 W/mK at 43.2°C

Insulation Thickness		Mean Temp. 40°F (4.4°C)				Mean Temp. 85°F (29.4°C)				Mean Temp. 110°F (43.2°C)			
		English Units		Metric Units		English Units		Metric Units		English Units		Metric Units	
Inches	Meters	R	U	R_m	U_m	R	U	R_m	U_m	R	U	R_m	U_m
2	0.051	7.09	0.141	1.249	0.800	6.57	0.152	0.158	0.864	6.22	0.161	1.095	0.912
2 1/2	0.064	8.34	0.120	1.470	0.680	7.71	0.130	1.359	0.736	7.28	0.137	1.283	0.779
3	0.076	9.59	0.104	1.690	0.592	8.85	0.112	1.559	0.641	8.34	0.119	1.469	0.680
3 1/2	0.088	10.84	0.092	1.910	0.524	9.98	0.100	1.758	0.569	9.41	0.106	1.658	0.603
4	0.102	12.09	0.083	2.130	0.469	11.12	0.090	1.959	0.510	10.47	0.096	1.844	0.542

R = °F, ft², hr, Btu R_m = km²W U = Btu per ft² per hr U_m = W/m²K

TABLE 106

FLAT ROOFS

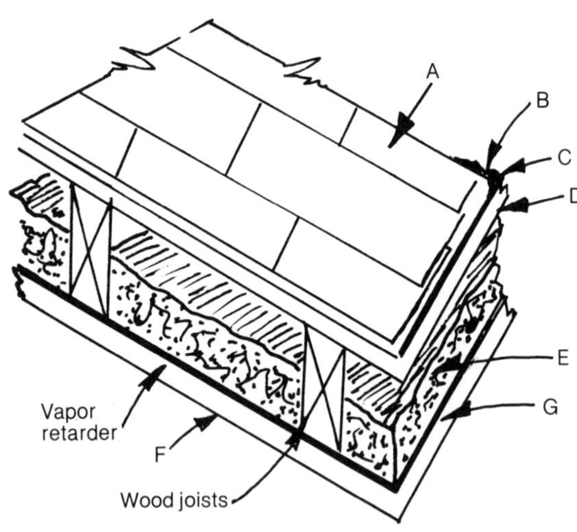

Heated and/or cooled area below

CONSTRUCTION ASSEMBLY IX

Note: Thickness of insulation may be limited by height dimension of joists.

NO THERMAL INSULATION
THERMAL RESISTANCE R and R_m

Mean Temperature 40° (4.4°C), Wind 10 mph (16.1 km/hr)

Components	Eng. Units		Metric Units
A, B, C, D and F same as Assem. VIII	2.09		0.369
E Insulation	—		—
G Gypsum board or plaster	0.45		0.079
(TOTAL) RESISTANCE R =	2.54	R_m =	0.448
TRANSMITTANCE U =	0.394	U_m =	2.232

Mean Temperature 85° (29.4°C), Wind 5 mph (8 km/hr)

Components	Eng. Units		Metric Units
A, B, C, D and F same as Assem. VIII	2.03		0.114
E Insulation	—		—
G Gypsum board or plaster	0.41		0.072
(TOTAL) RESISTANCE R =	2.44	R_m =	0.429
TRANSMITTANCE U =	0.410	U_m =	2.331

Mean Temperature 110° (43.2°C), No Wind

Components	Eng. Units		Metric Units
A, B, C, D and F same as Assem. VIII	1.96		0.343
E Insulation	—		—
G Gypsum board or plaster	0.38		0.066
(TOTAL) RESISTANCE R =	2.34	R_m =	0.066
TRANSMITTANCE U =	0.427	U_m =	2.445

18. GLASS FIBER BATT OR BLANKET INSULATION
THERMAL RESISTANCE AND TRANSMITTANCE OF ASSEMBLY IX WITH INSULATION:

k = 0.24 Btu, in./ft^2, hr °F at 40°F, k_m = 0.036 W/mK at 4.4°C, k = 0.28 Btu, in./ft^2, hr °F at 85°F, k_m = 0.040 W/mK at 29.4°C, k = 0.34 Btu, in./ft^2, hr °F at 110°F, k_m = 0.049 W/mK at 43.2°C

Insulation Thickness		Mean Temp. 40°F (4.4°C)				Mean Temp. 85°F (29.4°C)				Mean Temp. 110°F (43.2°C)			
		English Units		Metric Units		English Units		Metric Units		English Units		Metric Units	
Inches	Meters	R	U	R_m	U_m	R	U	R_m	U_m	R	U	R_m	U_m
1	0.025	7.68	0.130	1.353	0.739	6.85	0.146	1.207	0.829	5.66	0.176	0.997	1.003
2	0.051	11.75	0.085	2.070	0.483	10.42	0.096	1.836	0.545	9.10	0.110	1.603	0.623
3	0.076	15.92	0.063	2.805	0.356	13.99	0.073	2.407	0.415	12.04	0.083	2.121	0.471
4	0.102	20.09	0.049	3.539	0.282	17.57	0.057	3.096	0.323	14.98	0.067	2.639	0.379
5	0.127	24.23	0.041	4.273	0.234	21.13	0.047	3.723	0.268	17.84	0.056	3.143	0.318
6	0.152	28.43	0.035	5.009	0.199	24.71	0.040	4.354	0.230	20.86	0.048	3.676	0.272
7	0.178	32.63	0.031	5.749	0.174	28.28	0.035	4.983	0.201	23.81	0.042	4.195	0.238
8	0.203	36.75	0.027	6.475	0.154	31.85	0.031	5.612	0.178	26.76	0.037	4.715	0.212
9	0.229	40.92	0.024	7.210	0.139	35.42	0.028	6.241	0.160	29.70	0.034	5.233	0.191
10	0.254	44.09	0.023	7.769	0.129	38.99	0.026	6.870	0.145	32.63	0.080	5.749	0.174
11	0.279	49.25	0.020	8.678	0.115	42.56	0.023	7.499	0.133	35.58	0.028	6.269	0.159
12	0.305	58.33	0.019	9.397	0.106	45.44	0.022	8.007	0.125	37.81	0.026	6.662	0.150

R = °F, ft^2, hr, Btu R_m = km^{2}W U = Btu per ft^2 per hr U_m = W/m^{2}K

TABLE 107

PITCHED ROOF

19. MINERAL FIBER, RESIN BINDER, BATT OR BLANKET INSULATION
20. CELLULOSE FIBER BATT OR LOOSE FILL INSULATION
THERMAL RESISTANCE AND TRANSMITTANCE OF ASSEMBLY IX WITH CELLULOSE OR MINERAL WOOL INSULATION:

k = 0.30 Btu, in./ft^2, hr °F at 40 °F, k_m = 0.053 W/mK at 4.4 °C, k = 0.33 Btu, in./ft^2, hr °F at 85 °F, k_m = 0.058 W/mK at 29.4 °C, k = 0.37 Btu, in./ft^2, hr °F at 110 °F, k_m = 0.065 W/mK at 43.2 °C

Insulation Thickness		Mean Temp. 40 °F (4.4 °C)				Mean Temp. 85 °F (29.4 °C)				Mean Temp. 110 °F (43.2 °C)			
		English Units		Metric Units		English Units		Metric Units		English Units		Metric Units	
Inches	Meters	R	U	R_m	U_m	R	U	R_m	U_m	R	U	R_m	U_m
1	0.025	5.87	0.170	1.034	0.966	5.47	0.183	0.964	1.037	5.04	0.198	0.888	1.126
2	0.051	9.21	0.109	1.623	0.616	8.50	0.117	1.498	0.667	7.75	0.129	1.366	0.732
3	0.076	12.54	0.080	2.209	0.453	11.53	0.087	2.031	0.492	10.45	0.096	1.841	0.543
4	0.102	15.87	0.063	2.796	0.357	14.56	0.069	2.565	0.389	13.15	0.076	2.317	0.432
5	0.127	19.21	0.052	3.385	0.295	17.59	0.057	3.099	0.323	15.85	0.063	2.793	0.358
6	0.152	22.54	0.044	3.972	0.252	20.62	0.048	3.633	0.275	18.55	0.053	3.268	0.306
7	0.178	25.87	0.039	4.558	0.219	23.65	0.042	4.167	0.240	21.25	0.047	3.744	0.267
8	0.203	29.21	0.034	5.147	0.194	26.68	0.037	4.701	0.213	23.96	0.042	4.222	0.237
9	0.229	32.54	0.031	5.734	0.174	29.71	0.034	5.235	0.191	26.66	0.037	4.697	0.213
10	0.254	35.87	0.028	6.320	0.158	32.74	0.031	5.769	0.173	29.40	0.034	5.180	0.193
11	0.279	39.21	0.026	6.909	0.145	35.77	0.027	6.303	0.159	32.06	0.031	5.649	0.177
12	0.305	42.54	0.024	7.496	0.133	38.80	0.025	6.837	0.146	34.77	0.029	6.126	0.163

R = °F, ft^2, hr, Btu R_m = km^{2}W U = Btu per ft^2 per hr U_m = W/m^{2}K

TABLE 108

21. VERMICULITE (EXPANDED) LOOSE INSULATION
THERMAL RESISTANCE AND TRANSMITTANCE OF ASSEMBLY IX WITH VERMICULITE INSULATION:

k = 0.47 Btu, in./ft^2, hr °F at 40 °F, k_m = 0.068 W/mK at 4.4 °C, k = 0.48 Btu, in./ft^2, hr °F at 85 °F, k_m = 0.069 W/mK at 29.4 °C, k = 0.50 Btu, in./ft^2, hr °F at 110 °F, k_m = 0.072 W/mK at 43.2 °C

Insulation Thickness		Mean Temp. 40 °F (4.4 °C)				Mean Temp. 85 °F (29.4 °C)				Mean Temp. 110 °F (43.2 °C)			
		English Units		Metric Units		English Units		Metric Units		English Units		Metric Units	
Inches	Meters	R	U	R_m	U_m	R	U	R_m	U_m	R	U	R_m	U_m
1	0.025	4.66	0.214	0.821	1.218	4.52	0.221	0.796	1.256	4.34	0.230	0.765	1.308
2	0.051	6.79	0.147	1.196	0.836	6.61	0.151	1.165	0.859	6.34	0.158	1.117	0.895
3	0.076	8.92	0.112	1.571	0.636	8.69	0.115	1.531	0.653	8.34	0.120	1.469	0.680
4	0.102	11.05	0.090	1.947	0.513	10.77	0.093	1.897	0.527	10.34	0.125	1.822	0.529
5	0.127	13.18	0.076	2.308	0.433	12.85	0.078	2.264	0.442	12.34	0.081	2.174	0.460
6	0.152	15.31	0.065	2.697	0.371	14.94	0.067	2.632	0.380	14.34	0.070	2.527	0.396
7	0.178	17.43	0.057	3.071	0.326	17.02	0.059	2.999	0.333	16.34	0.061	2.879	0.347
8	0.203	19.56	0.051	3.446	0.290	19.11	0.052	3.367	0.297	18.34	0.054	3.232	0.309
9	0.229	21.69	0.046	3.821	0.262	21.19	0.047	3.733	0.268	20.34	0.049	3.584	0.279
10	0.254	23.81	0.042	4.195	0.238	23.27	0.043	4.100	0.243	22.34	0.045	3.936	0.254
11	0.279	25.94	0.038	4.571	0.218	25.36	0.039	4.468	0.223	24.34	0.041	4.289	0.233
12	0.305	28.07	0.036	4.946	0.202	27.44	0.036	4.835	0.207	26.34	0.040	4.641	0.215

Note: Vermiculite loose insulation must be installed from top level of the roof structure so as to completely fill cavity. Thickness is determined by size of cavity.

R = °F, ft^2, hr, Btu R_m = km^{2}W U = Btu per ft^2 per hr U_m = W/m^{2}K

TABLE 109

PITCHED ROOFS

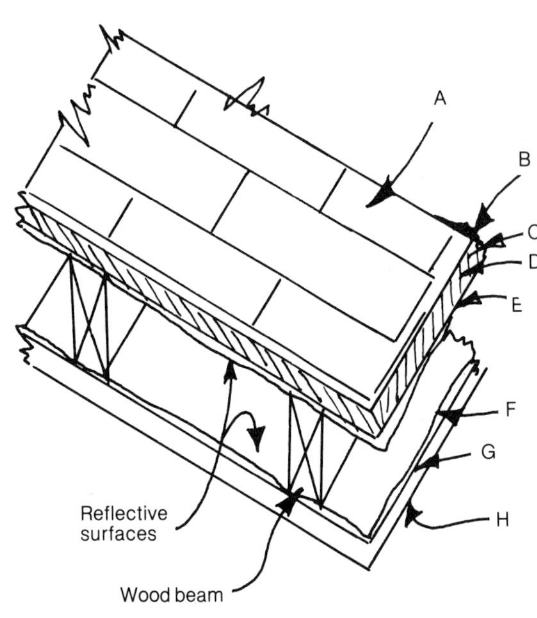

CONSTRUCTION ASSEMBLY X

Note: Mean temperature has less effect on thermal resistance than mass insulation. Direction of heat flow has greater influence.

NO THERMAL INSULATION
THERMAL RESISTANCE R and R_m
Heat Flow Up — Heating Season

Components	R		R_m
A Outside surface	0.20		0.035
B Shingles	0.44		0.078
C Building paper	0.05		0.009
D Plywood deck	0.80		0.141
E Reflective insulation sheet	—		—
*F Reflective insulation sheet	—		—
G Gypsum board or plaster	0.45		0.079
H Inside surface	0.60		0.106
(TOTAL) RESISTANCE R =	2.54	R_m =	0.448
TRANSMITTANCE U =	0.394	U_m =	2.232

Heat Flow Down — Summer

Components	R		R_m
A Outside surface	0.17		0.029
B Shingles	0.36		0.063
C Building paper	0.03		0.005
D Plywood deck	0.70		0.123
E Reflective insulation sheet	—		—
*F Reflective insulation sheet	—		—
G Gypsum board or plaster	0.45		0.076
H Inside surface	0.70		0.123
(TOTAL) RESISTANCE R =	2.41	R_m =	0.419
TRANSMITTANCE U =	0.414	U_m =	2.387

*Aluminum foil faced gypsum board can be used as lower reflective surface.

22. REFLECTIVE ALUMINUM SURFACE (OR SURFACES)
Heat flow up, resistance one surface R = 3.46, R_m = 0.609
Heat flow down, resistance one surface R = 3.24, R_m = 0.571
THERMAL RESISTANCE AND TRANSMITTANCE OF ASSEMBLY X WITH REFLECTIVE SURFACES FACING CAVITY

Number of	Heat Flow Up (Heating Season)				Heat Flow Down (Cooling Season)			
Reflective	English Units		Metric Units		English Units		Metric Units	
Surfaces	R	U	R_m	U_m	R	U	R_m	U_m
One	6.00	0.166	1.057	0.946	5.65	0.177	1.000	1.000
Two	9.46	0.106	1.666	0.600	8.89	0.112	1.566	0.638

R = °F, ft², hr, Btu R_m = km²W   U = Btu per ft² per hr U_m = W/m²K

TABLE 110

PITCHED ROOFS

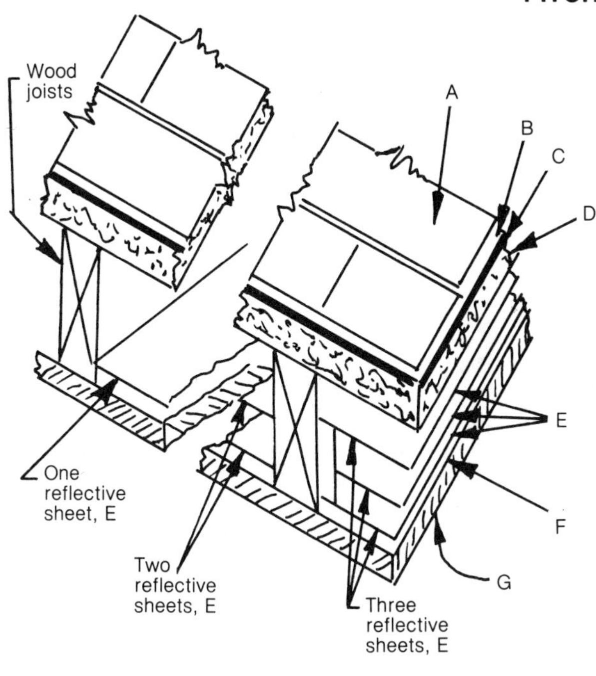

CONSTRUCTION ASSEMBLY X

Note: The reflective sheets shown are aluminum faced both sides.

NO THERMAL INSULATION
THERMAL RESISTANCE R and R_m

Heat Flow Up (Heating Season)

	Components	R		R_m
A	Outside surface	0.20		0.035
B	Shingles	0.44		0.078
C	Building paper	0.03		0.009
D	Plywood deck	0.80		0.141
E	Reflective insulation	—		—
F	Gypsum board or plaster	0.45		0.079
G	Inside surface	0.60		0.106
	(TOTAL) RESISTANCE R =	2.54	R_m =	0.448
	TRANSMITTANCE U =	0.394	U_m =	2.232

Heat Flow Down (Cooling Season)

	Components	R		R_m
A	Outside surface	0.17		0.029
B	Shingles	0.36		0.063
C	Building paper	0.03		0.005
D	Plywood deck	0.70		0.123
E	Reflective insulation	—		—
F	Gypsum board or plaster	0.47		0.083
G	Inside surface	0.65		0.114
	(TOTAL) RESISTANCE R =	2.38	R_m =	0.417
	TRANSMITTANCE U =	0.420	U_m =	2.398

23. REFLECTIVE ALUMINUM INSULATION (SHEETS OF FOIL) THERMAL RESISTANCE AND TRANSMITTANCE OF ROOF ASSEMBLY X WITH REFLECTIVE INSULATION SHEETS AS SHOWN: (ε = 0.05)

Reflective Space		Mean Temp. 40°F (4.4°C) Heat Flow Up (Heating)				Mean Temp. 85°F (29.4°C) Heat Flow Down (Cooling)			
		English Units		Metric Units		English Units		Metric Units	
Number of Enclosed Air Spaces	Aluminum Reflective Sheets (Reflective on Both Sides)	R	U	R_m	U_m	R	U	R_m	U_m
2	1 (between joists)	7.84	0.128	1.381	0.724	10.06	0.099	1.773	0.564
3	2 (between joists)	13.24	0.076	2.332	0.489	17.64	0.057	3.108	0.322
4	3 (between joists)	18.64	0.054	3.284	0.304	25.20	0.040	4.440	0.225

TABLE 111

CEILINGS UNDER VENTED SPACE

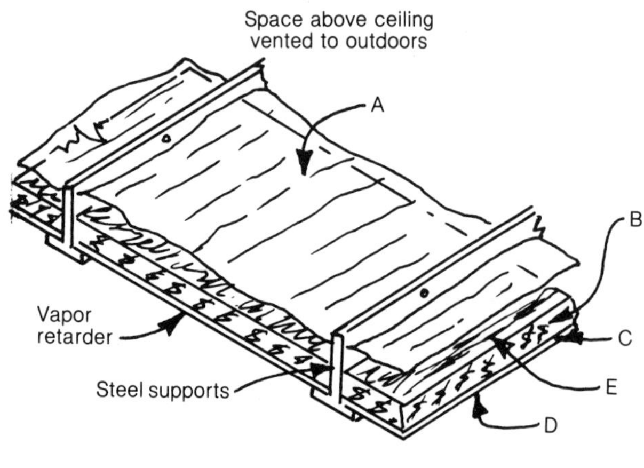

SUSPENDED CEILING

CONSTRUCTION ASSEMBLY XI

Note: As ceiling under vented space is shaded by roof above it is not exposed to direct solar heat and resistances of materials at 110°F (43.2°C) is not needed.

NO THERMAL INSULATION THERMAL RESISTANCE R and R_m

Heat Flow Up (Heating Season)

	Components	R		R_m
A	Top surface	0.35		0.062
B	Glass fiber ceiling panel	1.98		0.349
C	Panel facing	0.02		0.003
D	Bottom surface	0.60		0.106
E	Insulation	—		—
	(TOTAL) RESISTANCE R =	2.15	R_m =	0.377
	TRANSMITTANCE U =	0.46	U_m =	2.652

Heat Flow Down (Cooling Season)

	Components	R		R_m
A	Top surface	0.25		0.044
B	Glass fiber ceiling panel	1.98		0.349
C	Panel facing	0.02		0.003
D	Bottom surface	0.60		0.106
E	Insulation	—		—
	(TOTAL) RESISTANCE R =	2.85	R_m =	0.502
	TRANSMITTANCE U =	0.351	U_m =	1.992

24. GLASS FIBER, LIGHT DENSITY BATT OR BLANKET INSULATION THERMAL RESISTANCES OF ASSEMBLY WITH GLASS FIBER INSULATION:

k = 0.24 Btu, in./ft^2, hr °F at 40°F, k_m = 0.036 W/mK at 4.4°C,
k = 0.28 Btu, in./ft^2, hr °F at 85°F, k_m = 0.040 W/mK at 29.4°

Insulation Thickness		Mean Temp. 40°F (4.4°C)				Mean Temp. 85°F (29.4°C)			
		English Units		Metric Units		English Units		Metric Units	
Inches	Meters	R	U	R_m	U_m	R	U	R_m	U_m
1	0.025	7.11	0.141	1.253	0.798	6.42	0.156	1.539	0.650
2	0.051	11.28	0.089	1.988	0.503	9.99	0.100	1.760	0.568
3	0.076	15.45	0.065	2.722	0.367	13.56	0.074	2.389	0.418
4	0.102	19.62	0.051	3.457	0.289	17.14	0.058	3.020	0.331
5	0.127	23.78	0.042	4.190	0.239	20.71	0.048	3.649	0.274
6	0.152	27.95	0.036	4.924	0.203	24.28	0.041	4.278	0.233

R = °F, ft^2, hr, Btu R_m = km^{2}W   U = Btu per ft^2 per hr U_m = W/m^{2}K

TABLE 112

CEILINGS
UNDER VENTED SPACE

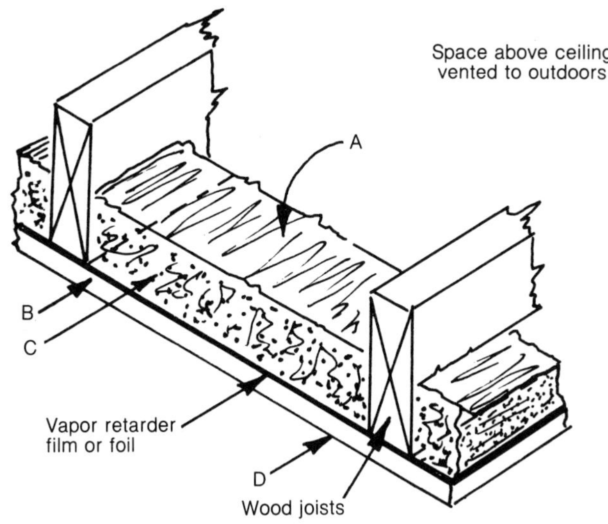

CONSTRUCTION ASSEMBLY XII

NO THERMAL INSULATION
THERMAL RESISTANCE R and R_m

Heat Flow Up (Heating Season)

	Components	R		R_m
A	Top surface	0.35		0.062
B	Gypsum board or plaster	0.45		0.079
C	Insulation	—		—
D	Inside surface	0.60		0.106
	(TOTAL) RESISTANCE R =	1.40	R_m =	0.247
	TRANSMITTANCE U =	0.714	U_m =	4.048

Heat Flow Down (Cooling Season)

	Components	R		R_m
A	Top surface	0.25		0.044
B	Gypsum board or plaster	0.45		0.079
C	Insulation	—		—
D	Inside surface	0.60		0.106
	(TOTAL) RESISTANCE R =	1.30	R_m =	0.229
	TRANSMITTANCE U =	0.769	U_m =	4.367

25. GLASS FIBER LIGHT DENSITY BATT OR BLANKET INSULATION THERMAL RESISTANCE AND TRANSMITTANCE OF ASSEMBLY XII WITH GLASS FIBER INSULATION:

k = 0.24 Btu, in./ft^2, hr °F at 40 °F, k_m = 0.036 W/mK at 4.4 °C, k = 0.28 Btu, in./ft^2, hr °F at 85 °F, k_m = 0.040 W/mK at 29.4 °C,

Insulation Thickness		Mean Temp. 40 °F (4.4 °C)				Mean Temp. 85 °F (29.4 °C)			
		English Units		Metric Units		English Units		Metric Units	
Inches	Meters	R	U	R_m	U_m	R	U	R_m	U_m
1	0.025	5.57	0.179	1.981	1.019	4.87	0.205	0.858	1.165
2	0.051	9.73	0.103	1.714	0.583	8.44	0.118	1.487	0.672
3	0.076	13.90	0.072	2.449	0.408	12.01	0.083	2.116	0.473
4	0.102	18.07	0.055	3.184	0.314	15.59	0.064	2.747	0.364
5	0.127	20.83	0.048	3.670	0.272	19.16	0.052	3.376	0.296
6	0.152	26.40	0.038	4.652	0.214	22.73	0.043	4.005	0.250
7	0.178	30.57	0.033	5.386	0.186	26.30	0.038	4.634	0.216
8	0.203	34.73	0.029	6.119	0.163	29.87	0.033	5.263	0.190
9	0.229	38.90	0.026	6.854	0.146	33.44	0.029	5.892	0.170
10	0.254	41.67	0.024	7.342	0.136	37.01	0.027	6.521	0.153
11	0.297	47.23	0.021	8.322	0.120	40.58	0.024	7.150	0.140
12	0.305	51.40	0.019	9.057	0.110	44.16	0.023	7.781	0.129

The thickness of insulation used may be limited to joists height.

R = °F, ft^2, hr, Btu R_m = km^{2}W   U = Btu per ft^2 per hr U_m = W/m^{2}K

TABLE 113

CEILINGS
UNDER VENTED SPACE

26. MINERAL FIBER, RESIN BINDER, BATT OR BLANKET INSULATION
27. CELLULOSE FIBER BATT OR LOOSE FILL INSULATION

THERMAL RESISTANCE AND TRANSMITTANCE OF ASSEMBLY XII WITH MINERAL FIBER OR CELLULOSE INSULATION: k = 0.30 Btu, in./ft^2, hr °F at 40°F, k_m = 0.053 W/mK at 4.4°C, k = 0.33 Btu, in./ft^2, hr °F at 85°F, k_m = 0.058 W/mK at 29.4°C

Insulation Thickness		Mean Temp. 40°F (4.4°C) Heat Flow Up				Mean Temp. 85°F (29.4°C) Heat Flow Down			
		English Units		Metric Units		English Units		Metric Units	
Inches	Meters	R	U	R_m	U_m	R	U	R_m	U_m
1	0.025	4.73	0.211	0.833	1.199	4.33	0.231	0.763	1.311
2	0.051	8.07	0.123	1.422	0.703	7.36	0.136	1.296	0.771
3	0.076	11.40	0.088	2.009	0.498	10.39	0.096	1.831	0.546
4	0.102	14.73	0.068	2.595	0.385	13.42	0.074	2.365	0.422
5	0.127	18.07	0.055	3.184	0.314	16.45	0.061	2.898	0.345
6	0.152	21.40	0.047	3.771	0.265	19.48	0.051	3.432	0.291
7	0.178	24.73	0.040	4.357	0.229	22.51	0.044	3.966	0.252
8	0.203	28.07	0.036	4.946	0.202	25.54	0.039	4.500	0.222
9	0.229	31.40	0.031	5.533	0.181	28.57	0.035	5.034	0.198
10	0.254	34.70	0.028	6.114	0.164	31.60	0.032	5.568	0.180
11	0.279	38.07	0.026	6.708	0.149	34.63	0.028	6.102	0.164
12	0.305	41.40	0.024	7.224	0.138	37.66	0.027	6.636	0.151

R = °F, ft^2, hr, Btu R_m = km^{2}W   U = Btu per ft^2 per hr U_m = W/m^{2}K

TABLE 114

27. VERMICULITE (EXPANDED) LOOSE INSULATION

THERMAL RESISTANCE AND TRANSMITTANCE OF ASSEMBLY XII WITH VERMICULITE LOOSE FILL INSULATION: k = 0.47 Btu, in./ft^2, hr °F at 40°F, k_m = 0.068 W/mK at 4.4°C, k = 0.48 Btu, in./ft^2, hr °F at 85°F, k_m = 0.069 W/mK at 29.4°C

Insulation Thickness		Mean Temp. 40°F (4.4°C) Heat Flow Up				Mean Temp. 85°F (29.4°C) Heat Flow Down			
		English Units		Metric Units		English Units		Metric Units	
Inches	Meters	R	U	R_m	U_m	R	U	R_m	U_m
1	0.025	3.52	0.284	0.620	1.612	3.48	0.287	0.613	1.630
2	0.051	5.65	0.177	0.995	1.004	5.57	0.179	0.981	1.019
3	0.076	7.78	0.128	1.371	0.729	7.65	0.131	1.348	0.742
4	0.102	10.91	0.092	1.922	0.520	9.73	0.103	1.714	0.583
5	0.127	12.04	0.083	2.121	0.471	11.81	0.085	2.080	0.480
6	0.152	14.17	0.071	2.497	0.400	13.90	0.072	2.449	0.408
7	0.178	16.29	0.061	2.870	0.348	15.98	0.063	2.816	0.355
8	0.203	18.42	0.054	3.246	0.308	18.07	0.055	3.178	0.314
9	0.229	20.55	0.049	3.621	0.276	20.14	0.049	3.549	0.282
10	0.254	22.67	0.044	3.994	0.250	22.23	0.045	3.917	0.255
11	0.279	24.80	0.040	4.360	0.229	24.32	0.041	4.285	0.233
12	0.305	26.93	0.037	4.745	0.210	26.40	0.037	4.652	0.214

The thickness of insulation used may be limited to joist height.

R = °F, ft^2, hr, Btu R_m = km^{2}W   U = Btu per ft^2 per hr U_m = W/m^{2}K

TABLE 115

CEILINGS
UNDER VENTED SPACE

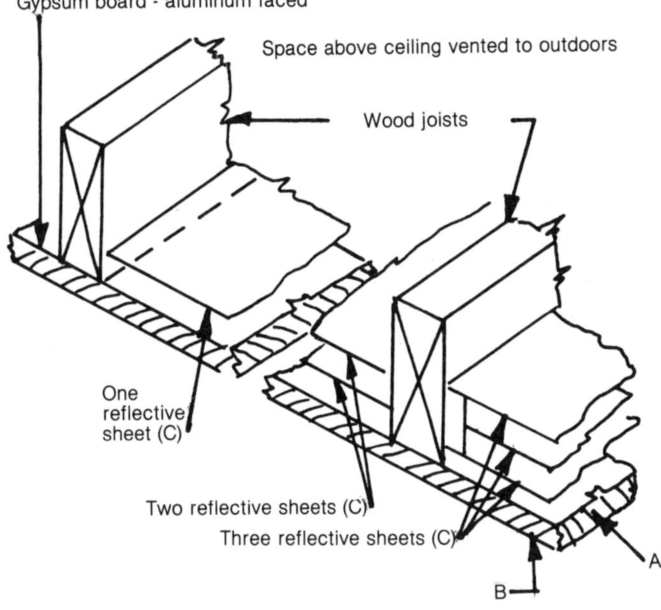

NO THERMAL INSULATION
THERMAL RESISTANCE R and R_m

Heat Flow Up — Heating Season

	Components	R		R_m
A	Gypsum board or plaster	0.45		0.079
B	Inside surface	0.60		0.106
C	Reflective insulation	—		—
	(TOTAL) RESISTANCE R =	1.05	R_m =	0.185
	TRANSMITTANCE U =	0.952	U_m =	5.405

Heat Flow Down — Cooling Season

	Components	R		R_m
A	Gypsum board or plaster	0.45		0.079
B	Inside surface	0.65		0.114
C	Reflective insulation	—		—
	(TOTAL) RESISTANCE R =	1.10	R_m =	0.193
	TRANSMITTANCE U =	0.909	U_m =	5.181

CONSTRUCTION ASSEMBLY XIII

28. ALUMINUM FACED PLASTER BOARD — Reflective Side Facing Up
Heat flow up $R = 3.46, R_m = 0.609$
Heat flow down $R = 3.24, R_m = 0.571$
THERMAL RESISTANCE AND TRANSMITTANCE OF ASSEMBLY XIII WITH ONE ALUMINUM FOIL ON PLASTER BOARD FACING UP

Number of Reflective Surfaces	Heat Flow Up (Heating Season) English Units R	U	Metric Units R_m	U_m	Heat Flow Down (Cooling Season) English Units R	U	Metric Units R_m	U_m
One-facing up	4.51	0.222	0.796	1.256	4.34	0.230	0.764	1.307

R = °F, ft^2, hr, Btu R_m = km^2W U = Btu per ft^2 per hr U_m = W/m^2K

TABLE 116

29. ALUMINUM REFLECTIVE LAMINATED SHEET (OR SHEETS)
THERMAL RESISTANCE AND TRANSMITTANCE OF ASSEMBLY XIII WITH REFLECTIVE INSULATION SHEETS AS SHOWN (ε = 0.05)

Number of Enclosed Air Spaces	Aluminum Reflective Sheets (Reflective on Both Sides)	Mean Temp. 40°F (4.4°C), Heat Flow Up English Units R	U	Metric Units R_m	U_m	Mean Temp. 85°F (29.4°C), Heat Flow Down English Units R	U	Metric Units R_m	U_m
1	1 (between joists)	6.35	0.157	1.119	0.894	8.78	0.113	1.547	0.646
2	2 (between joists)	11.75	0.085	2.070	0.483	16.36	0.061	2.883	0.347
3	3 (between joists)	17.15	0.058	3.022	0.331	23.92	0.042	4.215	0.237

TABLE 117

WALLS

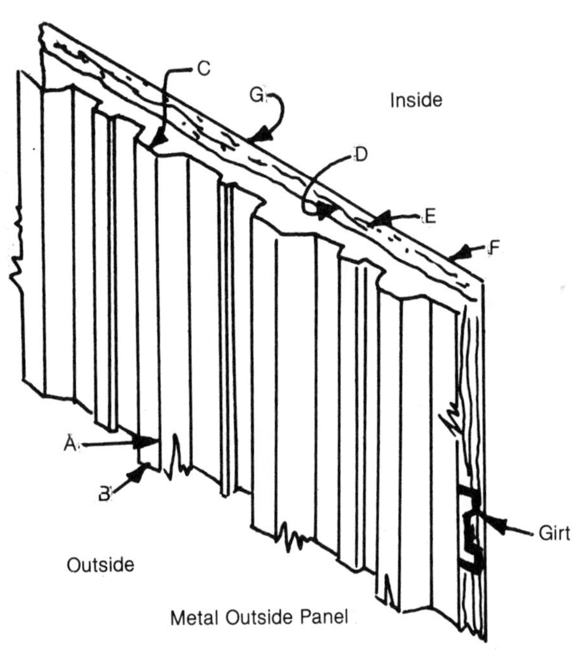

CONSTRUCTION ASSEMBLY XIV

Note: Outside and inside panels available in many configurations. Due to differential temperatures between inside and outside, expansion-contraction joints must be provided at ends of panels.

NO THERMAL INSULATION
THERMAL RESISTANCE R and R_m

Mean Temperature 40° (4.4°C), Wind 10 mph (16.1 km/hr)

	Components	Eng. Units		Metric Units
A	Outside surface	0.18		0.032
B	Metal panel	0.01		0.002
C	Inner surface of panel	0.40		0.070
D	Inner cavity surface	0.45		0.079
E	Insulation	—		—
F	Liner panel	0.01		0.002
G	Inside surface	0.60		0.106
	(TOTAL) RESISTANCE R =	1.65	R_m =	0.292
	TRANSMITTANCE U =	0.606	U_m =	3.425

Mean Temperature 85° (29.4°C), Wind 5 mph (8 km/hr)

	Components	Eng. Units		Metric Units
A	Outside surface	0.17		0.030
B	Metal panel	0.01		0.002
C	Inner surface of panel	0.36		0.063
D	Inner cavity surface	0.40		0.070
E	Insulation	—		—
F	Liner panel	0.01		0.002
G	Inside surface	0.52		0.092
	(TOTAL) RESISTANCE R =	1.47	R_m =	0.259
	TRANSMITTANCE U =	0.680	U_m =	3.861

Mean Temperature 110° (43.2°C), No Wind

	Components	Eng. Units		Metric Units
A	Outside surface	0.14		0.025
B	Metal panel	0.01		0.002
C	Inner surface of panel	0.32		0.056
D	Inner cavity surface	0.40		0.070
E	Insulation	—		—
F	Liner panel	0.01		0.002
G	Inside surface	0.50		0.088
	(TOTAL) RESISTANCE R =	1.38	R_m =	0.243
	TRANSMITTANCE U =	0.725	U_m =	4.113

30. LONG GLASS FIBER, PHENOLIC RESIN BINDER TEXTILE BLANKET THERMAL RESISTANCE AND TRANSMITTANCE OF ASSEMBLY XIV WITH LONG GLASS FIBER BLANKET: k = 0.34 Btu, in./ft², hr °F at 40°F, k_m = 0.058 W/mK at 4.4°C, k = 0.37 Btu, in./ft², hr °F at 85°F, k_m = 0.065 W/mK at 29.4°C, k = 0.41 Btu, in./ft², hr °F at 110°F, k_m = 0.072 W/mK at 43.2°C

Insulation Thickness		Mean Temp. 40°F (4.4°C)				Mean Temp. 85°F (29.4°C)				Mean Temp. 110°F (43.2°C)			
		English Units		Metric Units		English Units		Metric Units		English Units		Metric Units	
Inches	Meters	R	U	R_m	U_m	R	U	R_m	U_m	R	U	R_m	U_m
1	0.025	4.59	0.217	0.809	1.236	4.17	0.239	0.735	1.361	3.81	0.262	0.671	1.490
1 1/2	0.037	6.06	0.164	1.068	0.937	5.52	0.181	0.973	1.028	5.03	0.198	0.886	1.128
2	0.051	7.53	0.132	1.327	0.734	6.88	0.145	1.212	0.825	6.25	0.159	1.101	0.908
3	0.076	10.47	0.095	1.845	0.542	9.58	0.104	1.690	0.592	8.70	0.115	1.533	0.652
4	0.102	13.41	0.075	2.363	0.423	12.28	0.081	2.164	0.462	11.14	0.089	1.963	0.509

R = °F, ft², hr, Btu R_m = km²W U = Btu per ft² per hr U_m = W/m²K

TABLE 118

WALLS

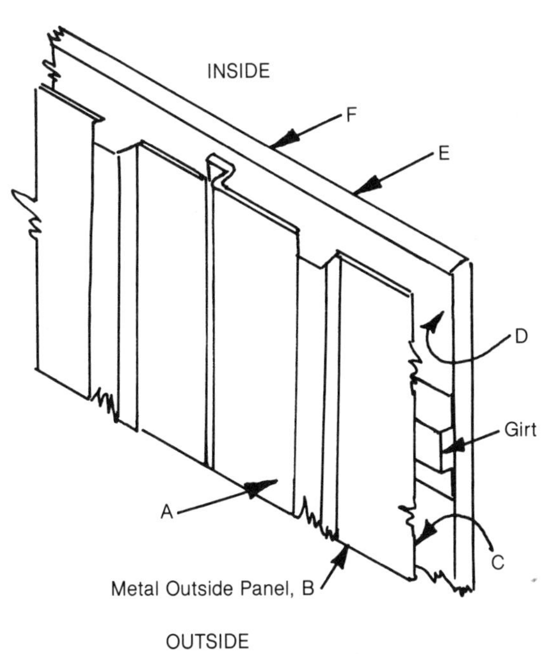

CONSTRUCTION ASSEMBLY XV

Note: Outside and inside panels available in many configurations.

Due to differential temperatures between inside and outside, expansion-contraction joints must be provided at ends of panels

NO THERMAL INSULATION
THERMAL RESISTANCE R and R_m

Mean Temperature 40° (4.4°C), Wind 10 mph (16.1 km/hr)

	Components	Eng. Units		Metric Units
A	Outside surface	0.18		0.032
B	Metal panel (outside)	0.01		0.002
C	Inner surface of panel	0.40		0.070
D	Inner cavity surface	0.45		0.079
E	Insualted panel	—		—
F	Inside surface	0.60		0.106
	(TOTAL) RESISTANCE R =	1.64	R_m =	0.290
	TRANSMITTANCE U =	0.610	U_m =	3.443

Mean Temperature 85° (29.4°C), Wind 5 mph (8 km/hr)

	Components	Eng. Units		Metric Units
A	Outside surface	0.17		0.030
B	Metal panel (outside)	0.01		0.002
C	Inner surface of panel	0.36		0.063
D	Inner cavity surface	0.40		0.070
E	Insulated panel	—		—
F	Inside surface	0.52		0.092
	(TOTAL) RESISTANCE R =	1.46	R_m =	0.267
	TRANSMITTANCE U =	0.685	U_m =	3.745

Mean Temperature 110° (43.2°C), No Wind

	Components	Eng. Units		Metric Units
A	Outside surface	0.14		0.025
B	Metal panel (outside)	0.01		0.002
C	Inner surface of panel	0.32		0.056
D	Inner cavity surface	0.40		0.070
E	Insulated panel	—		—
F	Inside surface	0.50		0.088
	(TOTAL) RESISTANCE R =	1.37	R_m =	0.241
	TRANSMITTANCE U =	0.729	U_m =	4.149

31. PRE-INSULATED LINER PANEL (MINERAL FIBER INSULATION) THERMAL RESISTANCE AND TRANSMITTANCE OF ASSEMBLY XV WITH PREINSULATED PANEL

CONDUCTIVITY OF INSULATION ONLY: k = 0.34 Btu, in./ft^2, hr °F, k_m = 0.058 W/mK at 4.9°C, k = 0.37 Btu, in./ft^2, hr °F at 85°F, k_m = 0.065 W/mK at 29.4°C, k = 0.41 Btu, in./ft^2, hr °F at 110°F, k_m = 0.072 W/mK at 43.2°C

Insulation Thickness		Mean Temp. 40°F (4.4°C)				Mean Temp. 85°F (29.4°C)				Mean Temp. 110°F (43.2°C)			
		English Units		Metric Units		English Units		Metric Units		English Units		Metric Units	
Inches	Meters	R	U	R_m	U_m	R	U	R_m	U_m	R	U	R_m	U_m
1 1/2	0.037	6.05	0.165	1.066	0.938	5.51	0.181	0.970	1.030	5.02	0.199	0.884	1.130
2	0.051	7.52	0.133	1.325	0.755	6.86	0.146	1.209	0.827	6.24	0.160	1.099	0.909

R = °F, ft^2, hr, Btu R_m = km^{2}W U = Btu per ft^2 per hr U_m = W/m^{2}K

TABLE 119

WALLS

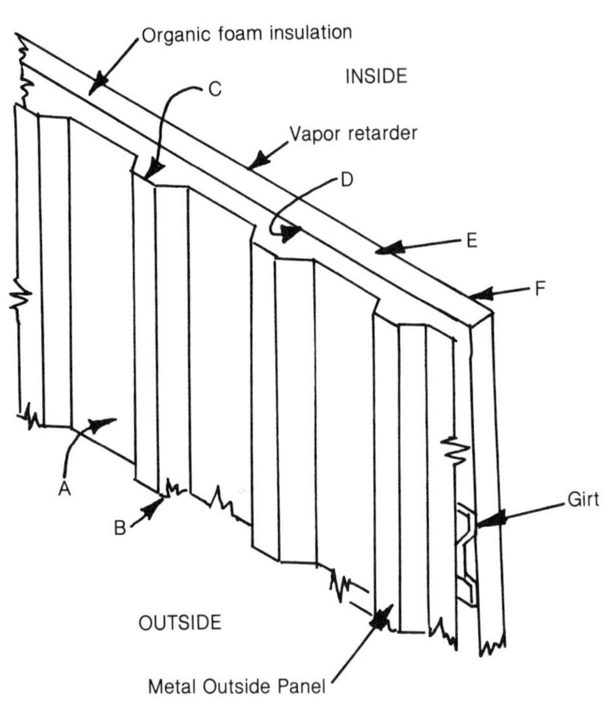

CONSTRUCTION ASSEMBLY XVI

Note: If organic foam is faced with film or foil, the inner face must be the vapor retarder. Any foil on outer face must be perforated. Outer metal panel or use of inner metal panel will not provide fire protection of organic foam.

NO THERMAL INSULATION
THERMAL RESISTANCE R and R_m

Mean Temperature 40° (4.4°C), Wind 10 mph (16.1 km/hr)

	Components	Eng. Units		Metric Units
A	Outside surface	0.18		0.032
B	Metal panel	0.01		0.002
C	Inner surface of panel	0.40		0.070
D	In insulation calculations	—		—
E	In insulation calculations	—		—
F	In insulation calculations	—		—
	(TOTAL) RESISTANCE R =	0.59	R_m =	0.104
	TRANSMITTANCE U =	1.695	U_m =	9.615

Mean Temperature 85° (29.4°C), Wind 5 mph (8 km/hr)

	Components	Eng. Units		Metric Units
A	Outside surface	0.17		0.030
B	Metal panel	0.01		0.002
C	Inner surface of panel	0.36		0.063
D	In insulation calculations	—		—
E	In insulation calculations	—		—
F	In insulation calculations	—		—
	(TOTAL) RESISTANCE R =	0.54	R_m =	0.095
	TRANSMITTANCE U =	1.85	U_m =	10.526

Mean Temperature 110° (43.2°C), No Wind

	Components	Eng. Units		Metric Units
A	Outside surface	0.14		0.025
B	Metal panel	0.01		0.002
C	Inner surface of panel	0.32		0.056
D	In insulation calculations	—		—
E	In insulation calculations	—		—
F	In insulation calculations	—		—
	(TOTAL) RESISTANCE R =	0.47	R_m =	0.083
	TRANSMITTANCE U =	2.127	U_m =	12.048

32. RIGID POLYURETHANE OR POLYISOCYANURATE FOAM SHEETING THERMAL RESISTANCE AND TRANSMITTANCE OF ASSEMBLY XVI WITH URETHANE (OR ISOCYANURATE) RIGID SHEETING: k = 0.24 Btu, in./ft², hr °F at 40°F, k_m = 0.042 W/mK at 4.4°C, k = 0.30 Btu, in./ft², hr °F at 85°F, k_m = 0.053 W/mK at 29.4°C, k = 0.33 Btu, in./ft², hr °F at 110°F, k_m = 0.058 W/mK at 43.2°C

Insulation Thickness		Mean Temp. 40°F (4.4°C)				Mean Temp. 85°F (29.4°C)				Mean Temp. 110°F (43.2°C)			
		English Units		Metric Units		English Units		Metric Units		English Units		Metric Units	
Inches	Meters	R	U	R_m	U_m	R	U	R_m	U_m	R	U	R_m	U_m
3/8	0.009	3.21	0.311	0.566	1.768	2.72	0.368	0.479	2.086	2.52	0.397	0.444	2.252
3/4	0.018	4.77	0.209	0.840	1.169	3.97	0.252	0.699	1.429	3.65	0.274	0.643	1.554
1	0.025	5.82	0.172	1.025	0.975	4.80	0.208	0.846	1.182	4.41	0.227	0.777	1.287
1 1/2	0.037	7.90	0.127	1.392	0.718	6.47	0.155	1.140	0.877	5.92	0.169	1.043	0.959
2 1/4	0.056	11.03	0.091	1.943	0.514	8.97	0.111	1.581	0.623	8.20	0.122	1.445	0.692

R = °F, ft², hr, Btu R_m = km²W   U = Btu per ft² per hr U_m = W/m²K

TABLE 120

WALLS

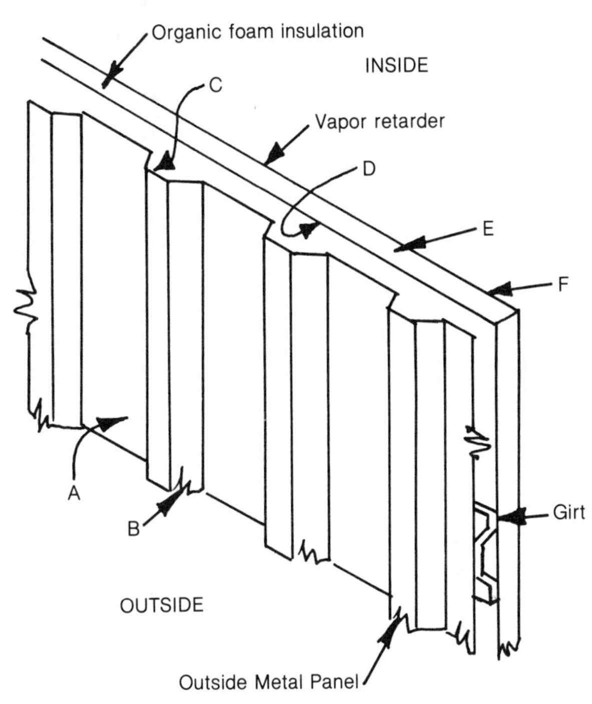

CONSTRUCTION ASSEMBLY XVI

Note: If organic foam is faced with film or foil, the inner face must be the vapor retarder. Any foil on outer face must be perforated. Outer metal panel or use of inner metal panel will not provide fire protection of organic foam.

NO THERMAL INSULATION THERMAL RESISTANCE R and R_m

Mean Temperature 40° (4.4°C), Wind 10 mph (16.1 km/hr)

	Components	Eng. Units		Metric Units
A	Outside surface	0.18		0.032
B	Metal panel	0.01		0.002
C	Inner surface of panel	0.40		0.070
D	In insulation calculations	—		—
E	In insulation calculations	—		—
F	In insulation calculations	—		—
	(TOTAL) RESISTANCE R =	0.59	R_m =	0.104
	TRANSMITTANCE U =	1.695	U_m =	9.615

Mean Temperature 85° (29.4°C), Wind 5 mph (8 km/hr)

	Components	Eng. Units		Metric Units
A	Outside surface	0.17		0.030
B	Metal panel	0.01		0.002
C	Inber surface of panel	0.36		0.063
D	In insulation calculations	—		—
E	In insulation calculations	—		—
F	In insulation calculations	—		—
	(TOTAL) RESISTANCE R =	0.54	R_m =	0.095
	TRANSMITTANCE U =	1.85	U_m =	10.526

Mean Temperature 110° (43.2°C), No Wind

	Components	Eng. Units		Metric Units
A	Outside surface	0.14		0.025
B	Metal panel	0.01		0.002
C	Inner surface of panel	0.32		0.056
D	In insulation calculations	—		—
E	In insulation calculations	—		—
F	In insulation calculations	—		—
	(TOTAL) RESISTANCE R =	0.47	R_m =	0.083
	TRANSMITTANCE U =	2.127	U_m =	12.048

33. POLYSTYRENE FOAM BOARD THERMAL RESISTANCE AND TRANSMITTANCE OF ASSEMBLY XVI WITH POLYSTYRENE FOAM BOARD INSULATION:

k = 0.20 Btu, in./ft², hr °F at 40°F, k_m = 0.035 W/mK at 4.4°C, k = 0.24 Btu, in./ft², hr °F at 85°F, k_m = 0.044 W/mK at 29.4°C, k = 0.28 Btu, in./ft², hr °F at 110°F, k_m = 0.049 W/mK at 43.2°C

Insulation Thickness		Mean Temp. 40°F (4.4°C)				Mean Temp. 85°F (29.4°C)				Mean Temp. 110°F (43.2°C)			
		English Units		Metric Units		English Units		Metric Units		English Units		Metric Units	
Inches	Meters	R	U	R_m	U_m	R	U	R_m	U_m	R	U	R_m	U_m
1 1/2	0.037	9.15	0.109	1.612	0.620	7.72	0.129	0.360	0.735	6.74	0.148	1.188	0.842
2	0.051	11.65	0.086	2.053	0.487	9.80	0.102	1.727	0.579	8.52	0.117	1.501	0.666
2 1/2	0.064	14.15	0.071	2.493	0.401	11.89	0.084	2.095	0.477	10.31	0.097	1.817	0.550
3	0.076	16.65	0.060	2.934	0.341	13.97	0.072	2.462	0.406	12.09	0.082	2.130	0.469

R = °F, ft², hr, Btu R_m = km²W U = Btu per ft² per hr U_m = W/m²K

TABLE 121

WALLS

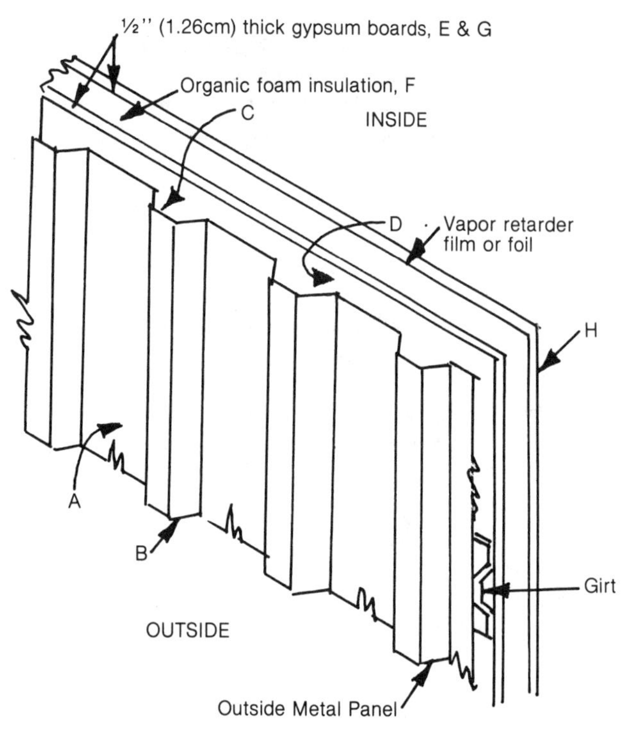

CONSTRUCTION ASSEMBLY XVII

Note: To be effective for fire protection from the combustible organic foam the foam must be completely enclosed — sides, ends and edges by a minimum of 1/2" (1.27 cm) thick gypsum board.

NO THERMAL INSULATION
THERMAL RESISTANCE R and R_m

Mean Temperature 40° (4.4°C), Wind 10 mph (16.1 km/hr)

Components	Eng. Units		Metric Units
A Outside surface	0.18		0.032
B Metal panel	0.01		0.002
C Inner surface of panel	0.40		0.070
D, E, F, G and H in insulation cal.	—		—
(TOTAL) RESISTANCE R =	0.59	R_m =	0.104
TRANSMITTANCE U =	1.695	U_m =	9.615

Mean Temperature 85° (29.4°C), Wind 5 mph (8 km/hr)

Components	Eng. Units		Metric Units
A Outside surface	0.17		0.030
B Metal panel	0.01		0.002
C Inner surface of panel	0.36		0.063
D, E, F, G and H in insulation cal.	—		—
(TOTAL) RESISTANCE R =	0.54	R_m =	0.095
TRANSMITTANCE U =	0.000	U_m =	00.000

Mean Temperature 110° (43.2°C), No Wind

Components	Eng. Units		Metric Units
A Outside surface	0.14		0.025
B Metal panel	0.01		0.002
C Inner surface of panel	0.32		0.056
D, E, F, G and H in insulation cal.	—		—
(TOTAL) RESISTANCE R =	0.47	R_m =	0.088
TRANSMITTANCE U =	2.127	U_m =	12.048

34. RIGID POLYURETHANE OR POLYISCYANURATE FOAM AND GYPSUM BOARD THERMAL RESISTANCE AND TRANSMITTANCE OF ASSEMBLY XVII WITH URETHANE (OR ISCYANURATE) RIGID SHEETING: k = 0.24 Btu, in./ft², hr °F at 40°F, k_m = 0.042 W/mK at 4.4°C, k = 0.30 Btu, in./ft², hr °F at 85°F, k_m = 0.053 W/mK at 29.4°C, k = 0.33 Btu, in./ft², hr °F at 110°F, k_m = 0.058 W/mK at 43.2°C
GYPSUM BOARD: k = 3.00 Btu, in./ft², hr °F at 40°F, k_m = 0.529 W/mK at 4.4°C, k = 3.06 Btu, in./ft², hr °F at 85°F, k_m = 0.539 W/mK at 29.4°C, k = 3.11 Btu, in./ft², hr °F at 110°F, k_m = 0.578 W/mK at 43.2°C

Insulation Thickness*		Mean Temp. 40°F (4.4°C)				Mean Temp. 85°F (29.4°C)				Mean Temp. 110°F (43.2°C)			
		English Units		Metric Units		English Units		Metric Units		English Units		Metric Units	
Inches	Meters	R	U	R_m	U_m	R	U	R_m	U_m	R	U	R_m	U_m
3/8	0.009	3.54	0.282	0.624	1.603	3.04	0.329	0.536	1.867	2.84	0.352	0.500	2.000
3/4	0.018	5.10	0.196	0.899	1.113	4.29	0.233	0.756	1.323	3.97	0.252	0.699	1.429
1	0.025	6.15	0.163	1.084	0.923	5.12	0.195	0.902	1.108	4.73	0.211	0.833	1.200
1 1/2	0.037	8.23	0.115	1.450	0.690	6.79	0.147	1.196	0.836	6.24	0.160	1.099	0.909
2 1/4	0.056	11.36	0.088	2.002	0.499	9.29	0.108	1.637	0.611	8.52	0.117	1.501	0.666

*Insulation is enclosed by 1/2" (1.27 cm) gypsum board.

R = °F, ft², hr, Btu R_m = km²W   U = Btu per ft² per hr U_m = W/m²K

TABLE 122

WALLS

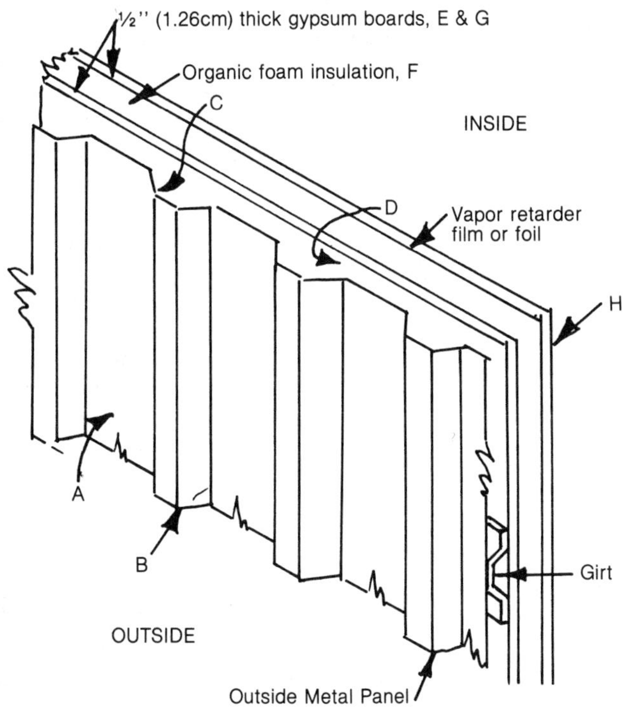

CONSTRUCTION ASSEMBLY XVII

Note: To be effective for fire protection from the combustible organic foam the foam must be completely enclosed — sides, ends and edges by a minimum of 1/2" (1.27 cm) thick gypsum board.

NO THERMAL INSULATION
THERMAL RESISTANCE R and R_m

Mean Temperature 40° (4.4°C), Wind 10 mph (16.1 km/hr)

Components	Eng. Units		Metric Units
A Outside surface	0.18		0.032
B Metal panel	0.01		0.002
C Inner surface of panel	0.40		0.070
D, E, F, G and H in insulation cal.	—		—
(TOTAL) RESISTANCE R =	0.59	R_m =	0.104
TRANSMITTANCE U =	1.695	U_m =	9.615

Mean Temperature 85° (29.4°C), Wind 5 mph (8 km/hr)

Components	Eng. Units		Metric Units
A Outside surface	0.17		0.030
B Metal panel	0.01		0.002
C Inner surface of panel	0.36		0.063
D, E, F, G and H in insulation cal.	—		—
(TOTAL) RESISTANCE R =	0.54	R_m =	0.095
TRANSMITTANCE U =	0.000	U_m =	00.000

Mean Temperature 110° (43.2°C), No Wind

Components	Eng. Units		Metric Units
A Outside surface	0.14		0.025
B Metal panel	0.01		0.002
C Inner surface of panel	0.32		0.056
D, E, F, G and H in insulation cal.	—		—
(TOTAL) RESISTANCE R =	0.47	R_m =	0.088
TRANSMITTANCE U =	2.127	U_m =	12.048

35. POLYSTYRENE FOAM BOARD AND GYPSUM BOARD
THERMAL RESISTANCE AND TRANSMITTANCE OF ASSEMBLY XVII WITH POLYSTYRENE

FOAM SHEETING: k = 0.20 Btu, in./ft^2, hr °F at 40°F, k_m = 0.035 W/mK at 4.4°C, k = 0.24 Btu, in./ft^2, hr °F at 85°F, k_m = 0.044 W/mK at 29.4°C, k = 0.28 Btu, in./ft^2, hr °F at 110°F, k_m = 0.049 W/mK at 43.2°C

GYPSUM BOARD: k = 3.00 Btu, in./ft^2, hr °F at 40°F, k_m = 0.529 W/mK at 4.4°C, k = 3.06 Btu, in./ft^2, hr °F at 85°F, k_m = 0.539 W/mK at 29.4°C, k = 3.11 Btu, in./ft^2, hr °F at 110°F, k_m = 0.578 W/mK at 43.2°C

Insulation Thickness*		Mean Temp. 40°F (4.4°C)				Mean Temp. 85°F (29.4°C)				Mean Temp. 110°F (43.2°C)			
		English Units		Metric Units		English Units		Metric Units		English Units		Metric Units	
Inches	Meters	R	U	R_m	U_m	R	U	R_m	U_m	R	U	R_m	U_m
1 1/2	0.037	9.48	0.105	1.670	0.598	8.04	0.124	1.417	0.706	7.06	0.142	1.244	0.804
2	0.051	11.98	0.083	2.111	0.474	10.12	0.099	1.783	0.561	8.84	0.113	1.558	0.642
2 1/2	0.064	14.48	0.069	2.551	0.391	12.21	0.082	2.151	0.465	10.63	0.094	1.873	0.533
3	0.076	16.98	0.058	2.992	0.334	14.29	0.070	2.158	0.397	12.41	0.081	2.187	0.457

*Insulation is enclosed by 1/2" (1.27 cm) gypsum board.

R = °F, ft^2, hr, Btu R_m = km^{2}W   U = Btu per ft^2 per hr U_m = W/m^{2}K

TABLE 123

WALLS

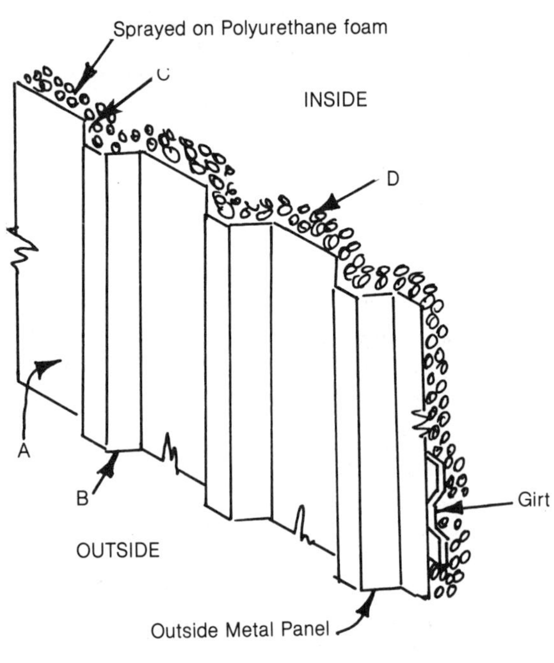

CONSTRUCTION ASSEMBLY XVIII

Note: Inner surface should be coated with vapor retarder - fire retardant finish

NO THERMAL INSULATION
THERMAL RESISTANCE R and R_m

Mean Temperature 40° (4.4°C), Wind 10 mph (16.1 km/hr)

Components	Eng. Units		Metric Units
A Outside surface	0.18		0.032
B Metal panel	0.01		0.002
C Sprayed on insulation	—		—
D Inside surface	0.60		0.106
(TOTAL) RESISTANCE R =	0.79	R_m =	0.140
TRANSMITTANCE U =	1.266	U_m =	7.143

Mean Temperature 85° (29.4°C), Wind 5 mph (8 km/hr)

Components	Eng. Units		Metric Units
A Outside surface	0.17		0.030
B Metal panel	0.01		0.002
C Sprayed on insulation	—		—
D Inside surface	0.52		0.092
(TOTAL) RESISTANCE R =	0.70	R_m =	0.124
TRANSMITTANCE U =	1.424	U_m =	8.065

Mean Temperature 110° (43.2°C), No Wind

Components	Eng. Units		Metric Units
A Outside surface	0.16		0.025
B Metal panel	0.01		0.002
C Sprayed on insulation	—		—
D Inside surface	0.50		0.089
(TOTAL) RESISTANCE R =	0.67	R_m =	0.115
TRANSMITTANCE U =	1.493	U_m =	8.696

35. SPRAYED ON POLYURETHANE INSULATION
THERMAL RESISTANCE AND TRANSMITTANCE OF ASSEMBLY XVIII SPRAYED ON POLYURETHANE INSULATION

k = 0.24 Btu, in./ft², hr °F at 40°F, k_m = 0.035 W/mK at 4.4°C, k = 0.25 Btu, in./ft², hr °F at 85°F, k_m = 0.036 W/mK at 29.4°C, k = 0.28 Btu, in./ft², hr °F at 110°F, k_m = 0.040 W/mK at 43.2°C

Insulation Thickness		Mean Temp. 40°F (4.4°C)				Mean Temp. 85°F (29.4°C)				Mean Temp. 110°F (43.2°C)			
		English Units		Metric Units		English Units		Metric Units		English Units		Metric Units	
Inches	Meters	R	U	R_m	U_m	R	U	R_m	U_m	R	U	R_m	U_m
1	0.025	4.95	0.202	0.872	1.146	4.70	0.213	0.828	1.207	4.24	0.236	0.747	1.339
1 1/2	0.038	7.04	0.142	1.240	0.806	6.70	0.149	1.181	0.847	6.03	0.166	1.062	0.941
2	0.051	9.12	0.109	1.607	0.622	8.70	0.115	1.533	0.652	7.81	0.128	1.376	0.727
2 1/2	0.064	11.21	0.089	1.975	0.506	10.70	0.093	1.885	0.530	9.60	0.104	1.692	0.591
3	0.076	13.23	0.075	2.331	0.429	12.70	0.079	2.234	0.447	11.38	0.087	2.005	0.499
3 1/2	0.088	15.37	0.065	2.708	0.369	14.70	0.068	2.590	0.386	13.17	0.076	2.320	0.431
4	0.102	17.46	0.057	3.076	0.325	16.70	0.060	2.943	0.340	14.96	0.067	2.635	0.379

Note: Due to the high combustibility of polyurethane foam all welding or fire must be prohibited close to the insulated area. All electrical circuits should be kept away from the organic foam insulation.

R = °F, ft², hr, Btu R_m = km²W U = Btu per ft² per hr U_m = W/m²K

TABLE 124

WALLS

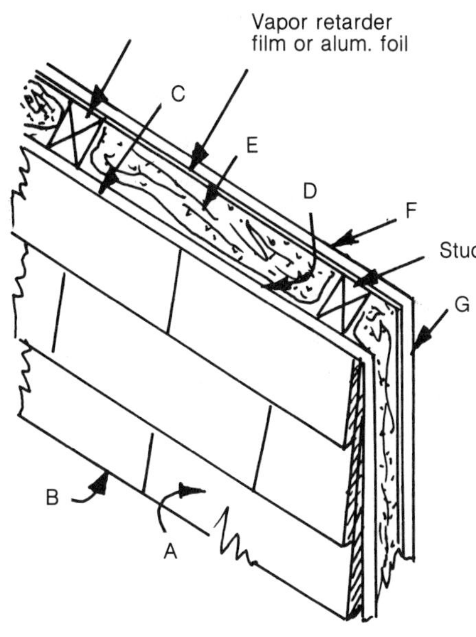

CONSTRUCTION ASSEMBLY XIX

Note: Do not put vapor retarder film or foil on sheathing or under siding (or shingles).

When facing is other than wood shingles or siding subtract the following from resistance calculation Tables 125 through 128.

FACING	SUBTRACT FROM R	SUBTRACT FROM R_m
Aluminum siding	0.20	0.035
Stucco	0.61	0.107
4″ brick	0.37	0.065

NO THERMAL INSULATION
THERMAL RESISTANCE R and R_m

Mean Temperature 40° (4.4°C), Wind 10 mph (16.1 km/hr)

	Components	Eng. Units		Metric Units
A	Outside surface	0.17		0.030
B	Shingles or siding	0.81		0.143
C	Sheathing	0.90		0.159
D	Inner surface of sheathing	0.45		0.079
E	Insulation	—		—
F	Gypsum wallboard	0.45		0.079
G	Inside surface	0.60		0.106
	(TOTAL) RESISTANCE R =	3.38	R_m =	0.596
	TRANSMITTANCE U =	0.296	U_m =	1.678

Mean Temperature 85° (29.4°C), Wind 8 mph (8 km/hr)

	Components	Eng. Units		Metric Units
A	Outside surface	0.17		0.030
B	Shingles or siding	0.75		0.132
C	Sheathing	0.85		0.150
D	Inner surface of sheathing	0.41		0.072
E	Insulation	—		—
F	Gypsum wallboard	0.43		0.075
G	Inside surface	0.58		0.102
	(TOTAL) RESISTANCE R =	3.19	R_m =	0.561
	TRANSMITTANCE U =	0.313	U_m =	1.783

Mean Temperature 110° (43.2°C), No Wind

	Components	Eng. Units		Metric Units
A	Outside surface	0.14		0.025
B	Shingles or siding	0.71		0.125
C	Sheathing	0.81		0.143
D	Inner surface of sheathing	0.38		0.067
E	Insulation	—		—
F	Gypsum wallboard	0.43		0.075
G	Inside surface	0.58		0.102
	(TOTAL) RESISTANCE R =	3.05	R_m =	0.537
	TRANSMITTANCE U =	0.327	U_m =	1.862

36. GLASS FIBER LIGHT DENSITY BATT OR BLANKET INSULATION THERMAL RESISTANCE AND TRANSMITTANCE OF ASSEMBLY XIX WITH GLASS FIBER INSULATION:

k = 0.24 Btu, in./ft², hr °F at 40°F, k_m = 0.035 W/mK at 4.4°C, k = 0.28 Btu, in./ft², hr °F at 85°F, k_m = 0.04 W/mK at 29.4°C, k = 0.34 Btu, in./ft², hr °F at 110°F, k_m = 0.049 W/mK at 43.2°C

Insulation Thickness		Mean Temp. 40°F (4.4°C)				Mean Temp. 85°F (29.4°C)				Mean Temp. 110°F (43.2°C)			
		English Units		Metric Units		English Units		Metric Units		English Units		Metric Units	
Inches	Meters	R	U	R_m	U_m	R	U	R_m	U_m	R	U	R_m	U_m
1	0.025	6.71	0.149	1.182	0.846	6.39	0.156	1.126	1.888	5.40	0.185	0.951	1.051
1 1/2	0.037	8.38	0.119	1.477	0.677	7.99	0.125	1.407	0.710	6.57	0.152	1.158	0.864
2	0.051	10.01	0.100	1.769	0.565	9.59	0.104	1.690	0.592	7.75	0.129	1.366	0.732
2 1/2	0.064	11.71	0.085	2.063	0.485	11.19	0.089	1.972	0.507	8.93	0.112	1.573	0.656
3	0.076	13.38	0.075	2.357	0.424	12.79	0.078	2.254	0.444	10.10	0.099	1.780	0.562
3 1/2	0.088	15.05	0.066	2.651	0.377	14.39	0.069	2.536	0.394	11.38	0.089	1.988	0.503
5 1/2*	0.138	21.71	0.046	3.825	0.261	20.79	0.048	3.663	0.273	16.02	0.062	2.823	0.354

*For walls with 2″ x 6″ studs.

R = °F, ft², hr, Btu R_m = km²W U = Btu per ft² per hr U_m = W/m²K

TABLE 125

WALLS

37. MINERAL FIBER, RESIN BINDER, BATT OR BLANKET INSULATION
38. CELLULOSE FIBER BATT INSULATION

THERMAL RESISTANCE AND TRANSMITTANCE OF ASSEMBLY XIX WITH MINERAL CELLULOSE FIBER BATT INSULATION: k = 0.30 Btu, in./ft^2, hr °F at 40 °F, k_m = 0.053 W/mK at 4.4 °C, k = 0.33 Btu, in./ft^2, hr °F at 85 °F, k_m = 0.058 W/mK at 29.4 °C, k = 0.37 Btu, in./ft^2, hr °F at 110 °F, k_m = 0.065 W/mK at 43.2 °C

Insulation Thickness		Mean Temp. 40 °F (4.4 °C)				Mean Temp. 85 °F (29.4 °C)				Mean Temp. 110 °F (43.2 °C)			
		English Units		Metric Units		English Units		Metric Units		English Units		Metric Units	
Inches	Meters	R	U	R_m	U_m	R	U	R_m	U_m	R	U	R_m	U_m
1	0.025	6.04	0.165	1.064	0.939	5.61	0.178	0.988	1.012	5.21	0.192	0.918	1.089
1 1/2	0.037	7.38	0.136	1.300	0.769	6.82	0.146	1.202	0.832	6.29	0.158	1.108	1.902
2	0.051	8.71	0.115	1.535	0.651	8.04	0.124	1.417	0.706	7.37	0.136	1.298	0.770
2 1/2	0.064	10.04	0.100	1.769	0.565	9.25	0.108	1.629	0.618	8.45	0.118	1.489	0.672
3	0.076	11.33	0.088	1.996	0.501	10.46	0.096	1.843	0.542	9.53	0.105	1.679	0.596
3 1/2	0.088	12.71	0.079	2.239	0.447	11.67	0.085	2.056	0.486	10.61	0.094	1.869	0.535
5 1/2*	0.138	18.05	0.055	3.180	0.314	16.52	0.061	2.911	0.344	14.94	0.066	2.632	0.380

*For walls with 2″ x 6″ studs.

R = °F, ft^2, hr, Btu R_m = km^{2}W U = Btu per ft^2 per hr U_m = W/m^{2}K

TABLE 126

39. VERMICULITE FLAKES FILL INSULATION

THERMAL RESISTANCE AND TRANSMITTANCE OF ASSEMBLY XIX WITH VERMICULITE FLAKE INSULATION FILLING THE STUD SPACE: k = 0.47 Btu, in./ft^2, hr °F at 40 °F, k_m = 0.0823 W/mK at 4.4 °C, k = 0.48 Btu, in./ft^2, hr °F at 85 °F, k_m = 0.085 W/mK at 29.4 °C, k = 0.50 Btu, in./ft^2, hr °F at 110 °F, k_m = 0.088 W/mK at 43.2 °C

Width of Stud Space		Mean Temp. 40 °F (4.4 °C)				Mean Temp. 85 °F (29.4 °C)				Mean Temp. 110 °F (43.2 °C)			
		English Units		Metric Units		English Units		Metric Units		English Units		Metric Units	
Inches	Meters	R	U	R_m	U_m	R	U	R_m	U_m	R	U	R_m	U_m
3 1/2	0.088	8.88	0.113	1.565	0.639	8.61	0.116	1.517	0.659	8.27	0.121	1.457	0.686
5 1/2	0.138	12.29	0.081	2.165	0.462	11.94	0.084	2.103	0.475	11.47	0.087	2.021	0.495

R = °F, ft^2, hr, Btu R_m = km^{2}W U = Btu per ft^2 per hr U_m = W/m^{2}K

TABLE 127

40. CELLULOSE FIBER LOOSE FILL INSULATION
41. MINERAL FIBER LOOSE FILL INSULATION
42. PERLITE LOOSE FILL INSULATION

THERMAL RESISTANCE AND TRANSMITTANCE OF ASSEMBLY XIX WITH CELLULOSE FIBER, MINERAL FIBER OR PERLITE INSULATION FILLING THE STUD SPACE:

k = 0.19 Btu, in./ft^2, hr °F at 40 °F, k_m = 0.027 W/mK at 4.4 °C, k = 0.24 Btu, in./ft^2, hr °F at 85 °F, k_m = 0.035 W/mK at 29.4 °C, k = 0.28 Btu, in./ft^2, hr °F at 110 °F, k_m = 0.040 W/mK at 43.2 °C

Width of Stud Space		Mean Temp. 40 °F (4.4 °C)				Mean Temp. 85 °F (29.4 °C)				Mean Temp. 110 °F (43.2 °C)			
		English Units		Metric Units		English Units		Metric Units		English Units		Metric Units	
Inches	Meters	R	U	R_m	U_m	R	U	R_m	U_m	R	U	R_m	U_m
3 1/2	0.088	9.29	0.108	1.638	0.611	8.87	0.113	1.562	0.640	8.50	0.118	1.498	0.667
5 1/2	0.138	12.93	0.077	2.278	0.439	12.34	0.081	2.175	0.460	11.83	0.084	2.086	0.479

Note: Calculations in Tables 125, 126, 127 and 128 include compensation for studs in walls.

R = °F, ft^2, hr, Btu R_m = km^{2}W U = Btu per ft^2 per hr U_m = W/m^{2}K

TABLE 128

WALLS

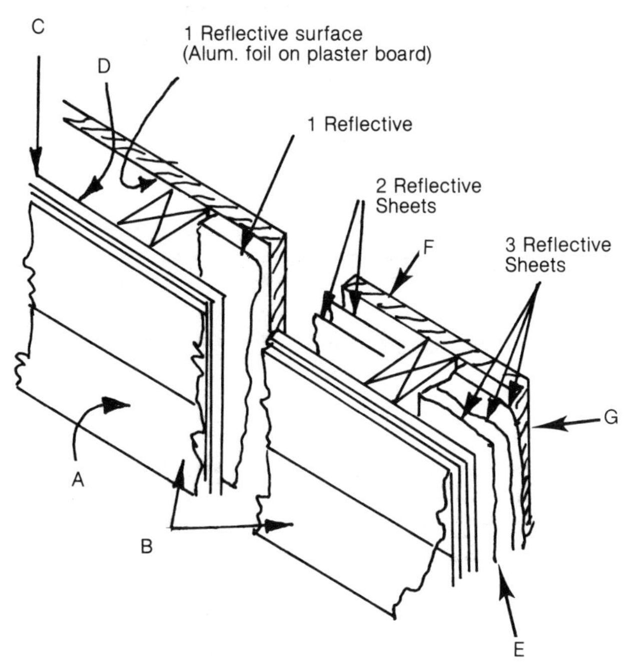

CONSTRUCTION ASSEMBLY XIX

Note: Do not put vapor retarder film or foil on sheathing or under siding (or shingles). Seal reflective sheets to studs to obtain vapor retarder to prevent vapor migration. All sheets show dividing stud space are reflective on both sides.

NO THERMAL INSULATION
THERMAL RESISTANCE R and R_m

Mean Temperature 40° (4.4°C), Wind 10 mph (16.1 km/hr)

	Components	Eng. Units		Metric Units
A	Outside surface	0.17		0.030
B	Shingles or siding	0.81		0.143
C	Sheathing	0.90		0.159
D	Inner surface of sheathing	0.45		0.079
E	Insulation, reflective	—		—
F	Gypsum wallboard	0.45		0.079
G	Inside surface	0.60		0.106
	(TOTAL) RESISTANCE R =	3.38	R_m =	0.596
	TRANSMITTANCE U =	0.296	U_m =	1.678

Mean Temperature 85° (29.4°C), Wind 5 mph (8 km/hr)

	Components	Eng. Units		Metric Units
A	Outside surface	0.17		0.030
B	Shingles or siding	0.75		0.132
C	Sheathing	0.85		0.150
D	Inner surface of sheating	0.41		0.072
E	Insulation, reflective	—		—
F	Gypsum wallboard	0.43		0.075
G	Inside surface	0.58		0.102
	(TOTAL) RESISTANCE R =	3.19	R_m =	0.561
	TRANSMITTANCE U =	0.313	U_m =	1.788

Mean Temperature 110° (43.2°C), No Wind

	Components	Eng. Units		Metric Units
A	Outside surface	0.14		0.025
B	Shingles or siding	0.71		0.125
C	Sheathing	0.81		0.143
D	Inner surface of sheathing	0.38		0.067
E	Insulation, reflective	—		—
F	Gypsum wallboard	0.43		0.075
G	Inside surface	0.58		0.102
	(TOTAL) RESISTANCE R =	3.05	R_m =	0.537
	TRANSMITTANCE U =	0.327	U_m =	1.862

43. REFLECTIVE ALUMINUM INSULATION (SHEETS OR FOIL) THERMAL RESISTANCE AND TRANSMITTANCE OF ASSEMBLY XIX WITH REFLECTIVE SHEETS AS SHOWN. $\varepsilon = 0.05$

No. of Air Spaces	Aluminum Reflective Sheets	Mean Temperature 40°F (4.4°C) English Units R	U	Metric Units R_m	U_m	Mean Temperature 85°F (29.4°) English Units R	U	Metric Units R_m	U_m	Mean Temperature 110°F (43.2°C) Engllish Units R	U	Metric Units R_m	U_m
1	faced by stud space divided by	6.02	0.166	1.061	0.943	5.72	0.175	1.001	0.992	5.49	0.182	0.967	1.033
2	1 sheet	6.60	0.151	1.163	0.860	6.31	0.158	1.111	0.899	6.08	0.164	1.071	0.933
3	2 sheets	10.04	0.099	1.769	0.565	9.44	0.106	1.663	0.601	8.93	0.112	1.573	0.636
4	3 sheets	13.48	0.074	2.375	0.421	12.57	0.075	2.215	0.452	11.78	0.085	2.076	0.481

TABLE 129

WALLS

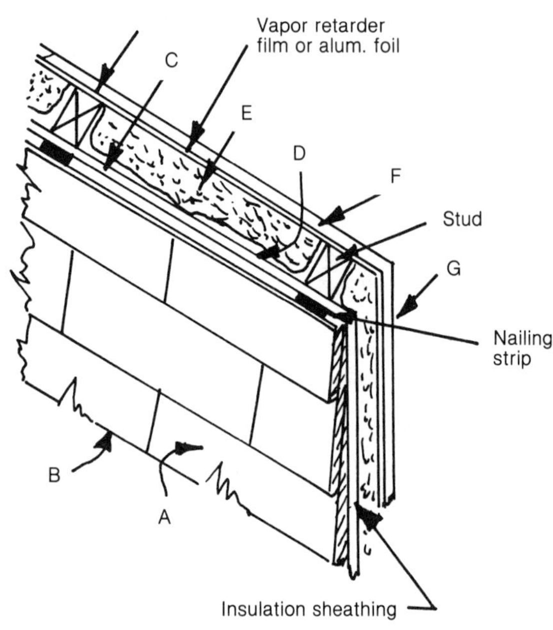

CONSTRUCTION ASSEMBLY XIX

Note: Do not put vapor retarder film or foil on sheathing or under siding (or shingles).

When facing is other than wood shingles or siding subtract the following from calculation Tables 125 through 128.

FACING	SUBTRACT R	FROM R_m
Aluminum siding	0.20	0.035
Stucco	0.61	0.107
4″ brick	0.37	0.065

NO THERMAL INSULATION
THERMAL RESISTANCE R and R_m

Mean Temperature 40° (4.4°C), Wind 10 mph (16.1 km/hr)

	Components	Eng. Units		Metric Units
A	Outside surface	0.17		0.030
B	Shingles or siding	0.81		0.143
C	Sheathing 1″ (2.54 cm) thk glass fiber	4.00		0.704
D	Inner surface of sheathing	0.45		0.079
E	Insulation	—		—
F	Gypsum wallboard	0.45		0.079
G	Inside surface	0.60		0.106
	(TOTAL) RESISTANCE R =	6.48	R_m =	1.141
	TRANSMITTANCE U =	0.154	U_m =	0.876

Mean Temperature 85° (29.4°C), Wind 5 mph (8 km/hr)

	Components	Eng. Units		Metric Units
A	Outside surface	0.17		0.030
B	Shingles or siding	0.75		0.132
C	Sheathing 1″ (2.54 cm) thk glass fiber	3.80		0.670
D	Inner surface of sheathing	0.41		0.072
E	Insulation	—		—
F	Gypsum wallboard	0.43		0.075
G	Inside surface	0.58		0.102
	(TOTAL) RESISTANCE R =	6.14	R_m =	1.081
	TRANSMITTANCE U =	0.163	U_m =	0.925

Mean Temperature 110° (43.2°C), No Wind

	Components	Eng. Units		Metric Units
A	Outside surface	0.14		0.025
B	Shingles or siding	0.71		0.125
C	Sheathing 1″ (2.54 cm) thk glass fiber	3.60		0.634
D	Inner surface of sheathing	0.38		0.067
E	Insulation	—		—
F	Gypsum wallboard	0.43		0.075
G	Inside surface	0.58		0.102
	(TOTAL) RESISTANCE R =	5.84	R_m =	1.028
	TRANSMITTANCE U =	0.171	U_m =	0.971

44. GLASS FIBER BATT INSULATION

THERMAL RESISTANCE AND TRANSMITTANCE OF ASSEMBLY XIX WITH GLASS FIBER BATT INSULATION: k = 0.24 Btu, in./ft², hr °F at 40°F, k_m = 0.035 W/mK at 4.4°C, k = 0.28 Btu, in./ft², hr °F at 85°F, k_m = 0.04 W/mK at 29.4°C, k = 0.34 Btu, in./ft², hr °F at 110°F, k_m = 0.049 W/mK at 43.2°C

Insulation Thickness		Mean Temp. 40°F (4.4°C)				Mean Temp. 85°F (29.4°C)				Mean Temp. 110°F (43.2°C)			
		English Units		Metric Units		English Units		Metric Units		English Units		Metric Units	
Inches	Meters	R	U	R_m	U_m	R	U	R_m	U_m	R	U	R_m	U_m
1	0.025	9.81	0.102	1.728	0.578	9.34	0.107	1.646	0.608	8.19	0.122	1.443	0.693
1 1/2	0.037	11.43	0.087	2.013	0.496	10.94	0.091	1.927	0.519	9.36	0.107	1.649	0.606
2	0.051	13.14	0.076	2.315	0.432	12.54	0.080	2.209	0.453	10.54	0.095	1.857	0.538
2 1/2	0.064	14.81	0.068	2.610	0.383	14.14	0.071	2.491	0.401	11.72	0.085	2.065	0.484
3	0.076	16.48	0.061	2.904	0.344	15.74	0.064	2.773	0.361	12.89	0.078	2.271	0.440
3 1/2	0.088	18.15	0.055	3.198	0.313	17.34	0.058	3.091	0.324	14.07	0.071	2.479	0.403
5 1/2*	0.138	24.81	0.040	4.372	0.229	23.74	0.042	4.183	0.239	18.81	0.053	3.314	0.302

*For walls with 2″ x 6″ studs.

R = °F, ft², hr, Btu R_m = km²W U = Btu per ft² per hr U_m = W/m²K

TABLE 130

WALLS

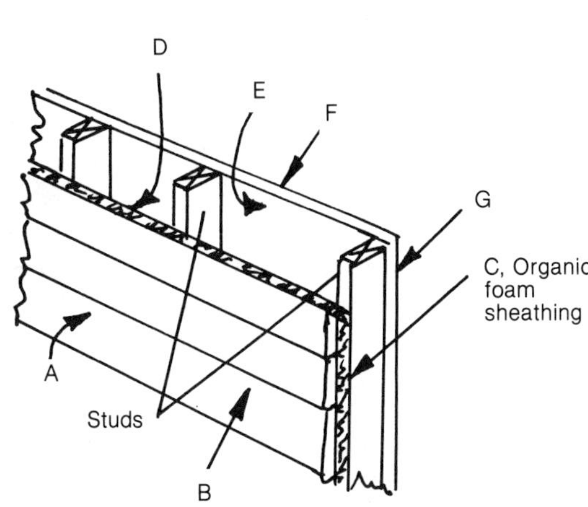

NO THERMAL INSULATION IN STUD SPACE
THERMAL RESISTANCE R and R_m

Mean Temperature 40° (4.4°C), Wind 10 mph (16.1 km/hr)

	Components	Eng. Units		Metric Units
A	Outside surface	0.17		0.030
B	Siding (wood)	0.81		0.140
C	Organic foam insulation sheathing	—		—
D	Inner surface of sheathing	0.45		0.079
E	Space surf. of gypsum wallboard	0.45		0.079
F	Gypsum wallboard	0.43		0.075
G	Inside surface	0.60		0.106
	(TOTAL) RESISTANCE R =	2.91	R_m =	0.509
	TRANSMITTANCE U =	0.344	U_m =	1.965

Mean Temperature 85° (29.4°C), Wind 5 mph (8 km/hr)

	Components	Eng. Units		Metric Units
A	Outside surface	0.16		0.028
B	Siding (wood)	0.75		0.132
C	Organic foam insulation sheathing	—		—
D	Inner surface of sheathing	0.41		0.072
E	Space surf. of gypsum wallboard	0.45		0.079
F	Gypsum wallboard	0.43		0.075
G	Inside surface	0.59		0.104
	(TOTAL) RESISTANCE R =	2.79	R_m =	0.490
	TRANSMITTANCE U =	0.858	U_m =	2.041

Mean Temperature 110° (43.2°C), No Wind

	Components	Eng. Units		Metric Units
A	Outside surface	0.14		0.025
B	Siding (wood)	0.71		0.125
C	Organic foam insulation sheathing	—		—
D	Inner surface of sheathing	0.38		0.067
E	Space surf. of gypsum wallboard	0.45		0.079
F	Gypsum wallboard	0.43		0.075
G	Inside surface	0.58		0.103
	(TOTAL) RESISTANCE R =	2.69	R_m =	4.74
	TRANSMITTANCE U =	0.372	U_m =	2.110

CONSTRUCTION ASSEMBLY XX

Note: Due to combustibility of organic foam sheathing electrical circuits should not be placed in stud spaces unless foam is protected by 1/2″ gypsum sheeting.

If gypsum board is used on both sides of organic foam sheeting for fire protection add 0.9 to R or 0.158 R_m in Tables 130.

45. STYRENE FOAM INSULATION SHEATHING
THERMAL RESISTANCE AND TRANSMITTANCE OF ASSEMBLY XX WITH POLYSTYRENE INSULATION SHEATHING:

k = 0.25 Btu, in./ft^2, hr °F at 40°F, k_m = 0.027 W/mK at 4.4°C, k = 0.29 Btu, in./ft^2, hr °F at 85°F, k_m = 0.051 W/mK at 29.4°C, k = 0.31 Btu, in./ft^2, hr °F at 110°F, k_m = 0.055 W/mK at 43.2°C

Sheathing Thickness		Mean Temp. 40°F (4.4°C)				Mean Temp. 85°F (29.4°C)				Mean Temp. 110°F (43.2°C)			
		English Units		Metric Units		English Units		Metric Units		English Units		Metric Units	
Inches	Meters	R	U	R_m	U_m	R	U	R_m	U_m	R	U	R_m	U_m
1	0.025	6.91	0.145	1.217	0.821	6.23	0.160	1.099	0.910	5.91	0.169	1.042	0.959

TABLE 131

46. POLYURETHANE FOAM INSULATION SHEATHING
THERMAL RESISTANCE AND TRANSMITTANCE OF ASSEMBLY XX WITH POLYURETHANE INSULATION SHEATHING:

k = 0.24 Btu, in./ft^2, hr °F at 40°F, k_m = 0.042 W/mK at 4.4°C, k = 0.25 Btu, in./ft^2, hr °F at 85°F, k_m = 0.044 W/mK at 29.4°C, k = 0.28 Btu, in./ft^2, hr °F at 110°F, k_m = 0.049 W/mK at 43.2°C

Sheathing Thickness		Mean Temp. 40°F (4.4°C)				Mean Temp. 85°F (29.4°C)				Mean Temp. 110°F (43.2°C)			
		English Units		Metric Units		English Units		Metric Units		English Units		Metric Units	
Inches	Meters	R	U	R_m	U_m	R	U	R_m	U_m	R	U	R_m	U_m
1	0.025	7.07	0.141	1.247	0.802	6.96	0.148	1.226	0.816	6.26	0.160	1.104	0.906

TABLE 132

WALLS

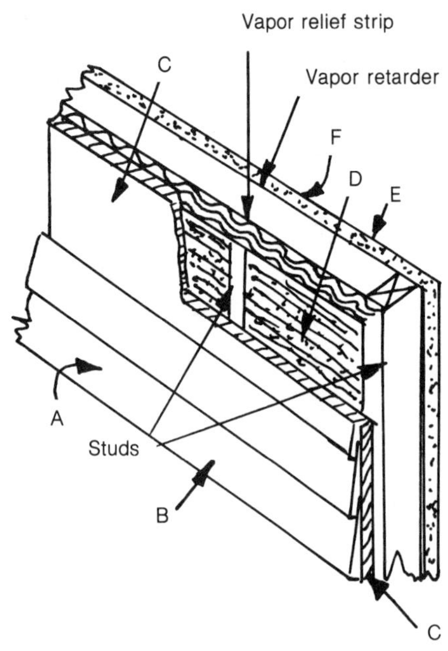

CONSTRUCTION ASSEMBLY XXI

Note: Due to high vapor resistance of faced organic foam sheathing and the lowering of its surface temperature, when batt insulation is installed, venting of vapor is required. This venting to outside air reduces the effectiveness of components A, B, C to provide thermal resistance.

NO THERMAL INSULATION OTHER THAN INSULATION SHEATHING
THERMAL RESISTANCE R and R_m

Mean Temperature 40° (4.4°C), Wind 10 mph (16.1 km/hr)

	Components	Eng. Units		Metric Units
A	Outside surface	0.08		0.014
B	Siding (wood)	0.25		0.044
C	Organic foam insulation sheathing	1.35		0.238
D	Insulation	—		—
E	Gypsum wall board	0.45		0.075
F	Inside surface	0.60		0.106
	(TOTAL) RESISTANCE R =	2.73	R_m =	0.477
	TRANSMITTANCE U =	0.366	U_m =	2.079

Mean Temperature 85° (29.4°C), Wind 5 mph (8 km/hr)

	Components	Eng. Units		Metric Units
A	Outside surface	0.07		0.012
B	Siding (wood)	0.23		0.041
C	Organic foam insulation sheathing	1.30		0.229
D	Insulation	—		—
E	Gypsum wallboard	0.45		0.075
F	Inside surface	0.60		0.106
	(TOTAL) RESISTANCE R =	2.79	R_m =	0.490
	TRANSMITTANCE U =	0.858	U_m =	2.041

Mean Temperature 110° (43.2°C), No Wind

	Components	Eng. Units		Metric Units
A	Outside surface	0.06		0.011
B	Siding (wood)	0.22		0.038
C	Organic foam insulation sheathing	1.25		0.221
D	Insulation	—		—
E	Gypsum wallboard	0.45		0.075
F	Inside surface	0.60		0.106
	(TOTAL) RESISTANCE R =	2.63	R_m =	0.451
	TRANSMITTANCE U =	0.380	U_m =	2.217

47. GLASS FIBER BATT INSULATION

THERMAL RESISTANCE AND TRANSMITTANCE OF ASSEMBLY XXI WITH GLASS FIBER BATT INSULATION: k = 0.24 Btu, in./ft^2, hr °F at 40°F, k_m = 0.035 W/mK at 4.4°C, k = 0.28 Btu, in./ft^2, hr °F at 85°F, k_m = 0.04 W/mK at 29.4°C, k = 0.34 Btu, in./ft^2, hr °F at 110°F, k_m = 0.047 W/mK at 43.2°C

Insulation Thickness		Mean Temp. 40°F (4.4°C)				Mean Temp. 85°F (29.4°C)				Mean Temp. 110°F (43.2°C)			
		English Units		Metric Units		English Units		Metric Units		English Units		Metric Units	
Inches	Meters	R	U	R_m	U_m	R	U	R_m	U_m	R	U	R_m	U_m
3 1/2	0.088	17.31	0.057	3.051	0.327	15.15	0.066	2.669	0.375	12.92	0.077	2.277	0.439
5 1/2*	0.138	25.64	0.039	4.518	0.221	22.29	0.045	3.928	0.255	18.81	0.053	3.314	0.302

TABLE 133

48. MINERAL WOOL OR CELLULOSE FIBER BATT INSULATION

THERMAL RESISTANCE AND TRANSMITTANCE OF ASSEMBLY XXI WITH MINERAL WOOL OR CELLULOSE FIBER BATT INSULATION: k = 0.30 Btu, in./ft^2, hr °F at 40°F, k_m = 0.053 W/mK at 4.4°C, k = 0.33 Btu, in./ft^2, hr °F at 85°F, k_m = 0.058 W/mK at 29.4°C, k = 0.37 Btu, in./ft^2, hr °F at 110°F, k_m = 0.065 W/mK at 43.2°C

Insulation Thickness		Mean Temp. 40°F (4.4°C)				Mean Temp. 85°F (29.4°C)				Mean Temp. 110°F (43.2°C)			
		English Units		Metric Units		English Units		Metric Units		English Units		Metric Units	
Inches	Meters	R	U	R_m	U_m	R	U	R_m	U_m	R	U	R_m	U_m
3 1/2	0.088	14.39	0.069	2.537	0.394	12.65	0.079	2.229	0.449	12.09	0.083	2.130	0.469
5 1/2*	0.138	21.06	0.047	3.711	0.269	18.34	0.055	3.232	0.309	17.49	0.057	3.083	0.324

*For walls with 2″ x 6″ studs.

TABLE 134

WALLS

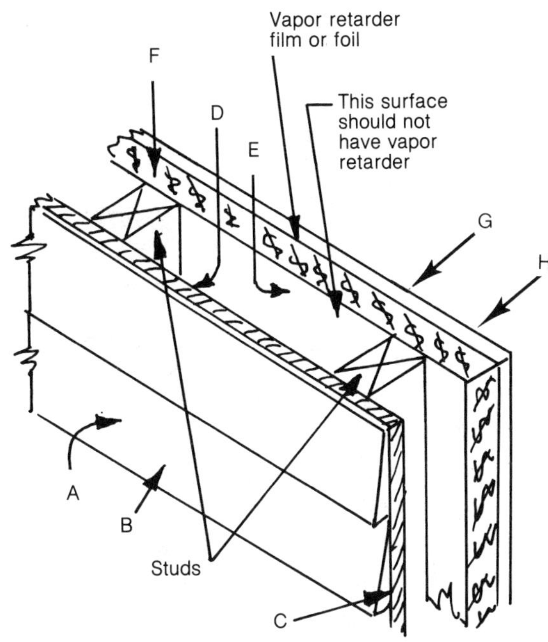

CONSTRUCTION ASSEMBLY XXII

Note: When organic foam is exposed to stud space no electrical circuits should be installed in the space.

Where potential ignition is possible from exterior or electric wires 1/2″ (1.27 cm) thick gypsum board should be installed on studs prior to application of foam insulation board.

Where gypsum wallboards are on both sides of organic foam insulation board add 0.43 to R and 0.106 to R_m in Tables 135 and 136.

When facing is other than wood shingles or siding subtract the following from Tables 135 and 136.

FACING	SUBTRACT FROM R	R_m
Aluminum siding	0.20	0.035
Stucco	0.61	0.107
4″ Face brick	0.37	0.065

NO THERMAL INSULATION
THERMAL RESISTANCE R and R_m

Mean Temperature 40° (4.4°C), Wind 10 mph (16.1 km/hr)

	Components	Eng. Units		Metric Units
A	Outside surface	0.17		0.030
B	Shingles or siding	0.81		0.143
C	Sheathing	0.90		0.155
D	Inner surface of sheathing	0.45		0.079
E	Inner surface of board	0.45		0.079
F	Organic foam insulation board	—		—
G	Gypsum wallboard	0.45		0.079
H	Inside surface	0.60		0.106
	(TOTAL) RESISTANCE R =	3.83	R_m =	0.671
	TRANSMITTANCE U =	0.261	U_m =	1.49

Mean Temperature 85° (29.4°C), Wind 5 mph (8 km/hr)

	Components	Eng. Units		Metric Units
A	Outside surface	0.17		0.030
B	Shingles or siding	0.75		0.132
C	Sheathing	0.85		0.150
D	Inner surface of sheathing	0.41		0.072
E	Inner surface of board	0.41		0.072
F	Organic foam insulation board	—		—
G	Gypsum wallboard	0.43		0.075
H	Inside surface	0.58		0.106
	(TOTAL) RESISTANCE R =	3.60	R_m =	0.637
	TRANSMITTANCE U =	0.277	U_m =	1.570

Mean Temperature 110° (43.2°C), No Wind

	Components	Eng. Units		Metric Units
A	Outside surface	0.14		0.025
B	Shingles or siding	0.71		0.125
C	Sheathing	0.81		0.143
D	Inner surface of sheathing	0.38		0.067
E	Inner surface of board	0.38		0.067
F	Organic foam insulation board	—		—
G	Gypsum wallboard	0.043		0.075
H	Inside surface	0.058		0.106
	(TOTAL) RESISTANCE R =	3.46	R_m =	0.608
	TRANSMITTANCE U =	0.289	U_m =	1.645

49. POLYSTYRENE FOAM INSULATING BOARD
THERMAL RESISTANCE AND TRANSMITTANCE OF ASSEMBLY XXII WITH POLYSTYRENE FOAM INSULATION BOARD:

$k = 0.25$ Btu, in./ft², hr °F at 40°F, $k_m = 0.044$ W/mK at 4.4°C, $k = 0.29$ Btu, in./ft², hr °F at 85°F, $k_m = 0.051$ W/mK at 29.4°C, $k = 0.31$ Btu, in./ft², hr °F at 110°F, $k_m = 0.055$ W/mK at 43.2°C

Insulation Board Thickness		Mean Temp. 40°F (4.4°C)				Mean Temp. 85°F (29.4°C)				Mean Temp. 110°F (43.2°C)			
		English Units		Metric Units		English Units		Metric Units		English Units		Metric Units	
Inches	Meters	R	U	R_m	U_m	R	U	R_m	U_m	R	U	R_m	U_m
1 1/2	0.038	9.83	0.102	1.732	0.577	8.77	0.114	1.546	0.646	8.30	0.121	1.462	0.684
2	0.051	11.83	0.085	2.084	0.480	10.49	0.095	1.849	0.541	9.91	0.101	1.746	0.573
2 1/2	0.064	13.83	0.072	2.437	0.410	12.22	0.082	2.153	0.464	11.52	0.086	2.031	0.492
3	0.076	15.83	0.063	2.789	0.359	13.94	0.071	2.457	0.407	13.14	0.076	2.315	0.432

R = °F, ft², hr, Btu R_m = km²W U = Btu per ft² per hr U_m = W/m²K

TABLE 135

WALLS
CONSTRUCTION ASSEMBLY XXII

50. POLYURETHANE OR POLYISOCYANURATE FOAM INSULATING BOARD

THERMAL RESISTANCE AND TRANSMITTANCE OF ASSEMBLY XXII WITH POLYURETHANE OR POLYISOCYANURATE INSULATION BOARD: k = 0.19 Btu, in./ft^2, hr °F at 40 °F, k_m = 0.027 W/mK at 4.4 °C, k = 0.24 Btu, in./ft^2, hr °F at 85 °F, k_m = 0.035 W/mK at 29.4 °C, k = 0.28 Btu, in./ft^2, hr °F at 110 °F, k_m = 0.040 W/mK at 43.2 °C

Insulation Board Thickness		Mean Temp. 40 °F (4.4 °C)				Mean Temp. 85 °F (29.4 °C)				Mean Temp. 110 °F (43.2 °C)			
		English Units		Metric Units		English Units		Metric Units		English Units		Metric Units	
Inches	Meters	R	U	R_m	U_m	R	U	R_m	U_m	R	U	R_m	U_m
3/8	0.009	5.80	0.172	1.023	0.977	5.16	0.193	0.910	1.099	4.76	0.208	0.846	1.183
3/4	0.019	7.77	0.128	1.370	0.730	6.75	0.149	1.185	0.844	6.13	0.163	1.082	0.925
1	0.025	9.09	0.110	1.602	0.624	7.76	0.129	1.368	0.731	7.03	0.142	1.239	0.807
1 1/2	0.038	11.72	0.085	2.065	0.484	9.85	0.102	1.736	0.576	8.82	0.113	1.554	0.644
2	0.051	14.40	0.070	2.530	0.395	11.93	0.083	2.103	0.476	10.60	0.094	1.868	0.535
2 1/2	0.064	16.99	0.059	2.993	0.334	14.02	0.071	2.469	0.404	12.38	0.081	2.183	0.458
3	0.076	19.62	0.051	3.457	0.289	16.10	0.062	2.837	0.352	14.35	0.070	2.529	0.395

R = °F, ft^2, hr, Btu R_m = km^{2}W U = Btu per ft^2 per hr U_m = W/m^{2}K

TABLE 136

MASONRY WALL ABOVE GRADE

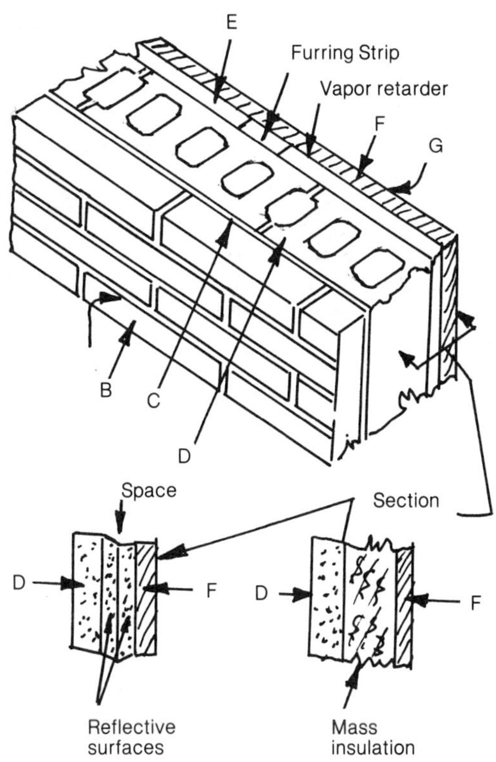

TABLE 137

CONSTRUCTION ASSEMBLY XXIII

Note: If aluminum siding is used instead of face brick add 0.17 to R and 0.029 to R_m. If common brick or gravel block is used instead of of cinder block subtract from:

4″ Block	8″ Block
0.61 from R	0.62 from R
0.107 from R_m	0.109 from R_m

NO THERMAL INSULATION
THERMAL RESISTANCE R and R_m

Mean Temperature 40° (4.4°C), Wind 10 mph (16.1 km/hr)

	Components	English Units 4″ Blk.	English Units 8″ Blk.		Metric Units 10.15 cm Blk.	Metric Units 20.3 Cm Blk.
A	Outside surface	0.13	0.13		0.023	0.023
B	Face Brick 4″ (10.15 cm)	0.33	0.33		0.058	0.058
C	Cement mortar 1/2″ (1.27 cm)	0.10	0.10		0.017	0.017
D	Cinder block or clay tile	1.11	1.72		0.196	0.303
E	Air space	0.97	0.97		0.171	0.171
F	Gypsum wallboard	0.45	0.45		0.079	0.079
G	Inside surface	0.60	0.60		0.106	0.106
	RESISTANCE R =	3.76	4.30	R_m =	0.650	0.757
	TRANSMITTANCE U =	0.266	0.233	U_m =	1.538	1.321

Mean Temperature 85°F (29.4°C), Wind 5 mph (8 km/hr)

	Components	English Units 4″ Blk.	English Units 8″ Blk.		Metric Units 10.15 cm Blk.	Metric Units 20.3 cm Blk.
A	Outside surface	0.11	0.11		0.019	0.019
B	Face brick 4″ (10.15 cm)	0.30	0.30		0.052	0.052
C	Cement mortar 1/2″ (1.27 cm)	0.09	0.09		0.016	0.016
D	Cinder block or clay tile	1.03	1.65		0.181	0.291
E	Air space	0.97	0.97		0.171	0.171
F	Gypsum wallboard	0.43	0.43		0.075	0.075
G	Inside surface	0.58	0.58		0.102	0.102
	RESISTANCE R =	3.51	4.12	R_m =	0.616	0.726
	TRANSMITTANCE U =	0.284	0.243	U_m =	1.623	1.377

Mean Temperature 110°F (43.2°C), No Wind

	Components	English Units 4″ Blk.	English Units 8″ Blk.		Metric Units 10.15 cm Blk.	Metric Units 20.3 cm Blk.
A	Outside surface	0.09	0.09		0.016	0.016
B	Face brick 4″ (10.15 cm)	0.27	0.27		0.048	0.048
C	Cement mortar	0.08	0.08		0.014	0.014
D	Cinder block or clay tile	0.98	1.60		0.173	0.282
E	Air space	0.97	0.97		0.171	0.171
F	Gypsum wallboard	0.43	0.43		0.075	0.075
G	Inside surface	0.58	0.58		0.102	0.102
	RESISTANCE R =	3.40	4.02	R_m =	0.599	0.708
	TRANSMITTANCE U =	0.294	0.249	U_m =	1.669	1.412

51. REFLECTIVE INSULATION: Aluminum foil faced gypsum board, R of space = 3.25, R_m = 0.573
One sheet reflective sheet dividing space (reflective both sides), R of space = 6.21, R_m 1.094

THERMAL RESISTANCE AND TRANSMITTANCE OF ASSEMBLY XXIII WITH REFLECTIVE INSULATION

Block Width	Reflective Insulation	Mean Temp. 40°F (4.4°C) English Units R	U	Metric Units R_m	U_m	Mean Temp. 85°F (29.4°C) English Units R	U	Metric Units R_m	U_m	Mean Temp. 110°F (43.2°C) English Units R	U	Metric Units R_m	U_m
	One — facing space	6.04	0.166	1.064	0.940	5.79	0.173	1.020	0.980	5.88	0.170	1.000	1.000
4″	Two — 1 on each surface	9.29	0.107	1.637	0.611	9.04	0.111	1.593	0.628	8.91	0.112	1.569	0.637
10.15 cm	Divided, both sides reflective	9.00	0.111	1.568	0.631	8.75	0.114	1.542	0.649	8.64	0.116	1.522	0.657
	One + divided reflec. surface	12.25	0.082	2.158	0.463	12.00	0.083	2.114	0.473	11.89	0.084	2.095	0.477
	One — facing space	6.58	0.152	1.159	0.863	6.40	0.156	1.127	0.887	6.30	0.159	1.110	0.901
8″	Two — 1 on each surface	9.83	0.102	1.732	0.577	9.65	0.104	1.700	0.588	9.55	0.105	1.683	0.594
20.3 cm	Divided, both sides reflective	9.54	0.105	1.681	0.595	9.36	0.106	1.649	0.606	9.26	0.108	1.632	0.613
	One + divided reflec. surface	12.79	0.078	2.254	0.444	12.61	0.073	2.222	0.450	12.51	0.079	2.202	0.454

R = °F, ft², hr, Btu R_m = km²W U = Btu per ft² per hr U_m = W/m²K

TABLE 137

MASONRY WALLS — ABOVE GRADE
CONSTRUCTION ASSEMBLY XXIII — Cont'd

52. POLYSTYRENE FOAM INSULATING BOARD
THERMAL RESISTANCE AND TRANSMITTANCE OF ASSEMBLY XXIII WITH POLYSTYRENE FOAM INSULATING BOARD:

k = 0.25 Btu, in./ft^2, hr °F at 40 °F, k_m = 0.044 W/mK at 4.4 °C, k = 0.29 Btu, in./ft^2, hr °F at 85 °F, k_m = 0.051 W/mK at 29.4 °C, k = 0.31 Btu, in./ft^2, hr °F at 110 °F, k_m = 0.055 W/mK at 43.2 °C

Block Thk.	Insulation Thickness		Mean Temp. 40 °F (4.4 °C)				Mean Temp. 85 °F (29.4 °C)				Mean Temp. 110 °F (43.2 °C)			
			English Units		Metric Units		English Units		Metric Units		English Units		Metric Units	
	Inches	Meters	R	U	R_m	U_m	R	U	R_m	U_m	R	U	R_m	U_m
4″	1 1/2	0.038	9.77	0.102	1.721	0.581	8.68	0.115	1.529	0.624	8.24	0.121	1.452	0.689
	2	0.051	11.77	0.084	2.074	0.482	10.40	0.096	1.832	0.545	9.85	0.102	1.736	0.576
10.15 cm	2 1/2	0.064	13.77	0.073	2.426	0.412	12.13	0.082	2.137	0.468	11.46	0.087	2.019	0.495
	3	0.076	15.77	0.063	2.779	0.360	13.85	0.072	2.940	0.410	13.08	0.076	2.305	0.433
8″	1 1/2	0.038	10.17	0.098	1.792	0.358	9.39	0.106	1.655	0.604	8.86	0.112	1.561	0.641
	2	0.051	12.17	0.082	2.144	0.466	11.11	0.090	1.957	0.510	10.47	0.095	1.845	0.542
20.3 cm	2 1/2	0.064	14.17	0.071	2.497	0.400	12.84	0.078	2.262	0.442	12.08	0.083	2.119	0.472
	3	0.076	16.17	0.062	2.849	0.351	14.56	0.068	2.565	0.390	13.70	0.073	2.414	0.414

R = °F, ft^2, hr, Btu R_m = km^{2}W U = Btu per ft^2 per hr U_m = W/m^{2}K

TABLE 138

53. POLYURETHANE OR POLYISOCYANURATE FOAM INSULATING BOARD
THERMAL RESISTANCE AND TRANSMITTANCE OF ASSEMBLY XXIII WITH POLYURETHANE OR POLYISOCYANURATE INSULATING BOARD:

k = 0.19 Btu, in./ft^2, hr °F at 40 °F, k_m = 0.027 W/mK at 4.4 °C, k = 0.24 Btu, in./ft^2, hr °F at 85 °F, k_m = 0.035 W/mK at 29.4 °C, k = 0.28 Btu, in./ft^2, hr °F at 110 °F, k_m = 0.040 W/mK at 43.2 °C

Block Thk.	Insulation Thickness		Mean Temp. 40 °F (4.4 °C)				Mean Temp. 85 °F (29.4 °C)				Mean Temp. 110 °F (43.2 °C)			
			English Units		Metric Units		English Units		Metric Units		English Units		Metric Units	
	Inches	Meters	R	U	R_m	U_m	R	U	R_m	U_m	R	U	R_m	U_m
	3/8	0.009	5.74	0.174	1.011	0.989	5.07	0.197	0.893	1.119	4.70	0.213	0.828	1.207
	3/4	0.019	7.71	0.129	1.358	0.736	6.66	0.150	1.173	0.852	6.07	0.165	1.069	0.935
4″	1	0.025	9.03	0.111	1.591	0.628	7.67	0.130	1.351	0.740	6.97	0.143	1.228	0.814
	1 1/2	0.038	11.66	0.086	2.054	0.487	9.76	0.102	1.719	0.581	8.76	0.114	1.543	0.648
10.15 cm	2	0.051	14.34	0.070	2.527	0.396	11.87	0.084	2.091	0.478	10.54	0.095	1.857	0.538
	2 1/2	0.064	16.93	0.059	2.983	0.335	13.96	0.072	2.459	0.407	12.32	0.081	2.171	0.461
	3	0.076	19.56	0.051	3.446	0.290	16.04	0.062	2.826	0.353	14.29	0.070	2.518	0.397
	3/8	0.009	6.34	0.158	1.117	0.895	5.78	0.173	1.018	0.982	5.32	0.188	0.937	1.067
	3/4	0.019	8.31	0.120	1.464	0.683	6.92	0.144	1.219	0.820	6.69	0.149	1.179	0.848
8″	1	0.025	10.44	0.095	1.839	0.544	7.93	0.126	1.397	0.715	7.59	0.132	1.337	0.747
	1 1/2	0.038	12.26	0.081	2.160	0.463	10.02	0.100	1.766	0.566	9.38	0.107	1.653	0.605
20.3 cm	2	0.051	14.94	0.067	2.632	0.380	12.10	0.083	2.132	0.469	11.16	0.090	1.966	0.508
	2 1/2	0.064	17.53	0.057	3.089	0.324	14.64	0.068	2.579	0.387	12.94	0.077	2.280	0.439
	3	0.076	20.16	0.049	3.552	0.282	16.72	0.060	2.946	0.339	14.91	0.067	2.627	0.381

Note: The aluminum foil facing on the organic foam must be placed to face the inside of the building, under the gypsum plaster board. If gypsum plaster board has aluminum foil this vapor retarder can be used instead of vapor retarder on the organic foam insulation. No electrical wiring should be placed where it might cause the organic foam to burn.

R = °F, ft^2, hr, Btu R_m = km^{2}W U = Btu per ft^2 per hr U_m = W/m^{2}K

TABLE 139

MASONRY WALL ABOVE GRADE

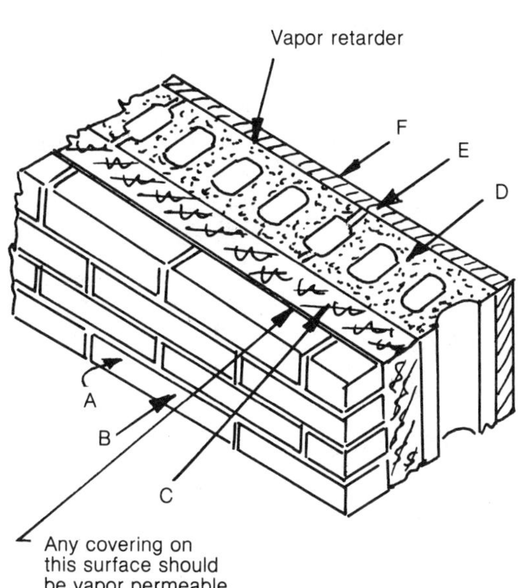

NO THERMAL INSULATION
THERMAL RESISTANCE R and R_m

Mean Temperature 40° (4.4°C), Wind 10 mph (16.1 km/hr)

	Components	English Units 4″ Blk.	8″ Blk.		Metric Units 10 cm Blk.	20 Cm Blk.
A	Outside surface	0.13	0.13		0.023	0.023
B	Face Brick 4″ (10.15 cm)	0.33	0.33		0.058	0.058
C	Insulation	—	—		—	—
D	Cinder block or clay tile	1.11	1.72		0.196	0.303
E	Gypsum wallboard	0.45	0.45		0.079	0.079
F	Inside surface	0.60	0.60		0.106	0.106
	RESISTANCE R =	2.62	3.23	R_m =	0.462	0.569
	TRANSMITTANCE U =	0.382	0.310	U_m =	2.165	1.757

Mean Temperature 85°F (29.4°C), Wind 5 mph (8 km/hr)

	Components	English Units 4″ Blk.	8″ Blk.		Metric Units 10 cm Blk.	20 cm Blk.
A	Outside surface	0.11	0.11		0.019	0.019
B	Face brick 4″ (10.15 cm)	0.30	0.30		0.052	0.052
C	Insulation	—	—		—	—
D	Cinder block or clay tile	1.03	1.65		0.181	0.291
E	Gypsum wallboard	0.43	0.43		0.075	0.075
F	Inside surface	0.58	0.58		0.102	0.102
	RESISTANCE R =	2.45	3.07	R_m =	0.429	0.539
	TRANSMITTANCE U =	0.408	0.326	U_m =	2.331	1.855

Mean Temperature 110°F (43.2°C), No Wind

	Components	English Units 4″ Blk.	8″ Blk.		Metric Units 10 cm Blk.	20 cm Blk.
A	Outside surface	0.09	0.09		0.016	0.016
B	Face brick 4″ (10.15 cm)	0.27	0.27		0.048	0.048
C	Insulation	—	—		—	—
D	Cinder block or clay tile	0.43	0.43		0.075	0.075
E	Gypsum wallboard	0.43	0.43		0.075	0.075
F	Inside surface	0.58	0.58		0.102	0.102
	RESISTANCE R =	2.35	2.97	R_m =	0.414	0.523
	TRANSMITTANCE U =	0.426	0.337	U_m =	2.415	1.912

CONSTRUCTION ASSEMBLY XXIV

Note: If aluminum siding is used instead of face brick add 0.17 to R and 0.029 to R_m. If common brick or gravel — sand block is used instead of of cinder block subtract from:

4″ Block	8″ Block
0.61 from R	0.62 from R
0.107 from R_m	0.109 from R_m

1. CELLULAR GLASS BLOCK INSULATION

THERMAL RESISTANCE AND TRANSMITTANCE OF ASSEMBLY XXIV WITH CELLULAR GLASS INSULATION: k = 0.33 Btu, in./ft^2, hr °F at 40°F, k_m = 0.047 W/mK at 4.4°C, k = 0.35 Btu, in./ft^2, hr °F at 85°F, k_m = 0.050 W/mK at 29.4°C, k = 0.36 Btu, in./ft^2, hr °F at 110°F, k_m = 0.052 W/mK at 43.2°C

Block Width	Insulation Thickness Inches	Meters	Mean Temp. 40°F (4.4°C) English Units R	U	Metric Units R_m	U_m	Mean Temp. 85°F (29.4°C) English Units R	U	Metric Units R_m	U_m	Mean Temp. 110°F (43.2°C) English Units R	U	Metric Units R_m	U_m
	1 1/2	0.038	7.19	0.139	1.267	0.787	6.72	0.149	1.184	0.847	6.47	0.155	1.140	0.877
4″	2	0.051	8.72	0.115	1.536	0.649	8.15	0.123	1.436	0.694	7.84	0.128	1.381	0.725
10.15 cm	2 1/2	0.064	10.24	0.098	1.804	0.556	9.57	0.104	1.686	0.592	9.22	0.108	1.625	0.617
	3	0.070	11.77	0.207	2.074	0.483	11.00	0.091	1.938	0.515	10.59	0.094	1.866	0.535
	1 1/2	0.038	7.80	0.128	1.374	0.730	7.34	0.136	1.293	0.775	7.09	0.141	1.249	0.800
8″	2	0.051	9.33	0.107	1.644	0.610	8.77	0.114	1.545	0.645	8.46	0.118	1.491	0.671
20.3 cm	2 1/2	0.064	10.85	0.092	1.912	0.524	10.19	0.098	1.795	0.556	9.84	0.102	1.734	0.578
	3	0.076	12.38	0.081	2.181	0.459	11.62	0.086	2.047	0.488	11.21	0.089	1.975	0.505

R = °F, ft^2, hr, Btu R_m = km^{2}W U = Btu per ft^2 per hr U_m = W/m^{2}K

TABLE 140

MASONRY WALLS — ABOVE GRADE
CONSTRUCTION ASSEMBLY XXIV — Cont'd

55. POLYSTYRENE FOAM INSULATION SHEETS
THERMAL RESISTANCE AND TRANSMITTANCE OF ASSEMBLY XXIV WITH POLYSTYRENE FOAM INSULATION SHEETS: k = 0.25 Btu, in./ft^2, hr °F at 40°F, k_m = 0.044 W/mK at 4.4°C, k = 0.29 Btu, in./ft^2, hr °F at 85°F, k_m = 0.051 W/mK at 29.4°C, k = 0.31 Btu, in./ft^2, hr °F at 110°F, k_m = 0.055 W/mK at 43.2°C

Block Width	Insulation Thickness		Mean Temp. 40°F (4.4°C) English Units		Metric Units		Mean Temp. 85°F (29.4°C) English Units		Metric Units		Mean Temp. 110°F (43.2°C) English Units		Metric Units	
	Inches	Meters	R	U	R_m	U_m	R	U	R_m	U_m	R	U	R_m	U_m
	1	0.025	6.62	0.151	1.166	0.857	5.90	0.169	1.039	0.961	5.57	0.179	0.981	1.019
4″	1 1/2	0.038	8.62	0.116	1.514	0.658	7.62	0.131	1.343	0.745	7.19	0.139	1.267	0.789
	2	0.054	10.62	0.094	1.871	0.534	9.34	0.107	1.646	0.608	8.80	0.114	1.551	0.645
10.15 cm	2 1/2	0.064	12.62	0.079	2.224	0.450	11.07	0.090	1.950	0.513	10.41	0.096	1.834	0.545
	3	0.076	14.62	0.068	2.576	0.388	12.79	0.078	2.253	0.444	12.03	0.083	2.119	0.472
	1	0.025	7.23	0.138	1.274	0.785	6.51	0.153	1.147	0.871	6.19	0.162	1.091	0.917
8″	1 1/2	0.038	9.23	0.108	1.626	0.614	8.24	0.121	1.452	0.688	7.81	0.128	1.376	0.727
	2	0.054	11.23	0.089	1.979	0.505	9.96	0.100	1.755	0.670	9.42	0.106	1.659	0.602
20.3 cm	2 1/2	0.064	13.23	0.076	2.331	0.429	11.63	0.085	2.049	0.488	11.03	0.091	1.942	0.515
	3	0.076	15.23	0.066	2.683	0.373	13.41	0.075	2.363	0.423	12.65	0.079	2.229	0.449

R = °F, ft^2, hr, Btu R_m = km^{2}W U = Btu per ft^2 per hr U_m = W/m^{2}K

TABLE 141

56. POLYURETHANE OR POLYISOCYANURATE FOAM INSULATION SHEETS
THERMAL RESISTANCE AND TRANSMITTANCE OF ASSEMBLY XXIV WITH POLYURETHANE OR POLYISOCYANURATE INSULATION SHEETS: k = 0.19 Btu, in./ft^2, hr °F at 40°F, k_m = 0.029 W/mK at 4.4°C, k = 0.24 Btu, in./ft^2, hr °F at 85°F, k_m = 0.035 W/mK at 29.4°C, k = 0.28 Btu, in./ft^2, hr °F at 110°F, k_m = 0.040 W/mK at 43.2°C

Block Width	Insulation Thickness		Mean Temp. 40°F (4.4°C) English Units		Metric Units		Mean Temp. 85°F (29.4°C) English Units		Metric Units		Mean Temp. 110°F (43.2°C) English Units		Metric Units	
	Inches	Meters	R	U	R_m	U_m	R	U	R_m	U_m	R	U	R_m	U_m
	1	0.025	7.88	0.127	1.388	0.720	6.61	0.151	1.165	0.859	5.92	0.168	1.043	0.959
4″	1 1/2	0.038	10.51	0.095	1.852	0.540	8.70	0.115	1.533	0.652	7.70	0.130	1.357	0.737
	2	0.054	13.15	0.076	2.317	0.432	10.78	0.093	1.899	0.526	9.49	0.105	1.672	0.598
10.15 cm	2 1/2	0.064	15.17	0.063	2.673	0.374	12.87	0.078	2.267	0.441	11.27	0.088	1.986	0.503
	3	0.076	18.41	0.054	3.244	0.308	14.95	0.067	2.634	0.380	13.06	0.077	2.301	0.435
	1	0.025	8.49	0.118	1.496	0.668	7.23	0.138	1.274	0.785	6.54	0.152	1.152	0.868
8″	1 1/2	0.038	11.12	0.090	1.959	0.510	9.32	0.107	1.642	0.609	8.33	0.120	1.467	0.681
	2	0.054	13.75	0.073	2.423	0.413	11.40	0.088	2.009	0.497	10.11	0.098	1.781	0.561
20.3 cm	2 1/2	0.064	16.33	0.061	2.877	0.348	13.49	0.074	2.377	0.421	11.90	0.084	2.097	0.477
	3	0.076	19.02	0.053	3.351	0.298	15.57	0.064	2.814	0.355	13.68	0.073	2.410	0.415

Note: Polyurethane foamed in place could be used to fill the cavity space between block and face brick, however, the brick requires reinforcing and ties to block to resist pressure caused by the expansion of the material as it foams. Pressures up to 5 lbs per square inch (3515.5 kg/square meter) can be exerted by expanding organic foam.

R = °F, ft^2, hr, Btu R_m = km^{2}W U = Btu per ft^2 per hr U_m = W/m^{2}K

TABLE 142

MASONRY WALL BELOW GRADE

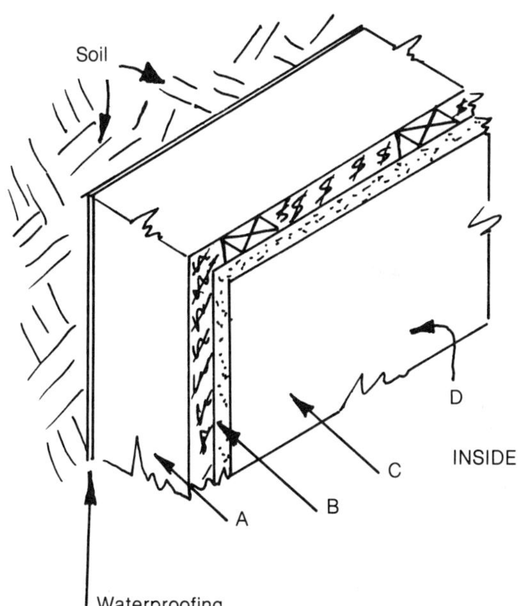

CONSTRUCTION ASSEMBLY XXV

Note: Mean temperature of 50 °F (10 °C) was selected for walls below grade as they are not exposed to high air temperature or direct solar heat.

NO THERMAL INSULATION
THERMAL RESISTANCE R and R_m

Sand and Gravel Concrete

No Plaster Board

Components	English Units 8″ Thk.	English Units 12″ Thk.		Metric Units 20.3 cm Thk.	Metric Units 30.4 cm Thk.
A Concrete	0.66	1.00		0.116	0.176
D Inside surface	0.50	0.50		0.088	0.088
RESISTANCE R =	1.16	1.50	R_m =	0.204	0.264
TRANSMITTANCE U =	0.866	0.666	U_m =	4.902	3.788

With Plaster Board

Components	English Units 8″ Thk.	English Units 12″ Thk.		Metric Units 20.3 cm Thk.	Metric Units 30.4 cm Thk.
A Concrete	0.66	1.00		0.116	0.176
B Air space*	0.97	0.97		0.171	0.171
C Gypsum wallboard	0.43	0.43		0.075	0.075
D Inside surface	0.58	0.58		0.102	0.102
RESISTANCE R =	2.64	2.98	R_m =	0.464	0.524
TRANSMITTANCE U =	0.379	0.336	U_m =	2.155	1.905

Cinder Block or Clay Tile

No Plaster Board

Components	English Units 8″ Thk.	English Units 12″ Thk.		Metric Units 20.3 cm Thk.	Metric Units 30.4 cm Thk.
A Cinder block or tile	1.72	1.90		0.303	0.325
B Inside surface	0.50	0.50		0.088	0.088
RESISTANCE R =	2.22	2.40	R_m =	0.391	0.423
TRANSMITTANCE U =	0.450	0.417	U_m =	2.558	2.364

With Plaster Board

Components	English Units 8″ Thk.	English Units 12″ Thk.		Metric Units 20.3 cm Thk.	Metric Units 30.4 cm Thk.
A Cinder block or tile	1.72	1.90		0.303	0.325
B Air space*	0.97	0.97		0.171	0.171
C Gypsum wallboard	0.43	0.43		0.075	0.075
D Inside surface	0.58	0.58		0.102	0.102
RESISTANCE R =	3.70	3.88	R_m =	0.651	0.683
TRANSMITTANCE U =	0.270	0.258	U_m =	1.536	1.464

*Insulation replaces air space in Table 143.

57. CELLULAR GLASS INSULATION
THERMAL RESISTANCE AND TRANSMITTANCE OF ASSEMBLY XXV WITH CELLULAR GLASS INSULATION: k = 0.36 Btu, in./ft², hr °F at 50 °F, k_m = 0.063 W/mKat 10 °C,

Masonry Wall	Insulation Thickness		8″ (20.3 cm) Wall Thickness				12″ (30.4 cm) Wall Thickness			
			English Units		Metric Units		English Units		Metric Units	
	Inches	Meters	R	U	R_m	U_m	R	U	R_m	U_m
Sand and	1 1/2	0.038	6.72	0.149	1.184	0.847	6.90	0.145	1.216	0.758
Gravel	2	0.051	8.23	0.122	1.450	0.690	8.41	0.119	1.482	0.676
Concrete	2 1/2	0.064	9.73	0.103	1.714	0.585	9.91	0.101	1.746	0.571
	3	0.076	11.23	0.089	1.979	0.505	11.41	0.088	2.010	0.498
Cinder Blk.	1 1/2	0.038	8.20	0.122	1.445	0.690	8.34	0.119	1.470	0.676
or	2	0.051	9.71	0.103	1.711	0.585	9.89	0.101	1.743	0.575
Clay	2 1/2	0.064	11.21	0.089	1.975	0.505	11.39	0.088	2.007	0.497
	3	0.076	12.71	0.079	2.239	0.446	12.89	0.079	2.271	0.441

R = °F, ft², hr, Btu R_m = km²W U = Btu per ft² per hr U_m = W/m²K

TABLE 143

MASONRY WALLS — BELOW GRADE
CONSTRUCTION ASSEMBLY XXV (Cont'd)

58. POLYSTYRENE FOAM INSULATION SHEETS
THERMAL RESISTANCE AND TRANSMITTANCE OF ASSEMBLY XXV WITH POLYSTYRENE FOAM INSULATION: $k = 0.26$ Btu, in./ft^2, hr °F at 50 °F, $k_m = 0.045$ W/mK at 10 °C,

Masonry Wall	Insulation Thickness		8″ (20.3 cm) Wall Thickness				12″ (30.4 cm) Wall Thickness			
			English Units		Metric Units		English Units		Metric Units	
	Inches	Meters	R	U	R_m	U_m	R	U	R_m	U_m
	1	0.025	5.52	0.181	0.973	1.028	6.58	0.152	1.159	0.862
Sand and	1 1/2	0.038	7.44	0.134	1.311	0.763	8.50	0.118	1.498	0.668
Gravel	2	0.051	9.36	0.107	1.649	0.606	10.42	0.096	1.836	0.545
Concrete	2 1/2	0.064	11.29	0.089	1.989	0.503	12.34	0.081	2.174	0.460
	3	0.076	13.21	0.076	2.328	0.429	14.26	0.070	2.513	0.398
	1	0.025	5.86	0.171	1.033	0.968	6.75	0.148	1.189	0.841
Cinder	1 1/2	0.038	7.78	0.128	1.371	0.729	8.68	0.115	1.529	0.654
Block or	2	0.051	9.70	0.103	1.692	0.591	10.60	0.094	1.868	0.535
Clay Tile	2 1/2	0.064	11.63	0.086	2.049	0.488	12.53	0.079	2.208	0.453
	3	0.076	13.55	0.074	2.388	0.418	14.45	0.069	2.546	0.393

R = °F, ft^2, hr, Btu R_m = km^{2}W   U = Btu per ft^2 per hr U_m = W/m^{2}K

TABLE 144

59. POLYURETHANE OR POLYISOCYANURATE FOAM INSULATION SHEETS
THERMAL RESISTANCE AND TRANSMITTANCE OF ASSEMBLY XXV WITH POLYURETHANE OR POLYISOCYANURATE FOAM INSULATION: $k = 0.20$ Btu, in./ft^2, hr °F at 50 °F, $k_m = 0.035$ W/mK at 10 °C,

Masonry Wall	Insulation Thickness		8″ (20.3 cm) Wall Thickness				12″ (30.4 cm) Wall Thickness			
			English Units		Metric Units		English Units		Metric Units	
	Inches	Meters	R	U	R_m	U_m	R	U	R_m	U_m
	1	0.025	6.67	0.150	1.175	0.851	7.01	0.142	1.235	0.809
Sand and	1 1/2	0.038	9.17	0.109	1.616	0.619	9.51	0.105	1.676	0.597
Gravel	2	0.051	11.67	0.086	2.056	0.486	12.01	0.083	2.116	0.473
Concrete	2 1/2	0.064	14.17	0.071	2.591	0.386	14.51	0.069	2.557	0.391
	3	0.076	16.67	0.060	2.937	0.340	17.01	0.059	2.997	0.334
	1	0.025	7.73	0.129	1.356	0.734	7.91	0.126	1.394	0.717
Cinder	1 1/2	0.038	10.23	0.098	1.803	0.555	10.41	0.096	1.834	0.545
Block or	2	0.051	12.73	0.079	2.243	0.446	12.91	0.077	2.275	0.440
Clay Tile	2 1/2	0.064	15.23	0.066	2.683	0.373	15.41	0.065	2.715	0.368
	3	0.076	17.73	0.056	3.124	0.320	17.91	0.056	3.156	0.317

R = °F, ft^2, hr, Btu R_m = km^{2}W   U = Btu per ft^2 per hr U_m = W/m^{2}K

Note: Do not run any electric wires between masonry wall and the inner wall board in the space insulated by polystyrene, polyurethane or polyisocyanurate foam insulation.

TABLE 145

MASONRY WALL BELOW GRADE
CONSTRUCTION ASSEMBLY XXV — Cont'd

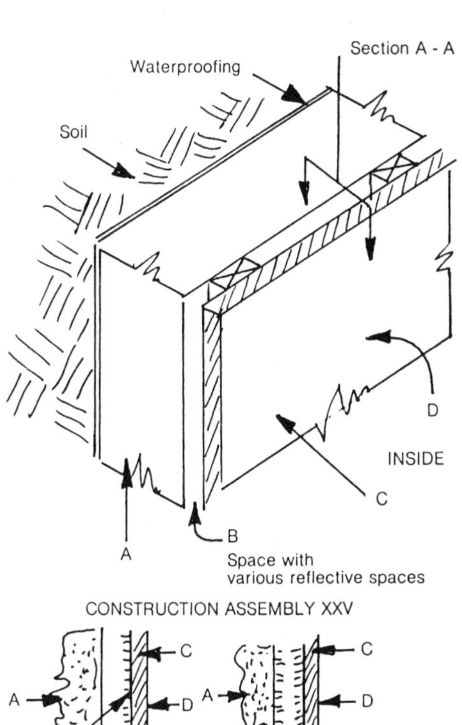

NO THERMAL INSULATION
THERMAL RESISTANCE R and R_m

Sand and Gravel Concrete
With Interior Wallboard

Components	English Units 8″ Thk.	12″ Thk.		Metric Units 20.3 cm Thk.	30.4 cm Thk.
A Concrete	0.66	1.00		0.116	0.176
B Space (without insul.)	0.97	0.97		0.171	0.171
C Gypsum wallboard	0.43	0.43		0.075	0.075
D Inside surface	0.58	0.58		0.102	0.102
RESISTANCE R =	2.64	2.98	R_m =	0.464	0.524
TRANSMITTANCE U =	0.379	0.336	U_m =	2.155	1.908

Cinder Block or Clay Tile
With Interior Wallboard

Components	English Units 8″ Thk.	12″ Thk.		Metric Units 20.3 cm Thk.	30.4 cm Thk.
A Cinder block or tile	1.72	1.90		0.303	0.325
B Space (without insul.)	0.97	0.97		0.171	0.171
C Gypsum wallboard	0.43	0.43		0.075	0.075
D Inside surface	0.58	0.58		0.102	0.102
RESISTANCE R =	3.70	3.88	R_m =	0.651	0.673
TRANSMITTANCE U =	0.270	0.258	U_m =	1.536	1.486

Note: Resistance of space (B) is part of resistance of reflective surfaces in that space.

60. REFLECTIVE INSULATION: Aluminum foil faced wallboard R = 3.25, R_m = 0.573 (1 reflective surface)
One sheet (reflective on both sides dividing space) R = 6.21, R_m = 1.094

THERMAL RESISTANCE AND TRANSMITTANCE OF ASSEMBLY XXV WITH REFLECTIVE INSULATION

Wall Thickness	No. of Spaces	Reflective Insulation	Sand & Gravel Masonry English Units R	U	Metric Units R_m	U_m	Cinder Block or Tile Wall English Units R	U	Metric Units R_m	U_m
	1	Reflective — one side	4.92	0.203	0.867	1.154	5.98	0.167	1.054	0.949
8″	1	Reflective — both sides	8.17	0.122	1.439	0.695	9.19	0.109	1.619	0.617
20.3 cm	2	Divided — reflective sheets	7.88	0.126	1.388	0.720	8.94	0.112	1.575	0.635
	2	Above — plus one	10.23	0.098	1.806	0.555	11.29	0.089	1.989	0.503
	1	Reflective — one side	5.26	0.190	0.927	1.079	6.16	0.162	1.085	0.921
12″	1	Reflective — both sides	8.51	0.118	1.499	0.666	9.41	0.106	1.658	0.603
30.4 cm	2	Divided — reflective sheets	8.22	0.122	1.448	0.690	9.12	0.109	1.607	0.622
	2	Above — plus one side	10.47	0.955	1.845	0.542	11.47	0.087	2.021	0.495

TABLE 146

MASONRY WALL BELOW GRADE

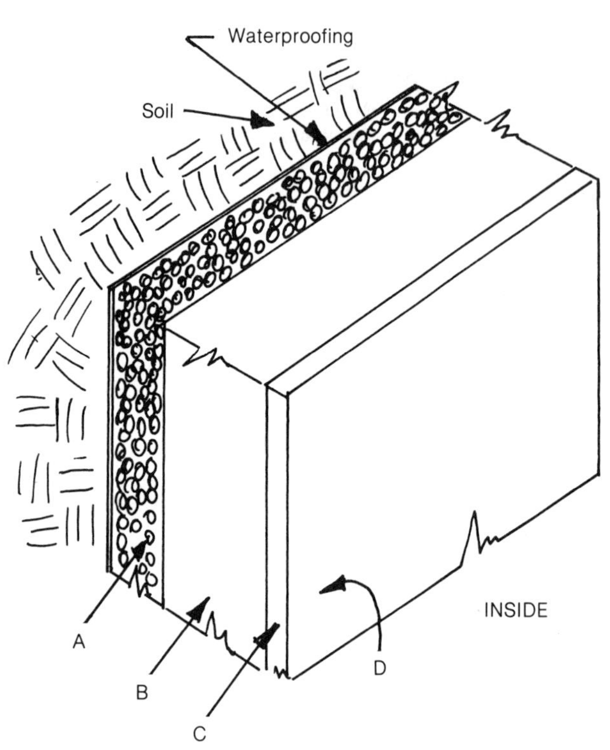

NO THERMAL INSULATION
THERMAL RESISTANCE R and R_m

Sand and Gravel Concrete
With Interior Wallboard

	Components	English Units 8″ Thk.	English Units 12″ Thk.		Metric Units 20 cm Thk.	Metric Units 30 cm Thk.
A	Insulation	—	—		—	—
B	Concrete	0.66	1.00		0.116	0.176
C	Gypsum wallboard	0.43	0.43		0.075	0.075
D	Inside surface	0.58	0.58		0.102	0.102
	RESISTANCE R =	1.67	2.01	R_m =	0.193	0.253
	TRANSMITTANCE U =	0.599	0.498	U_m =	5.181	3.953

Cinder Block or Clay Tile
With Interior Wallboard

	Components	English Units 8″ Thk.	English Units 12″ Thk.		Metric Units 20 cm Thk.	Metric Units 30 cm Thk.
A	Insulation	—	—		—	—
B	Cinder block or tile	1.72	1.90		0.303	0.325
C	Gypsum wallboard	0.43	0.43		0.075	0.075
D	Inside surface	0.58	0.58		0.102	0.102
	RESISTANCE R =	2.73	2.91	R_m =	0.480	0.502
	TRANSMITTANCE U =	0.367	0.344	U_m =	2.053	1.992

CONSTRUCTION ASSEMBLY XXVI

Note: If wallboard is not used subtract 0.43 from R and 0.075 from R_m.

61. CELLULAR GLASS INSULATION
THERMAL RESISTANCE AND TRANSMITTANCE OF ASSEMBLY XXVI WITH CELLULAR GLASS INSULATION: $k = 0.33$ Btu, in./ft^2, hr °F at 45°F, $k_m = 0.047$ W/mK at 7.2°C,

Masonry Wall	Insulation Thickness		8″ (20.3 cm) Wall Thickness English Units		8″ (20.3 cm) Wall Thickness Metric Units		12″ (30.4 cm) Wall Thickness English Units		12″ (30.4 cm) Wall Thickness Metric Units	
	Inches	Meters	R	U	R_m	U_m	R	U	R_m	U_m
	1 1/2	0.038	6.20	0.161	1.092	0.917	6.54	0.153	1.152	0.870
Sand and	2	0.051	7.71	0.130	1.359	0.735	8.05	0.124	1.418	0.704
Gravel	2 1/2	0.064	9.22	0.108	1.625	0.617	9.56	0.105	1.684	0.595
Concrete	3	0.076	10.73	0.093	1.891	0.529	11.73	0.085	2.067	0.483
	3 1/2	0.089	12.24	0.082	2.157	0.463	12.58	0.079	2.217	0.450
	4	0.102	13.75	0.073	2.423	0.413	14.09	0.071	2.483	0.403
	1 1/2	0.038	7.26	0.138	1.279	0.781	7.44	0.134	1.311	0.763
Cinder Blk.	2	0.051	8.77	0.114	1.545	0.667	8.95	0.112	1.577	0.633
or	2 1/2	0.064	10.28	0.092	1.811	0.552	10.46	0.096	1.843	0.543
Clay Tile	3	0.076	11.79	0.085	2.077	0.481	11.97	0.084	2.109	0.474
	3 1/2	0.085	13.30	0.075	2.349	0.427	13.48	0.074	2.375	0.420
	4	0.102	14.81	0.068	2.609	0.383	14.99	0.067	2.641	0.379

R = °F, ft^2, hr, Btu R_m = km^{2}W U = Btu per ft^2 per hr U_m = W/m^{2}K

TABLE 147

MASONARY FLOOR
ABOVE AREA OPEN TO OUTSIDE

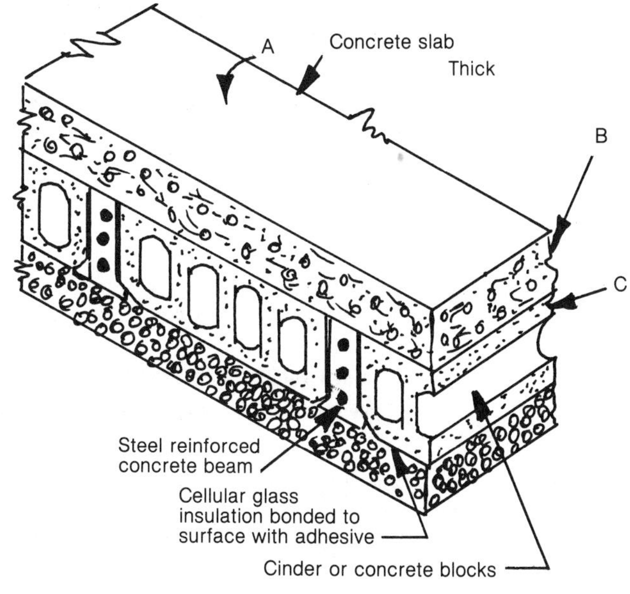

CONSTRUCTION ASSEMBLY XXVII

NO THERMAL INSULATION
THERMAL RESISTANCE R and R_m

Mean Temperature 40° (4.4°C) — Heating

	Components	Eng. Units		Metric Units
A	Upper surface	0.25		0.044
B	Concrete slab 6″ (15 cm)	0.48		0.085
C	Cinder blocks and beam	1.61		0.284
D	Cellular glass insulation	—		—
E	Lower surface	0.23		0.040
	(TOTAL) RESISTANCE R =	2.57	R_m =	0.453
	TRANSMITTANCE U =	0.397	U_m =	2.252

Mean Temperature 85° (29.4°C) — Cooling

	Components	Eng. Units		Metric Units
A	Upper surface	0.20		0.035
B	Concrete slab 6″ (15 cm)	0.48		0.085
C	Cinder block and beam	1.50		0.264
D	Cellular glass insulation	—		—
E	Lower surface	0.19		0.033
	(TOTAL) RESISTANCE R =	2.37	R_m =	0.417
	TRANSMITTANCE U =	0.422	U_m =	2.395

62. CELLULAR GLASS INSULATION

THERMAL RESISTANCE AND TRANSMITTANCE OF ASSEMBLY XXVII WITH CELLULAR GLASS INSULATION: k = 0.33 Btu, in./ft^2, hr °F at 40°F, k_m = 0.047 W/mK at 4.4°C, k = 0.35 Btu, in./ft^2, hr °F at 85°F, k_m = 0.050 W/mK at 29.4°C

Insulation Thickness		Mean Temp. 40°F (4.4°C)				Mean Temp. 85°F (29.4°C)			
		English Units		Metric Units		English Units		Metric Units	
Inches	Meters	R	U	R_m	U_m	R	U	R_m	U_m
1 1/2	0.038	7.14	0.140	1.258	0.793	6.64	0.151	0.170	1.855
2	0.051	8.67	0.115	1.528	0.624	8.07	0.124	1.422	0.704
2 1/2	0.064	10.19	0.098	1.794	0.556	9.49	0.105	1.672	0.599
3	0.076	11.72	0.085	2.065	0.483	10.91	0.092	1.922	0.521
3 1/2	0.089	13.24	0.076	2.333	0.429	12.34	0.081	2.174	0.461
4	0.102	14.77	0.068	2.602	0.385	13.77	0.073	2.462	0.412
4 1/2	0.114	16.29	0.061	2.870	0.348	15.19	0.066	2.676	0.373
5	0.127	17.81	0.056	3.138	0.318	16.62	0.061	2.928	0.341
6	0.152	20.86	0.048	3.676	0.272	19.46	0.051	3.429	0.292

R = °F, ft^2, hr, Btu R_m = km^{2}W   U = Btu per ft^2 per hr U_m = W/m^{2}K

TABLE 148

FLOOR
ABOVE AREA OPEN TO OUTSIDE

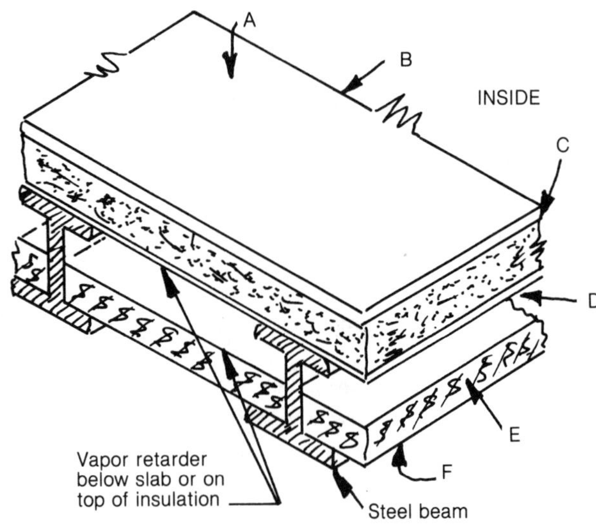

CONSTRUCTION ASSEMBLY XXVIII

Note: The air space D is not in the assembly before the insulation is installed. After installation of insulation its resistance is R = 0.85, R_m = 0.149.

The metal joists act as a thermal bipass around the insulation reducing its thermal resistance.

NO THERMAL INSULATION
THERMAL RESISTANCE R and R_m

Mean Temperature 40° (4.4°C)

	Components	Eng. Units		Metric Units
A	Inside surface	0.50		0.088
B	Linoleum or tile	0.05		0.008
C	Concrete slab 6″ (15.24 cm)	0.48		0.085
D	Air space	—		—
E	Insulation	—		—
F	Lower surface	0.23		0.040
	(TOTAL) RESISTANCE R =	1.26	R_m =	0.221
	TRANSMITTANCE U =	0.794	U_m =	4.504

Mean Temperature 85° (29.4°C)

	Components	Eng. Units		Metric Units
A	Inside surface	0.45		0.079
B	Linoleum or tile	0.05		0.008
C	Concrete slab 6″ (15.24 cm)	0.48		0.085
D	Air space	—		—
E	Insulation	—		—
F	Lower surface	0.19		0.033
	(TOTAL) RESISTANCE R =	1.17	R_m =	0.205
	TRANSMITTANCE U =	0.855	U_m =	4.878

63. CELLULAR GLASS INSULATION
THERMAL RESISTANCE AND TRANSMITTANCE OF ASSEMBLY XXVIII WITH CELLULAR GLASS INSULATION:

k = 0.33 Btu, in./ft^2, hr °F at 40°F, k_m = 0.047 W/mK at 4.4°C, k = 0.35 Btu, in./ft^2, hr °F at 85°F, k_m = 0.050 W/mK at 29.4°C

Insulation Thickness		Mean Temp. 40°F (4.4°C)				Mean Temp. 85°F (29.4°C)			
		English Units		Metric Units		English Units		Metric Units	
Inches	Meters	R	U	R_m	U_m	R	U	R_m	U_m
1 1/2	0.038	5.83	0.172	1.027	0.971	5.44	0.184	0.959	1.040
2	0.051	7.36	0.136	1.297	0.769	6.87	0.146	1.210	0.826
2 1/2	0.064	8.88	0.113	1.565	0.641	8.29	0.121	1.461	0.685
3	0.076	10.41	0.096	1.834	0.546	9.72	0.103	1.713	0.585
3 1/2	0.089	11.93	0.084	2.102	0.476	11.14	0.090	1.963	0.510
4	0.102	13.46	0.074	2.372	0.422	12.57	0.080	2.215	0.450

R = °F, ft^2, hr, Btu R_m = km^{2}W U = Btu per ft^2 per hr U_m = W/m^{2}K

TABLE 149

FLOOR
ABOVE AREA OPEN TO OUTSIDE
CONSTRUCTION ASSEMBLY XXVIII

63. GLASS FIBER BOARD, BLANKET OR BATT INSULATION THERMAL RESISTANCE AND TRANSMITTANCE OF ASSEMBLY XXVIII WITH GLASS FIBER INSULATION:

k = 0.24 Btu, in./ft^2, hr °F at 40 °F, k_m = 0.035 W/mK at 4.4 °C, k = 0.28 Btu, in./ft^2, hr °F at 85 °F, k_m = 0.040 W/mK at 29.4 °C

Insulation Thickness		Mean Temp. 40 °F (4.4 °C)				Mean Temp. 85 °F (29.4 °C)			
		English Units		Metric Units		English Units		Metric Units	
Inches	Meters	R	U	R_m	U_m	R	U	R_m	U_m
1	0.025	5.23	0.191	0.922	1.085	4.69	0.212	0.826	1.211
2	0.051	8.36	0.120	1.473	0.679	7.38	0.136	1.300	0.769
3	0.076	11.49	0.087	2.024	0.494	10.06	0.099	1.773	0.564
4	0.102	14.61	0.068	2.574	0.388	12.73	0.079	2.243	0.446
5	0.127	17.74	0.056	3.126	0.320	15.41	0.064	2.715	0.368
6	0.152	20.86	0.047	3.676	0.272	18.09	0.055	3.187	0.314
7	0.178	23.98	0.041	4.225	0.237	20.77	0.048	3.660	0.273
8	0.203	27.11	0.037	4.777	0.209	23.45	0.042	4.132	0.242

TABLE 150

64. MINERAL FIBER HEAVY DENSITY BOARD, BLANKET OR BATT INSULATION THERMAL RESISTANCE AND TRANSMITTANCE OF ASSEMBLY XXVIII WITH MINERAL INSULATION:

k = 0.36 Btu, in./ft^2, hr °F at 40 °F, k_m = 0.052 W/mK at 4.4 °C, k = 0.39 Btu, in./ft^2, hr °F at 85 °F, k_m = 0.056 W/mK at 29.4 °C

Insulation Thickness		Mean Temp. 40 °F (4.4 °C)				Mean Temp. 85 °F (29.4 °C)			
		English Units		Metric Units		English Units		Metric Units	
Inches	Meters	R	U	R_m	U_m	R	U	R_m	U_m
1	0.025	4.33	0.231	0.763	1.311	4.07	0.246	0.717	1.394
2	0.051	6.55	0.153	1.154	0.866	6.12	0.163	1.078	0.927
3	0.076	8.78	0.114	1.547	0.647	8.17	0.122	1.439	0.695
4	0.102	11.00	0.091	1.938	0.516	10.25	0.098	1.806	0.554
5	0.127	13.22	0.076	2.329	0.429	12.28	0.081	2.164	0.462
6	0.152	15.44	0.065	2.720	0.368	14.32	0.070	2.523	0.396
7	0.178	17.67	0.057	3.113	0.321	16.38	0.061	2.886	0.346
8	0.203	19.89	0.050	3.505	0.285	18.43	0.054	3.247	0.308

Note: Batt or blanket insulation may require hexagonal wire netting or sheets secured to bottom of steel member for support.

TABLE 151

FLOOR
ABOVE AREA OPEN TO OUTSIDE

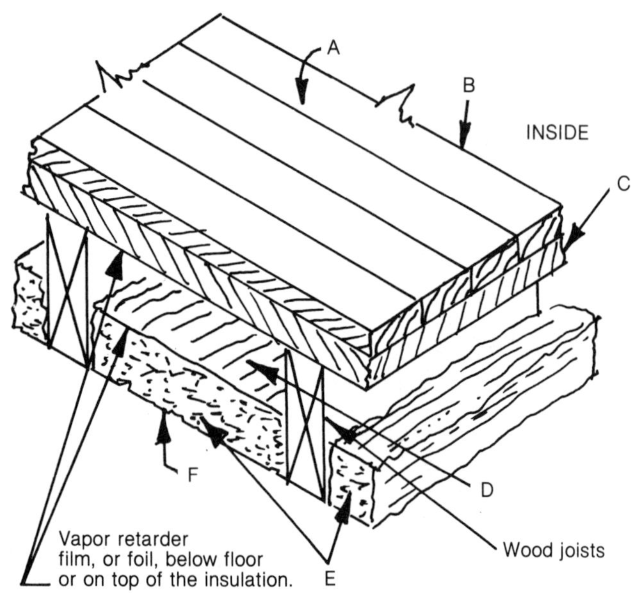

CONSTRUCTION ASSEMBLY XXIX

Note: The air space D is not in the assembly before the insulation is installed. After installation its resistance is R = 0.85, R_m = 0.149.

NO THERMAL INSULATION
THERMAL RESISTANCE R and R_m

Mean Temperature 40° (4.4°C)

	Components	Eng. Units		Metric Units
A	Inside surface	0.50		0.088
B	Wood floor 3/4″ (1.9 cm)	0.80		0.140
C	Plywood subfloor 3/4″ (1.9 cm)	0.86		0.151
D	Air space	—		—
E	Insulation	—		—
F	Lower surface	0.23		0.040
	(TOTAL) RESISTANCE R =	2.39	R_m =	0.419
	TRANSMITTANCE U =	0.418	U_m =	2.387

Mean Temperature 85° (29.4°C)

	Components	Eng. Units		Metric Units
A	Inside surface	0.45		0.079
B	Wood floor 3/4″ (1.9 cm)	0.80		0.140
C	Plywood subfloor 3/4″ (1.9 cm)	0.86		0.151
D	Air space	—		—
E	Insulation	—		—
F	Lower surface	0.19		0.033
	(TOTAL) RESISTANCE R =	2.30	R_m =	0.403
	TRANSMITTANCE U =	0.435	U_m =	2.481

65. GLASS FIBER BOARD, BLANKET OR BATT INSULATION
THERMAL RESISTANCE AND TRANSMITTANCE OF ASSEMBLY XXIX WITH GLASS FIBER INSULATION:

k = 0.24 Btu, in./ft², hr °F at 40°F, k_m = 0.035 W/mK at 4.4°C,
k = 0.28 Btu, in./ft², hr °F at 85°F, k_m = 0.040 W/mK at 29.4°C

Insulation Thickness		Mean Temp. 40°F (4.4°C)				Mean Temp. 85°F (29.4°C)			
		English Units		Metric Units		English Units		Metric Units	
Inches	Meters	R	U	R_m	U_m	R	U	R_m	U_m
1	0.025	7.40	0.135	1.304	0.767	6.72	0.149	1.184	0.845
2	0.051	11.57	0.086	2.039	0.490	10.29	0.097	1.813	0.551
3	0.076	15.74	0.064	2.773	0.361	13.86	0.072	2.442	0.409
4	0.102	19.91	0.050	3.508	0.285	17.44	0.057	3.073	0.325
5	0.127	24.07	0.042	4.241	0.236	21.01	0.048	3.702	0.270
6	0.152	28.24	0.035	4.976	0.201	24.58	0.041	4.331	0.231
7	0.178	32.41	0.031	5.710	0.175	28.15	0.036	4.960	0.202
8	0.203	36.57	0.027	6.444	0.155	31.72	0.032	5.589	0.179

R = °F, ft², hr, Btu R_m = km²W   U = Btu per ft² per hr U_m = W/m²K

TABLE 152

FLOOR
ABOVE AREA OPEN TO OUTSIDE
CONSTRUCTION ASSEMBLY XXIX

66. MINERAL FIBER HEAVY DENSITY BOARD, BLANKET OR BATT INSULATION THERMAL RESISTANCE AND TRANSMITTANCE OF ASSEMBLY XXIX WITH MINERAL FIBER INSULATION:

INSULATION: k = 0.36 Btu, in./ft^2, hr °F at 40 °F, k_m = 0.052 W/mK at 4.4 °C, k = 0.39 Btu, in./ft^2, hr °F at 85 °F, k_m = 0.056 W/mK at 29.4 °C

Insulation Thickness		Mean Temp. 40 °F (4.4 °C)				Mean Temp. 85 °F (29.4 °C)			
		English Units		Metric Units		English Units		Metric Units	
Inches	Meters	R	U	R_m	U_m	R	U	R_m	U_m
1	0.025	6.01	0.166	1.059	0.944	5.71	0.175	1.006	0.994
2	0.051	8.79	0.114	1.549	0.646	8.28	0.121	1.459	0.685
3	0.076	11.57	0.086	2.038	0.490	10.84	0.092	1.910	0.524
4	0.102	14.35	0.070	2.528	0.395	13.41	0.075	2.363	0.423
5	0.127	17.13	0.058	3.018	0.331	15.97	0.063	2.814	0.355
6	0.152	19.91	0.050	3.508	0.285	18.53	0.054	3.265	0.306
7	0.178	22.63	0.044	3.987	0.251	21.10	0.047	3.718	0.269
8	0.203	25.46	0.039	4.486	0.223	23.66	0.042	4.169	0.240

R = °F, ft^2, hr, Btu R_m = km^{2}W U = Btu per ft^2 per hr U_m = W/m^{2}K

TABLE 153

FLOOR
ABOVE AREA OPEN TO OUTSIDE

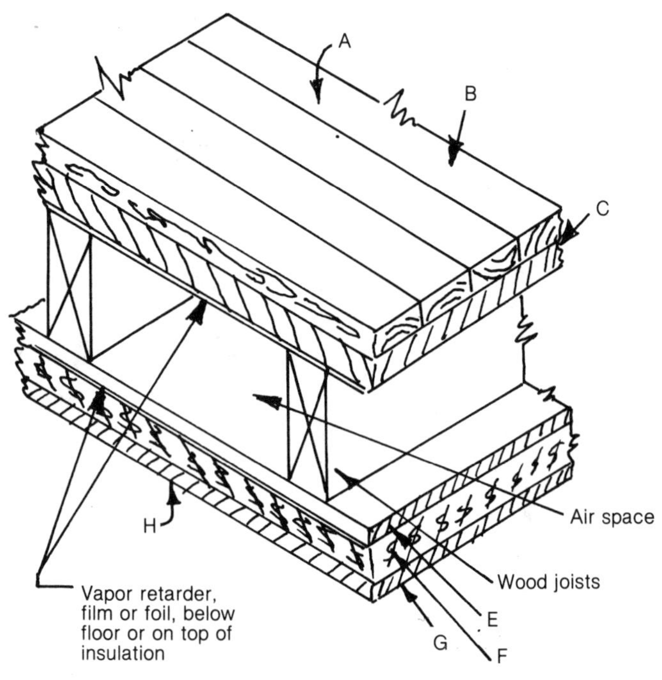

CONSTRUCTION ASSEMBLY XXX

Note: When organic foam boards are installed below wood floor they should be installed with 1/2″ gypsum boards on both sides to prevent rapid flame spread in case of fire below.

NO THERMAL INSULATION*
THERMAL RESISTANCE R and R_m

Mean Temperature 40° (4.4°C)

	Components	Eng. Units		Metric Units
A	Inside surface	0.50		0.088
B	Wood floor 3/4″ (1.9 cm)	0.80		0.140
C	Plywood subfloor 3/4″ (1.9 cm)	0.86		0.151
D	Air space	0.85		0.149
E	Gypsum board 1/2″ (1.27 cm)	—		—
F	Insulation	—		—
G	Gypsum board 1/2″ (1.27 cm)	0.45		0.079
H	Lower surface	0.23		0.040
	(TOTAL) RESISTANCE R =	3.69	R_m =	0.647
	TRANSMITTANCE U =	0.271	U_m =	1.546

Mean Temperature 85° (29.4°C)

	Components	Eng. Units		Metric Units
A	Inside surface	0.45		0.079
B	Wood floor 3/4″ (1.2 cm)	0.80		0.140
C	Plywood subfloor 3/4″ (1.2 cm)	0.86		0.151
D	Air space	0.85		0.149
E	Gypsum board 1/2″ (1.27 cm)	—		—
F	Insulation	—		—
G	Gypsum board 1/2″ (1.27 cm)	0.45		0.079
H	Lower surface	0.19		0.033
	(TOTAL) RESISTANCE R =	3.60	R_m =	0.631
	TRANSMITTANCE U =	0.278	U_m =	1.585

*Calculations include gypsum board on bottom.

67. POLYSTYRENE FOAM INSULATION SHEETS
THERMAL RESISTANCE AND TRANSMITTANCE OF ASSEMBLY XXX WITH POLYSTYRENE INSULATION

$k = 0.20$ Btu, in./ft^2, hr °F at 40°F, $k_m = 0.035$ W/mK at 4.4°C, $k = 0.24$ Btu, in./ft^2, hr °F at 85°F, $k_m = 0.044$ W/mK at 29.4°C

Insulation Thickness		Mean Temp. 40°F (4.4°C)				Mean Temp. 85°F (29.4°C)			
		English Units		Metric Units		English Units		Metric Units	
Inches	Meters	R	U	R_m	U_m	R	U	R_m	U_m
1	0.025	9.41	0.106	1.658	0.603	8.21	0.122	1.447	0.691
1 1/2	0.037	11.91	0.084	2.099	0.477	10.30	0.097	1.815	0.551
2	0.051	14.41	0.069	2.539	0.394	12.38	0.081	2.181	0.458
2 1/2	0.064	16.91	0.059	2.979	0.336	14.47	0.069	2.550	0.392
3	0.076	19.41	0.052	3.420	0.292	16.55	0.060	2.916	0.343

R = °F, ft^2, hr, Btu R_m = km^{2}W U = Btu per ft^2 per hr U_m = W/m^{2}K

TABLE 154

68. POLYURETHANE OR POLYISOCYANURATE FOAM INSULATION SHEETS THERMAL RESISTANCE AND TRANSMITTANCE OF ASSEMBLY XXX WITH URETHANE (OR ISOCYANURATE) FOAM INSULATION:

k = 0.24 Btu, in./ft^2, hr °F at 40 °F, k_m = 0.042 W/mK at 4.4 °C, k = 0.30 Btu, in./ft^2, hr °F at 85 °F, k_m = 0.053 W/mK at 29.4 °C

Insulation Thickness		Mean Temp. 40 °F (4.4 °C)				Mean Temp. 85 °F (29.4 °C)			
		English Units		Metric Units		English Units		Metric Units	
Inches	Meters	R	U	R_m	U_m	R	U	R_m	U_m
3/4	0.018	7.26	0.138	1.279	0.781	6.54	0.152	1.152	0.868
1	0.025	8.30	0.120	1.462	0.684	7.37	0.136	1.298	0.770
1 1/2	0.087	10.39	0.096	1.831	0.546	9.04	0.111	1.593	0.628
2	0.051	12.47	0.080	2.197	0.455	10.70	0.093	1.885	0.530
2 1/2	0.064	14.56	0.069	2.565	0.390	12.37	0.081	2.179	0.459
3	0.076	16.64	0.060	2.932	0.341	14.04	0.071	2.474	0.404

R = °F, ft^2, hr, Btu R_m = km^{2}W U = Btu per ft^2 per hr U_m = W/m^{2}K

TABLE 155

FLOOR
ABOVE AREA OPEN TO OUTSIDE

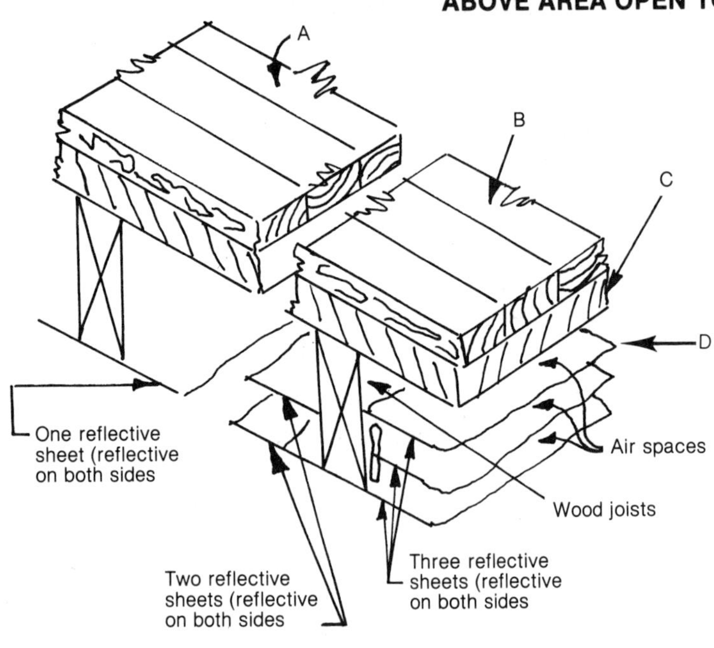

CONSTRUCTION ASSEMBLY XXXI

NO THERMAL INSULATION
THERMAL RESISTANCE R and R_m

Mean Temperature 40° (4.4°C)

	Components	Eng. Units		Metric Units
A	Inside surface	0.50		0.088
B	Wood floor 3/4" (1.9 cm)	0.80		0.140
C	Plywood subfloor	0.86		0.151
D	Insulation, reflective (Number of reflective sheets and spaces as shown) below)	—		—
	(TOTAL) RESISTANCE R =	2.16	R_m =	0.379
	TRANSMITTANCE U =	0.474	U_m =	2.639

Mean Temperature 85° (29.4°C)

	Components	Eng. Units		Metric Units
A	Inside surface	0.45		0.079
B	Wood floor 3/4" (1.9 cm)	0.80		0.140
C	Plywood subfloor	0.86		0.151
D	Insulation, reflective	—		—
	(TOTAL) RESISTANCE R =	2.11	R_m =	0.370
	TRANSMITTANCE U =	0.474	U_m =	2.703

69. REFLECTIVE ALUMINUM FOIL INSULATION
THERMAL RESISTANCE AND TRANSMITTANCE OF ASSEMBLY XXXI WITH REFLECTIVE INSULATION SHEETS. SURFACE EMITTANCE $\varepsilon = 0.05$

Horizontal Reflective Spaces		Mean Temp. 40°F (4.4°C) Heat Flow Up (Heating)				Mean Temp. 85°F (29.4°C) Heat Flow Down (Cooling)			
		English Units		Metric Units		English Units		Metric Units	
Number of Enclosed Air Spaces	Aluminum Reflective Sheets (Reflective on Both Sides)	R	U	R_m	U_m	R	U	R_m	U_m
1	1 (attached to bottom)	8.71	0.115	1.535	0.652	7.67	0.130	1.351	0.740
3	2 (between joists)	15.32	0.065	2.699	0.370	13.07	0.077	2.303	0.434
4	3 (between joists)	21.90	0.046	3.859	0.259	18.47	0.054	3.254	0.307

TABLE 156

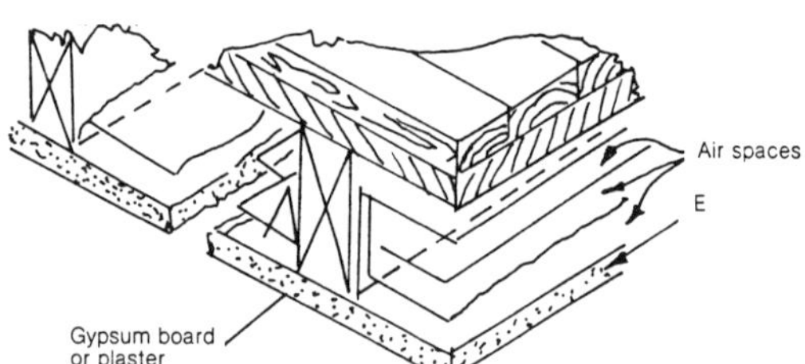

CONSTRUCTION ASSEMBLY XXXI

THERMAL RESISTANCE AND TRANSMITTANCE OF ASSEMBLY XXXI WITH CEILING BOARD BELOW THE REFLECTIVE INSULATION SHEETS.

Horizontal Reflective Spaces		Mean Temp. 40°F (4.4°C) Heat Flow Up (Heating)				Mean Temp. 85°F (29.4°C) Heat Flow Down (Cooling)			
		English Units		Metric Units		English Units		Metric Units	
Number of Enclosed Air Spaces	Aluminum Reflective Sheets (Reflective on Both Sides)	R	U	R_m	U_m	R	U	R_m	U_m
2	1 (between joists)	10.03	0.100	1.767	0.566	9.02	0.111	1.589	0.629
3	2 (between joists)	16.67	0.060	2.937	0.340	14.42	0.069	2.541	0.394
4	3 (between joists)	24.23	0.041	4.270	0.234	19.82	0.050	3.492	0.284

TABLE 157

FLOOR SLAB ON GRADE

PERIMETER INSULATION

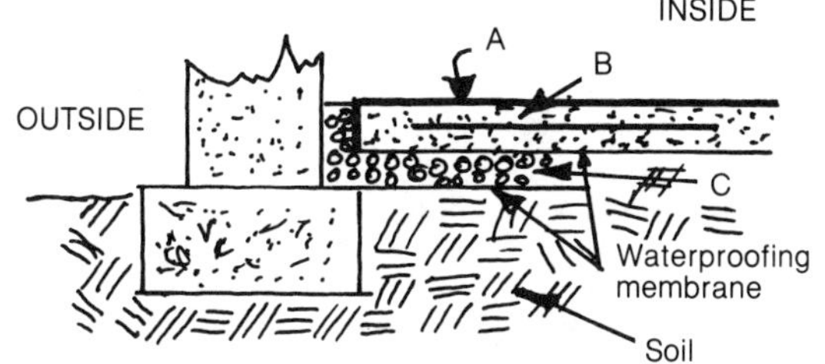

CONSTRUCTION ASSEMBLY XXXII

Note: The temperature difference depends upon location. The outermost Δt is determined by temperature between soil temperature and air temperature. Further inward the temperature difference is determined by ground temperature.

NO THERMAL INSULATION
THERMAL RESISTANCE R and R_m
Mean Temperature 50° (10°C)

	Components	English Units 4″ Slab	English Units 6″ Slab		Metric Units 10.1 cm Slab	Metric Units 15.2 cm Slab
A	Top surface	0.50	0.50		0.088	0.088
B	Concrete floor slab	0.33	0.50		0.058	0.088
C	Insulation	—	—		—	—
	RESISTANCE R =	0.83	1.00	R_m =	0.156	0.176
	TRANSMITTANCE U =	1.205	1.000	U_m =	6.849	5.682

70. CELLULAR GLASS INSULATION
THERMAL RESISTANCE AND TRANSMITTANCE OF ASSEMBLY XXXII WITH CELLULAR GLASS INSULATION: k = 0.33 Btu, in./ft², hr °F at 50°F, k_m = 0.000 W/mK at 10°C,

Insulation Thickness Inches	Meters	4″ (10.1 cm) Thk. Concrete Slab English Units R	U	Metric Units R_m	U_m	6″ (15.2 cm) Thk. Concrete Slab English Units R	U	Metric Units R_m	U_m
1 1/2	0.036	5.38	0.186	0.948	1.050	5.50	0.182	0.969	1.030
2	0.051	6.89	0.145	1.214	0.826	7.00	0.142	1.233	0.813
2 1/2	0.064	8.39	0.119	1.478	0.676	8.51	0.118	1.499	0.667
3	0.076	9.89	0.101	1.743	0.575	10.00	0.100	1.762	0.568
3 1/2	0.089	11.39	0.088	2.007	0.498	11.51	0.087	2.028	0.493
4	0.102	12.89	0.077	2.271	0.441	13.01	0.077	2.292	0.437

R = °F, ft², hr, Btu R_m = km²W U = Btu per ft² per hr U_m = W/m²K

TABLE 158

71. HEAVY DENSITY POLYURETHANE OR POLYSTYRENE SHEETS
THERMAL RESISTANCE AND TRANSMITTANCE OF ASSEMBLY XXXII WITH ORGANIC FOAM SHEETS: k = 0.26 Btu, in./ft², hr °F at 50°F, k_m = 0.045 W/mK at 10°C,

Insulation Thickness Inches	Meters	4″ (10.1 cm) Thk. Concrete Slab English Units R	U	Metric Units R_m	U_m	6″ (15.2 cm) Thk. Concrete Slab English Units R	U	Metric Units R_m	U_m
1	0.025	4.73	0.212	0.833	1.200	4.83	0.206	0.851	1.175
1 1/2	0.036	6.64	0.150	1.169	0.855	6.77	0.140	1.193	0.838
2	0.051	8.57	0.117	1.510	0.662	8.69	0.115	1.531	0.653
2 1/2	0.064	10.50	0.095	1.850	0.540	10.62	0.094	1.871	0.534
3	0.076	12.42	0.081	2.188	0.460	12.53	0.080	2.208	0.453

R = °F, ft², hr, Btu R_m = km²W U = Btu per ft² per hr U_m = W/m²K

TABLE 159

FLOOR SLAB BELOW GRADE

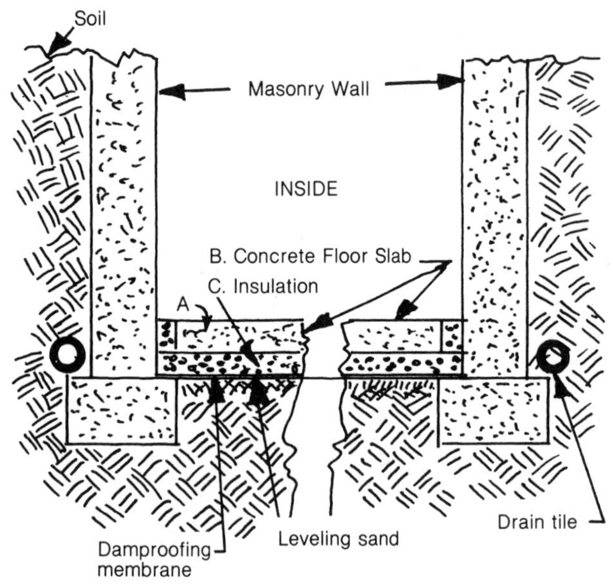

CONSTRUCTION ASSEMBLY XXXIII

Note: Due to low compressive strength organic foams are not recommended for insulation under entire slab.

NO THERMAL INSULATION THERMAL RESISTANCE R and R_m

Mean Temperature 50° (10°C)

	Components	English Units 4″ Slab	English Units 6″ Slab		Metric Units 10.1 cm Slab	Metric Units 20.3 cm Slab
A	Top surface	0.50	0.50		0.088	0.088
B	Concrete floor slab	0.33	0.50		0.058	0.088
C	Insulation	—	—		—	—
	RESISTANCE R =	0.83	1.00	R_m =	0.146	0.176
	TRANSMITTANCE U =	1.136	1.000	U_m =	6.849	5.682

71. CELLULAR GLASS INSULATION

THERMAL RESISTANCE AND TRANSMITTANCE OF ASSEMBLY XXXIII WITH CELLULAR GLASS INSULATION: k = 0.36 Btu, in./ft^2, hr °F at 50°F, k_m = 0.063 W/mK at 10°C,

Insulation Thickness		4″ (10.1 cm) Thk. Concrete Slab				6″ (15.2 cm) Thk. Concrete Slab			
		English Units		Metric Units		English Units		Metric Units	
Inches	Meters	R	U	R_m	U_m	R	U	R_m	U_m
1 1/2	0.036	5.38	0.186	0.948	1.050	5.50	0.182	0.969	1.030
2	0.051	6.89	0.145	1.214	0.826	7.00	0.142	1.233	0.813
2 1/2	0.064	8.39	0.119	1.478	0.676	8.51	0.118	1.499	0.667
3	0.076	9.89	0.101	1.743	0.575	10.00	0.100	1.762	0.568
3 1/2	0.089	11.39	0.088	2.007	0.498	11.51	0.087	2.028	0.493
4	0.102	12.89	0.077	2.271	0.441	13.01	0.077	2.292	0.437

R = °F, ft^2, hr, Btu R_m = km^{2}W U = Btu per ft^2 per hr U_m = W/m^{2}K

TABLE 160

72. PERLITE OR VERMICULITE LIGHT WEIGHT INSULATING CONCRETE

THERMAL RESISTANCE AND TRANSMITTANCE OF ASSEMBLY XXXIII WITH INSULATING CONCRETE: k = 0.50 Btu, in./ft^2, hr °F at 50°F, k_m = 0.216 W/mK at 10°C,

Insulation Thickness		4″ (10.1 cm) Thk. Concrete Slab				6″ (15.2 cm) Thk. Concrete Slab			
		English Units		Metric Units		English Units		Metric Units	
Inches	Meters	R	U	R_m	U_m	R	U	R_m	U_m
2	0.051	2.16	0.462	0.381	2.627	2.33	0.429	0.411	2.436
4	0.102	3.50	0.286	0.617	1.622	3.66	0.273	0.645	1.550
6	0.158	4.83	0.207	0.851	1.175	5.00	0.200	0.881	1.135

R = °F, ft^2, hr, Btu R_m = km^{2}W U = Btu per ft^2 per hr U_m = W/m^{2}K

TABLE 161

WINDOWS — GLASS

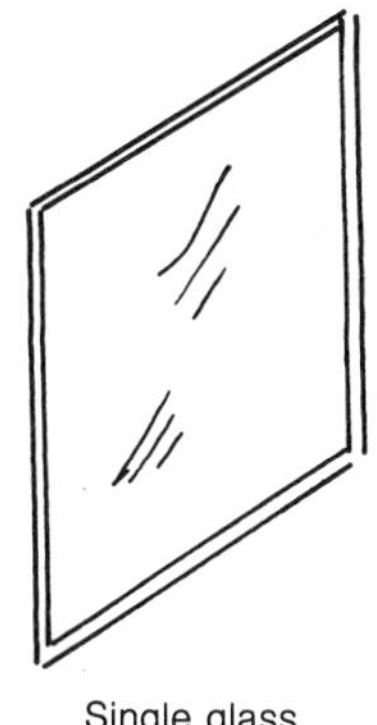

Single glass

Glass with storm window

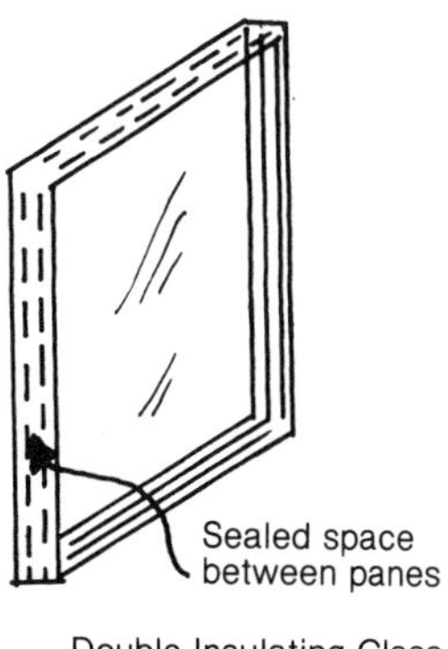

Double Insulating Glass

CONSTRUCTION ASSEMBLY XXXIV

73. GLASS, PANELS AND WINDOWS THERMAL RESISTANCE AND TRANSMITTANCE OF ASSEMBLY XXXIV

Window in Vertical Position	Frame and % Area	Mean Temperature 40°F (4.4°C) English Units R	U	Metric Units R_m	U_m	Mean Temperature 85°F (29.4°C) English Units R	U	Metric Units R_m	U_m
Single glass	Panel	0.91	1.099	0.160	6.237	0.94	1.064	0.166	6.038
Single glass	Wood (20%)	1.01	0.990	0.178	5.619	1.04	0.961	0.183	5.457
Single glass	Wood (40%)	1.13	0.885	0.199	5.022	1.16	0.862	0.204	4.893
Single glass	Metal	0.91	1.099	0.160	6.237	0.94	1.064	0.166	6.038
Glass & storm window	Wood (20%)	1.78	0.562	0.314	3.188	1.84	0.543	0.324	3.084
Glass & storm window	Metal	1.61	0.621	0.284	3.525	1.67	0.599	0.294	3.398
Double insul. glass	Panel	1.49	0.671	0.263	3.809	1.63	0.613	0.287	3.482
Double insul. glass	Wood (20%)	1.59	0.629	0.280	3.569	1.73	0.578	0.305	3.281
Double insul. glass	Metal	1.49	0.671	0.263	3.809	1.63	0.613	0.287	3.482

Window In Horizontal Position	Frame and % Area	Mean Temperature 40°F (4.4°C) English Units R	U	Metric Units R_m	U_m	Mean Temperature 85°F (29.4°) English Units R	U	Metric Units R_m	U_m
Single glass	Panel	0.82	1.219	0.144	6.921	1.20	0.833	0.211	4.729
Single glass	Wood (20%)	0.92	1.087	0.162	6.169	1.30	0.769	0.229	4.366
Single glass	Wood (40%)	1.01	0.990	0.178	5.619	1.39	0.719	0.245	4.083
Single glass	Metal	0.82	1.219	0.144	6.921	1.20	0.833	0.211	4.729
Double insul. glass	Panel	1.52	0.657	0.268	3.734	2.18	0.459	0.384	2.603
Double insul. glass	Wood (20%)	1.62	0.617	0.285	3.503	2.24	0.446	0.395	2.534
Double insul. glass	Metal	1.52	0.657	0.268	3.734	2.18	0.459	0.384	2.603

Note: The above resistances and transmittance includes surfaces on both sides. The above is for heat transfer from air to air on other side, it *does not* include heat transfer by solar radiation.

TABLE 162

GLASS — PLAIN & REFLECTIVE

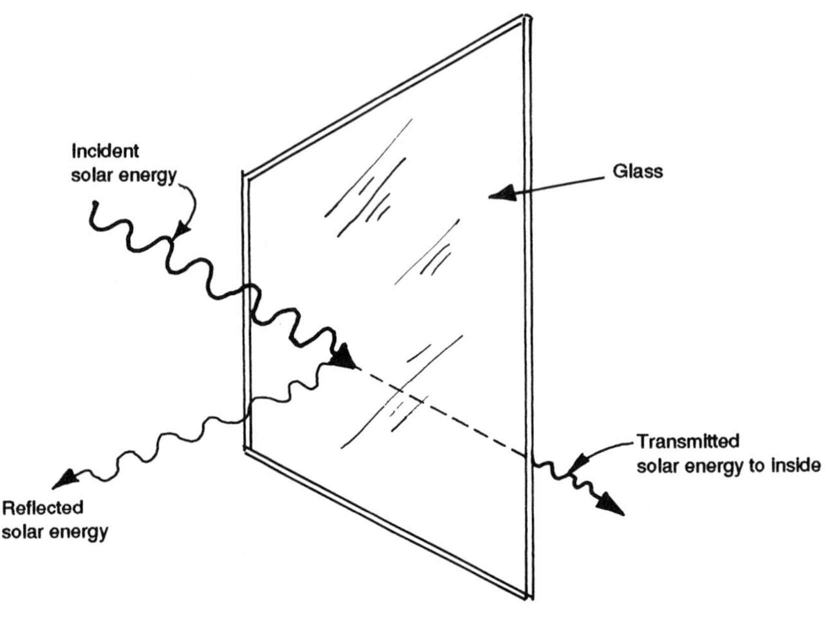

CONSTRUCTION ASSEMBLY XXXV

74. GLASS RADIATION TRANSMISSION

Thickness and Type of Glass	Treatment	Transmission Solar Energy Radiation	Ultra Violet Radiation	Visible Light	Overall Resistance and Transmission English Units R	U	Metric Units R_m	U_m
1/8″ (0.3 cm) clear	Untreated	86%	80%	90%	0.91	1.099	0.160	6.237
1/4″ (0.6 cm) clear	Untreated	74%	68%	87%	0.962	1.040	0.169	5.902
1/4″ (0.6 cm)	Green tint	50%	31%	65%	0.943	1.060	0.166	6.015
1/8″ (0.3 cm)	Solar control film	14%	25%	17%	1.087	0.920	0.192	5.221
1/4″ (0.6 cm)	Solar control film	12%	19%	16%	1.098	0.910	0.194	5.164
1/8″ (0.3 cm)	Glare control film	30%	45%	37%	1.052	0.950	0.185	5.391
1/4″ (0.6 cm)	Glare control film	27%	38%	36%	1.075	0.930	0.189	5.278

Note: See Table 38 to obtain solar heat to glass based on location, date, time and position.

TABLE 163

DOORS

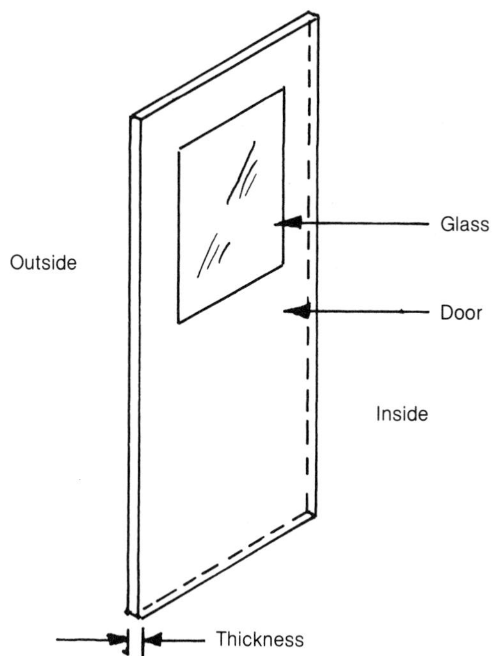

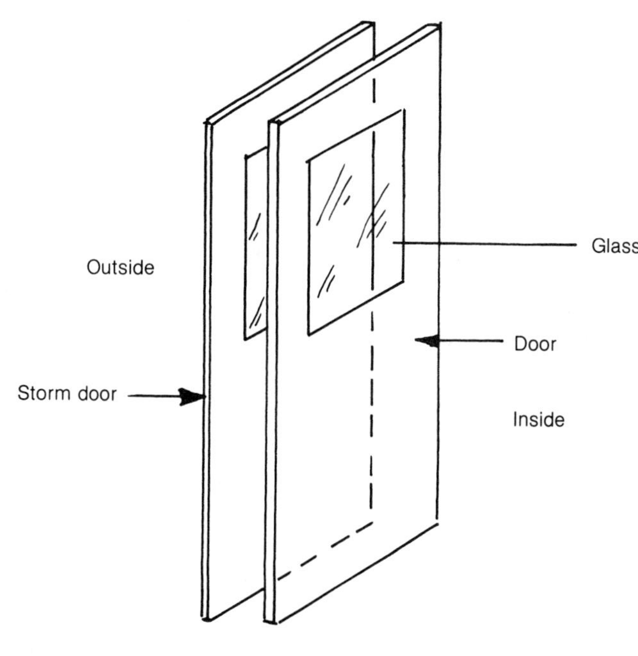

CONSTRUCTION ASSEMBLY XXXVI

75 DOORS

Description	Door Thickness Inches	Door Thickness cm	Storm Door With 50% Glass Area	Mean Temperature 40 °F (4.4 °C) English Units R	U	Metric Units R_m	U_m	Mean Temperature 85 °F (29.4 °C) English Units R	U	Metric Units R_m	U_m
Wood panel	1 (ave.)	2.5 (ave.)	None	1.56	0.640	0.275	3.638	1.60	0.610	0.282	3.547
Wood panel	1 (ave.)	2.5 (ave.)	Wood door	3.33	0.300	0.587	1.704	3.37	0.297	0.594	1.685
Wood panel	1 (ave.)	2.5 (ave.)	Metal door	2.56	0.390	0.451	2.217	2.61	0.383	0.460	2.174
Panel with glass	1 (ave.)	2.5 (ave.)	None	1.25	0.800	0.220	4.540	1.28	0.781	0.226	4.434
Panel with glass	1 (ave.)	2.5 (ave.)	Wood door	3.02	0.331	0.532	1.879	3.06	0.327	0.539	1.855
Panel with glass	1 (ave.)	2.5 (ave.)	Metal door	2.35	0.426	0.414	2.415	2.39	0.418	0.421	2.375
Solid wood door	1	2.54	None	1.56	0.641	0.275	3.638	1.60	0.625	0.282	3.547
Solid wood door	1	2.54	Wood door	3.33	0.300	0.587	1.704	3.37	0.297	0.594	1.684
Solid wood door	1	2.54	Metal door	2.56	0.391	0.451	2.217	2.61	0.383	0.459	2.174
Solid wood door	1 1/2	3.81	None	2.04	0.490	0.359	2.782	2.12	0.470	0.374	2.677
Solid wood door	1 1/2	3.81	Wood door	3.70	0.270	0.652	1.534	3.75	0.267	0.661	1.513
Solid wood door	1 1/2	3.81	Metal door	3.03	0.330	0.534	1.873	3.07	0.326	0.541	1.848
Solid wood door	2	5.08	None	2.32	0.431	0.409	2.446	2.36	0.482	0.416	2.405
Solid wood door	2	5.08	Wood door	4.16	0.240	0.733	1.364	4.21	0.237	0.742	1.348
Solid wood door	2	5.08	Metal door	3.44	0.291	0.606	1.650	3.49	0.287	0.615	1.626

Note: Above does not include heat loss (or gain) due to air infiltration.

TABLE 164

AIR INFILTRATION PER LENGTH OF WINDOW TO FRAME JOINTS

Type of Window Frame Material Type of Wall	Weather Stripping	Cubic Feet Per Foot Per Hour Wind Velocity, Miles Per Hour					Cubic Meters Per Meter Per Hour Wind Velocity, km Per Hour				
		5	**10**	**15**	**20**	**25**	**8.05**	**16.09**	**24.14**	**32.13**	**40.23**
Double hung — wood	Not stripped	7	21	39	59	80	0.198	0.594	1.104	1.670	2.26
Average Fitted	Stripped	4	13	24	36	49	0.113	0.368	0.679	1.019	1.387
Double hung — wood	Not stripped	27	69	111	154	249	0.764	1.954	3.140	4.359	7.049
Poorly fitted	Stripped	6	19	32	46	60	0.169	0.538	0.905	1.302	1.694
Double hung — metal	Not stripped	20	47	74	104	170	0.566	1.331	2.095	2.944	4.813
	Stripped	6	19	32	46	60	0.199	0.538	0.906	1.302	1.694
Industrial Pivoted	Not stripped	52	108	176	244	304	1.500	3.057	4.983	6.908	8.606
Residential Casement	Not stripped	10	25	43	62	80	0.283	0.708	1.217	1.755	2.265
Heavy Casement	Not stripped	6	19	30	41	53	0.199	0.538	0.906	1.302	1.500

Note: Storm windows will reduce the air infiltration to approximately one-half of the above listed values.

TABLE 165

WEIGHT OF AIR PER UNIT OF VOLUME

Temp. °F	Pounds Per Cubic Foot Percent Relative Humidity			Temp. °C	kg Per Cubic Meter Percent Relative Humidity		
	20%	**60%**	**100%**		**20%**	**60%**	**100%**
40	0.0658	0.0656	0.0654	14.4	1.054	1.051	1.048
50	0.0645	0.0643	0.0641	10.0	1.033	1.030	1.027
60	0.0633	0.0629	0.0625	15.6	1.014	1.007	1.001
70	0.0617	0.0613	0.0610	21.1	0.989	0.983	0.977
80	0.0606	0.0599	0.0588	26.7	0.971	0.959	0.942
90	0.0595	0.0588	0.0571	32.2	0.953	0.942	0.919
100	0.0580	0.0562	0.0555	37.8	0.929	0.899	0.890

TABLE 166

These tables provide the means of estimating the quantities of infiltration air which enters through the joints between windows and frames. The amount of heating or cooling each unit of air depends the thermodynamic properties of moist air at particular temperature and relative humidity conditions.*

*See ASHRAE GUIDE AND DATA BOOK.

DUCTS

Duct may be metal duct lined with insulation or insulation with outer reinforced aluminum foil.

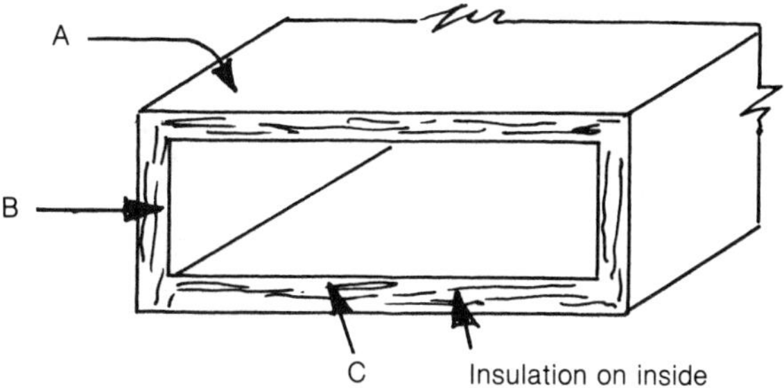

NO THERMAL INSULATION
THERMAL RESISTANCE R and R_m
Mean Temperature 60° (15.6°C) — Cooling

	Components	Eng. Units		Metric Units
A	Outside surface	0.50		0.088
B	Duct	0.01		0.002
C	Insulation	—		—
	(TOTAL) RESISTANCE R =	0.51	R_m =	0.090
	TRANSMITTANCE U =	1.961	U_m =	11.111

Mean Temperature 90° (32.2°C) — Heating — Same as Above

Note: No inner surface resistance is included as it would change with air velocity and due to the movement it would be small.

CONSTRUCTION ASSEMBLY XXXVII

Note: On ducts operating below dew-point temperature the outer casing acts as the vapor retarder.

76. GLASS FIBER INSULATING BOARD OR BLANKET INSULATION THERMAL RESISTANCE AND TRANSMITTANCE OF ASSEMBLY XXXVII WITH GLASS FIBER INSULATION:

k = 0.23 Btu, in./ft^2, hr °F at 60°F, k_m = 0.033 W/mKat 15.6°C, k = 0.28 Btu, in./ft^2, hr °F at 90°F, k_m = 0.040 W/mK at 32.2°C,

Insulation Thickness		Mean Temp. 60°F (15.6°C)				Mean Temp. 90°F (32.2°C)			
		English Units		Metric Units		English Units		Metric Units	
Inches	Meters	R	U	R_m	U_m	R	U	R_m	U_m
1	0.025	4.85	0.206	0.855	1.170	4.08	0.245	0.719	1.391
1 1/2	0.038	7.03	0.142	1.239	0.807	5.87	0.170	1.034	0.967
2	0.051	9.20	0.109	1.621	0.617	7.65	0.131	1.347	0.741

R = °F, ft^2, hr, Btu R_m = km^{2}W U = Btu per ft^2 per hr U_m = W/m^{2}K

TABLE 167

ACOUSTICAL ATTENUATION: Glass fiber insulation also acts as a sound absorber. The following table provides information as to sound attenuation.

H_2 = Inside Perimeter in Inches Divided by Cross Section in Sq Inches	Attenuation Db Per Linear Foot Octave Band Center Frequencies					
	125	250	500	1000	2000	4000
0.6	8.0	2.0	4.0	4.0	6.0	5.0
0.5	8.0	1.8	4.0	3.0	5.0	4.0
0.4	8.0	1.3	3.0	3.0	5.0	4.0
0.3	8.0	1.1	3.0	3.0	4.0	3.0
0.2	8.0	1.0	2.0	2.0	1.0	2.0
0.15	8.0	1.0	2.0	2.0	1.0	2.0
0.125	6.0	1.0	1.7	2.0	1.0	1.0

TABLE 168

DUCTS

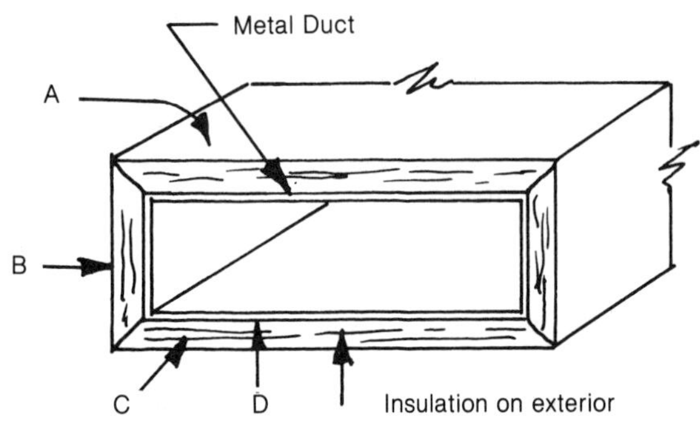

CONSTRUCTION ASSEMBLY XXXVIII

NO THERMAL INSULATION
THERMAL RESISTANCE R and R_m
Mean Temperature 60° (15.6°C) — Cooling

	Components	Eng. Units		Metric Units
A	Outside surface	0.50		0.088
B	Exterior jacked	0.02		0.003
C	Insulation	—		—
D	Metal duct	0.01		0.002
	(TOTAL) RESISTANCE R =	0.53	R_m =	0.093
	TRANSMITTANCE U =	1.876	U_m =	10.753

Mean Temperature 90° (32.2°C) — Heating

Same As Above

77. GLASS FIBER BOARD, BLANKET OR BATT INSULATION
THERMAL RESISTANCE AND TRANSMITTANCE OF ASSEMBLY XXXVIII WITH GLASS FIBER INSULATION (NOT SUITABLE IF DUCT TEMPERATURE IS BELOW AMBIENT AIR DEW POINT TEMPERATURE):

k = 0.23 Btu, in./ft², hr °F at 60°F, k_m = 0.033 W/mK at 15.6°C, k = 0.28 Btu, in./ft², hr °F at 90°F, k_m = 0.040 W/mK at 32.2°C

Insulation Thickness		Mean Temp. 60°F (15.6°C)				Mean Temp. 90°F (32.2°C)			
		English Units		Metric Units		English Units		Metric Units	
Inches	Meters	R	U	R_m	U_m	R	U	R_m	U_m
1	0.025	4.88	0.205	0.860	1.163	4.10	0.244	0.722	1.384
1 1/2	0.038	7.05	0.142	1.242	0.805	5.89	0.170	1.038	0.964
2	0.051	9.22	0.108	1.624	0.616	7.67	0.130	1.351	0.740

R = °F, ft², hr, Btu R_m = km²W U = Btu per ft² per hr U_m = W/m²K

TABLE 169

78. CELLULAR GLASS INSULATION
THERMAL RESISTANCE AND TRANSMITTANCE OF ASSEMBLY XXXVIII WITH CELLULAR GLASS INSULATION:

k = 0.32 Btu, in./ft², hr °F at 30°F, k_m = 0.04 W/mK at −1.1°C, k = 0.34 Btu, in./ft², hr °F at 60°F, k_m = 0.049 W/mK at 15.6°C, k = 0.36 Btu, in./ft², hr °F at 90°F, k_m = 0.052 W/mK at 32.2°C

Insulation Thickness		Mean Temp. 30°F (−1.1°C)				Mean Temp. 60°F (15.6°C)				Mean Temp. 90°F (32.2°C)			
		English Units		Metric Units		English Units		Metric Units		English Units		Metric Units	
Inches	Meters	R	U	R_m	U_m	R	U	R_m	U_m	R	U	R_m	U_m
1 1/2	0.038	5.21	0.192	0.918	1.089	5.07	0.197	0.893	1.119	4.70	0.213	0.828	1.206
2	0.051	6.78	0.147	1.195	0.837	6.59	0.152	1.161	0.861	6.08	0.164	1.071	0.933
2 1/2	0.061	8.34	0.120	1.470	0.680	8.10	0.123	1.427	0.701	7.47	0.134	1.316	0.760
3	0.076	9.91	0.100	1.746	0.573	9.62	0.104	1.695	0.590	8.86	0.113	1.561	0.641
3 1/2	0.089	11.46	0.087	2.019	0.495	11.13	0.090	1.961	0.510	10.25	0.098	1.806	0.554
4	0.102	13.03	0.077	2.296	0.436	12.65	0.079	2.229	0.449	11.64	0.086	2.051	0.488
5	0.127	16.16	0.062	2.847	0.351	15.68	0.064	2.763	0.362	14.42	0.069	2.541	0.394
6	0.152	19.28	0.052	3.397	0.294	18.71	0.053	3.297	0.303	17.20	0.058	3.031	0.330

R = °F, ft², hr, Btu R_m = km²W U = Btu per ft² per hr U_m = W/m²K

TABLE 170

DUCTS

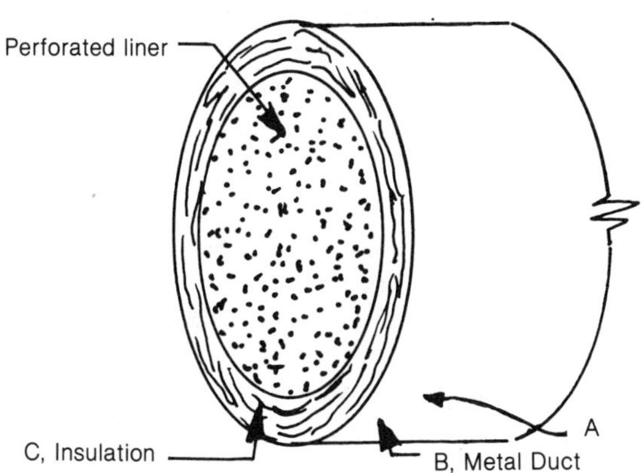

CONSTRUCTION ASSEMBLY XXXIX

Note: On ducts operating below ambient dew-point temperature the metal outer duct acts as vapor retarder.

NO THERMAL INSULATION
THERMAL RESISTANCE R and R_m

Mean Temperature 60° (15.6°C) — Cooling

	Components	Eng. Units		Metric Units
A	Outside surface	0.50		0.088
B	Metal duct	0.01		0.002
C	Insulation	—		—
	(TOTAL) RESISTANCE R =	0.51	R_m =	0.090
	TRANSMITTANCE U =	1.961	U_m =	11.111

Mean Temperature 90° (32.2°C) — Heating

Same as above.

Note: No inner surface resistance is included as it would change with air velocity and due to air movement it would be small.

79. GLASS FIBER INSULATION

THERMAL RESISTANCE AND TRANSMITTANCE OF ASSEMBLY XXXXIII WITH GLASS FIBER INSULATION: k = 0.27 Btu, in./ft^2, hr °F at 60°F, k_m = 0.039 W/mK at 15.6°C, k = 0.29 Btu, in./ft^2, hr °F at 90°F, k_m = 0.042 W/mK at 32.2°C

Insulation Thickness		Mean Temp. 60°F (15.6°C)				Mean Temp. 90°F (32.2°C)			
		English Units		Metric Units		English Units		Metric Units	
Inches	Meters	R	U	R_m	U_m	R	U	R_m	U_m
1	0.025	4.21	0.237	0.742	1.348	3.95	0.253	0.696	1.437
2	0.051	7.92	0.126	1.396	0.717	7.41	0.135	1.306	0.766

R = °F, ft^2, hr, Btu R_m = km^{2}W   U = Btu per ft^2 per hr U_m = W/m^{2}K

TABLE 171

ACOUSTICAL ATTENUATION

Duct Diameter		Attenuation Db Per Linear Foot Octave Band Center Frequencies					
Inches	Meters	125	150	500	1000	2000	4000
4	0.102	0.82	1.15	1.88	2.21	2.17	2.02
6	0.152	0.54	0.96	1.83	2.21	2.17	1.96
8	0.203	0.38	0.84	1.72	2.21	2.17	1.82
10	0.254	0.32	0.76	1.60	2.21	2.17	1.64
12	0.305	0.29	0.71	1.50	2.21	2.17	1.45
16	0.406	0.25	0.63	1.28	2.00	1.68	1.08
20	0.508	0.22	0.56	1.14	1.83	1.22	0.78
24	0.610	0.21	0.54	1.04	1.68	0.92	0.54

TABLE 172

DUCTS

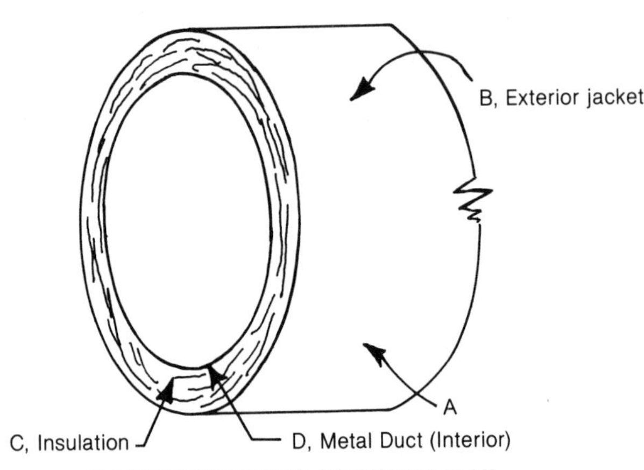

CONSTRUCTION ASSEMBLY XL

NO THERMAL INSULATION
THERMAL RESISTANCE R and R_m
Mean Temperature 60° (15.6°C) — Cooling

	Components	Eng. Units		Metric Units
A	Outside surface	0.50		0.038
B	Exterior jacket	0.02		0.003
C	Insulation	—		—
D	Metal duct	0.01		0.002
	(TOTAL) RESISTANCE R =	0.53	R_m =	0.093
	TRANSMITTANCE U =	1.887	U_m =	10.753

Mean Temperature 90° (32.2°C) — Heating

Same As Above

80. GLASS FIBER DUCT SECTIONS, BATT OR BLANKET INSULATION THERMAL RESISTANCE AND TRANSMITTANCE OF ASSEMBLY XL WITH GLASS FIBER INSULATION (NOT SUITABLE IF DUCT TEMPERATURE IS BELOW AMBIENT AIR DEW POINT TEMPERATURE):

k = 0.27 Btu, in./ft^2, hr °F at 60°F, k_m = 0.039 W/mK at 15.6°C, k = 0.29 Btu, in./ft^2, hr °F at 90°F, k_m = 0.042 W/mK at 32.2°C

Insulation Thickness		Mean Temp. 60°F (15.6°C)				Mean Temp. 90°F (32.2°C)			
		English Units		Metric Units		English Units		Metric Units	
Inches	Meters	R	U	R_m	U_m	R	U	R_m	U_m
1	0.025	4.23	0.236	0.745	1.342	3.98	0.251	0.701	1.426
2	0.051	7.94	0.126	1.399	0.715	7.42	0.135	1.307	0.765

R = °F, ft^2, hr, Btu R_m = km^{2}W U = Btu per ft^2 per hr U_m = W/m^{2}K

TABLE 173

81. CELLULAR GLASS INSULATION (SUITABLE FOR REFRIGERATION DUCTS) THERMAL RESISTANCE AND TRANSMITTANCE OF ASSEMBLY XL WITH CELLULAR GLASS INSULATION:

k = 0.32 Btu, in./ft^2, hr °F at 30°F, k_m = 0.040 W/mK at −1.1°C, k = 0.34 Btu, in./ft^2, hr °F at 60°F, k_m = 0.049 W/mK at 15.6°C, k = 0.36 Btu, in./ft^2, hr °F at 90°F, k_m = 0.052 W/mK at 32.2°C

Insulation Thickness		Mean Temp. 30°F (−1.1°C)				Mean Temp. 60°F (15.6°C)				Mean Temp. 90°F (32.2°C)			
		English Units		Metric Units		English Units		Metric Units		English Units		Metric Units	
Inches	Meters	R	U	R_m	U_m	R	U	R_m	U_m	R	U	R_m	U_m
1 1/2	0.038	5.21	0.192	0.918	1.089	5.07	0.197	0.893	1.119	4.70	0.213	0.828	1.206
2	0.051	6.78	0.147	1.195	0.837	6.59	0.152	1.161	0.861	6.08	0.164	1.071	0.933
2 1/2	0.061	8.34	0.120	1.470	0.680	8.10	0.123	1.427	0.701	7.47	0.134	1.316	0.760
3	0.076	9.91	0.100	1.746	0.573	9.62	0.104	1.695	0.590	8.86	0.113	1.561	0.641
3 1/2	0.089	11.46	0.087	2.019	0.495	11.13	0.090	1.961	0.510	10.25	0.098	1.806	0.554
4	0.102	13.03	0.077	2.296	0.436	12.65	0.079	2.229	0.449	11.64	0.086	2.051	0.488
5	0.127	16.16	0.062	2.847	0.351	15.68	0.064	2.763	0.362	14.42	0.069	2.541	0.394
6	0.152	19.28	0.052	3.397	0.294	18.71	0.053	3.297	0.303	17.20	0.058	3.031	0.330

R = °F, ft^2, hr, Btu R_m = km^{2}W U = Btu per ft^2 per hr U_m = W/m^{2}K

TABLE 174

TABLE 175 DEW-POINT TEMPERATURE

Dry-bulb temp. °F	Percent relative humidity 10	15	20	25	30	35	40	45	50	55	60	65	70	75	80	85	90	95	100
5	−35	−30	−25	−21	−17	−14	−12	−10	−8	−6	−5	−4	−2	−1	1	2	3	4	5
10	−31	−25	−20	−16	−13	−10	−7	−5	−3	−2	0	2	3	4	5	7	8	9	10
15	−28	−21	−16	−12	−8	−5	−3	−1	1	3	5	6	8	9	10	12	13	14	15
20	−24	−16	−11	−8	−4	−2	2	4	6	8	10	11	13	14	15	16	18	19	20
25	−20	−15	−8	−4	0	3	6	8	10	12	15	16	18	19	20	21	23	24	25
30	−15	−9	−3	2	5	8	11	13	15	17	20	22	23	24	25	27	28	29	30
35	−12	−5	1	5	9	12	15	18	20	22	24	26	27	28	30	32	33	34	35
40	−7	0	5	9	14	16	19	22	24	26	28	29	31	33	35	36	38	39	40
45	−4	3	9	13	17	20	23	25	28	30	32	34	36	38	39	41	43	44	45
50	−1	7	13	17	21	24	27	30	32	34	37	39	41	42	44	45	47	49	50
55	3	11	16	21	25	28	32	34	37	39	41	43	45	47	49	50	52	53	55
60	6	14	20	25	29	32	35	39	42	44	46	48	50	52	54	55	57	59	60
65	10	18	24	28	33	38	40	43	46	49	51	53	55	57	59	60	62	63	65
70	13	21	28	33	37	41	45	48	50	53	55	57	60	62	64	65	67	68	70
75	17	25	32	37	42	46	49	52	55	57	60	62	64	66	69	70	72	74	75
80	20	29	35	41	46	50	54	57	60	62	65	67	69	72	74	75	77	78	80
85	23	32	40	45	50	54	58	61	64	67	69	72	74	76	78	80	82	83	85
90	27	36	44	49	54	58	62	66	69	72	74	77	79	81	83	85	87	89	90
95	30	40	48	54	59	63	67	70	73	76	79	82	84	86	88	90	91	93	95
100	34	44	52	58	63	68	71	75	78	81	84	86	88	91	92	94	96	98	100
105	38	48	56	62	67	72	76	79	82	85	88	90	93	95	97	99	101	103	105
110	41	52	60	66	71	77	80	84	87	90	92	95	98	100	102	104	106	108	110
115	45	56	64	70	75	80	84	88	91	94	97	100	102	105	107	109	111	113	115
120	48	60	68	74	79	85	88	92	96	99	102	105	107	109	112	114	116	118	120
125	52	63	72	78	84	89	93	97	100	104	107	109	111	114	117	119	121	123	125

TABLE 176 DEW-POINT TEMPERATURE °C METRIC UNITS

Dry-bulb temp. °C	Percent relative humidity 10	15	20	25	30	35	40	45	50	55	60	65	70	75	80	85	90	95	100
−15.0	−37.2	−34.4	−31.7	−29.4	−27.2	−26.5	−24.4	−23.3	−22.2	−21.1	−20.6	−20.0	−18.9	−18.3	−17.2	−16.7	−16.1	−15.6	−15.0
−12.2	−35.0	−31.7	−28.9	−26.7	−25.0	−23.3	−21.7	−20.6	−19.4	−18.9	−17.8	−16.7	−16.1	−15.6	−15.0	−13.9	−13.3	−12.8	−12.2
−9.4	−33.3	−29.4	−26.7	−24.4	−22.2	−20.6	−19.4	−18.3	−17.2	−16.1	−15.0	−14.4	−13.3	−12.8	−12.2	−11.1	−10.6	−10.6	−9.4
−6.7	−31.1	−26.7	−23.9	−22.2	−20.0	−18.9	−16.7	−15.6	−14.4	−13.3	−12.2	−11.7	−10.6	−10.0	−9.4	−8.9	−7.8	−7.2	−6.7
−3.9	−28.9	−26.1	−22.2	−20.0	−17.8	−16.1	−14.4	−13.3	−12.2	−11.1	−9.4	−8.9	−7.8	−7.2	−6.7	−6.1	−5.0	−4.4	−3.9
−1.1	−26.1	−22.8	−19.4	−16.7	−15.0	−13.3	−11.7	−10.6	−9.4	−8.3	−6.7	−5.6	−5.0	−4.4	−3.9	−2.8	−2.2	−1.7	−1.1
1.7	−24.4	−20.6	−17.2	−15.0	−12.8	−11.1	−9.9	−7.8	−6.7	−5.6	−4.4	−3.3	−2.8	−2.2	−1.1	0	0.6	1.1	1.7
4.4	−21.7	−17.8	−15.0	−12.8	−10.0	−8.9	−7.2	−5.6	−4.4	−3.3	−2.2	−1.7	−0.6	0.6	1.7	2.2	3.3	3.9	4.4
7.2	−20.0	−16.1	−12.8	−10.6	−8.3	−6.7	−5.0	−3.9	−2.2	−1.1	0	1.1	2.2	3.2	4.1	5.0	5.8	6.7	7.2
10.0	−18.3	−13.9	−11.1	−8.3	−6.1	−4.4	−2.8	−1.1	0	1.0	2.0	3.0	4.0	5.0	6.0	7.0	8.0	9.0	10.0
12.8	−16.1	−11.7	−8.9	−6.1	−3.9	−2.2	0	1.1	2.8	3.9	5.0	6.1	7.2	8.3	9.4	10.0	11.1	11.7	12.8
15.6	−14.4	−10.0	−6.7	−3.9	−1.7	0	1.7	3.9	5.6	6.7	7.8	8.9	10.0	11.1	12.2	18.3	13.9	15.0	15.6
18.3	−12.2	−7.8	−4.4	−2.2	0.6	3.3	4.4	6.1	7.8	9.4	10.6	11.7	12.8	13.9	15.0	15.6	16.7	17.2	18.3
21.1	−10.6	−6.1	−2.2	0.6	2.8	5.0	7.2	8.9	10.0	11.7	12.8	13.9	15.6	16.7	17.8	18.3	19.4	20.0	21.1
23.9	−8.3	−3.9	0	2.8	5.6	7.8	9.4	11.1	12.8	13.9	15.6	16.7	17.8	18.9	20.6	21.1	22.2	23.3	23.9
26.7	−6.7	−1.7	1.7	5.0	7.8	10.0	12.2	13.9	15.6	16.7	18.3	19.4	20.6	22.2	23.3	24.2	25.0	25.6	26.7
29.4	−5.0	0	4.4	7.2	10.0	12.2	14.4	16.1	17.8	19.4	20.6	22.2	23.3	24.4	25.6	26.7	27.8	28.3	32.4
32.2	−2.8	2.2	6.7	9.4	12.2	14.4	16.7	18.9	20.6	22.2	23.3	25.0	26.1	27.2	28.3	29.4	30.6	31.7	32.2
35.0	−1.1	4.4	8.9	12.4	15.0	17.2	19.4	21.1	22.8	24.4	26.1	27.8	28.9	30.0	31.1	32.2	32.8	33.9	35.0
37.8	1.1	6.7	11.1	14.4	17.2	20.0	21.7	23.9	25.6	27.2	28.9	30.0	31.1	32.8	33.3	34.4	35.6	36.7	37.8
40.6	3.3	8.9	13.3	16.7	19.4	22.2	24.4	26.1	27.8	29.4	31.1	32.2	33.9	35.0	36.1	37.2	38.3	39.4	40.6
43.3	5.0	11.1	15.6	18.9	21.7	25.0	26.7	28.9	30.6	32.2	33.3	35.0	36.7	37.8	38.9	40.0	41.1	42.2	43.3
46.1	7.2	13.3	17.8	21.1	23.9	26.7	28.9	31.1	32.8	34.4	36.1	37.8	38.9	40.6	41.7	42.8	43.9	45.0	46.1
48.9	8.9	15.6	20.0	23.3	26.1	29.4	31.1	33.3	35.6	37.2	38.9	40.6	41.7	42.8	44.4	45.6	46.7	47.8	48.9
51.7	11.1	17.2	22.2	25.6	28.9	31.7	33.9	36.1	37.8	40.0	41.7	42.8	43.9	45.6	47.2	48.3	49.4	50.6	51.7

CONTROL OF WATER DRIP FROM SURFACE OF PIPE AND DUCTS USING INSULATING MASTICS

PERCENT RELATIVE HUMIDITY ABOVE WHICH CONDENSATION WILL OCCUR ON SURFACE

Pipe or Equipment Temperature	Insulated with 1/4" thickness of mastic (0.635 cm thickness)												Pipe or Equipment Temperature
	Ambient Temperature °C												
°F	7.2	10.0	12.8	15.6	18.3	21.1	23.9	26.7	29.4	32.2	35.0	37.8	°C
40	86	81	74	68	63	59	55	52	49	47	43	41	4.4
45		91	82	75	69	64	60	57	54	50	48	45	7.2
50			91	82	75	69	65	61	58	55	51	48	10.0
55				91	83	77	71	66	62	59	54	52	12.8
60					91	83	78	72	67	63	59	56	15.6
65						92	84	78	73	68	64	60	18.3
70							92	84	79	73	68	64	21.1
75								92	86	79	74	69	23.9
80									92	86	79	77	26.7
85										92	86	80	29.4
90											92	87	32.2
95												92	35.0
	45	50	55	60	65	70	75	80	85	90	95	100	
	Ambient Temperature °F												

Pipe or Equipment Temperature	Insulated with 3/8" thickness of mastic (0.935 cm thickness)												Pipe or Equipment Temperature
	Ambient Temperature °C												
°F	7.2	10.0	12.8	15.6	18.3	21.1	23.9	26.7	29.4	32.2	35.0	37.8	°C
40	92	84	79	76	69	66	63	59	57	55	53	50	4.4
45		92	85	80	74	70	67	64	61	58	56	53	7.2
50			92	85	80	76	71	68	64	62	59	57	10.0
55				92	86	80	76	72	69	66	63	60	12.8
60					92	86	81	77	73	69	67	63	15.6
65						92	87	82	78	73	70	68	18.3
70							93	88	82	78	74	71	21.1
75								93	90	82	79	74	23.9
80									93	88	83	79	26.7
85										93	88	84	29.4
90											93	88	32.2
95												93	35.0
	45	50	55	60	65	70	75	80	85	90	95	100	
	Ambient Temperature °F												

TABLE 177

TABLE 178 NPS PIPE

EQUIVALENT THICKNESS – INCHES – ENGLISH

PIPE		NOMINAL INSULATION THICKNESS – INCHES																			
NPS	Diam."	½	1	1½	2	2½	3	3½	4	4½	5	5½	6	6½	7	7½	8	8½	9	9½	10
¼	0.540	0.81	1.98	3.32	4.84	6.45	8.13	9.05	11.70	13.70	15.55	17.66	19.69	21.80	23.92	26.11	28.28	30.52	32.82	35.07	37.38
⅜	0.675	0.76	1.84	3.12	4.53	6.04	7.64	9.32	11.06	12.87	14.74	16.64	18.57	20.65	22.60	24.69	26.76	28.90	31.00	33.15	33.33
½	0.840	0.72	1.73	2.92	4.23	5.66	7.18	8.74	10.39	12.10	13.82	15.69	17.53	19.38	21.30	23.28	25.26	27.29	29.29	31.35	33.02
¾	1.030	0.69	1.63	2.73	3.96	5.29	6.73	8.21	9.73	11.36	12.98	14.64	16.44	18.19	20.02	21.91	23.78	25.63	27.62	29.57	31.57
1	1.315	0.66	1.52	2.57	3.72	4.96	6.29	7.65	9.12	10.62	12.16	13.79	15.44	17.11	18.76	20.56	22.33	24.08	25.98	27.83	29.73
1¼	1.660	0.63	1.45	2.40	3.45	4.63	5.86	7.14	8.50	9.91	11.37	12.85	14.41	15.90	17.54	19.24	20.93	22.58	24.28	26.03	27.83
1½	1.900	0.62	1.40	2.30	3.33	4.45	5.65	6.85	8.17	9.54	10.89	12.38	13.83	15.35	16.85	18.42	20.05	21.74	23.38	25.08	26.83
2	2.375	0.59	1.33	2.21	3.16	4.17	5.28	6.47	7.62	8.88	10.21	11.50	12.93	14.39	15.80	17.24	18.83	20.34	21.90	23.51	25.03
2½	2.875	0.58	1.27	2.09	2.99	3.95	4.98	6.07	7.23	8.42	9.66	10.96	12.20	13.57	14.93	16.35	17.74	19.18	20.66	22.20	23.16
3	3.500	0.56	1.24	2.02	2.85	3.78	4.75	5.78	6.84	7.94	9.11	10.30	11.54	12.79	14.09	15.36	16.77	18.14	19.46	20.93	22.33
3½	4.000	0.53	1.22	1.96	2.76	3.64	4.60	5.56	6.60	7.67	8.75	9.90	11.12	12.32	13.50	14.82	16.10	17.43	18.70	20.15	21.48
4	4.500	0.53	1.19	1.91	2.71	3.56	4.46	5.41	6.38	7.42	8.47	9.61	10.73	11.90	13.04	14.53	15.58	16.77	18.11	19.39	20.70
4½	5.000	0.54	1.18	1.88	2.66	3.45	4.35	5.28	6.24	7.21	8.25	9.28	10.37	11.52	12.73	13.90	15.12	16.28	17.60	18.84	20.13
5	5.563	0.54	1.17	1.84	2.58	3.38	4.28	5.15	6.03	6.99	8.01	9.03	10.10	11.14	12.32	13.47	14.55	15.79	16.96	18.30	19.43
6	6.625	0.53	1.13	1.78	2.50	3.25	4.10	4.91	5.78	6.72	7.65	8.64	9.59	10.70	11.65	12.76	13.91	15.00	16.12	17.30	18.50
7	7.625	0.53	1.12	1.75	2.44	3.22	3.95	4.75	5.62	6.48	7.40	8.29	9.22	10.31	11.24	12.33	13.35	14.41	15.50	16.64	17.82
8	8.625	0.53	1.11	1.74	2.40	3.13	3.88	4.61	5.49	6.26	7.17	8.05	8.97	9.95	10.86	11.93	12.93	13.97	15.04	16.02	17.17
9	9.625	0.53	1.10	1.70	2.38	3.07	3.75	4.57	5.29	6.15	6.97	7.84	8.76	9.62	10.63	11.66	12.56	13.58	14.50	15.60	16.59
10	10.750	0.52	1.09	1.69	2.34	3.05	3.66	4.50	5.22	6.09	6.81	7.69	8.61	9.22	10.35	11.28	12.25	13.13	14.18	15.12	16.24
12	12.750	0.52	1.08	1.68	2.33	3.00	3.63	4.32	5.06	5.85	6.57	7.50	8.26	8.98	9.99	10.78	11.75	12.60	13.49	14.56	15.50
14	14.000	0.52	1.07	1.65	2.26	2.90	3.56	4.26	4.97	5.71	6.47	7.25	8.05	8.86	9.70	10.30	11.43	12.25	13.22	14.19	15.07
16	16.000	0.52	1.06	1.63	2.23	2.85	3.50	4.17	4.86	5.57	6.31	7.06	7.83	8.62	9.42	10.23	11.09	12.05	12.81	13.65	14.59
18	18.000	0.51	1.05	1.62	2.21	2.82	3.45	4.10	4.78	5.47	6.19	6.91	7.66	8.42	9.21	9.90	10.81	11.55	12.48	13.32	14.19
20	20.000	0.51	1.05	1.61	2.19	2.80	3.41	4.05	4.71	5.44	6.08	6.80	7.52	8.16	9.02	9.80	10.58	11.47	12.20	13.06	13.86
24	24.000	0.51	1.04	1.59	2.16	2.74	3.35	3.96	4.60	5.25	5.92	6.58	7.30	7.99	8.73	9.56	10.22	11.07	11.74	12.47	13.29
28	28.000	0.51	1.03	1.58	2.14	2.71	3.30	3.91	4.52	5.15	5.80	6.46	7.12	7.82	8.51	9.25	9.94	10.80	11.32	12.22	12.93
32	32.000	0.51	1.03	1.57	2.12	2.68	3.27	3.85	4.46	5.07	5.71	6.34	7.01	7.66	8.35	9.17	9.73	10.54	11.15	11.98	12.62
36	36.000	0.51	1.03	1.56	2.11	2.66	3.24	3.81	4.41	5.02	5.63	6.25	6.90	7.53	8.11	8.92	9.56	10.34	10.94	11.83	12.37
40	40.000	0.51	1.02	1.55	2.10	2.65	3.21	3.79	4.37	4.97	5.58	6.20	6.82	7.46	8.10	8.80	9.42	10.26	10.78	11.80	12.16
48	48.000	0.49	1.02	1.55	2.08	2.62	3.18	3.74	4.31	4.89	5.48	6.08	6.69	7.30	7.92	8.51	9.20	9.75	10.54	11.06	11.84

Note: Tabular values are those of the Equivalent Thickness $L = r_2 \log_e \frac{r_2}{r_1}$ for the Nominal Insulation Thicknesses given in inches.

r_1 = Inside Radius of Insulation in inches r_2 = Outside Radius of Insulation in inches $r_2 - r_1$ = Actual Thickness (ℓ) in inches
L = Equivalent Thickness in inches

TABLE 179 NPS PIPE Equivalent Thickness – Meters (Metric)

Pipe Size		NOMINAL INSULATION THICKNESS – mm																			
NPS inches	Dia. mm	13	25	38	51	64	76	89	102	114	127	140	152	165	178	190	203	216	229	241	250
1/4	13.7	0.021	0.050	0.084	0.123	0.164	0.206	0.229	0.297	0.348	0.395	0.448	0.500	0.553	0.607	0.663	0.718	0.775	0.833	0.891	0.949
3/8	17.1	0.019	0.047	0.079	0.115	0.153	0.194	0.236	0.281	0.327	0.374	0.422	0.471	0.524	0.574	0.627	0.680	0.734	0.787	0.842	0.846
1/2	21.3	0.018	0.045	0.074	0.107	0.144	0.182	0.221	0.264	0.307	0.351	0.398	0.447	0.492	0.541	0.591	0.641	0.693	0.744	0.796	0.839
3/4	26.7	0.017	0.041	0.069	0.101	0.134	0.171	0.208	0.247	0.288	0.329	0.372	0.417	0.462	0.508	0.556	0.604	0.651	0.701	0.751	0.802
1	33.4	0.017	0.039	0.065	0.094	0.126	0.160	0.194	0.232	0.270	0.309	0.350	0.392	0.434	0.476	0.522	0.567	0.611	0.660	0.707	0.753
1 1/4	42.2	0.016	0.037	0.061	0.088	0.117	0.149	0.181	0.216	0.251	0.288	0.326	0.366	0.403	0.445	0.488	0.531	0.573	0.617	0.661	0.707
1 1/2	48.3	0.016	0.036	0.058	0.085	0.113	0.143	0.174	0.231	0.242	0.276	0.314	0.351	0.390	0.428	0.468	0.509	0.552	0.594	0.637	0.681
2	60.3	0.015	0.034	0.056	0.080	0.106	0.134	0.164	0.215	0.225	0.259	0.292	0.328	0.365	0.401	0.438	0.478	0.516	0.556	0.597	0.636
2 1/2	73.0	0.015	0.032	0.053	0.076	0.100	0.126	0.154	0.183	0.214	0.245	0.278	0.309	0.344	0.379	0.415	0.450	0.487	0.524	0.564	0.588
3	88.9	0.014	0.031	0.051	0.073	0.096	0.121	0.147	0.173	0.201	0.231	0.261	0.293	0.324	0.357	0.390	0.426	0.460	0.494	0.531	0.567
3 1/2	101.6	0.013	0.031	0.050	0.070	0.092	0.117	0.141	0.167	0.195	0.222	0.251	0.282	0.312	0.342	0.376	0.408	0.443	0.474	0.512	0.545
4	114.3	0.013	0.030	0.049	0.069	0.090	0.113	0.137	0.162	0.188	0.215	0.244	0.272	0.302	0.331	0.369	0.396	0.426	0.460	0.492	0.526
4 1/2	127.0	0.014	0.030	0.048	0.068	0.088	0.110	0.134	0.158	0.183	0.209	0.235	0.263	0.292	0.323	0.353	0.384	0.413	0.447	0.478	0.511
5	141.3	0.014	0.030	0.048	0.065	0.086	0.108	0.131	0.153	0.177	0.203	0.229	0.256	0.282	0.312	0.342	0.369	0.401	0.431	0.464	0.493
6	168.3	0.013	0.029	0.045	0.063	0.082	0.104	0.124	0.146	0.170	0.194	0.219	0.242	0.271	0.295	0.324	0.353	0.381	0.409	0.439	0.470
7	193.7	0.013	0.028	0.044	0.062	0.082	0.100	0.120	0.142	0.164	0.188	0.210	0.234	0.261	0.285	0.313	0.339	0.366	0.394	0.422	0.452
8	219.1	0.013	0.028	0.044	0.061	0.079	0.098	0.117	0.139	0.159	0.182	0.204	0.228	0.252	0.276	0.303	0.328	0.355	0.382	0.407	0.436
9	244.5	0.013	0.028	0.043	0.060	0.078	0.095	0.116	0.134	0.156	0.177	0.199	0.222	0.244	0.270	0.296	0.319	0.345	0.368	0.396	0.421
10	273.5	0.013	0.028	0.043	0.059	0.077	0.093	0.114	0.132	0.154	0.172	0.195	0.218	0.234	0.263	0.286	0.311	0.333	0.360	0.384	0.412
12	323.8	0.013	0.027	0.043	0.059	0.076	0.092	0.109	0.128	0.148	0.167	0.190	0.210	0.228	0.253	0.274	0.298	0.320	0.342	0.370	0.394
14	355.6	0.013	0.027	0.042	0.057	0.074	0.090	0.108	0.126	0.145	0.164	0.184	0.204	0.225	0.246	0.261	0.290	0.311	0.336	0.360	0.382
16	406.4	0.013	0.027	0.041	0.056	0.072	0.089	0.106	0.123	0.141	0.160	0.179	0.199	0.218	0.239	0.260	0.281	0.306	0.325	0.347	0.370
18	457.2	0.013	0.027	0.041	0.056	0.071	0.087	0.104	0.121	0.139	0.157	0.175	0.194	0.214	0.234	0.251	0.274	0.293	0.317	0.338	0.360
20	508.0	0.013	0.027	0.041	0.055	0.071	0.086	0.100	0.119	0.138	0.154	0.172	0.191	0.207	0.229	0.248	0.269	0.291	0.310	0.332	0.352
24	609.6	0.013	0.026	0.040	0.054	0.069	0.085	0.100	0.116	0.133	0.150	0.167	0.185	0.203	0.221	0.242	0.259	0.281	0.298	0.317	0.337
28	711.2	0.013	0.026	0.040	0.054	0.068	0.084	0.099	0.114	0.131	0.147	0.164	0.180	0.198	0.216	0.235	0.252	0.274	0.287	0.310	0.328
32	812.8	0.013	0.026	0.040	0.054	0.068	0.083	0.097	0.113	0.128	0.145	0.161	0.178	0.194	0.212	0.233	0.247	0.268	0.283	0.304	0.320
36	914.4	0.013	0.026	0.039	0.053	0.067	0.082	0.096	0.112	0.127	0.143	0.158	0.175	0.191	0.206	0.226	0.242	0.262	0.278	0.300	0.314
40	1016.0	0.013	0.026	0.039	0.053	0.067	0.081	0.096	0.111	0.126	0.141	0.157	0.173	0.189	0.205	0.223	0.239	0.260	0.274	0.299	0.309
48	1219.2	0.013	0.026	0.039	0.053	0.066	0.080	0.095	0.110	0.124	0.139	0.154	0.170	0.185	0.201	0.216	0.234	0.248	0.268	0.280	0.301

Note: Tabular values are those of the Equivalent Thickness $L_m = r_{2m} \log_e \frac{r_{2m}}{r_{1m}}$ for the Nominal Insulation Thickness given.

r_{1m} = Inside radius of insulation in metres

$r_{1m} - r_{2m}$ = Actual thickness (ℓ_m) in metres.

r_{2m} = Outside radius of insulation in metres.

L_m = Equivalent thickness in metres.

TABLE 180 TUBE SIZES EQUIVALENT THICKNESS – INCHES – ENGLISH

TUBE SIZE		NOMINAL INSULATION THICKNESS – INCHES																			
Nom."	Diam.	½	1	1½	2	2½	3	3½	4	4½	5	5½	6	6½	7	7½	8	8½	9	9½	10
¼	0.250	1.04	2.47	4.17	6.29	7.99	10.04	13.21	14.82	16.70	18.78	21.41	23.84	25.88	28.81	31.34	33.91	35.49	39.15	40.76	44.49
½	0.500	0.82	2.01	3.40	4.93	6.58	8.33	10.15	12.04	13.98	15.98	18.03	20.11	22.25	24.44	26.58	28.86	31.15	33.39	35.68	38.07
1	1.000	0.69	1.65	2.77	4.02	5.34	6.81	8.32	9.89	11.51	13.18	14.90	16.67	18.67	20.31	22.16	24.08	26.01	27.97	30.00	31.97
2	2.000	0.61	1.39	2.29	3.30	4.38	5.55	6.77	8.05	9.38	10.75	12.17	13.62	15.21	16.63	18.19	19.77	21.38	23.03	24.68	26.38
3	3.000	0.57	1.28	2.08	2.97	3.92	4.94	6.02	7.15	8.32	9.53	10.78	12.07	13.34	14.73	16.21	17.48	19.00	20.43	21.89	23.32
4	4.000	0.56	1.22	1.96	2.77	3.64	4.58	5.56	6.59	7.71	8.76	9.91	11.09	12.30	13.54	14.82	16.10	17.33	18.75	20.13	21.50
5	5.000	0.55	1.18	1.88	2.65	3.47	4.34	5.25	6.21	7.21	8.24	9.31	10.40	11.53	12.67	13.90	15.07	16.28	17.55	18.84	20.12
6	6.000	0.54	1.15	1.82	2.55	3.32	4.16	5.03	5.93	6.87	7.85	8.84	9.88	10.95	12.04	13.13	14.29	15.41	16.64	17.88	19.06
7	7.000	0.53	1.13	1.78	2.49	3.23	4.02	4.85	5.72	6.61	7.54	8.50	9.49	10.50	11.54	12.54	13.68	14.76	15.91	17.03	18.22
8	8.000	0.53	1.12	1.75	2.43	3.16	3.92	4.73	5.55	6.41	7.30	8.22	9.16	10.13	11.13	12.08	13.18	14.25	15.32	16.47	17.54
9	9.000	0.53	1.10	1.72	2.39	3.10	3.83	4.60	5.40	6.24	7.09	7.99	8.90	9.83	10.79	11.76	12.77	13.78	14.83	15.82	16.97
10	10.000	0.52	1.09	1.70	2.35	3.04	3.76	4.51	5.29	6.10	6.93	7.79	8.67	9.58	10.51	11.50	12.42	13.37	14.41	15.37	16.47
12	12.000	0.52	1.08	1.67	2.30	2.95	3.65	4.36	5.11	5.98	6.65	7.48	8.32	9.17	10.05	10.93	11.86	12.76	13.74	14.73	16.18
14	14.000	0.52	1.07	1.65	2.26	2.90	3.56	4.26	4.97	5.71	6.47	7.25	8.05	8.86	9.70	10.55	11.43	12.25	13.22	14.19	15.07
16	16.000	0.52	1.06	1.63	2.23	2.85	3.50	4.17	4.86	5.57	6.31	7.06	7.83	8.62	9.42	10.23	11.09	12.05	12.81	13.65	14.59
18	18.000	0.51	1.05	1.62	2.21	2.82	3.45	4.10	4.78	5.47	6.19	6.91	7.66	8.42	9.21	9.90	10.81	11.65	12.48	13.32	14.19
20	20.000	0.51	1.05	1.61	2.19	2.80	3.41	4.05	4.71	5.44	6.08	6.80	7.52	8.16	9.02	9.80	10.58	11.47	12.20	13.06	13.86
24	24.000	0.51	1.04	1.59	2.16	2.74	3.35	3.96	4.60	5.25	5.92	6.59	7.30	7.99	8.73	9.56	10.22	11.07	11.75	12.47	13.29
28	28.000	0.51	1.03	1.58	2.14	2.71	3.30	3.91	4.52	5.15	5.80	6.46	7.12	7.82	8.51	9.25	9.94	10.80	11.32	12.22	12.93
32	32.000	0.51	1.03	1.57	2.12	2.68	3.27	3.85	4.46	5.07	5.71	6.34	7.01	7.66	8.35	9.17	9.73	10.54	11.15	11.98	12.62
36	36.000	0.51	1.03	1.56	2.11	2.66	3.24	3.81	4.41	5.02	5.63	6.25	6.90	7.53	8.21	8.92	9.56	10.34	10.94	11.83	12.37
40	40.000	0.51	1.02	1.55	2.10	2.65	3.21	3.79	4.37	4.97	5.58	6.20	6.82	7.46	8.10	8.80	9.42	10.26	10.78	11.80	12.16
48	48.000	0.50	1.02	1.55	2.08	2.62	3.18	3.74	4.31	4.90	5.48	6.08	6.69	7.30	7.92	8.51	9.20	9.75	10.51	11.06	11.84
56	56.000	0.50	1.02	1.54	2.06	2.60	3.15	3.71	4.27	4.82	5.42	5.99	6.60	7.18	7.81	8.42	9.05	9.67	10.31	10.88	11.60
64	64.000	0.50	1.02	1.54	2.04	2.59	3.14	3.67	4.24	4.78	5.36	5.95	6.51	7.11	7.72	8.30	8.92	9.57	10.15	10.79	11.40
72	72.000	0.50	1.02	1.53	2.03	2.56	3.11	3.65	4.21	4.76	5.33	5.91	6.48	7.07	7.62	8.22	8.82	9.43	10.04	10.65	11.28

Note: Tabular values are those of the Equivalent Thickness $L = r_2 \log_e \frac{r_2}{r_1}$ for the Nominal Insulation Thicknesses given in inches.

r_1 = Inside Radius of Insulation in inches r_2 = Outside Radius of Insulation in inches $r_2 - r_1$ = Actual Thickness (ℓ) in inches
L = Equivalent Thickness in inches

TABLE 181 TUBE SIZES Equivalent Thickness – Meters (Metric)

Tube Size		NOMINAL INSULATION THICKNESS – mm																			
Dia. inches	Dia. mm	13	25	38	51	64	76	89	102	114	127	140	152	165	178	190	203	216	229	241	250
1/4	6.4	0.026	0.063	0.106	0.160	0.202	0.255	0.335	0.376	0.424	0.477	0.544	0.605	0.657	0.731	0.796	0.861	0.901	0.994	1.035	1.130
1/2	15.9	0.021	0.051	0.086	0.125	0.167	0.211	0.258	0.306	0.355	0.406	0.458	0.511	0.565	0.620	0.675	0.733	0.791	0.848	0.906	0.967
1	22.2	0.017	0.042	0.070	0.102	0.135	0.172	0.211	0.251	0.292	0.334	0.378	0.423	0.474	0.516	0.562	0.612	0.661	0.710	0.762	0.812
2	59.0	0.015	0.035	0.058	0.084	0.111	0.141	0.172	0.204	0.238	0.273	0.309	0.346	0.386	0.422	0.462	0.502	0.543	0.585	0.627	0.670
3	79.4	0.014	0.032	0.053	0.075	0.099	0.125	0.153	0.182	0.211	0.242	0.274	0.306	0.339	0.374	0.412	0.444	0.482	0.519	0.556	0.592
4	104.8	0.014	0.031	0.050	0.070	0.092	0.116	0.141	0.167	0.195	0.222	0.252	0.282	0.312	0.344	0.376	0.409	0.440	0.476	0.511	0.546
5	130.2	0.014	0.030	0.048	0.067	0.088	0.110	0.133	0.158	0.183	0.209	0.236	0.264	0.293	0.322	0.353	0.382	0.413	0.446	0.478	0.511
6	155.7	0.014	0.029	0.046	0.065	0.084	0.105	0.128	0.150	0.174	0.199	0.224	0.251	0.278	0.306	0.333	0.363	0.391	0.422	0.454	0.484
7	177.8	0.013	0.029	0.045	0.063	0.082	0.102	0.123	0.145	0.168	0.191	0.216	0.241	0.267	0.293	0.318	0.347	0.375	0.404	0.432	0.463
8	203.2	0.013	0.028	0.044	0.062	0.080	0.099	0.120	0.141	0.163	0.185	0.209	0.233	0.257	0.282	0.307	0.335	0.362	0.389	0.418	0.445
9	228.6	0.013	0.028	0.044	0.061	0.079	0.097	0.117	0.137	0.158	0.180	0.203	0.226	0.250	0.274	0.299	0.324	0.350	0.377	0.401	0.431
10	254.0	0.013	0.028	0.043	0.060	0.077	0.095	0.115	0.134	0.155	0.176	0.198	0.220	0.243	0.266	0.292	0.315	0.340	0.366	0.390	0.418
12	304.8	0.013	0.027	0.042	0.058	0.075	0.093	0.111	0.130	0.152	0.169	0.190	0.211	0.233	0.255	0.277	0.290	0.324	0.349	0.374	0.411
14	355.6	0.013	0.027	0.042	0.057	0.079	0.090	0.108	0.126	0.145	0.164	0.184	0.204	0.225	0.246	0.261	0.290	0.311	0.336	0.360	0.382
16	406.4	0.013	0.027	0.041	0.056	0.072	0.089	0.106	0.123	0.141	0.160	0.179	0.199	0.218	0.239	0.260	0.281	0.306	0.325	0.347	0.370
18	437.2	0.013	0.027	0.041	0.056	0.071	0.087	0.104	0.121	0.139	0.157	0.175	0.194	0.214	0.234	0.251	0.274	0.293	0.317	0.338	0.360
20	508.0	0.013	0.027	0.041	0.055	0.071	0.086	0.100	0.119	0.138	0.159	0.172	0.191	0.207	0.229	0.248	0.269	0.291	0.310	0.332	0.352
24	609.6	0.013	0.026	0.040	0.054	0.069	0.085	0.100	0.116	0.133	0.150	0.167	0.185	0.203	0.221	0.242	0.259	0.281	0.298	0.317	0.337
28	711.2	0.013	0.026	0.040	0.054	0.068	0.084	0.099	0.118	0.131	0.197	0.164	0.180	0.198	0.216	0.235	0.252	0.279	0.297	0.310	0.328
32	812.8	0.013	0.026	0.040	0.054	0.068	0.083	0.097	0.113	0.128	0.145	0.161	0.178	0.194	0.212	0.233	0.247	0.268	0.283	0.304	0.320
36	914.4	0.013	0.026	0.039	0.053	0.067	0.082	0.096	0.112	0.127	0.143	0.158	0.175	0.191	0.206	0.226	0.242	0.262	0.278	0.300	0.314
40	1016.0	0.013	0.026	0.039	0.053	0.067	0.081	0.096	0.111	0.126	0.141	0.157	0.173	0.189	0.205	0.223	0.239	0.260	0.274	0.299	0.309
48	1219.2	0.012	0.026	0.039	0.053	0.066	0.080	0.095	0.110	0.124	0.139	0.154	0.170	0.185	0.201	0.216	0.234	0.248	0.268	0.280	0.301
56	1422.4	0.012	0.026	0.039	0.052	0.066	0.080	0.094	0.108	0.122	0.138	0.152	0.168	0.182	0.198	0.214	0.230	0.245	0.262	0.276	0.295
64	1625.6	0.012	0.026	0.039	0.052	0.066	0.080	0.093	0.108	0.121	0.136	0.151	0.165	0.180	0.196	0.210	0.226	0.243	0.259	0.274	0.289
72	1828.8	0.012	0.026	0.039	0.052	0.065	0.079	0.093	0.107	0.121	0.135	0.150	0.164	0.179	0.194	0.209	0.224	0.239	0.255	0.270	0.286

Note: Tabular values are those of the Equivalent Thickness $L_m = r_{2m} \log_e \frac{r_{2m}}{r_{1m}}$ for the Nominal Insulation Thickness given.

r_{1m} = Inside radius of insulation in metres
$r_{1m} - r_{2m}$ = Actual thickness (ℓ_m) in metres.
r_{2m} = Outside radius of insulation in metres.
L_m = Equivalent thickness in metres.

THERMAL CONDUCTIVITIES OF PREFORMED PIPE INSULATIONS – ENGLISH UNITS

Pipe Insulation Material	Temperature Limits		Thermal Conductivity K in Btu, in./ft^2, hr, °F								
			Mean Temperature °F								
	Min. °F	Max. °F	0° F	50° F	100° F	150° F	200° F	250° F	300° F	350° F	400° F
Calcium Silicate	200	1200	–	–	0.35	0.39	0.43	0.46	0.50	0.52	0.55
Cellular Glass	–300	800	0.31	0.33	0.36	0.39	0.42	0.45	0.49	0.52	0.57
Glass Fiber	Dew Point*	500	–	0.26	0.28	0.30	0.32	0.35	0.39	0.43	0.48
Mineral Fiber	Dew Point*	1200	–	0.32	0.34	0.37	0.41	0.44	0.48	0.51	0.55
Perlite–Expanded	Dew Point*	1200	–	0.36	0.40	0.42	0.45	0.47	0.50	0.52	0.55
Polyiscyanurate	Dew Point*	200	–	0.21	0.27	–	–	–	–	–	–
Polystyrene	Dew Point*	165	–	0.22	0.27	–	–	–	–	–	–
Polyurethane	Dew Point*	200	–	0.21	0.27	–	–	–	–	–	–
Rubber–Foamed	Dew Point*	200	–	0.27	0.28	–	–	–	–	–	–
Silica–Expanded	Dew Point*	1500	–	0.38	0.40	0.41	0.42	0.43	0.45	0.47	0.50

*Due to vapor pressure moisture will migrate to low temperature pipe, or tube, if surface temperature is below ambient air dew point for an extended period of time.

TABLE 182

THERMAL CONDUCTIVITIES OF PREFORMED PIPE INSULATIONS – METRIC UNITS

Pipe Insulation Material	Temperature Limits		Thermal Conductivities K_m in W/mK								
			Mean Temperature °C								
	Min. °C	Max. °C	–17.8°C	10.0°C	37.8°C	65.6°C	93.3°C	121°C	149°C	177°C	204°C
Calcium Silicate	93.3	649	–	–	0.050	0.056	0.062	0.066	0.072	0.075	0.079
Cellular Glass	–18.4	427	0.044	0.047	0.052	0.056	0.061	0.065	0.071	0.075	0.082
Glass Fiber	Dew Point*	260	–	0.037	0.040	0.043	0.046	0.050	0.056	0.062	0.069
Mineral Fiber	Dew Point*	649	–	0.046	0.049	0.053	0.059	0.063	0.069	0.073	0.079
Perlite–Expanded	Dew Point*	649	–	0.052	0.058	0.061	0.065	0.068	0.072	0.075	0.079
Polyiscyanurate	Dew Point*	93	–	0.030	0.039	–	–	–	–	–	–
Polystyrene	Dew Point*	74	–	0.032	0.039	–	–	–	–	–	–
Polyurethane	Dew Point*	93	–	0.030	0.039	–	–	–	–	–	–
Rubber–Foamed	Dew Point*	93	–	0.039	0.040	–	–	–	–	–	–
Silica–Expanded	Dew Point*	816	–	0.054	0.058	0.059	0.061	0.062	0.065	0.068	0.072

*Due to vapor pressure moisture will migrate to low temperature pipe, or tube, if surface temperature is below ambient air dew point for an extended period of time.

TABLE 183

HIGH TEMPERATURE SERVICE
CALCULATED AIR FILM SURFACE RESISTANCE R_s and R_{ms}
STILL AIR – NATURAL CONVECTION – AIR TEMPERATURE 70°F

Insulation Jacket or Cover	Surface Emittance ϵ	Insulation Resistance R_c	Surface Resistance R_s (English Units) at Pipe Operating Temperature °F					
			100°F	200°F	300°F	400°F	500°F	600°F
Bright Aluminum (unpainted)	0.05	1.0 to 2.5	1.46	1.24	1.10	0.99	0.94	0.89
		2.5 to 5.0	1.56	1.37	1.22	1.11	1.05	1.00
		5.0 to 10.0	1.81	1.55	1.36	1.24	1.17	1.12
		10.0 to 15.0	1.92	1.68	1.50	1.37	1.28	1.22
		15.0 to 20.0	1.98	1.74	1.60	1.46	1.36	1.31
		20.0 to 25.0	2.02	1.83	1.66	1.52	1.42	1.36
Light Colored Mastic Pastel Painted Jacket	0.50	1.0 to 2.5	0.91	0.82	0.74	0.67	0.64	0.61
		2.5 to 5.0	0.95	0.88	0.81	0.76	0.71	0.68
		5.0 to 10.0	1.07	0.98	0.89	0.83	0.79	0.77
		10.0 to 15.0	1.17	1.06	0.97	0.89	0.85	0.82
		15.0 to 20.0	1.23	1.10	1.00	0.92	0.87	0.85
		20.0 to 25.0	1.28	1.14	1.03	0.95	0.90	0.87
Dark Colored Mastic or Painted Jackets	0.90	1.0 to 2.5	0.74	0.65	0.58	0.53	0.50	0.48
		2.5 to 5.0	0.77	0.68	0.61	0.56	0.54	0.52
		5.0 to 10.0	0.80	0.73	0.68	0.62	0.60	0.58
		10.0 to 15.0	0.83	0.77	0.72	0.67	0.65	0.64
		15.0 to 20.0	0.87	0.80	0.75	0.70	0.68	0.67
		20.0 to 25.0	0.90	0.83	0.78	0.72	0.70	0.69

TABLE 184

HIGH TEMPERATURE SERVICE
CALCULATED AIR FILM SURFACE RESISTANCE R_s and R_{ms}
STILL AIR – NATURAL CONVECTION – AIR TEMPERATURE 21.1°C

Insulation Jacket or Cover	Surface Emittance ϵ	Insulation Resistance R_{mc}	Surface Resistance R_{ms} (Metric Units) at Pipe Operating Temperature °C					
			37.8°C	98.3°C	149°C	204°C	260°C	316°C
Bright Aluminum (unpainted)	0.05	0.176 to 0.441	0.257	0.218	0.193	0.174	0.165	0.157
		0.441 to 0.881	0.275	0.241	0.215	0.195	0.185	0.176
		0.881 to 1.762	0.318	0.273	0.240	0.218	0.206	0.197
		1.762 to 2.643	0.338	0.296	0.264	0.241	0.226	0.214
		2.643 to 3.524	0.349	0.307	0.282	0.257	0.240	0.231
		3.524 to 4.405	0.356	0.322	0.292	0.267	0.250	0.240
Light Colored Mastic Pastel Painted Jacket	0.50	0.176 to 0.441	0.107	0.144	0.130	0.118	0.113	0.107
		0.441 to 0.881	0.167	0.155	0.143	0.134	0.125	0.120
		0.881 to 1.762	0.189	0.172	0.157	0.143	0.139	0.136
		1.762 to 2.643	0.206	0.187	0.171	0.157	0.150	0.144
		2.643 to 3.524	0.217	0.193	0.176	0.162	0.153	0.150
		3.524 to 4.405	0.226	0.201	0.181	0.167	0.159	0.153
Dark Colored Mastic or Painted Jackets	0.90	0.176 to 0.441	0.130	0.115	0.102	0.093	0.068	0.085
		0.441 to 0.881	0.136	0.120	0.107	0.098	0.095	0.092
		0.881 to 1.762	0.141	0.129	0.120	0.109	0.106	0.102
		1.762 to 2.643	0.146	0.136	0.127	0.118	0.114	0.113
		2.643 to 3.524	0.154	0.141	0.132	0.123	0.120	0.118
		3.524 to 4.405	0.158	0.146	0.137	0.127	0.123	0.121

TABLE 185

LOW TEMPERATURE SERVICE
CALCULATED AIR FILM SURFACE RESISTANCE R_s and R_{ms}
NATURAL CONVECTION – AIR TEMPERATURE 70°F

Insulation Jacket or Cover	Surface Emittance ϵ	Insulation Resistance R_c	Surface Resistance R_s (English Units) at Pipe Operating Temperature °F					
			−40°F	−20°F	0°F	20°F	40°F	60°F
Bright Aluminum (unpainted)	0.05	1.0 to 2.5	1.90	1.80	1.73	1.67	1.60	1.53
		2.5 to 5.0	2.09	1.92	1.84	1.80	1.76	1.72
		5.0 to 10.0	2.17	2.13	2.06	1.97	1.90	1.86
		10.0 to 15.0	2.24	2.17	2.11	2.06	2.00	1.96
		15.0 to 20.0	2.30	2.26	2.20	2.15	2.10	2.09
		20.0 to 25.0	2.35	2.30	2.25	2.20	2.15	2.11
Light Colored Mastic Pastel Painted Jacket	0.50	1.0 to 2.5	1.06	1.04	1.02	1.00	0.98	0.96
		2.5 to 5.0	1.08	1.06	1.04	1.03	1.02	1.01
		5.0 to 10.0	1.17	1.15	1.13	1.11	1.09	1.08
		10.0 to 15.0	1.28	1.26	1.24	1.22	1.20	1.18
		15.0 to 20.0	1.39	1.37	1.35	1.32	1.29	1.26
		20.0 to 25.0	1.49	1.46	1.43	1.40	1.37	1.34
Dark Colored Mastic or Painted Jackets	0.90	1.0 to 2.5	0.88	0.86	0.84	0.82	0.80	0.78
		2.5 to 5.0	0.91	0.89	0.87	0.85	0.83	0.81
		5.0 to 10.0	0.93	0.91	0.89	0.87	0.86	0.84
		10.0 to 15.0	0.95	0.93	0.92	0.90	0.88	0.87
		15.0 to 20.0	0.97	0.96	0.94	0.93	0.91	0.90
		20.0 to 25.0	0.99	0.98	0.96	0.95	0.93	0.92

TABLE 186

LOW TEMPERATURE SERVICE
CALCULATED AIR FILM SURFACE RESISTANCE R_s and R_{ms}
NATURAL CONVECTION – AIR TEMPERATURE 21.1°C

Insulation Jacket Cover	Surface Emittance ϵ	Insulation Resistance R_{mc}	Surface Resistance R_{ms} (Metric Units) at Pipe Operating Temperature °C					
			–40°C	–28.9°C	–17.8°C	–6.7°C	4.4°C	15.6°C
Bright Aluminum (unpainted)	0.05	0.176 to 0.441	0.335	0.317	0.304	0.294	0.282	0.270
		0.441 to 0.881	0.368	0.338	0.324	0.317	0.310	0.303
		0.881 to 1.762	0.382	0.375	0.362	0.347	0.334	0.327
		1.762 to 2.643	0.395	0.382	0.372	0.362	0.352	0.345
		2.643 to 3.524	0.405	0.398	0.387	0.379	0.370	0.368
		3.524 to 4.405	0.414	0.405	0.396	0.387	0.379	0.372
Light Colored Mastic Pastel Painted Jacket	0.50	0.176 to 0.441	0.187	0.183	0.180	0.176	0.173	0.169
		0.441 to 0.881	0.190	0.187	0.183	0.181	0.180	0.178
		0.881 to 1.762	0.206	0.202	0.199	0.195	0.192	0.190
		1.762 to 2.643	0.226	0.222	0.218	0.215	0.211	0.208
		2.643 to 3.524	0.244	0.241	0.238	0.233	0.227	0.222
		3.524 to 4.405	0.263	0.257	0.252	0.247	0.242	0.236
Dark Colored Mastic or Painted Jackets	0.90	0.176 to 0.441	0.155	0.151	0.148	0.144	0.140	0.137
		0.441 to 0.881	0.159	0.157	0.153	0.150	0.146	0.143
		0.881 to 1.762	0.164	0.159	0.157	0.153	0.151	0.148
		1.762 to 2.643	0.167	0.163	0.162	0.159	0.155	0.153
		2.643 to 3.524	0.171	0.169	0.166	0.163	0.160	0.159
		3.524 to 4.405	0.174	0.173	0.169	0.167	0.164	0.162

TABLE 187

TABLE 188 SURFACE AREAS OF PIPE INSULATION

English Units

Based on NPS pipe and ASTM dimensional standard/pipe insulation—square feet per linear foot													Pipe Diameter	
Nom. pipe size	Bare	Nominal insulation thickness – inches											Inches	mm
		1	1½	2	2½	3	3½	4	4½	5	5½	6		
1/8	0.106	0.62	0.92	1.18	1.46	1.73	2.00	2.26					0.405	10.3
1/4	0.141	0.75	1.05	1.30	1.46	1.73	2.00	2.26					0.504	12.8
3/8	0.177	0.75	1.05	1.30	1.46	1.73	2.00	2.26					0.695	17.1
1/2	0.220	0.75	1.05	1.30	1.73	2.00	2.25	2.52					0.840	21.3
3/4	0.275	0.75	1.05	1.30	1.73	2.00	2.25	2.52					1.050	26.7
1	0.344	0.92	1.18	1.46	1.74	2.00	2.26	2.52					1.315	33.4
1 1/4	0.435	0.92	1.31	1.46	1.74	2.00	2.26	2.52					1.660	42.2
1 1/2	0.498	1.05	1.31	1.74	2.00	2.26	2.52	2.81					1.900	48.3
2	0.622	1.18	1.46	1.73	2.00	2.26	2.52	2.81					2.375	60.3
2 1/2	0.753	1.31	1.73	2.00	2.26	2.52	2.81	3.08					2.875	73.0
3	0.917	1.46	1.73	2.00	2.26	2.52	2.81	3.08	3.34				3.500	88.9
3 1/2	1.047	1.73	2.00	2.26	2.52	2.81	3.08	3.34	3.67				4.000	101.6
4	1.178	1.73	2.00	2.26	2.52	2.81	3.08	3.34	3.67				4.500	114.3
4 1/2	1.309	2.00	2.26	2.52	2.81	3.08	3.34	3.67	3.67				5.000	127.0
5	1.456	2.00	2.26	2.52	2.81	3.08	3.34	3.67	3.93	4.19			5.563	141.3
6	1.734	2.26	2.52	2.81	3.08	3.34	3.67	3.93	4.18	4.45			6.625	168.3
7	1.996		2.81	3.08	3.34	3.67	3.93	4.19	4.45	4.71	4.97		7.625	198.7
8	2.258		3.08	3.34	3.67	3.93	4.19	4.45	4.71	4.97	5.24		8.625	219.1
9	2.520		3.34	3.67	3.93	4.19	4.45	4.71	4.97	5.24	5.50		9.625	244.5
10	2.814		3.67	3.93	4.19	4.45	4.71	4.97	5.24	5.50	5.76		10.750	273.0
11	3.076		3.93	4.19	4.45	4.71	4.97	5.24	5.50	5.76	6.02		11.750	298.4
12	3.338		4.19	4.45	4.71	4.97	5.24	5.50	5.76	6.02	6.28	6.54	12.750	323.8
14	3.665		4.45	4.71	4.97	5.24	5.50	5.76	6.02	6.28	6.54	6.81	14.750	855.6
16	4.189		4.97	5.24	5.50	5.76	6.02	6.28	6.54	6.81	7.07	7.33	16.000	406.4
18	4.712		5.50	5.76	6.02	6.28	6.54	6.81	7.07	7.33	7.59	7.85	18.000	459.2
20	5.236		6.02	6.28	6.54	6.81	7.07	7.33	7.59	7.85	8.12	8.38	20.000	508.0
22	5.759		6.54	6.81	7.07	7.33	7.59	7.85	8.12	8.38	8.64	8.90	22.000	558.8
24	6.283		7.07	7.33	7.59	7.85	8.12	8.38	8.64	8.90	9.16	9.42	24.000	609.5
26	6.807		7.59	7.85	8.12	8.38	8.64	8.90	9.16	9.42	9.69	9.95	26.000	6.604
28	7.331		8.12	8.38	8.64	8.90	9.16	9.42	9.69	9.95	10.21	10.47	28.000	711.2
30	7.854		8.64	8.90	9.16	9.42	9.69	9.95	10.21	10.47	10.73	11.00	30.000	762.0
		Nominal insulation thickness – mm												
			25	38	51	64	76	89	102	127	140	152		

TABLE 189 SURFACE AREAS OF NPS PIPE INSULATION

Based on NPS pipe and ASTM standard C-585 square meters per linear meter.

Metric Units

Pipe			Nominal Insulation Thickness										
Nom Size Inches	Dia mm	Bare Area	1" nom	1½" nom	2" nom	2½" nom	3" nom	3½" nom	4" nom	4½" nom	5" nom	5½" nom	6" nom
			25 mm	38 mm	51 mm	64 mm	76 mm	89 mm	102 mm	114 mm	127 mm	140 mm	152 mm
1/8	10.3	0.032	0.189	0.280	0.360	0.445	0.527	0.610	0.689				
1/4	13.7	0.043	0.229	0.320	0.399	0.445	0.527	0.610	0.689				
3/8	17.1	0.054	0.229	0.320	0.399	0.445	0.527	0.610	0.689				
1/2	21.3	0.068	0.229	0.320	0.399	0.527	0.610	0.689	0.768				
3/4	26.7	0.084	0.229	0.320	0.399	0.527	0.610	0.689	0.768				
1	33.4	0.105	0.280	0.360	0.445	0.527	0.610	0.689	0.768				
1 1/4	42.2	0.133	0.280	0.399	0.445	0.527	0.610	0.689	0.768				
1 1/2	48.3	0.152	0.320	0.399	0.527	0.610	0.689	0.768	0.857				
2	60.3	0.190	0.360	0.445	0.527	0.610	0.689	0.768	0.857				
2 1/2	73.0	0.230	0.399	0.527	0.610	0.689	0.768	0.857	0.939				
3	88.9	0.210	0.445	0.527	0.610	0.689	0.768	0.857	0.939	1.018			
3 1/2	101.6	0.319	0.527	0.610	0.689	0.768	0.857	0.939	1.018	1.119			
4	114.3	0.359	0.527	0.610	0.689	0.768	0.857	0.939	1.018	1.119			
4 1/2	127.0	0.399	0.610	0.689	0.768	0.857	0.939	1.018	1.119	1.119			
5	141.3	0.444	0.610	0.689	0.768	0.857	0.939	1.018	1.119	1.198	1.277		
6	168.3	0.529	0.689	0.768	0.857	0.939	1.018	1.119	1.198	1.277	1.356		
7	193.7	0.608		0.857	0.939	1.018	1.119	1.198	1.277	1.356	1.436	1.515	
8	219.1	0.688		0.939	1.018	1.119	1.198	1.277	1.356	1.436	1.515	1.600	
9	244.3	0.768		1.018	1.119	1.198	1.277	1.356	1.436	1.515	1.600	1.676	
10	273.0	0.858		1.119	1.198	1.277	1.356	1.436	1.515	1.600	1.676	1.760	
11	298.4	0.938		1.198	1.277	1.356	1.436	1.515	1.600	1.676	1.760	1.834	
12	323.8	1.017		1.277	1.356	1.436	1.515	1.600	1.676	1.760	1.834	1.914	1.993
14	355.6	1.117		1.356	1.436	1.515	1.600	1.676	1.760	1.834	1.194	1.993	2.076
16	406.4	1.277		1.515	1.600	1.676	1.760	1.834	1.914	1.993	2.076	2.155	2.234
18	457.2	1.436		1.676	1.760	1.834	1.914	1.993	2.076	2.155	2.234	2.313	2.393
20	508.0	1.596		1.834	1.914	1.993	2.076	2.155	2.234	2.313	2.393	2.475	2.554
22	558.6	1.755		1.993	2.076	2.155	2.234	2.313	2.393	2.475	2.554	2.634	2.713
24	609.6	1.915		2.155	2.234	2.313	2.393	2.475	2.554	2.634	2.713	2.792	2.871
26	660.4	2.074		2.313	2.393	2.475	2.554	2.634	2.713	2.792	2.871	2.954	3.033
28	711.2	2.234		2.475	2.554	2.634	2.713	2.792	2.871	2.954	3.033	3.112	3.191
30	762.0	2.394		2.634	2.713	2.792	2.871	2.954	3.033	3.112	3.191	3.271	3.353

TABLE 190 SURFACE AREAS OF TUBE INSULATION

Based on tube size insulation in accordance with ASTM standard C-585 square feet per linear foot.

English Units

Tube				Nominal Insulation Thickness — Inches						
Nom Size Inches	Dia		Bare Area	1" nom	1½" nom	2" nom	2½" nom	3" nom	3½" nom	4" nom
	inches	mm								
3/8	0.500	12.7	0.131	0.62	0.92	1.18	1.46	1.73	2.00	2.26
1/2	0.625	15.9	0.164	0.75	0.92	1.18	1.46	1.73	2.00	2.26
3/4	0.875	22.2	0.229	0.75	1.05	1.31	1.73	2.00	2.26	2.52
1	1.125	28.6	0.295	0.75	1.05	1.31	1.73	2.00	2.26	2.52
1 1/4	1.375	34.9	0.360	0.92	1.18	1.46	1.73	2.00	2.26	2.52
1 1/2	1.625	41.3	0.425	0.92	1.18	1.46	1.73	2.00	2.26	2.52
2	2.125	54.0	0.556	1.05	1.31	1.73	2.00	2.26	2.52	2.81
2 1/2	2.625	66.7	0.687	1.18	1.46	1.73	2.00	2.26	2.52	2.81
3	3.125	79.4	0.818	1.31	1.73	2.00	2.26	2.52	2.81	3.08
3 1/2	3.265	92.1	0.949	1.46	1.73	2.00	2.26	2.52	2.81	3.08
4	4.125	104.8	1.079	1.73	2.00	2.26	2.52	2.81	3.08	3.34
5	5.125	130.2	1.342	2.00	2.26	2.52	2.81	3.08	3.34	3.67
6	6.125	155.6	1.603	2.26	2.52	2.81	3.08	3.34	3.67	3.93

TABLE 191 SURFACE AREA OF TUBE INSULATION

Based on tube size insulation in accordance with ASTM C-585 square meters per linear meter.

Metric Units

Tube			Nominal Insulation Thickness						
Nom Size Inches	Dia mm	Bare Area	1" nom 25 mm	1½" nom 38 mm	2" nom 51 mm	2½" nom 64 mm	3" nom 76 mm	3½" nom 89 mm	4" nom 102 mm
3/8	12.7	0.040	0.189	0.280	0.360	0.445	0.527	0.610	0.689
1/2	15.9	0.050	0.229	0.280	0.360	0.445	0.527	0.610	0.689
3/4	22.2	0.070	0.070	0.320	0.399	0.527	0.610	0.689	0.768
1	23.6	0.090	0.229	0.320	0.399	0.527	0.610	0.689	0.768
1 1/4	34.9	0.110	0.280	0.360	0.445	0.527	0.610	0.689	0.768
1 1/2	41.3	0.130	0.280	0.360	0.445	0.527	0.610	0.689	0.768
2	54.0	0.169	0.320	0.399	0.527	0.610	0.689	0.768	0.857
2 1/2	66.7	0.209	0.360	0.445	0.527	0.610	0.689	0.768	0.857
3	79.4	0.249	0.399	0.527	0.610	0.689	0.768	0.857	0.939
3 1/2	92.1	0.289	0.445	0.527	0.610	0.689	0.768	0.857	0.939
4	104.8	0.329	0.527	0.610	0.689	0.768	0.857	0.939	1.018
5	130.2	0.409	0.610	0.689	0.768	0.857	0.939	1.018	1.112
6	155.6	0.489	0.689	0.768	0.857	0.939	1.018	1.112	1.198

HEAT LOSSES FROM BARE FLAT SURFACES

Still air conditions—air temperature 70° F, (21.1° C) black surface $\epsilon = 1.0$

Losses in Btu per hour, per square foot, or watts per hour per square meter, of surface for *total* temperature difference between hot surface and air

Hot Surface Temp. °F	Temperature Diff. °F	Btu/ft², hr Loss	Hot Surface Temp.°C	Temperature Diff. °C	W/m² Loss
100	30	50	38	17	157
200	130	295	93	72	930
300	230	654	149	128	2061
400	330	1145	204	183	3609
500	430	1806	260	239	5693
600	530	2666	316	295	8404
700	630	3765	371	350	11869
800	730	5126	427	406	16160

TABLE 192

HEAT LOSSES FROM BARE NPS PIPE IN BTU PER HR PER LINEAR FT AND W PER HR PER LINEAR METER
STILL AIR CONDITIONS – AIR TEMPERATURE 70°F, 21.1°C, BLACK SURFACE ON PIPE

English and Metric Units

Pipe size	Pipe temp °F	Btu/lin ft hr loss	Pipe temp °C	W/lin m loss
1/2" NPS	100	13	38	12
	200	75	93	72
	300	165	149	158
	400	287	204	275
outside	500	444	260	426
dia	600	649	316	624
0.084"	700	901	371	866
21.3 mm	800	1218	427	1170
3/4" NPS	100	15	38	14
	200	92	93	88
	300	204	149	196
	400	353	204	339
outside	500	547	260	525
dia	600	801	316	765
1.050"	700	1113	371	1069
26.7 mm	800	1508	427	1449
1" NPS	100	19	38	18
	200	113	93	108
	300	250	149	240
	400	433	204	416
outside	500	674	260	647
dia	600	989	316	950
1.315"	700	1379	371	1325
33.4 mm	800	1865	427	1792
1 1/4" NPS	100	24	38	23
	200	141	93	135
	300	312	149	299
	400	540	204	518
outside	500	843	260	810
dia	600	1237	316	1188
1.66"	700	1728	371	1660
42.2 mm	800	2342	427	2250
1 1/2" NPS	100	27	38	26
	200	159	93	152
	300	352	149	338
	400	612	204	588
outside	500	955	260	914
dia	600	1403	316	1348
1.90"	700	1960	371	1883
48.3 mm	800	2661	427	2557
2" NPS	100	33	38	32
	200	195	93	187
	300	431	149	414
	400	753	204	723
outside	500	1176	260	1130
dia	600	1732	316	1664
2.375"	700	2423	378	2328
60.3 mm	800	3295	427	3166
2 1/2" NPS	100	39	38	37
	200	232	93	222
	300	515	149	494
	400	899	204	863
outside	500	1408	260	1352
dia	600	2077	316	1995
2.875"	700	2907	371	2793
73.0 mm	800	3956	427	3801

Pipe size	Pipe temp °F	Btu/lin ft hr loss	Pipe temp °C	W/lin m loss
3" NPS	100	48	38	46
	200	280	93	269
	300	619	149	594
	400	1083	204	1040
outside	500	1695	260	1629
dia	600	2505	316	2407
3.50"	700	3511	371	3373
88.9 mm	800	4784	427	4597
4" NPS	100	60	38	57
	200	354	93	340
	300	783	149	752
	400	1370	204	1316
outside	500	2149	260	2064
dia	600	3179	316	3054
4.5"	700	4467	371	4292
114.3 mm	800	6094	427	5856
5" NPS	100	73	38	70
	200	428	93	411
	300	952	149	194
	400	1674	204	1608
outside	500	2633	260	2530
dia	600	3893	316	3741
5.563"	700	5473	371	5259
141.3 mm	800	7482	427	7189
6" NPS	100	86	38	82
	200	503	93	493
	300	1122	149	1078
	400	1976	204	1898
outside	500	3105	260	2583
dia	600	4604	316	4424
6.625"	700	6479	371	6225
168.3 mm	800	8858	427	8512
8" NPS	100	111	38	106
	200	643	93	617
	300	1436	149	1379
	400	2530	204	2431
outside	500	3988	260	3832
dia	600	5927	316	5695
8.625"	700	8356	371	8029
219.1 mm	800	11433	427	10986
10" NPS	100	136	38	130
	200	790	93	759
	300	1769	149	1700
	400	3114	204	2992
outside	500	4918	260	4725
dia	600	7312	316	7026
10.75"	700	10325	371	9921
273.0 mm	800	14147	427	13594
12" NPS	100	159	38	152
	200	930	93	894
	300	2076	149	1994
	400	3664	204	3520
outside	500	5809	260	5582
dia	600	8627	316	8290
12.75"	700	12194	371	11717
323.8 mm	800	16721	427	16067

Pipe size	Pipe temp °F	Btu/lin ft hr loss	Pipe temp °C	W/lin m loss
14" NPS	100	173	38	166
	200	1009	93	970
	300	2258	149	2170
	400	3989	204	3833
outside	500	6316	260	6069
dia	600	9404	316	9036
14.0"	700	13301	371	12781
355.6 mm	800	18236	427	17523
16" NPS	100	197	38	189
	200	1142	93	1097
	300	2560	149	2460
	400	4529	204	4352
outside	500	7160	260	6880
dia	600	10697	316	10278
16.0"	700	15124	371	14532
406.4 mm	800	20769	427	19957
18" NPS	100	220	38	211
	200	1282	93	1231
	300	2875	149	2763
	400	5074	204	4876
outside	500	8032	260	7718
dia	600	12010	316	11540
18.0"	700	16992	371	16328
457.2 mm	800	23346	427	22433
20" NPS	100	243	38	233
	200	1416	93	1360
	300	3168	149	3044
	400	5618	204	5398
outside	500	8897	260	8549
dia	600	13270	316	12751
20.0"	700	18777	371	18043
508.0 mm	800	25810	427	24801
24" NPS	100	288	38	276
200	200	1683	93	1617
	300	3772	149	3624
	400	6679	204	6418
outside	500	10595	260	10180
dia	600	15825	316	15206
24.0"	700	22375	371	21500
609.6 mm	800	30836	427	29630
30" NPS	100	351	38	337
	200	2042	93	1962
	300	4642	149	4460
	400	8242	204	7920
outside	500	13103	260	12590
dia	600	19606	316	18840
30.0"	700	27758	371	26673
762.0 mm	800	38299	427	36802

TABLE 193

TABULATION OF RESULTS – EXAMPLE 50 and 50a

EXAMPLE 50 – English Units						EXAMPLE 50a – Metric Units					
Covering or Jacket	Insul. Thk. Inches	Heat Gain Btu Per Hour		Surface Temp. °F	Max. % RH Without Surface Condensation	Covering or Jacket	Insul. Thk. Metres	Heat Gain W Per Hour		Surface Temp. °C	Max. % RH Without Surface Condensation
		ft^2	lin. ft					m^2	lin. m		
Mastic	1	12.96	18.9	59	68%	Mastic	0.025	41.10	18.29	15	68%
Mastic	1½	8.60	14.8	62	76%	Mastic	0.039	26.80	14.12	17	76%
Mastic	2	6.29	12.6	65	80%	Mastic	0.051	19.46	11.87	18	80%
Mastic	2½	4.85	10.9	66	86%	Mastic	0.066	15.12	10.42	19	86%
Aluminum	1	10.75	15.7	51	52%	Alum.	0.025	34.30	15.26	11	52%
Aluminum	1½	7.98	12.9	56	63%	Alum.	0.039	23.70	12.50	13	63%
Aluminum	2	5.60	11.2	59	68%	Alum.	0.051	17.45	10.64	15	68%
Aluminum	2½	4.43	10.0	61	71%	Alum.	0.066	13.81	9.52	16	71%

INSIDE AIR TEMPERATURE 75°F (23.9°C)
RELATIVE HUMIDITY 20%

CONDENSATION PREVENTION INSULATION THICKNESSES
OUTER SURFACE EMITTANCE NOT LESS THAN $\epsilon = 0.05$
THICKNESSES IN NOMINAL INCHES

CELLULAR GLASS INSULATION
(ALUMINUM JACKETED)

Nominal Pipe Sizes Inches	OPERATING TEMPERATURES °C											Nominal Pipe Sizes mm
	19 to 10	9 to 5	4 to 0	−1 to −5	−6 to −10	−11 to −15	−16 to −20	−21 to −25	−26 to −30	−31 to −35	−36 to −40	
1/2	1.0	1.0	1.0	1.0	1.0	1.0	1.0	1.0	1.0	1.0	1.0	21
3/4	1.0	1.0	1.0	1.0	1.0	1.0	1.0	1.0	1.0	1.0	1.0	27
1	1.0	1.0	1.0	1.0	1.0	1.0	1.0	1.0	1.0	1.0	1.0	33
1 1/2	1.0	1.0	1.0	1.0	1.0	1.0	1.0	1.0	1.0	1.0	1.0	48
2	1.0	1.0	1.0	1.0	1.0	1.0	1.0	1.0	1.0	1.0	1.0	60
2 1/2	1.0	1.0	1.0	1.0	1.0	1.0	1.0	1.0	1.0	1.0	1.0	73
3	1.0	1.0	1.0	1.0	1.0	1.0	1.0	1.0	1.0	1.0	1.0	89
4	1.0	1.0	1.0	1.0	1.0	1.0	1.0	1.0	1.0	1.0	1.0	114
5	1.0	1.0	1.0	1.0	1.0	1.0	1.0	1.0	1.0	1.0	1.0	141
6	1.0	1.0	1.0	1.0	1.0	1.0	1.0	1.0	1.0	1.0	1.0	168
8	1.5	1.5	1.5	1.5	1.5	1.5	1.5	1.5	1.5	1.5	1.5	219
10	1.5	1.5	1.5	1.5	1.5	1.5	1.5	1.5	1.5	1.5	1.5	273
12	1.5	1.5	1.5	1.5	1.5	1.5	1.5	1.5	1.5	1.5	1.5	324
14	1.5	1.5	1.5	1.5	1.5	1.5	1.5	1.5	1.5	1.5	1.5	356
16	1.5	1.5	1.5	1.5	1.5	1.5	1.5	1.5	1.5	1.5	1.5	406
18	1.5	1.5	1.5	1.5	1.5	1.5	1.5	1.5	1.5	1.5	1.5	457
20	1.5	1.5	1.5	1.5	1.5	1.5	1.5	1.5	1.5	1.5	1.5	508
24	1.5	1.5	1.5	1.5	1.5	1.5	1.5	1.5	1.5	1.5	1.5	610
28	1.5	1.5	1.5	1.5	1.5	1.5	1.5	1.5	1.5	1.5	1.5	711
32	1.5	1.5	1.5	1.5	1.5	1.5	1.5	1.5	1.5	1.5	1.5	813
36	1.5	1.5	1.5	1.5	1.5	1.5	1.5	1.5	1.5	1.5	1.5	914
FLAT	1.5	1.5	1.5	1.5	1.5	1.5	1.5	1.5	1.5	1.5	1.5	FLAT
	67 to 50	49 to 41	40 to 32	31 to 23	22 to 14	13 to 5	4 to −4	−5 to −13	−14 to −22	−23 to −31	−32 to −40	

OPERATING TEMPERATURES °F

Courtesy of Pittsburgh Corning Corp., Pittsburgh, Pa.

TABLE 194

INSIDE AIR TEMPERATURE 75°F (23.9°C)
RELATIVE HUMIDITY 40%

CONDENSATION PREVENTION INSULATION THICKNESSES
OUTER SURFACE EMITTANCE NOT LESS THAN ϵ = 0.05
THICKNESSES IN NOMINAL INCHES

CELLULAR GLASS INSULATION
(ALUMINUM JACKETED)

Nominal Pipe Sizes Inches	OPERATING TEMPERATURES °C											Nominal Pipe Sizes mm
	19 to 10	9 to 5	4 to 0	−1 to −5	−6 to −10	−11 to −15	−16 to −20	−21 to −25	−26 to −30	−31 to −35	−36 to −40	
1/2	1.0	1.0	1.0	1.0	1.0	1.0	1.0	1.0	1.0	1.0	1.0	21
3/4	1.0	1.0	1.0	1.0	1.0	1.0	1.0	1.0	1.0	1.5	1.5	27
1	1.0	1.0	1.0	1.0	1.0	1.0	1.0	1.0	1.0	1.0	1.5	33
1 1/2	1.0	1.0	1.0	1.0	1.0	1.0	1.0	1.0	1.0	1.5	1.5	48
2	1.0	1.0	1.0	1.0	1.0	1.0	1.0	1.0	1.5	1.5	1.5	60
2 1/2	1.0	1.0	1.0	1.0	1.0	1.0	1.0	1.0	1.5	1.5	1.5	73
3	1.0	1.0	1.0	1.0	1.0	1.0	1.0	1.5	1.5	1.5	1.5	89
4	1.0	1.0	1.0	1.0	1.0	1.0	1.0	1.5	1.5	1.5	1.5	114
5	1.0	1.0	1.0	1.0	1.0	1.0	1.0	1.5	1.5	2.0	2.0	141
6	1.0	1.0	1.0	1.0	1.0	1.0	1.5	1.5	1.5	2.0	2.0	168
8	1.5	1.5	1.5	1.5	1.5	1.5	1.5	1.5	1.5	2.0	2.0	219
10	1.5	1.5	1.5	1.5	1.5	1.5	1.5	1.5	1.5	2.0	2.0	273
12	1.5	1.5	1.5	1.5	1.5	1.5	1.5	1.5	2.0	2.0	2.0	324
14	1.5	1.5	1.5	1.5	1.5	1.5	1.5	1.5	2.0	2.0	2.5	356
16	1.5	1.5	1.5	1.5	1.5	1.5	1.5	2.0	2.0	2.0	2.5	406
18	1.5	1.5	1.5	1.5	1.5	1.5	1.5	2.0	2.0	2.0	2.5	457
20	1.5	1.5	1.5	1.5	1.5	1.5	1.5	2.0	2.0	2.5	2.5	508
24	1.5	1.5	1.5	1.5	1.5	1.5	1.5	2.0	2.0	2.5	2.5	610
28	1.5	1.5	1.5	1.5	1.5	1.5	1.5	2.0	2.0	2.5	2.5	711
32	1.5	1.5	1.5	1.5	1.5	1.5	1.5	2.0	2.0	2.5	2.5	813
36	1.5	1.5	1.5	1.5	1.5	1.5	1.5	2.0	2.0	2.5	2.5	914
FLAT	1.5	1.5	1.5	1.5	1.5	1.5	1.5	2.0	2.0	2.5	2.5	FLAT
	67 to 50	49 to 41	40 to 32	31 to 23	22 to 14	13 to 5	4 to −4	−5 to −13	−14 to −22	−23 to −31	−32 to −40	

OPERATING TEMPERATURES °F

Courtesy of Pittsburgh Corning Corp., Pittsburgh, Pa.

TABLE 194 (Cont'd)

INSIDE AIR TEMPERATURE 75°F (23.9°C)
RELATIVE HUMIDITY 60%

CONDENSATION PREVENTION INSULATION THICKNESSES
OUTER SURFACE EMITTANCE NOT LESS THAN ϵ = 0.05
THICKNESSES IN NOMINAL INCHES

CELLULAR GLASS INSULATION
(ALUMINUM JACKETED)

Nominal Pipe Sizes Inches	OPERATING TEMPERATURES °C											Nominal Pipe Sizes mm
	19 to 10	9 to 5	4 to 0	−1 to −5	−6 to −10	−11 to −15	−16 to −20	−21 to −25	−26 to −30	−31 to −35	−36 to −40	
1/2	1.0	1.0	1.0	1.0	1.5	1.5	1.5	1.5	2.0	2.0	2.0	21
3/4	1.0	1.0	1.0	1.5	1.5	1.5	1.5	2.0	2.0	2.0	2.5	27
1	1.0	1.0	1.0	1.0	1.5	1.5	1.5	2.0	2.0	2.0	2.5	33
1 1/2	1.0	1.0	1.0	1.5	1.5	1.5	2.0	2.0	2.0	2.0	2.5	48
2	1.0	1.0	1.0	1.5	1.5	2.0	2.0	2.0	2.5	2.5	3.0	60
2 1/2	1.0	1.0	1.0	1.5	1.5	2.0	2.0	2.0	2.0	2.5	2.5	73
3	1.0	1.0	1.0	1.5	2.0	2.0	2.0	2.5	2.5	3.0	3.0	89
4	1.0	1.0	1.5	1.5	2.0	2.0	2.5	2.5	3.0	3.0	3.5	114
5	1.0	1.0	1.5	1.5	2.0	2.5	2.5	2.5	3.0	3.5	3.5	141
6	1.0	1.0	1.5	2.0	2.0	2.5	2.5	3.0	3.0	3.5	3.5	168
8	1.5	1.5	1.5	2.0	2.0	2.5	2.5	3.0	3.5	3.5	4.0	219
10	1.5	1.5	1.5	2.0	2.0	2.5	3.0	3.0	3.5	4.0	4.0	273
12	1.5	1.5	1.5	2.0	2.5	2.5	3.0	3.5	3.5	4.0	4.0	324
14	1.5	1.5	1.5	2.0	2.5	3.0	3.0	3.5	4.0	4.0	4.5	356
16	1.5	1.5	1.5	2.0	2.5	3.0	3.0	3.5	4.0	4.5	4.5	406
18	1.5	1.5	1.5	2.0	2.5	3.0	3.5	3.5	4.0	4.5	5.0	457
20	1.5	1.5	1.5	2.0	2.5	3.0	3.5	3.5	4.0	4.5	5.0	508
24	1.5	1.5	2.0	2.0	2.5	3.0	3.5	4.0	4.0	4.5	5.0	610
28	1.5	1.5	2.0	2.5	2.5	3.0	3.5	4.0	4.0	4.5	5.0	711
32	1.5	1.5	2.0	2.5	2.5	3.0	3.5	4.0	4.5	4.5	5.0	813
36	1.5	1.5	2.0	2.5	3.0	3.0	3.5	4.0	4.5	4.5	5.0	914
FLAT	1.5	1.5	2.0	2.5	3.0	3.5	4.0	4.5	5.0	5.5	5.5	FLAT
	67 to 50	49 to 41	40 to 32	31 to 23	22 to 14	13 to 5	4 to −4	−5 to −13	−14 to −22	−23 to −31	−32 to −40	
	OPERATING TEMPERATURES °F											

Courtesy of Pittsburgh Corning Corp., Pittsburgh, Pa.

TABLE 194 (Cont'd)

INSIDE AIR TEMPERATURE 75° F (23.9° C)
RELATIVE HUMIDITY 80%

CONDENSATION PREVENTION INSULATION THICKNESSES
OUTER SURFACE EMITTANCE NOT LESS THAN $\epsilon = 0.05$
THICKNESSES IN NOMINAL INCHES

CELLULAR GLASS INSULATION
(ALUMINUM JACKETED)

Nominal Pipe Sizes Inches	OPERATING TEMPERATURES °C											Nominal Pipe Sizes mm
	19 to 10	9 to 5	4 to 0	−1 to −5	−6 to −10	−11 to −15	−16 to −20	−21 to −25	−26 to −30	−31 to −35	−36 to −40	
1/2	1.5	1.5	2.0	2.5	2.5	3.0	3.5	4.0	4.0	4.5	5.0	21
3/4	1.5	2.0	2.5	2.5	3.0	3.5	3.5	4.0	4.5	5.0	5.5	27
1	1.5	2.0	2.5	3.0	3.5	4.0	4.0	4.5	5.0	5.5	5.5	33
1 1/2	1.5	2.0	2.5	3.0	3.5	4.0	4.5	5.0	5.0	5.0	5.5	48
2	1.5	2.0	3.0	3.5	4.0	4.5	5.0	5.5	5.5	6.0	6.5	60
2 1/2	1.5	2.0	2.5	3.5	4.0	4.5	5.0	5.0	6.0	6.5	7.0	73
3	1.5	2.5	3.0	4.0	4.5	5.0	5.5	6.0	6.5	7.0	7.5	89
4	2.0	2.5	3.5	4.0	4.5	5.5	5.5	6.5	7.0	7.5	8.0	114
5	2.0	3.0	3.5	4.0	5.0	5.5	6.0	6.5	7.5	8.0	8.5	141
6	2.0	3.0	3.5	4.5	5.0	6.0	6.5	7.0	8.0	8.5	9.0	168
8	2.0	3.0	4.0	5.0	5.5	6.5	7.0	7.5	8.5	9.0	9.5	219
10	2.0	3.5	4.0	5.0	6.0	7.0	7.5	8.0	9.0	9.5	10.0	273
12	2.5	3.5	4.5	5.5	6.5	7.0	7.5	8.5	9.0	10.0	11.0	324
14	2.5	3.5	4.5	5.5	6.5	7.5	8.0	8.5	9.0	10.0	11.0	356
16	2.5	3.5	5.0	5.5	6.5	7.5	8.0	8.5	9.0	10.0	11.0	406
18	2.5	4.0	5.0	6.0	7.0	7.5	8.0	9.0	10.0	11.0	12.0	457
20	2.5	4.0	5.0	6.0	7.0	8.0	8.5	9.0	10.0	11.0	12.0	508
24	3.0	4.0	5.0	6.0	7.0	8.0	8.5	9.5	11.0	11.0	12.0	610
28	3.0	4.0	5.0	6.0	7.0	8.0	9.0	9.5	11.0	12.0		711
32	3.0	4.0	5.0	6.5	7.5	8.5	9.0	10.0	11.0	12.0		813
36	3.0	4.0	5.0	6.5	7.5	8.5	9.0	10.0	11.0	12.0		914
FLAT	3.0	4.5	6.0	7.5	8.5	10.0	11.0	12.0				FLAT
	67 to 50	49 to 41	40 to 32	31 to 23	22 to 14	13 to 5	4 to −4	−5 to −13	−14 to −22	−23 to −31	−32 to −40	
	OPERATING TEMPERATURES °F											

Courtesy of Pittsburgh Corning Corp., Pittsburgh, Pa.

TABLE 194 (Cont'd)

INSIDE AIR TEMPERATURE 75°F (23.9°C)
RELATIVE HUMIDITY 20%

CONDENSATION PREVENTION INSULATION THICKNESSES
OUTER EMITTANCE NOT LESS THAN $\epsilon = 0.9$
THICKNESSES IN NOMINAL INCHES

CELLULAR GLASS INSULATION
(COLORED MASTIC OR MEDIUM COLOR PAINTED METAL JACKET)

Nominal Pipe Sizes Inches	OPERATING TEMPERATURES °C											Nominal Pipe Sizes mm
	19 to 10	9 to 5	4 to 0	−1 to −5	−6 to −10	−11 to −15	−16 to −20	−21 to −25	−26 to −30	−31 to −35	−36 to −40	
1/2	1.0	1.0	1.0	1.0	1.0	1.0	1.0	1.0	1.0	1.0	1.0	21
3/4	1.0	1.0	1.0	1.0	1.0	1.0	1.0	1.0	1.0	1.0	1.0	27
1	1.0	1.0	1.0	1.0	1.0	1.0	1.0	1.0	1.0	1.0	1.0	33
1 1/2	1.0	1.0	1.0	1.0	1.0	1.0	1.0	1.0	1.0	1.0	1.0	48
2	1.0	1.0	1.0	1.0	1.0	1.0	1.0	1.0	1.0	1.0	1.0	60
2 1/2	1.0	1.0	1.0	1.0	1.0	1.0	1.0	1.0	1.0	1.0	1.0	73
3	1.0	1.0	1.0	1.0	1.0	1.0	1.0	1.0	1.0	1.0	1.0	89
4	1.0	1.0	1.0	1.0	1.0	1.0	1.0	1.0	1.0	1.0	1.0	114
5	1.0	1.0	1.0	1.0	1.0	1.0	1.0	1.0	1.0	1.0	1.0	141
6	1.0	1.0	1.0	1.0	1.0	1.0	1.0	1.0	1.0	1.0	1.0	168
8	1.5	1.5	1.5	1.5	1.5	1.5	1.5	1.5	1.5	1.5	1.5	219
10	1.5	1.5	1.5	1.5	1.5	1.5	1.5	1.5	1.5	1.5	1.5	273
12	1.5	1.5	1.5	1.5	1.5	1.5	1.5	1.5	1.5	1.5	1.5	324
14	1.5	1.5	1.5	1.5	1.5	1.5	1.5	1.5	1.5	1.5	1.5	356
16	1.5	1.5	1.5	1.5	1.5	1.5	1.5	1.5	1.5	1.5	1.5	406
18	1.5	1.5	1.5	1.5	1.5	1.5	1.5	1.5	1.5	1.5	1.5	457
20	1.5	1.5	1.5	1.5	1.5	1.5	1.5	1.5	1.5	1.5	1.5	508
24	1.5	1.5	1.5	1.5	1.5	1.5	1.5	1.5	1.5	1.5	1.5	610
28	1.5	1.5	1.5	1.5	1.5	1.5	1.5	1.5	1.5	1.5	1.5	711
32	1.5	1.5	1.5	1.5	1.5	1.5	1.5	1.5	1.5	1.5	1.5	813
36	1.5	1.5	1.5	1.5	1.5	1.5	1.5	1.5	1.5	1.5	1.5	914
FLAT	1.5	1.5	1.5	1.5	1.5	1.5	1.5	1.5	1.5	1.5	1.5	FLAT
	67 to 50	49 to 41	40 to 32	31 to 23	22 to 14	13 to 5	4 to −4	−5 to −13	−14 to −22	−23 to −31	−32 to −40	
	OPERATING TEMPERATURES °F											

Courtesy of Pittsburgh Corning Corp., Pittsburgh, Pa.

TABLE 194 (Cont'd)

INSIDE AIR TEMPERATURE 75°F (23.9°C)
RELATIVE HUMIDITY 40%

CONDENSATION PREVENTION INSULATION THICKNESSES
OUTER SURFACE EMITTANCE NOT LESS THAN $\epsilon = 0.9$
THICKNESSES IN NOMINAL INCHES

CELLULAR GLASS INSULATION
(COLORED MASTIC OR MEDIUM COLOR PAINTED METAL JACKET)

Nominal Pipe Sizes Inches	OPERATING TEMPERATURES°C											Nominal Pipe Sizes mm
	19 to 10	9 to 5	4 to 0	−1 to −5	−6 to −10	−11 to −15	−16 to −20	−21 to −25	−26 to −30	−31 to −35	−36 to −40	
1/2	1.0	1.0	1.0	1.0	1.0	1.0	1.0	1.0	1.0	1.0	1.0	21
3/4	1.0	1.0	1.0	1.0	1.0	1.0	1.0	1.0	1.0	1.0	1.0	27
1	1.0	1.0	1.0	1.0	1.0	1.0	1.0	1.0	1.0	1.0	1.0	33
1 1/2	1.0	1.0	1.0	1.0	1.0	1.0	1.0	1.0	1.0	1.0	1.0	48
2	1.0	1.0	1.0	1.0	1.0	1.0	1.0	1.0	1.0	1.0	1.0	60
2 1/2	1.0	1.0	1.0	1.0	1.0	1.0	1.0	1.0	1.0	1.0	1.0	73
3	1.0	1.0	1.0	1.0	1.0	1.0	1.0	1.0	1.0	1.0	1.0	89
4	1.0	1.0	1.0	1.0	1.0	1.0	1.0	1.0	1.0	1.0	1.0	114
5	1.0	1.0	1.0	1.0	1.0	1.0	1.0	1.0	1.0	1.0	1.0	141
6	1.0	1.0	1.0	1.0	1.0	1.0	1.0	1.0	1.0	1.0	1.0	168
8	1.5	1.5	1.5	1.5	1.5	1.5	1.5	1.5	1.5	1.5	1.5	219
10	1.5	1.5	1.5	1.5	1.5	1.5	1.5	1.5	1.5	1.5	1.5	273
12	1.5	1.5	1.5	1.5	1.5	1.5	1.5	1.5	1.5	1.5	1.5	324
14	1.5	1.5	1.5	1.5	1.5	1.5	1.5	1.5	1.5	1.5	1.5	356
16	1.5	1.5	1.5	1.5	1.5	1.5	1.5	1.5	1.5	1.5	1.5	406
18	1.5	1.5	1.5	1.5	1.5	1.5	1.5	1.5	1.5	1.5	1.5	457
20	1.5	1.5	1.5	1.5	1.5	1.5	1.5	1.5	1.5	1.5	1.5	508
24	1.5	1.5	1.5	1.5	1.5	1.5	1.5	1.5	1.5	1.5	1.5	610
28	1.5	1.5	1.5	1.5	1.5	1.5	1.5	1.5	1.5	1.5	1.5	711
32	1.5	1.5	1.5	1.5	1.5	1.5	1.5	1.5	1.5	1.5	1.5	813
36	1.5	1.5	1.5	1.5	1.5	1.5	1.5	1.5	1.5	1.5	1.5	914
FLAT	1.5	1.5	1.5	1.5	1.5	1.5	1.5	1.5	1.5	1.5	1.5	FLAT
	67 to 50	49 to 41	40 to 32	31 to 23	22 to 14	13 to 5	4 to −4	−5 to −13	−14 to −22	−23 to −31	−32 to −40	
	OPERATING TEMPERATURES°F											

Courtesy of Pittsburgh Corning Corp., Pittsburgh, Pa.

TABLE 194 (Cont'd)

INSIDE AIR TEMPERATURE 75°F (23.9°C)
RELATIVE HUMIDITY 60%

CONDENSATION PREVENTION INSULATION THICKNESSES
OUTER SURFACE EMITTANCE NOT LESS THAN $\epsilon = 0.9$
THICKNESSES IN NOMINAL INCHES

CELLULAR GLASS INSULATION
(COLORED MASTIC OR MEDIUM COLOR PAINTED METAL JACKET)

Nominal Pipe Sizes Inches	OPERATING TEMPERATURES ° C											Nominal Pipe Sizes mm
	19 to 10	9 to 5	4 to 0	−1 to −5	−6 to −10	−11 to −15	−16 to −20	−21 to −25	−26 to −30	−31 to −35	−36 to −40	
1/2	1.0	1.0	1.0	1.0	1.0	1.0	1.0	1.0	1.0	1.0	1.0	21
3/4	1.0	1.0	1.0	1.0	1.0	1.0	1.0	1.0	1.0	1.0	1.5	27
1	1.0	1.0	1.0	1.0	1.0	1.0	1.0	1.0	1.0	1.0	1.0	33
1 1/2	1.0	1.0	1.0	1.0	1.0	1.0	1.0	1.0	1.0	1.0	1.5	48
2	1.0	1.0	1.0	1.0	1.0	1.0	1.0	1.0	1.0	1.5	1.5	60
2 1/2	1.0	1.0	1.0	1.0	1.0	1.0	1.0	1.0	1.0	1.5	1.5	73
3	1.0	1.0	1.0	1.0	1.0	1.0	1.0	1.0	1.5	1.5	1.5	89
4	1.0	1.0	1.0	1.0	1.0	1.0	1.0	1.0	1.5	1.5	1.5	114
5	1.0	1.0	1.0	1.0	1.0	1.0	1.0	1.5	1.5	1.5	1.5	141
6	1.0	1.0	1.0	1.0	1.0	1.0	1.0	1.5	1.5	1.5	1.5	168
8	1.5	1.5	1.5	1.5	1.5	1.5	1.5	1.5	1.5	1.5	1.5	219
10	1.5	1.5	1.5	1.5	1.5	1.5	1.5	1.5	1.5	1.5	1.5	273
12	1.5	1.5	1.5	1.5	1.5	1.5	1.5	1.5	1.5	1.5	1.5	324
14	1.5	1.5	1.5	1.5	1.5	1.5	1.5	1.5	1.5	1.5	2.0	356
16	1.5	1.5	1.5	1.5	1.5	1.5	1.5	1.5	1.5	1.5	2.0	406
18	1.5	1.5	1.5	1.5	1.5	1.5	1.5	1.5	1.5	1.5	2.0	457
20	1.5	1.5	1.5	1.5	1.5	1.5	1.5	1.5	1.5	2.0	2.0	508
24	1.5	1.5	1.5	1.5	1.5	1.5	1.5	1.5	1.5	2.0	2.0	610
28	1.5	1.5	1.5	1.5	1.5	1.5	1.5	1.5	1.5	2.0	2.0	711
32	1.5	1.5	1.5	1.5	1.5	1.5	1.5	1.5	1.5	2.0	2.0	813
36	1.5	1.5	1.5	1.5	1.5	1.5	1.5	1.5	1.5	2.0	2.0	914
FLAT	1.5	1.5	1.5	1.5	1.5	1.5	1.5	1.5	1.5	2.0	2.0	FLAT
	67 to 50	49 to 41	40 to 32	31 to 23	22 to 14	13 to 5	4 to −4	−5 to −13	−14 to −22	−23 to −31	−32 to −40	
	OPERATING TEMPERATURES ° F											

Courtesy of Pittsburgh Corning Corp., Pittsburgh, Pa.

TABLE 194 (Cont'd)

INSIDE AIR TEMPERATURE 75°F (23.9°C)
RELATIVE HUMIDITY 80%

CONDENSATION PREVENTION INSULATION THICKNESSES
OUTER SURFACE EMITTANCE NOT LESS THAN ϵ = 0.9
THICKNESSES IN NOMINAL INCHES

CELLULAR GLASS INSULATION
(COLORED MASTIC OR MEDIUM COLOR PAINTED METAL JACKET)

Nominal Pipe Sizes Inches	OPERATING TEMPERATURES ° C											Nominal Pipe Sizes mm
	19 to 10	9 to 5	4 to 0	−1 to −5	−6 to −10	−11 to −15	−16 to −20	−21 to −25	−26 to −30	−31 to −35	−36 to −40	
1/2	1.0	1.0	1.0	1.0	1.5	1.5	1.5	1.5	2.0	2.0	2.0	21
3/4	1.0	1.0	1.0	1.5	1.5	1.5	2.0	2.0	2.0	2.0	2.5	27
1	1.0	1.0	1.0	1.5	1.5	1.5	1.5	2.0	2.0	2.0	2.5	33
1 1/2	1.0	1.0	1.0	1.5	1.5	2.0	2.0	2.0	2.0	2.0	2.5	48
2	1.0	1.0	1.0	1.5	1.5	2.0	2.0	2.0	2.5	2.5	2.5	60
2 1/2	1.0	1.0	1.5	1.5	1.5	1.5	2.0	2.0	2.0	2.5	2.5	73
3	1.0	1.0	1.5	1.5	2.0	2.0	2.0	2.5	2.5	3.0	3.0	89
4	1.0	1.0	1.5	1.5	2.0	2.0	2.5	2.5	2.5	3.0	3.0	114
5	1.0	1.0	1.5	1.5	2.0	2.0	2.5	2.5	3.0	3.0	3.0	141
6	1.0	1.0	1.5	2.0	2.0	2.0	2.0	2.5	3.0	3.0	3.5	168
8	1.5	1.5	1.5	2.0	2.0	2.5	2.5	2.5	3.0	3.0	3.5	219
10	1.5	1.5	1.5	2.0	2.0	2.5	2.5	3.0	3.0	3.5	3.5	273
12	1.5	1.5	1.5	2.0	2.0	2.5	2.5	3.0	3.0	3.5	3.5	324
14	1.5	1.5	1.5	2.0	2.5	2.5	3.0	3.0	3.5	3.5	4.0	356
16	1.5	1.5	1.5	2.0	2.5	2.5	3.0	3.0	3.5	3.5	4.0	406
18	1.5	1.5	1.5	2.0	2.5	2.5	3.0	3.0	3.5	4.0	4.0	457
20	1.5	1.5	1.5	2.0	2.5	2.5	3.0	3.0	3.5	4.0	4.0	508
24	1.5	1.5	1.5	2.0	2.5	3.0	3.0	3.5	3.5	4.0	4.0	610
28	1.5	1.5	1.5	2.0	2.5	3.0	3.0	3.5	3.5	4.0	4.0	711
32	1.5	1.5	1.5	2.0	2.5	3.0	3.0	3.5	3.5	4.0	4.5	813
36	1.5	1.5	2.0	2.0	2.5	3.0	3.0	3.5	3.5	4.0	4.5	914
FLAT	1.5	1.5	2.0	2.0	2.5	3.0	3.0	3.5	4.0	4.5	4.5	FLAT
	67 to 50	49 to 41	40 to 32	31 to 23	22 to 14	13 to 5	4 to −4	−5 to −13	−14 to −22	−23 to −31	−32 to −40	
	OPERATING TEMPERATURES ° F											

Courtesy of Pittsburgh Corning Corp., Pittsburgh, Pa.

TABLE 194 (Cont'd)

PIPE		APPROXIMATE WALL THICKNESS													
		Nominal thickness													
Nom. size	Outer dia	1		1 1/2		2		2 1/2		3		3 1/2		4	
inches	inches	Dec."	Fract."	Dec."	Fract."	Dec."	Fract."	Dec."	Fract."	Dec."	Fract."	Dec."	Fract."	Dec."	Fract."
1/2	0.840	1.01	1 1/64	1.57	1 9/16	2.07	2 1/16	2.88	2 7/8	3.38	3 3/8	3.88	3 7/8	4.38	4 3/8
3/4	1.050	0.90	29/32	1.46	1 15/32	1.96	1 31/32	2.78	2 25/32	3.28	3 11/32	3.78	3 25/32	4.28	4 11/32
1	1.315	1.08	1 5/64	1.58	1 37/64	2.12	2 1/8	2.64	2 41/64	3.14	3 9/64	3.64	3 41/64	4.14	4 9/64
1 1/4	1.660	0.91	29/32	1.66	1 21/32	1.94	1 15/16	2.47	2 15/32	2.97	2 31/32	3.47	3 15/32	3.97	3 31/32
1 1/2	1.900	1.04	1 3/64	1.54	1 35/64	2.35	2 23/64	2.85	2 27/32	3.35	3 23/64	3.85	3 27/32	4.42	4 27/64
2	2.375	1.04	1 3/64	1.58	1 37/64	2.10	2 7/64	2.60	2 39/64	3.10	3 7/64	3.60	3 39/64	4.17	4 11/64
2 1/2	2.875	1.04	1 3/64	1.86	1 55/64	2.36	2 23/64	2.86	2 55/64	3.36	3 23/64	3.92	3 59/64	4.42	4 27/64
3	3.500	1.02	1 1/64	1.54	1 35/64	2.04	2 3/64	2.54	2 35/64	3.04	3 3/64	3.61	3 39/64	4.11	4 7/64
3 1/2	4.000	1.30	1 19/64	1.80	1 51/64	2.30	2 19/64	2.80	2 51/64	3.36	3 23/64	3.86	3 55/64	4.36	4 3/64
4	4.500	1.04	1 3/64	1.54	1 35/64	2.04	2 3/64	2.54	2 35/64	3.11	3 7/64	3.61	3 39/64	4.11	4 7/64
4 1/2	5.000	1.30	1 19/64	1.80	1 51/64	2.30	2 19/64	2.86	2 55/64	3.36	3 23/64	3.86	3 55/64	4.48	4 31/64
5	5.563	0.99	1	1.49	1 1/2	1.99	2	2.56	2 9/16	3.06	3 1/16	3.56	3 9/16	4.18	4 3/16
6	6.625	0.96	31/32	1.46	1 15/32	2.02	2 1/64	2.52	2 33/64	3.02	3 1/64	3.65	3 21/32	4.15	4 5/32
7	7.625			1.52	1 33/64	2.02	2 1/64	2.52	2 33/64	3.15	3 5/32	3.65	3 21/32	4.15	4 5/32
8	8.625			1.52	1 33/64	2.02	2 1/64	2.65	2 21/32	3.15	3 5/32	3.65	3 21/32	4.15	4 5/32
9	9.625			1.52	1 33/64	2.15	2 5/32	2.65	2 21/32	3.15	3 5/32	3.65	3 21/32	4.15	4 5/32
10	10.750			1.58	1 37/64	2.08	2 5/64	2.58	2 37/64	3.08	3 5/64	3.58	3 37/64	4.08	4 5/64
11	11.750			1.58	1 37/64	2.08	2 5/64	2.58	2 37/64	3.08	3 5/64	3.58	3 37/64	4.08	4 5/64
12	12.750			1.58	1 37/64	2.08	2 5/64	2.58	2 37/64	3.08	3 5/64	3.58	3 37/64	4.08	4 5/64
14[a]	14.000[a]			1.46	1 15/32	1.96	1 31/32	2.46	2 15/32	2.96	2 31/32	3.46	3 15/32	3.96	3 31/32

[a] Sizes 14" through 36" in 1 inch increments

Fraction conversions to nearest 64th of an inch

TABLE 195 APPROXIMATE WALL THICKNESS OF INSULATION FOR NPS PIPES IN INCHES (DEC. AND FRACT.) – ENGLISH UNITS

Pipe		Insulation Dimensions									
Nom.	Outer	Inner Diameter			Approximate Wall Thickness in Millimeters						
size	dia	Dia	Tolerance		1" nom.	1 1/2" nom.	2" nom.	2 1/2" nom.	3" nom.	3 1/2" nom.	4" nom.
inches	mm	mm	Minus mm	Plus mm	25 mm	38 mm	51 mm	64 mm	76 mm	89 mm	102 mm
1/2	21.3	22	0	1.6	26	40	53	73	86	99	111
3/4	26.7	27	0	1.6	23	37	50	71	83	96	109
1	33.4	34	0	1.6	27	40	54	67	80	92	105
1 1/4	42.2	43	0	1.6	23	42	49	63	75	88	101
1 1/2	48.3	49	0	1.6	26	39	60	72	85	98	112
2	60.3	61	0	2.4	26	40	53	66	79	91	106
2 1/2	73.0	74	0	2.4	26	47	60	73	85	100	112
3	88.9	90	0	2.4	26	39	52	65	77	92	104
3 1/2	101.6	102	0.8	2.4	33	46	58	71	85	98	111
4	114.3	115	0.8	2.4	26	39	52	65	79	92	104
4 1/2	127.0	128	0.8	2.4	33	46	58	73	85	98	114
5	141.3	143	0.8	2.4	25	38	51	65	78	90	106
6	168.3	180	0.8	2.4	24	37	51	64	77	93	105
7	193.7	196	0.8	2.4		39	51	64	80	93	105
8	219.1	221	0.8	2.4		39	51	67	80	93	105
9	244.5	246	0.8	2.4		39	55	67	80	93	105
10	278.0	275	0.8	2.4		40	53	66	78	91	104
11	298.4	300	0.8	2.4		40	53	66	78	91	104
12	323.8	326	1.6	2.4		40	53	66	78	91	104
14	355.6	358	1.6	4.0		37	50	62	75	88	101

*Larger sizes 355.6 mm through 914.4 mm in 25.4 increments.

TABLE 196 APPROXIMATE WALL THICKNESS OF INSULATION FOR NPS PIPE – METRIC UNITS

TUBE		APPROXIMATE WALL THICKNESS													
		Nominal thickness													
Nom. size inches	Outer dia inches	1		1 1/2		2		2 1/2		3		3 1/2		4	
		Dec."	Fract."	Dec."	Fract."	Dec."	Fract."	Dec."	Fract."	Dec."	Fract."	Dec."	Fract."	Dec."	Fract."
3/8	0.500	0.93	15/16	1.49	1 31/64	1.99	2	2.52	2 33/64	3.05	3 3/64				
1/2	0.625	1.12	1 1/8	1.43	1 7/16	1.93	1 15/16	2.46	2 15/32	2.99	3				
3/4	0.875	1.00	1	1.56	1 9/16	2.06	1 1/16	2.86	2 55/64	3.36	3 23/64	3.87	3 7/8	4.37	4 3/8
1	1.125	0.87	7/8	1.43	1 7/16	1.93	1 15/16	2.74	2 47/64	3.24	3 15/64	3.74	3 47/64	4.24	4 15/64
1 1/4	1.375	1.06	1 1/16	1.56	1 9/16	2.08	1 5/64	2.62	2 5/8	3.12	3 1/8	3.62	3 5/8	4.12	4 1/8
1 1/2	1.625	0.93	15/16	1.43	1 7/16	1.96	1 31/32	2.49	2 1/2	2.99	3	3.49	3 1/2	3.99	4
2	2.125	0.93	15/16	1.43	1 7/16	2.23	2 15/64	2.73	2 47/64	3.23	3 15/64	3.73	3 47/64	4.30	4 19/64
2 1/2	2.625	0.92	59/64	1.45	1 29/64	1.98	1 63/64	2.48	2 31/64	2.98	2 63/64	3.48	3 31/64	4.04	4 3/64
3	3.125	0.92	59/64	1.73	1 47/64	2.23	2 15/64	2.73	2 47/64	3.23	3 15/64	3.80	3 51/64	4.30	4 19/64
3 1/2	3.625	0.95	61/64	1.48	1 31/64	1.98	1 63/64	2.48	2 31/64	2.98	2 63/64	3.54	3 35/64	4.04	4 3/64
4	4.125	1.23	1 15/64	1.73	1 47/64	2.23	2 15/64	2.73	2 47/64	3.30	3 19/64	3.80	3 51/64	4.30	4 19/64
5	5.125	1.23	1 15/64	1.73	1 47/64	2.23	2 15/64	2.80	2 51/64	3.30	3 19/64	3.80	3 51/64	4.42	4 19/64
6	6.125	1.21	1 7/32	1.71	1 23/32	2.28	2 9/32	2.78	2 25/32	3.28	3 9/32	3.90	3 29/32	4.40	4 7/16

Fraction conversions to nearest 64th of an inch

TABLE 197 APPROXIMATE WALL THICKNESS OF INSULATION FOR TUBES IN INCHES (DEC. and FRACT.) – ENGLISH UNITS

Tube		Insulation Dimensions – Millimeters										
Nom. size inches	Outer dia mm	Inner Diameter			Approximate Wall Thickness in Millimeters							
		Dia mm	Tolerance		1" nom.	1 1/2" nom.	2" nom.	2 1/2" nom.	3" nom.	3 1/2" nom.	4" nom.
			Minus mm	Plus mm	25 mm	38 mm	51 mm	64 mm	76 mm	89 mm	102 mm
3/8	12.7	13	0	1.6	24	38	51	64	77	90	103
1/2	15.9	16	0	1.6	28	36	49	62	76	89	102
3/4	22.2	23	0	1.6	25	40	52	73	85	98	111
1	28.6	29	0	1.6	22	36	49	70	82	95	108
1 1/4	34.9	35	0	1.6	27	40	53	67	79	92	105
1 1/2	41.3	42	0	1.6	24	36	50	63	76	89	101
2	54.0	55	0	1.6	23	36	57	60	82	95	109
2 1/2	66.7	68	0	1.6	23	37	50	63	76	88	103
3	79.4	80	0	1.6	23	44	57	69	82	97	109
3 1/2	92.1	93	0	1.6	24	38	50	63	76	90	103
4	104.8	106	0.8	2.4	31	44	57	69	84	97	109
5	130.2	131	0.8	2.4	31	44	57	71	84	97	112
6	155.6	157	0.8	2.4	31	43	58	71	83	99	112

TABLE 198 APPROXIMATE WALL THICKNESSES OF INSULATION FOR TUBES – METRIC UNITS

PIPE		INSULATION DIMENSIONS																	
		Inner diameter				Outer diameter for nominal thickness of:													
Nom. size	Outer dia	Diameter		Tolerance		1		1 1/2		2		2 1/2		3		3 1/2		4	
inches	inches	Dec."	Fract."	Minus	Plus	Dec."	Fract."	Dec."	Fract."	Dec."	Fract."	Dec."	Fract."	Dec."	Fract."	Dec."	Fract."	Dec."	Fract."
1/2	0.840	0.86	55/64	0	1/16	2.88	2 7/8	4.00	4	5.00	5	6.62	6 5/8	7.62	7 5/8	8.62	8 5/8	9.62	9 5/8
3/4	1.050	1.07	1 1/64	0	1/16	2.88	2 7/8	4.00	4	5.00	5	6.62	6 5/8	7.62	7 5/8	8.62	8 5/8	9.62	9 5/8
1	1.315	1.33	1 21/64	0	1/16	3.50	3 1/2	4.50	4 1/2	5.56	5 9/16	6.62	6 5/8	7.62	7 5/8	8.62	8 5/8	9.62	9 5/8
1 1/4	1.660	1.68	1 11/64	0	1/16	3.50	3 1/2	5.00	5	5.56	5 9/16	6.62	6 5/8	7.62	7 5/8	8.62	8 5/8	9.62	9 5/8
1 1/2	1.900	1.92	1 59/64	0	1/16	4.00	4	5.00	5	6.62	6 5/8	7.62	7 5/8	8.62	8 5/8	9.62	9 5/8	10.75	10 3/4
2	2.375	2.41	2 13/32	0	3/32	4.50	4 1/2	5.56	5 9/16	6.62	6 5/8	7.62	7 5/8	8.62	8 5/8	9.62	9 5/8	10.75	10 3/4
2 1/2	2.875	2.91	2 29/32	0	3/32	5.00	5	6.62	6 5/8	7.62	7 5/8	8.62	8 5/8	9.62	9 5/8	10.75	10 3/4	11.75	11 3/4
3	3.500	3.53	3 17/32	0	3/32	5.56	5 9/16	6.62	6 5/8	7.62	7 5/8	8.62	8 5/8	9.62	9 5/8	10.75	10 3/4	11.75	11 3/4
3 1/2	4.000	4.03	4 1/32	1/32	3/32	6.62	6 5/8	7.62	7 5/8	8.62	8 5/8	9.62	9 5/8	10.75	10 3/4	11.75	11 3/4	12.75	12 3/4
4	4.500	4.53	4 17/32	1/32	3/32	6.62	6 5/8	7.62	7 5/8	8.62	8 5/8	9.62	9 5/8	10.75	10 3/4	11.75	11 3/4	12.75	12 3/4
4 1/2	5.006	5.03	5 1/32	1/32	3/32	7.62	7 5/8	8.62	8 5/8	9.62	9 5/8	10.75	10 3/4	11.75	11 3/4	12.75	12 3/4	14.00	14
5	5.563	5.64	5 41/64	1/32	3/32	7.62	7 5/8	8.62	8 5/8	9.62	9 5/8	10.75	10 3/4	11.75	11 3/4	12.75	12 3/4	14.00	14
6	6.625	6.70	6 45/64	1/32	3/32	8.62	8 5/8	9.62	9 5/8	10.75	10 3/4	11.75	11 3/4	12.75	12 3/4	14.00	14	15.00	15
7	7.625	7.70	7 45/64	1/32	3/32			10.75	10 3/4	11.75	11 3/4	12.75	12 3/4	14.00	14	15.00	15	16.00	16
8	8.625	8.70	8 45/64	1/32	3/32			11.75	11 3/4	12.75	12 3/4	14.00	14	15.00	15	16.00	16	17.00	17
9	9.625	9.70	9 45/64	1/32	3/32			12.75	12 3/4	14.00	14	15.00	15	16.00	16	17.00	17	18.00	18
10	10.750	10.83	10 53/64	1/32	3/32			14.00	14	15.00	15	16.00	16	17.00	17	18.00	18	19.00	19
11	11.750	11.83	11 53/64	1/32	3/32			15.00	15	16.00	16	17.00	17	18.00	18	19.00	19	20.00	20
12	12.750	12.84	12 27/32	1/16	3/32			16.00	16	17.00	17	18.00	18	19.00	19	20.00	20	21.00	21
14[a]	14.000	14.09	14 3/32	1/16[b]	5/32[b]			17.00[a]	17[a]	18.00[a]	18[a]	19.00[a]	19[a]	20.00[a]	20[a]	21.00[a]	21[a]	22.00[a]	22[a]

[a] Sizes 16 through 36 in 1 inch increments
[b] Sizes 16 through 36 same tolerance as for 14 in NPS pipe, fraction conversions to nearest 64th of an inch

TABLE 199 DIMENSIONS OF INSULATION FOR NPS PIPE IN INCHES (DEC. and FRACT.) – ENGLISH UNITS

Pipe		Insulation Dimensions – Millimeters									
Nom.	Outer	Inner Diameter			Outer Diameter for Nominal Thickness in Millimeters						
size	dia	Dia	Tolerance		1" nom.	1 1/2" nom.	2" nom.	2 1/2" nom.	3" nom.	3 1/2" nom.	4" nom.
inches	mm	mm	Minus mm	Plus mm	25 mm	38 mm	51 mm	64 mm	76 mm	89 mm	102 mm
1/2	21.3	22	0	1.6	73.15	101.6	127.0	168.1	193.5	218.9	244.3
3/4	26.7	27	0	1.6	73.15	101.6	127.0	168.1	193.5	218.9	244.3
1	33.4	34	0	1.6	88.9	114.3	141.2	168.1	193.5	218.9	244.3
1 1/4	42.2	43	0	1.6	88.9	127.0	141.2	168.1	193.5	218.9	244.3
1 1/2	48.3	49	0	1.6	101.6	127.0	168.1	193.5	218.9	244.3	273.1
2	60.3	61	0	2.4	114.3	141.2	168.1	193.5	218.9	244.3	273.1
2 1/2	73.0	74	0	2.4	127.0	168.1	193.5	218.9	244.3	273.1	298.5
3	88.9	90	0	2.4	141.2	168.1	193.5	218.9	244.3	273.1	298.5
3 1/4	101.6	102	0.8	2.4	168.1	193.5	218.9	244.3	273.1	298.5	323.5
4	114.3	115	0.8	2.4	168.1	193.5	218.9	244.3	273.1	298.5	323.5
4 1/2	127.0	128	0.8	2.4	193.5	218.9	244.3	273.1	298.5	323.9	355.6
5	141.3	143	0.8	2.4	193.5	218.9	244.3	273.1	298.5	323.9	355.6
6	168.3	180	0.8	2.4	218.9	244.3	273.1	298.5	323.9	355.6	381.0
7	193.7	196	0.8	2.4		273.1	298.5	323.9	355.6	381.0	406.4
8	219.1	221	0.8	2.4		298.5	323.9	355.6	381.0	406.4	431.8
9	244.5	246	0.8	2.4		323.9	355.6	381.0	406.4	431.8	457.2
10	278.0	275	0.8	2.4		355.6	381.0	406.4	431.8	457.2	482.6
11	298.4	300	0.8	2.4		381.0	406.4	431.8	457.2	482.6	508.0
12	323.8	326	1.6	2.4		406.4	431.8	457.2	482.6	508.0	533.4
14*	355.6	358	1.6	4.0		431.8	457.2	482.6	508.0	533.4	558.8

*Larger sizes 355.6 mm through 914.4 mm in 25.4 increments.

TABLE 200 DIMENSIONS OF INSULATION FOR NPS PIPE – METRIC UNITS

TUBE		INSULATION																	
		Inner diameter				Outer diameter for nominal thickness of:													
Nom. size	Outer dia	Diameter		Tolerance		1		1 1/2		2		2 1/2		3		3 1/2		4	
inches	inches	Dec."	Fract."	Minus	Plus	Dec."	Fract."	Dec."	Fract."	Dec."	Fract."	Dec."	Fract."	Dec."	Fract."	Dec."	Fract."	Dec."	Fract."
3/8	0.500	0.52	33/64	0	1/16	2.38	2 7/8	3.50	3 1/2	4.50	4 1/2	5.56	5 9/16	6.62	6 5/8	7.62	7 5/8	8.62	8 5/8
1/2	0.625	0.64	41/64	0	1/16	2.88	2 7/8	3.50	3 1/2	4.50	4 1/2	5.56	5 9/16	6.62	6 5/8	7.62	7 5/8	8.62	8 5/8
3/4	0.875	0.89	57/64	0	1/16	2.88	2 7/8	4.00	4	5.00	5	6.62	6 5/8	7.62	7 5/8	8.62	8 5/8	9.62	9 5/8
1	1.125	1.14	1 9/64	0	1/16	2.88	2 7/8	4.00	4	5.00	5	6.62	6 5/8	7.62	7 5/8	8.62	8 5/8	9.62	9 5/8
1 1/4	1.375	1.39	1 25/64	0	1/16	3.50	3 1/2	4.50	4 1/2	5.56	5 9/16	6.62	6 5/8	7.62	7 5/8	8.62	8 5/8	9.62	9 5/8
1 1/2	1.625	1.64	1 41/64	0	1/16	3.50	3 1/2	4.50	4 1/2	5.56	5 9/16	6.62	6 5/8	7.62	7 5/8	8.62	8 5/8	9.62	9 5/8
2	2.125	2.16	2 5/32	0	1/16	4.00	4	5.00	5	6.62	6 5/8	7.62	7 5/8	8.62	8 5/8	9.62	9 5/8	10.75	10 3/4
2 1/2	2.625	2.66	2 21/32	0	1/16	4.50	4 1/2	5.56	5 9/16	6.62	6 5/8	7.62	7 5/8	8.62	8 5/8	9.62	9 5/8	10.75	10 3/4
3	3.125	3.16	3 5/32	0	1/16	5.00	5	6.62	6 5/8	7.62	7 5/8	8.62	8 5/8	9.62	9 5/8	10.75	10 3/4	11.75	11 3/4
3 1/2	3.625	3.66	3 21/32	0	1/16	5.56	5 9/16	6.62	6 5/8	7.62	7 5/8	8.62	8 5/8	9.62	9 5/8	10.75	10 3/4	11.75	11 3/4
4	4.125	4.16	4 5/32	1/32	3/32	6.62	6 5/8	7.62	7 5/8	8.62	8 5/8	9.62	9 5/8	10.75	10 3/4	11.75	11 3/4	12.75	12 3/4
5	5.125	5.16	5 5/32	1/32	3/32	7.62	7 5/8	8.62	8 5/8	9.62	9 5/8	10.75	10 3/4	11.75	11 3/4	12.75	12 3/4	14.00	14
6	6.125	6.20	6 13/64	1/32	3/32	8.62	8 5/8	9.62	9 5/8	10.75	10 3/4	11.75	11 3/4	12.75	12 3/4	14.00	14	15.00	15

Fraction conversions to nearest 64th of an inch

TABLE 201 DIMENSIONS OF INSULATION FOR TUBES IN INCHES (DEC. and FRACT.) – ENGLISH UNITS

Tube		Insulation Dimensions – Millimeters									
Nom.	Outer	Inner Diameter			Outer Diameter for Nominal Thicknesses in Millimeters						
size	dia	Dia	Tolerance		1" nom.	1 1/2" nom.	2" nom.	2 1/2" nom.	3" nom.	3 1/2" nom.	4" nom.
inches	mm	mm	Minus mm	Plus mm	25 mm	38 mm	51 mm	64 mm	76 mm	89 mm	102 mm
3/8	12.7	13	0	1.6	60.5	88.9	114.3	141.2	168.1	193.5	218.9
1/2	15.9	16	0	1.6	73.2	88.9	114.3	141.2	168.1	193.5	218.9
3/4	22.2	23	0	1.6	73.2	101.6	127.0	168.1	193.5	218.9	244.3
1	28.6	29	0	1.6	73.2	101.6	127.0	168.1	193.5	218.9	244.3
1 1/4	34.9	35	0	1.6	88.9	114.3	141.2	168.1	193.5	218.9	244.3
1 1/2	41.3	42	0	1.6	88.9	114.3	141.2	168.1	193.5	218.9	244.3
2	54.0	55	0	1.6	101.6	127.0	168.1	193.5	218.9	244.3	273.1
2 1/2	66.7	68	0	1.6	114.3	141.2	168.1	193.5	218.9	244.3	273.1
3	79.4	80	0	1.6	127.0	168.1	193.5	218.9	244.3	273.1	298.5
3 1/2	92.1	93	0	1.6	141.2	168.1	193.5	218.9	244.3	273.1	298.5
4	104.8	106	0.8	2.4	168.1	193.5	218.9	244.3	273.1	298.5	323.9
5	130.2	131	0.8	2.4	193.5	218.9	244.3	273.1	298.5	323.9	355.6
6	155.6	157	0.8	2.4	218.9	244.3	273.1	298.5	323.9	355.6	381.0

TABLE 202 DIMENSIONS OF INSULATION FOR TUBES – METRIC UNITS

12 Energy Loss and Economics

The preceding chapter established thermal resistances, or transmittances, for typical building materials and constructions. This provides an easier method of calculating heat transfer than dividing temperature differences by a sum of thicknesses divided by conductivities. Of course, it is impossible to calculate resistances for all potential combinations of building materials but where these have been precalculated, resistances can be used. They serve as an easy way to calculate the heat loss or gain into a house or building.

However, more than a single calculation as to heat loss and heat gain is necessary to provide the information for designing a heating or air conditioning system.

To determine needed heating capacity it is necessary to know the heat loss of roof, wall, floors, doors and windows under the coldest expected weather conditions.

To determine cost, the average mean temperature for the month is used to find the heat loss that occurs during the month. The total loss per month can then be used to calculate dollar costs.

Before calculating heat gain, it is also necessary to determine needed capacity by knowing the maximum temperatures and the sun load. Heat gain into a building during summer is actually an energy loss necessary to remove it. As with heating, monthly cost is determined by the average heat gain over the month.

In addition to heat transfer through the components of a building, all outside air entering the structure must be added to the heat loss. This air requires energy for heating or refrigeration. Figures were provided in Table 165 as to infiltration of air around windows. However, besides the air entering by infiltration, make up outside air is required for combustion of oil, coal or gas burning boilers or furnaces. More outside air is required for occupancy, vent fans to remove fumes over stoves and moist air in bathrooms etc. All this added energy loss is determined by the number of vent fans, combustion air intakes, and the particular design of the heating and refrigeration systems. Although this outside air must be added to the energy loss, it is beyond the scope of this book.

Another factor beyond the scope of this book is the determination of the efficiency of the heating or cooling systems. To illustrate: (1) Where heat is obtained by combustion, all heat which goes up the stack is wasted heat. However, flue temperature must be kept above dew point of vaporized acids. (2) All heat generated by electrical motors or compressors in refrigeration systems is waste heat. Thus, the losses by all building components are further increased by the lack of 100% efficiency of the heating or cooling system.

Since these last two energy losses are a part of the mechanical design of the air conditioning and heating systems, they generally are a concern of the Air Conditioning Engineer, whereas the thermal insulation as a part of the structure is the concern of the Architect or Structural Designer, even though it affects the sizing of air conditioning and heating equipment.

The total energy loss is the loss of energy through the roof, walls, windows, doors and floor divided by the efficiency of the heating (or conditioning) system plus the energy of heating or cooling of infiltration and make up air divided by the efficiency of the heating (or conditioning) system.

Expressed in algebraic terms: Total Energy Loss is,

when:

U_t = total energy loss per hour
U_r = energy loss — roof
U_w = energy loss — walls
U_f = energy loss — floors
U_{wd} = energy loss — windows
U_{dr} = energy loss — doors
U_{ia} = energy loss — infiltrated air
U_{va} = energy loss — vent and make up air
M_e = Mechanical efficiency — heating or cooling system-percent.

$$U_t = U_r + U_w + U_f + U_{wd} + U_{dr} + \frac{U_{ia} + U_{va}}{M_e} \tag{50}$$

This book assists in determining the values of U_r, U_w, U_f, U_{wd}, U_{dr} and the means to reduce them by using thermal insulation. Lack of knowledge as to the energy losses of roofs, walls, floors (slabs), doors and windows results in thermally inefficient residences and buildings such as the all glass office buildings.

Because the temperature difference is smaller between ground temperature and indoor temperature, little attention has been given to insulation under concrete slab floors on grade.

Two major factors make losses greater than would be assumed. One is that concrete is a very poor thermal insulation. The other is that since the ground temperature is constant and does not fluctuate like the outdoor environment, the heat loss is consistant over the entire time.

Another often overlooked factor is that when people walk or stand on a floor their only thermal protection is the soles of their shoes. This direct contact causes individuals to feel cold and to thus raise the temperature in the room excessively to compensate.

To determine heat loss, ground temperature must be established. The following tables provide ground temperatures at 40 degrees North latitude in relatively moderate elevations some distance from the western coast. Table 203 contains this information in English Units and Table 204 in Metric Units. This information is necessary to calculate U_f for concrete floors.

GROUND TEMPERATURES
40 DEGREES NORTH LATITUDE
MONTHLY AVERAGE

ENGLISH UNITS

Depth Below Surface	Ground Temperatures °F											
	Jan.	**Feb.**	**Mar.**	**Apr.**	**May**	**June**	**July**	**Aug.**	**Sept.**	**Oct.**	**Nov.**	**Dec.**
1 ft.	33	27	34	47	50	54	57	60	63	54	43	35
2 ft.	35	31	35	45	48	52	56	60	62	54	46	38
3 ft.	37	33	36	44	46	50	55	57	60	54	48	40
4 ft.	41	35	38	43	45	50	55	57	59	54	50	43
5 ft.	43	37	39	42	45	49	54	56	58	54	51	45
6 ft.	44	39	40	42	44	47	52	54	57	54	52	49
7 ft.	45	41	42	42	43	45	49	52	55	54	52	50
8 ft.	46	43	43	42	44	46	49	51	54	54	52	51
9 ft.	47	44	44	42	44	46	49	51	53	53	52	52
10 ft.	49	46	45	43	45	47	48	50	52	53	52	52
11 ft.	50	48	47	44	45	45	47	49	51	53	52	52
12 ft.	50	49	48	45	45	45	47	48	50	52	52	52

TABLE 203

GROUND TEMPERATURES
40 DEGREES NORTH LATITUDE
MONTHLY AVERAGE

METRIC UNITS

Depth Below Surface	Ground Temperatures °C											
	Jan.	**Feb.**	**Mar.**	**Apr.**	**May**	**June**	**July**	**Aug.**	**Sept.**	**Oct.**	**Nov.**	**Dec.**
0.30 m	0.6	−2.8	1.1	8.3	10.0	12.2	13.9	15.6	17.2	12.2	6.1	1.7
0.61 m	1.7	−0.6	1.7	7.2	8.9	11.1	13.3	15.6	16.7	12.2	7.8	3.3
0.91 m	2.8	0.6	2.2	6.7	7.8	10.0	12.8	13.9	15.6	12.2	8.9	4.4
1.22 m	5.0	1.7	3.3	6.1	7.2	10.0	12.8	13.9	15.0	12.2	10.0	6.1
1.52 m	6.1	2.8	3.9	5.6	7.2	9.4	12.2	13.3	14.4	12.2	10.6	7.2
1.82 m	6.7	3.9	4.4	5.6	6.7	8.3	11.1	12.2	13.9	12.2	11.1	9.4
2.13 m	7.2	5.0	5.3	5.6	6.1	7.2	9.4	11.1	12.8	12.2	11.1	10.0
2.44 m	7.8	6.1	6.1	5.6	6.7	7.8	9.4	10.6	12.2	12.2	11.1	10.6
2.74 m	8.3	6.7	6.7	5.6	6.7	7.8	9.4	10.6	11.7	11.7	11.1	11.1
3.05 m	9.4	7.8	7.2	6.1	7.2	8.3	8.9	10.0	11.1	11.7	11.1	11.1
3.35 m	10.0	8.9	8.5	6.7	7.2	7.2	8.3	9.4	10.6	11.7	11.1	11.1
3.66 m	10.0	9.4	8.9	7.2	7.2	7.2	8.3	8.9	10.0	11.1	11.1	11.1

TABLE 204

The following examples are provided to determine the energy savings possible by adding thermal insulation. The energy losses — maximum and average — will be calculated for a typical house, with no insulation and with insulation, to show the contrast.

The following examples using a typical frame house will show (a) the maximum heat loss to size the heating equipment, (b) the average heat per year, (c) the maximum heat gain to size the cooling equipment, and (d) the average heat gain per year. These examples present the house with and without thermal insulation in order to establish the energy savings which can be achieved through the use of thermal insulation.

Example 52—*English Units*

Determine the maximum and yearly averages of heat loss, uninsulated and insulated, for the house shown in Figure 163. Also, determine the maximum and yearly average heat gains. The house is located in Springfield, Mo. The inside temperature is maintained at 70°F.

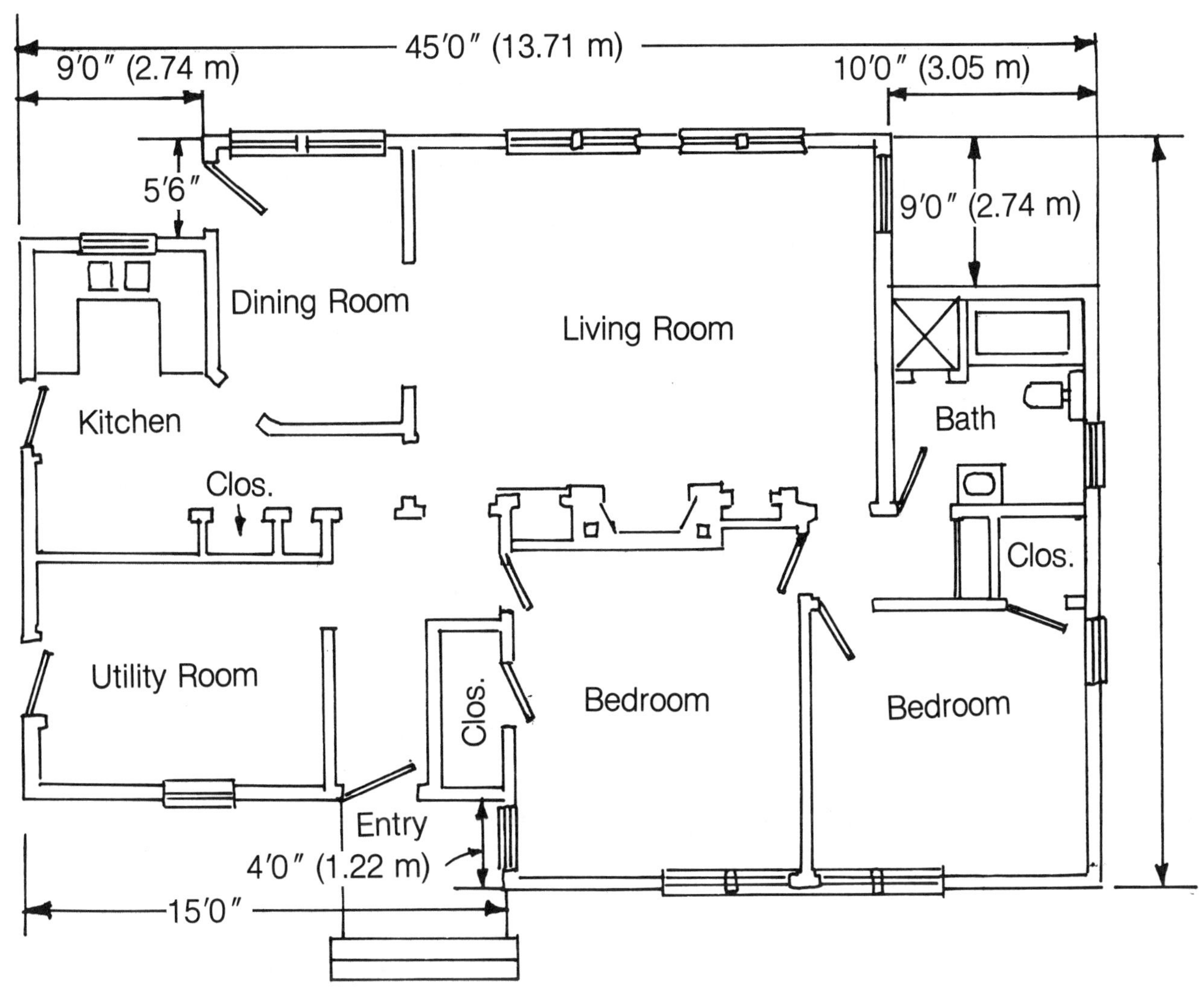

Ceiling Height 8′0″ (2.41 m)

FLOOR PLAN

WINDOWS: 14 3′0″ × 4′0″ (0.91 m × 1.21 m)
2 3′0″ × 3′0″ (0.91 m × 0.91 m)
DOORS: 4 3′0″ × 7′0″ (0.91 m × 2.13 m)
CONSTRUCTION: Roof — pitched, wood beams and deck, building paper, shingles
Ceiling — wood joists and 1/2″ (1.27 cm) gypsum board
Walls, Frame — siding, sheathing, 2″ × 4″ (5.1 cm × 10.2 cm) studs with 1/2″ (1.27 cm) gypsum board
Floor — 4″ (0.102 m) thick concrete slab on grade
Inside Height of Room — 8′0″ (2.44 m)
Attic Space Above Ceiling Vented to Outside Air

Figure 163

The weather data for Springfield, Mo. (from Table 24) is: Design Temperature, Winter 0 °F (−18 °C), Summer 97 °F (36 °C); Winter Heating: Degree-Hours 112,680°F (62,600°C); Summer Cooling Data: Degree-Hours 23,160 °F (12,867 °C). The location of this house, as shown, has the major window space facing South. From the map in Figure 45 Springfield is approximately 38° North Latitude (From Table 36 — use solar heat gain listed for 40° North Latitude). Maximum heat gain: at 11 am is 108.17 Btu/ft², hr (340.17 W/m²); at 12 noon it is 114 Btu/ft², hr (359.38 W/m²); and at 1:00 pm the maximum heat gain is 108.17 Btu/ft², hr (340.17 W/m²). Minimum ground temperature under the slab is 45 °F (7.2°C); average for winter is 50°F (10°C).

Areas—English Units

Window Area = 186 ft²

Wall Area = [(45′0″ + 33′0″) × 8 × 2] − (186 ft²)
= 1248 ft² − 186 ft² = 1,062 ft²

Ceiling Area = (33′0″) − [(9′0″) + (9′0″ × 10′0″) + (15′0″)]
= 1485 − [49.5 + 90 + 60] = 1485 − 199.5
= 1285.5 ft²

Floor Area = (same as ceiling) = 1,285.5 ft²

Maximum Heat Losses — Based on 70 °F inside temperature and 0° outdoor design temperature — per sq ft and total area.

Ceiling Uninsulated — (from Construction Assembly XII) R = 1.4

$$U_c = \frac{70-0}{1.4} = 50 \text{ Btu/ft}^2\text{, hr} \times 1285.5 \text{ ft}^2 \text{ (area)} = 64{,}275 \text{ Btu/hr}$$

Ceiling Insulated 6″ thick Glass Fiber — (from Table 113) R = 26.4

$$U_{ci} = \frac{70-0}{26.4} = \frac{70}{26.4} = \text{Btu/ft}^2\text{, hr} \times 1285.5 \text{ ft}^2 \text{ (area)} = 3{,}408 \text{ Btu/hr}$$

This is a savings of 60,867 Btu/hr (94%).

Walls 4″ Stud Wall (3½″ space) — Uninsulated — (from Construction Assembly XIX) R = 3.38

$$U_w = \frac{70-0}{3.38} = 18.04 \text{ Btu/ft}^2\text{, hr} \times 1062 \text{ ft}^2 \text{ (area)} = 21{,}994 \text{ Btu/hr}$$

Walls Insulated 3½″ thick Glass Fiber — (from Table 125) R = 15.05

$$U_{wi}\ \frac{70-0}{15.05} = 4.52 \text{ Btu/ft}^2\text{, hr} \times 1062 \text{ ft}^2 \text{ (area)} = 4{,}796 \text{ Btu/hr}$$

A savings of 17,198 Btu/hr (77%).

Windows single pane 20% wood frame — (from Table 162) R = 1.01

$$U_{wd} = \frac{70-0}{1.01} = 69.3 \text{ Btu/ft}^2\text{, hr} \times 186 \text{ ft}^2 \text{ (area)} = 12{,}891 \text{ Btu/hr}$$

Windows single pane 20% wood plus storm window — (from Table 162) R = 1.78

$$U_{wds} = \frac{70-0}{1.78} = 39.3 \text{ Btu/ft}^2 \times 186 \text{ ft}^2 \text{ (area)} = 7{,}314 \text{ Btu/hr}$$

A savings of 5,577 Btu/hr (43%).

Floor Uninsulated 4″ thick sand gravel concrete. Minimum temperature of ground 50 °F.

4″ Concrete Floor Slab — (from Construction Assembly XXXIII) R = 0.83

$$U_f = \frac{70-50}{0.83} = \frac{20}{0.83} = 24.1 \text{ Btu/ft}^2\text{, hr} \times 1285.5 \text{ ft}^2 \text{ (area)}$$
$$= 30{,}975 \text{ Btu/hr}$$

Floor Insulated 2″ thick cellular glass — (from Table 160) R = 6.89

$$U_{fi} = \frac{70-50}{6.89} = \frac{20}{6.89} = 2.9 \text{ Btu/ft}^2\text{, hr} \times 1285.5 \text{ ft}^2 \text{ (area)}$$
$$= 3{,}731 \text{ Btu/hr}$$

A savings of 27,244 Btu/hr (87%).

Construction		Maximum Heat Loss of House Structure		
Assembly	**No.**	**Uninsulated Loss Btu/hr**	**Insulation**	**Insulated Btu/hr**
Ceiling	XII	64,275	6″ Glass Fiber	3,408
Walls	XIX	21,994	3½″ Glass Fiber	4,796
Windows		12,891	Storm Window	7,314
Floor	XXXIII	30,975	2″ Cellular Glass	3,731
Totals		130,135 Btu/hr		19,249 Btu/hr

Insulation provides a savings of 110,886 Btu/hr in the total capacity required of the heating equipment. In other words, the insulation saves 84% of the required heating capacity for sizing equipment. Should the thermal efficiency of the heating equipment be 80% the savings in required capacity is

$$\frac{110{,}886}{0.8} = 138{,}607 \text{ Btu/hr.}$$

This saving is only on the energy loss of the structural components of the residences. Heating capacity for infiltration and fresh air for ventilation must be added to the amount of the heat loss of the building itself. Using the same thermal resistances, *yearly* heat loss can be calculated as follows:

WINTER SEASON HEAT LOSS — Based on average degree hours. From Construction Assembly XII, R = 1.4. From Table 24, Degree-hours (°F) = 112,680.

Ceiling — Uninsulated

$$U_{cdh} = \frac{112,680}{1.4} = 80,464 \text{ Btu/ft}^2 \times 1285.5 \text{ ft}^2 \text{ (area)}$$
$$= 103,464,385 \text{ Btu/yr}$$

Ceiling Insulated 6″ glass fiber — (from Table 113) R = 26.4

$$U_{cidh} = \frac{112,680}{26.4} = 4,268 \text{ Btu/ft}^2 \times 1285.5 \text{ ft}^2 \text{ (area)}$$
$$= 5,486,747 \text{ Btu/yr}$$

A savings of 97,977,638 Btu/yr (94%).

Walls — Uninsulated

$$U_{wdh} = \frac{112,680}{3.38} = 33,337 \text{ Btu/ft}^2 \times 1062 \text{ ft}^2 \text{ (area)}$$
$$= 35,404,189 \text{ Btu/yr}$$

Walls Insulated 3½″ glass fiber — (from Table 125) R = 15.05

$$U_{widh} = \frac{112,680}{15.05} = \text{ Btu/ft}^2 \times 1062 \text{ ft}^2 \text{ (area)}$$
$$= 7,951,239 \text{ Btu/yr}$$

A savings of 27,452,950 Btu/yr (77%).

Windows single pane 20% wood frame — (from Table 162) R = 1.01

$$U_{wddh} = \frac{112,680}{1.01} = 111,564 \text{ Btu/ft}^2 \times 186 \text{ ft}^2 \text{ (area)}$$
$$= 20,750,970 \text{ Btu/yr}$$

Windows single pane 20% wood frame plus storm door R = 1.78

$$U_{wddih} = \frac{112,680}{1.78} = 63,300 \text{ Btu/ft}^2 \times 186 \text{ ft}^2 \text{ (area)}$$
$$= 11,774,426 \text{ Btu/yr}$$

A savings of 8,976,544 Btu/yr (43%).

Floor Uninsulated 4″ thick gravel, sand concrete

In case the floors are on (or below) grade, the temperature is always below inside air temperature for the entire heating period. In many areas where heating is required in the United States this period is Nov., Dec., Jan., Feb., and March (151 days or 3624 hours). The average temperature under the entire slab for these months is estimated to be 50°F. Using the 65°F inside air temperature criteria, the degree hours would be (65 − 50) × 3624 – 54,360 degree-hours.

Uninsulated — (from Construction Assembly XXXIII) R = 0.83

$$U_{fdh} = \frac{54,360}{0.83} = 65,493 \text{ Btu/ft}^2 \times 1285.5 \text{ ft.}^2 \text{ (area)}$$
$$= 84,192,506 \text{ Btu/yr}$$

Floor Insulated 2″ thick cellular glass — (from Table 160) R = 6.89

$$U_{fidh} = \frac{54,360}{6.89} = 7,890 \text{ Btu/ft}^2 \times 1285.5 \text{ ft}^2 \text{ (area)}$$
$$= 10,142,203 \text{ Btu/yr}$$

A savings of 74,050,302 Btu/yr (87%).

Construction		**Yearly Heat Loss — Btu**			
Assembly	**No.**	**Uninsulated Loss**	**Insulation**	**Insulated Loss**	**Savings by Insulation**
Ceiling	XII	103,464,385	6″ Glass Fiber	5,486,747	97,977,638
Walls	XIX	30,841,793	3½″ Glass Fiber	7,949,239	22,890,554
Windows		20,750,970	Storm Windows	11,774,426	8,976,544
Floor	XXXIII	84,192,506	2″ Cellular Glass	10,142,203	74,050,302
Totals		239,249,654		35,352,615	203,897,039

This represents an 85% savings on energy. Should the efficiency of the heating equipment be 80%, the energy savings would be $\frac{203,897,034}{0.8} = 254,863,798$ Btu/yr.

MAXIMUM HEAT GAIN — Which must be removed by air conditioning equipment. Based on 75°F indoors temperature.

Ceiling Uninsulated — (from Construction Assembly XII) R = 1.30

Note: Even though outside design temperature is taken to be 97°F, solar radiation on the roof will cause the vented space between roof and ceiling to reach a temperature of 130°F.

$$U_{ca} = \frac{130 - 75}{1.3} = \frac{55}{1.3} = 42 \text{ Btu/ft}^2 \times 1285.5 \text{ ft}^2 \text{ (area)}$$
$$= 54,386 \text{ Btu/hr}$$

Ceiling Insulated 6″ thick glass fiber — (from Table 113 R = 22.73

$$U_{cai} = \frac{130 - 75}{22.73} = \frac{55}{22.73} = 2.4 \text{ Btu/ft}^2 \times 1285.5 \text{ ft}^2 \text{ (area)}$$
$$= 3,110 \text{ Btu/hr}$$

A savings of 51,276 Btu/hr (94%).

Walls Uninsulated 3½″ — Uninsulated — (from Construction Assembly XIX) R = 3.19 (at 85°mean temperature). R = 3.05 (at 110°F mean temperature — solar heat).

The maximum temperature will occur when the South wall is heated by solar radiation and the balance of the wall area is heated by outside ambient air. The area of the South wall is 45′ x 8′ = 360 ft². Window Area (6 x 12) + 9 = 81 or 360 − 81 = 279 ft².

The shingles have a surface emittance ε of 0.75. From Table 44 the maximum temperature increase of a wall surface facing south is 37°F. Therefore, the maximum surface temperature reached by this wall is 97 + 37 = 134°F. The maximum temperature of the other walls, while shaded, is 97°F. Heat Gain due to radiation and air temperature outside, wall:

Facing sun — (from Construction Assembly XIX) R = 3.05.

$$U_{wsr} = \frac{134 - 75}{3.05} = \frac{59}{3.05} = 19.34 \text{ Btu/ft}^2\text{, hr} \times 279 \text{ ft}^2 \text{ (area)} = 5{,}395 \text{ Btu/hr}$$

Shaded Wall — (from Construction Assembly XIX) R = 3.19

$$U_{wsh} = \frac{97 - 75}{3.19} = \frac{22}{3.19} = 6.9 \text{ Btu/ft}^2\text{, hr} \times (1062 \text{ ft}^2 - 279 \text{ ft}^2) = 5{,}402 \text{ Btu/hr}$$

Walls Total Maximum Gain = 10,797 Btu/hr.

Insulated Walls 3½° thick glass fiber insulation. Heat Gain due to radiation and air temperature outside, wall:

Facing sun — (from Table 125) R = 11.28

$$U_{wsri} = \frac{134 - 75}{11.28} = \frac{59}{11.28} = 5.23 \text{ Btu/ft}^2\text{, hr} \times 279 \text{ ft}^2 \text{ (area)} = 1{,}459 \text{ Btu/hr}$$

Shaded Wall — (from Table 125) R = 14.39

$$U_{wshi} = \frac{97 - 75}{14.39} = \frac{22}{14.39} = 1.52 \text{ Btu/ft}^2\text{, hr} \times (1062 \text{ ft}^2 - 279 \text{ ft}^2) = 1{,}190 \text{ Btu/hr}$$

Insulated Walls Maximum Gain = 2,649 Btu/hr.

Total of Walls — Maximum Heat Gain Per Hour
Heat capacity requirement

Uninsulated Walls = 10,797 Btu/hr
Insulated Walls = 2,649 Btu/hr
Savings by using insulation = 8,148 Btu/hr

SOLAR RADIATION — MAXIMUM HEAT GAIN PER HOUR

Windows — Single Pane 20% Wood Frame

When solar radiation is on windows, the maximum heat gain can be obtained from Table 36. For windows facing south, the maximum radiation is at 12 noon with radiation of 132 Btu/ft², hr.* The heat transferred through single unshaded glass is, for 1/8″ thick glass, 86%.

$$U_{wdso} = 132 \times 0.86 = 113 \times \text{area south windows, which is } 81 \text{ ft}^2 = 113 \times 81 = 9{,}195 \text{ Btu/hr}$$

Windows — Single pane 20% Wood Frame with Storm Window

*Spring, 40° North Latitude.

$$U_{ndsow} = 132 \times 0.86 \times 0.86 = 97 \text{ Btu/ft}^2\text{, hr} \times 81 = 7{,}857 \text{ Btu/hr}$$

A savings of 1,338 Btu/hr.

HEAT GAIN FROM OUTSIDE AIR TEMPERATURE — MAXIMUM PER HOUR

Windows — Single pane 20% wood — (from Table 162) R = 1.01

$$U_{wdair} = \frac{97 - 75}{1.01} = \frac{22}{1.01} = 21.7 \text{ Btu/ft}^2\text{, hr} \times \text{window area of } 186 \text{ ft}^2 = 4{,}051 \text{ Btu/hr}$$

Windows — Single pane 20% Wood with Storm Windows — (from Table 162) R = 1.78

$$U_{wdairs} = \frac{97 - 75}{1.78} = \frac{22}{1.78} = 12.4 \text{ Btu/ft}^2\text{, hr} \times \text{window area of } 186 \text{ ft}^2 = 2{,}298 \text{ Btu/hr}$$

A savings of 1,753 Btu/hr.

Total Windows Heat Gain Per Hour

Single pane 9,195 + 4,051 = 13,246 Btu/hr
Single pane with storm window = 7,857 − 2,298 = 10,155
A savings of 3,091 Btu/hr (23%)

Floors

Since the floors on a slab do not reach temperatures above that of the inside air, their insulation has no effect on the cooling energy load.

Construction		Maximum Heat Gain of House Structure		
Assembly	**No.**	**Uninsulated Btu/hr**	**Insulation**	**Insulated Btu/hr**
Ceiling	XII	54,386	6″ Glass Fiber	3,110
Walls	XIX	10,797	3½″ Glass Fiber	2,649
Windows		13,246	Storm Windows	10,155
Totals		78,429		15,914

The insulation provides a savings of 62,515 Btu/hr of required capacity of the cooling equipment. Should the thermal efficiency of the cooling equipment be 75% the savings in needed capacity is $\frac{62{,}515}{0.75} = 83{,}353$ Btu/hr.

YEARLY HEAT GAIN — Based on average degree hours from Table 24 for Springfield, Mo. Summer Cooling Degree hours = 23,160 °F.

Ceiling Uninsulated (from Construction Assembly XII) R = 1.3

$$U_{cdc} = \frac{23{,}160}{1.3} = 17{,}815 \text{ Btu/ft}^2\text{, yr} \times 1{,}285.5 \text{ ft}^2 \text{ (area)} = 22{,}901{,}676 \text{ Btu/yr}$$

Ceiling Insulated 6″ thick glass fiber — (from Table 113 R = 22.73

$$U_{cidc} = \frac{23{,}160}{22.73} = 1{,}018 \text{ Btu/ft}^2\text{, yr} \times 1{,}285.5 \text{ ft}^2 \text{ (area)}$$

$$= 1{,}309{,}818 \text{ Btu/yr}$$

A savings of 21,591,858 Btu/yr (94%).

Walls — 4″ Stud Uninsulated — (from Construction Assembly XIX) R = 3.05

$$U_{wc} = \frac{23{,}160}{3.05} = 7{,}593 \text{ Btu/ft}^2\text{, yr} \times 1{,}062 \text{ ft}^2 \text{ (area)}$$

$$= 8{,}064{,}236 \text{ Btu/yr}$$

Walls — 4″ Stud Insulated 3½″ Glass Fiber — (from Table 125) R = 11.28

$$U_{wic} = \frac{23{,}160}{11.28} = 2{,}053 \text{ Btu/ft}^2\text{, yr} \times 1{,}062 \text{ ft}^2 \text{ (area)}$$

$$= 2{,}180{,}489 \text{ Btu/yr}$$

A savings of 5,883,747 Btu/yr (72%).

Windows single pane 20% Wood Frame — (from Table 162) R = 1.04

$$Y_{wdc} = \frac{23{,}160}{1.04} = 22{,}269 \text{ Btu/ft}^2\text{, yr} \times 186 \text{ ft}^2 \text{ (area)}$$

$$= 4{,}142{,}076 \text{ Btu/yr}$$

Windows single pane 20% Wood Frame with Storm Windows — (from Table 162) R = 1.84

$$U_{wdic} = \frac{23{,}160}{1.84} = 12{,}587 \text{ Btu/ft}^2\text{, yr} \times 186 \text{ ft}^2 \text{ (area)}$$

$$= 2{,}341{,}173 \text{ Btu/yr}$$

A savings of 2,241,549 Btu/yr (51%)

Construction		**Yearly Heat Gain — Btu**			
Assembly	**No.**	**Uninsulated Gain**	**Insulation**	**Insulated Gain**	**Savings**
Ceiling	XII	22,901,676	6″ Glass Fiber	1,309,818	21,591,858
Walls	XIX	8,064,236	3½″ Glass Fiber	2,180,489	5,883,747
Windows		4,142,076	Storm Windows	2,341,173	1,800,901
Totals		35,107,988		5,831,480	29,276,506

If the thermal efficiency of the cooling equipment were 75% then total savings per year — cooling — by insulation would be:

$$\frac{29{,}276{,}508}{.75} = 39{,}035{,}345 \text{ Btu/yr.}$$

The total savings in energy is the savings obtained by the insulation in energy heat loss which is made up by energy supplied by the heating system plus the energy used to remove the heat entering during cooling season.

The following is a tabulation of these factors as calculated in Example 52.

Construction Assembly	**No.**	**Season**	**Uninsulated Season Loss, Btu's**	**Insulated Season Loss Insulation**	**Btu's**	**Savings Btu by Insulation**
Ceiling	XII	Heating	103,464,385	6″ Glass Fiber	5,486,747	97,977,638
Walls	XIX	Heating	30,841,793	3½″ Glass Fiber	7,949,239	22,892,554
Windows		Heating	20,750,970	Storm Windows	11,774,426	8,976,544
Floor		Heating	84,192,506	2″ Cell. Glass	10,142,203	74,050,303
Ceiling	XII	Cooling	22,901,676	6″Glass Fiber	1,309,818	21,591,858
Walls	XIX	Cooling	8,064,236	3½″ Glass Fiber	2,180,489	5,883,747
Windows		Cooling	4,142,076	Storm Windows	2,341,175	1,800,901
Totals			274,357,642		41,184,097	233,173,545

Total yearly savings is 233,173,545 Btu, accomplished by the insulation and storm windows. This represents a savings of 84% of energy requirements due to the building structure. Stated another way. — If building is not insulated it will cost 6.66 times more money to heat and cool.

Example 52a—*Metric Units*

Determine the maximum and yearly averages of heat loss, uninsulated and insulated, for the house shown in Figure 163. The house is located in Springfield, Mo. The inside temperature is maintained at 21.1 °C.

The weather data for Springfield, Mo. (from Table 24) is: Design temperature, Winter −18 °C, Summer 36 °C, Winter Heating: Degree Hours 62,600 °C. Summer cooling data: Degree Hours 12,867 °C. The location of the house, as shown, has the major window space facing South. From the map in Figure 45 the degrees North Latitude of Springfield, Mo. is approximately 38° (From Table 36 — use solar heat gain listed for 40° North Latitude) Maximum heat gain is: at 11 am 340.17 W/m², at 12 noon 359.88 W/m² and at 1:00 pm 340.17 W/m². Minimum ground temperature under slab is 7.2 °C. Winter average is 10 °C at depth of slab below grade.

Areas—Metric Units

$$\text{Windows} = 14 \times (0.9144_m \times 1.219_m) = 15.61 \text{ m}^2$$
$$2 \times (0.9144_m \times 0.9144_m) = 1.67 \text{ m}^2$$
$$\text{Total} \quad 17.28 \text{ m}^2$$

$$\text{Wall} = [(13.716 \text{ m} + 10.058 \text{ m}) \times 2.436 \text{ m} \times 2] - 17.28 \text{ m}^2$$
$$= [(23.76 \text{ m}) \times 2.436 \text{ m}] - 17.28 \text{ m}^2 = 115.75 \text{ m}^2 - 17.28 \text{ m}^2$$
$$= 98.478 \text{ m}^2$$

$$\text{Ceiling} = [(10.058 \text{ m} \times 13.716 \text{ m})] - [(2{,}743 \text{ m} \times 1.67 \text{ m})$$
$$+ (2.743 \text{ m} \times 3.048 \text{ m}) + (4.57 \text{ m} \times 1.219 \text{ m})] = [(137.55)] - [4.592$$
$$+ 8.336 + 5.571] = 137.55 \text{ m}^2 - 18.498 \text{ m}^2 = 119.05 \text{ m}^2$$

$$\text{Floor} = \text{(same as ceiling)} = 119.05 \text{ m}^2$$

MAXIMUM HEAT LOSSES — Based on 21.1 °C inside temperature and −17.8 °C outdoors design temperature — per m² and total area m².

Ceiling Uninsulated — (from Construction Assembly XII R_m = 0.247

$$U_{mc} = \frac{21.1-(-17.8)}{0.247} = \frac{38.9}{0.248} = 156.85 \text{ W/m}^2 \times 119.4 \text{ m}^2 \text{ (area)}$$
$$= 18{,}727 \text{ W}$$

Ceiling Insulated 0.1524 m thick Glass Fiber — (from Table 113) R = 4.652

$$U_{mci} = \frac{21.1-(-17.8)}{4.652} = \frac{38.9}{4.652} = 8.362 \text{ W/m}^2 \times 119.4 \text{ m}^2 \text{ (area)}$$
$$= 996 \text{ W}$$

A savings of 17,729 W (94%).

Walls 10 cm Stud Walls (8.89 cm air space) Uninsulated — (from Construction Assembly XIX) R_m = 0.596

$$U_{mw} = \frac{21.1-(-17.8)}{0.596} = \frac{38.9}{0.596} = 65.27 \text{ W/m}^2 \times 98.418 \text{ m}^2 \text{ (area)}$$
$$= 6423.6 \text{ W}$$

Walls Insulated 0.088 m thickness Glass Fiber — (from Table 125) R_m = 2.651

$$U_{mwi} = \frac{21.1-(-17.8)}{2.651} = \frac{38.9}{2.651} = 14.67 \text{ W/m}^2 \times 98.418 \text{ m}^2 \text{ (area)}$$
$$= 1444.2 \text{ W}$$

A savings of 4979.4 W (77%).

Windows single pane 20% Wood Frame — (from Table 162) R_m = 0.178

$$U_{mwd} = \frac{21.1-(-17.8)}{0.178} = \frac{38.9}{0.178} = 218 \text{ W/m}^2 \times 17.28 \text{ m}^2 \text{ (area)}$$
$$= 3{,}767 \text{ W}$$

Windows single pane 20% Wood Frame with Storm Window — (from Table 162) R_m = 0.314

$$U_{mwds} = \frac{21.1-(-17.8)}{0.314} = \frac{38.9}{0.314} = 123.9 \text{ W/m}^2 \times 17.28 \text{ m}^2 \text{ (area)}$$
$$= 2140 \text{ W}$$

A savings of 1,627 W (43%).

Concrete Floor Uninsulated 10.16 cm sand gravel concrete. Minimum temperature of ground 10°C —(from Construction Assembly XXXIII) R_m = 0.146

$$U_{mf} = \frac{21.1-10}{0.146} = \frac{11.1}{0.146} = 76 \text{ W/m}^2 \times 119 \text{ m}^2 \text{ (area)} = 9051 \text{ W}$$

Concrete Floor with 0.058 m thickness of cellular glass insulation — (from Table 160) R_m = 1.214

$$U_{mfi} = \frac{21.1-10}{1.214} = \frac{11.1}{1.214} = 9 \text{ W/m}^2 \times 119 \text{m}^2 \text{ (area)} = 1088 \text{ W}$$

A savings of 7,963 W (87%).

Construction		Maximum Heat Loss of House Structure		
Assembly	No.	Uninsulated Loss W	Insulation	Insulated Loss W
Ceiling	XII	18,727	0.152 m Glass Fiber	996
Walls	XIX	6,424	0.088 m Glass Fiber	1,444
Windows		3,767	Storm Window	2,140
Floor		9,051	0.058 Cellular Glass	1,088
Totals		36,971 W		5,568 W

Insulation realizes a savings of 31,403 W in needed capacity of the heating equipment. In other words, the insulation saves 84% of needed heating capacity as compared to uninsulated house. Should the thermal efficiency of the heating equipment be 80% the savings in needed capacity would then be

$$\frac{31{,}403}{.8} = 39{,}253 \text{ W.}$$

This savings is only on the energy loss of the structural components of the residence.

WINTER'S SEASON HEAT LOSS — Based on degree hours. From Construction Assembly XII R = 0.247. From Table 24 Degree hours °C = 62,600.

Ceiling Uninsulated — (from Construction Assembly XII) R_m = 0.247

$$U_{mcdh} = \frac{62{,}600}{0.247} = 253{,}441 \text{ W-hr/m}^2 \times 119.4 \text{ m}^2 \text{ (area)}$$
$$= 30{,}266{,}671 \text{ W-hr}$$

Ceiling Insulated 0.152 glass fiber insulation — (from Table 113) R_m = 4.652

$$U_{mcidh} = \frac{62{,}600}{0.247} = 13{,}456 \text{ W-hr/m}^2 \times 119.4 \text{ m}^2 \text{ (area)}$$
$$= 1{,}606{,}655 \text{ W-hr}$$

A savings of 28,660,015 W-hr (94%).

Walls 10 cm Stud Walls (8.89 cm air space) Uninsulated — (from Construction Assembly XIX) R_m = 0.596

$$U_{mw} = \frac{62{,}600}{0.596} = 105{,}033 \text{ W-hr/m}^2 \times 98.418 \text{ m}^2 \text{ (area)}$$
$$= 10{,}337{,}192 \text{ W-hr}$$

Walls Insulated 0.088 m thick glass fiber — (from Table 125) R_m = 2.651

$$U_{mwi} = \frac{62{,}600}{2.651} = 23{,}612 \text{ W-hr/m}^2 \times 98.418 \text{ m}^2 \text{ (area)}$$
$$= 2{,}324{,}016 \text{ W-hr}$$

A savings of 8,013,175 W-hr (77%)

Windows single pane 20% wood frame — (from Table 162) — $R_m = 0.178$

$$U_{mwd} = \frac{62{,}600}{0.178} = 351{,}685 \text{ W-hr/m}^2 \times 1728 \text{ m}^2 \text{ (area)}$$
$$= 6{,}077{,}123 \text{ W-hr}$$

Windows single pane 20% wood frame plus storm window — (from Table 162) $R_m = 0.314$

$$U_{mwdi} = \frac{62{,}600}{0.314} = 199{,}363 \text{ W-hr/m}^2 \times 17.28 \text{ m}^2 \text{ (area)}$$
$$= 3{,}444{,}992 \text{ W-hr}$$

A savings of 2,632,130 W-hr (43%).

Floor Uninsulated 10.16 cm thick sand-gravel concrete.

As previously stated, below grade temperatures are always below the temperature to which the house is heated in winter. If the inside surface temperature of the concrete were 18.33 °C and the average soil temperature under the slab were 10.0 °C, then the degree C hours would be 3624 hours × 8.33 = 30,200 degree C hours.

Uninsulated — (from Construction Assembly XXXIII) $R_m = 0.146$

$$U_{mfdh} = \frac{30{,}200}{0.146} = 206{,}849 \text{ W-hr/m}^2 \times 119.05 \text{ m}^2 \text{ (area)}$$
$$= 24{,}615{,}031 \text{ W-hr}$$

Floor Insulated 0.015 cm thick cellular glass — (from Table 160) $R_m = 1.214$

$$U_{mfdhi} = \frac{30{,}200}{1.214} = 24{,}876 \text{ W} \times 119.05 \text{ m}^2 \text{ (area)}$$
$$= 2{,}961{,}540 \text{ W-hr}$$

A savings of 21,653,490 W-hr per year (87%).

Construction		Yearly Heat Loss — Watt - Hours			
Assembly	No.	Uninsulated Loss	Insulation	Insulated Loss	Savings by Insulation
Ceiling	XII	30,266,671	0.152 m Glass Fiber	1,606,655	28,660,016
Walls	XIX	10,337,192	0.088 m Glass Fiber	2,324,016	8,013,176
Window		6,077,123	Storm Windows	3,444,992	2,632,131
Floor		24,615,031	0.051 Cellular Glass	2,961,540	21,653,490
Totals		71,296,017		10,337,283	60,958,764

This represents 85% savings of energy. Should the efficiency of the heating equipment be 80% the total savings of energy would be $\frac{60{,}958{,}764}{0.8}$ = 76,198,456 W-hr or 76,198 kW-hr per year.

MAXIMUM HEAT GAIN — Which must be removed by the capacity of the air conditioning equipment at 23.9°C indoor temperature.

Ceiling Uninsulated — (from Construction Assembly XII) $R_m = 0.229$

Note: Even though outside design temperature is given as 36.1 °C due to solar radiation, the space between the roof and ceiling — even vented — will reach a temperature of 54.4 °C.

$$U_{mca} = \frac{54.4 - 23.9}{0.229} = \frac{30.5}{0.229} = 133.2 \text{ W/m}^2 \times 119.4 \text{ m}^2 \text{ (area)}$$
$$= 15{,}902 \text{ W}$$

Ceiling Insulated 0.152 m Glass Fiber Insulation — (from Table 113) $R_m = 4.005$

$$U_{mcai} = \frac{54.4 - 23.9}{4.005} = \frac{30.5}{4.005} = 7.62 \text{ W/m}^2 \times 119.4 \text{ m}^2 \text{ (area)}$$
$$= 909 \text{ W}$$

A savings of 14,993 W (94%).

Walls Uninsulated 10 cm Stud Wall — (from Construction Assembly XIX) $R_m = 1.081$ at 27.4 °C mean temperature — $R_m = 1.028$ at 43.2 °C mean temperature

The maximum temperature will occur when the South wall is heated by solar radiation and the balance of the wall area is heated only by outside ambient air. The area of the South wall is 13.72 m × 2.43 m = 33.33 minus the window area of (1.82 × 3.66) + 0.83 = 6.65 + 0.83 = 7.48 m² or 33.33 − 7.48 = 25.85 m².

The shingles have a surface emittance ε of 0.75 (from Table 44) the maximum temperature that the wall surface will be increased is 20.55 °C. Therefore the maximum surface temperature reached by this wall is 36.1 °C + 20.6 °C = 56.7 °C.

The maximum temperature of the other walls, while shaded, is 36.1 °C.

Heat Gain due to radiation and air temperature outside wall:

Facing Sun — (from Construction Assembly XIX) $R_m = 0.537$

$$U_{mwsr} = \frac{56.7 - 23.9}{0.537} = \frac{32.8}{0.537} = 61 \text{ W/m}^2 \times 25.9 \text{ m}^2 \text{ (area)}$$
$$= 1{,}583 \text{ W}$$

Shaded Wall — (from Construction Assembly XIX) $R_m = 0.561$

$$U_{mwsh} = \frac{36.1 - 23.9}{0.561} = \frac{12.2}{0.561} = 21.7 \text{ W/m}^2 \times (98.6 \text{ m}^2 - 25.9 \text{ m}^2)$$
$$= 21.7 \text{ W} \times 72.7 \text{ m}^2 = 1{,}578 \text{ W}$$

Walls total maximum gain = 3,161 W.

Insulated Wall 0.88 m thick glass fiber. Heat gain due to radiation and air temperature outside wall:

Facing Sun — (from Table 125) $R_m = 1.988$

$$U_{mwsr} = \frac{56.7 - 23.9}{1.988} = \frac{32.8}{1.988} = 16.5 \text{ W/m}^2 \times 25.9 \text{ m}^2 \text{ (area)}$$
$$= 427.6 \text{ W}$$

Shaded Wall — (from Table 125) $R_m = 2.536$

$$U_{mwshi} = \frac{36.1 - 23.9}{2.536} = \frac{12.2}{2.536} = 4.8\ W/m^2 \times 7.27\ m^2$$

349 W

Insuated Walls total maximum gain per hour = 776.5 W

Total of Walls — Maximum Heat Gain

Heat capacity requirements

Uninsulated walls = 3,161 W
Insulated walls = 776 W
Savings by using insulation = 2,385 W

SOLAR RADIATION

Windows — Single Pane 20% Wood Frame

The maximum radiation due to solar radiation is given in Table 36. For windows facing south the maximum radiation* is at 12 noon, with radiation at 416.13 W/m². The heat transferred through single unshaded glass for 0.317 cm thick glass is 86%.

$$U_{mwdso} = 416.13 \times 0.86 = 357\ W/m^2 \times \text{area of south windows}$$
$$= 357\ W/m^2 \times 7.52\ (\text{area}) = 2{,}693\ W$$

Window Single Pane 20% Wood Frame with Storm Window

$$U_{mwdsow} = 416.13 \times 0.86 \times 0.86 = 307\ W/m^2 \times 7.52$$
$$= 2{,}314\ W$$

A savings of 378 W

HEAT GAIN FROM OUTSIDE AIR TEMPERATURE

Windows Single Pane 20% Wood Frame — (from Table 162) $R_m = 0.178$

$$U_{mwdir} = \frac{36.1 - 23.9}{0.178} = \frac{12.2}{0.178} = 66\ W/m^2 \times 17.28\ m^2\ (\text{area})$$
$$= 1{,}140\ W$$

Windows Single Pane 20% Wood Frame with Storm Window — (from Table 162) $R_m = 0.314$

$$U_{mwdirs} = \frac{36.1 - 23.9}{0.314} = \frac{12.2}{0.314} = 38\ W/m^2 \times 17.28\ m^2\ (\text{area})$$
$$= 671\ W$$

A savings of 469 W

Total Window Heat Gain, Maximum Per Hour

Single pane 2693 + 1140 = 3,833 watts maximum

Single pane with storm window 2314 + 671 = 2,985 watts maximum

A savings of 848 W — maximum.

*Spring, 40° North Latitude.

Floors

As the floors on a slab do not reach a temperature above that of the controlled inside air temperature, they have no effect on maximum heating load.

Construction		Maximum Heat Gain of House Structure		
Assembly	**No.**	**Uninsulated Loss W**	**Insulation**	**Insulated Loss W**
Ceiling	XII	15,902	0.152 m Glass Fiber	909
Walls	XIX	3,161	0.088 m Glass Fiber	776
Windows		3,833	Storm Window	2,985
Totals		22,896 W		4,670 W

The insulation and storm windows provide a savings of 18,226 W per hour of required capacity of the cooling equipment. Should the thermal efficiency of the cooling equipment be 75% the savings in needed capacity is $\frac{18{,}226}{0.75} = 24{,}301$ W or 24.3 kW.

YEARLY HEAT GAIN — Based on the average degree C hours from Table 24. For Springfield, Mo. the summer cooling degree hours is 12,867 °C.

Ceiling Uninsulated — (from Construction Assembly XII) $R_m = 0.229$

$$U_{mcdc} = \frac{12{,}867}{0.229} = 56{,}187\ W\text{-}hr/m^2 \times 119.4\ m^2\ (\text{area})$$
$$= 6{,}708{,}727\ W\text{-}hr \text{ per year}$$

Ceiling Insulated 0.152 m Glass Fiber — (from Table 113) $R_m = 4.005$

$$U_{mcdci} = \frac{12{,}867}{4.005} = 3212\ W\text{-}hr/m^2 \times 119.4\ m^2\ (\text{area})$$
$$= 383{,}600\ W\text{-}hr \text{ per year}$$

A savings of 6,325,127 W-hr or 6,325 kW-hr (94%).

Walls 10 cm Stud Walls (8.89 cm air space) Uninsulated — (from Construction Assembly XIX) $R_m = 0.537$

$$U_{mwc} = \frac{12{,}867}{0.537} = 23{,}960\ W\text{-}hr/m^2 \times 98.418\ m^2\ (\text{area})$$
$$= 2{,}363{,}888\ W\text{-}hr \text{ per year}$$

Walls 10 cm Stud Wall, 0.088 m Glass Fiber — from Table 125) $R_m = 1.988$

$$U_{mwci} = \frac{12{,}867}{1.988} = 6{,}472\ W\text{-}hr/m^2 \times 98.418\ m^2\ (\text{area})$$
$$= 636{,}994\ W\text{-}hr \text{ per year}$$

A savings of 1,726,894 W-hr/yr or 1726.8 kW-hr/yr

Windows Single Pane 20% Wood Frame — (from Table 162) $R_m = 0.183$

$$U_{mwdc} = \frac{12,867}{0.183} = 70,311 \text{ W-hr/m}^2 \times 17.28 \text{ m}^2 \text{ (area)}$$
$$= 1,214,931 \text{ W-hr per year}$$

Windows Single Pane 20% Wood Frame — (from Table 162) $R_m = 0.325$

$$U_{mwdcs} = \frac{12,867}{0.324} = 39,712 \text{ W-hr/m}^2 \times 17.28 \text{ m}^2 \text{ (area)}$$
$$= 686,239 \text{ W-hr per year}$$

A savings of 528,692 Watts per year or 528.6 kW-hr /yr.

Construction		Yearly Heat Gain — Watt - Hours			
		Uninsulated		Insulated	Savings by
Assembly	No.	Gain W	Insulation	Gain	Insulation
Ceiling	XII	6,708,727	0.152 m Glass Fiber	383,600	6,325,127
Walls	XIX	2,363,888	0.088 m Glass Fiber	636,994	1,726,894
Windows		1,214,931	Storm Windows	686,239	528,692
Totals		10,287,546		1,706,833	8,580,713 or 8580.7 kW-hr per year

If the thermal efficiency of the cooling equipment is 75% then the total savings per year - for cooling - would be, by insulation and storm windows, $\frac{8,580,713}{.75}$ = 11,440,950 W-hr per year or 11,440 kW-hr per year.

The total savings in energy is the savings obtained by the insulation in energy heat loss which is made up by the heating system plus the energy used to remove heat entering during the cooling season.

The cooling is a tabulation of these factors as calculated in Example 52a.

Construction			Uninsulated			Savings W-hr
Assembly	No.	Season	Season Loss, W-hr	Insulated Season Loss Insulation	W-hr	by Insulation
Ceiling	XII	Heating	30,266,671	0.152 m Glass Fiber	1,606,655	28,660,016
Walls	XIX	Heating	10,337,192	0.088 m Glass Fiber	2,324,016	8,013,176
Windows		Heating	6,077,123	Storm Windows	3,444,992	2,632,131
Floor	XXXIII	Heating	24,615,031	0.051 m Cell. Glass	2,961,590	21,653,441
Ceiling	XII	Cooling	6,708,727	0.152 m Glass Fiber	383,600	6,325,127
Walls	XIX	Cooling	2,363,888	0.088 m Glass Fiber	636,994	1,726,894
Windows		Cooling	1,214,931	Storm Windows	686,239	528,692
Totals			81,583,563		12,049,036	69,534,527 or 69,534 kW-hr per year

If the thermal efficiency of the cooling equipment is 75% then the total savings per year heating and cooling by insulation and storm windows would be:

$$\frac{69,534,527}{0.75} = 92,712,702 \text{ W-hr per year or } 92,712 \text{ kW-hr per year}$$

This represents a savings of 84% of energy requirements. Stated another way, if the building is not insulated it will cost 6.66 times more money to heat and cool.

Examples 52 and 52a illustrated the method to determine the energy loss of a house. These same methods can be used to determine the energy losses of any homes, office buildings or industrial buildings.

In all construction a major factor is the relationship between costs and savings. Unfortunately an easy conversion from energy to money or money to energy does not exist. This is because the value of money is always changing and the purchase of fuel energy is usually given in the purchase cost of cubic feet of gas, cost per ton of coal, cost per gallon of oil. Thus to establish a basis of economic evaluation it is necessary to establish the approximate energy content of commonly used fuels.

Using these values of heat content it is then possible to determine the quantity of fuel directly lost for individual constructions. The fuel quantities determined must be divided by the % efficiency factor of converting from fuel to heat energy to obtain the total required.

For illustration this will be converted to dollars based on 1982 prices at a particular location. Since pertinent information for this time period was available, values typical of 1982 will be used in these examples. These dollar values change with location and time and must not be considered constant.

In the determination of the quantities of fuel used, difficulty may be experienced due to the numerous units (of quantity or volume and heat) in common use. Frequently, the heat contents of fuels are given in units other than those used for calculating heat losses. Table 205 is provided so that the units of heat can be converted to the one that was used in the calculation of the heat transmissions of a building or residence. This table also provides conversions of weight units and volume units.

As fuels are most frequently sold by weight, volume, or liquid measure, the heat required must be converted to the units of fuel necessary to supply that heat. Again these values are given in many units. Table 206 provides the equivalent of energy in the various units by which it might be purchased.

The heat values of fuels such as coal, oil, or gas are typical values. For example natural gas was given as being 970 to 1010 Btu/cu ft. which is the value of most gas sold to consumers, yet there are some areas that natural gas may be as high as 1200 Btu/cu ft heat content. For this reason it is well to check the heat content of fuel which is available in the area in which the building is constructed.

HEAT CONVERSION TABLE

English Units	Conversion Factor	Metric Units	Conversion Factor	English Units
British Thermal Unit, Btu	0.252	kilogram calories	3.968	Btu
British Thermal Unit, Btu	1055	joules J	0.0009478	Btu
Calorie	4.1868	joules J	0.0002399	calorie
Therm (100,000 Btu)	105500000	joules J	9.478×10^{-9}	therm
Btu/hr	0.2931	watts W	3.41215	Btu/hr
Ton refrigeration (12,000 Btu/hr)	3517.2	watts W	0.000284	ton refrigeration
Watt-hr	3600	joules	0.000277	watt
British Thermal Unit, Btu	0.293	Watt-hours	3.1413	Btu
Btu/lb	0.1323	Watt-hours/kg	7.5585	Btu/lb
Btu/ft^3	37,260	joules/m^3	0.0000268	Btu/ft^3
Btu/lb	2326	joules/kg	0.0004299	Btu/lb
Btu/gal	278	joules/liter	0.003587	Btu/gal

WEIGHT CONVERSION TABLE

English Units	Conversion Factor	Metric Units	Conversion Factor	English Units
Pound per gallon, lb/gal	9.978	kg/m^3	0.10013	lb/gal
Pound per cubic foot, lb/ft^3	16.02	kg/m^3	0.06242	lb/ft^3
Pound per cubic foot, lb/ft^3	16.02	gram/liter	0.06242	lb/ft^3
Pound, lb	0.4536	kilogram kg	2.205	lb
Ton	907.2	kilogram kg	0.00102	ton

VOLUME CONVERSION TABLE

English Units	Conversion Factor	Metric Units	Conversion Factor	English Units
Cubic foot, ft^3	0.02832	cubic meters m^3	35.31	ft^3
Cubic foot, ft^3	28.33	liter ℓ	0.035	ft^3
Quart	0.009463	cubic meters m^3	105.67	quart
Quart	0.9463	liter ℓ	1.057	quart
Gallon	0.0037854	cubic meters m^3	264.09	gallon
Gallon	3.7854	liter ℓ	0.2642	gallon
Cubic foot, ft^3	7.481	gallons	0.1337	ft^3

TABLE 205

HEAT CONTENT OF FUELS

SOLIDS

Fuel	Energy Unit Per Unit of Weight		
	Btu/lb	Joules/kg	Watt-hr/kg
Anthracite Coal	12,900	30,005,400	1714
Low Volatile Bituminous Coal	14,300	33,261,800	1900
Medium Volatile Bituminous Coal	13,500	31,401,000	1786
High Volatile Bituminous Coal	12,000	27,912,000	1594
Coke	13,200	30,703,200	1754

LIQUIDS

Fuel	Energy Unit Per Unit of Weight or Volume				
	Btu/lb	Btu/gal	Joules/kg	Joules/liter	Watt-hr/kg
Grade 1 Fuel Oil	19,700	134,000	45,961,760	37,250,000	2626
Grade 2 Fuel Oil	19,600	138,000	45,589,600	38,364,000	2604
Grade 4 Fuel Oil	19,400	145,500	45,124,400	40,449,000	2578
Grade 5 Fuel Oil	18,980	149,000	44,147,430	41,422,000	2522
Grade 6 Fuel Oil	18,600	153,000	43,263,600	42,534,000	2472

GASES

Fuel	Energy Unit Per Unit of Weight or Volume				
	Btu/lb	Btu/cu ft	Joules/kg	Joules/m^3	Watt-hr/kg
Manufactured Fuel Gas	18,100 to	850 to	45,370,090 to	31,671,000 to	2405 to
Manufactured Fuel Gas	19,200	900	44,659,200	33,534,000	2551
Natural Gas	19,500	970 to	45,357,000 to	36,142,200 to	2591 to
Natural Gas	20,000	1010	46,520,000	37,632,600	2658
Commercial Butane	19500	3260	45,357,000	121,467,600	2591
Commercial Propane	19800	2500	46,050,800	93,150,000	2631

Example 53

Calculate the quantity of fuel used to heat the uninsulated house and the insulated house as based on the calculated heat losses in Examples 52 and 52a. Determine these amounts for coal, oil, gas and electricity.

ENGLISH UNITS

COAL QUANTITIES — Tons Used, Heating

COAL USED — HEATING SEASON* (Based on Coal Energy of 13,500 Btu/lb)

Construction Assembly	Heat Loss — Btu		Coal Used — Tons		Coal Saved by Insulation	
	Uninsulated	Insulated	Uninsulated	Insulated	Tons	%
Ceiling	103,464,385	5,486,747	3.832	0.203	3.629	94
Walls	30,841,793	7,949,239	1.142	0.294	0.848	74
Windows	20,750,970	11,774,426	0.768	0.436	0.332	43
Floor	84,192,506	10,142,303	3.118	0.876	2.742	88
Totals	239,249,654	35,352,615	8.860	1.329	7.548	85

*Not including the thermal efficiency of the furnace or boiler.

METRIC UNITS

COAL QUANTITIES — Kilograms Used, Heating

COAL USED — HEATING SEASON* (Based on Coal Energy of 31,401,000 J/kg)

Construction Assembly	Heat Loss — Watt/Hrs		Coal Used — Kilograms		Coal Saved by Insulation	
	Uninsulated	Insulated	Uninsulated	Insulated	Kilograms	%
Ceiling	30,266,671	1,606,655	3,467	184	3,292	94
Walls	10,337,192	2,324,016	1,036	266	769	74
Windows	6,077,123	3,444,992	696	395	301	43
Floor	24,615,031	2,961,540	2,828	341	2,487	88
Totals	71,296,017	12,049,036	8,037	1,205	6,847	85

*Not including the thermal efficiency of the furnace or boiler.

ENGLISH UNITS

OIL QUANTITIES — Gallons Used, Heating

OIL USED — HEATING SEASON* (Based on 145,000 Btu per gallon)

Construction Assembly	Heat Loss — Btu		Oil Used — Gallons		Oil Saved by Insulation	
	Uninsulated	Insulated	Uninsulated	Insulated	Gallons	%
Ceiling	103,464,385	5,486,747	713.5	37.8	675.7	94
Walls	30,841,793	7,949,239	212.7	54.8	157.9	74
Windows	20,750,970	11,774,426	143.1	81.2	61.9	43
Floor	84,192,506	10,142,303	580.6	69.9	510.7	87
Totals	239,249,654	35,352,615	1,649.9	243.7	1,406.2	85

*Not including the thermal efficiency of the furnace or boiler.

METRIC UNITS

OIL QUANTITIES — Liter Used, Heating

OIL USED — HEATING SEASON* (Based on 40,310,000 Joules per liter)

Construction Assembly	Heat Loss — Watt/Hrs		Oil Used — Liters		Oil Saved by Insulation	
	Uninsulated	Insulated	Uninsulated	Insulated	Liters	%
Ceiling	30,266,671	1,606,655	2,700.9	143.1	2,557.8	94
Walls	10,337,192	2,324,016	805.1	207.4	597.7	74
Windows	6,077,123	3,444,992	541.7	307.4	234.3	43
Floor	24,615,031	2,961,540	2,197.8	264.5	1,933.2	87
Totals	71,296,017	12,049,036	6,245.5	922.4	5,323.0	85

*Not including the thermal efficiency of the furnace or boiler.

ENGLISH UNITS

GAS QUANTITIES — Cubic Feet Used, Heating

GAS USED — HEATING SEASON* (Based on 1010 Btu/cubic foot)

Construction Assembly	Heat Loss — Btu Uninsulated	Heat Loss — Btu Insulated	Gas Used — Cubic Feet Uninsulated	Gas Used — Cubic Feet Insulated	Gas Saved by Insulation Cubic Feet	Gas Saved by Insulation %
Ceiling	103,464,385	5,486,747	102,439	5,432	97,007	94
Walls	30,841,793	7,949,239	30,536	7,870	22,666	74
Windows	20,750,970	11,774,426	20,545	11,657	8,888	43
Floor	84,192,506	10,142,303	83,358	10,041	73,317	88
Totals	239,249,654	35,352,615	236,878	35,000	201,878	85

*Not including the thermal efficiency of the furnace or boiler.

METRIC UNITS

GAS QUANTITIES — Cubic Meters Used, Heating

GAS USED — HEATING SEASON* (Based on 37,632,600 Joules/cub metre)

Construction Assembly	Heat Loss — Watt·Hours Uninsulated	Heat Loss — Watt·Hours Insulated	Gas Used — Cubic Meters Uninsulated	Gas Used — Cubic Meters Insulated	Gas Saved by Insulation Cubic m	Gas Saved by Insulation %
Ceiling	30,266,671	1,606,655	2,901	153	2,747	94
Walls	10,337,192	2,324,016	864	221	643	74
Windows	6,077,123	3,444,992	581	329	252	43
Floor	24,615,031	2,961,540	2,360	285	2,075	88
Totals	71,296,017	12,049,036	6,708	991	5,717	85

*Not including the thermal efficiency of the furnace or boiler.

ENGLISH AND METRIC UNITS

ELECTRIC QUANTITIES — Used, Heating

ELECTRICITY USED — HEATING SEASON* (kW·hrs)

Construction Assembly	Heat Loss — Watt·Hours Uninsulated	Heat Loss — Watt·Hours Insulated	Electricity Used — kW·hr Uninsulated	Electricity Used — kW·hr Insulated	Electricity Saved kW·hr	Electricity Saved %
Ceiling	30,266,671	1,606,655	30,266	1,606	28,660	94
Walls	10,337,192	2,324,016	10,337	2,324	8,013	74
Windows	6,077,123	3,444,992	6,077	3,444	2,633	43
Floor	24,615,931	2,961,540	24,615	2,961	21,654	88
Totals	71,296,017	12,049,036	71,295	10,335	60,960	85

*Not including the thermal efficiency of the furnace or boiler.

ENGLISH AND METRIC UNITS **ELECTRIC QUANTITIES — Used, Cooling**

ELECTRICITY USED — COOLING SEASON* (kW-hrs)

Construction Assembly	Heat Gain — Watt-Hours Uninsulated	Insulated	Electricity Used — kW-hr Uninsulated	Insulated	kW-hr Saved by Insulation kW-hr	%
Ceiling	6,708,727	383,600	6,709	384	6,325	94
Walls	2,363,888	636,994	2,364	637	1,727	73
Windows	1,214,931	686,239	1,215	686	8,581	43
Totals	10,287,546	1,076,833	10,288	1,707	8,581	83

*Not including the thermal efficiency of the heating equipment.

ENGLISH AND METRIC UNITS **ELECTRIC QUANTITIES — Capacity, Maximum**

ELECTRICAL CAPACITY REQUIRED — COOLING SYSTEM* (kW-hr)

Construction Assembly	Heat Gain — Watt-Hours Uninsulated	Insulated	Electric Demand kW Uninsulated	Insulated	Capacity Reduction in kW kW	%
Ceiling	15,902	909	15.09	0.90	15.00	94
Walls	3,161	776	3.16	0.78	2.38	75
Windows	3,833	2,985	3.83	2.99	0.84	21
Totals	22,860	4,670	22.86	4.67	18.22	79

*Not including the thermal efficiency of the cooling equipment.

All the losses and savings have been presented either in units of heat or fuel. As these are fixed technical units the values are fixed. Caculations of thermal heat transfer, heat losses, and quantities of fuel use also remain constant. Unfortunately, the money cost of materials, construction cost, and fuel costs are not constant.

CHARGES FOR 500 KILOWATT HOURS
OF ELECTRICITY IN U.S. CITIES
(From most to least expensive in September 1981)

City	In July 1979	In Sept. 1981	Percentage of Increase
New York	$45.83	$62.96	37.3%
Honolulu	$30.58	$62.22	103.4%
San Diego	$28.75	$47.77	66.1%
Chicago	$30.69	$44.44	44.8%
Cleveland	$32.06	$40.91	27.6%
Boston	$28.80	$39.84	38.3%
Detroit	$27.88	$39.24	40.7%
Philadelphia	$25.44	$39.12	53.7%
Miami	$26.59	$38.50	44.7%
Denver	$23.85	$37.10	55.5%
Kansas City, Mo.	$31.04	$36.25	16.7%
Baltimore	$27.24	$34.98	28.4%
Los Angeles	$25.05	$34.56	37.3%
Milwaukee	$25.12	$34.22	36.2%
Washington, D.C.	$26.64	$34.39	29.0%
Pittsburgh	$26.70	$33.53	25.5%
Dallas	$23.44	$32.71	39.5%
Buffalo, N.Y.	$23.59	$31.47	33.4%
St. Louis	$23.08	$30.79	33.4%
Houston	$24.66	$30.57	23.9%
Minneapolis	$25.08	$29.99	19.5%
San Jose	$18.23	$29.47	61.6%
Anchorage, Alaska	$20.36	$28.02	37.6%
Cincinnati	$22.24	$26.60	19.6%
Atlanta	$21.72	$26.52	22.0%
Charleston, W. Va.	$19.43	$25.07	29.0%
Portland, Ore.	$14.90	$22.82	53.1%
Seattle	$ 9.45	$13.14	39.0%
U.S. Cities Average	$25.30	$35.26	39.4%

TABLE 207

The following is presented to provide dollar values of fuel in 1982. Solid and liquid fuels were relatively constant in price over the entire United States at this time.

Prices of electric power vary with geographical location due to varying production and transmission costs. Price also varies as quantity used per month varies (see Table 207).

Gas prices vary as to location and type of gas (see Table 208).

AVERAGE GAS PRICES FOR THIRD QUARTER 1981
DOLLARS PER MILLION Btu (105,506,000,000 joule)

Area of United States	Type of Consumer Residential	Commercial	Industrial
New England	8.01	5.36	4.34
Middle Atlantic	6.48	4.70	3.96
East North Central	4.57	3.79	3.41
West North Central	4.66	3.17	3.03
South Atlantic	5.87	4.25	3.61
East South Central	4.82	3.92	3.42
West South Central	4.41	3.57	3.18
Mountain	4.32	3.55	3.00
Pacific	4.07	4.33	3.95
United States Average	4.95	3.99	3.42

Note: The above gas prices were based on information listing third quarter 1981 prices. Due to energy shortage these prices will probably increase, thus for calculating heating cost prevailing local cost should be determined.

TABLE 208

COSTS

Typical is the following electric charges for residential users.

Monthly Rate

First	30 kW/hr	$0.08 24	per kW·hr
Next	40 kW/hr	$0.07 70	per kW·hr
Next	130 kW/hr	$0.05 51	per kW·hr
Next	300 kW/hr	$0.04 27	per kW·hr
Next	1000 kW/hr	$0.03 87	per kW·hr
All Over	1500 kW/hr	$0.03 59	per kW·hr

Using 500 kW·hr as an average monthly usage, the charges are:

HEATING SEASON

COAL COSTS (at $75 a ton)

Construction Assembly	Uninsulated* Ton	Uninsulated* $ Cost	Insulated Ton	Insulated $ Cost	Savings $ Yr.	Insulation $ Cost	% Return on Investment
Ceiling	5.109	383.18	0.270	20.25	362.93	501.35	72
Wall	1.523	114.22	0.392	29.40	84.82	297.40	28
Windows	1.024	76.80	0.581	43.58	33.22	680.50	5
Floor	4.157	311.75	0.501	37.58	274.20	2,185.35	13
Totals	11.813	885.95	1.744	130.81	755.14	3,664.60	21

*Based on thermal efficiency of furnace of 75%.

HEATING SEASON

OIL COSTS (at $1.15 a gallon)

Construction Assembly	Uninsulated* Gallon	Uninsulated* $ Cost	Insulated Gallon	Insulated $ Cost	Savings $ Yr.	Insulation $ Cost	% Return on Investment
Ceiling	951.3	1093.99	50.4	57.96	1036.03	501.35	206
Wall	283.6	326.14	73.1	84.06	199.54	297.40	67
Windows	109.8	219.42	108.3	124.54	94.88	680.50	14
Floor	774.1	890.21	93.2	107.18	783.03	2185.35	35
Totals	2199.8	2529.76	325.0	373.74	2156.02	3664.60	58

*Based on thermal efficiency of furnace of 80%.

HEATING SEASON

GAS COSTS (Based on $4.85 per MCF or 0.485 per therm)

Construction Assembly	Uninsulated* MCF	Uninsulated* $ Cost	Insulated MCF	Insulated $ Cost	Savings $ Yr.	Insulation $ Cost	% Return on Investment
Ceiling	128.03	620.98	6.79	32.92	588.06	501.35	117
Wall	38.15	185.03	9.83	47.71	137.32	297.40	46
Windows	25.69	124.58	14.38	70.69	53.89	680.50	8
Floor	104.20	505.37	12.55	60.87	444.50	2185.35	20
Totals	296.07	1435.96	43.75	212.19	1223.77	3664.60	33

*Based on thermal efficiency of furnace of 80%.

HEATING SEASON

ELECTRIC COSTS (Based on $0.046 average per kW·hr)

Construction Assembly	Uninsulated* kW	Uninsulated* $ Cost	Insulated kW	Insulated $ Cost	Savings $ Yr.	Insulation $ Cost	% Return on Investment
Ceiling	33,629	1546.93	1785.11	82.12	1464.81	501.35	292
Wall	11,530	530.38	2582.22	118.78	384.60	297.40	129
Windows	6,752	310.59	3100.41	142.62	167.97	680.50	24
Floor	27,350	1258.10	3290.55	151.36	1106.74	2185.35	51
Totals	79,261	3646.00	10758.29	494.88	3123.62	3664.60	85

*Based on thermal efficiency of furnace of 90%.

The price of fuel gas varies in price depending upon location and type. Presently residential users of natural gas are charged $4.25 to $5.85 per MCF; this is $0.370 to $0.49 per therm(1 therm equals 100,000 Btu).

INSULATION COST

The following is based on the contractor installing the insulation as the house is being constructed.

TOTAL SEASONS COST

COAL FOR HEATING — ELECTRICITY FOR COOLING

Construction Assembly	Heating Cost $ Uninsulated	Heating Cost $ Insulated	Cooling Cost $ Uninsulated	Cooling Cost $ Insulated	Total Cost $ Uninsulated	Total Cost $ Insulated	Savings $/Yr.	Insulation $ Cost	% Return on Investment
Ceiling	383.18	20.25	308.61	17.66	691.79	37.91	653.88	501.35	130
Walls	114.22	29.40	108.70	29.30	222.90	58.70	164.20	297.40	55
Windows	76.80	43.58	55.66	31.56	132.40	75.14	57.26	680.50	8
Floor	311.75	37.58	none	none	311.75	37.58	274.17	2185.35	12
Totals	885.95	130.81	472.97	78.52	1358.84	209.33	1149.51	3664.60	31

TOTAL SEASONS COST

OIL FOR HEATING — ELECTRICITY FOR COOLING

Construction Assembly	Heating Cost $ Uninsulated	Heating Cost $ Insulated	Cooling Cost $ Uninsulated	Cooling Cost $ Insulated	Total Cost $ Uninsulated	Total Cost $ Insulated	Savings $/Yr.	Insulation $ Cost	% Return on Investment
Ceiling	1093.99	57.96	308.61	17.66	1402.60	75.62	1326.98	501.35	264
Walls	326.14	84.06	108.70	29.30	434.84	113.86	321.41	297.40	108
Windows	219.42	124.54	55.66	31.56	275.08	156.10	118.98	680.50	17
Floor	890.21	107.18	none	none	890.21	37.58	852.63	2185.35	39
Totals	2529.76	373.74	472.97	78.52	3002.73	382.66	2620.07	3664.60	71

Typical 1982 costs for fuel and materials will be used in the following examples. Natural gas costs in the eastern part of the United States were as follows. A moderate size commercial user was charged approximately $4.45 per MCF. The average residential user was charged $5.03 per MCF.*

The cost of coal delivered to a relatively moderate user was $70.00 to $80.00 per ton.

The cost of fuel oil delivered to a relatively moderate user was $1.15 per gallon.

Installed cost of 6″ thick glass fiber insulation with vapor retarder sealed to ceiling beams is $0.39 per sq ft as described in Example 51 and 51a.

Installed cost of 3 1/2″ thick glass fiber insulation with vapor barrier sealed to wall studs is $0.27 per sq ft.

Installed cost of 2″ thick cellular glass under slab floor is $1.90 per sq ft.

Installed total cost of the storm windows for the house as shown in Figure 163 is $680.00.

TOTAL SEASONS COST

GAS FOR HEATING — ELECTRICITY FOR COOLING

Construction Assembly	Heating Cost $ Uninsulated	Heating Cost $ Insulated	Cooling Cost $ Uninsulated	Cooling Cost $ Insulated	Total Cost $ Uninsulated	Total Cost $ Insulated	Savings $/Yr.	Insulation $ Cost	% Return on Investment
Ceiling	620.98	32.92	308.61	17.66	929.59	50.58	879.01	501.35	175
Walls	185.03	47.71	108.70	29.30	293.73	77.01	216.72	297.40	73
Windows	124.58	70.69	55.66	31.56	180.24	102.25	77.99	680.50	11
Floor	505.37	60.87	none	none	505.37	60.87	442.50	2185.35	20
Totals	1435.97	212.19	472.97	78.52	1906.93	290.71	1616.22	3664.60	44

TOTAL SEASONS COST

ELECTRICITY FOR HEATING AND COOLING

Construction Assembly	Heating Cost $ Uninsulated	Heating Cost $ Insulated	Cooling Cost $ Uninsulated	Cooling Cost $ Insulated	Total Cost $ Uninsulated	Total Cost $ Insulated	Savings $/Yr.	Insulation $ Cost	% Return on Investment
Ceiling	1546.93	82.12	308.61	17.66	1855.54	99.78	1755.76	501.35	350
Walls	530.38	118.78	108.70	29.30	639.08	148.08	491.00	297.40	165
Windows	310.59	142.62	55.66	31.56	366.25	174.18	192.07	680.50	28
Floor	1258.10	151.36	none	none	1258.10	151.36	1106.74	2185.35	51
Totals	3646.00	494.88	472.97	78.52	4118.97	573.40	3545.57	3664.60	97

Using the cost figures for fuel and the cost of insulation, the economics of thermal insulation can be determined. In this case Metric Units will not be used as at present all the dollar values for energy cost and insulation costs are in English Units.

INSULATION ECONOMICS

Example 54

Not considering the savings in investment of heating and cooling equipment, calculate return on investment of insulation and storm windows of the house as described in Example 52. Determine these returns on investment for various fuels for the heating season only.

The ceiling is insulated with vapor retarder and 6″ glass fiber at $0.39 per sq ft × area of 1285.5 ft² is total cost of $501.35.

The wall is insulated with vapor retarder and 3 1/2″ glass fiber at $0.28 per sq ft × area 1062 ft² at a total cost of $297.00.

The cost of insulating under slab floor is $1.70 a sq ft × 1285.5 ft² = $2442.45.

The cost of the storm windows was given as $680.00.

The total cost of insulation and storm windows is $3920.00.

The previously calculated savings and return on investment were for the heating season only. If the house was air conditioned additional savings and greater return on insulation would result.

Example 54 — continued

Calculate the dollar savings obtained by insulation as used in house described in Example 52. The cost of insulation is the same as that stated in Example 53.

In this case the energy source is electricity and the cost will be the same as used in Example 53.

Cooling Season
Electric Costs
(Based on $0.046 average per kW-hr)

Const. Assem.	Uninsulated kW-hr	Uninsulated $Cost	Insulated kW-hr	Insulated $Cost	Savings $ Per Year
Ceiling	6,709	308.61	384	17.66	290.95
Wall	2,363	108.70	637	29.90	79.40
Windows	1,210	55.66	686	31.56	24.10
Totals	10,282	472.97	1707	78.52	394.45

These savings must be added to savings of heating season to obtain total yearly savings.

Example 54 — continued

Tabulate the total dollar savings for the heating season for the various fuels used for heating and cooling of the house as described in Example 52, using results obtained in Examples 53 and 54.

*Average cost of gas per million Btu's given in Table 208, based on prices third quarter 1981.

In addition to saving energy and its costs, thermal insulation also reduces the size of heating and cooling units. If the house is being constructed, or when units must be replaced, the insulation saves capital investment in the heating and cooling equipment.

In most instances the savings in ductwork is relatively small because air flow must be maintained even though heating and cooling requirements are less. However, this would not be true for large commercial buildings.

In all cases the installed costs of the heating and cooling equipment must be determined for the particular time, installation and location. However, to illustrate the importance of this capital dollar savings a representative installed cost of heating and air conditioning equipment is presented below:

Heating Equipment — Installed Cost
(not including duct work)

Maximum Heat Output English Units	Metric Units	Gas Furnace Installed Dollar Cost
60,000 Btu/hr	17,586 watts	$ 680.00
80,000 Btu/hr	23,448 watts	$ 700.00
100,000 Btu/hr	29,310 watts	$ 790.00
120,000 Btu/hr	35,172 watts	$ 880.00
140,000 Btu/hr	41,034 watts	$ 970.00
160,000 Btu/hr	46,896 watts	$1,050.00

The cost of air conditioning cooling equipment, compressor, air cooled heat exchanger, cooling coils etc. is:

Cooling Equipment — Installed Cost
(not including duct work)

Maximum Cooling English Units	Metric Units	Air Cooling Equipment Installed Dollar
1 Ton 12,000 Btu/hr	3,517 watts	$1,050.00
2 Tons 24,000 Btu/hr	7,034 watts	$1,375.00
3 Tons 36,000 Btu/hr	10,551 watts	$1,875.00
4 Tons 48,000 Btu/hr	14,069 watts	$2,325.00
5 Tons 60,000 Btu/hr	17,586 watts	$2,800.00
6 Tons 72,000 Btu/hr	21,103 watts	$3,100.00
8 Tons 96,000 Btu/hr	28,137 watts	$3,500.00

Example 54 — continued

Calculate the savings of capital investment in heating and cooling equipment obtained by the thermal insulation of the house as shown in Figure 163, based on the location and temperature conditions as given in Example 52. Also recalculated the return on investment of the insulation when this capital investment savings is subtracted from the capital cost of the insulation. Base calculations on gas heating equipment and gas costs.

As the heating unit must provide sufficient heat for the coldest weather the unit must be sized by the maximum heating requirements. Likewise the cooling unit must be determined by the maximum cooling requirement. The maximum heating requirements for this house, both uninsulated and insulated, were calculated in Example 52. These were:

Total Maximum Heat Loss

Uninsulated	130,135 Btu/hr (38,142 W)
Insulated	19,249 Btu/hr (5,642 W)

Thus the uninsulated house would require a heating unit of 140,000 Btu capacity costing $1,050.00. The insulated house would only require a 20,000 Btu heating capacity, or the smallest unit costing $680.00. *This represents a savings of $370.00 in capital investment.**

On the air cooling equipment, similar capital savings are obtained. The maximum cooling equipment for the house, both insulated and uninsulated, was calculated in Example 52.

Total Maximum Heat Loss Per Hour

Uninsulated	74,429 Btu/hr
Insulated	15,914 Btu/hr

The uninsulated house would require cooling equipment of 8 tons, 96,000 Btu/hr (28,137 watts), costing $3,500.00. The insulated house would require only 2 tons (24,000 Btu/hr) unit costing $1,375.00, *representing a savings of $2,125.00 in capital investment*.

The total savings in capital investment of heating and cooling equipment is thus $370.00 + $2,125.00 = $3,895.00.

Since it was estimated that the total cost of the installed insulation and storm windows in this house was $3,660.00, the correctly insulated house costs $234.00 less than the uninsulated house.

In addition the insulation saves gas and electrical energy costs totaling *$1616.22 per year*.

Thus there is no cost for the insulation, only a savings of capital investment of $230.00 and a fuel cost savings of $1616.22 per year. This is a $1616.00 yearly return on *investment. Thermal insulation saves money.*

COMMERCIAL BUILDINGS

Commercial and office buildings can also obtain major savings by using thermal insulation in roofs, walls and floors. This is illustrated in the following examples.

Example 55—*English and Metric Units*

Calculate the quantity of fuel oil used to heat and electrical energy to cool a room having an uninsulated and insulated flat roof. The roof is 100′0″ (30.48 m) × 100′0″ (30.48 m). The building is located in New York City. The construction is roof deck board on metal beams with 1_0 plaster board on bottom of beams. (See Construction Assembly IV). The insulation is 4″ (10.15 cm) thick cellular glass.

*A minimum unit of 60,000 Btu/hr is necessary to provide for losses due to door openings and closings and ventilation requirements.

Maximum Hourly Heat Loss — Uninsulated
Thermal Resistance R — Mean Temperature 40° (4.4°C)

Components	English Units	Metric Units
Outside surface	0.20	0.035
Built up roofing	0.33	0.050
Deck board	0.76	0.133
Metal deck	0.01	0.002
Gypsum board 1″ (2.54 cm) thick	0.87	0.155
Bottom surface	0.85	0.149
Thermal Resistance	3.02	0.532

Resistance of 4" (10.15 cm) thick cellular glass (k = 0.33 Btu in/ft² hr) $R = \frac{4}{0.33}$ = 12.12 English Units

In Metric Units thickness = 0.1015 m and k_m = 0.047; $R_m = \frac{0.1015}{0.047}$ = 2.159 Metric Units.

With the addition of the thermal resistance R for the roof is 15.14 English Units and R_m = 2.691 Metric Units.

MAXIMUM HOURLY HEAT LOSS

The maximum heat transferred per hour is based on heating/cooling data given in Table 24 for New York, N.Y. The winter design temperature is 12°F or − 11°C; the indoor temperature is 75°F (23.9°C) at Δt = 63°F or Δt = 34.9°C.

The uninsulated hourly roof heat loss, in English Units, is $\frac{63}{3.02}$ = 20.88 Btu per sq ft, hr and for total area 100′ × 100′ = 208,800 Btu/hr.

The uninsulated hourly roof heat loss, in Metric Units, is $\frac{34.9}{0.532}$ = 65.6 W/m², hr and for total area 30.48 m × 30.48 m = 60,945 W-hr.

The insulated hourly roof heat loss, in English Units, is $\frac{63}{15.14}$ = 4.16 Btu per sq ft, hr and for total area 100′ × 100′ = 41,600 Btu/hr.

This is a savings in the required heat capacity of 167,200 Btu/hr. The insulated hourly roof heat loss, in Metric Units, is $\frac{34.9}{2.159}$ = 16 W/m², hr and for total area 30.48 m × 30.48 m = 15,017 W-hr.

A savings of 45,928 W-hr by the insulation.

At today's cost this reduces the size of heating equipment resulting in approximately *$501.00 savings in capital investment.**

YEARLY HEAT LOSS

The yearly heat loss is based on degree hours. From Table 24 the degree F hours for New York is 121,200 and degree C hours is 67,333.

*Based on $300.00 per 100,000 Btu/hr output of furnace.

The uninsulated roof total yearly heat loss, in English Units, is $\frac{121,200}{3.02} = 40,132$ Btu/ft² and for total area 100′ × 100′ = 401,320,000 Btu.

The uninsulated roof total yearly heat loss, in Metric Units, is $\frac{67,332}{0.532} = 126,563$ W-hr/m² and for total area 30.48 m × 30.48 m = 117,581,719 W-hr.

The insulated roof total yearly heat loss, in English Units, is $\frac{121,200}{15.14} = 8,005$ Btu/ft² and for total area 100′ × 100′ = 80,050,000 Btu.

A savings of 321,270,000 Btu.

The insulated roof total yearly heat loss, in Metric Units, is $\frac{67,332}{2.691} = 25,021$ W-hr/m² and for total area 30.48 m × 30.48 m = 23,245,438 W-hr.

DOLLAR SAVINGS

It was stated that this building is heated with oil. If this oil has a heat content of 145,000 Btu/gal and the furnace is 80% efficient, then the available heat would be 116,00 Btu per gallon. The insulation of the roof (only) would save 321,270,000 ÷ 116,000 = 2,770 gallons of oil.

If the oil sold for $1.15 a gallon (delivered), *the saving in dollars per year would be $3,185.00.*

MAXIMUM HOURLY HEAT GAIN

From Table 24 the high temperature design of New York City is 93 °F (34 °C). However a horizontal flat surface will increase in temperature, at 1.0 emittance, an average of 88 °F (31.1 °C) between the hours of 12 noon to 3 pm. (Table 44). If the surface emittance is 0.92, then the surface temperature at the top surface of asphaltic roofing would be:

In English Units:

$$T = 93 + (88 \times 0.92) = 93 + 81 = 174\,°F$$

In Metric Units:

$$T_m = 33.9 + [(88 \times \frac{5}{9}) \times 0.92] = 33.9 + 45.0 = 78.9\,°C$$

The heat gain maximum per hour is based on the difference of surface temperature of the roof, as shown above, and indoor temperature of 75°F (23.9°C). The temperature difference is

174 °F − 75 °F = 99 °F (78.9 °C − 23.9 °C = 55 °C)

The uninsulated roof maximum hourly heat gain, in English Units is $\frac{(174-75)}{3.02} = \frac{99}{3.02} = 32.78$ Btu/ft², hr and for total area of 100′ × 100′ = 327,800 Btu/hr.

The uninsulated roof maximum hourly heat gain, in Metric Units, is $\frac{(78.9-23.9)}{0.532} = \frac{55}{0.532} = 103.38$ W/m², hr and for total area 30.48 m × 30.48 m = 96,046 W-hr.

The insulated roof maximum hourly heat gain, in English Units, is $\frac{99}{15.14} = 6.54$ Btu/ft², hr and for total area 100′ × 100′ = 65,400 Btu/hr.

A savings of needed air conditioning capacity of 262,400 Btu/hr or 21 tons.

The insulated roof maximum hourly heat gain, in Metric Units, is $\frac{55}{2.159} = 25.47$ W/m², hr and for the total area 30.48 m × 30.48 m = 23.666 W-hr.

A savings of needed air conditioning capacity of 72,379 W-hr.

At today's cost, this reduces the size of cooling equipment by 21 tons. This represents approximately a savings of $8,400.00 capital investment.*

YEARLY HEAT GAIN

From Table 24 the °F degree hours of cooling per year for New York, N.Y. is 15,552 °F-hr or 8,640 °C-hr. This does not take solar effects into account.

The uninsulated roof yearly heat gain, in English Units, is $\frac{15,552}{3.02} = 5,150$ Btu/ft² and for total area of 100′ × 100′ = 51,500,000 Btu.

The uninsulated roof yearly heat gain, in Metric Units, is $\frac{8,640}{0.532} = 16,240$ W-hr/m² and for total area of 30.48 × 30.48 m = 15,088,012 W-hr.

The insulated roof yearly heat gain, in English Units, is $\frac{15,552}{15.14} = 1,027$ Btu/ft² and for total area 100′ × 100′ = 10,270,000 Btu.

A savings of 41,227,873 Btu's.

The insulated roof heat gain, in Metric Units, is $\frac{8,640}{2.159} =$ 4,001 W/m² or for 30.48 m × 30.48 m area = 3,717,842 W-hr.

A savings of 11,370,169 W-hr

A cost of $0.046 per kW the savings in electrical energy is $523.00 and if the unit was 80% efficient the savings would be $523.00 ÷ 0.8 = $653.00 per year.

The savings obtained by the 4″ cellular glass roof insulation is:

$ Capital Investment Savings		**$ Savings Per Year — Energy**
Heating	$ 501.00	$3,185.00
Cooling	$3,185.00	$ 653.00
Total	$3,686.00	$3,838.00

*Based on estimated cost of $400.00 per ton.

ECONOMIC EVALUATION

In the past economic evaluations of thermal efficiency of buildings and residences have not been done. The reason was the energy in any form was very cheap and readily available. This is no longer true. Thus, buildings and homes must be designed to be more thermally efficient, otherwise the costs associated with their heating and cooling will cause them to be a financial failure.

To illustrate: The most efficient double insulating glass, used as building outside walls, has a thermal resistance R value of 1.73. (At 85 °F — Table 162) whereas a metal wall with 4″ of glass fiber insulation with inner 1/2″ gypsum board has a thermal resistance of 12.28 (at 85 °F — Table 118). Thus the economics of the percentage of glass area as compared to well insulated panel area should be considered in office and commercial building designs.

RETROFIT

At least 98% of all residences and buildings are underinsulated, based on today's cost of energy. Wherever possible insulation should be added to the roof structure, walls or floors. Because of the wide variations in types of construction, energy costs, types of fuels used, locations, and labor costs it is impossible to provide general recommendations for retrofitting.

However, caution should be taken with insulation added to existing structures. These are:

Be sure that insulation is not packed around electrical wires and outlet boxes. It is essential that electrical circuits be allowed to dissipate the heat generated by electric flow.

Where combustible insulations are used, such as organic foams or fibers these should be covered by fire protective materials such as gypsum board to prevent the rapid spread of accidental fire. However, non-combustible insulations of inorganic fibers or foam can serve not only as insulation but also as fire protective barriers.

Where fiberous insulation is added to existing outside walls all possible means should be used to vapor seal the *inside* wall. Vapor resistant paint and sealing caulking may be used to accomplish this vapor sealing. The outside wall should be *vented* to provide a means for vapor to escape to the outside wall. One common error is to seal all siding on the outside walls. This traps moisture in the wall causing it to rot and the moisture vapor that penetrates the wood siding causes the paint to blister. The ends of siding should be sealed but the *joint below the overlap should not be vapor sealed*. See Figure 21.

Residences or buildings of solid masonry construction present a major problem for adding insulation to reduce energy losses.

If the structure has an outer facing of brick or stone it is not practical to attempt to add the thermal insulation on this outer decorative surface. Thus, the insulation can only be added on the inside of the walls.

An ordinary brick faced 8″ cinder block (or clay tile) wall with plaster board on the inside has a thermal resistance R of 2.79 (English Units) or R_m = 0.491 (Metric Units). This is very inefficient for the conservation of energy.

To add insulation inside is not easy as the space for the added insulation must come from the inside of the room. For this reason the addition of 2″ to 3″ or more insulation within a space to be provided is difficult to provide. Yet, that is what would be desirable for energy efficient walls. However, reflective aluminum sheets can provide good thermal resistance in a relative thin space.

The existing walls (after window, door facings and baseboards have been removed) can be covered with aluminum foil sheathing with reflective surface facing inward toward room area. Over this 3/4″ × 2″ (1.9 cm × 5.00 cm) wood furring strips on 18″ (45.72 cm) centers shall be installed. A reflective sheathing with foil facing on both sides shall be installed so as to divide the air space. Plaster or gypsum board with reflective surface facing outward should then be secured to the furring strips. Facings and baseboards then are reinstalled. See Figure 164.

This construction forms a 3/4″ (1.9 cm) air space divided with a sheet-reflective on both sides and a reflective surface on both sides and reflective surface on each side of the space. With the 1/2″ gypsum plaster board it reduces the inside dimension of the room only 1 1/4″ (3.175 cm).

The 3/4″ space with reflective surfaces on both sides and divided by a reflective sheet provides a thermal resistance of 8.46 which is 2.5 times more resistance than the original masonry wall. The total thermal resistance of wall with this reflective shields, R = 12.25 (English Units) or R_m = 2.158 (Metric Units).

To illustrate the importance of this added thermal resistance, consider the following:

On a cold day at 0 °F:

The uninsulated wall would have a heat loss of $\frac{75 - 0}{2.79}$ = 26.9 Btu/ft², hr or 84.8 W/m²

The reflective insulated wall would have a heat loss of $\frac{75 - 0}{12.25}$ = 6.1 Btu/ft², hr or 19.23 W/m²

On a hot day with a 95 °F (shade side):

The uninsulated wall will allow energy entry at the rate of $\frac{95 - 75}{3.05} = \frac{20}{3.05}$ = 6.56 Btu/ft², hr or 20.68 W/m²

The insulated wall will allow energy entry at the rate of $\frac{95 - 75}{12.0} = \frac{20}{12}$ = 1.66 Btu/ft² or 5.25 W/m²

On the wall, or walls, facing the sun during summer, these energy savings will be 2 to 4 times greater.

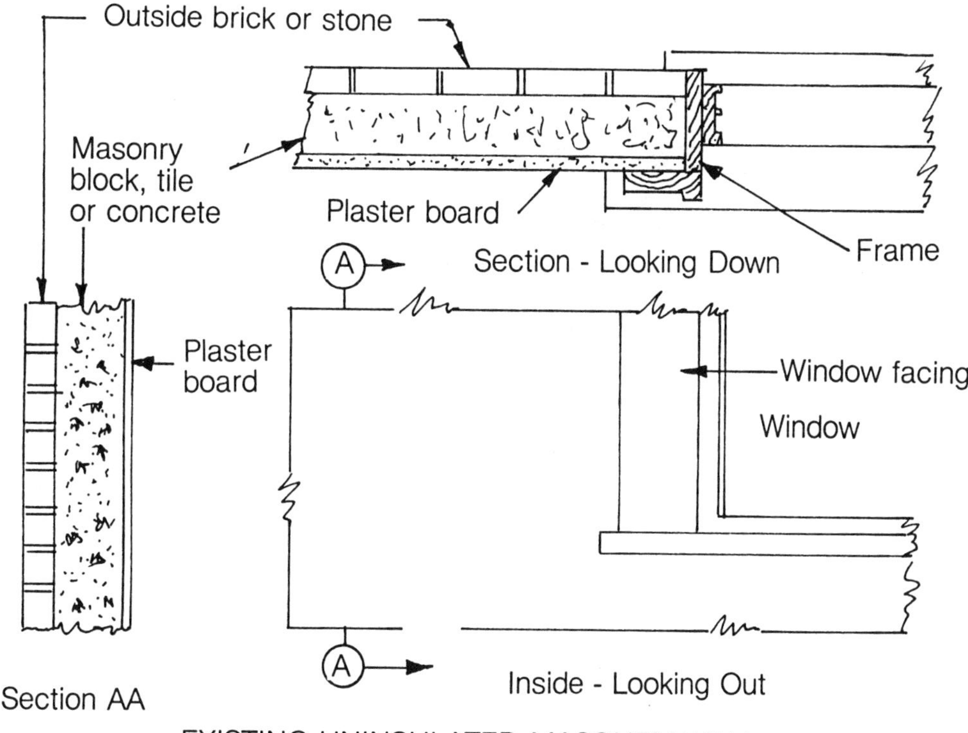

EXISTING UNINSULATED MASONRY WALL

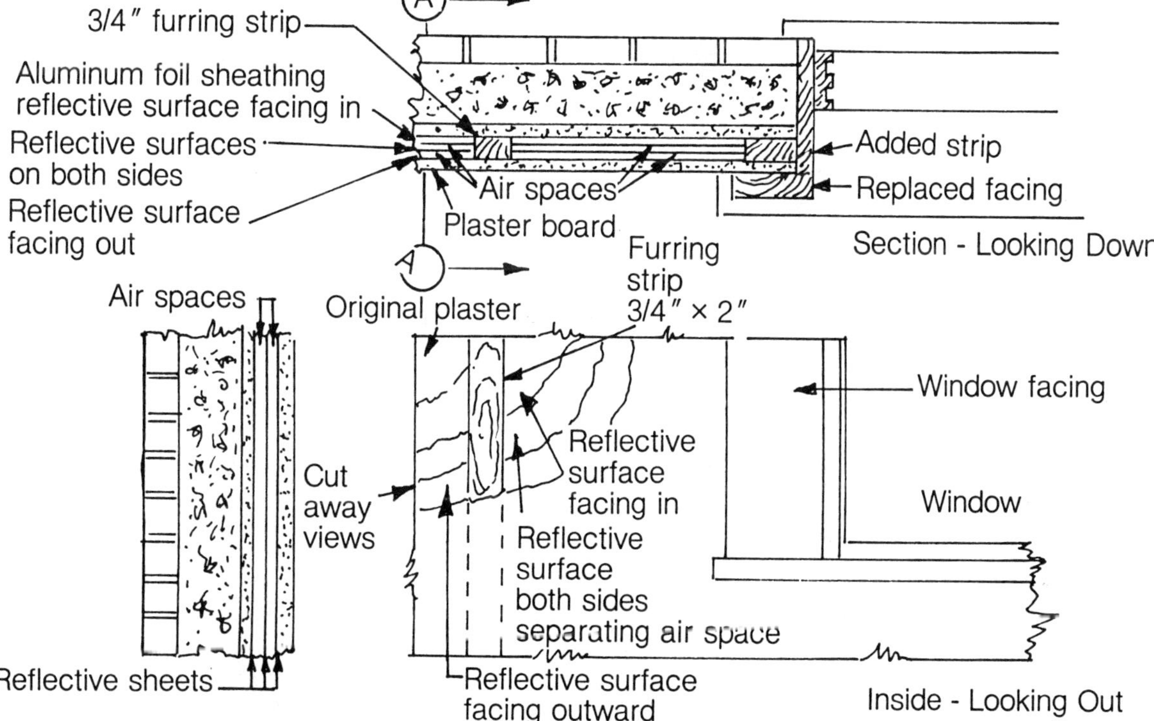

REFLECTIVE INSULATION ADDED TO EXISTING MASONRY WALL

Figure 164

GENERAL

Due to the rapidly rising energy cost the old solid masonry uninsulated residences are becoming too expensive to heat and cool.

Likewise, the all glass (so called modern) office building is already obsolete.

Correctly insulated structures in many cases cost less than the same structures not insulated, due to savings possible in capital investment of heating and cooling equipment.

In almost any residence or building thermal insulation provides a return on investment of 20% to 80%. As energy costs continue to increase, the return on investment of insulation in existing buildings will become greater.

Appendix A — Glossary and Definitions

References

List of Contributors

The following definitions sometimes are not as strict as those which would be applied by a physicist, but are phrased in the terms familiar to and commonly used by engineers and people in the construction industry. Users of this Glossary should note that, in many cases, dictionary definitions are not closely followed, but, instead, words are used in the connotations familiar to the industry, and the definitions are restricted to this limited usage.

Abrasion Force: A force caused by the rubbing together of two abutting objects, or by an external object rubbing against the surface of one of them.

Abrasion Resistance: The ability of a material to withstand abrasion without wearing away.

Absolute Humidity: The mass of water vapor present in a unit volume of atmospheric air.

Absolute Zero: The point at which all molecular motion ceases, with the resultant complete absence of heat. This point is −459.6° Fahrenheit, and −273.2° Celsius.

Absorbency: That property of a material which measures, in unit terms, its capacity to take up and assimilate liquids (from either the liquid or vapor form).

Absorptance: That property of a material which measures its total capacity to take up and assimilate liquids (from either the liquid or vapor form).

Absorption: That property of a material which enables it to take up liquids (in either the liquid or vapor form), especially by suction, and to assimilate them.

Acrylic Resin: Resins made by the polymerization of acrylic monomers, such as ethyl acrylate and methyl methacrylate.

Adherend: A body which is held to another body by an adhesive.

Adhesion—Dry: The property of a material which indicates its ability to bond to the surface to which it is applied and remain in place in service.

Adhesion—Wet: The property of a material which indicates its ability to stick to the surface to which it has been applied without sliding or falling off.

Adhesive: A substance capable of holding materials together by surface attachment.

Adsorption: That property of a material which enables it to retain liquids (in either liquid or vapor form) upon its surfaces—both internal and external.

Afterglow: The incandescence in a material after removal of an external flame or fire, or after an integral flaming has been extinguished.

Aggregate Size: The size of the coarsest particles in a cement, concrete, loose fill insulation, or similar product. The size is usually given in a percentage range of those particles which will or will not (with a maximum) pass through the mesh of a given size screen.

Alkalinity: The tendency of a material to have a basic (alkaline) reaction. The tendency is measured on the pH scale, with all readings above 7.0 alkaline, and below 7.0 acidic.

Alkyd Resins: Resins composed principally of polymeric esters (polyesters) in which the recurring ester groups are an integral part of the main polymer chain, and in which ester groups occur in most cross-links that may be present between chains.

Alligatoring: A term describing the action of a coating or mastic when it cracks into large segments. When the action is fine and incomplete it is usually referred to as "checking."

Ambient: Surrounding.

Ambient Temperature: The temperature of the medium, usually air, surrounding the object under consideration.

Anti-Abrasive Coating: A coating used on both of the mating surfaces of an insulation and its substrate, to prevent or retard the wearing away of the surface.

Appearance Covering: A material, or materials, used over insulation, the weather-barrier, or indoor covering, to provide the desired color or texture, for decorative purposes.

Application Temperature Limits: Temperatures between which it is usually safe to apply finishes.

Asbestos (Asbestos Fiber): A group of fibrous minerals which occur as small veins in the massive body of natural hydrous silicates of serpentine or amphobole, and have heat-, fire-, and solvent-resistant properties. Used as a reinforcement in the manufacture of mastics.

Asperigillus niger: One of the most common mold growths found on vegetable tanning vats and on leather, usually greenish or blackish in color.

Asphalt: A dark brown to black cementitious material, solid or semisolid in consistency, in which the predominating constituents are bitumens which occur in nature as such, or are obtained as residue in refining petroleum. The principal ingredient in asphalt mastics.

Asphalt Emulsion: A colloidal dispersion of petroleum asphalt in water. The emulsifying agent may be a colloidal clay or a chemical soap.

Batt: A piece of insulation, of the flexible type, cut into easily handleable sizes, square or rectangular in shape, usually 24" or 48" long and with a vapor-barrier on one side, and with, or without, a container sheet on the other side.

Bedding Compound: A plastic material, composed of various ingredients, spread on the substrate and used as a medium in which to embed the insulation layer. The compound acts as a cushion, anti-abrasive, and adhesive.

Binder: The cementing material used to bond fibers, flakes, or granular materials together.

Bitumen: Hydrocarbon material of natural or pyrogenous origin, or combinations of both, which may be liquid, semisolid, or solid, and which is completely soluble in carbon disulfide.

Blackbody: An ideal, perfect emitter and absorber of thermal radiated energy.

Blanket: Insulation, of the flexible type, formed into sheets or rolls, usually with a vapor-barrier on one side and with or without a container sheet on the other side.

Blanket Insulation: A flat flexible type of insulation formed into sheets or rolls. These blankets may be faced or coated on one or both sides.

Blanket Insulation—Metal Mesh: Blanket insulation covered with flexible metal mesh on one or both sides. Metal mesh secured to blanket with tie wires or other temperature resistant securements.

Bleeding: The diffusion of coloring matter through a coating from the substrate. (Such as bleeding of asphalt mastic through a topcoat of paint.)

Blister: Undesirable rounded elevation of the surface of a mastic whose boundaries may be either more or less sharply defined, somewhat resembling, in shape, a blister on the human skin. A blister may burst and become flattened.

Block: Rigid or semi-rigid insulation formed into sections, rectangular both in plan and cross section, usually 36"-48" long, 6"-24" wide, and 1"-6" thick.

Board: Rigid or semi-rigid insulation formed into sections, rectangular both in plan and cross section, usually more than 48" long, 24"-30" wide and up to 4" thick.

Boardy: Adjective applied to stiff inflexible mastic or coating resembling a board.

Body: The degree of consistency and internal cohesion. An increase in "body" indicates an increase in consistency, internal cohesion or both.

Bond: The union of materials by adhesives.

Bond Age: Time period elapsed since bonding specimens prior to testing.

Bond Strength: The unit load applied in tension, compression, peel, impact, cleavage, or shear required to break an adhesive assembly with failure occurring in or near the plane of the bond.

Bonding Time: Time period after application of adhesive, during which the adherends may be combined.

Brattice Cloth: A coarse plain weave heavy fabric of cotton or jute fiber used as an air curtain to control mine ventilation.

Breaking Load: That load, concentrated in the middle of the span, which will just break a measured sample of insulation under test, according to ASTM C-203 or C-446.

Bridging Ability: The ability of an insulation to span a gap in the substrate to which it is applied, or the ability of a weather-barrier, or vapor-barrier, to span a gap in the insulation.

British Thermal Unit: Originally the amount of heat necessary to raise one pound of water one degree Fahrenheit at standard atmospheric pressure. Now by international agreement, the Btu has been established as 778.26 ft lbs.

Bubble: An internal void or a trapped globule of air or other gas in a mastic application.

Build—Dry: The dry thickness attained by a coating or cement by the application of a given number of coats.

Build—Wet: The wet thickness attained by a coating or cement by the application of a given number of coats, combined with its ability to be applied to that wet thickness on a vertical surface without slipping, sliding, or sagging.

Canvas: A light, plain weave, coarse, cotton cloth with hard twisted yarns, usually not more than 8 oz per square yard. See "Duck."

Capillarity: That property of a material which will enable it to suck a liquid up into, or through itself, with the driving force of the liquid being its surface tension.

Cast Film: A film made by depositing a layer of a coating or adhesive onto a surface, stabilizing this form, and removing the film from the surface.

Caulking Compound: A soft, plastic material, consisting of pigment and vehicle, used for sealing joints in buildings, and other structures, where normal structural movement may occur.

Celsius: The temperature measuring scale (formerly Centigrade) in which the ice point of water is taken at 0° and the steam point at 100°. The absolute zero on this scale is −273.2°.

Cellular Plastic: A plastic whose apparent density is decreased substantially by the presence of numerous cells disposed throughout its mass.

Cement: See Adhesive.

Cement—Insulating: See Insulating Cement.

Cement—Fabrication: See Fabrication Cement.

Cement—Finishing: See Finishing Cement.

Cement—Finishing and Insulation: See Finishing and Insulation Cement.

Centigrade: See Celsius.

Chaetomium globosum: A type of mold growth found chiefly on cellulosic materials, and particularly on paper, usually black, and characterized by long stiff hairs variously straight, branched or curled.

Chalking: Dry, chalk-like appearance or deposit on the surface of a weathered finish.

Checking: A defect in a coated surface characterized by the appearance of fine fissures in all directions. Designated as "surface checking" if superficial, or "through checking" if extending deeply into, or through to an adjoining surface.

Chemical Reaction: The property of a material which measures its tendency to chemically combine (or react) with other materials which may come into contact with, or be absorbed by it.

Chemical Resistance: Capability of withstanding limited exposure to designated acids, alkalies, and salts and their solutions.

Chemically Foamed Plastic: A cellular plastic produced by gases generated from chemical interaction of constituents.

Chlorinated Solvent: An organic chemical liquid characterized by a high chlorine content and used in coating products to impart non-flammability.

Closed-Cell Foamed Plastic: A cellular plastic in which there is a predominance of non-interconnecting cells.

Coal Tar: Tar produced by the destructive distillation of bituminous coal.

Coating: A liquid, or semi-liquid, protective finish capable of application to thermal insulation or other surfaces, usually by brush or spray, in moderate thickness, less than 30 mils (0.030").

Cobwebbing: A phenomenon observed during spray application, characterized by the formation of web-like threads in addition to the usual droplets leaving the nozzle of the spray gun.

Coefficient of Expansion (Contraction): The increase (decrease) in length of a material, one unit long, due to the increase (decrease) of its temperature one degree. In the English System the unit is usually one foot, and the temperature Fahrenheit.

Cold: The absence of heat.

Color: The aspect of appearance dependent upon the spectral composition of the incident light, the spectral reflectance or transmittance of the object, and the spectral response of the observer.

Hue: The attribute by which a perceived color is distinguished, as red, yellow, green, blue, purple, or a combination of these. (White, gray, and black colors possess no hue.)

Lightness: The attribute by which a perceived color is judged to be equivalent to a member of the continuous series of grays ranging from black to white.

Saturation: The attribute by which a perceived color is judged to depart from gray of equal lightness toward a pure hue.

Combustible: Capable of uniting with air or oxygen in a reaction initiated by heating, accompanied by the subsequent evolution of heat and light. Capable of burning.

Combustibility: That property of a material which measures its tendency to burn. It is normally expressed in the arbitrary terms of "Flame Spread Index" and "Smoke Density Index," according to ASTM Test E-84.

Combustion: A chemical process, usually involving oxygen, which produces light and heat, either as glow or flames.

Compaction or Settling: The property of the blankets, or batts, which measures their change in density and thickness resulting from loading, or vibration, with a resultant change of thermal efficiency.

Compaction Resistance: That property of a fibrous or loose fill material to resist compaction under load or vibratory conditions.

Compressive Force: A force tending to cause the fibers of the material to be pressed or squeezed into more intimate contact, as opposed to a tensile force which tends to pull them apart.

Compressive Strength: That property of a material which enables it to resist any change in dimensions when acted upon by a force tending to squeeze or shorten it.

Concentrated Load: A load applied at a point, or over a very small percentage of the possible load bearing area.

Condensate-Barrier: A material, normally used as an inner lining for the metal jacket weather-barrier of an insulation installation, which will bar the alkaline condensate, which normally tends to form on the inner surface of the metal jacket, from contact with it.

Condensation: The act of water vapor turning into liquid water upon contact with a surface at a lower temperature than the dew point of the vapor.

Conditioning: The exposure of a material to the influence of a prescribed atmosphere for a stipulated period of time, or until a stipulated relation is reached between material and atmosphere.

Conductance: See Thermal Conductance.

Conduction: The transfer of energy within a body, or between two bodies in physical contact, from a higher temperature region to a lower temperature region by tangible contact.

Conductivity: See Thermal Conductivity.

Consistency: The resistance of a non-Newtonian material to deformation. Note: Consistency is not a fundamental property, but is comprised of viscosity, plasticity, and other phenomena.

Contact Adhesive: An adhesive which is apparently dry to the touch, and which will adhere to itself instantaneously upon contact; also called contact bond adhesive, or dry bond adhesive.

Convection—Forced: See Forced Convection.

Convection—Natural: See Natural Convection.

Copolymer: See Polymer.

Corrosion Effect: The wearing away, or destruction, of a substrate caused by acid or alkaline reactions between materials contained in the insulation and the substrate.

Covering—Appearance: See Appearance Covering.

Covering Capacity—Dry: The volume occupied in square area, expressed as thickness times the area of the material after being dried.

Covering Capacity—Wet: For materials which are mixed with water before application—the volume occupied in square area, expressed as thickness, for a unit of the dry material before being mixed with water. For materials which are pre-mixed (in cans)—the volume occupied in square area expressed as thickness of unit volume as received.

Crater: Small, shallow, crater-like surface imperfection in finishes.

Crawl Space: The space, usually 2-4' between the original undisturbed grade level and the bottom of the floor construction above, or the space between the top of the ceiling joists and the bottom of the roof rafters above. Neither space is ever high enough for a man to stand upright, hence the name "crawl" space.

Creep: The dimensional change with time of a material under load apart from, and following, the initial instantaneous elastic or rapid deformation.

Cryogenic: Pertaining to the extremely low temperatures, such as the liquefaction points of gaseous elements, usually below −150° F (−101 ° C) on down to absolute zero.

Cure: To change the properties of a plastic or resin by chemical reaction, which for example may be condensation, polymerization, or addition; usually accomplished by the action of either heat, catalyst, or both, and with or without pressure.

Curing Agent: An additive incorporated in a coating or adhesive resulting in increased chemical activity between the components, with an increase or decrease in the rate of cure.

Curing Time: The length of time necessary to affect a cure of a plastic or resin by chemical reaction. See Cure.

Curved Segmental Block: A piece of rigid insulation, rectangular in plan, and the sector of a tube, in cross section, molded, or cut from block of the proper thickness.

Cut-Back Products: Petroleum or tar residuums which have been blended with distillate solvents.

Cutting Force (Shearing Force): A force tending to cleave the object to which it is applied.

Dauber: A spatula-like instrument (such as the Foster Gooper) for smear application of adhesives.

Deflection: The distance by which an object deviates from its original axis or shape after the application of a load.

Delamination: The separation of the layers of material in a laminate.

Density—Apparent: The weight of a unit volume of a material in its manufactured state, including all voids. Usually expressed in pounds per cubic foot.

Density—Real: The weight of a unit volume of a material, excluding all voids. Usually expressed in pounds per cubic foot.

Deterioration: A permanent change in physical properties evidenced by impairment of these properties.

Dew Point: The temperature at which the quantity of water vapor in a material would cause saturation, with resultant condensation of the vapor into liquid water by any further reduction of temperature.

Dielectric Strength: The rate of electric stress (volts per cm) required to puncture a film of coating.

Diffusivity: The time rate of temperature change within a body, or between two of its surfaces.
Dilatancy: The property of some highly filled coating materials of increasing in viscosity when the system is subjected to a distortional stress.
Dimensional Stability: That property of a material which enables it to hold its original size, shape, and dimensions when subjected to aging, load, heat, cold, moisture, or cutting.
Dimensional Trueness: That property of a material which enables it to be manufactured or fabricated, without internal stresses causing it to depart from size or shape with the passage of time.
Discoloration: Any change from the initial color. A lack of uniformity in color, where it should be uniform over the whole area.
Dispersion: A heterogeneous system in which a finely divided material is distributed in another material. Note: A dispersion is usually the distribution of a finely divided solid in a liquid or a solid; for example, pigments or fillers in coatings. A dispersion of a solid in a liquid only is a suspension.
Doctor Blade: A device consisting of a fixed metal blade or blades with which coatings or adhesives may be applied with a scraping action.
Dropping Resistance: That property of a material which enables it to withstand being dropped without fracture, breaking, or crumbling.
Dry: To change the physical state by the loss of solvent constituents by evaporation, absorption, oxidation, or a combination of these factors.
Drying Oil: An oil which possesses, to a marked degree, the property of readily taking up oxygen from the air and changing to a relatively hard, tough, elastic substance when exposed to the air in a thin film.
Drying Time (Adhesives): Time elapsed since bonding at the optimum time when no further increase in bond strength is realized.
Drying Time (Finishes): Time elapsed after which no further significant changes take place in appearance or performance properties, due to drying.
Duck: A compact, firm, heavy, plain weave cotton fabric.
Ductility: That property of a material which enables it to undergo large deformations without rupture.
Durability: As applied to finishes, it is the lasting quality or permeance in service with particular reference to deterioration. May be related directly to an exposure condition.

Efflorescence (Bloom): A white powdery substance occurring on the surface of coated insulation products, caused by the migration of soluble salts from the insulation, followed by precipitation and carbonation.
Elasticity: The tendency of a material to recover its original size and shape after deformation.
Elastomer: A material which at room temperature can be stretched repeatedly to at least twice its original length and, immediately upon release of the stress, will return with force to its approximate original length.
Elongation (Extensibility): The extension between bench marks produced by a tension force applied to a specimen. It is expressed by a percentage of the original distance between the marks on the unstretched specimen (also known as Stretch).
Emissivity: The total heat lost per unit of time through a unit area of the surface of a body.
Emittance: The ratio of the total heat lost per unit of time through a unit area of the surface of a body to the total heat which would be lost in the same unit of time through the same unit area of a perfect blackbody.
Emittance—Directional: The ratio of the total heat transferred per unit of time and unit of surface area in a particular direction to that from a unit of blackbody surface of same area, temperature and conditions.
Emittance—Hemispherical: The ratio of the total heat radiant flux density from a body to that of a blackbody under same temperature conditions.
Emittance—Spectral: The emittance ratio based on energy emitted per unit wavelength interval.
Emulsion: Strictly stated, a colloidal suspension of one liquid in another. Often loosely used for "dispersion."
Energy: The measure of the amount of work a body (or system of bodies) can do, by virtue of its motion, or position, against forces applied to it. It is also a measure of the work it can do by virtue of its chemical composition, or as a result of having been heated. See Mechanical Potential Energy, Mechanical Kinetic Energy, and Internal Energy.
Epoxy Resins: Resins made by the reaction of epoxides or oxiranes with other materials such as amines, alcohols, phenols, carboxylic acids, acid anhydrides, and unsaturated compounds. Epoxy compounds may be cured at ambient temperature to form finishes which are highly resistant to solvents and chemicals.
Evaporation: The loss, from a liquid, through the transformation of a portion of it to its vapor state, caused by the application of heat.
Expandable Plastic: A plastic which can be made cellular by thermal, chemical, or mechanical means.
Expansion Ratio: The ratio of the Coefficients of Expansion of any two abutting materials.
Expansion—Wet State to Cured State: The property of a material which measures the difference in volumetric change in poured, sprayed, or foamed in place organics.
Exposure: The action by which a protective finish is exposed to the weather elements.

Facing: A thin layer, usually factory applied, on the surface of an insulating panel, variously acting as a vapor-barrier, weather-barrier, protector from damage, and a decorative coating.
Fading: Any lightening of an initial color.
Fahrenheit: The temperature scale of the English system of units in which the ice point of water is assigned the value of 32° and the steam point the value of 212°, with 180 even divisions between, and corresponding divisions above and below. Absolute zero on this scale is −459.6°.
False Body: Thixotropic flow property of a suspension or dispersion coating. When a compound "thins down" on stirring, or "builds up" on standing, it is said to exhibit false body.
Fatigue Resistance: That property of a material which enables it to be flexed back and forth, with reversal of stress each time, without rupture. Usually expressed as the number of cycles without rupture.
Felt: An insulation material composed of fibers of one or more kinds, in which they are interlocked, and have been compacted under pressure.
Fiber (Fibrated): See Asbestos.
Fill (in Fabrics): Yarn running from selvage to selvage at right angles to the warp yarn. See Pick Count.
Fill Insulation: An insulating material consisting of loose granules, fibers, beads, flakes, etc., which must be contained, and is usually placed in cavities of some description.
Filler: A relatively inert material added to a mastic or coating to modify its strength, permeance, working properties, or other qualities.
Fillet: That portion of an adhesive, mastic coating, or sealant which fills the corner, or angle, where two adherends or surfaces are joined.
Film: An optional term for sheeting having nominal thickness not greater than 0.010".
Film—Wet: The freshly applied layer of mastic, coating, or adhesive before curing or drying has occurred.
Finishing Cement: A mixture of fibrous or powdery materials, or both, with suitable binders, that, when mixed with suitable proportion of water will develop a plastic consistency and can be used on the surface of insulations to provide a medium-hard to hard, even finish.
Finishing and Insulating Cement: A mixture of fibers and binders, water-mixed to a plastic mass on the job, and used as a finishing cement, and for an insulation for situations where only a small insulating effect is desired.
Fire Endurance: That property of a material which measures the elapsed time during which it continues to exhibit resistance to fire, under specified conditions of test and performance.
Fire Point Temperature: The lowest temperature of a material at which it gives off vapor, which, when combined with air near its surface, forms an ignitable mixture at a rate sufficient to support combustion continuously after the external ignition source is removed.

Fire Resistance: That property of a material which enables it to resist decomposition or deterioration when exposed to a fire.

Fire Resistive: Having fire resistance.

Fire Retardance: That property of a material which delays the spread of fire, either through or over itself.

Fire Retardant: Having fire retardance.

Fish-Mouth: A transverse gap between layers of sheet materials caused by warping or bunching of one or both layers.

Flame Spread: The rate, expressed in distance-time, at which a material will propagate flame on its *surface.* As this is a difficult property to measure in time and distance, the measure is now by *flame spread index* to enable the comparison of materials by test methods.

Flame Spread Requirements G.S.A.: United States General Services Administration, Public Buildings Service, Guide Specification, Section 301-1,

"For all concealed horizontal and vertical ductwork, and for all concealed piping, covering materials and accessories shall have a fire hazard rating not to exceed 25 for flame spread and 50 for fuel contributed and smoke developed. Ratings shall be determined by Underwriters' Laboratories, Inc. Method of Test of Surface Burning Characteristics of Building Materials or by the method in Interim Federal Standard No. 00136, Flame-Spread Properties of Materials."

Flame Resistance: That property of a material which enables it to resist decomposition or deterioration when exposed to flame.

Flame Retardance: That property of a material which delays the spread of flame either through or over itself.

Flammable: That property of a material which permits it to oxidize rapidly and release heat of combustion when exposed to flame or fire, and allows continuous burning after the external ignition source is removed.

Flammable (or Explosive) Limits: In the case of solvent vapors which form flammable mixtures with air or oxygen, there is a minimum concentration of vapor in air or oxygen below which propagation of flame does not occur on contact with a source of ignition. There is also a maximum proportion of vapor or gas in air, above which propagation of flame does not occur. These boundary-line mixtures of vapor with air, which, if ignited, will just propagate flame, are known as the "lower and upper flammable," usually expressed in terms of percentage by volume of vapor in air.

Flammable (or Explosive) Range: The range of combustible vapor and air mixtures between the upper and lower flammable limits is known as the "flammable range," sometimes referred to as the "explosive range."

Flammability Index: The comparison of the flammability of a material with that of an arbitrarily chosen standard material. Expressed as an *index.*

Flash-Ignition Temperature (Flash Point): The lowest temperature of a material at which it gives off vapor, which, when combined with air near its surface, forms an ignitable mixture.

Flashing: A thin strip of material, usually, but not always, metal, inserted at the junction of two materials, or two parts of a material, to divert liquid water to a specific direction.

Flexibility: That property of a material which allows it to be bent (flexed) without loss of strength.

Flexural Resistance: That property of a material which enables it to resist bending (flexure).

Flexural Strength: That property of a material which measures its resistance to bending (flexing). Usually expressed in pounds per square inch.

Flow: Movement of an adhesive, coating, or sealant during the application process before set has occurred. See Sag.

Fly Ash: Extremely fine inorganic dust from combustion of solid fuel in large power boilers.

Force: The potential for movement, or relative movement, of physical mass or temperature.

Forced Convection: The movement of a body, with its associated energy, from one location to another, with the rate of this movement increased by some outside influence such as wind, or a fan.

Freezing Point: The temperature at which a liquid will change from the liquid to the solid stage, with a simultaneous loss of energy.

Freeze-Thaw Resistance: The property of a material which permits it to be alternately frozen and thawed—through many cycles—without damage from rupture or cracking.

Friction: The resistance to relative motion between two bodies in contact.

Fuel Contribution: Flammable by-products of fire generated by, and emitted from, a burning object.

Gallon Weight: Weight of the standard U.S. volumetric unit of 231 cubic inches.

Galvanic Corrosion: Pitting or eating away of one of the metals when two metals of different electric potential are in direct contact, or electrically connected by an electrolyte.

Gel: A semisolid system consisting of a network of solid aggregates in which liquid is held. Note: Gels have very low strength and do not flow like a liquid. They are soft, flexible and will rupture under their own weight unless supported externally.

Geodesic: Of, or pertaining to the shape of the earth; therefore, cut or made in the shape of a sphere.

Gloss: A term used to express the shine, sheen, or luster of a dried film.

"Gooper": A special disposable type of dauber designed for easy smear application of adhesives. Registered trade mark, Benjamin Foster Co.

Graybody: A body having the same spectral emittance (less than unity) at all wave lengths.

Grit: Hard, relatively large inclusions in a coating composition.

Hanger (Insulation): A device to carry the weight of insulation in which the load carrier is positioned *above* the insulation. See Support.

Hardness: See Indentation Hardness.

Hazard—Fire: The susceptibility of a material to ignition and consequent potential for spread of flame and release of toxic gases and smoke.

Heat: The result of molecular motion and interacting forces. Energy in transient form.

Heat Capacity: The amount of heat required to raise a unit mass of a material 1 degree in temperature.

Heat—Latent: See Latent Heat.

Heat Conduction: See Conduction.

Heat Flux: The time rate of heat flow per unit area, in a direction perpendicular to the isothermal shield and spacer surfaces.

Heat Convection: See Convection.

Heat Radiation: See Radiation.

Heat Retardance: That property of a material which enables it to delay the flow of heat from a hot surface to a cold surface of a body.

Heat Transfer Cement: A soft, plastic material, which under use quickly solidifies to a rock-like hardness, having a high coefficient of heat transfer, which is used to bond tubes, or other heat-conveying devices, to the pipe or equipment to which it is desired to transfer the heat.

Heat Resistance: That property of a material which enables it to withstand heat without deterioration or failure.

Hexagonal Wire Mesh: Generic term for poultry netting, chicken wire, etc., usually made from pregalvanized wire woven in 1" mesh size. Also available in post-galvanized and rustless metal alloys.

Holiday: In a coating application a place not covered by coating compound.

Humidity: The condition of the atmosphere in respect to water vapor. See also Humidity—Absolute, Humidity—Relative.

Humidity—Absolute: See Absolute Humidity.

Humidity—Relative: See Relative Humidity.

Hydrocarbon Resins: Resins composed of carbon and hydrogen alone.

Hygroscopicity: That property of a material which enables it to readily absorb and retain water in either its liquid or vapor state.

Ignition: The initiation of combustion as evidenced by glow, flame, or explosion.

Ignition Temperature: The minimum temperature to which a solid, liquid, or gas must be heated in order to initiate or cause self-sustained combustion independently of the heating element.

Impact: The single instantaneous stroke of a body in motion against another either in motion or at rest.

Impact Force: The force resulting from an impact of one body upon another.

Impact Resistance: Capability of a finish to withstand mechanical or physical abuse under severe service conditions. Resistance to blows, bumps, and shocks incident to plant operation.

Indentation Hardness: That property of a body which enables it to resist, to a measurable degree, the tendency of a moving body, or a force, to dent its surface.

Indoor Covering: A material which serves to protect insulation from mechanical damage and wear and tear on indoor applications.

Insulating Mastic: A premixed soft, plastic material of various consistencies, applied by spray, trowel, brush or palm, which possesses some insulating value in addition to its other vapor- or weather-barrier characteristics.

Insulation Finish: A material, or materials, applied to the insulation to provide the final contour and a smooth even finish, and to strengthen the outer surface.

Insulation—Thermal: See Thermal Insulation.

Insulating Cement: A mixture of various fibers and binders, to be mixed with water to form a soft, plastic mass, and used to insulate small irregular surfaces and fill the cracks and crevices between the units insulating larger surfaces. (See Chapter 4 for types and properties.)

Insulation Cover: The cover for a flange, pipe fitting, or valve, composed of the specified thickness insulating material, and preformed into its proper shape before application.

Insulation—Fill: See Fill Insulation.

Insulation—Loose: See Loose Insulation.

Internal Energy: Energy stored within a body such as a gas, liquid, or solid, or within any material from which it can be released by chemical reaction.

Intumescence: The process of swelling or expanding on fire exposure to form a cellular charred layer which insulates and retards flaming.

Isocyanate Resins: Resins made by the condensation of organic isocyanates with other compounds. See Urethane.

Jacket: A covering placed around an insulation to protect it from mechanical damage, and, insofar as it is intrinsically able, from weather, water, ultra violet light, etc.

Joint: The location at which two mating surfaces are in juxtaposition.

"k" Factor: See Thermal Conductivity.

Kelvin: A temperature scale on which the absolute zero point is taken as 0°, the ice point of water 273.2°, and the steam point 373.2°, with corresponding divisions on up the scale. The degree divisions of Kelvin are the same as those of Celsius.

Lag: A long, narrow piece of rigid insulation, rectangular in plan, trapezoidal in cross section, molded, or cut from block of the proper thickness.

Lagging n.: An insulation layer, on a cylindrical surface, composed of lags.

Lagging v.: Action of covering something, as a boiler with insulation, now generally used to mean covering installed pipe insulation with canvas, using a combination adhesive-coating. This is a common, but incorrect usage of the word.

Laminate: A product made by bonding together two or more layers of material or materials.

Lap Adhesive (Lap Cement): The adhesive material used to seal the side and end laps of insulation jackets.

Latent Heat: The energy which must be added to a liquid at its boiling point to change its state from liquid to gas at a constant pressure. Conversely, it is the energy which must be removed from a gas at the boiling point of a liquid to change its state to a liquid, at a constant pressure. The latent heat to or from a liquid or gas causes a change in state without change in temperature.

Leno Weave: See Weave.

Linear Expansion: Increase in length in an axial or lengthwise direction, due to increase in temperature of the material.

Load: The force exerted on a material, or a support, by other material placed upon it, or from some external object.

Load—Breaking: See Breaking Load.

Load—Superimposed: See Superimposed Load.

Loose Insulation: Insulation in the form of loose granules, fibers, beads, flakes, etc., which must be contained and is usually placed in cavities of some description.

Low Temperature Bending: That property of a material which allows it to be bent (flexed) without rupture at low temperatures, with complete recovery to line and shape upon removal of the force causing the bending.

m.a.c. (Maximum Allowable Concentration): See Threshold Limit Value.

Mastic: A relatively thick consistency protective finish capable of application to thermal insulation or other surfaces, usually by spray or trowel, in thick coats, greater than 30 mils (0.030").

Mat: A piece of insulation, of the semi-flexible type, cut into easily handled sizes, usually square or rectangular in shape, composed of fibers of one or more kinds, in which the fibers are in random arrangement and are compacted and held together by an adhesive. This material is used both as a reinforcement and an insulation, thus, the word is often used indiscriminately. Care should be taken to use the proper qualifying adjective, i.e., "reinforcing" or "insulating," when using the term.

Mean: The arithmetical average of a set of numbers.

Mechanical Abuse: See Impact Resistance.

Mechanically Foamed Plastic: A cellular plastic in which the cells are formed by the physical incorporation of gases.

Mechanical Kinetic Energy: Energy possessed by a body by virtue of the relative motion between it and other parts of a system.

Mechanical Potential Energy: Energy possessed by a body by virtue of its vertical distance above a horizontal plane.

Median: If the numerical values for a given property are arranged in ascending order, the median is (1) the middle value of the series if the number of values is odd, or (2) the mean of the two middle values if the number of values is even.

Melting Point: The temperature at which a material will change from the solid to the liquid state by the application of additional heat.

Membrane Reinforcement: Woven or non-woven fabrics used for saturation and embedment in mastic and coating applications to provide strength, continuity, and impact resistance.

Mil: A unit used in measuring thickness, being 0.001". (British equivalent: thou.)

Mildew: Any discoloration caused by parasitic fungi on vegetable matter or other substances. See Mold.

Mineral Spirits (Petroleum Spirits): A refined petroleum distillate with volatility, flash point, and other properties making it suitable as a thinner and solvent in coatings, mastics, and similar products.

Moisture Vapor: See Water Vapor.

Mold: A growth produced by fungi on various forms of organic matter, especially when damp or decaying. See Mildew.

Mold and Mildew Resistance: That property of a material which enables it to resist the formation of fungus growths under unfavorable conditions of temperature and humidity.

Monomer: A relatively simple compound which can react to form a polymer.

Mud Cracking: A form of alligatoring, or stress cracking, which may occur during drying in thick applications of water-base mastics or coatings, usually caused by shrinkage from excessive volatile content.

Muslin: Any of various plain weave coarse cotton fabrics having a weight per square yard usually not more than 2 oz.

Natural Convection: The movement of a body, with its associated energy, from one location to another.

Naphtha (Petroleum): A generic term applied to refined, partly refined, or unrefined petroleum products and liquid products of natural gas, not less than 10 percent of which distills below 347° F (175° C) and not less than 95 percent of which distills below 464° F (240° C), when subjected to distillation in accordance with ASTM Method D86, Test for Distillation of Petroleum Products. The "naphthas" used for specific purposes, such as manufacture of rubber cements, paints and varnishes, etc., are made to conform to specifications which may require products of considerably greater volatility than set by the limits of this generic definition.

Netting: Interwoven wires of some metal, usually either galvanized steel or Monel, woven into either a rectangular, square, or hexagonal pattern, used as a reinforcement in the application of insulation on large surfaces.

Newtonian (Simple) Liquid: A liquid in which the rate of sheer is proportional to the shearing stress.

Non-Newtonian (Complex) Liquid: A liquid in which the rate of shear is not proportional to the shearing stress.

Noncombustible: A material which will not contribute fuel or heat to a fire to which it is exposed.

Noncombustibility: That property of a material which prevents it from contributing fuel or heat to a fire to which it is exposed.

Nonflammable: That property of a material which prevents it from oxidizing rapidly and releasing heat of combustion when exposed to fire or flame.

Non-Volatile Content: That portion of a material which does not evaporate at ordinary temperatures.

Nylon Resins: Resins composed principally of a long-chain synthetic polymeric amide which has recurring amide groups as an integral part of the main polymer chain.

Odor Emission: That property of a material which indicates its relative scent, smell, or fragrance.

Opacity: The degree of obstruction to the transmission of visible light.

Open-Cell Foamed Plastic: A cellular plastic in which there is a predominance of interconnected cells.

Open Time Maximum (Adhesives): That open time which corresponds to 90 percent of the optimum strength after the maximum value has been reached.

Open Time Minimum (Adhesives): That open time which corresponds to 90 percent of the optimum strength prior to reaching the maximum value.

Open Time Optimum (Adhesives): That open time which gives the optimum strength at a bond age of 24 hours.

Open Time Range (Adhesives): Time spread between minimum open time and maximum open time.

Orange-Peel: Uneven surface of a spray-applied coating, somewhat resembling an orange peel.

Panel: An insulation, prefabricated at the factory into a rectangular shape, of relatively thin material, usually with a facing of some material on one of its rectangular surfaces, erected and secured as one piece. Panels are made in almost any erectable size, and in thicknesses up to about 2″.

Peel Resistance: That property of a material which imparts to it the maximum bond to its substrate, to enable it to resist any external forces tending to peel them apart.

Penetration: The consistency of a mastic material, expressed as the distance that a standard cone vertically penetrates a sample of the material under known conditions of loading, time, and temperature. The units of penetration indicate hundredths of a centimeter.

Penetrometer: The instrument used for determinating penetration values of consistency.

Perm: The accepted unit of Water Vapor Permeance. Is expressed as 1 grain per square foot, hr, inch of mercury.

Perm-Inch: The accepted unit of Water Vapor Permeability. Is expressed as 1 grain per square foot, hour, inch of mercury, inch of thickness.

Permanence: The property of a finish describing its resistance to appreciable changes in characteristics with time and environment.

Permeability: See Water Vapor Permeability.

Permeance: See Water Vapor Permeance.

pH: The negative logarithm of the hydrogen ion concentration. A solution at pH 7 is neutral; lower numbers indicate increasing acidity, higher numbers, increasing alkalinity.

Phenolic Resins: Resins made by the condensation of phenols, such as phenol and cresol, with aldehydes.

Phoma pigmentovora: A form of mold which attacks freshly applied coating surfaces, characterized by pink to purple spots up to 1″ or 2″ in diameter.

Pick Count (Fabrics): The number of fill yarns per inch of fabric.

Pigment: The fine solid color particles used in the preparation of colored coatings and substantially insoluble in the vehicle.

Pinhole: Very small hole through a mastic or coating.

Pipe: A circular conduit for the convenience of liquids or semi-solids. Nominal Pipe Sizes (NPS) are expressed as the *nominal inside* diameter of the conduit through 12″ NPS, and the actual outside diameter and Schedule from 14″ on up. See Tube.

Pipe Insulation: Rigid or semi-rigid preformed thermal insulation to fit over NPS pipe and tubing. Thickness and dimensions for pipe insulation should be in accordance with ASTM Standard C-585.

Pit: Small regular or irregular crater in the surface of a plastic, usually with its width approximately of the same order of magnitude as its depth.

Plastic: A material that contains, as an essential ingredient, an organic substance of large molecular weight, is solid in its finished state, and, at some state in its manufacture or in its processing into finished articles, can be shaped by flow.

Plastic Foam: See Cellular Plastic.

Plasticizer: A material incorporated in a plastic, or coating, to increase its flexibility or distensibility.

Plasticizer Migration: The transfer of a constituent from a plastic body to other contracting solids.

Polyamid Resins: See Nylon Resins.

Polyester Resins: Synonymous with alkyd resins.

Polymer: A compound formed by the reaction of simple molecules having functional groups that permit their combination to proceed to high molecular weights under suitable conditions. Polymers may be formed by polymerization (addition polymer) or polycondensation (condensation polymer). When two or more monomers are involved, the product is called a copolymer.

Polymer Emulsion: Colloidal dispersion of a polymer in a water base, with or without other ingredients.

Polystyrene: A resin made by polymerization of styrene as the sole monomer.

Polyurethane: See Urethane Resins.

Polyvinyl Acetate Resins: Resins made by the polymerization of vinyl acetate or copolymerization of vinyl acetate with minor amounts (not over 50 percent) of other unsaturated compounds.

Pot Life: See Working Life.

Power: The time rate of doing work.

Primer: The first of two or more coats of a finish system.

Pullularia pullulans: A fairly common mold occurring particularly on damp cellulosic materials, such as cotton or paper, varying in color from dirty white to greenish and eventually becoming black and leathery all over.

Puncture Resistance: That property of a material which enables it to resist a tendency to puncture or perforate under blows or pressure from sharp objects.

Punking: The incandescence, or glow, which lingers in some materials after any flame, or other evidence of fire, has departed.

Radiant Flux Density: The rate of radiant energy emissions from a unit area of a source in all the radial directions of the overspreading hemisphere.

Radiation: The transfer of energy from a higher temperature body, through space, to another body, or bodies, some distance away at a lower temperature, without raising the temperature of the medium through which the energy passes.

Rankine: A temperature scale on which the absolute zero point is assigned the value of 0° the ice point of water 491.6, and the steam point 671.6°, with corresponding divisions on up the scale. The degree divisions of Rankine are the same as those of Fahrenheit.

Recovery of Thickness After Compression: This is a vital factor of blankets or batts in the use of these materials as cushion blankets, or in expansion joints.

Reflectance: The ration of the radiant energy reflected by a body to that incident upon it.

Reflective Insulation: Insulation, composed of closely spaced sheets of either aluminum or stainless steel, which obtains its insulating value from the ability of the sheets to reflect a large part of the radiant energy incident on them.

Reinforcing Cloth or Fabric: A loosely woven cloth or fabric of glass or resilient fibers, placed approximately in the center of the vapor- or weather-barrier to act as reinforcing to the mastic of the barrier.

Relative Humidity: The ratio of the actual pressure of existing water vapor in the atmosphere at the same temperature, expressed as a percentage. (See Dew Point.)

Resin: A solid, semisolid, or pseudo-solid organic material which has an indefinite and often high molecular weight, exhibits a tendency to flow when subjected to stress, usually has a softening or melting range, and fractures conchoidally.

Resistance (Thermal): See Thermal Resistance.

Resistance to Abrasion: Ability to withstand scuffing, scratching, rubbing, or wind-driven particles without loss of mechanical protection properties.

Resistance to Air Flow: The quotient of the air pressure difference across the surface divided by the volume velocity of air flow across the surface.

Resistance to Air Movement: The property which indicates the ability of a blanket type material to resist erosion by air currents over its surface.

Resistance to Freeze-Thaw: The resistance to change in application, thermal and/or mechanical properties from exposure to alternate cycles of freezing and thawing.

Resistance to Impact: The ability to withstand mechanical blows without loss of physical integrity and protective properties.

Resistance to Mold and Mildew: The ability to resist deterioration by fungi.

Resistance to Plastic Flow: That property of a material which opposes flow (deformation) beyond the elastic range of the material.

Right Angle Test: A method designed by DuPont Engineering to determine cracking tendencies of emulsion mastics when applied over an interior right angle, formed by fastening together two sections of insulation board.

Rigidity: That property of a material which opposes any tendency for it to bend (flex) under load.

Room Temperature: A temperature in the range of 20°C-30°C (68°F-85°F).

Rubber Cement: A natural or synthetic elastomeric material suitably compounded to form an effective adhesive for specified uses.

Rust Blush: The earliest stage of rusting characterized by the orange or red color ferric hydroxide (common rust). Occurs frequently on freshly sandblasted steel if allowed to stand too long before coating.

Sag: Excessive flow in material after application to a surface, resulting in "curtaining" or running.

Self-Ignition Temperature (Autogenous Ignition): The lowest temperature of a material which will cause it to ignite without other ignition source.

Self-Extinguishing: That property of a material which enables it to stop its own ignition after external ignition sources are removed.

Sealer: A putty-like substance, composed of various materials, used as a barrier to the passage of water vaport or liquid water into the joint formed by the mating surfaces of jackets and water- and vapor-barriers over insulation. A good sealer will possess relatively little shrinkage. There are several types of sealers, such as nonsetting, setting, and heat resisting.

Securements—Insulation: Any device, wire, strap or adhesive used to fasten insulation into its service position and hold it there.

Self-Heating: The process whereby, due to exothermic reactions, heat is liberated within a material at a rate sufficient to raise its temperature.

Self-Ignition: Ignition resulting from self-heating.

Service Temperature Limits: The limiting temperatures at a coated surface, within which limits the applied coating will have satisfactory service performance.

Set: To convert into a fixed or hardened state by chemical or physical action.

Shade: A term descriptive of a lightness difference between surface colors, the other attributes of color being essentially constant. A lighter shade of a color is one that has higher lightness, but approximately the same hue and saturation. A darker shade is one that has a lower lightness.

Shear Strength: The maximum stress tending to cleave a material which it is capable of sustaining without destruction. Shear strength is calculated from the maximum load during a shear or torsion test, and is based on the original dimensions of the cross-section of the specimen.

Sheen: Shiny or lustrous appearance at, or near, grazing incidence of a surface that appears to have no gloss for near perpendicular incidence; also, the 85-degree specular gloss of such a surface.

Sheet: A piece of material which is very thin in relation to its length and breadth.

Shelf Life: See Storage Life.

Shrinkage: The property of a material which causes its dimensions to become smaller. The material may become smaller due to thermal contraction due to reduction in temperature or it may become smaller due to reduction of moisture content.

Shrinkage—Wet to Dry: The property of a material which measures the difference in volumetric and linear change which occurs in the drying of insulating cements and mastics.

Silicone Resins: Resins in which the main polymer chain consists of alternating silicon and oxygen atoms, with carbon-containing side groups.

Sizing: Any of various glutinous materials, used to fill the pores in the surface of a paper, fiber, or cloth.

Skinning: The formation of a relatively dense film on the surface of a mastic or coating material while stored in containers.

Smoke Density: The Smoke Density Factor is the amount of smoke given off by the burning material compared to the amount of smoke given off by the burning of a standard material.

Smoke Toxicity: The degree of hazard to health of the smoke.

Smoldering: The combustion of solid materials without the accompaniment of flame.

Soaking Heat Stability: That property of a material which enables it to endure a soaking heat over an appreciable length of time, without materially changing its dimensions or properties.

Softening Point: That temperature at which a material will change its property from firm or rigid to soft or malleable.

Solar Resistance: 1. The resistance of a material to decomposition by the ultra-violet rays from the sun. 2. The resistance of an insulation to the passage of radiant heat from the sun.

Solids Content: The percentage of the non-volatile matter. Note: The determined value of non-volatile matter in any adhesive, coating, or sealant will vary somewhat, according to the analytical procedure used. A standard test method must be used to obtain consistent results.

Solvent: Any substance, usually a liquid, which dissolves other substances. In the coatings industry, normally a liquid organic compound used to make a coating work more freely. See Thinner.

Specific Gravity—Apparent: The ratio of the weight of a unit volume (usually a cubic foot) of the material as manufactured—including all voids—to the weight of a unit volume (usually a cubic foot) of water.

Specific Gravity—Real: The ratio of the weight of a unit volume (usually a cubic foot) of the actual material—excluding all voids—to the weight of a unit volume (usually a cubic foot) of water.

Specific Heat (at Constant Pressure): The ratio of the amount of heat required to raise a unit mass of a material 1 degree, to that required to raise a unit mass of water 1 degree at some specified temperature. In the English system the unit of measurement is Btu/lb degree F; in the Metric system the unit of measurement is Cal/gg degree C. The numerical value for specific heat in these units of measurement is the same.

Specimen: A portion of a unit taken for a single measurement of a given property or characteristic.

Sprayed-On Insulation: Insulation of the fibrous, or foam, type which is applied to the substrate by means of any one of a large number of powered spray devices, and which is secured to its substrate by its own properties of adhesion.

Stability Under Edge Compression: That property of a material which enables it to retain its shape and dimensions during the application of a compressive load on its edge, or edges. This is usually applicable only to rigid reflective insulation.

Standard Deviation: The square root of the mean square of the deviations of a set of values from their mean. It is a measure of the dispersion of the data.

Standard Time—Temperature Curve: A curve depicting the allowable temperature rise in a material, as related to time. The Standard Time—Temperature Curve is published by ASTM.

Static: Motionless. At rest or in equilibrium.

Static Load Test: The application of a constant load at rest, in testing.

Storage Life: The period of time during which a packaged adhesive, coating, or sealant can be stored under specified temperature conditions and remain suitable for use. Sometimes called shelf life.

Storage Stability: The ability of a material to retain its shape, dimensions, and properties while in storage.

Strain: A measure of the change, due to force, in the size or shape of a body referred to its original size or shape. Strain is a nondimensional quantity, but it is frequently expressed in inches per inch, etc.

Strength—Dry: The strength of an adhesive joint determined immediately after drying under specified conditions, or after a period of conditioning in a standard atmosphere.

Strength—Wet: The strength of an adhesive joint determined immediately after bonding adherends under specified conditions.

Strength—Compressive: See Compressive Strength.

Strength—Flexural: See Flexural Strength.

Strength—Tensile: See Tensile Strength.

Strength—Transverse: See Transverse Strength.

Strength—Tear: See Tear Strength.

Stress: The intensity, at a point in a body, of the internal forces or components of force that act on a given plane through the point. Stress is expressed in force per unit of area (pounds per square inch, kilograms per square millimeter, etc.)

Stress Corrosion: Intergranular corrosion and cracking in a metallic material, caused by the combination of a minimum temperature, tensile stress, and a specific corrodent (the chloride ion in the case of Austenitic stainless steel).

Stress-Crack: External or internal cracks, caused by tensile stresses less than those concerned in the short-time mechanical strength. Note: The development of such cracks is frequently accelerated by the environment. The stresses which cause cracking may be present internally or externally or may be combinations of these stresses. The appearance of a network of fine cracks is called crazing.

Structural Insulation: Insulation which is used as a part of the loadcarrying frame of a structure such as the walls of a cold room, and which has the necessary physical properties to perform its structural function.

Substrate: A material upon the surface of which an adhesive or coating is spread. A broader term than Adherend.

Suction: The absorptive effect exerted on coating materials by a highly porous substrate.

Superimposed Load: A load on, and carried by, an insulation, over and above the dead weight of the insulation itself.

Support—Insulation: A device to carry the weight of insulation in which the load carrier is positioned *under* the insulation. See Hanger.

Surface Wetting: The property of a material applied to a substrate which enables it to thoroughly wet the substrate to produce a good bond.

Tack: The property of an adhesive that enables it to form a bond of measurable strength immediately after adhesive and adherend are brought into contact under low pressure.

Tape: A narrow strip or band of any flexible material.

Tar: Brown or black bituminous material, liquid or semi-solid in consistency, in which the predominating constituents are bitumens obtained as condensates in the destructive distillation of coal, petroleum, oil-shale, wood, or other organic materials, and which yields substantial quantities of pitch when distilled.

Tear: To divide, disrupt, or pull apart by the action of opposing forces.

Tear Strength: That property of a material which enables it to resist being pulled apart by opposing forces.

Temperature: The level of the thermal state as indicated on a designated scale. (For the English system—Fahrenheit.)

Temperature Limits: The upper and lower temperatures at which a material will experience no essential change in its properties.

Temperature Retardance: That property of a material which delays any change of the temperature of one of its surfaces as related to any change in the temperature of its other surface.

Temperature Rise—Internal: The rise in the internal temperature of a material, due to application of heat, to fire, or to contamination.

Tensile Strength: The force per unit of the original cross sectional area (of an unstretched specimen) which is applied at the time of rupture of the specimen. It is calculated by dividing the breaking force, in pounds by the cross-section of the unstretched specimen, in square inches.

Test Area: The designated location from which specimens for physical and chemical testing shall be taken.

Test Result: A single numerical quantity determined by measuring a test specimen for a given property. In the case of a physical characteristic, the test result shall be considered representative of the test unit from which the specimen was taken. In the case of a chemical property, the test result, being obtained from a composite sample, is not associated with any particular test unit.

Thermal Conductance: The amount of heat transferred through a unit area of a material in a unit time, through its *total* thickness, with a unit of temperature difference between the surfaces of the two opposite sides.

Thermal Conductivity: The amount of heat transferred through a unit area of a material in a unit time, through a *unit* thickness, with a unit of temperature difference between the surfaces of the two opposite sides.

Thermal Diffusivity: See Diffusivity.

Thermal Insulation: Material having air- or gas-filled pockets, void spaces, or heat-reflective surfaces, which, when properly applied, will retard the transfer of heat with reasonable effectiveness under ordinary conditions.

Thermal Insulation—Forms of:

Blanket Thermal Insulation: A flat, flexible thermal insulation which can be placed over flat, curved, or irregular surfaces. May or may not be faced, coated, or reinforced on one or both sides.

Block Thermal Insulation: Rigid thermal insulation preformed into rectangular units.

Board Thermal Insulation: Rigid and semi-rigid thermal insulation preformed into relatively large area rectangular units.

Cement Thermal Insulation: A prepared composition, in dry form, composed of granular, flaky, fibrous or powdery materials, which when mixed with correct proportion of water develops into a mix of plastic consistance. After application and becoming dry in place it forms a coherent semi-rigid insulation.

Loose Fill Thermal Insulation: Material in granular, nodular, fibrous, or powder form suitable for installation, dry, by pouring, blowing or hand placement into confined spaces or areas.

Pipe Thermal Insulation: Rigid or semi-rigid insulation preformed for application to pipe or tubing.

Reflective Thermal Insulation: Sheets or preformed shapes to provide spaces, and reflective sheets which obtain their insulating resistance from the ability of the sheets to reflect a large part of the radiant energy incident on them.

Foamed Thermal Insulation: Liquids sprayed or mixed together which foam and set into cellular or semicellular organic rigid or semi-rigid insulation.

Thermal Resistance: That property of a material which enables it to withstand the passage of heat through it, due to a temperature difference between its two opposite surfaces.

Thermal Shock Resistance: That property of a material which enables it to retain its shape and not distort, crack, or shatter, due to a sudden change in its temperature.

Thermal Transference: The steady-state heat flow from (or to) a body through applied thermal insulation and to (or from) the external surroundings by conduction, convection, and radiation. It is expressed as the time rate per unit area of the body surface per unit temperature difference between body surface and the external surroundings.

Thermal Transmission, Heat: The quantity of heat flowing due to all modes of heat transfer under the prevailing conditions.

Thermal Transmittance: The ratio of the steady flow of heat energy from ambient air on one side of a body, through the body, to the external surroundings on the opposite side of the body, due to the temperature difference between the two surroundings.

Thermally Foamed Plastic: A cellular plastic, produced by applying heat to effect gaseous decomposition or volatilization of a constituent.

Thermoplastic: Capable of being repeatedly softened by increase of temperature, and hardened by decrease of temperature. Note: Thermoplastic applies to those materials whose change upon heating is substantially physical.

Thermoset: A plastic or coating which, when cured by application of heat or chemical means, changes into a substantially infusible and insoluble product.

Thinner: Volatile organic liquid used to adjust consistency, or to modify other properties of mastics and coatings, which volatilizes during the drying process.

Thixotropy: The property of decreasing in consistency upon being sheared or worked, followed by a gradual recovery of consistency when the shearing stress is removed.

Threshold Limit Value: The maximum allowable concentration (m.a.c.) of vapor to which nearly all industrial workers may be repeatedly exposed, day after day, without adverse physiological effect.

Tint: A color produced by a mixture of white pigment or coating in predominating amount with a colored pigment or coating, not white. The tint of a color is, therefore, much lighter and much less saturated than the color itself.

Tolerance: The allowable variation from given dimensions in the manufacturing or fabrication of an object.

Total Solids: See Solids Content.

Toxicity: The degree of hazard to health.

Traced: The supplying of auxiliary heat (or refrigeration) to a line, or piece of equipment, by means of a comparison line containing a hot (or cold) liquid or gas, thermally bonded to the line or equipment, with the resultant assembly completely encased by the insulation.

Translucent: Allowing the passage of some light, but not a clear view of any object.

Transverse Strength: That property of a material which enables it to resist loads or impact, normal to its primary axis.

Tube: A circular conduit for the conveyance of liquids or semi-solids. Tube sizes are expressed as the *outside* diameter and wall thickness. See Pipe.

Ultimate Strength: That load in pounds per square inch (in the English system) at which a material will rupture.

Ultimate Strength (as applied to Adhesives): Bond strength after the established drying time has been determined.

Uniformity: The state of being unvarying in form or composition.

Urethane Resins: Resins made by the condensation of organic isocyanates with compounds or resins that contain hydroxol groups. Note: Urethanes are a type of isocyanate resins.

Vapor-Barrier: A material, or materials, which, when installed on the high vapor pressure side, retards the passage of the moisture vapor to the lower vapor pressure side.

Vapor Density: The relative density of a vapor or gas (with no air present) as compared with air. A figure less than 1 indicates that a vapor is lighter than air, and a figure greater than 1 that a vapor is heavier than air.

Vapor Migration: That property of a material which measures the rate at which water vapor will penetrate it, due to vapor pressure differences between its surfaces.

Vapor Pressure: The gas pressure exerted by the water vapor present in the air.

Vehicle: The liquid portion of a mastic or coating. Anything that is dissolved in the liquid portion is a part of the vehicle.

Vibration Resistance: That property of a material which enables it to stay whole and not disintegrate when subject to vibration.

Vinyl Resins: Resins made from vinyl monomers, except those specifically covered by other classifications such as acrylic and styrene resins. Typical vinyl resins are polyvinyl chloride, polyvinyl alcohol, and polyvinyl butyral as well as copolymers, or vinyl monomers, with unsaturated compounds.

Vinyl Chloride Resins: Resins made by the polymerization of vinyl chloride or copolymerization of vinyl chloride with minor amounts (not over 50 percent) of other unsaturated compounds.

Viscometer (viscosimeter): An instrument for measuring viscosity or consistency.

Viscosity: The property of resistance to flow exhibited within the body of a material. Note: This property can be expressed in terms of the relationship between applied shearing stress and resulting rate of strain in shear. Viscosity is usually taken to mean "Newtonian Viscosity," in which case the ratio of shearing stress to the rate of shearing strain is constant. In non-Newtonian behavior, which is the usual case with adhesives, coatings, and sealants, the ratio varies with the shearing stress. Such ratios are often called the "apparent viscosities" at the corresponding shearing stresses. See Consistency.

Volatile Content: The proportional part (usually expressed as a percentage) of a material comprised of volatile (readily vaporizable) compounds.

Volatile Loss: Weight loss by vaporization.

Warp: The yarn running lengthwise in a woven fabric.

Warpage: The change in dimension of one surface of insulation as compared to that of another surface, due to differences in temperature of the two surfaces.

Water Absorption: The increase in weight of a test specimen, expressed as a percentage of its dry weight after immersion in water for a specified time.

Waterproof: Impervious to prolonged exposure to water.

Water-Repellency: That property of a material which prevents it from adhering to, or mixing with, water.

Water-Repellent: Having the property of water repellency.

Water-Resistant: Capable of withstanding limited exposure to water, or wet conditions, without failure.

Water Vapor: Water in a gaseous state.

Water-Vapor-Barrier: A material, or materials, which serves both as a weather-barrier and a vapor-barrier.

Water Vapor Permeability: The water vapor permeability of a homogeneous material is a property of the substance. This property may vary with conditions of exposure. The average permeability of a specimen is the product of its permeance and thickness. An accepted unit of permeability is a perm inch, or 1 grain per square foot, hour, inch of mercury per inch of thickness. The test conditions must be stated.

Water Vapor Permeance: The water vapor permeance of a body between two specified parallel surfaces is the ratio of its WVT to the vapor pressure difference between the two surfaces. An accepted unit of permeance is a perm, or 1 grain per square foot, hour, inch of mercury. The test conditions must be stated.

Water Vapor-Retarder (Barrier): The correct term is "Retarder," as these materials or systems retard the transmission of water vapor. It should be noted that no material or system can completely stop water vapor transmission when there is a vapor pressure difference between each of its surfaces.

Water Vapor Diffusion: The process by which water vapor spreads or moves through materials caused by a difference in water vapor pressure.

Water Vapor Transmission (WVT): The rate of water vapor transmission of a body between two specified parallel surfaces is the time rate of water vapor flow normal to the surfaces under steady condition through unit area, under the conditions of test. An accepted unit of WVT is 1 grain per square foot, hour (with test conditions stated).

Weather-Barrier (Weathercoat): A material or materials, which, when installed on the outer surface of thermal insulation, protects the insulation from the ravages of weather, such as rain, snow, sleet, wind, solar radiation, atmospheric contamination, and mechanical damage.

Weathering: The exposure of mastics or coatings outdoors.

Weather-Ometer:* A machine device used to determine the life of coatings by subjection to ultraviolet light, water, and sometimes other conditions.

Weather-Vapor-Barrier: A material which combines the properties of a weather-barrier and a vapor-barrier.

Weave—Leno: A fabric pattern in which warp yarns are arranged in pairs, twisting around one another between picks of filling yarns.

Weave—Plain: A fabric pattern in which each yarn of the filling passes alternately over and under a yarn of warp, and each yarn of the warp passes alternately over and under a yarn of the filling.

Weight: The force (pounds in the English system) with which a body is attracted toward the earth.

Wet: In the freshly applied state before drying.

Work: Energy in transient form. Expressed as the force applied to an object multiplied by the distance through which the object moves. $W = F \times D$.

Working Life (Pot Life): The period of time during which an adhesive or coating, after mixing with catalyst, solvent, or other compounding ingredients remains suitable for use.

*T.M. Reg. Atlas Electric Devices Co.

REFERENCES

Adding Insulation to House Attics, Paper No. 3
By C. J. Shirtliff, Div. of Building Research, Research Council of Canada

A.S.H.R.A.E. Guide and Data Book
By American Society of Heating, Refrigeration and Air Conditioning Engineers

A.S.H.R.A.E. 90-75 Standard "Energy Conservation in New Building Design"
By American Society of Heating, Refrigeration and Air Conditioning Engineers

A.S.T.M. Annual Book of Standards, Part 18-1981
Insulation materials and testing specifications.
By American Society for Testing and Materials

Blown Cellulose Fiber Thermal Insulations DBR Paper No. 820
By Division of Building Research, Research Council of Canada

Building Materials and Structures — Effect of Ceiling on Summer Comfort, Report BMS 52
By T. D. Phillips, National Bureau of Standards

Building Materials and Structures, Stability of Fiber Building Boards, Report BMS 50
By D. A. Jessup, C. G. Weber, and S. G. Weissberg, National Bureau of Standards

Comparison of Thermal Performance of Three Insulating Materials Commonly Used to Retrofit Exterior Residential Walls
By D. M. Burch, C. Siu and F. J. Powell

Condensation Problems in Your House: Prevention and Solution, Agriculture Information Bulletin No. 373
U.S. Dept. of Agriculture, Forest Service

Field Measurements of Heat Flow Through a Roof with Saturated Thermal Insulation and Covered with Black and White Granules
By E. C. Shuman

Fire and Plastic Foam Insulations
By G. W. Shorter and J. H. McGuire, Division of Building Research, National Research Council of Canada

Heat
By The Architectural Forum

Heat Loss Calculations, Technical Circular No. 7
Federal Housing Administration

House Basements, Paper No. CBD 13
By C. R. Croker, Division of Building Research
National Research Council of Canada

Insulation to Prevent Ground Freezing, Paper No. 119
By D. G. Stephenson, Division of Building Research
National Research Council of Canada

Internal Condensation and Building Problems in Non-Vented Flat Roofs
By Dr. I. A. Hens and K. U. Leuven, Belgium

Journal of Thermal Insulations
C. B. Hilado, Editor, Technomic Publishing Co.

Let's Heat People Instead of Houses
By E. T. Hall, Human Nature, Jan. 1974

Moisture Condensation, Bulletin Vol. 44, Number 34
University of Illinois

Moisture Condensation in Building Walls
By H. W. Woolley, National Bureau of Standards

Moisture Conditions in Walls and Ceilings of a Simulated Older House During Winter
UDA Forest Service Research Paper EPL 290
U.S. Dept. of Agriculture, Forest Service

Moisture Transfer in Insulated Metal Deck
By I. Samelson, Lund Institute of Technology, Lund, Sweden

Net Annual Heat Loss, Factor Method for Estimating Heat Requirements of Buildings, Paper No. 117
By G. P. Mitala, Division of Building Research
National Research Council of Canada

Plastic No. CBD 154

Plastic Foams No. CBD 166

Properties and Behavior of Plastics No. CBD 177

Rigid Thermoplastic Foams No. CBD 169

REFERENCES (Continued)

Rigid Thermosetting Plastic Foams No. CBD 168
By A. Blaga, Division of Building Research
National Research Council of Canada

Polyurethane Foam as a Thermal Insulation, A Critical Examination No. 124
By C. J. Shirtliff, Division of Building Research
National Research Council of Canada

Principles Applied to Insulated Masonry Wall, No. CBD 50
By J. B. Hutchen, Division of Building Research
National Research Council of Canada

Reflective Insulating Blinds
By J. B. Shapira and P. R. Banes, Energy Division
Oak Ridge National Laboratory

Relation Between Thermal Resistance and Heat Storage in Buildings
By C. P. Mitalas, Paper No. 126, Division of Building Research
National Research Council of Canada

Testing Cellulose Fiber Insulation for Horizontal Applications, Paper No. 147
By M. Bomberg and C. J. Shirtliff, Division of Building Research
National Research Council of Canada

Thermal Conductivity of Polyurethane
By G. Bigolaro, F. D. Deponte and E. Fornaieri

Thermal Insulation Design Economics
By W. C. Turner and J. F. Malloy
Krieger Publishing Co., P.O. Box 9542, Melbourne, FL 32901

Thermal Insulation Handbook
By W. C. Turner and J. F. Malloy
Krieger Publishing Co., P.O. Box 9542, Melbourne, FL 32901 and McGraw-Hill Publishing Co., 1221 Avenue of the Americas, New York, N.Y. 10020

Thermal Insulating Values of Airspaces, Paper No. 32
Division of Housing Research

Thermal Insulation Systems, NASA SP 5027
National Aeronautics and Space Administration

Thermal Resistances of Building Insulation
By C. J. Shirtliff, Division of Building Research
National Research Council of Canada

Thermal Resistances of Airspaces and Fibrous Insulations Bonded by Reflective Surfaces
Building Materials and Structures Report 151
United States Dept. of Commerce

Thermoplastics Paper No. CBD 158

Thermosetting Plastics Paper No. CBD 159
By A. Blaga, Division of Building Research
National Research Council of Canada

Vapour Barriers in Home Construction
By C. G. Hangord, Division of Building Research
National Research Council of Canada

Vapour Barriers Under Concrete Slab on Ground, Use of
By R. G. Turene, Division of Building Research
National Research Council of Canada

Warm-Air Peremeter Heating
By J. R. Jamieson, R. W. Roose and S. Knozoff, Urbana, IL

HONOR ROLL OF ASSOCIATIONS AND FIRMS THAT HAVE CONTRIBUTED TO THERMAL INSULATION FOR RESIDENCES AND BUILDINGS

Alfoil Inc.
P.O. Box 7024
Charlotte, No. Carolina 28217

Compac Corp.
Old Flanders Road
Netcong, N.J. 07836

Dynaporte
195 Sweet Hollow Road
Old Bethplace, N.Y. 11804

Eneron
2320 I-35 South
San Marcos, Texas 78666

Fiberglass Canada Inc.
3080 Younge Street
Toronto Ontario, N4N3N1, Canada

Foilpleat Products Corp.
2020 W. 139th Street
Gardena, Ca. 80217

Manville Products Corp.
Ken-Caryl Ranch
Denver, Co. 80217

National Consumers League
1625 I Street, N.W., Suite 923
Washington, D.C. 20006

Owens Corning Fiberglass Co.
Fiberglass Tower
Toledo, Ohio 43659

Pamrod
P.O. Box 335
McQueeney, Tex. 78123

Pittsburgh Corning Corp.
800 Presque Isle Drive
Pittsburgh, Pa. 15239

Reflecto Shield
64 New Industrial Road
Woburn, Mass.

Thermon Mfg. Co.
100 Thermon Drive
San Marcos, Tex. 78666

TIMA
7 Kirgy Plaza
Mt. Kisco, N.Y. 10549

Vimasco Corp.
P.O. Box 516
Nitro, W. Va. 25143

The Wiremold Co.
West Hartford, Conn. 06110

Appendix B — Tables

UNITS OF MEASURE

The names English Units and Metric Units have been used in this book to designate the two different systems used in the world for measurement.

By English Units the system referred to is the system used in the United States known as United States Customary Units, and a few years ago renamed inch-pound Unit system. It also has been designated as British or UK Units. Of all these names English Units was chosen as being the one most used.

By Metric Units the system referred to is that system of measuring commonly used by most nations, other than the United States, for about 200 years. Recently, it has been renamed the International System of Units and called SI Units. Of these names Metric Unit was selected to designate the system of these names.

To establish the relationship of these systems, Table B-1 provides a table of energy, work and heat conversion factors; Table B-2a provides temperature conversion; Table B-2b the relationship of temperature scales; and Table B-2c the conversion factors of temperature scales. Table B-3 provides a general conversion table for measurements most used in the thermal insulation field. Table B-4 provides a direct relationship between the units used to measure energy.

Additional tables regarding units and heat transfer are contained in Appendix B.

Table B-1 Energy, work and heat conversion factors

Multiply number of → / To obtain ↓ By	BTU	Centimeter-grams	Ergs or centimeter-dynes	Foot-pounds	Horsepower-hours	Joules or watt-seconds	Kilo-calories	Kilowatt-hours	Meter-kilograms	Watt-hours
BTU	1	9.297×10^{-8}	9.480×10^{-11}	1.285×10^{-3}	2545	9.480×10^{-4}	3.969	3413	9.297×10^{-3}	3.413
Centimeter-grams	1.076×10^{7}	1	1.020×10^{-3}	1.383×10^{4}	2.737×10^{10}	1.020×10^{4}	4.269×10^{7}	3.671×10^{10}	10^{5}	3.671×10^{7}
Ergs or centimeter-dynes	1.055×10^{10}	980.7	1	1.356×10^{7}	2.684×10^{13}	10^{7}	4.186×10^{10}	3.6×10^{13}	9.807×10^{7}	3.6×10^{10}
Foot-pounds	778.3	7.233×10^{-3}	7.367×10^{-8}	1	1.98×10^{6}	0.7376	3087	2.655×10^{6}	7.233	2655
Horsepower-hours	3.929×10^{-4}	3.654×10^{-11}	3.722×10^{-14}	5.050×10^{-7}	1	3.722×10^{-7}	1.559×10^{-3}	1.341	3.653×10^{-6}	1.341×10^{-3}
Joules or watt-seconds	1054.8	9.807×10^{-6}	10^{-7}	1.356	2.684×10^{6}	1	4186	3.6×10^{6}	9.807	3600
Kilo-calories	0.2520	2.343×10^{-8}	2.389×10^{-11}	3.239×10^{-4}	641.3	2.389×10^{-4}	1	860.0	2.343×10^{-3}	0.8600
Kilowatt-hours	2.930×10^{-4}	2.724×10^{-11}	2.778×10^{-14}	3.766×10^{-7}	0.7457	2.778×10^{-7}	1.163×10^{-3}	1	2.724×10^{-6}	0.001
Meter-kilograms	107.6	10^{-5}	1.020×10^{-8}	0.1383	2.737×10^{5}	0.1020	426.9	3.671×10^{5}	1	367.1
Watt-hours	0.2930	2.724×10^{-8}	2.778×10^{-11}	3.766×10^{-4}	745.7	2.778×10^{-4}	1.163	1000	2.724×10^{-3}	1

Table is designed to convert quantities on the top row into quantities listed in the vertical column. For example, to convert kilowatt-hours (top row) to joules (vertical column), multiply by 3.6×10^{6}; that is, number of kilowatt-hours X (3.6×10^{6}) = Number of joules.

Table B-2 Temperature conversion

FAHRENHEIT SCALE LISTED IN EVEN NUMBERS

KELVIN K	CELSIUS C	FAHREN-HEIT F	RANKINE R	KELVIN K	CELSIUS C	FAHREN-HEIT F	RANKINE R	KELVIN K	CELSIUS C	FAHREN-HEIT F	RANKINE R
0	−273.2	−459.7	0	38.6	−234.4	−390	69.7	77.6	−195.6	−320	139.7
0.4	−272.8	−459	0.7	39.3	−233.9	−389	70.7	78.2	−195.0	−319	140.7
1.0	−272.2	−458	1.7	39.9	−233.3	−388	71.7	78.8	−194.4	−318	141.7
1.5	−271.7	−457	2.7	40.4	−232.8	−387	72.7	79.3	−193.9	−317	142.7
2.1	−271.1	−456	3.7	41.0	−232.2	−386	73.7	79.9	−193.3	−316	143.7
2.6	−270.6	−455	4.7	41.5	−231.7	−385	74.7	80.4	−192.8	−315	144.7
3.2	−270.0	−454	5.7	42.1	−231.1	−384	75.7	81.0	−192.2	−314	145.7
3.8	−269.4	−453	6.7	42.6	−230.6	−383	76.7	81.5	−191.7	−313	146.7
4.3	−268.9	−452	7.7	43.2	−230.0	−382	77.7	82.1	−191.1	−312	147.7
4.9	−268.3	−451	8.7	43.8	−229.4	−381	78.7	82.6	−190.6	−311	148.7
5.4	−267.8	−450	9.7	44.3	−228.9	−380	79.7	83.2	−190.0	−310	149.7
6.0	−267.2	−449	10.7	44.9	−228.3	−379	80.7	83.8	−189.4	−309	150.7
6.5	−266.7	−448	11.7	45.4	−227.8	−378	81.7	84.3	−188.9	−308	151.7
7.1	−266.1	−447	12.7	46.0	−227.2	−377	82.7	84.9	−188.3	−307	152.7
7.6	−265.6	−446	13.7	46.5	−226.7	−376	83.7	85.4	−187.8	−306	153.7
8.2	−265.0	−445	14.7	47.1	−226.1	−375	84.7	86.0	−187.2	−305	154.7
8.8	−264.4	−444	15.7	47.6	−225.6	−374	85.7	86.5	−186.7	−304	155.7
9.3	−263.9	−443	16.7	48.2	−225.0	−373	86.7	87.1	−186.1	−303	156.7
9.9	−263.3	−442	17.7	48.8	−224.4	−372	87.7	87.6	−185.6	−302	157.7
10.4	−262.8	−441	18.7	49.3	−223.9	−371	88.7	88.2	−185.0	−301	158.7
11.0	−262.2	−440	19.7	49.9	−223.3	−370	89.7	88.8	−184.4	−300	159.7
11.5	−261.7	−439	20.7	50.4	−222.8	−369	90.7	89.3	−183.9	−299	160.7
12.1	−261.1	−438	21.7	51.0	−222.2	−368	91.7	89.9	−183.3	−298	161.7
12.6	−260.6	−437	22.7	51.5	−221.7	−367	92.7	90.4	−182.8	−297	162.7
13.2	−260.0	−436,	23.7	52.1	−221.1	−366	93.7	91.0	−182.2	−296	163.7
13.8	−259.4	−435	24.7	52.6	−220.6	−365	94.7	91.5	−181.7	−295	164.7
14.3	−258.9	−434	25.7	53.2	−220.0	−364	95.7	92.1	−181.1	−294	165.7
14.9	−258.3	−433	26.7	53.8	−219.4	−363	96.7	92.6	−180.6	−293	166.7
15.4	−257.8	−432	27.7	54.3	−218.9	−362	97.7	93.2	−180.0	−292	167.7
16.0	−257.2	−431	28.7	54.9	−218.3	−361	98.7	93.8	−179.4	−291	168.7
16.5	−256.7	−430	29.7	55.4	−217.8	−360	99.7	94.3	−178.9	−290	169.7
17.1	−256.1	−429	30.7	56.0	−217.2	−359	100.7	94.9	−178.3	−289	170.7
17.6	−255.6	−428	31.7	56.5	−216.7	−358	101.7	95.4	−177.8	−288	171.7
18.2	−255.0	−427	32.7	57.1	−216.1	−357	102.7	96.0	−177.2	−287	172.7
18.8	−254.4	−426	33.7	57.6	−215.6	−356	103.7	96.5	−176.7	−286	173.7
19.3	−253.9	−425	34.7	58.2	−215.0	−355	104.7	97.1	−176.1	−285	174.7
19.9	−253.3	−424	35.7	58.8	−214.4	−354	105.7	97.6	−175.6	−284	175.7
20.4	−252.8	−423	36.7	59.3	−213.9	−353	106.7	98.2	−175.0	−283	176.7
21.0	−252.2	−422	37.7	59.9	−213.3	−352	107.7	98.8	−174.4	−282	177.7
21.5	−251.7	−421	38.7	60.4	−212.8	−351	108.7	99.3	−173.9	−281	178.7
22.1	−251.1	−420	39.7	61.0	−212.2	−350	109.7	99.9	−173.3	−280	179.7
22.6	−250.6	−419	40.7	61.5	−211.7	−349	110.7	100.4	−172.8	−279	180.7
23.2	−250.0	−418	41.7	62.1	−211.1	−348	111.7	101.0	−172.2	−278	181.7
23.8	−249.4	−417	42.7	62.6	−210.6	−347	112.7	101.5	−171.7	−277	182.7
24.3	−248.9	−416	43.7	63.2	−210.0	−346	113.7	102.1	−171.1	−276	183.7
24.9	−248.3	−415	44.7	63.8	−209.4	−345	114.7	102.6	−170.6	−275	184.7
25.4	−247.8	−414	45.7	64.3	−208.9	−344	115.7	103.2	−170.0	−274	185.7
26.0	−247.2	−413	46.7	64.9	−208.3	−343	116.7	103.8	−169.4	−273	186.7
26.5	−246.7	−412	47.7	65.4	−207.8	−342	117.7	104.3	−168.9	−272	187.7
27.1	−246.1	−411	48.7	66.0	−207.2	−341	118.7	104.9	−168.3	−271	188.7
27.6	−245.6	−410	49.7	66.5	−206.4	−340	119.7	105.4	−167.8	−270	189.7
28.2	−245.0	−409	50.7	67.1	−206.1	−339	120.7	106.0	−167.2	−269	190.7
28.8	−244.4	−408	51.7	67.6	−205.6	−338	121.7	106.5	−166.7	−268	191.7
29.3	−243.9	−407	52.7	68.2	−205.0	−337	122.7	107.1	−166.1	−267	192.7
29.9	−243.3	−406	53.7	68.8	−204.4	−336	123.7	107.6	−165.6	−266	193.7
30.4	−242.8	−405	54.7	69.3	−203.9	−335	124.7	108.2	−165.0	−265	194.7
31.0	−242.2	−404	55.7	69.9	−203.3	−334	125.7	108.8	−164.4	−264	195.7
31.5	−241.7	−403	56.7	70.4	−202.8	−333	126.7	109.3	−163.9	−263	196.7
32.1	−241.1	−402	57.7	71.0	−202.2	−332	127.7	109.9	−163.3	−262	197.7
32.6	−240.6	−401	58.7	71.5	−201.7	−331	128.7	110.4	−162.8	−261	198.7
33.2	−240.0	−400	59.7	72.1	−201.1	−330	129.7	111.0	−162.2	−260	199.7
33.8	−239.4	−399	60.7	72.6	−200.6	−329	130.7	111.5	−161.7	−259	200.7
34.3	−238.9	−398	61.7	73.2	−200.0	−328	131.7	112.1	−161.1	−258	201.7
34.9	−238.3	−397	62.7	73.8	−199.4	−327	132.7	112.6	−160.6	−257	202.7
35.4	−237.8	−396	63.7	74.3	−198.9	−326	133.7	113.2	−160.0	−256	203.7
36.0	−237.2	−395	64.7	74.9	−198.3	−325	134.7	113.8	−159.4	−255	204.7
36.5	−236.7	−394	65.7	75.4	−197.8	−324	135.7	114.3	−158.9	−254	205.7
37.1	−236.1	−393	66.7	76.0	−197.2	−323	136.7	114.9	−158.3	−253	206.7
37.6	−235.6	−392	67.7	76.6	106.7	−322	137.7	115.4	−157.8	−252	207.7
38.2	−235.0	−391	68.7	77.1	−196.1	−321	138.7	116.0	−157.2	−251	208.7

Table B-2 (continued)

KELVIN K	CELSIUS C	FAHREN-HEIT F	RANKINE R	KELVIN K	CELSIUS C	FAHREN-HEIT F	RANKINE R	KELVIN K	CELSIUS C	FAHREN-HEIT F	RANKINE R
116.5	−156.7	−250	209.7	156.5	−116.7	−178	281.7	196.5	−76.7	−106	353.7
117.1	−156.1	−249	210.7	157.1	−116.1	−177	282.7	197.1	−76.1	−105	354.7
117.6	−155.6	−248	211.7	157.6	−115.6	−176	283.7	197.6	−75.6	−104	355.7
118.2	−155.0	−247	212.7	158.2	−115.0	−175	284.7	198.2	−75.0	−103	356.7
118.8	−154.4	−246	213.7	158.8	−114.4	−174	285.7	198.8	−74.4	−102	357.7
119.3	−153.9	−245	214.7	159.3	−113.9	−173	286.7	199.3	−73.9	−101	358.7
119.9	−153.3	−244	215.7	159.9	−113.3	−172	287.7	199.9	−73.3	−100	359.7
120.4	−152.8	−243	216.7	160.4	−112.8	−171	288.7	200.4	−72.8	−99	360.7
121.0	−152.2	−242	217.7	161.0	−112.2	−170	289.7	201.0	−72.2	−98	361.7
121.5	−151.7	−241	218.7	161.5	−111.7	−169	290.7	201.5	−71.7	−97	362.7
122.1	−151.1	−240	219.7	162.1	−111.1	−168	291.7	202.1	−71.1	−96	363.7
122.6	−150.6	−239	220.7	162.6	−110.6	−167	292.7	202.6	−70.6	−95	364.7
123.2	−150.0	−238	221.7	163.2	−110.0	−166	293.7	203.2	−70.0	−94	365.7
123.8	−149.4	−237	222.7	163.8	−109.4	−165	294.7	203.8	−69.4	−93	366.7
124.3	−148.9	−236	223.7	164.3	−108.9	−164	295.7	204.3	−68.9	−92	367.7
124.9	−148.3	−235	224.7	164.9	−108.2	−163	296.7	204.9	−68.3	−91	368.7
125.4	−147.8	−234	225.7	165.4	−107.8	−162	297.7	205.4	−67.8	−90	369.7
126.0	−147.2	−233	226.7	166.0	−107.2	−161	298.7	206.0	−67.2	−89	370.7
126.5	−146.7	−232	227.7	166.5	−106.7	−160	299.7	206.5	−66.7	−88	371.7
127.1	−146.1	−231	228.7	167.1	−106.1	−159	300.7	207.1	−66.1	−87	372.7
127.6	−145.6	−230	229.7	167.6	−105.6	−158	301.7	207.6	−65.6	−86	373.7
128.2	−145.0	−229	230.7	168.2	−105.0	−157	302.7	208.2	−65.0	−85	374.7
128.8	−144.4	−228	231.7	168.8	−104.4	−156	303.7	208.8	−64.4	−84	375.7
129.3	−143.9	−227	232.7	169.3	−103.9	−155	304.7	209.3	−63.9	−83	376.7
129.9	−143.3	−226	233.7	169.9	−103.3	−154	305.7	209.9	−63.3	−82	377.7
130.4	−142.8	−225	234.7	170.4	−102.8	−153	306.7	210.4	−62.8	−81	378.7
131.0	−142.2	−224	235.7	171.0	−102.2	−152	307.7	211.0	−62.2	−80	379.7
131.5	−141.7	−223	236.7	171.5	−101.7	−151	308.7	211.5	−61.7	−79	380.7
132.1	−141.1	−222	237.7	172.1	−101.1	−150	309.7	212.1	−61.1	−78	381.7
132.6	−140.6	−221	238.7	172.6	−100.6	−149	310.7	212.6	−60.6	−77	382.7
133.2	−140.0	−220	239.7	173.2	−100.0	−148	311.7	213.2	−60.0	−76	383.7
133.8	−139.4	−219	240.7	173.8	−99.4	−147	312.7	213.8	−59.4	−75	384.7
134.3	−138.9	−218	241.7	174.3	−98.9	−146	313.7	214.3	−58.9	−74	385.7
134.9	−138.3	−217	242.7	174.9	−98.3	−145	314.7	214.9	−58.3	−73	386.7
135.4	−137.8	−216	243.7	175.4	−97.8	−144	315.7	215.4	−57.8	−72	387.7
136.0	−137.2	−215	244.7	176.0	−97.2	−143	316.7	216.0	−57.2	−71	388.7
136.5	−136.7	−214	245.7	176.5	−96.7	−142	317.7	216.5	−56.7	−70	389.7
137.1	−136.1	−213	246.7	177.1	−96.1	−141	318.7	217.1	−56.1	−69	390.7
137.6	−135.6	−212	247.7	177.6	−95.6	−140	319.7	217.6	−55.6	−68	391.7
138.2	−135.0	−211	248.7	178.2	−95.0	−139	320.7	218.2	−55.0	−67	392.7
138.8	−134.4	−210	249.7	178.8	−94.4	−138	321.7	218.8	−54.4	−66	393.7
139.3	−133.9	−209	250.7	179.3	−93.9	−137	322.7	219.3	−53.9	−65	394.7
139.9	−133.3	−208	251.7	179.9	−93.3	−136	323.7	219.9	−53.3	−64	395.7
140.4	−132.8	−207	252.7	180.4	−92.8	−135	324.7	220.4	−52.8	−63	396.7
141.0	−132.2	−206	253.7	181.0	−92.2	−134	325.7	221.0	−52.2	−62	397.7
141.5	−131.7	−205	254.7	181.5	−91.7	−133	326.7	221.5	−51.7	−61	398.7
142.1	−131.1	−204	255.7	182.1	−91.1	−132	327.7	222.1	−51.1	−60	399.7
142.6	−130.6	−203	256.7	182.6	−90.6	−131	328.7	222.6	−50.6	−59	400.7
143.2	−130.0	−202	257.7	183.2	−90.0	−130	329.7	223.2	−50.0	−58	401.7
143.8	−129.4	−201	258.7	183.8	−89.4	−129	330.7	223.8	−49.4	−57	402.7
144.3	−128.9	−200	259.7	184.3	−88.9	−128	331.7	224.3	−48.9	−56	403.7
144.9	−128.3	−199	260.7	184.9	−88.3	−127	332.7	224.9	−48.3	−55	404.7
145.4	−127.8	−198	261.7	185.4	−87.8	−126	333.7	225.4	−47.8	−54	405.7
146.0	−127.2	−197	262.7	186.0	−87.2	−125	334.7	226.0	−47.2	−53	406.7
146.5	−126.7	−196	263.7	186.5	−86.7	−124	335.7	226.5	−46.7	−52	407.7
147.1	−126.1	−195	264.7	187.1	−86.1	−123	336.7	227.1	−46.1	−51	408.7
147.6	−125.6	−194	265.7	187.6	−85.6	−122	337.7	227.6	−45.6	−50	409.7
148.2	−125.0	−193	266.7	188.2	−85.0	−121	338.7	228.2	−45.0	−49	410.7
148.8	−124.4	−192	267.7	188.8	−84.4	−120	339.7	228.8	−44.4	−48	411.7
149.3	−123.9	−191	268.7	189.3	−83.9	−119	340.7	229.3	−43.9	−47	412.7
149.9	−123.3	−190	269.7	189.9	−83.3	−118	341.7	229.9	−43.3	−46	413.7
150.4	−122.8	−189	270.7	190.4	−82.8	−117	342.7	230.4	−42.8	−45	414.7
151.0	−122.2	−188	271.7	191.0	−82.2	−116	343.7	231.0	−42.2	−44	415.7
151.5	−121.7	−187	272.7	191.5	−81.7	−115	344.7	231.5	−41.7	−43	416.7
152.1	−121.1	−186	273.7	192.1	−81.1	−114	345.7	232.1	−41.1	−42	417.7
152.6	−120.6	−185	274.7	192.6	−80.6	−113	346.7	232.6	−40.6	−41	418.7
153.2	−120.0	−184	275.7	193.2	−80.0	−112	347.7	233.2	−40.0	−40	419.7
153.8	−119.4	−183	276.7	193.8	−79.4	−111	348.7	233.8	−39.4	−39	420.7
154.3	−118.9	−182	277.7	194.3	−78.9	−110	349.7	234.3	−38.9	−38	421.7
154.9	−118.3	−181	278.7	194.9	−78.3	−109	350.7	234.9	−38.3	−37	422.7
155.4	−117.8	−180	279.7	195.4	−77.8	−108	351.7	235.4	−37.8	−36	423.7
156.0	−117.2	−179	280.7	196.0	−77.2	−107	352.7	236.0	−37.2	−35	424.7

Table B-2 Temperature conversion (continued)

KELVIN K	CELSIUS C	FAHRENHEIT F	RANKINE R	KELVIN K	CELSIUS C	FAHRENHEIT F	RANKINE R	KELVIN K	CELSIUS C	FAHRENHEIT F	RANKINE R
236.5	−36.7	−34	425.7	276.5	3.3	38	497.7	316.5	43.3	110	569.7
237.1	−36.1	−33	426.7	277.1	3.9	39	498.7	317.1	43.9	111	570.7
237.6	−35.6	−32	427.7	277.6	4.4	40	499.7	317.6	44.4	112	571.7
238.2	−35.0	−31	428.7	278.2	5.0	41	500.7	318.2	45.0	113	572.7
238.8	−34.4	−30	429.7	278.8	5.6	42	501.7	318.8	45.6	114	573.7
239.3	−33.9	−29	430.7	279.3	6.1	43	502.7	319.3	46.1	115	574.7
239.9	−33.3	−28	431.7	279.9	6.7	44	503.7	319.9	46.7	116	575.7
240.4	−32.8	−27	432.7	280.4	7.2	45	504.7	320.4	47.2	117	576.7
241.0	−32.2	−26	433.7	281.0	7.8	46	505.7	321.0	47.8	118	577.7
241.5	−31.7	−25	434.7	281.5	8.3	47	506.7	321.5	48.3	119	578.7
242.1	−31.1	−24	435.7	282.1	8.9	48	507.7	322.1	48.9	120	579.7
242.6	−30.6	−23	436.7	282.6	9.4	49	508.7	322.6	49.4	121	580.7
243.2	−30.0	−22	437.7	283.2	10.0	50	509.7	323.2	50.0	122	581.7
243.8	−29.4	−21	438.7	283.8	10.6	51	510.7	323.8	50.6	123	582.7
244.3	−28.9	−20	439.7	284.3	11.1	52	511.7	324.3	51.1	124	583.7
244.9	−28.3	−19	440.7	284.9	11.7	53	512.7	324.9	51.7	125	584.7
245.4	−27.8	−18	441.7	285.4	12.2	54	513.7	325.4	52.2	126	585.7
246.0	−27.2	−17	442.7	286.0	12.8	55	514.7	326.0	52.8	127	586.7
246.5	−26.7	−16	443.7	286.5	13.3	56	515.7	326.5	53.3	128	587.7
247.1	−26.1	−15	444.7	287.1	13.9	57	516.7	327.1	53.9	129	588.7
247.6	−25.6	−14	445.7	287.6	14.4	58	517.7	327.6	54.4	130	589.7
248.2	−25.0	−13	446.7	288.2	15.0	59	518.7	328.2	55.0	131	590.7
248.8	−24.4	−12	447.7	288.8	15.6	60	519.7	328.8	55.6	132	591.7
249.3	−23.9	−11	448.7	289.3	16.1	61	520.7	329.3	56.1	133	592.7
249.9	−23.3	−10	449.7	289.9	16.7	62	521.7	329.9	56.7	134	593.7
250.4	−22.8	−9	450.7	290.4	17.2	63	522.7	330.4	57.2	135	594.7
251.0	−22.2	−8	451.7	291.0	17.8	64	523.7	331.0	57.8	136	595.7
251.5	−21.7	−7	452.7	291.5	18.3	65	524.7	331.5	58.3	137	596.7
252.1	−21.1	−6	453.7	292.1	18.9	66	525.7	332.1	58.9	138	597.7
252.6	−20.6	−5	454.7	292.6	19.4	67	526.7	332.6	59.4	139	598.7
253.2	−20.0	−4	455.7	293.2	20.0	68	527.7	333.2	60.0	140	599.7
253.8	−19.4	−3	456.7	293.8	20.6	69	528.7	333.8	60.6	141	600.7
254.3	−18.9	−2	457.7	294.3	21.1	70	529.7	334.3	61.1	142	601.7
254.9	−18.3	−1	458.7	294.9	21.7	71	530.7	334.9	61.7	143	602.7
255.4	−17.8	0	459.7	295.4	22.2	72	531.7	335.4	62.2	144	603.7
256.0	−17.2	1	460.7	296.0	22.8	73	532.7	336.0	62.8	145	604.7
256.5	−16.7	2	461.7	296.5	23.3	74	533.7	336.5	63.3	146	605.7
257.1	−16.1	3	462.7	297.1	23.9	75	534.7	337.1	63.9	147	606.7
257.6	−15.6	4	463.7	297.6	24.4	76	535.7	337.6	64.4	148	607.7
258.2	−15.0	5	464.7	298.2	25.0	77	536.7	338.2	65.0	149	608.7
258.8	−14.4	6	465.7	298.8	25.6	78	537.7	338.8	65.6	150	609.7
259.3	−13.9	7	466.7	299.3	26.1	79	538.7	339.3	66.1	151	610.7
259.9	−13.3	8	467.7	299.9	26.7	80	539.7	339.9	66.7	152	611.7
260.4	−12.8	9	468.7	300.4	27.2	81	540.7	340.4	67.2	153	612.7
261.0	−12.2	10	469.7	301.0	27.8	82	541.7	341.0	67.8	154	613.7
261.5	−11.7	11	470.7	301.5	28.3	83	542.7	341.5	68.3	155	614.7
262.1	−11.1	12	471.7	302.1	28.9	84	543.7	342.1	68.9	156	615.7
262.6	−10.6	13	472.7	302.6	29.4	85	544.7	342.6	69.4	157	616.7
263.2	−10.0	14	473.7	303.2	30.0	86	545.7	343.2	70.0	158	617.7
263.8	−9.4	15	474.7	303.8	30.6	87	546.7	343.8	70.6	159	618.7
264.3	−8.9	16	475.7	304.3	31.1	88	547.7	344.3	71.1	160	619.7
264.9	−8.3	17	476.7	304.9	31.7	89	548.7	344.9	71.7	161	620.7
265.4	−7.8	18	477.7	305.4	32.2	90	549.7	345.4	72.2	162	621.7
266.0	−7.2	19	478.7	306.0	32.8	91	550.7	346.0	72.8	163	622.7
266.5	−6.7	20	479.7	306.5	33.3	92	551.7	346.5	73.3	164	623.7
267.1	−6.1	21	480.7	307.1	33.9	93	552.7	347.1	73.9	165	624.7
267.6	−5.6	22	481.7	307.6	34.4	94	553.7	347.6	74.4	166	625.7
268.2	−5.0	23	482.7	308.2	35.0	95	554.7	348.2	75.0	167	626.7
268.8	−4.4	24	483.7	308.8	35.6	96	555.7	348.8	75.6	168	627.7
269.3	−3.9	25	484.7	309.3	36.1	97	556.7	349.3	76.1	169	628.7
269.9	−3.3	26	485.7	309.9	36.7	98	557.7	349.9	76.7	170	629.7
270.4	−2.8	27	486.7	310.4	37.2	99	558.7	350.4	77.2	171	630.7
271.0	−2.2	28	487.7	311.0	37.8	100	559.7	351.0	77.8	172	631.7
271.5	−1.7	29	488.7	311.5	38.3	101	560.7	351.5	78.3	173	632.7
272.1	−1.1	30	489.7	312.1	38.9	102	561.7	352.1	78.9	174	633.7
272.6	−0.6	31	490.7	312.6	39.4	103	562.7	352.6	79.4	175	634.7
273.2	0	32	491.7	313.2	40.0	104	563.7	353.2	80.0	176	635.7
273.8	0.6	33	492.7	313.8	40.6	105	564.7	353.8	80.6	177	636.7
274.3	1.1	34	493.7	314.3	41.1	106	565.7	354.3	81.1	178	637.7
274.9	1.7	35	494.7	314.9	41.7	107	566.7	354.9	81.7	179	638.7
275.4	2.2	36	495.7	315.4	42.2	108	567.7	355.4	82.2	180	639.7
276.0	2.8	37	496.7	316.0	42.8	109	568.7	356.0	82.8	181	640.7

Table B-2 (continued)

KELVIN K	CELSIUS C	FAHRENHEIT F	RANKINE R	KELVIN K	CELSIUS C	FAHRENHEIT F	RANKINE R	KELVIN K	CELSIUS C	FAHRENHEIT F	RANKINE R
356.5	83.3	182	641.7	594	321	610	1070	983	710	1310	1770
357.1	83.9	183	642.7	600	327	620	1080	989	716	1320	1780
357.6	84.4	184	643.7	605	332	630	1090	994	721	1330	1790
358.2	85.0	185	644.7	611	338	640	1100	1000	727	1340	1800
358.8	85.6	186	645.7	616	343	650	1110	1005	732	1350	1810
359.3	86.1	187	646.7	622	349	660	1120	1011	738	1360	1820
359.9	86.7	188	647.7	627	354	670	1130	1016	743	1370	1830
360.4	87.2	189	648.7	633	360	680	1140	1022	749	1380	1840
361.0	87.8	190	649.7	639	366	690	1150	1027	754	1390	1850
361.5	88.3	191	650.7	644	371	700	1160	1033	760	1400	1860
362.1	88.9	192	651.7	650	377	710	1170	1039	766	1410	1870
362.6	89.4	193	652.7	655	382	720	1180	1044	771	1420	1880
363.2	90.0	194	653.7	661	388	730	1190	1050	777	1430	1890
363.8	90.6	195	654.7	666	393	740	1200	1055	782	1440	1900
364.3	91.1	196	655.7	672	399	750	1210	1061	788	1450	1910
364.9	91.7	197	656.7	677	404	760	1220	1066	793	1460	1920
365.4	92.2	198	657.7	683	410	770	1230	1072	799	1470	1930
366.0	92.8	199	658.7	689	416	780	1240	1077	804	1480	1940
366.5	93.3	200	659.7	694	421	790	1250	1083	810	1490	1950
367.1	93.9	201	660.7	700	427	800	1260	1089	816	1500	1960
367.6	94.4	202	661.7	705	432	810	1270	1094	821	1510	1970
368.2	95.0	203	662.7	711	438	820	1280	1100	827	1520	1980
368.8	95.6	204	663.7	716	443	830	1290	1105	832	1530	1990
369.3	96.1	205	664.7	722	449	840	1300	1111	838	1540	2000
369.9	96.7	206	665.7	727	454	850	1310	1116	843	1550	2010
370.4	97.2	207	666.7	733	460	860	1320	1122	849	1560	2020
371.0	97.8	208	667.7	739	466	870	1330	1127	854	1570	2030
371.5	98.3	209	668.7	744	471	880	1340	1133	860	1580	2040
372.1	98.9	210	669.7	750	477	890	1350	1139	866	1590	2050
372.6	99.4	211	670.7	755	482	900	1360	1144	871	1600	2060
373.2	100.0	212	671.7	761	488	910	1370	1150	877	1610	2070
377	104	220	680	766	493	920	1380	1155	882	1620	2080
383	110	230	690	772	499	930	1390	1161	888	1630	2090
389	116	240	700	777	504	940	1400	1166	893	1640	2100
394	121	250	710	783	510	950	1410	1172	899	1650	2110
400	127	260	720	789	516	960	1420	1177	904	1660	2120
405	132	270	730	794	521	970	1430	1183	910	1670	2130
411	138	280	740	800	527	980	1440	1189	916	1680	2140
416	143	290	750	805	532	990	1450	1194	921	1690	2150
422	149	300	760	811	538	1000	1460	1200	927	1700	2160
427	154	310	770	816	543	1010	1470	1205	932	1710	2170
433	160	320	780	822	549	1020	1480	1211	938	1720	2180
439	166	330	790	827	554	1030	1490	1216	943	1730	2190
444	171	340	800	833	560	1040	1500	1222	949	1740	2200
450	177	350	810	839	566	1050	1510	1227	954	1750	2210
455	182	360	820	844	571	1060	1520	1233	960	1760	2220
461	188	370	830	850	577	1070	1530	1239	966	1770	2230
466	193	380	840	855	582	1080	1540	1244	971	1780	2240
472	199	390	850	861	588	1090	1550	1250	977	1790	2250
477	204	400	860	866	593	1100	1560	1255	982	1800	2260
483	210	410	870	872	599	1110	1570	1261	988	1810	2270
489	216	420	880	877	604	1120	1580	1266	993	1820	2280
494	221	430	890	883	610	1130	1590	1272	999	1830	2290
500	227	440	900	889	616	1140	1600	1277	1004	1840	2300
505	232	450	910	894	621	1150	1610	1283	1010	1850	2310
511	238	460	920	900	627	1160	1620	1289	1016	1860	2320
516	243	470	930	905	632	1170	1630	1294	1021	1870	2330
522	249	480	940	911	638	1180	1640	1300	1027	1880	2340
527	254	490	950	916	643	1190	1650	1305	1032	1890	2350
533	260	500	960	922	649	1200	1660	1311	1038	1900	2360
539	266	510	970	927	654	1210	1670	1316	1043	1910	2370
544	271	520	980	933	660	1220	1680	1322	1049	1920	2380
550	277	530	990	939	666	1230	1690	1327	1054	1930	2390
555	282	540	1000	944	671	1240	1700	1333	1060	1940	2400
561	288	550	1010	950	677	1250	1710	1339	1066	1950	2410
566	293	560	1020	955	682	1260	1720	1344	1071	1960	2420
572	299	570	1030	961	688	1270	1730	1350	1077	1970	2430
577	304	580	1040	966	693	1280	1740	1355	1082	1980	2440
583	310	590	1050	972	699	1290	1750	1361	1088	1990	2450
589	316	600	1060	977	704	1300	1760	1366	1093	2000	2460

Table B-2a Temperature conversion

CELSIUS SCALE LISTED IN EVEN NUMBERS

KELVIN K	CELSIUS C	FAHREN-HEIT F	RANKINE R	KELVIN K	CELSIUS C	FAHREN-HEIT F	RANKINE R
0	−273.16	−459.60	0	77.16	−196	−320.8	138.9
1.16	−272	−457.6	2.1	78.16	−195	−319.0	140.7
2.16	−271	−455.8	3.9	79.16	−194	−317.2	142.5
3.16	−270	−454.0	5.7	80.16	−193	−315.4	144.3
4.16	−269	−452.2	7.5	81.16	−192	−313.6	146.1
5.16	−268	−450.4	9.3	82.16	−191	−311.8	147.9
6.16	−267	−448.6	11.1	83.16	−190	−310.0	149.7
7.16	−266	−446.8	12.9	84.16	−189	−308.2	151.5
8.16	−265	−445.0	14.7	85.16	−188	−306.4	153.3
9.16	−264	−443.2	16.5	86.16	−187	−304.6	155.1
10.16	−263	−441.4	18.3	87.16	−186	−302.8	156.9
11.16	−262	−439.6	20.1	88.16	−185	−301.0	158.7
12.16	−261	−437.8	21.9	88.16	−185	−301.0	158.7
13.16	−260	−436.0	23.7	89.16	−184	−299.2	160.5
14.16	−259	−434.2	25.5	90.16	−183	−297.4	162.3
15.16	−258	−432.4	27.3	91.16	−182	−295.6	164.1
16.16	−257	−430.6	29.1	92.16	−181	−293.8	165.9
17.16	−256	−428.8	30.9	93.16	−180	−292.0	167.7
18.16	−255	−427.0	32.7	94.16	−179	−290.2	169.5
19.16	−254	−425.2	34.5	95.16	−178	−288.4	171.3
20.16	−253	−423.4	36.3	96.16	−177	−286.6	173.1
21.16	−252	−421.6	38.1	97.16	−176	−284.8	174.9
22.16	−251	−419.8	39.9	98.16	−175	−283.0	176.7
23.16	−250	−418.0	41.7	99.16	−174	−281.2	178.5
24.16	−249	−416.2	43.5	100.16	−173	−279.4	180.3
25.16	−248	−414.4	45.3	101.16	−172	−277.6	182.1
26.16	−247	−412.6	47.1	102.16	−171	−275.8	183.9
27.16	−246	−410.8	48.9	103.16	−170	−274.0	185.7
28.16	−245	−409.0	50.7	104.16	−169	−272.2	187.5
29.16	−244	−407.2	52.5	105.16	−168	−270.4	189.3
30.16	−243	−405.4	54.3	106.16	−167	−268.6	191.1
31.16	−242	−403.6	56.1	107.16	−166	−266.8	192.9
32.16	−241	−401.8	57.9	108.16	−165	−265.0	194.7
33.16	−240	−400.0	59.7	109.16	−164	−263.2	196.5
34.16	−239	−398.2	61.5	110.16	−163	−261.4	198.3
35.16	−238	−396.4	63.3	111.16	−162	−259.6	200.1
36.16	−237	−394.6	65.1	112.16	−161	−257.8	201.9
37.16	−236	−392.8	66.9	113.16	−160	−256.0	203.7
38.16	−235	−391.0	68.7	114.16	−159	−254.2	205.5
39.16	−234	−389.2	70.5	115.16	−158	−252.4	207.3
40.16	−233	−387.4	72.3	116.16	−157	−250.6	209.1
41.16	−232	−385.6	74.1	117.16	−156	−248.8	210.9
42.16	−231	−383.8	75.9	118.16	−155	247.0	212.7
43.16	−230	−382.0	77.7	119.16	−154	−245.2	214.5
44.16	−229	−380.2	79.5	120.16	−153	−243.4	216.3
45.16	−228	−378.4	81.3	121.16	−152	−241.6	218.1
46.16	−227	−376.6	83.1	122.16	−151	−239.8	219.9
47.16	−226	−374.8	84.9	123.16	−150	−238.0	221.7
48.16	−225	−373.0	86.7	124.16	−149	−236.2	223.5
49.16	−224	−371.2	88.5	125.16	−148	−234.4	225.3
50.16	−223	−369.4	90.3	126.16	−147	−232.6	227.1
51.16	−222	−367.6	92.1	127.16	−146	−230.8	228.9
52.16	−221	−365.8	93.9	128.16	−145	−229.0	230.7
53.16	−220	−364.0	95.7	129.16	−144	−227.2	232.5
54.16	−219	−362.2	97.5	130.16	−143	−225.4	243.3
55.16	−218	−360.4	99.3	131.16	−142	−223.6	236.1
56.16	−217	−358.6	101.1	132.16	−141	−221.8	237.9
57.16	−216	−356.8	102.9	133.16	−140	−220.0	239.7
58.16	−215	−355.0	104.7	134.16	−139	−218.2	241.5
59.16	−214	−353.2	106.5	135.16	−138	−216.4	243.3
60.16	−213	−351.4	108.3	136.16	−137	−214.6	245.1
61.16	−212	−349.6	110.1	137.16	−136	−212.8	246.9
62.16	−211	−347.8	111.9	138.16	−135	−211.0	248.7
63.16	−210	−346.0	113.7	139.16	−134	−209.2	250.5
64.16	−209	−344.2	115.5	140.16	−133	−207.4	252.3
65.16	−208	−342.4	117.3	141.16	−132	−205.6	254.1
66.16	−207	−340.6	119.1	142.16	−131	−203.8	255.9
67.16	−206	−338.8	120.9	143.16	−130	−202.0	257.7
68.16	−205	−337.0	122.7	144.16	−129	−200.2	259.5
69.16	−204	−335.2	124.5	145.16	−128	−198.4	261.3
70.16	−203	−333.4	126.3	146.16	−127	−196.6	263.1
71.16	−202	331.6	128.1	147.16	−126	−194.8	264.9
72.16	−201	−329.8	129.9	148.16	−125	−193.0	266.7
73.16	−200	−328.0	131.7	149.16	−124	−191.2	268.5
74.16	−199	−326.2	133.5	150.16	−123	−189.4	270.3
75.16	−198	324.4	135.3	151.16	−122	−187.6	272.1
76.16	−197	−322.6	107.1	152.16	−121	−185.8	273.9

Table B-2a (continued)

KELVIN K	CELSIUS C	FAHREN-HEIT F	RANKINE R	KELVIN K	CELSIUS C	FAHREN-HEIT F	RANKINE R
153.16	−120	−184.0	275.7	231.16	− 42	− 43.6	416.1
154.16	−119	−182.2	277.5	232.16	− 41	− 41.8	417.9
155.16	−118	−180.4	279.3	233.16	− 40	− 40.0	419.7
156.16	−117	−178.6	281.1	234.16	− 39	− 38.2	421.5
157.16	−116	−176.8	282.9	235.16	− 38	− 36.4	423.3
158.16	−115	−175.0	284.7	236.16	− 37	− 34.6	425.1
159.16	−114	−173.2	286.5	237.16	− 36	− 32.8	426.9
160.16	−113	−171.4	288.3	238.16	− 35	− 31.0	428.7
161.16	−112	−169.6	290.1	239.16	− 34	− 29.2	430.5
162.16	−111	−167.8	291.9	240.16	− 33	− 27.4	432.3
163.16	−110	−166.0	293.7	241.16	− 32	− 25.6	434.1
164.16	−109	−164.2	295.5	242.16	− 31	− 23.8	435.9
165.16	−108	−162.4	297.3	243.16	− 30	− 22.0	437.7
166.16	−107	−160.6	299.1	244.16	− 29	− 20.2	439.5
167.16	−106	−158.8	300.9	245.16	− 28	− 18.4	441.3
168.16	−105	−157.0	302.7	246.16	− 27	− 16.6	443.1
169.16	−104	−155.2	304.5	247.16	− 26	− 14.8	444.9
170.16	−103	−153.4	306.3	248.16	− 25	− 13.0	446.7
171.16	−102	−151.6	308.1	249.16	− 24	− 11.2	448.5
172.16	−101	−149.8	309.9	250.16	− 23	− 9.4	450.3
173.16	−100	−148.0	311.7	251.16	− 22	− 7.6	452.1
173.16	−100	−148.0	311.7	252.16	− 21	− 5.8	453.9
174.16	− 99	−146.2	313.5	253.16	− 20	− 4.0	455.7
175.16	− 98	−144.4	315.3	254.16	− 19	− 2.2	457.5
176.16	− 97	−142.6	317.1	255.16	− 18	− 0.4	459.3
177.16	− 96	−140.8	318.9	256.16	− 17	1.4	461.1
178.16	− 95	−139.0	320.7	257.16	− 16	3.2	462.9
179.16	− 94	−137.2	322.5	258.16	− 15	5.0	464.7
180.16	− 93	−135.4	324.3	258.16	−15	5.0	464.7
181.16	− 92	−133.6	326.1	259.16	−14	6.8	466.5
182.16	− 91	−131.8	327.9	260.16	−13	8.6	468.3
183.16	− 90	−130.0	329.7	261.16	−12	10.4	470.1
184.16	− 89	−128.2	331.5	262.16	−11	12.2	471.9
185.16	− 88	−126.4	333.3	263.16	−10	14.0	473.7
186.16	− 87	−124.6	335.1	264.16	− 9	15.8	475.5
187.16	− 86	−122.8	336.9	265.16	− 8	17.6	477.3
188.16	− 85	−121.0	338.7	266.16	− 7	19.4	479.1
189.16	− 84	−119.2	340.5	267.16	− 6	21.2	480.9
190.16	− 83	−117.4	342.3	268.16	− 5	23.0	482.7
191.16	− 82	−115.6	344.1	269.16	− 4	24.8	484.5
192.16	− 81	−113.8	345.9	270.16	− 3	26.6	486.3
193.16	− 80	−112.0	347.7	271.16	− 2	28.4	488.1
194.16	− 79	−110.2	349.5	272.16	− 1	30.2	489.9
195.16	− 78	−108.4	351.3	273.16	0	32.0	491.7
196.16	− 77	−106.6	353.1	274.16	1	33.8	493.5
197.16	− 76	−104.8	354.9	275.16	2	35.6	495.3
198.16	− 75	−103.0	356.7	276.16	3	37.4	497.1
199.16	− 74	−101.2	358.5	277.16	4	39.2	498.9
200.16	− 73	− 99.4	360.3	278.16	5	41.0	500.7
201.16	− 72	− 97.6	362.1	279.16	6	42.8	502.5
202.16	− 71	− 95.8	363.9	280.16	7	44.6	504.3
203.16	− 70	− 94.0	365.7	281.16	8	46.4	506.1
204.16	− 69	− 92.2	367.5	282.16	9	48.2	507.9
205.16	− 68	− 90.4	369.3	283.16	10	50.0	509.7
206.16	− 67	− 88.6	371.1	284.16	11	51.8	511.5
207.16	− 66	− 86.8	372.9	285.16	12	53.6	513.3
208.16	− 65	− 85.0	374.7	286.16	13	55.4	515.1
209.16	− 64	− 83.2	376.5	287.16	14	57.2	516.9
210.16	− 63	− 81.4	378.3	288.16	15	59.0	518.7
211.16	− 62	− 79.6	380.1	289.16	16	60.8	520.5
212.16	− 61	− 77.8	381.9	290.16	17	62.6	522.3
213.16	− 60	− 76.0	383.7	291.16	18	64.4	524.1
214.16	− 59	− 74.2	385.5	292.16	19	66.2	525.9
215.16	− 58	− 72.4	387.3	293.16	20	68.0	527.7
216.16	− 57	− 70.6	389.1	294.16	21	69.8	529.5
217.16	− 56	− 68.8	390.9	295.16	22	71.6	531.3
218.16	− 55	− 67.0	392.7	296.16	23	73.4	533.1
219.16	− 54	− 65.2	394.5	297.16	24	75.2	534.9
220.16	− 53	− 63.4	396.3	298.16	25	77.0	536.7
221.16	− 52	− 61.6	398.1	299.16	26	78.8	538.5
222.16	− 51	− 59.8	399.9	300.16	27	80.6	540.3
223.16	− 50	− 58.0	401.7	301.16	28	82.4	542.1
224.16	− 49	− 56.2	403.5	302.16	29	84.2	543.9
225.16	− 48	− 54.4	405.3	303.16	30	86.0	545.7
226.16	− 47	− 52.6	407.1	304.16	31	87.8	547.5
227.16	− 46	− 50.8	408.9	305.16	32	89.6	549.3
228.16	− 45	− 49.0	410.7	306.16	33	91.4	551.1
229.16	− 44	− 47.2	412.5	307.16	34	93.2	552.9
230.16	− 43	− 45.4	414.3	308.16	35	95.0	554.7

Table B-2a Temperature conversion (continued)

KELVIN K	CELSIUS C	FAREN-HEIT F	RANKINE R	KELVIN K	CELSIUS C	FAREN-HEIT F	RANKINE R
309.16	36	96.8	556.5	513	240	464	923
310.16	37	98.6	558.3	523	250	482	941
311.16	38	100.4	560.1	533	260	500	959
312.16	39	102.2	561.9	543	270	518	977
313.16	40	104.0	563.7	553	280	536	995
314.16	41	105.8	565.5	563	290	554	1013
315.16	42	107.6	567.3	573	300	572	1031
316.16	43	109.4	569.1	583	310	590	1049
317.16	44	111.2	570.9	593	320	608	1067
318.16	45	113.0	572.7	603	330	626	1085
319.16	46	114.8	574.5	613	340	644	1103
320.16	47	116.6	576.3	623	350	662	1121
321.16	48	118.4	578.1	633	360	680	1139
322.16	49	120.2	579.9	643	370	698	1157
323.16	50	122.0	581.7	653	380	716	1175
324.16	51	123.8	583.5	663	390	734	1193
325.16	52	125.6	585.3	673	400	752	1211
326.16	53	127.4	587.1	683	410	770	1229
327.16	54	129.2	588.9	693	420	788	1247
328.16	55	131.0	590.7	703	430	806	1265
329.16	56	132.8	592.5	713	440	824	1283
330.16	57	134.6	594.3	723	450	842	1301
331.16	58	136.4	596.1	733	460	860	1319
332.16	59	138.2	597.9	743	470	878	1337
333.16	60	140.0	599.7	753	480	896	1355
334.16	61	141.8	601.5	763	490	914	1373
335.16	62	143.6	603.3	773	500	932	1391
336.16	63	145.4	605.1	783	510	950	1409
337.16	64	147.2	606.9	793	520	968	1427
338.16	65	149.0	608.7	803	530	986	1445
339.16	66	150.8	610.5	813	540	1004	1463
340.16	67	152.6	612.3	823	550	1022	1481
341.16	68	154.4	614.1	833	560	1040	1499
342.16	69	156.2	615.9	843	570	1058	1517
343.16	70	158.0	617.7	853	580	1076	1535
343.16	70	158.0	617.7	863	590	1094	1553
344.16	71	159.8	619.5	873	600	1112	1571
345.16	72	161.6	621.3	883	610	1130	1589
346.16	73	163.4	623.1	893	620	1148	1607
347.16	74	165.2	624.9	903	630	1166	1625
348.16	75	167.0	626.7	913	640	1184	1643
349.16	76	168.8	628.5	923	650	1202	1661
350.16	77	170.6	630.3	923	650	1202	1661
351.16	78	172.4	632.1	933	660	1220	1679
352.16	79	174.2	633.9	943	670	1238	1697
353.16	80	176.0	635.7	953	680	1256	1715
354.16	81	177.8	637.5	963	690	1274	1733
355.16	82	179.6	639.3	973	700	1292	1751
356.16	83	181.4	641.1	983	710	1310	1769
357.16	84	183.2	642.9	993	720	1328	1787
358.16	85	185.0	644.7	1003	730	1346	1805
359.16	86	186.8	646.5	1013	740	1364	1823
360.16	87	188.6	648.3	1023	750	1382	1841
361.16	88	190.4	650.1	1033	760	1400	1859
362.16	89	192.2	651.9	1043	770	1418	1877
363.16	90	194.0	653.7	1053	780	1436	1895
364.16	91	195.8	655.5	1063	790	1454	1913
365.16	92	197.6	657.3	1073	800	1472	1931
366.16	93	199.4	659.1	1083	810	1490	1949
367.16	94	201.2	660.9	1093	820	1508	1967
368.16	95	203.0	662.7	1103	830	1526	1985
369.16	96	204.8	664.5	1113	840	1544	2003
370.16	97	206.6	666.3	1123	850	1562	2021
371.16	98	208.4	668.1	1133	860	1580	2039
372.16	99	210.2	669.9	1143	870	1598	2057
373.16	100	212.0	671.7	1153	880	1616	2075
383	110	230	689	1163	890	1634	2093
393	120	248	707	1173	900	1652	2111
403	130	266	725	1183	910	1670	2129
413	140	284	743	1193	920	1688	2147
423	150	302	761	1203	930	1706	2165
433	160	320	779	1213	940	1724	2183
443	170	338	797	1223	950	1742	2201
453	180	356	815	1233	960	1760	2219
463	190	374	833	1243	970	1778	2237
473	200	392	851	1253	980	1796	2255
483	210	410	869	1263	990	1814	2273
493	220	428	887	1273	1000	1832	2291
503	230	446	905	1283	1010	1850	2309

Table B-2a (continued)

KELVIN K	CELSIUS C	FAREN-HEIT F	RANKINE R	KELVIN K	CELSIUS C	FAREN-HEIT F	RANKINE R
1293	1020	1868	2327	1543	1270	2318	2777
1303	1030	1886	2345	1553	1280	2336	2795
1313	1040	1904	2363	1563	1290	2354	2813
1323	1050	1922	2381	1573	1300	2372	2831
1333	1060	1940	2399	1583	1310	2390	2849
1343	1070	1958	2417	1593	1320	2408	2867
1353	1080	1976	2435	1603	1330	2426	2885
1363	1090	1994	2453	1613	1340	2444	2903
1373	1100	2012	2471	1623	1350	2462	2921
1383	1110	2030	2489	1633	1360	2480	2939
1393	1120	2048	2507	1643	1370	2498	2957
1403	1130	2066	2525	1653	1380	2516	2975
1413	1140	2084	2543	1663	1390	2534	2993
1423	1150	2102	2561	1673	1400	2552	3011
1433	1160	2120	2579	1683	1410	2570	3029
1443	1170	2138	2597	1693	1420	2588	3047
1453	1180	2156	2615	1703	1430	2606	3065
1463	1190	2174	2633	1713	1440	2624	3083
1473	1200	2192	2651	1723	1450	2642	3101
1483	1210	2210	2669	1733	1460	2660	3119
1493	1220	2228	2687	1743	1470	2678	3137
1503	1230	2246	2705	1753	1480	2696	3155
1513	1240	2264	2723	1763	1490	2714	3173
1523	1250	2282	2741	1773	1500	2732	3191
1533	1260	2300	2759				

Table B-2b Relationship of temperature scales

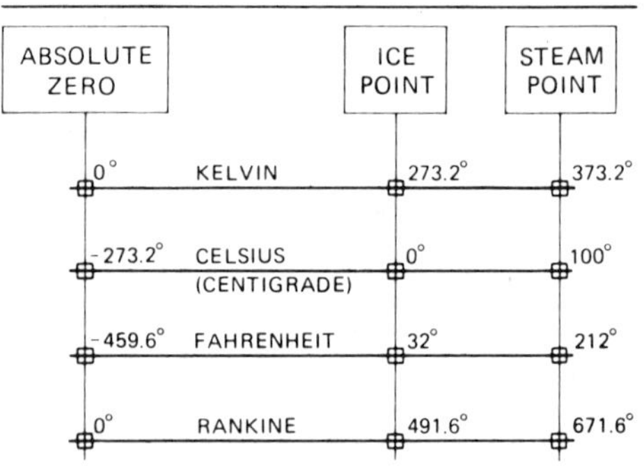

Table B-2c Temperature Conversion Factors

Temp. (t)	Fahrenheit (°F)	Rankine (°R)	Celsius (°C)	Kelvin (°K)
°F =	1	t°R − 459.6	1.8 t°C + 32	1.8(t°K − 273.16) + 32
°R =	t°F + 459.69	1	1.8 t°C + 491.69	1.8 t°K
°C =	(t°F − 32) X 5/9	5/9(t°R) − 273.16	1	t°K − 273.16
°K =	(t°F − 32) X 5/9 + 273.16	5/9(t°R)	t°C + 273.16	1

Table B-3 English to Metric to English Conversion Table

Physical or Thermal Property	English System Quantity	English System Units Symbol	Conversion Factors Multiply by →	Metric System Quantity	Metric System Units Symbol	Conversion Factor Multiply by →	English System Quantity	English System Units Symbol
Heat	calorie	cal	4.1868	joule	J	0.238846	calorie	cal
Heat	British Thermal Unit	Btu	1055.056	joule	J	0.0009478	British Thermal Unit	Btu
Heat	therm (100,000 Btu)	Btu x 10^5	105505600	joule	J	0.00000000948	therm	Btu x 10^5
Heat Capacity	Btu per degree F	Btu/°F	1899.0	joule per degree C	J/°C	0.00052659	Btu per degree F	Btu/°F
Heat Enthalpy	Btu per pound	Btu/lb	2326.0	joule per kilogram	J/kg	0.0004299	Btu per pound	Btu/lb
Specific Heat	Btu per pound per °F	Btu/lb, °F	4187.0	joule per kilogram °C	J/kg°C	0.000238846	Btu per pound °F	Btu/lb,°F
Entropy	Btu per degree Rankine	Btu/°R	1899.0	joule per degree K	J/°K	0.00052659	Btu per °R	Btu/°R
Specific Entropy	Btu per pound per °R	Btu/lb, °R	4187.0	joule per kilogram °K	J/kg°K	0.000238846	Btu per pound °R	Btu/lb,°R
Latent Heat	Btu per pound	Btu/lb	2326.0	joule per kilogram	J/kg	0.0004299	Btu per pound	Btu/lb
Vol. Heat	Btu per cubic foot	Btu/ft^3	37260.0	joule per cubic meter	J/m^3	0.00002684	Btu per ft^3	Btu/ft^3
Spec. Heat, Vol.	Btu per cubic foot °F	Btu/ft^3, °F	67010.0	joule per cubic meter C	J/m^3C	0.000014923	Btu per ft^3, °F	Btu/ft^3, °F
Spec. Heat, Wt.	Btu per pound per °F	Btu/lb, °F	1.163	watt hour per kilogram K	W, hr/kgK	0.859845	Btu per pound °F	Btu/lb, °F
Spec. Heat, Wt.	Btu per pound per °F	Btu/lb, °F	4186.8	joule per kilogram kelvin	J/kgK	0.0002388	Btu per pound °F	Btu/lb, °F
Heat Flow	calorie per second	cal/s	4.184	watt	W	0.239006	calorie/sec	cal/s
Heat Flow	calorie per minute	cal/min	0.06973	watt	W	14.34103	calorie/minute	cal/min
Heat Flow	Btu per second	Btu/s	1055.056	watt	W	0.0009478	Btu per second	Btu/s
Heat Flow	Btu per minute	Btu/min	17.58426	watt	W	0.056869	Btu per minute	Btu/min
Heat Flow	Btu per hour	Btu/hr	0.2930711	watt	W	3.141214	Btu per hour	Btu/hr
Heat/area, time	Btu per sq foot per hr	Btu/ft^2, hr	3.152481	watts per sq meter	W/m^2	0.31721	Btu per sq ft per hr	Btu/ft^2, hr
Heat/lin, time	Btu per lin foot per hr	Btu/lin, ft, hr	0.9609097	watts per lin meter	W/lin, m	1.04068	Btu per lin ft per hr	Btu/lin, ft, hr
Heat/area, time °	Btu per sq ft per hr °F	Btu/ft^2, hr°F	5.67826	watts per sq m per °K	W/m^2K	0.17611	Btu ft^2 per hr °F	Btu/ft^2, hr°F
Conductance	Btu per sq ft per hour °F	Btu/ft^2hr°F	5.678	watts per sq m per °K	W/m^2K	0.17611835	Btu per sq ft per °F	Btu/ft^2hr°F
Conductivity	Btu inch per ft^2, hr, °F	Btu in./ft^2 hr°F	0.1442	watts m per sq m per °K	W/mK	6.9348	Btu in. per ft^2, hr°F	Btu in./ft^2hr°F
Conductivity	Btu ft per ft^2, hr, °F	Btu/ft/ft^2, hr°F	1.731	watts m per sq m per °K	W/mK	0.5777	Btu ft per ft^2, hr°F	Btu ft/ft^2 hr°F
Resistance	Degree F, sq ft hr/Btu	°F, ft^2, hr/Btu	0.1761	°K meter^2/watt	K m^2W	5.6786	Degree, F, sq ft hr/Btu	°F, ft^2, hr/Btu
MOISTURE								
Content	Grains per pound		1.428	gram per kilogram	g/kg	0.70028	Grains per pound	
Vapor trans.	Perm-inch		1.8	Perm-centimeter		0.555555	Perm-inch	
Permeability	Pound foot/hr pound force		0.00000862	Kilogram, m/Newton sec		116.009.8	lb, ft/hr, lb	
Perm (23°C)	Grain/ft^2, hr, inch of Hg		5.74525 E-11*	Kilogram/pascal, sec sq m		17406 E + 10	Grain/ft^2, hr, inch of Hg	
Perm-in (23°C)	Grain/ft^2, hr, in. of Hg, in.		1.45929 E-12*	Kilogram/pascal, sec meter		.68526 E + 11	Grain/ft^2, hr, inch Hg, inch	
Perm, British	English Units		0.659	Metric Per m		1.517	Per m, British	

Note: This table of conversions from English Units to Metric Units and back to English Units is not a complete English-Metric conversion table. The units listed are those units which are likely to be used in the thermal insulation technology. Complete tabulation is available as ASTM Standard for Metric Practice E380-76.

*As defined by ASTM. Test methods E-11 and E-12.

Table B-3 English to Metric to English Conversion Table (Continued)

Physical or Thermal Property	English System Quantity	English System Units Symbol	Conversion Factors Multiply by →	Metric System Quantity	Metric System Units Symbol	Conversion Factor Multiply by →	English System Quantity	English System Units Symbol
Force	ounce force		0.2780139	Newton	N	3.596942	ounce force	
Force	pound force	lbf	4.448222	Newton	N	0.224808	pound force	lbf
Force	pound thrust/pound	lbf/lb	9.80665	Newton per kilogram	N/kg	0.1019716	pound thrust/lb	lbf/lb
Force	poundal		0.138255	Newton	N	7.233	poundal	
Force	dyne		0.00001	Newton		100000.0	dyne	
Force/length	pounds foot per foot	lbf/ft	14.5939	Newton per metre	N/m	0.068522	pound foot/foot	lbf/ft
Force/length	pounds foot per inch	lbf/in.	175.1268	Newton per metre	N/m	0.0057101	pound foot/inch	lbf/in.
Torque	pound force per foot		1.356	Newton metre	N/m	0.737463	pound force per ft	
Pressure	pound force per sq inch	$lbf/in.^2$	6895.0	Newton per square metre	N/m^2	0.000145	pound/square inch	psi
Pressure	pound force per sq inch	$lbf/in.^2$	6895.0	pascal	Pa	0.000145	pound/square inch	psi
Pressure	pound force per sq inch		703.1	kilogram per square metre	kgs/m^2	0.0014227	pound per sq ft	lb/ft^2
Pressure	pound force per sq ft	lbf/ft^2	47.88026	$Newton/m^2$ or pascal	Pa	0.020885	pound/square foot	lbf/ft^2
Pressure	inch H_2O (at 40° F)	in.H_2O	249.082	$Newton/m^2$ or pascal	Pa	0.0040147	in. of H_2O (40° F)	in.H_2O
Pressure	inch of mercury (at 40° F)	1″ Hg	3386.0	$Newton/m^2$ or pascal	Pa	0.000029	in. of Hg (40° F)	in. Hg
Pressure	foot of H_2O (at 40° F)	1′ H_2O	2988.98	$Newton/m^2$ or pascal	Pa	0.0033457	foot of H_2O (40° F)	ft H_2O
Pressure	torr (millimetre of Hg)	torr	133.32	$Newton/m^2$ or pascal	Pa	0.0075007	torr (mm of Hg)	torr
Pressure	atmosphere (standard)		101325.	$Newton/m^2$ or pascal	Pa	0.00000987	atmosphere, std.	
Press./Drop	in. of H_2O per 100 feet		8.170	Newton per cubic meter	N/m^2m	0.122309	inch H_2O/100 lin ft	
Press./Drop	foot H_2O per 100 feet		98.1	Newton per cubic meter	N/m^2m	0.0101938	foot H_2O/100 lin ft	
Dynamic Viscosity	pound force second/$foot^2$		47.88026	Newton, second/meter	Ns/m	0.020885	pound force sec/ft^2	
Dynamic Viscosity	pound force second/$foot^2$		47880.0	Centipoise	cP	0.0000209	pound force sec/ft^2	
Dynamic Viscosity	pound force hour/$foot^2$		172400.	Newton, second/meter	Ns/m	0.0000058	pound force hour/ft^2	
Energy*	foot pound force	ftlbf	1.355818	joule	J	0.737562	pound force hour/ft^2	
Energy	foot poundal		0.04214	joule	J	23.73042	foot poundal	
Energy	horsepower hour		2684.000	joule	J	0.000000372	horsepower hr	
Energy	kilowatt hour	kWh	3600000	joule	J	0.000000277	kilowatt hr	
Energy	watt per second	Ws	1.0000	joule	J	1.000	watt per second W/s	
Energy	calorie		4.1868	joule	J	0.238846	calorie	
Energy Flow	horsepower		745.7	watts	W	0.001341	horsepower	
Energy Flow	horsepower		0.7457	kilowatts	kW	1.341022	horsepower	
Energy Flow	kilocalorie-hour		1.163	watts	W	0.859845	kilocalorie/hr	

*Other than thermal energy

Note: This table of conversions from English Units to Metric Units and back to English Units is not a complete English-Metric conversion table. The units listed are those units which are likely to be used in the thermal insulation technology. Complete tabulation is available as ASTM Standard for Metric Practice E380-76.

Table B-3 English to Metric to English Conversion Table (Continued)

Physical or Thermal Property	English System Quantity	Units Symbol	Conversion Factors Multiply by →	Metric System Quantity	Units Symbol	Conversion Factor Multiply by →	English System Quantity	Units Symbol
Length	inch	"	25.4	millimeter	mm	0.03937	inch	"
Length	inch	"	0.0254	meter	m	39.37	inch	"
Length	foot	'	3.048	meter	m	0.32808	foot	'
Length	yard	yd.	0.9144	meter	m	1.09361	yard	yd.
Length	mile	mi.	1.6093	kilometer	km	0.62139	mile	mi.
Area	square inch	in.^{2}	645.16	square, millimeter	mm^2	0.00155	sq inch	in.^{2}
Area	square inch	in.^{2}	0.00064516	square meter	m^2	1549.148	sq inch	in.^{2}
Area	square foot	ft^2	0.0929	square meter	m^2	10.76426	sq ft	ft^2
Area	square yard	yd^2	0.8361	square meter	m^2	1.196	sq yd	yd^2
Area	square mile	mi^2	2589998.0	square meter	m^2	0.000000386	sq mi	mi^2
Area	square mile	mi^2	2.589998	square kilometer	km^2	0.386103	sq mi	mi^2
Area	acre	acre	4046.873	square meter	m^2	0.00024698	acre	acre
Area	acre	acre	0.004046	square kilometer	km^2	246.98	acre	acre
Volume	cubic inch	in.^{3}	16.39	cubic milli meter	mm^2	0.061013	cubic in.	in.^{3}
Volume	cubic inch	in.^{3}	0.00001639	cubic meter	m^3	61013.0	cubic in.	in.^{3}
Volume	cubic foot	ft^3	0.02832	cubic meter	m^3	35.31	cubic ft	ft^3
Volume	cubic foot	ft^3	28.32	liter	ℓ	0.03531	cubic ft	ft^3
Volume	cubic yard	yd^3	0.7646	cubic meter	m^3	1.30787	cubic yd	yd^3
Volume	cubic inch	in.^{3}	0.01639	liter	ℓ	61.0128	cubic in.	in.^{3}
Volume	pint (liquid)	pt.	0.0004732	cubic meter	m^2	2113.271	pint (liquid)	pt.
Volume	pint (liquid)	pt.	0.4732	liter	ℓ	2.113271	pint (liquid)	pt.
Volume	quart (liquid)	qt.	0.0009463	cubic meter	m^3	1056.688	quart (liquid)	qt.
Volume	quart (liquid)	qt.	0.9463529	liter	ℓ	1.056688	quart (liquid)	qt.
Volume	gallon (liquid)	gal.	0.0037854	cubic meter	m^3	264.1728	gallon (liquid)	gal.
Volume	gallon (liquid)	gal.	3.785412	liter	ℓ	0.2641728	gallon (liquid)	gal.
Mass	ounce	oz.	0.029349	kilogram	kg	35.274612	ounce	oz.
Mass	pound	lb.	0.4535924	kilogram	kg	2.204622	pound	lb.
Mass	ton	ton	1016.047	kilogram	kg	0.0009941	ton	ton
Mass	ton (short)	short ton	907.1847	kilogram	kg	0.0011023	ton (short)	short ton
Mass	ton (metric)	metric ton	1000.0	kilogram	kg	0.001	ton (metric)	metric ton
Mass	grain	grain	0.0000648	kilogram	kg	15432.36	grain	grain
Length Expan.	length, foot per° F	ft/° F	1.600	meter per ° K	m/° K	0.625	length, ft/° F	ft/° F

Note: This table of conversions from English Units to Metric Units and back to English Units is not a complete English-Metric conversion table. The units listed are those units which are likely to be used in the thermal insulation technology. Complete tabulation is available as ASTM Standard for Metric Practice E380-76.

Table B-3 English to Metric to English Conversion Table (Continued)

Physical or Thermal Property	English System Quantity	Units Symbol	Conversion Factors Multiply by →	Metric System Quantity	Units Symbol	Conversion Factor Multiply by →	English System Quantity	Units Symbol
Mass/Length	pound, per inch	lb/in.	17.85797	kilogram/meter	kg/m	0.055997	pound per inch	lb/in
Mass/Length	pound per foot	lb/ft	1.488164	kilogram/meter	kg/m	0.671968	pound per foot	lb/ft
Mass/area	ounce per sq foot	oz/ft^2	0.3051517	kilogram per meter^2	kg/m^2	3.277058	ounce/sq ft	oz./ft^2
Mass/area	pound per sq foot	lb/ft^2	4.882428	kilogram per meter^2	kg/m^2	0.204816	pound/sq ft	lb/ft^2
Specific Vol.	cubic foot per pound	ft^3/lb	0.06243	cubic meter per kilogram	m^3/kg	16.01794	cubic ft/lb	ft^3/lb
Specific Wt.	pound per cubic inch	lb/in.^{3}	27680.0	kilogram per cubic meter	kg/m^3	0.00003613	pound/cubic inch	lb/in.^{3}
Specific Wt.	pound per cubic foot	lb/ft^3	1.602	kilogram per cubic meter	kg/m^3	0.6242197	pound/cubic foot	lb/ft^3
Specific Wt.	pound per gallon	lb/gal	9.978	kilogram per cubic meter	kg/m^3	0.1002205	pound/gal	lb/gal
Mass Flow	pound per hour	lb/hr	0.000126	kilogram per second	kg/s	7936.507	pound/hr	lb/hr
Mass Flow/Area	pound/inch^2 per hour	lb/in.^{2}, hr	0.1953	kilogram/sq meter, second	kg/m^{2}s	5.120327	pound/inch^2, hr	lb/in.^{2}, hr
Mass Flow/Area	pound/foot^2 per hour	lb/ft^2, hr	0.001356	kilogram/sq meter, second	kg/m^{2}s	737.463127	pound/ft^2, hr	lb/ft^2, hr
Vol. Flow Rate	cubic foot/second	ft^3/sec	0.02832	cubic meter, per second	m^3/s	35.31073	cubic ft/sec	ft^3/sec
Vol. Flow Rate	cubic foot/second	ft^3/sec	2.832	liter per second	ℓ/s	0.353107	cubic ft/sec	ft^3/sec
Vol. Flow Rate	cubic foot/minute	ft^3/min	0.0094719	cubic meter per second	m^2/s	2119.093	cubic ft/min	ft^3/min
Vol. Flow Rate	cubic foot/minute	ft^3/min	0.4719	liber per second	ℓ/s	2.119093	cubic ft/min	ft^3/min
Vol. Flow Rate	gallon/minute	gal/min.	0.00006309	cubic meter per second	m^3/s	15850.372	gallon/minute	gal/min
Vol. Flow Rate	gallon/minute	gal/min	0.06309	liter per second	ℓ/s	15.850372	gallon/minute	gal/min
Voi. Flow Rate	gallon/hour	gal/hr	0.000001052	cubic meter per second	m^1/s	950570.34	gallon/hour	gal/hr
Vol. Flow Rate	gallon/hour	gal/hr	0.001052	liter per second	ℓ/s	950.57034	gallon/hour	gal/hr
MOTION								
Velocity	feet per sec	ft/sec	0.3048	meter per second	m/s	3.28084	feet/sec	ft/sec
Velocity	feet per min	ft/min	0.00508	meter per second	m/s	196.85039	feet/min	ft/min
Velocity	miles per hour	mph	0.447	meter per second	m/s	2.237136	miles/hour	mph
Velocity	miles per hour	mph	1.609	kilometer per hour	kms/hr	0.621504	miles/hour	mph
Velocity	miles per hour	mph	0.8684	knots	knots	1.151543	miles/hour	mph
Velocity	knots	knot	0.5144	meter per second	m/s	1.9440	knots	knot
Accelerotion	feet per second^2	ft/sec^2	0.3048	meter per second^2	m/s^2	3.28084	feet/sec^2	ft/sec^2
Frequency	cycle per second	cy/sec	1.00	hertz		1.00	cycle/sec	cy/sec
Frequency	sq ft per hour	ft^2/hr	0.0000258	sq meter per second	m^2/s	38750.0	sq ft per hour	ft^2/hr
Mass/area	gal/100 ft^2	gal/100 ft^2	0.228	liters per meter^2	ℓ/m^2	64.3859	gal per 100 sq ft	gal/100 ft^2
Mass/Volume	pound/gal	lb/gal	0.0978	kg per liter	kg/ℓ	8.34724	pound/gal	lb/gal
Area/Volume	sq ft/gal	ft^2/gal	0.024462	sq meters per liter	m^2/litre	40.8795	sq ft per gal	ft^2/gal

Note: This table of conversions from English Units to Metric Units and back to English Units is not a complete English-Metric conversion table. The units listed are those units which are likely to be used in the thermal insulation technology. Complete tabulation is available as ASTM Standard for Metric Practice E380-76.

Table B-4

CONVERSIONS: UNITS OF ENERGY PER UNIT OF AREA
Btu/Sq Ft to *Langleys (Gram-Calorie/Cm²) and Watt/Sq Meter

UNITS			UNITS			UNITS			UNITS			UNITS		
English	Metric		English	Metric		English	Metric		English	Metric		English	Metric	
Btu Per Sq Ft	*Gram Calorie Per Cm²	Watts Per Sq Meter	Btu Per Sq Ft	*Gram Calorie Per Cm²	Watts Per Sq Meter	Btu Per Sq Ft	*Gram Calorie Per Cm²	Watts Per Sq Meter	Btu Per Sq Ft	*Gram Calorie Per Cm²	Watts Per Sq Meter	Btu Per Sq Ft	*Gram Calorie Per Cm²	Watts Per Sq Meter
5	1	16	205	56	646	405	110	1277	605	164	1907	805	218	2538
10	3	32	210	57	662	410	111	1293	610	165	1923	810	220	2554
15	4	47	215	58	677	415	113	1308	615	167	1939	815	221	2569
20	5	63	220	59	694	420	114	1324	620	168	1955	820	222	2585
25	7	79	225	61	709	425	115	1340	625	169	1970	825	224	2601
30	8	95	230	62	725	430	117	1356	630	171	1986	830	225	2617
35	9	110	235	64	741	435	118	1371	635	172	2002	835	226	2632
40	11	126	240	65	757	440	119	1387	640	174	2018	840	228	2648
45	12	142	245	66	772	445	121	1403	645	175	2033	845	229	2664
50	14	158	250	67	788	450	122	1419	650	176	2049	850	231	2680
55	15	173	255	69	804	455	123	1434	655	178	2065	855	232	2695
60	16	189	260	71	820	460	125	1450	660	179	2081	860	233	2711
65	18	205	265	72	835	465	126	1465	665	180	2096	865	234	2727
70	19	221	270	73	851	470	127	1482	670	182	2112	870	236	2743
75	20	236	275	75	867	475	129	1497	675	183	2128	875	237	2758
80	22	252	280	76	883	480	130	1513	680	184	2144	880	239	2774
85	23	268	285	77	898	485	131	1529	685	186	2159	885	240	2790
90	24	284	290	78	914	490	133	1545	690	187	2175	890	241	2806
95	26	299	295	80	930	495	134	1560	695	188	2191	895	243	2821
100	27	315	300	81	946	500	136	1576	700	190	2207	900	244	2837
105	28	331	305	82	962	505	137	1592	705	191	2222	905	245	2853
110	30	347	310	84	977	510	138	1608	710	192	2238	910	247	2869
115	31	363	315	85	993	515	140	1624	715	194	2254	915	248	2885
120	33	378	320	87	1009	520	141	1639	720	195	2270	920	250	2900
125	34	394	325	88	1025	525	142	1655	725	197	2286	925	251	2916
130	35	410	330	89	1040	530	143	1671	730	198	2301	930	252	2932
135	37	426	335	91	1056	535	145	1687	735	199	2317	935	254	2948
140	38	441	340	92	1072	540	146	1702	740	201	2332	940	255	2963
145	39	457	345	94	1088	545	147	1718	745	220	2349	945	256	2979
150	41	473	350	95	1103	550	149	1734	750	203	2364	950	258	2995
155	42	489	355	96	1119	555	151	1750	755	205	2380	955	259	3011
160	43	504	360	98	1135	560	152	1765	760	206	2396	960	260	3020
165	45	520	365	99	1151	565	153	1781	765	207	2412	965	262	3042
170	46	536	370	100	1166	570	155	1797	770	209	2427	970	263	3058
175	47	552	375	102	1182	575	156	1813	775	210	2443	975	264	3074
180	49	567	380	103	1198	580	157	1828	780	212	2459	980	266	3089
185	50	583	385	104	1214	585	158	1844	785	213	2475	985	267	3105
190	51	599	390	106	1229	590	160	1860	790	214	2400	990	269	3121
195	52	615	395	107	1245	595	161	1876	795	216	2506	995	270	3137
200	54	630	400	108	1261	600	163	1891	800	217	2522	1000	271	3152

*1Btu per square foot = 0.27122 Langley (gram calorie per square centimeter)
= 3.152481 watt per square meter

Table B-4 (Continued)

CONVERSIONS: UNITS OF ENERGY PER UNIT OF AREA

*Langleys to Watts Per Sq Meter also to Btu/Sq Ft

UNITS			UNITS			UNITS			UNITS			UNITS		
Metric		English	Metric		English	Metric		English	Metric		English	Metric		English
*Gram Calorie Per Cm^2	Watts Per Sq Meter	Btu Per Sq Ft	*Gram Calorie Per Cm^2	Watts Per Sq Meter	Btu Per Sq Ft	*Gram Calorie Per Cm^2	Watts Per Sq Meter	Btu Per Sq Ft	*Gram Calorie Per Cm^2	Watts Per Sq Meter	Btu Per Sq Ft	*Gram Calorie Per Cm^2	Watts Per Sq Meter	Btu Per Sq Ft
5	58	18	205	2383	756	405	4707	1493	605	7032	2231	805	9357	2968
10	116	37	210	2441	774	410	4765	1512	610	7090	2249	810	9415	2986
15	174	55	215	2499	793	415	4824	1530	615	7148	2268	815	9473	3005
20	232	74	220	2557	811	420	4881	1549	620	7206	2286	820	9531	3023
25	291	92	225	2615	830	425	4939	1567	625	7264	2304	825	9589	3042
30	349	111	230	2673	848	430	4998	1585	630	7323	2323	830	9647	3060
35	407	129	235	2731	866	435	5056	1604	635	7380	2341	835	9705	3079
40	465	147	240	2790	885	440	5114	1622	640	7439	2350	840	9763	3097
45	523	166	245	2848	903	445	5172	1641	645	7497	2378	845	9822	3116
50	581	184	250	2906	922	450	5230	1659	650	7555	2397	850	9880	3134
55	639	203	255	2964	940	455	5289	1678	655	7613	2415	855	9938	3152
60	697	221	260	3022	959	460	5347	1696	660	7671	2433	860	9996	3171
65	756	240	265	3080	977	465	5405	1714	665	7729	2452	865	10054	3189
70	814	258	270	3138	995	470	5463	1733	670	7788	2470	870	10112	3208
75	872	277	275	3196	1014	475	5521	1751	675	7846	2489	875	10170	3226
80	930	295	280	3254	1032	480	5579	1770	680	7904	2507	880	10228	3245
85	988	313	285	3313	1051	485	5637	1788	685	7962	2526	885	10287	3263
90	1046	332	290	3371	1069	490	5695	1807	690	8020	2544	890	10234	3281
95	1104	350	295	3429	1088	495	5753	1825	695	8078	2562	895	10403	3299
100	1162	368	300	3487	1106	500	5812	1843	700	8136	2581	900	10461	3318
105	1220	387	305	3545	1125	505	5870	1862	705	8194	2599	905	10519	3337
110	1279	406	310	3603	1143	510	5928	1880	710	8252	2618	910	10577	3355
115	1337	424	315	3661	1161	515	5986	1899	715	8311	2636	915	10635	3374
120	1395	442	320	3719	1180	520	6044	1917	720	8369	2655	920	10693	3392
125	1453	461	325	3778	1198	525	6102	1936	725	8427	2673	925	10751	3410
130	1511	479	330	3836	1217	530	6160	1954	730	8485	2692	930	10810	3429
135	1569	498	335	3894	1235	535	6218	1973	735	8543	2710	935	10868	3447
140	1627	516	340	3952	1254	540	6277	1991	740	8601	2728	940	10926	3466
145	1685	535	345	4010	1272	545	6334	2009	745	8659	2747	945	10984	3484
150	1743	553	350	4068	1290	550	6393	2028	750	8717	2765	950	11042	3503
155	1802	571	355	4126	1309	555	6451	2046	755	8776	2784	955	11100	3521
160	1860	586	360	4184	1327	560	6509	2065	760	8834	2802	960	11158	3540
165	1918	608	365	4242	1346	565	6567	2083	765	8891	2820	965	11216	3558
170	1976	627	370	4300	1364	570	6625	2102	770	8550	2839	970	11274	3576
175	2034	645	375	4359	1383	575	6683	2120	775	9008	2857	975	11333	3595
180	2092	664	380	4417	1401	580	6741	2138	780	9066	2876	980	11391	3613
185	2150	682	385	4475	1419	585	6800	2157	785	9124	2864	985	11449	3532
190	2208	701	390	4533	1438	590	6858	2175	790	9182	2913	990	11507	3650
195	2267	719	395	4591	1456	595	6916	2194	795	9240	2931	995	11565	3669
200	2324	737	400	4649	1475	600	6974	2212	800	9299	2950	1000	11623	3687

*1 Langley = 1 gram calorie per square centimeter
= 11.6232 watts per square meter
= 3.687 Btu per square foot

(Sheet 2 of 2)

Table B-5 Water vapor transmission units

English and Metric Units

Water Vapor Transmission

Water Vapor Transmission

$$= \frac{\text{Weight Change in Grains}}{\text{Time x Area}}$$

In English Units:

In Unit Time
Water Vapor Transmission, per Hour

$$= \frac{\text{Weight Change in Grains per Hour}}{\text{Area in Square Feet}}$$

In Unit Time, Unit Area

WVT = Rate of water vapor transmission in number of grains per hour per square foot, or

= Grains per hour square foot

$$= \frac{G}{\text{Hr, Sq Ft}}$$

When G = Grains

In Metric Units:

Water vapor transmission is expressed in unit time of 24 hours. Thus:

WVT = Rate of vapor transmission in number of grams per hour per square meter

= Grams per 24 hours square meter

$$= \frac{g}{\text{24 hrs, sq m}}$$

When g = Grams

Conversion Between the Two Systems:

Grams per 24 hours, sq m x 0.0598
= Grains per hour, sq ft.

Grains per hour, sq ft x 16.7
= Grams per 24 hour, sq m

Permeance of a Material

Permeance

$$= \frac{\text{Water Vapor Transmission}}{\text{Vapor Pressure Difference Between the Two Sides of the Material}}$$

$$= \frac{WVT}{\Delta P}$$

When ΔP = Vapor pressure difference between the two sides of the material

In English Units:

$$\text{Permeance (Perms)} = \frac{G}{\Delta P\text{, Hr, Sq Ft}}$$

In Metric Units:

$$\text{Permeance (Metric Perms)} = \frac{g}{\Delta p\text{, Hr, Sq Ft}}$$

When Δp = Vapor pressure difference between the two sides of the material

Conversion Between the Two Systems:

Metric Perms x 1.52 = Perms
Perms x 0.659 = Metric Perms

Permeability of a Material

For thick materials permeability is given in perm-inches for *English Units* and in Metric Perm-Centimeters for *Metric Units*

Conversion Between the Two Systems:

Metric Perm-Centimeters x 0.598 = Perm-inches
Perm-inches x 1.67 = Metric Perm Centimeters

Note: Permeability and Permeance as used in this Manual do not follow the pattern used in defining other terms in the Manual ending in "ance" and "ility," but follow the definitions as given by ASTM.

Units of Measure: In this book the commonly used term English Unit has been used to designate the unit of measure now used in the United States, although the correct designation, very seldom used, is British Units. Thus English Units and British Units are the same. Presently an International System of Units (S.I.) is being adopted throughout the world. This is based on Metric Units with some very minor changes. Therefore, S.I. Units and Metric Units are the same. As the term S.I. Units is not well known in the United States at this time, the term Metric Units was used in this book.

Thus: English Units are the same as British Units.
Metric Units are the same as S.I. Units.

Table B-6 Hyperbolic logarithms

These two pages give the natural (hyperbolic, or Napierian) logarithms ($\log_e$) of numbers between 1 and 10, correct to four places. Moving the decimal point n places to the right [or left] in the number is equivalent to adding n times 2.3026 [or n times 3.6974] to the logarithm. Base e = 2.71828+

n	n(2.3026)	n(0.6974−3)
1	2.3026	0.6974−3
2	4.6052	0.3948−5
3	6.9078	0.0922−7
4	9.2103	0.7897−10
5	11.5129	0.4871−12
6	13.8155	0.1845−14
7	16.1181	0.8819−17
8	18.4207	0.5793−19
9	20.7233	0.2767−21

Number	0	1	2	3	4	5	6	7	8	9	Avg. diff.
1.0	0.0000	0100	0198	0296	0392	0488	0583	0677	0770	0862	95
1.1	0953	1044	1133	1222	1310	1398	1484	1570	1655	1740	87
1.2	1823	1906	1989	2070	2151	2231	2311	2390	2469	2546	80
1.3	2624	2700	2776	2852	2927	3001	3075	3148	3221	3293	74
1.4	3365	3436	3507	3577	3646	3716	3784	3853	3920	3988	69
1.5	0.4055	4121	4187	4253	4318	4383	4447	4511	4574	4637	65
1.6	4700	4762	4824	4886	4947	5008	5068	5128	5188	5247	61
1.7	5306	5365	5423	5481	5539	5596	5653	5710	5766	5822	57
1.8	5878	5933	5988	6043	6098	6152	6206	6259	6313	6366	54
1.9	6419	6471	6523	6575	6627	6678	6729	6780	6831	6881	51
2.0	0.6931	6981	7031	7080	7129	7178	7227	7275	7324	7372	49
2.1	7419	7467	7514	7561	7608	7655	7701	7747	7793	7839	47
2.2	7885	7930	7975	8020	8065	8109	8154	8198	8242	8286	44
2.3	8329	8372	8416	8459	8502	8544	8587	8629	8671	8713	43
2.4	8755	8796	8838	8879	8920	8961	9002	9042	9083	9123	41
2.5	0.9163	9203	9243	9282	9322	9361	9400	9439	9478	9517	39
2.6	9555	9594	9632	9670	9708	9746	9783	9821	9858	9895	38
2.7	0.9933	9969	*0006	*0043	*0080	*0116	*0152	*0188	*0225	*0260	36
2.8	1.0296	0332	0367	0403	0438	0473	0508	0543	0578	0613	35
2.9	0647	0682	0716	0750	0784	0818	0852	0886	0919	0953	34
3.0	1.0986	1019	1053	1086	1119	1151	1184	1217	1249	1282	33
3.1	1314	1346	1378	1410	1442	1474	1506	1537	1569	1600	32
3.2	1632	1663	1694	1725	1756	1787	1817	1848	1878	1909	31
3.3	1939	1969	2000	2030	2060	2090	2119	2149	2179	2208	30
3.4	2238	2267	2296	2326	2355	2384	2413	2442	2470	2499	29
3.5	1.2528	2556	2585	2613	2641	2669	2698	2726	2754	2782	28
3.6	2809	2837	2865	2892	2920	2947	2975	3002	3029	3056	27
3.7	3083	3110	3137	3164	3191	3218	3244	3271	3297	3324	27
3.8	3350	3376	3403	3429	3455	3481	3507	3533	3558	3584	26
3.9	3610	3635	3661	3686	3712	3737	3762	3788	3813	3838	25
4.0	1.3863	3888	3913	3938	3962	3987	4012	4036	4061	4085	25
4.1	4110	4134	4159	4183	4207	4231	4255	4279	4303	4327	24
4.2	4351	4375	4398	4422	4446	4469	4493	4516	4540	4563	23
4.3	4586	4609	4633	4656	4679	4702	4725	4748	4770	4793	23
4.4	4816	4839	4861	4884	4907	4929	4951	4974	4996	5019	22
4.5	1.5041	5063	5085	5107	5129	5151	5173	5195	5217	5239	22
4.6	5261	5282	5304	5326	5347	5369	5390	5412	5433	5454	21
4.7	5476	5497	5518	5539	5560	5581	5602	5623	5644	5665	21
4.8	5686	5707	5728	5748	5769	5790	5810	5831	5851	5872	20
4.9	5892	5913	5933	5953	5974	5994	6014	6034	6054	6074	20
5.0	1.6094	6114	6134	6154	6174	6194	6214	6233	6253	6273	20
5.1	6292	6312	6332	6351	6371	6390	6409	6429	6448	6467	19
5.2	6487	6506	6525	6544	6563	6582	6601	6620	6639	6658	19
5.3	6677	6696	6715	6734	6752	6771	6790	6808	6827	6845	18
5.4	6864	6882	6901	6919	6938	6956	6974	6993	7011	7029	18

Table B-6 Hyperbolic logarithms (continued)

Number	0	1	2	3	4	5	6	7	8	9	Avg. diff.
5.5	1.7047	7066	7084	7102	7120	7138	7156	7174	7192	7210	18
5.6	7228	7246	7263	7281	7299	7317	7334	7352	7370	7387	18
5.7	7405	7422	7440	7457	7475	7492	7509	7527	7544	7561	17
5.8	7579	7596	7613	7630	7647	7664	7681	7699	7716	7733	17
5.9	7750	7766	7783	7800	7817	7834	7851	7867	7884	7901	17
6.0	1.7918	7934	7951	7967	7984	8001	8017	8034	8050	8066	16
6.1	8083	8099	8116	8132	8148	8165	8181	8197	8213	8229	16
6.2	8245	8262	8278	8294	8310	8326	8342	8358	8374	8390	16
6.3	8405	8421	8437	8453	8469	8485	8500	8516	8532	8547	16
6.4	8563	8579	8594	8610	8625	8641	8656	8672	8687	8703	15
6.5	1.8718	8733	8749	8764	8779	8795	8810	8825	8840	8856	15
6.6	8871	8886	8901	8916	8931	8946	8961	8976	8991	9006	15
6.7	9021	9036	9051	9066	9081	9095	9110	9125	9140	9155	15
6.8	9169	9184	9199	9213	9228	9242	9257	9272	9286	9301	15
6.9	9315	9330	9344	9359	9373	9387	9402	9416	9430	9445	14
7.0	1.9459	9473	9488	9502	9516	9530	9544	9559	9573	9587	14
7.1	9601	9615	9629	9643	9657	9671	9685	9699	9713	9727	14
7.2	9741	9755	9769	9782	9796	9810	9824	9838	9851	9865	14
7.3	1.9879	9892	9906	9920	9933	9947	9961	9974	9988	*0001	13
7.4	2.0015	0028	0042	0055	0069	0082	0096	0109	0122	0136	13
7.5	2.0149	0162	0176	0189	0202	0215	0229	0242	0255	0268	13
7.6	0281	0295	0308	0321	0334	0347	0360	0373	0386	0399	13
7.7	0412	0425	0438	0451	0464	0477	0490	0503	0516	0528	13
7.8	0541	0554	0567	0580	0592	0605	0618	0631	0643	0656	13
7.9	0669	0681	0694	0707	0719	0732	0744	0757	0769	0782	12
8.0	2.0794	0807	0819	0832	0844	0857	0869	0882	0894	0906	12
8.1	0919	0931	0943	0956	0968	0980	0992	1005	1017	1029	12
8.2	1041	1054	1066	1078	1090	1102	1114	1126	1138	1150	12
8.3	1163	1175	1187	1199	1211	1223	1235	1247	1258	1270	12
8.4	1282	1294	1306	1318	1330	1342	1353	1365	1377	1389	12
8.5	2.1401	1412	1424	1436	1448	1459	1471	1483	1494	1506	12
8.6	1518	1529	1541	1552	1564	1576	1587	1599	1610	1622	12
8.7	1633	1645	1656	1668	1679	1691	1702	1713	1725	1736	11
8.8	1748	1759	1770	1782	1793	1804	1815	1827	1838	1849	11
8.9	1861	1872	1883	1894	1905	1917	1928	1939	1950	1961	11
9.0	2.1972	1983	1994	2006	2017	2028	2039	2050	2061	2072	11
9.1	2083	2094	2105	2116	2127	2138	2148	2159	2170	2181	11
9.2	2192	2203	2214	2225	2235	2246	2257	2268	2279	2289	11
9.3	2300	2311	2322	2332	2343	2354	2364	2375	2386	2396	11
9.4	2407	2418	2428	2439	2450	2460	2471	2481	2492	2502	11
9.5	2.2513	2523	2534	2544	2555	2565	2576	2586	2597	2607	10
9.6	2618	2628	2638	2649	2659	2670	2680	2690	2701	2711	10
9.7	2721	2732	2742	2752	2762	2773	2783	2793	2803	2814	10
9.8	2824	2834	2844	2854	2865	2875	2885	2895	2905	2915	10
9.9	2925	2935	2946	2956	2966	2976	2986	2996	3006	3016	10
10.0	2.3026										

$\log_e x = (2.3026)\log_{10} x \qquad \log_{10} x = (0.4343)\log_e x$

where $2.3026 = \log_e 10$ and $0.4343 = \log_{10} e$

Moving the decimal point n places to the right [or left] in the number requires adding n times 2.3026 [or n times (0.6974−3)] in the body of the table. See auxiliary table of multiples on top of the preceding page.

Table B-7 Decimals of a foot – English Units

0"	.0000	1"	.0833	2"	.166667	3"	.2500
1/16"	.0052	1-1/16"	.0885	2-1/16"	.171875	3-1/16"	.2552
1/8"	.0104	1-1/8"	.09375	2-1/8"	.1771	3-1/8"	.2604
3/16"	.015625	1-3/16"	.0990	2-3/16"	.1823	3-3/16"	.265625
1/4"	.0208	1-1/4"	.1042	2-1/4"	.1875	3-1/4"	.2708
5/16"	.0260	1-5/16"	.109375	2-5/16"	.1927	3-5/16"	.2760
3/8"	.03125	1-3/8"	.1146	2-3/8"	.1979	3-3/8"	.28125
7/16"	.0365	1-7/16"	.1198	2-7/16"	.203125	3-7/16"	.2865
1/2"	.0417	1-1/2"	.1250	2-1/2"	.2083	3-1/2"	.2917
9/16"	.046875	1-9/16"	.1302	2-9/16"	.2135	3-9/16"	.296875
5/8"	.0521	1-5/8"	.1354	2-5/8"	.21875	3-5/8"	.3021
11/16"	.0573	1-11/16"	.140625	2-11/16"	.2240	3-11/16"	.3075
3/4"	.0625	1-3/4"	.1458	2-3/4"	.2292	3-3/4"	.3125
13/16"	.0677	1-13/16"	.1510	2-13/16"	.234375	3-13/16"	.3177
7/8"	.0729	1-7/8"	.15625	2-7/8"	.2396	3-7/8"	.3229
15/16"	.078125	1-15/16"	.1615	2-15/16"	.2448	3-15/16"	.328125
4"	.3333	5"	.416667	6"	.5000	7"	.5833
4-1/16"	.3385	5-1/16"	.421875	6-1/16"	.5052	7-1/16"	.5885
4-1/8"	.34375	5-1/8"	.4271	6-1/8"	.5104	7-1/8"	.59375
4-3/16"	.3490	5-3/16"	.4323	6-3/16"	.515625	7-3/16"	.5990
4-1/4"	.3542	5-1/4"	.4375	6-1/4"	.5208	7-1/4"	.6042
4-5/16"	.359375	5-5/16"	.4427	6-5/16"	.5260	7-5/16"	.609375
4-3/8"	.3646	5-3/8"	.4479	6-3/8"	.53125	7-3/8"	.6146
4-7/16"	.3698	5-7/16"	.453125	6-7/16"	.5365	7-7/16"	.6198
4-1/2"	.3750	5-1/2"	.4583	6-1/2"	.5417	7-1/2"	.6250
4-9/16"	.3802	5-9/16"	.4635	6-9/16"	.546875	7-9/16"	.6302
4-5/8"	.3854	5-5/8"	.46875	6-5/8"	.5521	7-5/8"	.6354
4-11/16"	.390625	5-11/16"	.4740	6-11/16"	.5573	7-11/16"	.640625
4-3/4"	.3958	5-3/4"	.4792	6-3/4"	.5625	7-3/4"	.6458
4-13/16"	.4010	5-13/16"	.484375	6-13/16"	.5677	7-13/16"	.6510
4-7/8"	.40625	5-7/8"	.4896	6-7/8"	.5729	7-7/8"	.65625
4-15/16"	.4115	5-15/16"	.4948	6-15/16"	.578125	7-15/16"	.6615
8"	.666667	9"	.7500	10"	.8333	11"	.916667
8-1/16"	.671875	9-1/16"	.7552	10-1/16"	.8385	11-1/16"	.921875
8-1/8"	.6771	9-1/8"	.7604	10-1/8"	.84375	11-1/8"	.9271
8-3/16"	.6823	9-3/16"	.765625	10-3/16"	.8490	11-3/16"	.9323
8-1/4"	.6875	9-1/4"	.7708	10-1/4"	.8542	11-1/4"	.9375
8-5/16"	.6927	9-5/16"	.7760	10-5/16"	.859375	11-5/16"	.9427
8-3/8"	.6979	9-3/8"	.78125	10-3/8"	.8646	11-3/8"	.9479
8-7/16"	.703125	9-7/16"	.7865	10-7/16"	.8698	11-7/16"	.953125
8-1/2"	.7083	9-1/2"	.7917	10-1/2"	.8750	11-1/2"	.9583
8-9/16"	.7135	9-9/16"	.796875	10-9/16"	.8802	11-9/16"	.9635
8-5/8"	.71875	9-5/8"	.8021	10-5/8"	.8854	11-5/8"	.96875
8-11/16"	.7240	9-11/16"	.8073	10-11/16"	.890625	11-11/16"	.9740
8-3/4"	.7292	9-3/4"	.8125	10-3/4"	.8958	11-3/4"	.9792
8-13/16"	.734375	9-13/16"	.8177	10-13/16"	.9010	11-13/16"	.984375
8-7/8"	.7396	9-7/8"	.8229	10-7/8"	.90625	11-7/8"	.9896
8-15/16"	.7448	9-15/16"	.828125	10-15/16"	.9115	11-15/16"	.9948

Reproduced with permission, from Heat Insulation Manual, Papco Industrial Product Div., Fiberboard Corp.

Table B-8

Decimal Equivalents

Fraction	Decimal	Fraction	Decimal
1/64	.015625	33/64	.515625
1/32	.03125	17/32	.53125
3/64	.046875	35/64	.546875
1/16	.0625	9/16	.5625
5/64	.078125	37/64	.578125
3/32	.09375	19/32	.59375
7/64	.109375	39/64	.609375
1/8	.125	5/8	.625
9/64	.140625	41/64	.640625
5/32	.15625	21/32	.65625
11/64	.171875	43/64	.671875
3/16	.1875	11/16	.6875
13/64	.203125	45/64	.703125
7/32	.21875	23/32	.71875
15/64	.234375	47/64	.734375
1/4	.25	3/4	.75
17/64	.265625	49/64	.765625
9/32	.28125	25/32	.78125
19/64	.296875	51/64	.796875
5/16	.3125	13/16	.8125
21/64	.328125	53/64	.828125
11/32	.34375	27/32	.84375
23/64	.359375	55/64	.859375
3/8	.375	7/8	.875
25/64	.390625	57/64	.890625
13/32	.40625	29/32	.90625
27/64	.421875	59/64	.921875
7/16	.4375	15/16	.9375
29/64	.453125	61/64	.953125
15/32	.46875	31/32	.96875
31/64	.484375	63/64	.984375
1/2	.5	1	1.0

Table B-9 Radiation heat transfer insulation to air for surface emittances $\epsilon = 1.0$ to 0.1 in Btu/ft, hr

English Units

Temp. difference surface to air °F or °R	Emittance of outer weather barrier or jacket in reference to thermal black body										Temp. difference surface to air °C or °K
	1.0	0.9	0.8	0.7	0.6	0.5	0.4	0.3	0.2	0.1	
2	1.8	1.6	1.4	1.3	1.1	0.9	0.7	0.5	0.4	0.2	1.1
4	3.5	3.2	2.8	2.5	2.1	1.8	1.4	1.1	0.7	0.4	2.2
6	5.6	5.0	4.5	3.9	3.4	2.8	2.2	1.7	1.1	0.6	3.3
8	7.5	6.8	6.0	5.2	4.5	3.8	3.0	2.3	1.5	0.8	4.4
10	9.5	8.6	7.6	6.6	5.7	4.8	3.8	2.9	1.9	1.0	5.6
15	14.4	13.0	11.5	10.1	8.6	7.2	5.8	4.3	2.9	1.4	8.3
20	19.6	17.6	15.7	13.7	11.8	9.8	7.8	5.9	3.9	2.0	11.1
25	24.7	22.2	19.8	17.3	14.8	12.4	9.9	7.4	4.9	2.5	13.3
30	30.2	27.2	24.2	21.1	18.1	15.1	12.1	9.1	6.1	3.0	16.7
35	35.8	32.2	28.6	25.1	21.5	17.9	14.4	10.8	7.2	3.6	19.4
40	41.5	37.4	33.2	28.0	24.9	20.8	16.6	12.4	8.3	4.2	22.2
45	47.4	42.6	37.9	31.5	28.4	23.7	19.0	14.2	9.5	4.7	25.0
50	53.1	47.8	42.4	37.2	31.9	26.6	21.2	15.9	10.6	5.3	27.7
60	65.9	59.4	52.7	46.1	39.5	33.0	26.4	19.8	13.2	6.6	33.3
70	79.2	71.3	63.3	55.5	47.5	39.6	31.6	23.8	15.8	7.9	33.9
80	93.1	83.9	74.4	65.2	55.8	46.6	37.2	27.9	18.6	9.3	44.4
90	107.8	97.2	86.3	75.5	64.7	53.9	43.2	32.2	21.6	10.8	50.0
100	123.2	111.1	95.5	87.2	73.9	61.6	49.3	37.0	24.6	12.3	55.5
110	139.4	125.6	111.5	95.5	83.6	69.7	55.8	41.8	27.9	13.9	61.1
120	156.4	141.0	125.1	109.0	93.2	78.2	62.6	46.9	31.3	15.6	66.6
130	174.2	157.1	139.3	122.0	104.5	87.1	69.8	52.4	34.8	17.4	72.2
140	192.9	173.7	154.0	135.0	115.8	96.5	77.2	57.8	38.6	19.3	77.7
150	212.4	191.3	169.9	148.7	127.7	106.2	85.1	63.8	42.6	21.2	88.3
160	232.9	209.0	185.5	163.0	139.8	116.5	93.3	70.0	46.7	23.3	88.8
170	254.3	229.1	203.5	178.0	152.6	127.2	101.9	76.4	50.8	25.4	94.4
180	276.7	249.3	221.1	192.9	166.0	138.4	110.8	83.1	55.4	27.7	100.0
190	300.1	270.0	240.0	210.0	180.0	150.0	120.0	90.0	60.0	30.0	105.5
200	324.5	292.5	259.5	227.5	194.8	162.3	130.0	97.5	64.8	32.4	111.1

Note: To convert to Metric Units, multiply listed Btu/ft^2, hr by 3.152591 to obtain W/m^2.

Table B-10 Thermal expansion of pipes in inches per 100 linear feet

Temp. °F	Cast Iron Pipe	Steel Pipe	Wrought Iron Pipe	Copper Pipe	Temp. °C
−20	0	0	0	0	−28.9
0	0.127	0.145	0.152	0.204	−17.8
20	0.255	0.293	0.306	0.442	−6.7
40	0.390	0.430	0.465	0.655	4.4
60	0.518	0.593	0.620	0.888	15.6
80	0.649	0.725	0.780	1.100	26.7
100	0.787	0.898	0.939	1.338	37.8
120	0.926	1.055	1.110	1.570	48.9
140	1.051	1.209	1.265	1.794	60.0
160	1.200	1.368	1.427	2.008	71.1
180	1.345	1.528	1.597	2.255	82.2
200	1.495	1.691	1.778	2.500	93.3
220	1.634	1.852	1.936	2.720	104
240	1.780	2.020	2.110	2.960	116
260	1.931	2.183	2.279	3.189	127
280	2.085	2.350	2.465	3.422	138
300	2.233	2.519	2.630	3.665	149
320	2.395	2.690	2.800	3.900	160
340	2.543	2.862	2.988	4.145	171
360	2.700	3.029	3.175	4.380	182
380	2.859	3.211	3.350	4.628	193
400	3.008	3.375	3.521	4.870	204
420	3.182	3.566	3.720	5.118	216
440	3.345	3.740	3.900	5.358	227
460	3.511	3.929	4.096	5.612	238
480	3.683	4.100	4.280	5.855	249
500	3.847	4.296	4.477	6.110	260
520	4.020	4.487	4.677	6.352	271
540	4.190	4.670	4.866	6.614	282
560	4.365	4.860	5.057	6.850	293
580	4.541	5.051	5.268	7.123	304
600	4.725	5.247	5.455	7.388	316
620	4.896	5.437	5.660	7.636	327
640	5.082	5.627	5.850	7.893	338
660	5.260	5.831	6.067	8.153	349
680	5.442	6.020	6.260	8.400	360
700	5.269	6.229	6.481	8.676	371
720	5.808	6.425	6.673	8.912	382
740	6.006	6.635	6.899	9.203	393
760	6.200	6.833	7.100	9.460	404
780	6.389	7.046	7.314	9.736	416
800	6.587	7.250	7.508	9.992	427
820	6.779	7.464	7.757	10.272	438
840	6.970	7.662	7.952	10.512	449
860	7.176	7.888	8.195	10.814	460
880	7.375	8.098	8.400	11.175	471
900	7.579	8.313	8.639	11.360	482
920	7.795	8.545	8.867	11.625	493
940	7.989	8.755	9.089	11.911	504
960	8.200	8.975	9.300	12.180	516
980	8.406	9.916	9.547	12.473	527
1000	8.617	9.421	9.776	12.747	538

To obtain the amount of expansion between any two temperatures, take the proportionate difference between the values given for those temperatures.
To convert to Metric Units multiply by 8.2025 to obtain thermal expansion of pipes in millimeters per 10 meters in length.

Table B-11 Linear coefficients of expansion for one degree

Substance	Coefficient, n	
	Celsius	Fahrenheit
Metals and Alloys		
Aluminum, wrought	.0000231	.0000128
Brass	.0000188	.0000104
Brass wire	.0000193	.0000107
Bronze	.0000181	.0000101
Copper	.0000168	.0000093
German Silver	.0000183	.0000102
Gold	.0000150	.0000083
Iron, cast, gray	.0000106	.0000059
Iron, wrought	.0000120	.0000067
Iron, wire	.0000124	.0000069
Lead	.0000286	.0000159
Nickel	.0000126	.0000070
Platinum	.0000090	.0000050
Platinum-Iridium, 15% Ir.	.0000081	.0000045
Silver	.0000192	.0000107
Steel, cast	.0000110	.0000061
Steel, hard	.0000132	.0000073
Steel, medium	.0000120	.0000067
Steel, soft	.0000110	.0000061
Tin	.0000210	.0000117
Zinc, rolled	.0000311	.0000173
Miscellaneous Solids		
Glass	.0000085	.0000047
Graphite	.0000079	.0000044
Gutta-percha	.0005980	.0003322
Paraffin	.0002785	.0001547
Porcelain	.0000036	.0000020
Stone and Masonry		
Ashlar masonry	.0000063	.0000035
Brick masonry	.0000055	.0000031
Cement, Portland	.0000107	.0000059
Concrete	.0000143	.0000079
Concrete, masonry	.0000120	.0000067
Granite	.0000084	.0000047
Limestone	.0000080	.0000044
Marble	.0000100	.0000056
Plaster	.0000166	.0000092
Rubble masonry	.0000063	.0000035
Sandstone	.0000110	.0000061
Slate	.0000104	.0000058
Timber		
Fir, parallel to fiber	.0000037	.0000021
Maple, parallel to fiber	.0000064	.0000036
Oak, parallel to fiber	.0000049	.0000027
Pine, parallel to fiber	.0000054	.0000030
Fir, perpendicular to fiber	.000058	.000032
Maple, perpendicular to fiber	.000048	.000027
Oak, perpendicular to fiber	.000054	.000030
Pine, perpendicular to fiber	.000034	.000019
Liquid Substances	Volumetric Expan.	
Alcohol	.00104	.00058
Acid, nitric	.00110	.00061
Acid, sulphuric	.00063	.00035
Mercury	.00018	.00010
Oil, turpentine	.00090	.00050

Expansion of Water, Maximum Density = 1

C°	Volume	C°	Volume	C°	Volume	C°	Volume	C°	Volume	C°	Volume
0	1.000126	10	1.000257	30	1.004234	50	1.011877	70	1.022384	90	1.035829
4	1.000000	20	1.001732	40	1.007627	60	1.016954	80	1.029003	100	1.043116

Table B-12 Table of gauges and weights for wire, bands and flat sheet metal

English Units

Material	Gauge	Diameter or Thickness	Linear Feet Per Pound
Galvanized or B. A. Wire	W & M		
12 Gauge	W & M	0.1055″ Dia.	33.30
14 Gauge	W & M	0.0800″ Dia.	58.82
16 Gauge	W & M	0.0625″ Dia.	95.23
18 Gauge	W & M	0.0475″ Dia.	166.66
Soft Copper Wire	B & S		
12 Gauge	B & S	0.0808″ Dia.	52.63
14 Gauge	B & S	0.0640″ Dia.	83.33
16 Gauge	B & S	0.0508″ Dia.	142.85
18 Gauge	B & S	0.0403″ Dia.	250.00
Galvanized Steel Bands			
1/2″ Wide	–	.015″ thk.	39.26
1/2″ Wide	–	.020″ thk.	29.45
3/4″ Wide	–	.020″ thk.	19.63
1 1/4″ Wide	–	.035″ thk.	6.73
Stainless Steel Bands 18-8 Chrome-Nickel			
1/2″ Wide	–	.015″ thk	39.26
1/2″ Wide	–	.020″ thk.	29.45
3/4″ Wide	–	.015″ thk.	26.18
3/4″ Wide	–	.020″ thk.	19.63
Material	**Gauge**	**Thickness**	**Weight in Lbs Per Sq Ft**
Flat Black Sheet Metal			
18 Gauge	U. S. Standard	.05″	2.00
20 Gauge	U. S. Standard	.0375″	1.50
22 Gauge	U. S. Standard	.03125″	1.25
24 Gauge	U. S. Standard	.025″	1.00
26 Gauge	U. S. Standard	.01825″	0.75
28 Gauge	U. S. Standard	.015625″	0.625
30 Gauge	U. S. Standard	.0125″	0.50
Flat Galvanized Sheet Metal			
18 Gauge	U. S. Standard	.0540″	2.1563
20 Gauge	U. S. Standard	.0415″	1.6563
22 Gauge	U. S. Standard	.03535″	1.4063
24 Gauge	U. S. Standard	.0290″	1.1563
26 Gauge	U. S. Standard	.02275″	.9063
28 Gauge	U. S. Standard	.019625″	.7813
30 Gauge	U. S. Standard	.0165″	.6563
Aluminum Sheet			
16 Gauge	B & S	.051	.716
18 Gauge	B & S	.040	.568
20 Gauge	B & S	.032	.450
22 Gauge	B & S	.025	.357
24 Gauge	B & S	.020	.283
26 Gauge	B & S	.016	.225
28 Gauge	B & S	.012	.178
Stainless Steel Sheet			
18 Gauge	U. S. Standard	.050	2.10
20 Gauge	U. S. Standard	.037	1.58
22 Gauge	U. S. Standard	.031	1.31
24 Gauge	U. S. Standard	.025	1.05
26 Gauge	U. S. Standard	.018	.788
28 Gauge	U. S. Standard	.015	.66
30 Gauge	U. S. Standard	.013	.53
32 Gauge	U. S. Standard	.010	.43

Table B-13 Volumes of pipe insulation — English Units

Nom. NPS Pipe size	CUBIC FEET PER LINEAR FOOT — Nominal Insulation Thickness — Inches																	
	½	1	1½	2	2½	3	3½	4	4½	5	5½	6	6½	7	7½	8	8½	9
½	0.013	0.040	0.083	0.133	0.231	0.313	0.400	0.498										
¾	0.010	0.038	0.082	0.130	0.228	0.310	0.398	0.496										
1	0.018	0.056	0.100	0.154	0.225	0.307	0.394	0.493										
1¼	0.029	0.051	0.122	0.148	0.219	0.302	0.388	0.487										
1½	0.027	0.071	0.120	0.218	0.300	0.388	0.486	0.611										
2	0.038	0.082	0.137	0.208	0.289	0.377	0.474	0.600										
2½	0.044	0.093	0.191	0.273	0.360	0.458	0.583	0.709										
3	0.044	0.098	0.169	0.251	0.338	0.437	0.562	0.688	0.818	1.003	1.162	1.331						
3½	0.049	0.148	0.229	0.317	0.414	0.540	0.665	0.796	0.892	0.982	1.140	1.309						
4	0.055	0.126	0.208	0.294	0.393	0.518	0.643	0.774	0.960	1.118	1.288	1.468						
6	0.082	0.169	0.268	0.393	0.518	0.649	0.834	0.993	1.162	1.342	1.533	1.734	1.948	2.171	2.405	2.651	2.908	3.174
8		0.223	0.349	0.480	0.666	0.823	0.993	1.173	1.363	1.565	1.778	2.002	2.237	2.482	2.738	3.005	3.283	3.573
10		0.257	0.442	0.600	0.769	0.949	1.140	1.342	1.554	1.778	2.013	2.258	2.514	2.782	3.060	3.349	3.649	3.960
12		0.343	0.513	0.693	0.883	1.086	1.298	1.522	1.757	2.003	2.258	2.525	2.803	3.093	3.393	3.703	4.025	4.356
14		0.328	0.508	0.698	0.900	1.113	1.337	1.571	1.817	2.073	2.340	2.618	2.908	3.207	3.518	3.840	4.173	4.516
16		0.363	0.573	0.786	1.009	1.243	1.489	1.745	2.013	2.291	2.580	2.88	3.191	3.513	3.845	4.189	4.543	4.909
18		0.414	0.638	0.873	1.118	1.374	1.642	1.920	2.209	2.509	2.820	3.142	3.474	3.818	4.173	4.538	4.914	5.302
20		0.458	0.703	0.960	1.228	1.505	1.794	2.094	2.405	2.728	3.060	3.403	3.758	4.123	4.500	4.887	5.285	5.694
24		0.546	0.834	1.134	1.445	1.768	2.100	2.443	2.798	3.163	3.540	3.927	4.325	4.734	5.154	5.585	6.027	6.480
30		0.677	1.031	1.397	1.773	2.160	2.558	2.968	3.388	3,818	4.259	4.713	5.176	5.651	6.136	6.633	7.139	7.658
36		0.808	1.222	1.658	2.100	2.553	3.016	3.491	3.976	4.473	4.980	5.498	6.027	6.567	7.118	7.679	8.253	8.836

Metric Units: To convert cubic feet per linear foot to cubic meters per linear meter multiply by 0.0929.

Table B-14 Areas, sizes and capacities of standard NPS pipe (All dimensions and weights are nominal) English Units

Nom. Size in.	Diameter, in. External	Diameter, in. Internal	Thickness in.	Circumference in. External	Circumference in. Internal	Transverse areas, sq in. External	Transverse areas, sq in. Internal	External surface areas sq ft/lin ft of pipe	Length of pipe containing cu ft	wt of water per foot, lb
1/8	0.405	0.269	0.068	1.272	0.845	0.129	0.057	0.1060	2533.775	0.025
1/4	0.540	0.364	0.088	1.696	1.144	0.229	0.104	0.1414	1383.789	0.045
3/8	0.675	0.493	0.091	2.121	1.549	0.358	0.191	0.1767	754.360	0.083
1/2	0.840	0.622	0.109	2.639	1.954	0.554	0.304	0.220	473.906	0.132
3/4	1.050	0.824	0.113	3.299	2.589	0.866	0.533	0.275	270.034	0.231
1	1.315	1.049	0.133	4.131	3.296	1.358	0.861	0.344	166.618	0.375
1¼	1.660	1.380	0.140	5.215	4.335	2.164	1.495	0.435	96.275	0.65
1½	1.900	1.610	0.145	5.969	5.058	2.835	2.036	0.498	70.733	0.88
2	2.375	2.067	0.154	7.461	6.494	4.430	3.355	0.622	42.913	1.45
2½	2.875	2.469	0.203	9.032	7.757	6.492	4.788	0.753	30.077	2.07
3	3.500	3.068	0.216	10.996	9.638	9.621	7.393	0.917	19.479	3.20
3½	4.000	3.548	0.226	12.566	11.146	12.566	9.886	1.047	14.565	4.29
4	4.500	4.026	0.237	14.137	12.648	15.904	12.730	1.178	11.312	5.50
4½	5.000	4.506	0.247	15.708	14.156	19.635	15.947	1.3009	9.030	6.91
5	5.563	5.047	0.258	17.477	15.856	24.306	20.006	1.4586	7.198	8.67
6	6.625	6.065	0.280	20.813	19.054	34.472	28.891	1.7384	4.984	12.51
7	7.625	7.023	0.301	23.955	22.063	45.664	38.738	1.996	3.717	16.80
8	8.625	8.071	0.277	27.096	25.356	58.426	51.161	2.2350	2.815	22.18
8	8.625	7.981	0.322	27.096	25.073	58.426	50.027	2.2058	2.878	21.70
9	9.625	8.941	0.342	30.238	28.089	72.760	62.786	2.5220	2.294	27.20
10	10.750	10.192	0.279	33.772	32.019	90.763	81.685	2.8174	1.765	35.37
10	10.750	10.136	0.307	33.772	31.843	90.763	80.691	2.8104	1.826	34.20
10	10.750	10.020	0.365	33.772	31.479	90.763	78.855	2.8104	1.826	34.20
11	11.750	11.000	0.375	36.914	34.558	108.434	95.033	3.076	1.515	41.20
12	12.750	12.090	0.330	40.055	37.892	127.676	114.800	3.338	1.254	49.70
12	12.750	12.000	0.375	40.055	37.699	127.676	113.097	3.338	1.273	49.00

Note: Conversions to metric units not given, as the above is specific information regarding standard NPS pipe which is based on English Units.

INDEX

Page numbers in boldface type refer to figures and tables.